Heinemann BIOLOGY 2

5TH EDITION

VCE Units 3 & 4

Philip Batterham
Krista Bayliss
Ian Bentley
Christina Bliss
Sally Cash
Sarah Edwards
Barbara Evans
Kate Gready
Samantha Hopley
Pauline Ladiges
Fran Maher
John McKenzie
Jonathan Meddings
Aline Poh
Yvonne Sanders
Helen Silvester
Rebecca Wood

Pearson Australia
(a division of Pearson Australia Group Pty Ltd)
707 Collins Street, Melbourne, Victoria 3008
PO Box 23360, Melbourne, Victoria 8012
www.pearson.com.au

First published 2016 by Pearson Australia
2020 2019 2018 2017
10 9 8 7 6 5 4 3 2

Publisher: Alicia Brown
Series Consultant: Malcolm Parsons
Development Editor: Zoe Hamilton
Project Lead: Caroline Williams
Project Managers: Shelly Wang, Hannah Turner, Susan Keogh
Series Designer: Anne Donald
Rights and Permissions Editors: Jenny Jones, Amirah Fatin
Senior Publishing Services Analyst: Rob Curulli
Illustrators: Bruce Rankin, DiacriTech, Anne Donald
Editor: Marta Veroni
Proof checker: Lorna Hendry
Indexer: Jon Jermey

Printed in Malaysia (CTP-VVP)

National Library of Australia Cataloguing-in-Publication entry

Title: Heinemann Biology 2: VCE units 3 & 4 / Philip Batterham [and sixteen others]
Edition: 5th edition.
ISBN: 978 1 4886 1123 0 (paperback)
Notes: Includes index.
Previous edition: 2005.
Target Audience: For year 12 students.
Subjects: Biology Textbooks.
Biology Study and teaching (Secondary)Victoria.
Dewey Number: 570

Pearson Australia Group Pty Ltd ABN 40 004 245 943

Heinemann

BIOLOGY 2

5TH EDITION

Writing and Development Team

We are grateful to the following people for their time and expertise in contributing to the **Heinemann Biology 2** project.

Malcolm Parsons
Publishing Consultant
Series Consultant

Philip Batterham
Professor, Educator
Author

Krista Bayliss
Teacher
Author

Ian Bentley
Educator
Reviewer and Author

Christina Bliss
Teacher
Contributor and Reviewer

Sally Cash
Teacher
Author

Sarah Edwards
Teacher
Author

Barbara Evans
Professor, Educator
Author

Kate Gready
Teacher
Author

Samantha Hopley
Science Educator
Author

Pauline Ladiges
Professor, Educator
Author

Louise Lennard
Teacher, Head of Department
Contributor

Fran Maher
Bioscience Educator
Author and Reviewer

John McKenzie
Educator
Author

Jonathan Meddings
Science Writer
Author and Reviewer

Aline Poh
Teacher
Author

Yvonne Sanders
Teacher, Head of Department
Author and Reviewer

Helen Silvester
Teacher, Head of Department
Author and Reviewer

Neil van Herk
Teacher
Author

Rebecca Wood
Educator
Author

Contents

Unit 4 How does life change and respond to challenges over time?

AREA OF STUDY 1

How are species related?

AREA OF STUDY 2

How do humans impact on biological processes?

AREA OF STUDY 3 *Heinemann Biology 2 5th Edition ProductLink* provides extensive support material for Unit 4 Area of Study 3 Practical Investigation.

How to use this book

Heinemann Biology 2 5th edition

Heinemann Biology 2 5th edition has been written to the new VCE Biology Study Design 2017–2021. The book covers Units 3 and 4 and is an easy-to-use resource. Explore how to use this book below.

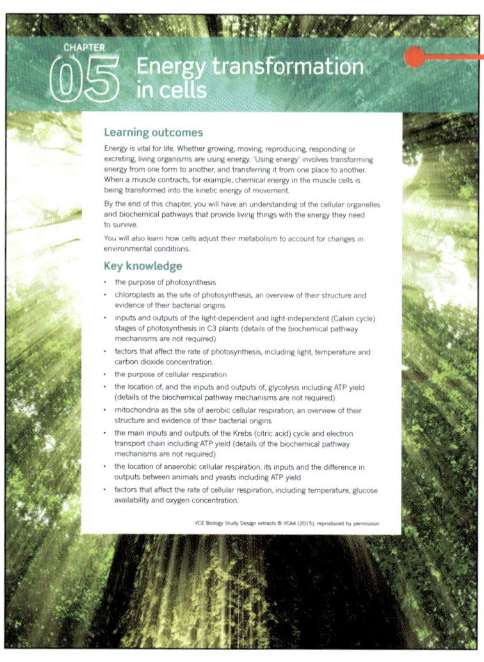
CHAPTER 05 Energy transformation in cells

Learning outcomes

Energy is vital for life. Whether growing, moving, reproducing, responding or excreting, living organisms are using energy. 'Using energy' involves transforming energy from one form to another, and transferring it from one place to another. When a muscle contracts, for example, chemical energy in the muscle cells is being transformed into the kinetic energy of movement.

By the end of this chapter, you will have an understanding of the cellular organelles and biochemical pathways that provide living things with the energy they need to survive.

You will also learn how cells adjust their metabolism to account for changes in environmental conditions.

Key knowledge

- the purpose of photosynthesis
- chloroplasts as the site of photosynthesis, an overview of their structure and evidence of their bacterial origins
- inputs and outputs of the light-dependent and light-independent (Calvin cycle) stages of photosynthesis in C3 plants (details of the biochemical pathway mechanisms are not required)
- factors that affect the rate of photosynthesis, including light, temperature and carbon dioxide concentration
- the purpose of cellular respiration
- the location of, and the inputs and outputs of, glycolysis including ATP yield (details of the biochemical pathway mechanisms are not required)
- mitochondria as the site of aerobic cellular respiration, an overview of their structure and evidence of their bacterial origins
- the main inputs and outputs of the Krebs (citric acid) cycle and electron transport chain including ATP yield (details of the biochemical pathway mechanisms are not required)
- the location of anaerobic cellular respiration, its inputs and the difference in outputs between animals and yeasts including ATP yield
- factors that affect the rate of cellular respiration, including temperature, glucose availability and oxygen concentration

VCE Biology Study Design extracts © VCAA (2015); reproduced by permission

Chapter opener

Chapter opening pages link the Study Design to the chapter content. Key knowledge addressed in the chapter is clearly listed.

Biology in Action

Biology in Action places biology in an applied situation or relevant context. Text and artwork refer to the nature and practice of biology, applications of biology and associated issues, and the historical development of biological concepts and ideas.

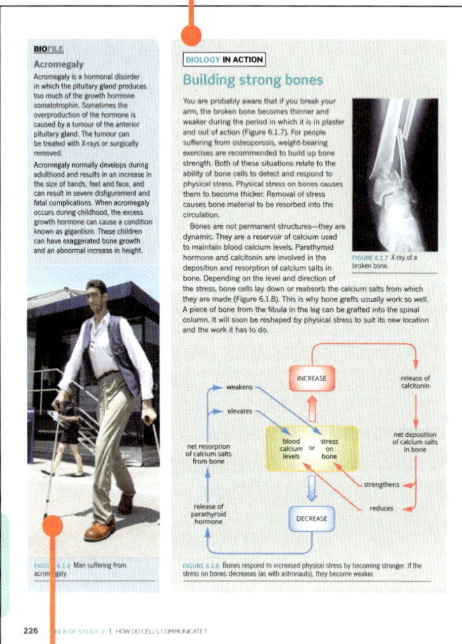
BIOFILE

Acromegaly

BIOLOGY IN ACTION

Building strong bones

226 AREA OF STUDY 2 | HOW DO CELLS COMMUNICATE?

EXTENSION

Treatment for allergic reactions

Medications

Allergen immunotherapy

An antihistamine is a drug that counteracts the effects of histamine by blocking histamine receptors and therefore suppressing some allergy symptoms.

CHAPTER 8 | IMMUNITY, IMMUNE MALFUNCTIONS AND IMMUNOTHERAPY 307

Highlight

Focus on important information such as key definitions and summary points.

BioFile

BioFiles include interesting information and real world examples.

Extension

Extension goes beyond the core content of the Study Design. Material is intended for students who wish to expand their depth of understanding.

Chapter review

Each chapter concludes with a set of higher-order questions to test students' ability to apply the knowledge gained from the chapter.

Section summary

Each section includes a summary to assist students consolidate key points and concepts.

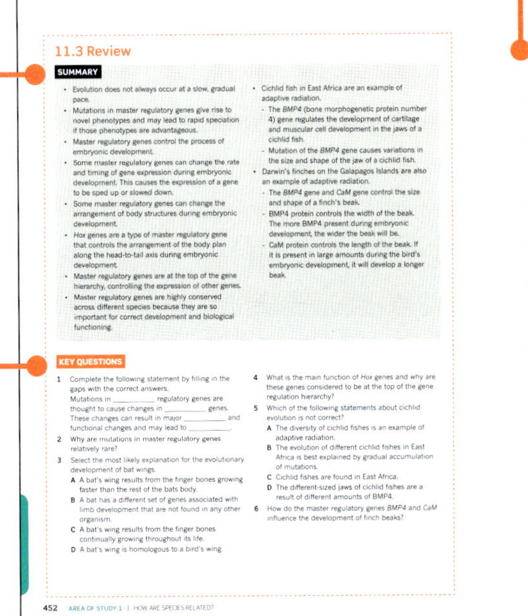

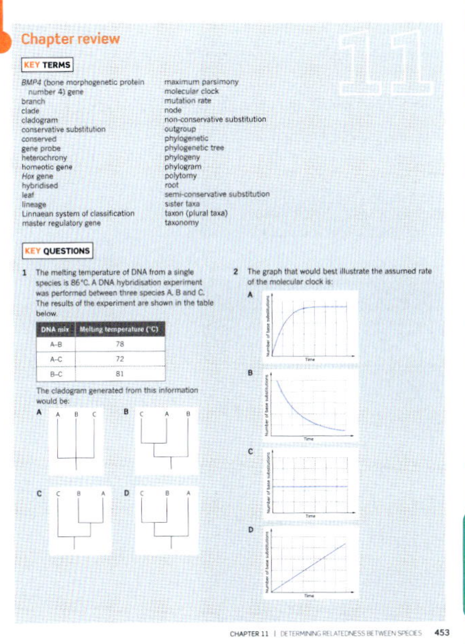

Section review

Each section concludes with questions to test students' understanding and ability to recall the section's key concepts.

Area of Study review

Each Area of Study concludes with a comprehensive set of exam-style questions, including multiple choice and extended response, that assist students in drawing together their knowledge and understanding and applying this to these question styles.

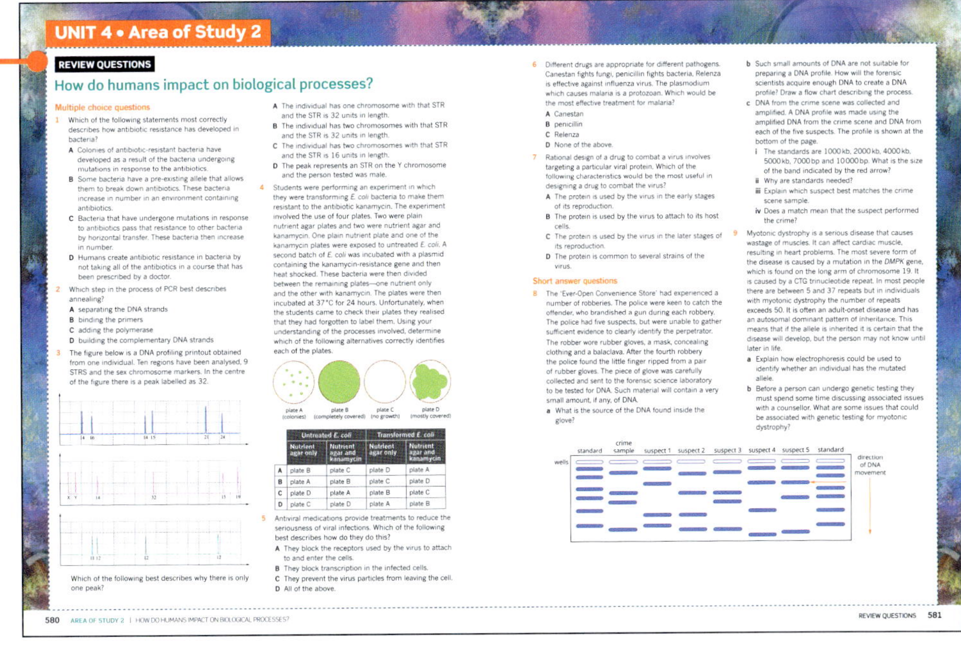

Answers

Comprehensive answers for all section review, chapter review and Area of Study review questions are provided via *Heinemann Biology 2 5th edition ProductLink*.

Glossary

Key terms are shown in **bold** throughout, and listed at the end of each chapter. A comprehensive glossary at the end of the book defines all key terms.

Heinemann Biology 2
5th edition

Student Book

Heinemann Biology 2 5th edition has been written to fully align with the VCE Biology Study Design 2017–2021. The series includes the very latest developments and applications of biology and incorporates best practice literacy and instructional design to ensure the content and concepts are fully accessible to all students.

Pearson eBook 3.0

Pearson eBook 3.0 is the digital version of the student book. It retains the integrity of the printed page and is available online or offline on PCs, Macs, Android tablets and iPads.

Student Workbook

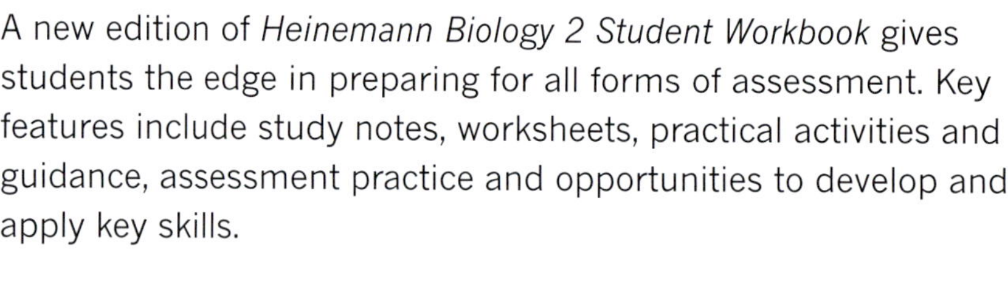

A new edition of *Heinemann Biology 2 Student Workbook* gives students the edge in preparing for all forms of assessment. Key features include study notes, worksheets, practical activities and guidance, assessment practice and opportunities to develop and apply key skills.

ProductLink

Heinemann Biology 2 5th edition ProductLink provides comprehensive support for teachers and students, completely free of charge. ProductLink includes comprehensive answers to all Student Book and Student Workbook questions, plus tests, quizzes, practice exams, teaching programs and risk assessments.

Heinemann Biology 2 5th edition ProductLink provides extensive support material for Unit 4 Area of Study 3 Practical Investigation. This includes teacher notes and advice, logbook template and sample logbook, poster template and sample poster, rubrics, checklists and more.

CHAPTER 01

Biology keystones — Foundation skills

Learning outcomes

The development of a set of key science skills is a core component of the study of VCE Biology and applies across Units 1 to 4 in all areas of study. Chapter 1 scaffolds the development of these skills. The opportunity to develop, use and demonstrate these skills in a variety of contexts is important ahead of undertaking investigations and when evaluating the research of others.

Although this chapter can be read as a whole, it is best to refer to it and use it when the need arises as you work through other chapters. For example, you may need a refresher on the process of the scientific method. It also contains useful checklists to assist when drawing scientific diagrams, graphing and completing aspects of your report. Similarly, when performing a first-hand investigation, refer to this chapter to make sure you collect data properly and that your data is of high quality.

Key skills

Develop aims and questions, formulate hypotheses and make predictions

- determine aims, hypotheses, questions and predictions that can be tested
- identify independent, dependent and controlled variables

Plan and undertake investigations

- determine appropriate type of investigation: conduct experiments (including use of controls); solve a scientific or technological problem; use of databases; simulations; access secondary data, including data sourced through the internet that would otherwise be difficult to source as raw or primary data through fieldwork, a laboratory or a classroom
- select and use equipment, materials and procedures appropriate to the investigation, taking into account potential sources of error and uncertainty

Comply with safety and ethical guidelines

- apply ethical principles when undertaking and reporting investigations
- apply relevant occupational health and safety guidelines while undertaking practical investigations, including following relevant bioethical guidelines when handling live materials

Conduct investigations to collect and record data

- work independently and collaboratively as appropriate and within identified research constraints
- systematically generate, collect, record and summarise both qualitative and quantitative data

KEY SKILLS CONTINUED

Analyse and evaluate data, methods and scientific models

- process quantitative data using appropriate mathematical relationships and units
- organise, present and interpret data using schematic diagrams and flow charts, tables, bar charts, line graphs, ratios, percentages and calculations of mean
- take a qualitative approach when identifying and analysing experimental data with reference to accuracy, precision, reliability, validity, uncertainty and errors (random and systematic)
- explain the merit of replicating procedures and the effects of sample sizes in obtaining reliable data
- evaluate investigative procedures and possible sources of bias, and suggest improvements
- explain how models are used to organise and understand observed phenomena and concepts related to biology, identifying limitations of the models

Draw evidence-based conclusions

- determine to what extent evidence from an investigation supports the purpose of the investigation, and make recommendations, as appropriate, for modifying or extending the investigation
- draw conclusions consistent with evidence and relevant to the question under investigation
- identify, describe and explain the limitations of conclusions, including identification of further evidence required
- critically evaluate various types of information related to biology from journal articles, mass media and opinions presented in the public domain
- discuss the implications of research findings and proposals

Communicate and explain scientific ideas

- use appropriate biological terminology, representations and conventions, including standard abbreviations, graphing conventions and units of measurement
- discuss relevant biological information, ideas, concepts, theories and models and the connections between them
- identify and explain formal biological terminology about investigations and concepts
- use clear, coherent and concise expression
- acknowledge sources of information and use standard scientific referencing conventions.

1.1 The scientific method

Biology is the study of living organisms. As scientists, biologists extend their understanding using the scientific method, which involves investigations that are carefully designed, carried out and reported (Figure 1.1.1). Well-designed research is based on a sound knowledge of what is already understood about a subject, as well as careful preparation and observation.

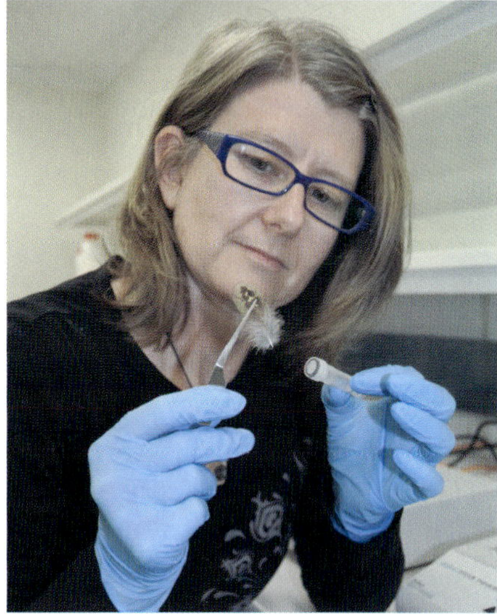

FIGURE 1.1.1 Biological research may employ diverse approaches and procedures, such as molecular biology. Analysis of DNA extracted from feathers by scientists at the Museum of Western Australia has confirmed that the night parrot (*Pezoporus occidentalis*) is not extinct, as previously thought.

OBSERVATION

Observation includes using all your senses and a wide variety of instruments and laboratory techniques to allow closer observation. Through careful inquiry and observation you can learn a lot about organisms, the ways they function, and their interactions with each other and the environment. For example, animals clearly function very differently from plants. Animals usually move around, take in nutrients and water, and often interact with each other in groups. Plants, however, are green, stationary, turn their leaves towards the light and grow. Many other distinguishing macroscopic structures and behaviours can be discerned from simple observation. Microscopic observation of cells reveals similarities and differences in the cellular structure of plant and animal cells, as well as the specialisations in the cells of a particular organism. Observational studies are a common research method, but they do not explain all the details of how organisms function.

First-hand investigations

The idea for a first-hand investigation of a complex problem arises from prior learning and observations that raise further questions. For example, indoor plants do not grow well in the long term without artificial lighting, which suggests light is required for photosynthesis in plants (Figure 1.1.2). This aspect of photosynthesis can be researched and the new knowledge applied to other applications, such as methods for growing plants in the laboratory for genetic selection and modification for crop improvement.

FIGURE 1.1.2 Laboratory methods such as plant tissue culture rely on careful observations and data collection about the requirements for growth of plants in natural conditions. Laboratory investigations then provide new information that can be applied to plants growing in the field.

Interpreting observations

How observations are interpreted depends on past experiences and knowledge, but to enquiring minds they will usually provoke further questions such as:

- How do organisms gain and expend energy?
- Are there differences between cellular processes in plants, animals, bacteria, fungi and protists?
- How do multicellular organisms develop specialised tissues?
- What are the molecular building blocks of cells?
- How do species change and evolve over time?
- How do cells communicate with each other?
- What is the molecular basis of heredity and how is this genetic information decoded?

Many of these questions cannot be answered by observation alone, but they can be answered through scientific investigations. Many great discoveries have been made when a scientist has been busy investigating another problem. Good scientists have acute powers of observation and enquiring minds, and they make the most of these chance opportunities, like Alexander Fleming did when he discovered penicillin.

● *You will now be able to answer Key Question 1.*

BIOLOGY IN ACTION

Observation and discovery

Scottish physician Alexander Fleming was growing cultures of *Staphylococcus* bacteria in his laboratory in the 1920s (Figure 1.1.3a). Some of the agar plates he was growing the bacteria on became contaminated with a fungus called *Penicillium notatum*. From his observation that the bacteria were unable to grow in the region around the contaminating fungus, Fleming inferred that the fungus was releasing a substance that killed the bacteria. Experiments followed that used extracts from the fungus, and when a paper disc was soaked in this extract and applied to an agar plate culture of *Staphylococcus*, a clear zone appeared around the disc (Figure 1.1.3b). The bacteria could not grow in this area, demonstrating the antibacterial properties of this substance. Fleming named it penicillin after the type of fungus producing the chemical.

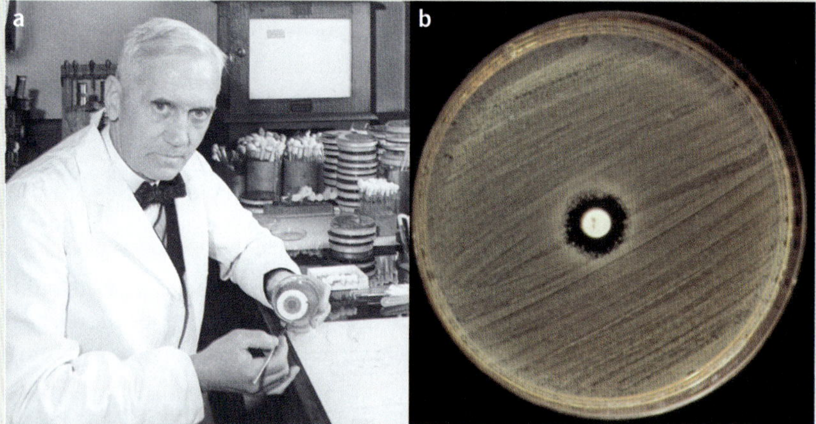

FIGURE 1.1.3 (a) Scottish biologist Alexander Fleming. (b) A culture of *Staphylococcus aureus* bacteria with a white disc containing penicillin placed at the centre. *Staphylococcus aureus* has not been able to grow near the penicillin disc.

After Fleming made the initial key observation that led to the discovery of naturally occurring antibiotics, the Australian scientist Howard Florey (then working at Oxford, England) and his colleagues further developed the methods for extracting penicillin on a large scale, and showed it was effective against staphylococcal and pneumococcal infections. Following the success of penicillin, pharmaceutical companies searched for other naturally occurring antibiotics, many of which were found in fungi (Figure 1.1.4).

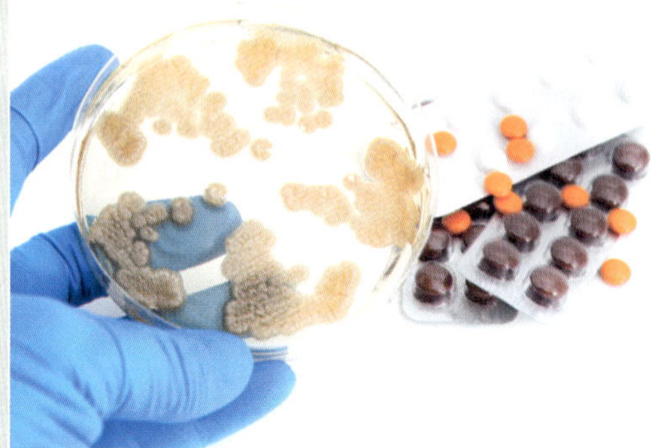

FIGURE 1.1.4 An agar plate with fungal colonies. Many naturally occurring antibiotics now used as medications were discovered by studying fungi.

LEARNING BY EXPERIMENTATION

Scientists observe, study what is already known, and then ask questions. Using their knowledge and experience, scientists suggest possible explanations for the things they observe. A **hypothesis** is a possible explanation to a research question that can be used to make predictions, which can often be tested experimentally. This is the basis of the **scientific method** (Figure 1.1.5).

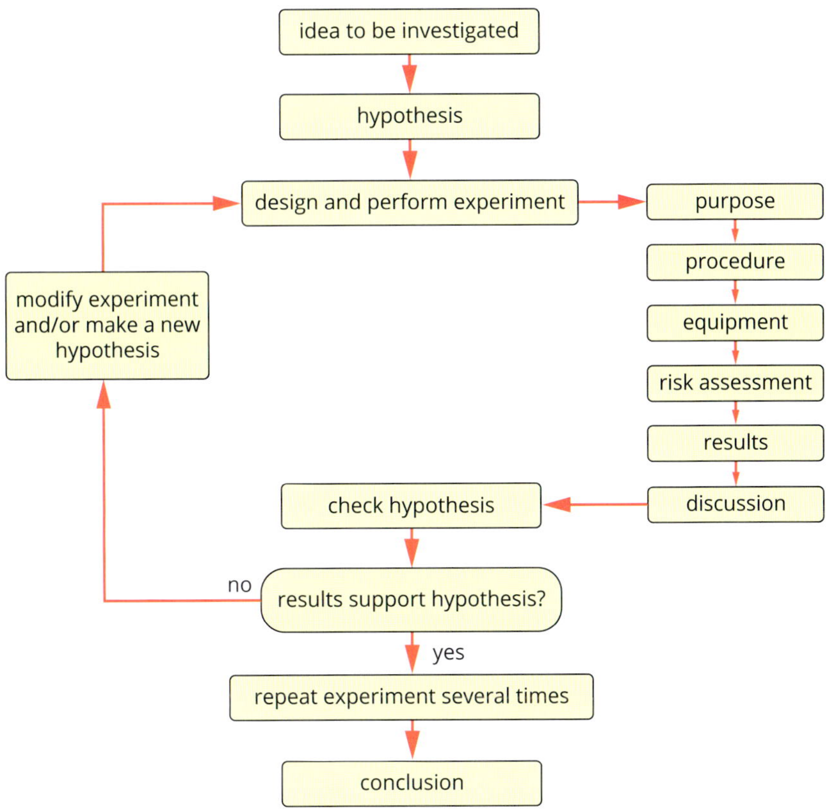

FIGURE 1.1.5 The scientific method.

Carefully designed experiments are carried out to determine whether the predictions are accurate or not. If the results of an experiment do not fall within an acceptable range, the hypothesis is rejected. If the predictions are found to be accurate, the hypothesis is supported. If, after many different experiments, one hypothesis is supported by all the results obtained so far, then this explanation can be given the status of a **theory** or **principle**.

There is nothing mysterious about the experimental approach to the study of science that is called the scientific method. You might use the same process to find out how an unfamiliar machine works if you had no instructions. Careful observation is usually the first step.

● *You will now be able to answer Key Question 2.*

ASKING THE RIGHT QUESTIONS

In science, there is little value in asking questions that cannot be answered. An experimental hypothesis must be testable, but your inability to test a particular hypothesis does not mean that the hypothesis cannot be correct.

Your ability to test a hypothesis may be limited by the resources and equipment you have available. If you ask a research question, form and test your hypothesis, and find your hypothesis is supported, that does not mean it is true in all circumstances. Likewise, if your hypothesis is not supported, that does not mean it is never true.

For example, you might hypothesise that 'Hydrogen peroxide is a toxic by-product of respiration that is broken down by catalases. As all eukaryotes undergo respiration they will all contain catalase'. However, there may be a eukaryote that lacks catalase, but testing every eukaryotic organism would be impossible, and just because a eukaryote without catalase hasn't been identified does not mean none exist.

- *You will now be able to answer Key Question 3.*

HAVING A GOOD METHOD

Methods must be described clearly and in sufficient detail to allow other scientists to repeat the experiment. If other scientists cannot obtain similar results when an experiment is repeated, then the experiment is considered unreliable. It is also important to avoid personal bias that might affect the collection of data or the analysis of results. A good scientist works hard to be objective (free of personal bias) rather than subjective (influenced by personal views). The results of an experiment must be clearly stated and must be separate from any discussion of the conclusions that are drawn from the results.

In science, doing an experiment once is not sufficient. You can have little confidence in a single result because you cannot be sure that the result was not due to some unusual circumstance that occurred at the time. The same experiment is usually repeated a number of times over a period of time and the combined results are then analysed statistically. If the statistics show that there is a low probability (usually less than 5%, referred to as $p < 0.05$) that the results could have occurred as a result of chance, then the result is accepted as being significant.

- *You will now be able to answer Key Question 4.*

THE NEED FOR EXPERIMENTAL CONTROLS

It is difficult—sometimes impossible—to eliminate all **variables** that might affect the outcome of an experiment. In biology, time of day, temperature, amount of light, humidity, and unidentified infections in organisms are examples of such variables. A way to eliminate the possibility that random factors affect results is to set up a second group within the experiment (called a **control group**) that is identical in every way to the first group (the **experimental group**) except for the single experimental variable that is being tested. This is a controlled experiment, because it allows you to examine one variable at a time. Controlled experiments are an important way of testing your hypothesis.

> The experimental conditions of the control group are identical to those of the experimental group, except that the variable of interest (the independent variable) is also kept constant.

The variable that the experimenter is testing is the **independent variable**.

The **dependent variable** is what is measured when the independent variable changes. All of the other factors that could vary but must be kept the same in all experimental groups are called **controlled variables**.

- *You will now be able to answer Key Question 5.*

> In an experiment, controlled (fixed) variables are kept constant; only one variable (the independent variable) is changed, and the dependent variable is measured to determine any effect of the change. Experiments and their results must be able to be repeated by other scientists to be validated.

When investigating antibacterial activity of compounds extracted from fungi or other sources, the variables to consider include the source, purity and concentration of the extract, the composition and consistency of the agar plates, the type of bacteria tested, the amount of substance on the test disc, the thickness of the discs and the incubation temperature. The independent variable would be the extract being tested. The dependent variable would be the presence and size of the zone of inhibition around the disc. The other variables listed above all need to be controlled. In Section 1.4 you will learn about setting up an investigation with controls.

MAKING VALID CONCLUSIONS

Conclusions are based on results and other knowledge. Making valid conclusions depends on the reliability of results and whether they are correctly interpreted. Speculation involves going beyond the results to make suggestions about what might be occurring. Conclusions are necessary, but speculation is interesting and thought-provoking. Both concluding and speculating are worthwhile, but you must be careful to keep them separate. It is also the usual practice of scientists to accept the simplest hypothesis that accounts for all the evidence available.

The conclusion made by Fleming, that *Penicillium notatum* produced a substance that can kill bacteria, was valid. It has been repeated many times and the principle generalised to the search for other antibiotics in a range of fungi and other organisms including bacteria and plants.

Experiments and their results must be able to be repeated by other scientists to be validated.

BIOFILE

Detecting antibiotic resistance

The conclusion made by Fleming was valid and led to the development of standard operating procedures for detecting antibiotic sensitivity and resistance in bacteria. If a bacterium is susceptible to an antibiotic, its growth around a disc containing the antibiotic is inhibited and observed as a clear zone on the agar plate, called the zone of inhibition. The greater the zone of inhibition, the more sensitive the bacteria are to the antibiotic. If a bacterium has developed resistance to an antibiotic it can grow around its antibiotic disc. Sometimes there is still a small zone of inhibition, but the bacteria are not sensitive enough for the antibiotic to be effective, so they are still considered to be resistant (Figure 1.1.6). The spread of antibiotic resistance by gene transfer between different species of bacteria is an important healthcare problem today. In 2015, the World Health Organization endorsed a global action plan to tackle antimicrobial resistance.

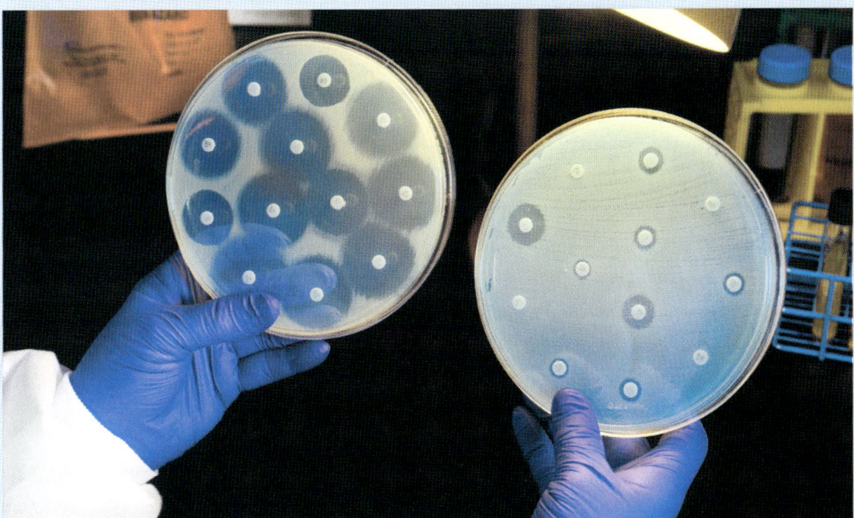

FIGURE 1.1.6 A microbiologist holding two Petri dish culture plates growing bacteria in the presence of discs containing various antibiotics. In the left plate bacteria are not growing around the discs because they are susceptible to the antibiotics on the discs. The plate on the right was inoculated with a carbapenem-resistant Enterobacteriaceae (CRE) bacterium that proved to be resistant to, and therefore able to grow well around, all of the antibiotics tested. Photographed at the Centers for Disease Control (CDC), USA.

LIMITATIONS OF THE SCIENTIFIC METHOD

The scientific method is not perfect; however, it remains the best way to understand our surroundings, and to constantly improve on that understanding. Even when the scientific method is strictly adhered to, there is still an element of chance in scientific discovery.

The scientific method can be applied only to hypotheses that can be tested, and to questions that can be answered. A hypothesis that is not testable can be neither supported nor disproved by the scientific method. Such hypotheses therefore remain as possible explanations. For example, Fleming's observation led to the hypothesis that certain fungi can produce chemicals that inhibit the growth of certain bacteria. This was testable for *Penicillium* and other fungi that can be grown on agar plates in the laboratory. If the hypothesis was broadened to 'All fungi produce antibiotics', this might not be testable, as it depends on being able to grow all fungi and all potential bacterial targets in the laboratory to test this hypothesis.

It is also important to understand that although science can prove a particular hypothesis wrong, it cannot prove that hypothesis to be true in all circumstances—only under the conditions that have been tested.

The scientific method cannot be used to test morality or ethics. These judgements belong to the fields of philosophy, history, politics and law. Science can, however, provide valuable information that people can take into account when making these judgements. For example, science can be used to predict the environmental consequences of pollution and the medical consequences of chemical weapons, but it cannot itself make value or moral judgements about either.

EXPERIMENTATION

Once you have a testable hypothesis, you are ready to conduct an experiment to test it. Every experiment has to be designed and planned carefully. You need to be sure that someone else can repeat your experiment exactly the way you did it and get similar results. In Section 1.2 you will learn how to formulate your hypothesis and design an experiment to test it.

● *You will now be able to answer Key Questions 6 and 7.*

MODELS

Scientific models are used to create and test theories and explain concepts. They may also be developed as prototypes for functional devices such as replacement organs. The introduction of computer technology, including two- and three-dimensional animations, has helped to create more detailed and realistic representations of biological processes. Different types of models can be used, but each model has limitations on the type of information it can provide.

Modelling concepts

Models are created to answer specific questions or demonstrate specific processes. How a model is designed will depend on its purpose. The two most familiar types of models are visual models and physical models, but mathematical models and computational models are also common and increasingly important in the biosciences. Models help to make sense of ideas by visualising:

- objects that are difficult to see because of their size (too big or too small) or position, such as ecosystems, organs such as the heart and pancreas, cells, molecules and atoms
- processes that cannot easily be seen directly, such as digestion, feedback loops, biochemical reactions, gene expression and protein folding
- abstract ideas such as energy transfer and the particulate nature of matter
- complex processes such as networks of biochemical reactions, genome organisation and regulation, evolution, and brain connectivity and function.

For example, models of all the connections between neurons in the human brain have been constructed from brain scanning technology. The models are used to predict and test signalling and communication between neurons (Figure 1.1.7).

Using digital modelling software to develop physical or mathematical models has enhanced our understanding in many areas. For example, dissection and surgical simulations can replace the practice of dissecting living organisms. As another example, computational modelling enables scientists to handle the huge amounts of data, such as that generated by genome sequencing.

A deeper understanding of concepts can be developed through models. However, you need to identify the benefits and limitations of using a particular model to represent a concept. Furthermore, the quality and validity of a model is limited by the depth and accuracy of the information used to construct the model.

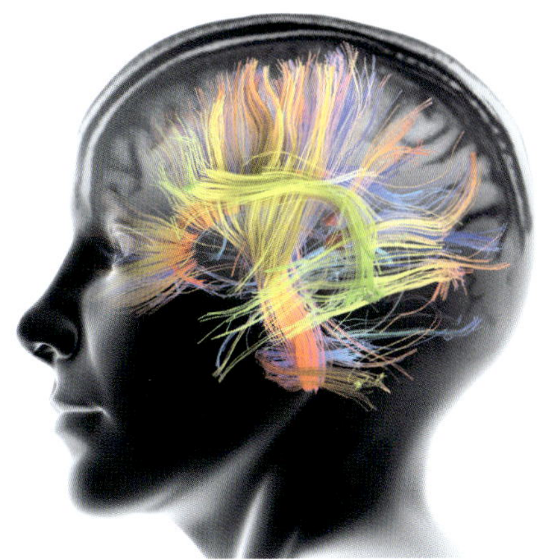

FIGURE 1.1.7 A model of the brain's wiring pattern explored in the Human Connectome Project.

Visual models

Visual models are used to represent concepts. Diagrams and flow charts are examples of visual models (Figure 1.1.8). Computer animations of these structures and processes can give a more dynamic view.

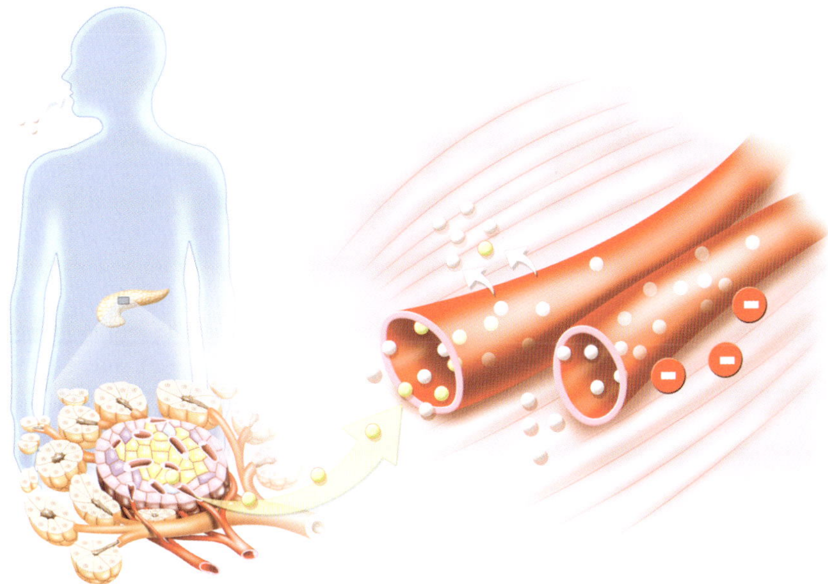

FIGURE 1.1.8 A diagram of a human body with a pancreas, showing aspects of internal structure and an indication of the organ's function, is an example of a visual model. This visual model of the human pancreas has multiple levels of detail. It illustrates the location in the body, the external structure of the organ and its internal cellular architecture, including the islets of Langerhans. It also represents functional elements, in this case release of the hormone insulin from islet beta cells into the blood. Insulin is a signalling molecule that promotes cellular glucose uptake and metabolism.

Physical models

Physical models can be scaled-up or scaled-down three-dimensional versions of reality. You have probably already used physical models many times in the classroom without being aware of it. The human skeleton is a physical model often seen in classrooms.

Although models help us to understand concepts, they are limited in how well they can represent what they are modelling. For example, although a plastic model of a lung does inflate and deflate, it does not take in oxygen and release carbon dioxide, and it is hard and solid instead of soft and flexible.

When making physical models (Figure 1.1.9), it is important to consider what materials are used to represent reality, so that the model has fewer limitations. The materials you use to construct your model should relate to what you are modelling.

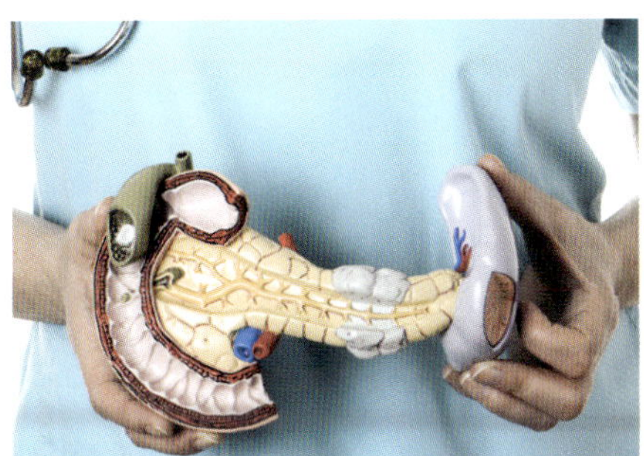

FIGURE 1.1.9 A physical model of a pancreas used in a clinical setting to help explain to a patient what is happening in that part of their body.

Computer and mathematical models

The complexity of biological systems has led to the use of mathematical modelling and computer simulations for testing hypotheses and conducting virtual experiments.

Computer simulations and mathematical models are being developed to model the complexity of whole cells, systems, organs or organisms, and allow virtual experimentation. Examples include:

- the bacterium *Mycoplasma genitalium* virtual cell
- connections between cells, for example all of the neural connections in the brain referred to as the connectome
- whole organs (virtual liver and heart)
- whole organisms such as the nematode worm *Caenorhabditis elegans*
- mathematical modelling of the way in which immune cells attack other cells
- gene interactions using data from the human genome project
- relationships between genotype and phenotype, using gene and protein sequence databases
- protein structure and function, using protein sequence databases and three-dimensional molecular modelling.

Bionic models

Physical models are often used as a prototype for developing replacement organs, such as prosthetic limbs. However, the complexity of biological systems limits the capacity of physical models to replace non-functioning organs. Research to make functional models focuses on single functions. Combining physical concepts with computer modelling of biochemical and physiological processes enables the development of models that mimic biological function.

For example, the pancreas is a complex organ with many different specialised cells and functions. Among these, it is the endocrine cells that detect blood glucose levels and release hormones to control blood glucose that are of interest when modelling diabetes. There is a bionic pancreas in development for the treatment of type 1 diabetes, which occurs when insulin-releasing beta cells are damaged. The bionic pancreas uses a glucose sensor to monitor blood glucose and a computer-controlled algorithm to direct the amount of insulin to be delivered by an insulin pump (Figure 1.1.10). Years of research and development are needed to gain enough understanding of the biological processes to develop such devices.

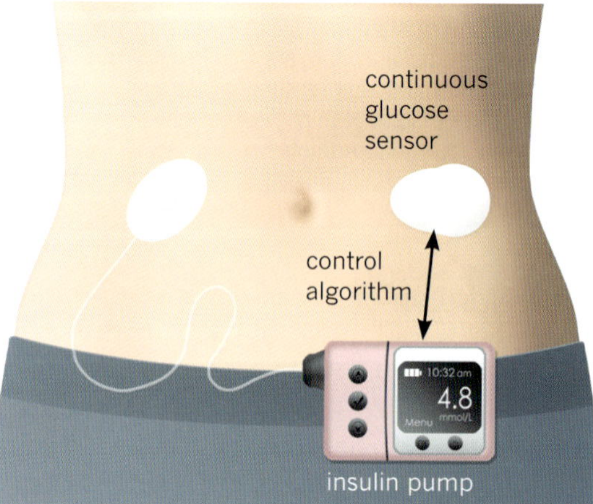

FIGURE 1.1.10 The components of a bionic pancreas developed for people with type 1 diabetes. A glucose sensor samples blood to measure blood glucose concentration. Computations determine whether a signal is sent to the insulin pump, which releases insulin when needed to maintain blood glucose in the normal range.

Model organisms

Biologists use live bacteria, animals and plants as model organisms for the investigation of cells and systems *in situ* and *in vivo*. It is possible to test hypotheses in animals that cannot be tested in humans for ethical reasons. Most of the advances in understanding animal and plant biology, genetics, pathology and medicine result from the use of model organisms. These organisms include the bacterium *Escherichia coli*, the nematode *Caenorhabditis elegans* (Figure 1.1.11), rats and mice, the plant *Arabidopsis thaliana*, and the vinegar fly *Drosophila melanogaster*.

Studies that are *in situ* are 'in position' or 'in place', such as when studying cells functioning within an intact organ, or molecules in their normal cellular location.

Studies that are *in vivo* are 'within the living', such as when cells are studied in a living organism.

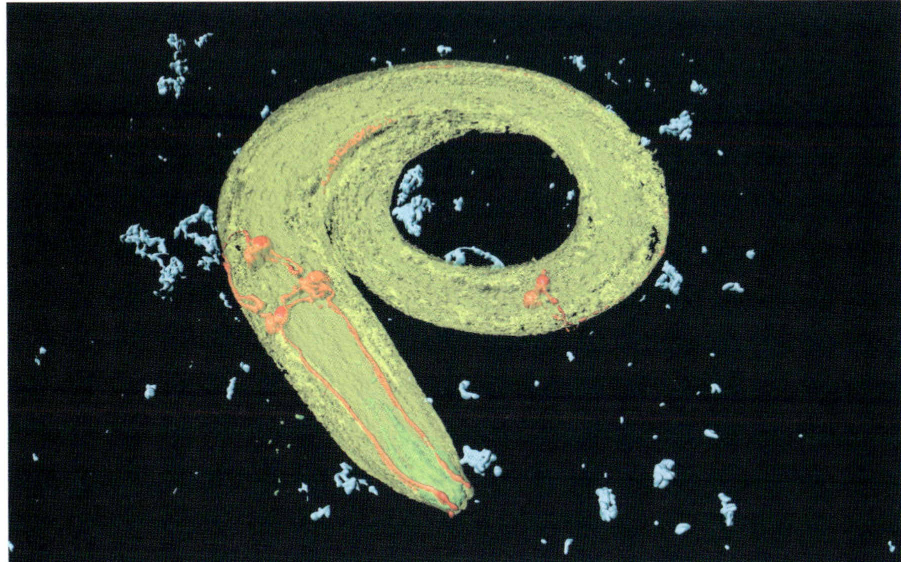

FIGURE 1.1.11 Model organism *Caenorhabditis elegans* worm. Confocal laser scanning micrograph of *C. elegans* with neurons stained green and the digestive tract stained red. *C. elegans* is a soil-dwelling nematode worm about 1 mm long and one of the most studied animals in biological and genetic research. A great deal is known about this organism, including its full genome, details of its life cycle, and the exact number of neurons in its nervous system (302) and how they form the nervous system.

Studies that are *in vitro* are 'in glass' or in a dish or test tube, such as when cells are removed from the organism and studied in a culture dish (it doesn't have to be glass).

Efforts are being made to reduce the number of animals used in research, and strict ethical guidelines must be followed in their use. Studies performed *in vitro*, and advances in computer simulation and 'virtual' cells and organisms that have made *in silico* studies possible, allow for a reduced reliance on live animals. But keep in mind that the value and validity of a virtual model or simulation is only as good as the data and information used to construct the model. This ultimately comes from living cells and organisms.

Studies that are *in silico* are 'in silicon', which refers to the silicon chips used in computers for computer simulations.

- *You will now be able to answer Key Questions 8–12.*

1.1 Review

SUMMARY

- Well-designed experiments are based on a sound knowledge of what is already understood or 'known' and careful observation.
- The scientific method is an accepted procedure for conducting experiments.
- A hypothesis is a possible explanation for a set of observations that can be used to make predictions, which can then be tested experimentally.
- Controlled experiments allow us to examine one factor at a time; they are the major means of testing hypotheses.
- Science can prove that a particular hypothesis is wrong, but it cannot prove it to be true in all circumstances.
- Science cannot be used to evaluate hypotheses that are not testable, nor can it make value or moral judgements.
- Models are useful tools that can be created and used to assist in a deeper understanding of concepts.

KEY QUESTIONS

1 The scientific method is a multistep process. Which two of the following are important parts of the method?

- **A** observations made by eye and with instrumentation
- **B** subjective decisions based on data collected
- **C** careful manipulation of results to fit your ideas
- **D** the use of prior knowledge to help objectively interpret new data

2 The following steps of the scientific method are out of order. Place a number (1–7) to the left of each point to indicate the correct sequence.

	Form a hypothesis
	Collect results
	Plan experiment and equipment
	Draw conclusions
	Question whether results support hypothesis
	State the biological question to be investigated
	Perform experiment

3 Scientists make observations and ask questions from which a testable hypothesis is formed.

- **a** Define 'hypothesis'.
- **b** Three statements are given below. One is a theory, one is a hypothesis and one is an observation. Identify which is which.
  - **i** If skin cells are exposed to UV light, cells will be damaged.
  - **ii** The skin burned when exposed to UV light.
  - **iii** Skin is formed from units called cells.

4 **a** What do 'objective' and 'subjective' mean?

b Why must experiments be carried out objectively?

5 Define 'independent', 'controlled' and 'dependent' variables.

6 **a** Explain what is meant by the term 'controlled experiment'.

b A student conducted an experiment to find out whether a bacterial species could use sucrose (cane sugar) as an energy source for growth. She already knew that these bacteria could use glucose for energy. Three components of the experiment are listed. Next to each one, indicate the type of variable described.

- **i** presence or absence of sucrose
- **ii** measurement of cell density after 24 h
- **iii** incubation temperature, volume of culture, size of flask

7 A scientist carries out a set of experiments, analyses the results and publishes them in a scientific journal. Other scientists in different laboratories repeat the experiment, but do not get the same results as the original scientist. Suggest several possible reasons that could explain this.

8 Explain what the visual model below represents.

9 The following diagram illustrates a body function involving a feedback loop. Describe what the model shows, and discuss the benefits and limitations of this diagram as a visual model of biological feedback.

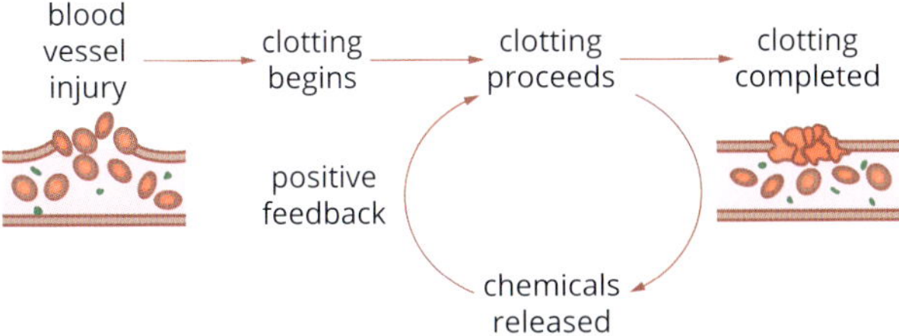

10 Below is a molecular model of the enzyme catalase, which converts hydrogen peroxide to water and oxygen. Suggest reasons why scientists construct molecular models in addition to simple diagrams or a written description of its molecular composition.

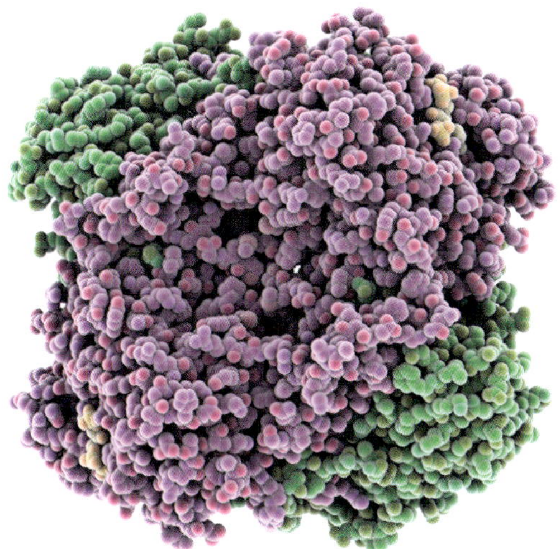

11 Suggest some limitations of using models. Include examples.

12 Discuss how computer modelling could assist in representing scientific concepts and advancing scientific knowledge.

1.2 Planning investigations

FIGURE 1.2.1 Scientists collecting grape vine samples for genetic research on the geographical origins of vines in the Mediterranean Basin.

First-hand investigations are those for which you gather the raw data yourself. These often take the form of experiments, activities, field trips or surveys (Figure 1.2.1). There are many elements to this type of practical investigation. A step-by-step approach will help you through the process and assist you in completing a solid and worthwhile investigation.

Taking the time to carefully plan and design an investigation before you begin will help you maintain a clear and concise focus throughout. Preparation is essential. In this section you will learn about some of the key steps to take when planning investigations:

- choosing a topic
- defining key terms
- sourcing information
- obtaining ethics approval
- ensuring occupational health and safety
- writing a protocol and schedule.

CHOOSING A TOPIC

Throughout this course you will conduct practical work (laboratory or field work) on a range of topics. For Unit 4 Area of Study 3 you are required to design and conduct a practical investigation related to cellular processes and/or biological change and continuity over time.

- 'Cellular processes' are any of the cell processes and biochemical pathways covered in Units 3 and 4, such as respiration, photosynthesis, enzyme regulation, cell signalling via hormones and neurotransmitters, immune responses and gene expression.
- 'Biological change and continuity over time' covers topics in Unit 4 such as changes in allele frequency in populations, impacts of mutation, environmental selection pressures, selective breeding and evolutionary processes.

When you choose a topic consider the following:

- Choose a research question you find interesting.
- Start with a topic about which you already have some background information, or some clues about how to perform the experiments.
- Check that your school laboratory has the resources for you to perform the experiments or investigate the topic.
- Choose a topic that can provide clear measurable data.

A number of topics that may be addressed in the course are suggested in Table 1.2.1. You will learn more about useful research techniques for topics like these in Section 1.3.

Before you start

The topics in Table 1.2.1 are only suggestions. Select your topic based on what resources are available to you. Before commencing your project, check that you have:

- the materials required to grow or culture an organism (e.g. plants, bacteria, yeast, protists or invertebrates)
- equipment such as microscopes, pH meters, spectrophotometers, centrifuges, and data loggers
- reagents needed to perform the experiments, such as biochemical test strips (glucose, protein), enzymes and substrates, acids and bases.

Also ensure that you:

- can order any materials needed that are not on hand
- have a solid understanding of the theory behind your investigation
- are trained to use the required equipment
- have a detailed plan for the practical components of your investigation
- are able to access the school laboratory when you need to.

Laboratory experiments may be used to investigate factors affecting cellular and/or biochemical processes.	Possible topics for laboratory investigation include: • phagocytosis or endocytosis in living cells • photosynthesis in plants, algae or cyanobacteria • respiration in plants, algae, bacteria, fungi or yeast • comparison of photosynthetic pigments by chromatography • enzyme activity in living cells or tissues, or purified enzymes • plant and animal response to infection • cell signalling mechanisms—phototaxis and chemotaxis • antibiotics—mode of action and biological effectiveness • enzymes and electrophoresis for DNA manipulation and analysis • transformation of bacteria by plasmid transfer.
Fieldwork may be used for an investigation on cellular processes or for investigating biological change and continuity over time.	Possible topics for fieldwork investigation include: • collecting samples (e.g. for photosynthetic pigment extraction) • surveying populations for phenotypes and phenotypic change • assessing impacts of selective breeding programs • investigating the role of geological change on populations and evolutionary processes.
The use of data from online databases may facilitate, or be central to, your investigation.	Possible uses of online databases include: • bioinformatics using DNA sequence data • comparison of protein structures with digital 3D protein models • global statistics on disease incidence and vaccination • species distribution • characteristics and images of hominin and other fossils • geological sites of fossil evidence.

TABLE 1.2.1 Potential areas for investigation in Units 3 and 4.

DEFINING KEY TERMS

When you begin a research investigation, you first have to develop and evaluate a research question, determine the associated variables, formulate a hypothesis and define the aims. It is important to understand that each of these can be refined as the planning of your investigation continues.

- The **research question** defines what is being investigated. For example: Is the rate of photosynthesis in plants dependent on temperature?
- The variables are the factors that change during your experiment. For example: Temperature is a variable.
- The hypothesis is a suggested outcome of the experiment based on previous knowledge, evidence or observations that attempts to answer the research question. For example: If the temperature increases from 20°C to 40°C then the rate of photosynthesis will increase.
- The **aim** is a statement describing in detail what will be investigated. For example: To investigate the effect of temperature on the rate of photosynthesis in plants at 20°C, 30°C and 40°C.

Determining your research question

Before conducting an experimental investigation you need a research question to address. Once you have come up with a topic or idea of interest, the first thing you need to do is conduct a search of the relevant literature; that is, reading scientific reports and other articles on the topic to find out what is already known, and what is not known or not yet agreed upon. The literature also gives you important information for the introduction to your report and ideas for experimental methods. Use this information to generate questions.

When you have defined the question, you are able to formulate a hypothesis, identify the measurable variables, proceed with designing your investigation and suggest a possible outcome of the experiment.

Stop to evaluate the question before you progress; it may need further refinement or even further investigation before it is suitable as a basis for an achievable and worthwhile investigation. Consider the following checklist:

- ☐ relevance—Your question must be related to your chosen topic. For your practical investigation decide whether your question will relate to cellular structure or organisation, or to structural, physiological or behavioural adaptations of an organism to an environment.
- ☐ clarity and measurability—Your question must be able to be framed as a clear hypothesis. If the question cannot be stated as a specific hypothesis, then it is going to be very difficult to complete your research.
- ☐ time frame—Make sure your question can be answered within a reasonable period of time. Ensure your question isn't too broad.
- ☐ knowledge and skills—Make sure you have a level of knowledge and a level of laboratory skills that will allow you to explore the question. Keep the question simple and achievable.
- ☐ safety and ethics—Consider the safety and ethical issues associated with the question you will be investigating. If there are issues, determine if these need to be addressed.
- ☐ advice—Seek advice from your teacher on your question. Their input may prove very useful. Their experience may lead them to consider aspects of the question that you have not thought about.

Defining your variables

The factors that can change during your experiment or investigation are called the variables. An experiment or investigation determines the relationship between variables. There are three categories of variables:

- independent—a variable that is controlled by the researcher (the one that is selected and changed)
- dependent—a variable that may change in response to a change in the independent variable, and is measured or observed
- controlled variables—the variables that are kept constant during the investigation.

You should have only one independent variable. Otherwise you could not be sure which independent variable was responsible for changes in the dependent variable.

Constructing your hypothesis

The hypothesis is an educated guess (based on evidence and prior knowledge) to answer your research question. It defines a proposed relationship between two variables. To do this, you will need to identify the dependent and independent variables.

A good hypothesis is written in terms of the dependent and independent variables:

If *x* happens, then *y* will happen. The 'if ' part of the hypothesis refers to the independent variable—the variable you alter in the experiment. The 'then' part relates to the dependent variable—the variable you measure or observe.

For example:

***If** yeast is grown in acidic conditions **then** the rate of respiration will decrease.*

A hypothesis does not need to include 'if ' and 'then' in its wording. For example, the previous hypothesis could also be stated the following way:

The rate of respiration in yeast will decrease when yeast cells are grown in acidic conditions.

A good hypothesis can be tested to determine whether it is true (verified), or false (falsified) by investigation. To be testable, your hypothesis should include variables that are measurable.

Writing a hypothesis from inference

Scientists often develop a hypothesis by **inference** (reasoning) based on preliminary observations. A **valid** inference is one that explains the observations and gives rise to testable hypotheses.

For example, stem cells are of great interest in biomedical science. Embryonic stem cells (ESCs) are pluripotent, meaning they can differentiate into a range of specialised cells. From the early days of ESC culture in 1981 scientists noticed that stem cells will only maintain their pluripotency if grown on a layer of 'feeder cells' such as fibroblasts (Figure 1.2.2).

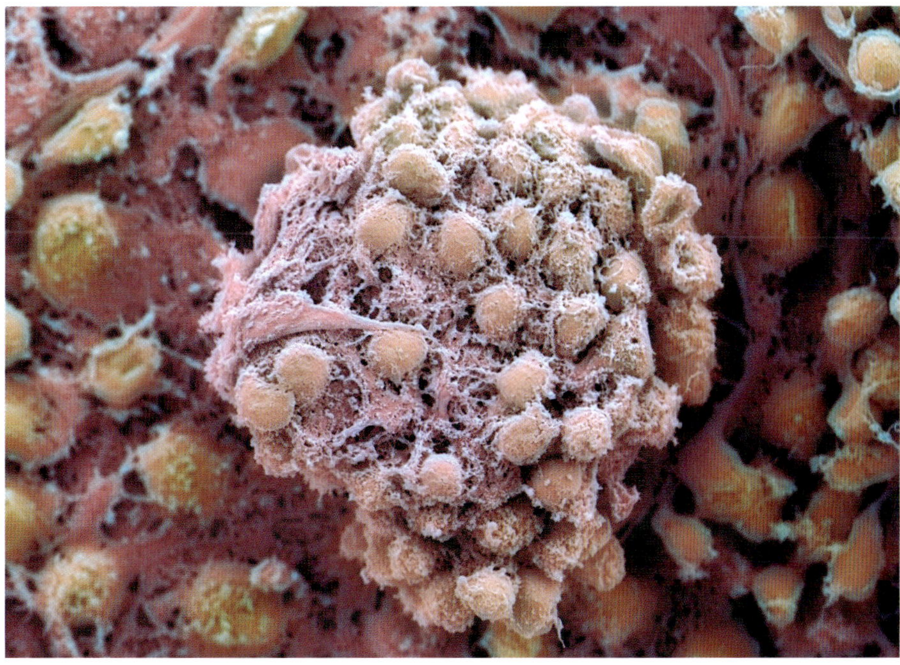

FIGURE 1.2.2 Coloured scanning electron micrograph of human embryonic stem cells that have formed themselves into a round clump. Behind this clump is the 'feeder cell layer' that provides growth signals for the embryonic cells. These cells were grown in 2005 at the Centre for Life, Newcastle Upon Tyne, UK.

Possible inferences about the role of feeder cells in maintaining pluripotent stem cells might be:

- Inference 1: They release molecules that promote cell growth and/or inhibit cell differentiation.
- Inference 2: They alter gene expression in the ESCs.
- Inference 3: Physical contact between the cells is necessary for pluripotency.

Hypotheses formed from these inferences might include:

- that purified molecules from feeder cell cultures will maintain ESC cultures
- that specific genes in ESCs will be turned on or off when grown with feeder cells.

Scientists tested these hypotheses and now know that feeder cells do release signalling molecules that keep the cells in a pluripotent state. You will learn how signalling molecules work in Chapter 6.

When you evaluate your research question, consider the variables, and think about different potential hypotheses; it helps to create a table that outlines them. For example, Table 1.2.2 outlines a research question, the variables, and a potential hypothesis that relates to the effect of glucose on the rate of respiration in yeast.

Research question	If I grow yeast cells in higher amounts of glucose, will they respire faster?
Independent variable	glucose concentration
Dependent variable	rate of respiration (measured as CO_2 release)
Controlled variables	yeast culture volume, temperature, light conditions
Potential hypothesis	The rate of respiration in yeast will increase as glucose concentration increases.

TABLE 1.2.2 Summary table of example research question, variables and potential hypothesis.

You will now be able to answer Key Questions 1 and 2.

Determining your aims

The aims are the key steps required to test your hypothesis. Each aim should directly relate to the variables in the hypothesis, describing how each will be studied or measured. The aims do not need to include the details of the method.

Example 1

- Hypothesis: If a person has a bacterial infection, then the number of neutrophils in the blood will be elevated.
- Aim: To compare microscope slides of blood samples (blood smears) from infected and healthy subjects.
- Variables: blood smears (independent) and number of neutrophils (dependent).

Example 2

- Hypothesis: If algae are exposed to low light levels, the rate of photosynthesis will decrease.
- Aim: To compare the rates of photosynthesis in algae at different distances from a light source.
- Variables: distance from light source, i.e. light intensity (independent) and rate of photosynthesis (dependent).

You will now be able to answer Key Questions 3–6.

SOURCING INFORMATION

When sourcing information during your search of the literature, researching experimental methods and investigating a broader issue, consider whether that information is from primary or secondary sources. You should also consider the advantages and disadvantages of using resources such as books or the internet.

Primary and secondary sources

Primary and secondary sources provide valuable information for research.

Sometimes the same type of resource may be classified as both a primary and a secondary source, depending on when and by whom it was written. For example, a scientist's journal article on a clinical trial of treatments for teenage obesity is a **primary source**, while a general magazine article about teenage obesity written by a journalist and referring to the scientific study is a **secondary source**. Table 1.2.3 compares primary and secondary sources.

Secondary sources of information may have a bias, so you need to determine if they are accurate, reliable and valid sources of information. You will learn about assessing the accuracy, reliability and validity of data in Section 1.4.

	Primary sources	Secondary sources
Characteristics	• first-hand records of events or experiences • written at the time the event happened • original documents	• interpretations of primary sources • written by people who did not see or experience the event • use information from original documents but rework it
Examples	• results of experiments • scientific journal/magazine articles • reports of scientific discoveries • photographs, specimens, maps and artefacts • interviews with experts • websites (if they meet the criteria above)	• textbooks • biographies • newspaper articles • magazine articles • radio and television documentaries • websites that interpret the scientific work of others • podcasts

TABLE 1.2.3 Summary of primary and secondary sources.

Using books and the internet

Peer-reviewed scientific journals are the best sources of information, but you are unlikely to have access to many of them, and much of the information is difficult to interpret if you are not an expert in the field. Good science magazines are more accessible to a non-expert, because they interpret the complex primary data and present it in a way that is easy to understand.

As books, magazines and internet searches will be your most commonly used resources for information, you should be aware of their limitations (Table 1.2.4). Reputable science magazines you might find in your school library include *New Scientist*, *Cosmos*, *Scientific American* and *The Helix* (Figure 1.2.3).

FIGURE 1.2.3 A reputable science magazine you might find in your school library.

	Book resources	Internet resources
Advantages	• written by experts • authoritative information • proofread, so information is accurate • logical, organised layout • content is relevant to the topic • contain a table of contents and index to help find relevant information	• quick and easy to access • allow access to hard-to-find information • access to the whole world; millions of websites • up-to-date information • may be interactive and use animations to enhance understanding
Disadvantages	• may not have been published recently • usable by only one person at a time	• time-consuming looking for relevant information • a lot of 'junk' sites and biased material • search engines may not display the most useful sites • cannot always tell how up-to-date information is • difficult to tell if information is accurate • hard to tell who has responsibility for authorship • information may not be well ordered • less than 10% of sites are educational

TABLE 1.2.4 Advantages and disadvantages of book and internet resources.

Searching online

Online sources include online magazines, the websites of print magazines such as those described above, the news and education sections of major journals such as *Science* magazine and *Nature-Scitable*, and podcasts and blogs (institutional, company and personal). Bioscience animations can also be found at the Walter and Eliza Hall Institute's WEHI-TV, the Tree of Life web project, and the education and resource sections of museum websites.

Often you will not be able to access journal papers without a journal subscription. 'Open access' sources are science papers and/or databases that are made available to everyone, without needing to purchase a subscription. Open access journals and databases are good sources of primary information. They include the Public Library of Science journals (e.g. PLOS Biology, PLOS Genetics), and the US National Center for Biotechnology Information (NCBI), a source of gene sequences and protein structures.

When searching for relevant information you need appropriate search terms to enter into a search engine. Here are some tips when searching online:

- Break your search statement into concepts and key words.
- Find synonyms, other related terms and concepts that apply to the topic.
- Create concepts of 1–3 words to enter into the search engine.
- Try different combinations of terms.
- Don't settle for the first sites on the list or your first attempt; look through the results for sites from science organisations and research institutions (e.g. CSIRO, WEHI, NIH; .gov, .org), universities (.edu) and science journals and magazines.

Evaluating websites

Remember that anyone can publish anything on the internet, so it is important to evaluate the credibility, currency and content of online information. To evaluate online information, follow this checklist:

- ☐ credibility—Consider who the author is, their qualifications and expertise; check for their contact information and for a trusted abbreviation in the web address, such as .gov or .edu; websites using .com may have a bias towards selling a product (but this product could be a reputable science magazine or journal), and .org sites might have a bias towards one point of view (although these sites can be a good starting points for general information).
- ☐ currency—Check the date the information you are using was last revised.
- ☐ content—Consider whether the information presented is fact or opinion; check for properly referenced sources; compare to other reputable sources, including books and science journals.

Evaluating books and journals

Your textbook should be your first source of reliable information. Other information should be consistent with this. Articles published in journals and magazines often present findings of new research, which may or may not be confirmed later, so be careful not to treat such sources of information as established fact. Scientific journals are **peer-reviewed** (critically reviewed by other specialist scientists), which gives them more credibility than other sources.

● *You will now be able to answer Key Questions 7 and 8.*

ETHICS APPROVAL

Ethics is a set of moral principles by which your actions can be judged as right or wrong. Every society or group of people has its own principles or rules of conduct. Scientists have to obtain approval from an ethics committee and follow ethical guidelines when conducting research that involves animals including, and especially, humans.

If you work with animals as part of your studies, you may need to obtain a licence. Check with your school, teacher or laboratory technician. All animal use should follow the Victorian Government's guidelines for the care and use of animals in schools. These guidelines recommend that schools consider the '3Rs rule':

- Replace the use of animals with other methods where possible.
- Reduce the number of animals used.
- Refine techniques to reduce the impact on animals.

You should treat animals with respect and care. The welfare of the animal must be the most important factor to consider when determining the use of animals in experiments. If at any time the animal being used in your experiment is distressed or injured, the experiment must stop.

● *You will now be able to answer Key Question 9.*

OCCUPATIONAL HEALTH AND SAFETY

While planning for an investigation in the laboratory or outside in the field, it is important for your safety and the safety of others that you consider the potential risks.

Everything we do has some risk involved. **Risk assessments** are performed to identify, assess and control hazards. A risk assessment should be performed for any situation, whether in the laboratory or out in the field, that could cause harm to people or animals. Always identify the risks and control them to keep everyone safe.

To identify risks, think about:

- the activity that you will be carrying out
- where in the environment you will be working (e.g. in a laboratory, the school grounds, or a natural environment)
- how you will use equipment, chemicals, organisms or parts of organisms that you will be handling
- what clothing you should wear.

The hierarchy of risk controls (Figure 1.2.4) is organised from most effective to least effective at reducing risk:

- elimination—Eliminate dangerous equipment, procedures or substances.
- substitution—Find different equipment, procedures or substances to use that will achieve the same result, but have less risk associated.
- isolation—Ensure there is a barrier between the person and the hazard. Examples include physical barriers such as guards in machines or fume hoods to work with volatile substances.
- engineering controls—Modify equipment to reduce risks.
- administrative controls—Provide guidelines, special procedures, warning signs and safe behaviours for any participants.
- personal protective equipment (PPE)—Wear safety glasses, lab coats, gloves, respirators and any other necessary safety equipment where appropriate, and provide these to other participants. As PPE can be damaged, it is considered the least effective control measure, but it remains an essential safety feature after other control measures are in place.

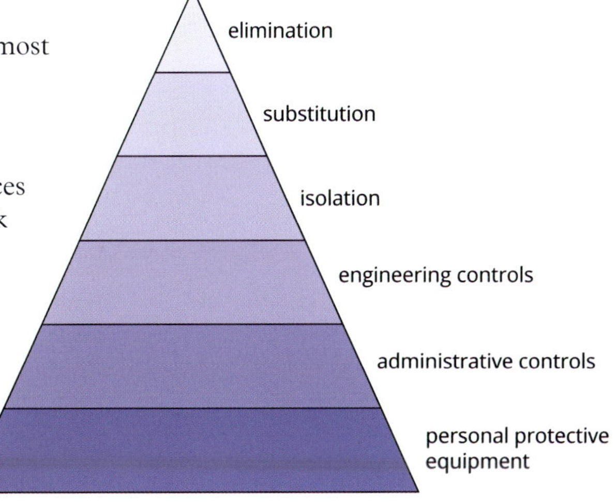

FIGURE 1.2.4 The hierarchy for hazard control is shown in this pyramid, marked from top to bottom in order of effectiveness at reducing risk.

Personal protective equipment

FIGURE 1.2.5 A lab coat, gloves and safety glasses are essential items of personal protective equipment in the laboratory.

Everyone who works in a laboratory wears clothing and equipment to improve safety (Figure 1.2.5). This is called personal protective equipment and includes:

- safety glasses
- shoes with covered tops
- disposable gloves when handling chemicals or organisms
- an apron or a lab coat to prevent spills from coming into contact with your clothes and skin
- ear protection if there is risk to your hearing.

Science outdoors

Your investigation may involve outdoor field work (Figure 1.2.6). All the potential risks, and ways to minimise them, must be considered when planning field work such as use of suitable protective clothing, knowledge of the geography, up to date maps, and checking predicted weather and fire risk.

FIGURE 1.2.6 Researchers excavating human fossils at a cave in Atapuerca, Spain. Hard hats, ropes, harnesses, strong clothing and footwear are essential during fossil research in the field.

Chemical safety

Some chemicals used in laboratories are harmful. When you are working with chemicals in the laboratory or at home, it is important you keep them away from your body. Laboratory chemicals can enter the body in three ways:

- ingestion—Chemicals that have been ingested (eaten) may be absorbed across cells lining the mouth or enter the stomach, and may then be absorbed into the bloodstream.
- inhalation—Chemicals that are breathed in (inhaled) can cross the thin cell layer of the alveoli in the lungs and enter the bloodstream.
- absorption—Some chemicals are able to pass through the skin in a process called absorption.

When working with any type of chemical you should:

- identify the chemical codes and be aware of the dangers they are warning about
- become familiar with the relevant Safety Data Sheet (SDS), formerly known as the Material Safety Data Sheets (MSDS)
- use personal protective equipment
- wipe up any spills
- wash your hands thoroughly after use.

Chemical codes

The chemicals in laboratories, supermarkets, pharmacies and hardware shops have warning symbols on their labels. These are a chemical code indicating the nature of the contents (Table 1.2.5).

Symbol	Meaning and examples	Symbol	Meaning and examples
DANGER BIOLOGICAL HAZARD	Biological hazards are living organisms that may pose a threat of infection or irritation, e.g. bacterial cultures. To dispose of, place in a biohazard bag ready for autoclaving (sterilisation at 121 °C), or soak contaminated paper towel in ethanol or bleach. Clean contaminated surfaces with 70% ethanol or bleach.	ORGANIC PEROXIDE 5.2	Organic peroxides, including hydrogen peroxide, are powerful bleaching agents that cause skin and hair to turn white. They can irritate and damage skin and eyes.
CORROSIVE	These can dissolve or eat away at substances, including tissues such as your skin or airways; examples include bleach, acids and bases, e.g. hydrochloric acid, acetic acid, sodium hydroxide, some stains used for microscopy, biochemical reagents for detecting protein and sugars.	IRRITANT	These cause discomfort, pain or itchiness; examples include urea, some microscopy stains, acetic acid.
POISON	Poisons can cause injury or death if ingested, inhaled or absorbed; examples include ninhydrin, methanol, Lugol's iodine, hydrochloric acid, formalin/formaldehyde.	FLAMMABLE LIQUID 3	Flammable liquids include alcohols such as ethanol, acetone, glacial acetic acid.

TABLE 1.2.5 Some of the different warning labels you might see on chemicals.

Safety data sheets

Every chemical substance used in a laboratory has a **safety data sheet** (SDS). This contains important information about the possible hazards in using the substance and how it should be handled and stored. An SDS states:

- the name of the hazardous substance
- the chemical and generic names of certain ingredients
- the chemical and physical properties of the hazardous substance
- health hazard information
- how to store the chemical safely
- precautions for safe use and handling
- how to dispose of the chemical safely
- the name of the manufacturer or importer, including an Australian address and telephone number.

An SDS contains important safety and first aid information for teachers and technicians about each chemical you commonly use in the laboratory.

The SDS provides employers, workers and emergency crews with the necessary information to safely manage the risk of hazardous substance exposure.

First aid

Minimising the risk of injury reduces the chance of requiring first aid assistance. However, it is still important to have someone with first aid training with you during practical investigations. Always tell your teacher or laboratory technician if an injury or accident happens.

● *You will now be able to answer Key Questions 10–13.*

WRITING A PROTOCOL AND SCHEDULE

Once you have determined your research question and aims, defined your variables, sourced information and considered the ethical and safety implications, you should write a detailed description (or protocol) of how you will conduct your experiment. You should also create a work schedule that outlines the time frame of your experiments, being sure to include sufficient time to repeat experiments if necessary. Check with your teacher that your protocol and schedule are appropriate, and that others will be able to repeat your experiment exactly by following the protocol you have written.

Test your protocol, and evaluate and modify it if necessary. You need to be able to perform your experiment independently, in the time available in the school laboratory, and with minimal support from your teachers and school laboratory staff. Quantitative results are preferable for high-quality reproducible science, so if possible use methods that enable you to count, measure or grade what you observe.

1.2 Review

SUMMARY

- A research question is a statement that broadly defines what is being investigated.
- A hypothesis:
  - is a possible outcome based on previous knowledge and evidence or observations, and addresses the research question
  - often takes the form of a proposed relationship between two or more variables in a cause and effect relationship
  - must be testable; that is, able to be supported (verified) or refuted (falsified) by investigation.
- A practical investigation determines the relationship between variables, measuring the results.
- The three types of variables are:
  - independent—a variable that is controlled by the researcher (the one that is selected and changed)
  - dependent—a variable that may change in response to a change in the independent variable, and is measured or observed
  - controlled variables—variables that are kept constant during the investigation.
- An aim is a statement describing in detail what will be investigated.
- Write a detailed protocol and schedule for your experiment before you start it.
- Ethical and safety considerations must be of the highest priority at all times during a practical investigation.

KEY QUESTIONS

1 Write each of the three inferences below as an 'if... then...' hypothesis that could be tested in an experiment.

a Fungi produce compounds that kill bacteria.

b An enzyme in stomach fluid causes meat to be digested.

c Acidic conditions are not good for respiration in eukaryotic cells.

2 Write a hypothesis for each of the following scenarios:

a A student investigating algal blooms wondered whether *Chlorella*, a unicellular eukaryotic alga, carries out photosynthesis faster than *Anabaena*, a cyanobacterium.

b A student on work placement at a dairy research station wondered whether dairy cattle with mastitis (a bacterial infection of the udder) would have more white blood cells such as neutrophils in their blood to fight the infection.

3 Which of these hypotheses is written in the correct manner? Explain why the other options are not good hypotheses.

A If light and temperature increase, the rate of photosynthesis increases.

B Respiration is affected by temperature.

C Light is related to the rate of photosynthesis.

D Light triggers a response in motile algae to move towards the light source.

4 a State the meaning of the term 'variable'.
b Copy and complete the table below with definitions of the types of variables.

Independent variable	Dependent variable	Controlled variables

5 Identify the independent, dependent and controlled variables that would be needed to investigate each of the following hypotheses:
a An increase in temperature will lead to an increase in the rate of respiration in plants.
b If a gene for salt tolerance is delivered to wheat plants, then the modified wheat will be able to grow in saline soils.
c If mice are exposed to an allergen, then histamine will be released from mast cells.
d Algae that live in shaded water have photosynthetic pigments that are different from those of algae that live in sunny water.

6 Consider the planning process for the following investigation of an enzyme that breaks down cellulose:

Aim:

Procedure:
1. Set up 6 equal-sized test tubes in a test tube rack.
2. Label 2 test tubes pH 5. Add 5 mL of pH 5 buffer to each tube.
3. Label 2 tubes pH 7. Add 5 mL of pH 7 buffer to each tube.
4. Label 2 tubes pH 9. Add 5 mL of pH 9 buffer to each tube.
5. Add 0.1 mL of cellulase enzyme solution to one tube at each pH.
6. Add 0.1 mL of the appropriate buffer (pH 5, 7 or 9) to the other tube at each pH.
7. Place the test tube rack, with all tubes, in a 37°C water bath.
8. Add 0.1 g shredded cellulose paper to each of the test tubes.
9. Incubate for 24 h.
10. Take 1 mL of solution from each tube and test for presence of glucose.

Experimental design

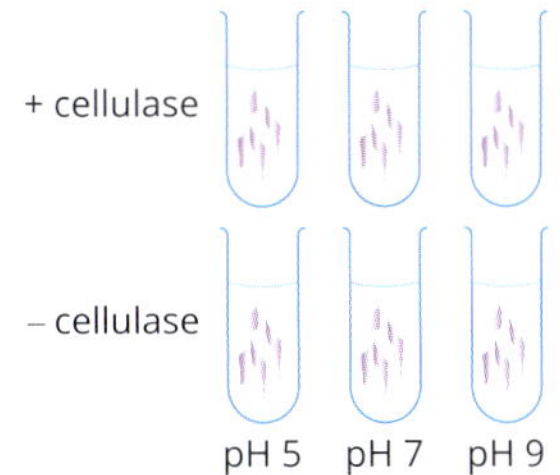

a Before conducting this experiment, what information would be researched as background for the introduction of the practical report?
b What information would you need to find out to conduct the experiment effectively?
c Identify the independent variable for the experiment.
d Identify the dependent variable for the experiment.
e List the controlled variables stated in the procedure.
f Write an aim for this experiment.
g Why was it important to use the set of test tubes without cellulase?
h Suggest improvements to the design of this experiment.

7 Decide whether each of the following is a primary or a secondary source of information.
a a newspaper article about genetically edited human embryos
b an experiment to investigate molecular changes within cells treated with hormones
c an interview with a fisheries molecular scientist about using DNA analysis for tracking tiger sharks
d a website with information about genetic engineering

8 You are learning about genetically inherited diseases and are searching for facts about cystic fibrosis. From the list below, which would be the best resource to use? Give reasons for your choice.
A the book *Cystic Fibrosis*, published in 1997
B the article 'Living with cystic fibrosis' published in the *Daily Mail* on 23 February 2008
C the website www.cysticfibrosis.org.au, accessed on 30 October 2015

9 List three things (the 3Rs rule) that should be considered in the care and use of animals in schools.

10 Complete the following table to list and describe the three ways a laboratory chemical could enter the body and how you might prevent this occurring.

Mode of entry	How the substance enters	Prevention

11 What does SDS stand for? Explain the reasons for having an SDS for each of the chemicals used in the laboratory.

12 If you spilled a chemical substance with the following label on yourself, what would be the appropriate thing to do?

13 If you spilled a live bacterial culture on the lab bench, you would use paper towel to soak up the liquid.
a Who would you consult about proper clean-up procedures?
b What personal protective equipment (PPE) would you wear during this clean up?
c What would you use to clean the bench top?

1.3 Research techniques

In this section you will learn about a number of research techniques useful in scientific research, including techniques that can be used to research cellular processes and biochemical pathways such as respiration and photosynthesis, as well as ways in which you can study biological changes and continuity over time.

MICROSCOPY

Your practical investigation may involve the study of live cells or prepared slides using microscopic methods. You may need to include cell size, number and cellular behaviour as part of your experimental evidence. You probably have access to light microscopes with magnifications up to ×400 and possibly ×1000 (oil immersion).

Field of view and size of specimens

Biological drawings should include a scale. Calculating the field of view under the microscope is required for estimating the size of specimens viewed. To calculate the field of view you use a minigrid. This is a 1 mm × 1 mm grid with a smaller microgrid of 100 µm × 100 µm in the centre (used with the ×40 objective) (Figure 1.3.1).

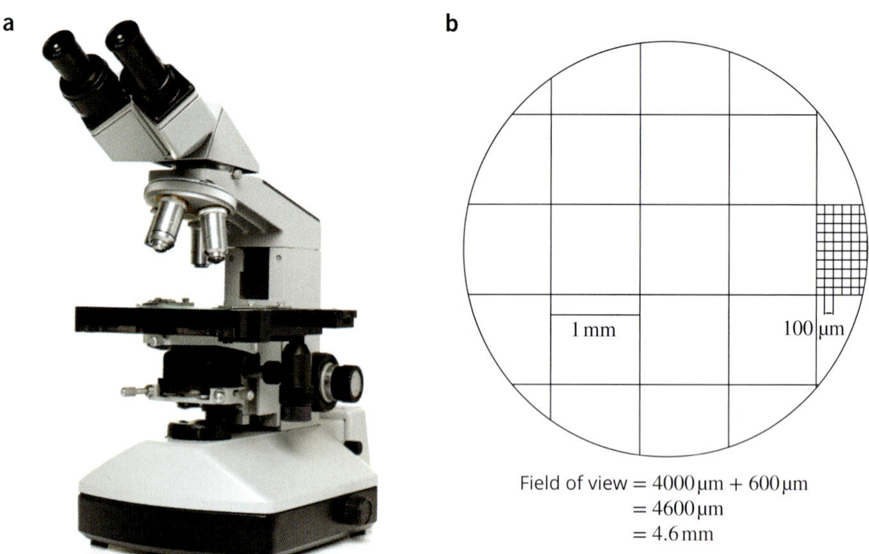

FIGURE 1.3.1 Light microscopy is used extensively in biology (a). Using a minigrid (b) allows you to measure the field of view and calculate cell size. The image on the right is a view of a minigrid at ×40 magnification. Each large square is 1 mm square, so the field of view is 4.6 mm (or 4600 µm).

Typical magnifications and fields of view in a school microscope are listed. Microscopes usually have ×10 eyepieces. The total magnification is the product of the eyepiece and objective lenses.

Objective lens	Total magnification	Field of view
×4	×40	4.5 mm
×10	×100	1.5 mm
×40	×400	450 µm
×100	×1000	150 µm

Once you have calculated your field of view for each lens, you can estimate the size of the cells. For example, you may be studying the processes of phagocytosis and lysosome action in *Amoeba* or *Paramecium* (Figure 1.3.2). If at ×400 magnification you estimate that one cell occupies half the diameter of the field of view (or two cells span the field of view) and the field of view is 450 μm, then the estimated size of each cell is 450 μm ÷ 2 cells = 225 μm.

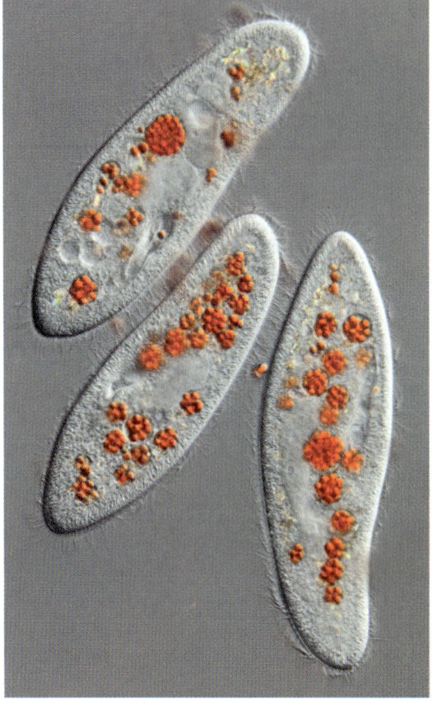

FIGURE 1.3.2 *Paramecium caudatum* viewed in the light microscope, bright field. *Paramecium* is a large unicellular ciliated protist. It consumes bacteria, small protists and algae. The process of phagocytosis and the fusion of food vacuoles with lysosomes for digestion can be studied in these live cells. In this image vacuoles have taken up yeast cells, stained red, that were engulfed by the cells.

CELL CULTURE

Cell culture is a core technique in biological sciences; unfortunately animal cell culture is restricted to laboratories with the specialised equipment and training, as well as the ethics and safety approvals. However, in your school lab you are able to grow cultures of eukaryotic cells including unicellular algae (e.g. *Chlorella*), protists (e.g. *Paramecium caudatum* and *Amoeba proteus*) and yeast (e.g. *Saccharomyces cerevisiae*, baker's yeast) (Table 1.3.1). Purchase a starter culture and culture medium to maintain a culture. Keep in mind that cells take time to grow, so plan early. You can also grow cultures of bacteria (low risk category 1) such as *Escherichia coli*, *Staphylococcus epidermidis* and *Bacillus subtilis* on agar plates or in broth cultures. Live cell cultures can be used to investigate factors affecting cellular processes that may be reflected in cell growth rates, cell responses and other cellular processes.

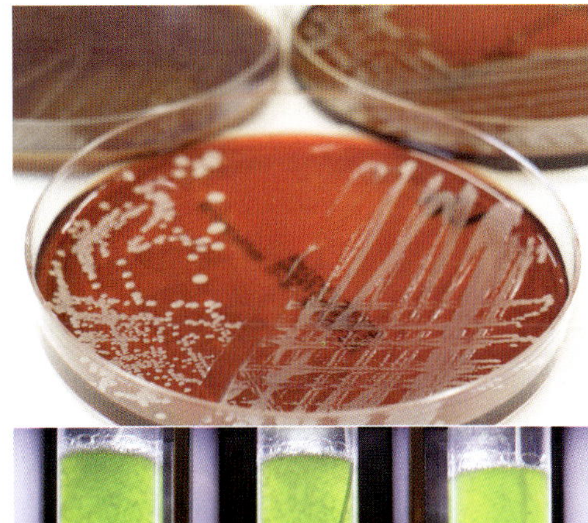	Bacteria and yeast are cultured in appropriate liquid nutrient broth or nutrient agar plates.
	Algae and protists can be grown in suitable protist medium in sterile glassware. Algae are grown in good light conditions. Protists prefer the dark.
	Plant tissue culture. Small segments of stem or leaf are surface sterilised to remove contaminants. Explants are cultured on plant nutrient agar over days or weeks.

TABLE 1.3.1 Growing cells for biology investigations.

- *You will now be able to answer Key Question 1.*

BIOLOGY IN ACTION

Growing body parts

The study of cell biology makes extensive use of cells grown in culture (Figure 1.3.3). Cells are removed from an animal or plant and grown *in vitro*, in a dish bathed in nutrient medium under sterile conditions. Cells are treated for investigations of cellular processes and analysed by methods such as immunofluorescence microscopy, and analysis of biochemical pathways and gene expression.

FIGURE 1.3.3 Cells attach to the surface of the culture dish. A medium containing nutrients such as glucose, amino acids, vitamins and proteins covers the cells. The red colour is a pH indicator.

Cell culture is essential for stem cell research and exploration of stem cells as a therapeutic tool. Stem cells differentiate, so they can be identified with fluorescent tags and sorted by a technique called flow cytometry (Figure 1.3.4).

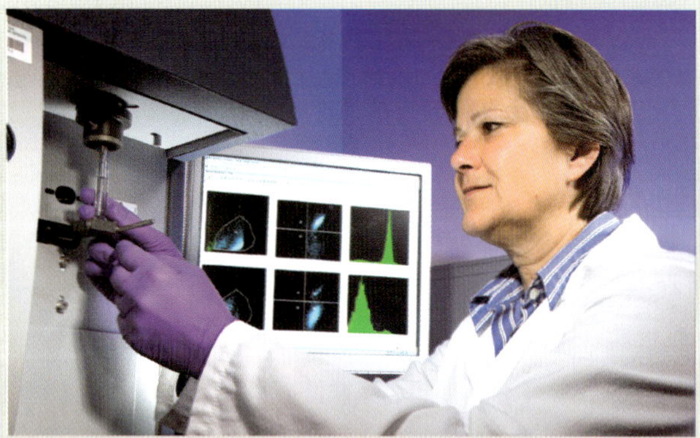

FIGURE 1.3.4 A scientist analyses cells with a flow cytometer. Cells are tagged with a fluorescent marker, sorted in the flow cytometer and visualised on a computer screen.

Replacement of cells and tissues lost through disease or injury is a challenge for medical research. Advances in cell culture methods, including stem cell technology, provide an opportunity. Tissue engineering combines cell culture and biopolymer scaffolds for the growth of specialised cells into the shape of body tissues or organs. Adult cells with the ability to replicate can be cultured directly in vitro; for example, skin cells are grown as sheets to replace skin damaged by burns. Alternatively, stem cells are cultured and differentiated into the cell types needed to reconstruct tissues. Skin, cartilage, heart valves and corneas are examples of tissues grown in the lab (Figure 1.3.5).

FIGURE 1.3.5 Corneal tissue grown in the laboratory. It was cultured from human epithelial cells that line the cornea of the eye.

Growing complex organs such as a urinary bladder has been done in the laboratory; however, achieving a functional organ in the body has so far proven challenging due to the difficulty of connecting all the blood vessels and nerves needed for a functional organ.

INVESTIGATING CELLULAR PROCESSES

Cellular respiration and photosynthesis can be detected in several ways, some of which provide qualitative results, and others provide quantitative data. A few examples of materials you might use in a laboratory to conduct an investigation are described here.

Cellular respiration

Cellular respiration may be studied in plant seedlings, yeast cultures, or in insect populations. Carbon dioxide (CO_2) is a product of respiration. CO_2 can be detected directly using a data logger with a CO_2 sensor, or indirectly by mixing the air from a growth chamber with calcium carbonate solution, commonly known as limewater (Figure 1.3.6). CO_2 dissolved in water forms carbonic acid, causing an acidic pH change, so respiration in water plants, algae and yeast cultures can be detected with a pH indicator, pH test strips or a pH meter. Yeast in broth culture or immobilised in alginate balls may be investigated for factors affecting respiration, such as temperature, nutrient concentration and inhibitors.

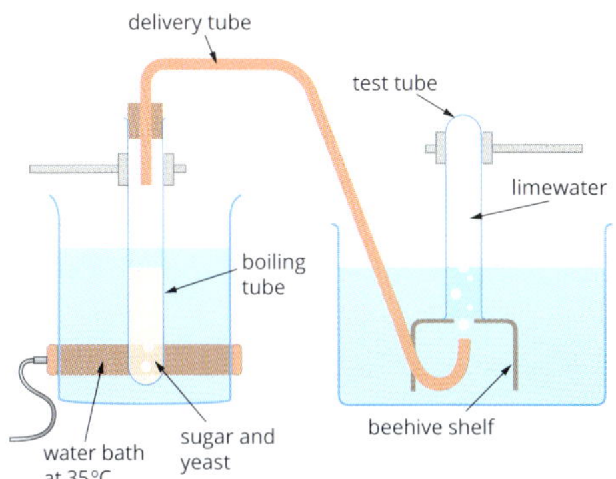

FIGURE 1.3.6 Yeast carbon dioxide test: A water bath being used to heat a stoppered test tube containing a yeast and sucrose solution (left), is connected by a delivery tube to a test tube of limewater (right). Cloudiness in limewater is a positive test for carbon dioxide, indicating fermentation in the yeast.

Photosynthesis

Photosynthesis can be studied in plants, growing seedlings, algae and cyanobacteria. In water, photosynthesis of plants and algae can be measured by oxygen production using a photosynthometer, which is a syringe connected by tubing to pond water surrounding a water plant, such as *Elodea*. Oxygen is collected and measured in the syringe (Figure 1.3.7a). Another approach is to measure the change in pH of the water as CO_2 is removed for photosynthesis, for example, by using hydrogen carbonate pH indicator (Figure 1.3.7b).

Photosynthesis can also be investigated in small leaf discs; the discs trap oxygen gas as they photosynthesise, become buoyant and float. Conditions that may affect the rate of photosynthesis, such as light intensity, temperature and chlorophyll concentration, can be investigated using these methods.

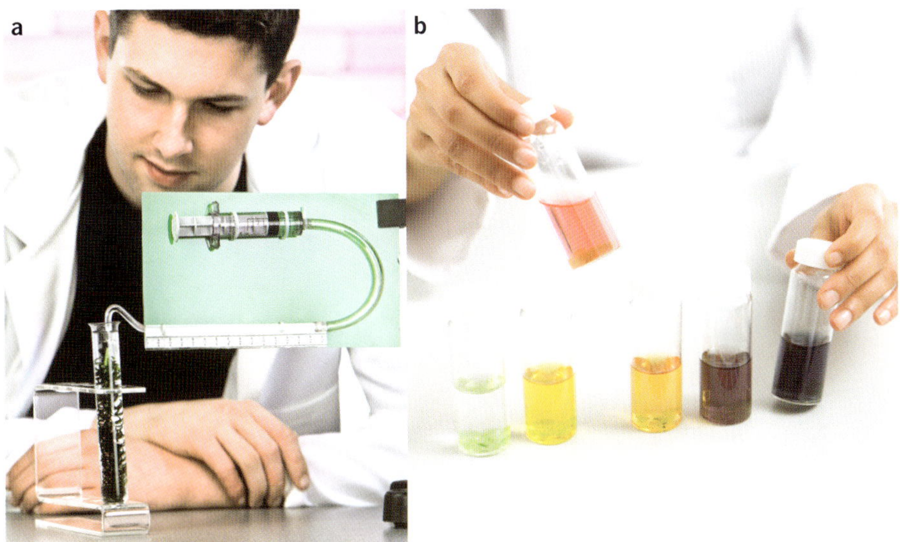

FIGURE 1.3.7 Students investigating photosynthesis by different methods: (a) Measuring oxygen produced by pond weed (in the test tube). As the pond weed photosynthesises it produces oxygen, which is collected and the volume measured in the photosynthometer (syringe connected by tubing to pond water). (b) Measuring pH change with hydrogen carbonate indicator. Algae are immobilised is small alginate balls in each tube. pH change reflects CO_2 intake for photosynthesis.

TOOLS TO SUPPORT YOUR PRACTICAL INVESTIGATIONS

Table 1.3.2 lists some tools you might use during your investigations.

Simple indicator of pH	Measuring pH or temperature	Measuring solutes
Tool: A dipstick test for the full pH range. A strip with pH-sensitive coloured pads is dipped into a solution then read against a reference colour chart after a defined time. **Purpose:** To measure the pH of a solution.	**Tool:** Electronic meters and probes. **Purpose:** To measure pH or temperature.	**Tool:** Strip tests for measuring glucose, protein and other solutes: • Multistix tests for several substances • Uriscan strips test glucose and protein • Glucostix tests glucose only **Purpose:** Usually designed for urine testing. Coloured pads on the strip are dipped into urine or other solutions; colour develops and is read against a reference chart. Detection is often based on an enzyme reaction within the pad.

Data loggers for a range of measurements	**Biochemical/chemical tests to detect molecules**	**Measuring absorbance, optical density or turbidity**
Tool: Common types of probes and capabilities in data loggers include: • pH • temperature • oxygen concentration • carbon dioxide concentration • absorption colorimeter • concentration of various compounds. **Purpose:** Data loggers enable data collection over significant time periods.	**Tools include:** **a** biuret reagent* for detecting protein (purple) **b** Benedict's reagent* for detecting reducing sugars such as glucose, maltose, fructose; not sucrose (red) **c** iodine–potassium iodide (IKI)* reagent for detecting starch (blue/purple). **Purpose:** To detect different biochemical reactions.	**Tool:** Colorimeter or spectrophotometer. **Purpose:** Used for quantitation of colour reactions, or turbidity for monitoring cell growth.
	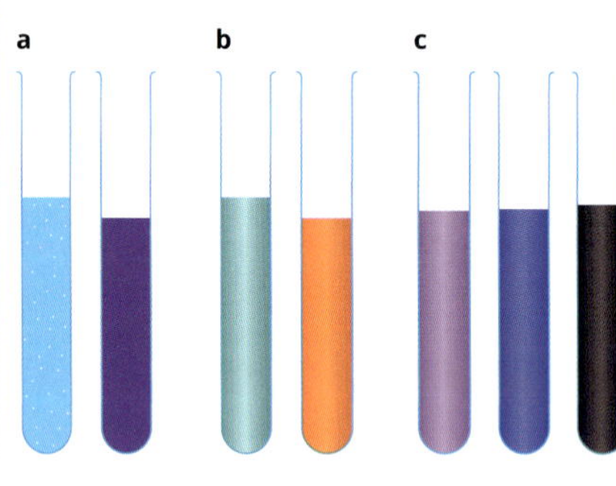 * Some tests are qualitative; quantitative or 'semi-quantitative' results may be achieved if combined with standards and absorbance readings.	

TABLE 1.3.2 Tools that may be available for practical investigations.

● *You will now be able to answer Key Questions 2 and 3.*

INVESTIGATING ENZYMATIC REACTIONS

Enzymes are biological catalysts that regulate biochemical processes in cells. Enzymes speed up the conversion of substrates to products. They are not consumed in the reaction. The general equation for enzymatic reactions is:

$$\text{substrate(s)} \xrightarrow{\text{enzyme}} \text{product(s)}$$

To measure enzymatic reactions, you can either measure the decrease in the amount of substrate or the increase in the amount of product. You will learn more about enzymes in Chapter 4.

Enzymes that could be used for your practical investigation may be present in biological material or purchased from a commercial supplier. To study the factors regulating an enzyme reaction you need an enzyme, its substrate, and a method to measure the change in amount of the substrate or product or both; Table 1.3.3 outlines some examples.

	Sources	Examples
The enzyme	present in fresh cells and tissues	• catalase in some bacteria, liver or potato • amylase in seeds
	present in commercial products	• proteases in meat tenderiser • lipases and proteases in washing powders • amylases in some wallpaper strippers
	purchased in purified form from a biochemicals supplier	• trypsin and pepsin • amylase • catalase • cellulase • restriction enzymes
The substrate	present in biological tissues	• protein in egg white or meat • starch in grain-based foods • cellulose in plant material
	purchased in purified form from a biochemicals supplier	• starch • albumin • hydrogen peroxide • DNA from plasmid or bacteriophage

TABLE 1.3.3 Useful enzymes for practical investigations.

Some reactions can be detected by visible changes in the mass of the starting material, a change in physical appearance or the production of bubbles from a gaseous product. Other reactions require a biochemical detection test that gives an obvious colour reaction, or a spectrophotometer to detect a colourless product. Colour reactions without measurement are examples of qualitative results. If you have equipment to measure absorbance of a colour reaction, such as a colorimeter, then it may be possible to quantify your results for more accurate and reliable data.

Enzymes are used extensively in DNA analysis. Restriction enzymes are a class of enzyme that recognise specific base sequences in the DNA. They cut the DNA at these sites in the DNA sequence. For a particular DNA molecule, different enzymes will produce different patterns of fragments. These enzymes have specific requirements, such as ion concentrations in the buffer. Specialised gel electrophoresis equipment is needed to visualise the results of restriction enzyme experiments.

Examples of enzymes experiments are shown in Figure 1.3.8.

FIGURE 1.3.8 (a) Some bacteria produce catalase. A bacterial colony is placed on a slide with a drop of hydrogen peroxide. If catalase is produced by the bacteria, oxygen is released and seen as bubbles. (b) Measuring the volume of oxygen released from a yeast culture after catalase enzyme is added. Catalase reacts with hydrogen peroxide, a by-product of cellular reactions. (c) Investigating the effect of pH on digestive proteases acting on muscle tissue. (d) Restriction enzymes recognise specific base sequences in the DNA and cut the DNA at this site. They are used extensively in recombinant DNA technology.

a

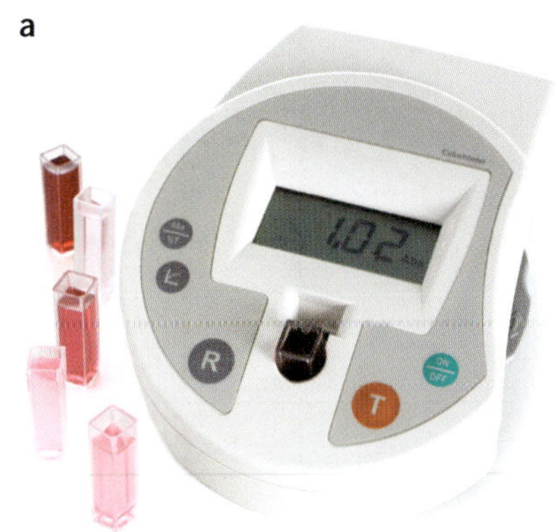

b

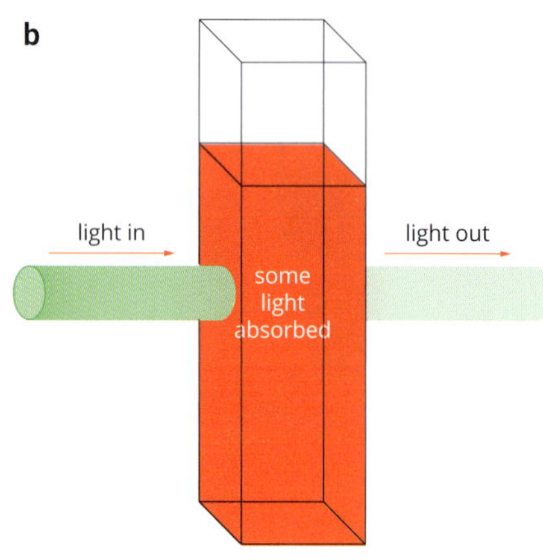

FIGURE 1.3.9 A colorimeter or spectrophotometer (a) reads absorbance of light. A sample is placed in a cuvette and placed in the instrument. Light of a particular wavelength is shone through the sample (b). The meter reads the amount of light absorbed by the sample. A sample with higher concentration gives a higher absorbance reading.

Measuring absorbance for quantifying reactions

If your school has a colorimeter or spectrophotometer, then you may be able to get quantitative results for experiments that use colour-based reactions, such as the detection of protein or starch. You can also use this instrument to measure turbidity or optical density of a bacterial or yeast broth culture when measuring their growth rates.

A sample is placed in a special tube called a cuvette, which is placed in the instrument. Light of a particular wavelength is shone through the sample, which absorbs some of the light (Figure 1.3.9). The appropriate wavelength of light to select is the one that is maximally absorbed by the sample, and this differs for each substance measured. For example, blue solutions absorb light around 600 nm, and red solutions absorb light around 490 nm (wavelength is measured in nanometres, nm, and represented by the symbol λ). The meter reads the amount of light absorbed by the sample. A sample with a high concentration of the substance will absorb more light and therefore give a higher reading.

CHROMATOGRAPHY

Chromatography methods available in your school laboratory may include paper or thin layer chromatography (Figure 1.3.10). Photosynthetic pigments vary in different organisms, such as different plants, algae and cyanobacteria. Different photosynthetic pigments have different properties of light absorbance, so are relevant to the rates of photosynthesis in different conditions.

Amino acids, the building block of proteins, can also be investigated by chromatography with the detection agent ninhydrin, which must be used safely in a fume hood.

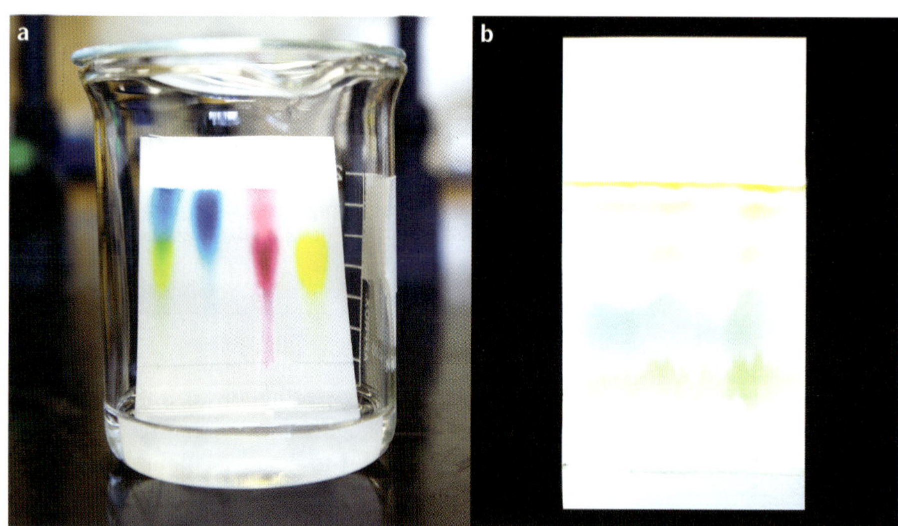

FIGURE 1.3.10 (a) Thin-layer chromatography (TLC) plate in a beaker, showing separated components (colours). TLC is performed on a sheet of glass, plastic or foil coated in a thin layer of adsorbent material. (b) An example of plant pigment molecules separated by paper chromatography. The sample is applied to the plate or paper and a solvent is drawn up the plate or paper via capillary action. Different components move up at different rates, causing them to separate.

ELECTROPHORESIS

Gel electrophoresis is a technique for separating proteins and DNA according to their size. Electrophoresis uses specialised equipment, which may not be available at your school. For DNA analysis, an agarose gel with DNA samples loaded in small wells is placed in an electrophoresis apparatus (Figure 1.3.11a). An electric current is applied, causing DNA fragments of different size to separate (Figure 1.3.11b). This method is used to investigate how different restriction enzymes cut DNA and to analyse plasmids for gene cloning. It is also used to view the products of a polymerase chain reaction (PCR), which is used for many applications, such as DNA barcoding.

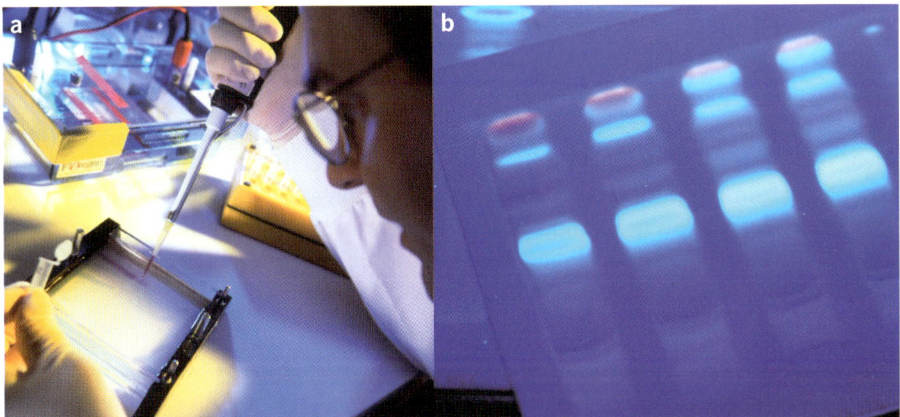

FIGURE 1.3.11 (a) A scientist loads DNA samples into the wells of an agarose gel for electrophoresis. Upper portion of photograph shows an operating electrophoresis chamber, with electrodes connected to red and black power cables. (b) DNA bands form when an electric current is applied because smaller fragments of DNA travel further than larger fragments.

IMMUNOLOGICAL INVESTIGATIONS

You may have access to prepared microscope slides of blood smears that can be used to investigate human responses to invading pathogens. (It is not safe to prepare your own blood smears, this should only be done in specialised labs or clinical settings by trained personnel.)

Different types of white blood cells (or leukocytes) in a stained blood smear can be identified under a light microscope by their size and shape (or morphology), particularly the morphology of their nucleus, as well as by the colour the cells stain (Figure 1.3.12). White blood cells are the cells of the immune system that are involved in protecting the body against both infectious disease and foreign invaders. The numbers of specific white blood cells can vary, depending on the presence of different types of infection, or in blood cancers such as leukaemia and myeloma.

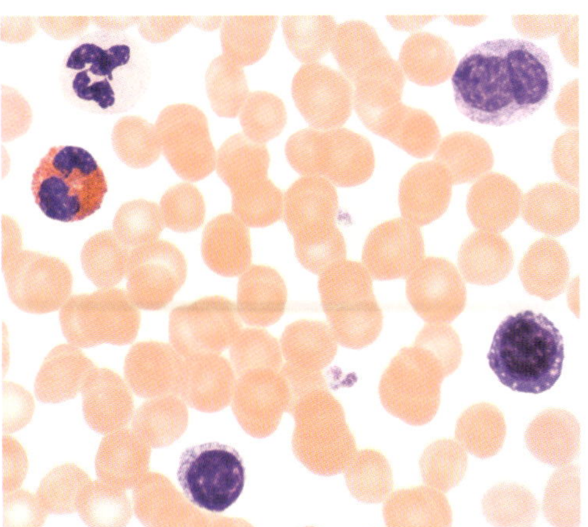

FIGURE 1.3.12 Light microscopy of a normal human blood smear showing the different types of white blood cells (neutrophil, monocyte, eosinophil, lymphocyte and basophil) and red blood cells (or erythrocytes). The mature human red blood cell is small, round, biconcave, and lacks a nucleus and organelles.

BIOFILE

Biotechnology from plant pigments

Researchers investigate the structure of the different photosynthetic pigments in plants and algae with instruments such as high performance liquid chromatography (Figure 1.3.13). Better understanding of the structure and function of photosynthetic pigments may lead to biotechnology applications such as silicon-based artificial photosynthesis systems for CO_2 capture, and genetic enhancement of photosynthetic organisms used as food sources and for pharmaceutical production.

FIGURE 1.3.13 Leaf pigment chromatography. Plant physiology researcher extracts a sample of pigments from leaf tissue (green liquid, centre right) for analysis in the high-performance liquid chromatography machine in the background. Photographed at the ARS (Agricultural Research Service) Natural Products Utilization Research Unit in Oxford, Mississippi, USA, which carries out research for the US Department of Agriculture.

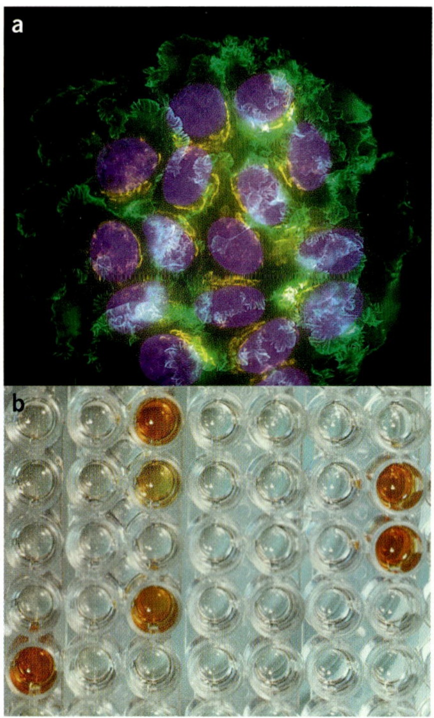

FIGURE 1.3.14 (a) Breast cancer cells (cell nuclei purple) detected by immunofluorescence staining and microscopy under UV light. Antibodies have been used to detect Golgi apparatus (yellow) and actin (green). (b) ELISA used to detect HIV in patient serum. Many samples can be tested at once in this multiwell plate. The brown colour is a positive result for HIV.

Other specialised laboratory tests you may learn about make use of proteins produced by immune cells, called antibodies, to detect other molecules. Antibodies may be tagged with fluorescent markers to detect specific cell types or processes, such as apoptosis or abnormal cell growth (Figure 1.3.14a). An enzyme-linked immunosorbent assay (or ELISA) uses antibodies tagged with an enzyme to generate a colour reaction. ELISAs are used in many laboratory and diagnostic applications, including snake venom detection kits and to screen blood for infectious agents (Figure 1.3.14b).

● *You will now be able to answer Key Questions 4 and 5.*

INVESTIGATING BIOLOGICAL CHANGE AND CONTINUITY OVER TIME

Many databases of biological information other than gene and protein sequences are available. They include databases for biochemical pathways and cellular signalling. Other open-access databases provide a large body of information for investigating the living world, biosciences and molecular biology. They include databases from museums and research institutions and include the records of specimens, fauna and flora, biodiversity and fossil collections (Table 1.3.4). They may include images, physical data and geographic distribution about samples that can be compared when investigating biological change and continuity over time (Figure 1.3.15).

Bioinformatics databases for phylogeny and palaeobiology	Type of data, information or applications
Encyclopedia of Life Tree of Life Web Project	species information, biodiversity, taxonomy, phylogeny
Museum Victoria	species data, classification, geographic distribution over time, skull image databases, biological data, fossils
Australia Museum—Learning Resources	evolution and extinction of Australian mammals; human evolution with 3D virtual skulls
American Museum of Natural History Smithsonian Museum of Natural History	research and collections with links to various resources, e.g. palaeobiology, bioinformatics
The Paleobiology Database Fossilworks	databases of fossils, geological distribution, timescales, analysis tools, construct maps

TABLE 1.3.4 Useful databases for investigating biological change and continuity over time.

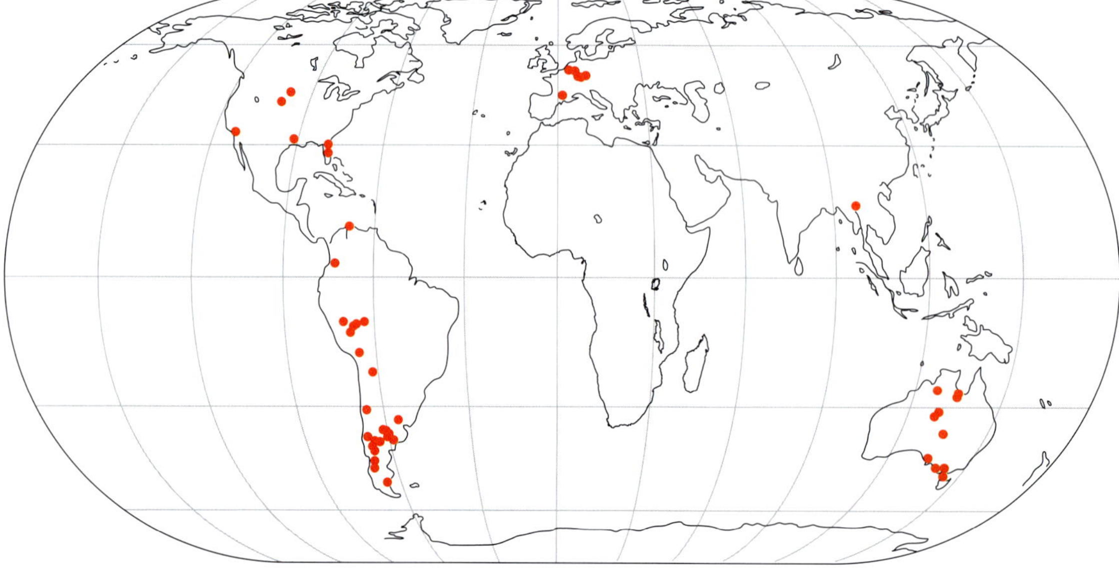

FIGURE 1.3.15 Map showing the distribution of marsupials in the Miocene geological period, constructed using a palaeontology database with search and mapping tools.

BIOLOGY IN ACTION

The rapid rise of bioinformatics

Bioinformatics is the use of mathematics, statistics and computer science to analyse and understand biological data. Bioinformatics computer programs can be used to organise raw biological data to visualise patterns, identify genes, model protein structures (Figure 1.3.16), compare DNA sequences (Figure 1.3.17), predict evolutionary relationships and discover and design drugs, along with many other applications.

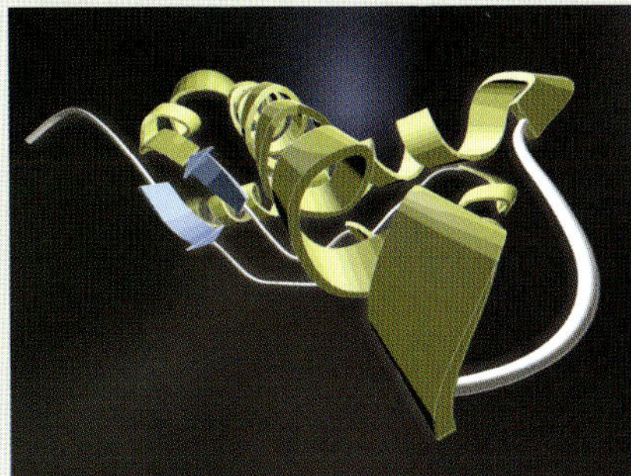

FIGURE 1.3.16 Bioinformatics tools enable you to view the secondary structure of proteins such as bovine prion protein, which misfolds and forms clumps in the brains of animals with 'mad cow disease'.

Bioinformatics is one of the fastest growing areas of biological science and is now integral to most research and development in biology. The global bioinformatics market is predicted to grow from US$4.1 billion in 2014 to US$12.5 billion by 2020. This rapid growth is primarily driven by the medical biotechnology sector, with demand for more comprehensive and efficient storage and access to personal medical data, and the development of personalised medicine.

One of the most well-known applications of bioinformatics is whole genome sequencing. The Human Genome Project is still known as the world's largest collaborative biological project. It began in 1990 and by 2003 the three billion nucleotide bases of the human genome had been sequenced. With the technology available at the time, this was an enormous undertaking, costing approximately US$3 billion dollars. With rapid advances in sequencing technology, the output of genome sequencing has skyrocketed, while the cost to sequence a genome has plummeted. It now costs just over $1000 to have your entire genome sequenced, making it affordable for many people.

Although sequencing technology is becoming more accessible, the pool of raw data for analysis is growing, requiring efficient data storage and management systems and the processing power to analyse it. The potential applications of whole genome data are vast but are limited by the bioinformatics tools, computational power and specialised knowledge of bioinformatics currently available to most biologists. As the demand for and capabilities of this technology grows, the scope of biological research is also shifting. Biologists are increasingly required to hone interdisciplinary skills in computer science, mathematics and statistics in order to keep pace with the rapid rise of bioinformatics.

A sample of online bioinformatics resources is listed in Table 1.3.5.

Centres providing bioinformatics databases	Type of data or information; applications
Biology Workbench, San Diego Supercomputer Center	search DNA and protein sequences; sequence alignment, construct evolutionary trees
US National Center for Biotechnology Information (NCBI) (GenBank) European Bioinformatics Institute (EMBL)	nucleotide (gene) sequences; protein sequences and protein structures, chromosome maps, genome maps, SNPs, epigenetics, molecular homology
NCBI–Cn3D OpenScience—Jmol	3D protein structure viewing—free downloads
Sanger Institute	bacterial, protozoan, virus and helminth (worm) genomes

TABLE 1.3.5 Bioinformatics resources.

```
B4F917.1   13 SIKLWPPSESTRIMLVDRMTNNLST..ESIFSRK..YRLLGKQEAHENAKTIEELCFALADE.....HFREEPDGDGSSAVQLYAKETSKMMLEVLK 100
A9S1V2.1   23 VFKLWPPSQGTREAVRQKMALKLSS..ACFESQS..FARIELADAQEHARAIEEVAFGAAQE......ADSGGDKTGSAVVMVYAKHASKLMLETLR 109
B9GSN7.1   13 SVKLWPPGQSTRLMLVERMTKNFIT..PSFISRK..YGLLSKEEAEEDAKKIEEVAFAAANQ.....HYEKQPDGDGSSAVQIYAKESSRLMLEVLK 100
Q8H056.1   30 SFSIWPPTQRTRDAVVRRLVDTLGG..DTILCKR..YGAVPAADAEPAARGIEAEAFDAAAA..SGEAAATASVEEGIKALQLYSKEVSRRLLDFVK 120
Q0D4Z3.2   44 SLSIWPPSQRTRDAVVRRLVQTLVA..PSILSQR..YGAVPEAEAGRAAAAVEAEAYAAVTES.SSAAAAPASVEDGIEVLQAYSKEVSRRLLELAK 135
B9MVW8.1   56 SFSIWPPTQRTRDAIISRLIETLST..TSVLSKR..YGTIPKEEASEASRRIEEEAFSGAST.......VASSEKDGLEVLQLYSKEISKRMLETVK 141
Q0IYC5.1   29 SFAVWPPTRRTRDAVVRRLVAVLSGDTTTALRKRYRYGAVPAADAERAARAVEAQAFDAASA....SSSSSSSVEDGIETLQLYSREVSNRLLAFVR 121
A9NW46.1   13 SIKLWPPSESTRLMLVERMTDNLSS..VSFFSRK..YGLLSKEEAAENAKRIEETAFLAAND.....HEAKEPNLDDSSVVQFYAREASKLMLEALK 100
Q9C500.1   57 SLRIWPPTQKTRDAVLNRLIETLST..ESILSKR..YGTLKSDDATTVAKLIEEEAYGVASN.......AVSSDDDGIKILELYSKEISKRMLESVK 142
Q2HRI7.1   25 NYSIWPPKQRTRDAVKNRLIETLST..PSVLTKR..YGTMSADEASAAAIQIEDEAFSVANA.......SSSTSNDNVTILEVYSKEISKRMIETVK 110
Q9M7N3.1   28 SFKIWPPTQRTREAVVRRLVETLTS..QSVLSKR..YGVIPEEDATSAARIIEEEAFSVASV.ASAASTGGRPEDEWIEVLHIYSQEIXQRVVESAK 119
Q9M7N6.1   25 SFSIWPPTQRTRDAVINRLIESLST..PSILSKR..YGTLPQDEASETARLIEEEAFAAAGS.......TASDADDGIEILQVYSKEISKRMIDTVK 110
Q9LE82.1   14 SVKMWPPSKSTRLMLVERMTKNITT..PSIFSRK..YGLLSVEEAEQDAKRIEDLAFATANK.....HFQNEPDGDGTSAVHVYAKESSKLMLDVIK 101
```

FIGURE 1.3.17 Bioinformatics tools allow comparison of many gene or protein sequences at once.

1.3 Review

SUMMARY

- Respiration and photosynthesis can be detected in several ways, some of which provide qualitative results, and others provide quantitative data.
- When investigating cellular respiration, you might perform a yeast carbon dioxide test to detect the presence of carbon dioxide.
- When investigating photosynthesis, you might use a photosynthometer to measure the amount of oxygen produced by a pond weed.
- Bioinformatics is the use of mathematics, statistics and computer science to analyse and understand biological data.
- Online databases such as the Encyclopaedia of Life and Fossilworks are useful for investigations of biological change and continuity over time.

KEY QUESTIONS

1 In a cell experiment you were viewing filamentous algae under the microscope. The following diagram represents what you saw in the field of view with the ×40 objective lens. The eyepiece lens on your microscope is ×10.

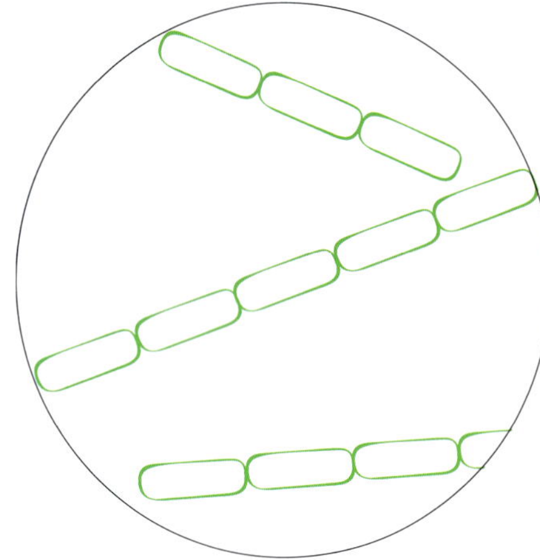

You have previously calculated that the field of view when using this lens is 450 μm.

a What is the total magnification?

b What is the size of each cell?

2 Suggest some methods you could use for detecting carbon dioxide generation during respiration in yeast, water plants or algae.

3 Suggest a method you could use for detecting photosynthesis in plants or algae.

4 Which materials or method(s) from list A–I could you use for the experiments listed in the table? Copy and complete the table by writing the letter(s) into the right-hand column.

A biochemical test
B bacterial culture
C glucose test strip
D pH meter, indicator or pH stick
E data logger—temperature probe
F plant tissue culture
G data logger—oxygen probe
H staining and microscopy
I spectrophotometer/colorimeter

	Materials or method(s)
i measure oxygen released in photosynthesis	
ii test the effectiveness of antibiotics on rate of bacterial growth	
iii quantitative measure of protein concentration in an enzymatic reaction	
iv identify phagocytosis in ciliate protozoa	
v measure glucose in an enzyme experiment	

5 The general formula for an enzymatic reaction is

$$\text{substrate(s)} \xrightarrow{\text{enzyme}} \text{product(s)}$$

a If you measured the amount of substrate, what would you expect to see if the reaction continued under the conditions tested?

b Suggest another way to measure the progress of the reaction, and the direction of change expected if the reaction occurs.

1.4 Data collection and quality

In this section you will learn about data collection, and how to identify and reduce sources of error that can affect data quality. You will also learn about the various factors that contribute to data quality, and the importance of controlled experiments in producing valid results.

FIGURE 1.4.1 A student recording the results of a photosynthesis experiment directly into a logbook.

KEEPING A LOGBOOK

Throughout Units 3 and 4, and during your practical investigation for Unit 4 Outcome 3, you must keep a logbook that includes every detail of your research (Figure 1.4.1). The following checklist will help you to remember what to include in your logbook:

- ☐ your ideas when planning the research
- ☐ clear protocols for each stage of your research experiments and observations (e.g. what standard procedures you will use and follow exactly each time)
- ☐ instructions noting exactly what needs to be recorded
- ☐ tables ready for data entry
- ☐ records of all materials, methods, experiments and raw data
- ☐ all notes, sketches, photographs and results; these should be recorded directly into your logbook, not on loose paper
- ☐ records of any incidents or errors that may influence the results

DATA COLLECTION

The measurements or observations that *you* collect during *your* investigation are your **first-hand data**. Keep in mind there are different types of data that can be collected in a scientific investigation, including **second-hand data** (data you have not collected yourself), so when planning your investigation, consider the type of data you will collect and how best to record it. Data can be raw or processed, and qualitative or quantitative.

First-hand data is data you collect yourself. Second-hand data is data that someone else has collected.

Raw and processed data

The data you record in your logbook is **raw data**. This data often needs to be **processed** or analysed before it can be presented. If an error occurs in processing the data, or you decide to present the data in a different format, you will always have the recorded raw data to refer back to.

Raw data that should be recorded includes:

- ☐ tables of results
- ☐ all observations and other notes
- ☐ diagrams and/or photographs of results.

Raw data is the data you record directly into your logbook.

Processed data is data obtained by applying a calculation or formula to raw data.

For example, you might want to study the effect of glucose concentration on respiration in yeast. To do this you might record two sets of raw data: the concentration of glucose added to each yeast culture flask and the amount of carbon dioxide produced by each culture (Table 1.4.1).

Carbon dioxide released by yeast cells in different glucose concentrations

Culture flask	Glucose concentration (g/L)	Amount of CO_2 released (ppm)	Amount of CO_2 released (ppm per 10^6 cells)
1	0.0	5	0.5
2	1.0	50	5.0
3	5.0	210	21.0
4	10.0	250	25.0

TABLE 1.4.1 An example of the kind of table that you might include in your logbook for first-hand data collection. Data tables should have a title and headings for each column and row, including units as required.

You can then process this data further. For example, to compare across different experiments in which the cell number may vary, it is useful to perform a cell count and express the result per cell (or per million cells). This value, shown in the last column in Table 1.4.1, is processed data.

● *You will now be able to answer Key Question 1.*

Qualitative data

Data collected about categorical variables is known as **qualitative data**. Categorical variables can be counted but not measured, and relate to a type or category, such as colour or gender, or states such as on/off or wet/dry. Categorical variables can be nominal or ordinal:

- **Nominal** (or unordered) variables are categorical variables in which there is no inherent order; they can be counted but not ordered. Examples are flower colour, gender, number of offspring and cattle breed.
- **Ordinal** (or ordered) variables are categorical variables in which there is an inherent order. They have a ranking or level, so they can be counted and also ordered. Examples include age group, position of a gene on a chromosome and position of a mutation in a gene sequence.

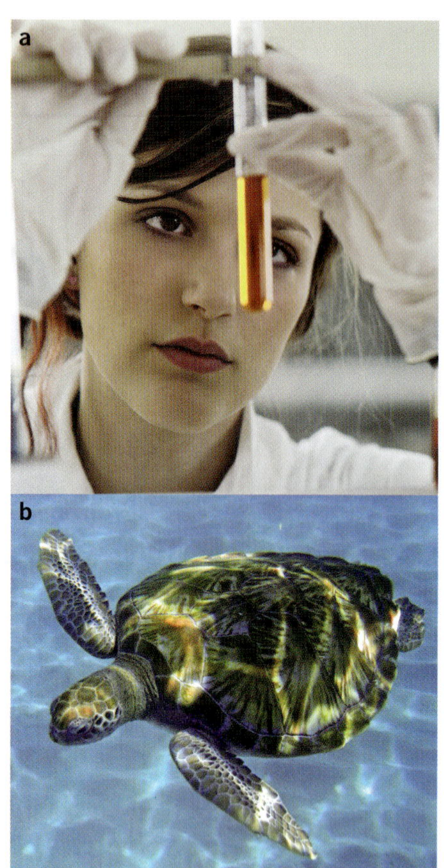

FIGURE 1.4.2 Colour-based biochemical reactions (a) and structural features (b) are examples of qualitative data. For example, different patterns on the top part of a turtle's shell (or carapace) can help distinguish one species from another. Reference keys and classification criteria are used to maintain consistency in recording qualitative data.

Recording qualitative data

Qualitative data can be represented as names, symbols or numbers. Observations of categorical variables can be descriptions or images. For example, dog breeds can be shown in a diagram, and textures of materials can be described using words such as brittle, coarse, crumbly, dense, flexible, rocky, rough, silky, slimy, smooth, spongy or velvety.

When you have to record qualitative data, think carefully about how each categorical variable will be defined. Creating a referencing system, such as assigning codes to different colours, allows you to quickly and easily record your data. For example, a photograph of reference colours with a scale (such as +++, ++, + for the colour reactions in Figure 1.4.2a) is a good way of maintaining consistency across experiments.

If you are recording details of structural features, such as when comparing variations in turtle shells (Figure 1.4.2b), make a key with diagrams to define your criteria for recording each feature. Samples may have both qualitative data, such as a particular pattern on the top of the shell (or carapace), and **quantitative data**, such as the number of clearly defined sections (or scutes) on the carapace.

Quantitative data

Data collected about numeric variables is quantitative data. Like categorical variables, numeric variables can be counted. Unlike categorical variables, numeric variables can also be measured, because they have a measurable quantity, such as length, mass or time. Numeric variables can be discrete or continuous:

- **Discrete variables** are values that can be counted or measured, but which can only have certain values. Examples are number of chromosomes in a karyotype, number of white blood cells on a slide, or the number of times a lever is pressed.
- **Continuous variables** may be any number value within a given range that can be measured. Examples are age, temperature, length, mass and wavelength.

Recording quantitative data

When you record quantitative data, remember to use scientific measuring units such as grams, centimetres, millimetres or degrees Celsius, and to use standards for quantification.

Sometimes qualitative data can become quantitative if accurate and consistent measurement is applied. For example, biochemical reactions based on a colour change, as in Figure 1.4.2a, can be prepared with known concentrations and a detailed grading system used to indicate colour intensity (such as ++, +, –; this might be considered semi-quantitative). If a colorimeter or spectrophotometer is available to read absorbance values, then you obtain quantitative data. A calibration curve or standard curve can then be prepared for reading the experimental values. You will learn about standard curves in Section 1.5.

Figure 1.4.3 summarises the different types of data and their variables.

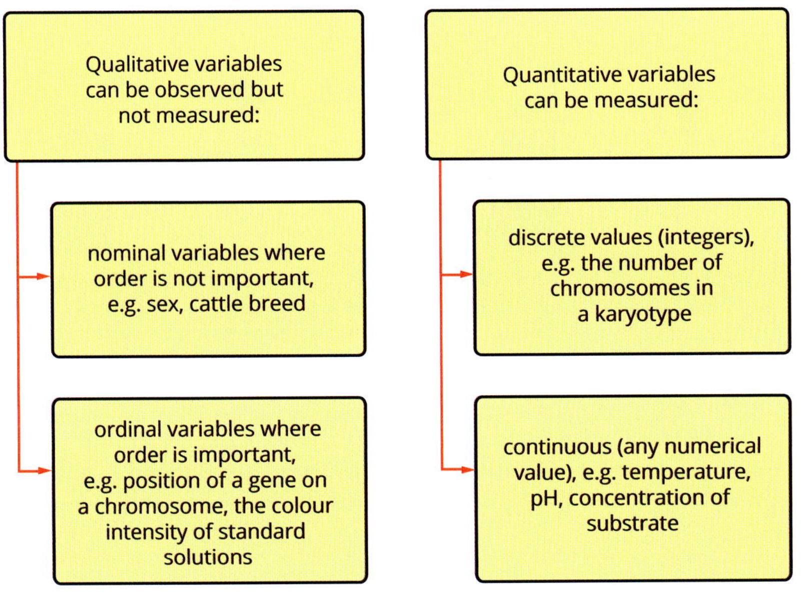

FIGURE 1.4.3 Summary of types of data and variables.

● *You will now be able to answer Key Questions 2–4.*

IDENTIFYING AND REDUCING ERRORS

When an instrument is used to measure a physical quantity and obtain a numerical value, the aim is to determine the true value. However, for a number of reasons the measured value is often not the true value. The difference between the true value and the measured value is called the **error**. This error in the measured value is the result of errors in the experiment. The two types of experimental errors are systematic errors and random errors.

Systematic errors

A systematic error (or bias) is a consistent error that occurs every time you take a measurement. Systematic errors are not easy to spot, because they do not appear as a single difference in the data set. Instead, repeated measurements give results that differ by the same amount from the true value. There are many different types of systematic errors, but the most common types are selection bias and measurement bias.

Selection bias

Selection bias occurs when your sample is not representative of the population being studied. This can have a number of different causes, including sampling bias, which is when your sample has not been selected randomly, and time-interval bias, which is when you stop your study too early because the results support your hypothesis.

> A meniscus is the curved upper surface of liquid in a tube.

Measurement bias

Measurement bias is usually a result of instruments that are faulty or not calibrated, or the incorrect use of instruments, which produces inaccurate results. For example, if a scale under-reads by 1%, a measurement of 99 mm will actually be 100 mm. Another example would be if you repeatedly used a piece of equipment incorrectly throughout your investigation, such as reading from the top of the **meniscus** instead of the bottom when using a measuring cylinder or graduated pipette (Figure 1.4.4).

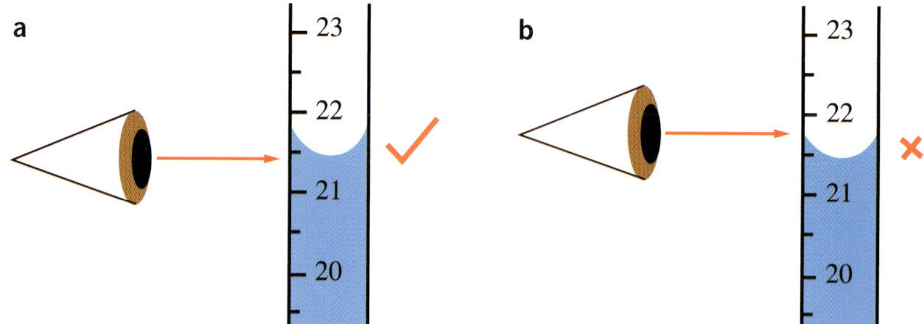

FIGURE 1.4.4 When measuring liquid levels in cylinders and pipettes, measure the value at the bottom of the meniscus of the liquid as in (a), not at the top as in (b).

Reducing systematic errors

The appropriate selection and correct use of calibrated equipment will help you reduce systematic errors. Because systematic errors are difficult to identify, it is also a good idea (if you have time) to repeat your measurements using different equipment.

Appropriate equipment

Use the equipment best suited to the data you need to collect. Determining the units and scale of the data you are collecting will help you to select the correct equipment. For example, if you need to measure 10 mL of a liquid, using a 10 mL graduated pipette or a 20 mL measuring cylinder will give more accurate readings than when using a 200 mL measuring cylinder, because the pipette or 20 mL cylinder will have a finer scale.

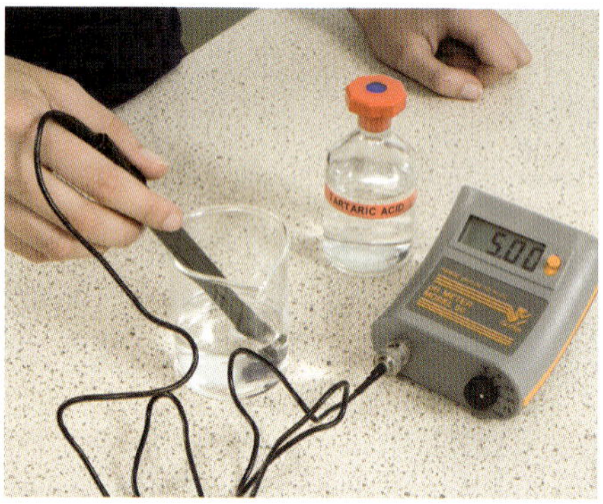

FIGURE 1.4.5 Measuring the pH level of tartaric acid with a pH meter. To ensure an accurate reading, the student would first have calibrated the meter using standard solutions of known pH.

Calibrated equipment

Accurate measurement requires properly calibrated equipment. Before you carry out your investigation, make sure your instruments or measuring devices are properly calibrated and functioning correctly (Figure 1.4.5). Your school laboratory may have a set of standard masses that can be used to calibrate a balance or scale. A pH meter should have a set of standard pH solutions (e.g. at pH 4, pH 7 and pH 9) to check the meter readings and adjust the meter if necessary.

Correct use of equipment

Use your equipment properly. Ensure you have been trained to use the equipment. Write the instructions in detail so you can follow them exactly each time, and practise using the equipment before you start your investigation. Improper use of equipment can result in inaccurate, imprecise data with large errors, and the validity of the data can be compromised. An example of incorrect use of a balance would be if it was not placed on a level surface, or it was used in a room with air currents or vibrations.

Random errors

Random errors (also called variability) are unpredictable variations that can occur with each measurement. Random errors can occur because instruments are affected by small variations in their surroundings, such as changes in temperature. All instruments have a limited precision, so the results they produce will always fall within a range of values.

Reducing random errors

To reduce random errors you need to take more measurements or increase your sample size. You can then calculate the average (the mean), which should be close to the true value.

More measurements

The impact of random errors can be minimised by taking more measurements and then calculating the average value. In general, more measurements will improve the accuracy of the measured value. The minimum number of measurements you should make is three. If one reading differs greatly from the rest, mention this in your results and discuss possible reasons for the difference. If you think it is the result of an error, do not include it in your results because it will skew (bias) the result.

Sample size

Increasing the sample size reduces the effect of random errors, which in turn makes your data more reliable. For example, if you are conducting an investigation into the effects of light intensity on the rate of photosynthesis in *Elodea*, do not test your hypothesis on just one stem. Test several stems (minimum three). If two stems photosynthesise and one does not, it is reasonable to conclude that one stem was unhealthy or the conditions incorrect. Using a large number of samples will reduce the likelihood of your results being skewed.

● *You will now be able to answer Key Questions 5 and 6.*

DATA QUALITY

The results of your data analysis will only be as good as the quality of the data. A well-designed scientific experiment should produce accurate, precise, reliable and valid results. You should consider all of these factors when collecting first-hand data in your practical investigations, and also when you assess the quality of second-hand data from other sources.

Accuracy and precision

In science and statistics the terms 'accuracy' and 'precision' have very specific and different meanings:

- **Accuracy** is the ability to obtain the correct measurement. To obtain accurate results, you must minimise systematic errors.
- **Precision** is the ability to consistently obtain the same measurement. To obtain precise results, you must minimise random errors.

To understand more clearly the difference between accuracy and precision, think about firing arrows at an archery target (Figure 1.4.6). Accuracy is being able to hit the bullseye, whereas precision is being able to hit the same spot every time you shoot. If you hit the bullseye every time you shoot, you are both accurate and precise (Figure 1.4.6a). If you hit the same area of the target every time but not the bullseye, you are precise but not accurate (Figure 1.4.6b). If you hit the area around the bullseye each time but don't always hit the bullseye, you are accurate but not precise (Figure 1.4.6c). If you hit a different part of the target every time you shoot, you are neither accurate nor precise (Figure 1.4.6d).

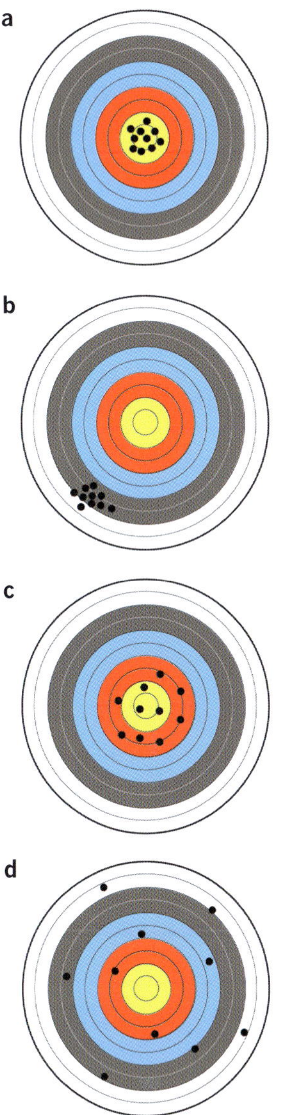

FIGURE 1.4.6 Examples of accuracy and precision: (a) both accurate and precise, (b) precise but not accurate, (c) accurate but not precise, and (d) neither accurate nor precise.

Recording numerical data

When using measuring instruments, the number of significant figures (or digits) and decimal places you use is determined by how precise your measurements are.

This depends on the scale, accuracy and precision of the instrument and technique you are using (Figure 1.4.7). For example, a beaker is used to measure volumes approximately and has limited accuracy, for example ±5%. A graduated pipette is more accurate, with accuracies of ±0.1% or ±0.2%. Your pipette may be accurate but if your technique using the pipette is variable, the overall accuracy and precision will be limited.

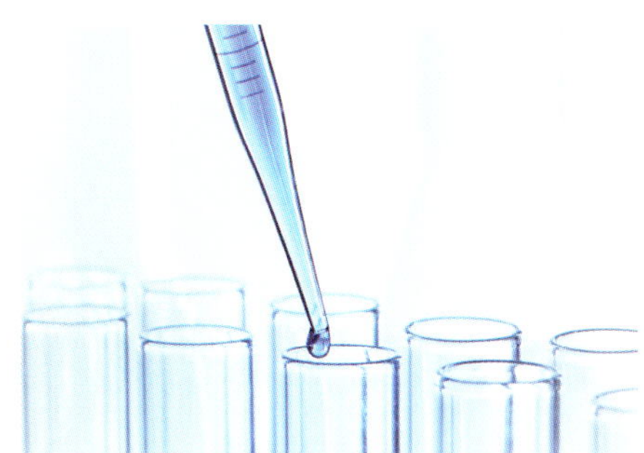

FIGURE 1.4.7 A 5 mL graduated pipette can measure volumes to an accuracy of one hundredth of a millilitre, or 5 mL ± 0.01 mL. The pipette has major divisions of 1 mL and minor divisions of 0.1 mL. You can estimate to 0.01 mL and record volumes to 2 decimal places, for example 3.80 mL or 4.52 mL.

When you record raw data and report processed data, use the number of significant figures available from your equipment or observation. Using either a greater or smaller number of significant figures can be misleading. For example, Table 1.4.2 shows measurements of 5 tissue samples taken using an electronic balance accurate to two decimal places. The data was entered into a spreadsheet to calculate the mean, which was displayed with 4 decimal places. You would record the mean as 20.83 g, not 20.8260 g, because two decimal places is the precision limit of the instrument. Recording 20.8260 g would be an example of false precision.

Sample	1	2	3	4	5	Mean
Mass (g)	20.13	20.62	21.22	20.99	21.17	20.8260

TABLE 1.4.2 An example of false precision in a data calculation.

Reliability

Reliability (sometimes called repeatability) is the ability to obtain the same results if an experiment is repeated (Figure 1.4.8). Because a single measurement or experimental result could be affected by errors, **replication** of samples within an experiment and **repeat trials** are key components of reliability. To improve reliability you should:

- specify the materials and methods in detail
- include replicate (several) samples within each experiment
- take repeat readings of each sample
- run the experiment or trial more than once.

FIGURE 1.4.8 If you can reproduce your results, they are reliable.

● *You will now be able to answer Key Questions 7 and 8.*

Validity

Validity refers to whether your results are sound and whether they can be generalised. Results are invalid, for example, if you think you have measured a variable but have actually measured something else. Factors influencing validity include:

- whether your experiment measures what it claims to measure. In other words, your experiment should test your hypothesis.

- the certainty that something observed in your experiment was the result of your experimental conditions and not some other cause that you did not consider. In other words, whether the independent variable influenced the dependent variable in the way you have concluded.
- the degree to which your findings can be generalised to the wider population from which your sample is taken, or to a different population, place or time.

Controls

To ensure your results are valid, carefully determine:

- the independent variable (the variable that you will change) and how you will change it
- the dependent variable (the variable that you will measure)
- the controlled variables (the variables that must remain constant) and how you will maintain them (see Section 1.2).

Your experiment should be designed so that only one independent variable is changed at a time. The remaining variables must be kept constant (or controlled) so that meaningful conclusions can be drawn in turn about the effect of each independent variable on the dependent variable you are measuring.

A control group is a comparison group. This means you need to set up two groups side by side within an experiment. Both groups are the same, except for the variable you are testing. This is the independent variable in the hypothesis, and is applied to your experimental group but not your control group. All the other variables have to stay the same. We do not want them to change, as these may affect the result of our experiment. For example, when testing a new medication (the variable being tested), two groups of patients are involved. The control group of patients is given a placebo (a blank capsule). The other group is given the actual medication, and the data collected from this group is compared to the data from the control group to see the effects of the medication.

Randomisation

Random selection of your sample reduces selection bias and improves validity. Selection bias occurs when your sample doesn't reflect the wider population you wish to generalise your results to. For example, if you were scoring phenotypes in large trials of genetically selected or genetically modified crop plants, a study design in which you choose locations at random throughout the plot would have better validity than one in which you choose only the edges of the plot.

USING AND EVALUATING SECOND-HAND DATA

In researching your topic for investigation you will find a range of sources of information. Not all will have valid or reliable information suitable for a scientific investigation. Determine whether the information source is reputable, such as a university, research and education organisation, or peer-reviewed journal. Other sources of information, such as interest groups or companies, may have a specific bias (as outlined in Table 1.4.3 on page 44). Current secondary school, university, and specialist area textbooks are good places to start. Sections of some specialist texts are available online. The best source of experimental detail and up to date information comes from peer-reviewed scientific journals no older than about 10 years.

As many peer-reviewed journals require a subscription, you may not have access to the original articles in full, but you can probably find the abstract (a summary of the study). Also, these original articles can be very complex and hard to interpret if you are not an expert in the field, so an alternative way to access current information about a topic is through print and online science magazines, such as *New Scientist* and *The Scientist*. Good science magazines and journalists provide the background, the results and the relevance of the study in a way that non-experts can understand. However, the methods will be in the original peer-reviewed report.

> Peer-reviewed means that other scientists have checked the information and have agreed that it is appropriate for publication.

Your investigation may use scientific data such as protein structures, DNA sequences, or fossil and biogeographic data from open-access databases. Most of these databases are linked to large research centres and are usually reliable sources of information.

DATA QUALITY SUMMARY

Table 1.4.3 summarises factors to consider when evaluating and using first-hand and second-hand data. Make sure you consider all the factors that might affect the quality of the data when you are doing your research and when you write a report of your investigation.

	First-hand data	Second-hand data
Accuracy	• Use appropriate and calibrated instruments. • Address systematic errors.	• Use reputable sources such as peer-reviewed journals and books. • Check that systematic errors have been addressed.
Precision	• Use an appropriate number of significant figures. • Address random errors.	• Check that random errors were addressed. • Check that any data analysis was appropriate.
Reliability	• Use replicates within the experiment. • Perform repeat readings. • Repeat your experiment.	• Check that the experimental method was relevant. • Were the results analysed and statistically valid? • Check that information is consistent with other reputable sources.
Validity	• Ensure your experiment tests your hypothesis. • Randomise your sample and use one or more controls.	• Check the study and information is current. • Check the information is based on scientific investigation, controlled trials or research. • Determine if the source is unbiased, or from a particular interest group, e.g. pharmaceutical company, religious group. • Check that the results relate to the hypothesis and aims.

TABLE 1.4.3 Summary of factors impacting quality of first- and second-hand data.

You will now be able to answer Key Question 9.

1.4 Review

SUMMARY

- Record all information objectively in your logbook including your data and method of investigation.
- Raw data is the data you collect in your logbook.
- Processed data is raw data that has been mathematically manipulated.
- Beware of potential errors when conducting an investigation:
  - Systematic errors are consistent errors that reduce accuracy.
  - Random errors are unpredictable errors that reduce precision.
- Reduce systematic errors by:
  - selecting appropriate equipment
  - properly calibrating equipment
  - using equipment correctly
  - repeating experiments.
- Reduce random errors by:
  - having a large sample size
  - repeating measurements.
- Accuracy is the ability to obtain a correct measurement.
- Precision is the ability to consistently obtain the same measurement.
- Reliability is the ability to reproduce your results.
- Validity refers to whether your results are sound.
- Controlled experiments are important for obtaining valid results.

KEY QUESTIONS

1 Explain the difference between raw and processed data.

2 Compare and contrast quantitative and qualitative data.

3 Identify which of the following pieces of information about plant material used in a plant hormone experiment are qualitative and which are quantitative. Place a tick in the appropriate column.

Information	Qualitative	Quantitative
leaf colour		
leaf smoothness		
length of stem		
number of leaves		
presence of roots		
change in internode length		

4 Using a Venn diagram, present the differences and similarities between discrete and continuous data.

5 Two sets of data are given below. Both sets contain errors. Identify which set is more likely to contain a systematic error and which is more likely to contain a random error.
Data set A: 11.4, 10.9, 11.8, 10.6, 1.5, 11.1
Data set B: 25, 27, 22, 26, 28, 23, 25, 27

6 What type of error is associated with:
- **a** inaccurate measurements?
- **b** imprecise measurements?

7 Describe the difference between replication and repeat trials.

8 Explain why replication and repeat trials are necessary.

9 A student conducted the following experiment on *Chlorella*, a unicellular fresh water alga. The student's research found a method to put the alga into small jelly-like alginate balls. Due to equipment limitations (only 2 oxygen sensors were available), only two sets of data could be collected at one time.
- **a** Suggest a hypothesis for this experiment.
- **b** Identify the independent variable.
- **c** Identify the dependent variable.
- **d** Identify the controlled variables.
- **e** Will the results be objective or subjective? Explain.
- **f** Suggest ways to increase the reliability and validity of this experiment.
- **g** What conclusion would you draw from this experiment?
- **h** Will the conclusion be valid for all algae? Explain.

Aim: To investigate the effect of light intensity on photosynthesis in *Chlorella*.

Hypothesis:

Materials:
- freshly prepared alginate balls with *Chlorella* (unicellular alga)
- two equal-sized tubes
- two dissolved oxygen sensors and data logger
- bright fibre optic light

Method:
1. Add 10 mL tap water to each tube.
2. Add 10 algae balls to each tube.
3. Place a probe for detecting dissolved oxygen into each tube and connect to the data logger.
4. Place tube one 30 cm from the light source. Place tube two 5 cm from the light source.
5. Turn on the light and measure the O_2 concentration over 24 h.

Experimental design

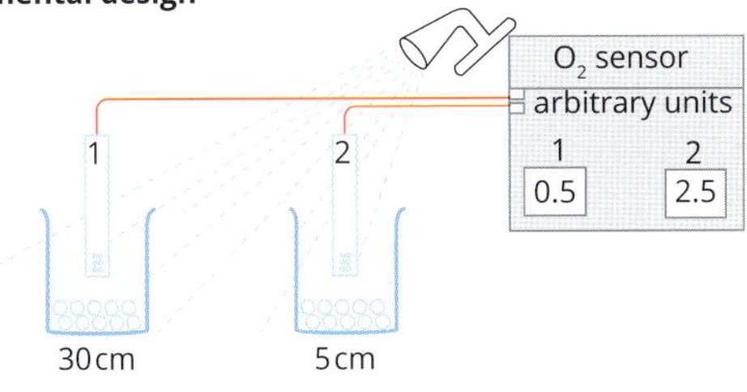

1.5 Data analysis and presentation

In Section 1.4 you learnt about different types of data and the factors that affect data quality. In this section you will learn about different descriptive statistics you can use to analyse your data, as well as how you can present your data in tables and graphs.

DESCRIPTIVE STATISTICS

Descriptive statistics can be used to analyse both quantitative and qualitative data. An important type of descriptive statistic is the measure of central tendency. It is good practice to use a measure of central tendency to provide a clearer understanding of the data.

Measures of central tendency

Measures of central tendency are single values that allow you to describe the central position in a set of data. Measures of central tendency are sometimes also called measures of central location. The mean, median and mode are all measures of central tendency.

Consider the following data set: 3, 5, 7, 8, 8, 8, 10.

- The **mean** (or average) is the sum of the values divided by the number of values, which in this case is $(3 + 5 + 7 + 8 + 8 + 8 + 10) \div 7 = 7$.
- The **median** is the 'middle' value in an ordered list of values, which in this case is the fourth value, which is 8.
- The **mode** is the value that occurs most often in a list of values, which in this case is 8. This measure is particularly useful for describing qualitative or discrete data.

The appropriate measure of central tendency to use depends on the type of data you are working with (Table 1.5.1).

Type of data	Mode	Median	Mean
nominal (qualitative)	✔	✘	✘
ordinal (qualitative)	✔	✔	maybe
discrete or continuous (quantitative)	✔	✔	✔

TABLE 1.5.1 When to use the different measures of central tendency.

Percentage change

Calculating the change in a variable is a helpful statistic because it provides a general trend or pattern, rather than listing a specific value, which will vary depending on the sample being studied. Percentage change applies to increases and decreases relative to the control or the starting point of the measurement.

For example, Table 1.5.2 shows data collected in an experiment that investigated the osmotic strength of different solutions. Four sets of dialysis tubing (a semipermeable membrane), each containing a different solution, were suspended in a beaker of physiological saline solution. The mass was measured at the start and after 24 hours.

	Mass (g) at 0 h	Mass (g) at 24 h	% change
sample 1	20.55	20.89	1.65
sample 2	20.01	21.94	9.65
sample 3	21.25	22.09	3.95
sample 4	20.55	20.32	−1.12

TABLE 1.5.2 Percentage change in mass of dialysis tubing over 24 h.

The percentage change in mass is calculated using the equation:

$$\text{percentage mass change} = \frac{\text{final mass} - \text{original mass}}{\text{original mass}} \times 100$$

Calculating percentage change accounts for variation and/or errors in the replicates within your experiment, or for the same experiment repeated by others. In Table 1.5.2, the starting mass is not identical in each sample, perhaps due to errors in measuring the volume put into the tubing. Although the final mass for sample 3 is the greatest, the percentage change is less than for sample 2 because the original mass was higher. Calculating percentage mass change shows that sample 2 has the greatest osmotic effect.

Percentage difference

The percentage difference (also often expressed as a fraction) is a measure of the precision of two measurements. It is calculated by working out the difference between the two measurements and dividing by the average of the two measurements:

$$\text{percentage difference} = \frac{\text{measurement 1} - \text{measurement 2}}{\text{average of measurements}} \times 100$$

For example, if your two measurements were 25 cm and 24 cm, you would calculate percentage difference as follows:

$$\text{percentage difference} = \frac{(25 - 24)}{(25 + 24) \div 2} \times 100 = \frac{1}{24.5} = 0.041 \times 100 = 4.1\%$$

Range

The **range** is simply the difference between the highest and lowest values in a data set. Table 1.5.3 shows the measurements taken for five different plants after treatment with a plant hormone.

Plant	1	2	3	4	5	Mean	Range
hormone-treated plants (mm)	158	378	320	377	363	319.2	378 – 158 = 220
untreated control plants (mm)	140	135	170	171	193	161.8	193 – 135 = 58

TABLE 1.5.3 Plant height in a hormone treatment experiment.

To determine the range for values in Table 1.5.3 you would subtract the smallest value from the largest value. Notice how an abnormally large or abnormally small value in the data set makes the variability appear high. If one value appears way out of range, such as plant 1 in the hormone-treated group, it is considered an **outlier** and can be deleted from the calculations. The range for the hormone-treated plants would then be 378 – 320 = 58. This illustrates the importance of having a sample size that is large enough to limit the impact of anomalies in the data set.

Uncertainty in measurement

When averaging repeat measurements, the **uncertainty** should be reported alongside your average. Uncertainty results from errors and represents a realistic range within which the true value is likely to be. A simple way to calculate the uncertainty is the range divided by 2:

$$\text{uncertainty} = \pm(\text{maximum value} - \text{minimum value}) \div 2$$

For example, if an experiment were conducted to measure the length of time it takes to convert a substrate to a product in an enzymatic reaction, and three replications of the experiment produced the times 2.50, 3.47 and 2.81 seconds, the average time taken would be 2.93 seconds. The uncertainty would be calculated as follows:

$$\text{uncertainty} = \pm(3.47 - 2.50) \div 2 = \pm 0.49$$

The result showing the mean and uncertainty is expressed as mean = 2.93 ± 0.49 seconds.

For the data set in Table 1.5.3 (page 47), in which we calculated the range, the uncertainties are as follows:

- control plants 161.8 ± 29.0
- hormone-treated plants 359.5 ± 29.0 (with the outlier removed).

● *You will now be able to answer Key Questions 1–3.*

PRESENTING DATA

When you have completed your experiment, you will need to organise and display the data. This makes it much easier to identify trends or patterns in the data. It also helps to identify any relationships that result from cause and effect between the independent and dependent variables, and helps you see if one variable has had any effect on another variable.

There are a number of ways to present data, including tables, graphs, flow charts and diagrams. The best way of visualising your data depends on its nature. Try several formats before you make a final decision to create the best possible presentation.

PRESENTING DATA IN TABLES

Tables record number values and allow you to organise your data.

Presenting raw data in tables

Tables organise data into rows and columns, and vary in complexity according to the nature of your data. Tables can be used to organise raw data and processed data, or to summarise results.

The simplest form of a table is a two-column chart. The first column should contain the independent variable (the one you control) and the second column should contain the dependent variable (the one that may change in response to a change in the independent variable).

As you can see in Figure 1.5.1, tables should have the following features:

- a descriptive title
- column headings (including the units)
- aligned figures (align the decimal points)
- the independent variable placed in the left column
- the dependent variable/s placed in the right column/s.

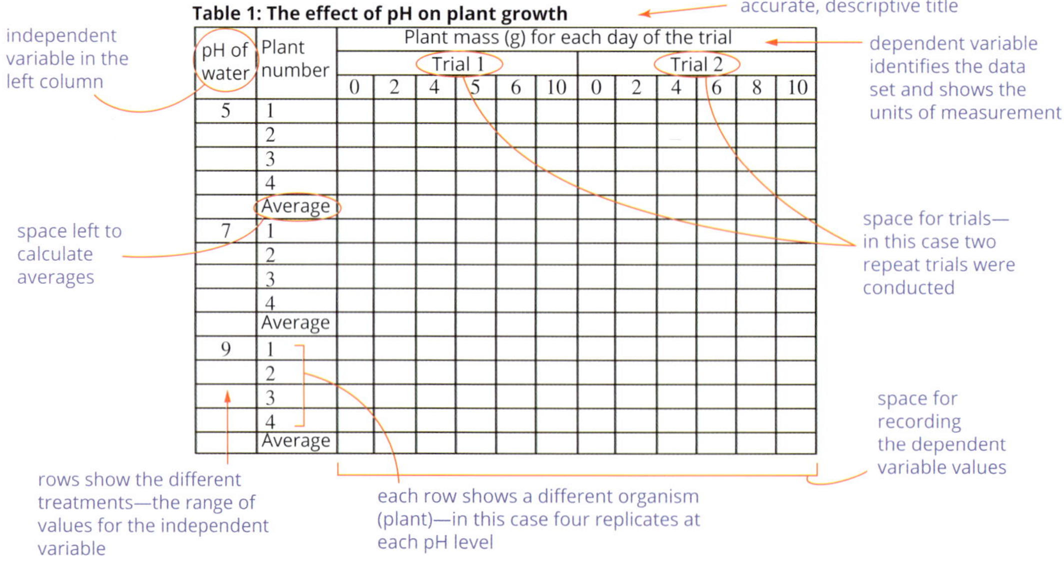

FIGURE 1.5.1 Features of a good table.

You should tailor the layout of your data table to suit your experiment. Table 1.5.4 is an example of a raw data table. It contains data from an experiment on the effect of temperature on the activity of enzyme X. A reaction between the enzyme and substrate was carried out for 10 minutes and the reaction product was measured. Three trials were performed.

	Product released (μg)		
Temperature (°C)	**Trial 1**	**Trial 2**	**Trial 3**
10	100	120	120
20	850	790	820
40	1350	1420	1390
60	1250	1210	1150
80	200	220	230

TABLE 1.5.4 Raw data table for the effect of temperature on reaction rate of enzyme X; measurement of reaction product.

Presenting processed data in tables

Table 1.5.5 also contains data on the relationship between temperature and mean enzyme reaction rate. It displays the data in a processed format; that is, the replicate values from Table 1.5.4 have been averaged to calculate the mean. The reaction rate per minute was also calculated (mean ÷ 10 min). The mean of the reaction rate per minute and its uncertainty are listed in Table 1.5.6.

Temperature (°C)	Mean (μg)	Mean rate (μg/min)
10	113.3	11.3
20	820.0	82.0
40	1386.7	138.7
60	1203.3	120.3
80	216.7	21.7

TABLE 1.5.5 Processed data table for the effect of temperature on enzyme X reaction rate; calculation of the mean product (μg) and rate (μg/min).

Temperature (°C)	Mean (μg/min)	Uncertainty
10	11.3	± 1.0
20	82.0	± 3.0
40	138.7	± 3.5
60	120.3	± 5.0
80	21.7	± 1.5

TABLE 1.5.6 Processed data table for the effect of temperature on enzyme X reaction rate; calculation of mean and uncertainty.

PRESENTING DATA IN GRAPHS

In general, tables provide more detailed data than graphs, but it is easier to observe trends and patterns in data in graph form than in table form. Graphs are used when two variables are being considered and one variable is dependent on the other.

There are several types of graphs, including line graphs, bar graphs, scatterplots and pie charts. The best one to use will depend on the nature of the data.

General rules to follow when preparing a graph include the following:

- ☐ Keep the graph simple and uncluttered.
- ☐ Use a descriptive title.
- ☐ Represent the independent variable on the *x*-axis.
- ☐ Represent the dependent variable on the *y*-axis.
- ☐ Start each axis at zero.
- ☐ Match the length of the axes to the data.
- ☐ Clearly label axes with both the variable and the unit in which it is measured.
- ☐ Use small symbols such as circles or squares for data points.
- ☐ Use different symbols for different data sets.

Line graphs

A **line graph** is a good way of representing continuous quantitative data. In a line graph, the values are plotted as a series of points on the graph. A line can then be drawn from each point to the next. The independent variable, which is set by the experimenter, is always shown on the *x*-axis. The dependent variable, which is the variable measured in the experiment, is always shown on the *y*-axis. Each point should be drawn in pencil as a small symbol, such as a circle, square or cross. Alternatively you can use a computer program to generate your graphs.

The data from the enzyme example in Table 1.5.6 is presented in Figure 1.5.2.

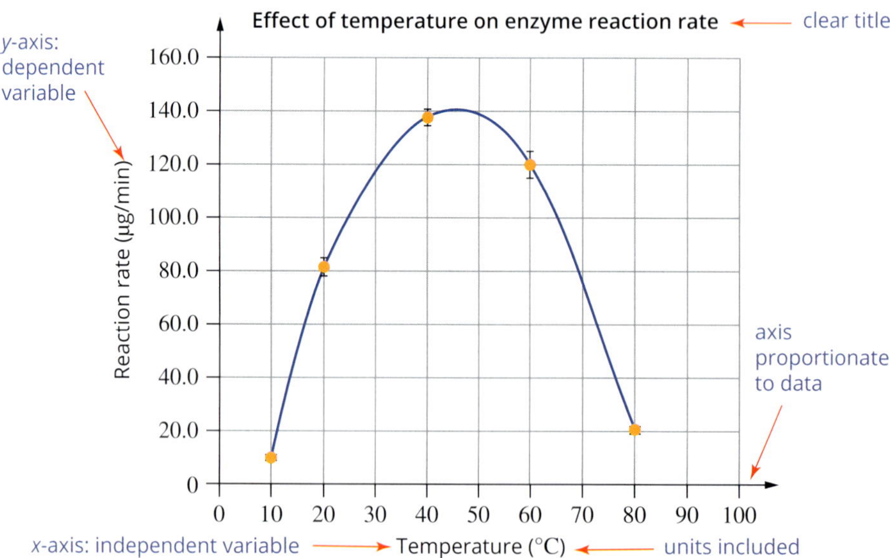

FIGURE 1.5.2 A line graph showing the relationship between two variables: temperature (independent variable) and reaction rate (dependent variable). The uncertainty values are represented as vertical bars above and below the mean at each data point.

For continuous data, the line does not need to join each data point. Rather a straight or curved line can be drawn to represent the overall trend of the data, as shown in the different graphs in Figure 1.5.3. This line is called a trend line or a line of best fit, and can be used to predict values between the data points. Its position can be estimated by eye or calculated mathematically from the data.

When graphing discrete quantitative data, a line shows the change in data from one point to the next, but does not predict the value of a point between the plotted data. An example is plotting the incidence of disease in a population (Figure 1.5.4). This type of data could also be represented as a bar graph, but lines may be better when multiple data sets are being compared on the same graph, such as when plotting the incidence of several different diseases.

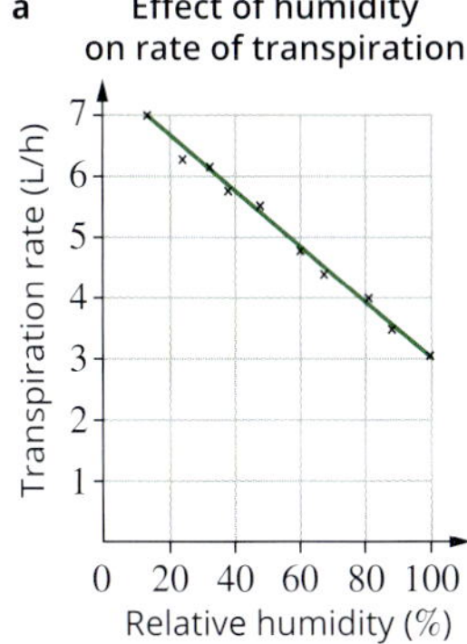

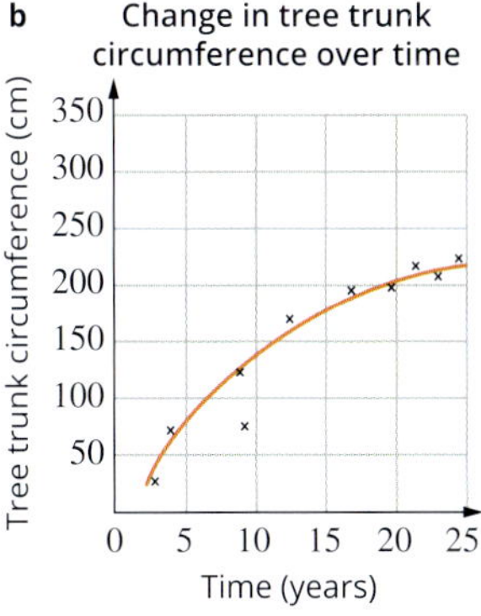

FIGURE 1.5.3 Graphs showing straight (a) and curved (b) trend lines or lines of best fit.

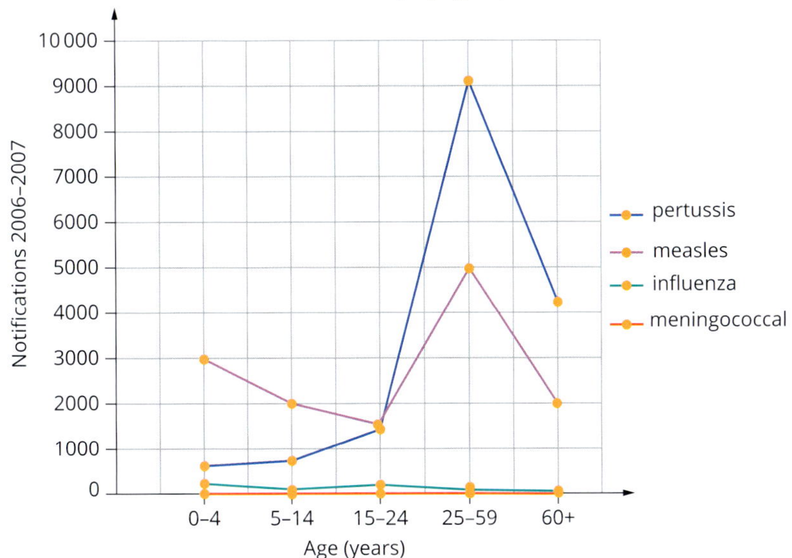

FIGURE 1.5.4 The Department of Health is notified of the number of new cases (or incidence) of a disease. The graph displays Australian Government Department of Health data for notifications of four infectious diseases (pertussis, measles, influenza and meningococcal) for the period 2006–2007. Vaccines are available for all of these infectious diseases.

You will now be able to answer Key Questions 4 and 5.

Scatterplots

Scatterplots are commonly used to display data when looking to see if there is a correlation or relationship between two variables. For example, in human evolution there is extensive interest in the relationship between brain size and time since early human ancestors, hominins, first appeared (Figure 1.5.5), as well as other changes over time, such as body size, bipedalism and tool use.

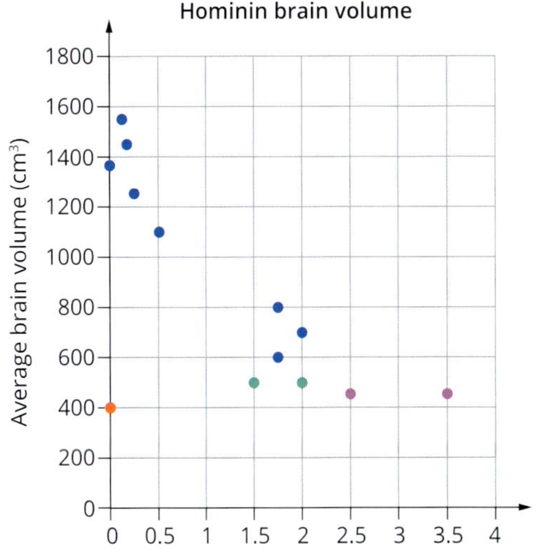

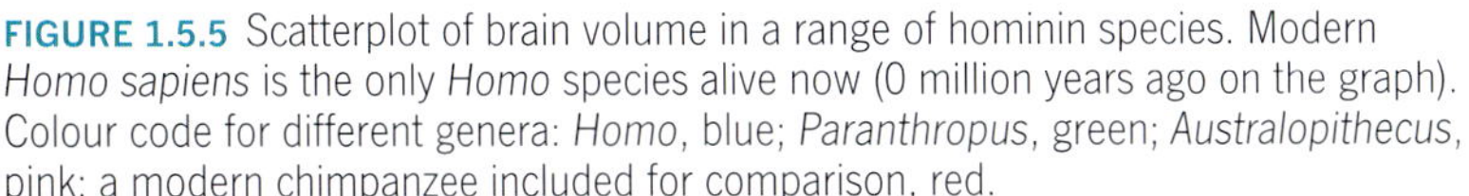

FIGURE 1.5.5 Scatterplot of brain volume in a range of hominin species. Modern *Homo sapiens* is the only *Homo* species alive now (0 million years ago on the graph). Colour code for different genera: *Homo*, blue; *Paranthropus*, green; *Australopithecus*, pink; a modern chimpanzee included for comparison, red.

Bar and column graphs

Bar and column graphs are used to show categories of data that have been counted.

- A **column graph** shows the value of the dependent variable by the height of the column; the categories are labelled across the *x*-axis.
- A **bar graph** shows the value of the dependent variable by the length of the horizontal bar; the categories are labelled up the *y*-axis.

Bar and column graphs are commonly used when the independent variable is categorical rather than numerical. The bars or columns are always the same width and the same distance apart. Bar and column graphs are very useful for graphing qualitative and discontinuous data, such as the number of base pairs and number of genes on each human chromosome (Figure 1.5.6).

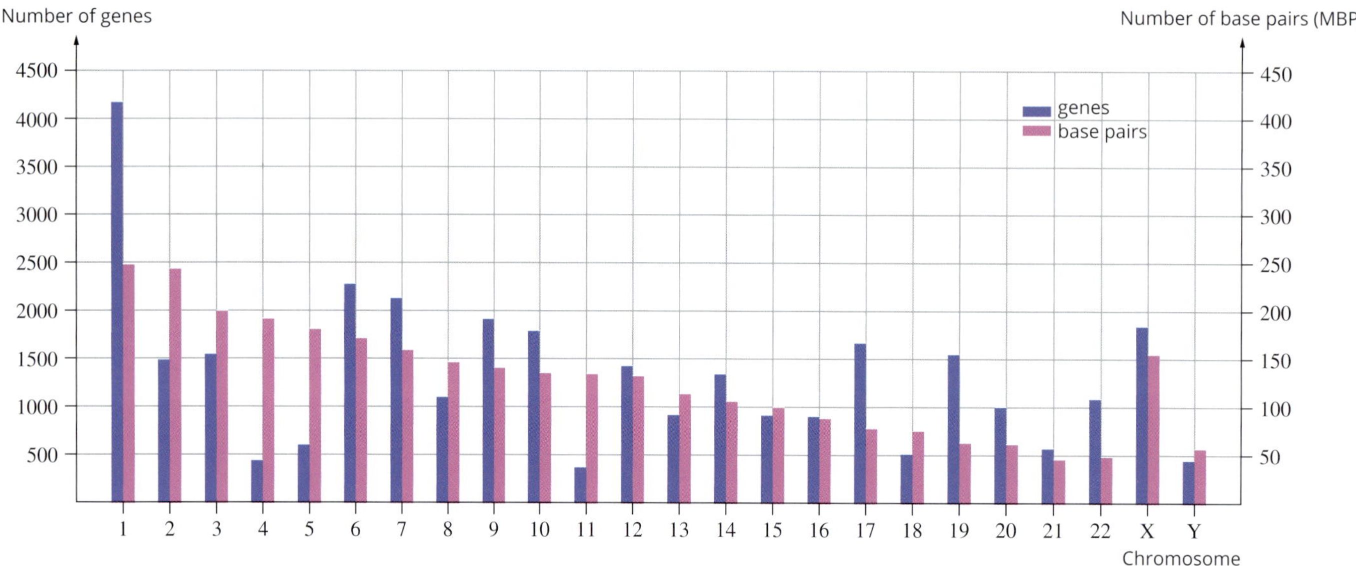

FIGURE 1.5.6 A column graph comparing the number of genes (blue, left axis) and the number of base pairs (pink, right axis) on human chromosomes. Note that two different vertical axes are used for the different data sets, which have very different scales.

When the labels of the variables are long, horizontal bar graphs can be used. Bar graphs are also used when the data ranges are variable and overlapping, such as genome sizes (Figure 1.5.7).

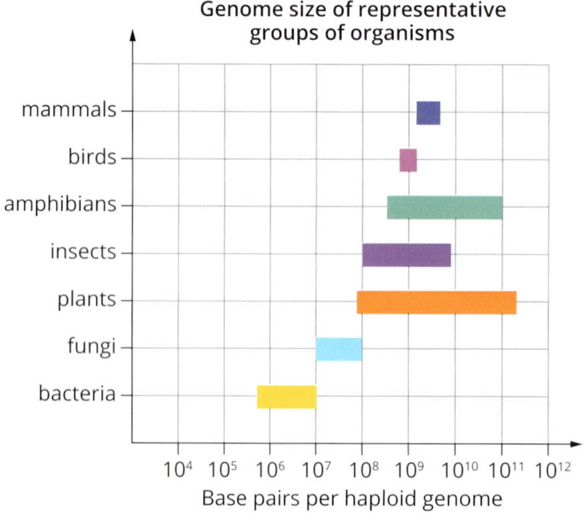

FIGURE 1.5.7 A horizontal bar graph comparing genome size of representative organisms from the different kingdoms of life.

Pie charts

A **pie chart** is a way of presenting qualitative and categorical data. It shows each category of data as a proportion of the total data. The chart is a circle divided into sections according to the proportions of each category, like slices of a pie (Figure 1.5.8). Each category is coloured or shaded differently so that it can be distinguished clearly from the other categories. Pie charts should only be used when there are few categories.

To draw a pie chart you must find how many degrees are needed for each category to fit within the 360° circle. This can be done as follows:

- ☐ Add the amounts in each category to find the total.
- ☐ Divide 360° by the total (this will tell you how many degrees of the circle one value is worth).
- ☐ Multiply this value by the amount in each category.
- ☐ This gives the degrees for each category that can then be marked, using a protractor, on the circle.

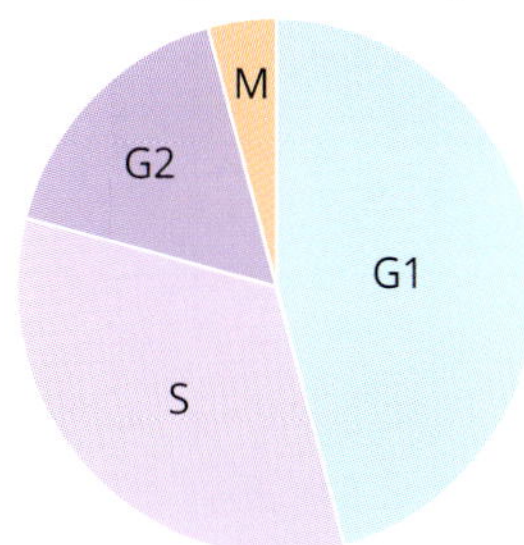

FIGURE 1.5.8 Pie chart presenting data on the length of time a population of mammalian cells spends in each stage of the cell cycle. These cells have a total cell cycle time of about 24 hours. Stage and time: G1, 11 hours; S (DNA synthesis), 8 hours; G2, 4 hours; M (mitosis), 1 hour.

Missing data

When you have missing data, leave a gap for it, as shown in Figure 1.5.9. Ensure that the axes are complete (do not skip values) and do not join data points that have data missing between them. Joining points could be misleading. For example, if the data in Figure 1.5.9 was collected to determine the need for a pertussis (whooping cough) booster vaccination program, it is important to know which age groups in the population are most at risk, so that the right age groups are targeted and public health dollars are well-directed. Try to predict the result for the missing data in the 25–59 year old age group in Figure 1.5.9, then compare it to Figure 1.5.4 to see how accurate your prediction was.

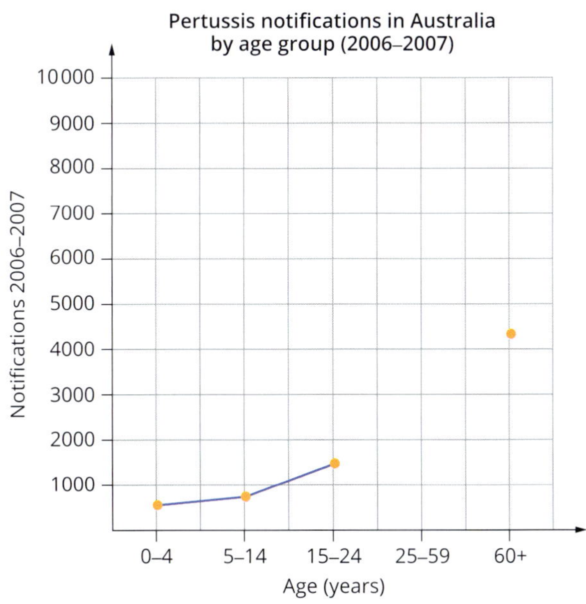

FIGURE 1.5.9 A line graph with missing data.

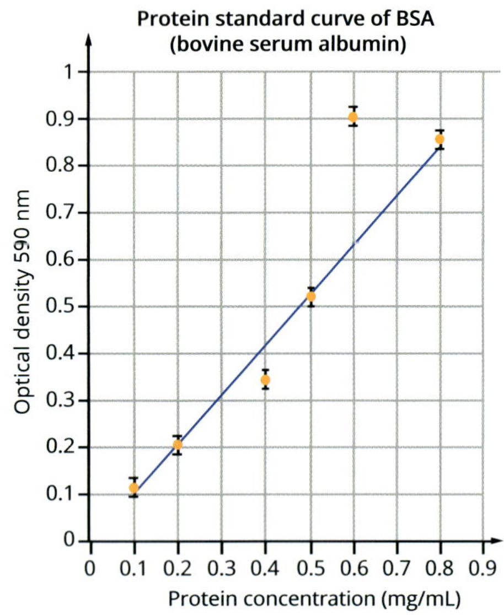

FIGURE 1.5.10 A line graph showing an outlier, which has been ignored when adding the line of best fit.

Outliers

Sometimes when you collect data, there may be one point that does not fit the trend and is clearly an error. This is called an outlier. An outlier is often caused by a mistake made in measuring or recording data, or from a random error in the measuring equipment. You should plot all of the data points on your graph, but if a data point in a continuous data series is clearly outside the trend line (an outlier) you can ignore it when drawing the line of best fit (Figure 1.5.10). Outliers can also be calculated mathematically.

● *You will now be able to answer Key Questions 6 and 7.*

Distorting the truth

Poorly constructed graphs can distort the truth. For example, in Figure 1.5.12 you can see two graphs that show the same data—the test results of two groups of students. One group of students did not eat breakfast before doing the test, and scored an average of 42 marks out of 50. The other group of students did eat breakfast and scored an average of 48 marks out of 50. One graph distorts the difference in marks between the two groups by using a scale of from 40 to 50 marks on the *y*-axis. It is important to make sure the graphs you create do not distort your data. You should also be wary of distorted data when interpreting graphs in other publications.

EXTENSION

Standard curves

Using standard curve graphs for quantification

A standard curve (or calibration curve) can be used to gain quantitative results for samples assayed with a biochemical colour-based assay. For example, you may be experimenting with an enzyme that breaks down protein (starting with a 1 mg/mL solution).

You can use the biuret test to detect changes in protein concentration at the end of a 24-hour reaction. This involves the use of biuret reagent and a protein such as bovine serum albumin (BSA). For the standard curve you would prepare a set of standard protein solutions of a known concentration (e.g. a range from 0.1 to 1.0 mg/mL), then perform the assay on the standards. Numeric values are then obtained for the standards by reading the absorbance (or optical density) with a colorimeter or spectrophotometer.

The absorbance values obtained can be used to plot a graph and draw a line of best fit. This is the standard curve. Keep in mind that a standard curve may not be linear if you are not working in the correct concentration range for the substance being measured.

Once you have your standard curve, you can use it to determine the protein concentrations at different optical densities. For example, for a protein assay on a sample that gave an optical density of 0.32, from the protein standard curve you find that the protein concentration is 0.3 mg/mL (red lines on Figure 1.5.11).

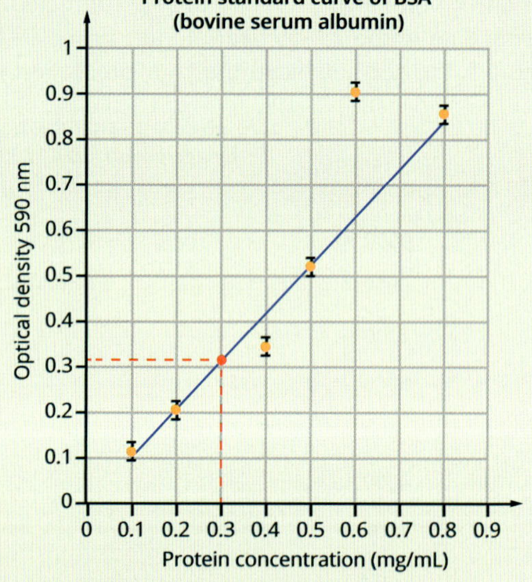

FIGURE 1.5.11 A protein standard curve. If an experimental sample has an optical density of 0.32, then go up the *y*-axis until you reach 0.32, trace a line horizontally to the right until you meet the line of best fit, then trace a line down to meet the *x*-axis, which tells you the protein concentration of the sample was 0.3 mg/mL.

If you do not have an instrument to read absorbance, you can make a visual grading scheme with colour photographs (for identification) and relative values (for quantification) assigned to every colour intensity (e.g. –, +, ++, +++ or 0, 1, 2, 3). This provides a semi-quantitative result.

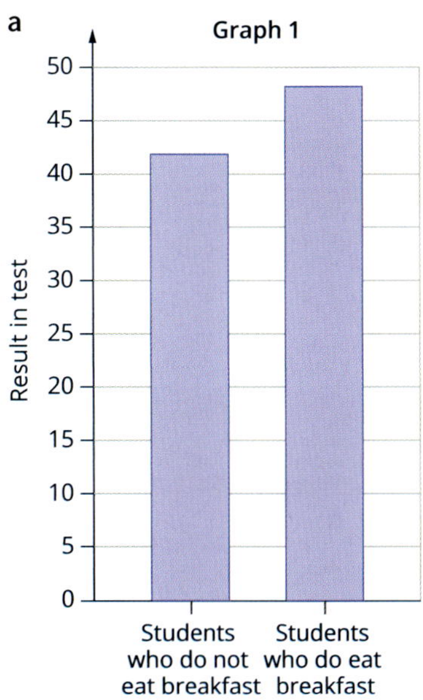

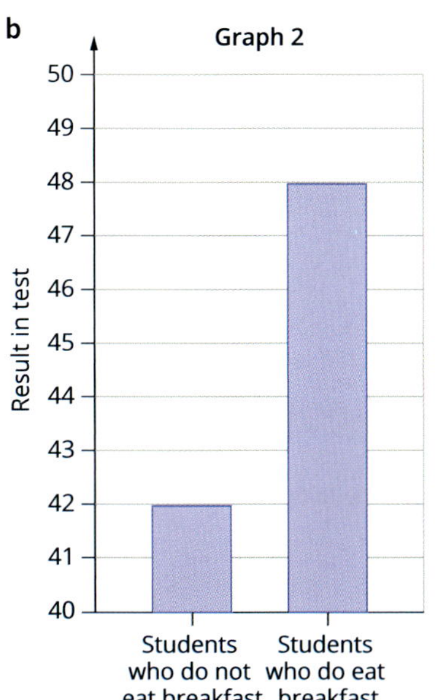

FIGURE 1.5.12 (a) A graph showing the difference between the test marks of two groups of students out of the total 50 marks on the *y*-axis. (b) A graph showing the difference between the test marks of the two groups within only a narrow range of marks on the *y*-axis, which distorts the difference and makes it appear larger than it really is.

PREPARING DIAGRAMS

It is important to learn how to draw and label diagrams of equipment and biological specimens in your studies of biology. There are certain rules you should follow in order to produce a diagram that will be acceptable in your reports and exams.

When drawing scientific equipment, diagrams should:

- ☐ be large, simple, two-dimensional pencil drawings
- ☐ have ruled lines where possible
- ☐ keep proportions realistic (Figure 1.5.13).

When drawing biological specimens follow these guidelines:

- ☐ Draw the whole diagram (including labels, lines, magnification, heading and scale if possible) in pencil.
- ☐ A diagram of microscopic objects does not require a circle representing the field of view.
- ☐ Draw one or a few cells to represent a sample; there is no need to try and draw every cell in a field of view.
- ☐ Draw your diagram with simple and clear lines (do not sketch).
- ☐ Use stippling rather than shading to indicate depth.
- ☐ Make your diagram as large as possible (at least 10 × 10 cm).
- ☐ If there are many features to show, it is useful to pair a photo with a detailed supporting diagram that shows cellular detail (Figure 1.5.14, page 56).
- ☐ Draw only the structures that you see, not things you think you should see, such as mitochondria (Figure 1.5.15, page 56).
- ☐ Include clear labels for the features you want to highlight.
- ☐ Place labels outside the drawing.
- ☐ Make sure label pointers do not cross over each other.
- ☐ Labels should line up on either side of the diagram where possible.
- ☐ Use straight lines without arrowheads that meet the features being labelled.
- ☐ Include a scale bar or scale (e.g. 1:100) in the diagram, or state the magnification (e.g. ×400) in the caption.

FIGURE 1.5.13 Diagram showing a dialysis tubing arrangement. Note the straight lines for labels that are horizontal where possible, and the realistic proportions of different parts in relation to each other.

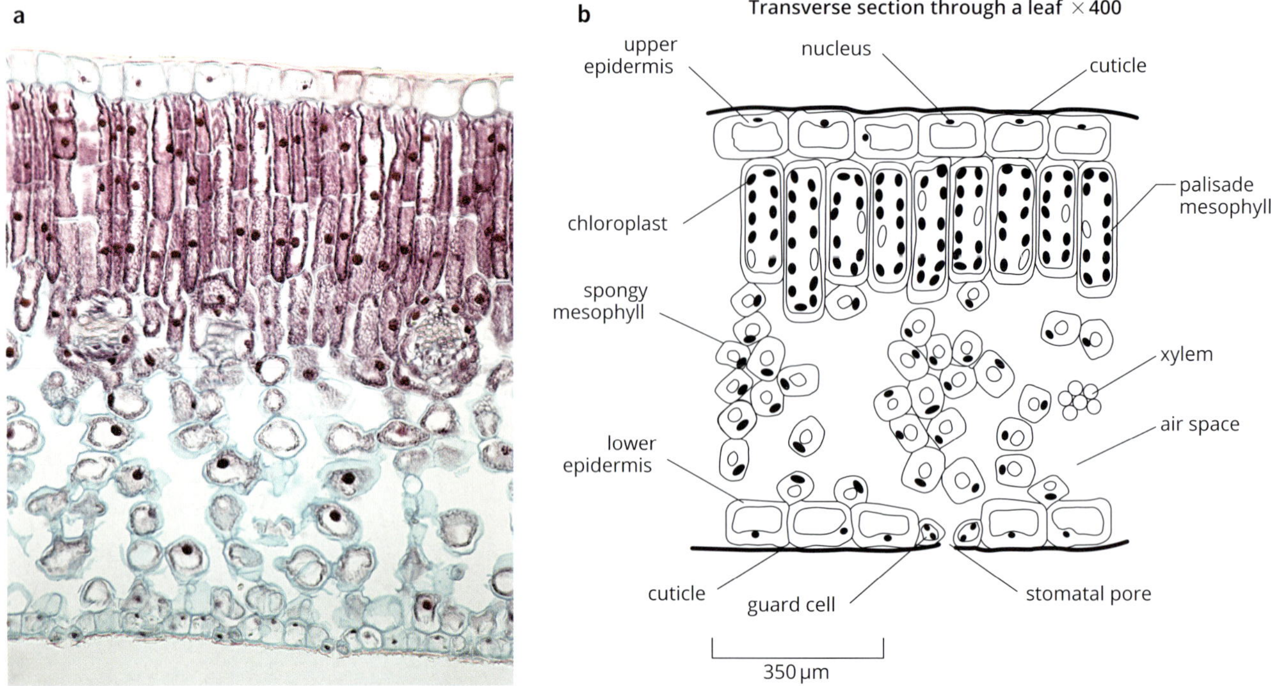

FIGURE 1.5.14 A photomicrograph (a) and a diagram (b) of a transverse section through a leaf.

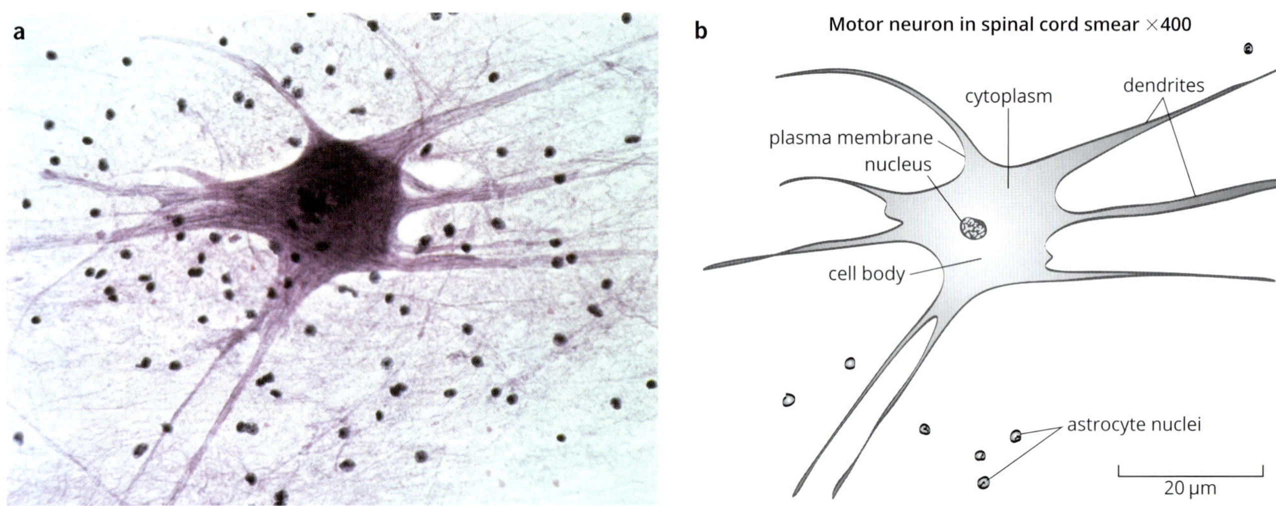

FIGURE 1.5.15 A photomicrograph (a) and a scientific diagram (b) showing a motor neuron in a spinal cord smear.

1.5 Review

SUMMARY

- Descriptive statistics can be used for qualitative and quantitative data.
- Descriptive statistics include three measures of central tendency:
  - the mean, which is the sum of the values divided by the number of values
  - the median, which is the 'middle' value in an ordered list of values
  - the mode, which is the value that occurs most often in a list of values.
- Other helpful descriptive statistics include:
  - percentage change, which applies to increases and decreases relative to the control or the starting point of the measurement
  - percentage difference, which is a measure of the precision of two measurements
  - range, which is simply the difference between the highest and lowest values in a data set
  - uncertainty, which results from errors and represents a realistic range within which the true value is likely to be.
- Tables are used to record raw and processed data.
- Tables allow the presentation of more detail, while graphs allow trends to be shown more clearly.
- When presenting the results of an investigation, do not distort the truth—this includes selecting appropriate scales on graph axes, including outliers in graphs, and including and explaining all errors.

KEY QUESTIONS

1 For the following data set, calculate and record (a) the median, (b) the mode and (c) the mean and uncertainty. Show your calculations.
Data: 21, 28, 19, 19, 25, 24, 20

2 Calculate the percentage change in mass for these plants exposed to light of different intensity.

Plant	Mass on day 1 (g)	Mass on day 2 (g)	% change
plant 1 (control)	12.3	12.5	
plant 2 (intense light)	12.4	12.7	
plant 3 (low light)	12.1	12.0	

3 Describe the advantages of calculating percentage change for the results of an experiment repeated by different groups of scientists.

4 Immunologists have measured the levels of antibodies in blood serum to gather background data on population responses to infection. They collected the following data on the concentration of two different types of antibody, IgG and IgA, from subjects ranging in age from 6 months to 20 years (the antibody levels are listed in order of increasing age of subject).

- Age of subject: 6 months, 1, 2, 4, 10, 20 years
- Concentration of IgG (mg/100 mL): 300, 600, 800, 1000, 1500, 1500
- Concentration of IgA (mg/100 mL): 50, 100, 100, 150, 200, 400

a Prepare a data table.
b Prepare a graph of the data.

5 Describe at least four ways the graph below could be improved.

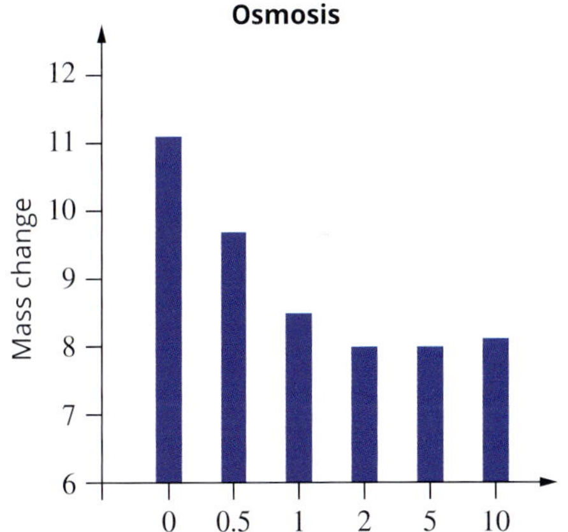

6 Distinguish between the times when a line of best fit should be drawn on a graph and when a line ruled from point to point is more appropriate.

7 What are outliers, and what is the statistical measurement most affected by them?

1.6 Reporting investigations

FIGURE 1.6.1 Posters at a scientific conference.

Now that you have thoroughly researched your topic, formulated a research question and hypothesis, conducted experiments, and collected data, it is time to bring it all together. The final part of an investigation involves summarising the findings in an objective, clear and concise manner for your audience.

Scientists report their findings in a number of ways: written peer-reviewed journal articles, on web pages, and at scientific conferences with short oral presentations or scientific posters (Figure 1.6.1). Regardless of the reporting and presentation method, the same key information is presented in the same order.

The coursework for Unit 3 will include your completed practical logbook (which you should be completing as you proceed with practical work) and either written reports or multimodal presentations of research and practical activities. Upon completion of the practical investigation for Unit 4, Area of Study 3, you are required to present the methodologies, findings and conclusions as a scientific poster. In this section you will learn how to present your findings effectively, discuss your investigation, and draw evidence-based conclusions in relation to your hypothesis and research question.

PRESENTATION FORMATS

All modes of presenting scientific research and experimental investigations have the same elements, but with different emphases on visual or textual components depending on the mode of delivery. Table 1.6.1 provides some guidelines for different presentation formats.

Format	Characteristics	General guidelines for the presentation format
poster presentation	• concise visual display of information • suitable for presenting information to many people • summary of ideas	• title that attracts attention • large headings that stand out • subheadings of a smaller size • attractive presentation • balance of written material and visual material such as diagrams, photographs, tables, graphs • writing large enough to be read from a distance
written report of a practical activity	• presents clear and detailed information on a topic • suitable for providing detailed and more comprehensive background information	• appropriate written style for introduction, materials and methods, results, discussion and conclusion • use subheadings to organise sections • text should be supported by tables, graphs, diagrams or photographs
oral presentation with supporting slides and/or handouts	• easy-to-follow format • good for presenting to a large audience • supporting slides can be printed as notes to be given to the audience • opportunity to answer questions from audience	• brief oral descriptions • use clear visuals that complement what is spoken • minimal text on each slide • consistent format on all slides—background, colours and text • images, diagrams and graphs are clear and large
online presentation e.g. website, blog	• can present visual and written information • accessible to a worldwide audience • easy to follow • easy to update with new information	• include hyperlinks to related information • include multimedia, such as video clips and audio, if appropriate • use the same format throughout—font, background, colours • use clear headings • list all hyperlinks on the main page • include your name, credentials and date of publication

TABLE 1.6.1 Characteristics of the main formats for presenting research work.

EFFECTIVE SCIENCE WRITING

Effective science writing is objective, clear and concise, and has a consistent narrative and visual support. If you have time, it is a good idea to put your finished writing aside for a few days and then go back and read it over again, fixing anything that is incorrect or poorly written. Checking the spelling is also an essential part of editing your writing. Do not rely only on computer programs to check spelling; they can make mistakes too, and often do not recognise scientific words. Make sure the spellchecker is set to Australian English; the default setting is usually American English.

Objective writing

Scientific reports should be written in an objective (unbiased) style. This is in contrast to literary writing, which often uses subjective (biased) techniques of persuasion (Table 1.6.2).

Unscientific writing examples	Scientific writing examples
Examples of biased and subjective language: • The results were weird/bad/atrocious/wonderful... • This produced a disgusting odour... • This is a major health crisis... • This breathtakingly beautiful golden bowerbird...	Examples of unbiased and objective language: • The results showed... • This produced a pungent odour... • This is a serious health issue... • The golden bowerbird...
Examples of exaggeration: • The object weighed a colossal amount... • No one has ever seen this phenomenon... • The magnesium exploded into flames... • Millions of ants swarmed over the next...	Examples of accurate language: • The object weighed about 250 kg... • This phenomenon has not been reported previously... • The magnesium burnt vigorously... • Ants swarmed over the next...
Examples of everyday language: • The bacteria passed away... • The results don't... • We guessed that ... • Previous researchers were slack and missed...	Examples of formal language: • The bacteria died... • The results do not... • We predicted/hypothesised/... • Previous researchers did not notice that...

TABLE 1.6.2 Examples of unscientific and scientific writing.

Qualified writing

It is best to avoid words that are absolute, such as always, never, shall, will, or proven. Instead, qualify your writing using words such as may, might, possible, probably, likely, suggests, indicates, appears, tends, can and could.

Concise writing

To be concise use short sentences with a simple structure. The opposite of being concise is being verbose (wordy). When editing your writing consider how you could say the same thing using fewer words (Table 1.6.3).

Verbose	Concise	Verbose	Concise
due to the fact that	because	is well known to be	is
Carlos undertook an investigation into	Carlos investigated	on an annual basis	yearly
It is possible that the cause could be	the cause may be	until such time as	until
a total of five experiments	five experiments	in the vicinity of	near
the end result	the result	while in the process of preparation	while preparing
in the event that	if	I am of the opinion that	I think that
at the time of writing	today	we took measurement readings	we measured

TABLE 1.6.3 Examples of verbose writing and concise alternatives.

Voice

'Voice' means whether the subject of the sentence is the 'doer' or 'receiver' of the action. In the active voice the subject is the doer; for example, 'We added 20 mL of sodium chloride to the beaker.' In the passive voice the subject is the receiver; for example, '20 mL of water was added to the solution.' Choose the voice that helps you communicate your ideas clearly. You will often find that a mixture of active and passive voice is best (Table 1.6.4).

Active voice	Passive voice
We recorded oxygen levels hourly.	The oxygen concentration was recorded every 60 minutes.
We used a pH meter to measure pH.	The pH was recorded with a pH meter.
A thermostat controlled the temperature in the water bath.	The temperature in the water bath was controlled by a thermostat.
We placed 50 g of solute in a conical flask containing distilled water and then slowly added 1 M hydrochloric acid drop by drop.	Fifty grams of solute was placed in a conical flask containing distilled water, and then 1 M hydrochloric acid was added dropwise.

TABLE 1.6.4 Examples of active and passive voice.

Tense

Use the past tense when describing your research, including the planning, the experiments and the results, as well as the work of previous researchers. For everything else (including describing facts and theories) you should use the present tense. Avoid using conditional verbs (could or would) and the future tense (unless you are talking about something that has not yet happened). Table 1.6.5 shows some examples of the correct and incorrect use of tenses in scientific writing.

Correct tense	Incorrect tense
Zhu (2013) described a similar phenomenon.	Zhu (2013) describes a similar phenomenon.
Hormone was added to the tips of coleoptiles.	Hormone will then be added to the tips of coleoptiles.
Enzyme B reacted best at pH 9.	Enzyme B reacts best at pH 9.
The DNA sequence comparison supported the conclusion that species A and B share a common ancestor.	The DNA sequence comparison supports the conclusion that species A and B share a common ancestor.

TABLE 1.6.5 Examples of correct and incorrect use of tense.

Visual support

Use graphs or diagrams to present complex concepts or information. This will reduce the number of words you need, and also make your research more accessible for your audience. Details of experimental methods can be presented as a diagram or flow chart. This can make it easier to see the procedure than to read through a series of steps. Flow charts use simple diagrams, small text boxes and connecting lines to represent the methods and sequence of steps in a scientific method. Diagrams should use clear outlines and labels—they are not works of art.

WRITING A SCIENTIFIC REPORT

Whether the investigation is presented as a poster, written report or oral presentation, the same key elements are included in the same sequence, as summarised in Figure 1.6.2.

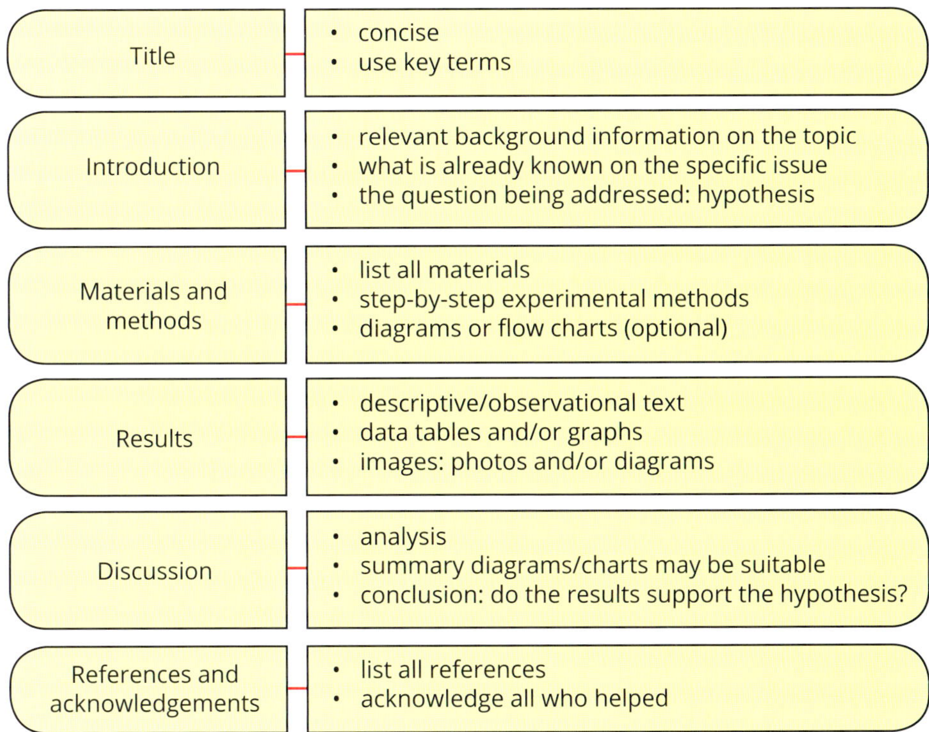

FIGURE 1.6.2 Elements of a scientific report or presentation.

Title

The title should give a clear idea of what the report is about, without being too long. It should include key terms that tell the reader what your study is about.

Introduction

The introduction sets the context of your report. It should outline relevant biological ideas, concepts, theories and models, and how they relate to your specific research question and hypothesis. It introduces the key terms, the specific question to be addressed, and states your hypothesis.

For example, consider a student investigating the cellular processes affected by a growth-promoting plant hormone. The research and introduction for this investigation might include the following points:

- the name and chemical nature of the hormone
- where the hormone is found (natural or synthetic)
- what is currently known about the actions of the hormone
- the specific question being addressed
- the hypothesis.

While researching this topic, the student found prior evidence that suggests this hormone increases the height of some dwarf plants, but the mechanism for this effect was not clear. There were some reports of increased cell division, while other studies reported a change in cell length. The student's hypothesis was that the hormone would increase the growth of dwarf peas by increasing the cell number.

Materials and methods

The materials and methods section lists all the materials that were used in the research, and describes in detail all the steps that were undertaken. For a poster presentation use step-wise lists, diagrams of specific methods, and/or flow charts of the overall experimental design. There should be enough detail for someone else to replicate your experiments. Therefore, your method needs to be in the correct sequence and include how you observed, measured, recorded and analysed the results. This section should also identify the independent, dependent and controlled variables (see Section 1.2).

Here is an example of a materials and methods section for an experiment on plant hormone action as it might be presented in a written report. For a poster presentation, the methods may be easier to follow in a step-wise list accompanied by large, clearly labelled diagrams. Alternatively, flow charts are a good way to clearly present experimental designs.

Materials

- 20 dwarf pea seeds
- 3 pots and potting mix
- plant hormone—gibberellic acid (GA) solutions, 0.01% and 0.1%, diluted from 1% stock solution in distilled water (dH_2O)
- small spray bottles
- scalpel blade and forceps
- toluidine blue stain (0.025%)
- 1 M HCl
- microscope slides and coverslips
- compound light microscope

Methods

Example experiment 1: Plant growth and treatment with GA

Dwarf pea seeds were germinated and transferred into 3 pots with potting mix, 5 plants per pot. After 1 week, when the seedlings were approximately 20 mm tall, plants were sprayed with either dH_2O, 0.01% GA or 0.1% GA (Figure 1). Plant height was measured 1 and 2 weeks later.

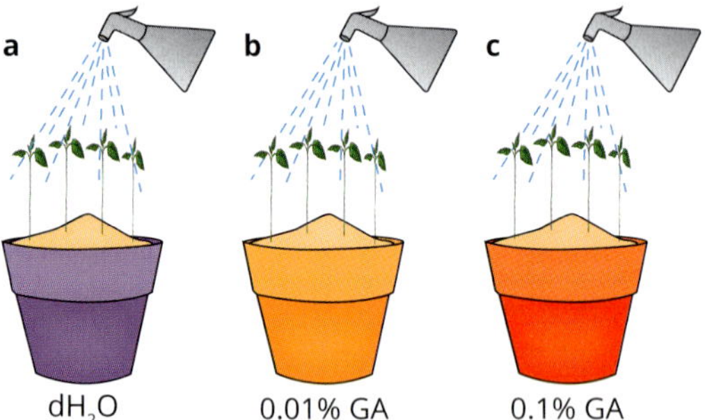

Figure 1: Seedlings were sprayed with dH_2O (a), 0.01% GA (b), and 0.1% GA (c).

The independent variable in this experiment was the GA concentration, and the dependent variable was plant height.

Example experiment 2: Microscopic analysis of internode cells

At week 3, a 5 mm section of internode was cut from the stem, placed on a microscope slide and sliced lengthwise. Three drops of 1 M HCl were added to the tissue and the slide placed on a 60 °C hotplate for 2 minutes. Excess HCl was soaked up with paper towel; 2 drops of toluidine blue stain were added for 2 minutes, then a coverslip was placed on the tissue and gently pressed down (Figure 2). The slide was viewed under the microscope at ×100 magnification. Two stems from each pot were stained and viewed in this manner.

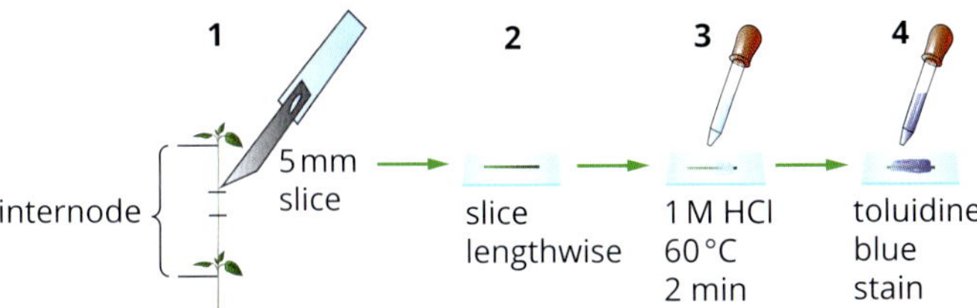

Figure 2: A 5 mm section of internode was cut from the stem (1) and placed lengthwise on a microscope slide (2), then three drops of 1 M HCl were added and the slide heated at 60 °C for 2 minutes (3). Finally, 2 drops of toluidine blue stain were added for 2 minutes, then a coverslip placed on top (4) before viewing by light microscopy.

The independent variable in this experiment was GA concentration, and the dependent variable was cell length and number.

Results

The results section is a record of your observations. It is where you present your data using graphs, diagrams, tables or photographs. In Section 1.5 you learnt tips on using graphs and tables appropriately.

For the plant hormone experiment described above, the results section might include the following table and figures:

Results

Example experiment 1: Effect of hormone on plant growth

	GA concentration		
Plant no.	**0**	**0.01%**	**0.10%**
1	23	117	158
2	20	210	378
3	22	240	320
4	30	211	377
5	31	198	363
mean	25	195	319

Table 1: Results of plant height (mm) at week 3 of GA treatment.

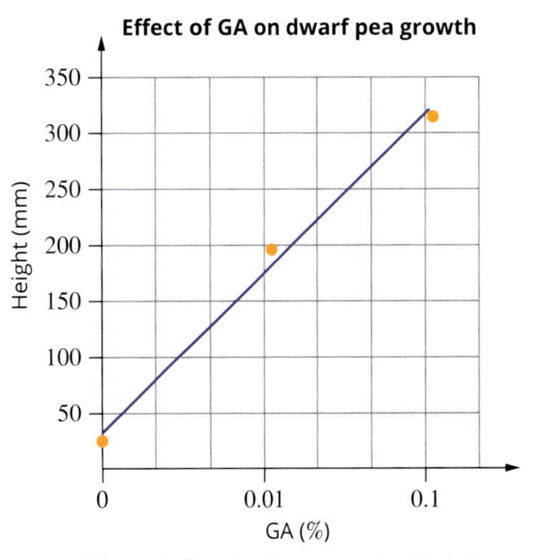

Figure 3: Graph of average plant height at week 3 (2 weeks after GA treatment).

Example experiment 2: Microscopy. The effect of GA on cell growth

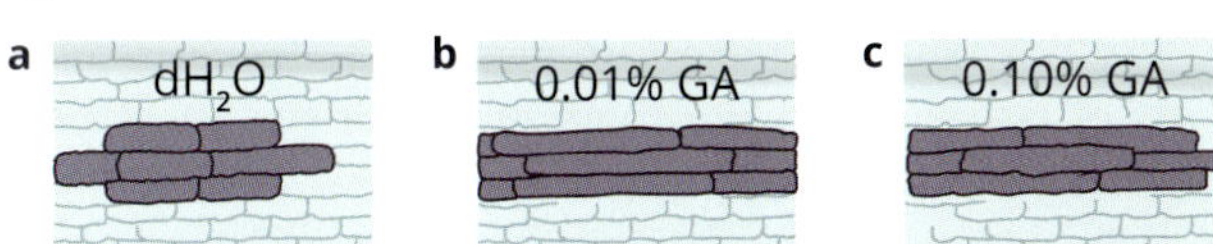

Figure 4: Diagrams of representative samples at week 3 (2 weeks after GA treatment), viewed at ×100 magnification. Estimated average cell length in control dH_2O, 60 μm (a); 0.01% GA, 100 μm (b); 0.10% GA, 100 μm (c).

Discussion

In the discussion you interpret your results, and discuss how your findings relate to your initial question and hypothesis, the research of others, and the biological concepts outlined in your introduction.

Interpret the results

When you interpret your results, you need to state clearly whether a pattern, trend or relationship was observed between the independent and dependent variables, describe what kind of pattern it was, and specify under what conditions it was observed.

In experiments with continuous variables, such as a range of concentrations, temperatures or pH, the types of relationships that may occur between variables are:

- **linear relationship**—variables that change in linear or direct proportion to each other produce a straight trend line (Figure 1.6.3a)
- **exponential relationship**—variables that change exponentially in proportion to each other produce a curved trend line (Figure 1.6.3b, c)
- **inverse relationship**—when there is an inverse relationship, one variable increases as the other variable decreases; this relationship may be linear or exponential (Figure 1.6.3d, e)
- none—when there is no relationship between two variables, one variable will not change even if the other does (Figure 1.6.3f).

More complex relationships might have to be evaluated mathematically to obtain a formula that describes the trend line.

Interpreting the result for the plant hormone experiment previously described, Figure 3 on page 63 shows that as the GA concentration increases, the height of the peas increases in a linear fashion.

● *You will now be able to answer Key Question 1.*

Evaluate investigative methods

Your discussion should evaluate your investigative methods and identify any issues that could have affected the validity, reliability, accuracy or precision of the data. Any possible sources of error in your experiment should be stated. Remember that controls are essential to the reliability and validity of your investigation, so if you have overlooked or were unable to control a variable that should have been controlled, this may explain unexpected results.

Identify any ways your experiment could be improved. In the example plant hormone experiment, the sources of error the experimenter should consider include whether there were enough replicates to obtain reliable data, whether microscopy was a reliable way of determining cell number and cell length, whether the microscope was calibrated, and whether enough cells were viewed. When writing your report, provide specific suggestions for improvements to the methodology based on what you have learnt.

It is also important to acknowledge contradictions in data and information. Again consider the example plant hormone experiment, in which the results of Experiment 2 indicated an increase in cell length in the GA-treated plants compared to the controls, but both GA concentrations had the same effect. This is not consistent with the concentration effect on plant height. So this raises several questions. Is it a limitation of the experimental design or methods? Are there more biological effects that are not being detected or measured? In your discussion, acknowledge these sorts of issues and make suggestions for further experiments to address them.

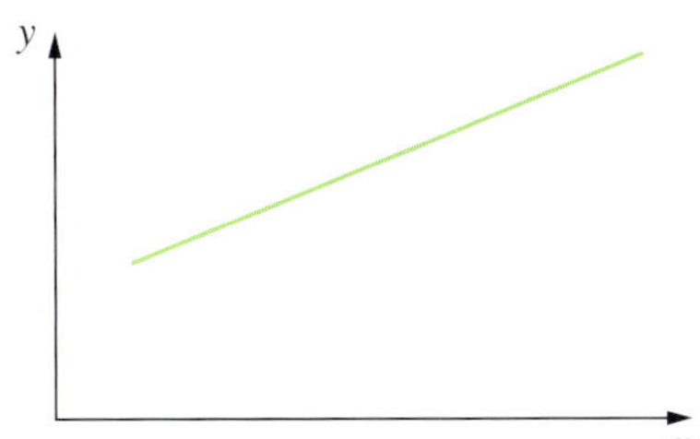

a Direct or linear proportional relationship

- Variables change at the same rate (graph line is straight, slope is constant)
- Positive relationship—as *x* increases, *y* increases

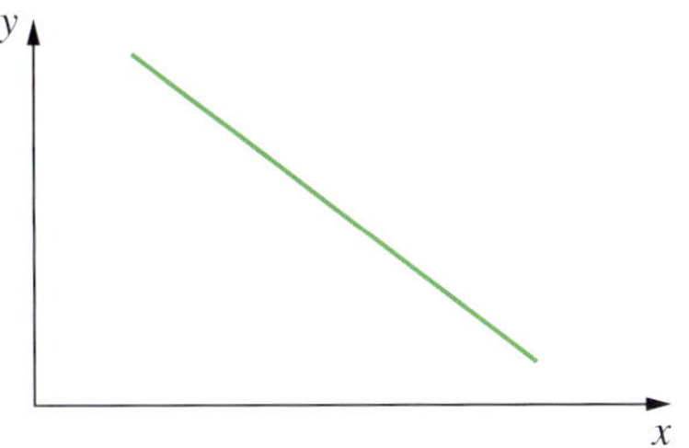

d Inverse direct or linear proportional relationship

- Variables change at the same rate (graph line is straight, slope is constant)
- Negative relationship—as *x* increases, *y* decreases

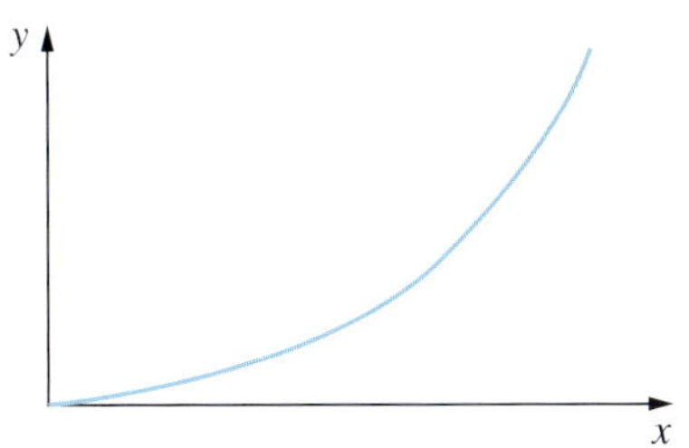

b Exponential relationship

- As *x* increases, *y* increases slowly, then more rapidly

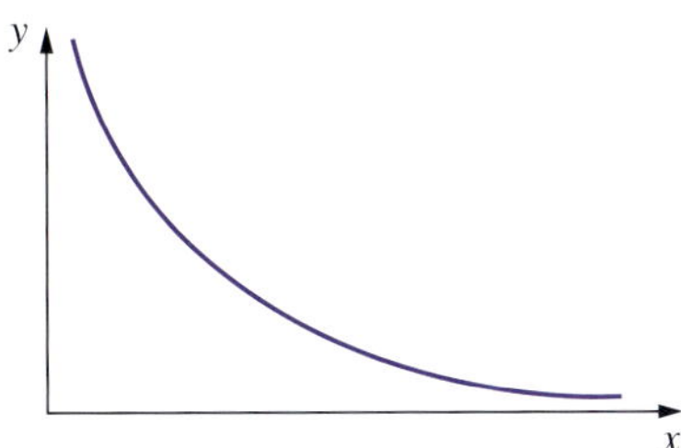

e Inverse exponential relationship

- As *x* increases, *y* decreases rapidly, then more slowly, until a minimum *y* value is reached

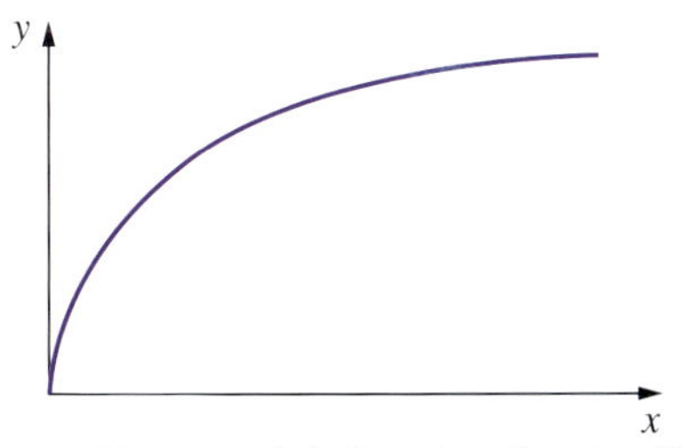

c Exponential rise, then levels off or plateaus (stops rising)

- As *x* increases, *y* increases rapidly at first, then slows, then finally does not increase at all—*y* reaches a maximum value

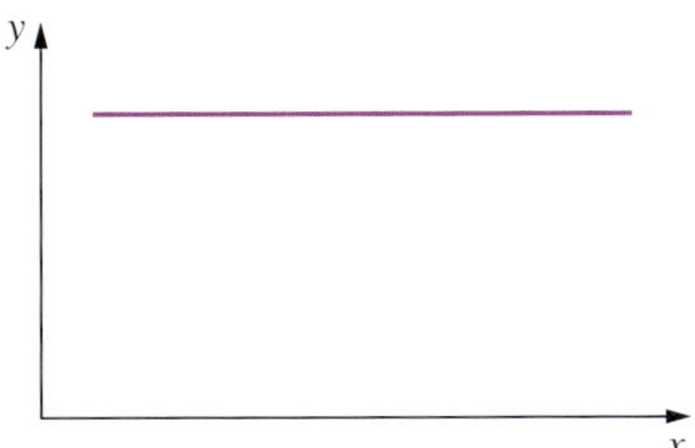

f No relationship between *x* and *y*

- As *x* increases, *y* remains the same

FIGURE 1.6.3 Line graphs illustrating common relationships between variables: (a) direct linear relationship, (b, c) exponential relationships, (d, e) inverse relationships, (f) no relationship.

Relate findings to biological concepts

In your introduction you established a context. Now you have a framework in which to discuss whether your data supports or refutes your hypothesis. Providing context also enables you to compare your results with existing research and knowledge. Use the points in the Figure 1.6.4 to help frame your discussion.

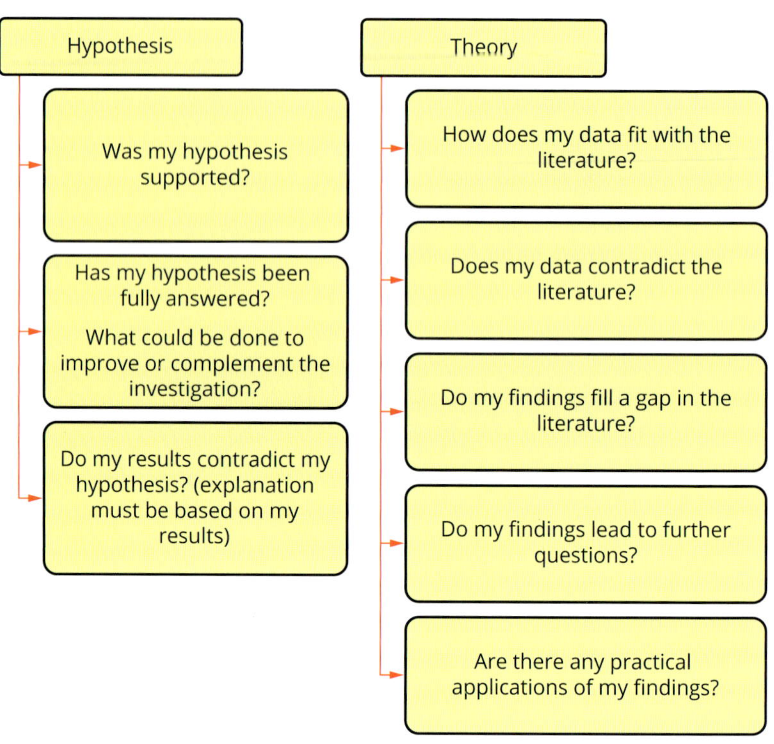

FIGURE 1.6.4 Points to help frame your discussion.

Conclusion

Your conclusion should be one or two paragraphs that link your evidence to your hypothesis. It should provide a carefully considered response to your research question based on your results and discussion. You should clearly state whether your hypothesis was supported or not. Draw your conclusions by identifying trends, patterns and relationships in the data.

It is important to appreciate your limitations and the limitations of the scientific method. Be careful not to overstate your conclusion. Your results will support or not support the hypothesis. They will not 'prove' something is true, as you can only ever provide evidence that indicates the probability of something being true.

Do not provide irrelevant information or introduce new information in your conclusion. Refer to the specifics of your hypothesis and research question, and do not make generalisations.

- *You will now be able to answer Key Questions 2 and 3.*

References

All the scientific papers and other sources that are mentioned in the report are to be listed at the end of your report. Cite any information you obtained from secondary sources in the text of your report, and provide a list of references at the end of your report. This demonstrates that you are aware of previous work in the area, and allows readers to locate sources of information if they want to study them further.

The usual approach is to give a short reference in the text, such as 'Hedden and Sponsel (2015)', and give the full reference in the reference list. If you are stating factual information from another source, you can either quote it word-for-word, or rewrite it in your own words. However, if you rewrite it you must make it clear that the information is not your own. Plagiarism (claiming that another person's work is your own) is not tolerated in scientific research.

Table 1.6.6 shows examples of ways to reference the three most common sources of information: journal articles, books and web pages. Use a consistent format for all references.

Source of information and Example of reference in text	**Format for listing references** and Example of a reference as written in the reference list
Research article or review article in a scientific journal GA is well established as a naturally occurring plant growth regulator with effects on... (Hedden and Sponsel, 2015)	**Author(s), date, article title, journal, volume number, page numbers** Hedden P and Sponsel V. (2015) A Century of Gibberellin Research *J Plant Growth Regul* 34:740–760
Book The molecular mechanisms of plant growth regulators is being studied by many groups (Karssen et al. 2012)	**Author(s) or Editor(s), date (and page number of referenced text), title of book, publisher, publisher's location** Karssen CM, van Loon LC, and D. Vreugdenhil D. (eds), (2012) *Progress in Plant Growth Regulation* Springer Science & Business Media, NY
Online article or page Plant hormones play many roles in plant growth and development, and sensing and responding to environment, possibly even by 'hearing' (Coghlan, 1998)	**Author(s), title of article, URL address, date accessed** Coghlan A. (1998) *Sensitive Flower Magazine*, issue 2153, published 26 September 1998 https://www.newscientist.com/article/mg15921534-900-sensitive-flower/ (accessed 1 January 2016)

TABLE 1.6.6 Examples of references for three common information sources.

Acknowledgements

Finally, it is important to acknowledge anyone who has assisted you in your investigation. This includes people who helped you find appropriate literature and references, learn to use equipment, prepare solutions, set up the experiments, find and navigate online databases, edit your report or prepare graphs and images.

1.6 Review

SUMMARY

- Your reports should include the following sections:
  - title
  - introduction
  - materials and methods
  - results
  - discussion
  - conclusion
  - references
  - acknowledgements.
- The title should give a clear idea of what the report is about, without being too long.
- The introduction sets the context of your report. It should outline relevant biological ideas, concepts, theories and models, and how they relate to your specific question and hypothesis.
- A materials and methods sections should:
  - clearly state the materials required and the method used to conduct your study
  - be presented in a clear, logical order that accurately reflects how you conducted your study.
- A results section should state your results and display them using graphs, figures and tables, but not interpret the results.
- A discussion should:
  - interpret data (identifying patterns, discrepancies and limitations)
  - evaluate the investigative method (identifying any issues that may have affected validity, reliability, accuracy or precision), make recommendations for improving the method
  - explain the link between investigation findings and relevant biological concepts (defining concepts and investigation variables, discussing the investigation results in relation to the hypothesis, linking the investigation's findings to existing knowledge and literature, and discussing the implications and possible applications of the investigation's findings).
- A conclusion should succinctly link the evidence collected to the hypothesis and research question, indicating whether the hypothesis was supported or refuted.
- References and acknowledgements should be presented in an appropriate format.

KEY QUESTIONS

1 a Which of the graphs A–D shows that the rate of respiration increases in direct proportion to an increase in temperature?

b Which of the graphs A–D depicts the results of a mammalian cell culture experiment in which a hormone stimulates cells to multiply exponentially, and then slow down when the nutrient supply is depleted?

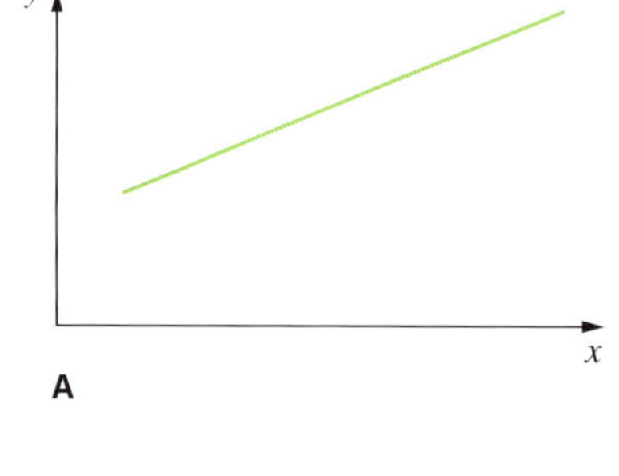

A

B

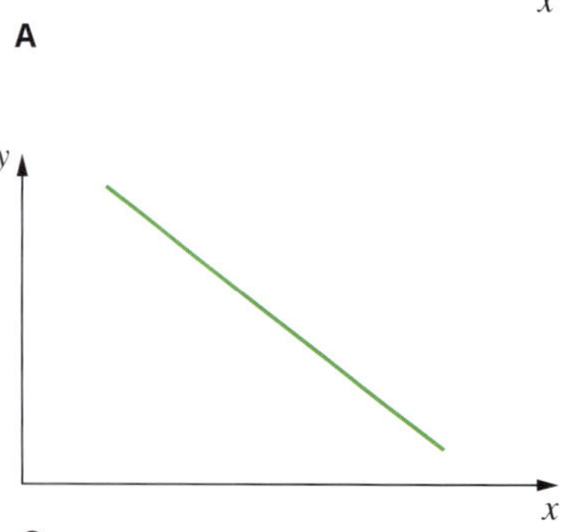

C

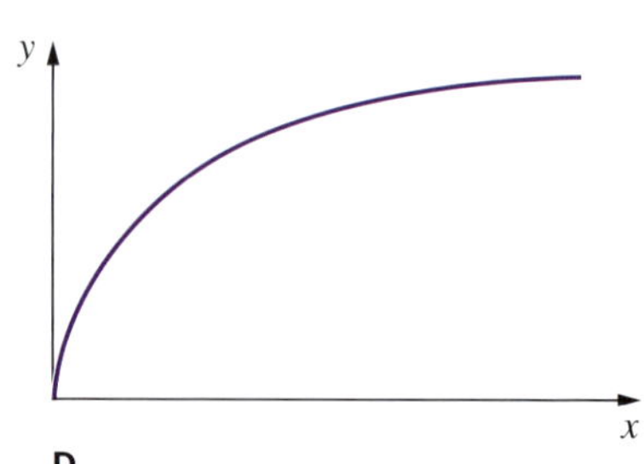

D

2 A scientist designed and conducted an experiment to test the following hypothesis: An increased consumption of fast food causes a decrease in the function of the liver.

a The discussion section of the scientist's report included comments on the reliability and validity of the investigation. Read each of the following statements and determine whether they relate to reliability or validity.

i Only teenage boys were tested.

ii Six boys were tested.

b The scientist then conducted the fast food study with 50 people in the experimental group and 50 people in the control group. In the experimental group, all 50 people gained weight. The scientist concluded all the subjects gained weight as a result of the experiment. Is this conclusion valid? Explain why or why not.

c What recommendations would you make to the scientist to improve the investigation?

3 Review the plant hormone experiment outlined in this section and answer the following questions.

a Discuss whether the experimental design, materials and methods were described clearly enough. For example, are there any missing experimental details needed to repeat the experiment? Would you suggest a different layout?

b How would you interpret the results?

c Write a conclusion for the experiment. Remember to state whether or not the results support the hypothesis.

KEY TERMS

accuracy
aim
bar graph
column graph
continuous variable
control group
controlled variable
dependent variable
discrete variable
error
experimental group
exponential relationship
first-hand data
hypothesis
independent variable
inference
inverse relationship
line graph
linear relationship
mean
median
meniscus
mode
nominal variable
observation
ordinal variable
outlier
peer-reviewed
pie chart
precision
primary source
principle
processed data
qualitative data
quantitative data
random selection
range
raw data
reliability
repeat trials
replication
research question
risk assessment
safety data sheet
scatterplot
scientific method
secondary source
second-hand data
theory
uncertainty
valid
validity
variable

UNIT 3 How do cells maintain life?

AREA OF STUDY 1

How do cellular processes work?

Outcome 1: On completion of this unit the student should be able to explain the dynamic nature of the cell in terms of key cellular processes including regulation, photosynthesis and cellular respiration, and analyse factors that affect the rate of biochemical reactions.

AREA OF STUDY 2

How do cells communicate?

Outcome 2: On completion of this unit the student should be able to apply a stimulus–response model to explain how cells communicate with each other, outline human responses to invading pathogens, distinguish between the different ways that immunity may be acquired, and explain how malfunctions of the immune system cause disease.

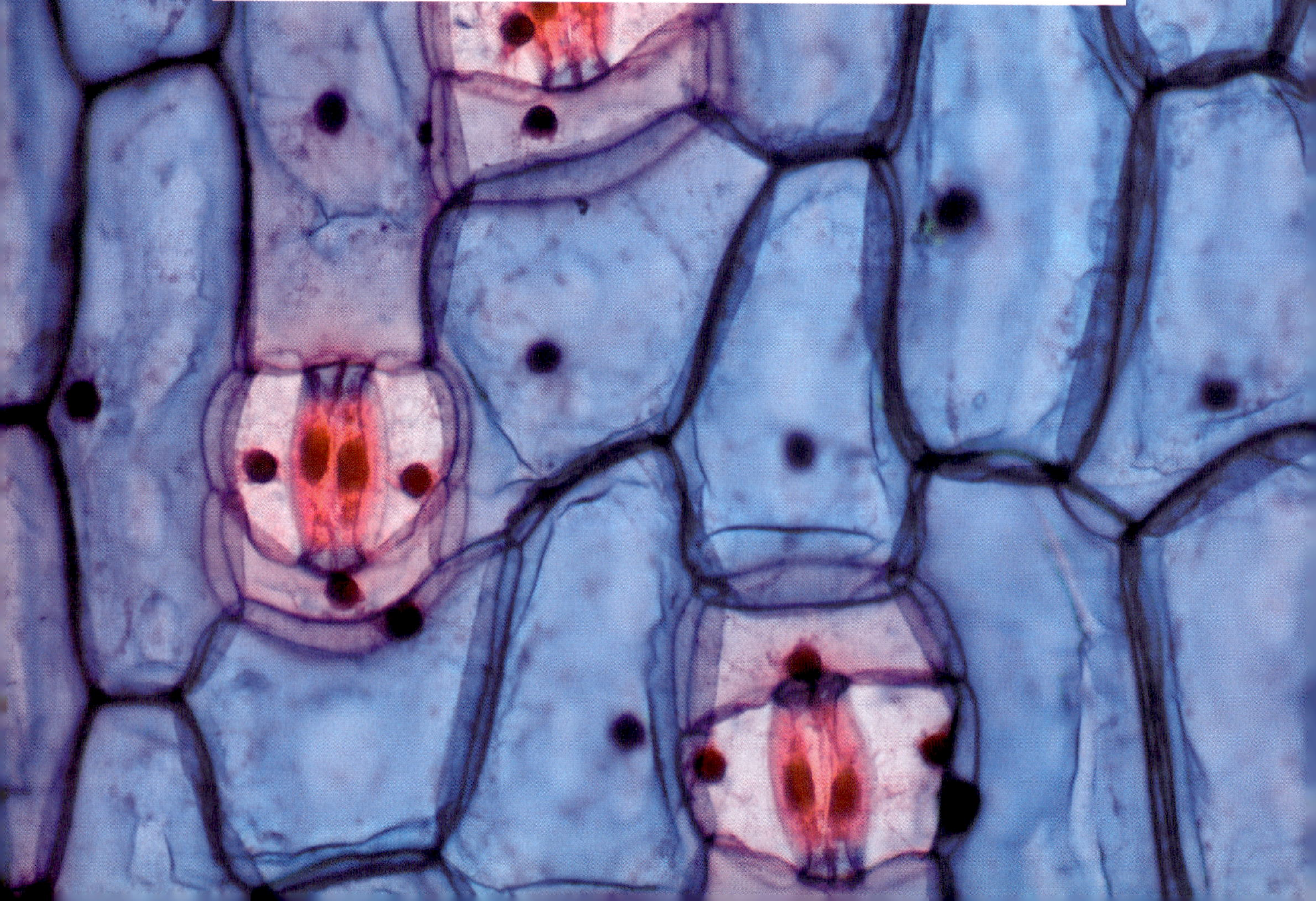

Cells and composition of cells

Learning outcomes

By the end of this chapter, you will have an understanding of the fluid mosaic model of the plasma membrane and of the different processes that a cell can use to move substances across this membrane. You will also be able to describe the various functions that proteins play in cells and the nature of the proteome. Finally, you will be able to explain how proteins are synthesised and to outline the importance of the four hierarchal levels of protein structure.

Key knowledge

- the fluid mosaic model of the structure of the plasma membrane and the movement of hydrophilic and hydrophobic substances across it based on their size and polarity
- the role of different organelles including ribosomes, endoplasmic reticulum, Golgi apparatus and associated vesicles in the export of a protein product from the cell through exocytosis
- cellular engulfment of material by endocytosis
- protein functional diversity and the nature of the proteome
- the functional importance of the four hierarchal levels of protein structure
- the synthesis of a polypeptide chain from amino acid monomers by condensation polymerisation.

2.1 Cells

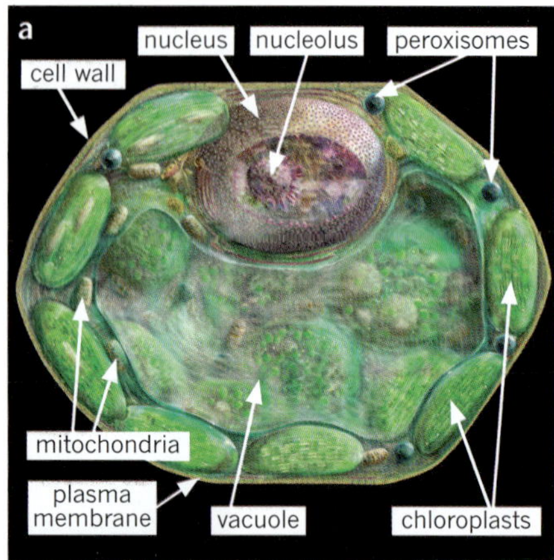

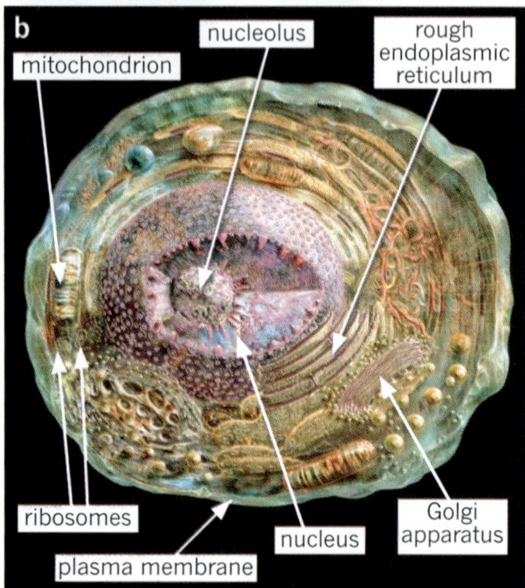

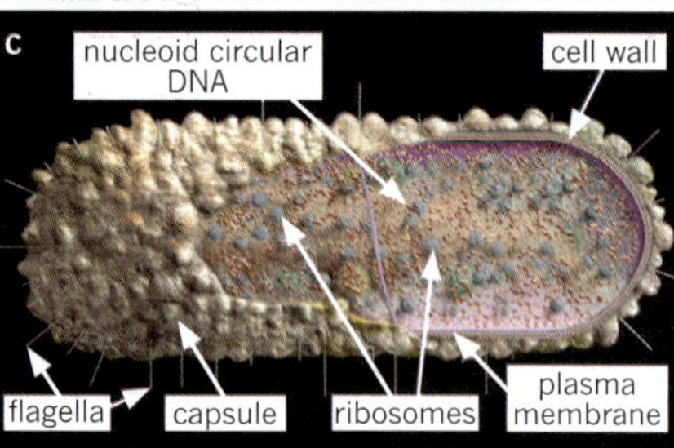

FIGURE 2.1.1 Cells can be classified as eukaryotic cells, which include (a) plant and (b) animal cells, or (c) prokaryotic cells.

An organelle is one of the specialised structures in a cell. Examples include the Golgi apparatus, mitochondria and vacuoles.

An enzyme is a protein molecule that acts as a biological catalyst. Enzymes speed up the rates of reactions that would otherwise take place much more slowly.

Cells are the basic structural and functional units of life on Earth. The cell theory is one of the fundamental principles of biology, and describes the properties of cells. Cells can be classified into two types, prokaryotic and eukaryotic cells (Figure 2.1.1). Each type of cell has many different structures in place to sustain life.

In this section you will learn about cell theory and the differences between prokaryotic and eukaryotic cells. The structure and function of organelles of cells will also be explored.

CELL THEORY

If you are to understand life you need to understand how cells work. **Cells** are the basic functional units of living organisms. The cell theory is based on detailed microscopic and experimental studies of tissues, from all types of organisms, carried out over the last 300 years.

The cell theory states that:

- all organisms are composed of cells (and the products of cells)
- all cells come from pre-existing cells
- the cell is the basic organisational unit of living things.

All types of cells perform similar basic processes and many also carry out highly specialised functions (Figure 2.1.2). The activities of cells require considerable energy, and produce a variety of biological molecules. These biological molecules, called biomolecules, are used to build new **organelles**, and used for repair or exported from the cell. All these processes are **catalysed** (sped up) by **enzymes** and are precisely regulated. Some biochemical processes involve hundreds of enzymes operating sequentially along a complex integrated chemical pathway in which each step is tightly controlled.

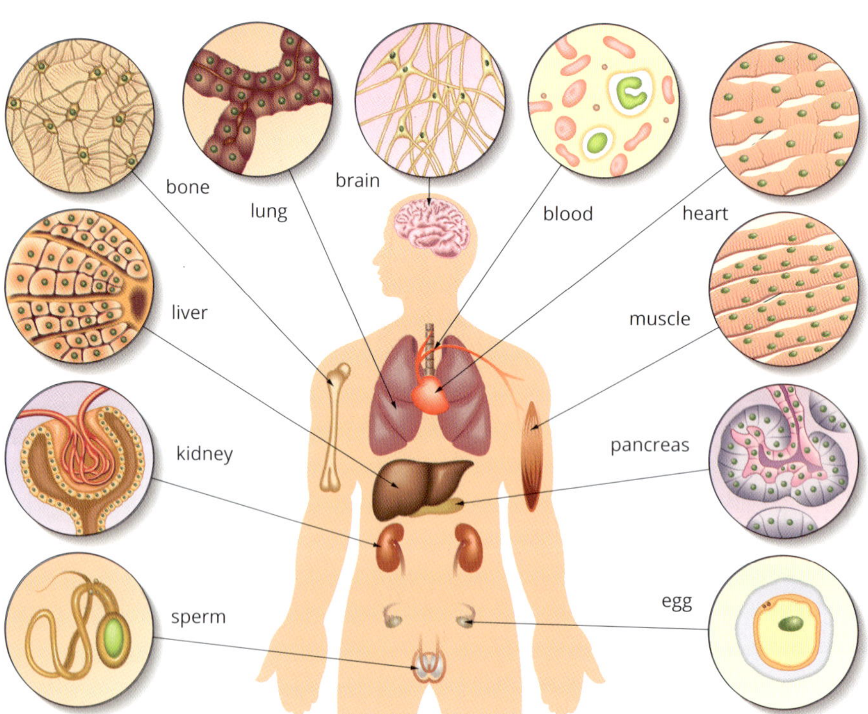

FIGURE 2.1.2 There are many different types of cells in the human body each carrying out specialised functions.

BIOLOGY **IN ACTION**

History of the cell theory

Our understanding that organisms are made of cells came about after the invention of the microscope at the turn of the 16th century. In 1590, Hans and Zacharias Janssen placed two convex lenses in a tube, thus making the first compound microscope.

Hooke: the discovery of cells

The first description of cells was made by Robert Hooke in his book *Micrographia*, published in 1665. Hooke took a thin slice of cork from the bark of a tree and examined it under a microscope he had made himself (Figure 2.1.3). He saw that the bark was made up of hundreds of little 'empty boxes' that gave it a honeycomb appearance. He called the boxes 'cells'.

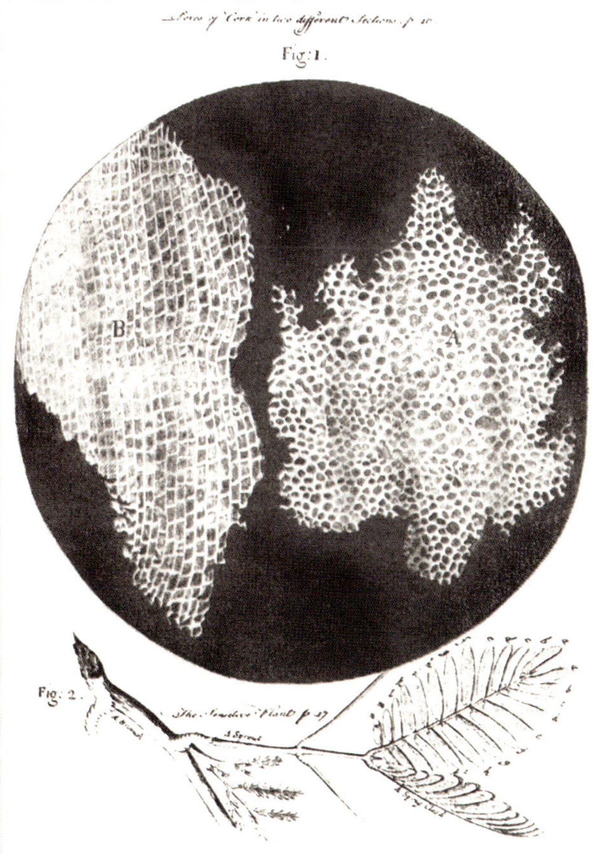

FIGURE 2.1.3 Robert Hooke's drawings of the cellular structure of cork from *Micrographia* (1665).

Hooke was actually looking at empty dead cells. When he later looked at fresh plant tissue, he noted the cells appeared to contain water. A few years later, Marcello Malpighi produced more detailed descriptions of plant cells.

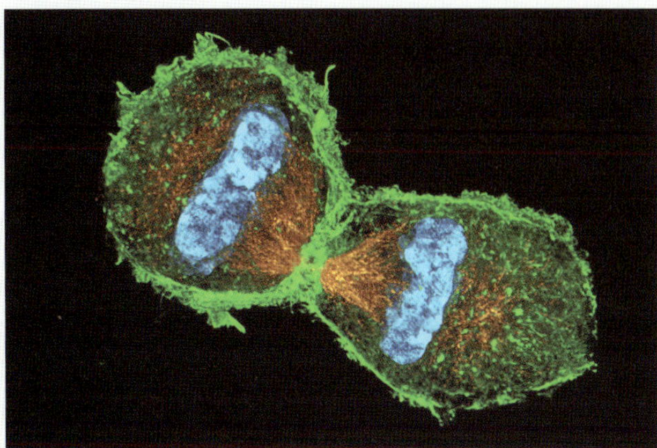

FIGURE 2.1.4 A cell divides to form a new cell.

Leeuwenhoek: first observations of living cells

In 1676 Anton van Leeuwenhoek observed many living cells under the microscope, including bacteria, blood cells and sperm. He was the first to describe the reproduction of unicellular organisms, which he called 'animalcules'.

Lamarck and Dutrochet: all living things are composed of cells

By the early 19th century the microscope had become a standard tool of biologists, and living animal and plant cells were easy to observe. In the early 19th century Jean Lamarck stated that all living things are a mass of cells, and that complex solutions move in and out of cells. René Dutrochet supported this idea, stating that: 'plants are composed entirely of cells, or of organs that are obviously derived from cells... the same is true for animals'.

Schleiden and Schwann: cells are organised into tissues

By the middle of the 19th century the fundamental principle that entire organisms are composed of highly organised groups of cells was broadly accepted. This was largely because of the work of Matthias Schleiden on plant tissues, and Theodor Schwann on animal tissues.

Remak and Virchow: the theory of biogenesis

Until the 1840s most biologists still believed that cells formed spontaneously from body fluids or from the nucleus, which they thought was the embryo of a new cell. Then Robert Remak discovered that new cells were formed by a single cell dividing in two, with the nucleus dividing at the same time (Figure 2.1.4). In the 1850s Rudolph Virchow used Remak's discovery to popularise the theory of biogenesis: that all cells come from pre-existing cells.

TYPES OF CELLS

There are two fundamentally different types of cells. Organisms are classified according to the type of cell they are composed of:

- **Prokaryotes** are composed of prokaryotic cells and include bacteria and archaea. Prokaryotic cells are usually unicellular and are generally smaller and less complex than eukaryotic cells
- **Eukaryotes** are composed of eukaryotic cells and include plants, animals, fungi and protists. Eukaryotic cells contain membrane-bound organelles.

Common features of cells

There is really no such thing as a typical cell. Cells are specialised for many different purposes and their structures reflect those purposes. However, there are some features that are shared by all cells:

- Plasma membrane (also called the cell membrane) is a semipermeable structure that separates the interior of the cell from the exterior environment.
- Cytoplasm comprises the cytosol (gel-like substance). In eukaryotes, the cytoplasm includes the cytosol and the organelles present in the cell except the **nucleus**.
- Deoxyribonucleic acid or DNA carries hereditary information, directs the cell's activities and is passed accurately from generation to generation. (In prokaryotes, the DNA is located in a region called the nucleoid; in eukaryotes, the DNA is located the nucleus.)
- **Ribosomes** are structures that assist in the synthesis of proteins. Although they are not membrane bound, ribosomes are often grouped with organelles.

Table 2.1.1 opposite shows some examples of the structures found inside different types of eukaryotic and prokaryotic cells.

BIOFILE

Enzymes: biological catalysts

All biological reactions within cells depend on enzymes. Enzymes as biological catalysts enable biological reactions to occur usually in milliseconds. Dr Richard Wolfenden, a scientist who has carried out extensive research on enzyme mechanisms, has reported that without enzymes a biological transformation essential in creating the building blocks of DNA and RNA would take 78 million years, and the biosynthesis of haemoglobin and chlorophyll would take 2.3 billion years, about half the age of Earth.

FIGURE 2.1.5 Dr Richard Wolfenden, Alumni Distinguished Professor of Biochemistry and Biophysics, and member of the National Academy of Sciences.

Prokaryotic cells

Prokaryotic organisms are unicellular and have a simple cell structure. Prokaryotic organisms can be found everywhere, even in extreme environments such as volcanoes.

The structure of a typical prokaryotic cell is shown in Figure 2.1.6. Prokaryotic cells are small and lack membrane-bound organelles, including a distinct nucleus. Their cytoplasm contains scattered ribosomes that are involved in the synthesis of proteins. The genetic material of prokaryotic cells is usually a single, circular DNA chromosome called the genophore, which is contained in an irregularly shaped region called the nucleoid. The nucleoid does not have a nuclear membrane, unlike the nucleus of eukaryotes.

This chromosomal DNA is attached to the plasma membrane by a region of the chromosome called the origin. In addition to this chromosomal DNA, many prokaryotic cells also contain small rings of double-stranded DNA called plasmids.

Protista are a large group of mostly single-celled organisms that cannot be classified as a plant, animal or fungus, but that have plant, animal and fungus-like characteristics. Table 2.1.1 shows an example of a photosynthetic protist that contains chloroplasts for photosynthesis, and a flexible, helical protein structure (the pellicle) that supports the cell membrane.

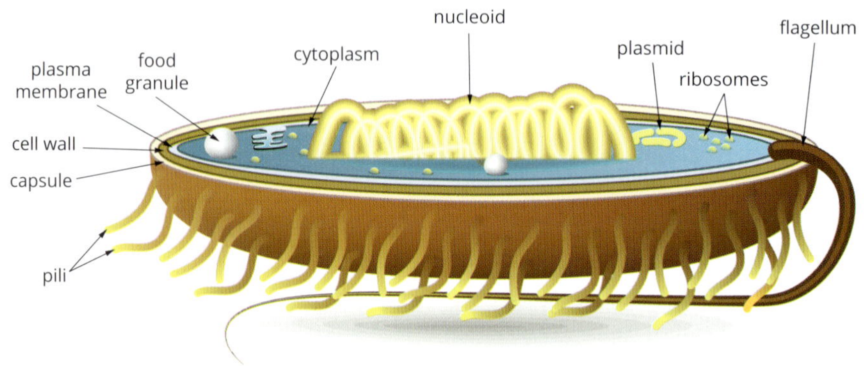

FIGURE 2.1.6 A cross-sectional diagram of a typical prokaryotic bacterial cell.

The plasma membrane of prokaryotic cells is surrounded by an outer **cell wall** of protein and complex carbohydrate (murein). Many bacteria also have a capsule outside the cell wall, which protects the cell from damage and dehydration. Many prokaryotes also have **flagella** (singular flagellum) that enable them to move freely, and small hair-like projections called pili (singular pilus), which are involved in the transfer of DNA between organisms and which can also help generate movement. Specialised pili that can attach to surfaces are called fimbriae.

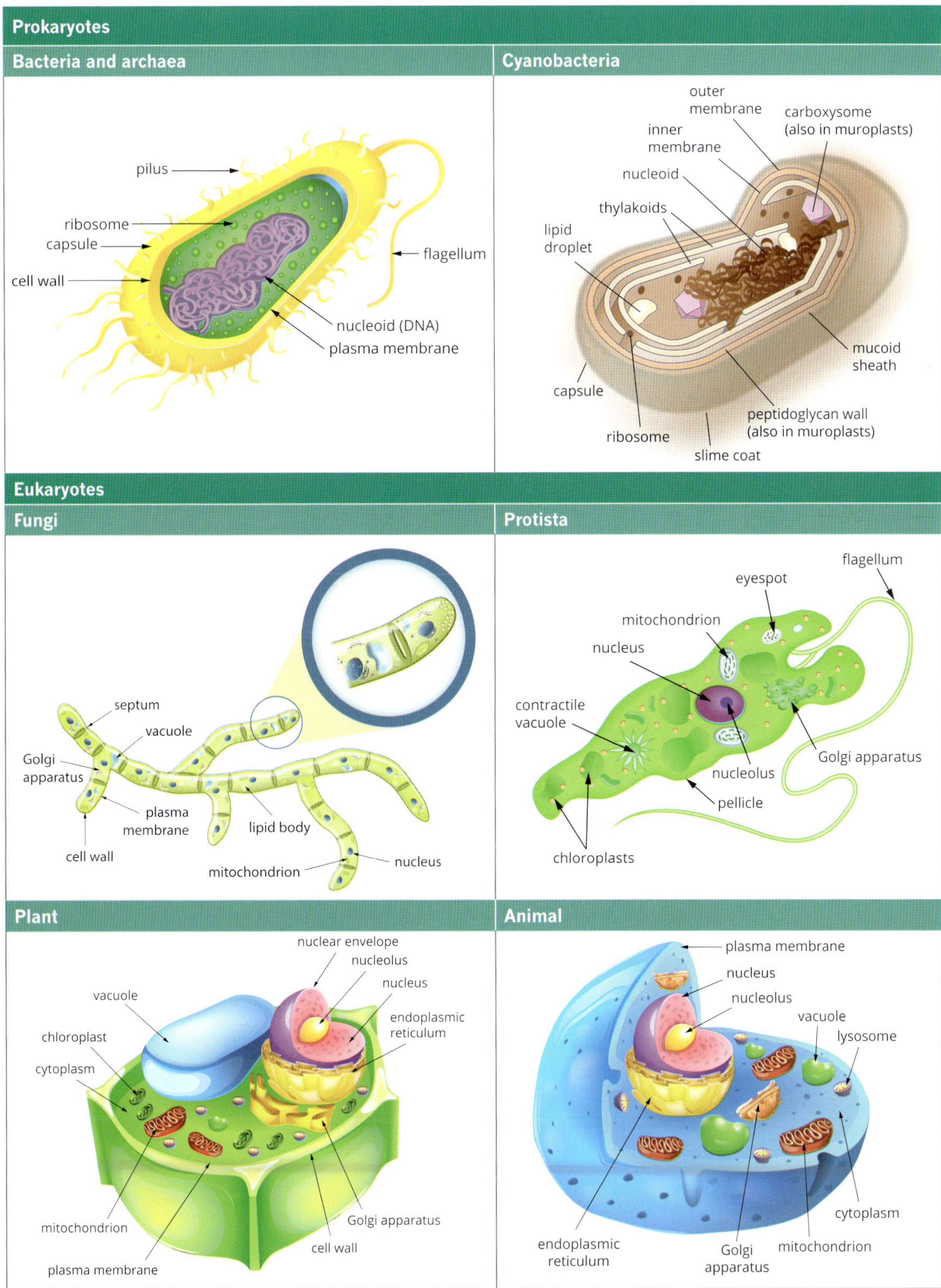

TABLE 2.1.1 These representative diagrams of bacterial, fungal, protistan, plant and animal cells show the organelles that are visible using the electron microscope.

Eukaryotic cells

Eukaryotic cells not only have a plasma membrane surrounding the cytoplasm, but also have internal (non-plasma) membranes that form specialised membrane-bound compartments within the cell. This is known as cell compartmentalisation. The membrane-bound structures are called organelles.

Organelles of eukaryotic cells

Organelles are subcellular structures that have a specific function. Because they have a specific function, their presence depends on the needs of the cell. In eukaryotes, some organelles are membrane-bound and some are not (Figure 2.1.7). Prokaryotic cells possess some non-membrane bound organelles such as ribosomes, flagella and a cell wall, although the structure and composition of these is usually different from that of equivalent eukaryotic organelles.

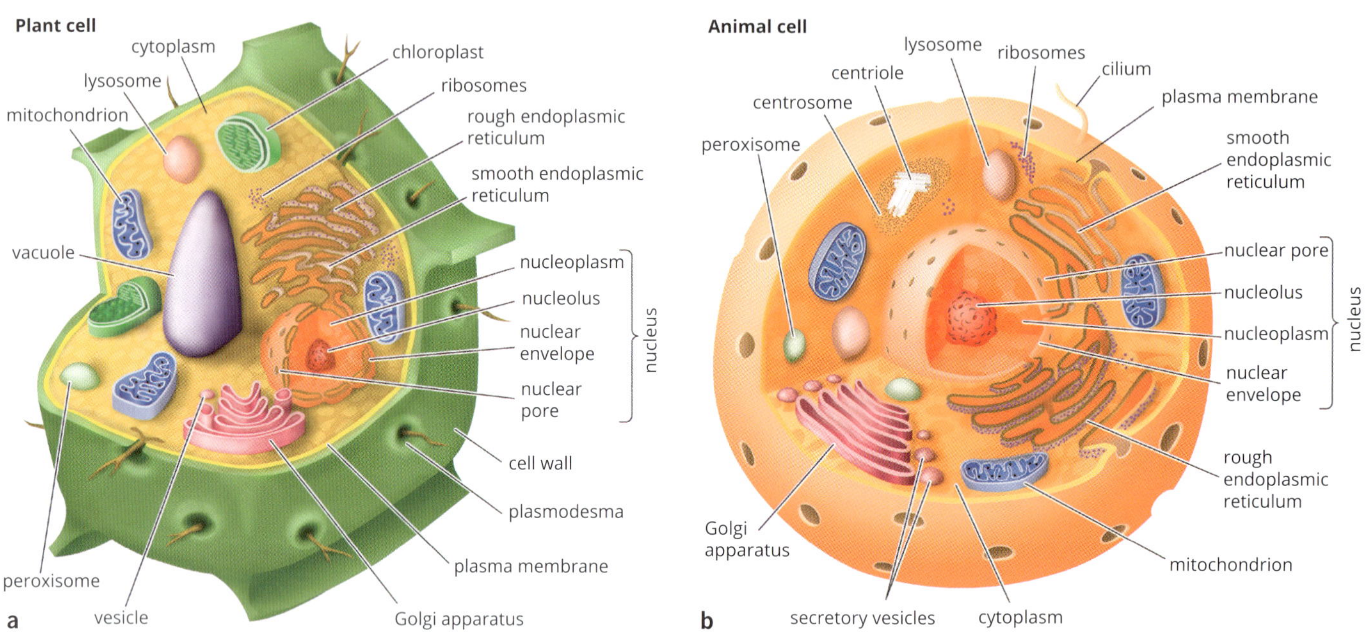

FIGURE 2.1.7 The many organelles of eukaryotic cells can be seen in these illustrations of (a) a plant and (b) an animal cell.

BIOFILE

Chloroplasts

Chloroplasts are one of the features that distinguish a plant cell from an animal cell. Chloroplasts contain chlorophyll, a green pigment that is needed to absorb energy from sunlight during photosynthesis. A chloroplast contains stacks (circular) of flattened membranes, called grana, that contain the chlorophyll. It is thought that, during cell evolution, chloroplasts originated as once-independent microorganisms.

Some of the most important organelles of eukaryotes are listed in Table 2.1.2 with a summary of their structure and function. Definitions for each of the organelles listed in Table 2.1.2 can also be found in the Glossary. Table 2.1.2 also indicates whether each organelle is present in animal and plant cells. Table 2.1.3 includes some other important structures that are not considered to be organelles.

TABLE 2.1.2 Structure and function of eukaryotic organelles.

Organelle: nucleus		Organelle: nucleolus
Structure: • membrane-bound: double membrane • contains DNA **Function:** • contains genetic information (used for the synthesis of proteins) • directs activities of the cell **Present in plants:** Yes **Present in animals:** Yes		**Structure**: • made up of proteins and RNA **Function:** • responsible for formation of incomplete ribosomes **Present in plants:** Yes **Present in animals:** Yes
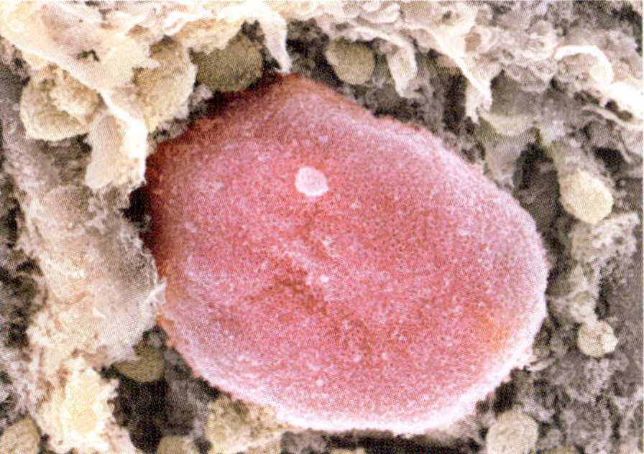 Coloured scanning electron micrograph (SEM) of a section through a liver cell showing the nucleus (pink) and the nuclear envelope with its many pores (tiny circles).	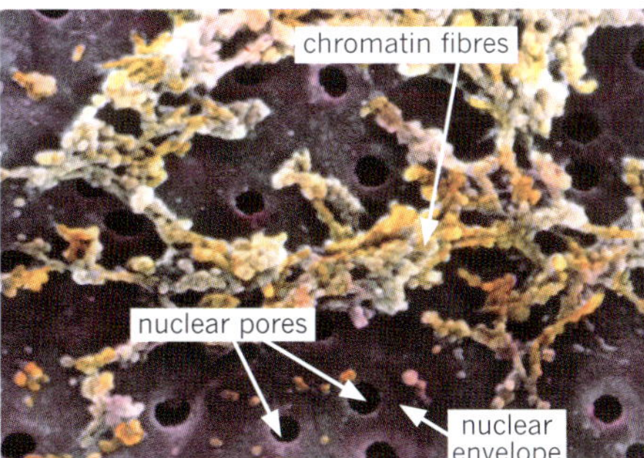 Coloured SEM of the external surface of a nuclear envelope in an onion root tip cell. The envelope consists of a double membrane (purple), with nuclear pores (black circles). Contained within the nucleus are the chromatin fibres (yellow and orange).	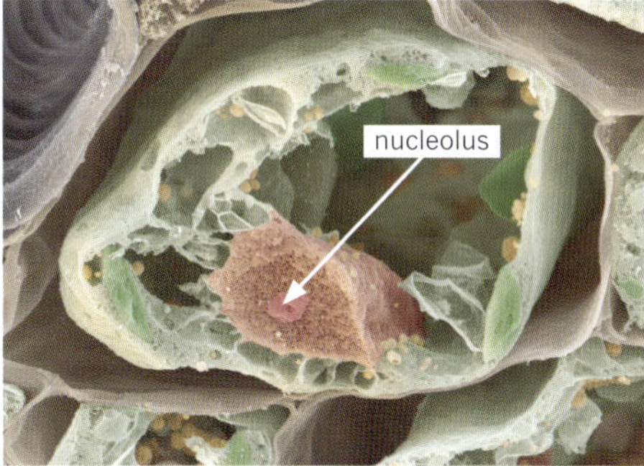 Coloured SEM of a section through cells in leaf tissue from an *Aconitum* sp. plant. At centre is a cell nucleus with its nucleolus (red).
Organelle: rough endoplasmic reticulum (RER)	**Organelle: ribosome**	**Organelle: Golgi apparatus (also known as Golgi body, Golgi complex)**
Structure: • membrane-bound • composed of a network of membranous tubules and sacs (called cisternae) • ribosomes bind to the membrane **Function:** • synthesises and processes proteins (often by adding carbohydrates to proteins produced by the ribosomes to form glycoproteins) **Present in plants:** Yes **Present in animals:** Yes	**Structure**: • composed of proteins and ribosomal RNA • found free in the cytoplasm or attached to endoplasmic reticulum **Function:** • synthesise proteins (translate messenger RNA into proteins) • RER-bound ribosomes synthesise proteins for export from the cell **Present in plants:** Yes **Present in animals:** Yes	**Structure**: • membrane-bound • stack of cisternae that are not connected to each other **Function:** • further processes and packages proteins into vesicles for export from the cell (except lysosomes, which remain in the cell) **Present in plants:** Yes **Present in animals:** Yes
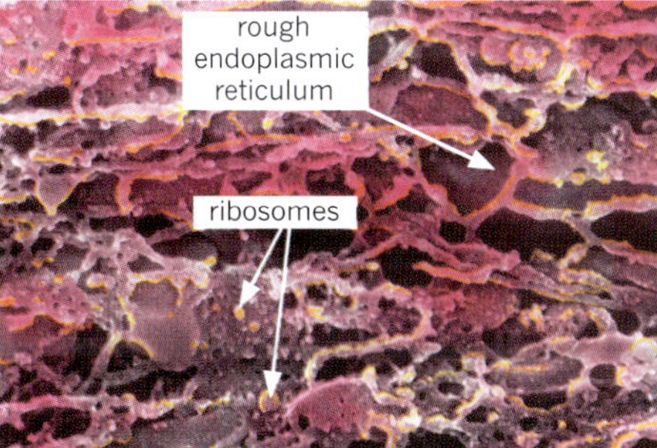 Coloured SEM of endoplasmic reticulum in an olfactory epithelium supporting cell. On the surface of some of the ER membranes are ribosomes (yellow spheres).	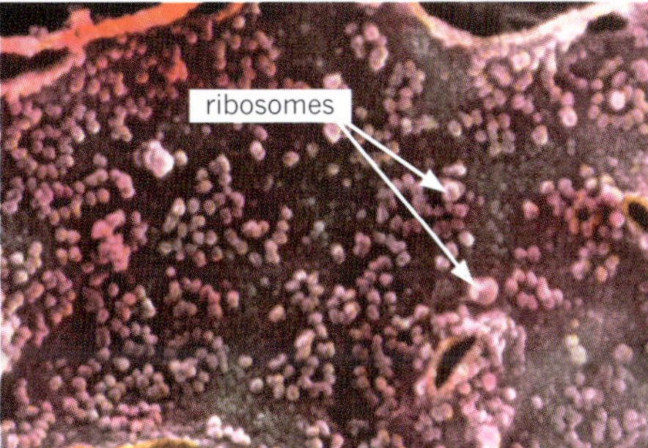 Coloured SEM of rough endoplasmic reticulum in an olfactory bulb mitral cell. On the surface of the ER membrane are numerous ribosomes (small spheres).	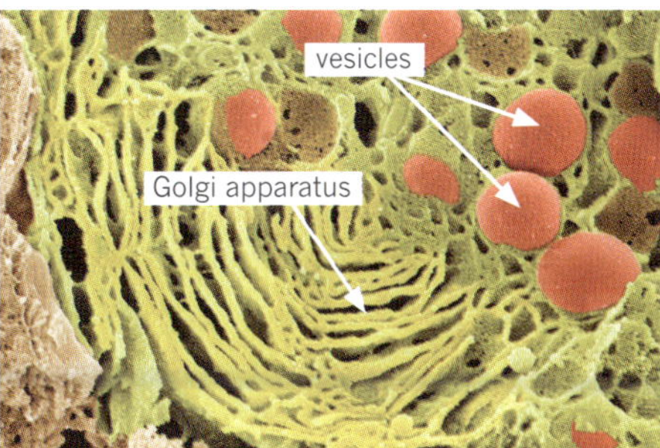 Coloured SEM of a pancreatic cell, showing the Golgi apparatus and vesicles.

TABLE 2.1.2 Structure and function of eukaryotic organelles (continued).

<table>
<tr><th>Organelle: lysosome</th><th>Organelle: smooth endoplasmic reticulum (SER)</th><th>Organelle: mitochondrion</th></tr>
<tr><td>Structure:
• membrane-bound
• vesicle containing digestive enzymes

Function:
• digest waste and foreign material

Present in plants: No

Present in animals: Yes</td><td>Structure:
• membrane-bound
• network of cisternae

Function:
• synthesis of lipids

Present in plants: Yes

Present in animals: Yes</td><td>Structure:
• membrane-bound: double membrane; the inner membrane is highly folded
• contains DNA

Function:
• release energy from organic compounds

Present in plants: Yes

Present in animals: Yes</td></tr>
<tr><td>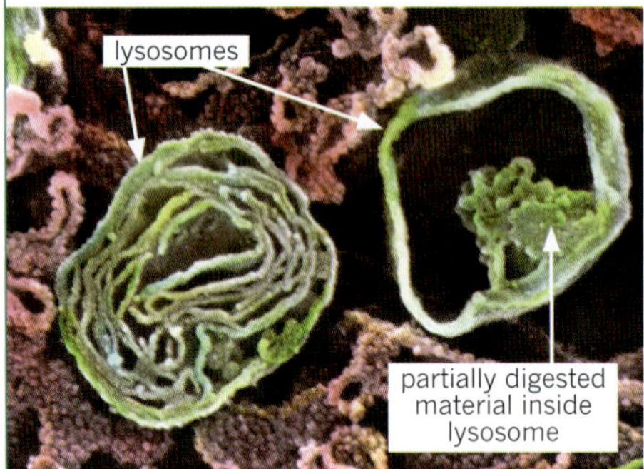

Coloured SEM of two lysosomes in a pancreatic cell. Lysosomes (green) are small spherical vesicles bound by a single membrane (clearest on right lysosome). Left lysosome is shown in cross-section.</td><td>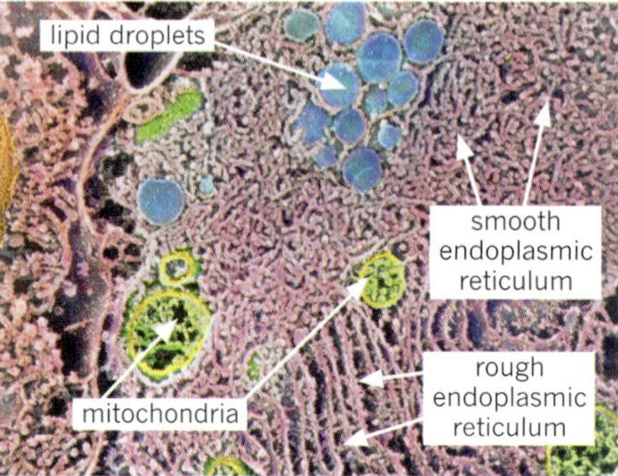

Coloured SEM showing smooth (top right) and rough (bottom centre) endoplasmic reticulum (light pink) inside a cell. Lipid droplets (round blue structures) and mitochondria can also be seen in this image.</td><td>

Coloured SEM of a single mitochondrion (pink, centre) in the cytoplasm of an intestinal epithelial cell.</td></tr>
<tr><th>Organelle: chloroplast</th><th colspan="2">Organelle: centriole</th></tr>
<tr><td>Structure:
• membrane-bound: double membrane
• contains thylakoids (disc-shaped, membranous sacs)
• contains DNA

Function:
• use light energy, carbon dioxide and water to produce glucose

Present in plants: Yes

Present in animals: No</td><td colspan="2">Structure:
• small structures composed of microtubules

Function:
• involved in cell division
• involved in the formation of cell structures such as flagella and cilia

Present in plants: Sometimes

Present in animals: Yes</td></tr>
<tr><td>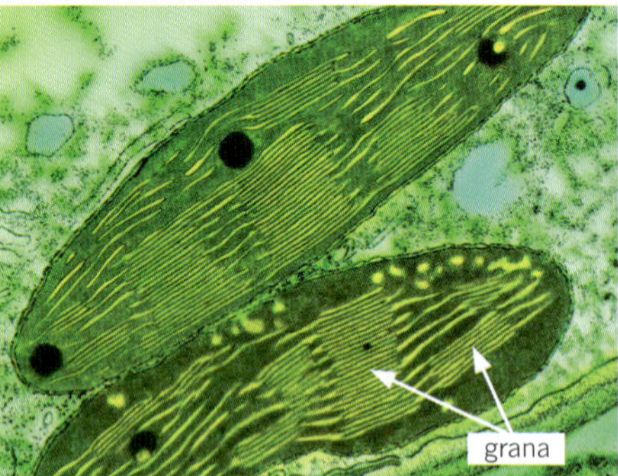

Coloured transmission electron micrograph (TEM) of two chloroplasts seen in the leaf of a pea plant Pisum sativum. Each chloroplast is seen cut lengthways and contains stacks of flattened membranes (yellow) known as grana.</td><td>

Coloured TEM of two centrioles.</td><td>
Diagram showing the barrel-shaped structure of centrioles within a human cell.</td></tr>
</table>

TABLE 2.1.2 Structure and function of eukaryotic organelles (continued).

<table>
<tr><th colspan="2">Organelle: cilia and flagella</th><th>Organelle: vacuoles</th></tr>
<tr><td colspan="2">Structure:
• external structures
• composed of microtubules
Function:
• involved in movement (of the cell or of things around the cell)
Present in plants: Yes
Present in animals: Yes</td><td>Structure:
• membrane-bound
• vesicles
Function:
• storage
• in plant cells, cell structure
Present in plants: Yes
Present in animals: Small or absent</td></tr>
<tr><td>
Coloured SEM of the trachea showing mucus-secreting goblet cells (yellow) and epithelial cells bearing hair-like cilia (green).</td><td>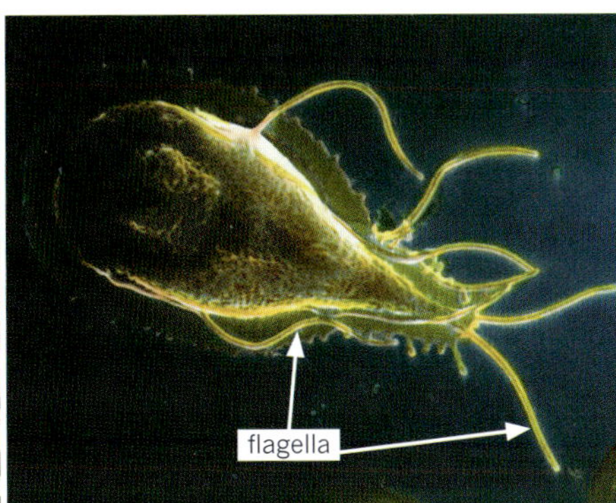

Coloured SEM of a Giardia parasite, which moves by using its long, hair-like flagella.</td><td>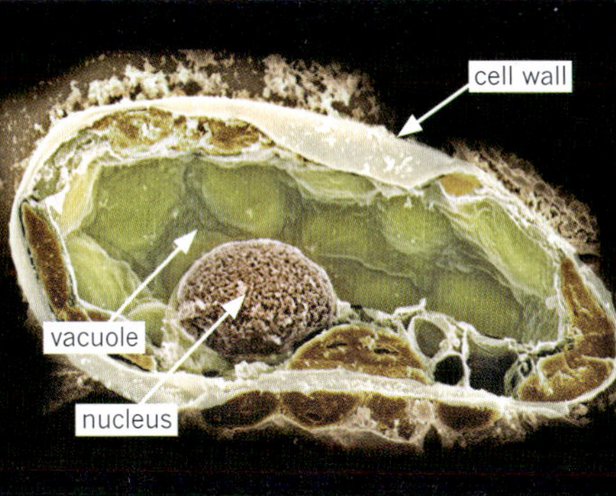

Coloured SEM of a section through a plant cell showing a large vacuole at the centre of the cell.</td></tr>
<tr><th>Organelle: plastids other than chloroplasts</th><th>Organelle: peroxisomes</th><th></th></tr>
<tr><td>Structure:
• small
• membrane-bound: double membrane
• contains DNA
Function:
• synthesis and storage of diverse organic compounds
Present in plants: Yes
Present in animals: No</td><td>Structure:
• small
• membrane-bound: double membrane
• contains DNA
Function:
• involved in a variety of metabolic reactions including oxidation of harmful materials
Present in plants: Yes
Present in animals: Yes</td><td></td></tr>
<tr><td>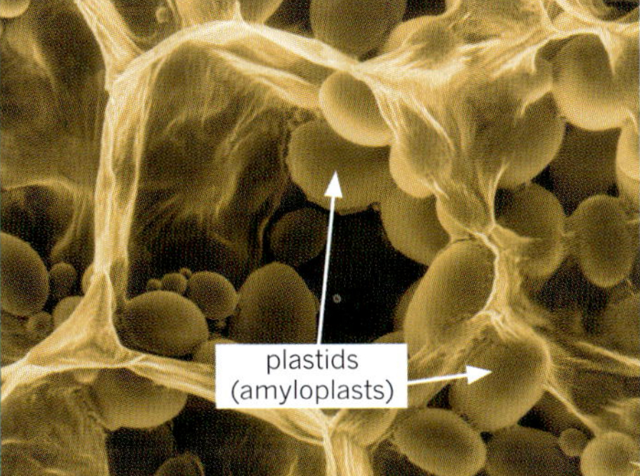

Coloured SEM of amyloplasts (oval) in the sectioned cells of a potato (Solanum tuberosum). Amyloplasts are starch-storing plastids.</td><td>
Close up view of plant cell structure showing the cell wall (yellow), the nucleus (pink), and two peroxisomes (blue spheres).</td><td></td></tr>
</table>

TABLE 2.1.3 Other important structures.

<table>
<tr><th>Cell wall *The cell wall is not an organelle, but it is an important structure of plant and bacterial cells.</th><th colspan="2">Cytoskeleton *The cytoskeleton is not an organelle, but it is an important structure of plant and bacterial cells.</th></tr>
<tr><td>Structure:
• external structure that surrounds the plasma membrane
• composed of cellulose (note that prokaryotic bacteria also have a cell wall composed of murein)

Function:
• structural support and protection

Present in plants: Yes

Present in animals: No</td><td colspan="2">Structure:
• three-dimensional internal support and transport network that fills the cytoplasm of eukaryotic cells

Function:
• provides both internal support and a transport network for movement and stability through microfilaments, intermediate filaments and microtubules

Present in plants: Yes

Present in animals: No</td></tr>
<tr><td>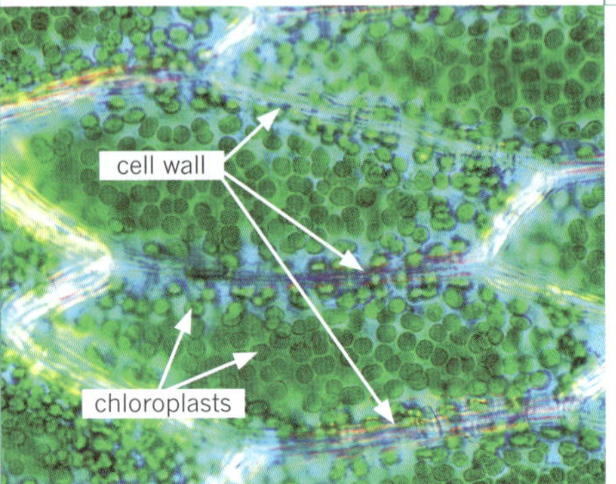
</td><td>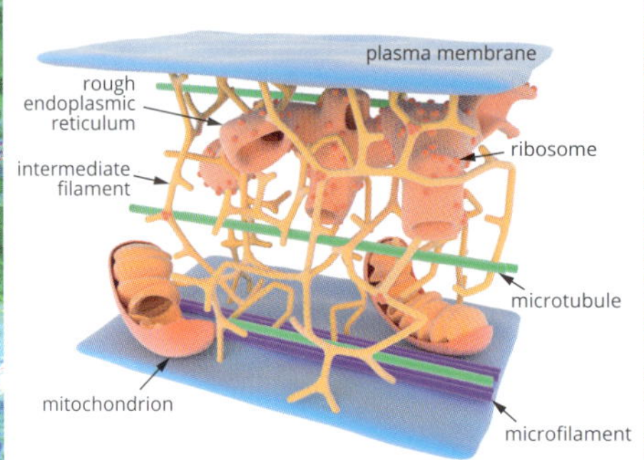
</td><td>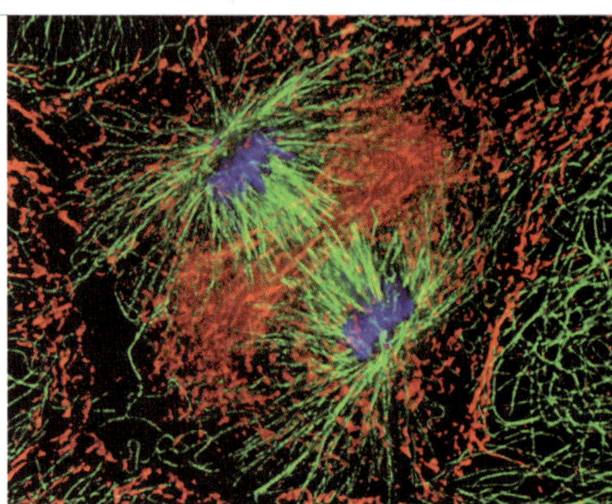</td></tr>
<tr><td>Polarised light micrograph of cells in a leaf of shining Hookeria moss (Hookeria lucens) showing cell wall (blue), which encloses each cell, and numerous chloroplasts.</td><td>The internal support and transport network found in cells. The cell membrane (blue) is partially supported by structures of the cytoskeleton such as microfilaments (purple), intermediate filaments (yellow), and microtubules (green).
(Also shown are cell organelles such as mitochondria (orange) and rough endoplasmic reticulum (pink), with ribosomes (red dots) on the rough endoplasmic reticulum).</td><td>During mitosis, the microfilaments (red) and tubulin microtubules (green) of the cytoskeleton maintain the cell's structure.</td></tr>
</table>

2.1 Review

SUMMARY

- Cells are the basic structural units of organisms.
- The cell theory states that:
  - all organisms are composed of cells (or the products of cells)
  - all cells come from pre-existing cells
  - the cell is the smallest living organisational unit.
- There are two fundamentally different types of cells: prokaryotic and eukaryotic.
- Prokaryotes include bacteria and archaea. Prokaryotic cells are small, with a relatively simple structure, and are usually unicellular.
- Eukaryotes include protists, fungi, plants and animals. Eukaryotic cells contain membrane-bound organelles in the cell cytoplasm.
- Some common features shared by all cells are the plasma membrane, cytoplasm, genetic material in the form of DNA, and ribosomes.
- Prokaryotic cells have a simple structure, with a nucleoid lacking a membrane, scattered ribosomes, and DNA mainly in a single-stranded loop in the nucleoid.
- Eukaryotic cells have a complex structure, a membrane-bound nucleus, many organelles in the cell cytoplasm, and DNA mainly in chromosomes in the nucleus.
- The main structures in a plant cell include the nucleus, vacuole, Golgi apparatus, rough and smooth endoplasmic reticulum, ribosomes, plastids, mitochondria and cell wall.
- The main structures in an animal cell include the nucleus, ribosomes, Golgi apparatus, rough and smooth endoplasmic reticulum, vacuoles, mitochondria, lysosomes, vesicles and centrioles.

KEY QUESTIONS

1 State the cell theory.
2 Distinguish a prokaryotic cell from a eukaryotic cell.
3 Compare plant and animal cells.
4 Label the parts of the animal cell in this diagram.

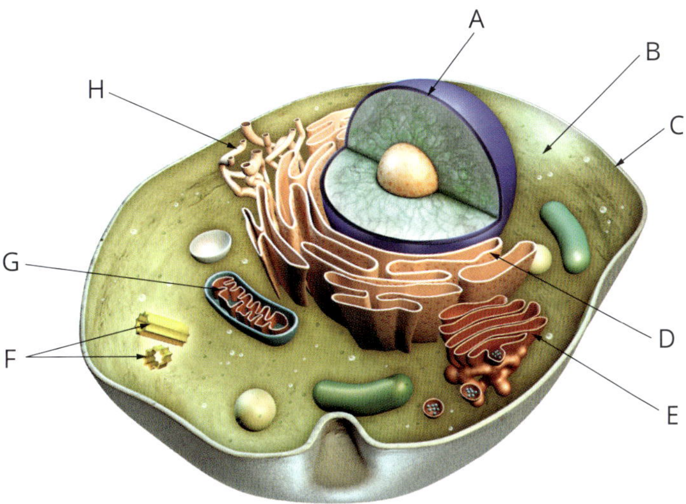

2.2 Molecular composition of organisms

> Elements are substances consisting of one type of atom.

> Compounds are made up of two or more different types of atoms.

> Molecules are made up of two or more atoms of the same or different types.

All life is composed of the same few elements. There are 92 naturally occurring elements. Only 11 of these are found in organisms in more than trace amounts, and four of these—carbon (C), hydrogen (H), oxygen (O) and nitrogen (N)—make up 99% of organisms by mass. The same elements are also found in rocks, soil and air; however, there is a difference in the way that these atoms are organised into larger **compounds** in living organisms (Figure 2.2.1). Organisms produce compounds that contain carbon and hydrogen known as **organic compounds**. All other compounds, whether in living or non-living things, are called **inorganic compounds**.

In this section, you will learn about the difference between organic and inorganic compounds. In addition, the four main types of organic molecules—nucleic acids, carbohydrates, lipids and proteins—will also be explored.

FIGURE 2.2.1 Aerial view of the Murray River wetlands lagoons and basin. The plants, water and animals all contain molecules. Only the plants and animals produce organic molecules.

ORGANIC MOLECULES

Organisms produce characteristic complex compounds that contain carbon and hydrogen (Figure 2.2.2). These are called organic compounds because the first compounds discovered were produced by or found in organisms. Most large organic molecules are composed of many smaller organic molecules linked together.

FIGURE 2.2.2 Some common molecules in organisms. Carbon atoms are coloured black, oxygen red, hydrogen white and nitrogen blue.

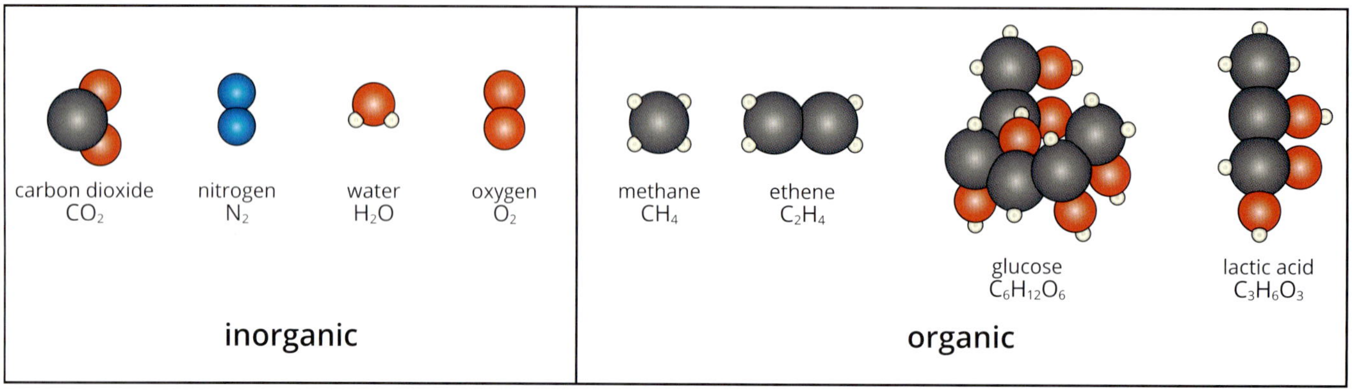

carbon dioxide CO_2
nitrogen N_2
water H_2O
oxygen O_2
inorganic

methane CH_4
ethene C_2H_4
glucose $C_6H_{12}O_6$
lactic acid $C_3H_6O_3$
organic

All other elements and compounds, whether in living or non-living things, are referred to as inorganic. Inorganic substances that are important for living organisms include water, oxygen, carbon dioxide, nitrogen and minerals.

The four main types of organic molecules are carbohydrates, proteins, nucleic acids and lipids (Figure 2.2.3). Carbohydrates, proteins and nucleic acids are huge and are also known as **biomacromolecules**. Biomacromolecules are chain-like molecules called **polymers** (*polys* meaning 'many' and *meros* meaning 'part'). Polymers are formed by joining together many smaller units (**monomers**) to form a chain.

In organisms, organic molecules can be converted from one form into another. Units may be linked together to form larger molecules. For example, glucose units may be linked together to form larger carbohydrates such as starch, glycogen or cellulose (Figure 2.2.3a). Other chemical groups may be attached to form molecules such as glycoproteins (proteins with sugars attached, Figure 2.2.4) and phospholipids (lipids with phosphate attached, Figure 2.2.5). When food is plentiful, carbohydrates are converted into fats for storage; when it is scarce, the reverse will occur and even proteins can be converted into small molecules to use for energy.

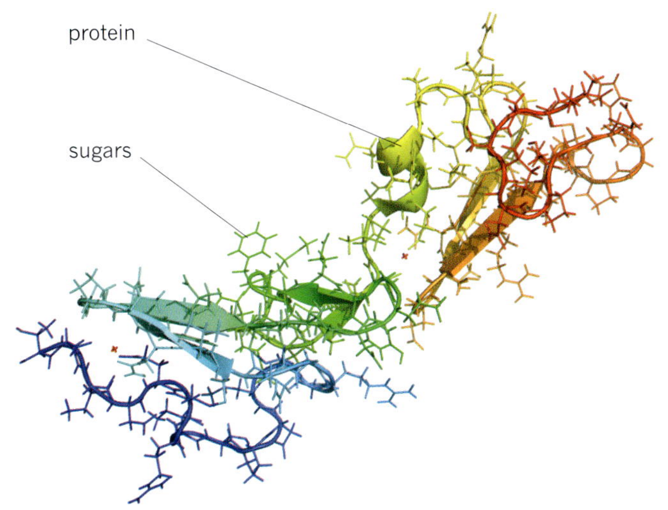

FIGURE 2.2.4 Computer model of a fibrillin glycoprotein. The protein component is represented by ribbons and the sugars are represented by the ring structures.

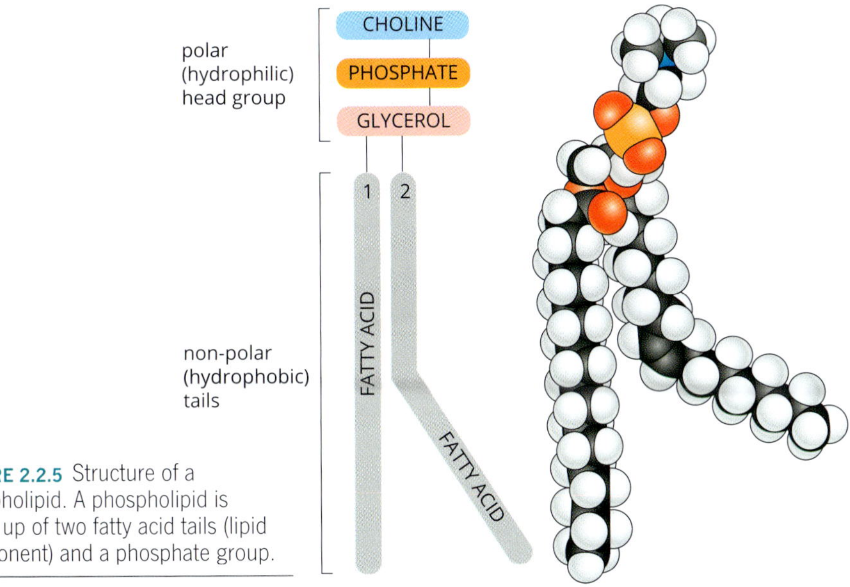

FIGURE 2.2.5 Structure of a phospholipid. A phospholipid is made up of two fatty acid tails (lipid component) and a phosphate group.

A molecule is made up of two or more atoms that are held together by chemical bonds. Any molecule that is found in a living organism is called a biomolecule. Examples of biomolecules are fatty acids, carbohydrates and hormones. Large biomolecules are called biomacromolecules. Biomacromolecules can be made up of thousands of atoms and include proteins and nucleic acids.

a Carbohydrates

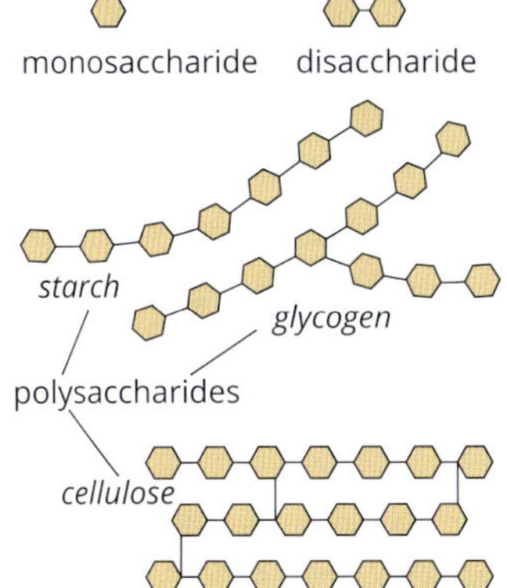

b Proteins

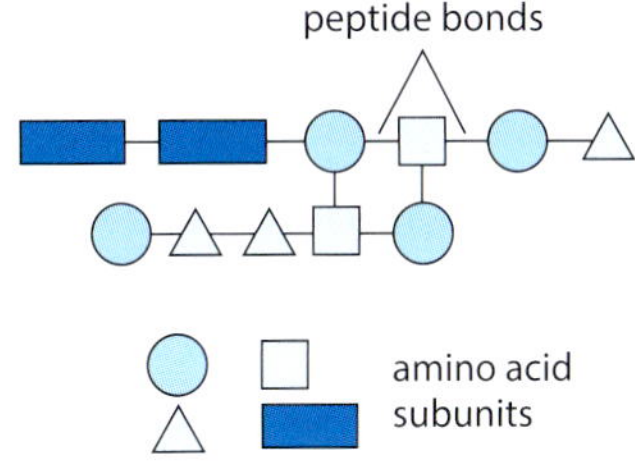

c Nucleic acids

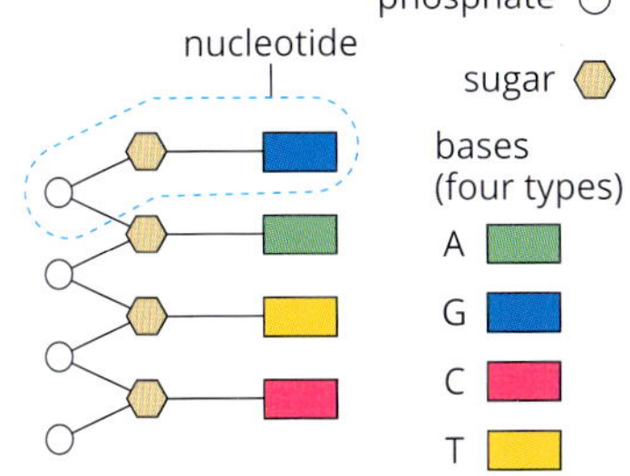

d Lipids

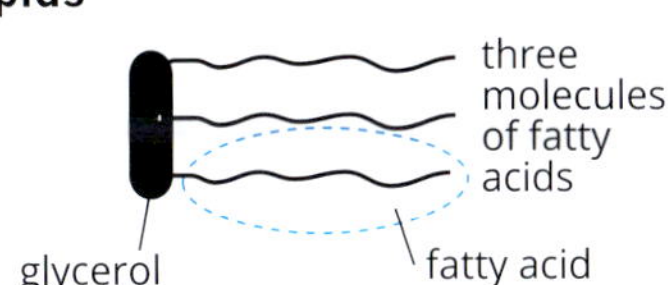

FIGURE 2.2.3 The structures of some organic molecules: (a) carbohydrates, (b) proteins, (c) nucleic acids and (d) lipids.

EXTENSION

A little chemistry

A little chemistry is useful to understand the activities of cells at the molecular level.

Elements are made of atoms

Atoms are the basic unit of all matter. Substances consisting of only one kind of atom are called elements. An atom has a nucleus (which is composed of positively charged protons and uncharged neutrons) and one or more negatively charged electrons in orbit around the nucleus. Atoms that have the maximum number of electrons in their outer 'shell' are most stable. The first shell (see the top row of the periodic table, Figure 2.2.6) can contain two electrons—hydrogen (H) has one electron and helium (He) has two. Helium has a full shell and is stable, hydrogen has not and this makes it likely to combine with other atoms (Figure 2.2.7).

The second and third shells (represented by the second and third rows in the periodic table) have a maximum of eight electrons. Neon (Ne) and argon (Ar) are the only stable atoms in these rows.

Chemical bonding of atoms makes molecules

Because atoms are more stable when their outer shells are filled, they tend to combine with other atoms to achieve this state, forming molecules. Molecules are two or more atoms held together by chemical bonds. Covalent bonds are created by sharing electrons between two atoms to achieve stability.

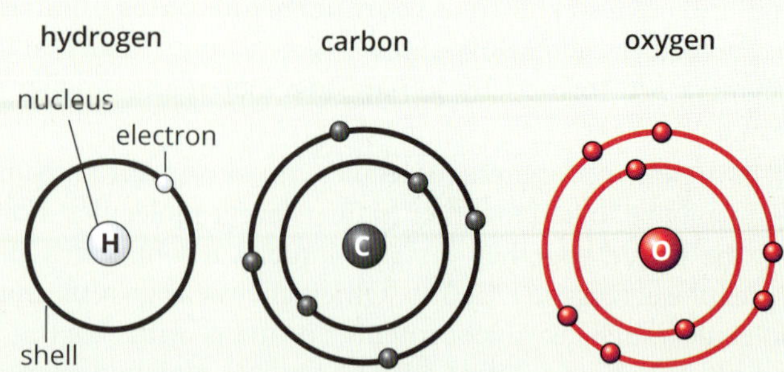

FIGURE 2.2.7 Hydrogen, carbon and oxygen showing electrons in shells.

Compound molecules, such as methane CH_4, involve the combination of different types of atoms (Figure 2.2.8). Covalent bonds between carbon and hydrogen are energy rich, which is why hydrocarbons (such as petrol and gas) make good fuels.

Sometimes in covalent bonds one atom attracts the shared electron more strongly than the other, resulting in a polar covalent bond. This is the case in water molecules where oxygen has a stronger attraction for the electrons, causing the molecules to be polar, meaning that they are slightly positive at the hydrogen atoms and slightly negative at the oxygen atom. Polar molecules have positive and negative regions (denoted by δ^+ and δ^- in Figure 2.2.9), whereas non-polar molecules have an even distribution of charge and are electrically neutral (Figure 2.2.8).

1 H																	2 He
3 Li	4 Be											5 B	6 C	7 N	8 O	9 F	10 Ne
11 Na	12 Mg											13 Al	14 Si	15 P	16 S	17 Cl	18 Ar
19 K	20 Ca	21 Sc	22 Ti	23 V	24 Cr	25 Mn	26 Fe	27 Co	28 Ni	29 Cu	30 Zn	31 Ga	32 Ge	33 As	34 Se	35 Br	36 Kr
37 Rb	38 Sr	39 Y	40 Zr	41 Nb	42 Mo	43 Tc	44 Ru	45 Rh	46 Pd	47 Ag	48 Cd	49 In	50 Sn	51 Sb	52 Te	53 I	54 Xe
55 Cs	56 Ba	57 La	72 Hf	73 Ta	74 W	75 Re	76 Os	77 Ir	78 Pt	79 Au	80 Hg	81 Tl	82 Pb	83 Bi	84 Po	85 At	86 Rn

FIGURE 2.2.6 This periodic table highlights the elements commonly found in living organisms: major elements (pink), other elements (yellow) and trace elements (blue).

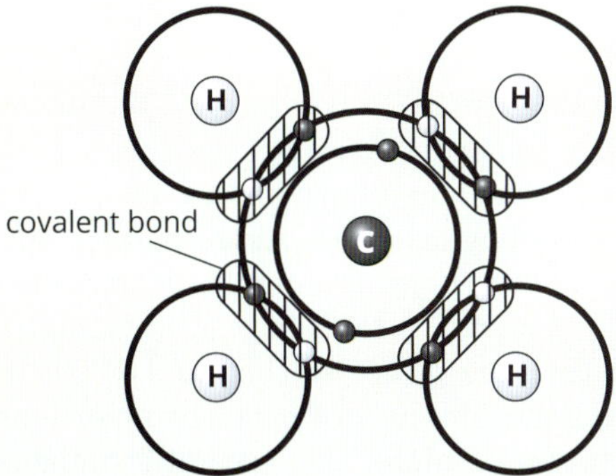

FIGURE 2.2.8 Hydrogen and carbon combine to form methane, a non-polar compound molecule held together by covalent bonds.

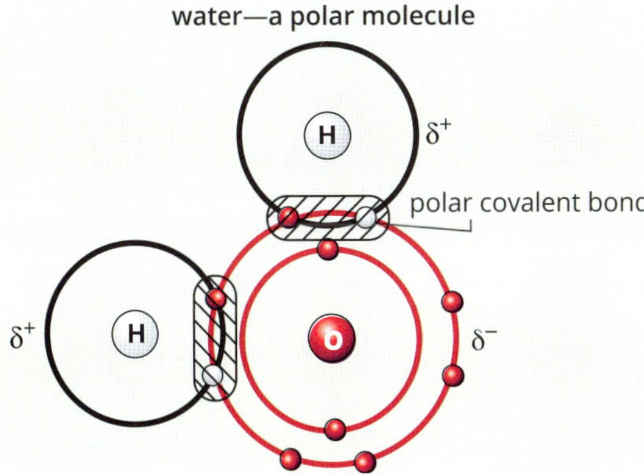

FIGURE 2.2.9 Oxygen and hydrogen combine to form water (a polar molecule held together by polar covalent bonds). Fractional charges on molecules are denoted as δ^+ (delta plus) and δ^- (delta minus).

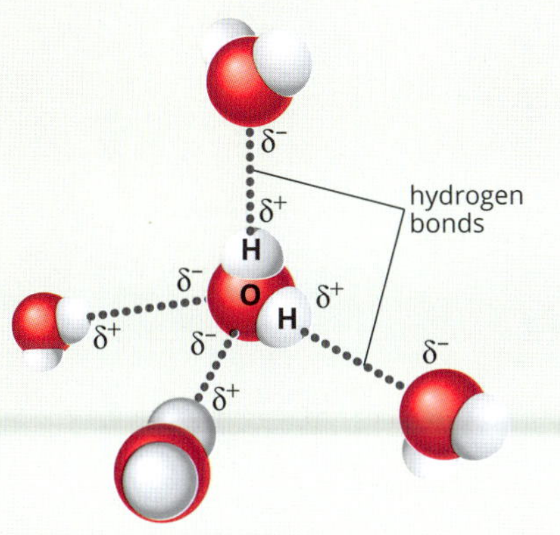

FIGURE 2.2.10 Hydrogen bonds form between hydrogen and oxygen atoms of polar water molecules.

Polarity of molecules is an important property in biology; for example, it governs the way that many molecules cross cell membranes. Individual water molecules are held together by hydrogen bonds. Hydrogen bonds are bonds between the slightly positive hydrogen atom of one polar molecule and the slightly negative region (usually an oxygen or nitrogen atom) of another polar molecule (Figure 2.2.10).

Sometimes the attraction for an electron is so strong that the electron actually leaves one atom to become part of another, resulting in the formation of ions (Figure 2.2.11). Ions are electrically charged atoms or group of atoms. The atom that loses electrons will be a positive ion (a cation) and the atom that gains electrons will be a negative ion (an anion). Positive and negative ions often come together and are held by weaker ionic bonds that can easily be broken in biological systems. The bonds formed in molecular recognition processes that are important to many biological functions usually include ionic bonds.

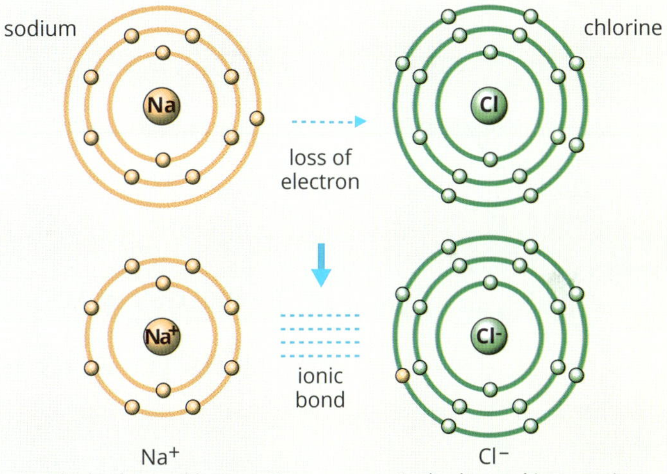

FIGURE 2.2.11 Sodium and chlorine form Na^+ and Cl^- held together by an ionic bond.

The special role of carbon

The periodic table (Figure 2.2.6) shows that carbon has four electrons in its outer shell (Figure 2.2.7). This allows each carbon atom to combine with up to four other atoms, as shown in Figure 2.2.8. This gives carbon the ability to form many different kinds and sizes of molecules with other atoms. This is why carbon is the key atom in organic molecules.

a glucose

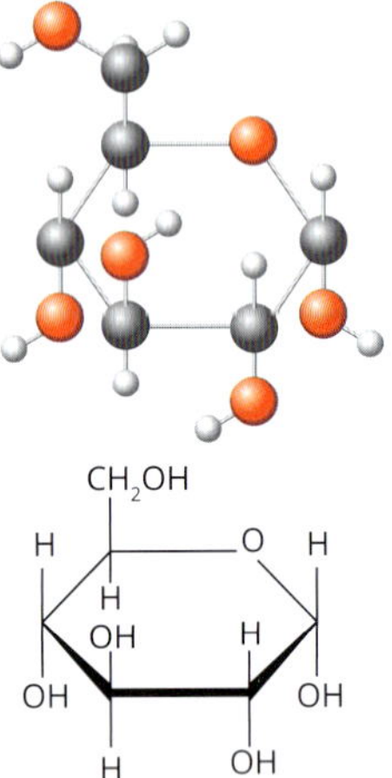

b fructose

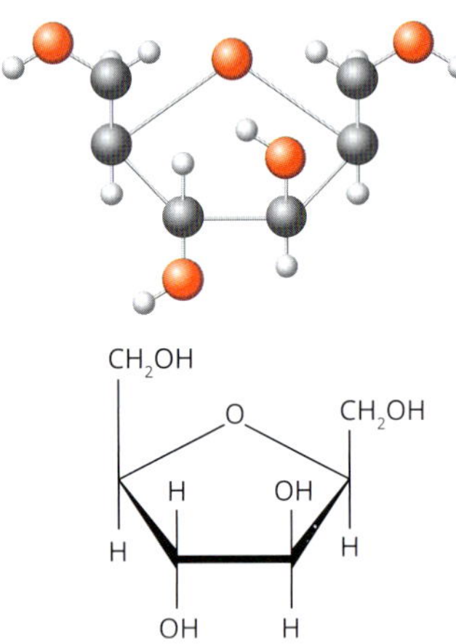

c galactose

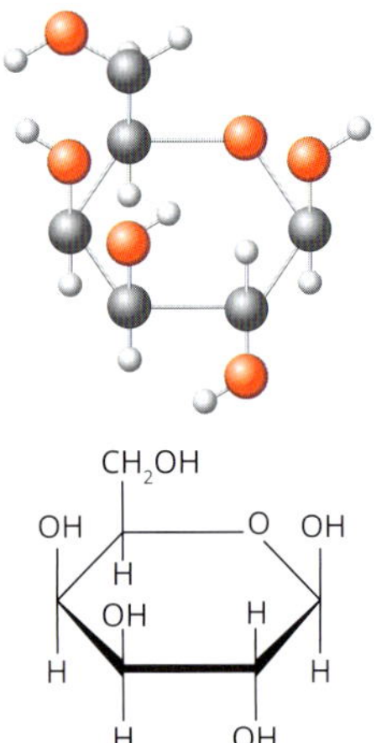

FIGURE 2.2.12 Structural chemical formula and model of three monosaccharides (a) glucose, (b) fructose and (c) galactose. In the models, grey spheres represent carbon atoms, white spheres represent hydrogen atoms and red spheres represent oxygen atoms.

Carbohydrates

Carbohydrates are the most abundant organic molecules in nature.

- They are an important source of chemical energy for living organisms (e.g. glucose).
- They are used as energy reserves in plants (e.g. starch) and animals (e.g. glycogen).
- They form structural components such as cell walls (e.g. cellulose in plants).
- They form part of both DNA and RNA.
- They combine with proteins and lipids to form glycoproteins and glycolipids, as in cell membranes

Carbohydrates are compounds made of carbon, hydrogen and oxygen. There are three main groups of carbohydrates: monosaccharides, disaccharides and polysaccharides. The basic subunits of carbohydrates are the simple sugars, called **monosaccharides**, meaning 'single sugar' (Figure 2.2.3a on page 83). Examples of monosaccharides include glucose, fructose and galactose (Figure 2.2.12). In monosaccharides, the hydrogen and oxygen are present in the same proportions as in water: two hydrogen atoms for each oxygen atom. The general formula is $C_n(H_2O)_n$ (e.g. glucose is $C_6H_{12}O_6$).

> **i** Carbohydrates are organic compounds, such as sugars, starch and cellulose, that are made of carbon, hydrogen and oxygen. Carbohydrates as a class includes monosaccharides, disaccharides and polysaccharides (complex carbohydrates). Only polysaccharides are polymers and hence also fall under the category of biomacromolecules.

When two sugars are joined together they form a **disaccharide** (meaning 'two sugars') and a molecule of water is removed in a chemical reaction known as condensation. Milk sugar (lactose) is made from glucose and galactose, whereas cane sugar (sucrose) is made from glucose and fructose (Figure 2.2.13).

BIOLOGY IN ACTION

Right- and left-handed molecules

Louis Pasteur was a French scientist famous for his work on microbes and the cause of disease. Before he was famous for his work on microbes, he identified a fundamental principle of chemistry that has important implications in biology. He found that the same molecule can exist in two forms that are mirror images of each other (Figure 2.2.15): right-handed and left-handed molecules (like a pair of gloves). Labels of chemicals or drugs denote these forms as dextro- or D-molecule and laevo- or L-molecule (from the Latin, *dextra* meaning 'turning to the right' and *laevus* meaning 'turning to the left'). These are referred to as optical or stereoisomers and their study is stereochemistry.

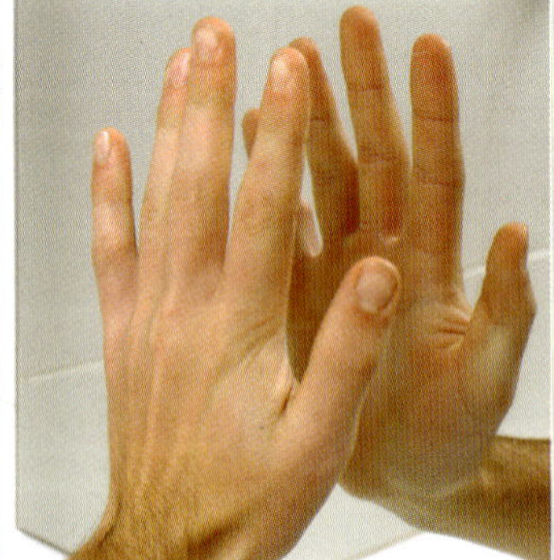

FIGURE 2.2.15 This left hand and its mirror image illustrate how the same parts of a structure, such as a hand or a molecule, can exist in two different shapes. In molecules these are known as stereoisomers.

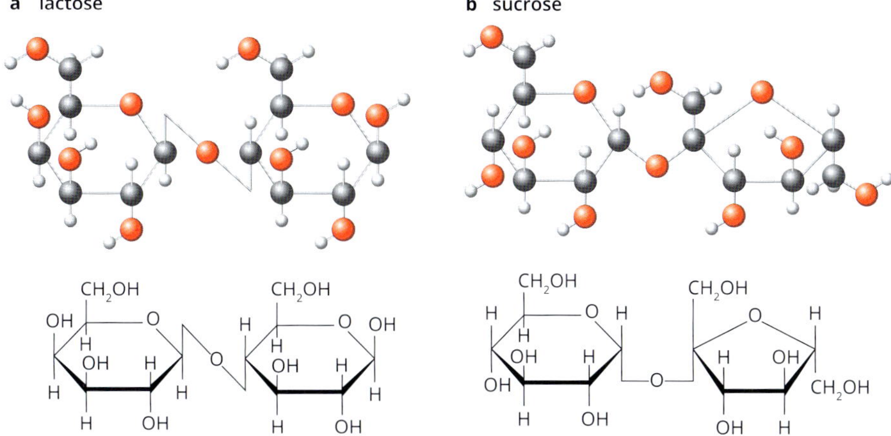

FIGURE 2.2.13 Structural chemical formula and model of the disaccharides (a) lactose and (b) sucrose. In the models, grey spheres represent carbon atoms, white spheres represent hydrogen atoms and red spheres represent oxygen atoms.

When many sugars are joined together they form biomacromolecules called **polysaccharides** ('many sugars'). Cellulose, the major component of plant cell walls, is the most abundant organic molecule on Earth. Starch is the polysaccharide used for energy storage in plants. In animals, the polysaccharide glycogen is used for energy storage. These three polysaccharides are each composed of glucose subunits, but they differ in a number of ways (Figure 2.2.3a on page 83). Starch is a long chain molecule, glycogen has a branching structure and cellulose has additional bonds forming cross-linking between the subunits of the chain. **Complex polysaccharides** are those that consist of different monosaccharide subunits in the same molecule, such as murein found in the cell walls of bacteria (Figure 2.2.14).

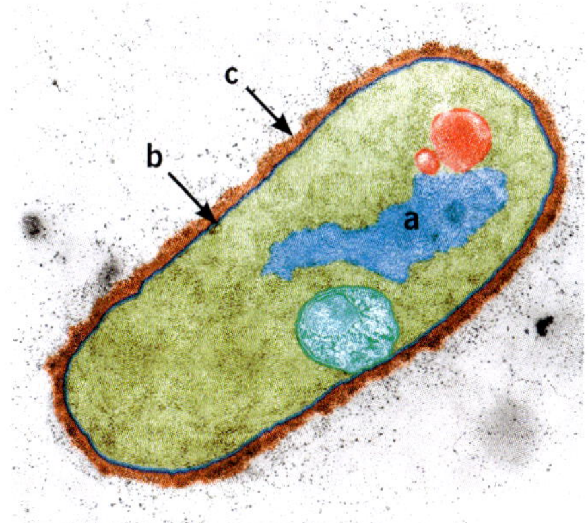

FIGURE 2.2.14 Coloured transmission electron micrograph of the bacterium *Bacillus megaterium*, showing the nucleoid (a), inner plasma membrane (b) and cell wall, which contains murein (c).

Stereochemistry is important in biology as molecules synthesised by organisms are usually of one form only and the organism can only utilise that form. Organisms usually make sugars in the D-configuration whereas proteins are made in the L-configuration. We can't digest 'wrong-handed' sugars. For example, there are two isomers of glucose, D-glucose and L-glucose (Figure 2.2.16). D-glucose occurs widely in nature but L-glucose only occurs in the laboratory. The isomers taste the same, but living organisms cannot use L-glucose as a source of energy.

Thalidomide, an anti-morning sickness drug, caused its terrible effects on unborn babies in the early 1960s because, while one of its isomeric forms was effective against morning sickness, the other was teratogenic—it caused serious damage in early embryonic growth.

When chemicals are manufactured in a laboratory, both right- and left-handed forms are produced in equal amounts, and they are hard to separate on a commercial scale.

Even if it were possible to administer only the correct form of thalidomide, it has been shown that these isomers are converted to each other in vivo (within the body), so both forms would be produced in the body and the teratogenic effect would still occur.

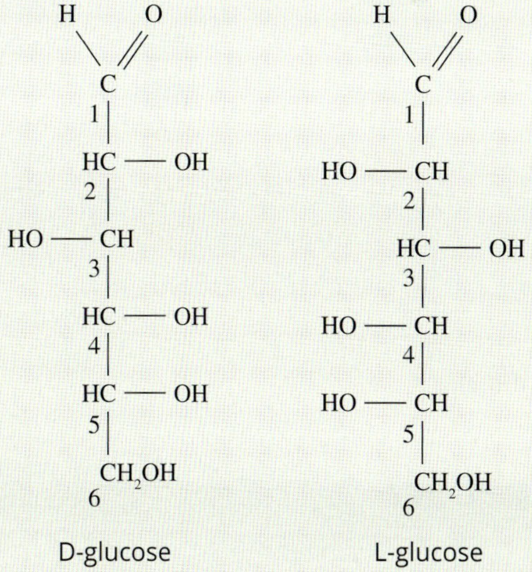

D-glucose: H–C(=O) (1) – HC—OH (2) – HO—CH (3) – HC—OH (4) – HC—OH (5) – CH_2OH (6)

L-glucose: H–C(=O) (1) – HO—CH (2) – HC—OH (3) – HO—CH (4) – HO—CH (5) – CH_2OH (6)

FIGURE 2.2.16 Structure of D-glucose and L-glucose.

Lipids

Lipids are 'fatty' substances that are composed of non-polar **hydrophobic** molecules. This means that lipids are insoluble in water and gives rise to their critical role in living organisms—they can form an effective barrier between two watery environments.

The roles of lipids in organisms include:

- as the main component of plasma membranes and organelle membranes (as **phospholipids**)
- storing energy (e.g. fats and oils are energy-storing molecules)
- playing an important role as hormones (e.g. steroids).

Lipids are relatively small molecules and vary widely in structure. There are two general forms of lipids—simple and compound. Simple lipids are composed of carbon, hydrogen and oxygen, but in different proportions to those of carbohydrates. Simple lipids contain a much smaller proportion of oxygen than do carbohydrates. Simple lipids include fats (composed of fatty acids and glycerol, see Figure 2.2.3d on page 83), and steroids such as cholesterol (Figure 2.2.17a) and the hormones cortisone and testosterone. Fatty acids may be saturated or unsaturated. A saturated fat has the maximum number of hydrogen atoms (it is 'saturated' with hydrogen atoms) and no double bonds (–C=C–) between carbon atoms in the chain (Figure 2.2.17b). An unsaturated fat is one in which there is at least one double bond between the carbon atoms in the chain (Figure 2.2.17c). Steroids have quite a different structure to fats but they are also insoluble in water.

Compound lipids contain fatty acids and glycerol, as well as other elements such as phosphorus and nitrogen. Phospholipids have a **hydrophilic** end (the 'phospho' end) and a hydrophobic end (the lipid end). Phospholipids are the main components of biological membranes and their fundamental role in membrane function is described in the next section.

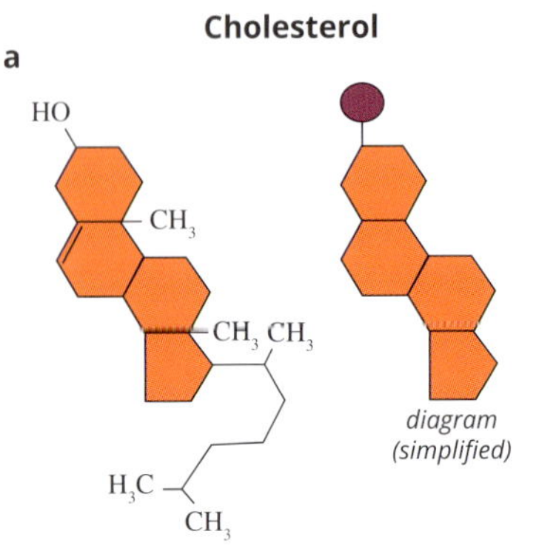

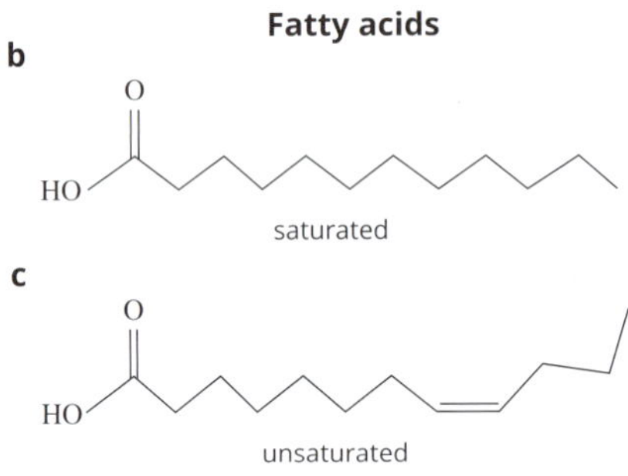

FIGURE 2.2.17 Structure of (a) cholesterol, (b) saturated and (c) unsaturated fatty acids.

Nucleic acids

Nucleic acids are the genetic material of all organisms, and they determine many of the features of an organism. Nucleic acids are biomacromolecules composed of long chains of subunits called nucleotides. A nucleotide consists of a phosphate, a sugar and a nitrogen base (Figure 2.2.18).

There are two types of nucleic acids:

- **Deoxyribonucleic acid (DNA)**—DNA carries the 'instructions' required to assemble proteins from amino acid subunits using a **genetic code**. It is passed accurately from cell to cell during cell division. The four bases in DNA are adenine (A), thymine (T), guanine (G) and cytosine (C).
- **Ribonucleic acid (RNA)**—RNA plays a major role in the manufacture of proteins within cells. The four bases in RNA are adenine (A), uracil (U), guanine (G) and cytosine (C).

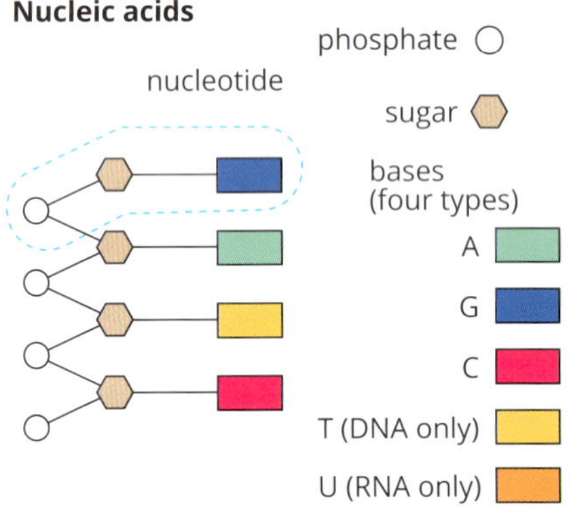

FIGURE 2.2.18 The structure of a nucleic acid.

Proteins

Proteins are more complex than carbohydrates or lipids. Proteins make up over 50% of the dry weight of cells. There are thousands of different kinds of proteins, and their functions vary widely. Although carbohydrates and lipids are similar in all plants and animals, organisms can have a variety of unique proteins that are specific to a particular species.

Protein functions vary widely. Proteins can:

- catalyse cellular reactions (e.g. enzymes such as amylase)
- play an important role as hormones (e.g. insulin)
- act as carrier molecules (e.g. haemoglobin)
- form structural components in organisms (e.g. collagen)
- play an important role in the immune system (e.g. antibodies and antigens).

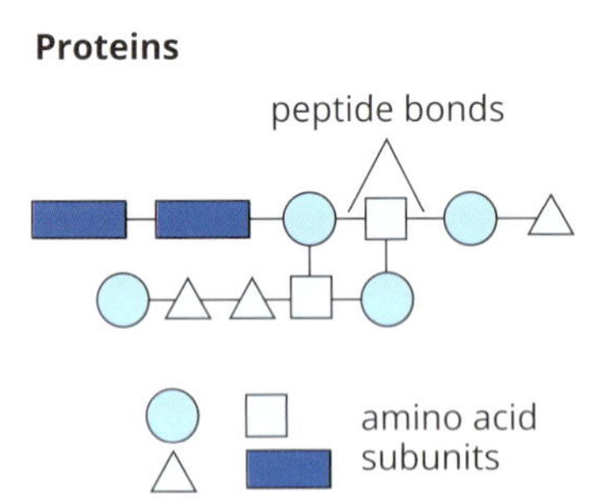

FIGURE 2.2.19 The structure of proteins.

All proteins contain carbon, hydrogen, oxygen and nitrogen; many also contain sulfur, and often phosphorus and other elements. Proteins are biomacromolecules composed of chains of smaller subunits called **amino acids** (Figure 2.2.19). Amino acids are linked by a particular kind of chemical bond, called a peptide bond, and form **polypeptides** or polypeptide chains. (Polypeptide means 'many peptide bonds'.) A protein is formed by one or more polypeptides arranged in a biologically functional way. In other words, 'protein' is the term used for a fully functioning molecule while 'polypeptide' refers to a non-functioning version.

There are 20 different amino acids commonly found in proteins. Nine of these are known as essential amino acids because they cannot be produced by humans. So, humans must obtain these amino acids by consuming other organisms.

The properties of many proteins are determined by their shape, which is determined by their amino acid sequence. You will learn more about proteins, including their shapes and functions, in the next section.

INORGANIC SUBSTANCES

Water

Life evolved in water. Most organisms are 70–90% water, and the chemical reactions that take place in cells take place in a watery medium. This is why the properties of water, such as pH, cohesiveness and heat capacity, are important in many biological processes. Water molecules are very **cohesive**, which means they have a strong tendency to stick together. This property allows thin columns of water to be pulled up tree trunks without breaking (Figure 2.2.20). Bonds between surface molecules also cause surface tension, which allows small insects to walk across the surface of water without breaking into the water molecules and sinking (Figure 2.2.21).

Water has a high **heat capacity**; that is, it can absorb a great deal of heat with very little increase in temperature. This is important for temperature regulation. When you exercise, the chemical reactions taking place in your cells produce heat. Much of this heat can be absorbed by the water in your body, without the cells heating up significantly. Because water has a high heat of vaporisation (a high amount of energy is required to transform one gram of liquid water into water vapour), evaporation of even small amounts of water will be effective in cooling that part of the body surface (Figure 2.2.22).

FIGURE 2.2.20 Mountain ash trees (*Eucalyptus regnans*) are the tallest of all flowering trees and can reach a height of up to 114 metres. Cohesion between water molecules holds the water together and allows for water to be drawn up the trunks of the trees.

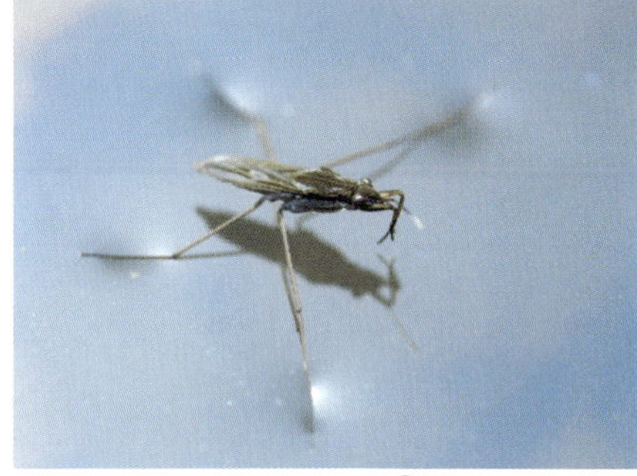

FIGURE 2.2.21 A common pond skater (*Gerris lacustris*) can walk on water. Notice that the legs of the pond skater seem to 'press' on the surface of the water.

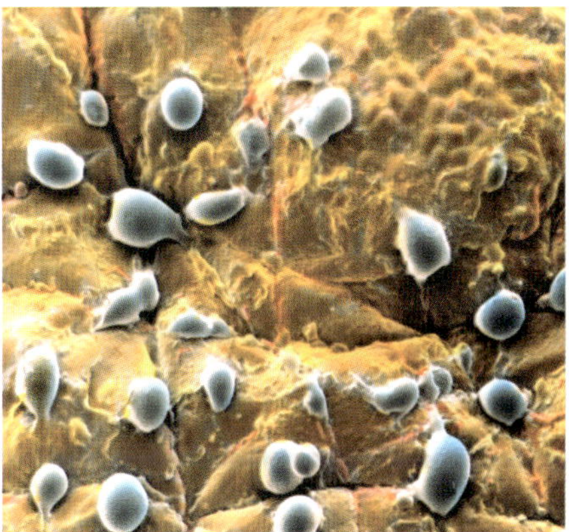

FIGURE 2.2.22 Coloured scanning electron micrograph of the skin surface of the back of a human hand, showing sweat droplets (blue). Evaporation of water on the skin surface is a method of controlling body temperature.

BIOFILE

Water bears

Imagine drying out from about 85% water to just 3% water. This tiny water bear *(Paramacrobiotus craterlaki)* can do this and survive, often for as long as 6–10 years. If the environment dries or freezes, the one-millimetre long tartigrade gradually dries out and lives in a state of suspended animation until water becomes available again.

FIGURE 2.2.23 Water bear (*Paramacrobiotus craterlaki*).

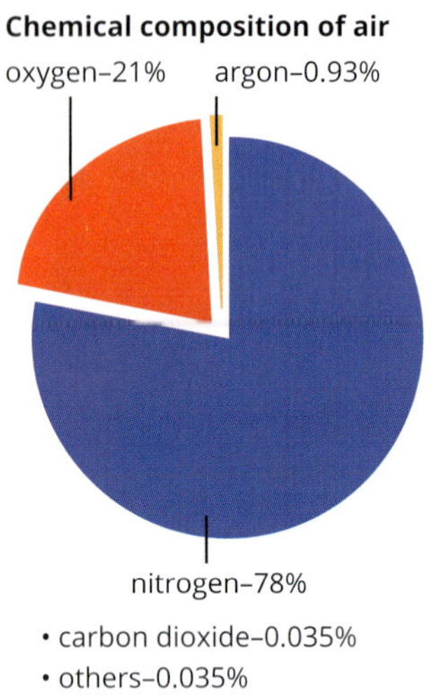

Oxygen and carbon dioxide

In most cells, oxygen is needed to release energy from food molecules in processes known collectively as **cellular respiration**. A constant supply of oxygen is therefore necessary to maintain the activity of these cells. This is usually easy for organisms that get their oxygen from air, because the atmosphere is 21% oxygen (Figure 2.2.24). However, oxygen is not very soluble in water, so organisms that get their oxygen from water are either small, flat and relatively inactive, or they have very efficient ventilation systems with large surface area for gaseous exchange, such as fish gills (Figure 2.2.25).

Carbon is the key atom in organic molecules. Carbon dioxide (CO_2) is taken from the atmosphere (which contains approximately 0.035% by volume of carbon dioxide, Figure 2.2.24) by plants, some bacteria and some protists. It is used in the process of photosynthesis to make sugars, some of which are eaten by animals. Carbon dioxide is returned to the atmosphere mainly by the decay of organic material and as an end-product of cellular respiration. This cycling of carbon through organisms and the atmosphere is critical to the survival of all organisms.

Chemical composition of air

oxygen–21%
argon–0.93%
nitrogen–78%
• carbon dioxide–0.035%
• others–0.035%

FIGURE 2.2.24 Air is a mixture that is mainly made up of nitrogen, oxygen and argon.

Nitrogen

Nitrogen is required by organisms in relatively large amounts because it is a key component of all proteins. There is plenty of nitrogen around because the atmosphere is about 78% nitrogen gas (Figure 2.2.24); however, most organisms are unable to use nitrogen in this form. Atmospheric nitrogen is converted by certain bacteria and cyanobacteria into compounds that can be used by plants in a process known as **nitrogen fixation**. The most important nitrogen-fixing bacteria are the symbiotic bacteria found in the roots of plants, including legumes, casuarinas and acacias (Figure 2.2.26). Nitrogen compounds produced by the bacteria in the soil are absorbed by plants and used to make amino acids. Heterotrophs obtain their amino acids by consuming plants and other organisms. They also produce nitrogen-rich waste (manure), which has traditionally been used as a plant fertiliser.

FIGURE 2.2.25 Coloured scanning electron micrograph of mackerel (*Scomber scombrus*) gills showing the large surface area for gaseous exchange.

FIGURE 2.2.26 (a) Nodules containing nitrogen-fixing bacteria on the roots of a garden pea (*Pisum sativum*). (b) Coloured scanning electron micrograph of nitrogen-fixing soil bacteria (*Rhizobium* species) in a root nodule of a bean plant. These bacteria (green) have a symbiotic relationship with the plant.

BIOFILE

The Australian lungfish

The Australian lungfish (*Neoceratodus forsteri*) is a unique fish that has a single lung in addition to its gills. When additional oxygen is required due to increased activity, drought or reduced oxygen levels in the water, the lungfish can rise to the surface and swallow air into its lung to supplement the oxygen supply through the gills.

FIGURE 2.2.27 The Australian lungfish (*Neoceratodus forsteri*).

Minerals

Mineral salts are naturally occurring inorganic compounds produced by the weathering of rocks. The water-soluble mineral salts produced are absorbed as ions into the roots of plants (Figure 2.2.28), making them available to be eaten by animals. Humans require more than 20 minerals. Biologically important minerals include phosphorus, potassium, calcium, magnesium, iron, sodium, iodine and sulfur. Many others are needed in small (trace) amounts.

Mineral ions are found in the cytosol of cells, in structural components (such as bone), and in the molecules of many enzymes and vitamins. They may also be incorporated into other important organic compounds in cells. Phosphorus is present in the phospholipids of cell membranes and in **ATP (adenosine triphosphate**—an important energy carrier in cells, see Chapter 5). Magnesium is an important constituent of chlorophyll, and iron is the central atom in every haemoglobin molecule in red blood cells (Figure 2.2.29). Calcium, potassium and sodium ions are important for the normal performance of cardiac muscle cells, and calcium and phosphorus are found in bones and teeth (Figure 2.2.30).

FIGURE 2.2.28 A soil profile showing the horizons (layers), which vary in colour depending on the mineral content in the soil. Plants absorb these minerals when they draw water out of the soil.

Molecules of ATP provide energy for immediate use by the cell and are produced during glycolysis and cellular respiration.

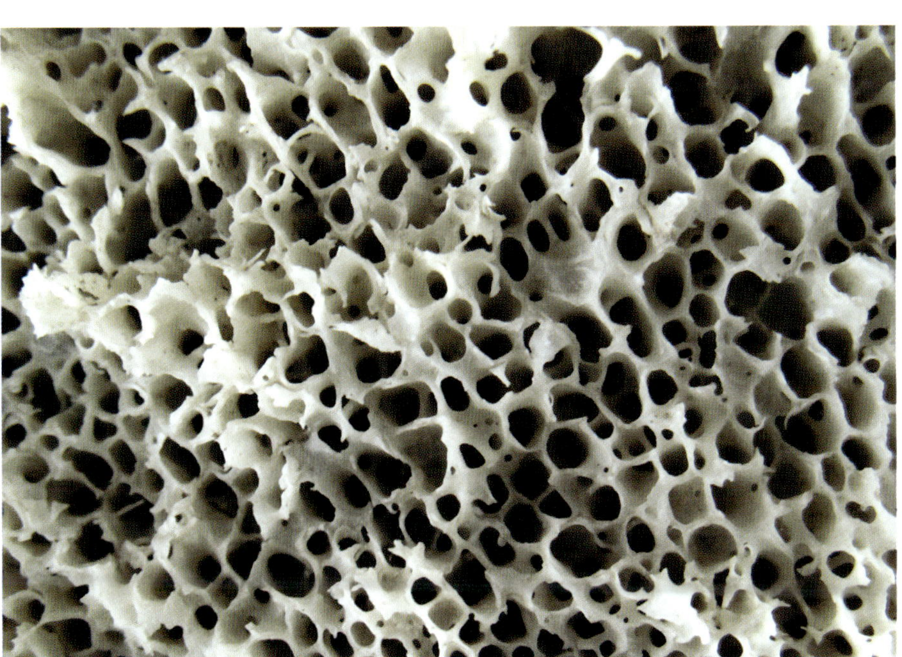

FIGURE 2.2.30 Bone matrix is made up of inorganic components including salts of calcium and phosphorus.

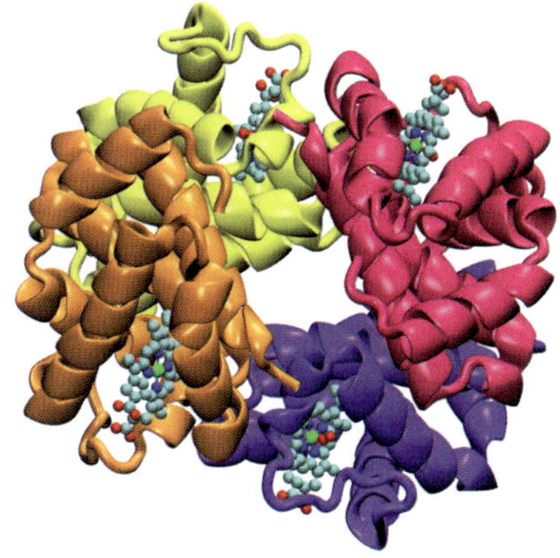

FIGURE 2.2.29 Haemoglobin is made up of four protein subunits (coloured ribbon structures). Each subunit has an oxygen-binding site, or haem group (turquoise). Within each haem group there is one atom of iron (green). Oxygen molecules are shown as paired red spheres.

BIOLOGY IN ACTION

Too much or too little copper

Copper is a cofactor in several enzymes, meaning that it is required for the enzyme to function efficiently. Because copper occurs widely in foods, a dietary deficiency of copper is rare in a balanced diet. However, too much or too little copper in the body can result from inherited copper management disorders, which cause serious problems.

Wilson disease is a copper toxicosis disorder. Normally the liver functions as a copper storage organ and any excess copper is excreted in bile. In people with Wilson disease, a protein needed to excrete copper is defective. Copper therefore accumulates in the liver, causing a very serious hepatitis-like disease (Figure 2.2.31) that may be accompanied by neurological effects in some patients. A clinical sign of an individual with Wilson disease is the appearance of a brownish ring overlying the outer rim of the iris of the eye and known as a Kayser–Fleischer ring (Figure 2.2.32). A Kayser–Fleischer ring is caused by a deposit of copper granules on the iris.

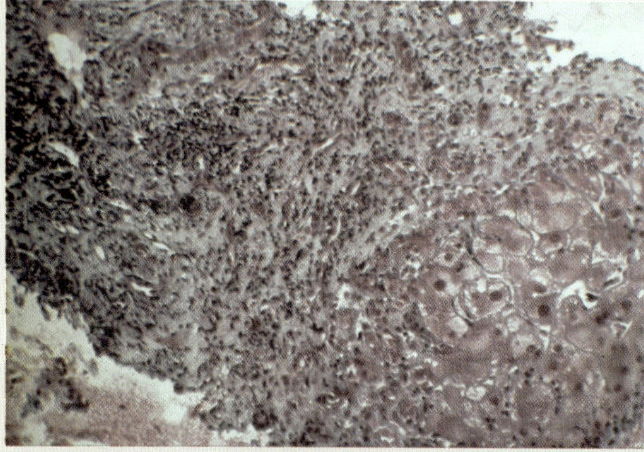

FIGURE 2.2.31 Light micrograph of a section through a liver affected by cirrhosis (scarring of the liver). The cirrhosis is caused by Wilson disease.

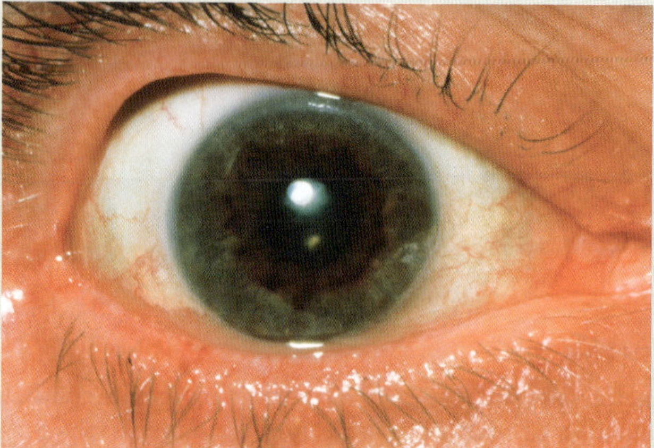

FIGURE 2.2.32 Kayser–Fleischer ring, a brownish ring overlying the outer rim of the iris of the eye.

The opposite problem occurs in Menkes disease—there is too little copper. Intestinal absorption of copper is defective, resulting in a low copper supply for the tissues and organs. Due to deficiencies in copper-dependent enzymes, the major symptoms are found in connective tissues (arteries and bone) and in the brain. It usually leads to death due to the rupture of a weak major artery. Babies with Menkes disease have unusual 'steely' or 'kinky' hair. Interestingly, it was an observant Australian scientist who noted the similarity between the kinky hair of Menkes sufferers and the wool of sheep grazing on copper-deficient soil. This astute observation led to the discovery that the cause of Menkes disease was too little copper.

Both Wilson disease and Menkes disease are caused by defects in molecular copper pumps. The two proteins (the pumps) are very closely related, even though they are functionally opposite.

2.2 Review

SUMMARY

- Organic compounds contain carbon and hydrogen and are found in living things.
- Inorganic substances are all elements and compounds other than organic compounds (e.g. oxygen, water and carbon dioxide).
- There are four main types of organic compounds: carbohydrates, proteins, nucleic acids and lipids.
- Biomacromolecules are large organic molecules formed by joining together many smaller units (monomers) to form a chain or polymer. Examples are polysaccharides, nucleic acids and proteins.
- Plants convert inorganic molecules into carbohydrates.
- Carbohydrates:
  - are an important source of chemical energy for living organisms (e.g. glucose)
  - are used as energy reserves in plants (e.g. starch) and animals (e.g. glycogen)
  - form structural components such as cell walls (e.g. cellulose in plants).
  - form part of both DNA and RNA
  - combine with proteins and lipids to form glycoproteins and glycolipids, as in cell membranes.
- Lipids are fatty substances that are not soluble in water. They:
  - are the main component of cell membranes
  - store energy
  - play an important role as hormones.
- Nucleic acids contain the genetic material of all organisms. There are two types:
  - DNA—carries the 'instructions' required to assemble proteins
  - RNA—plays a major role in the manufacture of proteins.
- The functions of proteins vary widely. Proteins can:
  - catalyse cellular reactions
  - play an important role as hormones
  - act as carrier molecules
  - form structural components in organisms
  - play an important role in the immune system (e.g. antibodies and antigens).
- The polar nature of water molecules explains many of water's biologically important properties.
- Important properties of water include cohesiveness, surface tension, heat capacity and pH.
- Oxygen is needed in most organisms to release energy from food molecules.
- Atmospheric carbon dioxide is the main source of carbon, which is the key atom in organic molecules.
- Nitrogen, which is a component of all proteins, is 'fixed' from the atmosphere by certain bacteria.
- Minerals are required in lesser amounts and form important parts of organic molecules such as enzymes and structural molecules.

KEY QUESTIONS

1 Define the terms 'organic compound' and 'inorganic substance'. Give an example of each.

2 Copy the following table and complete the summary of biologically important organic compounds.

Type of organic compound	Elements that make up compound	Role of compound in living organisms	Examples of compound
carbohydrate			glucose
protein			
			DNA, RNA
lipid			

3 a What is a polymer?
b List the four biomacromolecules. Indicate whether or not they are polymers and describe the subunits involved.

4 Distinguish between monosaccharides, disaccharides and polysaccharides.

5 List the two forms of nucleic acids and explain their roles.

6 Water molecules have a number of properties that make water important in biological processes. These include cohesion and high heat capacity. Explain what is meant by each term. Outline a specific example to explain its biological importance.

7 For living organisms, outline the role of:
a oxygen
b carbon dioxide.

8 a Outline the importance of nitrogen for living organisms.
b What are nitrogen-fixing bacteria? Why are they important?

9 a Define 'mineral salts'.
b Include specific examples to explain the significance of minerals for living organisms.

2.3 The plasma membrane

> Extracellular fluid is the body fluid outside the cell membranes. It includes blood plasma and interstitial fluid (fluid that surrounds and bathes the cells).

The plasma membrane encloses the contents of cells and controls the movement of substances between the exterior of the cell (the extracellular fluid) and the interior of the cell (the cytoplasm). The plasma membrane therefore assists in maintaining a composition within the cell that is different from that of the surrounding external environment. As well as controlling transport of molecules into and out of the cell, the plasma membrane also performs other important functions, such as cell recognition and communication with other cells.

In this section, you will learn about the fluid mosaic model of the structure of the plasma membrane.

THE FLUID MOSAIC MODEL

The **fluid mosaic model** that describes the structure of the plasma membrane was first proposed by Jonathan Singer and Garth Nicholson in 1972. It is now widely accepted as the basic model of all biological membranes. According to this model, plasma membranes consist of two layers of phospholipid molecules, with other molecules including proteins, carbohydrates and cholesterol scattered throughout the membrane (Figure 2.3.1).

The fluid mosaic model is a representation of our current knowledge of the plasma membrane. It is modified and updated as developments in cellular techniques and technology allow more information to be gathered.

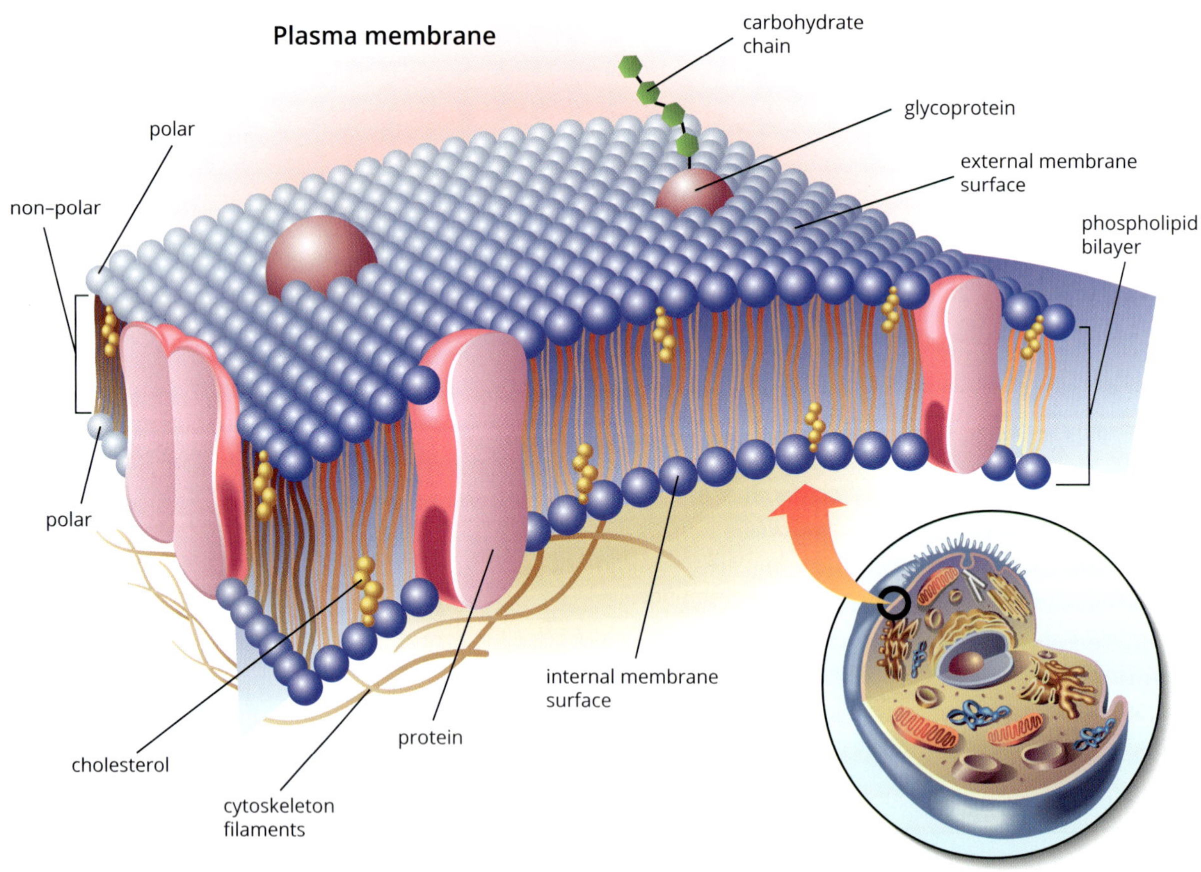

FIGURE 2.3.1 This illustration of the plasma membrane fluid mosaic model shows the phospholipid bilayer with proteins, glycoproteins and cholesterols embedded throughout and supported by the cytoskeletal filaments.

Structure of phospholipids and the formation of the phospholipid bilayer

The phospholipids found in the plasma membrane are composed of a hydrophilic head containing a phosphate group and glycerol, and two hydrophobic fatty acid tails. The tails can be saturated fatty acid tails or unsaturated fatty acid tails (Figure 2.3.2).

The presence of *cis* double bonds creates kinks in the unsaturated fatty acid chains of plasma membrane phospholipids, which makes it more difficult for the molecules to pack together, and results in the membrane remaining fluid at lower temperatures. As the temperature drops, cells synthesise fatty acids containing more *cis* double bonds to avoid a reduction in membrane fluidity.

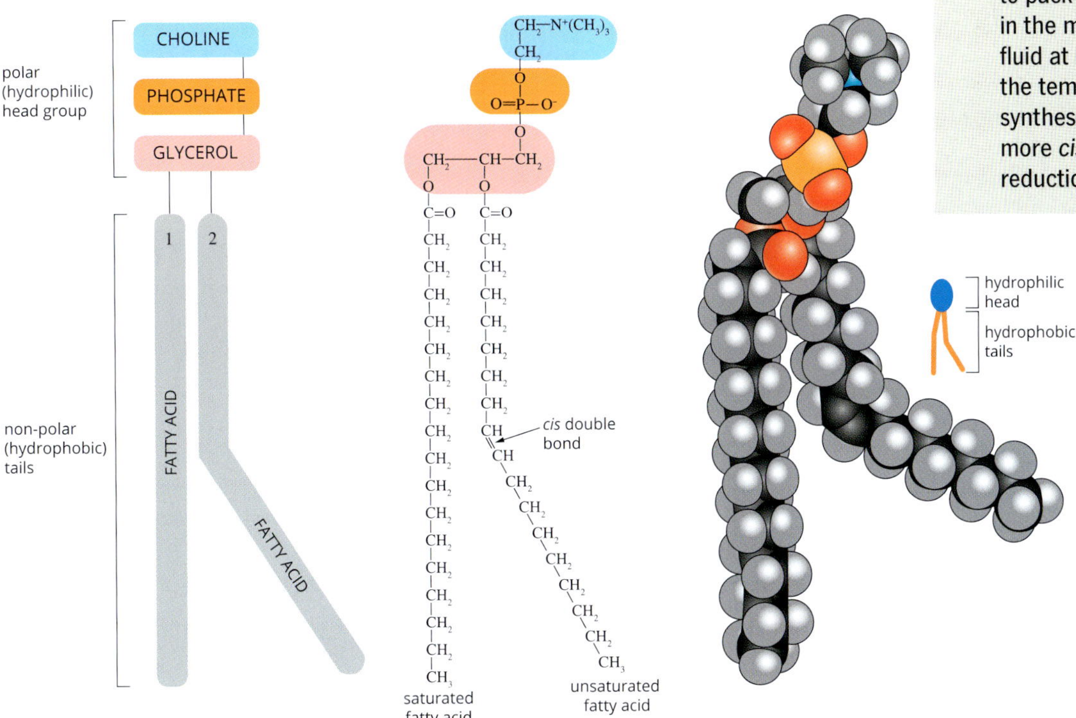

FIGURE 2.3.2 The structure of a phospholipid.

As phospholipids have a hydrophilic head and two hydrophobic tails, the phospholipids will react to the presence of water (Figure 2.3.3). If the phospholipids are in contact with water on one side and oil on the other side, all the hydrophilic heads will be embedded in water and the tails will be embedded in the oil. If the phospholipids are in contact with water on both sides, it would result in all hydrophilic heads being embedded in water, and the hydrophobic tails pointing towards each other with oil sandwiched in between. This bilayer arrangement shelters the hydrophobic tails of the phospholipids from water while exposing the hydrophilic heads to water (Figure 2.3.4).

The impermeability of membranes to water-soluble (polar) molecules is due to the phospholipid bilayer. Most other membrane functions are carried out by the proteins, which are located throughout the membrane.

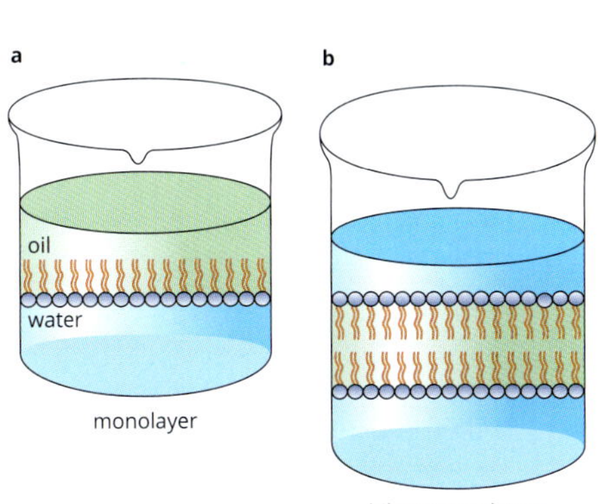

FIGURE 2.3.3 (a) Arrangement of phospholipids in the presence of water on one side and oil on the other side. (b) Arrangement of phospholipids in the presence of water on both sides forming a bilayer.

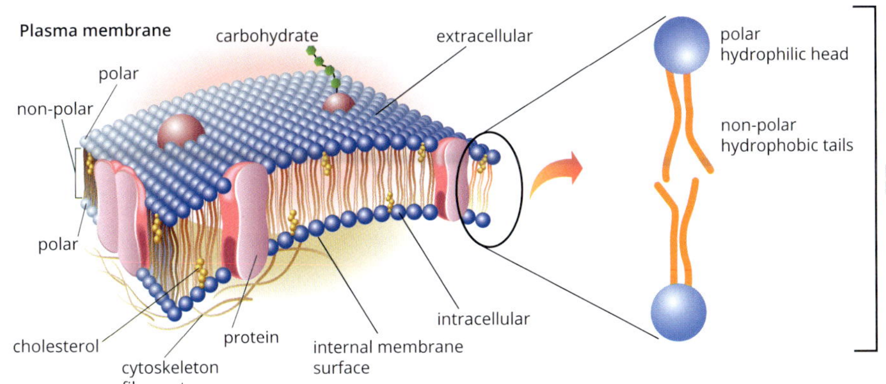

FIGURE 2.3.4 A phospholipid has a hydrophobic 'tail' and a hydrophilic 'head'.

BIOFILE

Fluidity of nerve cell membranes

Fluidity of nerve cell membranes is extremely important because membrane composition determines what is able to pass into and out of the cell. Oxygen, glucose and the nutrients that the cell needs to survive must all pass through the membrane and into the cell's interior. In addition to letting in essential nutrients and keeping out harmful substances, nerve cell membranes also contain proteins that act as receptors for some neurotransmitters.

In order for the receptors to be able to recognise neurotransmitters and send along the messages that they contain, the nerve cell membrane must be fluid. If the nerve cell membrane is too rigid, the receptors on the membrane become less capable of recognising neurotransmitters and passing along messages to the nerve cell. Thus, membrane composition is extremely important because it influences the ability of nerve cells to communicate with each other and, ultimately, to survive.

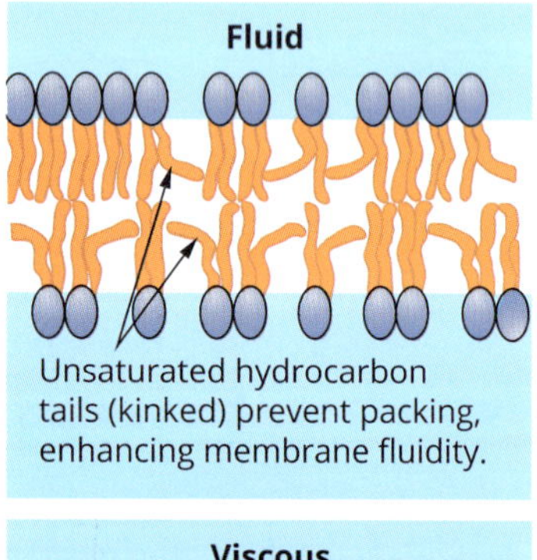

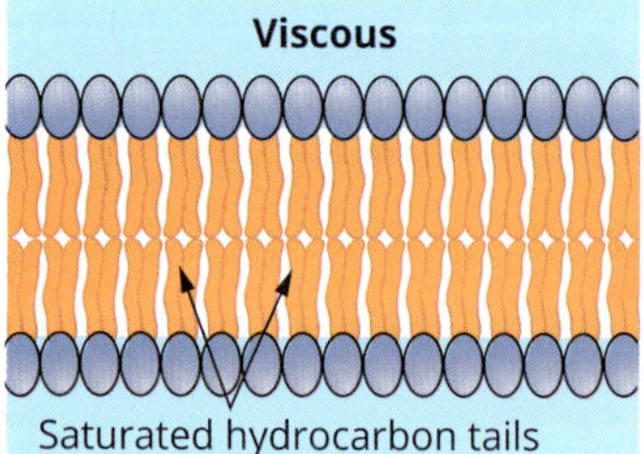

FIGURE 2.3.6 These two diagrams show the effects of unsaturated and saturated fatty acid tails on the fluidity of the plasma membrane.

Fluidity of the plasma membrane

Molecules of the plasma membrane are not fixed in place. Most of the phospholipids and some of the proteins can move about laterally, and sometimes some molecules are able to flip-flop transversely across the membrane (Figure 2.3.5). The rate at which the molecules move within a layer of the plasma membrane varies. Proteins in the membrane can move sideways throughout the plasma membrane, but they move at a much slower rate than the phospholipids. The ability of the phospholipids and proteins to move gives the plasma membrane its fluid nature.

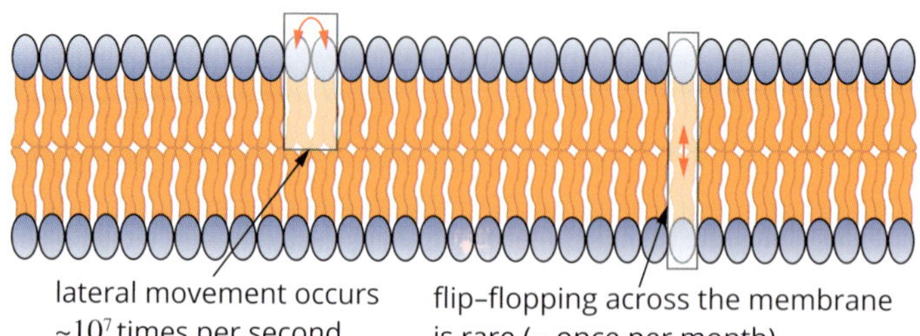

FIGURE 2.3.5 The movement of phospholipids in plasma membranes.

The fluidity of the plasma membrane is very important because it affects:

- the permeability of the membrane to substances
- the capacity for proteins to move within the membrane to particular areas where they are required to carry out their function.

Factors that alter the fluidity of the plasma membrane include:

- phospholipid composition and structure
- temperature
- the presence of cholesterol.

Phospholipid composition and structure

If a large proportion of phospholipids in a plasma membrane have saturated fatty acid tails, then that plasma membrane will be less fluid and more viscous than a plasma membrane in which a large proportion of phospholipids have unsaturated fatty acid tails. This is because saturated fatty acid tails are straight, allowing the phospholipids to pack tightly together. By contrast, phospholipids with unsaturated fatty acids have 'kinks' that prevent the tails from packing together closely, giving the plasma membrane greater fluidity (Figure 2.3.6).

Temperature

As temperature increases, the fluidity of plasma membranes increases. This is because the phospholipids become less closely packed together and are able to move more freely.

As temperature decreases, plasma membrane with a large proportion of saturated fatty acids may solidify at a certain point. This will not occur in plasma membrane with a large proportion of unsaturated fatty acids, because the kinks in the tails of these fatty acids prevent the phospholipids from becoming closely packed.

Cholesterol

In animal cells, the presence of cholesterol between phospholipid molecules alters the fluidity of the plasma membrane (Figure 2.3.7). The cholesterol molecules restrict the movement of the phospholipid molecules and prevent them from packing together as closely as they would otherwise.

Cholesterol acts as a buffer for changing temperatures:

- At higher temperatures, cholesterol stops the plasma membrane from becoming too fluid by restricting the movement of phospholipids.
- At lower temperatures, cholesterol prevents the plasma membrane from solidifying by restricting the tight packing of phospholipids.

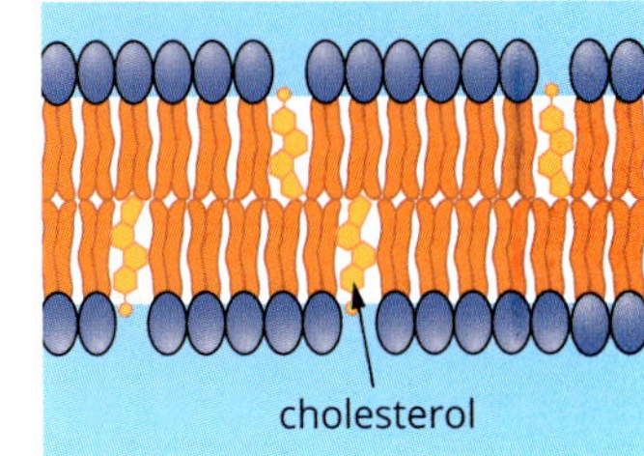

FIGURE 2.3.7 Cholesterol molecules are present in the plasma membrane.

BIOLOGY IN ACTION

Freeze-fracture electron microscopy

Freeze-fracture electron microscopy was one of the methods used to demonstrate that proteins are embedded in the plasma membrane. Using the freeze-fracture method, the plasma membrane can be split into its two phospholipid layers to reveal the structure of the plasma membrane's interior (Figure 2.3.8a).

The first step in freeze-fracture electron microscopy is to prepare the cell. A cell is frozen and fractured with a knife. When the frozen cell is fractured, the fracture plane often follows the hydrophobic tails of the phospholipid bilayer, splitting the phospholipid bilayer into two monolayers, an outer or E (extracellular) surface and an inner or C (cytoplasmic) surface (Figure 2.3.8b). Membrane protein does not get 'broken' in the technique and would be found on one of the layers.

Once the plasma membrane has been split into the two layers, scanning electron microscopy is used to take a picture of each layer. Membrane proteins are shown as 'bumps', demonstrating that proteins are embedded in the phospholipid bilayer.

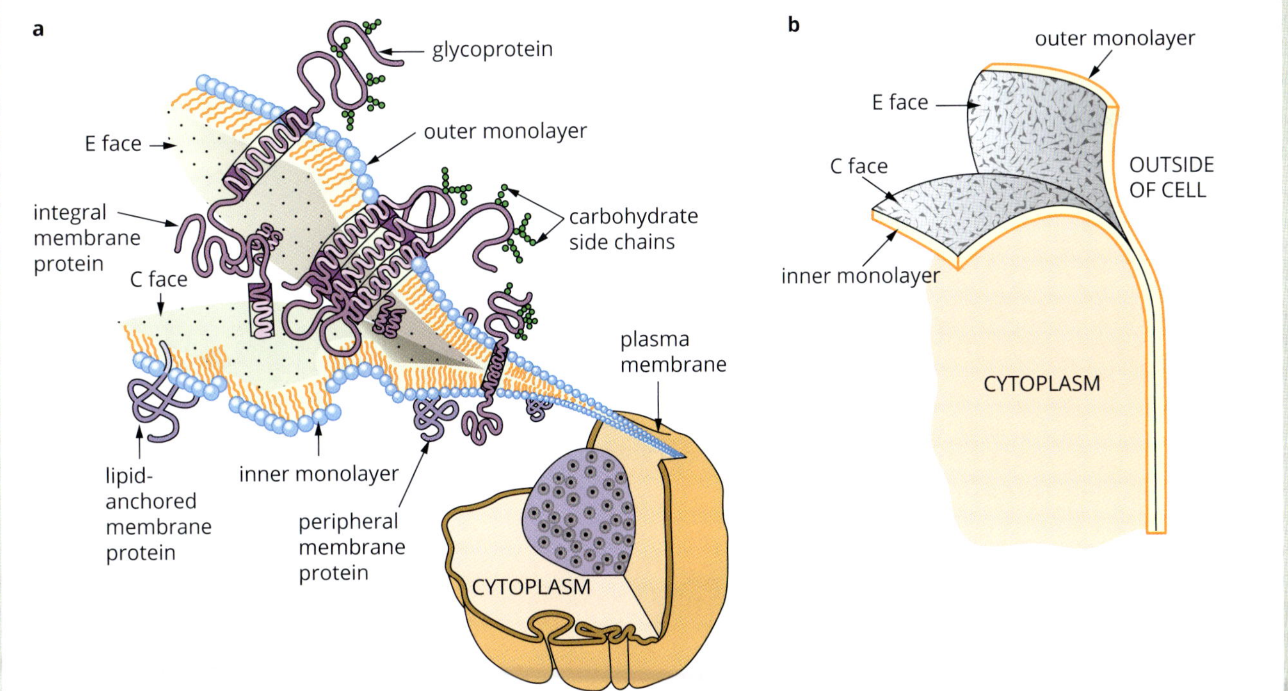

FIGURE 2.3.8 (a) The fracture plane has passed through the hydrophobic interior of the membrane, revealing the inner surfaces of the two monolayers. Integral membrane proteins remain embedded in one of the layers. (b) Surface view of the monolayers. The bumps on the surfaces are transmembrane proteins.

The mosaic nature of the plasma membrane

The plasma membrane consists of different proteins embedded in the phospholipid bilayer (Figure 2.3.9, Figure 2.3.10). The mosaic nature of the phospholipid bilayer is important for the particular functioning of a cell. Plasma membranes of different cells have different mosaic patterns depending on the proteins they require for their particular function.

The location of membrane proteins

The pattern of proteins within a membrane can also change periodically. Plasma membrane proteins can be found (Figure 2.3.9, Figure 2.3.10):

- clustered in groups at specific locations, in order to either carry out similar functions or because they are part of the same **biochemical pathways** (the long chains of chemical reactions that take place in cells)
- distributed randomly throughout the membrane.

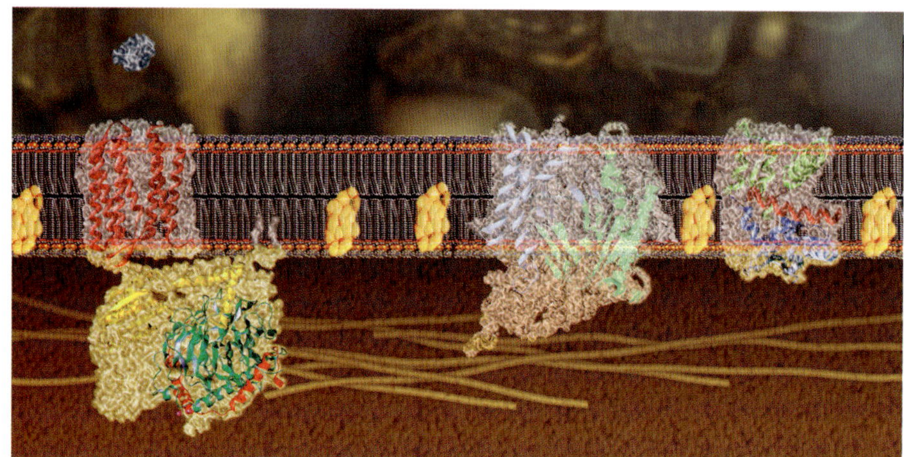

FIGURE 2.3.9 Biomedical illustration. Proteins (coloured ribbons) can be clustered together in groups (left) or distributed randomly throughout the membrane (middle and right).

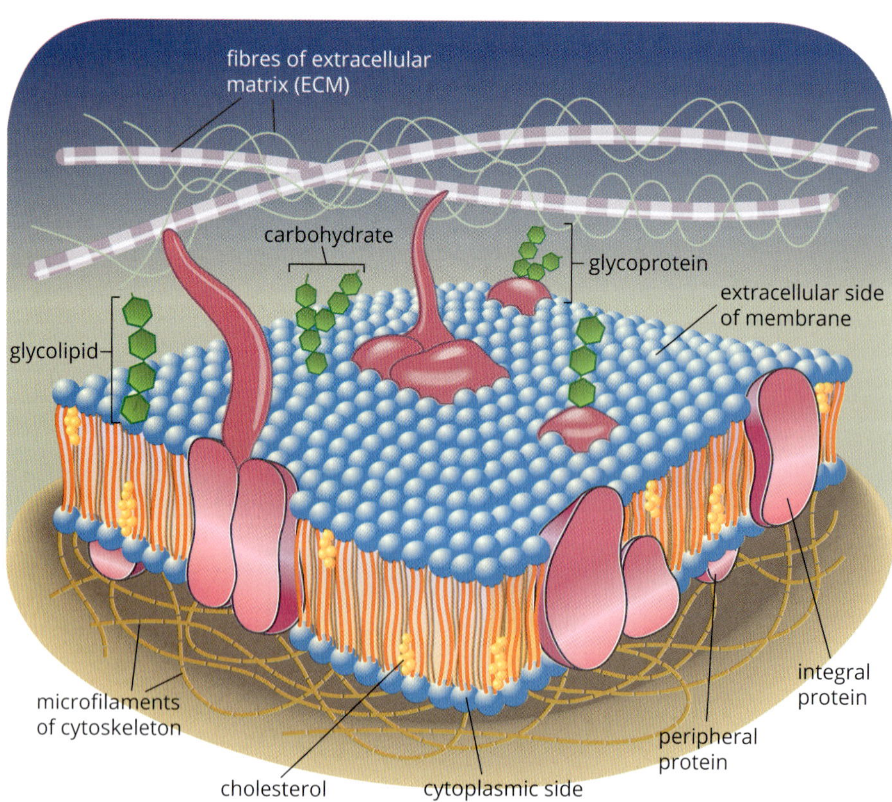

FIGURE 2.3.10 An updated model of an animal cell's plasma membrane showing the attachment of proteins to fibres of extracellular matrix (far left) and to microfilaments of cytoskeleton (second protein from right).

Depending on their role, membrane proteins can be:

- free to move within the bilayer (that is, not attached to other molecules)
- attached to the **cytoskeleton** on the interior surface or to the fibres of the **extracellular matrix** (ECM) so that their movement is directed and restricted to specific locations (Figure 2.3.10).

The cytoskeleton is the network of microtubules and microfilaments that provides a supporting framework for cells and attachments for various organelles.

The extracellular matrix, or ECM, is a collection of extracellular molecules secreted by cells that provides structural and biochemical support to the surrounding cells.

Types of membrane proteins

There are two types of proteins within the plasma membrane: integral proteins and peripheral proteins (Figure 2.3.10). Proteins may have both hydrophobic and hydrophilic regions, which determine their position and function in the plasma membrane. (You will learn more about proteins in Section 2.5.)

Integral proteins are proteins that are a permanent part of the plasma membrane. When integral proteins span both phospholipid layers they are also called **transmembrane proteins**.

Integral proteins have both hydrophilic regions and hydrophobic regions (Figure 2.3.11). The hydrophobic regions keep the protein embedded among the hydrophobic tails of the phospholipid bilayer (hydrophobic molecules are attracted to each other). Integral proteins can span the entire phospholipid bilayer, or extend only partway into the hydrophobic interior. The hydrophilic regions are exposed to the aqueous environment on one or both sides of the membrane. Additionally, hydrophilic regions may also form water-filled channels through the protein, through which various materials can be transported from one side of the cell to the other.

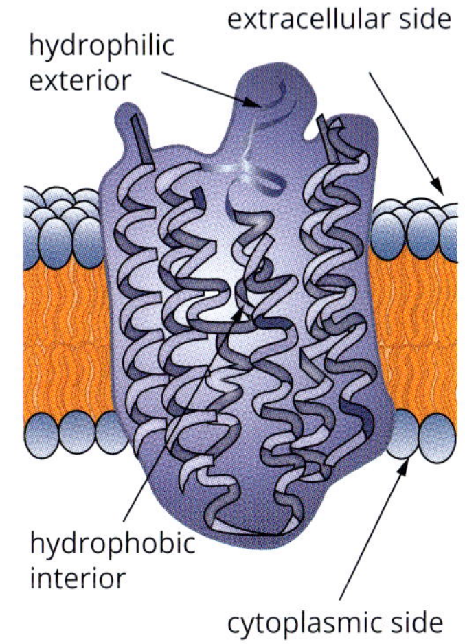

FIGURE 2.3.11 Integral transmembrane protein of the plasma membrane.

Integral proteins have many functions in the plasma membrane (Figure 2.3.12). For example, these proteins:

a can act as transport channels to transport molecules and ions through the membrane
b function as enzymes
c are involved in signal transduction
d function in cell–cell recognition
e connect cells to each other
f act as attachments to the cytoskeleton and the extracellular matrix.

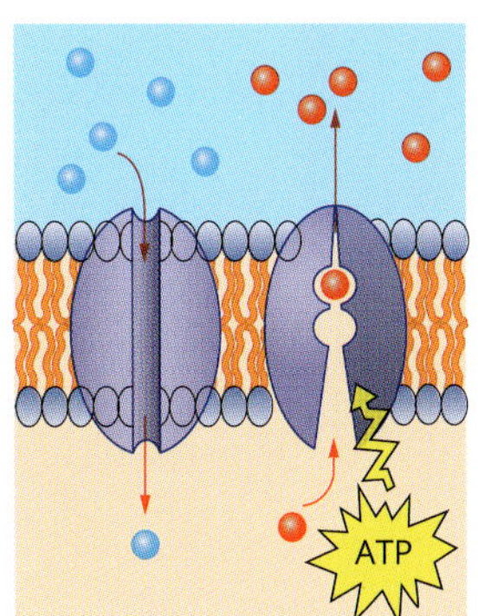

a Transport

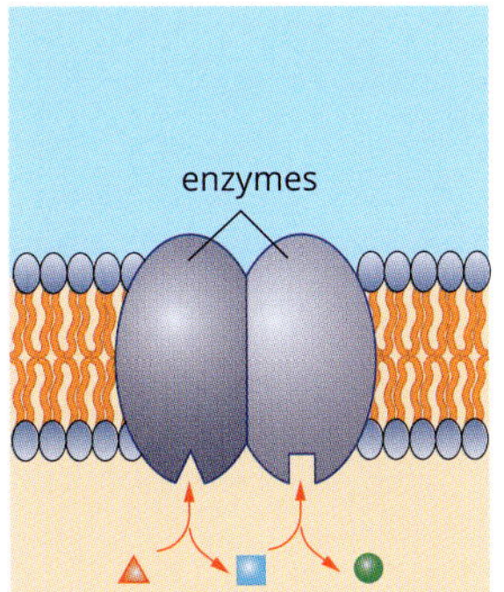

b Enzymatic activity

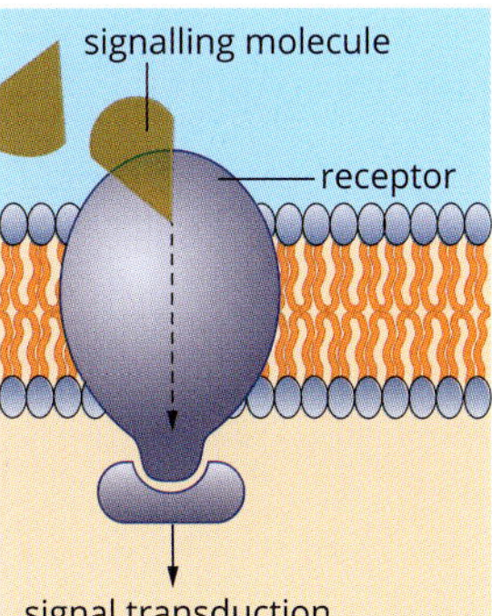

c Signal transduction

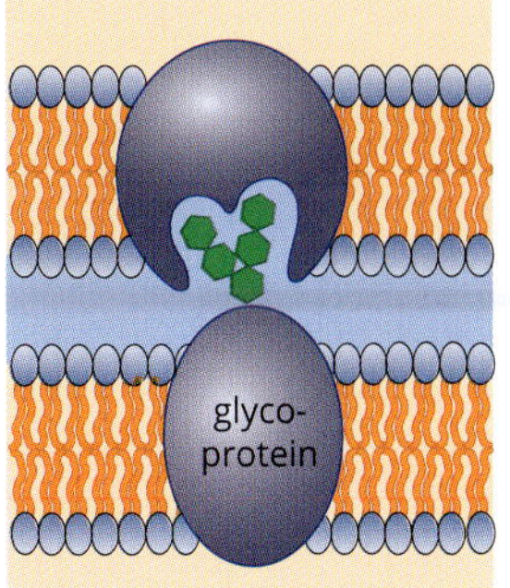

d Cell–cell recognition

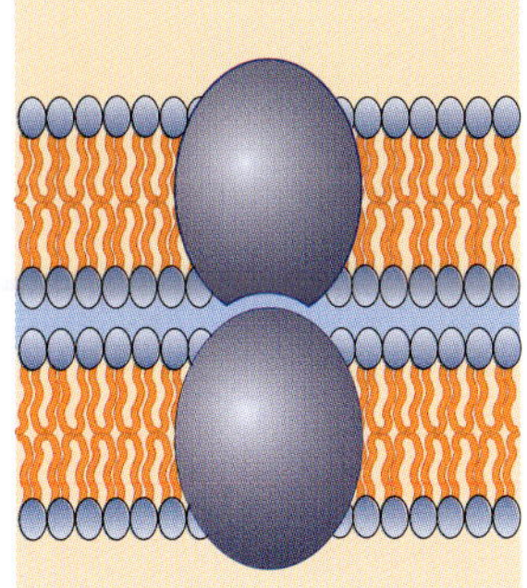

e Intercellular joining

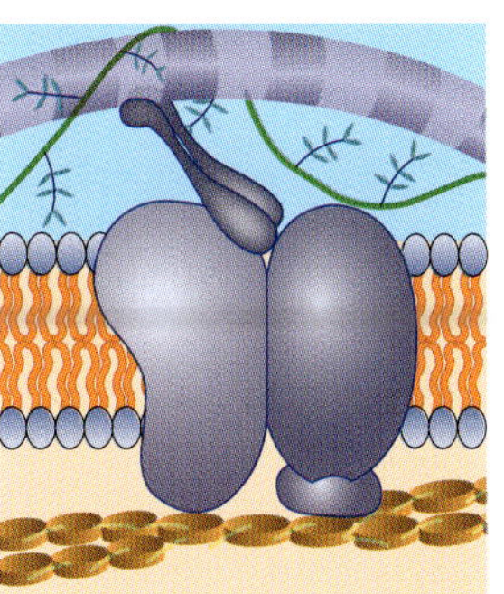

f Attachment to the cytoskeleton and extracellular matrix

FIGURE 2.3.12 Functions of integral proteins in the plasma membrane.

Proteins that are a temporary part of the plasma membrane are called **peripheral proteins**. Peripheral proteins are not embedded in the phospholipid bilayer and bind to integral proteins or penetrate into one surface of the plasma membrane. These proteins:

- can be attached either to phospholipid molecules or to integral proteins in either layer of the plasma membrane
- function as enzymes, receptors, structural attachment points and cellular recognition sites.

Carbohydrates

Carbohydrates associated with plasma membranes are usually linked to protruding proteins (forming glycoproteins) or to lipids (forming glycolipids) on the outer surface of the membrane (Figure 2.3.10, page 98). They play a role in recognition and adhesion between cells, and in the recognition of antibodies, hormones and viruses by cells.

Membranes in organelles

Eukaryotic organelles, such as the nucleus, endoplasmic reticulum, Golgi apparatus and vesicles, are surrounded by, or constructed from, membranes (Figure 2.3.13). These membranes establish compartments within the cell in which specific functions are carried out efficiently. The composition of these membranes is similar to that of the plasma membrane, enabling the two membranes to fuse. For example, vesicles are able to fuse with the plasma membrane, allowing the transport of materials into and out of a cell, as well as from one organelle to another.

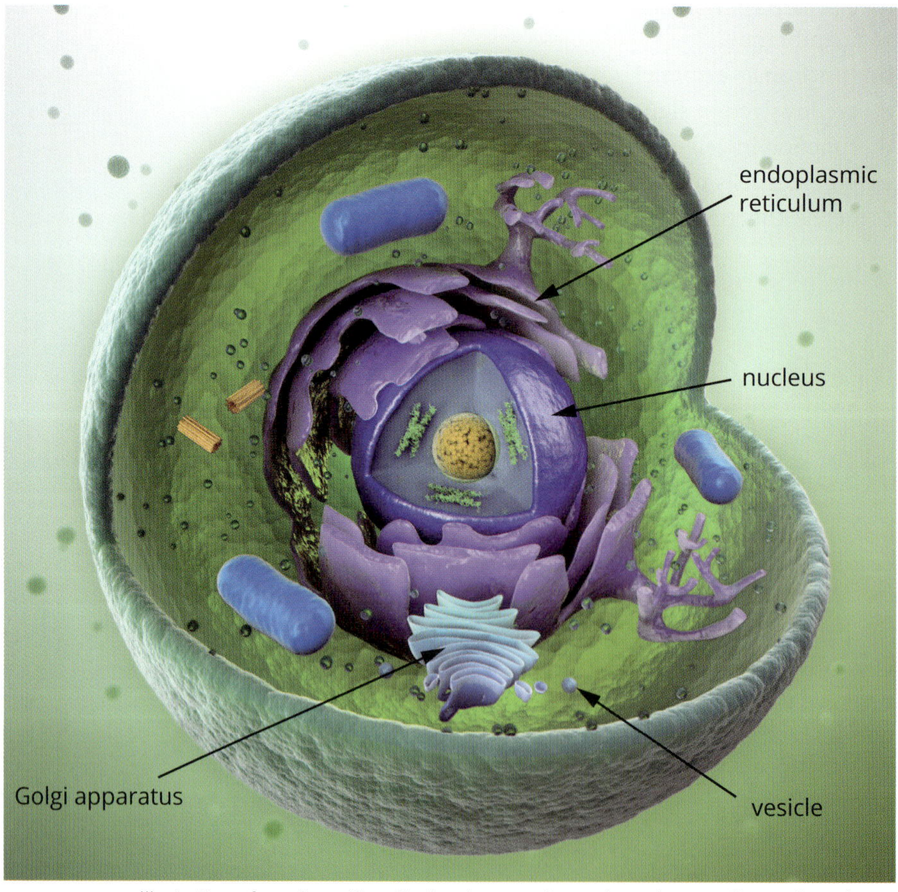

FIGURE 2.3.13 Illustration of a eukaryotic cell, showing membrane-bound organelles such as the nucleus, endoplasmic reticulum, Golgi apparatus and vesicles.

Compartmentalisation

Because each membrane-bound organelle has a different function, each organelle requires a different internal composition, including a high concentration of enzymes and reactants related to the organelle's particular function. The role of the membrane surrounding the organelle is to control the movement of substances between the organelle and the cell's cytosol, ensuring the organelle contains the substances it requires for its function.

Compartmentalisation in eukaryotic cells benefits the cell by:

- optimising the efficiency of processes by grouping enzymes and reactants for a particular function together in high concentrations and at the right conditions, such as at optimum pH levels
- enabling incompatible reactions to be performed simultaneously within a cell by separating the processes, or reactions, in space
- decreasing vulnerability to environmental change by separating processes inside the cell (as the environmental change would affect the cytosol much more so than the membrane-bound organelles).

2.3 Review

SUMMARY

- The plasma membrane is made up of a phospholipid bilayer with embedded proteins, cholesterol and glycoproteins.
- The fluid mosaic model describes the structure of the plasma membrane.
- The fluidity of the plasma membrane is a result of the free-moving phospholipids and proteins. This fluidity is dependent on phospholipid structure, temperature, and the presence of cholesterol.
  - Phospholipids are made up of hydrophilic phosphate heads and hydrophobic fatty acid tails. These tails can either be unsaturated (have double bonds and so have kinks in the tails) or saturated (have no double bonds and so have straight tails). Unsaturated tails give the membrane greater fluidity as the phospholipids are less tightly packed together.
  - Increases in temperature increase membrane fluidity.
  - Cholesterol in animal plasma membranes acts as a buffer for changing temperatures, decreasing fluidity at high temperatures and preventing the membrane solidifying at low temperatures.
- The fluidity of the plasma membrane affects its permeability and the movement of proteins within the membrane.
- The mosaic nature of a plasma membrane relates to the function and location of the integral and peripheral proteins within the membrane. The positioning of proteins within a membrane relates directly to their function.
  - Integral proteins are embedded within the membrane and span either part or the entire phospholipid bilayer.
  - Integral proteins that span the entire phospholipid bilayer are also known as transmembrane proteins.
  - Peripheral proteins are not embedded in the membrane but are attached to phospholipid molecules or to integral proteins.
- Lipid bilayer membranes also surround organelles and have a similar composition to the plasma membrane. This allows the two types of membranes, including those of vesicles, to fuse. This in turn enables the transport of substances between different organelles and into and out of the cell.
- Compartmentalisation in eukaryotic cells allows the concentration of enzymes and reactants in particular organelles of the cell and maintenance of the right conditions for their function. It also allows performance of incompatible chemical reactions simultaneously within the cell, and reduces vulnerability of the cell to environmental changes.

continued overleaf

2.3 Review *continued*

KEY QUESTIONS

1 List three functions of the plasma membrane.

2 Label the key components A–E of the plasma membrane shown in the diagram below.

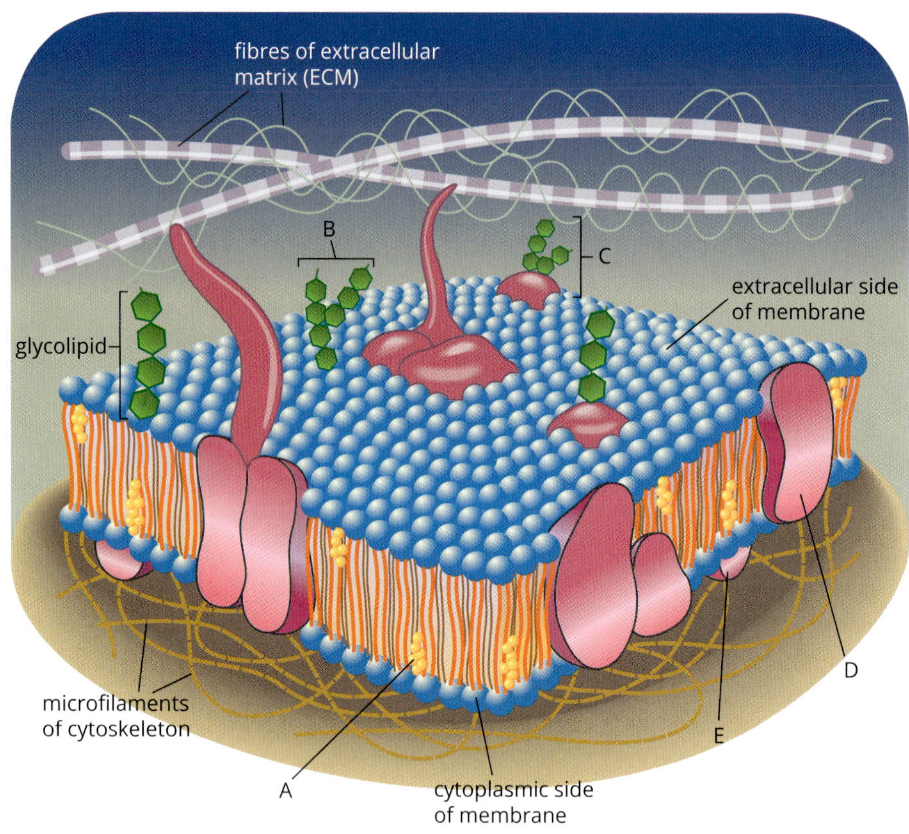

3 Outline how the properties of phospholipids forms the phospholipid bilayer.

4 Explain why the structure of the membrane is described as a 'fluid mosaic'?

5 Explain the factors that can affect the fluidity of a membrane.

6 What are the functions of membrane proteins?

2.4 Movement of material across membranes

One of the most important properties of membranes is their ability to regulate the transport of materials. Small molecules and water are constantly transported across the plasma membrane in both directions. Depending on their size and polarity, they either diffuse between the phospholipid molecules, or pass through channels created by proteins embedded within the plasma membrane. For larger molecules, such as proteins and polysaccharides, bulk transport across the plasma membrane is used—endocytosis if the molecules are to be transported into cells, or exocytosis if the molecules are to be transported out of cells.

In this section you will learn about the selective permeability of the plasma membrane. The various modes of transport to control the movement of molecules across the plasma membrane, including passive transport (simple diffusion, facilitated diffusion and osmosis) and active transport, will be explored. In addition, the role of different organelles, including the endoplasmic reticulum, Golgi apparatus and associated vesicles, in the export of a protein product from the cell through exocytosis, and the cellular engulfment of material by endocytosis will also be discussed.

PERMEABILITY OF THE PLASMA MEMBRANE

Many different types of molecules can move across the plasma membrane (Figure 2.4.1). The ways in which these molecules move across the plasma membrane depends on their chemical properties, such as size, charge and polarity, and whether or not the phospholipid bilayer is permeable to the substance (Table 2.4.1).

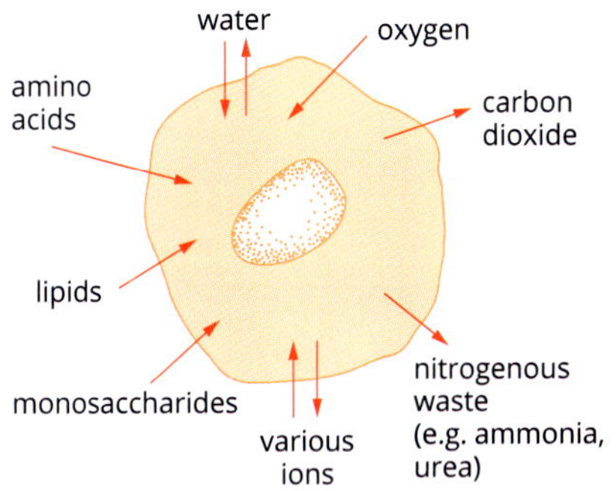

Example of molecule or ion	Chemical properties	Permeability of membrane to the substance
oxygen, carbon dioxide	small uncharged molecule	permeable
steroids, alcohol, chloroform	lipid-soluble, non-polar molecule	permeable
water, urea	small polar molecule	permeable or selectively permeable
potassium ion (K^+) sodium ion (Na^+), chloride ion (Cl^-)	small ion	non-permeable (ions pass through protein channels)
amino acid, glucose	large, polar, water-soluble molecule	non-permeable (molecule passes through protein channels)

TABLE 2.4.1 Permeability of plasma membrane to different molecules.

FIGURE 2.4.1 Cells exchange many substances with their environment across the plasma membrane.

The interior of the plasma membrane is made up of a lipid bilayer that is hydrophobic (Figure 2.4.2). Most hydrophobic molecules can dissolve in the lipid bilayer of the plasma membrane and move freely without the aid of membrane proteins.

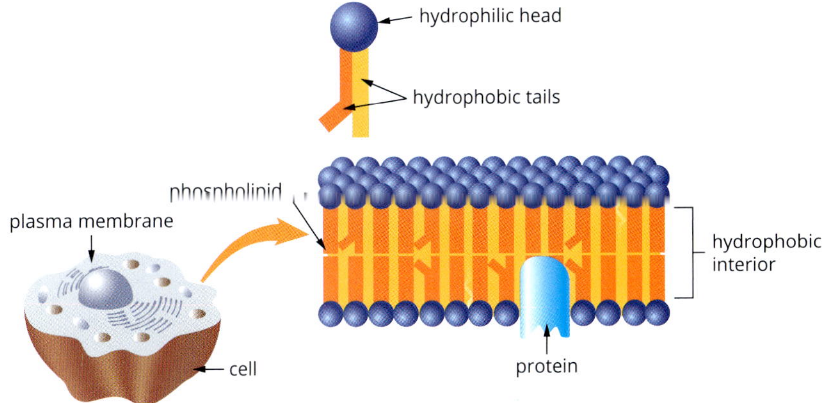

FIGURE 2.4.2 Structure of the plasma membrane.

However, the hydrophobic nature of the interior of the plasma membrane makes the plasma membrane impermeable to hydrophilic molecules including:

- most water-soluble molecules
- ions (atoms or groups of atoms with an overall positive or negative charge)
- polar molecules (molecules with charged regions but no overall charge).

When polar molecules such as glucose and amino acids try to pass through the plasma membrane, they pass through slowly. Extremely small polar molecules, such as water and urea, are able to pass between the phospholipid molecules, however it occurs slowly. A charged atom or molecule has a surrounding shell of water (Figure 2.4.3); hence, it is unable to penetrate the lipid bilayer of the plasma membrane. These hydrophilic substances must therefore pass through specific protein channels in the plasma membrane (Figure 2.4.4).

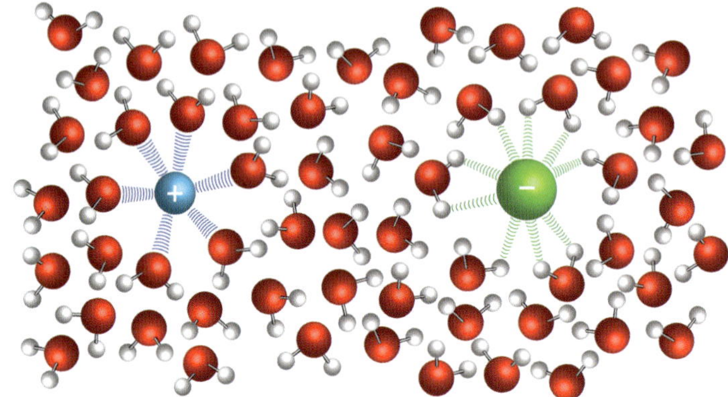

FIGURE 2.4.3 Water molecules surrounding ions.

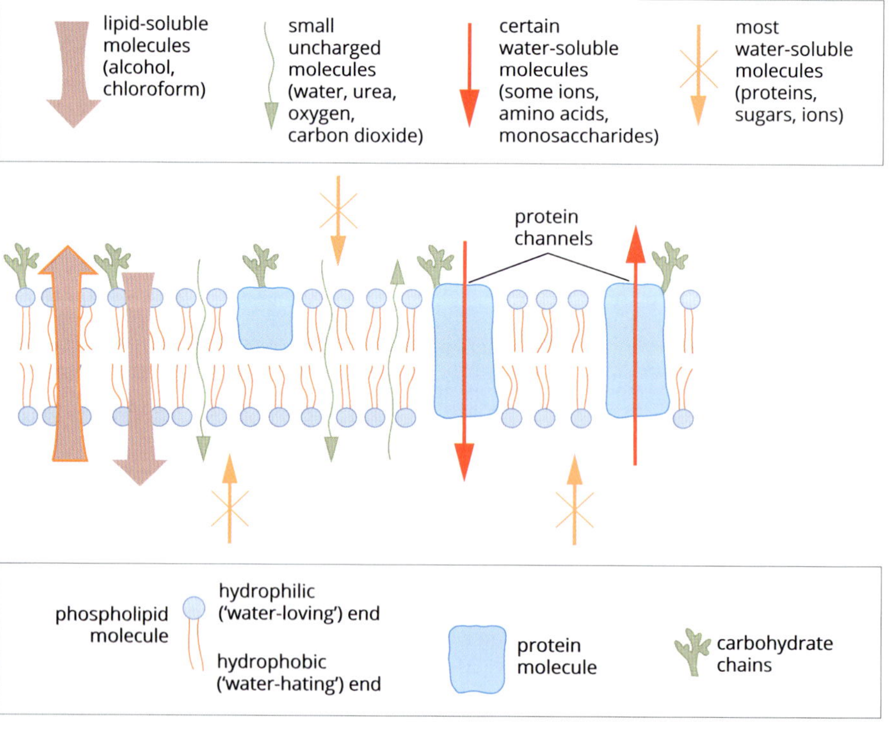

FIGURE 2.4.4 Protein channels are required for the movement of molecules for which the plasma membrane is impermeable.

PASSIVE TRANSPORT

Diffusion of molecules across the plasma membrane without the expenditure of energy is known as **passive transport**. The three types of passive transport across membranes are diffusion, facilitated diffusion and osmosis. The movement of a molecule across a biological membrane is called passive transport when the cell does not have to expend energy to make it happen.

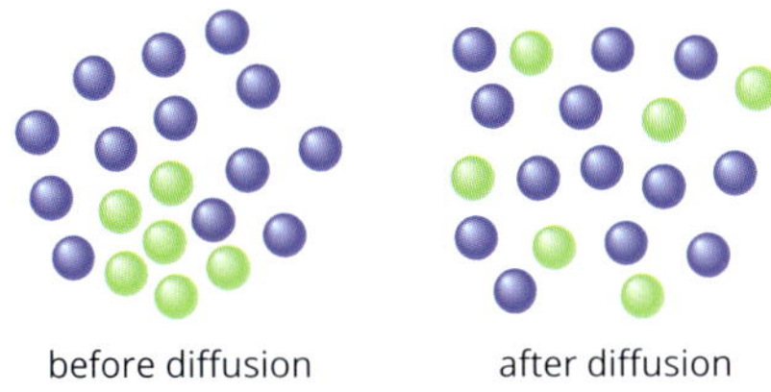

FIGURE 2.4.5 Diffusion of particles.

Diffusion

Particles are always in constant random motion. One result of this motion is diffusion by which particles move from an area of higher concentration to an area of lower concentration (Figure 2.4.5). As there are many particles colliding with each other during this process, the overall movement of particles is relatively slow.

Diffusion occurs in both gases and liquids, as the particles are able to move from place to place. Diffusion can be seen most prominently when a drop of ink (the **solute**), is placed in a jar of still water (the **solvent**). The dye particles in the ink move randomly through the water until the colour is homogenous (evenly spread). In other words, the solute particles move from an area of high solute concentration (the drop of ink) to the areas of low solute concentration (the rest of the jar). The solute particles are said to have moved down the **concentration gradient**.

Diffusion is called a passive process because it does not require energy. It occurs only because there is a concentration gradient.

BIOFILE

The diffusion of alcohol molecules

Alcohol enters the bloodstream quickly. It is a non-polar molecule and does not need to be broken down into smaller molecules by digestion. Alcohol molecules pass across membranes easily by simple diffusion and as a result they are absorbed rapidly by the mouth, stomach and small intestine, where absorption is fastest. Eating a meal before drinking reduces the efficiency and rate of alcohol absorption.

Diffusion across membranes

Diffusion can occur even in the presence of a membrane, but molecules can diffuse across a membrane only if the membrane is permeable to them (Figure 2.4.6).

If the membrane is permeable to the molecules, there will be a constant movement of molecules backwards and forwards across the membrane. If the concentration of the molecules is higher on one side of the membrane than the other, more molecules will cross from the area of higher concentration to the area of lower concentration (i.e. down its concentration gradient), as you can see in Figure 2.4.7a.

Molecules will still be moving backwards and forwards across the membrane, even when the concentration of molecules is the same on both sides of the membrane. However, about the same number move across in each direction, so there will be no net movement from one side to the other.

If the membrane is semipermeable—that is, it is impermeable to some molecules—there will be no movement of those molecules from the area of higher concentration to the area of lower concentration, as you can see in Figure 2.4.7b.

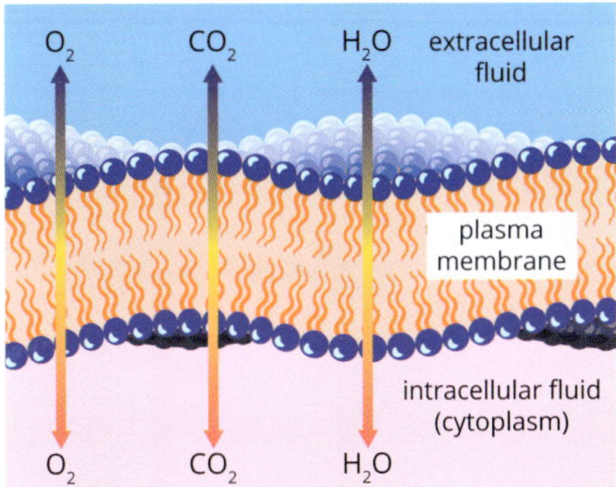

FIGURE 2.4.6 The transport of materials across the plasma membrane by simple diffusion.

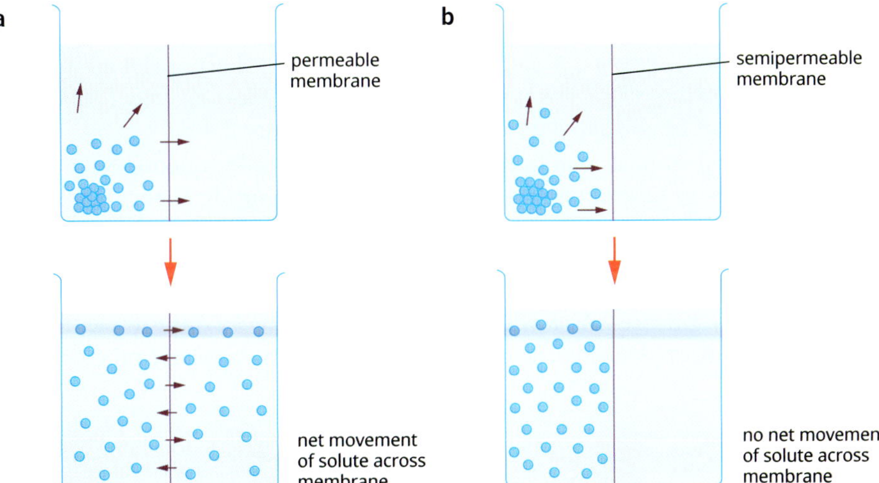

FIGURE 2.4.7 (a) If a membrane that is permeable to both the solute and the solvent is inserted in the liquid it does not affect the pattern of diffusion. (b) A membrane that allows the solvent molecules to pass through, but not the solute molecules, stops the solute from diffusing through the membrane.

Osmosis

Osmosis is a special case of diffusion that occurs across semipermeable membranes. Semipermeable (also sometimes called partially permeable, selectively permeable or differentially permeable) membranes allow the free passage of water molecules (and certain other molecules such as urea), but restrict the passage of most solutes. If a diluted and a concentrated solution are separated by a semipermeable membrane that allows the movement of free water molecules across the membrane, but not the movement of the solute molecules, the free water molecules will move across the membrane from the diluted to the concentrated solution. This is because water molecules bind to solute molecules in solution, so there are more free water molecules in a dilute solution than in a concentrated solution (Figure 2.4.8).

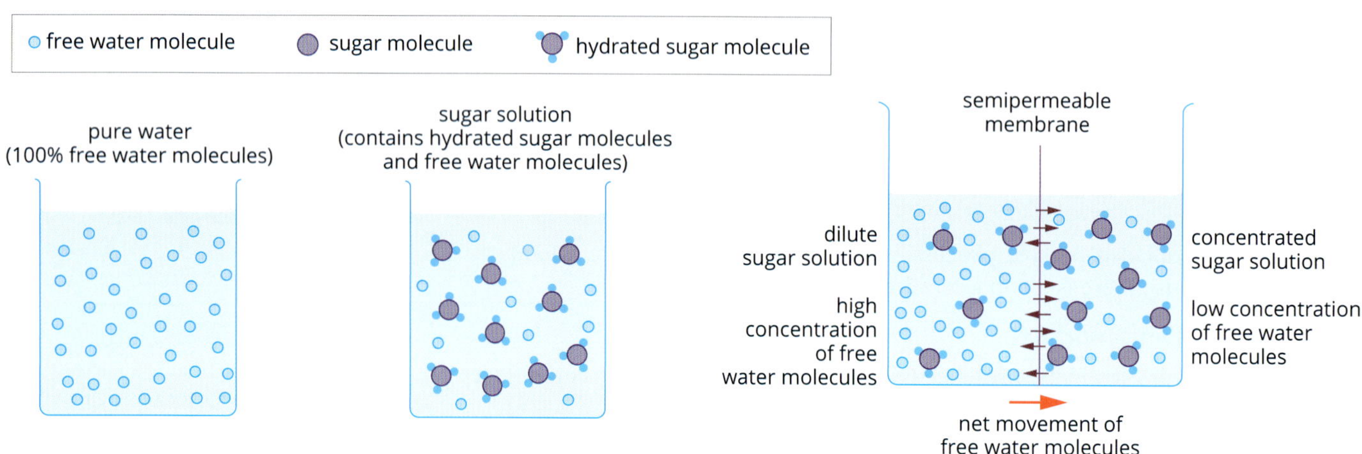

FIGURE 2.4.8 A net movement of water molecules from a dilute solution through a semipermeable membrane into a concentrated solution is osmosis.

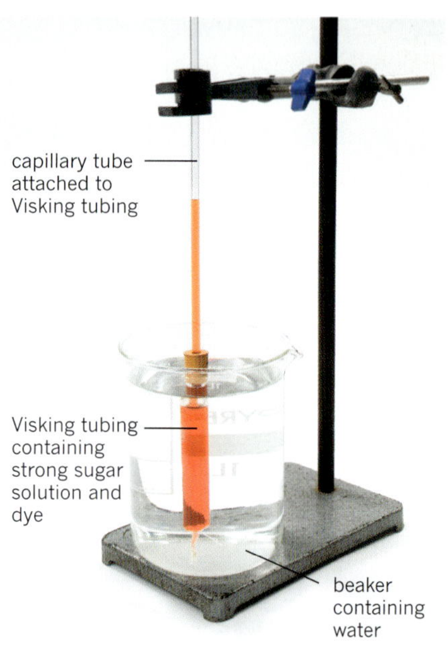

FIGURE 2.4.9 An experimental setup used for demonstrating osmosis.

In osmosis, net diffusion of water occurs through a semipermeable membrane from a diluted to a concentrated solution along the water's own concentration gradient, which is known as the **osmotic gradient**. The pressure causing the water to move along this gradient is called **osmotic pressure**.

Osmosis can be demonstrated using Visking tubing (containing a strong sugar solution coloured with food dye) attached to a clear capillary tube and submerged in a beaker of water (Figure 2.4.9). Visking tubing is a synthetic semipermeable membrane made from cellulose. It is also called dialysis or cellulose tubing. In the Visking tubing, water molecules will bind to the sugar molecules. As a result, there will be fewer free water molecules in the Visking tubing and a net movement of free water molecules into the tubing by osmosis will occur. This causes an increase in the volume of liquid in the tubing, increasing the pressure and forcing the coloured solution up the capillary tube as shown in Figure 2.4.9.

Diffusion and osmosis can occur together. For example, glucose is similar in size to the pores in Visking tubing. If the experiment setup in Figure 2.4.9 used a concentrated glucose solution inside the Visking tubing, water would still enter into the tubing by osmosis and cause the coloured solution up the capillary tube, as glucose does not easily diffuse out of the Visking tube. However, if the apparatus is left for 24 hours, glucose will slowly move out of the tubing by diffusion with water following by osmosis, causing the coloured solution in the capillary tube to fall.

In osmosis, we are always comparing solute concentration between two solutions. The terms isotonic, hypertonic and hypotonic solution are often used to describe the differences:

- Isotonic solutions: The solutions being compared have equal concentrations of solutes.
- Hypertonic solution: The solution with the higher concentration of solute (hence lower concentration of free water molecules).
- Hypotonic solution: The solution with the lower concentration of solute (hence higher concentration of free water molecules).

Facilitated diffusion

Recall that the phospholipid bilayer of the membrane is impermeable to certain particles (ions or molecules). However, transport proteins in the membrane allow the diffusion of these particles by a process called **facilitated diffusion**. In facilitated diffusion:

- the membrane transport proteins are specific for particular particles, so transport is selective; some particles are transported and others are not
- transport is more rapid than by simple diffusion
- the transport proteins can become saturated (fully occupied) as the concentration of the transported substances increases
- the transport of one particle may be inhibited by the presence of another particle that uses the same transport protein
- no energy is required; the particles move down their concentration gradient.

EXTENSION

Effect of osmosis on cells

The plasma membrane is permeable to water, so when cells are placed in pure water an osmotic gradient will draw water into the cells. This is because the cytosol is a concentrated solution containing many dissolved substances. For example, if red blood cells are placed in a solution that is more concentrated than their cytosol, water leaves the red blood cells by osmosis and causes them to shrink and become crenated (Figure 2.4.10a). Conversely, if red blood cells are placed in fresh water, the cells absorb so much water by osmosis that they swell and may eventually burst (lyse), releasing red pigment into the water (Figure 2.4.10c).

For cells with the presence of walls, such as plant cells and prokaryotes, the cell wall helps to maintain the cell's water balance. For example, if a plant cell loses water by osmosis, it will start to shrivel and the plasma membrane will start to pull away from the cell wall; the cell is said to have become plasmolysed (Figure 2.4.10d). However, if a plant cell absorbs water by osmosis, it swells to some extent but the relatively inelastic cell wall prevents the cell from bursting. The relatively inelastic wall will expand until it exerts a pressure back onto the cell, known as turgor pressure. Turgor pressure prevents further water uptake. At this point, the plant cell is turgid (Figure 2.4.10f).

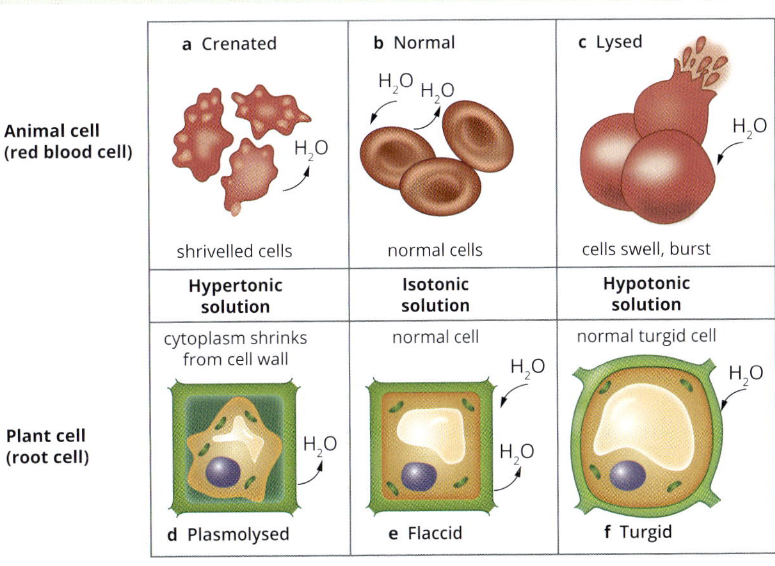

FIGURE 2.4.10 The effect of three different solution concentrations on an animal cell and plant cell.

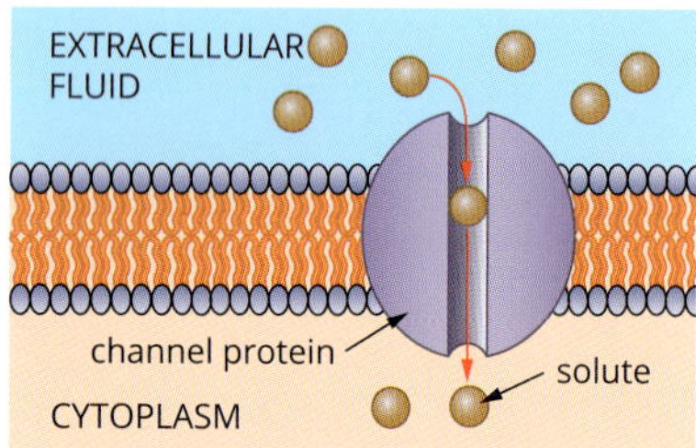

a A channel protein (purple) has a channel through which water molecules or a specific solute can pass.

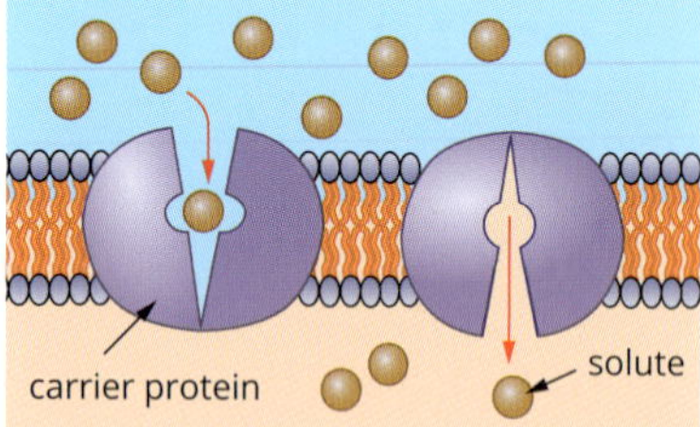

b A carrier protein alternates between two shapes, moving a solute across the membrane during the shape change.

FIGURE 2.4.11 Two types of transport proteins that carry out facilitated diffusion. In both cases, the protein can transport the solute in either direction, but the net movement is down the concentration gradient of the solute.

The two main types of membrane transport proteins in facilitated diffusion are **channel proteins** and **carrier proteins**. Membrane transport proteins provide channels for the passage of water-soluble (polar) molecules and ions across the phospholipid bilayer.

Channel proteins are specific for a substance. Channel proteins do not usually bind with the molecules being transported. They function like pores that open and close to allow the passage of specific molecules (Figure 2.4.11a).

Carrier proteins bind the molecules being transported, causing the protein to undergo changes in shape (conformation) that allow specific molecules to be transported across the membrane (Figure 2.4.11b). After the molecule has crossed the membrane, the protein is restored to its original shape.

Factors affecting rate of diffusion

There are three main factors that affect the rate of diffusion across a membrane:

- concentration—the greater the difference in concentration gradient, the faster the rate of diffusion. When the concentration is equal on both sides of the membrane the net diffusion is zero, even at high temperatures
- temperature—the higher the temperature, the higher the rate of diffusion. Increasing the temperature increases the speed at which molecules move
- particle size—the smaller the particles, the faster the rate of diffusion through a membrane.

ACTIVE TRANSPORT

Active transport involves the use of energy by the cell to transport particles across membranes (Figure 2.4.12). Because active transport uses energy, it can move substances against a concentration gradient (from low concentrations to high concentrations). Active transport enables a cell to maintain internal concentrations of small solutes that differ from concentrations in its environment. The membrane transport proteins in active transport are all carrier proteins.

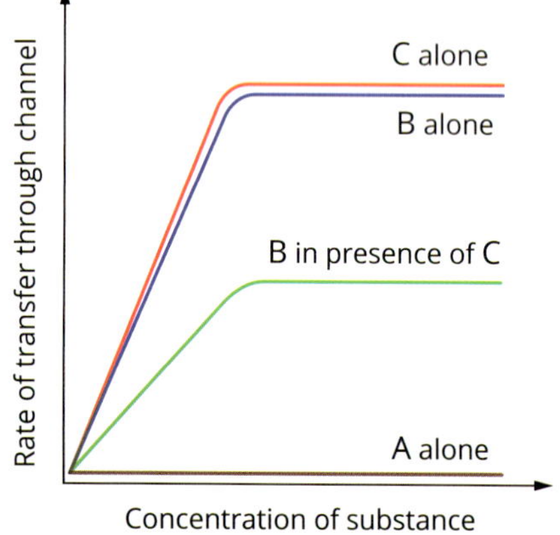

FIGURE 2.4.13 Theoretical transport rate vs concentration graph for the movement of three substances through a channel protein. Substances B and C are transported, but not substance A, demonstrating selectivity. The rate of transfer of substances B and C reaches a plateau when their concentrations reach a certain level, demonstrating saturation. The rate of transport of B is less when C is present, demonstrating competitive inhibition.

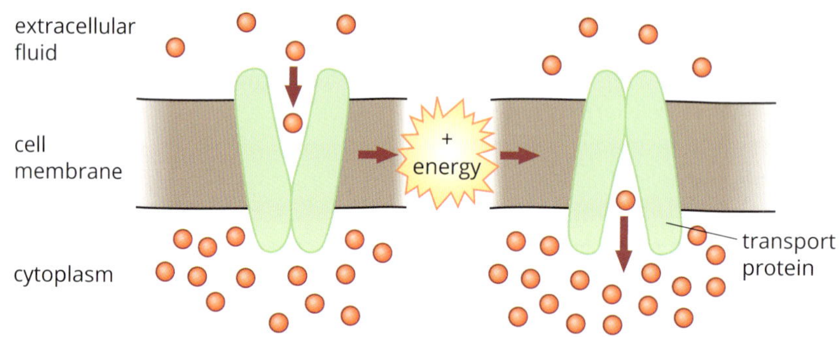

FIGURE 2.4.12 Active transport requires an energy source to result in a conformation change in the transport protein, so that molecules can be transported across the membrane.

Active transport has the same properties of selectivity, saturation and competitive inhibition as facilitated diffusion, because it also occurs through transport proteins (Figure 2.4.13). Selectivity means that some substances are transported but others are not. Saturation means that there is no increase in the rate of transfer when all transport proteins are open. Competitive inhibition means that one substance can inhibit the transport of another substance that uses the same transport protein.

In different situations, either facilitated diffusion or active transport may be used to transport a particular molecule. Whether a cell uses facilitated diffusion or active transport depends on the specific needs of the cell. For example, glucose is actively transported from the gut into epithelial cells lining the gut so it can enter the bloodstream. The regulation of this process is controlled by hormones, principally insulin and glucagon. If gut glucose levels are high, blood glucose levels will increase. If gut glucose levels are low, active transport makes sure that the little glucose that is in the gut gets pumped into the epithelium from where it can move to bloodstream via facilitated diffusion.

It is worth remembering that there are no mechanisms for the active transport of water molecules across cell membranes. The net movement of water across membranes always occurs by osmosis.

The comparison between passive transport and active transport can be seen in Figure 2.4.14 and Table 2.4.2.

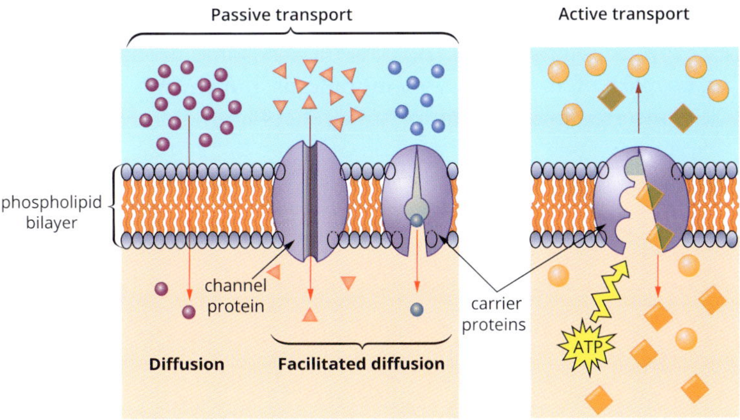

FIGURE 2.4.14 Methods of transport of different materials across the plasma membrane.

	Passive transport			Active transport
	Simple diffusion	Osmosis	Facilitated diffusion	
Type of substance	hydrophobic molecules, small polar molecules; examples: water, oxygen, carbon dioxide	water	hydrophilic molecules; examples: calcium ions, glucose, amino acids, sodium ions	hydrophilic molecules; examples: glucose, amino acids, sodium ions, potassium ions
Type of membrane protein required	none	none	channel protein carrier protein	carrier protein
Direction of movement of molecules	down concentration gradient	down concentration gradient	down concentration gradient	against concentration gradient
Energy requirement	none	none	none	energy in the form of ATP is required

TABLE 2.4.2 A comparison of passive and active transport.

BULK TRANSPORT

Large molecules, such as proteins and polysaccharides, as well as cellular waste products, nutrients and water, can be transported in bulk across the membrane by exocytosis and endocytosis. Like active transport, bulk transport requires energy.

Protein synthesis and exocytosis

Secretory proteins are proteins that are produced to be exported out of a cell. The movement of secretory proteins occurs by **exocytosis**, also known as secretion. Before reaching the plasma membrane for exocytosis, secretory proteins must first be synthesised and modified.

Ribosomes and endoplasmic reticulum

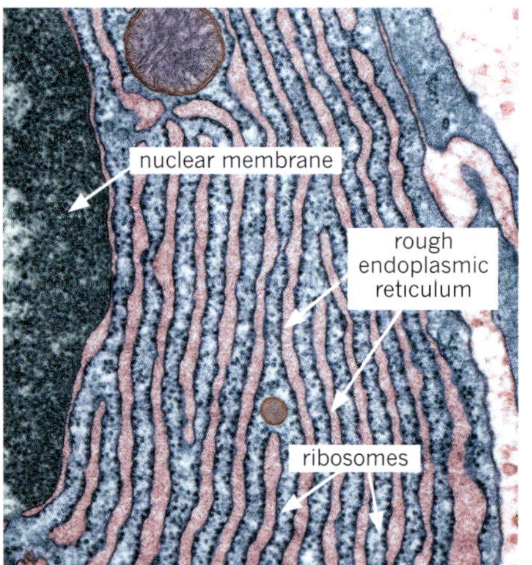

FIGURE 2.4.15 Coloured transmission electron micrograph of rough endoplasmic reticulum (blue) with ribosomes (black dots) on the outer surface. The outer layer of the nuclear membrane can be seen at the left edge (dark grey).

Proteins destined for use within the cell are synthesised by free ribosomes that are found in the cytosol. Proteins that are to be secreted are synthesised by ribosomes that stud the outer surface of the **rough endoplasmic reticulum** (Figure 2.4.15).

As it is produced by the ribosome, the polypeptide chain is inserted into the lumen (the fluid-filled space between the membranes) of the rough endoplasmic reticulum through a pore in the membrane. Once the secretory protein has been synthesised, it is transported through the tubules of the rough endoplasmic reticulum, where it is modified. For example, if the secretory protein is a glycoprotein (most are), carbohydrates are attached to the protein in the endoplasmic reticulum by enzymes that are present on its membranes.

When the secretory proteins reach the end of the tubules, they are wrapped in the membranes of vesicles that bud off from the endoplasmic reticulum. The vesicles are then transported to other parts of the cell. Vesicles that move from one part of the cell to another are called **transport vesicles**.

In addition to making secretory proteins, the rough endoplasmic reticulum also produces transmembrane proteins. As the ribosome produces polypeptides that are to be part of the plasma membrane as transmembrane proteins, the polypeptide is inserted into the endoplasmic reticulum membrane itself.

BIOLOGY IN ACTION

Carbon nanotubes used to make artificial pores in cell membranes

The plasma (or cell) membrane plays a vital role in the protection of the internal cellular environment and the transport of substances vital to cell growth, communication and survival. The selectively permeable nature of the plasma membrane enables tight regulation of what can and cannot enter and exit the cell.

Synthetic membranes are used widely in the laboratory and in industry for separation and purification purposes (e.g. water treatment), but, until recently, scientists have not been able to mimic the selective permeability of plasma membranes. Researchers at the University of California have successfully developed carbon nanotubes that can be inserted into both synthetic and natural membranes (Figure 2.4.16), creating artificial pores. (Carbon nanotubes are thin cylinders of carbon atoms.) The researchers used lipids to refine the structure of the nanotubes, ensuring that they were just the right size and composition to penetrate the membrane. Because plasma membranes consist of a lipid bilayer in which lipid and protein molecules are embedded, the nanotube structure needed to be consistent with this. The researchers coated the nanotubes with lipids, which allowed them to slide into the plasma membrane of a cell without causing damage.

FIGURE 2.4.16 Graphical representation of nanotube drug delivery. Nanotube technology provides the potential for drugs to be delivered to specific tissues and cells.

The researchers also observed the nanotubes opening and closing, just like biological ion channels. They discovered that the charge of the ends of the nanotubes shifted in response to changes in the ion concentration of the surrounding environment. These properties of nanotubes make it possible for scientists to artificially control what enters and exits cells, providing potential applications in areas such as drug delivery, where the selective transport of molecules could significantly improve the efficacy of treatments.

Golgi apparatus

After leaving the endoplasmic reticulum, the transport vesicles travel to and fuse with the Golgi apparatus at the ***cis* face** (Figure 2.4.17). The *cis* face is usually found near the endoplasmic reticulum.

The **Golgi apparatus** consists of flattened sacs called **cisternae** (Figure 2.4.17). The secretory protein enters the Golgi apparatus and moves from one cisternae to the next, carried by vesicles. As it moves through the Golgi it is progressively modified. For example, the Golgi may modify the carbohydrate on the glycoproteins by removing some sugar monomers and substituting them with others, producing a large variety of carbohydrates.

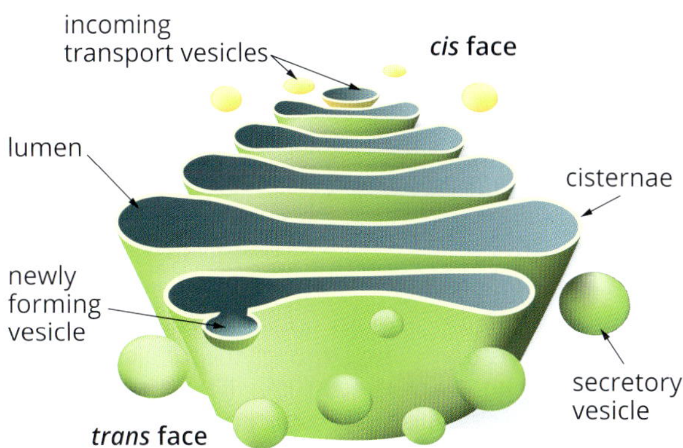

FIGURE 2.4.17 Structure of the Golgi apparatus.

When the secretory protein, such as a hormone or enzyme, is ready for secretion, **secretory vesicles** containing the protein bud off from the ***trans* face** end of the Golgi apparatus and move to the plasma membrane, where the product for secretion is released out of the cell via exocytosis.

Sometimes secretory vesicles are not transported to the plasma membrane. Instead, the vesicles are stored in the Golgi apparatus until the secretory protein is needed. For example, in pancreatic cells, digestive enzymes are stored in secretory vesicles within the Golgi apparatus until the presence of food in the stomach triggers a signal for their secretion (Figure 2.4.18).

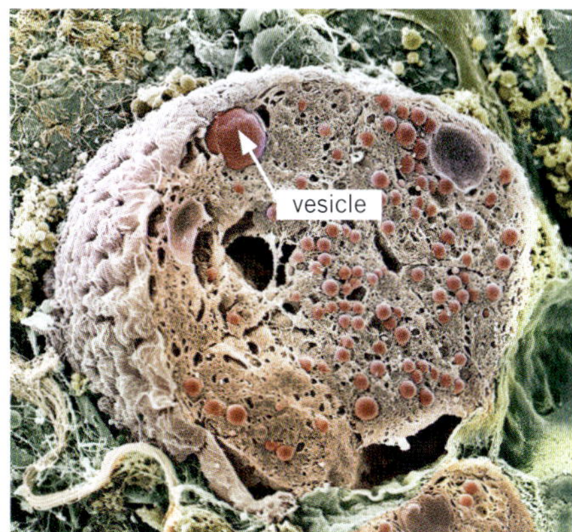

FIGURE 2.4.18 Coloured scanning electron micrograph of a pancreatic cell. Pancreatic cells produce and excrete digestive enzymes in vesicles (purple).

Exocytosis

When the secretory vesicle membrane and plasma membrane come into contact, specific proteins alter the arrangement of the phospholipids of the phospholipid bilayer, enabling the fusion of the two membranes. The fluid and dynamic nature of the plasma membrane enables this membrane fusion. Once the two membranes are fused, the contents of the secretory vesicle are released out of the cell. This is called exocytosis. The vesicle membrane becomes a permanent part of the plasma membrane (Figure 2.4.19). The plasma membrane is continually recycled as vesicles fuse during exocytosis and are conversely formed and released during endocytosis.

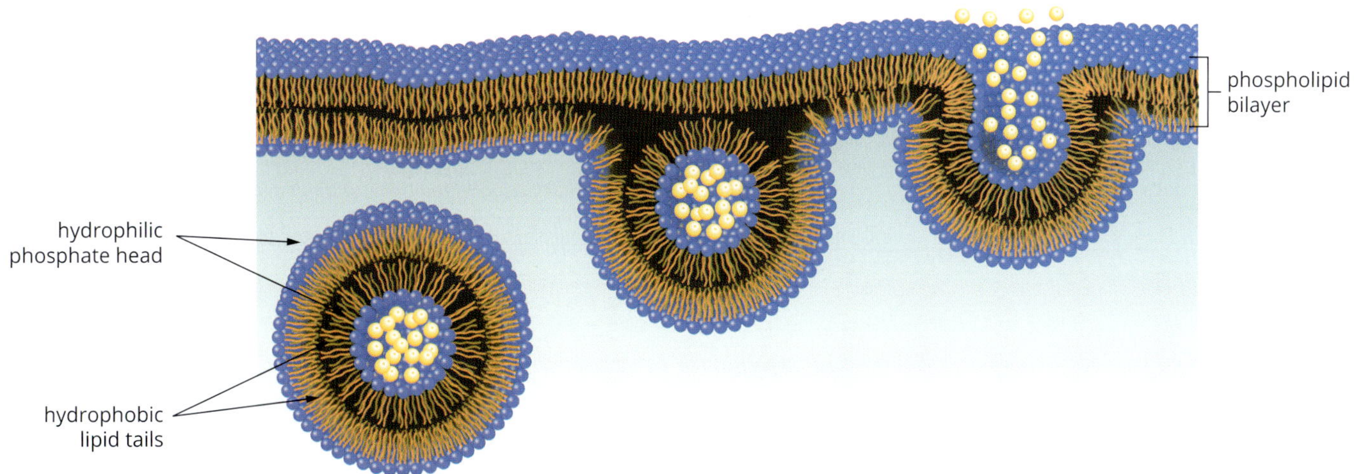

FIGURE 2.4.19 Exocytosis is the movement of a secretory vesicle towards the plasma membrane and the release of its contents.

In addition to being used for secreting proteins, exocytosis is also involved in the release of cellular waste and the breakdown products from **lysosomes**. For example, after phagocytic cells (such as unicellular protists or macrophages in multicellular organisms) engulf food or foreign matter and digest them with the aid of lysosomes, the waste products of this digestion are released by exocytosis.

A summary of the roles of the rough endoplasmic reticulum and Golgi apparatus in the exocytosis of proteins can be seen in Figure 2.4.20.

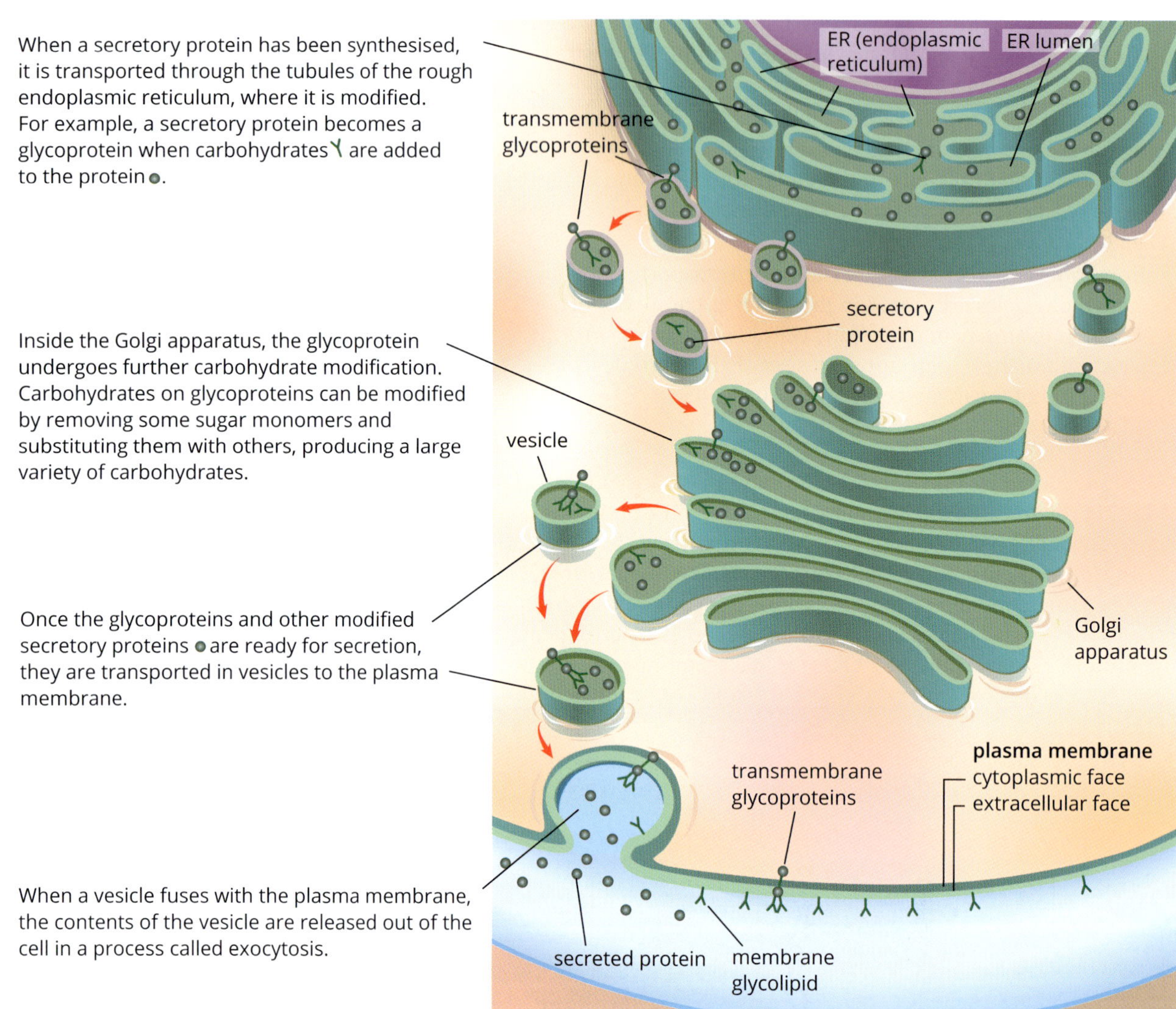

FIGURE 2.4.20 Roles of rough endoplasmic reticulum and Golgi apparatus in the production and exocytosis of proteins.

Endocytosis

In **endocytosis**, the cell takes in materials in bulk by forming new vesicles from the plasma membrane. During the process of endocytosis, a small area of the plasma membrane sinks inwards to form a pocket. As the pocket deepens, materials near the plasma membrane are enclosed by the membrane, which then pinches off to form a vesicle. The vesicle then transports the substance to where it is required within the cell.

There are three types of endocytosis (Figure 2.4.21):

- In **phagocytosis** a cell engulfs a solid material by wrapping **pseudopodia** around it, forming a phagosome (Figure 2.4.22). The material will be digested when the food vacuole fuses with a lysosome containing enzymes.
- In **pinocytosis** the plasma membrane engulfs liquid that contains dissolved molecules (Figure 2.4.23).
- **Receptor-mediated endocytosis** is a type of pinocytosis that engulfs specific substances. Protein **receptors** located on the surface of the plasma membrane respond to particular molecules, binding to the molecule and then triggering the engulfment of the substance into the cell.

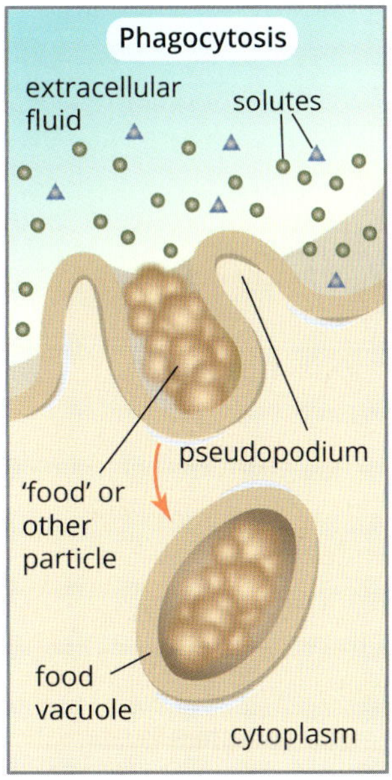

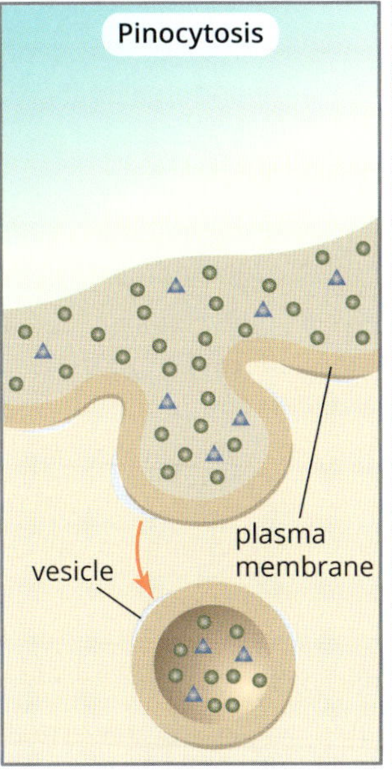

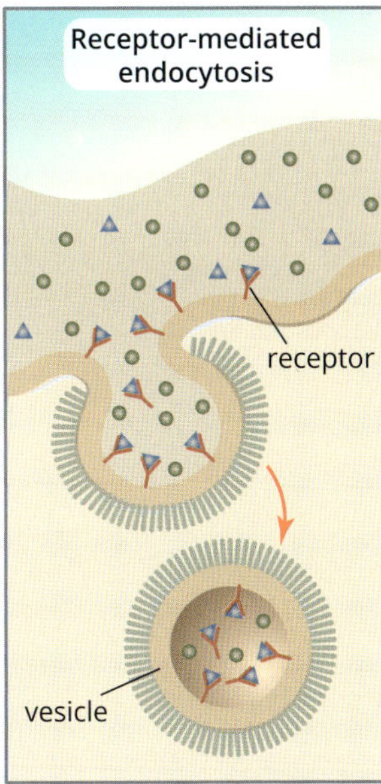

FIGURE 2.4.21 The three types of endocytosis showing the engulfment of different materials by the plasma membrane and the release of the vesicle within the cell.

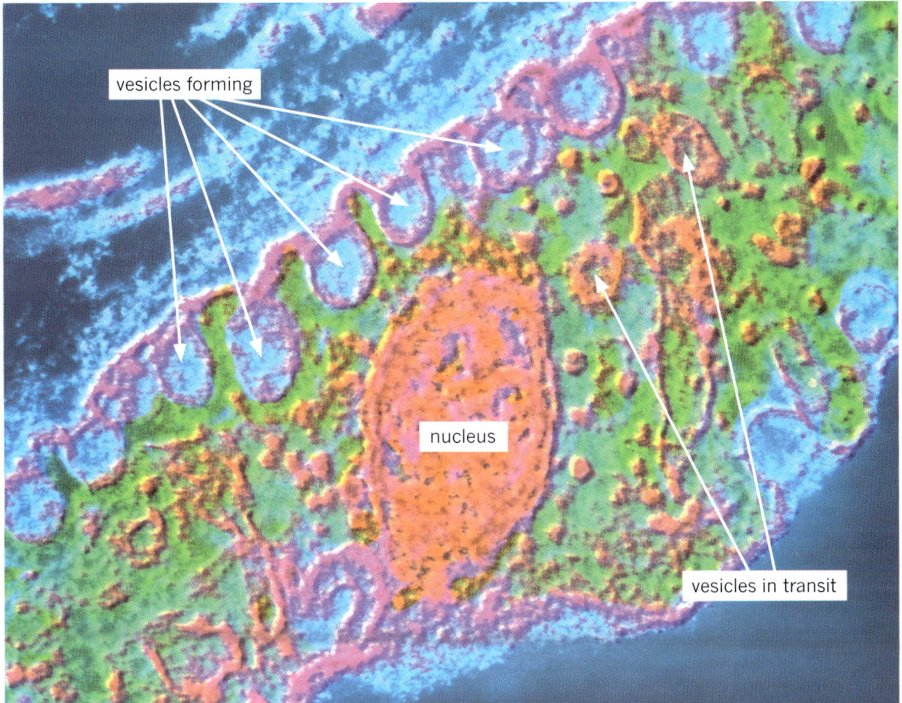

FIGURE 2.4.23 Transmission electron micrograph showing pinocytosis in a blood capillary. The lumen of the blood capillary is at top (blue), with one of the endothelial cells of the capillary lining below, the nucleus of which is red. A row of the vesicles formed during pinocytosis can be seen in the upper cell membrane (purple) and in transit towards the outside of the capillary (bottom).

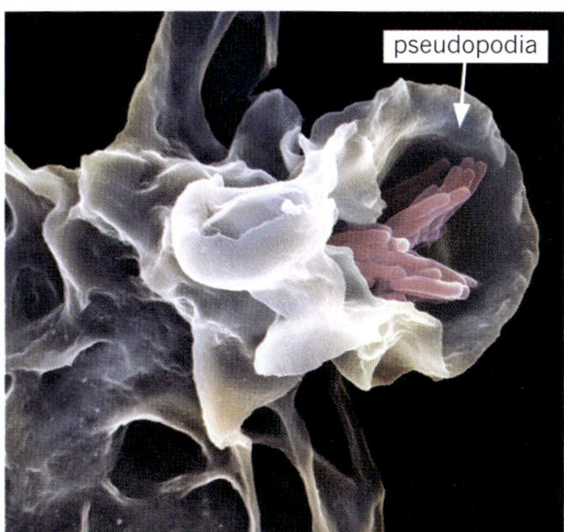

FIGURE 2.4.22 Coloured scanning electron micrograph of a macrophage white blood cell (purple) engulfing a tuberculosis bacterium (*Mycobacterium tuberculosis*) (pink) by phagocytosis.

2.4 Review

SUMMARY

- Transport of molecules across the membrane depends on their chemical and physical properties, such as size, charge and polarity, and whether or not the phospholipid bilayer is permeable to the substance.
- Passive transport is the movement of molecules without the expenditure of energy.
- Passive transport includes:
  - simple diffusion: movement of small, non-polar molecules and other uncharged molecules through a semipermeable membrane, down the concentration gradient
  - osmosis: the net diffusion of water molecules across a semipermeable membrane
  - facilitated diffusion: movement of polar molecules and ions through protein channels and carrier proteins down the concentration gradient.
- Active transport is the movement of molecules with the expenditure of energy, against the concentration gradient.
- Bulk transport is used for the movement of larger molecules and requires energy.
  - Movement of molecules out of the cell (for secretion) occurs through exocytosis.
  - Movement of materials into the cell occurs through endocytosis.
- The rough endoplasmic reticulum, Golgi apparatus and vesicles are involved in the production, modification, packaging and transport of proteins that are released from the cell by exocytosis.
  - Ribosomes are located on the surface of the rough endoplasmic reticulum and synthesise the proteins.
  - The rough endoplasmic reticulum then modifies the proteins.
  - A vesicle forms at the end of the rough endoplasmic reticulum, transporting the protein to the Golgi apparatus, which further modifies the protein.
  - A vesicle forms from the Golgi apparatus and transports the protein to the plasma membrane.
  - The vesicle fuses with the plasma membrane, releasing the contents out of the cell by exocytosis.
- Substances can enter a cell by endocytosis. The membrane surrounds and engulfs the substance, forming a vesicle, which enters the cell.
  - Solid particles enter by phagocytosis.
  - Liquid containing dissolved molecules enters by pinocytosis.
  - Receptors bind to the substance and enter by receptor-mediated endocytosis.

KEY QUESTIONS

1 Distinguish between simple diffusion and osmosis.

2 Distinguish between facilitated diffusion and active transport in the movement of substances.

3 State the function of ribosomes.

4 Outline how vesicles are used to transport materials secreted by a cell.

5 Outline the process of endocytosis.

6 Which of the following is a feature of exocytosis but not endocytosis?

A changes in the shape of the plasma membrane
B formation of vesicles
C use of ATP
D secretion

2.5 Proteins

Nearly every function of a living organism depends on proteins. Proteins have a large range of functions in living organisms including speeding up chemical reactions, playing a role in cell–cell recognition and cellular communication, movement, storage and even structural support. A human has tens of thousands of different proteins, and each protein has a specific sequence of amino acids, giving it a unique shape that enables it to carry out a particular function. Proteins have a large variety of functions, so they vary extensively in structure, with each type of protein having a unique three-dimensional shape (Figure 2.5.1).

In this section, you will learn about the diversity in the functions of proteins, and the nature of the proteome. You will also learn about the synthesis of a polypeptide chain from amino acid monomers by condensation polymerisation, and the functional importance of the four hierarchal levels of protein structure.

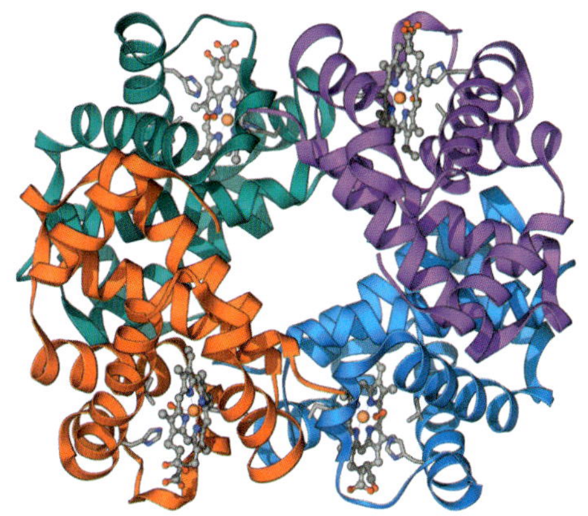

FIGURE 2.5.1 A ribbon diagram showing the three-dimensional structure of haemoglobin, a protein made up of four polypeptide chains.

THE FUNCTIONAL DIVERSITY OF PROTEINS

There are many different types of proteins within a particular organism. Each protein has a different function, and each plays a vital role in the regulation, functioning and maintenance of both individual cells and entire organisms. In fact, almost every function of living organisms depends on proteins. The specific structure of each protein enables it to carry out its function.

Some of the functional types of proteins are described in Table 2.5.1.

TABLE 2.5.1 An overview of protein function.

Function: enzymatic proteins	Function: hormonal proteins
Description: act as catalysts in cellular reactions (enzymes) **Examples:** • catabolic enzymes, such as lipase and amylase, that catalyse the breakdown of bonds (also known as hydrolysis) (Figure 2.5.2) • anabolic enzymes, such as DNA polymerase, that catalyse the formation of bonds (also known as condensation)	**Description:** coordinate an organism's activities by triggering a responses **Examples:** insulin, glucagon
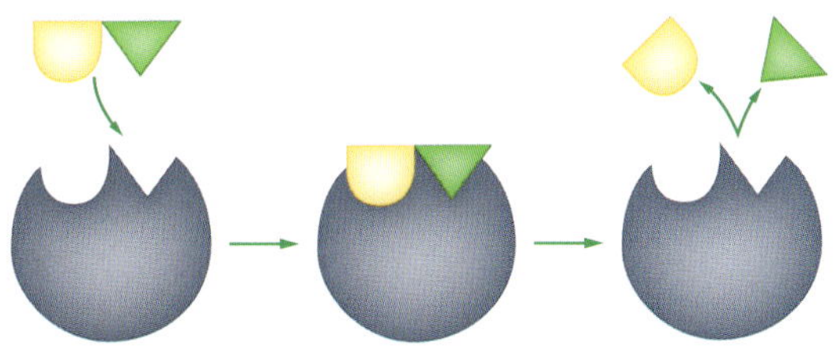 FIGURE 2.5.2 Catabolic enzymes, such as lipase, catalyse (speed up) reactions in which their specific substrate is broken into smaller products.	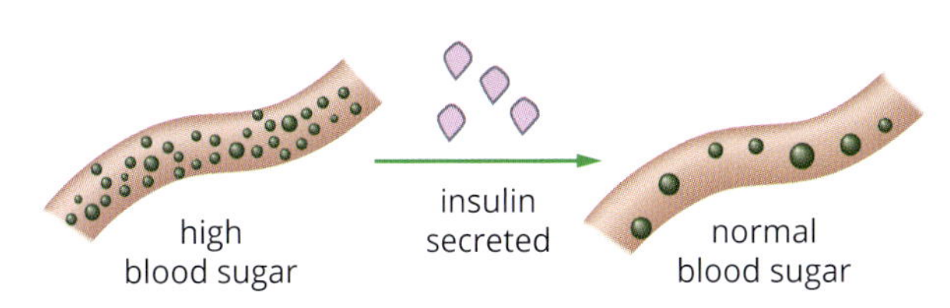 FIGURE 2.5.3 The hormone insulin regulates blood sugar levels..
Function: immunological proteins	**Function: contractile and motor proteins**
Description: • protect against disease by recognising foreign bodies and microbes • activate immune cells **Examples:** immunoglobulins (antibodies), complement, major histocompatibility complex proteins	**Description:** • contractile proteins aid muscle contraction • motor proteins are responsible for the movement of cilia and flagella **Examples:** myosin, actin, kinesin , dynein
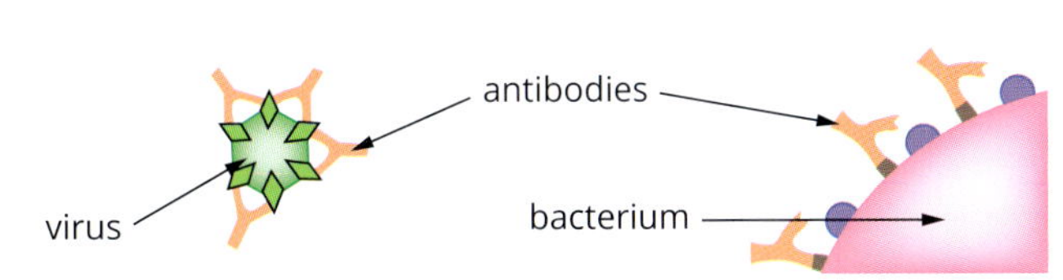 FIGURE 2.5.4 Antibodies help destroy viruses and bacteria.	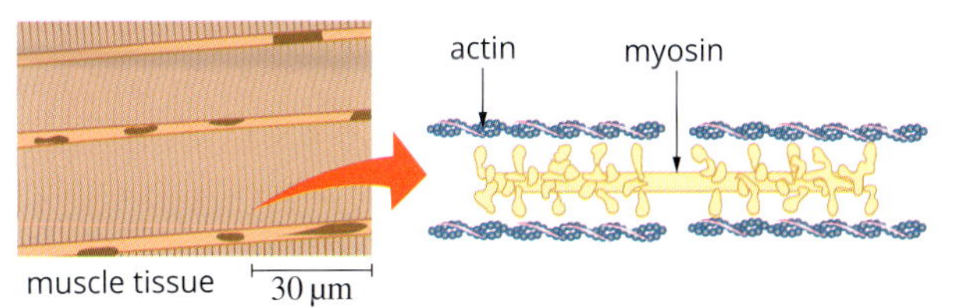 FIGURE 2.5.5 Actin and myosin are responsible for muscle contraction.

TABLE 2.5.1 An overview of protein function (continued).

Function: structural proteins	Function: transport proteins
Description: • provide support by forming the structural components of cells and organs • assist in contractile functions in tissue such as muscle **Examples:** collagen, keratin, actin, cytoskeleton 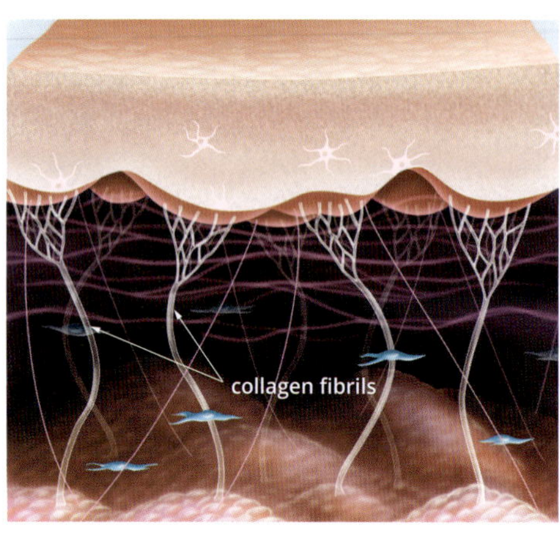 FIGURE 2.5.6 Collagen fibrils provide elasticity and support to the skin.	**Description:** • transport of substances by acting as carrier molecules within or between cells • act as membrane channel proteins **Examples:** haemoglobin, sodium–potassium pump, calcium channel 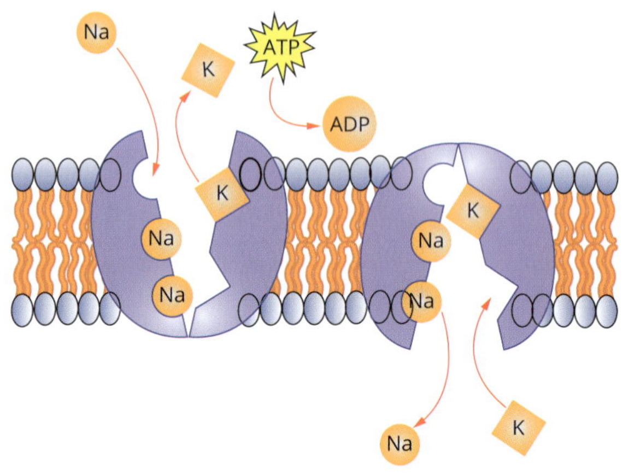 FIGURE 2.5.7 Sodium–potassium pump uses ATP to transport sodium ions and potassium ions across the plasma membrane.
Function: receptor proteins	**Function: storage proteins**
Description: assist the cell in responding to a chemical stimuli **Examples:** neurotransmitter receptors, hormone receptors 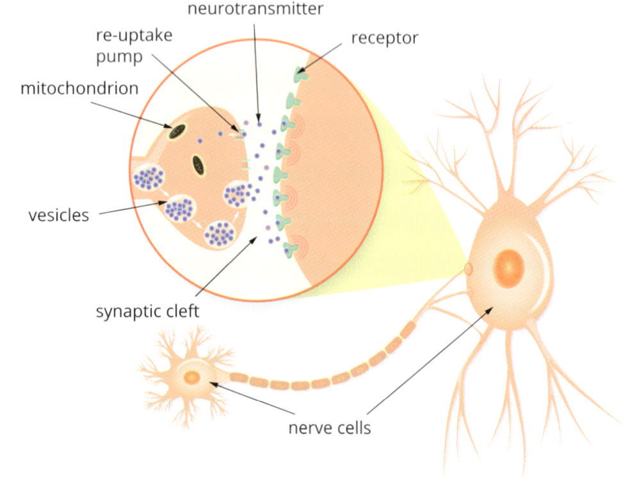 FIGURE 2.5.8 Receptors built into the plasma membrane of nerve cells detect neurotransmitters secreted by neighbouring nerve cells.	**Description:** storage of metal ions and amino acids **Examples:** ovalbumin and casein (to store amino acids), and ferritin (to store iron) 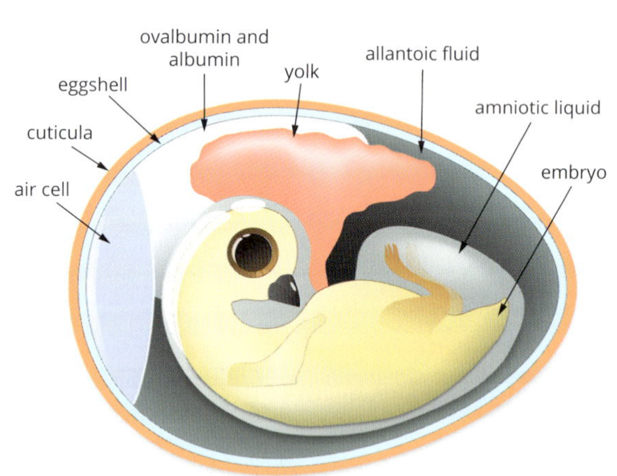 FIGURE 2.5.9 Ovalbumin is a protein found in egg white, used as an amino acid source for the developing embryo.

THE NATURE OF THE PROTEOME

The **proteome** is the complete set of proteins expressed by the **genome**—the complete set of genes or genetic material—of an individual cell or organism at a given time. The proteome varies between cell type, developmental stage and environmental conditions. Although a cell may contain the entire genome, only specific genes will be expressed, or 'switched on', at any given time. This ensures a cell produces only the proteins required for the specific functions it carries out.

For example, all human somatic cells (any body cell of an organism, apart from cells that give rise to eggs and sperm) in an individual contain identical genomes, but the array of proteins produced by a fibroblast cell is different from the array of proteins produced by a B lymphocyte. Fibroblasts are found in connective tissue and produce collagen to give the tissue strength and elasticity. B lymphocytes are found in the blood circulation and lymphatic system, and produce antibodies to help defend the body against infection or foreign, non-self, materials. A fibroblast does not produce antibodies and a B lymphocyte does not produce collagen. Each cell type only produces the proteins required to carry out its own specific functions.

Interestingly, there are many similarities between human and other proteomes, reflecting their common evolutionary origins. The human proteome contains proteins related by evolutionary descent (homologous) with 61% of the fruit fly proteome, 43% of the worm proteome and 46% of the proteome of baker's yeast.

Proteomics

Proteomics is the large-scale study of the structure, function and interactions of proteins. Proteomics is essential as it is proteins that actually carry out most of the activities of the cell, not the genes that encode them. By knowing when and where proteins are produced in an organism, as well as how proteins interact, we can better understand the functioning of cells and organisms.

One of the ways to determine changes in the proteome is by comparing the proteomes of cells under different conditions. For example, by comparing the protein expression of a diseased cell and a healthy cell, the proteins affected by the disease can be determined.

The research from proteomics can lead to the creation of protein biomarkers that can be used for screening individuals and populations for early detection of disease. The study of proteomics is also important in the production of drugs that interact with proteins involved in disease and alter their function.

BIOFILE

Rational drug design

Rational drug design uses high-speed computers to compare the three-dimensional structure of a faulty protein with a database containing many different chemical compounds. The compounds most likely to interact with the faulty protein are identified, and the interactions between these compounds and the faulty protein can then be tested in the laboratory to design drugs.

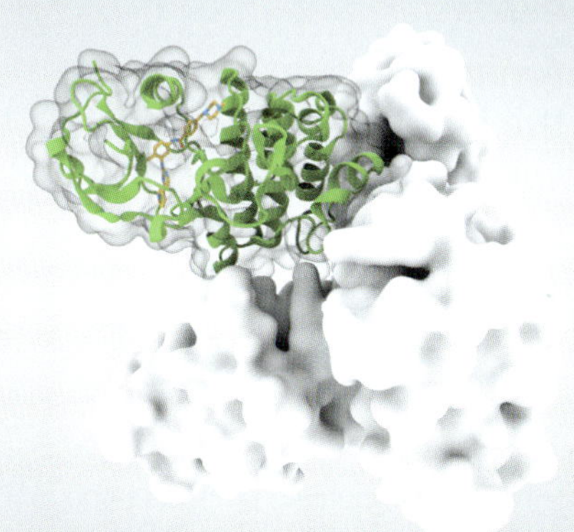

FIGURE 2.5.10 Illustration of Gleevec, a drug that has been created to interact with a faulty enzyme that leads to an overproduction of abnormal white blood cells in a rare type of leukaemia. Gleevec binds to the active site of the enzyme, altering its shape and preventing it from functioning.

SYNTHESIS OF PROTEINS

There are many steps involved in producing a functional protein. As you will recall from Section 2.2, although protein structure, size and function are quite diverse, all proteins are made up amino acids. These smaller subunits (or monomers) are joined together in a particular order to form polypeptide chains. The polypeptide chains are then folded and coiled into proteins.

Amino acid structure

All amino acids have the same basic structure (Figure 2.5.11):

- an **amine group** (NH_2)
- a **carboxyl group** (COOH)
- a **variable R group** (or side chain).

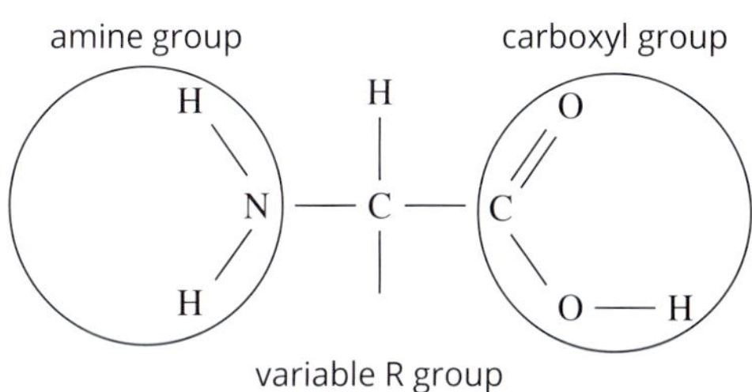

FIGURE 2.5.11 The basic structure of an amino acid, showing an amine, carboxyl and variable R group.

In the synthesis of proteins in organisms, there are 20 different standard (or canonical) amino acids, and each has a different R group (Figure 2.5.12). The variable properties of the R group (e.g. charged or uncharged, polar or non-polar, hydrophobic or hydrophilic) determine the type of protein that the amino acid will form. R groups can be as simple as a hydrogen atom (as in the amino acid glycine) or more complex; for example, -$CH(CH_3)_2$ in the amino acid valine.

Non-polar

glycine

leucine

tryptophan

alanine

isoleucine

methionine

valine

phenylalanine

proline

Acidic

aspartic acid

glutamic acid

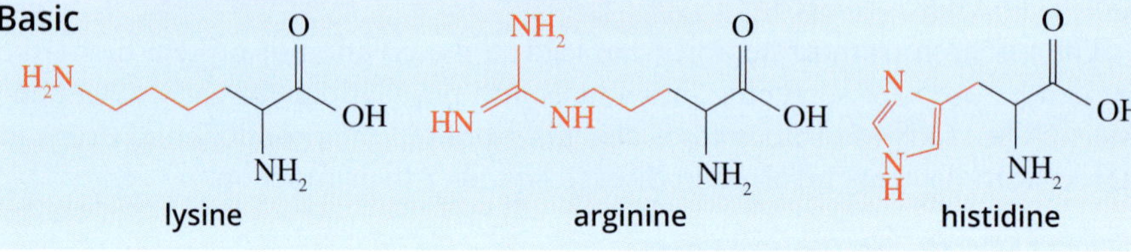

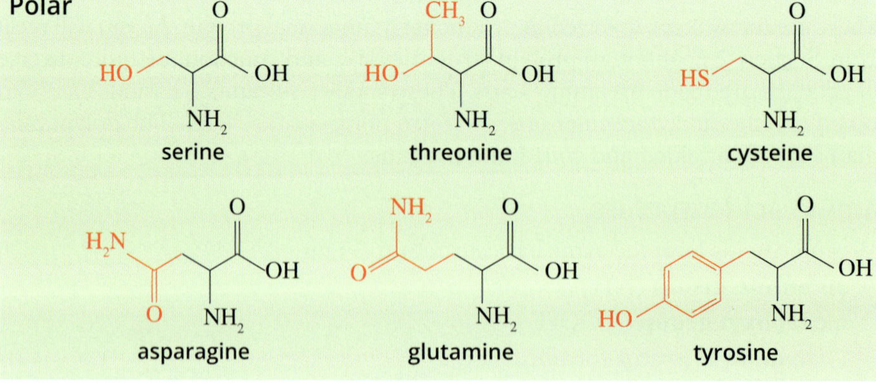

FIGURE 2.5.12 Chemical structure of the 20 standard amino acids. The R groups are coloured in red. Amino acids can be classified according to their chemical nature as non-polar, acidic, basic or polar.

BIOFILE

Amino acids in the human diet

In the human diet, amino acids can be classified into three main groups:

- Essential amino acids: These cannot be synthesised by the body and must be obtained from our diet. The nine essential amino acids are histidine, isoleucine, leucine, lysine, methionine, phenylalanine, threonine, tryptophan and valine.
- Non-essential amino acids: These can be produced by the body if is not obtained from the diet. The non-essential amino acids include alanine, asparagine, aspartic acid, glutamic acid and serine.
- Conditional amino acids: These are only required by the body in times of illness or stress. The conditional amino acids include arginine, cysteine, glycine, glutamine, proline and tyrosine.

Condensation polymerisation of amino acids

Amino acids are joined by peptide bonds in a **condensation polymerisation** reaction, which involves the removal of water (dehydration). A hydrogen and oxygen from the carboxyl group of one amino acid join with a hydrogen atom from the amine group of another amino acid to produce water. The water is released and a **dipeptide** is synthesised, with a peptide bond holding the two amino acids together. A chain of amino acids joined by peptide bonds is known as a polypeptide chain. The backbone of the polypeptide chain is formed by the repeats of the carboxyl and amine groups, with the R groups forming the side chains of the polypeptide chain (Figure 2.5.13).

A polypeptide chain forms the primary structure of a protein. With further folding and modification, a fully functional protein can be formed.

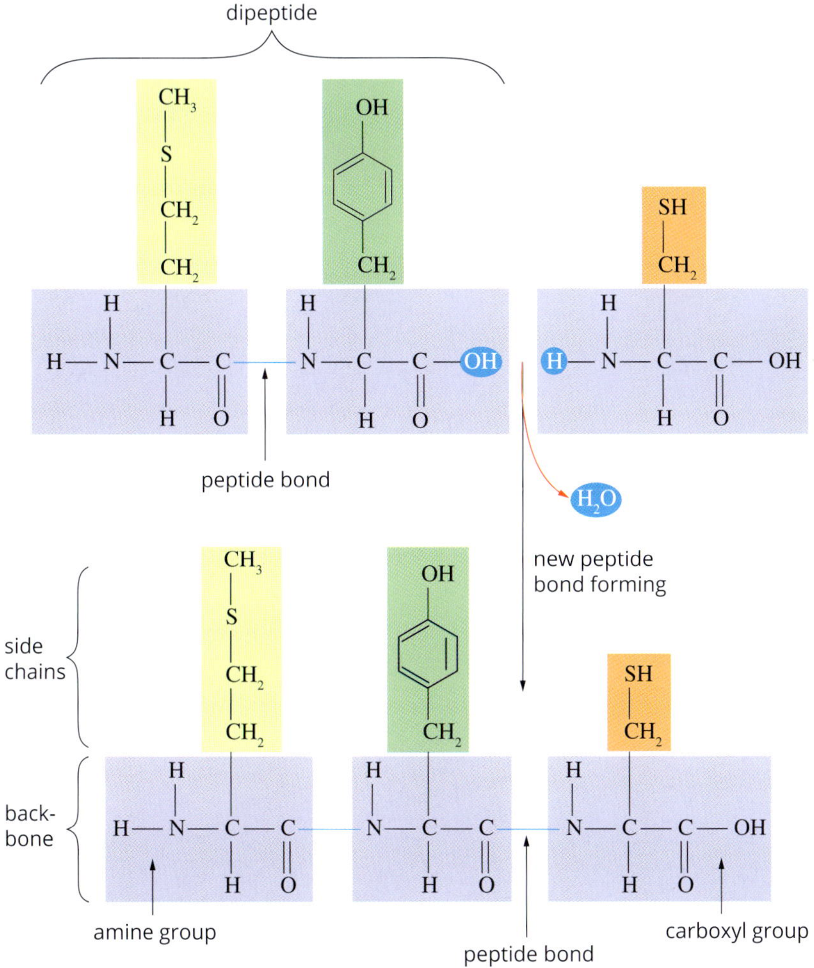

FIGURE 2.5.13 Formation of a polypeptide chain by a condensation polymerisation reaction.

BIOFILE

Protein size

Proteins come in vastly different sizes. The peptide hormones oxytocin and antidiuretic hormone are only nine amino acids long. Dystrophin is a large protein that is important for muscle cell structure. It has about 3600 amino acids. Defects in this protein lead to muscular dystrophy. Another muscle protein, titin, is involved in the elasticity of the muscle and is the largest known protein. It has about 30 000 amino acids!

BIOFILE

Protein folding and degenerative disease

A number of serious degenerative diseases, such as Alzheimer disease and Creutzfeldt–Jakob disease, appear to be the result of proteins not folding correctly. The incorrectly folded proteins are resistant to proteases and accumulate in brain tissue.

PROTEIN STRUCTURE

Proteins are large biomolecules that can contain thousands of amino acids and may be synthesised as one or several polypeptide chains. These polypeptide chains are folded and organised into specific shapes that are vital to the correct functioning of the protein. Most proteins are required to bind to other molecules. A single change to one amino acid within the sequence can alter the shape, and consequently the function, of a protein.

There are four different levels of organisation when describing protein structure (Figure 2.5.14):

- primary structure
- secondary structure
- tertiary structure
- quaternary structure.

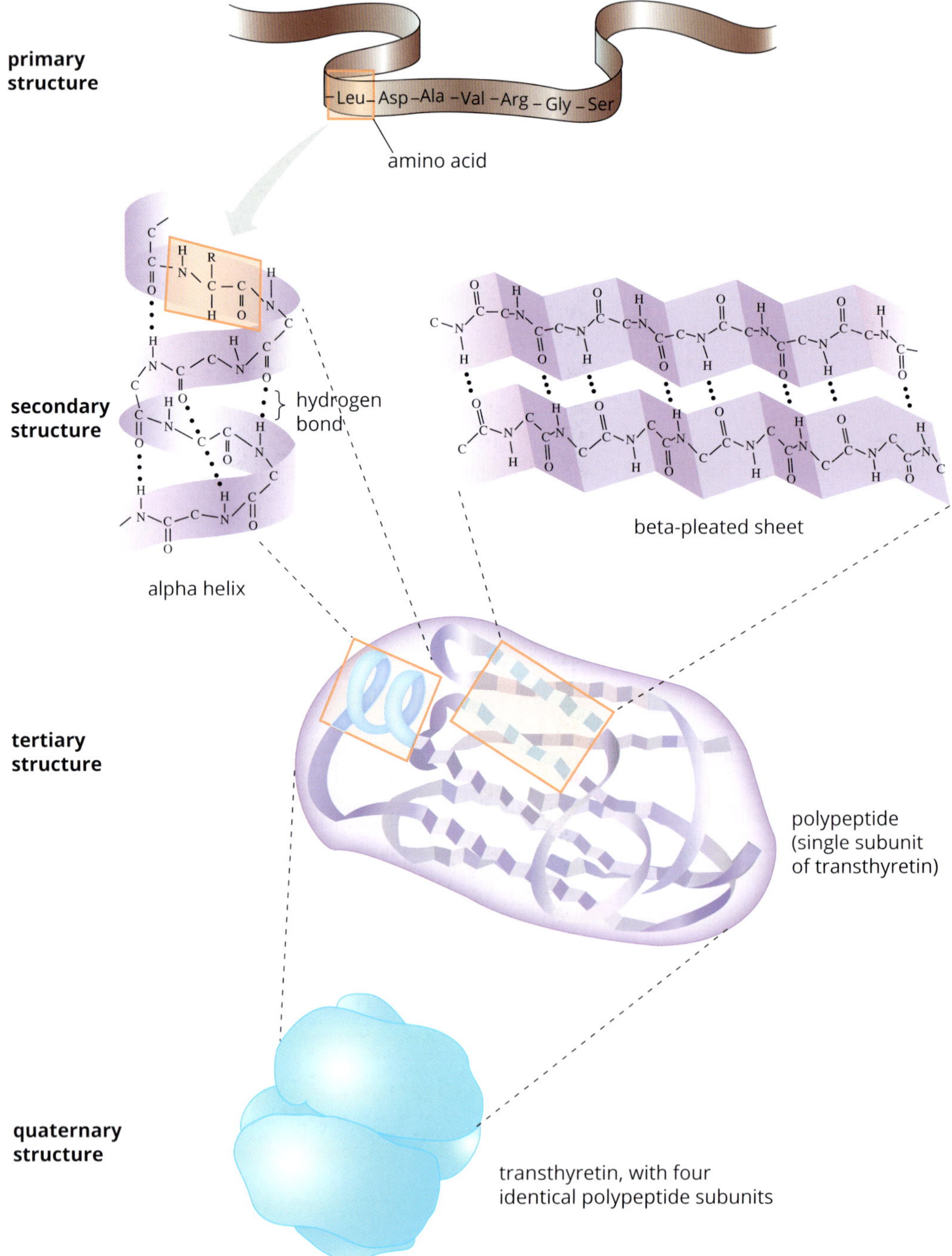

FIGURE 2.5.14 Overview of the four levels of protein structure—primary, secondary, tertiary and quaternary.

Primary structure

The linear sequence of amino acids in the polypeptide chain is referred to as the **primary structure** of a protein (Figure 2.5.15). It is unique to each protein. The primary structure of one polypeptide may consist of only 50 amino acids, whereas another may consist of 103 amino acids. Linear sequences shorter than 50 amino acids are known as **peptides**.

The linear sequence:

- provides information on how proteins will fold
- of functional and non-functional proteins can be compared in order to identify what changes to the sequence render the protein non-functional
- can be compared between proteins to determine the evolutionary history of a protein.

Secondary structure

The next step in the formation of a functional protein is the folding or coiling of the polypeptide chain—its **secondary structure**. Folding or coiling occurs due to the formation of hydrogen bonds between the amine and carboxyl groups of amino acids within a polypeptide chain that have come in close proximity to each other. This results in the formation of secondary structures. There are three types of secondary structures:

- **Alpha helix**—Hydrogen bonds form between adjacent amine and carboxyl groups within the polypeptide chain and the chain coils to form a helical shape (Figure 2.5.16a).
- **Beta-pleated sheets**—Hydrogen bonds form between amine and carboxyl groups in different parts of adjacent polypeptide chains, causing the chains to fold back on each other (Figure 2.5.16b).
- **Random coil**—Although parts of the polypeptide chain appear to have a random structure, the folding is not in fact random and the same folding occurs in all molecules of the same protein. For example, all insulin molecules will have the same random coils within their structure.

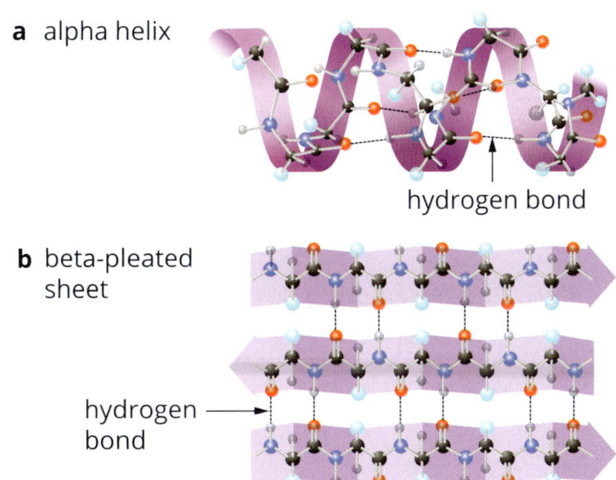

FIGURE 2.5.16 Regions stabilised by hydrogen bonds between atoms of the polypeptide chain result in the secondary structures (a) alpha helix and (b) beta-pleated sheets.

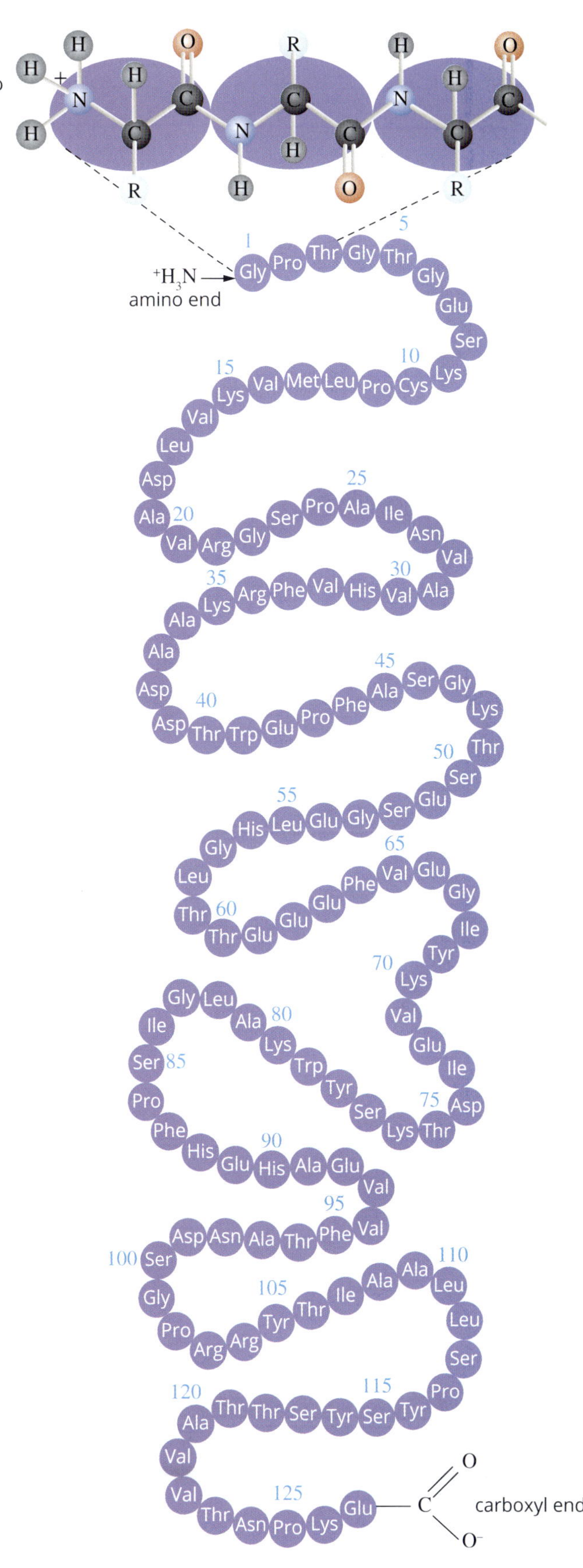

FIGURE 2.5.15 Primary structure of a protein showing the linear amino acid sequence of a polypeptide chain.

Tertiary structure

Polypeptides also fold further, forming more stable globular or fibrous three-dimensional shapes (Figure 2.5.17). This is known as the **tertiary structure**, and is usually the result of a combination of alpha helices and beta-pleated sheets along with other folded areas. The tertiary structure occurs due to different types of bonds, such as the disulfide bridge and the hydrogen bridge, between the R groups (side chains) of the amino acids (Figure 2.5.18).

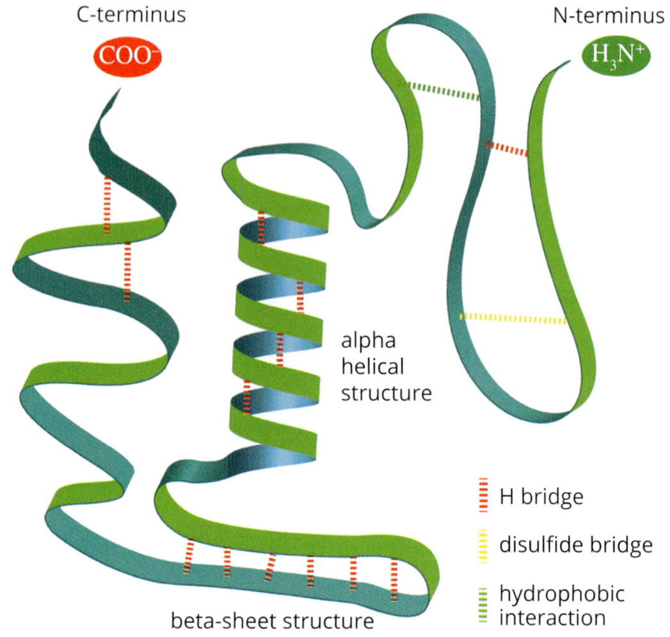

FIGURE 2.5.17 The tertiary structure of a protein is stabilised by the presence of different types of bonds.

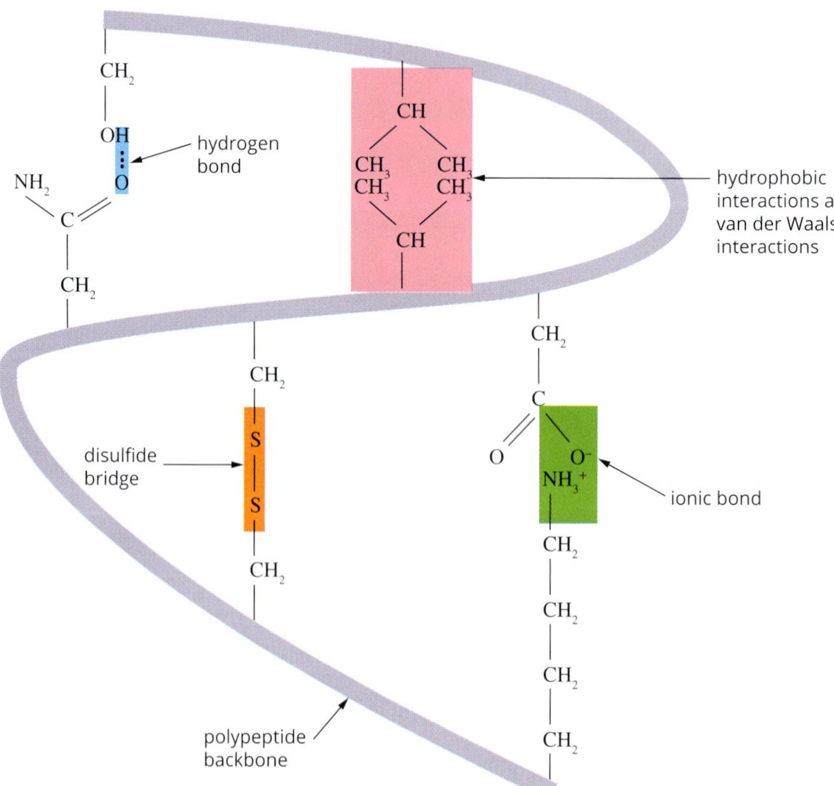

FIGURE 2.5.18 The different types of bonds between the R groups of the amino acids.

The three-dimensional structure of a protein is critical to its function. In some smaller polypeptides, this folding process occurs spontaneously due to its chemical environment. However, larger, more complex proteins require specialised proteins to help them fold correctly and, in some cases, to refold if they unravel and lose their native shape (**denature**).

The tertiary structure is the final structure for some proteins.

Quaternary structure

A **quaternary structure** is formed when two or more polypeptide chains or **prosthetic groups** (an inorganic compound that is involved in protein structure or function) join together to create a single functional protein. The polypeptides may be identical or different. Some proteins will not become active until they achieve their quaternary structure. A protein with a prosthetic group is known as a **conjugated protein**. Haemoglobin is an example of a conjugated protein with a quaternary structure (Figure 2.5.19).

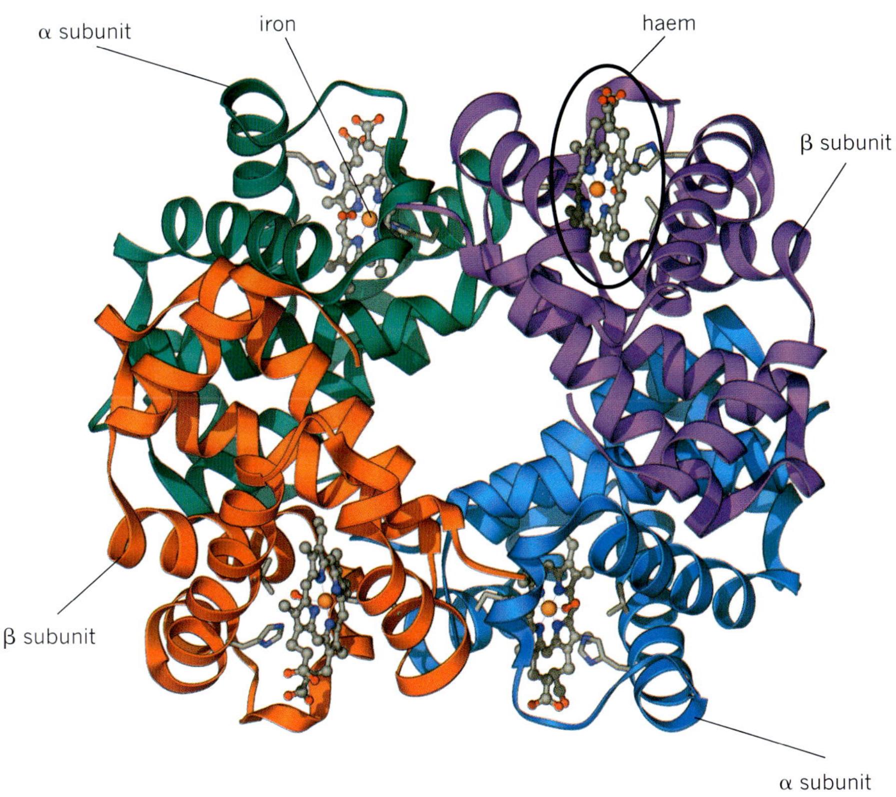

FIGURE 2.5.19 Quaternary structure of haemoglobin. Four polypeptides (two alpha (α) subunits of 141 amino acids and two beta (β) subunits of 146 amino acids) join together with haem prosthetic groups to form the functional haemoglobin molecule. Haemoglobin is a conjugated protein.

Chaperonins

Crucial to the folding process are **chaperonins** (or **chaperone proteins**). Chaperonins are protein molecules that assist in the proper folding of other proteins. Chaperonins do not specify the final structure of a polypeptide, but instead provide the polypeptides an area to fold in without the influences from the cytoplasmic environment, such as changes in pH (Figure 2.5.20). Another function of chaperonins is to prevent newly synthesised polypeptide chains and assembled subunits from becoming non-functional structures due to high temperatures.

> **BIOFILE**
>
> **Chaperonins and inherited disease**
>
> It is now thought that some inherited diseases associated with the lack of function of a particular protein may be due to a fault in chaperonins rather than a mutation in the gene for the protein itself. The sequence of amino acids in the polypeptide may be correct, but the protein is not correctly folded into its functional structure.

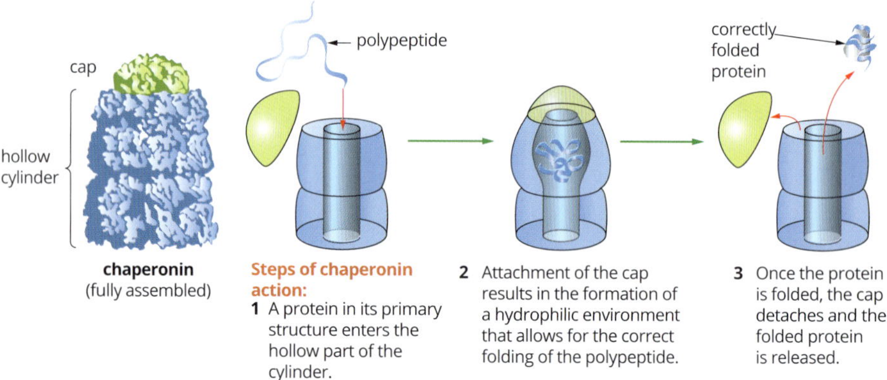

FIGURE 2.5.20 The computer graphic (left) shows a large chaperonin protein complex from *Escherichia coli*. It has an interior space that provides a shelter for the proper folding of newly made polypeptides.

PROTEIN CLASSIFICATION

Proteins can be classed as one of two types depending on their shapes:

- **Fibrous proteins** are typically elongated and insoluble (Figure 2.5.21a). Many have structural roles and have little or no tertiary folding (e.g. collagen found in connective tissue and keratin found in hair and nails).
- **Globular proteins** are compactly folded and coiled into spherical tertiary and quaternary structures (Figure 2.5.21b). Globular proteins are generally soluble. They have a core with hydrophobic properties and an outer hydrophilic region. Most enzymes and hormones are globular proteins.

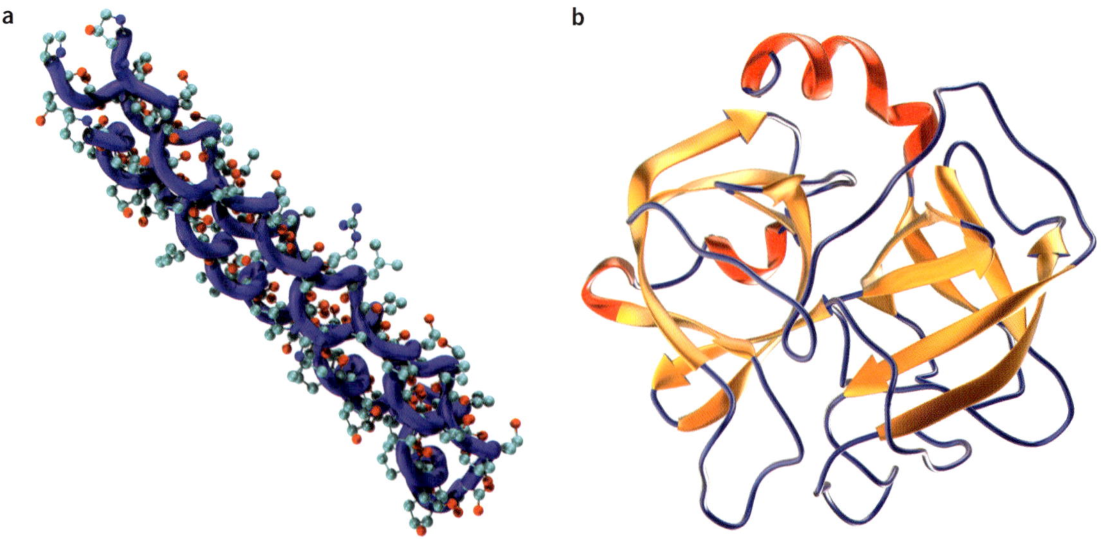

FIGURE 2.5.21 (a) The fibrous protein collagen. (b) The globular protein elastase (an enzyme that catalyses the hydrolysis of elastin in the pancreas).

FACTORS THAT AFFECT THE FUNCTION OF A PROTEIN

The environment surrounding proteins plays an important role in maintaining the structure and function of the protein. Usually the loss of function of the protein is due to **denaturation** of the protein. The factors in the environment affecting protein structure and function include:

- temperature
- pH
- concentration of ions or molecules that act as cofactors.

Denaturation and renaturation of proteins

A protein is said to have denatured when the hydrogen bonds, disulfide bridges, hydrophobic interactions and van der Waals forces that create the tertiary structure of the protein are broken and the shape of the protein is altered (Figure 2.5.22). As a result, the misshapen protein is biologically inactive. If a protein becomes fully denatured, the reaction is non-reversible and the protein remains non-functional. However, a protein that is partially denatured may be able to fold again (renature) when the appropriate conditions are present.

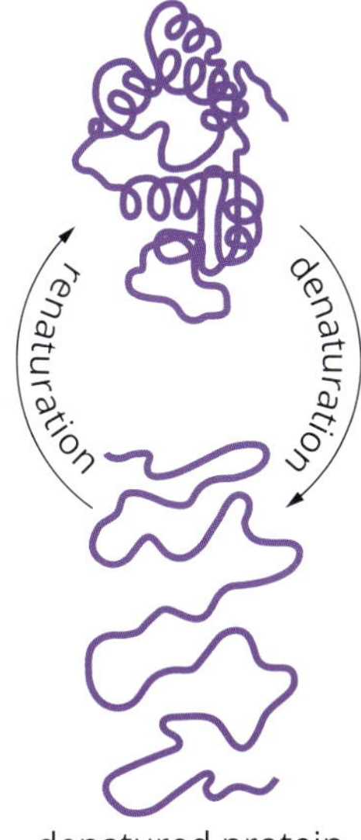

FIGURE 2.5.22 A denatured protein will lose its shape and hence its ability to function. Sometimes a protein can renature, when the chemical and physical aspects of its environment are restored to normal.

The effect of temperature on protein function

Proteins can be denatured at high temperatures due to the breaking of bonds. For example, hydrogen bonds break at temperatures above 40°C. However, at temperatures below 35°C, the bonds are not flexible enough to allow the necessary conformational changes.

The optimal temperature for proteins varies with the organism and its environment. In humans, the optimum temperature for proteins is 37°C, but proteins found in organisms living in extreme environments, such as hot springs or icy environments, tend to have different optimal temperatures.

BIOLOGY IN ACTION

How to unboil an egg

Professor Colin Raston from Flinders University, Australia, has been honoured with an Ig Nobel prize for creating a way to unboil an egg. The Ig Nobel is a humorous parody of the Nobel Prize and is almost as famous as the actual Nobel Prize.

When an egg is boiled, the proteins unfold and refold into a more tangled, disordered form. The vortex fluidic device invented by Professor Colin Raston (Figure 2.5.23) is capable of unravelling proteins, a process required to make the white of a cooked egg runny again.

Other then being used to unboil an egg, the vortex fluid device has applications in the treatment of cancer, the manufacture of pharmaceuticals, the production of biofuels and in food processing. For example, drug companies often make cancer antibodies in expensive hamster ovary cells, because they don't often create misfolded proteins. If, instead, these companies could use proteins from cheaper yeast or *E. coli* cells, it could make cancer treatments more affordable.

FIGURE 2.5.23 Professor Colin Raston and the vortex fluidic device.

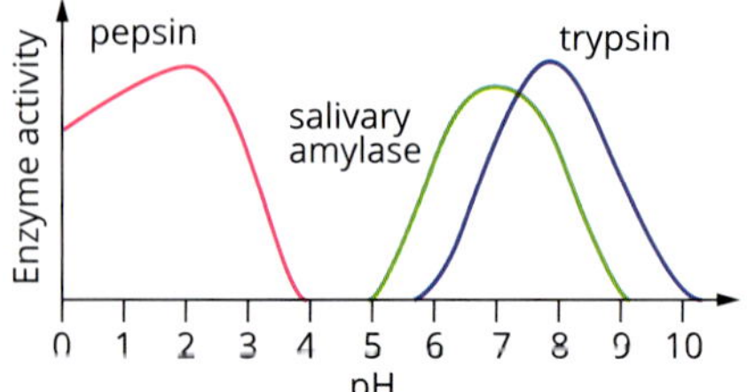

FIGURE 2.5.24 Graph showing the optimum pH for different digestive enzymes.

The effect of pH on protein function

Most proteins have a specific pH range in which their function is optimal, but this range can be quite different for each specific protein (whereas the range of optimal temperatures is similar for most proteins within an organism). In humans for example, the enzyme salivary amylase (which starts the digestion of starch in the mouth) has an optimum pH of about 7, the enzyme pepsin (a digestive enzyme found in the stomach) has an optimal pH of about 2, and the enzyme trypsin (a digestive enzyme found in the small intestines) has an optimal pH of about 8 (Figure 2.5.24).

If the pH reaches too far above or falls too far below the optimal pH, then the tertiary structure is affected. The interactions between the R groups of different amino acids are altered and the bonds between them are broken. As a result, the protein may be denatured, and in the case of enzymes the enzyme activity will decrease.

The effect of cofactors on protein function

Some proteins require non-protein chemical compounds known as **cofactors** for their biological function. The presence and concentration of cofactors such as salts, specific elements such as iron, magnesium and calcium ions, or organic molecules such as vitamins can play a significant role in the folding and function of proteins. For example, magnesium is essential for chlorophyll function in plants (Figure 2.5.25). A lack of magnesium ions causes the yellowing of leaves due to the plant's inability to synthesise chlorophyll.

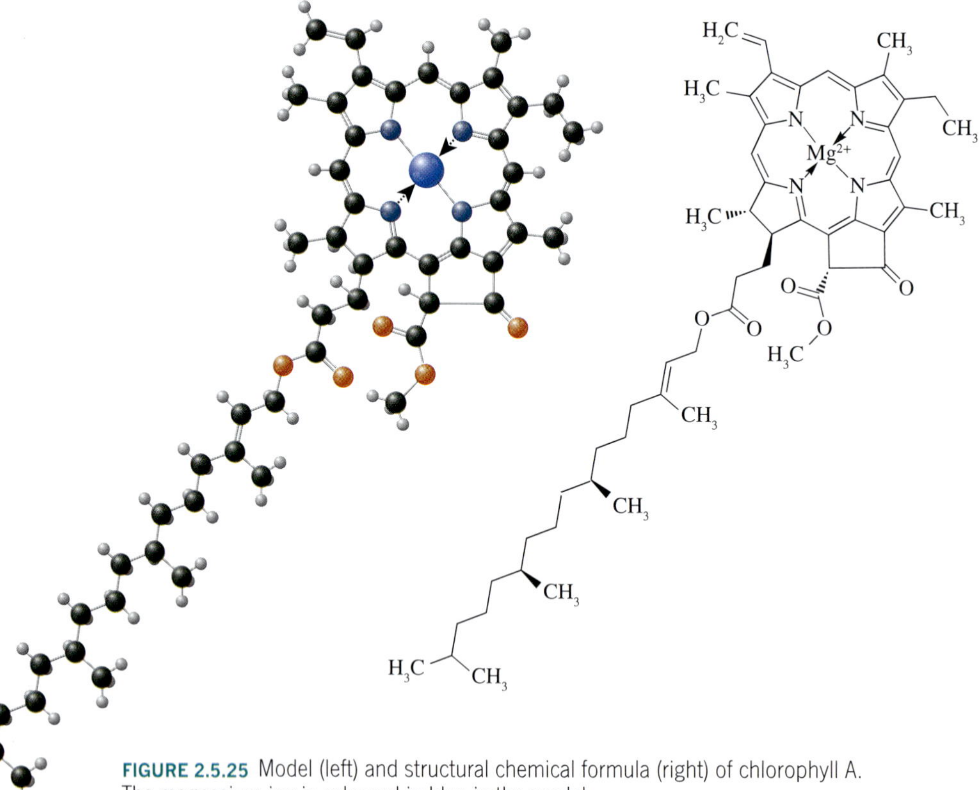

FIGURE 2.5.25 Model (left) and structural chemical formula (right) of chlorophyll A. The magnesium ion is coloured in blue in the model.

2.5 Review

SUMMARY

- Proteins have very diverse functions. The specific folding and final structure of proteins relate directly to their function.
- Functional types of proteins include:
  - enzymatic proteins
  - structural proteins
  - transport proteins
  - hormonal proteins
  - receptor proteins
  - immunological proteins
  - contractile and motor proteins
  - storage proteins.
- The proteome is the complete set of proteins expressed by the genome.
- Proteomics is the study of proteomes, including protein structure and function.
- Amino acids have an amine group, a carboxyl group and an R group. There are 20 standard amino acids. All have the same amine and carboxyl group, but differ in their R group.
- Amino acids join together to form polypeptide chains in a condensation polymerisation reaction. This reaction removes water between two amino acids and forms a peptide bond between them.
- Proteins are made up of one or more polypeptide chains, which are folded and organised into specific shapes that relate to their specific function.
  - Primary structure of a protein is the linear sequence of amino acids in the polypeptide chain.
  - Secondary structure of a protein is achieved with the folding or coiling of the polypeptide chains due to hydrogen bonds. There are three forms: alpha helices, beta-pleated sheets and random coils.
  - Tertiary structure of a protein is achieved by further folding, which creates more stable shapes. This structure occurs as a result of bonds forming between the R groups of the amino acids.
  - Quaternary structure of a protein is achieved when two or more polypeptide chains join to create a single functional protein.
- Proteins can be either fibrous or globular:
  - Fibrous proteins are elongated and insoluble.
  - Globular proteins are spherical and compact.
- Factors within the environment can have an impact on the structure and function of a protein, and can also lead to denaturation. These factors include temperature, pH, concentration of ions and molecules that act as cofactors.
- Proteins have optimal temperature and pH ranges within which they function most effectively.

KEY QUESTIONS

1 Proteins are key components of cells. Outline, with examples, at least five different roles carried out by proteins.

2 Distinguish between the proteome and proteomics.

3 Distinguish between peptides, dipeptides and polypeptides.

4 **a** Draw the structure of an amino acid, and label the groups that are used in peptide bond formation.
b Outline the production of a dipeptide by a condensation polymerisation reaction. Include the structure of a generalised dipeptide in your answer.

5 Use a single sentence and a simple diagram to explain what is meant by the following structures of a protein:
a primary
b secondary
c tertiary
d quaternary

6 Distinguish between fibrous and globular proteins.

7 **a** Explain what is meant by a protein becoming 'denatured'.
b Outline the factors that can cause a protein to become denatured.

8 **a** What are cofactors?
b How does the absence of cofactors affect protein function?

Chapter review

02

KEY TERMS

alpha helix
amine group
amino acid
ATP (adenosine triphosphate)
beta-pleated sheet
biochemical pathway
biomacromolecule
carbohydrate
carboxyl group
carrier protein
catalyse
cell
cell wall
cellular respiration
centriole
channel protein
chaperonin (chaperone proteins)
chloroplast
cilia
cis face
cisternae
cofactor
cohesive (cohesion)
complex polysaccharide
compound
concentration gradient
condensation polymerisation
conjugated protein
cytoskeleton
denature (denaturation)
deoxyribonucleic acid (DNA)
diffusion
dipeptide
disaccharide
endocytosis
enzyme
eukaryote
exocytosis
extracellular matrix
facilitated diffusion
fibrous protein
flagellum (plural flagella)
fluid mosaic model
genetic code
genome
globular protein
Golgi apparatus (also known as Golgi body, Golgi complex)
heat capacity
hydrophilic
hydrophobic
inorganic compound
integral protein
lipid
lysosome
mitochondrion
monomer
monosaccharide
nitrogen fixation
nucleic acid
nucleolus
nucleus
organelle
organic compound
osmosis
osmotic gradient
osmotic pressure
passive transport
peptide
peripheral protein
peroxisome
phagocytosis
phospholipid
pinocytosis
plastid
polymer
polypeptides
polysaccharide
primary structure
prokaryote
prosthetic group
protein
proteome
proteomics
pseudopodia
quaternary structure
random coil
receptor
receptor-mediated endocytosis
ribonucleic acid (RNA)
ribosome
rough endoplasmic reticulum (RER)
secondary structure
secretory protein
secretory vesicle
smooth endoplasmic reticulum (SER)
solute
solvent
tertiary structure
trans face
transmembrane protein
transport vesicle
vacuole
variable R group

KEY QUESTIONS

1 The cell theory states that:
- **A** all organisms are made up of cells
- **B** all cells arise from pre-existing cells
- **C** the cell is the smallest functional unit of living things
- **D** all of the above

2 Which of the following lists represents key organic compounds in cells?
- **A** carbohydrates, water, proteins, lipids
- **B** water, carbon dioxide, oxygen, enzymes
- **C** carbohydrates, proteins, lipids, nucleic acids
- **D** carbohydrates, proteins, lipids, nucleic acids, minerals

3 Look carefully at the diagram of a cell membrane. Which of the following statements is correct?

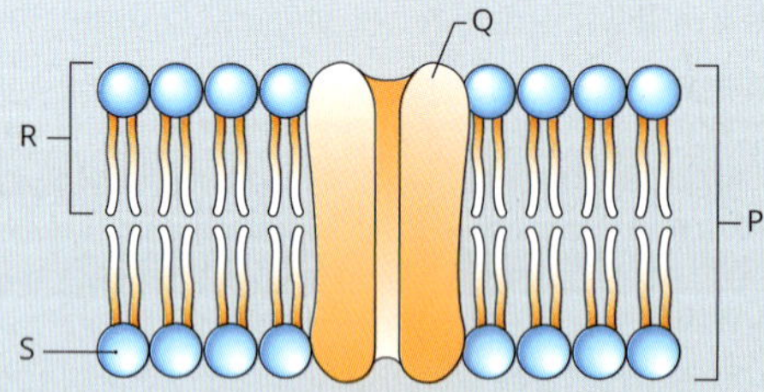

- **A** Structure P represents the phospholipid bilayer.
- **B** Structure R is hydrophobic.
- **C** Structure Q is involved in osmosis.
- **D** Structure S is largely composed of protein.

4 Which of the following will occur when red blood cells are placed in distilled water?

A Osmosis will result in a net movement of water out of the cell causing them to shrivel.

B Osmosis will result in a net movement of water into the cells causing them to burst.

C Active transport will move excess water molecules out of the cell against the concentration gradient.

D Facilitated diffusion will assist water molecules to move passively into the cells.

5 A compound is a molecule that is composed of:

A both carbon and hydrogen

B one type of atom

C two or more atoms

D more than one type of atom

6 Which of the following shows a beta-pleated sheet?

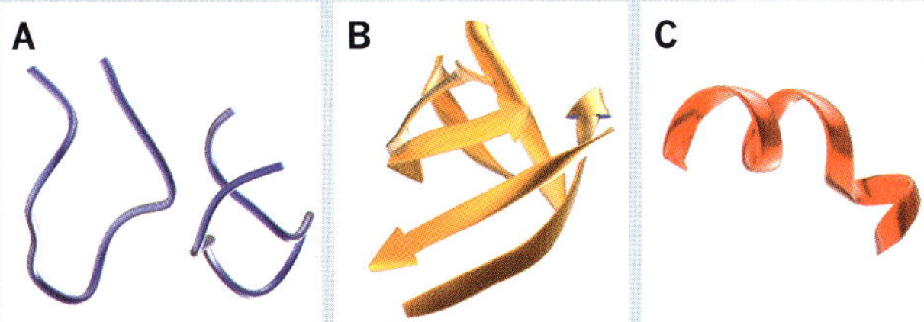

D None of the above

7 Consider the pictures of the following levels of protein structure.

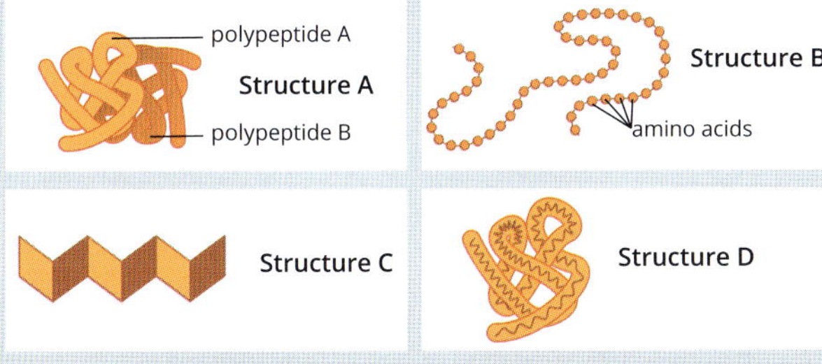

The diagram showing a quaternary structure is:

A structure A

B structure B

C structure C

D structure D

8 Integral proteins found in the plasma membrane include:

A protein carriers

B protein channels

C receptor proteins

D all of the above

9 Which of the following is not made of monomers and does not form a polymer?

A lipid

B protein

C nucleic acid

D carbohydrate

10 The internal environment of eukaryotic cells is divided into compartments by membranes. This makes eukaryotic cells more efficient than prokaryotic cells because:

A it allows tasks to be merged together

B it decreases the surface area of eukaryotic cells

C each compartment is entirely independent of the others

D it allows different areas of the cell to maintain different conditions

11 A proteome is defined as:

A the sum of all of the functional proteins that an individual organism produces

B a primitive, simple form of protein

C the kinds of proteins produced by prokaryotic organisms

D the kinds of proteins produced by eukaryotic organisms

12 Examine the following diagram of a cell.

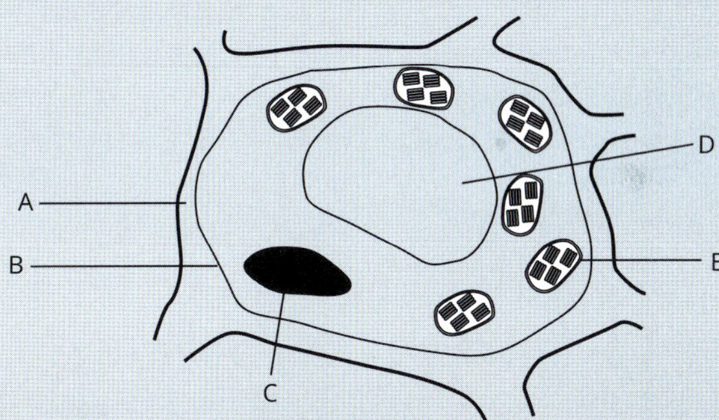

a Name the structures represented by each of the letters A–E.

b Is the cell from an animal or a plant? Give two reasons why you think so.

c Name a specialised function carried out by this cell. Describe the feature(s) of this cell that suggest the function you have named.

d Describe two key differences between plant and animal cells.

13 A scientist was interested in how serine (an amino acid) enters cells. The scientist established that serine is a polar molecule.

a The polar nature of the amino acid rules out one method of entry for serine. Explain which method of entry this is.

In order to further investigate the entry of serine, the scientist takes three cell cultures (A, B and C), adds radioactively labelled serine molecules in solution to each and then monitors the cells to determine whether the serine enters the cells. The results of the experiment were then tabulated.

Results of the ability of cell cultures to absorb serine following various treatments

Culture	Treatment	Result
A	radioactive serine solution only	radioactive serine found in the cells
B	mercury solution (damages protein carriers and channels) and radioactive serine solution	no radioactive serine in the cells
C	ATPase inhibitor and radioactive serine solution (ATPase catalyses the formation of ATP from ADP and P_i)	radioactive serine found in the cells

b What is the function of culture A in the experiment?

c What do the results of the experiment suggest about the method used by the cell to take in serine?

d What is the independent variable in this experiment?

14 Examine the cell below.

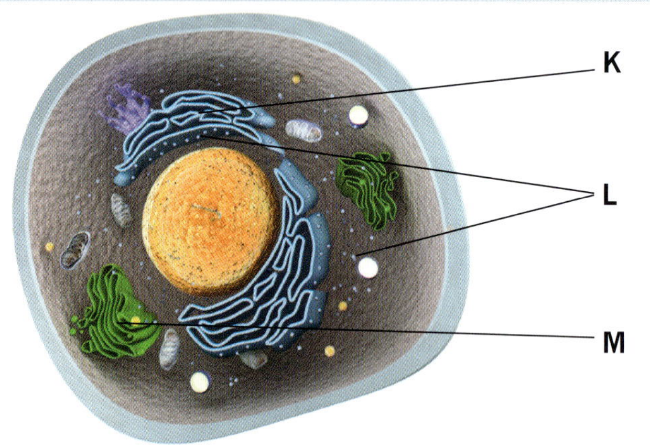

a Identify the organelles K and M.

b i Describe one similarity in function between K and M.

ii Describe one difference in function between K and M.

c A cell contains large numbers of organelle M. Suggest a possible function for the cell, giving a reason for your suggestion.

d How might organelle L be involved in the specialised function of the cell?

15 Students were observing bovine (cow) blood cells under the microscope. Bovine red blood cells are smaller than human blood cells. Their normal mean diameter is 5.84 µm. The students observed the cells under a light microscope in three different saline solutions and measured the mean diameters of the cells. Their results are presented in the table below.

Mean diameters of bovine blood cells in solutions of varying concentrations

Solution	Solution concentration	Mean red blood cell size
1	0.154 M	5.82 µm
2	0.146 M	5.63 µm
3	0.169 M	5.49 µm
4	0.138 M	5.89 µm

a What is the approximate concentration of the cytosol of the red blood cells? How do you know?

b i Draw a diagram of how you would expect the red blood cells to appear in solution 3.

ii Explain the reasoning that led to you drawing the diagram as you did.

c While observing the cells in solution 4 the students noticed that the cell was drying out, so one of them added some distilled water to the slide.

i Propose what happened to the cells on the slide.

ii Explain why this happened.

iii Would you have expected a different result if the cells under observation had been from a plant? Explain why or why not.

iv Why were the students using bovine rather than human blood?

v What safety procedures need to be implemented when using fresh biological samples?

16 Electron microscopy has greatly enhanced our understanding of cellular structure due to its ability to greatly magnify the internal structure of cells. Below are two cells observed under a transmission electron microscope.

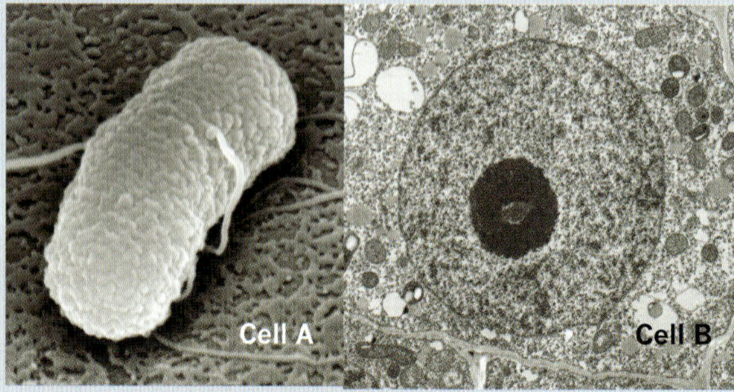

a One of the two cells is from a prokaryote. Explain which one.

b Is the eukaryotic cell from an animal or a plant? Give a reason for your choice.

17 Carbohydrates are composed of carbon, hydrogen and oxygen, which are always in the same ratio in a given carbohydrate molecule. The general formula for carbohydrates is $C_n(H_2O)_n$. Use the information provided to write the correct formula for each compound:

a glucose: $C_6H_{2n}O_n$

b maltose: $C_nH_{24}O_n$

18 An experiment was performed to determine the concentration of solutes in the cytosol of potato cells. Discs of potato were weighed and then placed in sucrose solutions of varying concentrations. After 24 hours the discs were re-weighed and any changes noted. The percentage change in mass of the discs was calculated for each concentration. The graph below shows the results of the experiment.

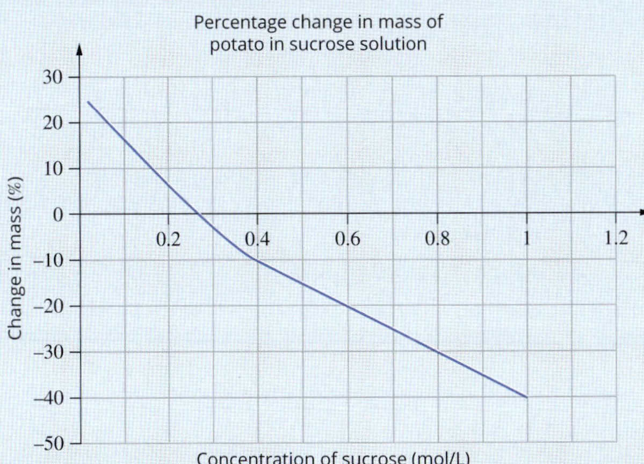

a i At what concentration of sucrose is the solution isotonic to the cytosol of the potato cells?

ii How do you know?

b i What is the percentage change in mass in a solution of 0.8 mol/L?

ii Is a solution of 0.8 mol/L isotonic, hypertonic or hypotonic? Explain your reasoning.

c Suggest a reason why percentage change in mass was used in this experiment rather than the actual number of grams in each set up.

19 The following diagram shows a cross-section through a typical plasma membrane. Explain why a plasma membrane is sometimes described as a 'fluid mosaic'.

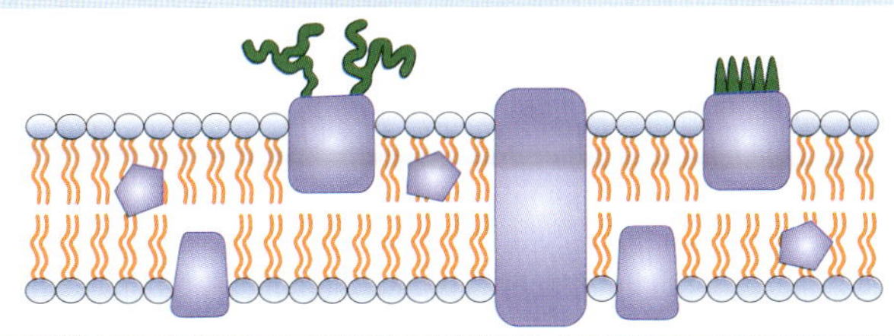

20 The diagram below shows two phospholipid molecules.

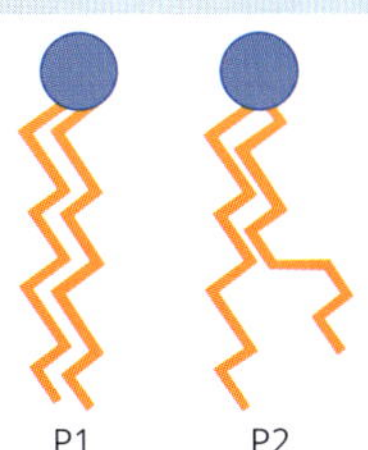

a Where in the cell are phospholipid molecules found?

b Redraw P1 and label the hydrophilic and hydrophobic parts.

c Why are P1 and P2 different shapes?

d How are the different shapes of P1 and P2 significant in determining the entry of materials into the cell?

e Many drugs are most effective when delivered into the cytoplasm of the target cells. If the drugs are hydrophilic it can be hard to get them into the cells. One delivery method that scientists have developed is the liposome. The drug to be delivered is encapsulated in the liposome. This facilitates its entry into the cell. A liposome is shown below.

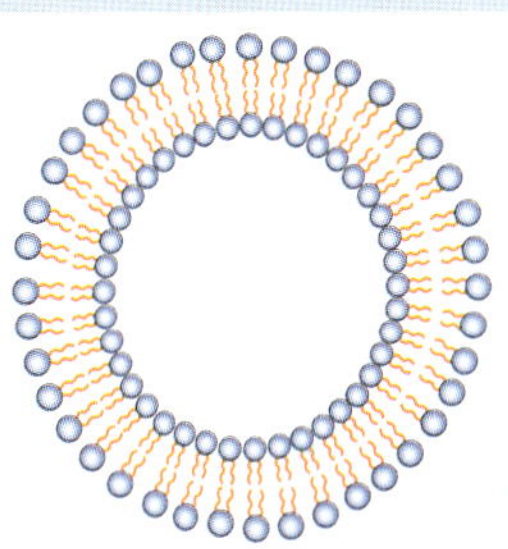

i Explain why the phospholipids maintain the liposome structure when introduced into the body.

ii How might liposomes assist with the delivery of the drugs into cells?

21 Explain why tadpoles living in a puddle of water may die well before the water has completely dried up.

22 In the experiments shown below, what were the original concentrations of solutions A, B, C and D? Explain your reasoning.

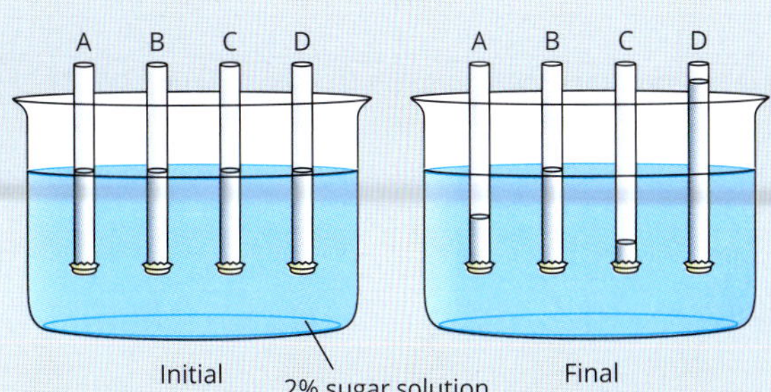

23 If a drowning person inhales sea water instead of fresh water, death occurs more slowly, taking about six to eight minutes. Use your understanding of osmosis to explain the difference between inhaling fresh water and sea water. You will need to consider the relative salt concentrations of sea water (1100 mOsm), blood (300 mOsm) and fresh water (0 mOsm).

24 A young teenager kept fish as a hobby. The marine aquarium was home to a variety of species of tropical fish. When asked to care for the freshwater goldfish of a friend the teenager put the fish into the marine aquarium along with the other fish. After some time the fish became sluggish and eventually died.

The teenager took the fish to the pet shop to seek advice. The pet-shop owner said that it had died in part because its body cells had become dehydrated.

a Explain how the body cells of the fish could have become dehydrated.

b Name the process involved.

25 A genome is the sum total of all of the genes contained in an individual organism. A proteome is the sum total of all of the proteins produced in the body of an individual organism.

a Outline the relationship between the genome and the proteome of an individual organism.

b Look carefully at the diagrams in boxes P, Q, R and S, and the text in boxes W, X, Y and Z below. Place the letter for each into the table so that they correspond to the correct level of protein structure.

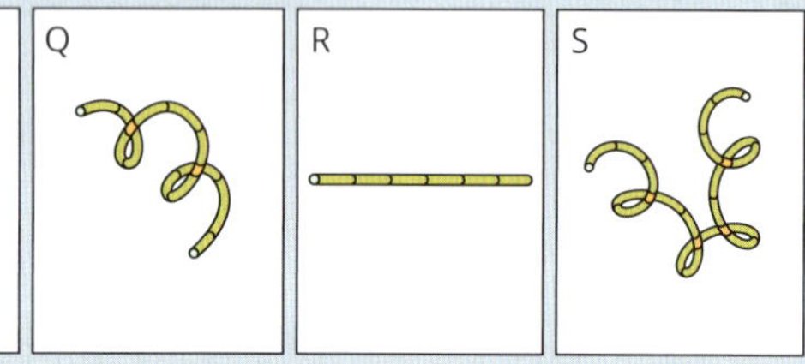

W	X	Y	Z
Polypeptide chain becomes coiled or pleated.	Amino acids become joined by peptide bonds to form a polypeptide.	Two or more polypeptide chains become entwined and chemically bonded together.	Polypeptide chain folds on itself to form a 3D structure.

Level of structure	Diagram (P–Q)	Description (W–Z)
primary		
secondary		
tertiary		
quaternary		

26 Bacteria live in a vast range of different environments. Conditions range from the ice sheets of Antarctica to the superheated water surrounding the undersea volcanoes of the mid-ocean ridges or the hot springs of Yellowstone. Bacteria living in the Antarctic ice are called cryophiles while those living in water at close to boiling point are hyperthermophiles. An example of a cryophilic bacterium is *Psychrobacter*, which thrives at temperatures between −10 °C and 42 °C. *Methanopyrus* is a hyperthermophilic bacterium. It has been shown to survive and reproduce at temperatures between 84 °C and 110 °C. Despite their extreme lifestyles these bacteria, like all living things, use proteins in the form of enzymes to regulate their metabolism.

An experiment was performed using both of these groups of bacteria. Cultures of *Psychrobacter* and *Methanopyrus* were incubated at a temperature of 60 °C for 3 hours. The bacterial cultures were then returned to their optimal temperature and the growth of the bacteria in each culture was monitored.

a What is meant by the optimal temperature for a protein?

b i Which of the cultures, if any, would you expect to show growth?

ii Explain your reasoning.

CHAPTER

03 Nucleic acids, gene structure and regulation

Learning outcomes

By the end of this chapter, you will be able to describe the structure of the nucleic acids DNA and RNA, and their synthesis through condensation polymerisation. You will also understand the role of these nucleic acids as information molecules that encode instructions for protein synthesis, and the steps in eukaryotic gene expression: transcription, RNA processing and translation.

You will have explored gene structure and analysed the distinction between structural and regulatory genes and have an understanding of the regulation of gene transcription by transcriptional factors.

You will learn about genetic changes due to transcriptional errors in Chapter 9.

Key knowledge

- nucleic acids as information molecules that encode instructions for the synthesis of proteins in cells
- the structure of DNA and the three forms of RNA, including similarities and differences in their subunits, and their synthesis by condensation polymerisation
- the genetic code as a degenerate triplet code and the steps in gene expression including transcription, RNA processing in eukaryotic cells and translation
- the functional distinction between structural genes and regulatory genes
- the structure of genes in eukaryotic cells including stop and start instructions, promoter regions, exons and introns
- use of the *lac* operon as a simple prokaryotic model that illustrates the switching off and on of genes by proteins (transcriptional factors) expressed by regulatory genes.

3.1 Nucleic acids—DNA and RNA

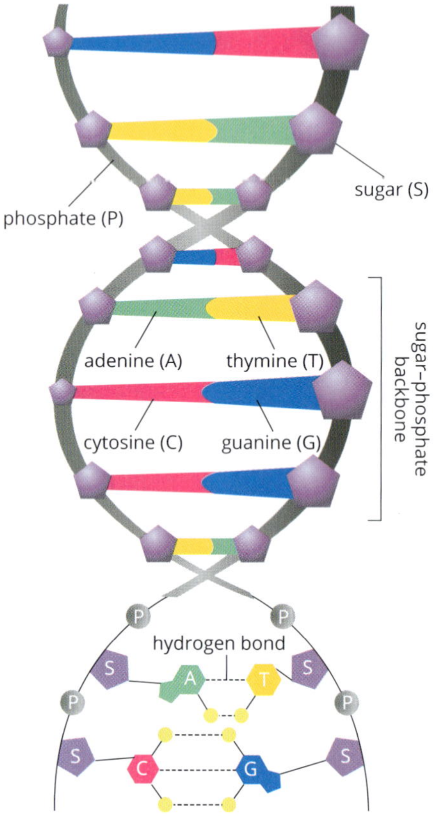

Nucleic acids are organic biomolecules that store and transmit inherited characteristics of organisms. There are two types of nucleic acids: deoxyribonucleic acid (DNA) and ribonucleic acid (RNA). Both DNA and RNA are made up of nitrogenous bases and a sugar–phosphate backbone. RNA is single-stranded and DNA has a double-stranded helix (spiral) structure with complementary pairing of its nitrogenous bases (Figure 3.1.1).

In this section, you will learn about the structure and function of DNA and RNA.

FIGURE 3.1.1 The structure of the DNA double helix biomolecule. The helical structure of DNA is formed by two strands of complementary nitrogenous bases that are joined by hydrogen bonds. Each side of the helix is comprised of deoxyribose sugar and phosphate molecules, known as the sugar–phosphate backbone.

NUCLEIC ACIDS

Nucleic acids are large **biomolecules** that store and transmit hereditary information. Specifically, nucleic acids encode instructions for the synthesis of proteins.

- Deoxyribonucleic acid (DNA) carries the instructions that code for the production of RNA, which may be functional (such as transfer RNA) or contain information for protein synthesis (messenger RNA). DNA is able to self-replicate.
- Ribonucleic acid (RNA) has different forms that perform different functions. It can carry a copy of a DNA sequence (messenger RNA) and it also has the ability to 'read' and translate the DNA information (transfer RNA). RNA plays a major role in the process of **protein synthesis**.

Nucleic acids are polymers, made up of repeated subunit monomers called **nucleotides**.

Nucleotides

A single nucleotide consists of three basic units:

- a phosphate group—the same in all nucleotides
- a five-carbon (pentose) sugar
  - **deoxyribose** in DNA nucleotides
  - **ribose** in RNA nucleotides
- a nitrogenous (nitrogen-containing) **base**.

The five carbon atoms in a pentose sugar molecule are labelled 1' to 5'. The phosphate is always attached to the 5' carbon and the base to the 1' carbon in a single nucleotide. You can see this structure in Figure 3.1.2.

> One end of the DNA and RNA strands has a free 3rd carbon atom. This is called the 3 prime (3') end. The other end has a free 5th carbon atom. This is called the 5 prime (5') end. DNA and RNA are synthesised in the 5' to 3' direction.

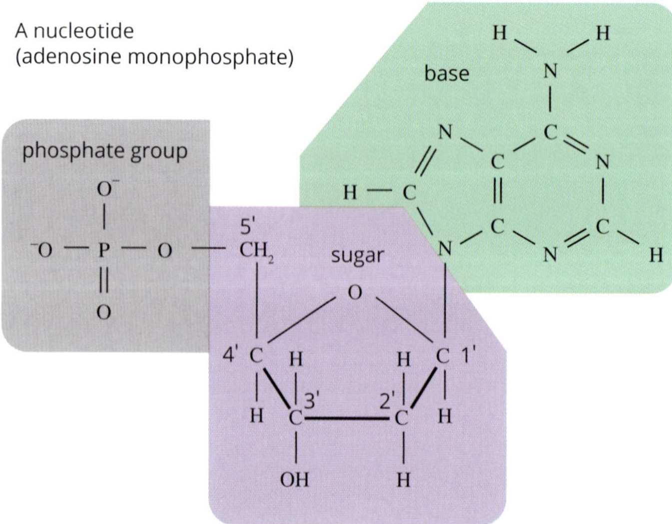

FIGURE 3.1.2 The basic structure of a DNA nucleotide, showing the phosphate group, the five-carbon sugar and the nitrogenous base adenine (A).

The nitrogenous bases

There are five different nitrogenous bases:

- **adenine (A)**
- **guanine (G)**
- **cytosine (C)**
- **thymine (T)**—in DNA only
- **uracil (U)**—in RNA only.

These five nitrogenous bases can be categorised into one of two groups based on their structure:

- **Purines** (A and G) have two rings in their structure.
- **Pyrimidines** (T, U and C) have one ring in their structure (see Figure 3.1.3).

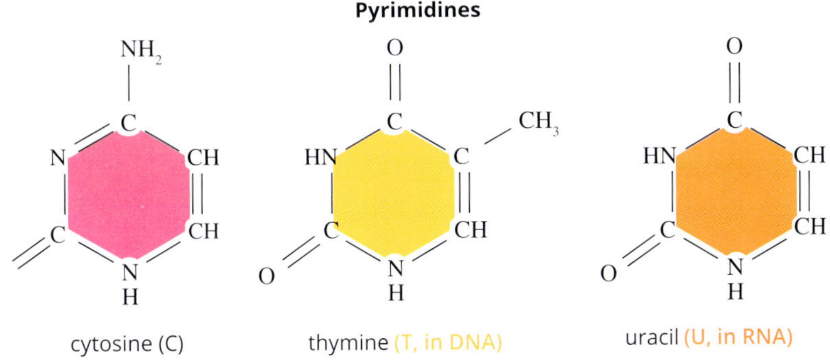

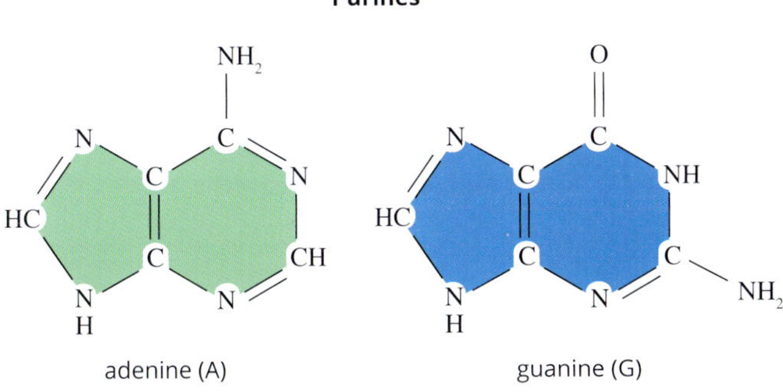

FIGURE 3.1.3 The structure of pyrimidines (cytosine, thymine and uracil) and purines (adenine and guanine). Pyrimidines have one ring and purines have two rings.

Condensation polymerisation of nucleotides

In Section 2.5, you learnt about condensation polymerisation in which single amino acids join together to form a polypeptide chain. Water molecules are produced as a by-product in this reaction.

Free nucleotides also link together to form strands through a condensation polymerisation reaction. This reaction occurs initially between two nucleotides, enabling them to join together to form a **dinucleotide**:

- The hydroxyl group (OH) on the 3' carbon atom of the sugar of one nucleotide joins with the phosphate on the fifth carbon of the pentose sugar (5') of the other nucleotide to form water, which is released.
- Free nucleotides can then be continuously added to the 3' carbons in this way, forming a long sugar–phosphate–sugar–phosphate backbone strand known as a **polynucleotide**.
- The nucleotides in the sugar–phosphate chain are joined by **phosphodiester bonds** (a type of strong covalent bond).
- In polynucleotide strands, one end has a free phosphate group on the 5' carbon; this is called the 5' end. The other end of the strand has a free hydroxyl on the 3' carbon; this is called the 3' end.

Both DNA and RNA are polynucleotides, formed through condensation polymerisation reactions (Figure 3.1.4).

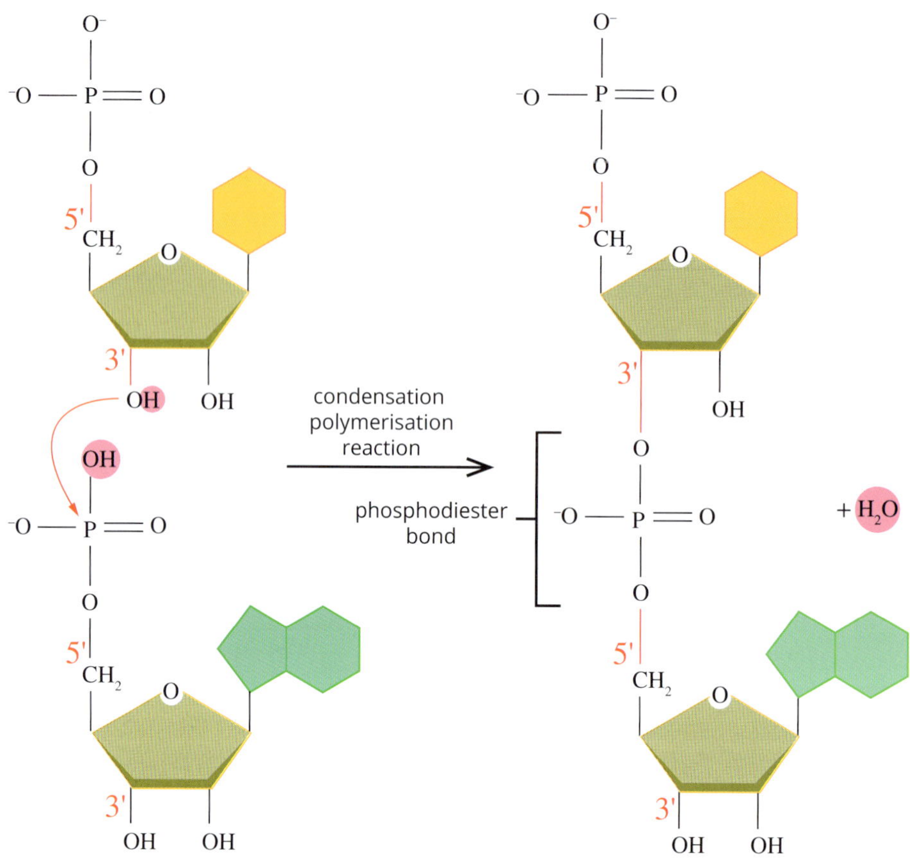

FIGURE 3.1.4 The condensation polymerisation reaction of two DNA nucleotides, forming a dinucleotide by the removal of water.

STRUCTURE OF DNA

> The diameter of the DNA double helix is approximately 2.0 nanometres and there are 10–10.5 pairs of nucleotide bases in each twist of the helix.

DNA is double-stranded, as it consists of two chains of nucleotides or 'strands' twisted into a **double helix** structure (Figure 3.1.5). The primary structure of DNA is the single strand of polynucleotides, which consists of a specific sequence of nitrogenous bases (A, T, C and G). The hydrogen bonds between pairs of nitrogenous bases stabilise the secondary structure of the DNA and form the double helix.

When DNA is uncoiled, the two strands can be represented as a ladder. The sides of the ladder consist of the sugar–phosphate backbone. The two strands of a DNA molecule are **antiparallel**, meaning that they run in opposite directions, with the 3' end of one strand matching with the 5' end of the other strand. The rungs of the ladder are the nitrogenous bases of each nucleotide.

Complementary base pairing occurs between the nitrogenous bases, forming the double-stranded DNA molecule. In complementary base pairing:

- the purine adenine (A) always pairs with the pyrimidine thymine (T), held together with two weak hydrogen bonds
- the purine guanine (G) always pairs with the pyrimidine cytosine (C), held together with three weak hydrogen bonds (Figure 3.1.5).

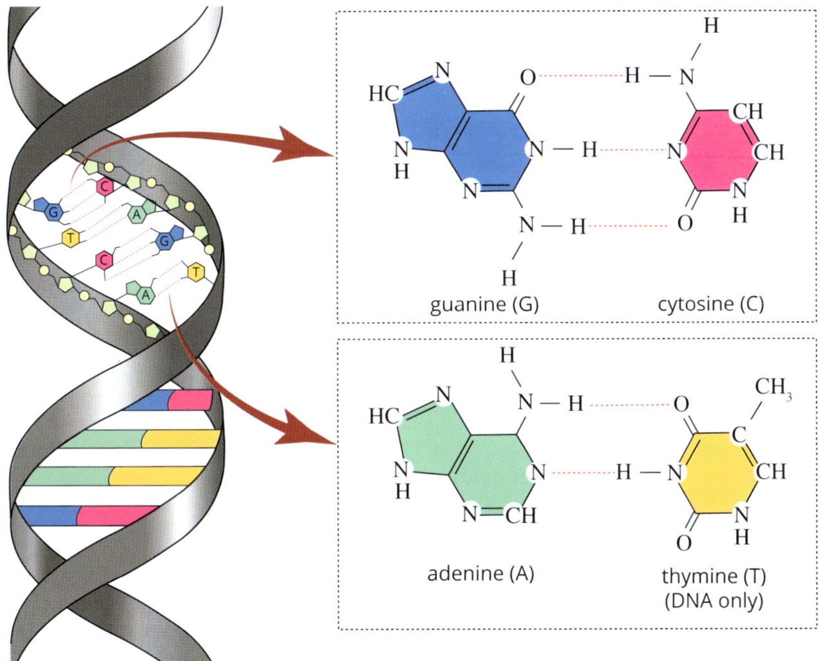

FIGURE 3.1.5 The helical structure of DNA. Two complementary strands form a double helix joined by base pairs guanine (G) and cytosine (C), and adenine (A) and thymine (T).

In the DNA double helix, hydrogen bonds hold the pairs of polynucleotides together. Note that there are two hydrogen bonds between adenine (A) and thymine (T) and three hydrogen bonds between cytosine (C) and guanine (G).

BIOFILE

Eukaryotic versus prokaryotic DNA

In eukaryotic cells, DNA is found in the nucleus, where it is tightly coiled up to form structures called chromosomes. In prokaryotic cells, DNA is found in the nucleoid region as a large circular chromosome and small circular molecules called plasmids (Figure 3.1.6).

Eukaryotic cells are more complex than prokaryotes (Figure 3.1.7 and Figure 3.1.8). Eukaryotes have an average of 25 times more DNA than prokaryotes. Because of this, the processes involved in DNA replication are much faster in prokaryotic cells. Some bacterial cells take just 40 minutes to replicate their DNA, while in some animals this can take up to 10 hours.

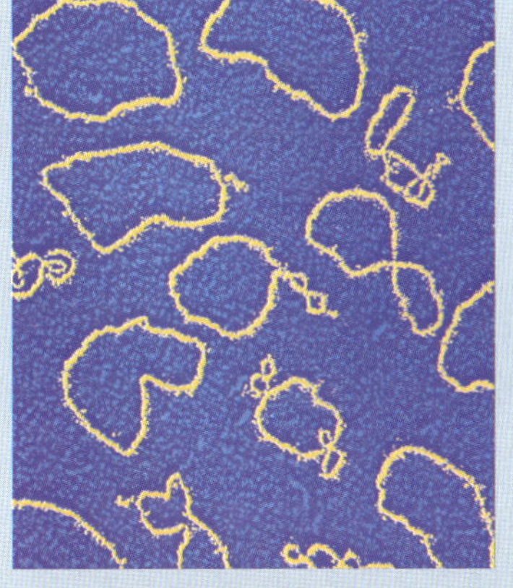

FIGURE 3.1.6 Coloured transmission electron micrograph of plasmids from the bacteria *Escherichia coli*. Plasmids are small, circular DNA molecules that are commonly found in bacteria.

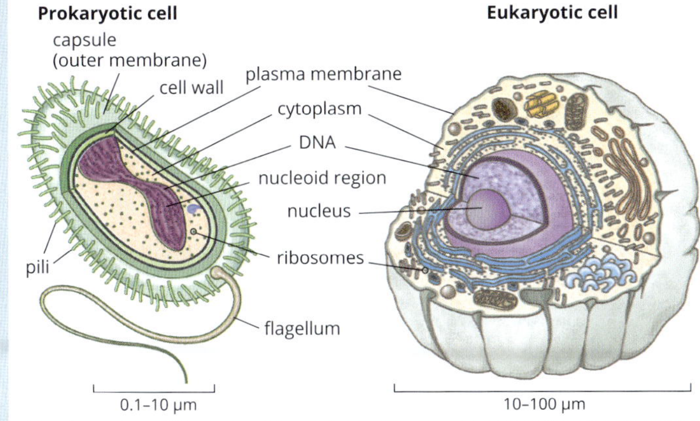

FIGURE 3.1.7 Prokaryotic cell compared to a eukaryotic cell. Prokaryotic cells and the processes involved in their DNA replication are generally much simpler than in eukaryotic cells.

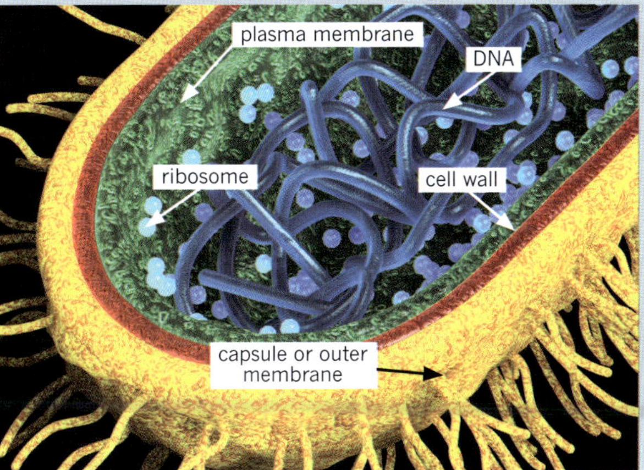

FIGURE 3.1.8 Artwork of the inner structures of an *E. coli* cell. The pili and capsule are shown in yellow, the cell wall in red, the plasma membrane in green, the ribosomes are the light blue structures and the DNA is shown in blue. You can see in this image that the DNA is not bound inside a nucleus as it is in eukaryotic cells.

STRUCTURE OF RNA

Unlike DNA, RNA is usually found as a single strand, sometimes folded onto itself. RNA molecules are usually much shorter than DNA molecules. The nucleotides of RNA have the same basic structure as those of DNA, with a few differences. DNA contains the sugar deoxyribose, while RNA contains ribose. The nitrogenous base thymine in DNA is replaced by uracil in RNA, both of which pair with adenine (Figure 3.1.9). Uracil is more stable in single-stranded polynucleotides.

There are three main forms of RNA: messenger RNA (mRNA) (Figure 3.1.10), ribosomal RNA (rRNA) (Figure 3.1.11) and transfer RNA (tRNA) (Figure 3.1.12). Table 3.1.1 summarises and highlights the differences between the structure of DNA and RNA.

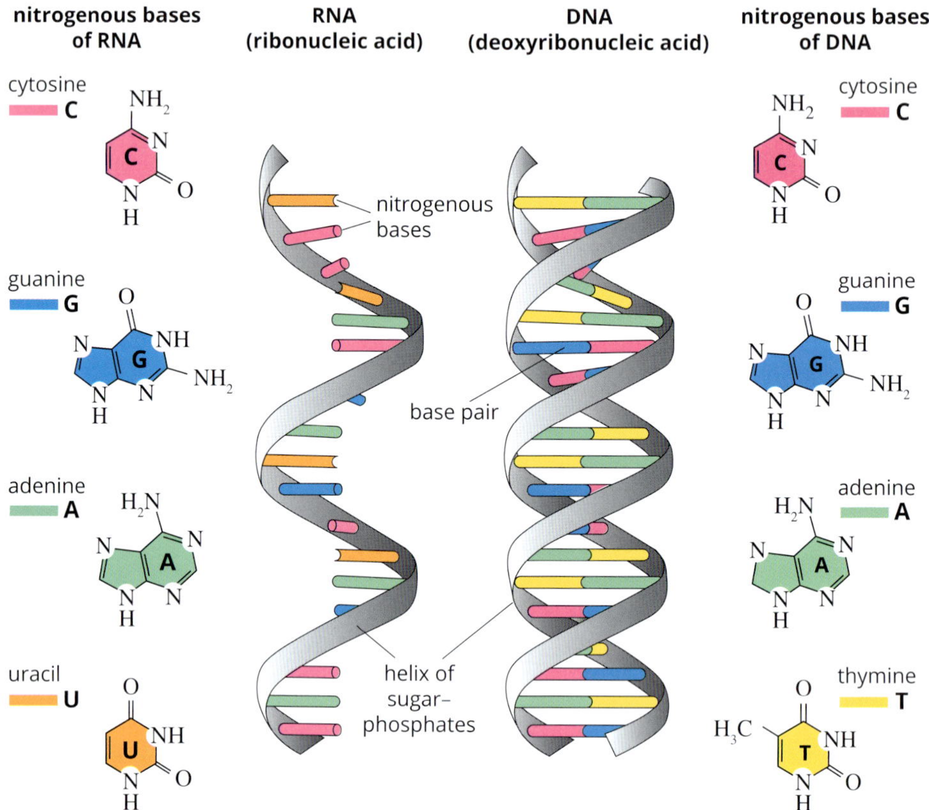

FIGURE 3.1.9 Comparison of the structures of RNA and DNA. Both molecules are made up of nitrogenous bases and a sugar–phosphate backbone, but in RNA ribose sugar replaces deoxyribose sugar and uracil replaces thymine. RNA is also single stranded; DNA is double stranded.

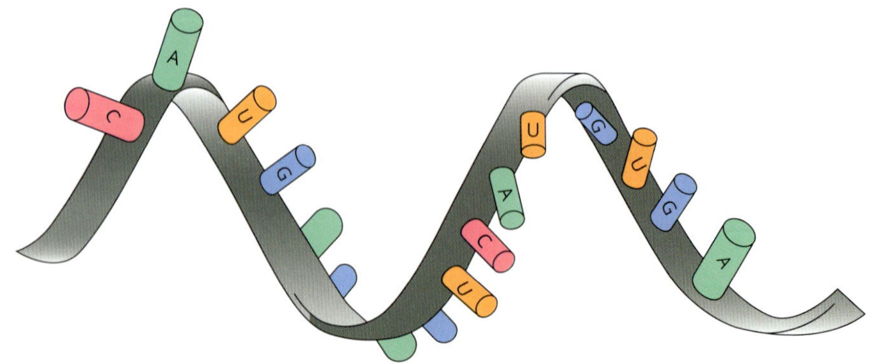

FIGURE 3.1.10 Messenger RNA (mRNA) carries a copy of the DNA's nucleotide sequence to be translated into proteins.

	DNA	RNA
relative length	long	short
sugar	deoxyribose deoxyribose in DNA	ribose ribose in RNA
bases	adenine cytosine guanine thymine thymine in DNA	adenine cytosine guanine uracil uracil in RNA
strands	double	usually single
base pairing	adenine–thymine cytosine–guanine	adenine–uracil cytosine–guanine

TABLE 3.1.1 A summary of the differences between DNA and RNA.

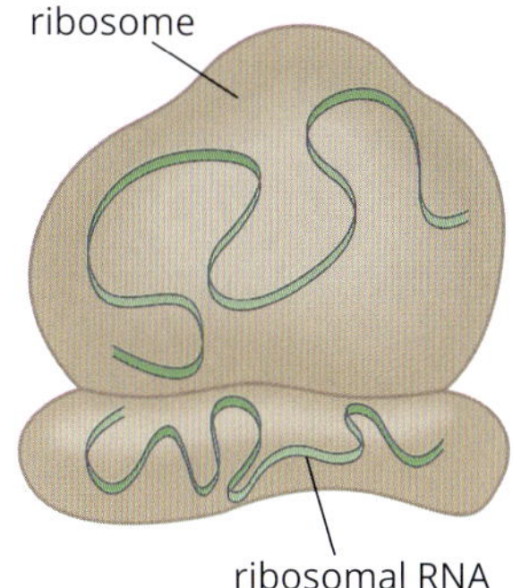

FIGURE 3.1.11 Ribosomal RNA (rRNA) forms ribosomes, the site of translation of the mRNA into proteins.

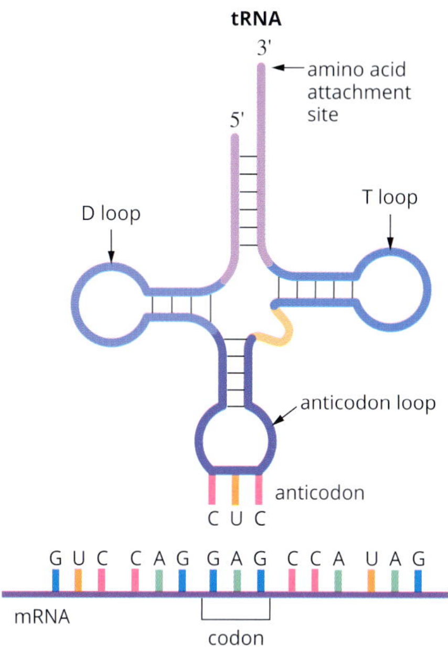

FIGURE 3.1.12 Transfer RNA (tRNA) carries amino acids to the appropriate positions on the mRNA by matching its anticodon sequence to the complementary sequence in the mRNA. The amino acids transferred by tRNA builds the polypeptide chain during translation.

ROLES OF DNA AND RNA IN PROTEIN SYNTHESIS

DNA and RNA both play vital roles in protein synthesis. DNA provides the instructions, which are translated by RNA into proteins that carry out all of the functions that are essential to life.

DNA and protein synthesis

DNA stores and transmits hereditary information. This information is stored as a sequence of nucleotides that encodes biological information. The order of the nucleotides in this biological code determines which products are synthesised. These products are mainly proteins, but they may also be functional RNA molecules. A **gene** is a region of DNA that contains the information to produce a protein or a functional RNA molecule.

BIOFILE

Viral RNA

Some viruses, such as rotavirus, which causes gastroenteritis in humans, have double-stranded RNA. Eukaryotic host cells have defence systems that detect and inactivate double-stranded RNA. However, some double-stranded RNA viruses avoid detection by creating another set of single-stranded mRNA (with their own RNA polymerase enzyme), and releasing this into the host cell instead of releasing their own double-stranded RNA.

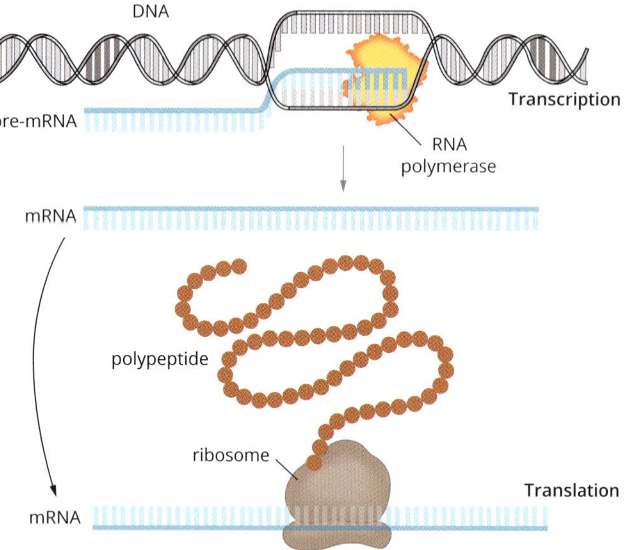

FIGURE 3.1.13 Transcription and translation are the processes by which genes are expressed as functional proteins. During transcription, RNA polymerase makes a copy of the DNA, which first becomes pre-mRNA and then mature mRNA. The mRNA is then transported to the ribosomes where it is translated into a chain of amino acids.

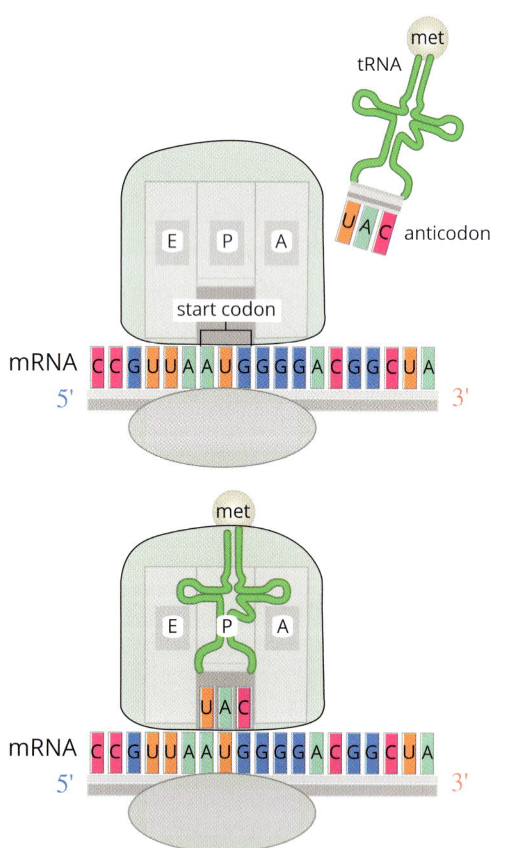

FIGURE 3.1.15 Translation synthesises a protein from mRNA. It is carried out in the ribosomes, where a tRNA finds codons in the mRNA that are complementary to its anticodon, binds to this site and transfers its amino acid. In this figure, a tRNA molecule carrying the amino acid methionine (met) recognises the codon on mRNA that is complementary to its anticodon sequence and transfers the amino acid in the correct position on the mRNA. Met is a start codon and the first amino acid to be incorporated into the polypeptide chain.

RNA and protein synthesis

RNA plays an important role in expressing the information contained in the sequence of a gene to synthesise proteins. (You will learn about gene expression in more detail in Section 3.2.)

Each type of RNA has a different role in the process of protein synthesis:

- **Messenger RNA** (mRNA) is formed in the nucleus by the process of **transcription**. mRNA carries a copy of the nucleotide sequence of DNA that specifies the amino acid sequence for a particular protein. During transcription, pre-mRNA is first formed by the enzyme RNA polymerase. Pre-mRNA is then processed (post-transcriptional modification) to form mature mRNA, which is a single-stranded copy of the coding DNA (gene) (Figure 3.1.13). The mature mRNA travels from the nucleus to the cytosol where it binds to ribosomes ready for **translation**.
- **Ribosomal RNA** (rRNA) is synthesised in the nucleolus of the cell nucleus and is based on the nucleotide sequence of the DNA. Together with proteins, rRNA forms a small organelle called a ribosome. Ribosomes are the sites where the information in the mRNA is translated into a chain of amino acids (Figure 3.1.14).
- **Transfer RNA** (tRNA) molecules transfer amino acids from the cytoplasm to the ribosomes, where they are joined to form a polypeptide chain based on the arrangement of nucleotides in the mRNA. There are 61 different tRNA molecules, each of which combines with only one particular amino acid at one end of its molecule. (There are 64 codons that each represent an amino acid, 3 of which are stop codons. There are no tRNA molecules that recognise the 3 stop codons and so translation is terminated.) There are three places for tRNA to bind to the ribosome: the exit site (E), the peptidyl site (P) and the aminoacyl (A) as shown in Figure 3.1.15. At the other end of the tRNA molecule, there is a sequence of nucleotides known as the **anticodon**. The anticodon recognises a particular sequence of nucleotides in the mRNA. This enables an amino acid to be positioned in the correct space on a protein.

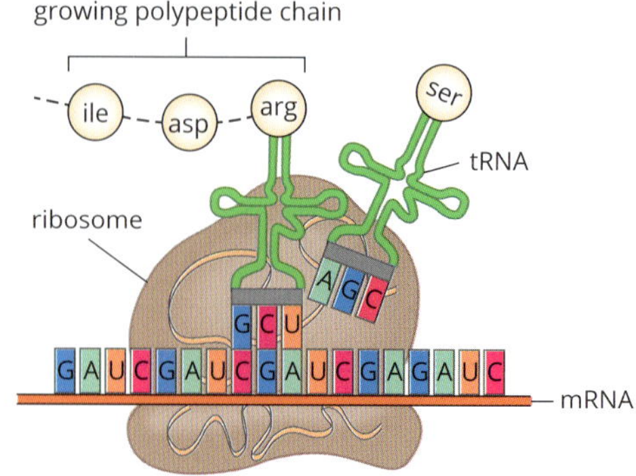

FIGURE 3.1.14 The three different types of RNA work together to use the information contained in a gene to synthesise a protein.

BIOLOGY IN ACTION

Protein folding and the need for speed

In Chapter 2 you learnt that the three-dimensional arrangement of proteins is vital to their biological functioning. Proteins display an extraordinary diversity in structural forms, with thousands of variations in size and folding configuration. Some examples are shown in Figure 3.1.16. However, it is not only the structure of the folds that is important for protein functioning, but also the speed at which they fold.

Researchers at the Heidelberg Institute for Theoretical Studies and the University of Illinois investigated the folding speed of all known proteins using computer analysis. The researchers combined all known protein structures and genomes to obtain a dataset of 92000 proteins and 989 genomes. By identifying protein sequences in the organisms' genomes that matched proteins for which the folding structure was known, the researchers were able to determine when different protein structures appeared in evolutionary history.

In order to determine the folding rates of different proteins, the researchers developed a mathematical model that used the known folding configurations of proteins. As proteins always fold at the same points, the speed and efficiency with which they fold is determined by how far apart these points are. The researchers found that over the course of evolution, from Archaea to multicellular organisms, protein-folding speed has increased. Amino acid chains that make up proteins have also become shorter over evolutionary time, contributing to the increased folding speeds. The researchers speculate that faster folding speeds of proteins may make them less susceptible to clumping and so increase their functional efficiency.

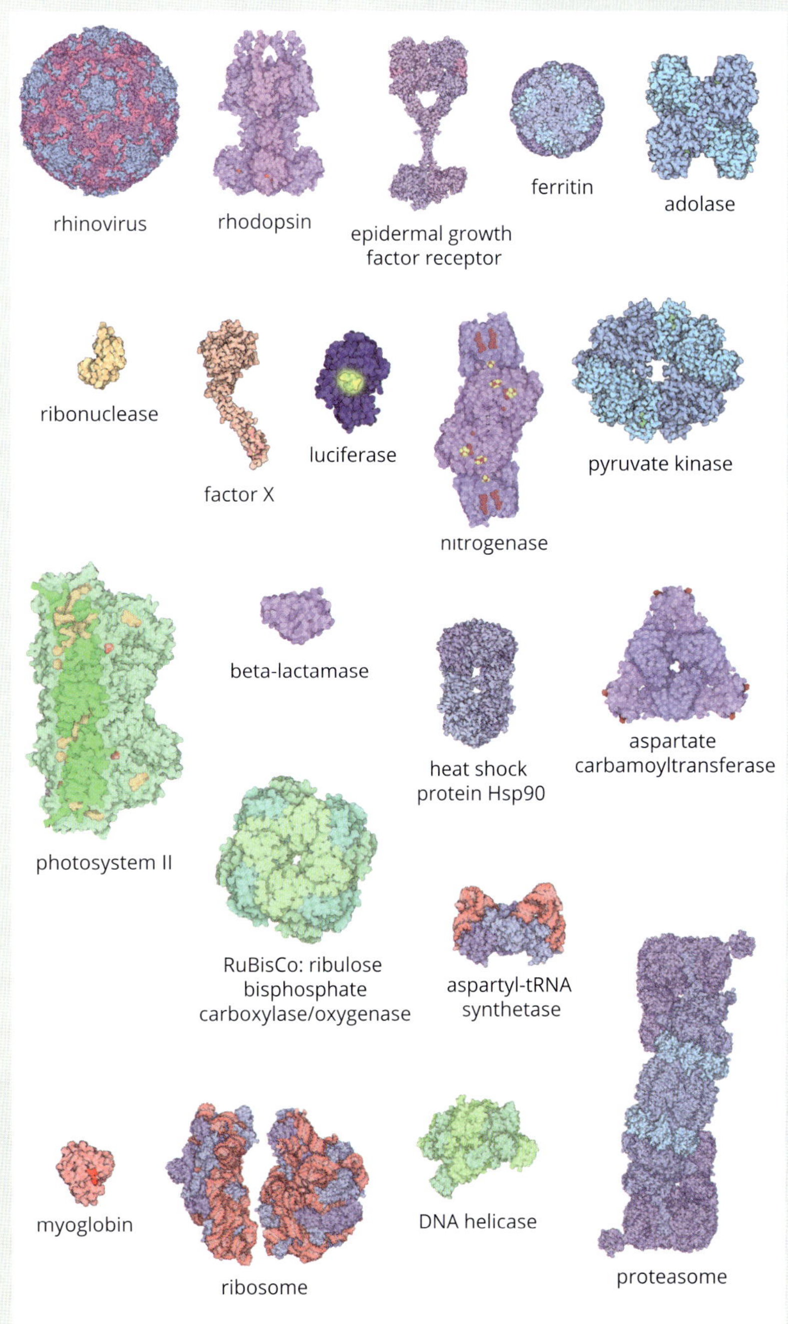

FIGURE 3.1.16 Examples of the diversity of structures and folding configurations of different proteins from the Worldwide Protein Data Bank.

3.1 Review

SUMMARY

- Nucleic acids are polymers of nucleotides (polynucleotides). There are two types of nucleic acids: DNA and RNA.
- Nucleotides are made up of a five-carbon sugar, a phosphate and a nitrogenous base.
- The nitrogenous bases adenine and guanine are purines; cytosine, thymine and uracil are pyrimidines.
- Nucleotides are joined in a condensation polymerisation reaction to form polynucleotides. Water is released and a covalent bond (phosphodiester bond) forms joining the nucleotides together.
- DNA is a long, coiled, double-stranded nucleic acid that forms a double helix.
  - Each strand of DNA contains nucleotides that are made up of deoxyribose sugar, a phosphate and one of four nitrogenous bases (adenine, cytosine, guanine and thymine).
  - The two strands of DNA are joined by complementary base pairing between the nitrogenous bases. Adenine joins with thymine by two weak hydrogen bonds, while cytosine joins with guanine by three hydrogen bonds.
  - The two strands of DNA are antiparallel. One runs in the 5' to 3' direction, while the other runs in the opposite direction.
- RNA is a short, usually single-stranded nucleic acid.
  - RNA contains nucleotides that are made up of ribose sugar, a phosphate and one of four nitrogenous bases (adenine, cytosine, guanine and uracil).
- DNA stores hereditary information, carrying the instructions that code for the production of mainly proteins but also functional RNA molecules, in a specific sequence of nucleotides. A gene is a region of DNA that codes for a protein or a functional RNA molecule.
- The role of RNA is to express the information contained in the nucleotide sequence of a gene to synthesise proteins:
  - mRNA is a single-stranded nucleic acid that carries a copy of the genetic sequence in DNA, specifying the amino acid sequence for a particular protein.
  - rRNA makes up part of a ribosome. Ribosomes are the sites where the information in the mRNA is translated into a chain of amino acids.
  - tRNAs carry specific amino acids to ribosomes in order to form polypeptide chains.

KEY QUESTIONS

1. What are the functions of DNA and RNA?
2. What are the three basic units of a nucleotide?
3. Complete the table by assigning 'purine or 'pyrimidine' to each nitrogenous base and filling in their complementary bases.

Base	Purine or pyrimidine	Complementary base	Purine or pyrimidine
adenine (A)			
guanine (G)			
cytosine (C)			
thymine (T)			
uracil (U)			

4. Explain how RNA differs from DNA, mentioning at least three features that differentiate them.
5. Name the three types of RNA and outline their basic functions.

3.2 Gene structure and expression

The genetic code represents the genetic information stored in DNA as a triplet code within sections called genes. This information is used to synthesise the amino acid sequences that form proteins through a process called **gene expression** (Figure 3.2.1).

In this section, you will learn about the roles of DNA and RNA in protein synthesis and the different steps of gene expression. You will learn about changes due to transcriptional errors in Chapter 9.

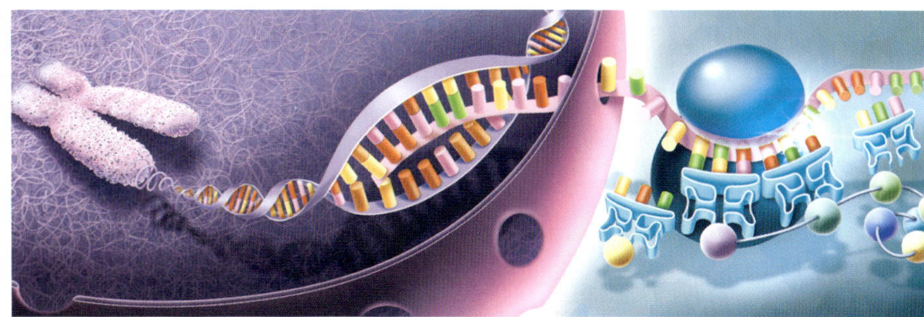

FIGURE 3.2.1 Overview of gene expression, showing mRNA being produced during transcription and a polypeptide being synthesised during translation.

GENES AND THE GENETIC CODE

As you learnt in the previous section, a gene is a region of DNA that may be translated into a polypeptide or an RNA molecule that can be functional, such as tRNA. When coding for a polypeptide, it is the sequence of nucleotides within a gene that contains the information for the protein to be synthesised. Genes can be millions of nucleotides in length. For example, the longest known gene, which codes for the protein dystrophin, is 2.5 megabase pairs long (2 500 000 base pairs).

The genetic code

The genetic code is a set of rules that defines how the information in nucleic acids (DNA and RNA) is translated into proteins and functional RNA molecules. The information in DNA and RNA is stored as a three-letter code of nucleotides. In DNA, this three-letter code is called a **triplet**. When a DNA triplet is transcribed into mature mRNA, the triplet is then called a **codon**. Each triplet or codon codes for one amino acid (Figure 3.2.2 and Figure 3.2.3 on page 144), and may also provide specific instructions, such as 'start translation' and 'stop translation'. So, the genetic code is the set of rules determining which sequences of nucleotides are translated into which amino acids. These amino acids form polypeptide chains and polypeptide chains form proteins. The genetic code is universal—the rules are the same for all organisms on Earth.

> Messenger RNA (mRNA) is produced during transcription and then translated to produce an amino acid chain (polypeptide).

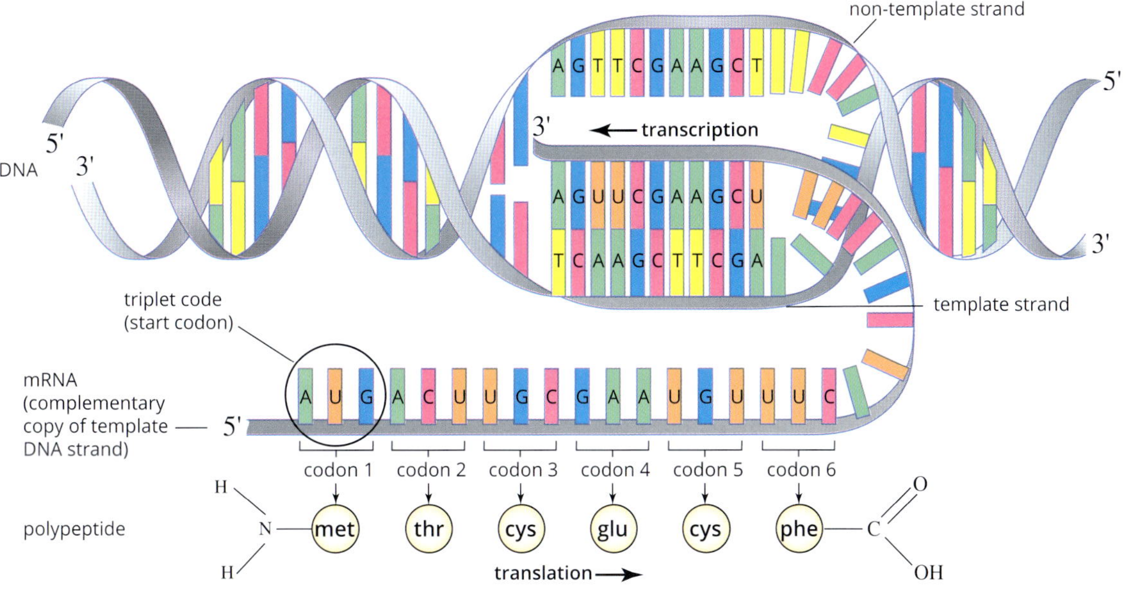

FIGURE 3.2.2 The DNA is transcribed into mRNA (messenger RNA). mRNA is read as codons. Each codon codes for a particular amino acid with the exception of stop codons, which end protein synthesis.

Degeneracy

The genetic code is said to be **degenerate** because more than one codon can code for the same amino acid (Figure 3.2.3). Differences in codons encoding the same amino acid usually occur at the second or third base. As the genetic code uses four nucleotides and three nucleotides code for an amino acid, the combinations of these nucleotides make a total of 64 possible codons ($4^3 = 64$), to code for the total 20 amino acids (Figure 3.2.3). The degeneracy of the code acts as a buffer for gene mutations in that a single change in one base may not necessarily lead to a change in the amino acid produced and therefore may not necessarily lead to a change in the structure of the protein produced.

First base of codon	Second base of codon: U		C		A		G		Third base of codon
	U		**C**		**A**		**G**		
U	UUU	phenylalanine (phe)	UCU	serine (ser)	UAU	tyrosine (tyr)	UGU	cysteine (cys)	U
	UUC		UCC		UAC		UGC		C
	UUA	leucine (leu)	UCA		UAA	STOP	UGA	STOP	A
	UUG		UCG		UAG		UGG	tryptophan (trp)	G
C	CUU	leucine (leu)	CCU	proline (pro)	CAU	histidine (his)	CGU	arginine (arg)	U
	CUC		CCC		CAC		CGC		C
	CUA		CCA		CAA	glutamine (gln)	CGA		A
	CUG		CCG		CAG		CGG		G
A	AUU	isoleucine (ile)	ACU	threonine (thr)	AAU	asparagine (asn)	AGU	serine (ser)	U
	AUC		ACC		AAC		AGC		C
	AUA		ACA		AAA	lysine (lys)	AGA	arginine (arg)	A
	AUG	methionine (met) START	ACG		AAG		AGG		G
G	GUU	valine (val)	GCU	alanine (ala)	GAU	aspartic acid (asp)	GGU	glycine (gly)	U
	GUC		GCC		GAC		GGC		C
	GUA		GCA		GAA	glutamic acid (glu)	GGA		A
	GUG		GCG		GAG		GGG		G

FIGURE 3.2.3 The genetic code for the 20 amino acids and stop codons.

THE STRUCTURE OF GENES

Eukaryotic genes all have a number of structural features in common, including:

- stop and start triplet sequences—regions where encoding DNA begins and ends for a specific gene
- **promoter** regions—an upstream binding region for the enzyme that is involved in the encoding process (which is **RNA polymerase**)
- **exons**—DNA regions that are the coding segments
- **introns** (or spacer DNA)—DNA regions that are non-coding segments.

These structural features are illustrated in Figure 3.2.4.

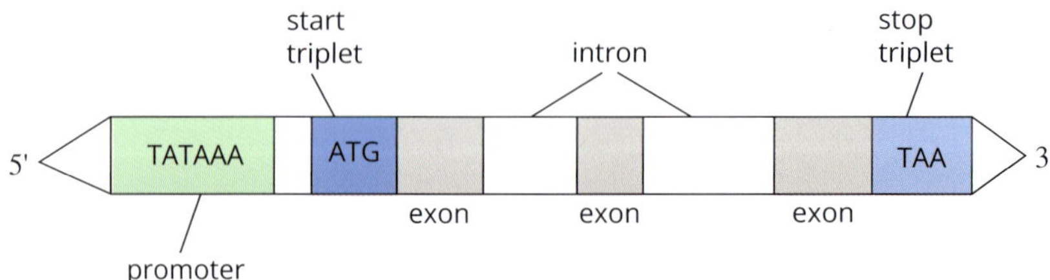

FIGURE 3.2.4 Eukaryotic genes have start and stop triplets, promoter regions, non-coding introns and coding exons. The information on the coding strand of DNA (shown here) is transcribed into mRNA ready for translation into proteins. AUG is the most common start codon in mRNA, but ATG is considered the start triplet in DNA. This is because when writing a DNA sequence, the scientific standard is to show only the coding strand of DNA and to write it in the 5' to 3' direction. The triplet that is complementary to AUG (TAC) is found on the non-coding (template) strand of DNA, which is read by RNA polymerase when building mRNA.

Start and stop instructions

A start triplet indicates where the first stage of gene expression will begin. When transcribed into mRNA this triplet will become the start codon AUG. This codon initiates translation and codes for the amino acid methionine. Most functional proteins start with the amino acid methionine (AUG), but there are some rare exceptions to this. For example, a protein in the fungus *Candida albicans* uses GUG as a start codon.

A stop triplet indicates where transcription will end. The stop triplet does not code for an amino acid. When the stop triplets are transcribed into mRNA they become the codons UAA, UAG and UGA.

Promoter region

Promoter regions are sections of a gene that are found before the start triplet, at the 5' end of the site where transcription will begin. A promoter region:

- is the location where the RNA polymerase (the enzyme that initiates transcription) attaches to the gene
- identifies which DNA strand will be transcribed
- identifies where transcription of the gene will start
- identifies in which direction transcription will occur.

In many eukaryotic genes, the promoter region is coded for by the sequence of bases TATAAA, which is sometimes called the **TATA box**.

Introns and exons

In eukaryotes, not all sections of a gene are translated:

- Exons are regions of a gene that are usually 'expressed' as proteins or RNA. Exons come together to make up mRNA, which is then translated into proteins.
- Introns are non-coding, or intervening, regions of a gene. Introns are spliced out of the mRNA during the stage of gene expression called **RNA processing**.

There are no rules about the number of exons and introns in a gene. In the dystrophin gene, for example, 99% of its length is made up of introns.

GENE EXPRESSION

Gene expression is the process by which the information stored in a gene is used to synthesise a functional gene product (protein or RNA) (Figure 3.2.5). This process is highly regulated so that proteins or RNA molecules are only produced if and when they are required by a cell. Multicellular organisms, in particular, can have specialised cells that require a specific set of proteins. For example, in humans, the cells in connective tissue and bone require the protein fibrillin to form elastic fibres, and skin cells require the enzyme tyrosinase to produce melanin and other pigments. The ability to regulate gene expression conserves energy and materials (nucleotides and amino acids) in the cell.

Gene expression leading to protein synthesis in eukaryotic cells occurs in three stages:

- transcription
- RNA processing
- translation.

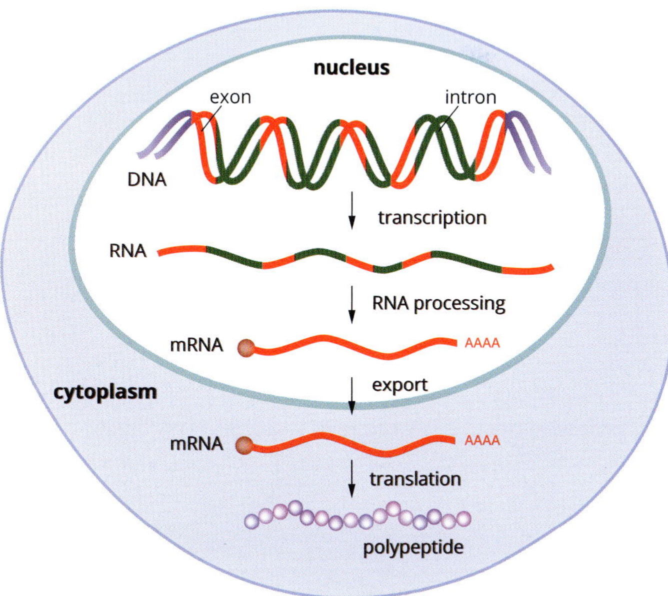

FIGURE 3.2.5 Transcription creates a primary RNA transcript from DNA. The introns are then spliced (cut out) during RNA processing to create a mature strand of mRNA. The mRNA exits the nucleus via a nuclear pore. A ribosome translates the mRNA into a polypeptide chain during translation.

BIOFILE

Gene switching

Trypanosomes (*Trypanosoma brucei*) are parasites that cause African sleeping sickness (Figure 3.2.6). These protozoan parasites are carried in the saliva of the blood-feeding tsetse fly (*Glossina fuscipes fuscipes*) (Figure 3.2.7). The tsetse fly transmits the *Trypanosoma* to humans and other animals when they feed on their blood. African sleeping sickness occurs in Sub-Saharan African, across 36 countries, and threatens the lives of millions of people. The disease causes about 9000 deaths per year. It causes damage to the central nervous system, leading to behavioural changes, confusion, loss of coordination and disturbance to sleep. Without treatment, the disease can be fatal.

T. brucei have up to 1000 genes, which code for proteins that will be positioned on their cell surface, but they can express only one of these genes at a time. When this occurs, one gene is transcribed and the rest are repressed. When a human is infected by a parasite, their immune system will usually recognise the proteins on the cell surface of the parasite and respond by producing antibodies as a defence. However, parasites such as trypanosome can switch from expressing one gene to another, thereby overcoming the human defence response and evading detection.

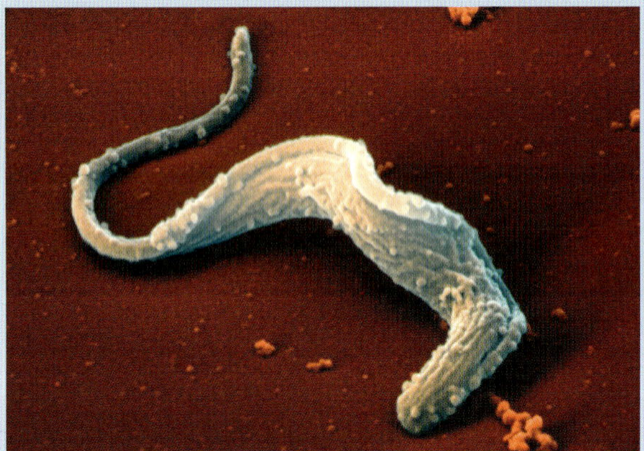

FIGURE 3.2.6 Coloured scanning electron micrograph of the parasite that causes African sleeping sickness, *Trypanosoma brucei*. The protozoan is carried in the saliva of the tsetse fly and is transmitted to humans when the fly feeds on their blood.

FIGURE 3.2.7 A tsetse fly (*Glossina fuscipes fuscipes*) feeding on blood. When the fly feeds on blood, it transmits the parasite *Trypanosoma*, which causes African sleeping sickness in humans.

Transcription

The production of single-stranded mRNA from DNA is called transcription and occurs within the nucleus of eukaryotic cells. The DNA segment that undergoes transcription is known as the transcription unit.

Transcription occurs in three steps:

1. Initiation: Transcription factors combine with the region at the start of the gene, known as the promoter. The promoter region contains specific nucleotide sequences (TATA box) that are recognised by an appropriate subunit of the enzyme RNA polymerase. In eukaryotic cells, transcription factors are required for RNA polymerase to attach to the DNA. RNA polymerase then attaches to the promoter, unwinding and unzipping the DNA molecule by breaking the weak hydrogen bonds between the two strands to expose the bases (see Figure 3.2.8a).
2. Elongation: During transcription, the RNA polymerase molecule covers a region of approximately 30 base pairs. Within this region, a segment of about 15 base pairs is uncoiled. This results in the formation of a transcription bubble. As the RNA polymerase moves along the gene, DNA strands located behind the transcription bubble are coiled again. The RNA polymerase moves along the DNA molecule, producing a strand of mRNA. It uses a strand of DNA as a template, attaching nucleotides (A, U, G, C) by complementary base pairing. mRNA is always synthesised in the 5' to 3' direction, with new nucleotides added to the 3' end. The initial mRNA molecule transcribed is called a primary RNA transcript (see Figure 3.2.8b). The primary RNA transcript will then be processed into mature mRNA.
3. Termination: Transcription ends when RNA polymerase reaches the termination site of the gene. This region contains a stop triplet code, which binds release factors that signal termination. The RNA polymerase detaches, releasing the mRNA and allowing the DNA molecule to reform (Figure 3.2.8c).

Many RNA polymerase molecules may attach to the gene being transcribed, producing many of the same mRNA molecules. The strand of DNA that is transcribed to the mRNA is known as the **template strand**, and the other complementary strand is known as the **coding strand**. The mRNA carries the same base sequence as the coding strand, except it contains uracil in place of thymine.

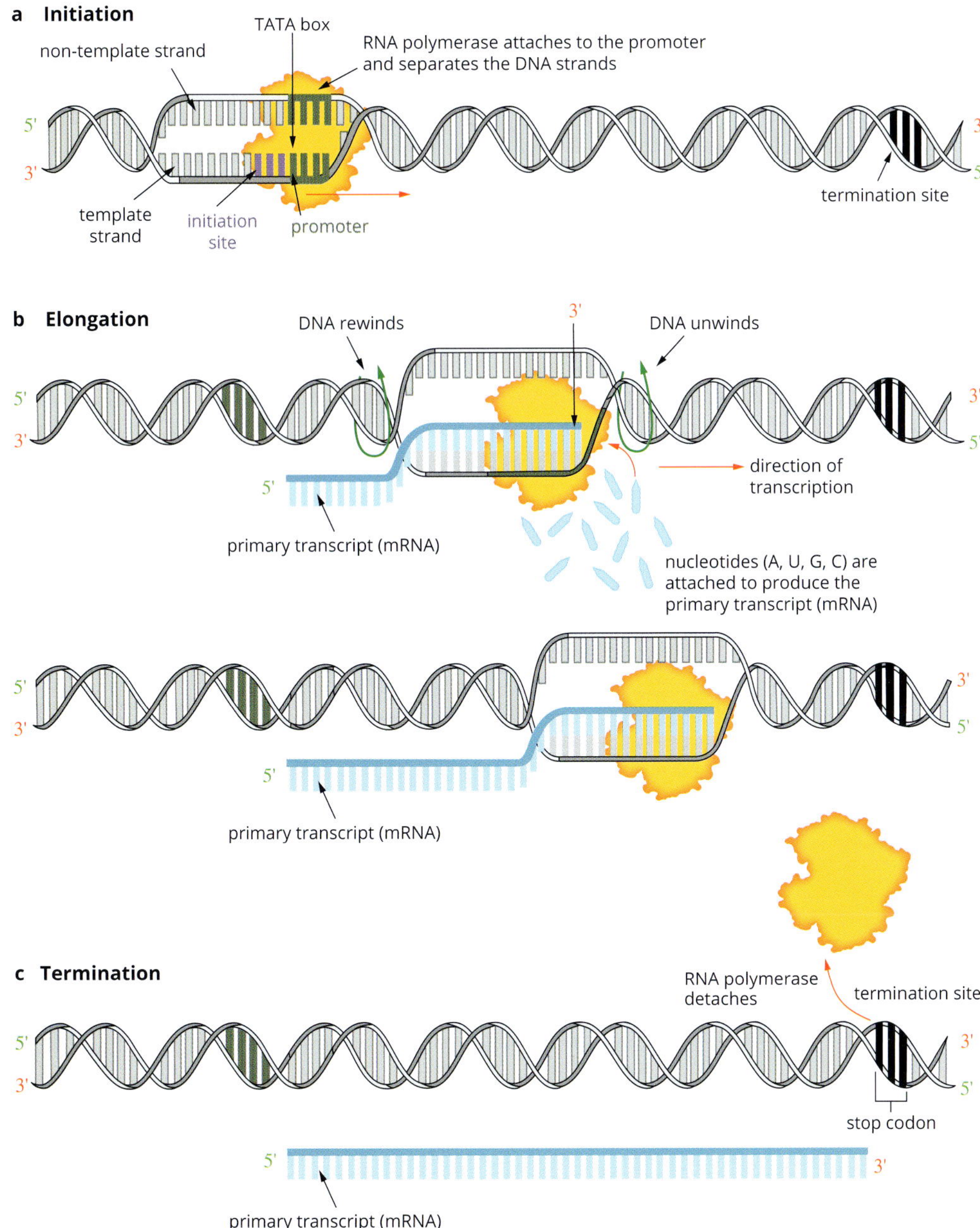

FIGURE 3.2.8 Transcription occurs in three stages: Initiation (a), elongation (b) and termination (c). The initiation (a) of transcription takes place when RNA polymerase attaches to the promoter region of the DNA. The enzyme RNA polymerase unzips the two DNA strands by breaking their hydrogen bonds, exposing the nitrogenous bases (A, T, G, C). During elongation (b), the RNA polymerase moves along the template DNA strand and produces a strand of mRNA by attaching complementary nucleotides (A, U, G, C). The mRNA strand is known as the primary transcript at this stage. Transcription is terminated (c) when the RNA polymerase reaches the stop triplet code (stop codon) at the termination site. The RNA polymerase then detaches and the two DNA strands come together.

RNA processing

After transcription, the primary RNA transcript undergoes processing before it is translated. This stage of gene expression is called RNA processing and includes:

- the addition of a **5' cap**
- the addition of a **poly-A tail**
- **splicing** (removal) of the introns (mRNA maturation).

5' cap and poly-A tail

A cap consisting of a methylguanosine triphosphate molecule, called a 5' cap, is added to the 5' end of the primary RNA transcript while it is being synthesised during transcription. Once transcription has finished, a chain of up to 250 adenine nucleotides is added to the 3' end of the primary RNA transcript. This chain is called a poly-A tail.

These modifications to either end of the primary RNA transcript increase its stability and prevent it from degrading. Additionally, the 5' cap aids the binding of the ribosome to the mRNA at the beginning of translation.

Splicing

Eukaryotic genes have protein-coding regions called exons and non-protein-coding regions called introns. Exons carry the code for the protein, while introns are 'intervening' regions. Most prokaryotes contain only exons and therefore the RNA processing described in this section does not occur in prokaryotes.

In eukaryotes, before a protein can be produced, the introns must be cut out of the primary RNA transcript to form the mature mRNA molecule. This process is known as splicing. During splicing, a complex molecule composed of protein and RNA molecules, called a **spliceosome**, removes the introns from the primary RNA transcript and joins the exon sections together to make mature mRNA (Figure 3.2.9). (Not all of the exons will necessarily be included, as you will see in the next section.)

The single-stranded mature mRNA molecule then exits the nucleus via a nuclear pore.

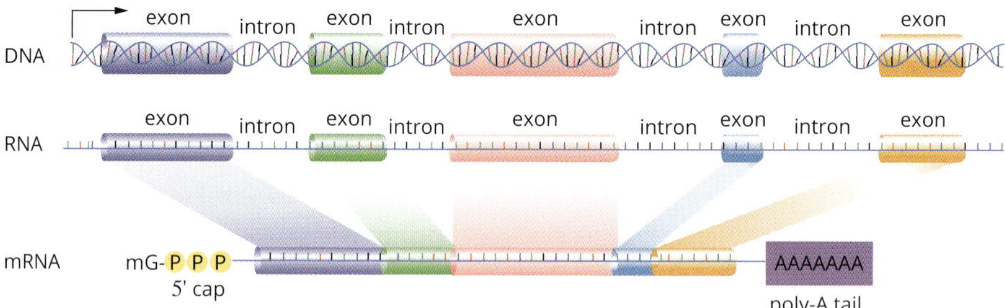

FIGURE 3.2.9 During RNA processing, the introns (non-protein-coding regions) are spliced from the primary RNA transcript, resulting in mature messenger RNA, which consists of only exons (protein-coding regions).

Alternative splicing

A primary transcript can be spliced in many different ways, resulting in alternative mature mRNA strands from a single gene and, thus, different proteins. This is the result of some exons being removed along with the introns during RNA processing. For example, a particular gene may result in a mature mRNA that contains all exons 1–5, but the same gene may result in another mature mRNA that contains only exons 1, 2, 4 and 5 (Figure 3.2.10). Alternative splicing is one reason why the 21 000 genes of humans can produce so many more than 21 000 proteins.

In early research on gene structure, introns were called 'junk DNA' because it was believed they had no role in protein production. It is now known that protein expression is much more complex than first thought and that by splicing mRNA during RNA processing, different proteins may be made from the same gene.

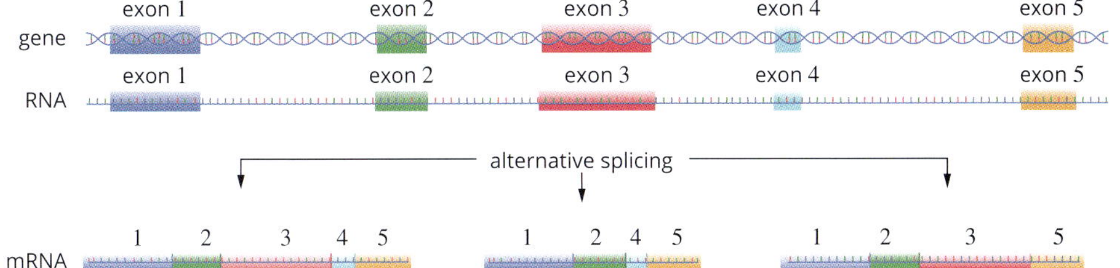

FIGURE 3.2.10 Alternative splicing of a single gene gives rise to alternative mRNA molecules, resulting in many different proteins.

Translation

Translation is the process in which the codons on mRNA are translated into a sequence of amino acids resulting in a polypeptide. This process occurs on ribosomes. Ribosomes bind to an mRNA molecule and act as docking stations for the tRNAs to deposit their specific amino acids. A part of the tRNA, called an anticodon, recognises and binds to the codon on the mRNA by complementary base pairing. Each tRNA carries a specific amino acid related to the codon to which it binds.

Translation occurs in a series of steps, as outlined below.

1 Initiation: To begin protein synthesis, a small ribosomal subunit attaches to the 5' end of an mRNA strand. It then moves along the mRNA until it reaches a start codon AUG. The sequence AUG, which codes for the amino acid methionine, is the most common initiation triplet of mRNA (there are some rare exceptions). A tRNA molecule with the anticodon UAC then brings the amino acid methionine to the mRNA. The tRNA molecule joins to the mRNA start codon, attaching by complementary base pairing between the codon and anticodon (Figure 3.2.11). A large ribosomal subunit also attaches to the tRNA and the small ribosomal subunit. The binding of both ribosomal subunits causes the formation of three special sites for tRNA to bind: the aminoacyl site (A site), the peptidyl site (T site) and the exit site (E site). The attachment of amino acids to their corresponding tRNA molecules occurs in the cytosol—a process that is catalysed by enzymes.

2 Elongation: Following the attachment of the amino acid methionine, another tRNA, with a complementary anticodon to the next codon on the mRNA attaches and adds its specific amino acid to the growing polypeptide chain (Figure 3.2.11). The deposited amino acid joins by a peptide bond to the first amino acid through a condensation polymerisation reaction. The ribosome then releases the tRNA and moves further along the mRNA strand. At each codon a new tRNA binds and adds another amino acid. The tRNA molecules can be reused, allowing them to pick up more of their specific amino acids and return to the mRNA molecule.

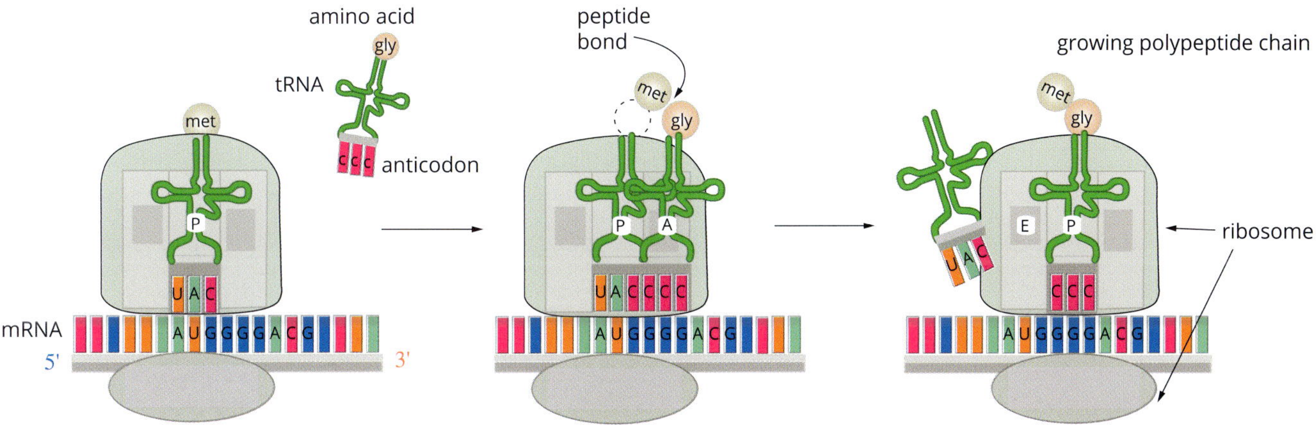

FIGURE 3.2.11 The process of translation on a ribosome. The ribosome moves along the mRNA one codon at a time, and tRNA molecules bring their specific amino acids to their complementary mRNA codon. The amino acids join together by peptide bonds to form a polypeptide chain.

3 Termination: Attachment of amino acids continues until a stop codon is reached. The polypeptide chain is then released from the ribosome into the cytoplasm or the endoplasmic reticulum. Some proteins consist of more than one polypeptide; the polypeptides of these proteins associate in the cytoplasm or the Golgi apparatus to form the fully functional protein.

Many ribosomes can translate the same, single strand of mRNA, enabling many polypeptide chains to be produced at the same time. Once the polypeptides are fully functional, they either remain in the cell for use, or are exported from the cell via exocytosis for use elsewhere in the organism.

EXTENSION

Protein synthesis in prokaryotes

Prokaryotes do not have membrane-bound organelles, so all cellular processes occur within the cytosol. This allows transcription and translation to be a continuous process rather than two separate stages. Ribosomes are able to attach to the mRNA while it is being transcribed, so translation can occur at the same time. Prokaryotes only contain exons, so splicing is not necessary prior to translation.

There are many differences between protein synthesis in prokaryotic and eukaryotic cells (Figure 3.2.12). Table 3.2.1 summarises the major differences. These differences have been used to develop drugs that target protein synthesis in prokaryotes only. For example, some antibiotics disrupt or inhibit the production of proteins.

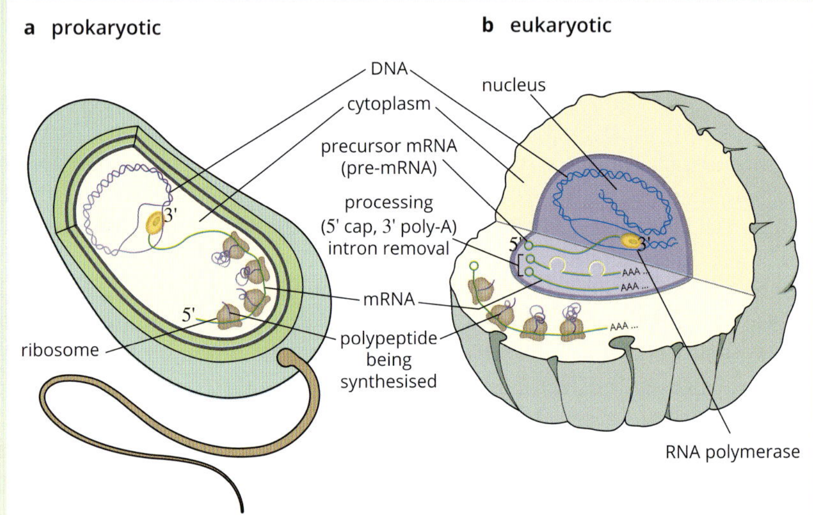

FIGURE 3.2.12 Comparison of protein synthesis in (a) prokaryotic and (b) eukaryotic cells. The structural differences between the cell types means that there are many differences in the way proteins are synthesised.

Prokaryotic protein synthesis	Eukaryotic protein synthesis
30S and 50S ribosomal subunits	40S and 60S ribosomal subunits
mRNA contains the coding sequences of several genes.	mRNA contains the coding sequence for one gene.
There is an overlap between transcription and translation, with protein synthesis beginning during transcription. This is known as coupled transcription–translation. This can occur because the DNA and ribosomes are in the cytosol together.	There is no overlap between transcription and translation. The transcription of DNA occurs in the nucleus and translation and protein synthesis occur in the cytoplasm.
Most prokaryotes contain only exons, so no mRNA processing is required.	Eukaryotes have introns and exons.
No further processing of mRNA is required after transcription.	The introns are spliced out of the mRNA before translation.
Prokaryotes have about 3 different initiation factors.	Eukaryotes have around 10 different initiation factors.
No 5' cap is added to mRNA.	A methylguanosine triphosphate molecule, called a 5' cap, is added to the 5' end of the mRNA.
No poly-A tail is added to mRNA.	A poly-A tail Is added to the 3' end of mRNA.

TABLE 3.2.1 A summary of the differences between prokaryotic and eukaryotic protein synthesis.

3.2 Review

SUMMARY

- The genetic code is the set of rules about how the instructions carried in nucleic acids are translated to synthesise proteins and functional RNA molecules. In DNA this information is stored as a three-letter code of nucleotides known as a triplet. When these triplets are transcribed into mature mRNA, they are then known as codons.
- The genetic code is universal and degenerate. There are 64 possible codons of three nucleotides (e.g. UAC) for the 20 amino acids.
- Eukaryotic genes have a number of structural features in common:
  - Stop and start instructions—These indicate where transcription starts and stops. Stop codons do not code for amino acids.
  - Promoter region—This is the site at which the RNA polymerase attaches to the gene to begin transcription (sometimes called the TATA box).
  - Exons—These are the DNA regions that are coding segments.
  - Introns—These are the DNA regions that are non-coding segments.
- Gene expression is the process in which the information stored in a gene is used to synthesise a functional gene product (protein or RNA). Gene expression is regulated so that it occurs if and when the particular protein or RNA is required by the cell.
- Protein synthesis in eukaryotes occurs in three stages: transcription, RNA processing and translation.
  - Transcription occurs in the nucleus and involves RNA polymerase transcribing the DNA into a primary RNA transcript.
  - During RNA processing in eukaryotes, a 5' cap is added to the 5' end of the primary RNA transcript and a poly-A tail is added to the 3' end. The cap and tail make the mRNA more stable and prevent it from degrading. Next, the primary RNA transcript is spliced to remove the introns, and sometimes some exons, resulting in mature mRNA. The mature mRNA then exits the nucleus.
  - Translation occurs on a ribosome. The codons on mRNA are translated into a sequence of amino acids, which are delivered by their specific tRNA molecules, to form a polypeptide chain.

KEY QUESTIONS

1 What is the genetic code?

2 What are the main structural features of eukaryotic genes and their functions?

3 What are the three stages of protein synthesis?

4 Transcription occurs in three stages: initiation, elongation and termination. Rearrange the table to match the transcription event with the correct stage at which it occurs.

Stage of transcription	Transcription event
initiation	• The RNA polymerase moves along the DNA molecule, producing a strand of mRNA. • The RNA polymerase detaches, releasing the mRNA and allowing the DNA molecule to reform.
elongation	• RNA polymerase uses a strand of DNA as a template, attaching nucleotides (A, U, G, C) by complementary base pairing. • Transcription factors combine with the region at the start of the gene, known as the promoter. • RNA polymerase reaches the termination site of the gene (stop codon) and translation ends.
termination	• RNA polymerase attaches to the promoter, unwinding and unzipping the DNA molecule by breaking the weak hydrogen bonds between the two strands to expose the bases.

5 What is splicing and alternative splicing?

6 Like transcription, translation occurs in three stages: initiation, elongation and termination. However, the events during each of these stages are different. Rearrange the table to match the translation event with the correct stage at which it occurs.

Stage of translation	Translation event
initiation	• Following the attachment of the amino acid methionine, another tRNA, with a complementary anticodon to the next codon on the mRNA attaches and adds its specific amino acid to the growing polypeptide chain. • The tRNA reaches a stop codon.
elongation	• A small ribosomal subunit attaches to the 5' end of an mRNA strand. It then moves along the mRNA until it reaches a start codon (AUG). • The polypeptide chain is released from the ribosome into the cytoplasm or the endoplasmic reticulum.
termination	• The ribosome then releases the tRNA and moves further along the mRNA strand. At each codon a new tRNA binds and adds another amino acid. • A tRNA molecule with an anticodon (UAC) brings the amino acid methionine to the mRNA. The tRNA molecule joins to the mRNA start codon, attaching by complementary base pairing between the codon and anticodon.

3.3 Gene regulation

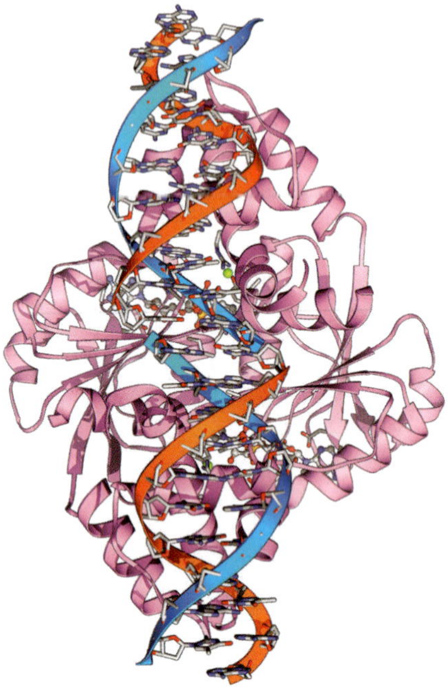

As you have learnt, the genome consists of many thousands of genes. A cell is able to express a selection of these genes at a given time. The genes expressed determine which proteins are produced, giving the cell its functionality and characteristics. Gene expression is the process through which information from a gene is used to synthesise a functional gene product—a protein or RNA. Gene expression in eukaryotes is tightly regulated by multiple mechanisms, at different points.

Although gene expression is controlled at many points, this section focuses on gene regulation at the transcription stage. You will learn the difference between structural and regulatory genes and understand how some proteins called transcription factors are able to regulate transcription by 'switching genes on and off' (Figure 3.3.1).

FIGURE 3.3.1 Transcription repressor protein (pink) bound to DNA (red and blue). The repressor protein physically blocks access to the DNA, preventing transcription of the underlying gene.

GENE REGULATION

Gene regulation is tightly controlled in both eukaryotes and prokaryotes. However, as the process of gene expression in eukaryotes is more complex, gene regulation occurs at a greater number of stages in eukaryotes than in prokaryotes.

Gene regulation in eukaryotes

In Section 3.2 you learnt that gene expression in eukaryotes comprises the processes of transcription, RNA processing and translation. Gene expression is highly controlled, and can be regulated at any of these stages (Figure 3.3.2). In eukaryotic cells, transcription and RNA processing occur within the nucleus, and translation occurs in the cytoplasm.

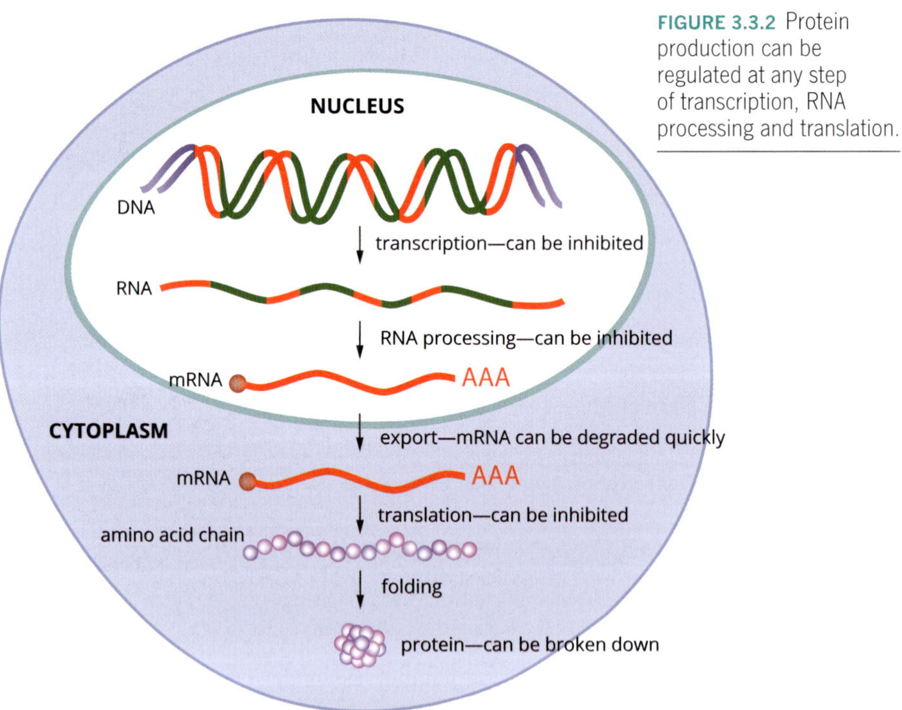

FIGURE 3.3.2 Protein production can be regulated at any step of transcription, RNA processing and translation.

Gene expression in prokaryotes consists only of transcription and translation, whereas in eukaryotes, it involves transcription, RNA processing and translation.

Gene regulation in prokaryotes

In prokaryotic cells, gene expression consists only of transcription and translation and occurs in the cytoplasm of cells. (Prokaryotic cells do not have a nucleus or any other membrane-bound organelles.) Here, transcription and translation occur at almost the same time. Gene expression in prokaryotes is regulated during transcription, which will be the focus of this section.

REGULATORY AND STRUCTURAL GENES

Some genes are expressed constitutively (continually), while other genes can be **induced** or **repressed**.

- **Constitutive genes** are always switched on: they are transcribed continually.
- For other genes, transcription may be induced or repressed by transcription factors as needed, depending on the cell type, stage or environmental conditions.

Regulatory genes code for transcription factors. **Transcription factors** are proteins that control gene expression at the transcription stage. They bind to DNA sequences close to the promoter region of a gene or to the RNA polymerase to induce or repress the expression of specific genes (Figure 3.3.3).

Structural genes code for proteins and RNAs that are not involved in gene regulation. For example, they can code for enzymes, protein channels, protein components for the cytoplasmic skeleton, or tRNA, among others.

Regulatory genes control the expression of structural genes via the production of transcription factors.

Structural genes code for any RNA or protein that are not involved in gene regulation.

Genes may be switched on all the time (constitutively expressed) or may be induced or repressed as needed by transcription factors.

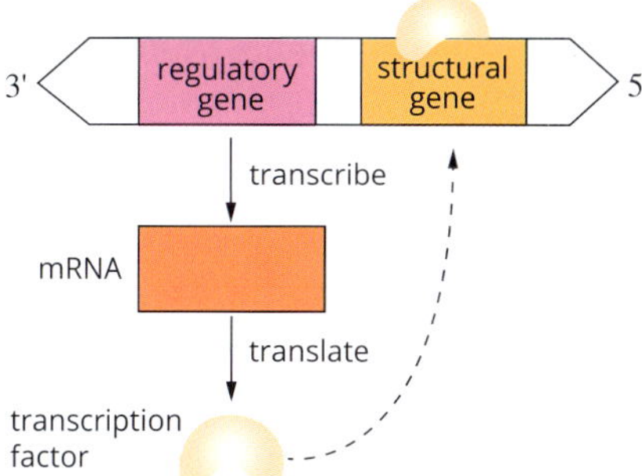

FIGURE 3.3.3 Regulatory genes code for transcription factors that induce or repress structural genes. This diagram shows an example of a transcription factor acting as an inhibitor of transcription.

THE *LAC* OPERON—A MODEL FOR GENE REGULATION

Gene regulation in eukaryotes is quite complex. Therefore, to illustrate how transcription factors regulate gene transcription, a simple prokaryotic model, the *lac* operon, is commonly used (Figure 3.3.4).

The ***lac* operon** is found in *Escherichia coli* (*E. coli*) and some other bacteria. In prokaryotes, an **operon** is a unit of DNA under the regulation of a single promoter that codes for several proteins. The *lac* operon is an example of an **inducible operon** (it can be switched on or off). It expresses three structural genes that code for three enzymes, but only when the sugar **lactose** is available. The enzymes break down lactose into the useable forms glucose and galactose.

Producing the enzymes that break down lactose constitutively would be a misuse of energy for the bacteria because lactose is not the preferred energy source for *E. coli*. Therefore, it is important that the *lac* operon genes are inducible and are expressed only in response to the presence of lactose in the environment.

The *lac* operon consists of:

- a promoter—the binding site of the RNA polymerase
- an **operator**—the binding site of the transcription factor, which, in this case, is a repressor
- three structural genes—*lacZ* (β-galactosidase), *lacY* (β-galactoside permease) and *lacA* (β-galactoside transacetylase)—that code for three different enzymes (Figure 3.3.4).

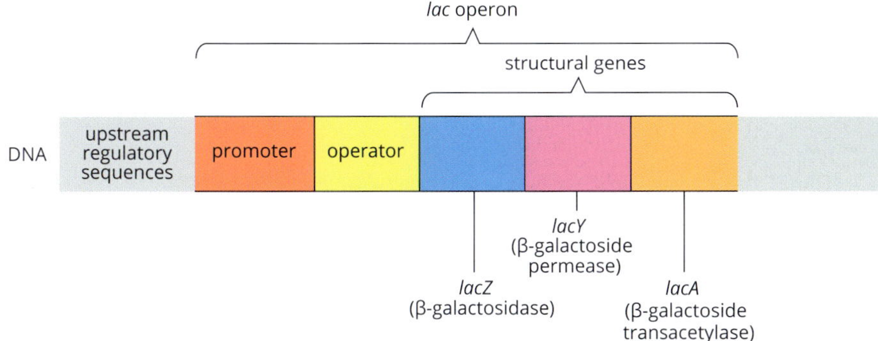

FIGURE 3.3.4 The prokaryotic *lac* operon. The structural genes in *E. coli* that encode for enzymes involved in lactose metabolism are under the control of one promoter.

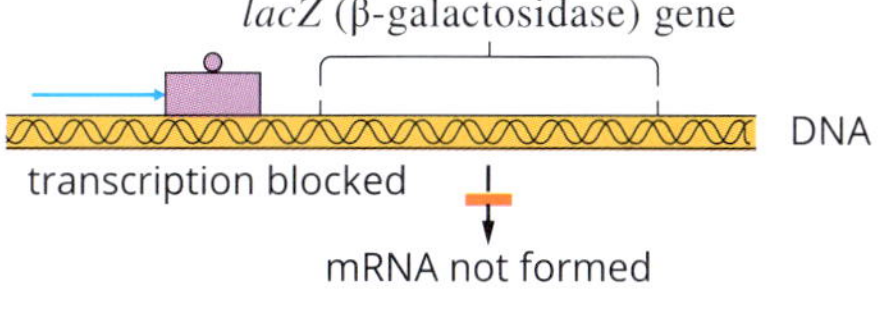

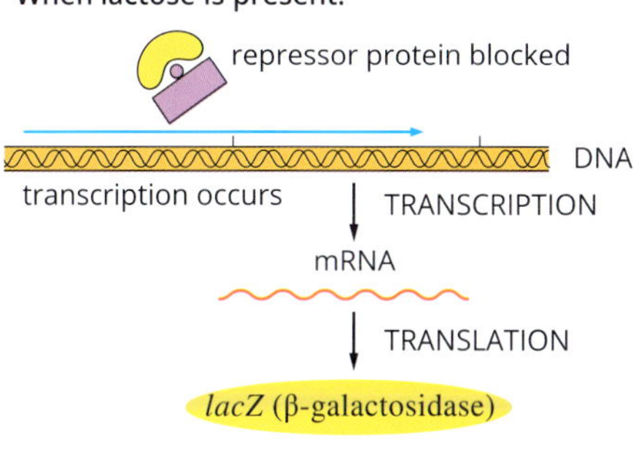

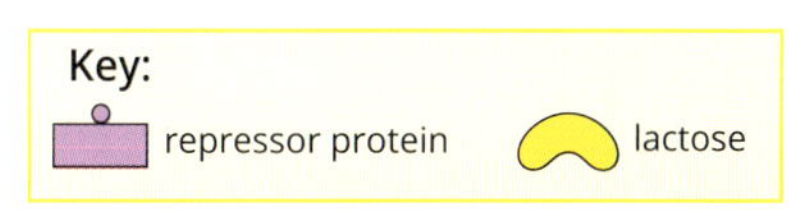

FIGURE 3.3.5 The genes of the *lac* operon are only expressed in the presence of lactose.

Adjacent to the *lac* operon is the regulatory gene ***lacI***. *LacI* codes for a transcription factor called the ***lac* repressor**. The *lacI* gene is expressed constitutively, so the *lac* repressor is always present. This transcription factor binds to the operator in the *lac* operon, blocking the RNA polymerase from binding to and transcribing the structural genes in the *lac* operon, thereby preventing the synthesis of the three enzymes involved in lactose metabolism (Figure 3.3.5).

However, when lactose is present, the lactose binds to the *lac* repressor, inhibiting the transcription factor from binding to the operator. This enables the RNA polymerase to attach to the promoter in the *lac* operon and transcribe the three structural genes, resulting in the production of the enzymes involved in lactose metabolism (Figure 3.3.5).

BIOLOGY IN ACTION

Master regulatory genes

The development of a complex, trillion-celled adult organism from a single fertilised cell occurs in a series of steps. Master regulatory genes code for transcription factors that turn genes on and off in different cells in the developing embryo. They can also start a sequence of events by turning other regulatory genes on and off, leading to the production of transcription factors that will, in turn, regulate other genes, and so on. A single master regulatory gene can, in this way, control the development of a complex structure such as an eye or nervous system, or a whole organism.

Some of the most important master regulatory genes are the *Hox* genes. These genes control the structure and organisation of body segments during embryonic development (Figure 3.3.6). The proteins of *Hox* genes bind to regulatory regions of target genes, which then activate or repress cellular activity, directing the development of the organism. This important gene family has been highly conserved throughout animal evolution. Because of the high degree of genetic similarity in these genes across animal species, researchers can use model organisms, such as *Drosophila*, to investigate human birth defects and diseases.

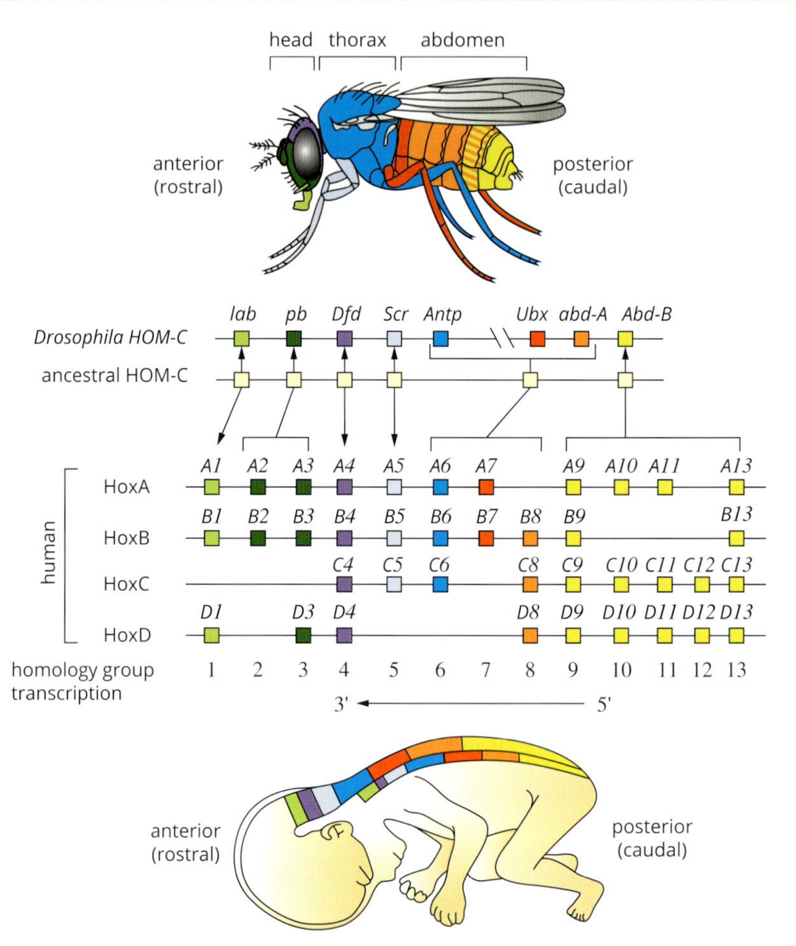

FIGURE 3.3.6 The master regulators of embryonic development in all animals are *Hox* genes. These genes code for the body plan of an embryo along the head to tail axis. This figure shows the genomic organisation and expression patterns of the *Hox* gene complex in *Drosophila* and humans.

BIOFILE

Tumour-suppressor proteins

Tumour-suppressor proteins function to prevent cancer. These proteins are produced in response to exposure to radiation and chemicals that cause damage to the structure of DNA.

An important tumour-suppressor protein is p53 (Figure 3.3.7). If DNA damage has occurred, the p53 protein binds to specific sites on the DNA to repress genes that play a role in the continuation of the cell cycle. This inhibits cell division and prevents damaged DNA from replicating. If the damage is minor, p53 activates genes that repair DNA. In cases where the damage cannot be repaired, p53 will initiate cell death (apoptosis).

p53 plays a major role in the prevention of cancers. If the gene coding for the p53 protein is deleted or mutated, the risk of cancer is greatly increased. In 50% of all cancers, p53 has been found to be inactive.

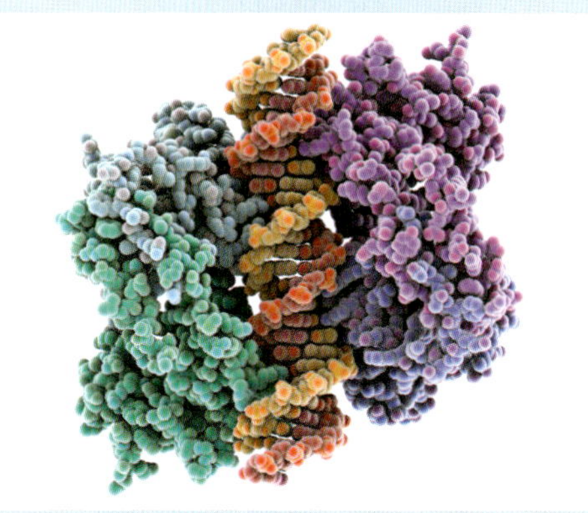

FIGURE 3.3.7 A molecular model of the tumour suppressor protein p53 (left and right) bound to a DNA molecule (centre).

EUKARYOTIC VS PROKARYOTIC GENETIC MATERIAL

Prokaryotic and eukaryotic genes are structurally different in many ways. These differences affect the way in which genetic information is transcribed, translated and expressed. Table 3.3.1 summarises the major differences between the structure of the genetic material of prokaryotes and eukaryotes.

Prokaryote	Eukaryote
There is one chromosome per cell.	There are multiple chromosomes per cell.
It has a circular chromosome without ends (no telomeres).	It has linear chromosomes with ends (with telomeres).
Contains plasmids—small, circular DNA.	Contains no plasmids but there are other sources of DNA apart from chromosomes—mitochondrial DNA and chloroplast DNA.
There is much less DNA than in eukaryotes (thousands to millions of bases, depending on species).	There is much more DNA than in prokaryotes (millions to billions of bases, depending on species).
There are fewer genes than in eukaryotes (thousands).	There are more genes than in prokaryotes (tens of thousands).
There is less non-coding DNA than in eukaryotes (greater number of genes per number of bases).	There is more non-coding DNA than in prokaryotes (fewer genes per number of bases).
DNA is not packaged into an organelle (less DNA to fit into the cell) (Figure 3.3.8).	DNA is tightly packaged—coiled around histones, which form nucleosomes, which are condensed into chromatin and packed as chromosomes into the nucleus (a lot of DNA to fit into a small space) (Figure 3.3.9, page 156).
Genes cluster into functional groups, known as operon regions (e.g. genes that code for enzymes in the same biochemical pathway are next to each other on the chromosome and are controlled by the same promoter, so all the genes for the pathway can be transcribed and expressed at once).	There is no operon region. (Genes that code for functionally similar enzymes can be physically far apart or located on different chromosomes. Eukaryotes have mechanisms to express these genes at the same time.)

TABLE 3.3.1 Summary of the major differences between the structure of the genetic material of prokaryotes and eukaryotes.

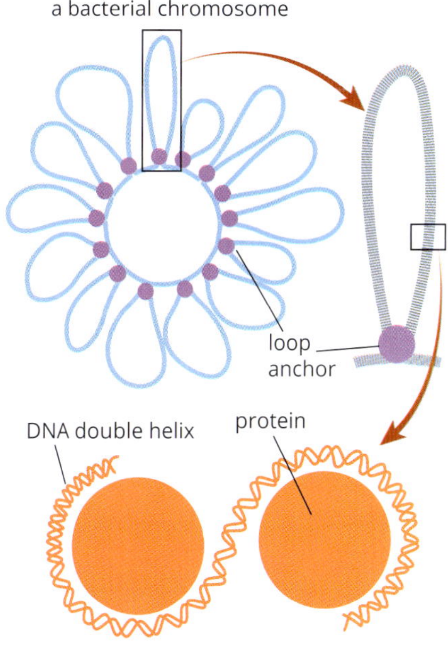

FIGURE 3.3.8 DNA is packaged into loop structures in prokaryotes. The DNA of prokaryotes does not have to be as tightly packaged as in eukaryotes because there is much less genetic material.

Histones are proteins found in eukaryotic cells that tightly package DNA into structures called nucleosomes.

Although there are many differences between the gene structures of prokaryotes and eukaryotes, shared evolutionary history means that there are also many fundamental similarities.

- Both prokaryotes and eukaryotes have double-stranded DNA that is made up of the nitrogenous bases adenine (A), thymine (T), cytosine (C) and guanine (G).
- Both have mRNA, which acts as an intermediate code to building proteins, with uracil (U) replacing thymine (T).
- Because the genetic information of prokaryotes and eukaryotes is composed of the same code, the way mRNA codons are translated into amino acids and proteins is also much the same.

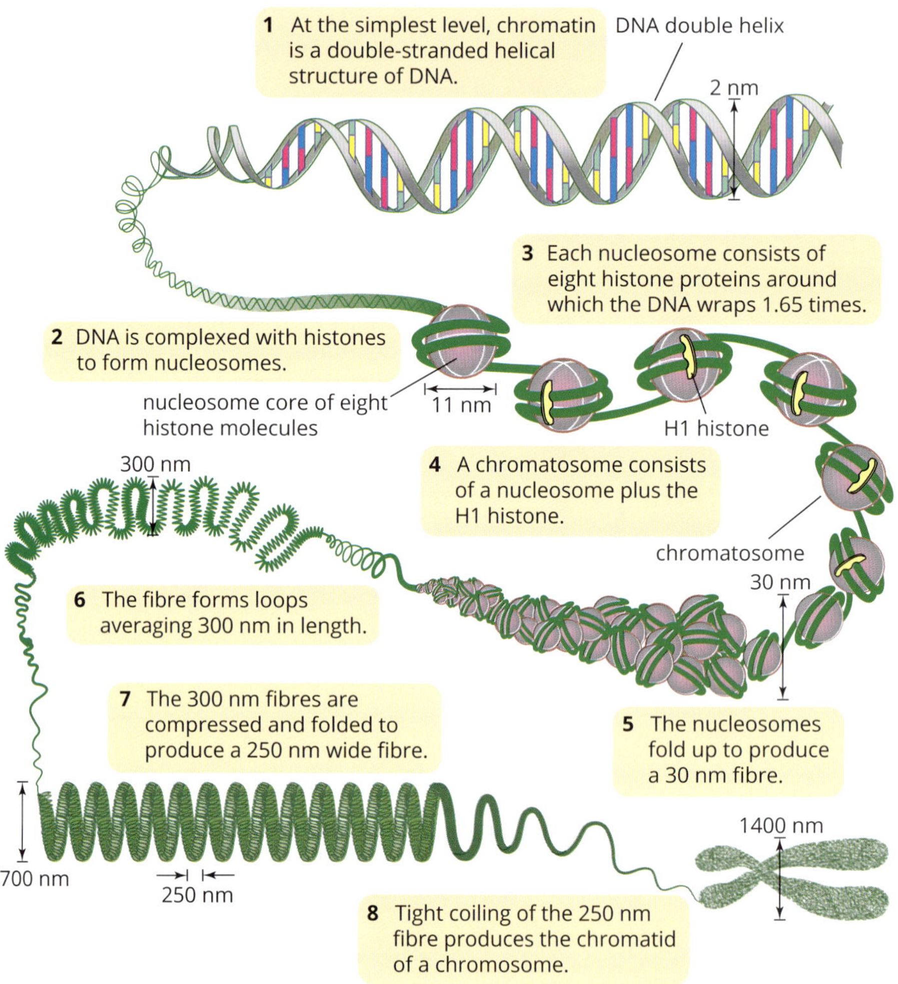

FIGURE 3.3.9 The packaging of DNA in eukaryotic cells. Because eukaryotes have large quantities of DNA to fit into a small space, the DNA needs to be tightly and efficiently packaged. Double-stranded DNA (1) is tightly coiled around histones to form nucleosomes (2 and 3). Nucleosomes and histones together form chromatosomes (4). The nucleosomes fold (5), loop (6) and compress (7) into chromatin. Tight coiling of the chromatin produces the chromatids of a chromosome (8).

3.3 Review

SUMMARY

- Gene expression in eukaryotes can be regulated at any of the three stages: transcription, RNA processing and translation.
- Gene regulation in prokaryotes occurs during transcription.
- Constitutive genes are expressed continually.
- Transcription of other genes can be induced or repressed as needed by transcription factors, which is a form of gene regulation.
- Regulatory genes code for the production of transcription factors.
- Transcription factors are proteins that control gene expression at the transcription stage. They induce or repress the expression of specific genes by binding to DNA sequences close to the promoter region of a gene or to the RNA polymerase.
- Structural genes code for proteins and RNAs that are not involved in gene regulation; for example, enzymes.
- The *lac* operon in *E. coli* provides an example of a unit of DNA for which transcription can be induced or repressed—or in other words, regulated. The regulatory gene *lacI* constitutively expresses a transcription factor called the *lac* repressor. The *lac* repressor binds to the operator of the *lac* operon, inhibiting transcription, unless lactose is present. In the presence of lactose, the lactose binds to the *lac* repressor, inhibiting it from binding to the *lac* operon, and enabling the RNA polymerase to bind to the *lac* operon and begin transcription.
- There are many differences in the structure of the genetic information of prokaryotes and eukaryotes, e.g. prokaryotes have much less DNA but more genes per number of bases than eukaryotes. Because eukaryotes have much more DNA, they have to package it more tightly. These structural differences affect the way in which genetic information is transcribed, translated and expressed.
- Although there are many differences, the basic genetic structures of prokaryotes and eukaryotes share many similarities, e.g. the same code of nitrogenous bases (A, T, C, G and U) translates into amino acids and proteins in much the same way in prokaryotes and eukaryotes.

KEY QUESTIONS

1 What are some of the main differences in gene regulation between prokaryotic and eukaryotic cells?

2 Define the following terms:

a transcription factors
b operator
c regulatory gene
d structural gene
e constitutive gene
f induced gene
g repressed gene

3 Complete the diagram of this prokaryotic gene by inserting the names of the gene regions in the correct locations.
structural genes, termination site, promoter, operator

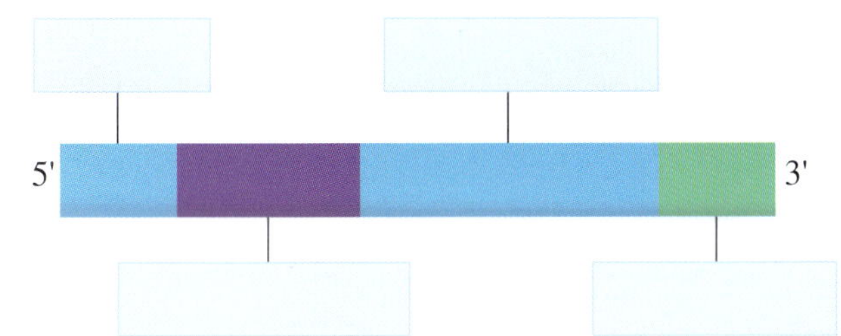

4 Complete the table by inserting in the correct order the processes that occur in *Escherichia coli* in the presence and absence of lactose.

Presence of lactose in the environment	Absence of lactose in the environment

- inhibition of gene expression
- production of the repressor
- intake of lactose into the cell
- repressor binds to the operator
- inhibition of repressor activity
- breakdown of lactose
- gene expression

5 List five basic differences between prokaryotic and eukaryotic gene structure.

Chapter review

03

KEY TERMS

5' cap
adenine (A)
anticodon
antiparallel
base
biomolecule
coding strand
codon
complementary base pairing
constitutive gene
cytosine (C)
degenerate
deoxyribose
dinucleotide
double helix
exon
gene
gene expression
gene regulation
guanine (G)
induced
inducible operon
intron
lac operon
lac repressor
lacI
lactose
messenger RNA (mRNA)
nucleotide
operator
operon
phosphodiester bond
poly-A tail
polynucleotide
promoter
protein synthesis
purine
pyrimidine
regulatory gene
repressed
ribose
ribosomal RNA (rRNA)
RNA polymerase
RNA processing
spliceosome
splicing
structural gene
TATA box
template strand
thymine (T)
transcription
transcription factor
transfer RNA (tRNA)
translation
triplet
uracil (U)

KEY QUESTIONS

1 What are the sequences that are included in the final mRNA product called?

A introns
B termons
C exons
D spliced codons

2 In polypeptide synthesis, the function of the ribosome is to:

A synthesise the required amino acids
B ensure that the DNA base sequence is complete
C provide the energy needed for the synthesis
D provide the site for the synthesis

3 DNA provides the code for the synthesis of polypeptides. Which one of the following statements is true?

A Every codon codes for its own exclusive amino acid.
B The code is read as sets of three bases called triplets.
C Each triplet codes for at least two different amino acids.
D There are 20 different amino acids therefore there are 20 different codons.

4 Which of the following statements about histones is false?

A All cells need histones to package their DNA.
B Chromatosomes are made from nucleosomes joined to histones.
C A nucleosome consists of 8 histones with DNA wrapped around them.
D Without histones the eukaryotic cell's DNA would be unable to fit into the nucleus.

5 Geneticists used DNA sequencing to discover the base sequence of a plasmid. If 27% of the bases in the plasmid were cytosine then the plasmid consisted of:

A 27% cytosine–guanine pairs
B 46% adenine–thymine pairs
C 46% cytosine–guanine pairs
D 23% adenine–thymine pairs

6 RNA comes in different forms, and each form has a specific role in the synthesis of proteins. What form of RNA is specifically involved with carrying amino acids to ribosomes for assembly into polypeptides?

A mRNA
B rRNA
C tRNA
D none of the above

7 A promoter is:

A a specific sequence of DNA to which a repressor may bind

B a specific sequence of DNA to which DNA polymerase may bind

C a specific sequence of DNA to which RNA polymerase may bind

D a specific sequence of RNA in mRNA

8 The nucleotide sequence A G U G A C C A A could represent:

A part of the DNA template of a particular gene

B the amino acid chain of a polypeptide

C a sequence of mRNA

D a section of double helix

To answer questions 9–11, refer to the genetic code in the table below. With a few rare exceptions, the genetic code is accepted as being universal.

First position (5' end)	Second position				Third position (3' end)
	U	C	A	G	
U	Phe Phe Leu Leu	Ser Ser Ser Ser	Tyr Tyr STOP STOP	Cys Cys STOP Trp	U C A G
C	Leu Leu Leu Leu	Pro Pro Pro Pro	His His Gin Gin	Arg Arg Arg Arg	U C A G
A	Ile Ile Ile Met	Thr Thr Thr Thr	Asn Asn Lys Lys	Ser Ser Arg Arg	U C A G
G	Val Val Val Val	Ala Ala Ala Ala	Asp Asp Glu Glu	Gly Gly Gly Gly	U C A G

9 a The following flow chart represents the production of a polypeptide chain as directed by the DNA template.

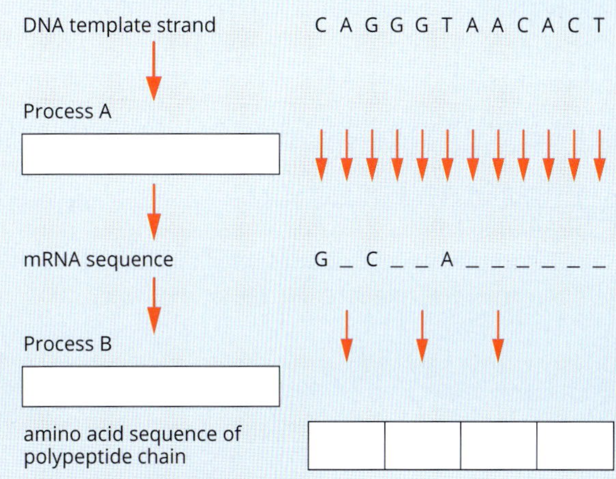

i Name processes A and B.

ii Complete the mRNA sequence.

iii Use the genetic code to determine the correct sequence of amino acids.

b i How is the fourth codon different from all of the others in this sequence?

ii Outline the significance of this codon in terms of polypeptide production.

10 The DNA sequence of a particular gene is shown below.

TAC - GGA - TCT - AGA	- ATA - AAA -	CGG - AAT - GCT	- GGG -	ACA - CGG - GTA - ACA
exon 1		exon 2		exon 3

a What do the bright blue sections represent?

b Show the strand of mature mRNA that would be produced using this gene.

c Decode the mRNA.

d What protein would be formed if an alternative splice, skipping of exon 2, were to occur?

11 The DNA sequence of the promoter and first exon of a gene are shown below.

G G G C T C T A T A A A G G G T A C C A C T T C A A T G C T

a Identify the TATA box.

b Explain where RNA polymerase will start transcription.

c The mRNA strand produced is shown below:
A U G G U G A A G U U A C G A

i Explain whether the DNA strand shown is the coding strand or the template strand.

ii Decode the strand.

d Suggest the likely consequence for the organism of a mutation that changes the sequence of the TATA box.

12 In order to acquire the nucleotides needed for the production of DNA and RNA in cells, humans must have nucleic acids in their diet. These nucleic acids must be broken down into nucleotides so that they can be absorbed from the digestive tract and into the bloodstream so they can be carried to the cells that need them.

The breakdown of the nucleic acids into nucleotides requires the input of water.

a Suggest why this is the case.

b A dinucleotide is shown at right.
Show the two nucleotides after digestion into single nucleotides. Indicate the fate of the water.

13 **a** Use the following terms to label the structures A–I on the diagram at right:
amino acid, anticodon, loaded tRNA, mRNA, polypeptide, pre-mRNA (primary mRNA transcript), ribosome, RNA polymerase, unloaded tRNA

b Describe the role of structure B in protein synthesis.

c Describe the process occurring in structure H. Ensure you include the significance of structure I in your discussion.

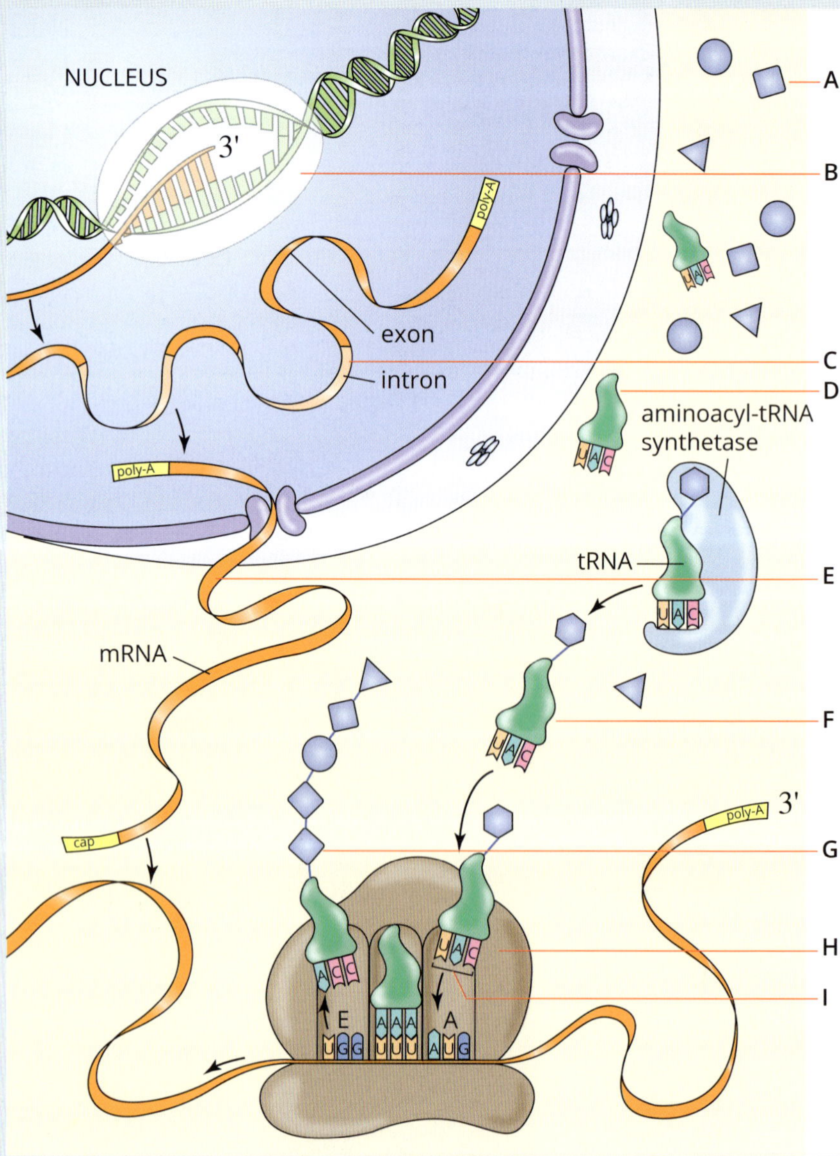

14 A nucleic acid strand is under investigation. It has been found to contain 29% A, 32% G and 17% C.

a Is the fourth base uracil or thymine?

b How do you know?

c Explain what percentage of the strand of nucleic acids is the fourth base?

15 A strand of nucleic acid is shown below.

AUG AAU CCU UAU GGU GGC UUU UAA

The protein produced as a result of the information encoded in the strand of nucleic acids is shown below.

met–asn–pro–phe

a Explain whether the strand given is DNA, pre-mRNA or mRNA.

b i During translation of the strand of nucleic acid shown, a tRNA having the anticodon UUA approached the ribosome. Which amino acid would the tRNA have been carrying?

ii Draw the tRNA molecule with its amino acid and anticodon.

16 In Section 3.2 you learnt that the longest known gene is the dystrophin gene which is 2.5 megabases long and that this gene is 99% introns.

a What is the maximum number of bases in the exons of the dystrophin gene?

b How many amino acids (approximately) make up the protein dystrophin?

17 Genetic engineering is used to transform bacteria by inserting human genes into their genome in order to produce human polypeptides, such as those that form insulin. Before the bacterium can be transformed, a copy of the human gene is required. A common method of acquiring the gene is to extract the appropriate mRNA from human cells and to use it as a template to make a DNA copy. This cDNA (complementary DNA) is then introduced into the bacterium, which then produces the required protein.

a Why can it be expected that a bacterium is able to decode a human gene and produce the correct protein?

b A gene made of cDNA is better for use in a bacterium that a gene cut directly from a human chromosome. Why?

c Another method of obtaining an appropriate gene is to analyse the protein needed, identify the amino acid sequence and construct a suitable section of DNA. Explain whether the gene created by this method is likely to be identical to the cDNA sequence made using mRNA as a template.

d Consider the amino acid sequence leu–pro–val. How many different DNA sequences would result in this amino acid chain? Explain how you arrived at your answer.

18 Tryptophan is an amino acid which prokaryotic cells are able synthesise when it is in short supply. Tryptophan production is controlled by a series of 5 enzymes that are coded for by 5 genes (*Trp A, B, C, D* and *E*), which form a single unit called the tryptophan operon. This operon has been extensively studied in the bacterium *Escherichia coli*.

a What is an operon?

The *trp* operon with some of its upstream region is shown below.

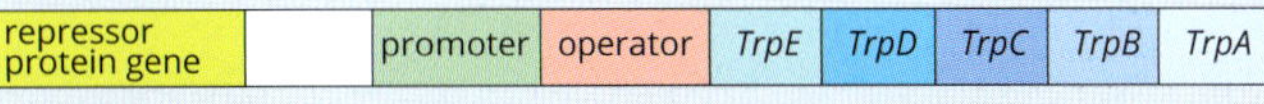

b Label the regulatory and structural genes on the diagram.

c The gene for the repressor protein is constitutively expressed. Explain the significance of this for the cell.

d The repressor protein is produced in an inactive form. It becomes activated when it binds to tryptophan. Explain how these features of the repressor protein benefit the organism.

e Once activated the repressor protein binds to the operator region. Propose how the binding of the repressor to the operator stops production of tryptophan.

f Identify a difference in the means of repression of the *lac* operon and the *trp* operon.

19 In recent genetics research, scientists replaced the gene controlling eye development in *Drosophila* flies with the gene that controls eye development in mice. The transgenic *Drosophila* developed normal compound fly eyes. What does this observation suggest about:

a the gene controlling eye development in *Drosophila* and mice?

b the factors that control eye development in these two vastly different species of insect and mammal?

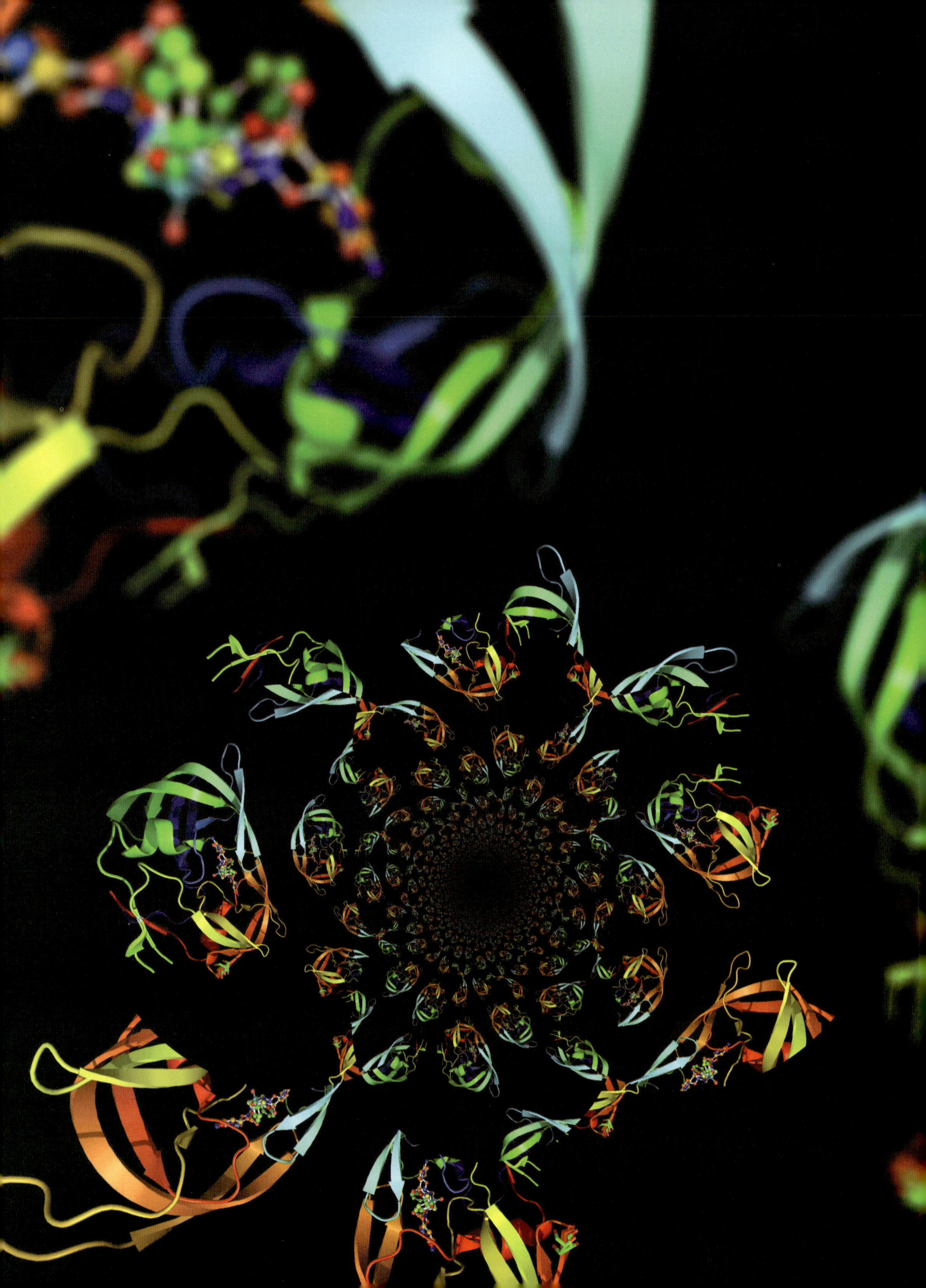

CHAPTER 04

Enzyme regulation in biochemical pathways

Learning outcomes

By the end of this chapter, you will understand the role and action of enzymes in catalysing biochemical reactions, how enzyme activity is regulated, and that enzyme regulation in turn regulates biochemical pathways. You will also understand coenzymes and their importance in moving energy, protons and electrons between reactions in the cell.

Key knowledge

- the role of enzymes as protein catalysts in biochemical pathways
- the mode of action of enzymes including reversible and irreversible inhibition of their action due to chemical competitors at the active site, and by factors including temperature, concentration and pH
- the cycling of coenzymes (ATP, NADH and NADPH) as loaded and unloaded forms to move energy, protons and electrons between reactions in the cell.

4.1 Enzymes and biochemical pathways

In this section, you will learn about the features of enzymes, including their specificity for particular substrates, and how they interact with substrates to catalyse biochemical reactions. You will also learn about the importance of enzymes in biochemical pathways.

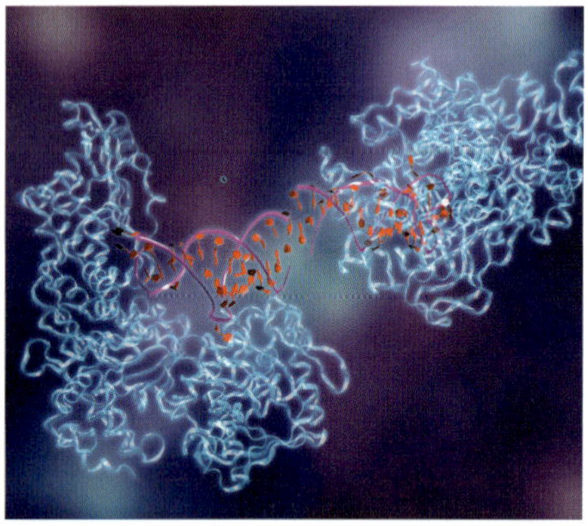

FIGURE 4.1.1 In this digital illustration, a metabolic enzyme called aconitase (blue) has formed a complex with ferritin messenger ribonucleic acid (mRNA, red). Aconitase is an important enzyme in the citric acid cycle, where it interconverts citrate and isocitrate. However, aconitase also helps to regulate iron levels, in which case it is known as iron regulatory protein 1 (IRP-1). By binding to ferritin mRNA (ferritin is a protein that stores iron), IRP-1 prevents translation and the production of ferritin.

ENZYME FEATURES

Most enzymes are globular proteins that have a tertiary or quaternary structure. The main features of enzymes are their specificity for a substrate and their **catalytic power**:

- Specificity—Different enzymes act as catalysts for different biochemical reactions by binding to a specific type of molecule called a **substrate**. Although many enzymes have evolved to be highly specific, and to act on a single substrate and catalyse one specific reaction, some enzymes are able to act on multiple substrates (Figure 4.1.1) and catalyse multiple reactions.
- Catalytic power—Enzymes do not make reactions occur that would not occur on their own; they only make reactions occur more quickly (sometimes over a million times more quickly).

ⓘ Catalytic power is the ability or potential of an enzyme to increase the rate of a biochemical reaction compared to the reaction occurring without the enzyme present.

ⓘ A substrate is a molecule upon which an enzyme acts.

ⓘ Enzymes are not consumed when they catalyse reactions.

Enzyme specificity

A key structure of enzymes is their **active site**. This is a pocket or groove-like part of the enzyme formed by the tertiary folding of the protein. Each enzyme's active site is a complex three-dimensional shape that interacts with a specific substrate to catalyse a specific reaction. When the active site binds to the substrate, it forms an **enzyme–substrate complex** (Figure 4.1.2).

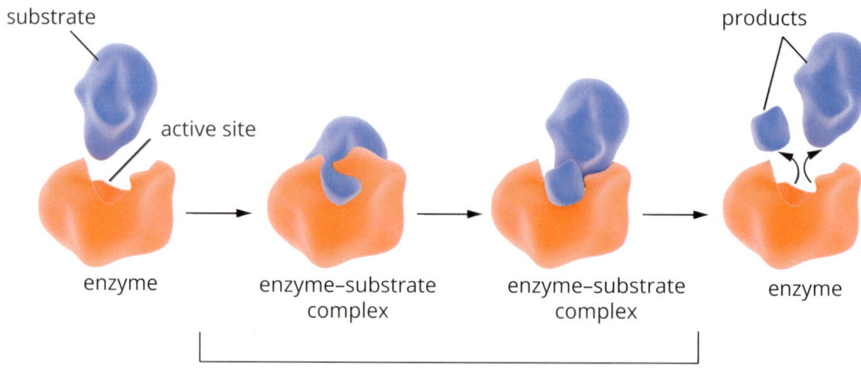

FIGURE 4.1.2 The stages of an enzyme–substrate interaction in a reaction. From left, the substrate molecule (blue) attaches to the active site on the enzyme (red), forming an enzyme–substrate complex (centre left). This complex goes through a transition state where the substrate molecule is strained, resulting in it being broken into two (centre right). The final stage (far right) is the release of the product molecules.

Enzyme–substrate interaction models

Multiple hydrogen bonds and hydrophobic interactions form between the substrate and the active site within the enzyme to stabilise the substrate in the active site. There are two models that describe how enzymes and their substrates interact:

- the lock and key model
- the induced-fit model.

ⓘ 'Hydrophobic' describes a non-polar molecule, or part of a molecule, that is unable to form energetically favourable reactions with water molecules, making it unable to dissolve in water.

The lock and key model describes the active site and the specific substrate as fitting together like a lock and key (Figure 4.1.3a). If the 'key' (the substrate), does not fit into the 'lock' (the active site), then no reaction occurs.

The induced-fit model states that when a substrate binds to the active site of an enzyme, a change in shape (or **conformational change**) of the active site occurs. This model is a more accurate representation of enzyme–substrate interactions, because we know that the active site is flexible and capable of changing its shape in order to conform to the shape of the substrate and achieve a tighter fit (Figure 4.1.3b).

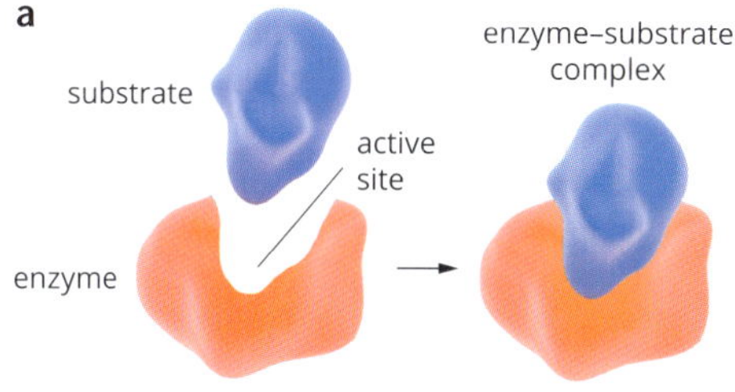

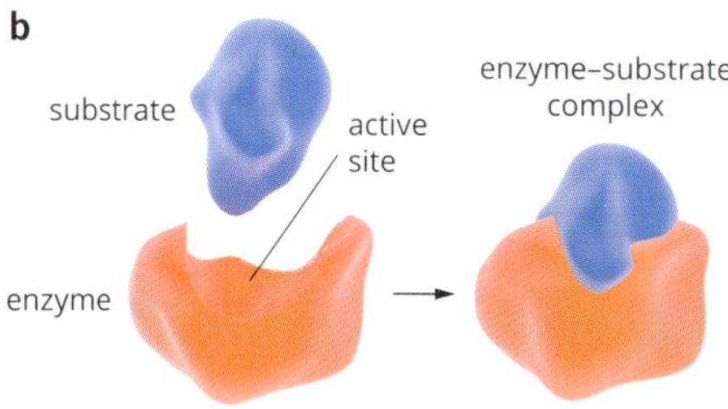

FIGURE 4.1.3 There are two major models of enzyme–substrate interaction: (a) The lock and key model describes the active site and substrate as fitting together like a lock and key. (b) The induced-fit model describes the active site as being more flexible and changing shape to fit the substrate more tightly.

Enzyme catalysis

Reactions are often reversible and so can often be catalysed in both directions (substrate → product, and product → substrate). However, this is not always the case. For example, some of the reactions in glycolysis are reversible but others are not.

Usually different enzymes catalyse a reaction in each direction. For example, DNA polymerase builds DNA and DNAase breaks it down. The direction of the reaction will depend on the concentration of substrates and products, as well as on the energy requirements.

Enzymes are not consumed when they catalyse reactions; they do not form part of the products of the reactions they catalyse. In other words, at the end of a reaction enzyme molecules are the same as they were at the beginning. This means that enzymes can be reused over and over again.

EXTENSION

Ribozymes

Although they are composed of only four nucleotides that are chemically similar, some ribonucleic acid (RNA) molecules can fold into three-dimensional structures that often serve as binding sites for proteins that function with the RNA molecules. In the 1980s, however, it was discovered that some RNA molecules catalyse biochemical reactions on their own, without the assistance of proteins.

The discovery of these RNA enzymes won Sidney Altman and Thomas Cech the Nobel Prize for Chemistry in 1989, signalled the end of the long-held belief that all enzymes are proteins, and lent support to the 'RNA world' hypothesis, which suggests that RNA appeared before DNA and proteins, and was crucial to the evolution of self-replicating systems.

Catalytic RNA molecules (or ribozymes) are considered enzymes because they act as catalysts and enhance the rate of reaction, and also they do not change as a result of the reaction. Since their discovery, several ribozymes have been identified, and most have been found to catalyse their own cleavage, or the cleavage of other RNAs.

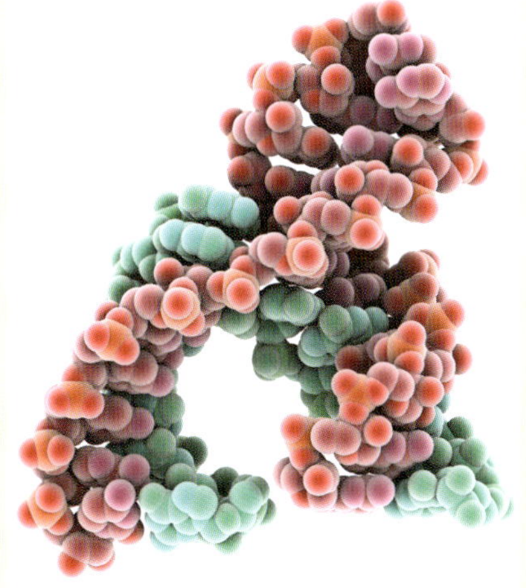

FIGURE 4.1.4 A molecular model of a hammerhead ribozyme.

For example, viroids are a type of self-cleaving circular ribozyme that has detrimental effects on plants. The damage to the plant comes from the proliferation of viroids, which uses up nucleotides that the plant needs, and from viroid bundles, which are like tumours and can interfere with the internal structures of plants. Researchers have identified the site of self-cleavage in viroids. The site is less than 30 nucleotides long and has three stems branching off from a central loop. Ribozymes with this characteristic self-cleaving secondary structure are called hammerheads (Figure 4.1.4).

Although viroids damage plants, other ribozymes may have therapeutic applications. Hammerheads as small as 19 nucleotides have now been synthesised to act as very specific catalysts, and similar ribozymes are being developed to break down RNA viruses such as HIV, or the RNA needed for the transcription and translation of DNA that contains mutations that cause genetic disorders.

Enzymes reduce activation energy

All reactions need an input of energy to start. This is called the **activation energy**. Whether a reaction releases or consumes energy, activation energy is needed for the reaction to start. The catalytic power of enzymes comes from their ability to reduce this activation energy, so that less energy is required for the reaction to occur (Figure 4.1.5).

Enzymes reduce the activation energy required for a reaction by influencing:

- proximity and orientation. Enzymes bring the parts of the molecules involved in the reaction closer to each other in the active site and position them where a reaction is more likely to occur.
- the micro-environment. Most active sites are hydrophobic. The absence of water results in a non-polar environment, allowing stabilising interactions such as hydrogen bonds, and hydrophobic and van der Waals interactions to occur.
- ion exchange. The amino acids in the active site can often take H^+ ions from, or donate them to, the substrate, to facilitate steps in certain reactions.

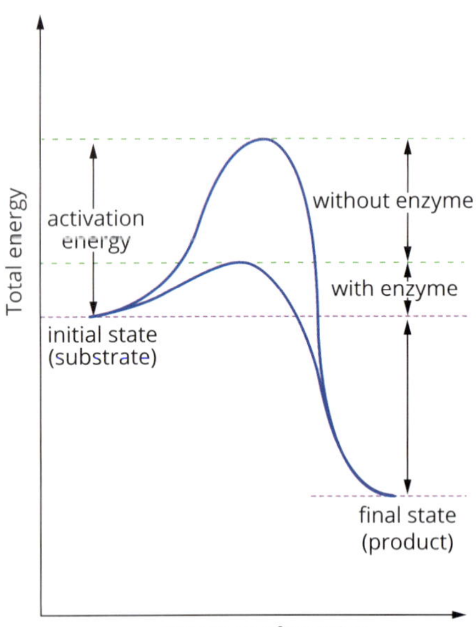

FIGURE 4.1.5 The addition of a catalyst reduces the amount of energy needed to initiate a reaction.

ENZYMES REGULATE BIOCHEMICAL PATHWAYS

Many chemical processes occur as a sequence of reactions, in which each reaction is catalysed by a specific enzyme and the product of one reaction becomes the substrate in the next reaction (Figure 4.1.6). Such sequences of biochemical reactions form biochemical pathways. Some biochemical pathways are linear, some are branched (leading to many final products), and others are cycles. An example of a cyclic pathway is shown in Figure 4.1.7.

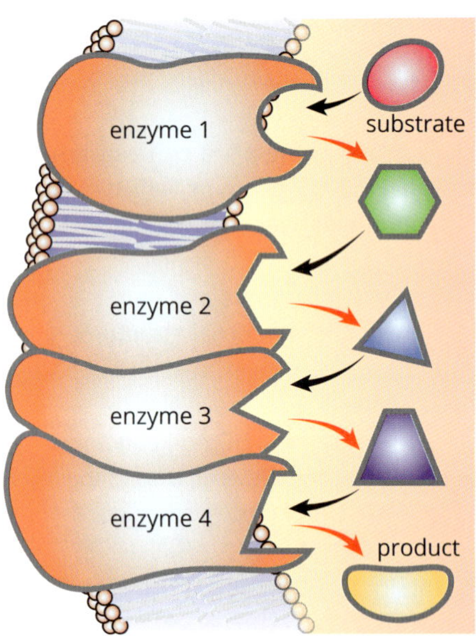

FIGURE 4.1.6 A biochemical pathway is a sequence of biochemical reactions catalysed by different enzymes. The product (indicated with the red arrow) of each reaction becomes the substrate (indicated by the black arrow) in the next reaction.

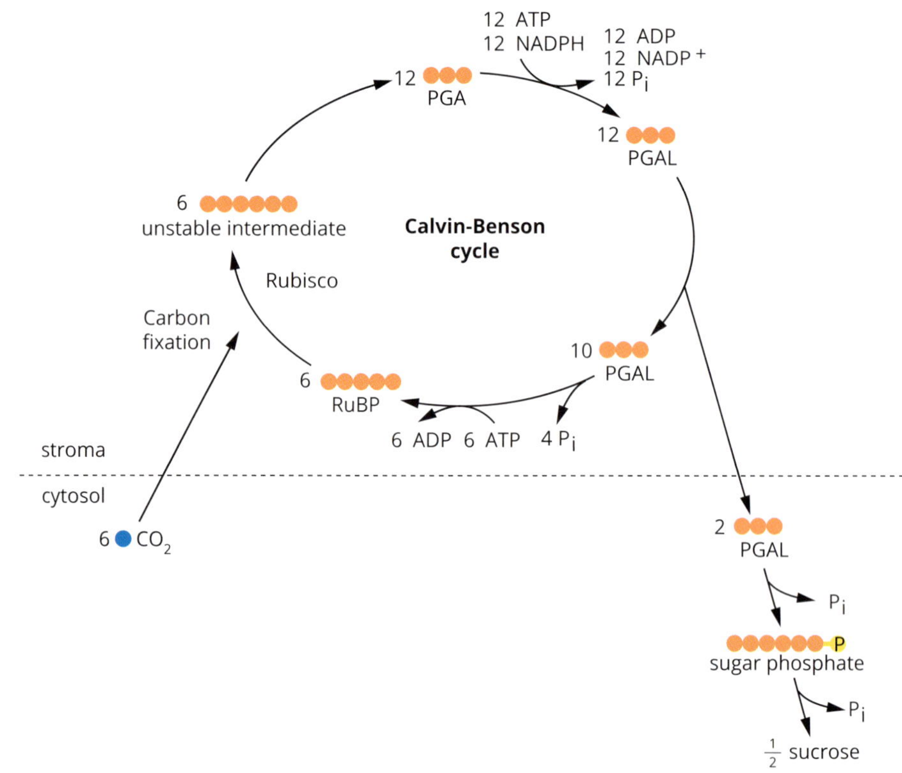

FIGURE 4.1.7 The Calvin-Benson cycle for carbon fixation is a biochemical pathway that occurs in the chloroplasts of plants and results in the production of organic compounds.

Metabolism

Metabolism is the collection of all the biochemical (or metabolic) reactions that occur in living cells. Metabolic reactions can be catabolic or anabolic:

- **Catabolic reactions** are reactions in which substrates are broken down and energy is released. Catabolic reactions are exergonic because they release energy (Figure 4.1.8a).
- **Anabolic reactions** are reactions that require an input of energy in order to produce larger molecules from smaller substrates. Anabolic reactions are endergonic because they require energy. The energy is required to form bonds between molecules (Figure 4.1.8b).

Some reactions are not energetically favourable and require an input of energy. To address this need for energy input, sometimes biochemical pathways are coupled, and a reaction that releases energy is coupled with a reaction that requires energy input so that the reaction proceeds.

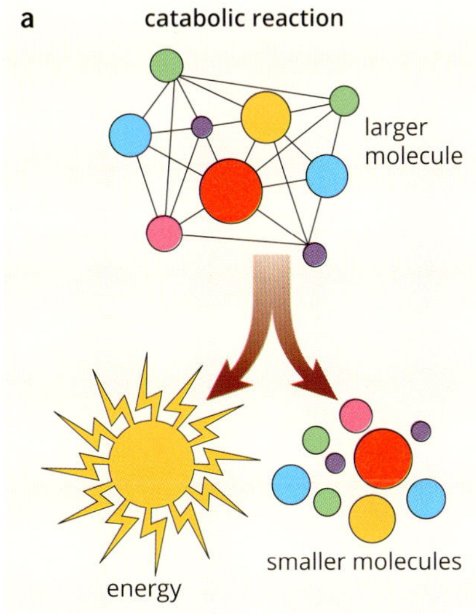

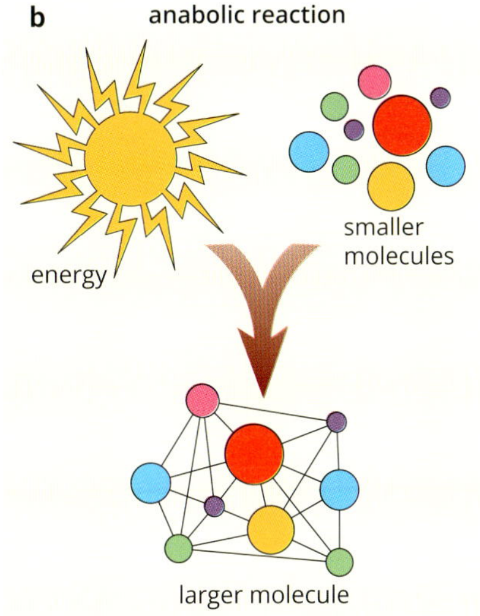

FIGURE 4.1.8 An overview of the reactions in metabolism. In catabolic reactions (a), a molecule is broken down into smaller molecules and there is a release of energy. In anabolic reactions (b), energy is required to combine small molecules into a larger molecule.

BIOFILE

Anabolic steroids

Anabolic steroids are synthetic variants of testosterone. More correctly, they are termed anabolic-androgenic steroids: 'anabolic' for their ability to stimulate anabolic reactions leading to muscle production (Figure 4.1.9), and 'androgenic' for their ability to increase male sexual characteristics. Anabolic steroids can be used to treat certain medical conditions. Some bodybuilders and sportspeople also abuse anabolic steroids to enhance their physical appearance or performance. They do this despite anabolic steroids being banned by many sports leagues and their side effects, which include irritability and aggression, and shrinkage of the testicles in men or growth of facial hair in women.

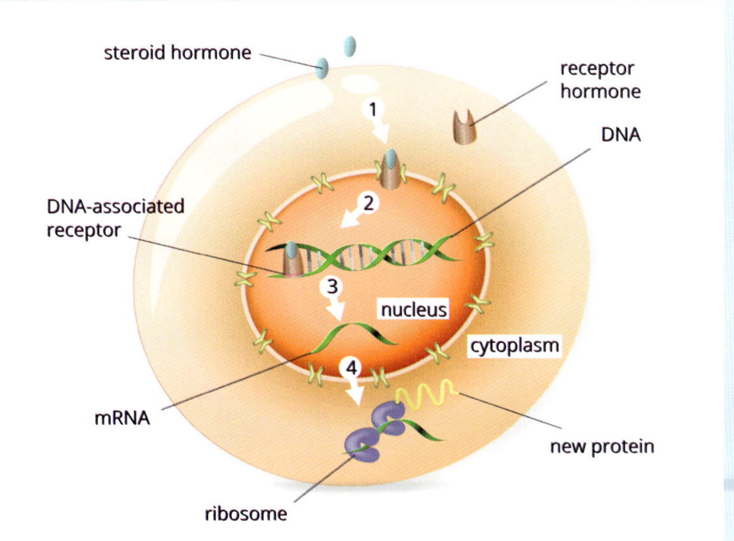

FIGURE 4.1.9 Once steroids are in the cytoplasm (1) they travel to the nucleus, where they bind to an intracellular receptor (2), which activates gene transcription (3) and causes proteins to be manufactured (4).

BIOLOGY IN ACTION

New enzyme may hold the key to beating obesity

Rising levels of sugar consumption in countries such as Australia (average consumption is 95.6 grams per person per day) and the United States (average consumption is 126.4 grams per person per day) are leading to epidemics of obesity, heart disease, tooth decay and type 2 diabetes. The World Health Organization recommends a maximum daily intake of 25 grams of sugar per day.

Although there are many health problems associated with excess sugar intake, sugar (in the form of glucose) is an essential source of energy for your cells. Sugars are a form of carbohydrate, which enzymes break down during glycolysis. After you eat, your blood-sugar levels rise, signalling the pancreas to release the hormone insulin. Insulin regulates blood-sugar levels either by allowing glucose to enter cells when it's needed or by storing excess glucose in the liver to be used later. Once the liver's storage capacity has been reached, the excess sugar is converted to fatty acids and stored in adipose fat cells. Figure 4.1.10 shows how the liver processes glucose. A lifestyle that includes a diet high in sugar with little to no exercise (that is, energy expenditure) will result in a build-up of stored fat, leading to obesity and the many health problems that are caused by it (e.g. heart disease and type 2 diabetes).

A compound essential to the metabolism of glucose in cells is glycerol-3-phosphate (Gro3P). Gro3P regulates lipid, energy and glucose metabolism, but high levels of glucose in the blood cause overproduction of this compound. Too much Gro3P can cause tissue damage and malfunction in the biochemical processes in which it is involved, ultimately leading to insulin resistance and type 2 diabetes.

Researchers investigating Gro3P have discovered a previously unknown metabolic pathway and the enzyme Gro3P phosphatase (G3PP), which directly converts Gro3P to glycerol, preventing the excess formation of fat and reducing the production of glucose in the liver. By regulating the amount of Gro3P present in cells, the enzyme, G3PP controls several important cellular processes, including glycolysis, ATP production and fatty acid oxidation in pancreatic cells.

The level of expression of the G3PP enzyme throughout the body varies depending on the health of the tissue. When G3PP was overexpressed in rat livers, the researchers observed lowered weight gain, lowered glucose production from glycerol in the liver and increased levels of fat-removing molecules in the blood plasma. The discovery of this enzyme and its important roles in lipid, energy and glucose metabolism opens up new possibilities for the development of treatment options for obesity, type 2 diabetes and metabolic heart disorders.

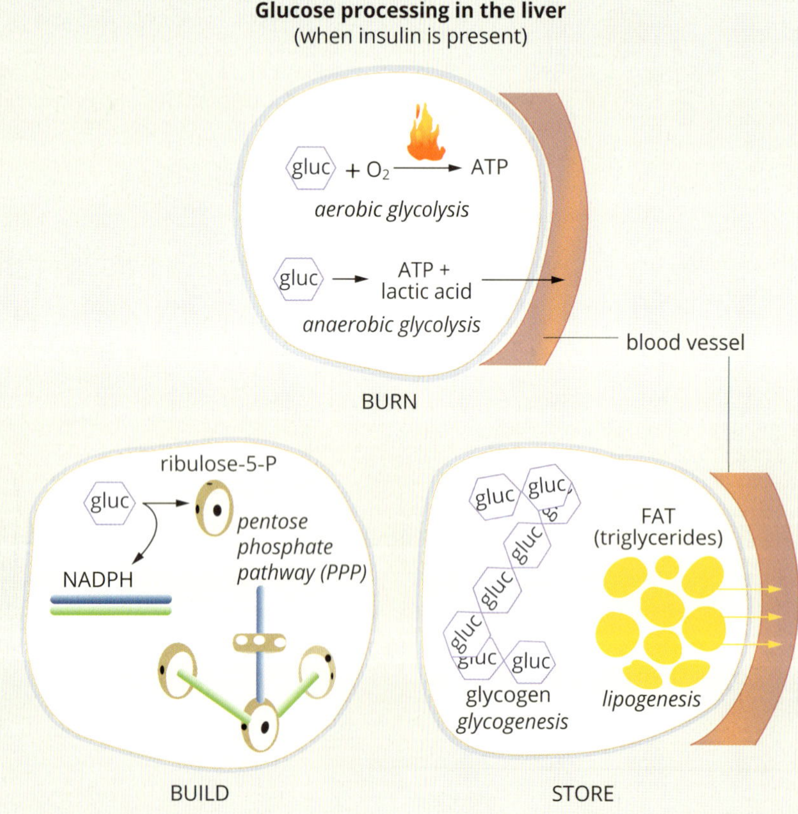

FIGURE 4.1.10 Glucose processing in the liver. When insulin is present, glucose can be used as an energy source (converted to ATP or lactic acid), to build new cellular components (e.g. DNA), or stored as glycogen in the liver or fats transported to other cells.

4.1 Review

SUMMARY

- Most enzymes are globular proteins that have a tertiary or quaternary structure.
- Enzymes are:
  - often specific to particular substrates, and therefore catalyse specific reactions
  - biological catalysts that increase the rate of reactions by lowering the activation energy needed for reactions to occur.
- Activation energy is the amount of energy input required for a reaction to start.
- As enzymes do not form part of the product of the reaction they catalyse, they are not used up and can be reused over and over again.
- The active site is a three-dimensional pocket-like part of the enzyme that is shaped to interact with its specific substrate.
- When enzymes and substrates interact, they form an enzyme–substrate complex.
- There are two models of enzyme–substrate interaction:
  - The lock and key model states that the substrate fits into the active site like a key fits into a lock.
  - The induced-fit model states that when a substrate binds to the active site of an enzyme, the active site can alter its shape to better fit the shape of the substrate.
- A biochemical pathway is a sequence of biochemical reactions catalysed by different enzymes, in which the product of each reaction becomes the substrate in the next reaction.
- Metabolism is the collection of all the biochemical (or metabolic) reactions that occur in living cells.
- Metabolic reactions can be catabolic or anabolic:
  - Catabolic reactions are reactions in which substrates are broken down and energy is released.
  - Anabolic reactions are reactions in which larger molecules are produced from smaller substrates. They require an input of energy.

KEY QUESTIONS

1 State the two main features of enzymes and explain their importance.

2 Describe the difference between the lock and key and induced-fit models of an enzyme–substrate interaction.

3 Does the following diagram showing an enzyme catalysing a reaction represent a catabolic or an anabolic reaction?

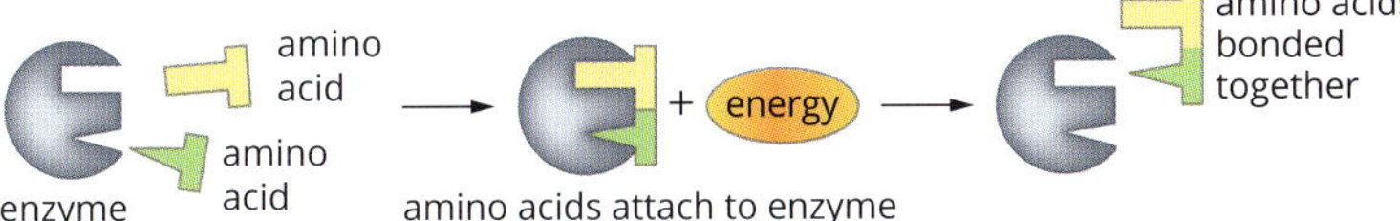

4 Describe the difference between catabolic and anabolic reactions.

4.2 Regulation of enzymes

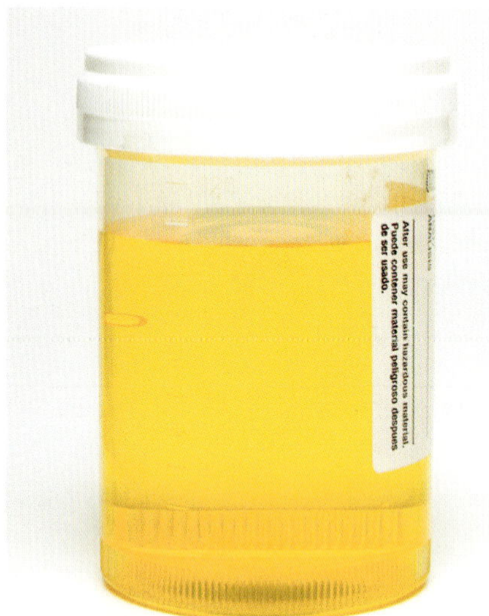
FIGURE 4.2.1 Maple syrup urine disease is caused by a defective group of enzymes that work together (an enzyme complex called the branched-chain ketoacid dehydrogenase complex). It results in a build-up of branched-chain amino acids that causes the urine to smell like maple syrup, and the blood to become acidic, which can result in death.

There can be a negative effect on the whole organism when cells produce too much or not enough of particular substances, or are unable to properly break some substances down (Figure 4.2.1). Cells that produce excess substances are also wasting energy and resources. To account for this, cells have mechanisms that regulate biochemical reactions to ensure the final product is not over- or under-produced.

As you learnt in the previous section, enzymes control metabolism through the regulation of biochemical reactions at every step of a biochemical (or metabolic) pathway. Because enzymes play a vital role in biochemical reactions, a whole biochemical pathway can be regulated through enzyme regulation.

In this section, you will learn how enzymes are regulated, including the different ways in which enzymes are inhibited, and other factors that affect the rate of the biochemical reactions enzymes catalyse. You will also learn about the coenzymes ATP, NADH and NADPH, which transport energy and electrons from one biochemical reaction or pathway to another.

FACTORS THAT REGULATE ENZYME ACTIVITY

The amounts of final products and the speed at which they are produced in a biochemical pathway can be controlled through the regulation of individual reactions that make up that pathway. As each reaction in a biochemical pathway uses the product from the previous reaction as a substrate, slowing down one reaction will have an effect on all subsequent reactions.

All enzymes have specific conditions in which they perform at their best. Factors such as temperature, pH and the concentration of the substrate and enzyme all affect the rate of enzymatic reactions. When these conditions are optimal, enzyme activity is at its highest, and the rate of reaction is at its fastest.

Other factors, such as inhibitors, phosphorylation, and cofactors and coenzymes, regulate the activation and inhibition of enzymes, and determine whether they can catalyse reactions at all, rather than the rate at which reactions occur.

Temperature

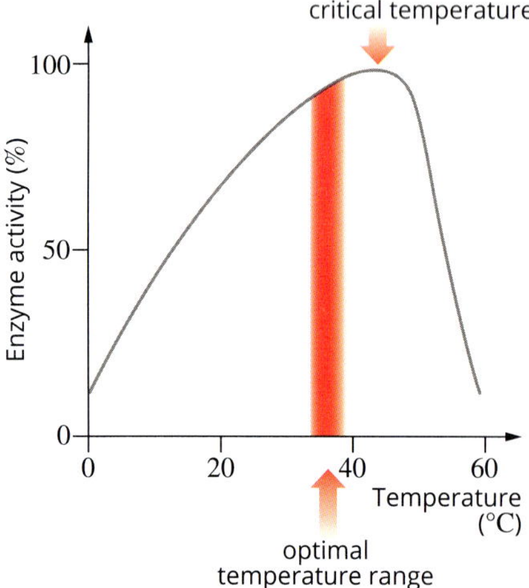

FIGURE 4.2.2 As the temperature increases, enzyme activity increases until a critical temperature is reached, at which point the protein denatures and the shape of the active site changes such that the substrate cannot bind. The rate of enzyme activity then decreases rapidly.

As with most chemical reactions, the rate of enzyme-catalysed reactions will generally increase as the temperature increases. This is because the warmer particles become during a reaction, the more rapidly they move, which makes successful collisions between them more likely to occur.

However, proteins, including enzymes, can be denatured at high temperatures. When this occurs, the hydrogen bonds and hydrophobic interactions that create the tertiary and quaternary structures of the enzyme are broken, and the shape of the enzyme is changed such that substrate cannot bind to it, and the reaction cannot occur.

Human enzymes have an optimum temperature of 36–38 °C, which matches our normal body temperature (approximately 37 °C). Indeed, many animal enzymes will begin to denature at temperatures above 40 °C (Figure 4.2.2). However, there are some enzymes that have optimum temperatures much higher than this. For example, Taq polymerase is an enzyme that was originally found in bacteria living in volcanic hot springs, and it has an optimum temperature of 70 °C.

If enzymes are cooled below their optimum temperature, the rate of reaction will slow down, because particles will move more slowly, making successful collisions less likely, and because the bonds are not as flexible at lower temperatures and conformational changes do not occur. However, if the enzyme is reheated, its activity and reaction rate will increase again.

BIOFILE

Psychrophiles

Bacteria and fungi that are adapted to the extreme cold (or psychrophiles) of Antarctica and the Arctic tend to have enzymes with a broader temperature range, or two optimum temperature zones, including an optimum range as low as 0–15 °C. This allows their cells to grow and metabolise at the constant low temperatures of their cold environments (Figure 4.2.3).

FIGURE 4.2.3 A glaciologist in Antarctica saws an ice core containing frozen psychrophiles.

BIOFILE

Cool surgery

Doctors can slow down enzyme activity when they perform complex operations on the heart by running the patient's blood through a heart–lung machine that cools it (Figure 4.2.4). The cooling of blood slows the rate of enzyme-controlled reactions, and reduces the impact a possible lack of oxygen during the operation could have on the heart, brain and other tissues.

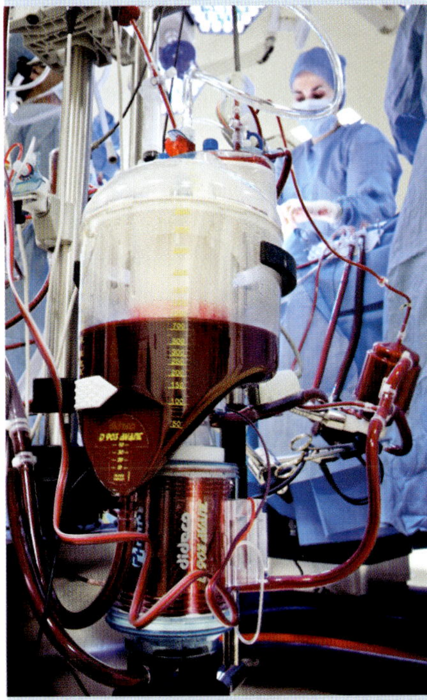

FIGURE 4.2.4 A heart–lung machine.

pH

The pH scale is used to measure acidity or alkalinity (Figure 4.2.5). Enzymes have a specific pH range at which they function best. If enzymes are taken too far above or below their optimum pH, then the tertiary structure is affected (the enzyme may become denatured) and the substrate may not be able to bind.

If the reaction occurs in an environment in which the pH is not ideal, the micro-environment of the active site may provide a different pH in order to create a specific environment for catalysis to take place.

The optimum pH range of enzymes can be quite different, and varies depending on the function of the enzyme and where it is located. Examples include the following digestive enzymes:

- Amylase starts the digestion of starch in the mouth and has an optimum pH of about 7.
- Pepsin is found in the stomach and has an optimum pH of about 2.
- Trypsin is found in the small intestine and has an optimum pH of about 8 (Figure 4.2.6).

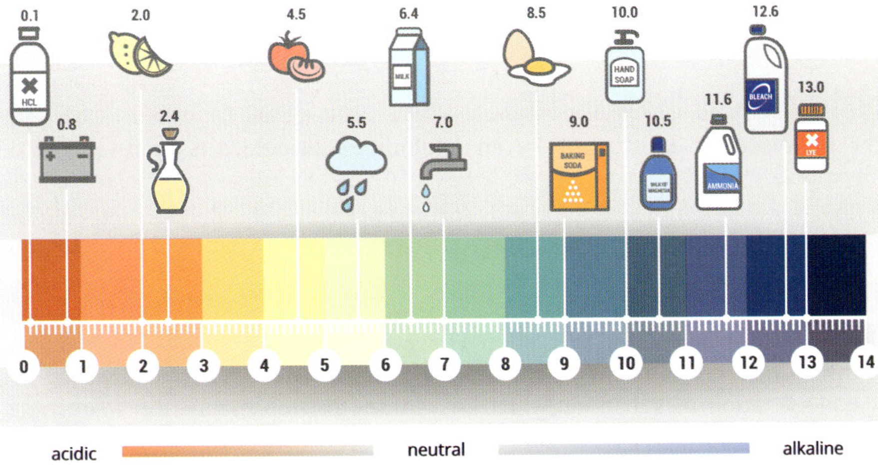

FIGURE 4.2.5 The pH scale.

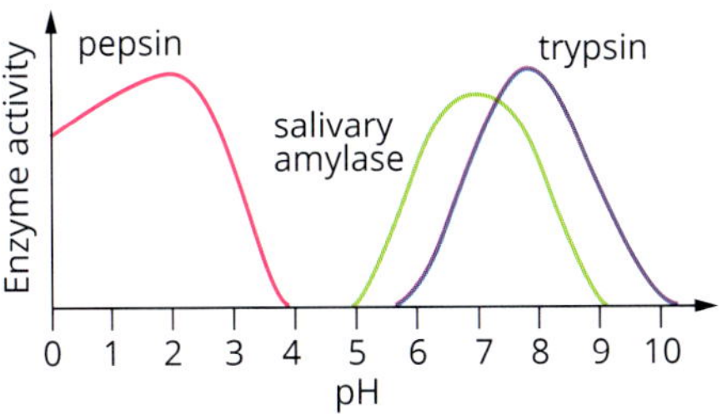

FIGURE 4.2.6 The rate of enzyme activity for three digestive enzymes in relation to pH values. At the optimum pH for an enzyme, its rate of reaction is at a maximum.

BIOFILE

Pepsin and ulcers

If the lining of the stomach is damaged, the action of acid and the stomach enzyme pepsin can contribute to the formation of ulcers (Figure 4.2.8). As the Nobel Prize-winning research by Australians Dr Barry Marshall and Dr Robin Warren showed, the bacterium *Helicobacter pylori* is a common cause of the damage that results in stomach ulcers. Antacid medications raise the pH of the stomach contents, which neutralises the acidity. As pepsin is only active at pH 1–4, with an optimal pH of around 2, raising the pH of the stomach contents above pH 4 reduces its activity.

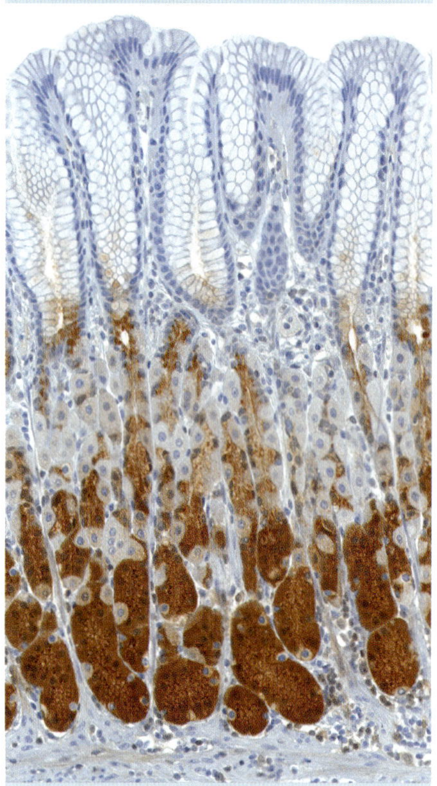

FIGURE 4.2.8 Inactive pepsin enzymes (or pepsinogens) appear brown in this stained section of stomach tissue.

Enzyme and substrate concentration

The concentration of enzyme compared to substrate affects the rate of reaction. If the enzyme concentration is high compared to that of the substrate, the reaction will occur over a short period of time. This is because the more enzyme molecules that are available, the more active sites there will be for the substrate to bind to, and so the rate of reaction will be faster (Figure 4.2.7).

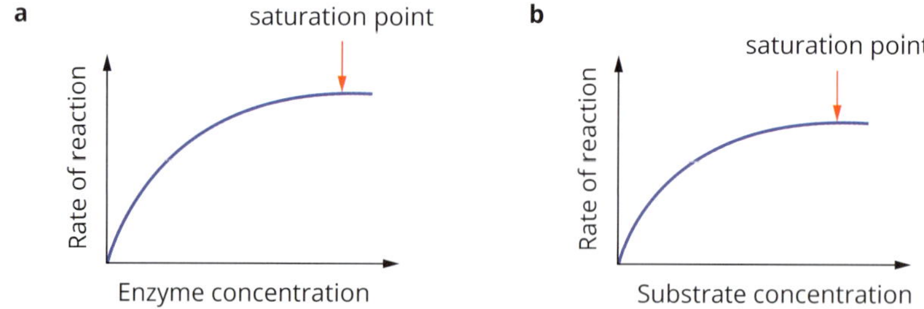

FIGURE 4.2.7 The rate of a reaction increases with increasing enzyme concentration (a), or substrate concentration (b). In each scenario, there is a point at which increasing the enzyme or substrate concentration further does not affect the reaction rate. The enzyme has reached the maximum rate of reaction and this is known as the saturation point.

If the enzyme concentration is low compared to the substrate concentration, then the reaction rate will be slow and will continue for longer. This is because there are few active sites compared to the number of substrate molecules, so the substrate molecules will have to 'wait' until there is a free active site.

Cellular compartmentalisation

Cellular compartmentalisation in eukaryotes helps to create an environment in which conditions are optimal for enzymes to catalyse reactions. Each organelle has a specific function. Within an organelle, the molecules involved in a particular biochemical pathway related to the organelle's function are brought together for the reaction to take place.

For example, lysosomes are specialised packages of digestive enzymes called acid hydrolases. These enzymes are delivered to lysosomes via vesicles that bud off the Golgi apparatus. Lysosomes contain about 40 types of enzyme that break down proteins, lipids, nucleic acids and more. These enzymes function optimally in the acidic lysosomal environment, around pH 5. Lysosomes digest cell components that are past their use-by date. Specialist phagocytic cells, such as macrophages, use lysosomal enzymes to destroy ingested foreign matter.

Inhibition of enzyme activity

The inhibition of an enzyme by an inhibiting molecule can be reversible or irreversible:

- In reversible inhibition, the bonds formed between the inhibitor and enzyme are weak (i.e. hydrogen bonds), so they are easily broken and the inhibition reversed. This means the inhibitor can move in and out of the active site, which reduces the activity of the enzyme because its active site will not be available for the substrate to bind to as often. However, because the binding is reversible, the reduction in enzyme activity can be partially overcome by increasing the concentration of substrate, which provides a greater chance for a substrate, rather than an inhibitor, to bind to the enzyme.
- In irreversible inhibition, the bonds formed between the inhibitor and enzyme are strong (i.e. covalent bonds), so the binding is irreversible. This means the inhibitor blocks the enzyme's active site permanently, so the enzyme will no longer be able to take part in reactions.

Enzyme inhibition is also classified as being **competitive inhibition** or **non-competitive inhibition**, depending on where the inhibiting molecule binds to the enzyme.

Competitive inhibition

Competitive inhibition occurs when the shape of the inhibitor is similar to the shape of the substrate that normally binds to the active sites of a particular enzyme. Due to their similar shapes, such inhibitors are able to bind to the active site of the enzyme, and block the substrate from binding to the site (Figure 4.2.9). Unlike a substrate, when an inhibitor binds to an enzyme's active site it does not trigger a catalytic reaction.

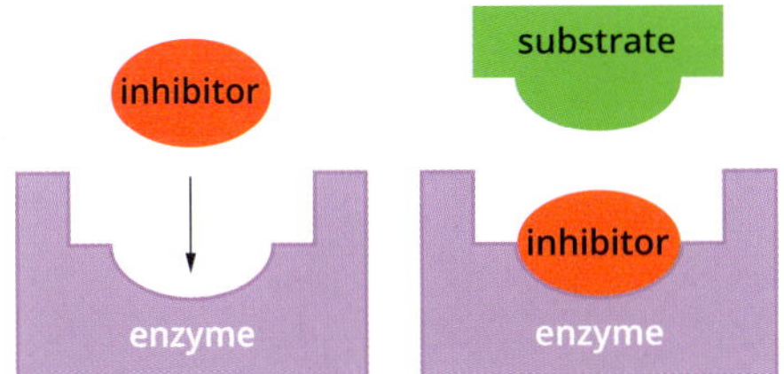

FIGURE 4.2.9 Competitive inhibition involves an inhibitor binding directly to the active site of the enzyme. The substrate is then unable to bind to the enzyme.

Non-competitive inhibition

Non-competitive inhibition (or allosteric inhibition) occurs when an inhibitor binds to an enzyme site other than the active site (an **allosteric site**). Binding to the allosteric site either changes the shape (or conformation) of the enzyme such that the substrate cannot bind to its active site (Figure 4.2.10), or it blocks the changes in shape needed for the reaction to progress once the substrate is bound.

Binding of molecules to an allosteric site does not always result in allosteric inhibition. Some molecules that bind to allosteric sites can cause conformational changes that allow reactions to occur (or allosteric activation). Whether binding to an allosteric site causes inhibition of activation, both are examples of allosteric regulation.

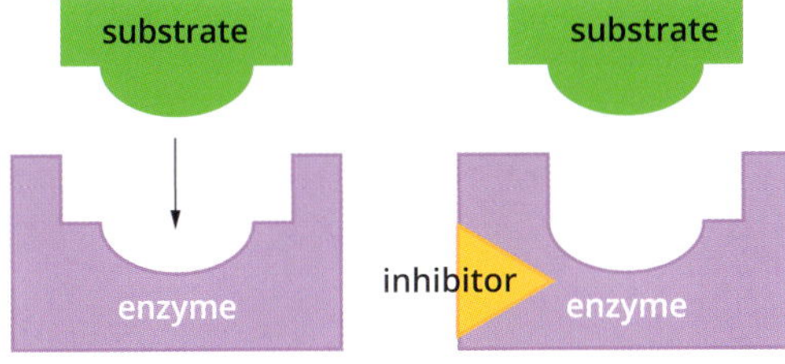

FIGURE 4.2.10 Non-competitive inhibition involves an inhibitor binding to the enzyme and causing a change in the shape of its active site, so that the substrate no longer fits.

Feedback inhibition

Feedback inhibition occurs when a product produced late in a pathway is also an inhibitor of an enzyme earlier in the pathway, such that as the amount of the inhibiting product increases, the number of enzyme molecules being inhibited also increases (Figure 4.2.11). This in turn reduces the amount of the inhibiting product. As the level of inhibiting product reduces, less of it will bind to the enzyme, allowing the enzyme to function again. Feedback inhibition is an important mechanism in controlling enzyme activity.

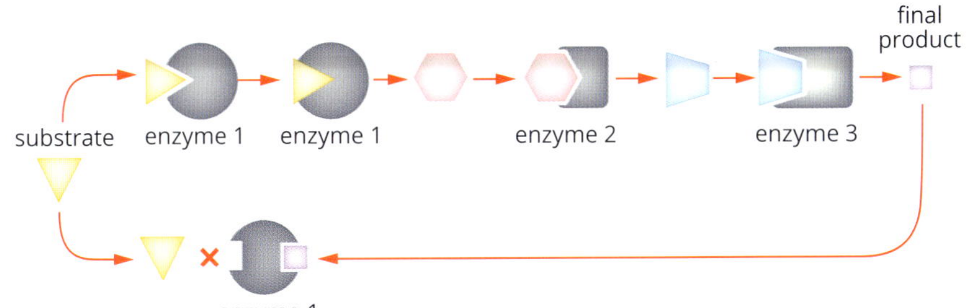

FIGURE 4.2.11 In feedback inhibition, when the amount of inhibiting product is high, the pathway slows down. When the amount of inhibiting product is low, the pathway speeds up.

BIOLOGY IN ACTION

Enzyme inhibition to treat Alzheimer's disease

Alzheimer's disease is characterised by a progressive degeneration of the brain that affects memory and cognitive function. A drug currently used in Australia to treat Alzheimer's disease inhibits the activity of the enzyme acetylcholinesterase in the central nervous system. The role of acetylcholinesterase is to break down the substrate acetylcholine (ACh), a neurotransmitter that is important in memory processes. Neurotransmitters are signalling molecules of the nervous system. When ACh is broken down, nerve cells (neurons) are able to return to their resting state.

In people who suffer from Alzheimer's disease, the level of ACh is low, so less memory-related signalling occurs. To compensate for the low levels of ACh, drugs that inhibit the activity of acetylcholinesterase can be used (Figure 4.2.12). These drugs can act by either competitive or non-competitive inhibition. Figure 4.2.13 shows an example of an acetylcholinesterase inhibitor that is used to treat Alzheimer's disease and acts through competitive inhibition.

FIGURE 4.2.12 Molecular model of acetylcholinesterase enzyme (purple) with a competitive inhibitor (green) bound to the active site preventing acetylcholine from binding.

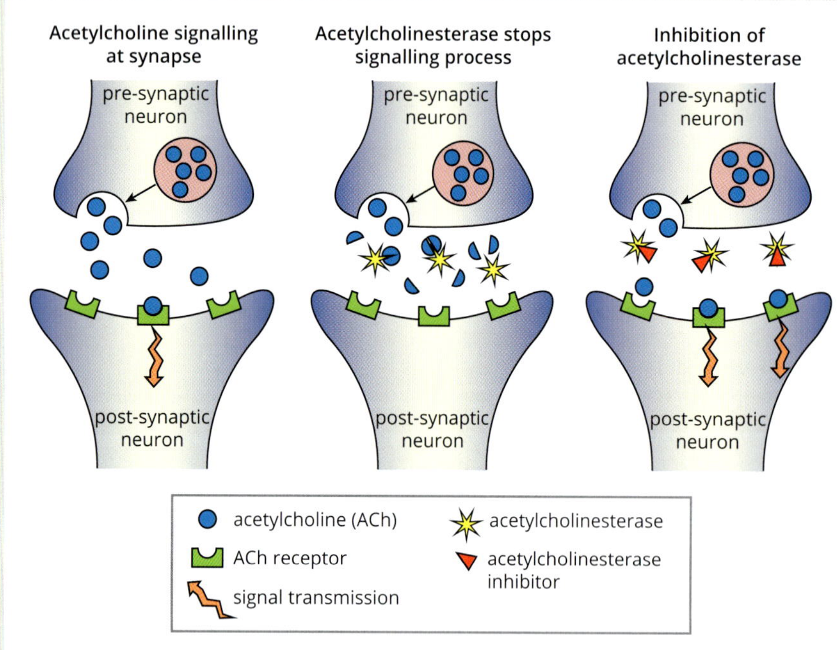

FIGURE 4.2.13 Inhibition of acetylcholinesterase blocks the enzyme's activity. Left: The normal signalling action of acetylcholine. Middle: Signal transmission is reduced when the acetylcholinesterase is present. Right: An acetylcholinesterase inhibitor blocks the enzyme from binding to acetylcholine.

Phosphorylation

The binding of a phosphate group to a protein is known as **phosphorylation**. The removal of a phosphate group from a protein is known as **dephosphorylation**. Phosphorylation is thought to be the most common regulatory mechanism of protein function, with as many as one third of all proteins in the human body being substrates for phosphorylation at some point.

Phosphorylation and dephosphorylation both change the structure (or conformation) of proteins, and therefore both can regulate enzymes. For example, the phosphorylation of glycogen synthase inhibits its ability to interact with glucose. The phosphorylation of glycogen synthase is performed by another enzyme called glycogen synthase kinase 3.

Cofactors and coenzymes

Some enzymes need additional components to enable them to catalyse a reaction. These components, called cofactors, bind to the enzyme. Cofactors can be inorganic ions such as iron (Fe^{2+}), magnesium (Mg^{2+}) and zinc (Zn^{2+}), or organic molecules such as proteins, vitamins and adenosine triphosphate (ATP).

Small, non-protein organic cofactors are known as **coenzymes** (Figure 4.2.14). For certain enzymes, a specific coenzyme is required to catalyse reactions. Often the coenzyme is structurally altered during the reaction, but afterwards reverts to its original form, which allows it to be reused. There are many types of coenzymes, and they include vitamins, ATP, nicotinamide adenine dinucleotide (NADH), flavin adenine dinucleotide ($FADH_2$) and nicotinamide adenine dinucleotide phosphate (NADPH).

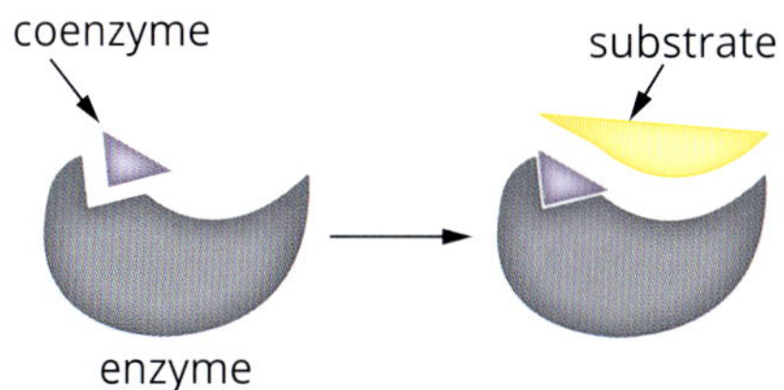

FIGURE 4.2.14 Coenzymes are non-protein organic cofactors that enable an enzyme to catalyse a reaction.

The cycling of coenzymes

Many of the reactions catalysed by enzymes involve the transfer of **chemical groups** (e.g. phosphates) from one molecule to another. Coenzymes such as ATP, NADH, $FADH_2$ and NADPH act as metabolic intermediates, and carry chemical groups between different reactions. These coenzymes are continuously recycled as they move a chemical group from one molecule to another molecule in different enzymatic reactions.

The coenzymes ATP, NADH, $FADH_2$ and NADPH play a major role in maintaining cellular processes such as cellular respiration and photosynthesis. These coenzymes can store and transport chemical groups, **protons** and **electrons** from one reaction to another. Energy is also transferred in this way, stored in the chemical bonds between the coenzymes and the chemical group, proton or electron they carry.

The **unloaded** form of a coenzyme is free to accept a proton, electron or a chemical group, and once it has accepted it, it is considered to be **loaded** (Table 4.2.1). The cycling between loaded and unloaded forms is referred to as the cycling of a coenzyme (Figure 4.2.15).

A chemical group is a group of covalently linked atoms, such as an amino group or a hydroxyl group, that has a characteristic chemical behaviour.

Cellular respiration is a general term applied to the breakdown of glucose in the cell to yield energy for the production of ATP.

Photosynthesis is a cellular process in which light energy is trapped by chlorophyll and used to combine carbon dioxide and water to make glucose.

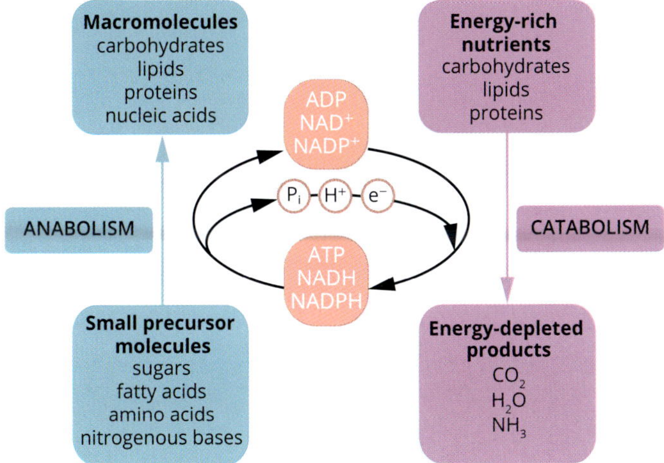

FIGURE 4.2.15 An overview of the cycling of coenzymes in anabolic and catabolic reactions in the cell. Phosphate (P_i), hydrogen ion (H^+), electron (e^-).

Unloaded	Loaded
ADP	ATP
NAD^+	NADH
FAD^+	$FADH_2$
$NADP^+$	NADPH

TABLE 4.2.1 Unloaded and loaded coenzymes.

ADP/ATP

Adenosine triphosphate (ATP) and adenosine diphosphate (ADP) are involved in the transfer of energy between reactions. ADP contains two phosphate molecules ('di' in diphosphate meaning 'two'). It accepts a third phosphate group by phosphorylation to become ATP ('tri' in triphosphate meaning 'three'). The breakdown of ATP to ADP releases the energy stored in the bond between the third phosphate and the ADP molecule.

ATP is the most useable form of energy that a cell can produce. In fact, the cells of a human body use the equivalent of 75 kg of ATP each day to transfer the energy they need. However, the body uses only a relatively small number of ATP molecules, because the molecules are continually being cycled—first being loaded to store energy and then unloaded to release energy. You will learn more about ATP in Chapter 5.

NAD^+/NADH and $NADP^+$/NADPH

NAD^+/NADH is involved in the transfer of electrons in many different processes, but the most important role of NADH is in cellular respiration. $NADP^+$/NADPH are very similar to NAD^+/NADH in structure and function, but have an additional phosphate group attached. NADPH is mainly used as a reducing agent that provides electrons in anabolic reactions, including the production of lipids, nucleic acids and carbohydrates. In photosynthesis, NADPH is involved in the reactions that convert carbon dioxide into glucose. You will learn more about photosynthesis and cellular respiration yield in Chapter 5.

EXTENSION

Oxidation and reduction of coenzymes

Coenzymes such as NAD^+ and $NADP^+$ are involved in the transfer of electrons (coupled with protons) between reactions or biochemical pathways. Electrons are transferred to and from coenzymes without the electrons losing their energy potential. When a coenzyme accepts electrons, the reaction is referred to as a reduction reaction. When the coenzyme donates electrons, the reaction is referred to as an oxidation reaction (Table 4.2.2, Figure 4.2.16).

Reduction	Oxidation
$NAD^+ + 2e^- + H^+ \rightarrow NADH$	$NADH \rightarrow NAD^+ + 2e^- + H^+$
$NADP^+ + 2e^- + H^+ \rightarrow NADPH$	$NADPH \rightarrow NADP^+ + 2e^- + H^+$

TABLE 4.2.2 Reduction of NAD^+ and $NADP^+$, and oxidation of NADH and NADPH.

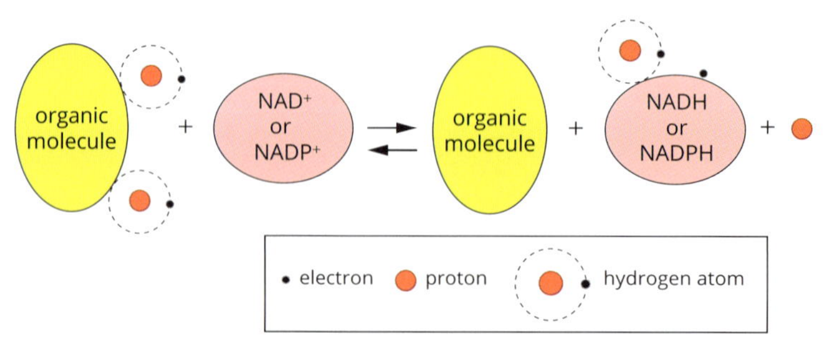

FIGURE 4.2.16 Coenzymes NAD^+ and $NADP^+$ accept an electron in a reduction reaction (left to right), and NADH and NADPH donate an electron in an oxidation reaction (right to left).

4.2 Review

SUMMARY

- Enzymes control metabolism through the regulation of biochemical reactions at every step of a biochemical (or metabolic) pathway.
- Through enzyme regulation, a whole biochemical pathway can be regulated.
- Factors affecting the rate of enzymatic reactions include:
  - temperature—High temperatures can alter an enzyme's three-dimensional structure and change the shape of the active site. This means the enzyme cannot act as a catalyst. This change is called denaturation.
  - pH—Enzymes have an optimum pH range and if the enzyme is exposed to a pH outside its tolerances, the enzyme may become denatured.
  - concentration—The concentrations of enzymes and their substrates will have an effect on the rate of a reaction, as both substrate and enzyme are required for the reaction to occur.
- Factors that determine whether enzymatic reactions can occur include inhibitors, phosphorylation, and cofactors and coenzymes.
- Inhibition can be reversible or irreversible:
  - reversible inhibition—The bonds formed between inhibitor and enzyme are weak and easily broken, so the inhibition is reversible. Reversible inhibition can be partially overcome by increasing the concentration of substrate.
  - irreversible inhibition—The bonds formed between inhibitor and enzyme are strong, so the binding is irreversible. If the inhibition is irreversible increasing substrate concentration will have no effect.
- Inhibition is classified as being competitive or non-competitive, depending on where an inhibiting molecule binds to an enzyme:
  - competitive inhibition—The inhibitor binds to the active site of the enzyme, blocking the substrate from binding.
  - non-competitive inhibition—The inhibitor binds to a site of the enzyme that is not the active site (an allosteric site) and changes the shape (or conformation) of the enzyme, which affects the binding of substrate to the active site, or it blocks the changes in shape needed for the reaction to progress once the substrate is bound.
- Phosphorylation and dephosphorylation both change the structure (or conformation) of proteins, and therefore both can regulate enzymes.
- Cofactors are additional components required by some enzymes to catalyse a reaction.
- Coenzymes are small, non-protein organic cofactors such as vitamins, ATP, NADH and NADPH.
- Coenzymes are needed for reactions in biochemical pathways, including cellular respiration and photosynthesis.
- Coenzymes carry chemical groups, protons and electrons between reactions in order to transfer energy between them.
- Coenzymes can be unloaded or loaded, and cyclically load and unload:
  - Unloaded coenzymes accept a chemical group, proton or electron.
  - Loaded coenzymes donate a chemical group, proton or electron.

KEY QUESTIONS

1 In non-competitive inhibition, the inhibitor binds to:

A a substrate

B an enzyme's active site

C an allosteric site on the enzyme

D another inhibitor

2 Explain the difference between reversible and irreversible inhibition. Include reference in your answer to how changing substrate concentration affects each.

3 Is the following diagram an example of competitive or non-competitive inhibition?

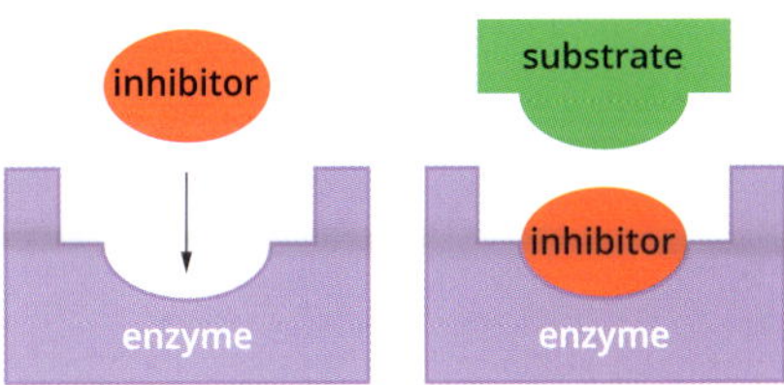

4 Explain what coenzymes are, and describe the difference between an unloaded and a loaded coenzyme.

Chapter review

KEY TERMS

activation energy
active site
allosteric site
anabolic reaction
catabolic reaction
catalytic power
chemical group
coenzyme
competitive inhibition
conformational change
dephosphorylation
electrons
enzyme–substrate complex
feedback inhibition
loaded coenzyme
metabolism
non-competitive inhibition
phosphorylation
protons
substrate
unloaded coenzyme

KEY QUESTIONS

1 An organic molecule required by an enzyme, in order for it to function, is best described as:
  A a cofactor
  B a coenzyme
  C a chemical group
  D an enzyme activator

2 Enzymes reduce the activation energy of a reaction by:
  A bringing the reactants close together so that a reaction is more likely to occur
  B orientating the reactants in the most favourable position for the reaction
  C providing a micro-environment favourable to the chemical reaction
  D all of the above

3 A student investigating the activity of the enzyme pepsin, which is found in the stomach of humans, observed the change in enzyme activity as the concentration of the substrate (protein) increased. The experiment was conducted at pH 3 and 37 °C. The student's data was presented in the graph shown.

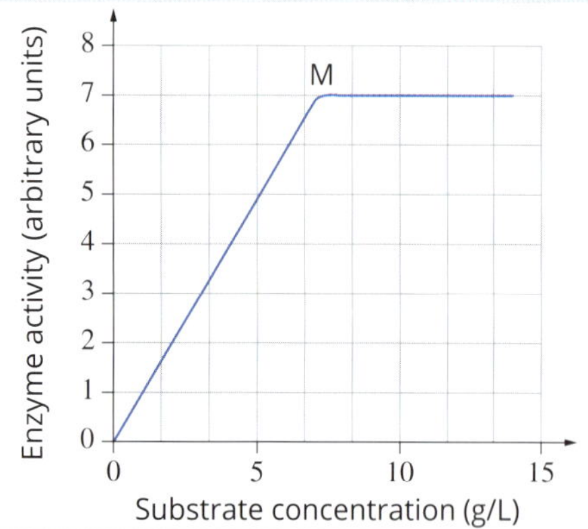

The student wanted to change the experiment so that the rate of reaction at point M was higher than that shown. In order to do this the student could:
  A increase the pH to 8
  B decrease the temperature
  C increase the amount of enzyme
  D increase the concentration of the substrate

4 Which of the statements about NAD^+ is not true?
  A NAD^+ can accept a hydrogen ion.
  B When NADH gives up a proton NAD^+ results.
  C Only a small amount of NAD^+ is needed in a cell.
  D NAD^+ can only be used once before it must be resynthesised.

5 An experiment was performed to investigate enzyme activity in three different species: the two-toed sloth (a mammal), an Arctic trout, and a bacterium from a thermal spring. The activities of the enzymes from each organism are plotted on the graph below.

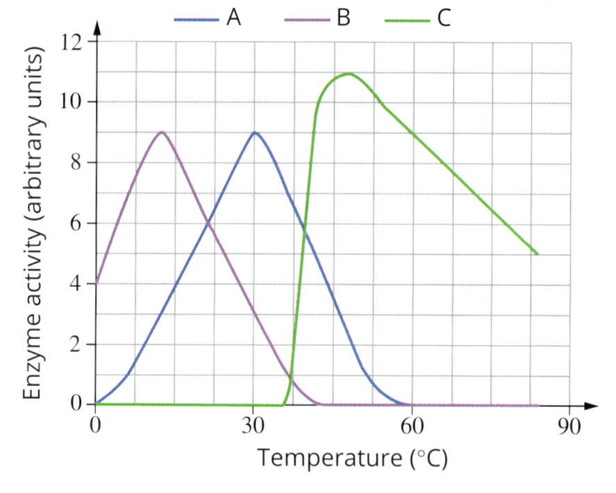

  a Explain which graph belongs to each animal.
  b Why was no activity observed at 60 °C for enzyme A?

6 Trypsin is a digestive enzyme secreted by the pancreas of vertebrate animals. A biologist studying enzyme activity isolated trypsin from a mammal and two different species of fish and tested the enzyme activity at different temperatures. The experimental results are illustrated in the graph.

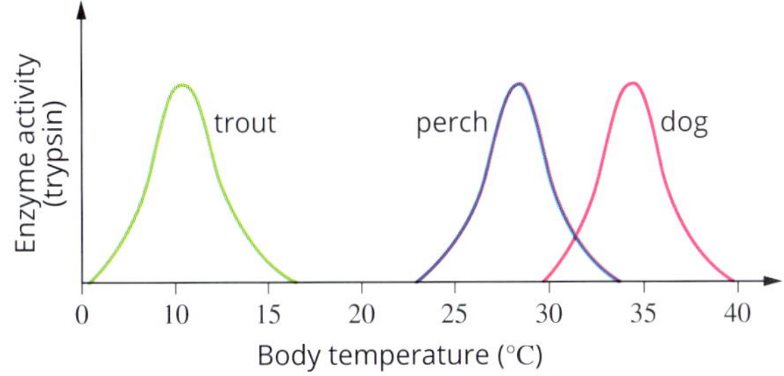

a Define 'optimum temperature of an enzyme'.
b What is the optimum temperature for trypsin in the:
 i trout?
 ii perch?
 iii dog?
c Describe what happens to the activity of trypsin after 10 °C in the trout, 28 °C in the perch and 34 °C in the dog. Explain why this occurs.
d Fish are ectotherms. Suggest a reason for the difference in the optimal temperatures of the enzyme in trout and perch.

7 The breaking or formation of bonds between atoms during biochemical reactions results in changes in the energy content of the molecules. Formation of new chemical bonds requires energy and the breaking of chemical bonds releases energy.

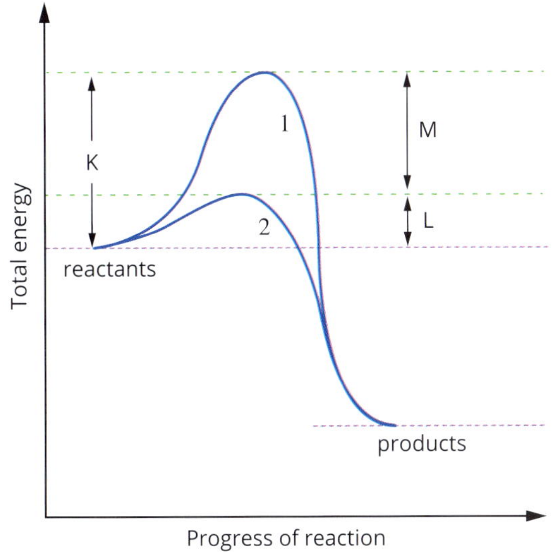

The graph above shows the energy changes during one particular chemical reaction with and without an enzyme.

a What energy change is represented by
 i K?
 ii L?
 iii M?
b Is the reaction exergonic or endergonic? Explain.
c Explain which line shows the enzyme-catalysed reaction.

8 The following diagram illustrates one model of enzyme activity.

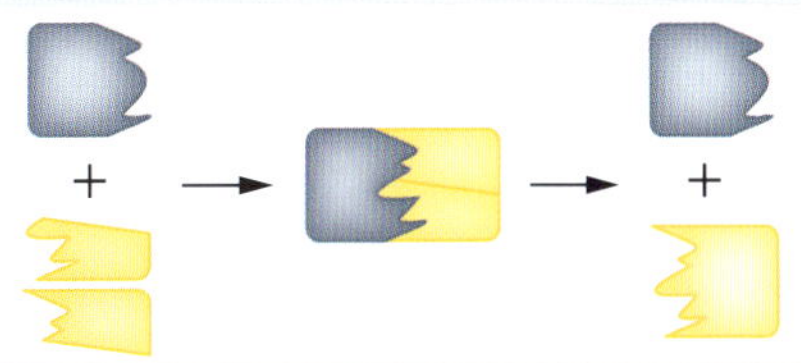

a Explain which of the two models of enzyme activity is being illustrated.
b Explain how this model of enzyme activity could be used to explain how some enzymes can act on multiple substrates.

9 Many chemical reactions in living things occur as part of a metabolic pathway. Metabolic pathways involve a series of reactions with a different enzyme catalysing each step in the pathway. In these pathways the product of one reaction is the substrate for the next reaction. The diagram below shows the enzymes involved in one such pathway.

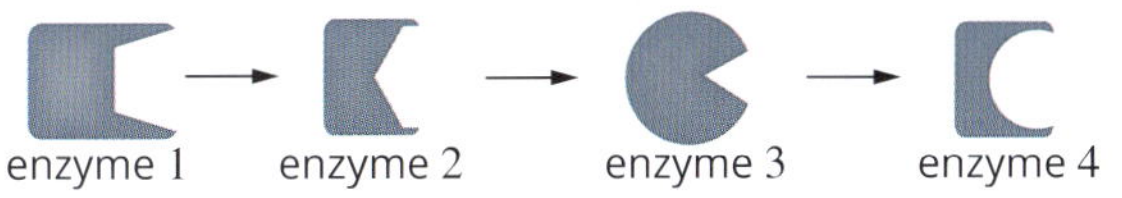

The enzyme substrates are shown below.

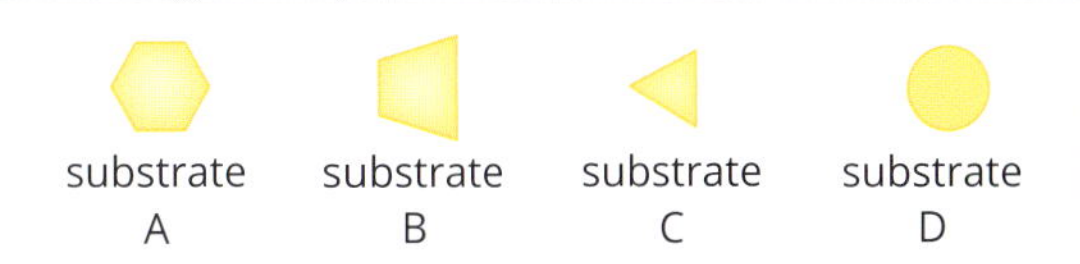

a i Match each substrate with its enzyme.
 ii What was basis of your decision?
b The following molecule is the final product of the reaction.

 i If the concentration of the final product builds up in the cell, the reaction will stop. Explain how the increase in the concentration of the final product stops the reaction proceeding.
 ii Is the product acting as a competitive inhibitor or an allosteric inhibitor? Explain your answer.

10 Consider the following incomplete graph for an enzyme-controlled reaction in which the enzyme is present at concentration x. Assume there is a fixed amount of substrate present.

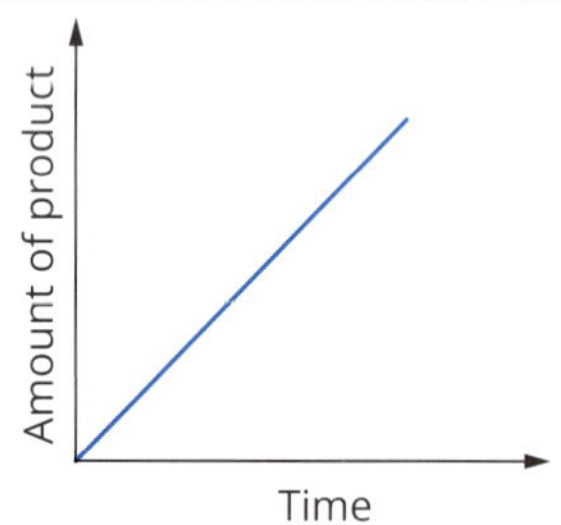

a Eventually the shape of the graph will change. Continue the line graph according to your expectations and explain what happens.

b Redraw the graph for an enzyme concentration of $2x$.

11 Pepsin is an enzyme that is released into the stomach of humans (pH 1.5–3.5, 37°C), where it breaks down proteins into polypeptides. Explain how you would expect the activity of pepsin to change as:

a the temperature is increased from 37°C to 45°C

b the pH is increased above 5.

12 The following graph illustrates the relationship between the concentration of an enzyme substrate and enzyme activity.

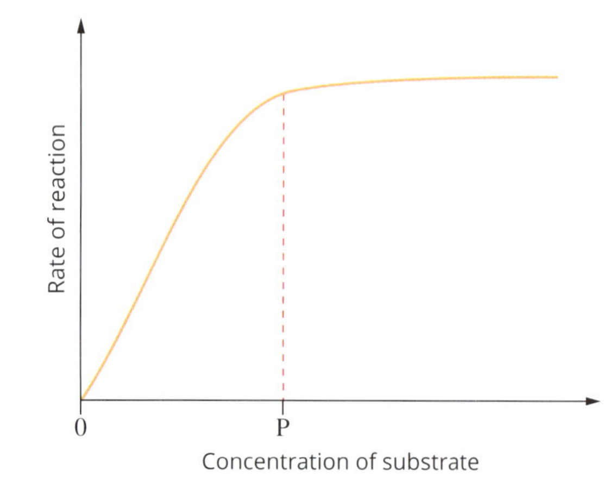

a Describe the relationship between the substrate concentration and the rate of reaction from 0 to P units of substrate concentration.

b What happens at and after point P? Explain.

c The concentration of the enzyme is described as being a limiting factor. Explain what this means.

13 Bile acids are processed in the liver to form bile salts. Bile salts are used in the small intestine to mechanically break down fats in the diet. The metabolic pathway for the formation of two bile acids is shown in the flow chart. The chart is simplified; some steps in the pathway between 3β,7α-dihydroxy-5-cholestenoic acid and chenodeoxycholic acid are not known, nor are some of the enzymes in the pathway.

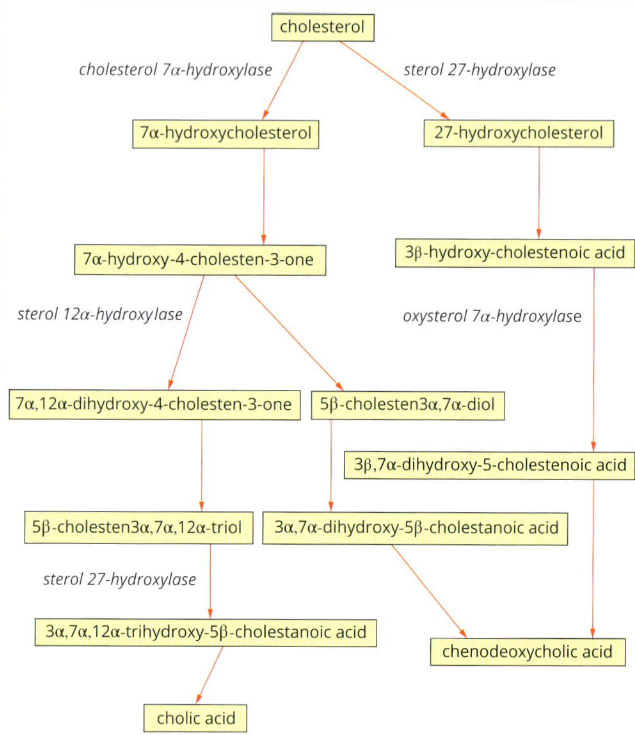

a One enzyme used in the pathway is sterol 27-hydroxylase. If an individual was unable to make functional sterol 27-hydroxylase would they be able to make bile acids? Explain.

b Explain whether it is likely that chenodeoxycholic acid acts as an inhibitor of sterol 27-hydroxylase.

c Suggest how sterol 27-hydoxylase can catalyse two different steps in the pathway.

14 During chemical reactions changes in free energy occur. The change of energy is given as ΔG (delta G). A positive energy change means that there had to be an input of energy for the reaction to occur and a negative energy change means the reaction released energy. Energy released by one step can be used in subsequent steps.

A particular metabolic pathway involves four enzymes (K, L, M and N) in a four-step pathway. Molecule A is the initial substrate of the pathway and molecule E is the final product.

$$\mathbf{A} \xrightarrow{\mathbf{K}} \mathbf{B} \xrightarrow{\mathbf{L}} \mathbf{C} \xrightarrow{\mathbf{M}} \mathbf{D} \xrightarrow{\mathbf{N}} \mathbf{E}$$

The energy requirements of each step (in joules) are shown in the table below.

Step	ΔG	Enzyme
A → B	+5	K
B → C	+2	L
C → D	−6	M
D → E	−4	N

a Identify the exergonic steps.

b Is the pathway anabolic or catabolic?

c The shape of enzyme K is shown below.

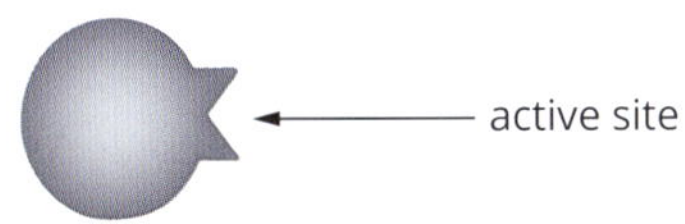

High concentrations of molecule E inhibit enzyme K.

i What is the substrate for enzyme K?

ii Draw a possible shape for molecule E.

iii Using annotated diagrams explain how molecule E inhibits enzyme K catalysing the formation of molecule B, thereby regulating the pathway.

15 One toxic product of respiration is hydrogen peroxide (H_2O_2). Hydrogen peroxide is so toxic that even though it spontaneously breaks down to form water and oxygen all living things contain an enzyme, called catalase, that speeds up this reaction.

The chemical breakdown occurs according to the following equation:

$$2H_2O_2 \rightarrow 2H_2O + O_2$$

A group of students performed an experiment to investigate the effects of varying substrate concentration on the activity of catalase. Potatoes were used as a source of catalase. The potatoes were pureed and then the puree was strained to collect the potato juice, which contains the enzyme catalase.

The equipment was set up to collect the oxygen gas as it was produced, as shown in the experimental setup below. The measuring cylinder was filled with water before being inverted into the tank, which was also filled with water. 20 mL of potato juice was placed in the flask and the flask was sealed as shown.

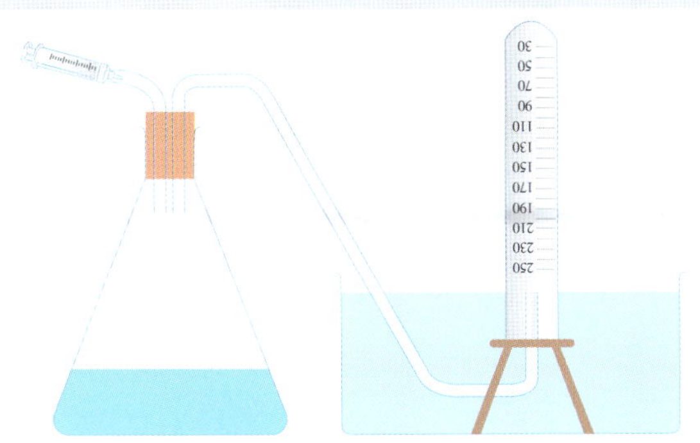

2 mL of hydrogen peroxide was drawn into the syringe and attached to the tube as shown. A 250 mL measuring cylinder and a 10 mL syringe having 1 mL graduations were used. The classroom clock was used to time the reaction. Timing began as soon as the syringe was depressed to mix the hydrogen peroxide and the enzyme mixture.

In the first test 2 mL of a 5% solution of H_2O_2 was used. Subsequent tests used 2 mL of 10%, 15%, 20% and 25% solutions of H_2O_2. The amount of oxygen produced in five minutes was measured. The results obtained by the students are shown in the table below.

Volume of oxygen produced by increasing concentrations of H_2O_2 when acted on by catalase from potato.

Concentration of H_2O_2	Volume of oxygen collected
5%	9.5 cm^3
10%	20 cm^3
15%	31 cm^3
20%	40.5 cm^3
25%	52 cm^3

a Why can the volume of oxygen produced be used as a measure of the activity of catalase?

b **i** Plot a graph of the students' results.

ii Consider the graph. What relationship appears to exist between the concentration of H_2O_2 and the volume of oxygen produced?

c **i** Identify possible sources of error in the experimental design and equipment.

ii For each source of error identified suggest a way of reducing the error or its impact on the results.

d How could the reliability and accuracy of the results be increased?

e Why must the H_2O_2 and the potato mixture be kept apart until timing starts?

f The breakdown of hydrogen peroxide is an exothermic reaction.

i What is an exothermic reaction?

ii Explain how this might impact on the results of the experiment.

iii How might you investigate the level of influence that the exothermic nature of the reaction is having on the outcome?

g Enzymes from plants often have a much greater range of temperatures over which they maintain their activity than mammalian enzymes. Why might this be the case?

16 Many junior science books state that 'All enzymes are proteins. Each enzyme is specific to one reaction'. Discuss the truth, or otherwise, of this statement.

CHAPTER

05 Energy transformation in cells

Learning outcomes

Energy is vital for life. Whether growing, moving, reproducing, responding or excreting, living organisms are using energy. 'Using energy' involves transforming energy from one form to another, and transferring it from one place to another. When a muscle contracts, for example, chemical energy in the muscle cells is being transformed into the kinetic energy of movement.

By the end of this chapter, you will have an understanding of the cellular organelles and biochemical pathways that provide living things with the energy they need to survive.

You will also learn how cells adjust their metabolism to account for changes in environmental conditions.

Key knowledge

- the purpose of photosynthesis
- chloroplasts as the site of photosynthesis, an overview of their structure and evidence of their bacterial origins
- inputs and outputs of the light-dependent and light-independent (Calvin cycle) stages of photosynthesis in C3 plants (details of the biochemical pathway mechanisms are not required)
- factors that affect the rate of photosynthesis, including light, temperature and carbon dioxide concentration.
- the purpose of cellular respiration
- the location of, and the inputs and outputs of, glycolysis including ATP yield (details of the biochemical pathway mechanisms are not required)
- mitochondria as the site of aerobic cellular respiration, an overview of their structure and evidence of their bacterial origins
- the main inputs and outputs of the Krebs (citric acid) cycle and electron transport chain including ATP yield (details of the biochemical pathway mechanisms are not required)
- the location of anaerobic cellular respiration, its inputs and the difference in outputs between animals and yeasts including ATP yield
- factors that affect the rate of cellular respiration, including temperature, glucose availability and oxygen concentration.

5.1 Photosynthesis

A biochemical pathway is a series of chemical reactions that can occur inside a cell.

FIGURE 5.1.1 Licuala fan palms from Northern Queensland stretch their broad leaves skywards to capture the sunlight they need for photosynthesis.

Living organisms need a constant supply of energy. The ultimate supply of energy for living things comes from the physical environment—in most cases, the light energy of the Sun (Figure 5.1.1).

In this section, you will learn about the biological structures and biochemical pathways involved in photosynthesis, which capture the Sun's light energy, transform and store it as chemical potential energy (Figure 5.1.2). Cyanobacteria, algae, phytoplankton and terrestrial plants all obtain their energy through the process of photosynthesis. This section will focus on plants because they are the most visible photosynthetic organisms in our everyday lives.

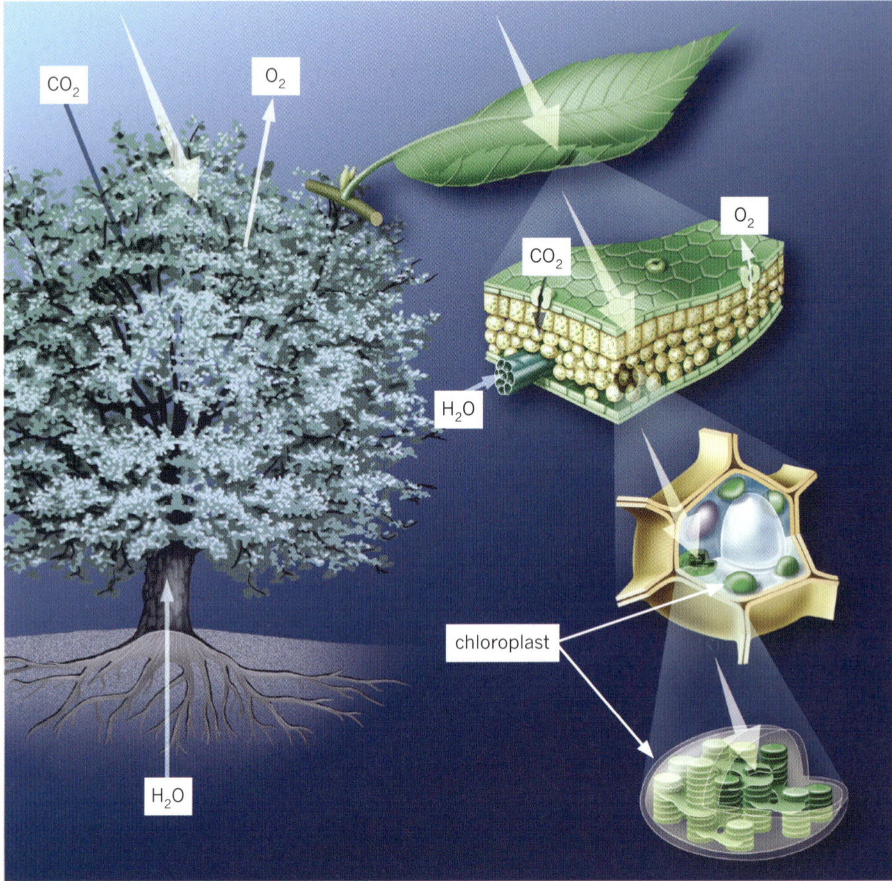

FIGURE 5.1.2 Leaves have an enlarged surface area to maximise the amount of light absorbed. Each leaf contains cells with specialised organelles called chloroplasts, that are capable of converting light energy into chemical potential energy.

During photosynthesis 12 H_2O are consumed and 6 H_2O are produced. A simpler equation for photosynthesis only shows the net water consumed:

$6CO_2 + 6H_2O \rightarrow C_6H_{12}O_6 + 6O_2$

Light energy is written above the arrow in the equation for photosynthesis, as it is not a reactant of photosynthesis, but it is required for the reaction to take place.

PHOTOSYNTHESIS—CONVERTING SOLAR ENERGY

Photosynthesis begins with the absorption of light energy by **chlorophyll** (Figure 5.1.3). This absorbed energy is then used to synthesise **glucose** from carbon dioxide and water.

Photosynthesis is a complex multistage process, but it is usually summarised in the following way:

INPUTS				OUTPUTS				
carbon dioxide	+	water	$\xrightarrow[\text{chlorophyll}]{\text{light energy}}$	glucose	+	water	+	oxygen
$6CO_2$	+	$12H_2O$	$\xrightarrow[\text{chlorophyll}]{\text{light energy}}$	$C_6H_{12}O_6$	+	$6H_2O$	+	$6O_2$

FIGURE 5.1.3 Chlorophyll is a pigment that gives leaves – both (a) terrestrial and (b, c) aquatic – their green appearance.

Figure 5.1.4 is a simple illustration of the process of photosynthesis. Each stage involves a series of biochemical reactions, often referred to as a biochemical pathway (or metabolic pathway). Each reaction in the pathway is catalysed (accelerated) by a particular enzyme.

The glucose formed in photosynthesis may be:

- used as an immediate source of energy by the plant
- stored by the plant as starch for later conversion back to glucose and use as a source of energy
- used as a chemical starting point for the synthesis of complex compounds, such as cellulose and proteins.

The oxygen formed in photosynthesis may be used for aerobic cellular respiration by the plant or released into the atmosphere.

FIGURE 5.1.4 Water is taken up through the roots of the plant towards the leaves. Carbon dioxide is taken in through the stomata in the leaves while oxygen is released. Light is absorbed by the chlorophyll in the leaves. Glucose (not shown) can be used immediately by the cells or stored for later use.

THE STRUCTURE OF CHLOROPLASTS

In the cells of eukaryotic **autotrophs**, photosynthesis occurs in organelles called chloroplasts. Each chloroplast has an outer and an inner membrane, which together regulate the movement of materials into and out of the organelle. Inside these membranes is a fluid matrix called **stroma** and a highly complex inner **thylakoid** membrane system. Figure 5.1.5 shows that the thylakoid membranes fold to form flat hollow discs, which form stacks called **grana**. Each **granum** looks like a stack of coins. Between the grana are flat membrane sheets called **thylakoid lamellae**.

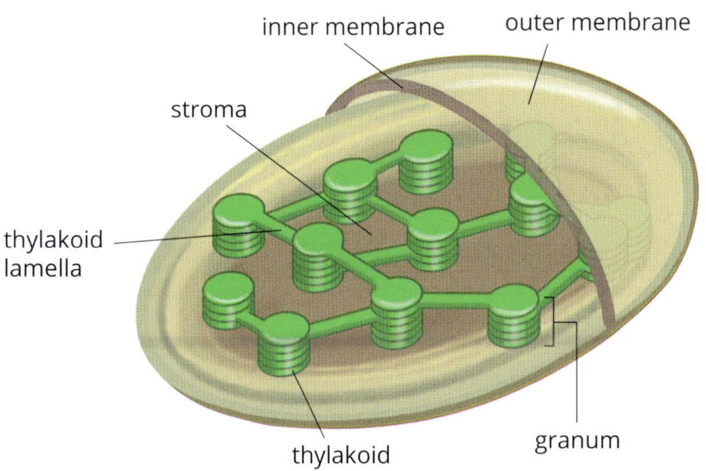

FIGURE 5.1.5 Diagram showing the main structures of a chloroplast.

Cellulose can be found in cell walls and woody tissue.

Plants conduct gas exchange through stomatal pores in their leaves.

The origin of chloroplasts

Cyanobacteria are photosynthetic prokaryotes. The cell membranes of these unicellular organisms are heavily in-folded to form a complex internal membrane structure that is much like the thylakoid membranes of chloroplasts. The cyanobacteria cells also contain chlorophyll and enzymes similar to those of chloroplasts.

It was the structural, functional and biochemical similarities between cyanobacteria and chloroplasts that first led scientists to propose the theory of **endosymbiosis** as an explanation for the origin of chloroplasts. This theory proposes that chloroplasts are the descendants of cyanobacteria that were engulfed by eukaryotic cells in ancient times (about 2 billion years ago) and lived in **symbiotic** relationship within the cells that had engulfed them. Over time, the cyanobacteria became permanent residents of the cells and progressively changed to become the chloroplasts we see today.

> Plastids are large double-membrane bound organelles found in plants and algae. For this reason, the DNA found in chloroplasts can be referred to as 'plastidial DNA'.

There is considerable additional evidence to support the endosymbiotic theory:

- In Figure 5.1.6 you can see that chloroplasts have their own circular DNA (sometimes called plastidial DNA) and replicate independently by binary fission. Bacteria also possess circular DNA.
- Chloroplasts have ribosomes that share similarities with bacterial ribosomes.
- If chloroplasts are removed from a plant or algal cell, the cell is unable to generate new chloroplasts.
- Studies of the DNA of modern-day cyanobacteria support the idea that they arose from a common ancestor of the chloroplast.
- The outer membranes of chloroplasts contain transport proteins called **porins**. Porins are membrane proteins also found in mitochondria. Otherwise porins are only found in prokaryotic organisms. This fact supports the idea that both chloroplasts and mitochondria are the descendants of prokaryotes once engulfed by eukaryotic cells.

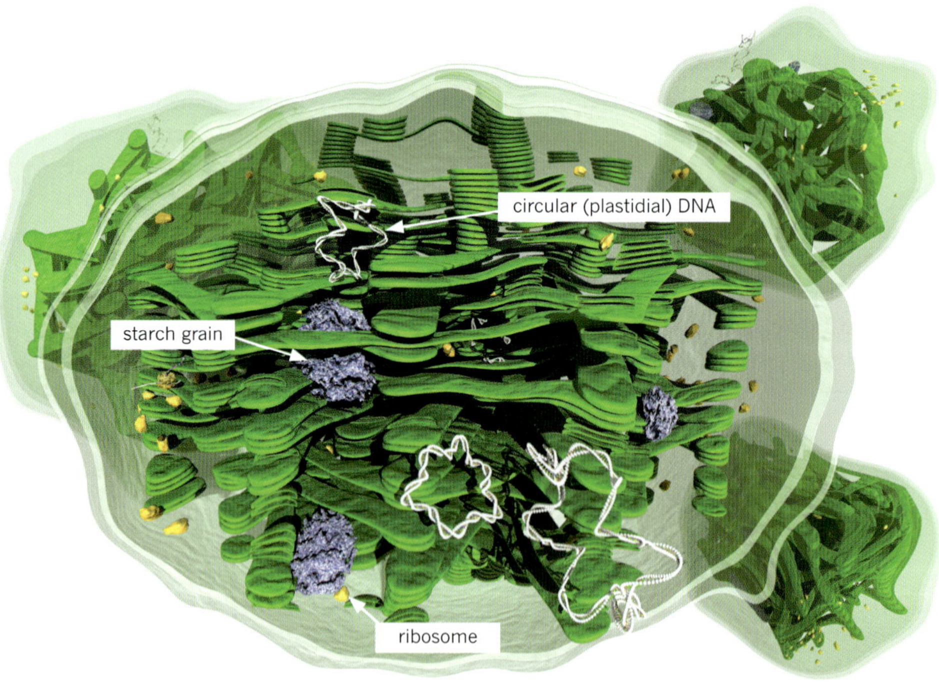

FIGURE 5.1.6 Chloroplasts contain their own circular (or plastidial) DNA (white) and ribosomes (yellow). The DNA codes for proteins found within the chloroplasts, including proteins found in the thylakoid membranes. The ribosomes are used to synthesise new proteins.

EXTENSION

Photosynthetic pigments

Photosynthetic pigments are coloured substances that collect the light energy that is used in photosynthesis. The green pigment in chloroplasts is chlorophyll. It is the most abundant and visible photosynthetic pigment found in plants.

Figure 5.1.7a shows us that the rate of photosynthesis is highest under red light (between 650 and 700 nm). The rate of photosynthesis is also high under blue and violet light (between 400 and 450 nm). The lowest rate of photosynthesis occurs under green light (between 500 and 600 nm). We can explain this using Figure 5.1.7b, as different types of chlorophyll absorb different wavelengths of light. Chlorophyll absorbs many of the colours that make up the spectrum of white light, but it reflects green light. The spectrum (Figure 5.1.7b) shows clearly that red, blue and violet light are all strongly absorbed, while yellow is absorbed to a lesser extent and green is not absorbed at all.

There are several types of chlorophyll found in photosynthetic organisms:

- Chlorophyll *a* is found in all photosynthetic organisms.
- Chlorophyll *b* is found in some plants.
- Chlorophyll c is found in algae.
- Chlorophyll *d* and *f* are found in cyanobacteria.

These different types of chlorophyll have slightly different molecular shapes and so have different light-absorption spectra.

Photosynthetic organisms also contain pigments called carotenoids, which are red, orange and yellow in colour and are also involved in capturing light. When chlorophyll is broken down in some plants in autumn, the colours of the carotenoids are no longer masked by the green chlorophyll and are exposed, giving autumn leaves their colours (Figure 5.1.8). Carotenoids are known as accessory pigments because they cannot pass their absorbed energy directly to photosynthesis but can transfer it to chlorophyll to then be used in photosynthesis. Accessory pigments broaden the range of wavelengths absorbed and therefore increase the amount of light that a plant can absorb for use in photosynthesis.

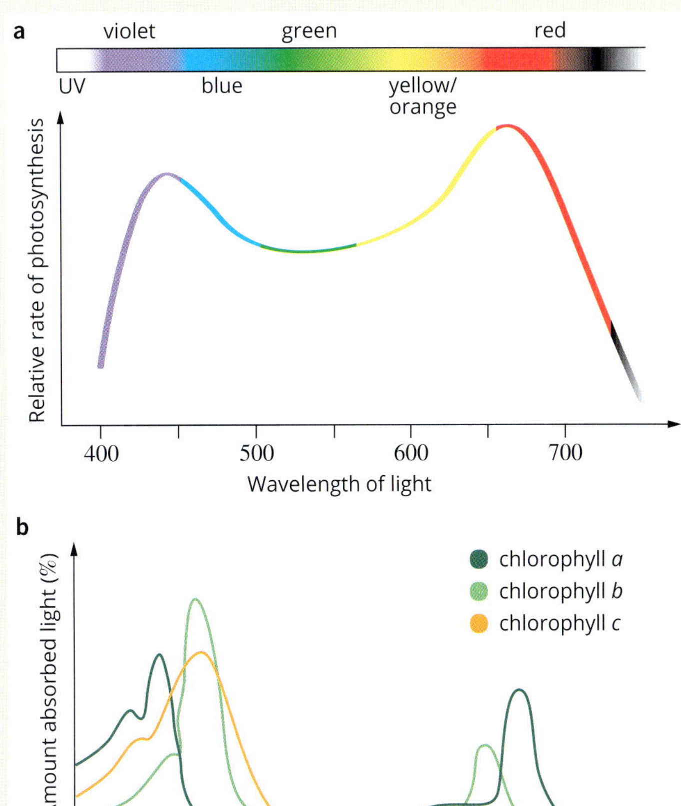

FIGURE 5.1.7 The rate of photosynthesis (a) occurs at different rates in light of different wavelengths. The pattern of photosynthetic activity is very similar to (b) the absorption spectrum of chlorophyll.

FIGURE 5.1.8 Deciduous plants such as the red maple decrease chlorophyll production in autumn, before losing their leaves. The red, yellow, orange and brown colours of carotenoids are visible when there is no green-coloured chlorophyll in the leaves.

THE LIGHT-DEPENDENT REACTIONS

The first stage of photosynthesis involves biochemical pathways known as the **light-dependent reactions**. These reactions occur on the thylakoid membranes of the chloroplast, where chlorophyll and the enzymes involved are located. The reactions can only take place in the presence of light.

First, light energy is absorbed by chlorophyll. The energy is then used to split water to produce oxygen, and to form the energy-carrying molecules ATP (adenosine triphosphate) and NADPH from ADP (adenosine diphosphate) and **NADP⁺ (nicotinamide adenine dinucleotide phosphate)**.

The light-dependent reactions are summarised in Figure 5.1.9.

NADP⁺ is a carrier molecule. It can receive hydrogen ions to become NADPH.

ADP can create a temporary bond with inorganic phosphate (P_i) to become ATP. ATP is an energy-carrying molecule.

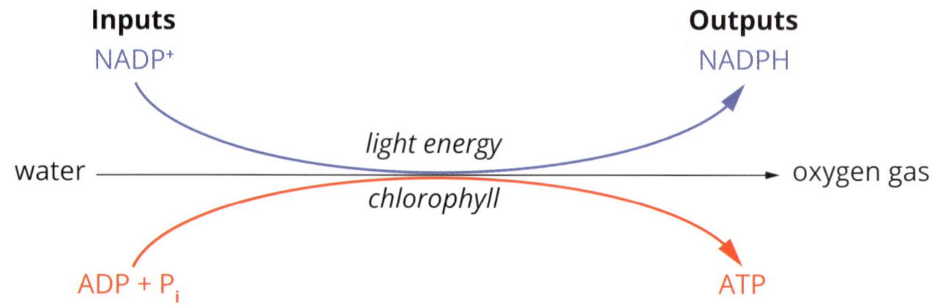

FIGURE 5.1.9 Water is split into oxygen and hydrogen. The oxygen is released as a gas. The NADPH and ATP are used in the second stage of photosynthesis.

EXTENSION

Photosystems I and II

Within the thylakoid membranes are a number of highly complex, linked chemical systems and a large number of enzymes. Two key chemical systems are photosystem I and photosystem II. Both systems contain chlorophyll and depend on light to function. Together with the enzyme ATP synthase these systems carry out the processes called the light-dependent reactions.

The chlorophyll in photosystem II absorbs light energy that is used to split water into oxygen, hydrogen ions and electrons. This can be summarised as:

$$2H_2O \rightarrow O_2 + 4H^+ + 4e^-$$

water → oxygen + hydrogen ions + electrons

The chlorophyll in photosystem I, absorbs light energy that is transferred together with hydrogen ions and electrons to form a coenzyme energy carrier called NADPH from NADP⁺ (nicotinamide adenine dinucleotide phosphate). This reaction is summarised as:

$$NADP^+ + H^+ + 2e^- \rightarrow NADPH$$

low energy → high energy

ATP synthase is the enzyme that uses energy generated by both photosystems to form ATP from ADP and inorganic phosphate (P_i).

THE LIGHT-INDEPENDENT REACTIONS

The products of the light-dependent reactions, NADPH and ATP, are released into the stroma of the chloroplast. There, these coenzymes provide energy to drive a biochemical pathway known as the **light-independent reactions**. The light-independent reactions are summarised in Figure 5.1.10.

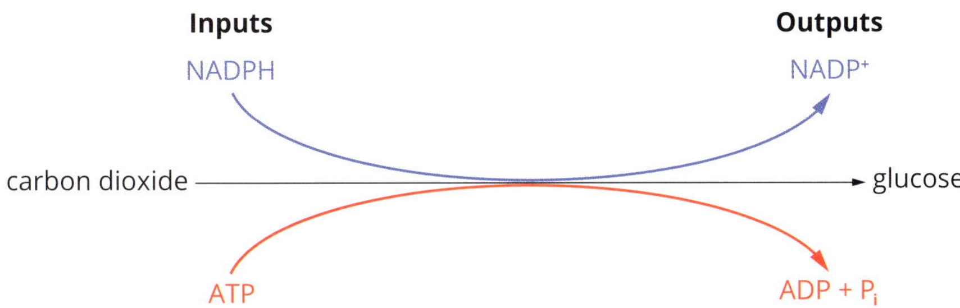

FIGURE 5.1.10 In the second stage of photosynthesis carbon dioxide is reduced to form the sugar glucose.

The second stage of photosynthesis does not require light for the reaction to be able to proceed. However, it does require the NADPH and ATP of the light-dependent reactions. If NADPH and ATP are present, the reaction can continue in the absence of light (Figure 5.1.11).

The main stage of the light-independent reactions is called the **Calvin cycle**. The individual steps of the Calvin cycle are outlined in the Extension box on page 191.

During the Calvin cycle carbon dioxide is reduced to form GAP (a three-carbon molecule). Once the GAP molecules are formed, they can leave the chloroplast and move into the cytosol of the cell, where two GAP molecules can join together to form one glucose unit.

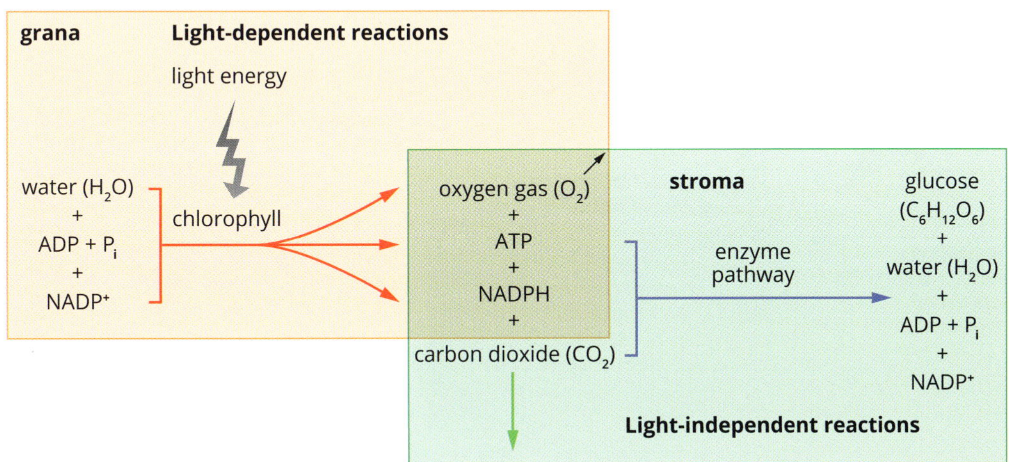

FIGURE 5.1.11 Photosynthesis occurs in two phases. The light-dependent reactions occur in the thylakoid membranes of the grana. The light-independent reactions occur in the stroma. The light-independent reactions require the carriers generated by the light-dependent reactions.

BIOLOGY IN ACTION

Plants and microbes provide electricity in the Amazon rainforest

When floodwater damaged cables supplying electricity to an indigenous community in the eastern Amazonian region of Ucayali, 65% of the community was left without power. Kerosene lamps were used as a source of light, but the smoke emitted by them caused breathing and vision problems.

A research team at the University of Engineering and Technology (UTEC) in Lima, Peru, wanted to provide a clean, renewable source of electrical light to the community. Using findings from research on plant–microbial fuel cells, a research team of seven professors and eight students developed a lamp powered by plants and microbes (Figure 5.1.12). Over four months, the researchers worked on the 'plant lamps', travelling to the remote community to conduct research and then later, in October 2015, to deliver ten prototypes of their lamps.

The 'plant lamp' technology takes advantage of a natural energy exchange between photosynthesis, plant roots and soil bacteria. During photosynthesis, excess energy that is generated by the plant is excreted through the root system into the soil in the form of organic matter. Microbes, known as *Geobacter*, inhabit the soil around the plant roots and break down the plant's organic matter, releasing electrons as a waste product. The 'plant lamp' uses a grid of electrodes to capture the free electrons in the soil and stores the energy in a battery, which is used to power a low energy-consumption, high-illumination LED light bulb.

This innovation has very real economic and educational benefits for people in the remote Ucayali community, lighting their homes and providing more hours in the day for work and study. The 'plant lamps' can generate a clean and safe source of light for two hours per day and have the potential to be used in many renewable energy applications.

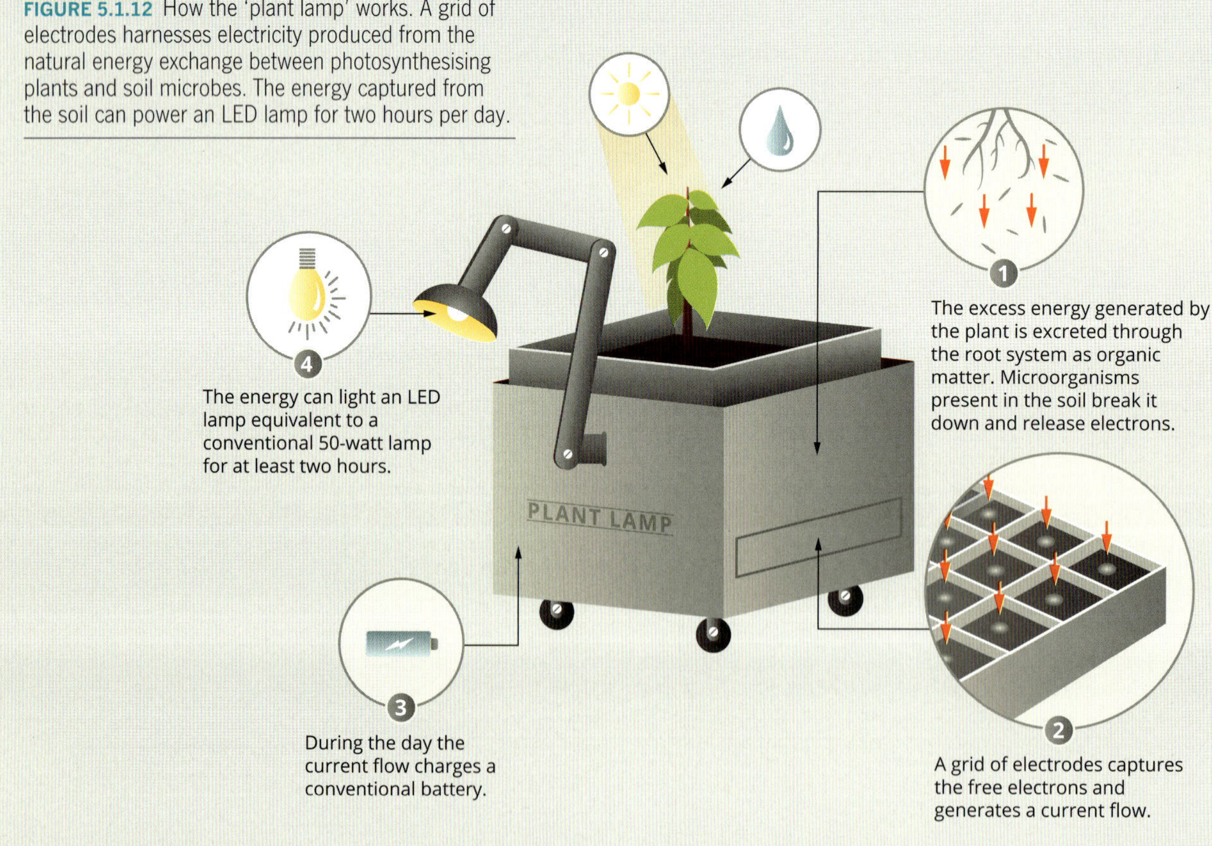

FIGURE 5.1.12 How the 'plant lamp' works. A grid of electrodes harnesses electricity produced from the natural energy exchange between photosynthesising plants and soil microbes. The energy captured from the soil can power an LED lamp for two hours per day.

EXTENSION

The Calvin cycle

The Calvin cycle comprises all the light-independent reactions. This biochemical pathway of enzyme-catalysed reactions uses carbon dioxide and the energy in the NADPH and ATP produced in the light-dependent reactions to make carbohydrates (Figure 5.1.13). To understand how this process works, it is necessary to realise that each cycle does not exist as a separate entity. There are many instances of the same reactions occurring simultaneously in the stroma and therefore many concurrent Calvin cycles occurring. If it were possible to isolate a single cycle, then one carbon atom would be added to the system with each 'turn' but no one-carbon product is actually produced. The identifiable product of the Calvin cycle is a three-carbon carbohydrate called glyceraldehyde-3-phosphate (GAP). In the cytoplasm two GAP molecules combine to produce the glucose ($C_6H_{12}O_6$) molecule that is normally identified as the product of photosynthesis. The remaining GAP molecules are recycled into ribulose 1,5 bisphosphate, the starting molecule of the light-independent reaction.

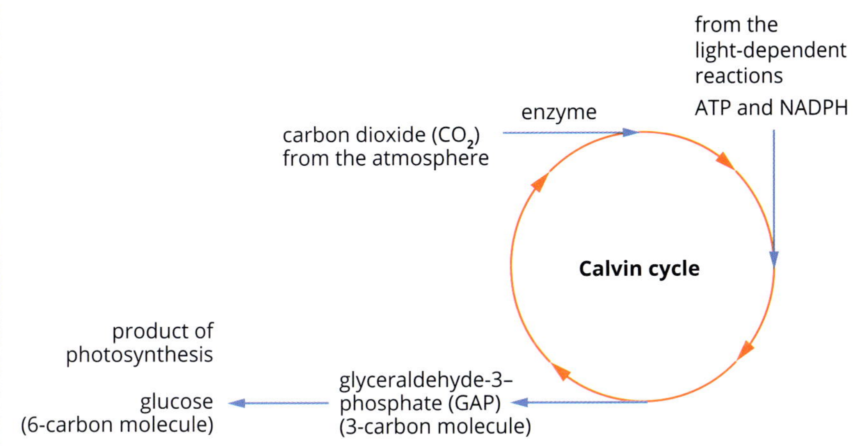

FIGURE 5.1.13 Carbon dioxide from the atmosphere feeds into the Calvin cycle, which uses energy carried by ATP and NADPH from the light-dependent reactions.

5.1 Review

SUMMARY

- Photosynthesis is carried out in two stages: the light-dependent reactions and the light-independent reactions.
- Chloroplasts contain thylakoid membranes and fluid called stroma. They also contain their own DNA and ribosomes.
- The light-dependent reactions occur on the grana. Light energy is trapped by chlorophyll, water is split and oxygen is released as a product.
- The light-independent reactions occur in the stroma. Carbon dioxide is reduced to form glucose.

KEY QUESTIONS

1 Write the balanced chemical equation for photosynthesis.

2 Which organelle is the site of photosynthesis?

3 Which three structural features are shared by bacteria and chloroplasts, thus providing evidence for the theory of endosymbiosis?

4 a Where in the chloroplasts do the light-dependent reactions occur?

b Name the reactant that is required for the light-dependent reactions.

c Identify two other inputs required for the first phase of photosynthesis and state their role in the process of photosynthesis.

d What process occurs in the stroma and what is the final product of this stage?

5.2 Factors that affect the rate of photosynthesis

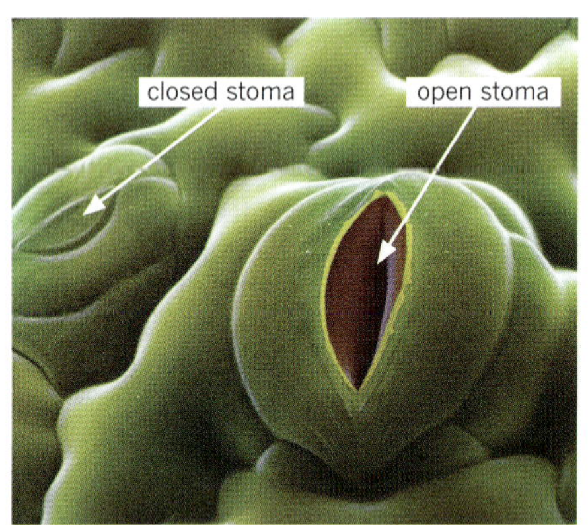

FIGURE 5.2.1 This coloured scanning electron micrograph image of a tobacco leaf shows one open stoma and one closed stoma. When the stoma is open gas exchange can occur and water molecules can escape.

As for all biochemical processes, the rate of photosynthesis varies according to the internal conditions of the cell, which in turn are often affected by the external environmental conditions.

Photosynthesis depends on variables that are interconnected; for example, the opening of **stomata** controls both loss of water and entry of carbon dioxide, as shown in Figure 5.2.1. For this reason, the only way of studying the effect of each factor on the rate of photosynthesis is in the laboratory. Laboratory studies using suspensions of isolated chloroplasts or green algae enable scientists to test the effect of varying different factors on the rate of photosynthesis under controlled conditions.

In this section, you will learn about the factors that affect the rate of photosynthesis in **C3 plants** (the most common type of plant).

C3 plants are so named because they rely on the C3 carbon fixation pathway (the Calvin cycle) to convert CO_2 into an organic compound.

INPUTS: CARBON DIOXIDE AND LIGHT ENERGY

If we look closely at the process of photosynthesis described in the overall equation, we can predict that the rate at which it occurs will be affected by a number of factors.

The main requirements of photosynthesis are carbon dioxide, water and light energy. If any one of these factors is in limited supply, it is reasonable to predict that the rate of photosynthesis will be also be limited. Because the amount of water used in photosynthesis is small compared with the amount needed to keep the cells alive, a living plant cell will normally have sufficient water for photosynthesis to occur. Water therefore does not have a direct effect on the rate of photosynthesis in nature. However, water does have an indirect effect, because when the plant is suffering from water stress (Figure 5.2.2), the stomata in the leaf close and reduce the availability of carbon dioxide. So, under normal circumstances we can say that photosynthesis will be affected by the availability of carbon dioxide and light.

FIGURE 5.2.2 This plant has wilted due to water loss. When water is not available, the plant cells lose water, reducing the turgor (pressure) inside the cell. Therefore the cells cannot maintain a rigid shape, and the plant collapses.

Carbon dioxide

Terrestrial plants receive their carbon dioxide from the air. Aquatic plants can utilise the carbon dioxide dissolved in the water.

The carbon dioxide level in the air remains relatively constant. The factors that affect the amount available for photosynthesis in most terrestrial plants are the number of stomata in the leaves and whether these stomata are open or closed. If the stomata are closed, photosynthesis will use up the carbon dioxide inside the leaf, therefore lowering the carbon dioxide concentration in the leaf. With less carbon dioxide available, the rate of photosynthesis, even in the presence of light, will be limited.

In the laboratory it is possible to control the concentration of carbon dioxide to which plants are exposed without changing other factors. Figure 5.2.3 shows a comparison of the rate of photosynthesis for a particular species of plant exposed to different concentrations of carbon dioxide at different light intensities.

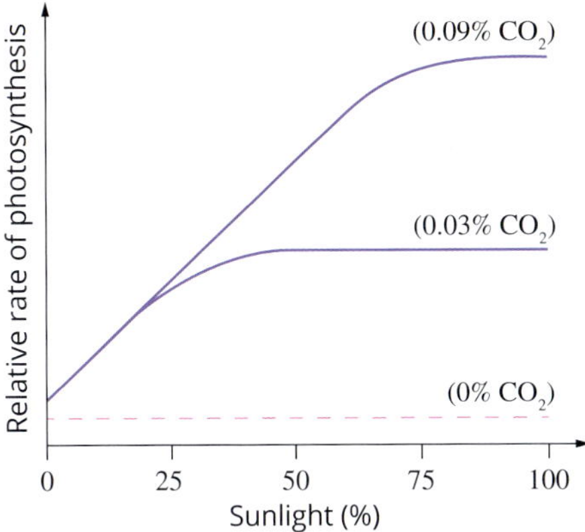

FIGURE 5.2.3 The rate of photosynthesis increases with increasing light intensity at two different concentrations of carbon dioxide. At 0.03% carbon dioxide the rate of photosynthesis increases until approximately 30% sunlight at which point the rate of photosynthesis is limited by the availability of carbon dioxide. At 0.09% carbon dioxide the rate of photosynthesis continues to increase up to about 80% sunlight. The concentration of carbon dioxide in the atmosphere is currently about 0.04%.

Light

In the laboratory, chloroplasts can be extracted from plant cells and tested in isolation. By varying the amount of light shining on these isolated chloroplasts, while keeping the carbon dioxide levels constant, it is possible to measure the rate at which photosynthesis occurs at different light levels. The results of this experiment are shown in Figure 5.2.4 and present what we refer to as a **light saturation curve**. The curve shows a steady increase in the photosynthesis rate with an increase in light intensity until a plateau is reached. The plateau indicates that there is a maximum rate at which photosynthesis can occur. Assuming unlimited amounts of carbon dioxide (and water), the limit will be the point at which all of the photosynthesis systems and enzymes in the chloroplasts are working at their optimum rate.

In the natural environment, the amount of light available to a plant for photosynthesis will be determined by the amount of sunlight in its environment. Sunlight will vary during the cycle of a day and will change with the seasons and the weather. Figure 5.2.5 shows how trees and taller plants will shade plants on the forest floor and the amount of light available to aquatic plants will be dependent on how far underwater they grow.

BIOFILE

The unexpected result of rising carbon dioxide levels

The rising atmospheric carbon dioxide levels and consequent global warming will lead to increased rates of photosynthesis in plants. This means plants will produce more glucose, and it has long been predicted that crops will therefore grow more quickly. Scientists studying the effects of elevated carbon dioxide levels on some important crop plants have, however, made some unexpected and alarming findings. In experiments in which crops were grown in an atmosphere with artificially elevated carbon dioxide levels, it was shown that although overall plant growth increased, levels of zinc, iron and protein in the plants decreased. People rely on food from crops for these important nutrients, and decreased levels may lead to serious malnutrition for people throughout the world.

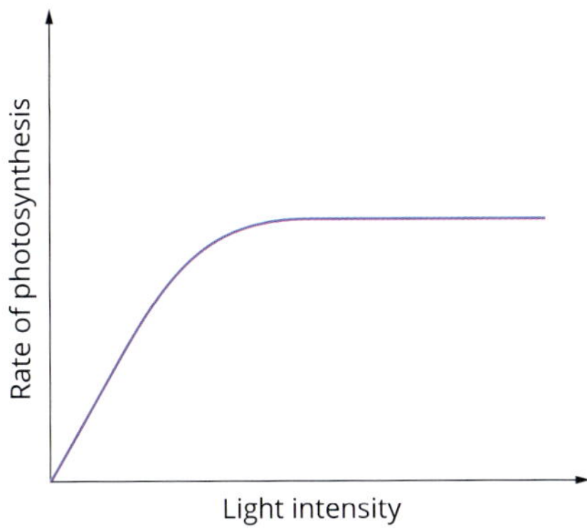

FIGURE 5.2.4 Light saturation curve. At a constant temperature and with unlimited carbon dioxide, the rate of photosynthesis will increase as the light intensity increases up to the point at which the photosynthetic processes are working at a maximum rate.

FIGURE 5.2.5 (a) Plants growing on the forest floor will receive less light than taller trees. (b) Aquatic photosynthetic organisms, such as this sea kelp, can only grow near the water's surface. As you move deeper underwater, there is less light available.

> When the temperature is below the optimum range of an enzyme, there is low kinetic energy (movement) in the molecules. Therefore the rate of photosynthesis is low, as the water and carbon dioxide molecules are moving slower.

Temperature

Many enzymes in chloroplasts catalyse the reactions of photosynthesis. Without these enzymes, photosynthesis would not occur. In Section 4.2 you learnt that the rate of an enzyme reaction is affected by temperature and so, accordingly, the reactions of photosynthesis are affected by temperature. The temperature that is optimum for the functioning of the enzymes of photosynthesis will also be the temperature at which the maximum rate of photosynthesis occurs. Similarly, as enzymes are denatured and no longer functional at particularly high temperatures, photosynthesis will cease at these high temperatures. This is seen in Figure 5.2.6.

> Plants from Arctic regions will have a lower optimum temperature range than plants found in tropical climates.

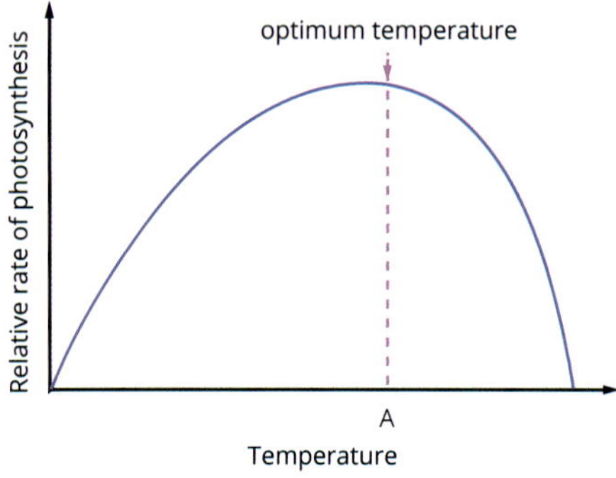

FIGURE 5.2.6 The rate of photosynthesis at different temperatures. The shape of the curve is typical of those shown for enzyme reactions. The rate rises with temperature to an optimum temperature and then falls to zero at high temperatures as the photosynthetic enzymes denature.

BIOFILE

Chlorophyll

Chlorophyll is required to absorb light for photosynthesis and so the amount of chlorophyll will affect the rate of photosynthesis.

There is variation in the amount of chlorophyll per chloroplast in plant cells, the number of chloroplasts per photosynthesising cell and the number of photosynthesising cells per leaf.

Light falling on a plant during its growth affects the amount of chlorophyll that is produced. The leaves of seedlings grown in the dark will be yellow in colour. If they are exposed to light, they will begin to turn green, as a result of the production of chlorophyll.

FIGURE 5.2.7 A comparison of runner bean plants (*Phaseolus coccineus*) grown in the light (left) with runner bean plants grown in the dark (right). Those grown in the light clearly show a darker green colour than the paler plants grown in the dark, which also show etiolated (long pale) stems.

5.2 Review

SUMMARY

- The factors affecting the rate of photosynthesis are interconnected.
- An increase in carbon dioxide levels can increase the rate of photosynthesis.
- An increase in light intensity can increase the rate of photosynthesis.
- The availability of chloroplasts to carry out photosynthesis can limit the rate of reaction.
- Plants have an optimum temperature range for photosynthesis. Too cold and the rate of reaction will be slow. Too hot and the enzymes in chloroplasts can denature.

KEY QUESTIONS

1 When a plant closes its stomata it can no longer exchange oxygen and carbon dioxide, therefore the rate of photosynthesis decreases. What is the benefit of closing its stomata?

2 Looking at the graph below, what is the overall effect of increased carbon dioxide levels on the rate of photosynthesis?

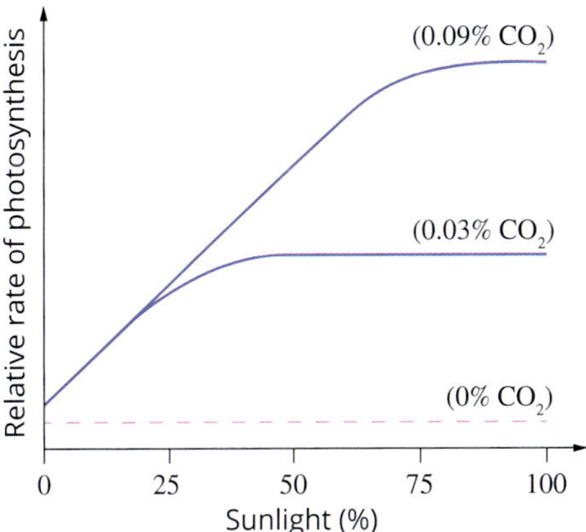

3 If a plant is given unlimited carbon dioxide and unlimited access to light, will the rate of photosynthesis increase? Explain your answer.

4 The graph below shows the rate of photosynthesis at different temperatures.

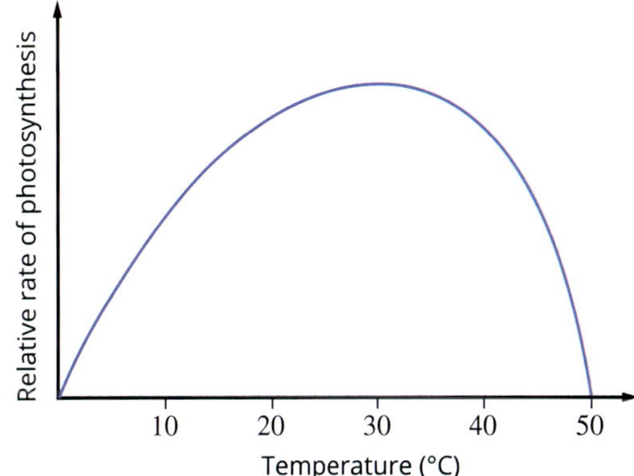

Use the graph to answer the following questions.

a Describe the rate of photosynthesis between 0 and 20°C and explain the trend.

b At what temperature does the rate of photosynthesis begin to decrease? Explain why the rate of reaction begins to decrease at this temperature.

5.3 Cellular respiration: glycolysis

Cellular respiration is the name given to the combination of biochemical pathways that together release energy from glucose.

Glycolysis is the first biochemical pathway in cellular respiration.

All cells need energy to function (Figure 5.3.1). They obtain this by releasing energy from organic compounds through a series of biochemical pathways; glycolysis is the first of these. Cellular respiration is the name given to the combination of biochemical pathways that together release energy from glucose.

In this section, you will learn about the first biochemical pathway in cellular respiration, glycolysis, and how the energy produced in glycolysis is transferred from glucose to coenzymes to be used in other processes in the cell.

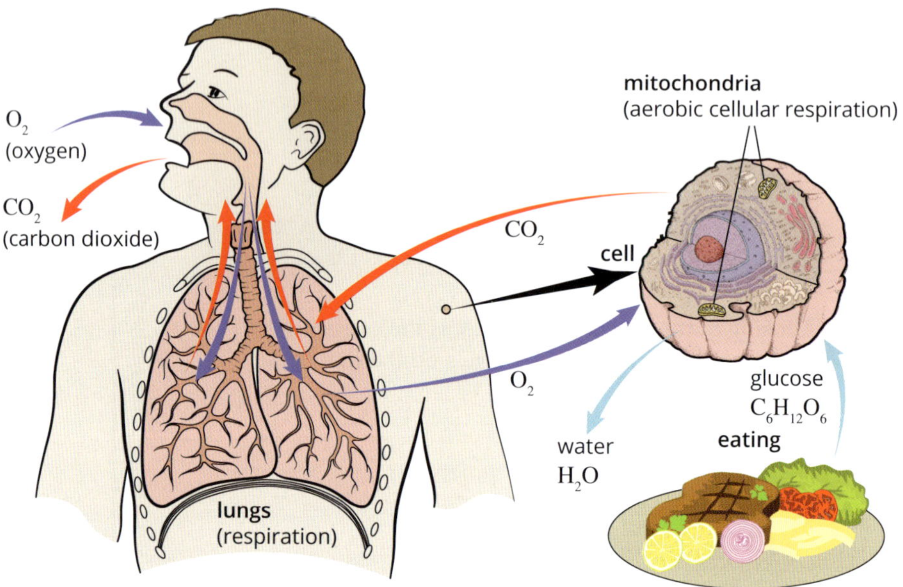

FIGURE 5.3.1 As animals, we take our energy from the food we eat. Plants generate their own energy using sunlight (through photosynthesis).

FIGURE 5.3.2 Complex carbohydrates are broken down by our digestive system into simple sugars, such as glucose. Glucose is the primary source of energy for most of our cells.

ENERGY FROM GLUCOSE

Cells can obtain energy from high-energy fats and other organic compounds including proteins. However, most cells use glucose as their immediate source of energy (Figure 5.3.2). To extract energy from other organic molecules, some must be converted into glucose first, others are broken down to small molecules that can enter the cellular respiration pathways.

Cellular respiration is the name given to the combination of biochemical pathways that occur together within a cell to release energy from glucose. The energy released from glucose through cellular respiration is used to generate the coenzyme adenosine triphosphate (ATP). The energy is transferred when ATP is formed from ADP and inorganic phosphate (P_i), and it is stored in the bond between ADP and phosphate. This can be represented as:

$$ADP + P_i \rightarrow ADP{\sim}P_i \text{ (ATP)}$$

The '~' represents the high-energy bond in which the energy is stored.

When the high-energy bond in ATP is broken, energy is released for use in the many energy-demanding processes that occur in the cell. The transfer of energy is summarised in Figure 5.3.3.

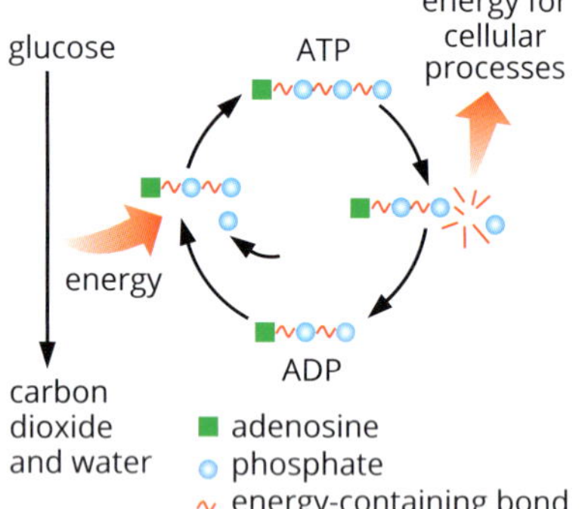

FIGURE 5.3.3 Energy from glucose is transferred in the synthesis of ATP from ADP and phosphate. The energy is stored in the phosphate bond. When the bond is broken, the energy is released to drive cellular processes. The ADP and phosphate are recycled.

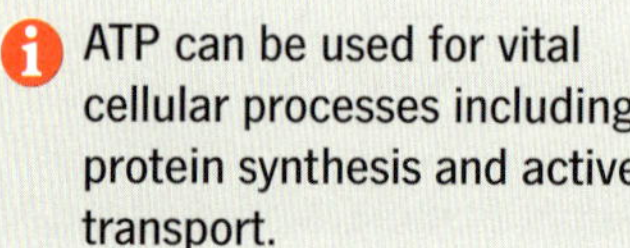

ATP can be used for vital cellular processes including protein synthesis and active transport.

Aerobic cellular respiration requires oxygen. When oxygen supplies are low, a cell can carry out anaerobic respiration.

Aerobic cellular respiration consists of the three interconnected biochemical pathways known as:

- glycolysis
- the Krebs cycle
- the electron transport chain.

Anaerobic cellular respiration also involves glycolysis. In plant and yeast cells, the product of glycolysis will undergo a fermentation pathway.

GLYCOLYSIS

Glycolysis is the first stage of cellular respiration and occurs in the cytosol of the cell. It results in the net production of 2 ATP molecules. During glycolysis, glucose (a six-carbon molecule) is broken down to two three-carbon molecules called pyruvate (or pyruvic acid).

During the process of glycolysis, a small amount of energy is released. This energy is transferred to the coenzymes ATP and NADH, which act as energy carriers. The overall reaction of glycolysis is:

INPUTS		OUTPUTS
Glucose + 2ADP + $2P_i$ + $2NAD^+$	→	2 pyruvate + 2ATP + 2NADH

NADH and $FADH_2$

An important coenzyme in cellular respiration is **nicotinamide adenine dinucleotide** (**NAD^+**). It acts as an energy carrier. There are various steps in cellular respiration in which energy is transferred in the making of NADH from NAD^+. During the final stage of aerobic cellular respiration (the electron transport chain) NADH is converted back to NAD^+ and the energy released is used in the formation of ATP.

Another important coenzyme is **flavin adenine dinucleotide** (**FAD**). It is also an energy carrier. During the second stage of aerobic cellular respiration (the Krebs cycle) energy is transferred, making $FADH_2$ from FAD. During the electron transport chain $FADH_2$ is converted back to FAD and the energy released is used in the formation of ATP.

BIOFILE

Glycolysis appeared early in evolution

Before the evolution of photosynthetic prokaryotes the Earth's atmosphere contained very little oxygen. Glycolysis was one of the first biochemical pathways to evolve in this oxygen-poor atmosphere.

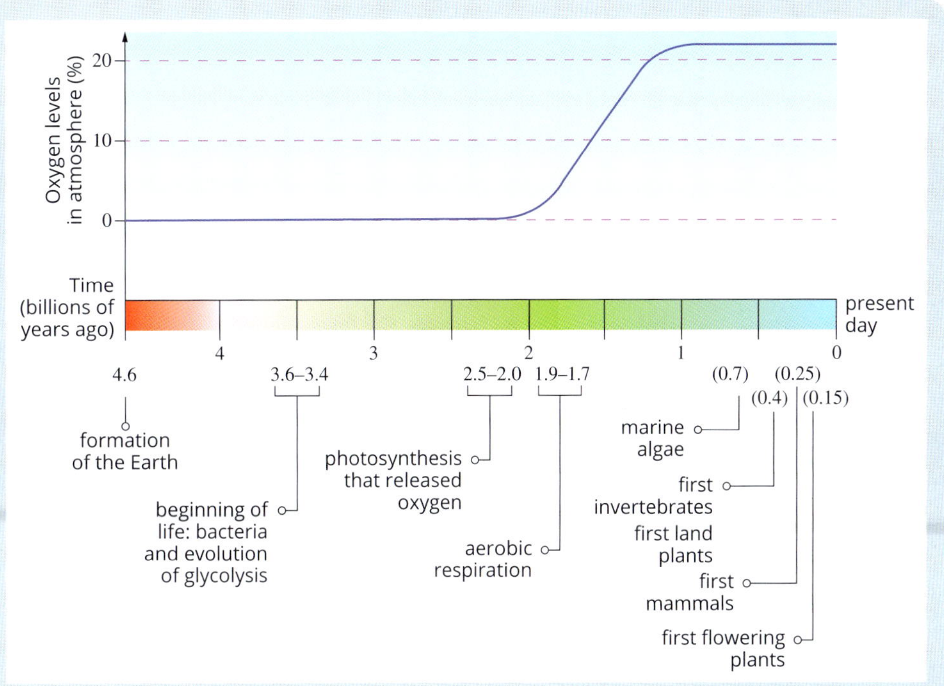

FIGURE 5.3.4 Glycolysis evolved in the low oxygen atmosphere that existed before the evolution of photosynthesis.

5.3 Review

SUMMARY

- Glucose is the primary energy source for cellular respiration.
- The energy released in the breakdown of glucose is carried by ATP.
- Two ATP (per glucose molecule) are created during glycolysis, which is the first stage of cellular respiration.

KEY QUESTIONS

1 Which organic molecule is used as the primary source of energy for most cells?

2 What is the useable form of energy for cellular processes?

3 Where does glycolysis occur?

4 How many ATP molecules can be created from one glucose molecule during glycolysis?

5.4 Cellular respiration: aerobic and anaerobic respiration

After glycolysis, cellular respiration can be diverted into one of two biochemical pathways depending on the availability of oxygen.

In this section you will learn about aerobic and anaerobic respiration, including some of the details of the biochemical reactions involved, where these processes occur and the different amounts of energy released in each (Figure 5.4.1).

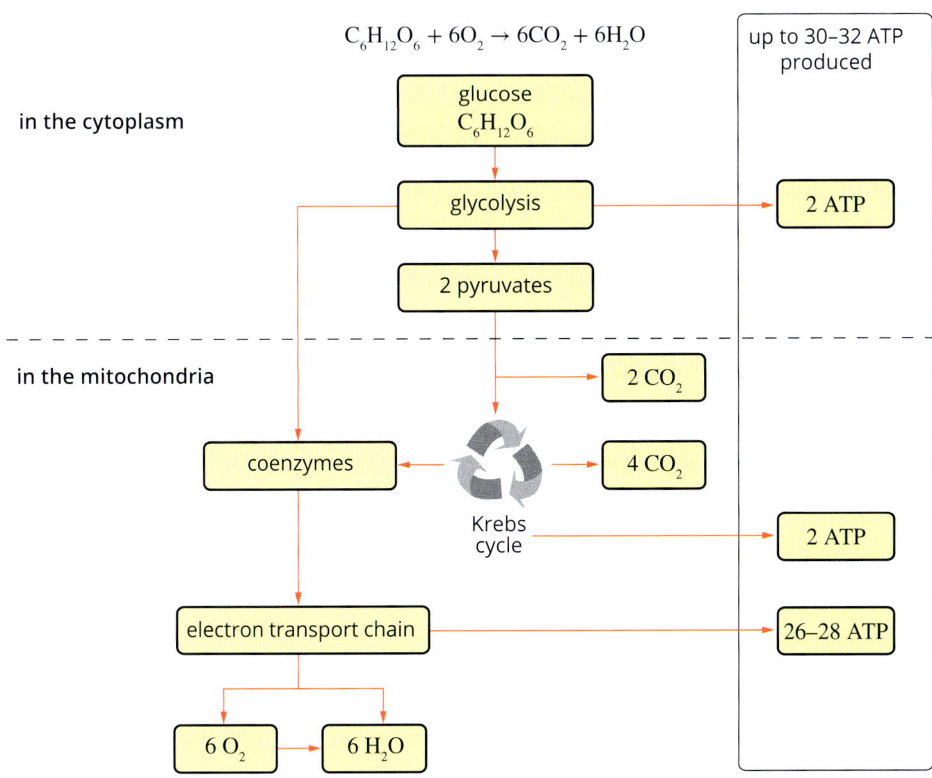

FIGURE 5.4.1 Overview of the stages of aerobic cellular respiration.

AEROBIC RESPIRATION

If oxygen is available, most eukaryotic organisms will use it to release energy from glucose in a process known as **aerobic respiration**. Aerobic respiration comprises three stages: glycolysis (see Section 5.3, page 196), the Krebs cycle and the electron transport chain. The latter two stages occur in the mitochondria. The balanced chemical equation for aerobic cellular respiration is shown below:

$$C_6H_{12}O_6 + 6O_2 \rightarrow 6CO_2 + 6H_2O \quad \text{(30–32 ATP produced)}$$

Mitochondria are about 3–5 microns in size. Chloroplasts are about 5–10 microns in size.

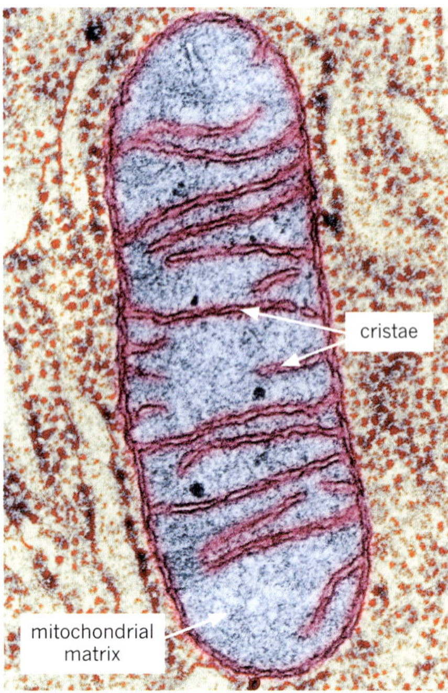

FIGURE 5.4.2 This coloured transmission electron micrograph of a mitochondrion shows the inner membrane (cristae) in pink and the matrix fluid in blue.

FIGURE 5.4.3 A mitochondrion contains many copies of circular DNA (mtDNA) and ribosomes within the matrix.

Mitochondria

The mitochondria are often referred to as the powerhouses of the cell. They are the sites of the Krebs cycle and the electron transport chain—two of the three processes that comprise aerobic respiration. If the cell has a supply of oxygen, the pyruvate formed from glycolysis passes into the mitochondria, where it is further broken down through a series of biochemical steps into carbon dioxide and water.

Mitochondria are slightly smaller than chloroplasts. Inside the outer membrane, there is a folded inner membrane structure. The folded structures of the inner membrane are called **cristae**. Inside the folded inner membrane is the **mitochondrial matrix** (Figure 5.4.2).

The origin of mitochondria

Like chloroplasts (on page 196), the origin of mitochondria can also be explained by the endosymbiotic theory. Evidence suggests that mitochondria are the result of endosymbiosis between prokaryotes and eukaryotic cells:

- Mitochondrial ribosomes and mitochondrial DNA (mtDNA) can be found in the mitochondrial matrix. The mtDNA contains the genetic code for the production of many of the proteins the mitochondria need to function (Figure 5.4.3). This mtDNA can be used as evidence for evolutionary relationships.
- Mitochondria reproduce by binary fission, independently of the cell.
- Mitochondrial DNA is circular, like bacterial DNA (eukaryotic nuclear DNA is linear).
- Mitochondrial ribosomes have similarities with bacterial ribosomes.

The Krebs cycle

The second important stage of aerobic respiration occurs after glycolysis and is called the **Krebs cycle** (or the citric acid cycle). It takes place in the mitochondrial matrix. The Krebs cycle is a series of eight reactions, each catalysed by a different enzyme. The pyruvate formed from glycolysis diffuses from the cytoplasm through the outer membrane of the mitochondria and is then moved by active transport through the inner membrane.

When the pyruvate, a three-carbon molecule, is in the mitochondrial matrix, it is converted into acetyl coenzyme A (acetyl CoA), a two-carbon molecule, which is the substrate for the first of a series of reactions that make up the Krebs cycle. In the formation of acetyl CoA from pyruvate, one carbon dioxide molecule is formed. In addition, in one turn of the Krebs cycle 2 carbon dioxide molecules are formed. That is a total of 3 molecules of carbon dioxide formed for every pyruvate molecule and 6 molecules of carbon dioxide for every glucose molecule metabolised.

During the reactions of the Krebs cycle, energy is transferred to energy-carrying coenzymes such as NADH, $FADH_2$ and ATP. From each turn of the Krebs cycle, one molecule of acetyl CoA is metabolised into 2 molecules of carbon dioxide, 3 molecules of NADH, one molecule of $FADH_2$ and one molecule of ATP (Figure 5.4.4).

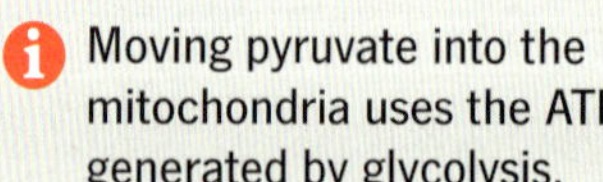

Moving pyruvate into the mitochondria uses the ATP generated by glycolysis.

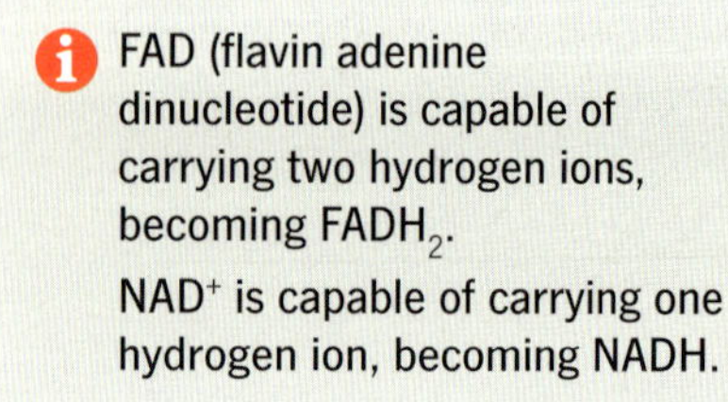

FAD (flavin adenine dinucleotide) is capable of carrying two hydrogen ions, becoming $FADH_2$.

NAD^+ is capable of carrying one hydrogen ion, becoming NADH.

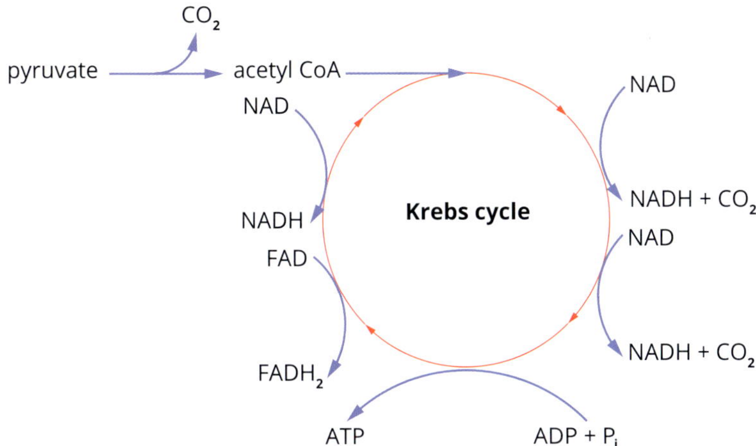

FIGURE 5.4.4 The three-carbon molecule pyruvate produced in glycolysis is converted to a two-carbon molecule acetyl CoA with the production of a carbon dioxide molecule. The acetyl CoA enters the Krebs cycle where it results in carbon dioxide, ATP, NADH and $FADH_2$ being formed.

The electron transport chain

The overall purpose of the electron transport chain is to move protons and electrons across a membrane as a system for generating ATP. Oxygen is essential because it picks up electrons at the end of the chain. If oxygen is not available, the electron transport chain stops.

Embedded in the inner membrane of the mitochondria are a number of protein complexes, including enzymes and cytochromes (Figure 5.4.5 and Figure 5.4.6). These complexes form an interconnected series that together make up the electron transport chain, which is the third stage of aerobic cellular respiration.

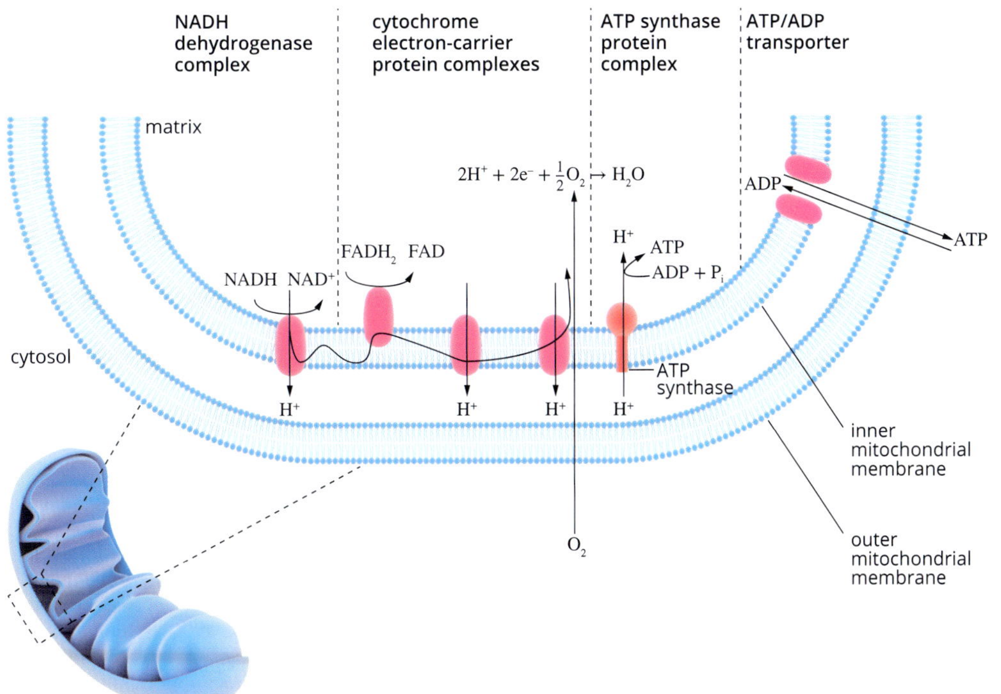

FIGURE 5.4.5 The inner mitochondrial membrane has several embedded proteins that pass along the energy and H^+ ions from carriers (NADH and $FADH_2$). The final enzyme in the pathway, ATP synthase, generates large amounts of ATP.

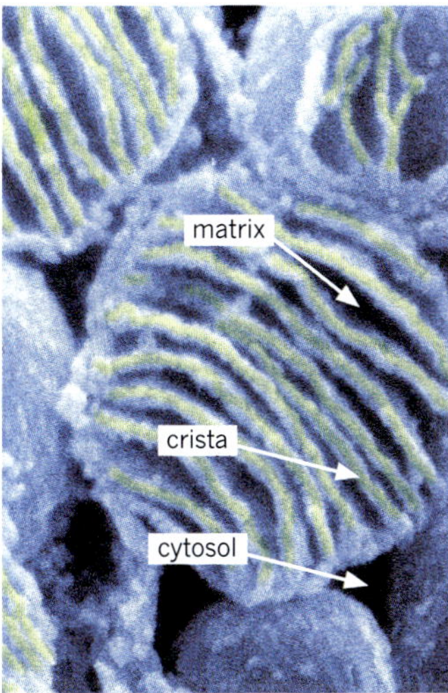

FIGURE 5.4.6 The first stage of aerobic cellular respiration (glycolysis) occurs in the cytosol. The second stage (the Krebs cycle) occurs in the matrix of the mitochondria. The last stage (the electron transport chain) occurs on the cristae of the mitochondria.

Energy-carrying molecules from the Krebs cycle feed into the electron transport chain. NADH is converted back to NAD^+ by interacting with the first complex at the beginning of the electron transport chain, and $FADH_2$ is converted back to FAD by interacting with the second complex.

The hydrogen ions (H^+) originating from the conversion of NADH and $FADH_2$ are moved into the intermembrane space and the electrons are transferred along the chain. The energy obtained in this process is used to make ATP. At the end of the electron transport chain, hydrogen ions and electrons combine with oxygen to form water:

$$2H^+ + 2e^- + \tfrac{1}{2}O_2 \rightarrow H_2O$$

The electron transport chain forms 26–28 molecules of ATP using the energy that was contained originally in each glucose molecule.

Adding up the ATP molecules formed at each stage of aerobic cellular respiration (Figure 5.4.7) we find that for every molecule of glucose metabolised, 30–32 ATP molecules can be formed:

- 2 ATP from glycolysis
- 2 ATP from the Krebs cycle
- 26–28 ATP from the electron transport chain.

Figure 5.4.7 summarises aerobic cellular respiration.

Scientists have not specified an exact number because the number of ATP molecules formed depends on a range of conditions that can differ. It is a very complex process.

BIOFILE

ATP yield in aerobic cellular respiration

Many sources state that 36–38 molecules of ATP are formed from the complete breakdown of one glucose molecule via aerobic cellular respiration, but it is now estimated that the overall production is closer to 30–32 ATP in the average cell.

It is a very complex process and scientists are still making new discoveries. Recent research shows that $FADH_2$ is not as efficient as first thought and generates less ATP than previously stated. In conjunction with this, the energy used during the final stages of aerobic cellular respiration is also higher than originally estimated. Due to the internal structure of mitochondria, more energy is required to convert and move substances within the mitochondrial membranes and matrix, reducing the overall yield of ATP. In this research, the results suggest that the electron transport chain produces 26–28 ATP per glucose molecule.

Some tissues and cells are more efficient than others and scientists have varying results in their research. There is research to suggest that some cells can make 38 ATP from one glucose molecule. The number (30–38 ATP) is a quoted as a range, as the exact number depends on a range of conditions that can differ.

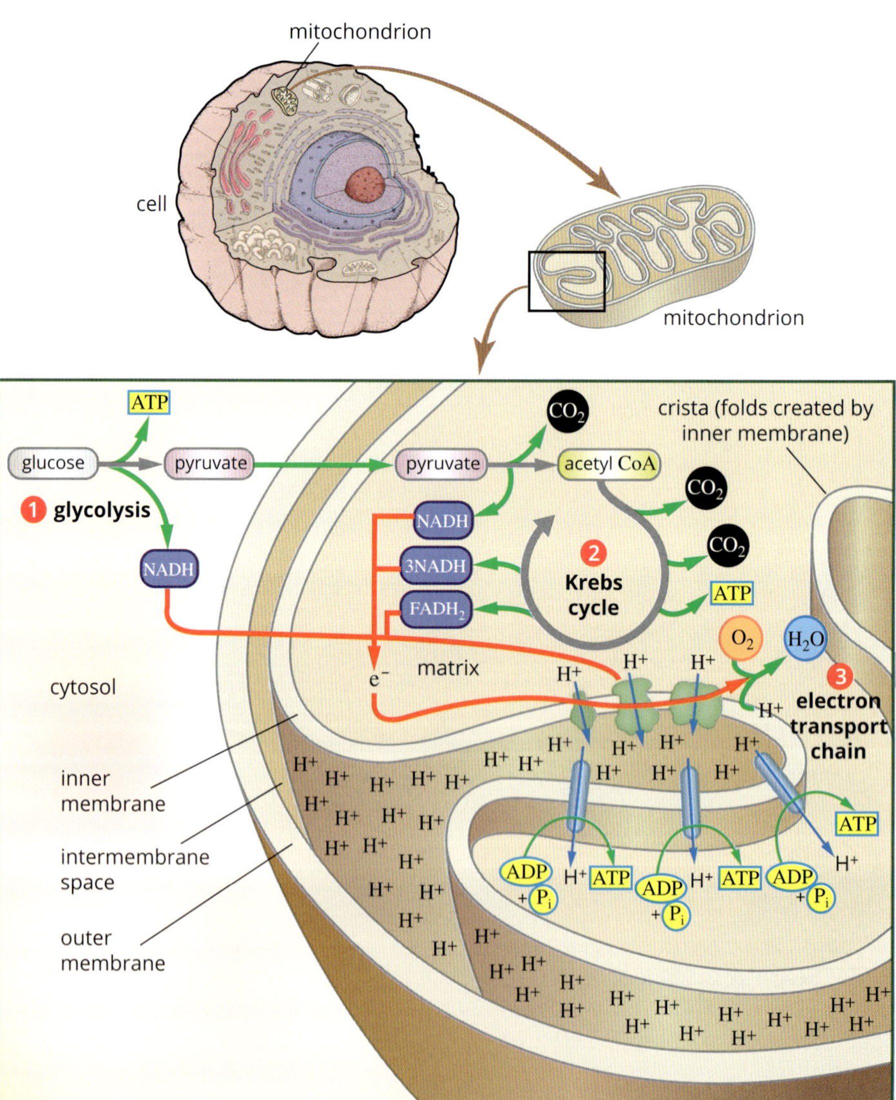

FIGURE 5.4.7 Summary of aerobic cellular respiration.

BIOFILE

ATP synthase—the cell's 'hydro' scheme

Hydroelectricity is generated using the potential energy in water stored in a dam. The potential energy is converted to kinetic energy and then to electrical energy by running the water downhill through water turbines that spin electricity generators.

Similarly, the electron transport chain builds up potential energy by pumping protons (hydrogen ions, H^+) released from NADH into the intermembrane space. This forms a concentration gradient across the inner membrane. The potential energy accumulated in the protons in the intermembrane space is transferred to ATP as the protons run through an ATP synthase enzyme complex embedded in the membrane and into the matrix.

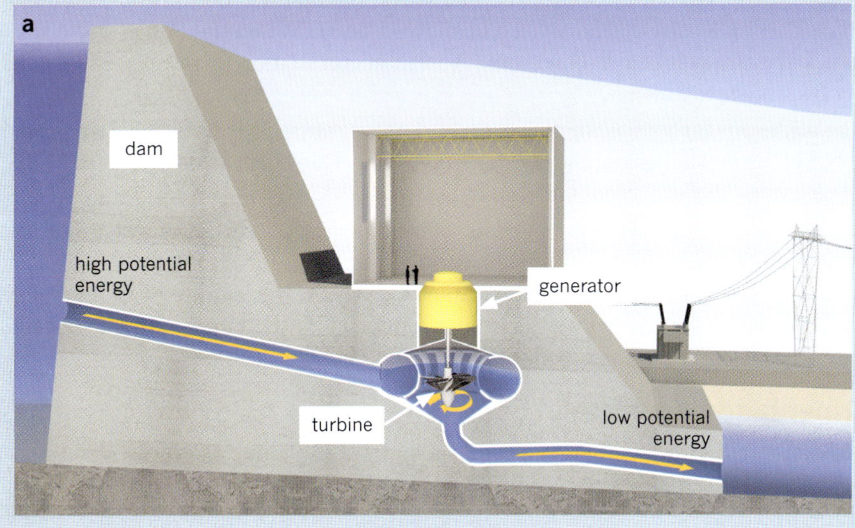

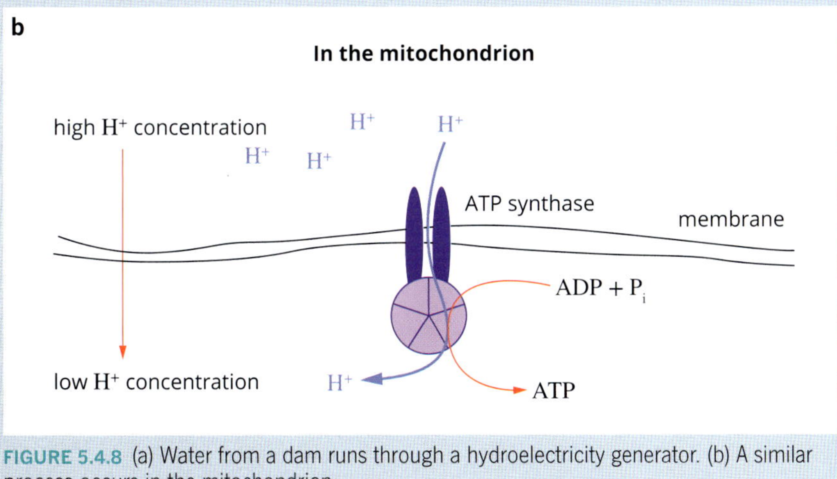

FIGURE 5.4.8 (a) Water from a dam runs through a hydroelectricity generator. (b) A similar process occurs in the mitochondrion.

ANAEROBIC RESPIRATION

If there is little or no oxygen available, a eukaryotic organism can still release some energy from glucose through the anaerobic pathway known as fermentation. In animals lactic acid **fermentation** is the name of the process that most often occurs in the cytoplasm of cells after glycolysis when oxygen is in short supply. In yeast a different type of fermentation occurs, ethanol fermentation (Figure 5.4.9).

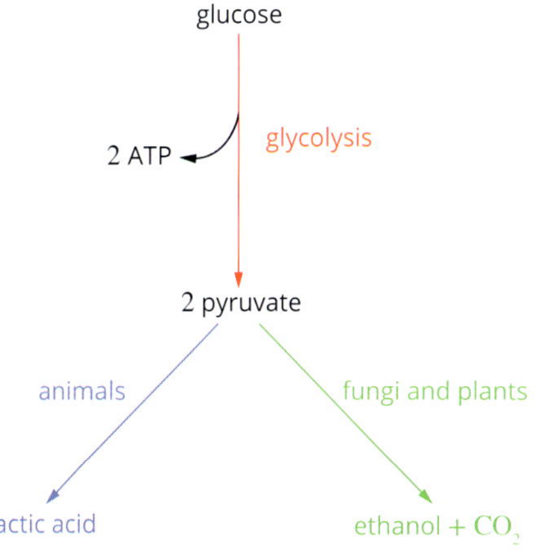

FIGURE 5.4.9 Anaerobic respiration occurs in the cytosol of cells when oxygen is not available.

Importance of fermentation

When the product of a biochemical reaction increases in concentration, it slows down the reaction that forms it, causing the reactants to accumulate. For a biochemical pathway to continue, the products of each reaction must be processed by the next reaction in the chain. If the product is not removed, it slows down the reaction, which causes the reaction before it to be slowed down and so on back through the pathway so that the whole pathway is slowed down or even stopped. Think of it like people moving in a queue. If the person at the head of the queue stops moving, the person behind must stop and so on back through the queue, until the whole queue stops moving.

The final reaction in the electron transport chain requires oxygen. Lack of oxygen will limit the rate of this final reaction, which in turn will limit the reaction before it and so on back through the pathways. NADH will not be converted back to NAD^+. The Krebs cycle will also be slowed down. When the Krebs cycle slows down, pyruvate will begin to accumulate. An accumulation of pyruvate will cause glycolysis to slow down.

In addition, for glycolysis to occur, a constant supply of NAD^+ is required. If the NAD^+ is not being regenerated, glycolysis will slow down or even stop.

Fortunately, to allow glycolysis to continue at low oxygen levels, some protists, fungi and some animal cells, such as muscle cells, contain an enzyme that can catalyse the conversion of pyruvate to lactic acid and NADH to NAD^+. This reaction solves both problems and enables glycolysis to continue. It first removes pyruvate so that the pathway is no longer blocked and then it provides a source of NAD^+ that is essential for glycolysis.

Fermentation in animals

> In animals, glycolysis and the final reaction that converts pyruvate to lactic acid are together called lactic acid fermentation (or lactate fermentation).

The lactic acid produced through fermentation can leave the cell by diffusion through the cell membrane and via a special membrane transport protein. This means the lactate level in the cell can remain low so that the pyruvate to lactate reaction can continue to occur, glycolysis can occur and ATP can continue to be produced to satisfy the cell's needs. In humans, lactic acid produced by muscle cells, once diffused out, can be circulated in the blood to other tissues in the body including liver and heart muscle, where it is converted back to pyruvate and enters aerobic cellular respiration pathways to produce ATP (Figure 5.4.10).

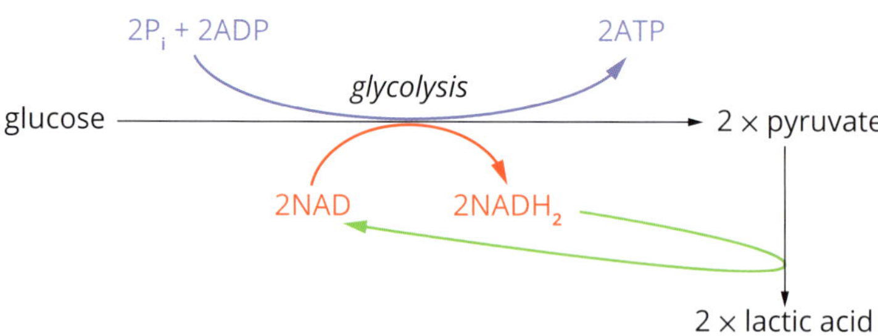

FIGURE 5.4.10 In animals, pyruvate is converted into lactic acid during anaerobic respiration to prevent a build-up of pyruvate. Later the lactic acid can be converted back into pyruvate for aerobic cellular respiration.

Fermentation in yeast

Yeast cells (Figure 5.4.11) carry out a different type of fermentation called ethanol fermentation (Figure 5.4.12). As in animal cells, one glucose molecule is broken down to two pyruvate molecules and NADH is formed from NAD^+. In yeast fermentation, the pyruvate is then broken down to ethanol (alcohol) and carbon dioxide, and NADH is converted back to NAD^+ through two reactions. In the first step, pyruvate is broken down to acetaldehyde and carbon dioxide. In the second step, the acetaldehyde is broken down to alcohol and NADH is converted to NAD^+.

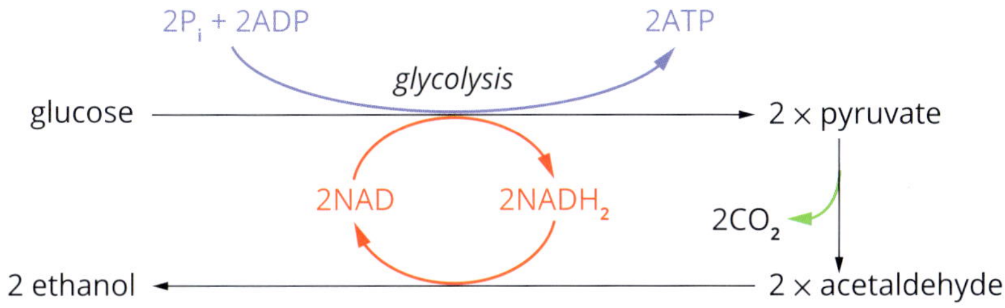

FIGURE 5.4.12 In yeast, pyruvate is converted into ethanol and carbon dioxide during anaerobic respiration.

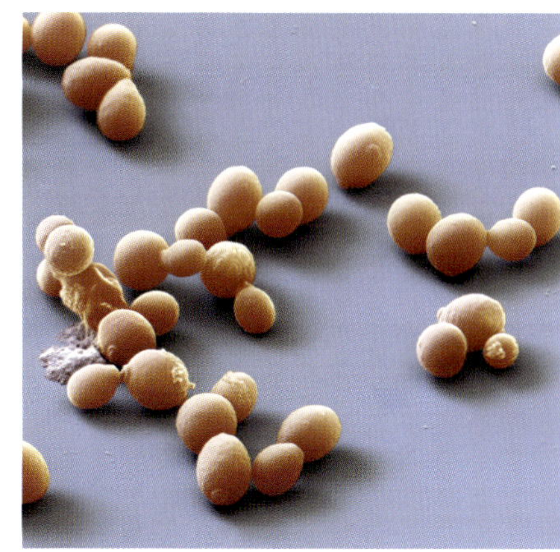

FIGURE 5.4.11 Coloured scanning electron micrograph of baker's yeast (*Saccharomyces cerevisiae*) cells.

The alcohol readily diffuses out of the cell by passive diffusion into the surrounding environment. The loss of alcohol from the cytoplasm means that the reactions of fermentation can continue until the alcohol builds up to a level in the external environment where alcohol no longer diffuses from the cell; that is, there is no longer a concentration gradient. Alcohol builds up in the cell to a point at which the fermentation reactions are blocked and they stop.

Fermentation can be conducted commercially on a large scale to produce fuels (such as bioethanol) and alcohol.

COMPARING AEROBIC AND ANAEROBIC RESPIRATION

Anaerobic respiration pathways in eukaryotes produce only two ATP molecules per molecule of glucose during glycolysis, whereas aerobic respiration produces 30–32 ATP molecules per molecule of glucose during glycolysis, the Krebs cycle and the electron transport chain. Therefore, aerobic respiration is much more efficient in supplying the cell with energy. The organic products of anaerobic respiration (lactic acid from animals and alcohol from yeast; Figure 5.4.13) still contain much energy and both can be further metabolised to release energy. Table 5.4.1 compares aerobic and anaerobic cellular respiration.

	Aerobic	Anaerobic
location	cytosol and mitochondria	cytosol
reactants	glucose and oxygen	glucose
products	carbon dioxide and water	lactic acid (animals) ethanol and CO_2 (plants/fungi)
energy output (per glucose molecule)	30–32 ATP total	2 ATP total

TABLE 5.4.1 Summary of cellular respiration.

FIGURE 5.4.13 Beer and wine are commercial products of yeast fermentation. Yoghurt and cheese are made using bacterial fermentation.

BIOFILE

The biochemistry of a hangover

A headache and nausea are the common symptoms of a 'hangover' experienced after drinking too much alcohol. Cells in the liver convert alcohol to acetyl CoA in three enzyme-catalysed reactions. First, the alcohol is converted to acetaldehyde by an enzyme called alcohol dehydrogenase (ADH), then to acetic acid and finally to acetyl CoA. The Krebs cycle metabolises the acetyl CoA to carbon dioxide and water. However, the enzyme reactions that form acetaldehyde and acetic acid are much faster than the one that forms acetyl CoA. The acetaldehyde and acetic acid therefore accumulate in the blood and circulate to the brain where they cause the toxic effect experienced as a hangover.

Other factors that contribute to a hangover include dehydration, which upsets the ADH metabolism, although drinking water only helps a bit. There are additional effects due to the diuretic effect of alcohol, which includes the depletion of electrolytes from the body.

FIGURE 5.4.14 It is not the alcohol that causes a hangover but the intermediate substances formed in its breakdown by the body.

5.4 Review

SUMMARY

- Aerobic cellular respiration has three stages: glycolysis, the Krebs cycle and the electron transport chain.
- Mitochondria contain folded membranes (cristae) and fluid (matrix). They also have their own circular DNA (mtDNA) and ribosomes.
- Glycolysis occurs in the cytosol and yields 2 ATP. Glucose is converted into two pyruvate molecules.
- The Krebs cycle occurs in the matrix of the mitochondria and yields 2 ATP. Pyruvate is broken down into carbon dioxide.
- The electron transport chain occurs in the cristae of the mitochondria and yields 26–28 ATP. Oxygen accepts the hydrogen ions that are used to generate a large amount of energy.
- In animals, the product of anaerobic respiration is lactic acid. In yeast, the products of anaerobic respiration are ethanol and carbon dioxide.
- Aerobic respiration is more efficient than anaerobic respiration. Aerobic respiration produces 30–32 ATP per glucose molecule; anaerobic respiration only yields 2 ATP per glucose molecule.

KEY QUESTIONS

1. Name each stage of aerobic cellular respiration. Identify the location of each stage.
2. What is the function of mtDNA?
3. What is the product of glycolysis?
4. Where does the Krebs cycle take place? What are the products of this process?
5. What is the main purpose for the electron transport chain?
6. In animals, what happens to lactic acid when oxygen becomes available?
7. What are the products of anaerobic respiration in yeast?
8. What roles do NAD and FAD have in aerobic respiration?

5.5 Factors that affect the rate of cellular respiration

Cellular respiration is affected by a number of factors including temperature, glucose availability and oxygen concentration (Figure 5.5.1).

In this section, you will learn how each of these factors affects aerobic cellular respiration.

FIGURE 5.5.1 Rebecca Clarke of New Zealand shows the exhaustion of completing a triathlon at the World Championship in Stockholm. During the race her glucose and oxygen levels would have been depleted. Her body would have worked hard to stop her body temperature from rising.

TEMPERATURE

You have learnt that cellular respiration is comprised of a number of interconnected biochemical pathways and that each pathway is a series of chemical reactions catalysed by specific enzymes. You have also learnt that the rate of an enzyme reaction is affected by temperature and that each enzyme has an optimum temperature.

As the temperature either rises above or falls below the optimum temperature, the rate of the enzyme reactions, and therefore the rate at which cellular respiration occurs, will slow down (Figure 5.5.2).

- When the temperature drops, the reactant molecules contain less kinetic energy and so do not react as quickly.
- When the temperature rises above the optimum level, the increased heat energy can disrupt the hydrogen bonds in the enzyme, causing the enzyme to denature. This means that the active site of the enzyme has lost its three-dimensional functional shape. This distortion in the shape means that the enzyme cannot bind to the substrate effectively, slowing down the rate of reaction.

Living organisms have particular temperature tolerance limits within which they will survive. More complex organisms such as birds and mammals control their body temperature at levels that are optimal for the functioning of their enzymes. Other organisms do not have the capacity to control the temperature of their cells, and the cellular respiration for these organisms is affected by the temperature of the external environment.

When an enzyme is denatured it cannot be repaired. The loss of shape is permanent and it will no longer catalyse the reaction.

BIOFILE

Birds use their beaks

Many animals have structural, physiological and behavioural adaptations to regulate body temperature. Toucans and many other birds, including Australian parrots, use their beaks as heat exchangers to lose excess heat to cool their bodies. When it is hot, the bird pumps more blood to its beak from which heat can be lost to the air. The ability to lose excess heat is of benefit to the bird because the enzymes that catalyse the reactions in the biochemical pathways of animals have an optimum temperature. The optimum temperature for enzyme reactions in birds is slightly higher than in humans at about 40 °C. At temperatures above the optimum, the bird's metabolism will begin to function poorly. Cellular respiration rate will slow down, and the supply of ATP to the cells will drop.

FIGURE 5.5.3 The large beak of a toucan is used as a heat exchanger to regulate body temperature.

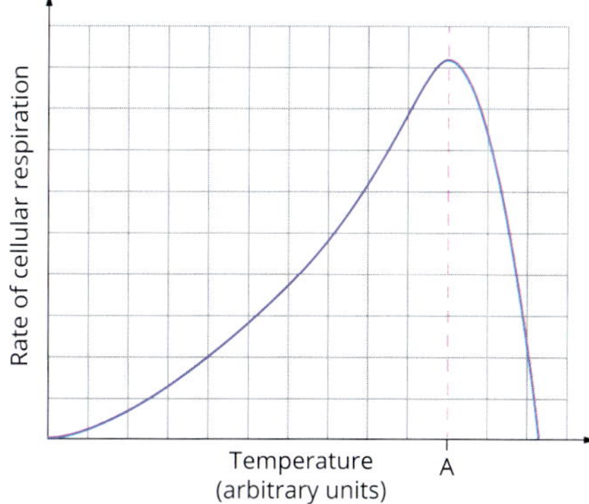

FIGURE 5.5.2 This graph shows the relationship between temperature and the rate of cellular respiration. As the temperature increases towards the optimum range, the rate of cellular respiration speeds up. At the optimum temperature (point A), cellular respiration will occur at the maximum rate. At temperatures above the optimum temperature, the rate of cellular respiration rapidly decreases.

GLUCOSE AVAILABILITY

All chemical reactions are limited by the concentration of reactants. An enzyme's reaction is limited by the availability of its substrate(s). Glucose is the substrate for glycolysis, and therefore it is the substrate for the first reaction in cellular respiration. The availability of glucose will affect the rate at which this first reaction occurs. The products of the first reaction become the substrates for the next and so on along the pathway. Hence the availability of glucose will affect the first and subsequent reactions in the cellular respiration biochemical pathways.

If the temperature remains constant, increasing the availability of glucose will increase the rate of cellular respiration up to a maximum level. This maximum level will be determined by the concentration of the enzymes and other cofactors required for cellular respiration.

Living cells require a constant supply of glucose from the surrounding environment. Simple organisms such as unicellular organisms have a range of adaptations that enable them to obtain glucose molecules from the food they capture in their environment. In the case of photosynthetic organisms, glucose is created by photosynthesis (Figure 5.5.4). The organs and systems of multicellular organisms work together to ensure that the cells have a constant supply of the substances they need for cellular respiration, including glucose.

The concentration of enzymes for cellular respiration is linked to the availability of mitochondria in the cell.

In photosynthetic organisms, the compensation point describes the point at which the rate of photosynthesis equals the rate of aerobic cellular respiration.

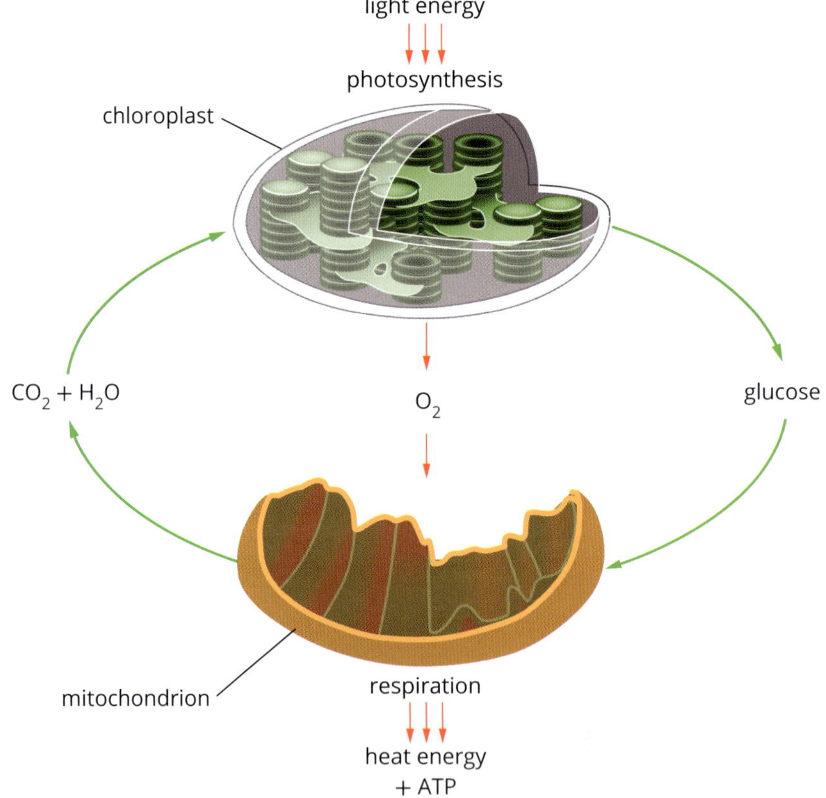

FIGURE 5.5.4 Photosynthetic organisms create their own supply of glucose (for cellular respiration) through photosynthesis. The products of photosynthesis are the reactants for aerobic cellular respiration (and vice versa).

BIOFILE

Glucose reserves in flowering plants

Plants often produce more glucose than they need and there is an obvious survival advantage in being able to store energy reserves. Energy reserves enable plants that were dormant over winter to grow rapidly in spring. Energy stores are also important for plants such as desert plants that must flower and fruit very rapidly after rain. Plants store carbohydrates mainly in the form of starch, a large molecule made up of glucose subunits. The starch molecules cluster into dense granules in cells. The seeds of many plants contain a store of oils and carbohydrates to provide energy for the growth of the new seedling (Figure 5.5.5).

FIGURE 5.5.5 A young plant growing from a seed.

Energy storage in animals

In winter, food generally becomes scarce for animals and their ability to store energy is necessary for survival. Animals store carbohydrates as **glycogen**, which, like starch, is a large molecule made from glucose subunits. In humans, about 100 g of glucose is stored as glycogen in the liver (Figure 5.5.6) and a further 200 g is stored in the muscles. This provides enough energy for about half a day at a moderate level of activity. The remainder of our energy reserves are stored as fats.

Animals use fats rather than carbohydrates as their main form of energy storage because:

- almost 25% more ATP is produced (per carbon atom) from fats than from carbohydrates
- fat is almost 50% lighter than carbohydrate
- stored carbohydrates attract and bind water molecules, increasing their weight between two and five times, whereas fats do not
- one gram of carbohydrate or protein can provide up to 17 kJ of energy, whereas one gram of fat provides 39 kJ of energy.

An average 70 kg male human stores about 11 kg of fat, which provides enough energy to last about one month without eating food, whereas the same amount of energy stored as carbohydrate could weigh more than 100 kg.

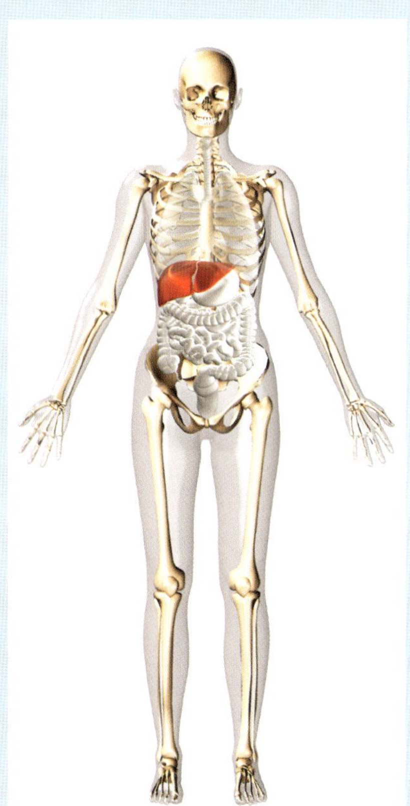

FIGURE 5.5.6 The liver (shown in red) is a large organ located just above the stomach. Excess glucose is stored (as glycogen) in the liver for later use.

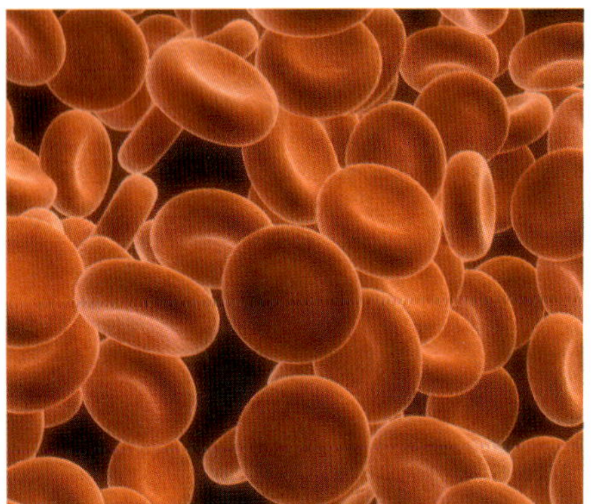

FIGURE 5.5.7 Red blood cells contain a red pigment (haemoglobin) that binds to oxygen. As blood moves through the capillaries in the lungs, the haemoglobin collects oxygen. As blood is pumped around the body, the red blood cells deliver oxygen to the cells in all of our tissues.

OXYGEN CONCENTRATION

In aerobic respiration, a constant supply of oxygen is necessary (Figure 5.5.7). Oxygen is the final reactant of the electron transport chain, so oxygen concentration affects the rate of aerobic cellular respiration. When the concentration of oxygen is low, the rate at which the electron transport chain can occur will be reduced. As you learnt in the previous section, when oxygen is in very short supply or absent, some cells will use fermentation reactions so that pyruvate does not accumulate and the glycolysis reactions can continue. This means ATP will continue to be produced, although at much lower rates, and the cell can remain alive.

BIOFILE

Myoglobin

Myoglobin is a protein that binds oxygen. It is present in muscle tissue of mammals and some other vertebrates. Diving mammals such as whales have large amounts of myoglobin in their muscles to enable them to store the oxygen needed for cellular respiration during a long dive.

FIGURE 5.5.8 Whales have large amounts of myoglobin in their muscles.

SUMMARY OF CELLULAR ENERGY TRANSFORMATIONS

Table 5.5.1 summarises what you have learnt about cellular energy transformations.

	Photosynthesis	Cellular respiration	
		Aerobic	Anaerobic
location	chloroplasts	cytosol and mitochondria	cytosol
reactants	water and carbon dioxide	glucose and oxygen	glucose
products	glucose, oxygen and water	water and carbon dioxide	• lactic acid (animals) • ethanol and carbon dioxide (plants/fungi)
carriers	NADPH	NADH and $FADH_2$	NADH

TABLE 5.5.1 Summary of cellular energy transformations.

5.5 Review

SUMMARY

- Aerobic cellular respiration is affected by temperature, glucose concentration and oxygen concentration.
- When the temperature is above or below the optimum range, the rate of cellular respiration is slower.
- Glucose is a substrate of glycolysis, therefore an increase in glucose availability will increase the rate of cellular respiration.
- Oxygen is a substrate of the electron transport chain, therefore an increase in oxygen concentration will increase the rate of aerobic cellular respiration.

KEY QUESTIONS

1. Which factors limit the rate of cellular respiration?
2. What would happen to the rate of cellular respiration at temperatures above the optimum range? Explain your answer.
3. Which stage of aerobic cellular respiration requires glucose as a reactant?
4. Which stage of aerobic cellular respiration requires oxygen as a reactant? What happens at low oxygen concentrations?

Chapter review

KEY TERMS

aerobic respiration
autotroph
C3 plant
Calvin cycle
chlorophyll
crista (plural cristae)
cyanobacteria
endosymbiosis
FAD (flavin adenine dinucleotide)
fermentation
glucose
glycolysis
granum (plural grana)
Krebs cycle
light-dependent reaction
light-independent reaction
light saturation curve
mitochondrial matrix
NAD^+ (nicotinamide adenine dinucleotide)
$NADP^+$ (nicotinamide adenine dinucleotide phosphate)
photosynthesis
porin
stoma (plural stomata)
stroma (plural stromata)
symbiotic
thylakoid
thylakoid lamella (plural thylakoid lamellae)

KEY QUESTIONS

1 Which of the following is true of light-independent reactions of photosynthesis in C3 plants?

A They occur on the inner membranes of chloroplasts.
B They involve the production of organic compounds containing three-carbon atoms in the Calvin cycle.
C They are adapted to maximise glucose production in hot, dry environments.
D They represent a special kind of photosynthesis that can produce organic compounds completely in the absence of light.

2 The following graph shows the rate of photosynthesis in a bean plant (a sun plant suited to high light intensities) and an oxalis plant (a shade plant suited to low light intensities). What is likely to be the limiting factor for photosynthesis for bean and oxalis respectively at light intensities A, B and C?

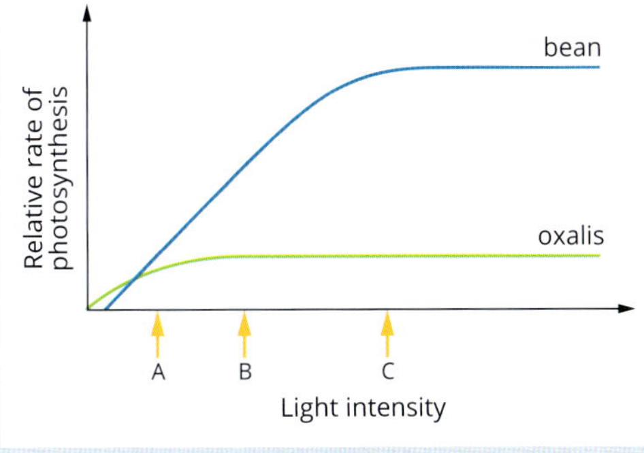

3 **a** Light availability is a significant limiting factor influencing the rate at which photosynthesis can occur.
i Define 'limiting factor'.
ii Why is light significant?

Photosynthetic organisms living in deep water have especially difficult challenges to survival. As shown on the graph below, light has limited ability to penetrate water. The graph shows light penetration in clear ocean water. The greater the material (e.g. mud) in the water the less light can penetrate.

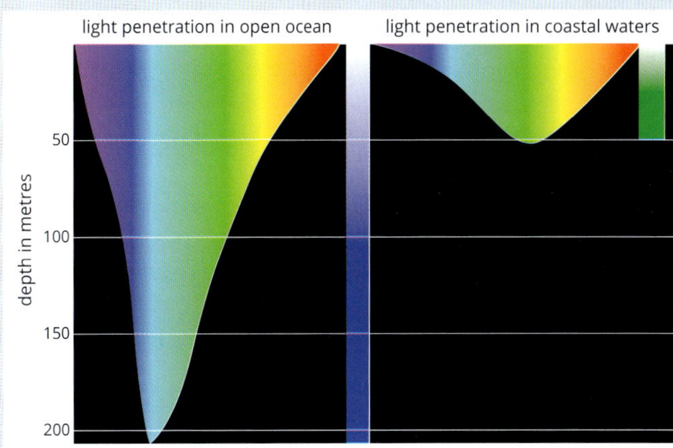

b What is the depth beyond which photosynthetic organism are unable to survive in the deep ocean?
c All photosynthetic organisms must contain chlorophyll, why?

As well as chlorophyll many photosynthetic organisms also contain a range of other pigments called

Question continued overleaf

accessory pigments. Together accessory pigments are able to absorb most wavelengths of light. The absorption spectrum for various photosynthetic pigments is shown in the graphs below.

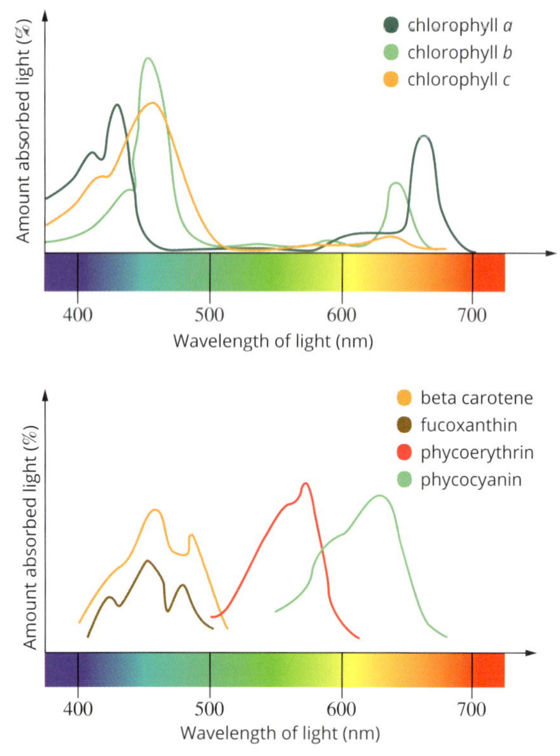

d Using the data presented in the graph above and your knowledge of the pigments used in photosynthesis, explain how having a range of accessory pigments could assist the survival of a photosynthetic organism.

Plants are one of the major groups of photosynthetic organisms, another is algae. Algae are eukaryotic organisms belonging to the kingdom Protista. Seaweeds are a type of alga. The diagram below shows the results of a chromatography experiment in which the pigments of various types of seaweed and an oak tree are compared.

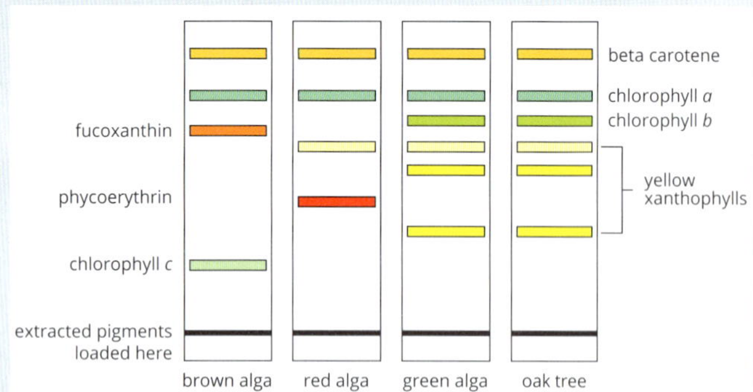

Chromatography is a method used to separate chemicals according to their solubility. The material to be separated is placed at one end of a special type of paper and that end is dipped into a solvent that is allowed to diffuse up the paper. The chemicals to be separated dissolve in the solvent and are carried up the paper. The chemicals separate according to their solubility. More soluble chemicals move further than less soluble chemicals.

e i Explain which type of seaweed is most likely to be found in the deepest waters in coastal regions.

ii The brown algae tested lack chlorophyll *b* and yet they are very successful, with some growing to 60 m in length. Explain how, despite the lack of chlorophyll *b*, the brown algae can grow so large.

4 An experiment was set up to investigate photosynthesis. A plant was placed in a sealed container at 20 °C. The air in the container was 0.09% CO_2 at the beginning of the experiment. (CO_2 concentration in room air is between 0.03 and 0.04%.) The experiment was undertaken at three different light intensities—dim, moderate and bright. The CO_2 concentration in the container was monitored over a 4-hour period. The results for each light intensity were collected and are shown in the table below.

Time (min)	Light intensity		
	Dim	Moderate	Bright
0	0.090	0.090	0.090
30	0.084	0.080	0.080
60	0.076	0.070	0.070
90	0.069	0.060	0.054
120	0.060	0.053	0.045
150	0.054	0.045	0.038
180	0.048	0.038	0.030
210	0.041	0.030	0.027
240	0.039	0.027	0.021

a Draw graphs of the information in the table. Use the same set of axes for the three light intensities. Ensure that the graphs comply with all of the graphing conventions.

b Describe the trends evident in the data.

c i Why can CO_2 uptake be used as a measure of the rate of photosynthesis?

ii Explain why the CO_2 decreased fastest in bright light.

d Why was the reduction in CO_2 the same for bright and moderate light until after 60 minutes had passed?

e Why were all of the experiments carried out at 20 °C?

f What hypothesis could this experiment have been testing?

5 Atrazine is a chemical commonly used as a weed killer.

It is absorbed by roots from the soil. In the leaves, it attaches to a protein, called D1, which is a part of the electron transport chain used to generate ATP during photosynthesis. It blocks the movement of electrons along this chain.

a Which phase of photosynthesis is disrupted by atrazine?

b ATP is not a final product of photosynthesis. What happens to the ATP normally produced in the electron transport chain?

c **i** Atrazine travels in the transpiration stream. What is the transpiration stream?

ii Hot weather enhances the effects of atrazine. Suggest why this would be the case.

d The advertising blurb on one particular brand of weed killer which has atrazine as its active ingredient says, '...this product works by starving the plant...'. Is this description of the action of atrazine accurate? Explain your answer.

6 The diagram below shows a cell from the palisade mesophyll of a leaf. A, B, C and D are molecules. Each letter may represent more than one molecule. The arrows show the direction of movement of these molecules.

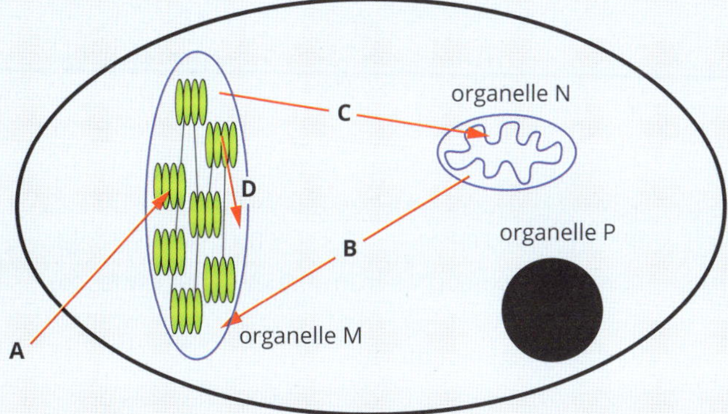

a Identify the position of the palisade mesophyll on the diagram below.

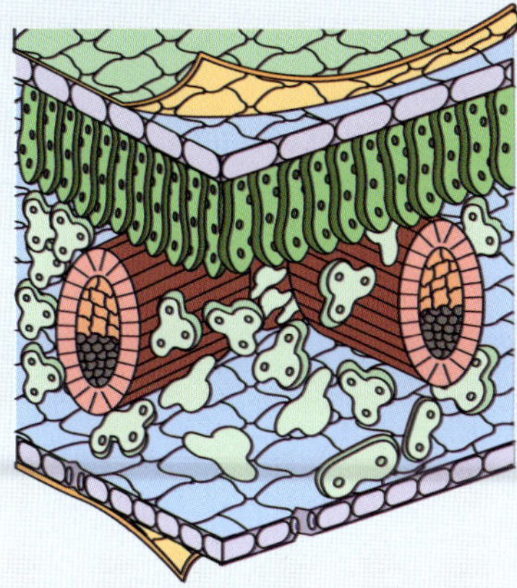

b **i** What is organelle M? How do you know?

ii Identify organelle N.

c Complete the table below by identifying molecules A, B, C and D.

	Molecules
A	
B	
C	
D	

d What is organelle P? How does it influence the events in organelle M?

7 The rate of photosynthesis for a particular species of plant was monitored under different wavelengths of light and varying temperature conditions.

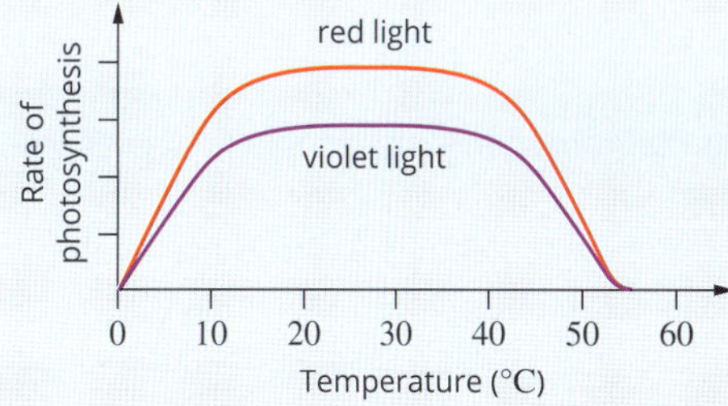

a Use information from the graph to name two factors that affect the rate of photosynthesis.

b Name two other environmental factors that can affect the rate of photosynthesis.

c Under which wavelength of light—red or violet—is the greater amount of oxygen gas produced? Explain.

d Plants grown under both red and violet light showed a sharp decline in the rate of photosynthesis after approximately 40 °C, until it ceased completely at approximately 55 °C.

i Explain this observation.

ii What does this suggest about the process of photosynthesis?

e Is photosynthesis an endergonic or an exergonic reaction? Explain.

8 An experiment was performed in which a plant was kept in a sealed container. The plant was supplied with air and water. The oxygen in the water was a radioactive isotope (^{18}O). This isotope can be used to identify the fate of the oxygen.

a The plant was kept in bright light for 2 hours and then the plant and it surroundings were tested to find out where the radioactive oxygen could be found. Explain where you would expect the radioactive oxygen to be discovered, giving reasons for your opinion.

b Later the experiment was repeated using normal water but the surrounding air contained carbon dioxide containing the radioactive oxygen. How would you expect the results of this second experiment to differ from those of the previous experiment? Explain why.

9 Study the diagram that summarises the processes of

energy release in cells.

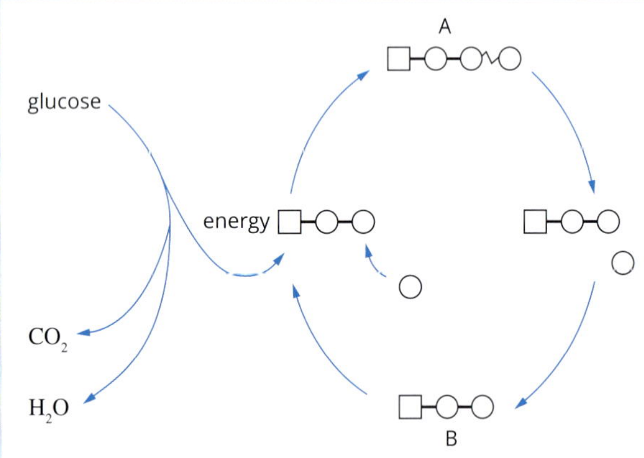

a In the diagram, name the molecule represented by each letter.

b Describe how molecules A and B are related.

c Glucose is the energy-rich molecule that enters glycolysis. In glycolysis a single glucose molecule is split to produce _______________________.

10 The function of aerobic respiration is to produce ATP. It does so by the following chemical reaction.

ADP + molecule X → ATP

Which of the following statements about this reaction in aerobic respiration is false?

A Molecule X is P_i.

B The electron transport chain is the major site of this reaction.

C The energy needed for the reaction to proceed can come from $FADH_2$.

D The energy needed for this reaction to proceed can come from NADP.

11 During aerobic respiration free oxygen:

A combines with water and helps produce ATP

B is produced as pyruvate is broken down

C is the final electron acceptor in the electron transport chain

D combines with carbon to produce carbon dioxide

12 $FADH_2$ is produced during:

A glycolysis in cellular respiration

B the Krebs cycle in aerobic respiration

C the electron transport chain of aerobic respiration

D the electron transport chain of the light-dependent reactions of photosynthesis

13 The Krebs cycle is an important stage of aerobic respiration. The diagram below shows the chemical changes that occur during the cycle for one molecule of pyruvate.

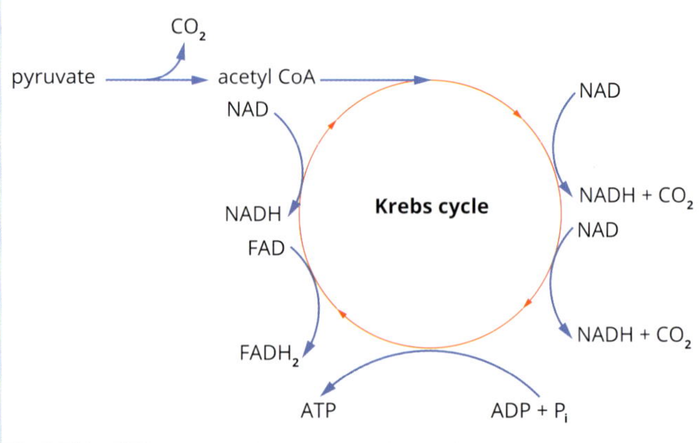

Use the information in the diagram above and your knowledge of cellular respiration to answer the following questions.

a How many molecules of ATP are formed in the Krebs cycle from one molecule of glucose?

b How many H^+ ions are loaded onto carriers during one turn of the Krebs cycle?

c NAD and FAD are sometimes described as proton carriers. Why?

d i How many molecules of CO_2 are produced during one turn of the Krebs cycle?

ii Which stage produces the remainder of the CO_2?

e Most of the ATP produced during the aerobic phase of respiration comes from the electron transport chain, yet this would be unable to occur without the Krebs cycle, explain why.

14 Mitochondria were extracted from some cells and isolated from the other cell contents. The mitochondria were suspended in a nutrient solution containing pyruvate in order to investigate respiration.

a What stages of respiration would be occurring in the mitochondria?

b Why did the nutrient solution contain pyruvate rather than glucose?

c Describe how the concentrations of carbon dioxide and oxygen would change throughout the experiment, assuming that the mitochondria were in a sealed container.

15 The enzyme cytochrome c oxidase is found embedded in the cristae of mitochondria. It is the last enzyme in the electron transport chain. It transfers electrons to oxygen and also binds protons (H^+) to oxygen to form water.

a What is the significance of the electron transport chain to living cells?

b Sodium azide is a pesticide that binds irreversibly to cytochrome c oxidase. Explain why this results in the death of the pests.

c An experiment was performed using human skin cells. Some skin cells were grown in culture so that the cells were all separated. The culture was divided in half and one half of the culture was placed in a test tube. The other half of the culture was treated to separate the mitochondria from the rest of the cell contents. The mitochondria were placed in one tube and the residue was placed in a third test tube. Samples from each of the tubes were grown in the following solutions:

A glucose + sodium azide

B pyruvate + sodium azide

C glucose alone

D pyruvate alone

The test tubes were supplied with oxygen. All other variables were kept the same. After 30 minutes the tubes were tested for the presence of CO_2 and lactic acid. The results are shown in the table below.

	A Glucose + sodium azide	B Pyruvate + sodium azide	C Glucose only	D Pyruvate only
whole cells	lactic acid	lactic acid	CO_2	CO_2
mitochondria	neither	neither	neither	CO_2
cell residue without mitochondria	lactic acid	neither	lactic acid	neither

i What is/are the control(s) in this experiment?

ii Explain why CO_2 was produced in the 'pyruvate only' tube but not in the 'glucose only' tube, of the tubes that contained only mitochondria.

iii The experiment could not be extended beyond 30 minutes because after that time the whole cells in series B died. The whole cells in series A continued to live for some time. Explain why the cells with pyruvate + sodium azide died but those with glucose + sodium azide did not.

iv Do the results of this experiment support the contention that sodium azide disrupts the electron transport chain?

v How would the results have differed if plant cells had been used for the experiment?

16 The graphs below show the contributions of the two energy-producing pathways to physical activity.

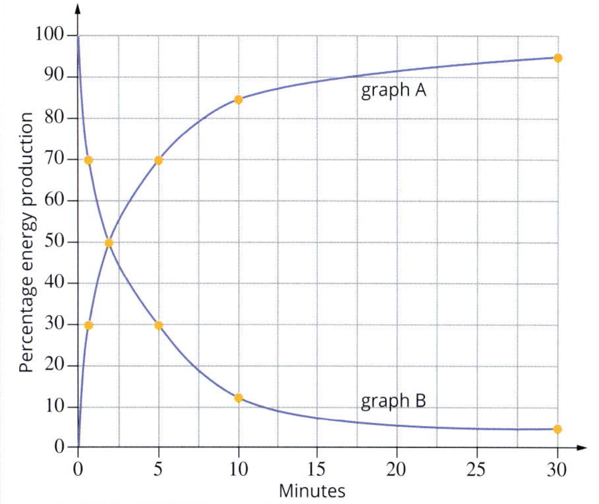

a Athletes competing in sports requiring short-term power output, such as sprinting, obtain most of their ATP from the anaerobic pathway, but athletes requiring sustained energy use aerobic respiration to meet most of their energy needs.

i Why do muscle cells need ATP?

ii What is the name of the process that produces ATP during anaerobic respiration?

iii Explain which graph (A or B) is most likely to represent ATP production by a sprinter.

iv Why can't anaerobic respiration supply the energy needs of athletes in events requiring energy over a sustained period of time?

b Lactic acid is produced during anaerobic respiration. What happens to this lactic acid?

c Animals make lactic acid during anaerobic respiration but yeasts and plants produce ethanol and CO_2. Suggest a reason why the products are different in animals and plants.

REVIEW QUESTIONS

How do cellular processes work?

Multiple choice questions

1 A student was interested in the actions of phospholipids and how they act in environments that are different from the cellular environment. The first experiment the student performed was to add some phospholipids to a test tube containing oil. Which of the following is the expected result?

A The phospholipids would dissolve in the oil.
B The phospholipids would line up in single file on the surface of the oil with the heads pointing into the oil.
C The phospholipids would form a bilayer with the heads pointing inwards and the tails pointing outwards.
D The phospholipids would form a bilayer with the heads pointing outwards and the tails pointing inwards.

2 What are transcription factors?

A promoters
B TATA boxes
C start and stop triplets
D proteins that attach to DNA to regulate gene expression

3 Upstream areas of the gene which regulate transcription are:

A promoters
B stop triplets
C start codons
D transcription factors

4 The graph below shows the rate of photosynthesis of two different plant species when the plants are experiencing the same environmental conditions.

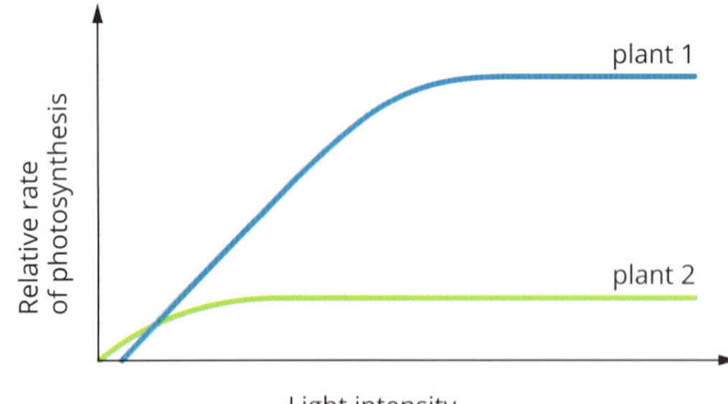

Which limiting factor is most likely to be causing the difference in photosynthetic rates between the two plants?

A oxygen availability
B chloroplast availability
C carbon dioxide availability
D water availability

Use the following diagram showing the structure of the *lac* operon to assist with answering Questions 5 and 6.

5 The regulatory gene in the *lac* operon is labelled:

A *lacI*
B promoter
C operator
D *lacZ*

6 An experiment is performed using a bacterial culture in which the cells have a mutation in both the *lacI* and *lacA* genes that causes these genes to be non-functional. The cells are exposed to both lactose and glucose. Which of the following shows the expected activity of the structural genes?

	lacZ	*lacY*	*lacA*
A	low	low	low
B	high	high	high
C	high	high	low
D	low	low	high

7 Which of the structures shown below is a tertiary protein structure?

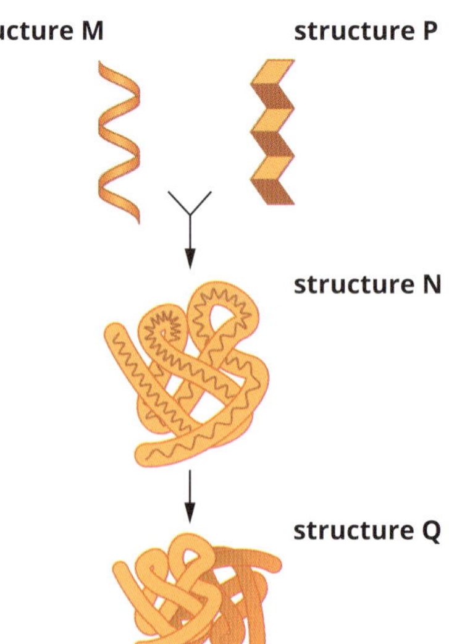

A structure M
B structure N
C structure P
D structure Q

Short answer questions

8 The electron micrograph below shows part of a cell that produces digestive enzymes.

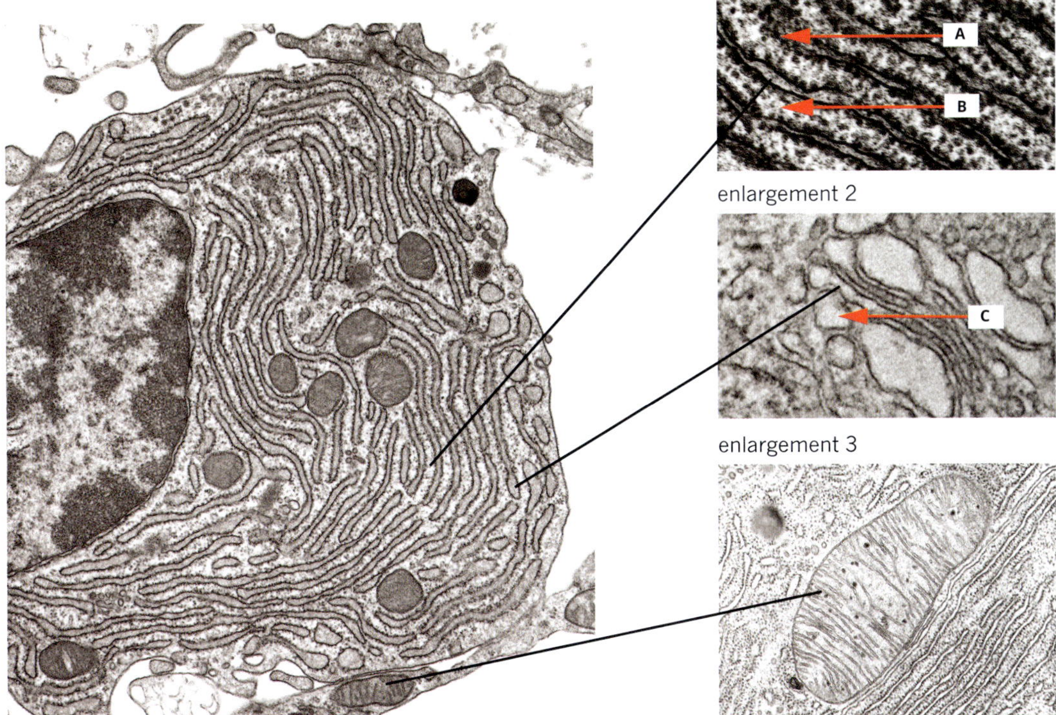

a i Identify the organelles illustrated in enlargements 2 and 3.

ii Identify the structures labelled A, B and C.

b What role does the organelle shown in enlargement 3 play in the making and secretion of the digestive enzymes?

c Draw a flow chart of the production and secretion of the enzymes. Refer to the relevant organelles and structures from the diagram above in your chart. Ensure you name the process by which the enzymes are secreted from the cell.

9 Aerobic respiration is a process that is essential for the survival of most life forms. The diagram below shows the stages of aerobic (cellular) respiration and the inputs and outputs of each stage.

a Complete the diagram by filling in the empty boxes.

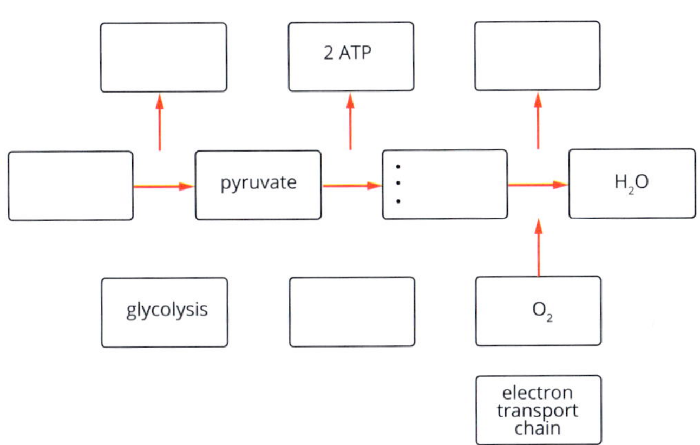

b DNP (2,4-dinitrophenol) is a chemical that was used during World War I to make explosives. It was noticed that the workers handling the DNP had extremely high body temperatures (up to 44 °C) and suffered from severe weight loss. Some workers died as a result of absorbing DNP through their skin, ingesting it or inhaling it. Research into DNP has shown that one of its actions is to block the movement of phosphate ions into the mitochondria.

i Normally, during aerobic respiration much of the energy released from the breakdown of glucose is used to build ATP. What happens to the rest of the energy?

ii Suggest how the action of DNP could lead to the very high body temperatures observed.

iii The very high body temperatures have been associated with a number of fatalities. Explain how high body temperatures can lead to cell death.

c Why would ingestion of DNP lead to weight loss?

d A student was investigating the effects of DNP on cellular respiration. A cell culture was supplied with glucose and monitored for the products of respiration. Part way through the experiment DNP was added to the culture. The graph below shows the results of the experiment.

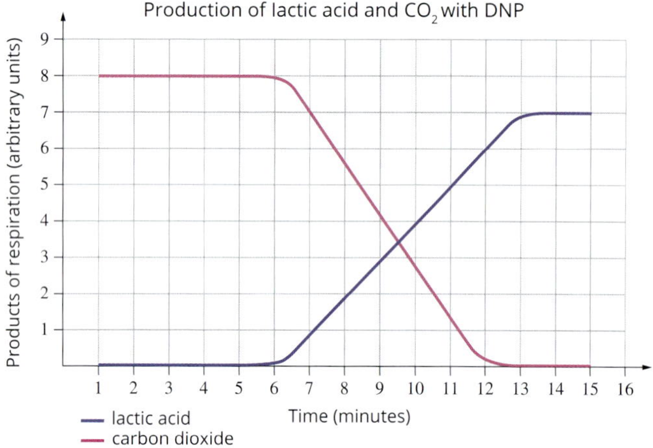

- **i** At what time was the DNP added to the culture? How do you know?
- **ii** Why is lactic acid being produced by the cells?
- **iii** DNP is highly toxic. Describe two safety precautions that the student should implement while performing this experiment.

10 The Calvin cycle is a series of chemical reactions. Each reaction is catalysed by its own enzyme. The first step in the Calvin cycle adds a CO_2 molecule to a ribulose 1,5 bisphosphate molecule. The new molecule (called 3-phosphoglycerate) is then further modified, ultimately forming glucose. The enzyme that catalyses the first step is called Rubisco.

$$\text{Ribulose 1,5 bisphosphate} + CO_2 \xrightarrow{\text{Rubisco}} \text{3-phosphoglycerate}$$

a Which stage of photosynthesis is the Calvin cycle part of?

b Examine the diagram of a chloroplast.

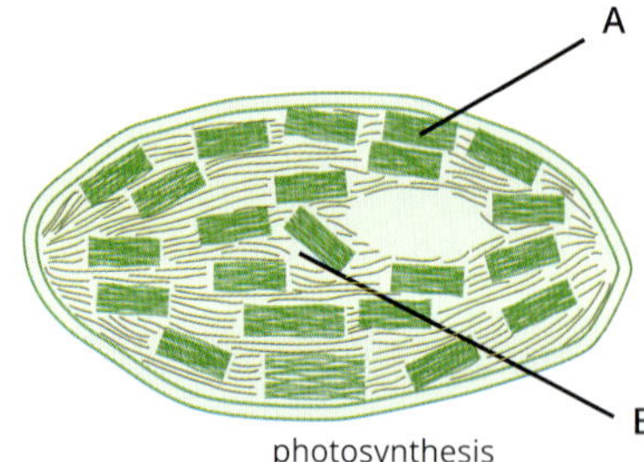

photosynthesis

- **i** Identify structures A and B.
- **ii** Describe the chemical reactions occurring at A and B.

c **i** What is/are the normal substrate(s) of Rubisco?
ii How does Rubisco facilitate the reaction?

d Experiments have shown that oxygen acts as a competitive inhibitor of the normal substrate(s) of Rubisco. How might this regulate the metabolic pathway that produces glucose?

11 A permanent alteration to the DNA of an organism is called a mutation. Many mutations result in the formation of proteins that do not function correctly. One such mutation leads to Tay-Sachs disease.

Tay-Sachs disease is caused by a mutation in the hexosaminidase A (*HEXA*) gene. This gene codes for a family of enzymes that are involved in the breakdown of fats, especially in neurons. Mutations in the *HEXA* gene result in a build-up of fats in the neurons. This causes damage to the neurons, causing muscle weakness and brain damage.

a How can one gene code for more than one enzyme?

b There are a number of different mutations of the *HEXA* gene and all result in Tay-Sachs disease. A section of one exon of the gene, showing one common mutation, is shown in the table below. Also shown is the normal sequence of the same section of the gene. The strands of DNA given are the coding strands.

normal *HEXA* allele	CGT ATA TCC TAT GCC CCT GAC ...
Tay-Sachs allele	CGT ATA TCT ATC CTA TGC CCC TGA C ...

- **i** Use Figure 3.2.3 on page 144 to work out the amino acid sequences of the exon shown for the normal *HEXA* gene and the Tay-Sachs variant.
- **ii** Explain why the mutant gene results in Tay-Sachs disease.

12 Enzymes are used to catalyse cellular reactions.

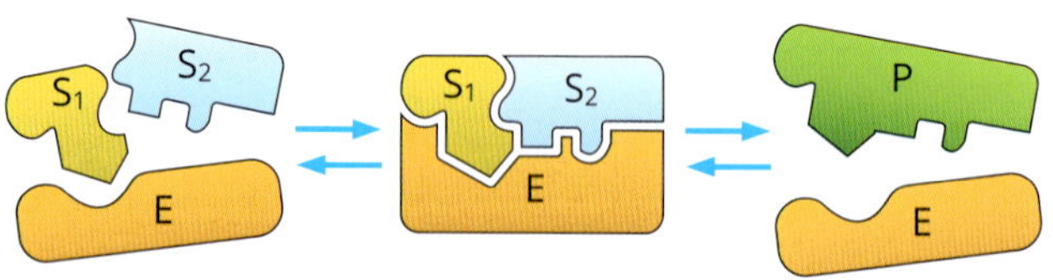

a What enzyme activation model does the diagram above represent?

b If heated above its critical temperature, an enzyme denatures. Define 'denature'.

c How does this affect the enzyme's activity?

13 The graph below shows the change in mass of potato pieces in solutions of sucrose.

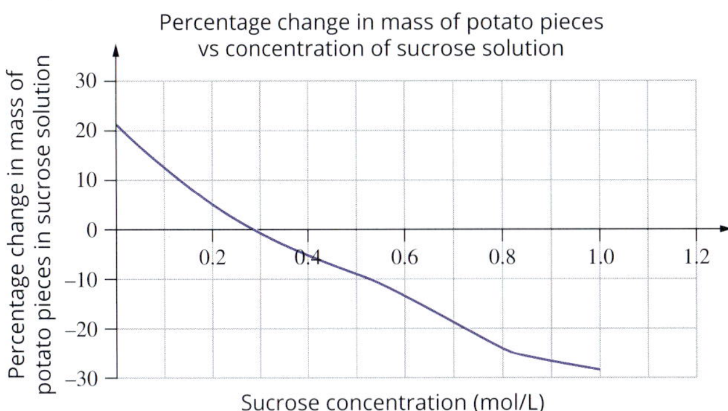

a What is the normal concentration of the cytosol of potato cells?

b i Potato cell membranes are impermeable to sucrose, explain why.

ii Explain why the potato pieces increased in mass when placed in a solution of 0.2 mol/L.

14 A student intended to investigate the processing of RNA in a cell. In order to do this the student decided to introduce a radioactively labelled RNA monomer to a culture of human skin cells.

a Which of the RNA monomers would be most appropriate to radioactively label for this experiment? Why?

b The radioactively labelled RNA was collected from the cell and X-ray diffraction was used to find the shapes of the various molecules. Two shapes that were identified are shown below.

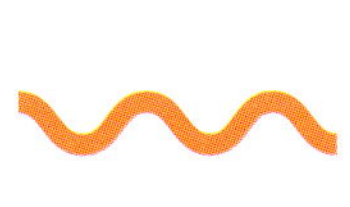

structure A

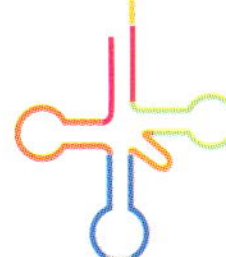

structure B

i Identify the two types of RNA shown.

ii Explain where in a cell each type of RNA would be discovered.

iii Describe the role of each type of RNA in the cell.

c A third type of RNA was extracted from the nucleus of the cell.

i What is this RNA?

ii Where else in the cell would this RNA be found?

iii What is its function?

15 a An experiment was performed in which muscle cells were incubated in an oxygen-free environment at 20 °C. The cumulative uptake of glucose was measured in grams. The results for the first 10 minutes are shown in the following table.

Glucose use in the absence of oxygen	
Time (minutes)	**Glucose uptake (g)**
2	5
4	10
6	15
8	20
10	25

After 10 minutes oxygen was infused into the culture and measurement of the uptake of glucose continued. The results for the next 10 minutes are tabulated below. Temperature was maintained at 20 °C.

Glucose use in the presence of oxygen	
Time (minutes)	**Glucose uptake (g)**
12	26
14	27
16	28
18	29
20	30

i Graph the uptake of glucose versus time for the 20 minutes of the experiment. Clearly mark the point at which oxygen was introduced into the culture.

ii Explain why glucose use declined so significantly after oxygen was added to the culture.

iii What is the independent variable in the experiment?

iv Why was it necessary to ensure that the temperature remained at 20 °C throughout the experiment?

b i Some mitochondrial diseases are caused by mutations in the genes needed for respiration. Mitochondrial diseases caused by these mutations are relatively common in comparison to diseases caused by nuclear chromosomal mutation. Why?

ii One mitochondrial disease is caused by a mutation in the gene that encodes the protein cytochrome *c* oxidase. Cytochrome *c* oxidase is the last of the cytochrome proteins forming the electron transport chain. Where in the mitochondria will this protein be found?

iii An experimenter investigating mitochondrial mutations performed the experiment from part a using cells with mitochondria possessing mutations. The scientist noted that when oxygen was added after 10 minutes to these cells no change in glucose use was observed. Explain this observation.

16 *Amoeba proteus* is a unicellular member of the kingdom Protista. It is frequently infected with bacteria.

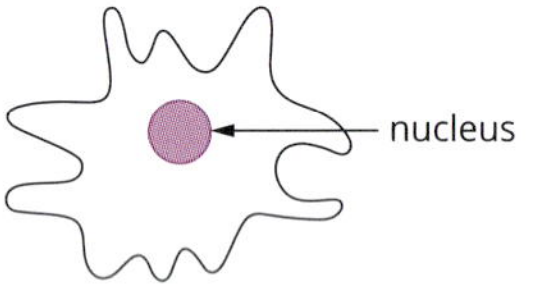

Amoeba proteus without bacteria

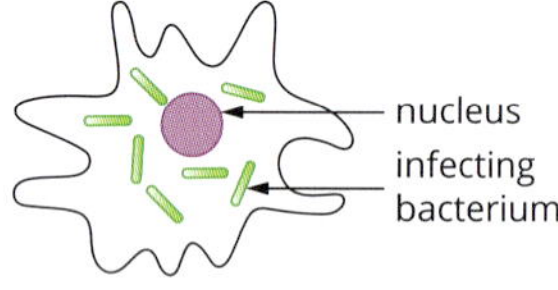

Amoeba proteus with infecting bacteria

a To which domain of life does *Amoeba proteus* belong?

b Describe how something as large as a bacterium would have entered *Amoeba proteus*.

c In one experiment the nucleus was removed from a number of infected cells and transferred to other *Amoeba proteus* cells that had had their nucleus removed (enucleated). Some of the receiving cells had bacteria and others did not.

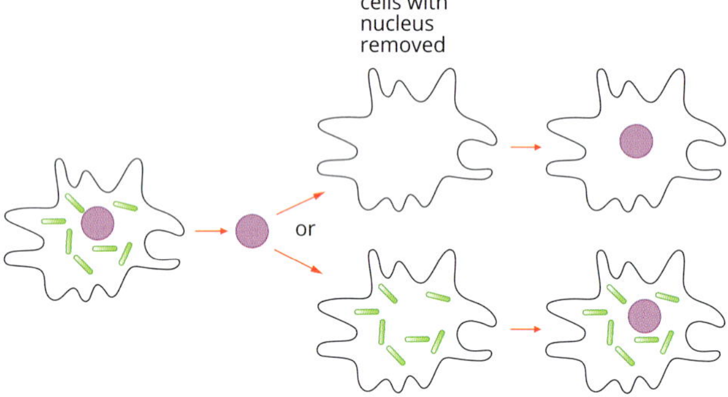

The *Amoeba proteus* were then provided with all of the nutrients they needed. The nuclei transplanted from a bacteria-infected cell to a non-bacteria-infected cell did not survive. The nuclei transplanted from a bacteria-infected cell to a bacteria-infected cell survived.

What conclusion can be drawn from this experiment?

d As part of the experiment, nuclei were taken from *Amoeba proteus* that had not been infected with the bacteria and transplanted into enucleated cells with and without bacteria.

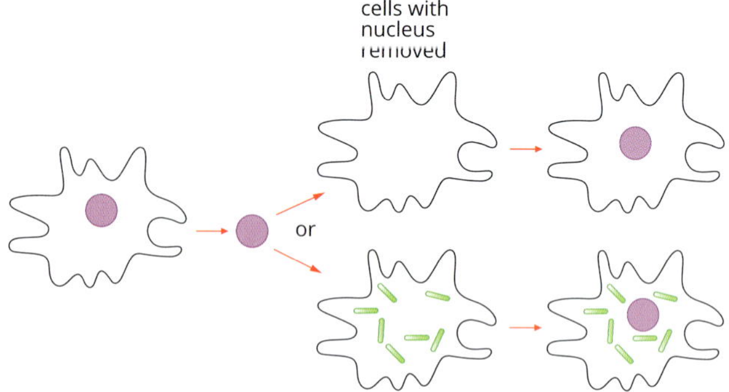

Most of the cells, both with and without the bacteria, survived.

i What conclusion can be drawn from this experiment?

ii Why was it necessary to perform this second experiment?

e Further research showed that the *Amoeba proteus* infected with the bacteria were unable to produce an enzyme that uninfected *Amoeba proteus* were still able to produce. It was shown that the bacteria were producing the enzyme being used by the infected *Amoeba proteus*.

i How does this experiment support the theory that mitochondria and chloroplasts are the result of endosymbiosis?

ii State two other pieces of evidence that support the theory that chloroplasts and mitochondria became organelles as a result of endosymbiosis.

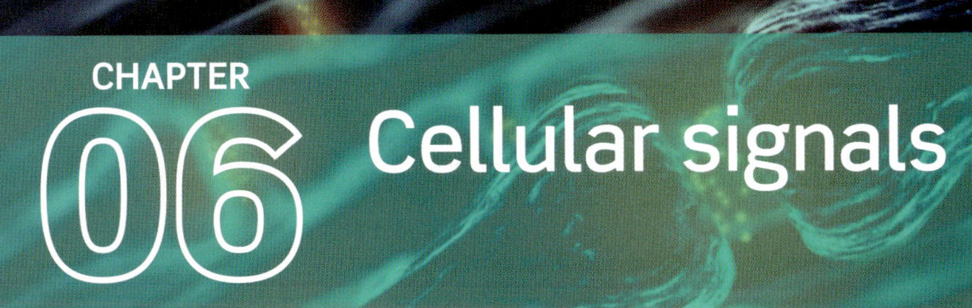

CHAPTER 06 Cellular signals

Learning outcomes

To survive, all living organisms must be able to detect and then respond appropriately to changes in their external and internal environments. In multicellular organisms, a key step in achieving these responses is the ability of cells within the individual to communicate with each other.

This chapter focuses on the ways in which cells send targeted signals that elicit specific responses in other cells. The stimulus–response model will be explored in terms of the different types of signalling molecules, the locations of their complementary receptors and the transduction of the signals in the target cells, which results in a response. Programmed cell death (or apoptosis) will be presented as an example of a process regulated by signalling molecules.

Key knowledge

- the sources and mode of transmission of various signalling molecules to their target cell, including plant and animal hormones, neurotransmitters, cytokines and pheromones
- the stimulus–response model when applied to the cell in terms of signal transduction as a three-step process involving reception, transduction and cellular response
- difference in signal transduction for hydrophilic and hydrophobic signals in terms of the position of receptors (on the membrane and in the cytosol) and initiation of transduction (details of specific chemicals, names of second messengers, G protein pathways, reaction mechanisms or cascade reactions are not required)
- apoptosis as a natural, regulatory process of programmed cell death, initiated after a cell receives a signal from inside (mitochondrial pathway) or from outside (death receptor pathway) the cell, resulting in the removal of cells that are no longer needed or that may be a threat to an organism, mediated by enzymes (caspases) that cleave specific proteins in the cytoplasm or nucleus (details of specific cytoplasmic or nuclear proteins are not required)
- malfunctions in apoptosis that result in deviant cell behaviour leading to diseases including cancer.

6.1 Signalling molecules

Homeostasis is the maintenance of a relatively stable internal environment in the face of changes in external or internal conditions.

Multicellular organisms survive, grow and reproduce because they are able to detect and respond to signals from their internal and external environments. Many of these responses involve negative feedback mechanisms through which homeostasis is maintained (refer to *Heinemann Biology 1*, Chapter 5). To achieve these responses, communication between cells is essential and there are many types of molecules that are involved (Figure 6.1.1). In this section, you will learn about some of the major groups of plant and animal signalling molecules, including hormones, cytokines, pheromones and neurotransmitters, as well as their sources and modes of transmission.

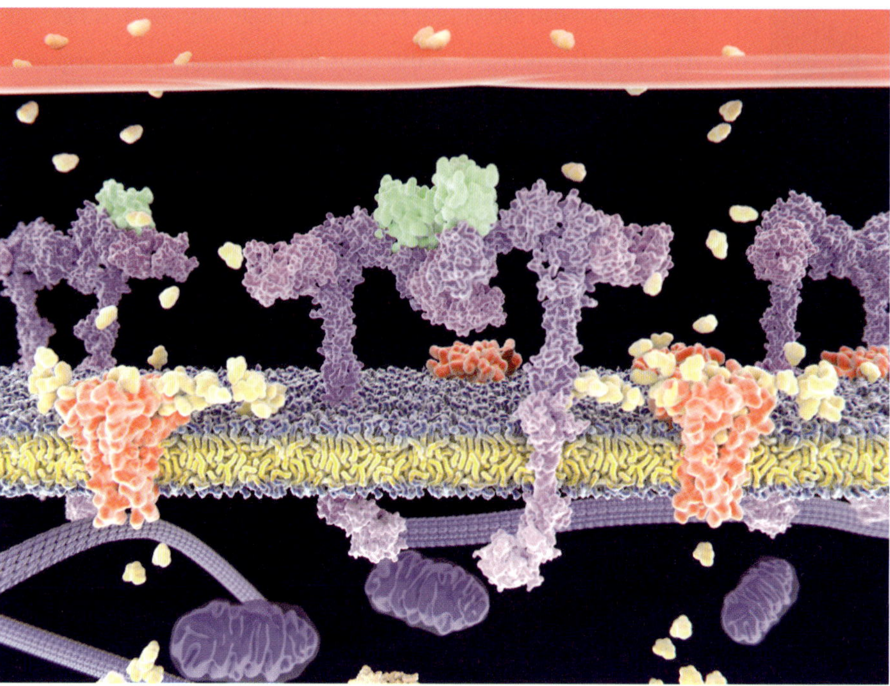

FIGURE 6.1.1 In this digital illustration, the signalling molecule insulin (green) is shown bound to the transmembrane insulin receptor (violet). Binding of signalling molecules to extracellular receptors triggers cellular responses inside the cell.

CELLULAR COMMUNICATION

Multicellular organisms have developed mechanisms to communicate between cells and convey messages about changes to internal and external environmental conditions. A change in conditions that elicits a **response** from a cell is referred to as a **stimulus**. Stimuli are varied in nature; changes in pressure, light, temperature or chemical molecules are all examples of stimuli.

Not all cells are capable of detecting all stimuli, and not all cells are able to respond to all stimuli. Cells that are able to detect stimuli can pass this information to other cells by producing and releasing **signalling molecules**. **Effector cells** respond to signalling molecules. The signal may be passed to a number of different cells before reaching the ultimate effector cells, and different signalling molecules may be released at each step. Signalling molecules can trigger a response even at very low concentrations.

An effector cell is a cell that responds to a stimulus.

Once the signal reaches the effector cells, various cellular processes may be activated, depending on the nature of the original stimulus. These changes in cellular activities form the ultimate response.

Mode of transmission of signalling molecules

Communication via signalling molecules can occur in nearby and distant environments. Signalling can be classified by the mode of transmission (Figure 6.1.3).

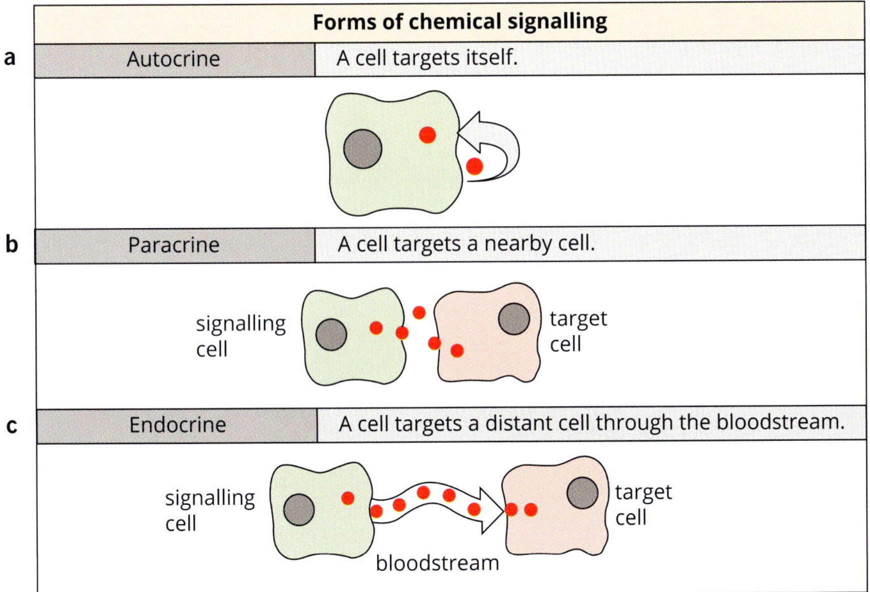

FIGURE 6.1.3 Signalling molecules can be classified according to the distance they need to travel to reach their target cells.

- **Autocrine signalling**: Signalling molecules act on the actual cell or the same type of cell that secreted them (Figure 6.1.3a). For example, activated T lymphocytes produce a type of cytokine for which they also express receptors, and binding of that cytokine with its receptors promotes cell division and proliferation.
- **Paracrine signalling**: Signalling molecules act on cells that are close to the secreting cell (Figure 6.1.3b). For example, neurotransmitters are secreted by a neuron and target the next neuron in the neural pathway or a neighbouring effector cell.
- **Endocrine signalling**: Signalling molecules act on cells that are far from the cell that secretes them (Figure 6.1.3c). For example, animal **hormones** are released from glands and organs and travel throughout the body, usually in the bloodstream, to act on distant target cells.

Nature and source of signalling molecules

Signalling molecules can be classified according to their chemical properties as hydrophobic or hydrophilic. Hydrophobic molecules are non-polar molecules such as lipid-based molecules that are relatively insoluble in water. Hydrophilic molecules are polar ions and molecules such as peptide-based molecules that dissolve easily in water. The main groups of signalling molecules and their sources, modes of transmission and chemical properties are summarised in Table 6.1.1 (page 224).

BIOFILE

Slowing cancer growth

Vascular endothelial growth factor (VEGF) is a chemical signalling molecule released by tissues to stimulate the proliferation of new blood vessels—a process known as angiogenesis (Figure 6.1.2). Cancerous tumours release VEGF to stimulate the growth of the blood vessels they need to feed their rapid growth. By making drugs that block VEGF receptors (angiogenesis inhibitors), doctors can prevent the rapid growth of the blood vessels and 'starve' the developing cancer. This dramatically reduces the growth rate of the tumour and is an effective anti-cancer therapy.

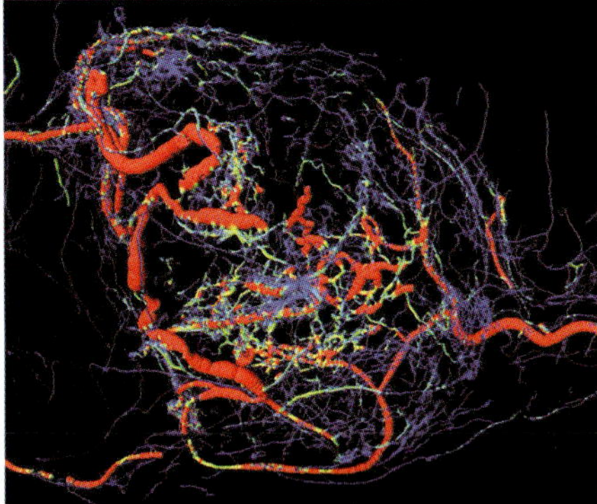

FIGURE 6.1.2 Coloured 3D computed tomography scan of blood vessels inside a tumour of the colon. The vessels are colour coded on the basis of their thickness. Red vessels are thickest, green and yellow vessels are of intermediate thickness, and blue vessels are the thinnest.

Group of signalling molecules	Source	Mode of transmission	Hydrophobic or hydrophilic
animal hormones	glands and organs	• autocrine • paracrine—through interstitial fluid between cells • endocrine—via the blood circulation	• hydrophobic steroid hormones • hydrophilic peptide hormones • hydrophobic or hydrophilic amino-acid-derived hormones
plant hormones	most plant cells are capable of producing a variety of plant hormones	• various, including diffusion	• hydrophobic phytohormones • hydrophilic phytohormones
neurotransmitters	neurons	• paracrine—through exocytosis into the synaptic gap and diffusion across the gap	• hydrophilic
cytokines	many cells, including immune cells such as macrophages, B- and T-lymphocytes and mast cells	• autocrine • paracrine • endocrine	• hydrophilic
pheromones	various cells depending on the species	• communication between organisms by diffusion outside the organism	• hydrophobic pheromones • hydrophilic pheromones

TABLE 6.1.1 Sources, modes of transmission and chemical properties of signalling molecules.

HORMONES

A hormone is a signalling molecule produced in tiny amounts that can have relatively long-lasting effects on target cells. Hormones help regulate the growth and activity of cells in most animals and plants (Figure 6.1.4).

FIGURE 6.1.4 Hormones are some of the signalling molecules responsible for human growth and development.

Animal hormones

In vertebrates, the **endocrine system** is responsible for coordinating many body functions, including growth, metabolism and reproduction. This system is made up of many glands and organs within the body that, along with some specialised tissues, synthesise and secrete hormones into the bloodstream. The blood flow transports the hormones to the target cells and tissues. The main glands and organs of the human endocrine system are shown in Figure 6.1.5.

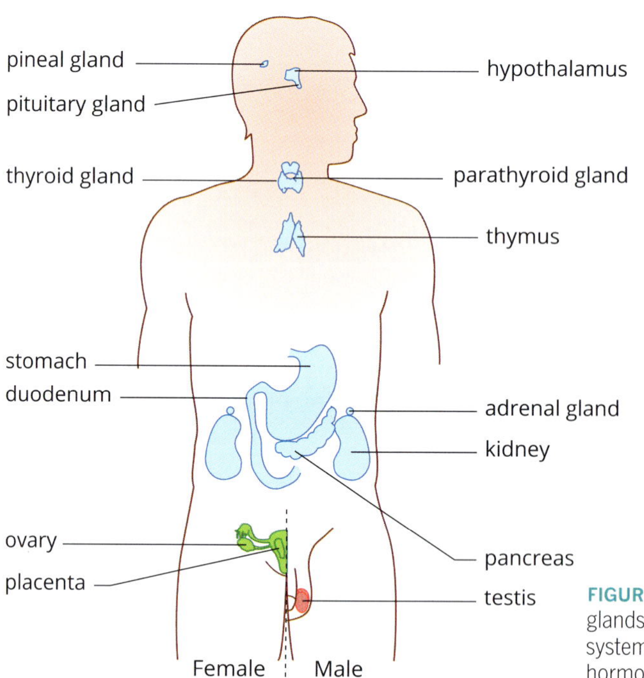

FIGURE 6.1.5 The organs and glands of the human endocrine system produce and secrete hormones into the bloodstream.

Hormones can be broadly grouped into three main classes:

- Lipid hormones are a class of hydrophobic signalling molecules derived from fatty acids (eicosanoids) or cholesterol (steroids). Eicosanoids include prostaglandins, which are involved in cell growth, fever and inflammation. **Steroid hormones** help to regulate metabolism, salt and water balance, inflammation and sexual function. Examples of steroid hormones include **testosterone**, **oestrogen** and **cortisol**.
- **Peptide and protein hormones** are a class of hydrophilic signalling molecules. An example of a peptide hormone is **insulin** and an example of a protein hormone is **growth hormone**.
- **Amino-acid-derived hormones** are a class of small signalling molecules derived from the amino acids tyrosine and tryptophan. They can be further divided into **catecholamines** and **thyroid hormones**. Thyroid hormones such as thyroxine are hydrophobic. Catecholamines are hydrophilic; examples include adrenaline and dopamine. (Dopamine acts as both a neurotransmitter and a hormone.)

A single hormone can trigger different responses in multiple target cells at the same time. An example is adrenaline, a hormone that is released from the adrenal glands. Adrenaline targets cardiac muscle cells, vascular smooth muscle cells and the various glands and organs of the digestive system. An increase of adrenaline in the bloodstream will result in an increase in heart rate and blood pressure and will simultaneously decrease digestive functions, preparing for a 'fight or flight' response.

Some common mammalian hormones, their sources, target tissues and functions are listed in Table 6.1.2.

Gland (source)	Hormone(s)	Hormone class	Hydrophobic or hydrophilic	Target	Function
adrenal cortex	glucocorticoids	steroid	hydrophobic	many cell types	regulate glucose metabolism and stimulate fat breakdown
	mineralocorticoids	steroid	hydrophobic	kidney tubule cells	regulate reabsorption of salts
hypothalamus	dopamine	amino-acid-derived	hydrophilic	anterior pituitary	inhibits release of prolactin
	growth hormone releasing hormone (GHRH)	peptide	hydrophilic	somatotroph cells in the pituitary gland	stimulates release of growth hormone
anterior pituitary	adrenocorticotrophic hormone (ACTH)	peptide	hydrophilic	adrenal cortex	promotes release of adrenal cortex hormones
	growth hormone (GH)	protein	hydrophilic	bone muscle	promotes protein synthesis and growth
	follicle stimulating hormone (FSH)	protein	hydrophilic	ovaries	promotes development of follicle and secretion of oestrogen
	luteinising hormone (LH)	protein	hydrophilic	ovaries	promotes ovulation, development of corpus luteum and secretion of progesterone
	prolactin	protein	hydrophilic	mammary glands	stimulates milk synthesis and secretion
	thyroid stimulating hormone (TSH)	protein	hydrophilic	thyroid	promotes production and release of thyroxine
pancreas	insulin	peptide	hydrophilic	most cells	regulates blood glucose levels
thyroid	thyroxine	amino-acid-derived	hydrophobic	most cells	regulates cellular metabolic rate

TABLE 6.1.2 Common mammalian hormones, their sources, target tissues and functions.

BIOFILE

Acromegaly

Acromegaly is a hormonal disorder in which the pituitary gland produces too much of the growth hormone somatotrophin. Sometimes the overproduction of the hormone is caused by a tumour of the anterior pituitary gland. The tumour can be treated with X-rays or surgically removed.

Acromegaly normally develops during adulthood and results in an increase in the size of hands, feet and face, and can result in severe disfigurement and fatal complications. When acromegaly occurs during childhood, the excess growth hormone can cause a condition known as gigantism. These children can have exaggerated bone growth and an abnormal increase in height.

FIGURE 6.1.6 Man suffering from acromegaly.

BIOLOGY IN ACTION

Building strong bones

You are probably aware that if you break your arm, the broken bone becomes thinner and weaker during the period in which it is in plaster and out of action (Figure 6.1.7). For people suffering from osteoporosis, weight-bearing exercises are recommended to build up bone strength. Both of these situations relate to the ability of bone cells to detect and respond to physical stress. Physical stress on bones causes them to become thicker. Removal of stress causes bone material to be resorbed into the circulation.

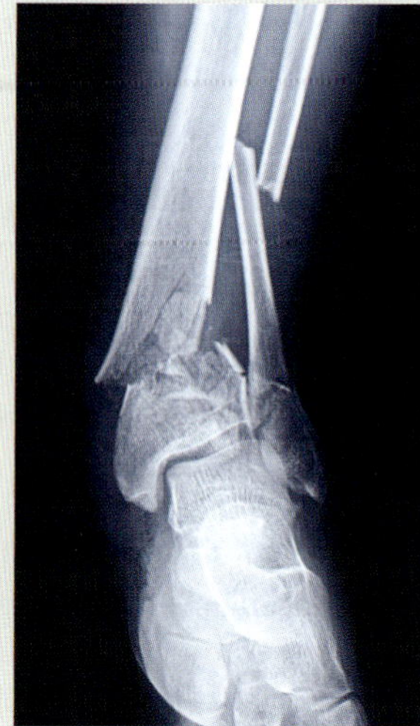

FIGURE 6.1.7 X-ray of a broken bone.

Bones are not permanent structures—they are dynamic. They are a reservoir of calcium used to maintain blood calcium levels. Parathyroid hormone and calcitonin are involved in the deposition and resorption of calcium salts in bone. Depending on the level and direction of the stress, bone cells lay down or reabsorb the calcium salts from which they are made (Figure 6.1.8). This is why bone grafts usually work so well. A piece of bone from the fibula in the leg can be grafted into the spinal column. It will soon be reshaped by physical stress to suit its new location and the work it has to do.

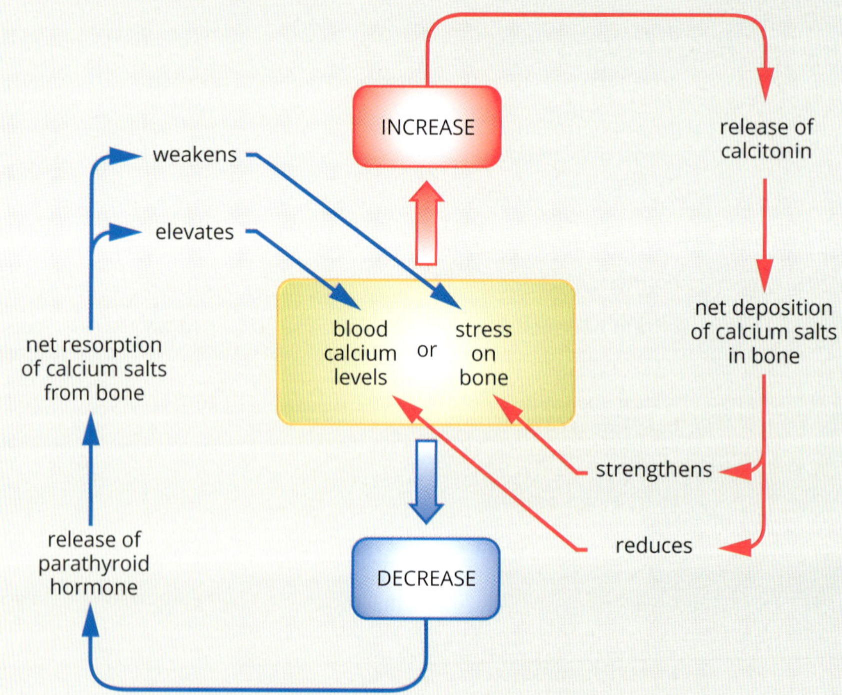

FIGURE 6.1.8 Bones respond to increased physical stress by becoming stronger. If the stress on bones decreases (as with astronauts), they become weaker.

EXTENSION

The pituitary gland

The pituitary gland is located at the base of the brain, just above the roof of the mouth. It is often called the 'master gland' of the endocrine system because it produces many of the body's hormones, a number of which help regulate the production of other hormones around the body. Hormones secreted by the pituitary gland are also involved in a range of cellular processes including growth, reproduction, lactation, kidney function, skin pigmentation and regulation of the activity of the thyroid and the adrenal glands.

The hypothalamus is above the pituitary gland, and connects directly to the pituitary gland (Figure 6.1.9). The hypothalamus is responsible for detecting internal stimuli from all over the body and determining whether or not optimal conditions are being maintained. These internal stimuli trigger the production of releasing hormones from the hypothalamus. Releasing hormones are those that control and regulate specific hormone production in the pituitary gland. The combined functions of the hypothalamus and pituitary gland are vital for homeostasis.

For example, when your body temperature drops, the hypothalamus secretes thyrotropin-releasing hormone (TRH). The pituitary gland responds to TRH by secreting thyroid-stimulating hormone (TSH), also known as thyrotropin. TSH in return stimulates the release of thyroid hormone by the thyroid gland (Figure 6.1.10). As thyroid hormone accumulates, it increases metabolic rate, resulting in the release of heat energy, which raises body temperature. When the blood concentration of thyroid hormones increases above a certain threshold, TRH-secreting neurons in the hypothalamus are inhibited and stop the secretion of TRH.

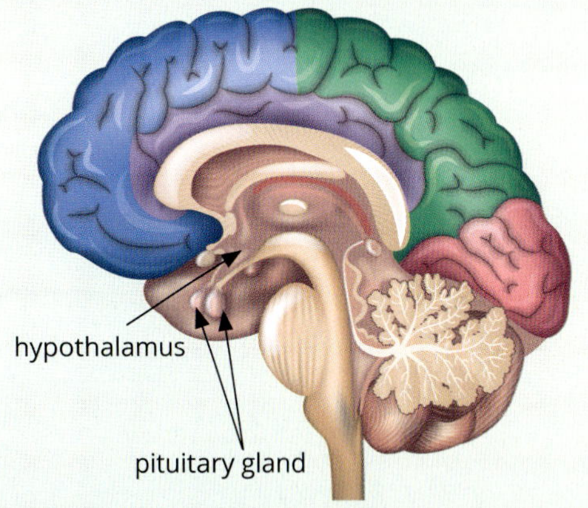

FIGURE 6.1.9 Cross-section of the human brain showing the location of the hypothalamus and pituitary gland.

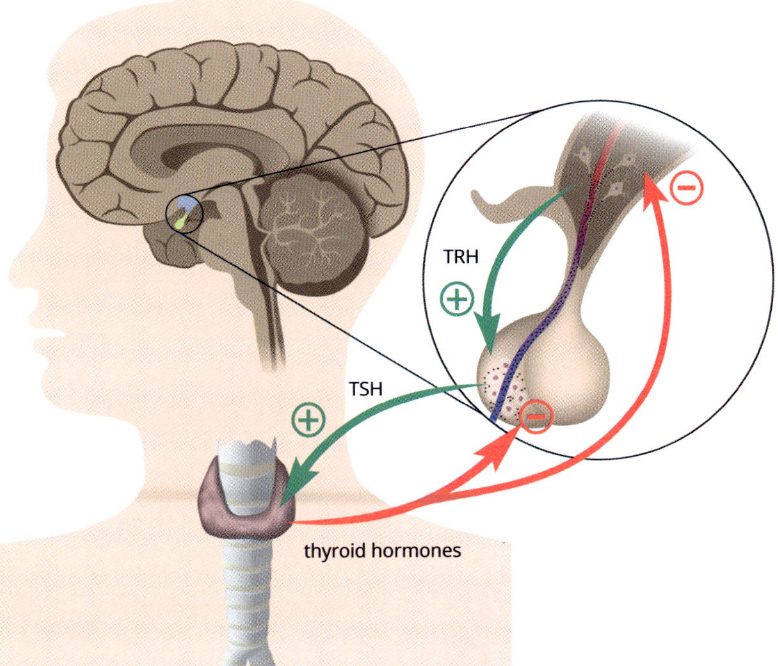

FIGURE 6.1.10 The secretion of TRH by the hypothalamus stimulates the release of TSH by the pituitary gland. TSH stimulates the thyroid gland to produce thyroid hormones (green). When the blood concentration of thyroid hormones increases above a certain threshold, the secretion of TRH in the hypothalamus is inhibited (red).

Plant hormones

Plant hormones, like all signalling molecules, are produced in low concentrations but have a significant effect on plant development and growth. In contrast to animal hormones, which are produced by specific glands and organs, each plant cell is able to produce many different types of hormone. Plant hormones also have a variety of modes of transmission.

There are five main types of plant hormones. These are sometimes called **phytohormones**. A summary of the biological roles of these hormones is provided in Table 6.1.3.

Type of plant hormone	abscisic acid
Source	leaves (chloroplasts), roots
Effector site	seeds, buds, guard cells, leaves and fruit
Mode of transmission	transported in the xylem from roots and phloem from leaves
Visible effect	seed and bud dormancy, drought tolerance and apical dominance
Type of plant hormone	**auxins**
Source	shoot tip (meristem), seeds
Effector site	growing region of shoots and roots, developing fruit
Mode of transmission	transported from cell to cell, often with directional transport, usually moving from shoots towards roots
Visible effect	shoot tip bends towards the light (phototropism), roots grow downwards (gravitropism), apical dominance
Type of plant hormone	**cytokinins**
Source	roots and developing fruits
Effector site	branch and leaf buds
Mode of transmission	transported in xylem
Visible effect	growth of lateral branches
Type of plant hormone	**ethene**
Source	ripening fruits and other parts of the plant
Effector site	most cells
Mode of transmission	diffusion (ethene is a gas)
Visible effect	increases sugar content of fruit, fruit and leaf drop
Type of plant hormone	**gibberellins**
Source	root and shoot apical meristems, growing leaves and seeds
Effector site	meristems, leaves, seeds and flowers
Mode of transmission	typically used in the cell that made it, otherwise transported cell-to-cell in xylem and phloem
Visible effect	elongation of stems, leaf expansion, seed germination, fruit and flower maturation

TABLE 6.1.3 The biological roles of the five main types of plant hormones.

BIOFILE

Using ethene at home

Ethene (or ethylene) is released from ripening fruit into the air. The gaseous hormone will also trigger nearby fruit to ripen. Commercial application of ethene allows farmers to pick and transport fruit before it ripens and induce ripening using ethene once the fruit arrives at its destination.

Placing a ripe banana in a paper bag will trap the ethene it releases. Placing an unripe fruit, such as an avocado, in the bag with the banana will cause the second fruit to ripen rapidly.

FIGURE 6.1.11 Ethene produced from ripe fruit can ripen surrounding fruit.

NEUROTRANSMITTERS

Neurotransmitters are a group of hydrophilic signalling molecules secreted by **neurons**. As you learnt in *Heinemann Biology 1*, Chapter 5, the neuron is the basic cellular unit of the nervous system. Neurons are specialised cells with structures that enable rapid transmission of information between cells. There are three different types of neurons: efferent (motor) neurons, interneurons and afferent (sensory) neurons. These different types of neurons have different structures but are made up of the same basic components: one or more **dendrites**, a cell body and an **axon** (Figure 6.1.13). Branching dendrites receive signals including neurotransmitters from other cells, and transmit these to the cell body. A single axon conducts a signal from the nerve cell body to nerve endings, which form synapses with the dendrites, cell body, or axon hillock of other neurons. Neurotransmitters are produced in the synaptic terminals (also known as synaptic buttons, synaptic knob or terminal buttons) of the neuron.

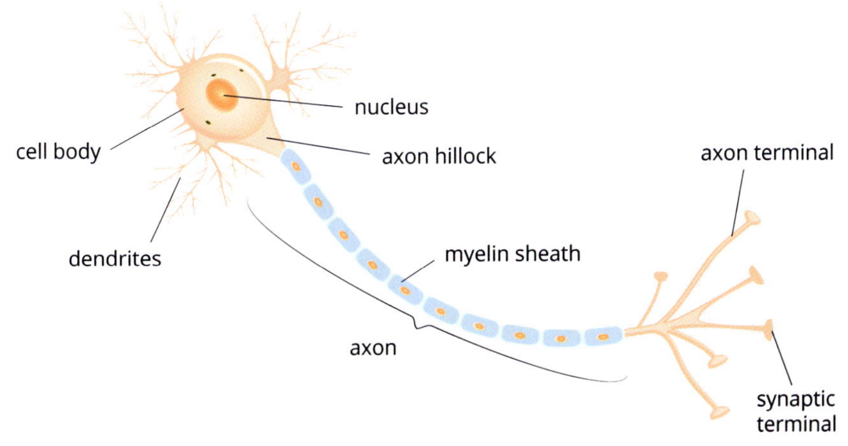

FIGURE 6.1.13 The key structures of a generalised neuron.

BIOFILE

Types of neurons

Structural classification: The dendrites of neurons don't always extend directly from the cell body, but when they do and the neuron has a single long process this is an example of a multipolar neuron (Figure 6.1.12a). Multipolar neurons are the most common type of neuron in vertebrates. Other types of neurons include bipolar neurons (two processes from the cell body, Figure 6.1.12b), and unipolar neurons (cell body appears to sit in the middle of the axon, Figure 6.1.12c).

Functional classification: Sensory (or afferent) neurons transmit impulses towards the central nervous system (CNS). Most sensory neurons are unipolar. Motor (or efferent) neurons transmit impulses away from the CNS. Motor neurons are multipolar. Interneurons (or association neurons) link motor and sensory neurons. Most interneurons are multipolar.

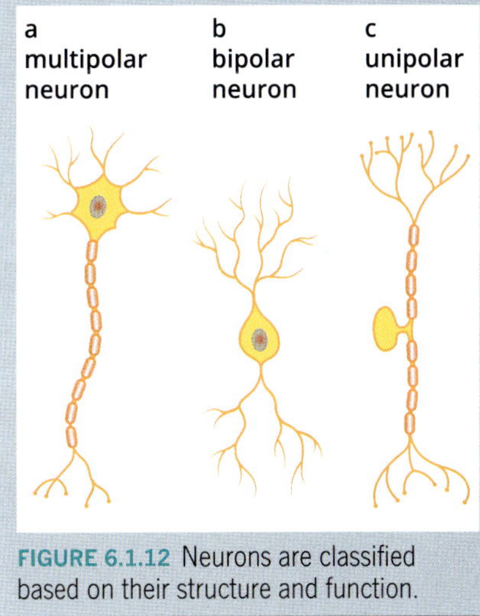

FIGURE 6.1.12 Neurons are classified based on their structure and function.

The axon hillock is where the axon joins the cell. It is from here that the electrical firing known as an action potential usually occurs.

BIOFILE

Weapons against the nervous system

Chemicals that cause nervous systems to malfunction have evolved as weapons in both plants and animals. Redback spider venom, for example, contains a neurotoxin that acts on both motor and sensory nerves, causing an excessive outpouring of neurotransmitters and may lead to paralysis. The Australian corkwood tree produces hyoscine, which blocks muscle receptors for the neurotransmitter acetylcholine. The paralysing actions of ingested corkwood leaves were understood by Aborigines. It has been suggested that they used it as an emu and fish poison.

FIGURE 6.1.15 The redback spider *Latrodectus hasseltii*.

The mode of transmission of neurotransmitters is paracrine signalling. Neurons secrete neurotransmitters into specialised junctions they form with local target cells, such as other neurons, gland cells or muscle cells. These specialised junctions are called **synapses**. When an electrical signal arrives at the synapse, it triggers the release of neurotransmitters into the intersynaptic space. The neurotransmitters will diffuse across the gap and bind to receptors on the surface of the **postsynaptic neuron**, which in turn cause the transmission of the nervous impulse from one neuron to another (Figure 6.1.14).

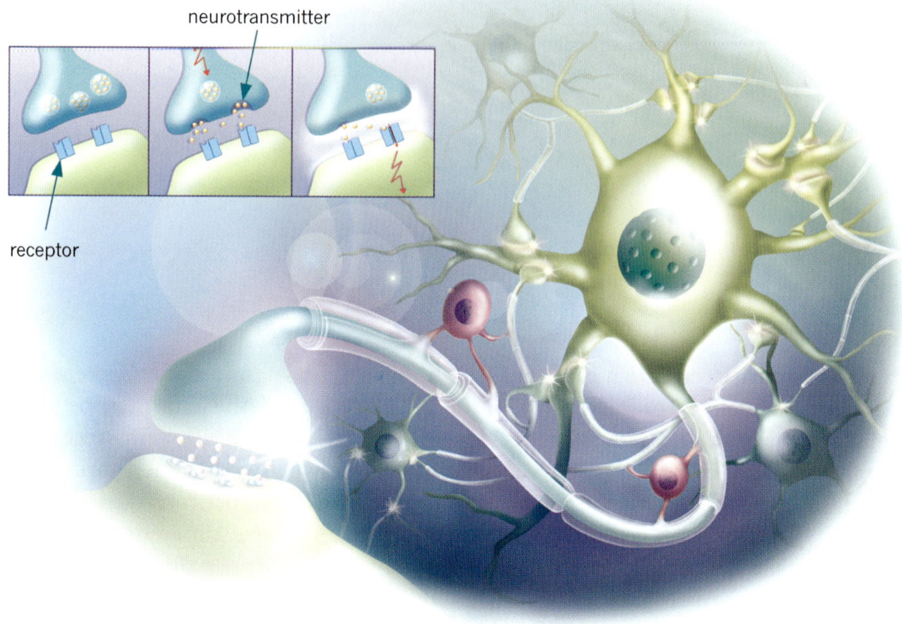

FIGURE 6.1.14 Transmission of neurotransmitters at the synapse.

Neurotransmitters are involved in many cellular responses including movement (Figure 6.1.16), regulating hormone production and organ function. Examples of neurotransmitters include serotonin (which influences mood, sleep and appetite) and dopamine (which is involved in reward-motivated behaviour).

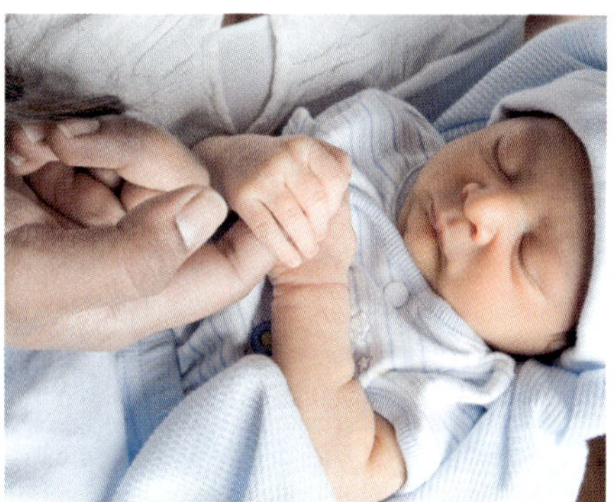

FIGURE 6.1.16 Neurotransmitters are responsible for the reflex action of grasping in a newborn baby.

CYTOKINES

Cytokines are a group of hydrophilic signalling molecules that are involved in communication between immune cells. Cytokines coordinate many aspects of the immune response and are released by body cells in response to damage or pathogens. Sources of cytokines include **macrophages**, **T lymphocytes** and **B lymphocytes**. Some cytokines, such as **interferons** and **interleukins**, are primarily involved in regulating inflammation and other immune system responses to infection (Figure 6.1.17).

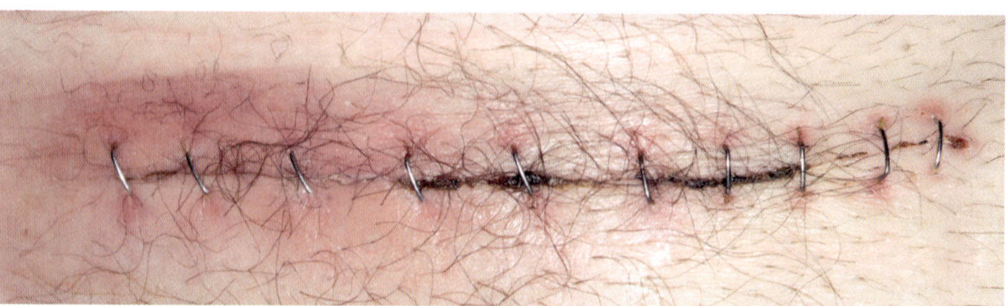

FIGURE 6.1.17 Inflammation of a wound is an immune response initiated by cytokines.

Cytokines bind to specific receptors on the surface of target cells, where they trigger a variety of cellular responses, depending on the receptor. Cytokines are typically involved in either paracrine or autocrine signalling (see page 223). Table 6.1.4 shows some examples of cytokines.

Cytokine and its function	Molecular structure
interleukin-6 (IL-6) IL-6 is produced by T lymphocytes and macrophages. It has many functions including raising body temperature during inflammation and supporting the growth and development of B lymphocytes, which produce antibodies.	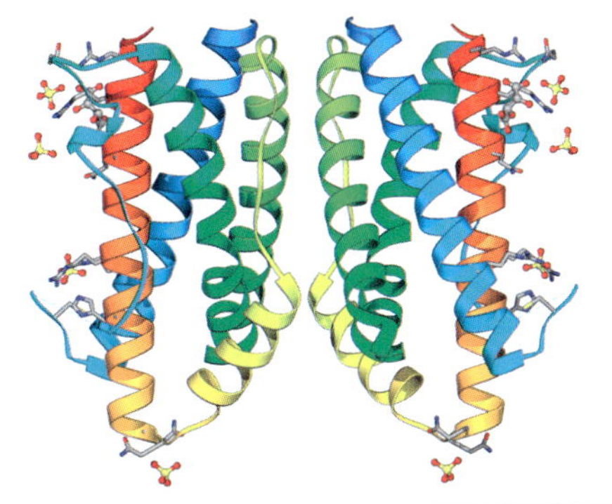
transforming growth factor beta 1 (TGF-b 1) It is produced by many different types of cells, including macrophages. Among its many functions, TGF-β 1 is involved in regulating the cell cycle and in some situations can trigger apoptosis.	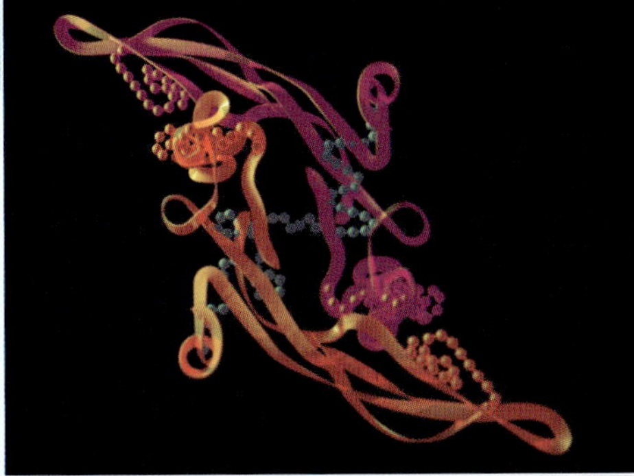
granulocyte-colony stimulating factor (G-CSF) This cytokine stimulates bone marrow stem cells to replicate and differentiate granulocytes. Other CSFs are involved in stimulating bone marrow stem cells to differentiate into the various blood cells. CSFs were discovered in 1977 by Australian scientist Professor Don Metcalf at the Walter and Eliza Hall Institute and are now used in cancer therapy.	

TABLE 6.1.4 Molecular structure and function of cytokines.

BIOFILE

Pheromones in tears

A study conducted by Noam Sobel, a cognitive neuroscientist at the Weizmann Institute of Science in Rehovot, Israel, discovered that human tears contain pheromones. In the study, women were shown a sad movie and the tears they shed were carefully collected. When men were given the tears to smell, it was found that the scent of the tears reduced testosterone levels and lowered the activity in parts of their brain involved in levels of sexual desire.

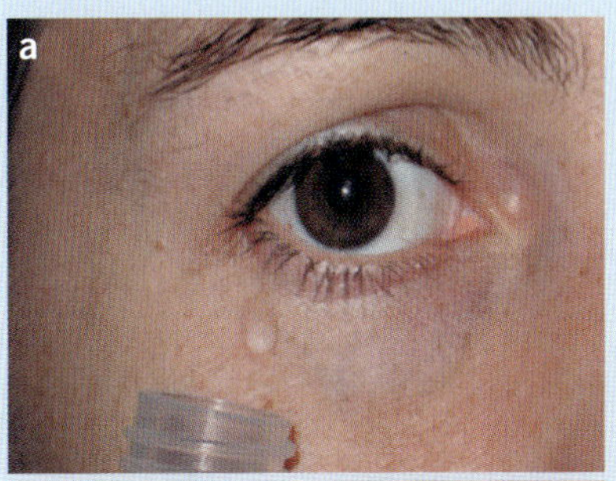

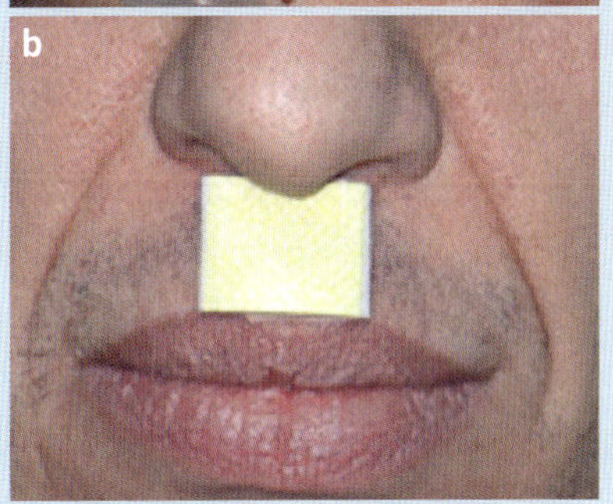

FIGURE 6.1.18 (a) Collection of tears from a woman participating in the study. (b) A pad soaked in either tears or saline solution (control) that is placed on the upper lip of male participants in the study.

PHEROMONES

Signalling molecules that are excreted into the external environment of an organism are called **pheromones**. Pheromones influence the behaviour or physiology of another individual, usually of the same species. Common functions of pheromones in animals include triggering alarm and aggressive responses, marking territory, marking food trails and attracting mates.

The use of pheromones for communication has been widely studied in insects (Figure 6.1.19), but many other animals and plants produce pheromones.

FIGURE 6.1.19 Many organisms, such as these ants, use pheromones to communicate with other individuals of the same species.

BIOLOGY IN ACTION

Using pheromones to control insect pests

Insect pests can destroy crops. One method of controlling insect pests involves the use of their own communication systems to disrupt their behaviour. Females of many moth species release pheromones to attract males. These substances are effective in extremely small quantities. In the silk moth (*Bombyx mori*), a single female contains enough pheromone in her abdomen to stimulate more than a hundred thousand million males—far more than are alive at any one time in her community (Figure 6.1.20).

An understanding of sexual communication in the oriental fruit moth (*Cydia molesta*), a major pest of peach and nectarine crops, led to the first commercial application of a pheromone for pest control in Australia. Normally the females release their pheromone into the air in the evening. Males, detecting just a few molecules, respond by flying in an upwind zigzag path, changing direction each time they detect a decrease in the amount of scent (Figure 6.1.21).

Flooding the area around crops with the female pheromone can prevent mating between moths. This confuses males and makes locating females by the usual zigzag tracking method impossible (Figure 6.1.22).

Pheromone-baited traps can be used to detect the presence of a pest species or to reduce the male pest population (Figure 6.1.23). Pheromone-baited traps are lined with an adhesive surface. As insects enter the trap, they become stuck to the inner surface. The traps can be used to detect the presence of a pest species and indicate when further action is required. Large numbers of pheromone traps can also be used to reduce the male pest population.

Pheromones offer three distinct advantages over conventional chemical insecticides:

- They tend to be more specific to the target pest.
- They leave no harmful residues in the environment.
- Insects are unlikely to develop a resistance to them, as insects produce pheromones naturally.

Almost all of Australia's commercial stone fruit growers now use the pheromone method to control pest insects, reducing their use of chemical insecticides by 90%.

FIGURE 6.1.20 A silkworm (*Bombyx mori*) feeding on a leaf. Silkworms are the larvae (caterpillars) of the silk moth and are responsible for damaging crops.

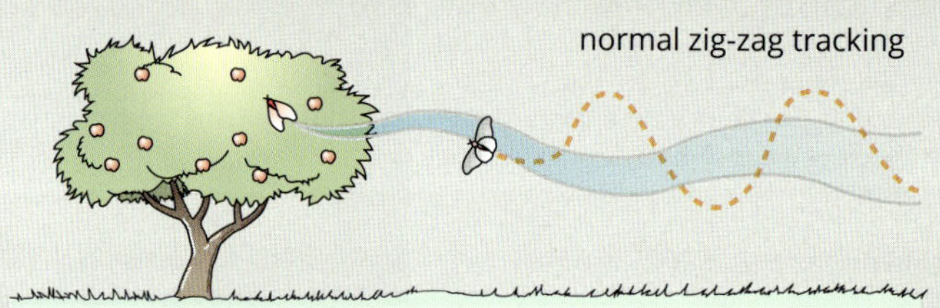

FIGURE 6.1.21 Male moths track female moths by the pheromones they release.

FIGURE 6.1.22 Pheromones can be used to interrupt mating behaviours of pest insects.

FIGURE 6.1.23 Pheromones can be used to trap pest insects and monitor their numbers. (a) Pheromone baits can lure moths into traps. (b) A few baited traps can be used to monitor the size of a moth population to determine whether further action is required.

6.1 Review

SUMMARY

- Multicellular organisms produce signalling molecules as a form of intercellular communication.
- Signalling molecules transmit information about an internal or external stimulus to other cells, including, ultimately, the effector cells, which enact the response.
- Signalling molecules can be classified according to their chemical structure, their source and their mode of transmission.
- Animal hormones:
  - are produced by organs and glands in animals
  - can be hydrophilic or hydrophobic
  - involve autocrine, paracrine or endocrine modes of transmission.
- Plant hormones:
  - are produced in a variety of cells in plants
  - can be hydrophilic or hydrophobic
  - are transported cell-to-cell or via the xylem and/or phloem.
- Neurotransmitters:
  - are produced by neurons in animals
  - are hydrophilic
  - involve paracrine transmission
  - transmit the signal to various types of cells that form synapses with neurons.
- Cytokines:
  - are produced by immune cells such as lymphocytes and macrophages
  - have an overall hydrophilic structure
  - involve autocrine, paracrine or endocrine modes of transmission
  - elicit a variety of immune cell responses such as inflammation, leukocyte differentiation and antiviral activity.
- Pheromones:
  - are produced by various specialised cells depending on the species
  - are excreted outside the organism
  - diffuse through the air to other individuals
  - typically only elicit a response in another individual of the same species.

KEY QUESTIONS

1. a What are 'signalling molecules'?
   b Give an example of how a signalling molecule may be used in an immune response in an animal.
2. Describe briefly the different modes of transmission for signalling molecules.
3. Describe the feature of a target cell that makes it receptive to a particular signalling molecule.
4. Hormones and neurotransmitters are both signalling molecules. Distinguish between the two.
5. Outline the difference in the source of plant and animal hormones.
6. What kind of signalling molecule is a pheromone? How does it reach its target cells?
7. Copy and complete the table below to show the mode of transmission for each type of signalling molecule.

Type of signalling molecule	Mode of transmission				
	Endocrine	Paracrine	Autocrine	Transported cell-to-cell or via the xylem and/or phloem	Diffusion by air
hormones					
neurotransmitters					
cytokines					
pheromones					

6.2 Signal transduction

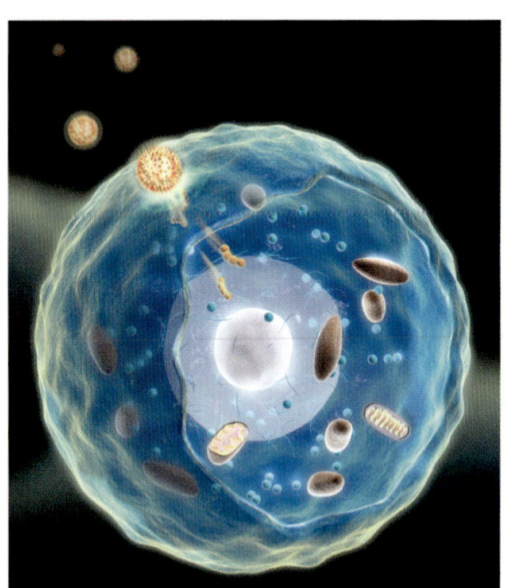

FIGURE 6.2.1 Signal transduction occurs when a signalling molecule from outside the cell (orange), activates a specific receptor located on the cell surface or inside the cell.

The process of converting the original stimulus signal into a response is called **signal transduction** (Figure 6.2.1). Signal transduction involves changing the form of the signal in some way. This may involve a change in the type of signalling molecule used, passing the signal into or out of a cell or converting the type of signal from one form to another (for example, from a chemical to an electrical signal). The specific processes involved in the transduction of a particular signal depend on the signalling molecules involved.

In this section, you will learn about the general characteristics of signal transduction involving hydrophobic and hydrophilic signalling molecules.

STIMULUS-RESPONSE MODEL

The ability of a multicellular organism to detect and respond to stimuli relies on cells communicating with each other. As you learnt in the previous section, cells communicate with each other through signalling molecules (Figure 6.2.2).

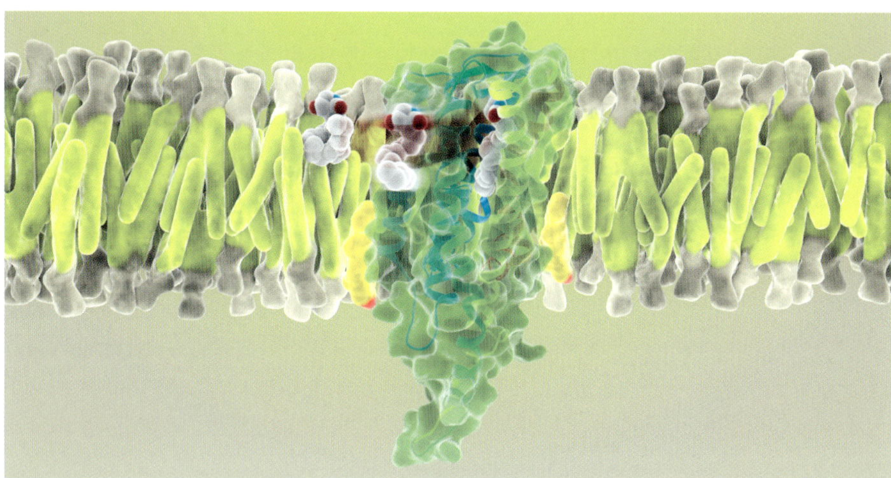

FIGURE 6.2.2 Digital illustration of a signalling molecule (mostly white, centre right) binding to a receptor (green).

The processes involved in a cell detecting and responding to a signalling molecule are together known as signal transduction. Signal transduction can be considered in terms of a **stimulus–response model** (Figure 6.2.3).

The stimulus–response model can be divided into a three-step process:

Step 1 reception—the detection of the signalling molecule by a receptor

Step 2 transduction—the relay of the signal into the cell

Step 3 cellular response—the activation of a cellular activity.

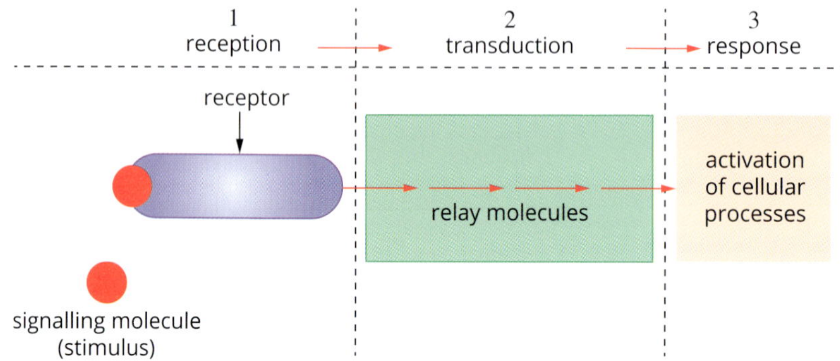

FIGURE 6.2.3 The stimulus–response model applied to the cell in terms of signal transduction.

BIOLOGY IN ACTION

CYBERNOSE biosensor

Biosensors are able to detect minute molecules to help monitor a person's health and their environment. Scientists are developing sensors that are sensitive enough to detect even the smallest presence of molecules, so that disease or dangerous environmental contaminants can quickly be assessed. A biosensor can consist of a tiny glass slide of sliver with protein receptors applied to it. When the receptors bind with their matching signal molecule, light waves traveling across the chip are perturbed into a recognisable interference pattern. Using this combination of light waves and biomolecules has resulted in a biochip a magnitude more sensitive than other current comparable sensors (Figure 6.2.4).

An example of a biosensor is the CYBERNOSE biosensor developed by CSIRO. DNA from nematode worms was used to make modified smell receptors, and the smell receptors were placed into the electronic device known as the CYBERNOSE sensor. Nematode worms were chosen as these animals have similar smell receptors to those in human noses. The receptors work by changing shape when an odour molecule binds to them. In the CYBERNOSE sensor, the modified smell receptors will emit a mixture of blue and green light that can be measured when an odour molecule binds to them. The detailed analysis of the light, using optical sensors, can indicate whether a particular substance is in the test sample.

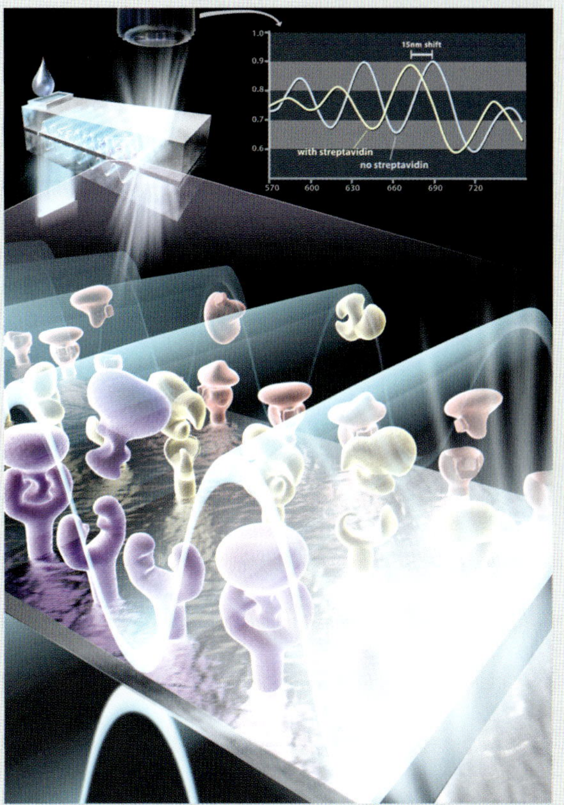

FIGURE 6.2.4 An illustration showing receptors on the surface of a biosensor device.

Reception

Reception involves the detection of a signalling molecule by a cell. The receptor that detects a signalling molecule can be located on the surface of the plasma membrane or in the cytosol or nucleus of the cell. The position of the receptor depends on whether the signalling molecule is hydrophobic or hydrophilic. Receptors are specific and will only bind to particular signalling molecules (Figure 6.2.5).

Not all cells are responsive to all signalling molecules. If a cell does not express the gene for a specific receptor, it will not have the receptors and hence that cell cannot respond to the complementary signalling molecule. For example, in the brain, neurons in certain neural pathways may express a defined set of receptors to be responsive only to the specific neurotransmitters used in that pathway.

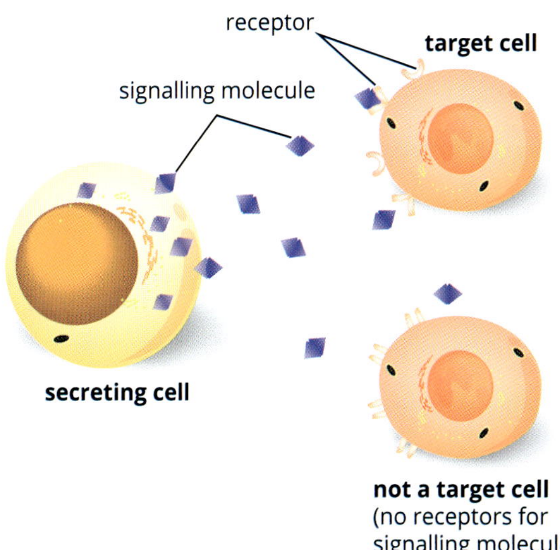

FIGURE 6.2.5 Each type of signalling molecule is designated for certain cells (target cells). Receptors are specific and will only bind to a particular signalling molecule.

Transduction

Transduction involves converting the signal into a form that can be relayed to reach its final destination within the cell and bring about a cellular response. Transduction may involve a one-step process in which a signalling molecule binds to one receptor and this complex produces a response. Alternatively, it may involve a multi-step process in which a signalling molecule binds to its receptor, leading to the sequential activation of different molecules in a chain of events. These multi-step transduction pathways are commonly called **cascades**.

> **BIOFILE**
>
> **Managing high blood pressure**
>
> One of the common groups of drugs used in the management of high blood pressure (hypertension) is an angiotensin receptor antagonist. When angiotensin receptors are activated, it results in large number of processes that increase blood pressure, such as vasoconstriction of the blood vessels. Angiotensin receptor antagonist acts by blocking the activation of angiotensin receptors, hence blood vessels undergo vasodilation and blood pressure is reduced.

Cellular response

Following transduction, a response is initiated. Cellular responses include any cellular activity such as gene transcription, the activation of enzymes or the secretion of signalling molecules by the cell. Responses can occur in the:

- nucleus
- cytosol
- plasma membrane.

SIGNAL TRANSDUCTION OF HYDROPHOBIC SIGNALLING MOLECULES

Hydrophobic signalling molecules are usually lipid-based molecules involved in gene regulation. Lipid-based molecules are lipid soluble so can easily diffuse through the plasma membrane. Inside the target cell, they bind to an intracellular receptor, either in the cytosol or in the nucleus.

Steroid hormones are examples of lipid-based hydrophobic signalling molecules that bind to receptors in the cytosol (Figure 6.2.6). Once bound, the signalling molecule–receptor complex moves from the cytosol through **nuclear pores** to its final destination in the nucleus (Figure 6.2.7).

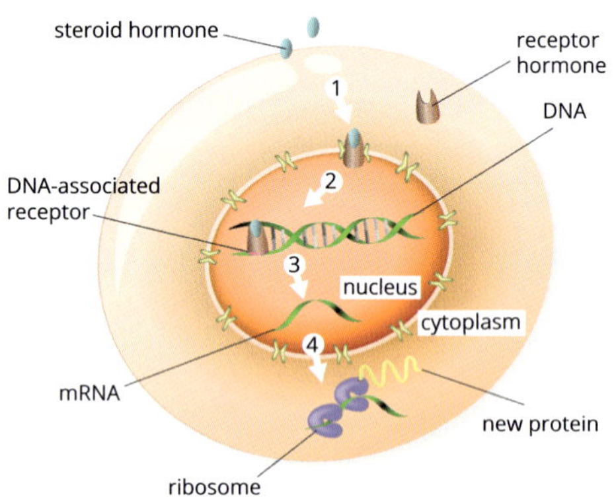

FIGURE 6.2.7 Once steroids are in the cytoplasm (1) they bind to receptors in the cytosol, then travel to the nucleus and move through pores in the nuclear membrane (2), which activates gene transcription (3) and causes proteins to be manufactured (4).

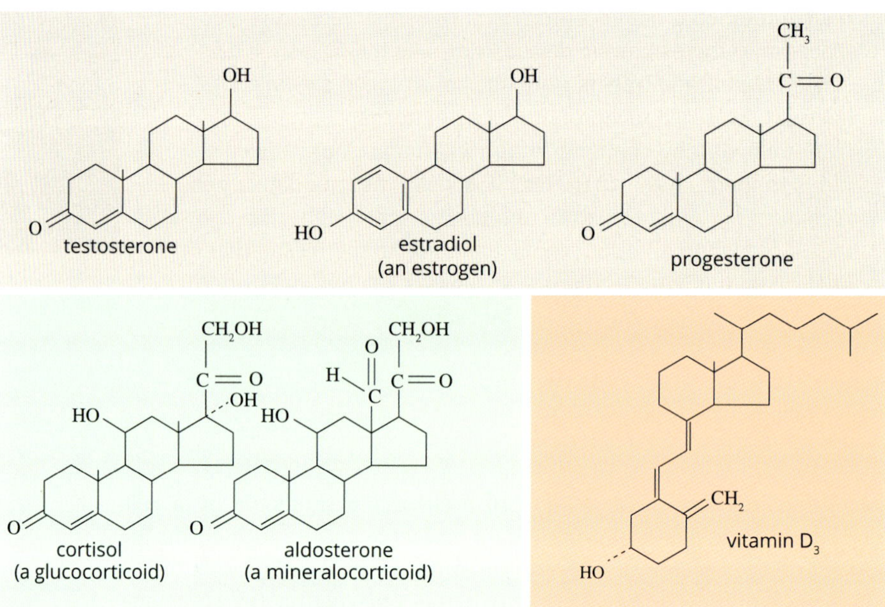

FIGURE 6.2.6 Examples of steroid hormones.

> Gene regulation refers to processes that control gene expression such as turning a gene on or off.

In the nucleus, the signalling molecule–receptor complex acts as a transcription factor. In Chapter 3 you learnt that transcription factors regulate gene expression by either inducing or repressing gene transcription. Figure 6.2.8 shows a molecular model of an androgen–receptor complex (signalling molecule–receptor complex) acting as a transcription factor and binding to DNA. The androgen receptor binds to hydrophobic signalling molecules called androgens, which are steroid hormones such as testosterone. Binding of the androgen–receptor complex switches on the genes involved in development of reproductive organs and secondary sexual characteristics.

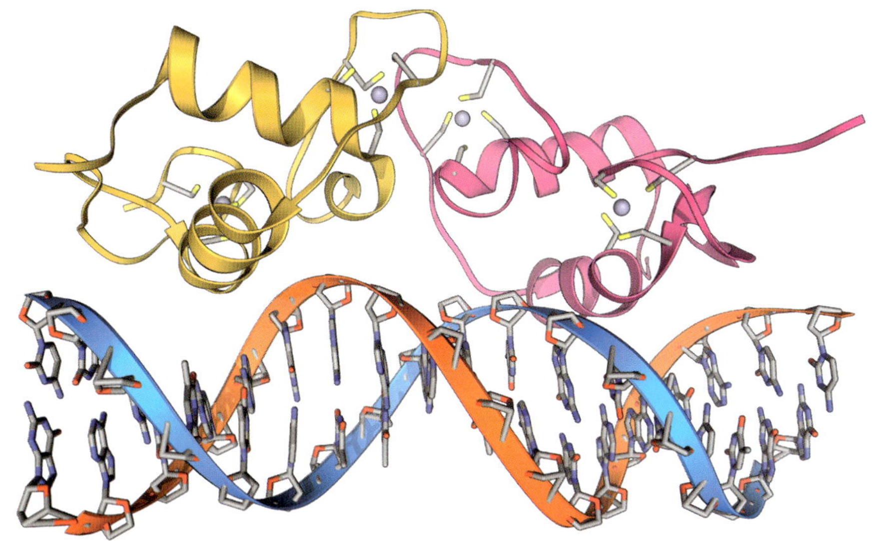

FIGURE 6.2.8 A molecular model of the DNA (red and blue) binding region of the androgen receptor (yellow and pink).

For example, testosterone is secreted by the cells of the testes in males. Testosterone travels in the bloodstream and enters all cells in the body. Only cells that contain androgen receptor molecules in the cytosol respond. In these cells, testosterone binds to the androgen receptor and activates it. The active form of the receptor then enters the nucleus and turns on specific genes that control male sex characteristics by binding onto a specific sequence of DNA known as hormone response elements.

> Although traditionally thought of as male hormones, androgens are sex hormones produced in the testes and adrenal glands in men and ovaries and adrenal glands in women.

SIGNAL TRANSDUCTION OF HYDROPHILIC SIGNALLING MOLECULES

Hydrophilic signalling molecules include hydrophilic peptide hormones, neurotransmitters and cytokines.

Reception

Hydrophilic signalling molecules are water soluble and so are unable to diffuse through plasma membranes. Therefore, the first step for all hydrophilic signalling molecules is to interact with receptors on the external surface of the plasma membrane. The receptors for these signalling molecules are transmembrane proteins, which are made up of one or more protein molecules and span both layers of the plasma membrane. An example of a transmembrane protein is the G-protein-coupled receptor. The G-protein-coupled receptor consists of seven membrane-spanning domains connected by extracellular loops. The receptor is activated by the binding of molecules on the surface of the cell to induce cellular responses inside the cell (Figure 6.2.9).

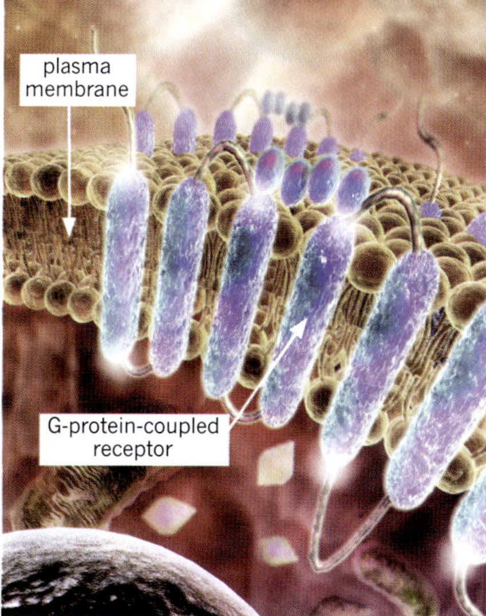

FIGURE 6.2.9 Digital illustration of a transmembrane protein, the G-protein-coupled receptor (purple), in a plasma membrane.

Transduction

Transmembrane receptors have an extracellular domain that acts as a binding site for the signalling molecule and an intracellular domain that transfers the signal into the cell (Figure 6.2.10 on page 238). When the signalling molecule binds to the extracellular part, the intracellular part of the receptor typically changes shape. This conformational change of the receptor results in the activation of molecules inside the cell (cellular responses).

The receptor may initiate a cellular response directly or indirectly by the activation of other molecules such as second messengers, G proteins or both.

> A conformational change is a change in shape of a macromolecule.

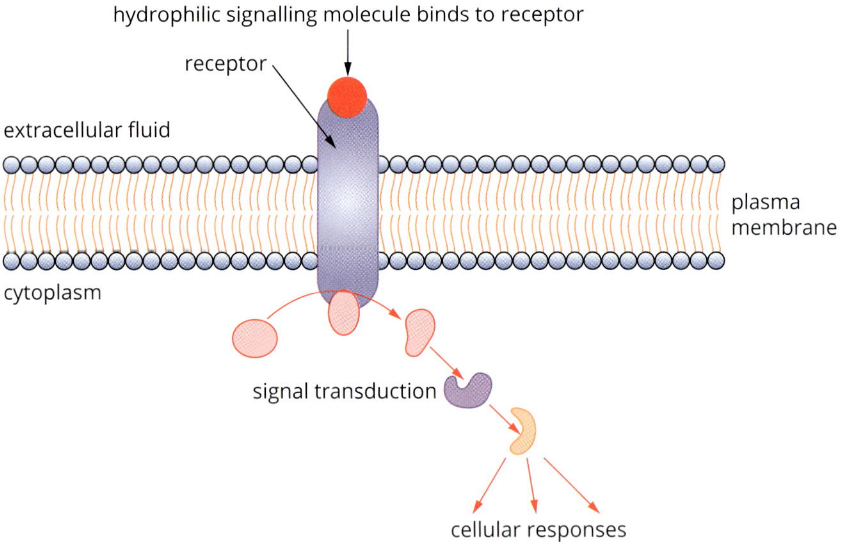

FIGURE 6.2.10 Hydrophilic signalling molecules interact with a specific membrane-bound receptor, causing the intracellular domain to change shape.

> Second messengers are intermediate signalling molecules that relay signals.

Second messengers

Signal transduction of hydrophilic signalling molecules typically uses intermediate proteins and other small non-protein molecules to relay the signal to its final destination. The small non-protein molecules are referred to as **second messengers**, and are usually water-soluble molecules. The size and solubility of second messengers enables them to diffuse quickly through the cytosol and to then trigger and amplify a response from multiple parts of the cell at once.

> Transduction cascades involve a series of events in which a change in one molecule causes a change in another, which in turn causes a change in yet another, and so on.

Transduction cascades

Whichever type of molecule that is activated by the conformational change of the receptor after binding to a hydrophilic signalling molecule, the result is the trigger of a transduction cascade inside the cell that eventually initiates the response.

Transduction cascades involve a series of events in which a change in one molecule causes a change in another, which in turn causes a change in yet another, and so on. The molecules may be enzymes, channel proteins or cell structure proteins. For example, when hydrophilic signalling molecules, such as peptide hormones, activate metabolic pathways, a second messenger might be produced that activates a series of enzymes in a set order, so that the activation of one enzyme causes the activation of the next in the sequence or cascade (Figure 6.2.11).

At the end of the cascade, molecules are activated that bring about a cellular response. A signalling molecule can initiate different responses in different cell types:

- Different receptors for the same signalling molecule exist in different cells, which may activate different transduction cascades.
- Proteins can be specific to particular cells and will lead to a particular response only in the cells in which they are present.

Even though signalling molecules are usually found in low concentrations, a cascade in a transduction pathway allows the **amplification** of the original signal that leads to enough response molecules to have an effect on the cell.

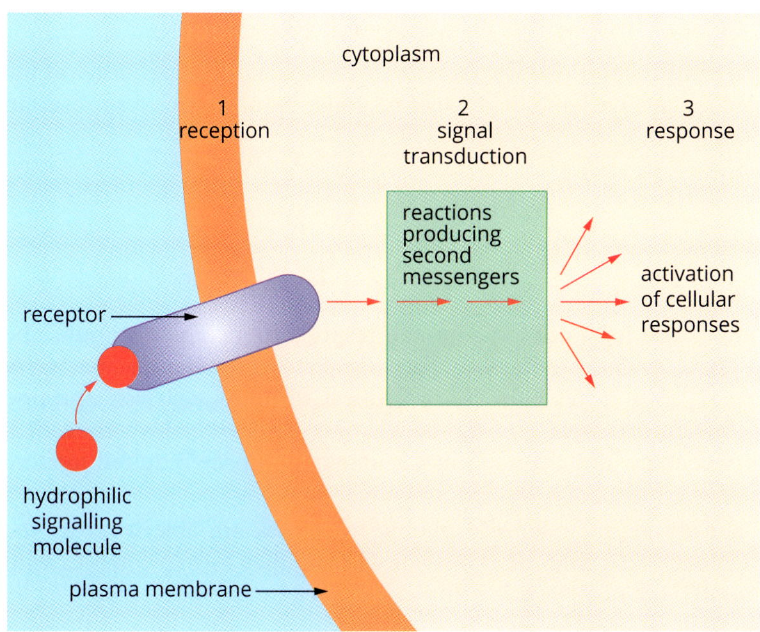

FIGURE 6.2.11 A hydrophilic signalling molecule binds to the receptor on the plasma membrane leading to a series of reactions that produce second messengers that, in turn, activate a cascade of enzymes or other molecules in the cell, leading to a range of possible cellular responses.

EXTENSION

G proteins

The G proteins are a family of intracellular proteins that are coupled to receptors in the plasma membrane. When they are activated, they modulate the activity of proteins such as enzymes, often using second messengers to relay the signal throughout the cell (Figure 6.2.12). Two molecules that affect G protein activity are guanosine triphosphate (GTP) and guanosine diphosphate (GDP). G proteins are active when they bind to GTP and are inactive when they bind to GDP.

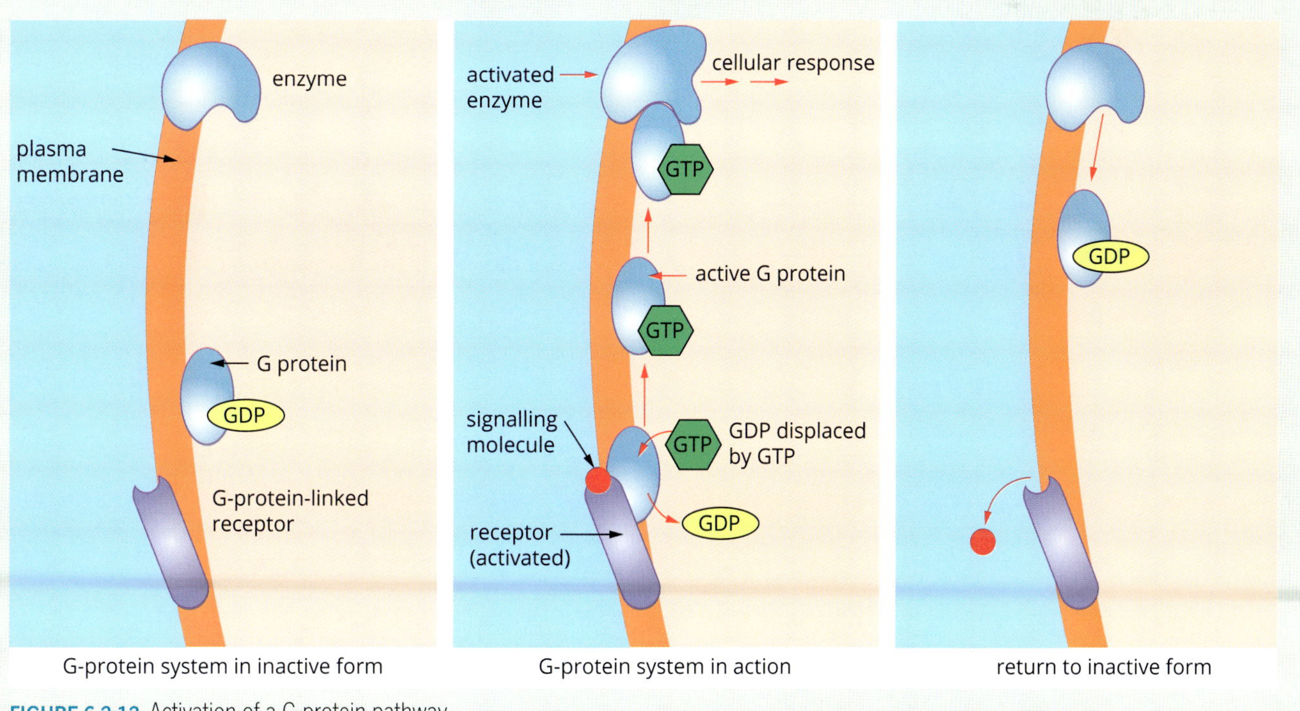

FIGURE 6.2.12 Activation of a G-protein pathway.

BIOFILE

Inactive precursors

Many enzymes are stored within a cell as inactive precursors, meaning they are stored with a non-functional 3D shape. Activation of a precursor molecule changes its 3D shape and makes its active site functional. Enzymes are activated only when required. For example, trypsin is released by the pancreas to break down proteins into smaller chains of amino acids. It is released in the inactive form trypsinogen and converted to trypsin in the small intestine by the action of the enzyme enterokinase.

a

b

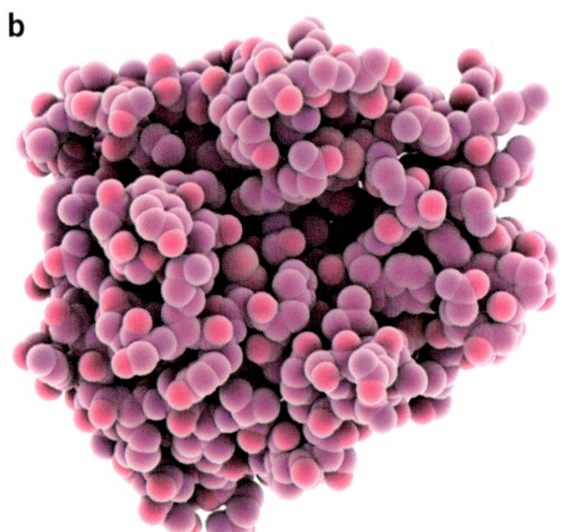

FIGURE 6.2.13 Molecular model of the digestive protease enzyme (a) trypsinogen and (b) trypsin.

Cellular responses for hydrophilic signalling molecules

At the end of a transduction pathway, a response is initiated. Depending on the signalling molecules involved and the type of response required, this response can occur in the:

- nucleus
- cytosol
- plasma membrane.

Responses in the nucleus

Gene regulation is vital for a cell to adapt and respond to incoming stimuli. The cell's need for enzymes and products such as signalling molecules fluctuates. By regulating gene expression to synthesise proteins as required, cells can optimise their functionality. For example, a hormone will only be produced when needed, so as to conserve energy and also to limit the amount of space required to store excess hormone.

You have already learnt that hydrophobic hormones are able to cross the plasma membrane. Once within the cytosol or nucleus, they form a complex with their receptor and in the nucleus this complex acts as a transcription factor.

Some hydrophilic hormones elicit a similar response but through a different transduction process. Peptide and protein hormones can activate protein receptors in the plasma membrane that, in turn, activate a signal transduction cascade. The cascade ends with a functional transcription factor, leading to gene regulation (Figure 6.2.14). One example of a peptide-based hormone that regulates gene expression is oxytocin. This hormone is produced in the pituitary gland and is involved in lactation, uterine contractions to expel the placenta after the birth of a baby and in social bonding.

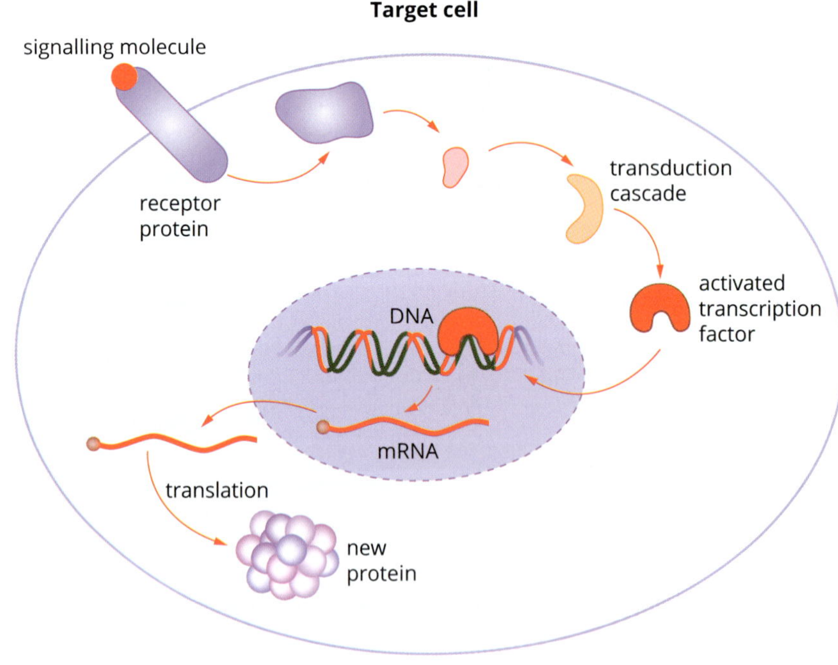

FIGURE 6.2.14 A hydrophilic signalling molecule interacts with its membrane-bound receptor, causing transduction of the signal into the cell through the plasma membrane.

Responses in the cytosol

Signal transduction can result in the **inhibition** or **activation** of enzymes in the cytosol of the target cell. Enzymes regulate cellular processes by catalysing chemical reactions, so inhibiting or activating enzymes will decrease or increase cellular functions respectively.

One example is the response of the liver cells to insulin. When glucose levels in the blood are high, insulin is released from the pancreas. Insulin binds to insulin receptors on the liver cell surface, which allows the receptor to activate second messenger molecules and initiate multiple transduction cascades. In liver cells, the pathway initiated by insulin leads to the activation of enzymes for glycogen synthesis, fatty acid synthesis, increased glycolysis and an increase in glucose transporters in the plasma membrane for the uptake of glucose into the cytoplasm (Figure 6.2.15).

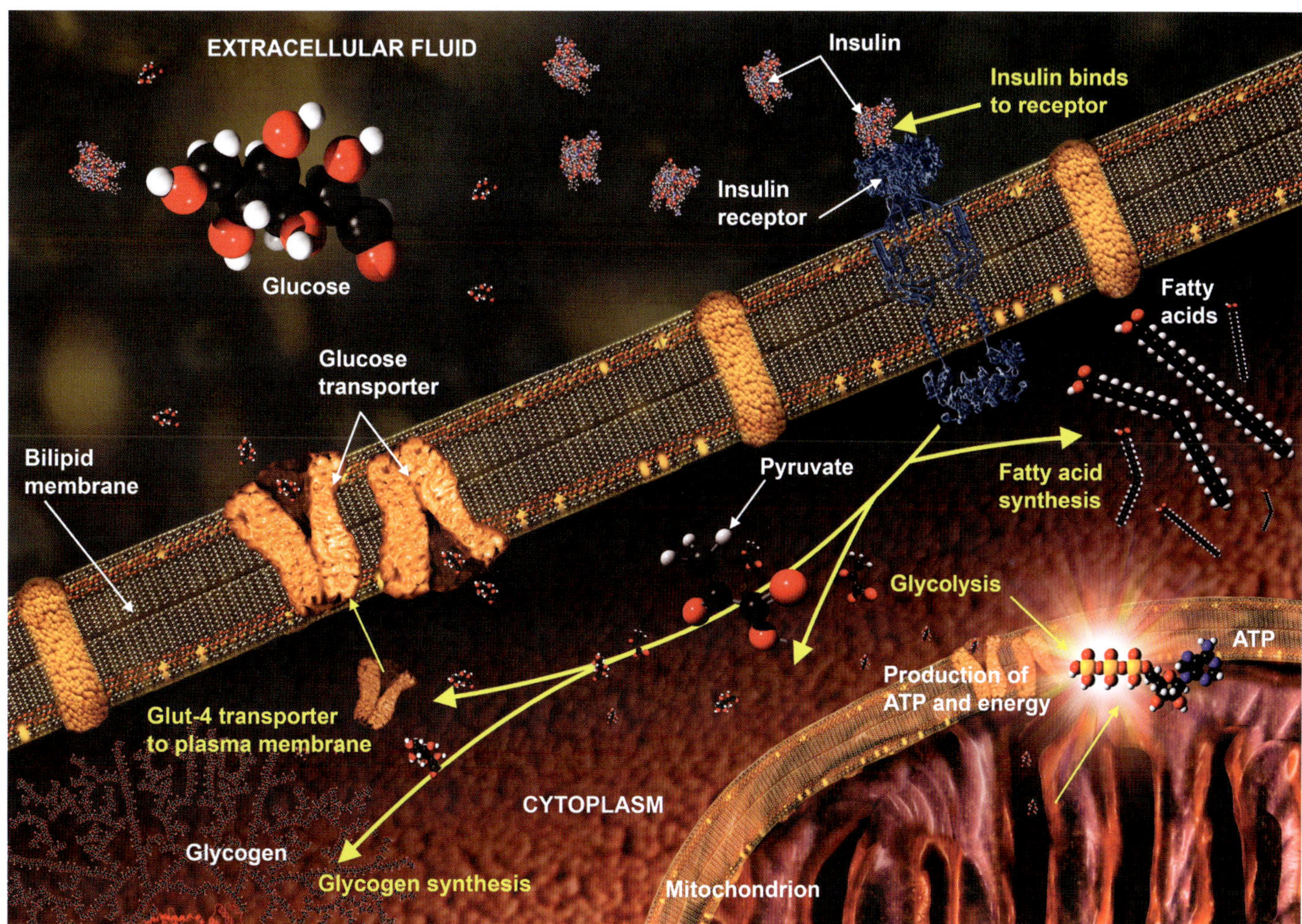

FIGURE 6.2.15 In this illustration the hydrophobic signalling molecule insulin interacts with its membrane-bound receptor (dark blue). This causes the intracellular domain receptor to trigger the activation of enzymes involved in the synthesis of glycogen and fatty acids, as well as glycolysis (yellow arrows).

Responses on the plasma membrane

The plasma membrane of a cell controls all the substances that move into or out of the cell. Not all substances can cross the membrane passively. Some cellular responses involve changes to the plasma membrane that allow certain substances to enter or exit the cell.

Ion channels

An **ion channel** is a type of transmembrane protein that allows specific ions to cross the plasma membrane. Some ion channels can be opened or closed in response to signalling molecules (Figure 6.2.16).

Ion channels are important in the process of converting chemical signals to electrical signals. Skeletal muscle cells are an example of cells that respond to signalling molecules by opening ion channels. When acetylcholine, a neurotransmitter released by neurons, binds to the receptors on the plasma membrane of a skeletal muscle cell, ion channels open that allow sodium ions (Na^+) to move into the cell. Sodium flow into the muscle cell leads to the cellular response, which in this case is contraction of the muscle.

In cellular communication, a receptor antagonist is a type of receptor ligand that blocks a biological response by binding to a receptor.

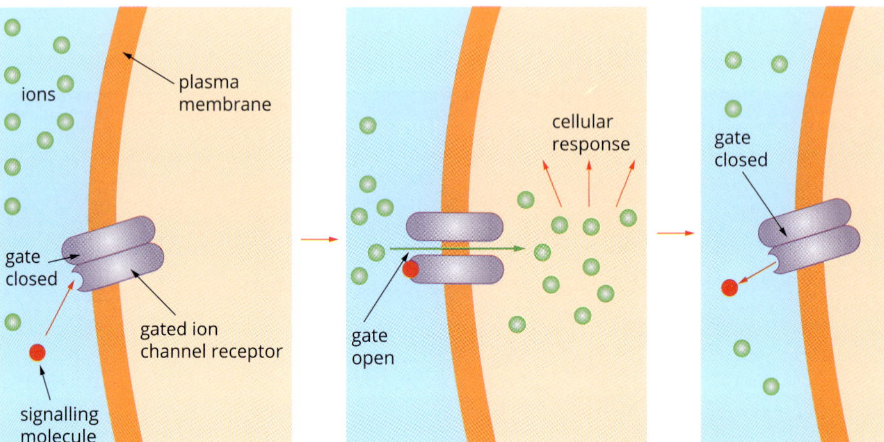

FIGURE 6.2.16 A gated ion channel receptor is both a receptor and an ion channel. It opens when a signalling molecule binds to the extracellular part of the channel, allowing ions to move through the channel from the extracellular fluid into the cytoplasm.

BIOFILE

Poisoned arrows

Hunters in South America place curare, a plant-based poison, on the tips of darts that are fired through blowguns to cause the targeted animal's muscles to relax so that it will, for example, fall out of the tree. Curare is an antagonist that blocks acetylcholine receptors on muscles causing paralysis and death. Curare has been adapted and used in some surgeries as a muscle relaxant.

FIGURE 6.2.17 The plant *Chondrodendron tomentosum* (a) is a source of arrow poison for hunters using a blow dart (b).

SIGNAL TRANSDUCTION IN NEURONS

Transduction of a signal carried by the nervous system uses a combination of electrical and chemical signalling involving the opening of gated ion channels and the release of neurotransmitters through exocytosis.

Signal transduction into a neuron

When a neuron is stimulated, gated sodium and potassium ion channels on its membrane are opened. The sudden movement of the ions into and out of the neuron initiates the **action potential**, which travels the length of the axon of the neuron in a wave-like sequence, opening other ion channels along the membrane. An action potential is the reversal of the normal potential difference across a cell membrane, or between the inside and the outside of a nerve fibre.

Cellular response of a neuron

The action potential eventually reaches the **synaptic terminals** of the axon, where it causes another ion, Ca^{2+}, to enter the cell. Vesicles containing the neurotransmitters are located at the synaptic terminals. The increased concentration of Ca^{2+} causes these vesicles to fuse with the nearby membrane (known as the presynaptic membrane) and to release the neurotransmitters into the synaptic gap via exocytosis.

The postsynaptic cell can be another neuron or another type of target cell such as a muscle or gland cell. The neurotransmitters released from the synaptic terminal diffuse across the synaptic gap and bind to specific receptors on the postsynaptic cell. If the postsynaptic cell is another neuron, binding of neurotransmitters to the receptors on its dendrites triggers the ion channels of the postsynaptic membrane to open, leading to release of neurotransmitters from the synaptic terminals (Figure 6.2.18). In this case, the signal has been transformed, or transduced, from an electrical impulse into chemical signalling molecules and back again.

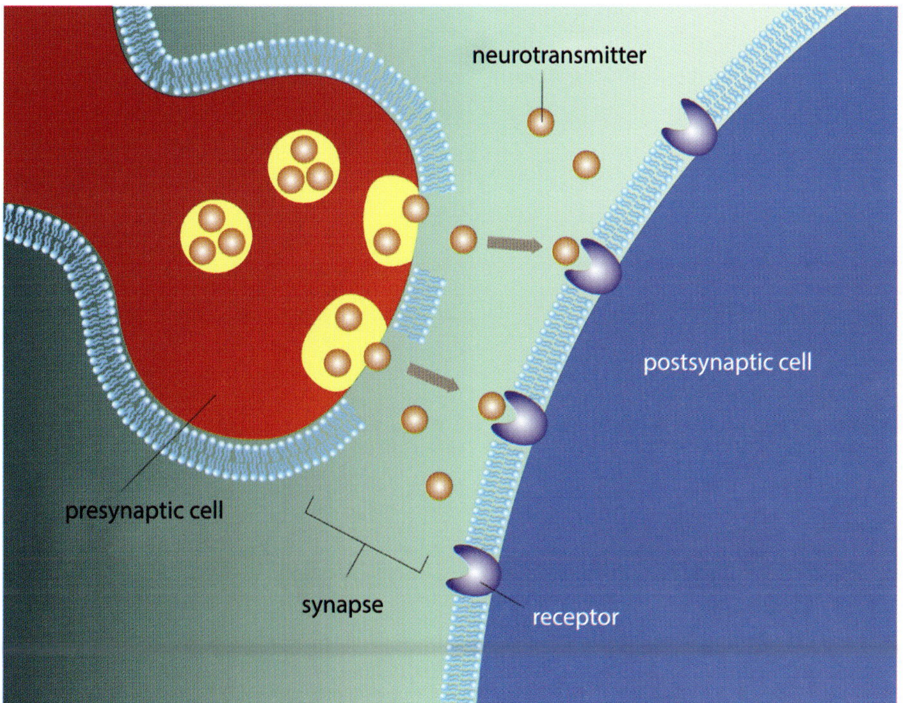

FIGURE 6.2.18 Secretory vesicles containing neurotransmitters fuse onto the presynaptic membrane, releasing the neurotransmitters into the synaptic gap. Neurotransmitters will diffuse through the gap and bind to receptors found on the postsynaptic cell.

BIOFILE

Speed

The main advantage of communication via electrical signals rather than other signalling molecules is speed. Electrical impulses can travel long distances within the body much more quickly than other signalling molecules can travel in the blood circulation system. A nerve impulse can travel through the whole body in a few milliseconds.

EXTENSION

The action potential

An action potential or nerve impulse is a wave of electrical change that passes rapidly along an axon membrane. When a neuron is not stimulated or is at rest, the membrane is polarised—the inside of the cell is more negative (Figure 6.2.19).

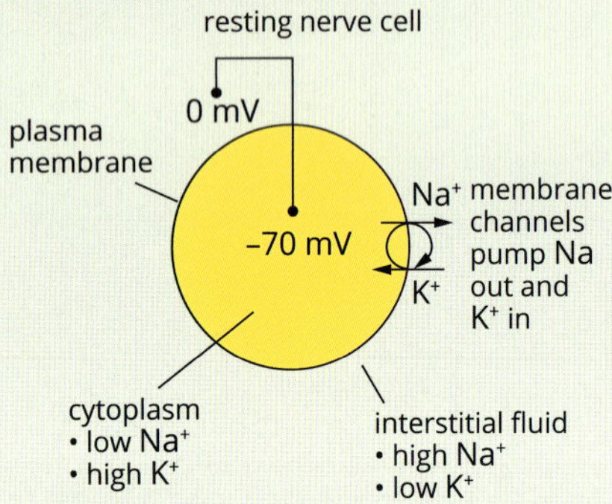

FIGURE 6.2.19 The inside of the neuron is negative with respect to the outside of the neuron, and there is an unequal distribution of Na^+ and K^+ across the plasma membrane.

A charge difference is maintained between the inside and outside of the cell largely by active transport using sodium–potassium pumps. The pumps actively transport sodium ions out of the cell and potassium ions into the cell.

An action potential begins when a stimulus disturbs the plasma membrane on a dendrite, causing sodium ion channels to open. As sodium ions flow into the neuron, it causes the inside of the neuron to become slightly less negative (the membrane is depolarised). If the depolarisation is sufficient to reach the threshold potential for the cell, Na^+ (sodium ion) channels in the membrane open and Na^+ floods into the cell along its concentration gradient, initiating an action potential (Figure 6.2.20). Because sodium ions are positive, the inside of the cell becomes briefly positive, which causes K^+ (potassium ion) channels to open, and K^+ diffuses out of the cell along its concentration gradient.

As the positive potassium ions leave, the inside of the cell becomes negative again. Therefore, the inside of the cell first becomes briefly positive and then negative again. These electrical changes are the action potential. Following the action potential, the Na^+ and K^+ channels close and the membrane returns intracellular ion concentrations to their initial values.

Once an action potential is generated, it sweeps along the neuron by conduction. During an action potential, when the sodium ions diffuse into the cell, some diffuse sideways inside the axon, depolarising the adjacent region of the axon membrane. As with the initial stimulus, this causes Na^+ channels in this part of the membrane to open and Na^+ floods into the cell, initiating an action potential in this region (Figure 6.2.21). This depolarises the next adjacent region of the axon membrane, and so the cycle continues, conducting the action potential along the length of the axon membrane.

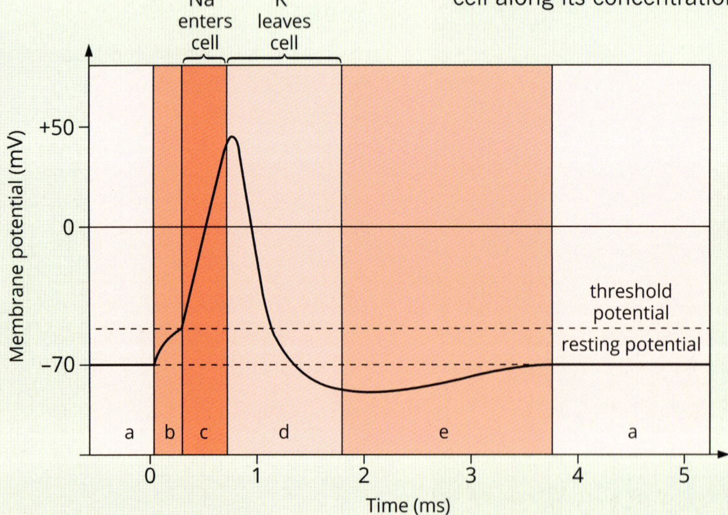

FIGURE 6.2.20 A simplified diagram illustrating the internal membrane potential of a nerve cell during an action potential: (a) resting membrane potential; (b) depolarisation reaches the threshold potential of the axon; (c) the membrane depolarises and the inside of the cell becomes briefly positive as sodium ions (Na^+) diffuse into the cell; (d) the potential becomes negative again as potassium ions (K^+) diffuse out of the cell; (e) the resting membrane potential and the original distribution of ions are re-established by active transport of Na^+ and K^+ across the membrane.

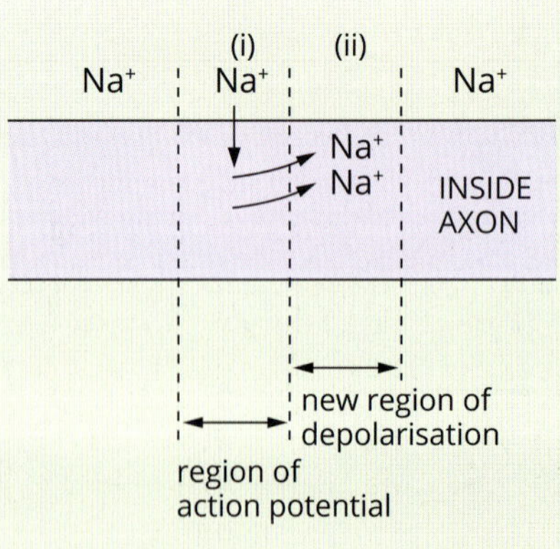

FIGURE 6.2.21 (i) As Na^+ enters the cell, some diffuses sideways, (ii) depolarising the adjacent region of membrane and causing an action potential in this region. In this way an action potential is conducted along the axon.

6.2 Review

SUMMARY

- The stimulus–response model outlines the three main steps involved in signal transduction in cells:
  - reception—the detection of a signalling molecule (the stimulus) by its specific receptor (including the physical binding of the signalling molecule to the receptor)
  - transduction—the transformation of the signal in terms of form, type of signalling molecule and the passage into and out of a cell
  - response—the change in cellular activity as a result of the initial stimulus.
- The transduction pathway of a signal depends on the cells and the signalling molecules involved.
- Transduction of hydrophobic signalling molecules includes:
  - passage of the signalling molecule through the plasma membrane into the cell
  - binding with a receptor
  - passage of the signalling molecule–receptor complex through nuclear pores into the nucleus if receptor is located in the cytosol
  - the complex acting as a transcription factor that can either induce or repress the transcription of a gene.
- Transduction of hydrophilic signalling molecules includes:
  - conformational change of the intracellular domain of the transmembrane receptor to activate molecules inside the cell
  - production of second messengers or the activation of G proteins
  - the cascade activation of transduction molecules.
- Cellular responses are dependent on the stimulus and the type of cell, the signalling molecules and response molecules involved.
- Cellular responses to hydrophilic signalling molecules can take place:
  - in the nucleus, where typically gene transcription is induced (switched on) or repressed (switched off)
  - in the cytosol, where the activation or inhibition of enzymes regulates most cellular activities
  - on the membrane, which includes the opening and closing of ion channels, initiating action potentials and the release of signalling molecules by exocytosis.
- Neuron signal transduction involves the conversion of the electrical impulse of the action potential into chemical signalling molecules, called neurotransmitters, and back again at the synapse between cells.

KEY QUESTIONS

1 a Define ‘signal transduction’.
  b Outline the stimulus–response model in signal transduction.
2 Briefly explain how a breakdown in any of the three steps of the stimulus–response model will prevent a response from occurring.
3 Outline the difference in how the two main types of signalling molecules interact with cells.
4 What are second messengers?
5 The table at right is a list of statements describing the stages of the mechanism of hormone activity. Match the statements to the correct hormone. Cortisol is a steroid hormone and melatonin is a peptide hormone.
6 What are some of the cellular responses to:
  a hydrophobic signalling molecules?
  b hydrophilic signalling molecules?
7 Outline how signal transduction occurs from one neuron to the next neuron.

Statement	Cortisol	Melatonin
Hormone passes freely through the cell surface membrane.		
Hormone binds with a receptor located in the cell surface membrane.		
Hormone combines with a receptor located in the cytoplasm.		
The hormone–receptor complex enters the nucleus.		
Formation of a hormone–receptor complex activates the enzymes involved in the synthesis of second messengers.		
The hormone–receptor complex binds with a specific space in DNA by means of a specific protein.		
Activation (or inhibition) of the transcription of a gene is dependent on the hormone.		
Activation or inhibition of multiple metabolic pathways occurs.		

6.3 Apoptosis

The growth and development of a multicellular organism requires careful regulation of cell division, differentiation and cell death. All three processes are controlled by cellular signals. In this section, you will learn about the regulatory process of programmed cell death, known as apoptosis (Figures 6.3.1 and 6.3.2).

PROGRAMMED CELL DEATH

As you learnt in *Heinemann Biology 1*, Section 8.3, billions of cells in your body die every hour as a result of programmed cell death, called **apoptosis**. This process of 'self-destruction' enables a multicellular organism to regulate the number of cells in the body, to remove cells that are no longer required and to remove cells that might contain damaged DNA that cannot be repaired. Apoptosis is a highly regulated process involving signalling molecules.

If you have recently had an infection, many of the lymphocytes called plasma cells that made the antibodies to fight a pathogen will die by apoptosis once the pathogen has been removed (Figure 6.3.3). During the menstrual cycle, the endometrium becomes progressively thicker, with more glands and blood vessels, in preparation for the implantation of an embryo. If this does not occur, much of the endometrium breaks down due to apoptosis and is lost during menstruation (Figure 6.3.4).

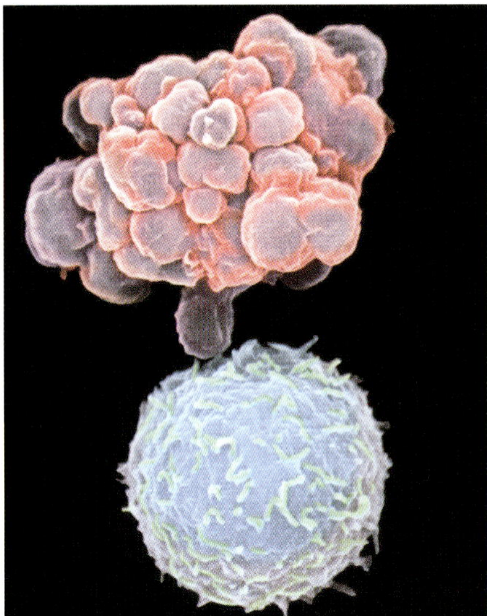

FIGURE 6.3.1 Coloured scanning electron micrograph of a healthy leucocyte (blue) and a leucocyte undergoing apoptosis (pink).

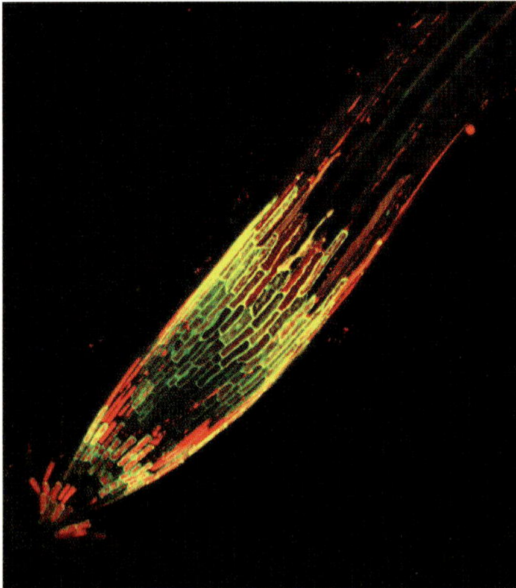

FIGURE 6.3.2 Fluorescent light micrograph of a thale cress (*Arabidopsis thaliana*) plant root. The root cap (very bottom left) is continually shed and the cells to be shed undergo apoptosis.

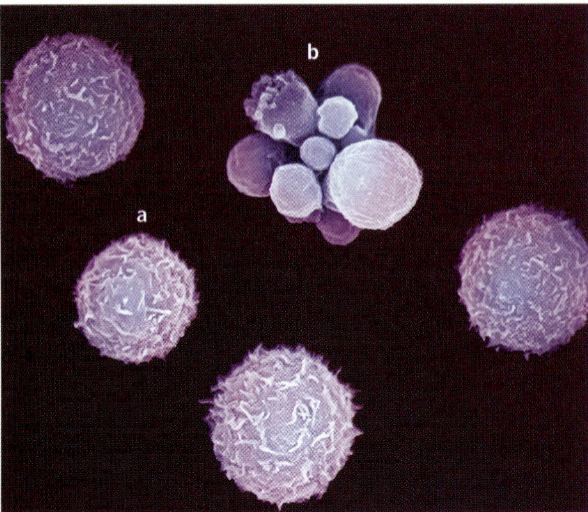

FIGURE 6.3.3 Coloured scanning electron micrograph showing healthy human lymphocytes (a) and a cell undergoing apoptosis (b).

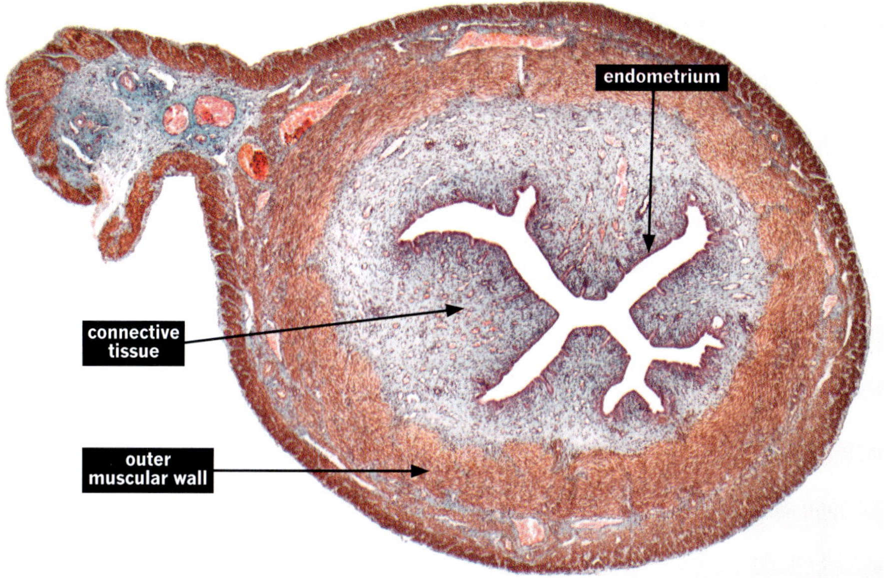

FIGURE 6.3.4 This light micrograph is of a section through a whole uterus. The endometrium (thin red inner layer) is surrounded by a thick layer of connective tissue (blue) and an outer muscular wall (red).

BIOFILE

Origin of the term 'apoptosis'

The term 'apoptosis' was first used in a journal article written by John F. Kerr, Andrew H. Wyllie and A. R. Currie in 1972. They created the term from the Greek prefix *apo-* meaning 'separate' or 'apart', and root word *-ptosis* meaning 'falling'.

Apoptosis was also central to your embryonic development; for example, in the removal of the skin webbing between the digits to form individual fingers and toes (Figure 6.3.5). Finally, we are born with many more neurons than will survive into our adulthood. During brain development, neurons that fail to make synaptic connections with other neurons are triggered to die by apoptosis (Figure 6.3.6).

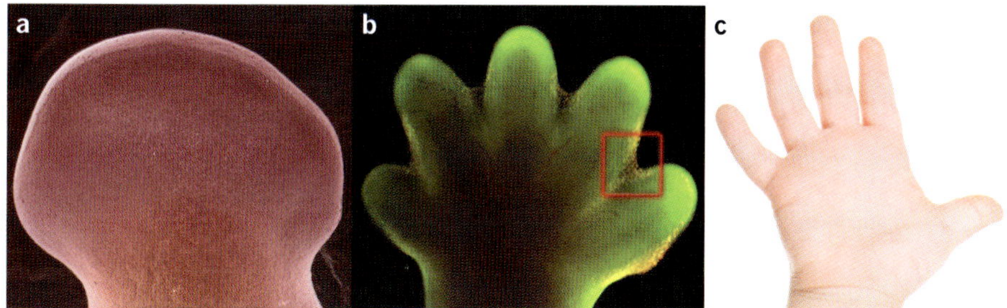

FIGURE 6.3.5 (a) Embryonic development of a hand starts with a small paddle-shaped limb bud. (b) As the embryo develops, the digits (fingers or toes) gradually emerge as the skin joining them dies by apoptosis. (c) Finally, fully separated digits are revealed.

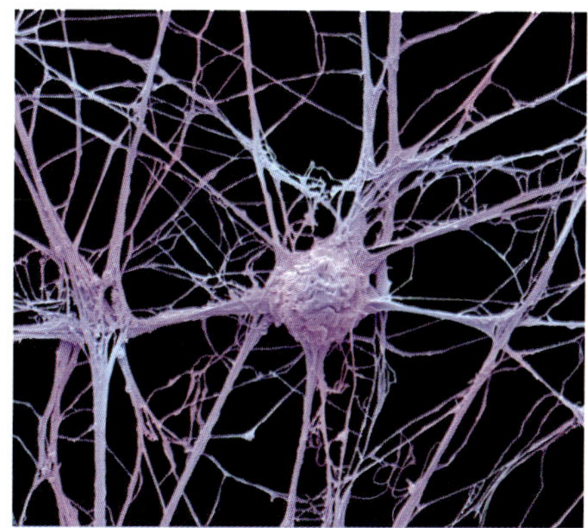

FIGURE 6.3.6 Coloured scanning electron micrograph of a human nerve cell (neuron).

CASPASES AND THE PROCESS OF APOPTOSIS

Apoptosis is a receptor-mediated response triggered by signalling molecules inside or outside the cell. Some common occurrences that promote the production of the signalling molecules that stimulate apoptosis include cell aging, cell obsolescence (cell is no longer needed) and damaged DNA. Once signalled, apoptosis occurs as a series of coordinated steps.

Enzymes called **caspases** are responsible for apoptosis. They are produced within the cell and stored as inactive precursors. When a signalling molecule is produced to stimulate apoptosis (e.g. when a cell ages), the signalling molecule binds to a receptor and this binding activates the caspase precursor. The caspase then stimulates the activation of other caspases, triggering a cascade that leads to cell death.

FIGURE 6.3.7 A molecular model of caspase-3, also known as apopain (pink and blue), complexed with an inhibitor (green).

An example of a caspase is caspase-3 (Figure 6.3.7). Caspase-3 is a protease that plays a role in apoptosis. It causes the fragmentation of actin filaments, part of a cell's cytoskeleton, and the inactivation of DNA repair. It also activates other caspases as part of the caspase cascade of apoptosis. Caspase-3 is the predominant caspase involved in Alzheimer's disease, processing the protein that is associated with neuron death.

There are many caspases involved in the cascade, which cause extensive damage to the cell. Each caspase is responsible for a particular step in the destruction of the cell. Some of the roles of caspases during apoptosis include:

- cleavage of DNA into characteristic fragments
- degradation of nuclear proteins, leading to condensation of chromatin
- cleavage of proteins of the nuclear membrane
- dismantling of the cytoskeleton and allowing the protrusion of the plasma membrane; this protrusion is called a **bleb**
- breakdown of proteins in the cytosol
- breakdown and fragmentation of organelles.

Apoptosis is an all-or-nothing process. Once triggered, it cannot be stopped or reversed.

The major steps in apoptosis include (Figure 6.3.8):

1. separation from adjacent cells
2. collapse of the cell's cytoskeleton
3. cell shrinkage
4. breakdown of organelles and nucleus
5. blebbing of the plasma membrane
6. budding of plasma-membrane-bound vesicles called apoptotic bodies (Figure 6.3.9), which prevent toxic or immunogenic substances from leaking when the apoptotic bodies are phagocytosed.
7. phagocytosis of the apoptotic bodies by specialised cells, typically macrophages, without spilling cell contents or triggering an **inflammatory response**.

Phagocytosis is the process by which a cell engulfs a solid particle.

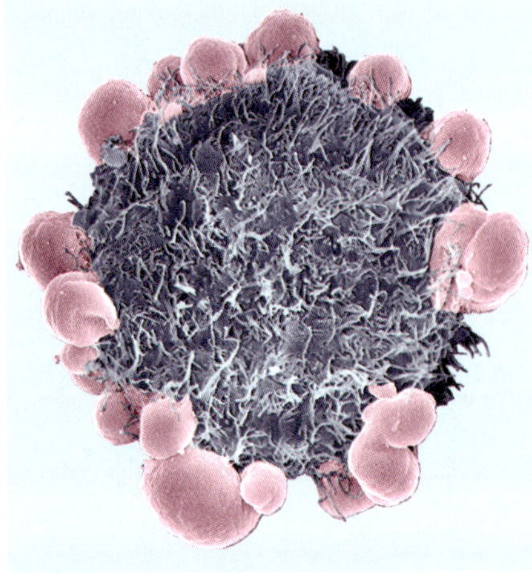

FIGURE 6.3.9 Coloured scanning electron micrograph of a HeLa cell undergoing apoptosis. Apoptotic bodies (in pink) form on its surface.

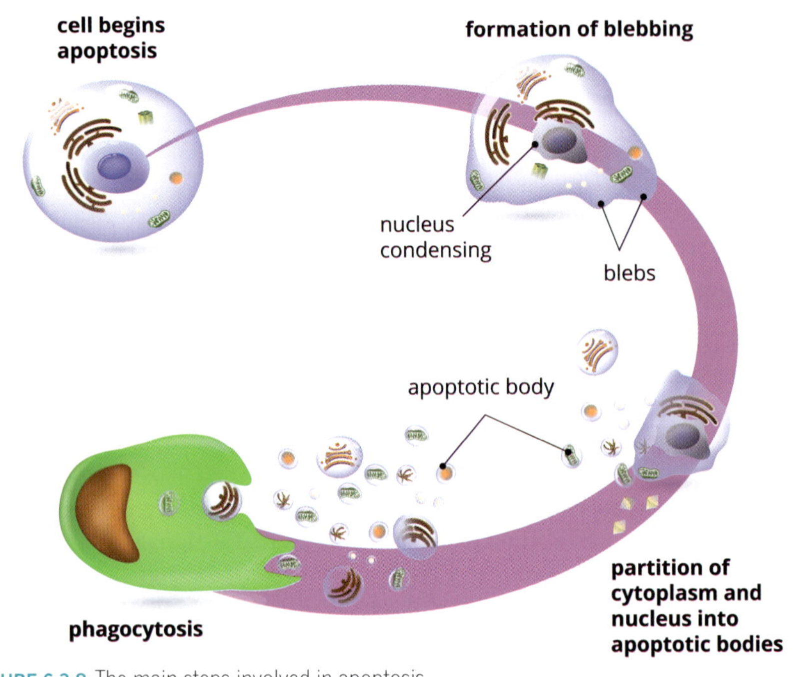

FIGURE 6.3.8 The main steps involved in apoptosis.

EXTENSION

Necrosis: another path to cell death

Not all cells die by apoptosis. Cells damaged by injury, infection, toxins or loss of blood supply die by a process called necrosis. In necrosis, the dying cells often lyse and spill their contents, triggering an inflammatory immune response that removes the debris (Figure 6.3.10). Depending on the cause, the necrosis can spread to neighbouring cells and, in some cases, the inflammation itself can cause further cellular damage.

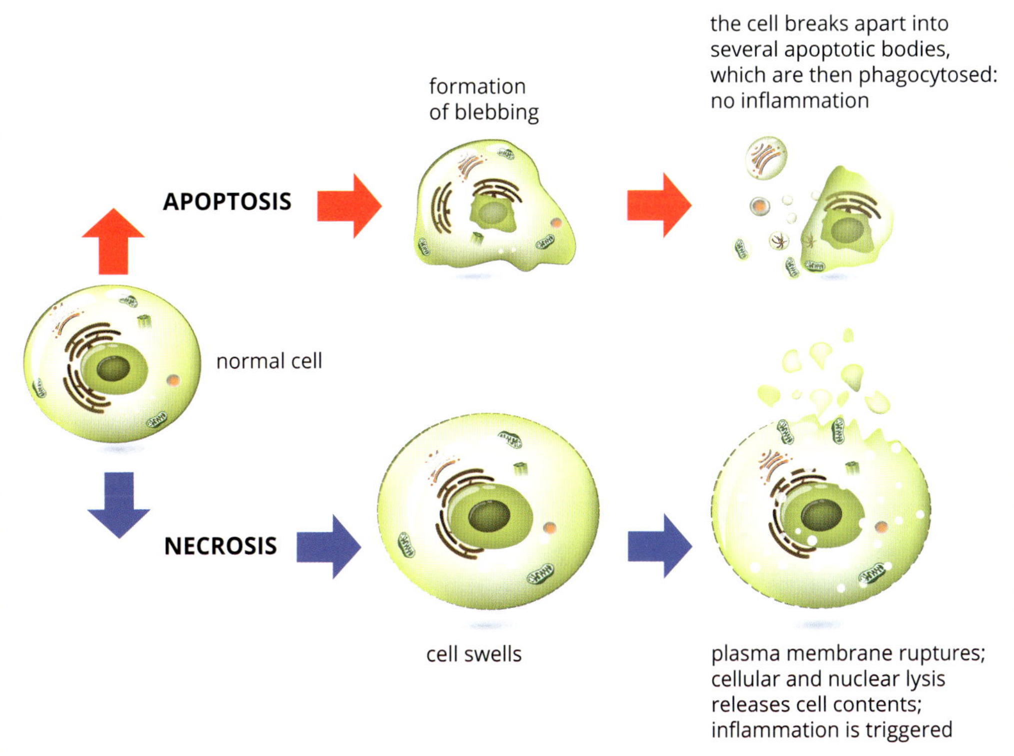

FIGURE 6.3.10 Necrosis, in contrast to apoptosis, produces an inflammatory response.

An example of a disease that can result in necrosis is peripheral vascular disease, which affects blood flow and can cause necrosis (Figure 6.3.11). Peripheral vascular disease is a complication of diabetes.

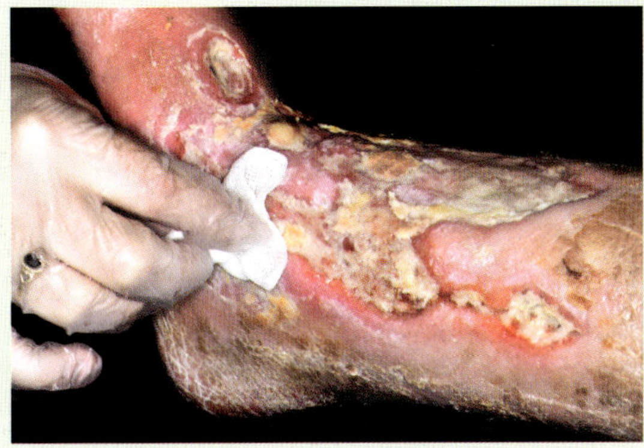

FIGURE 6.3.11 Close-up of a doctor cleaning deep ulcers on the leg of a 78-year-old female diabetic patient with peripheral vascular disease.

PATHWAYS OF APOPTOSIS

There are two different pathways of apoptosis depending on whether the signal comes from inside (intrinsic) or outside (extrinsic) the cell (Figure 6.3.12):

- intrinsic or mitochondrial pathway
- extrinsic or death receptor pathway.

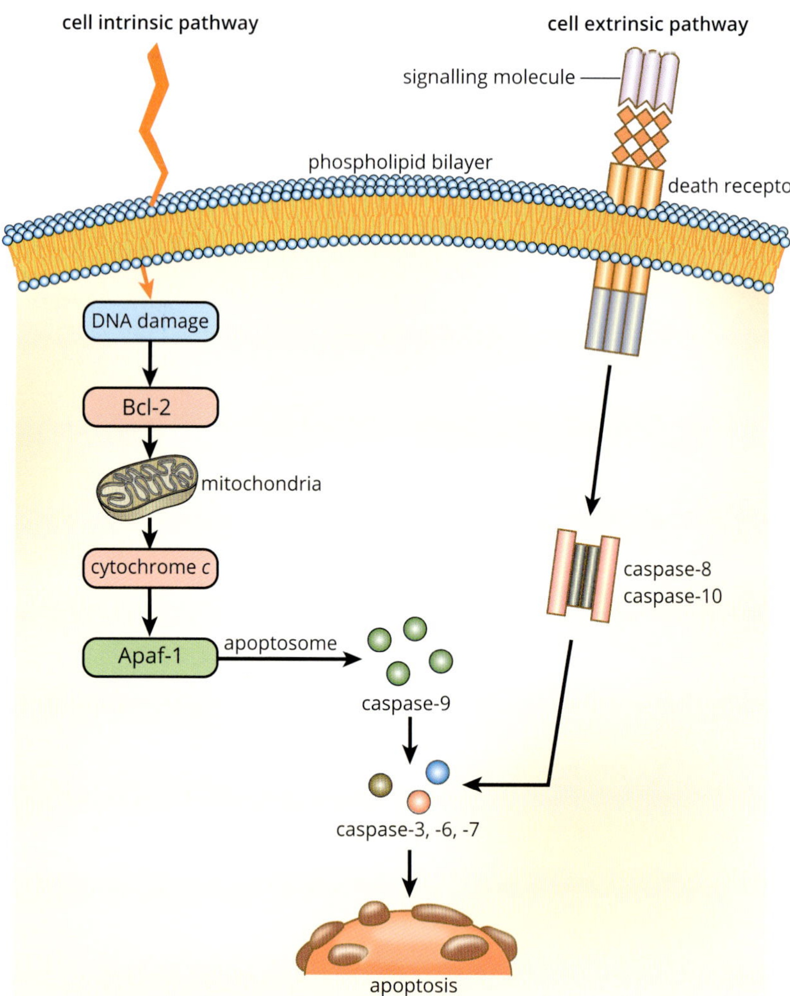

FIGURE 6.3.12 Summary of the two pathways of apoptosis.

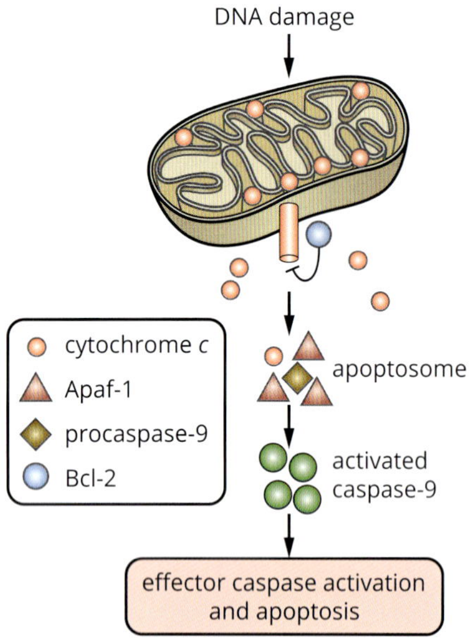

Intrinsic or mitochondrial pathway

Damage to the cell from radiation, viral infections, toxins or damaged DNA stimulates the transcription and translation (see Chapter 10) of genes into proteins (signalling molecules) that activate the mitochondrial pathway of apoptosis. In these cases, the signal for apoptosis is coming from within the cell.

The mitochondrial apoptosis signalling pathway is regulated by a number of proteins including a family of signalling proteins called **Bcl-2** (Figure 6.3.13). The Bcl-2 family of proteins is central to controlling the mitochondrial apoptotic pathway. These proteins come in different forms that promote (pro-) or inhibit (anti-) apoptosis. Following stress or cellular damage, Bcl-2 proteins that promote apoptosis relocate from the cytoplasm to the surface of the mitochondria causing the formation of pores. This leads to the release of **cytochrome *c*** from the mitochondria into the cytoplasm. Recall that cytochrome *c* is a protein involved in the electron transport chain. In the cytosol, cytochrome *c* forms a complex with a protein called Apaf-1 (apoptotic protease-activating factor). This complex is known as an **apoptosome** and it activates a cascade of 'executioner' caspases, resulting in apoptosis.

FIGURE 6.3.13 The mitochondrial or intrinsic pathway of apoptosis.

Extrinsic or death receptor pathway

All cells have transmembrane proteins on their plasma membrane called **death receptors**, which are specific to a number of different cytokine signalling molecules. Once the signalling molecule binds to the death receptor, signal transduction in the cell initiates a cascade of caspase activation that leads to apoptosis (Figure 6.3.14).

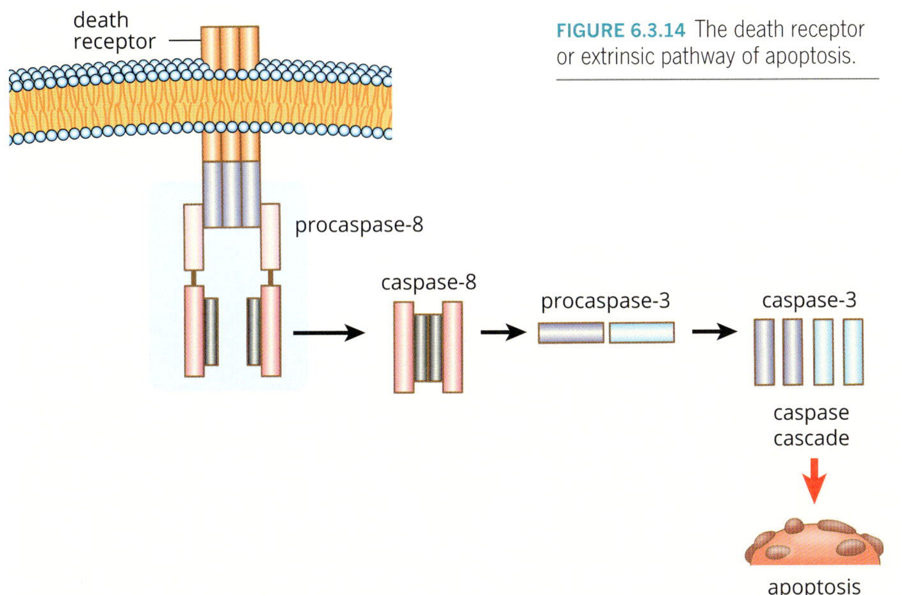

FIGURE 6.3.14 The death receptor or extrinsic pathway of apoptosis.

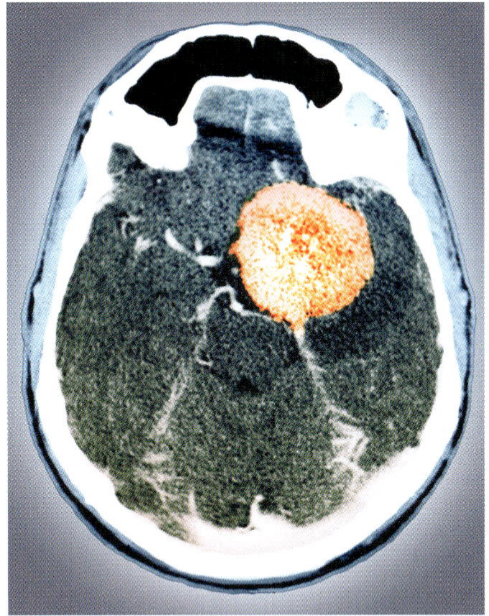

FIGURE 6.3.15 Coloured computed tomography scan of a section through the head of a 42-year-old patient with a benign (non-cancerous) meningioma (orange).

MALFUNCTIONS IN APOPTOSIS

Like many processes in the body, apoptosis is highly regulated. Too much cell death may result in a loss of vital tissue, while too little may result in tumours, cancer or other diseases. For example, meningioma is a tumour arising from the meninges, the membranes that surround the brain and spinal cord. The growth of the tumour is due to a breakdown in the apoptosis pathways, leading to uncontrolled proliferation of cells (Figure 6.3.15).

Excessive apoptosis

Too much apoptosis may be a cause of neurodegeneration as seen, for example, in Alzheimer's disease, a neurodegenerative disease that is characterised by a shrinking of the brain due to a loss of neurons associated with excessive apoptosis (Figure 6.3.16). As well as a decrease in brain volume, the surface of the brain is often more deeply folded. Tangled protein filaments (neurofibrillary tangles) occur within nerve cells. Alzheimer's disease accounts for most cases of senile dementia. Symptoms include memory loss, disorientation, personality change and delusion. It ultimately leads to death. Other degenerative diseases such as Parkinson's disease and motor neuron disease are also influenced by excessive apoptosis.

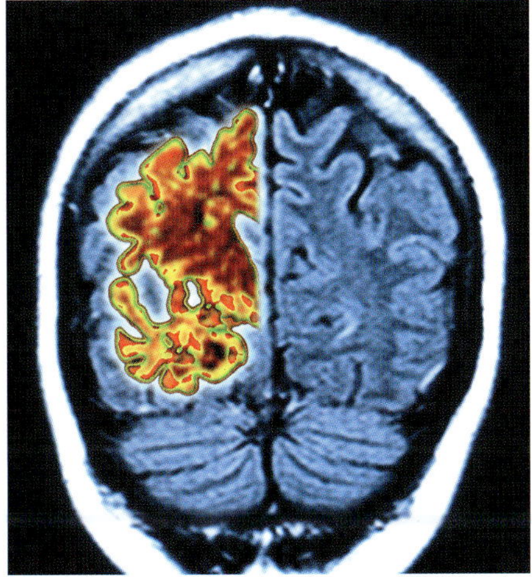

FIGURE 6.3.16 Composite image of a computer graphic of a vertical (coronal) slice through the brain of an Alzheimer's patient (orange) overlaid on a magnetic resonance imaging (MRI) scan of a normal brain (blue).

BIOFILE

Neuroblastoma

One of the common forms of childhood cancer is called neuroblastoma. Neuroblastoma arises from nerve cells in the adrenal glands. Embryonic nerve cells (neuroblasts) are responsible for nerve development. But a cancerous tumour may also develop from prolific divisions of these cells. Neuroblasts are normally deleted through programmed cell death before or soon after birth. If cell death does not occur then the cancer forms.

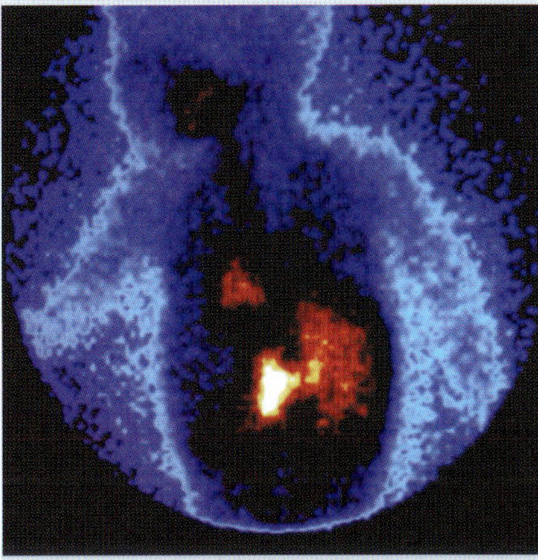

FIGURE 6.3.17 Coloured gamma scan of neuroblastoma tumours (white and red) in a 2-year-old's abdomen.

Inhibited apoptosis

An example of inhibited (too little) apoptosis is seen in a condition called syndactyly. Usually the skin between the digits is removed during embryonic development, but this apoptosis does not occur in patients with syndactyly (Figure 6.3.18).

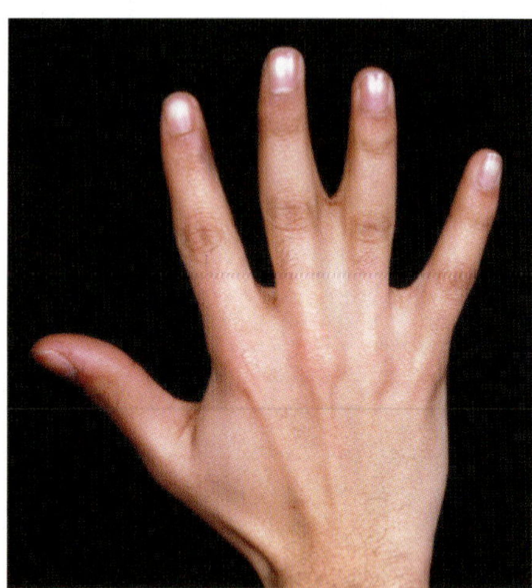

FIGURE 6.3.18 An example of syndactyly, in which there is congenital fusion of two fingers.

EXTENSION

Alzheimer's and apoptosis

Ongoing research into the causes of Alzheimer's disease has revealed a complex web of processes. Characteristic changes in the brain of Alzheimer's sufferers include the formation of lesions or 'plaques', clumps of a protein fragment called beta-amyloid. Beta-amyloid peptides are a product of the abnormal cleavage of a normal neuronal protein. These beta-amyloid peptides act as signalling molecules that activate the mitochondrial apoptotic pathway.

As you have already learnt, caspases in the apoptotic pathway break down many cellular proteins including the cytoskeleton. Another characteristic of Alzheimer-affected neurons is the presence of 'tangles' of a protein called tau. Tau protein is part of the cytoskeleton. Caspases degrade tau, causing it to form tangles (Figure 6.3.19).

Potential therapeutic drugs might include inhibitors of the mitochondrial apoptotic pathway to prevent apoptosis of the neurons.

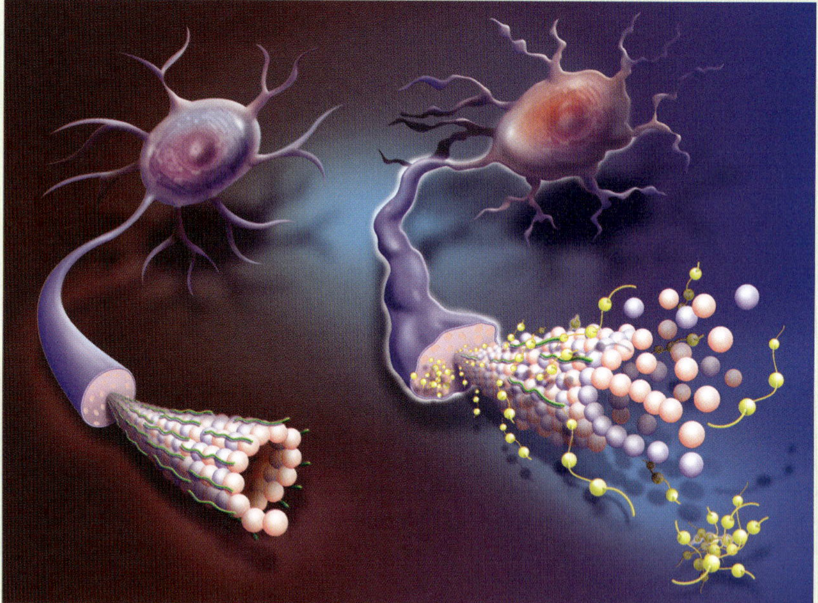

FIGURE 6.3.19 Illustration of a healthy neuron (left), showing a microtubule (pink) and tau proteins (green). On the right, neuronal degeneration is linked to the build up of 'tangles' of tau proteins (yellow) between cells.

Cancer

Cancers are a group of diseases that commonly involve unregulated and abnormal cell growth and division. Cancer can be caused by genetic mutations in the cells that either increase the rate of cell division and/or result in the suppression of apoptosis. Either case can lead to the growth of tumours.

Recall that the Bcl-2 family of proteins is central to controlling the mitochondrial apoptotic pathway. Bcl-2 is named for its discovery in *B*-*c*ell *l*ymphoma, a cancer of B lymphocytes. In B-cell lymphoma, excessive amounts of anti-apoptotic Bcl-2 proteins are produced. Anti-apoptotic Bcl-2 proteins inhibit the release of cytochrome *c* from mitochondria. Without cytochrome *c*, apoptosomes cannot form and apoptosis cannot continue. As a result, mutated B cells that would normally be removed by apoptosis instead survive, replicate and develop into cancer. Lymphomas most commonly occur in the lymph nodes and spleen, which are rich in tissue containing lymphocytes, and can spread to the liver and bone marrow (Figure 6.3.20).

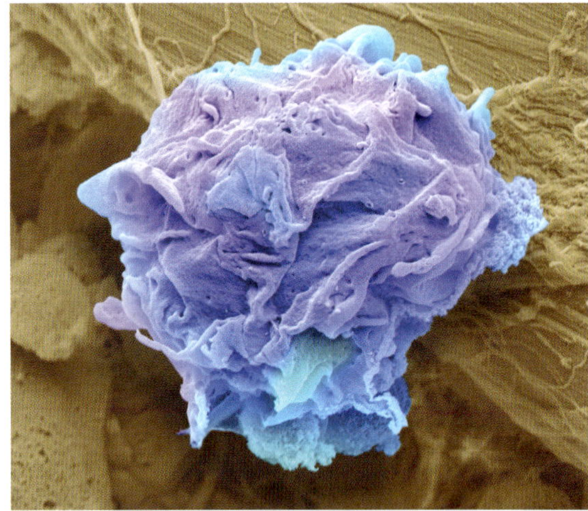

FIGURE 6.3.20 Coloured scanning electron micrograph of a lymphoma cell.

Defects in other aspects of the apoptotic pathways have been identified in various other cancers. Some of these defects include:

- mutations in caspases that prevent them from functioning, and therefore cause the cascade to stop
- defects in the adaptor (Apaf) proteins that prevent apoptosomes from forming
- defects in one or more components of the death receptor pathway.

In each case, apoptosis does not occur when it should, and defective cells survive and replicate, potentially developing into cancer.

Many cancer treatments involve the use of radiation or chemicals, which aim to induce the mitochondrial pathway of apoptosis (Figure 6.3.21). However, the reduced likelihood of apoptosis in cancerous cells also explains why these treatments are not always effective.

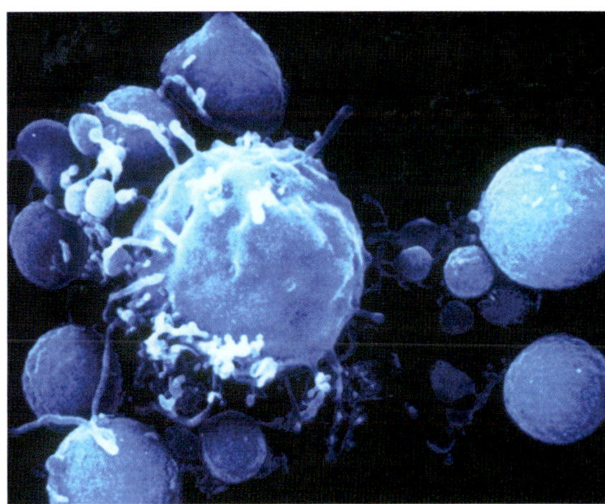

FIGURE 6.3.21 Coloured scanning electron micrograph showing apoptosis being induced in a cancer cell. The cell (at centre left) is becoming surrounded by blebs.

Tumour suppressor gene, p53

Not all mutations will lead to cancer. Most mutations are detected and repaired by enzymes in the nucleus. Recall that during the cell cycle, there are checkpoints through which cells must pass (see *Heinemann Biology 1*, Chapter 8). During the first phase of the cell cycle G1, there are several tumour suppressor proteins that act during the checkpoint, including a protein called p53 which checks for damaged DNA. If damage to DNA is detected, p53 activates the DNA repair system. During DNA repair, the cell cycle halts at the G1 checkpoint until the damage is repaired. If the DNA damage is irreparable, it will initiate apoptosis.

The mutation in the gene responsible for the formation of functional p53 is found in a wide variety of cancers (Figure 6.3.22). An example of a condition associated with a mutation of p53 is the Li–Fraumeni syndrome or sarcoma, breast, leukaemia and adrenal gland (SBLA) syndrome. Individuals with Li-Fraumeni syndrome have mutations in the p53 tumour suppressor gene in every cell of their body. As a consequence, these individuals often suffer from early onset of cancer, and develop multiple and diverse cancers throughout their lives.

FIGURE 6.3.22 Light micrograph showing proliferation of a cluster of lung cells from a rat embryo due to an induced mutation of the tumour suppressor gene p53. The inactivity of mutant p53 is believed to give rise to these circular groups of proliferating cells.

BIOLOGY IN ACTION

Cell death discovery helps protect female fertility

In Melbourne, a team of researchers from the Walter and Eliza Hall Institute, Monash University and Prince Henry's Institute of Medical Research (Figure 6.3.23), have found two proteins, PUMA and NOXA, which cause the death of egg cells in the ovaries. PUMA and NOXA trigger the death of eggs when the DNA of eggs is damaged after being exposed to radiation or chemotherapy. This egg death causes many female cancer patients to become infertile.

The team found that when primordial follicle oocytes (immature egg cells) were missing the PUMA or NOXA protein, the cells did not die after being exposed to radiation therapy. In addition, the egg cells were able to repair the DNA damage sustained and could potentially mature and be fertilised. This means that, in the future, there is a possibility of medication that could block the function of PUMA. This would stop the death of egg cells in patients undergoing chemotherapy or radiotherapy and maintain the fertility of cancer patients.

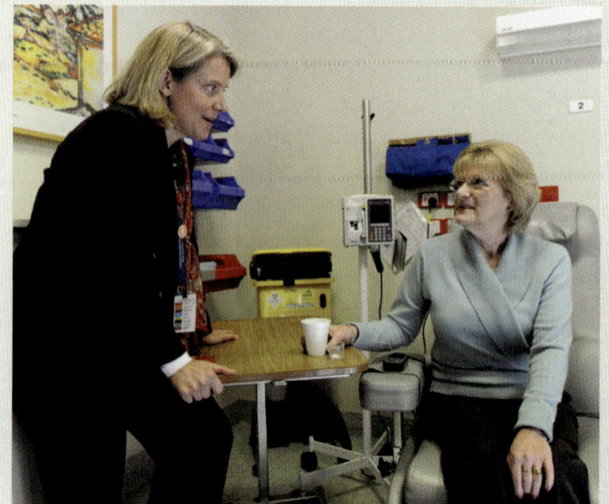

FIGURE 6.3.23 Associate Professor Clare Scott (left) and colleagues have identified a potential new way of protecting female fertility.

6.3 Review

SUMMARY

- Apoptosis is a natural process that regulates the growth and development of multicellular organisms through programmed cell death.
- There are two pathways of apoptosis:
  - the mitochondrial pathway (intrinsic), which is triggered by cell damage; signal comes from inside the cell
  - the death receptor pathway (extrinsic), which is triggered by cytokines; signal comes from outside the cell.
- Both apoptosis pathways involve a cascade of caspase activation that results in cell death.
- Caspases are enzymes that physically break down the components of the cell during apoptosis.
- Malfunctions in apoptosis result in deviant cell behaviour, which can cause a variety of diseases:
  - Excess apoptosis leads to a loss of tissue; e.g. Alzheimer's disease.
  - Inhibition of apoptosis leads to excess growth of tissue; e.g. cancer.

KEY QUESTIONS

1. a Explain what is meant by 'apoptosis'.
   b How is apoptosis important in embryo development?
2. a What are caspases?
   b Outline the functions of caspases.
3. Outline the functions of Bcl-2, cytochrome *c* and Apaf-1 in the intrinsic pathway of apoptosis.
4. Explain the function of death receptors.
5. Briefly explain why cytokines are required in extrinsic apoptosis.
6. Give an example of a disease caused by:
   a excessive apoptosis
   b inhibited apoptosis.
7. Suggest why some types of cancer are resistant to radiation and chemical treatments that would kill a healthy cell.

Chapter review

KEY TERMS

abscisic acid
action potential
activation
amino-acid-derived hormone
amplification
apoptosis
apoptosome
autocrine signalling
auxin
axon
B lymphocyte
Bcl-2
bleb
cancer
cascade
caspase
catecholamine
cortisol
cytochrome *c*
cytokine
cytokinin
death receptor
dendrite
effector cell
endocrine signalling
endocrine system
ethene
gibberellin
growth hormone
hormone
inflammatory response
inhibition
insulin
interferon
interleukin
ion channel
macrophage
neuron
neurotransmitter
nuclear pore
oestrogen
paracrine signalling
peptide and protein hormones
pheromone
phytohormone
postsynaptic neuron
response
second messenger
signal transduction
signalling molecule
steroid hormone
stimulus
stimulus–response model
synapse
synaptic terminal
T lymphocyte
testosterone
thyroid hormone

KEY QUESTIONS

1 Cytokines are signalling molecules that *always* act:
- **A** in an autocrine fashion
- **B** in an endocrine fashion
- **C** by binding to specific receptors
- **D** antagonistically with other cytokines

2 Interleukins are molecules produced by the immune system. Which of the following would it be true to say?
- **A** Interleukins are signalling molecules that allow cells to communicate with each other.
- **B** They are produced by many cells in the immune system.
- **C** They require a receptor on the target cell in order to stimulate a response.
- **D** All of the above are true.

3 Complete the table below to summarise the five main types of signalling molecules in terms of their sources, modes of transmission and their chemical properties.

Signalling molecule	Source	Mode of transmission	Chemical properties
animal hormone			
plant hormone			
neurotransmitter			
cytokine			
pheromone			

4 Blood glucose levels in normal and diabetic individuals, after eating similar meals, are shown in the following graph.

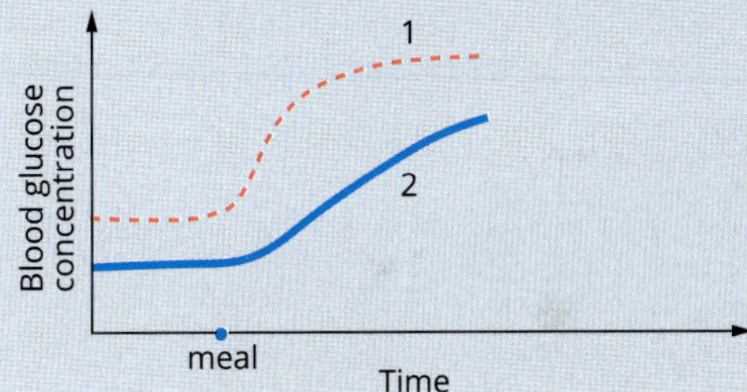

- **a** Which graph represents the diabetic individual? Explain your reasoning.
- **b** **i** Name the hormone that is produced in insufficient amounts in a diabetic individual.
  - **ii** In which organ is this hormone produced in a normal individual?
- **c** **i** Draw the negative feedback model for the control of blood glucose levels in the body.
  - **ii** Use the example of blood glucose level to explain what is meant by a 'feedback mechanism'.
- **d** Explain how the hormones glucagon and insulin work in the body to regulate blood glucose concentrations.

5 Examine the following diagram of a neuron.

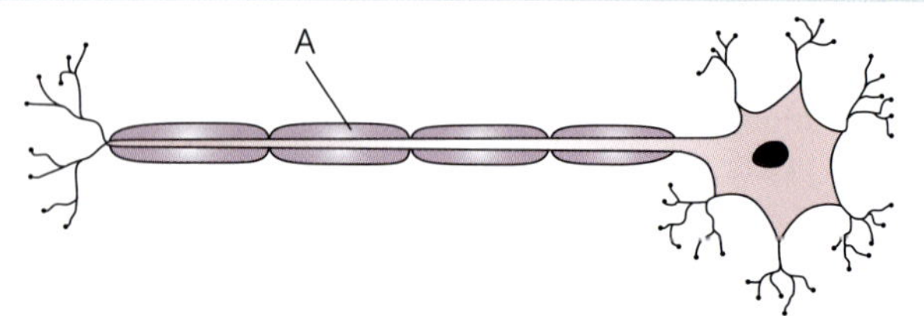

a In which direction will electrical impulses flow along the neuron?

b i Name the structure labelled A.

ii What is the main function of structure A?

c When an action potential passes into a nerve terminal, vesicles containing neurotransmitter molecules move to the nerve cell membrane and release their contents to the outside. Neurotransmitter molecules diffuse across the narrow gap and bind to specific receptors on the membrane of the subsequent neuron. Name the junction between a neuron and the cell it stimulates.

d Explain the similarity between neurotransmitters and hormones.

e Explain why nerve impulses can only be transmitted in one direction.

f Using a table, outline the differences in speed, nature of signal, method of transmission and cells affected between the nervous and endocrine systems.

6 Consider the diagram below:

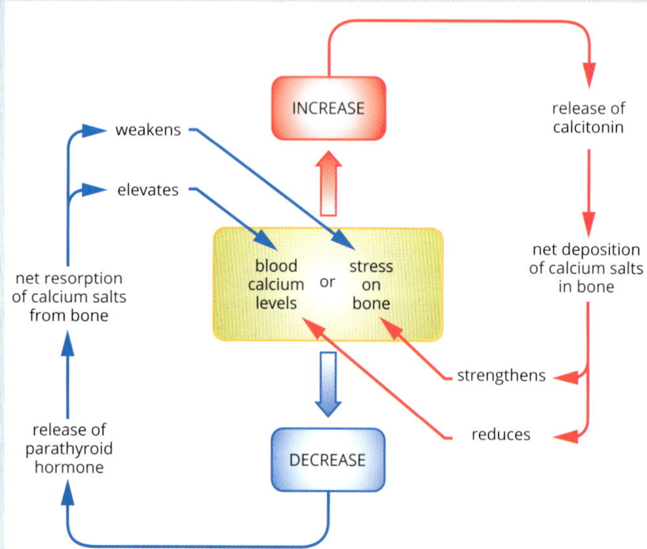

a Many people in Victoria become vitamin D deficient through the winter months due to the lack of sun exposure. In the absence of vitamin D dietary calcium is not absorbed efficiently from the digestive tract. Use the diagram to describe the likely effect that this will have on the relative concentrations of parathyroid hormone and calcitonin in the blood.

b Low levels of calcium can lead to problems at neuromuscular junctions, resulting in poor muscle control. Propose an explanation for this observation.

c i What type of signalling molecule is found in neuromuscular junctions?

ii Describe how this molecule stimulates a response in the muscle.

7 Briefly explain how cells communicate with each other, with reference to the stimulus–response model.

8 Guard cells, found in the leaf epidermis, are the entry point for carbon dioxide. Abscisic acid is a plant hormone that has been shown to be a significant contributor to guard cell opening and closing. When plants are under water stress guard cells lose turgor and close.

As soils dry out cells in the roots lose turgor. This stimulates these cells to produce abscisic acid.

a Suggest how abscisic acid produced in the roots causes the closing of stomata in the leaves.

b Explain whether abscisic acid is hydrophilic or hydrophobic.

c The figure below shows three types of cell signalling in humans. Explain which type of signalling is most like the action of abscisic acid.

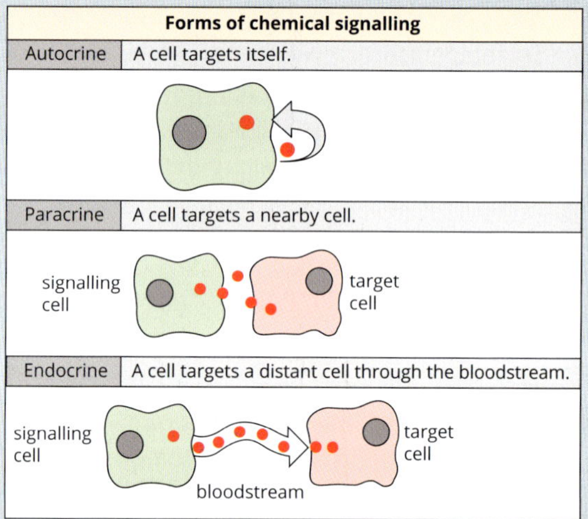

d The graph below shows the size of the stomatal pore of barley plants as the concentration of abscisic acid (ABA) is increased. The experiment was undertaken at three different temperatures.

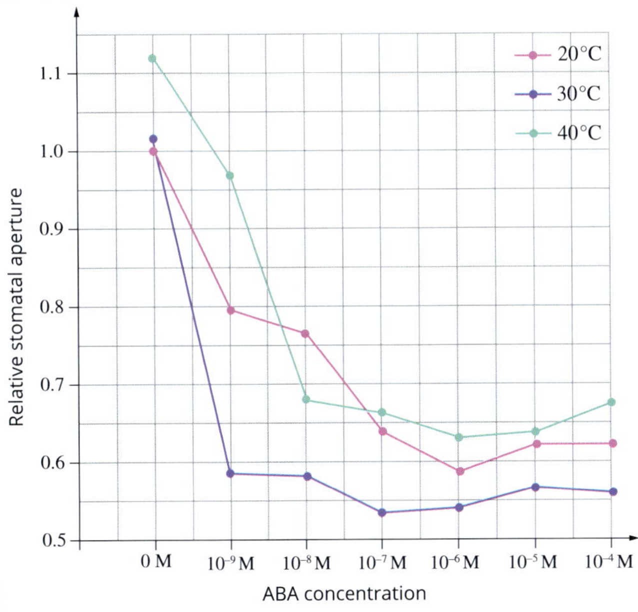

Use the information displayed on the graph to answer the following questions.

i At which temperature was stomatal aperture size most affected by an increase in ABA?

ii Describe the trends in stomatal aperture size as ABA concentration increases at 40 °C.

9 Distinguish between the key characteristics of typical hydrophobic and hydrophilic signalling molecules by completing the table below.

	Hydrophobic signalling molecules	**Hydrophilic signalling molecules**
example molecule	steroid hormones	neurotransmitters
passage through plasma membrane		
location of receptors		
transduction		
general responses		

10 Thyroxine is a very small signalling molecule. It is made of two tyrosine (amino acid) molecules joined together. Explain why the receptor for thyroxine is found inside the cell rather than on the outside.

11 *Clostridium botulinum* is a bacterium that grows in poorly preserved foods. It produces a toxin that binds to the membrane of vesicles containing acetylcholine. Acetylcholine is an important neurotransmitter. Among its other functions in the nervous system, it is the signalling molecule used at neuromuscular junctions with skeletal muscles.

a What is a neuromuscular junction?

b i Explain how botulinum toxin could disrupt nerve transmission.

ii What symptoms would you expect to see in a person with botulism (botulinum toxin poisoning)? Why?

12 Cortisol is an important human hormone. It plays a role in glucose regulation, immune system regulation and regulation of metabolic rate.

a Despite its role in many aspects of human physiology, not all cells respond to cortisol. Explain why some cells do not respond to cortisol.

b Receptors for cortisol are found in the cytosol of the cell. What does this indicate about the chemical nature of this hormone?

c Insulin is the hormone that stimulates the uptake of glucose by cells. Fat and muscle cells are generally particularly sensitive to insulin but cortisol is known to limit their response to this hormone. Propose how cortisol could reduce the normal response by fat and muscle cells.

13 Serotonin is an important neurotransmitter in the brain. It has been shown to influence mood. Depression has been linked to serotonin imbalances in the brain. It has been postulated that depression is caused by low levels of serotonin in the synapse.

In the brain serotonin is released from the presynaptic membrane. It diffuses across the synapse and attaches to receptors on the membrane of the next neuron. This stimulation results in an enhanced mood.

Unlike some neurotransmitters, which are broken down by enzymes in the synapse, serotonin is recycled back into the presynaptic neuron through special receptors called 5-HT re-uptake channels (or receptors).

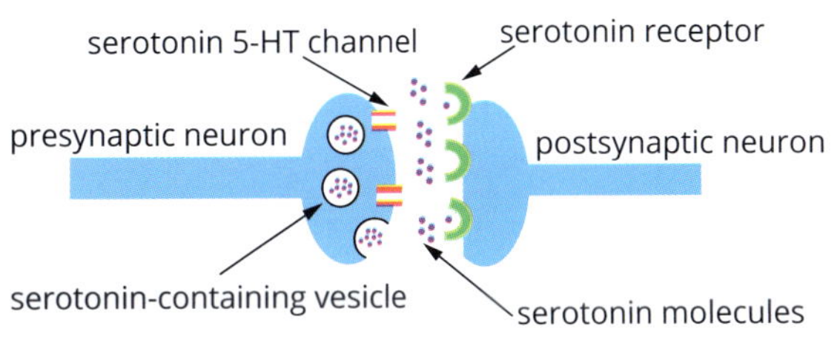

A group of drugs called selective serotonin re-uptake inhibitors (SSRIs) are one of the major treatments for clinical depression. The drugs work by increasing the amount of serotonin available in the synapse.

a What effect would higher concentrations of serotonin in the synapse have on the postsynaptic neuron? Why?

b Suggest how SSRIs inhibit the re-uptake of serotonin.

14 The Queensland fruit fly (*Bactrocera tryoni*) is a pest throughout Australia. It attacks a large variety of fruits and vegetables. Constant spraying with insecticides is expensive and harmful to the environment. The costs to agriculture of this pest have resulted in some innovative solutions. Pheromones are used by many species of insect, including *B. tryoni*, to communicate with each other. One form this communication takes is as an attractant of the opposite sex.

It is now common for orchardists to hang pheromone traps in their orchards. These traps are baited with chemicals that mimic the pheromones of the female Queensland fruit fly. Males are attracted into the traps, which also contain insecticide, and are killed. By monitoring the traps daily the farmer can tell when the orchard has been infested with fruit fly due to the increase in captures. The farmer can then spray the crop with insecticide. This reduces costs and environmental harm as spraying only occurs if and when the crop is under attack.

a How do pheromones reach their target cells?

b What feature must the chemicals of the pheromone trap have in order to make them effective?

c Scientists at a chemical company have been researching pheromones and claim that they have designed a new pheromone mimic that is more effective than those in current use. You are a farmer who has decided to test the truth of this claim in your next growing season. If the claims prove accurate you will switch to this new pheromone trap in the following year.

  i Design an experiment you could use in your orchards to test the claim of the chemical company.

  ii Identify the dependent and independent variables in your experiment.

  iii How will you know if the new pheromone trap is more effective than the old one?

  iv What might be some variables that would be hard to control in your experiment?

15 Compare and contrast the two pathways of apoptosis.

16 Apoptosis is a regulatory process in multicellular organisms. Propose another process that regulates cell numbers in multicellular organisms and briefly explain how in addition to apoptosis this process helps to maintain, increase or decrease cell numbers.

17 Mutations in the genes that encode the intracellular proteins of the Bcl-2 family of proteins have been identified as a common feature of cancer cells.

a i What is meant by the term 'intracellular'?

  ii Explain how mutations in Bcl-2 could lead to cancer.

b Crotamine is a protein found in rattle snake (*Crotalus atrox*) venom. This venom has been shown to have chemical features in common with another intracellular protein called BAX, which is a member of the Bcl-2 family of proteins and promotes apoptosis by the intrinsic pathway. Crotamine causes cell death in muscle cells using intracellular apoptotic pathways.

  i Hypothesise the pathway by which crotamine causes cell death.

  ii Crotamine has high affinity and toxicity for rapidly dividing cells. Explain why this has made this substance a target for cancer research.

18 Explain the consequence for the individual of a mutation during embryonic development in the cells between the fingers which has resulted in the absence of the extrinsic death receptors.

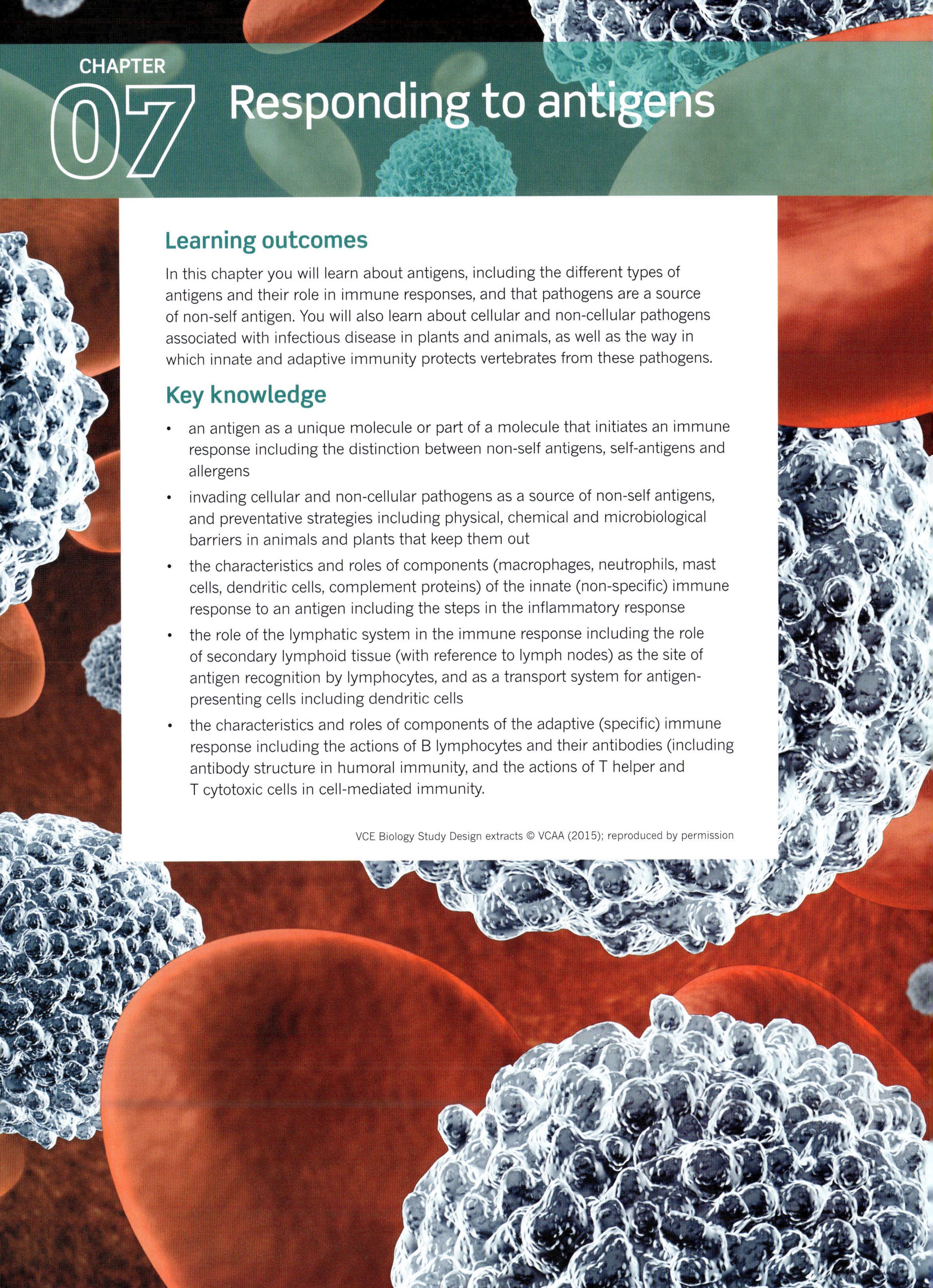

CHAPTER 07

Responding to antigens

Learning outcomes

In this chapter you will learn about antigens, including the different types of antigens and their role in immune responses, and that pathogens are a source of non-self antigen. You will also learn about cellular and non-cellular pathogens associated with infectious disease in plants and animals, as well as the way in which innate and adaptive immunity protects vertebrates from these pathogens.

Key knowledge

- an antigen as a unique molecule or part of a molecule that initiates an immune response including the distinction between non-self antigens, self-antigens and allergens
- invading cellular and non-cellular pathogens as a source of non-self antigens, and preventative strategies including physical, chemical and microbiological barriers in animals and plants that keep them out
- the characteristics and roles of components (macrophages, neutrophils, mast cells, dendritic cells, complement proteins) of the innate (non-specific) immune response to an antigen including the steps in the inflammatory response
- the role of the lymphatic system in the immune response including the role of secondary lymphoid tissue (with reference to lymph nodes) as the site of antigen recognition by lymphocytes, and as a transport system for antigen-presenting cells including dendritic cells
- the characteristics and roles of components of the adaptive (specific) immune response including the actions of B lymphocytes and their antibodies (including antibody structure in humoral immunity, and the actions of T helper and T cytotoxic cells in cell-mediated immunity.

7.1 Antigens

Antigens are unique molecules, or parts of molecules, that can often elicit an immune response, and so play a crucial role in immunity. Antigens can be classified as self or non-self antigens, and an organism's immune cells can usually differentiate between self and non-self antigens and respond accordingly.

This section will focus on the nature of antigens, the distinction between different types of antigens, and on pathogens as sources of non-self antigens (Figure 7.1.1).

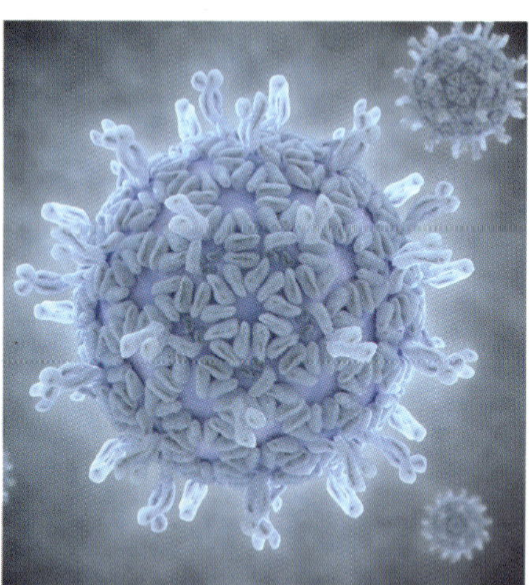

FIGURE 7.1.1 An artist's impression of rotavirus, a virus that is a common cause of gastroenteritis and diarrhoea in infants. Proteins on the surface of the virus act as antigens, which are recognised by the body's immune cells.

THE NATURE OF ANTIGENS

Antigens are unique molecules, or parts of molecules, that can be recognised by T lymphocytes or by **antibodies** produced by B lymphocytes (you will learn more about T and B lymphocytes in Section 7.3). Antibodies can be bound to the surface of, or secreted by, B lymphocytes. Antigens are important because they allow the body to recognise potentially harmful pathogens and mount an immune response against them. Although many antigens trigger an immune response, some do not. Antigens that elicit an immune response are more properly known as **immunogens**; however, in the context of an immune response it is still common to simply refer to them as antigens.

> Antibodies (also known as immunoglobulins) are proteins produced by B lymphocytes that bind to specific antigens.

Structure of antigens

Most antigens are protein-based and can be composed of one or more polypeptide chains. However, antigens can also be composed of carbohydrates, lipids and even nucleic acids. For example, the complex carbohydrates of the human ABO blood group are antigens.

It is the structure of the carbohydrate that makes the A antigen different from the B antigen. The presence or absence of A and B antigens on the surface of red blood cells determines whether the blood group is A, B or AB. Group O blood has neither A nor B antigens on the surface of red blood cells (Figure 7.1.2).

Blood type	Red blood cells	Antibodies present in plasma	Antigens present on cells
A		anti-B	A
B	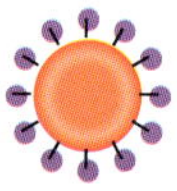	anti-A	B
AB	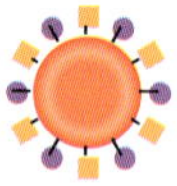	none	A and B
O		anti-A and anti-B	none

FIGURE 7.1.2 The A and B blood type antigens are carbohydrate molecules attached to proteins and lipids in the red blood cell membrane. If the blood type transfused into a patient is different from the patient's own blood type, an immune response will be elicited by the patient's immune system, which can lead to death.

Types of antigens

Antigens are expressed or presented on the surface of the plasma membrane of cells, where they act as recognition sites for the immune system. However, not all antigens are attached to a cell; for example, some antigens, such as toxins released by bacteria, circulate freely in body fluids. The immune system is normally able to distinguish antigens that are expressed by its own cells (**self-antigens**), from those that are not (**non-self antigens**), and respond accordingly.

Responding to antigens

Antigen recognition is dependent on the detection of antigens by receptors:

- The receptors on B lymphocytes are membrane-bound antibodies that recognise free antigens or antigens that are on the surface of a pathogen (Figure 7.1.3a). Antibodies can also be secreted by the B lymphocytes (Figure 7.1.3b).
- The receptors on T lymphocytes are different from the membrane-bound antibodies of B lymphocytes, and recognise antigens presented by the organism's own cells (Figure 7.1.3c).

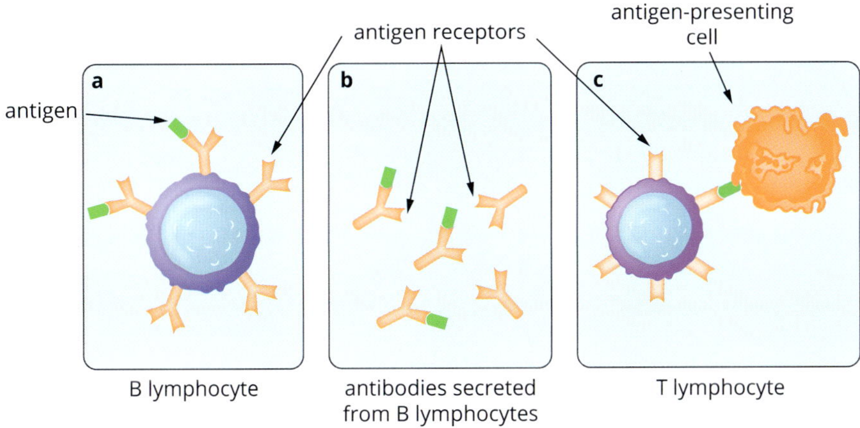

FIGURE 7.1.3 Antigen receptors on immune cells. Antibodies are antigen receptors that may be bound to the plasma membrane of B lymphocytes (a) or secreted from B lymphocytes (b). T lymphocytes have their own type of receptors that recognise antigen presented by specialised antigen-presenting cells (c).

BIOFILE

Antigens and antibodies

The term 'antigen' stands for 'antibody generator'. Antibodies are proteins made when immune cells called B lymphocytes become activated. Antibodies bind to specific antigens (Figure 7.1.4) and play an essential role in removing pathogens and preventing disease.

FIGURE 7.1.4 In this digital illustration, antibodies (pink) can be seen binding to specific antigens on the surface of influenza virus (yellow).

There are many different receptors and they are specific for particular antigens.

The **major histocompatibility complex (MHC) proteins**, also called **human leukocyte antigens** (HLA), are proteins on the surface of your body's cells that present self or non-self antigens to T lymphocytes. There are different classes of MHC proteins, which you will learn more about in Sections 7.2 and 7.3.

In the thymus, T lymphocytes undergo a maturation stage called positive selection, in which the T lymphocytes that do not interact with MHC proteins are destroyed by apoptosis. They then undergo a second stage of maturation, called negative selection, in which T lymphocytes that react with self-antigens in the thymus bind tightly to the cells in the thymus and eventually die. This two-stage process of selecting T lymphocytes that can recognise MHC proteins, and eliminating T lymphocytes that react to self-antigen is called clonal deletion.

The inability to respond to self-antigens is called tolerance, or **self-tolerance**. If self-tolerance breaks down and the immune system responds to self-antigens, it results in autoimmune diseases. You will learn more about autoimmune diseases in Chapter 8.

As mentioned earlier, not all antigens (including not all non-self antigens) elicit an immune response, and antigens that do elicit an immune response are called immunogens. In allergy certain antigens elicit an allergic immune response. Antigens that trigger an allergic response are called **allergens**. You will learn more about allergic reactions in Chapter 8.

PATHOGENS AS SOURCES OF NON-SELF ANTIGENS

Pathogens are agents that cause disease. Depending on their ability to cause disease, pathogens are divided into two groups:

- Primary pathogens—cause disease any time they are present.
- Opportunistic pathogens—only cause disease when the host's defences have been weakened, for example, by poor nutrition or stress.

Most pathogens contain unique antigens that can be recognised by the immune system. For example, the tuberculosis bacterium, the fungus that causes tinea, and the virus that causes influenza each have antigens that are unique to them. Toxins secreted by pathogens can also act as antigens.

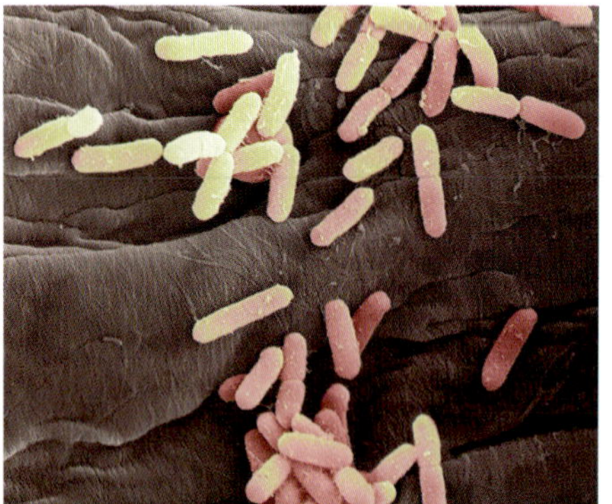

FIGURE 7.1.6 Coloured scanning electron micrograph of *Escherichia coli* bacteria (rod-shaped) found in a urine sample from a patient with a urinary tract infection.

Cellular pathogens

Pathogens may be cellular or non-cellular. **Cellular pathogens** of plants and animals include bacteria, fungi, oomycetes, protozoans, worms and arthropods.

Bacteria

Bacteria are prokaryotes that are almost everywhere, and exposure to pathogenic bacteria is a certainty (Figure 7.1.5). However, bacteria are not always pathogenic. The human body supports and relies on a range of bacteria that reside on and inside it. For example, humans benefit from the metabolic products of non-pathogenic *Escherichia coli*, an inhabitant of the intestine. However, it is possible for the same strain of *E. coli* that is beneficial in the intestine to cause infection if it enters the urinary tract (Figure 7.1.6).

FIGURE 7.1.7 Although mushrooms sold at the supermarket are safe to eat, many in the wild are poisonous.

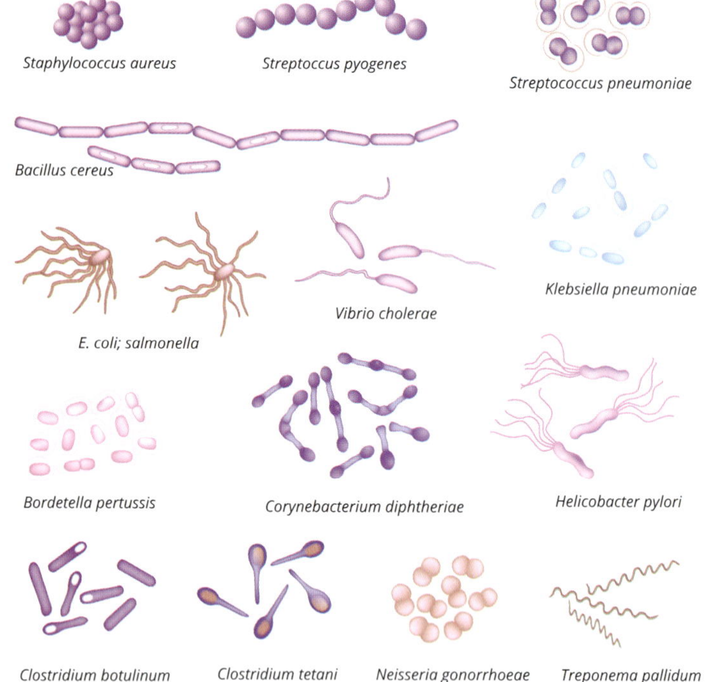

FIGURE 7.1.5 Common pathogenic bacteria.

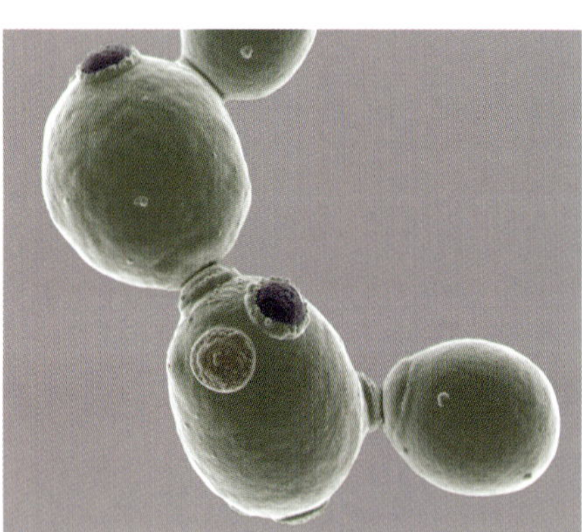

FIGURE 7.1.8 Coloured scanning electron micrograph of the yeast *Candida albicans*, which causes thrush (candidiasis). Depending on environmental conditions, *C. albicans* takes a unicellular yeast-like form or a multicellular filamentous form.

Fungi

Fungi are a diverse family ranging from the macroscopic (mushrooms, Figure 7.1.7) to the microscopic (moulds, unicellular yeasts and yeast-like fungi, Figure 7.1.8). Fungi can secrete digestive enzymes and other chemicals into their environment to break down organic matter, which can then be absorbed into the fungus. It is these secreted substances that are usually responsible for causing disease in animals and plants. Fungal cells produce surface glycoproteins and polysaccharides that act as antigens, allowing them to be identified by cells of the immune system.

Oomycetes

The **oomycetes** (phylum Oomycota) include organisms that cause blight and downy mildew on plants and life-threatening infections in animals. Originally thought of as fungi, the oomycetes, including *Phytophthora* (which means 'plant destroyer'), are now classified in the kingdom Protista.

Oomycetes have motile cells (with flagella), walls of cellulose, and many cellular processes that are not found in fungi. When spores of oomycetes are released on a leaf they may be carried in water droplets to other leaves, swim to a germination site, or germinate directly, sending out a hypha (fungal thread) that branches and invades plant tissue. These branching hyphae (haustoria) penetrate living cells and absorb nutrients or release enzymes that digest cytoplasm into molecules that can be absorbed. In plants it has been shown that ooymcetes release molecules that suppress their host's innate immune response and inhibit apoptosis.

There are about thirty-five species of *Phytophthora*, which infect many crops including potato, tomato, apple, tobacco plants and citrus trees. In Australia, *P. cinnamomi* has destroyed tens of thousands of hectares of valuable eucalypt timberland (Figure 7.1.9). The spores of *P. cinnamomi*, some of which can survive for years in moist soil, are attracted to the roots of the plants they infect by a chemical released from the roots.

FIGURE 7.1.9 A eucalypt forest in the Brisbane Ranges National Park infected with cinnamon 'fungus', *Phytophthora cinnamomi*, which causes dieback disease. Species of *Xanthorrhoea* (grass tree) are also very susceptible to this disease and rapidly turn brown and die.

Pythium insidiosum is the only oomycete known to infect mammals including humans.

Motile cells are capable of motion.

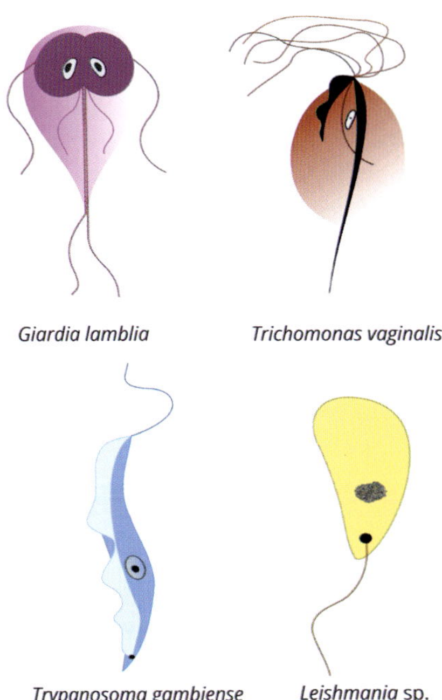

FIGURE 7.1.10 Common pathogenic protozoa.

Protozoans

Protozoans are unicellular eukaryotes (Figure 7.1.10). Some reproduce within their host's cells, while others, like *Giardia lamblia* (Figure 7.1.11), demonstrate extracellular reproduction. The life cycles of some protozoans include multiple stages in different hosts. Many of these protozoans express different proteins on the surface of their plasma membranes at different stages of their life cycle. These proteins act as antigens. The mechanism by which their surface antigens change is known as **antigenic variation**, and it assists protozoans in evading detection by the host.

FIGURE 7.1.11 Coloured scanning electron micrograph of a *Giardia lamblia* protozoan (pink), which undergoes asexual reproduction in the small intestine and causes diarrhoea.

BIOFILE

The changing face of malaria

The malarial protozoan *Plasmodium* infects red blood cells, where it resides to evade recognition by circulating immune cells (Figure 1.2.14). *Plasmodium* produces adhesion proteins, which it presents on the surface of the red blood cell. These proteins interfere with the cell's activities within capillaries.

The immune system recognises the adhesion proteins as non-self antigens, but before it can mount an effective immune response, the parasite replaces the adhesion protein with a different adhesion protein. *Plasmodium* has approximately 60 different adhesion proteins that it can continually interchange to remain a step ahead of the immune system. Scientists have recently discovered that an enzyme known as ribonuclease (RNase) causes this process of antigenic variation.

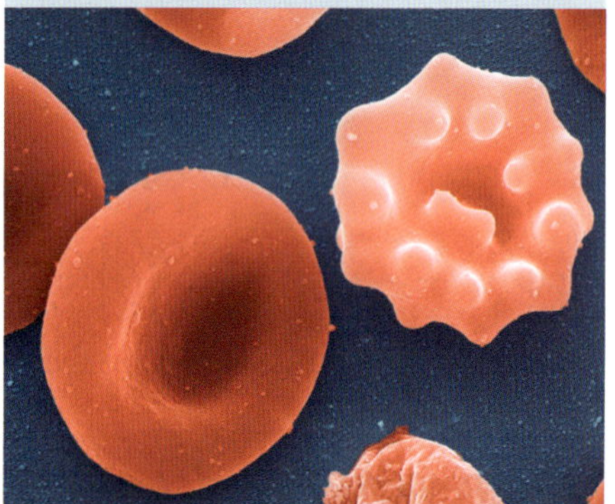

FIGURE 7.1.14 Scanning electron micrograph of red blood cells infected with *Plasmodium* parasite. Changes to the normal red blood cell membrane (left) are clearly seen in the infected red blood cell (top right).

Worms

Parasitic worms can infect plants and animals. They include flatworms such as tapeworms, and roundworms such as hookworms, pinworms and threadworms. In plants, roundworms infect roots and are major pests of orchard trees and crops. In animals, parasitic worms can regulate the immune system in a number of ways so that the immune response against them is suppressed. For example, some roundworms (Figure 7.1.12), such as *Nippostrongylus brasiliensis*, secrete inhibitors that block the action of enzymes needed for **antigen presentation**, which is an important part of the immune response you will learn more about in Section 7.2.

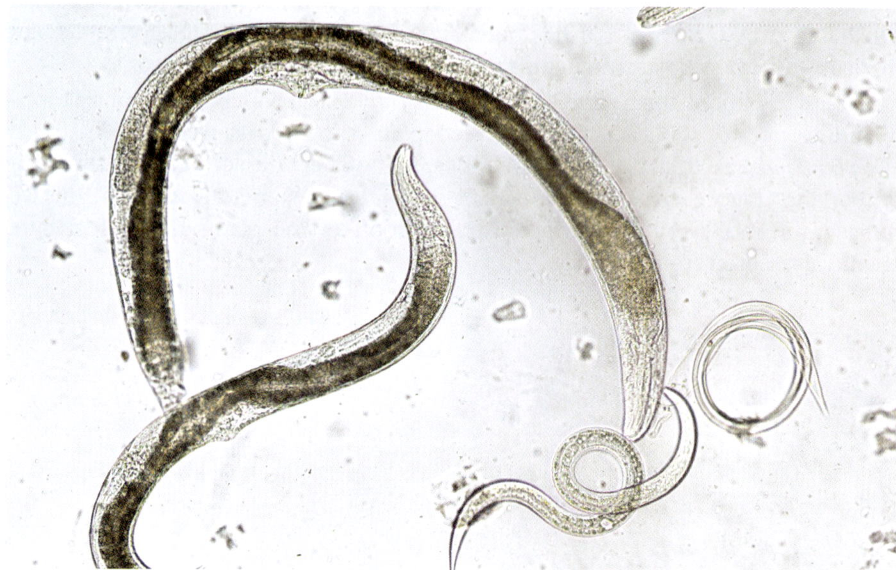

FIGURE 7.1.12 A light micrograph of female and juvenile roundworms (or nematodes).

Arthropods

Arthropods are invertebrates with external skeletons (or exoskeletons). Arthropods able to transmit or cause disease in humans include insects such as mosquitoes, ticks, lice and mites (Figure 7.1.13). Arthropod saliva contains molecules that modulate the host immune response and inhibit inflammation. These molecules create a favourable environment for pathogen transmission. Arthropod saliva also contains antigens that can trigger an adaptive immune response, and their immunogenic properties are being used to help develop vaccines against some vector-borne diseases.

Some arthropods can also damage plants. For example, psyllids (lerp insects) are small insects that in their larval stages induce the formation of swollen areas of leaf tissue known as galls. The larval stages of many psyllid species construct a covering (a lerp) under which they feed on the leaf surface. The saliva from feeding psyllids kills leaf tissue, causing extensive discolouration of the leaf.

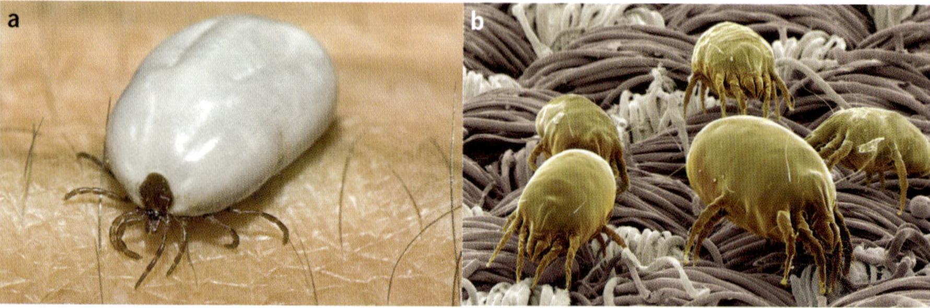

FIGURE 7.1.13 (a) Ticks feed on blood, so they can act as vectors, able to transmit a number of bacterial and viral pathogens when they bite. (b) Common household dust mites feed on flakes of dead skin in dust, and their waste products can cause skin and respiratory allergies.

Non-cellular pathogens

Viruses, **viroids** and **prions** are not living organisms, but they have the ability to cause disease.

Viruses

A single virus particle (or virion) is composed of genetic material, either DNA or RNA, enclosed in a protein coat. Some viruses also have a lipoprotein envelope (Figure 7.1.15).

In a process called antigenic drift, some viruses are constantly making minor changes to the antigens on their surface. These small genetic changes result in viruses that are very similar but not identical. They will usually be recognised by the immune system (if a similar virus has infected the host on a previous occasion). Eventually, though, the small genetic changes add up to make the virus significantly different.

Antigenic shift, by contrast, is a much more abrupt change in the genetic code of a virus due to re-assortment of genes from different viral strains, resulting in significantly different protein antigens on the coat of the virus.

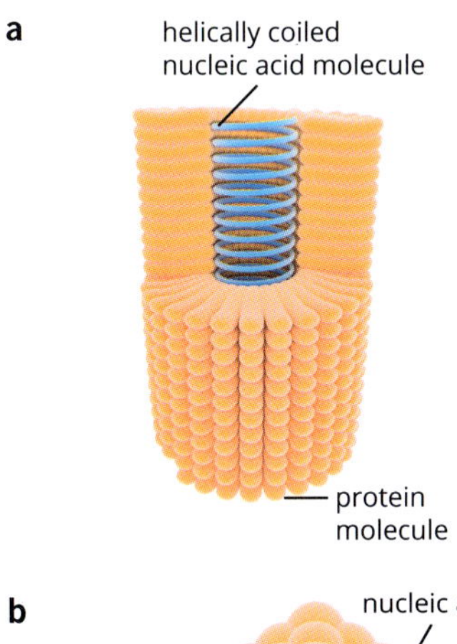

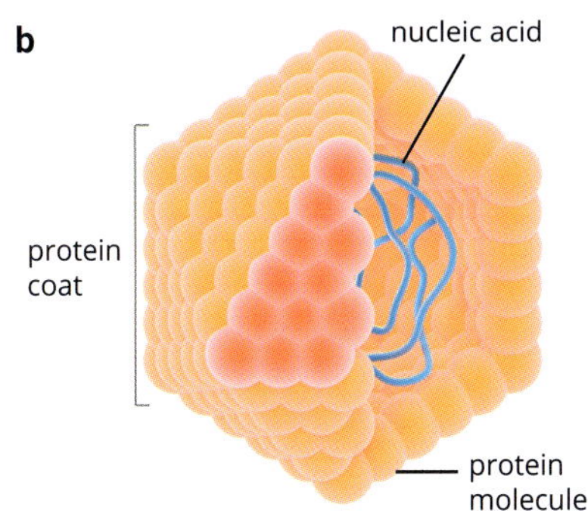

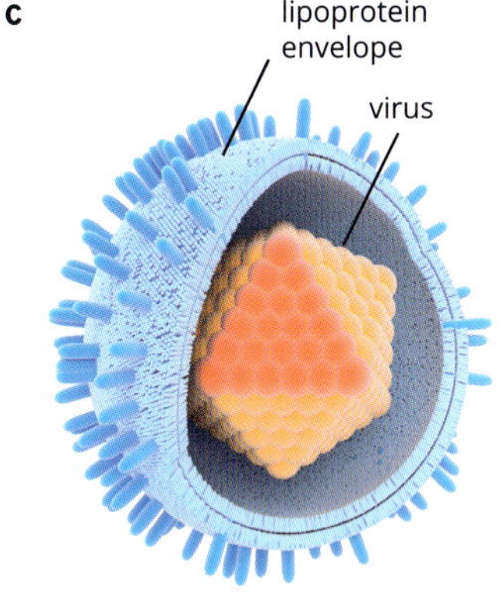

FIGURE 7.1.15 Different structures of virions. (a) A rod-shaped virion with proteins (orange) surrounding a helically coiled nucleic acid molecule (blue). (b) An isometric virion with an icosahedral protein coat surrounding a nucleic acid core (blue). (c) An icosahedral virion (orange) enclosed by a lipoprotein envelope (blue).

Viroids

A viroid is a type of self-cleaving RNA enzyme (or ribozyme) that is composed of short, circular strands of RNA that lack a protein coat. Viroids are only known to be pathogens of plants. They damage plants by competing for nucleotides and forming viroid bundles, which mechanically interfere with the internal structures of plants much like a tumour.

Prions

The only pathogens smaller than viroids are prions. Prions are the only known infectious agents that do not contain genetic material. Prions are proteins that are similar to normal cellular prion proteins (PrP), which are located mainly in the central nervous system of the organism. However, unlike PrP, prions have a shape (or conformation) that is abnormal.

Prions stimulate the organism's normal PrP to misfold into the infectious prion form. Prions are resistant to being denatured, as well as to being broken down by proteases, which normally break down proteins. Prions cause neurodegenerative diseases in mammals, the first of which to be discovered was scrapie in sheep (Figure 7.1.16).

In humans prions cause Creutzfeldt–Jakob disease (CJD). As in scrapie, in CJD prions cause vacuoles and misfolded proteins (or plaques) to form in the brain, which kills neurons and makes the brain appear 'spongy' under a microscope. Symptoms include dementia and sudden muscle contractions, leading to death. The equivalent disease in cattle, bovine spongiform encephalopathy (BSE), commonly known as mad cow disease, has been linked to human variant CJD through human consumption of BSE-contaminated beef.

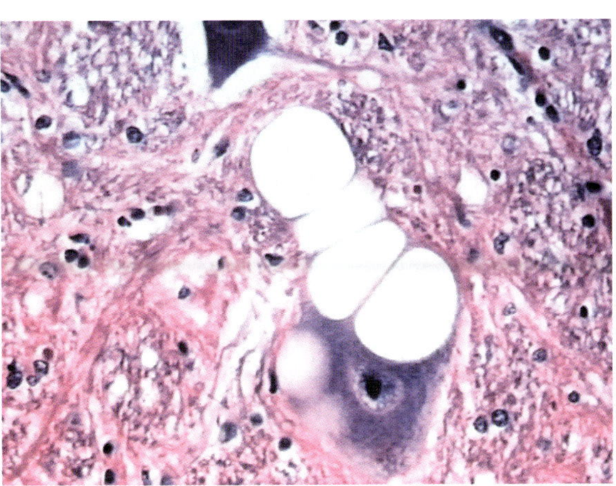

FIGURE 7.1.16 Scrapie is a disease in sheep caused by prions. This light micrograph shows a section through the brain of a sheep infected with scrapie. The large empty vacuoles (white) in the centre show the effects of the disease. As scrapie progresses, an increase in the number of empty vacuoles makes the brain tissue appear spongy and destroys neurons. Symptoms of the disease include glazed eyes and body tremors.

Prions elicit an ineffective innate immune response, and the adaptive immune system is unable to identify and respond to them. It has been suggested the reason the adaptive immune system cannot respond to prions is that, despite being abnormal in shape, prions remain very similar to the normal PrP, and any T lymphocytes that would have responded to normal PrP would have been destroyed to prevent an autoimmune reaction. Another reason could be that prions are unable to be broken down and presented by antigen-presenting cells. (You will learn more about antigen presentation in Section 7.2.)

BIOLOGY IN ACTION

Pioneering studies of disease

Perhaps the greatest health-related discoveries of the 19th century were those of Louis Pasteur who, among his many achievements, established the existence of microorganisms and showed that infectious diseases were caused by microbes. Prior to this time, deaths in hospitals due to post-operative infection were commonplace. This was not helped by the fact that many doctors would do post-mortems in the mornings and surgery in the afternoons, without changing their clothes! In some hospitals where this practice was followed, deaths following childbirth were very high in doctors' wards, and much lower in nurses' wards where mothers were cared for only by nurses who did not participate in post-mortems.

Joseph Lister, an English surgeon, had observed that when wounds were left open to the air, such as after amputations, almost half the patients died from infection. But with other wounds that were closed, infection was not nearly so severe. He had concluded that infection was due to 'something in the air'. When he heard of Pasteur's experiments showing microbes to be the cause of putrefaction (rotting) of food, he decided they might also be the cause of the infections in his patients. It was known that carbolic acid was highly poisonous to living organisms, so Lister decided to use it in the hospital wards in the hope that it would kill the 'invisible microbes'. He used it on patients, his own hands and around hospital rooms, and required nurses to use it also. The incidence of infection in his patients was dramatically reduced. This was the first practice of antiseptic surgery.

One of the earliest experimental studies that investigated the cause of disease was the work of a German doctor Robert Koch (Figure 7.1.17), and other scientists. Towards the end of the 19th century, with the improvement of the microscope, scientists were able to identify different species of bacteria and protozoa. In Koch's early work, he studied anthrax, a disease of cattle and sometimes humans.

Koch's experimental method involved examining blood samples taken from patients with different diseases, then growing microbes from the blood on nutrient plates. When he injected specific microbes into mice he found that they developed diseases similar to those of the original patient. As a result of these studies, specific microbes became recognised as the cause of particular diseases.

Koch formulated a set of criteria, known as *Koch's postulates*, which were used to establish whether a specific microorganism was the cause of a particular disease. Koch's postulates are:

FIGURE 7.1.17 Dr Robert Koch (1843–1910), German bacteriologist, in his laboratory in South Africa.

1 The microorganism should always be found in organisms suffering from the disease, but not in healthy organisms.
2 The microorganism must be cultivated in pure culture away from the body of the infected organism.
3 When this culture is inoculated into a susceptible organism, the organism should develop symptoms of the disease.
4 The microorganism should then be re-isolated from the experimentally infected organism and re-cultured in the laboratory. It should be the same as the original microorganism.

7.1 Review

SUMMARY

- Antigens are molecules, or parts of molecules, that interact with the receptors of T lymphocytes, B lymphocytes and with antibodies.
- Antigens:
  - have a unique molecular structure
  - are composed of one or more polypeptide chains but can also be composed of nucleic acids, carbohydrates or lipids
  - can identify cells as self or non-self
  - can be found on the surface of the plasma membrane of cells or circulating freely in body fluids (e.g. bacterial toxins).
- Some, but not all, non-self antigens elicit an immune response.
- Antigens that elicit an immune response are called immunogens.
- Antigens that elicit an allergic response in susceptible individuals are called allergens.
- Self-antigens do not normally elicit an immune response. This is known as self-tolerance.
- Under normal conditions, any foreign molecule is recognised by the immune system as a non-self antigen.
- Pathogens are sources of non-self antigens and agents that cause disease.
- Cellular pathogens include bacteria, fungi, oomycetes, protozoans and some worms and arthropods:
  - Bacteria are prokaryotes that exist almost everywhere. They are not always pathogenic and many support the functions of the human body, but they do always present non-self antigens.
  - Fungal pathogens include yeasts, moulds and yeast-like fungi. They produce glycoproteins and polysaccharides that act as antigens, allowing the immune system to identify them.
  - Oomycetes are fungus-like pathogens that mainly affect plants. They release molecules that suppress the innate immune response and inhibit apoptosis.
  - Protozoans are unicellular, eukaryotic organisms that have multiple stages in a complete life cycle. Many protozoans express different antigens at different stages in their life cycle. This is known as antigenic variation and assists in evading detection by their host's immune system.
  - Parasitic worms include flatworms such as tapeworms, and roundworms such as hookworms, pinworms and threadworms. In animals, parasitic worms can regulate the immune system in a number of ways so that the immune response against them is suppressed.
  - Arthropods are invertebrates with external skeletons (or exoskeletons). Some, such as dust mites, can trigger allergies; others, such as ticks, can transmit bacterial or viral pathogens when they bite.
- Non-cellular pathogens include viruses, viroids and prions:
  - Viruses are composed of genetic material enclosed in a protein coat. Viral antigens change as a virus evolves, which helps the virus evade detection by the host.
  - Viroids are a type of self-cleaving RNA enzyme (or ribozyme). They are composed of short, circular strands of RNA that lack a protein coat.
  - Prions are infectious agents that do not contain genetic material.

KEY QUESTIONS

1. Describe an antigen. Include the difference between an antigen, an immunogen and an allergen in your answer.
2. Are pathogens sources of self-antigens or non-self antigens? Explain the difference between self and non-self antigens in your answer.
3. What happens if an immune response is directed against a self-antigen?
4. Which of the following is true?
   A All immunogens are antigens.
   B All antigens are immunogens.
   C Antigens are produced by T lymphocytes to defend against non-self antigens.
   D Antigens are produced by B lymphocytes to defend against immunogens.

7.2 Innate immunity

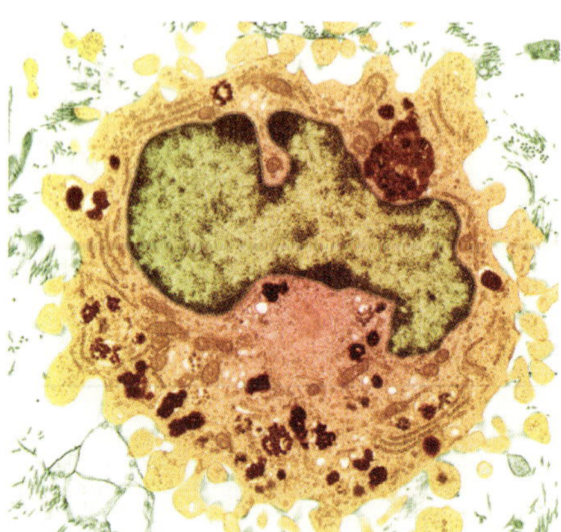

FIGURE 7.2.1 Coloured transmission electron micrograph of a macrophage. Macrophages are cells of the innate immune response in vertebrates that recognise and engulf foreign material.

There are several mechanisms by which the immune systems of organisms respond to non-self antigens and defend against pathogens. These defence mechanisms include barriers that help prevent infection, and immune responses to pathogens that breach these barriers. In vertebrates, immune responses are divided into innate (or non-specific) and adaptive (or specific) immune responses.

In this section you will learn about **innate immunity**, which consists of physical, chemical and microbiological barriers that provide innate resistance to infection, as well as the innate immune response to infection that occurs when these barriers are breached (Figure 7.2.1).

BARRIERS TO INFECTION

Organisms have a number of first-line defences (or barriers) that provide innate resistance against pathogens, including:

- physical barriers such as skin or bark
- chemical barriers such as the **lysozyme** enzymes in saliva and other body secretions
- microbiological barriers, namely **microflora**.

Physical barriers in plants

Physical barriers in plants largely involve cell walls that provide strength and flexibility. Cutin and waxes are fatty substances that make up the cuticle, which is found on the outer cell wall. A thicker cuticle generally prevents more pathogens from infecting the plant than a thinner cuticle. Likewise, in trees a thicker layer of bark is better able to prevent pathogens from entering the plant. Stomata create openings in the physical barriers of plants, providing an entry point for pathogens, but these openings can be closed when signalled (Figure 7.2.2).

In addition to the presence of physical barriers, the orientation of leaves can also play a role in defence. By positioning leaves vertically, water is unable to collect on the surface of leaves. This prevents infection by pathogens that are reliant on water for motility.

Stomata are tiny epidermal pores bound by two specialised guard cells; they are the main way in which gas exchange occurs in plants.

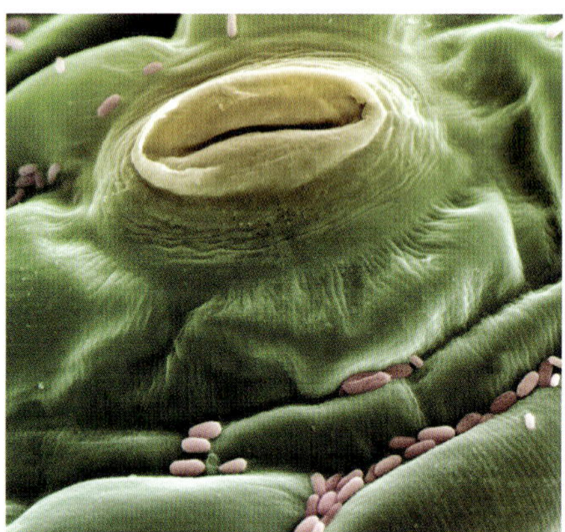

FIGURE 7.2.2 Scanning electron micrograph of a single stomata on the surface of the leaf of a tomato plant. Stomata are able to close to prevent bacteria (rod-shaped, pink) entering and infecting the plant.

Physical barriers in animals

In vertebrates, epithelial cells create a physical barrier that prevents pathogens from entering the organism. Epithelial cells line the skin, as well as the respiratory, gastrointestinal and urogenital tracts. They are joined tightly by specialised membrane proteins to form a continuous barrier against pathogens.

In addition to toughened (keratinised) skin, adaptations that provide physical barriers to pathogens in animals include mucus-secreting membranes that trap invading organisms in mucus and membranes lined with cilia that sweep foreign bodies away (e.g. those that line the airways) (Figure 7.2.3).

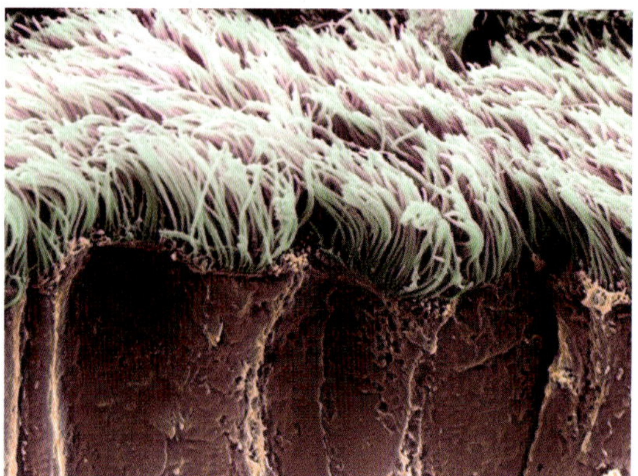

FIGURE 7.2.3 A coloured scanning electron micrograph of the mucus membrane (or bronchial epithelium) that lines the major airways of the lung. Mucus traps potential pathogens and foreign particles, and the rhythmic movement of hair-like cilia moves bacteria and other particles away from the lung and towards the throat.

Chemical barriers in plants

Plants have developed a vast array of chemical defences. Table 7.2.1 lists some of the chemicals produced by plants to defend against pathogens.

Chemical	Source/plant	Function
saponin	wheat	disrupt cell membranes of fungi
caffeine	coffee, tea and cocoa plants	toxic to insects and fungi
tannins	tea and grapes	toxic to insects
citronella	essential oil from lemongrass	repels insects
defensins	barley and wheat	toxic to microbes
chitinases	barley, tomato, banana	enzymes that disrupt cell membranes of fungi

TABLE 7.2.1 Chemical barriers in plants.

Chemical barriers in animals

External chemical barriers in vertebrates include lysozyme enzymes and toxic metabolites, for example lactic acid and fatty acids, which are found in secretions such as tears, sweat and saliva (Figure 7.2.4). Here, they have protective functions and provide a generalised defence, for example, by destroying bacterial cell walls.

Other chemical barriers include stomach acid and digestive enzymes, which are primarily involved in the digestion of food, but also kill many pathogens. The fluid in the lungs contains proteins that act as surfactants. Surfactants coat the pathogens, making it easier for the pathogens to be eliminated by macrophages. In female mammals, the lining of the vagina is coated in acidic secretions that serve several functions, including defence against pathogens.

Surfactants are 'detergent-like' substances found in lung secretions that lower the surface tension of lung fluids and prevent the alveoli from collapsing on exhalation. They are also antimicrobial.

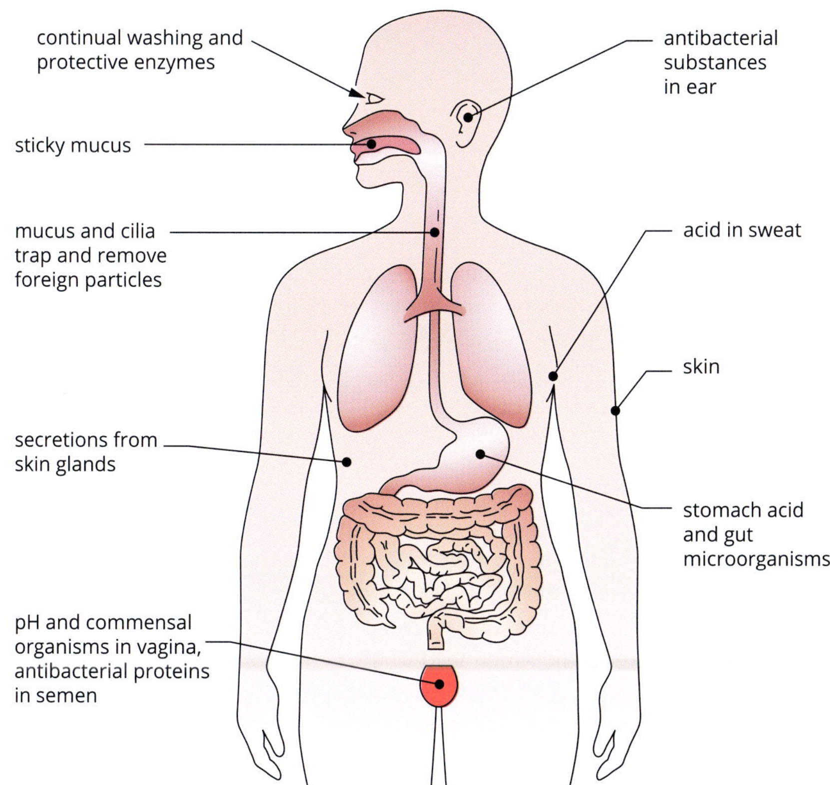

FIGURE 7.2.4 Some of the physical and chemical defence mechanisms that prevent foreign organisms from gaining access to the human body.

Microbiological barriers in animals

Non-pathogenic bacteria, referred to as normal flora, are found on the skin, and in the mouth, nose, throat, lower part of the gastrointestinal tract and the urogenital tract in healthy individuals. The presence of normal flora prevents the growth and colonisation of other bacteria because normal flora competes with other bacteria for space and resources, and produces chemicals that reduce the pH of the micro-environment.

Taking a course of antibiotics can disrupt the normal flora of the intestine, as the antibiotics do not discriminate between beneficial normal flora and harmful bacteria. This can disturb the normal gut function and predispose a person to various infections until the normal flora return to their pre-treatment levels.

Although not a problem in healthy individuals, in people with weakened immune systems normal flora can sometimes grow unchecked and cause disease.

THE INNATE IMMUNE RESPONSE

If pathogens manage to breach the barriers that act as a first line of defence, they are immediately met by attacking cells and molecules.

The innate immune response is found in all organisms, and its persistence over millions of years of evolution indicates its fundamental importance. Indeed, even when the innate immune response is unable to eliminate a pathogen, it remains critical for keeping infections under control until the adaptive immune response (see Section 7.3), which can take up to several days to develop, kicks in.

Innate immune responses in vertebrates:

- are non-specific—they do not target a specific antigen
- are rapid—they occur within hours
- are present in all animals
- are fixed responses—they do not adapt
- do not lead to an immunological memory of the pathogen that caused the infection.

Cells of the innate immune response

White blood cells (or **leukocytes**) are immune cells that are present in blood and other tissues. Leukocytes have pattern recognition molecules, also known as toll-like receptors (TLRs) on their surface, which are able to recognise microbial molecules called pathogen-associated molecular patterns (PAMPs). There are different TLRs that recognise different PAMPs. For example, TLR-2 recognises the lipoproteins and peptidoglycan of Gram-positive bacteria, TLR-4 recognises lipopolysaccharide of Gram-negative bacteria, TLR-5 recognises bacterial flagellin, and TLR-7 and 8 recognise single-stranded viral RNA.

PAMPs are common to a range of pathogens, meaning that the innate immune response to them is not specific to a particular pathogen. In contrast, the adaptive immune response is able to target a specific pathogen by the specific antigens it expresses.

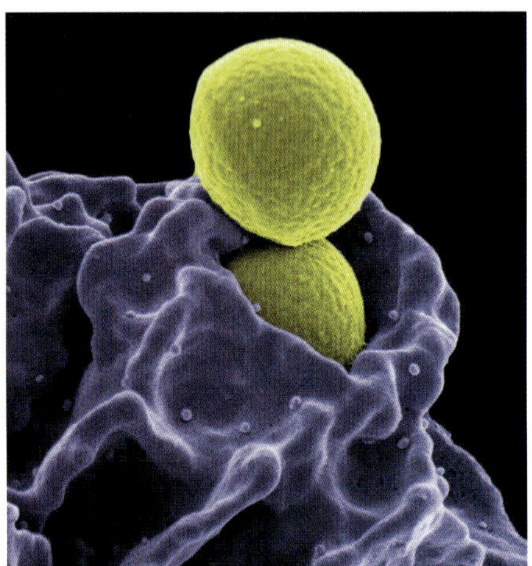

FIGURE 7.2.5 Phagocytes are a key part of the innate immune response. This neutrophil (purple) is engulfing *Staphylococcus aureus* bacteria (yellow), which it will then phagocytose.

Exocytosis is the release of substances enclosed within a vesicle to the outside of a cell. It occurs by fusion of the vesicle with the plasma membrane.

Phagocytes

Phagocytes are leukocytes that are able to engulf and break down pathogens in a process known as phagocytosis (Figure 7.2.5). Phagocytes include neutrophils, macrophages, monocytes and dendritic cells.

TLRs on a phagocyte interact with a microbe's PAMPs, causing signal transduction events to occur, which lead to the activation of the phagocyte. Once activated, the phagocyte engulfs the microbe, with the plasma membrane forming a vacuole called a **phagosome** around it. Then a lysosome containing digestive enzymes fuses with the phagosome, forming a phagolysosome, which breaks down the foreign material. The fragments can then be expelled from the cell by exocytosis.

Some phagocytes, namely macrophages and dendritic cells, also act as **antigen-presenting cells** (APCs). When APCs phagocytose a pathogen, fragments of digested antigen are linked to MHC-I proteins and displayed (or presented) on the surface of the membrane (Figure 7.2.6).

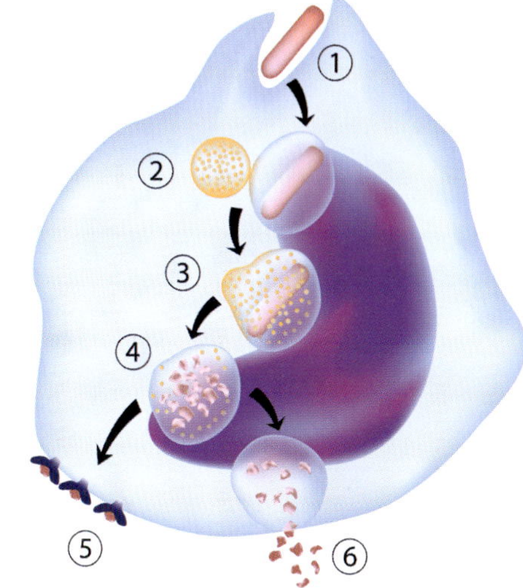

1 engulfing of foreign material
2 fusion of lysosome and phagosome
3 enzymes start to degrade foreign material
4 foreign material broken into small fragments
5 antigen fragments are bound to MHC-II and presented on the APC surface
6 leftover fragments released by exocytosis

FIGURE 7.2.6 Phagocytosis and antigen presentation in an antigen-presenting cell (APC). APCs communicate with other immune cells by presenting antigens or fragments of antigens on the cell surface.

Antigen presentation links the innate and adaptive immune responses

As you learnt in Section 7.1, there are major histocompatibility (MHC) proteins on the surface of your body's cells, which present self or non-self antigens to T lymphocytes. There are different classes of MHC proteins, including MHC class I and class II, which are both involved in antigen presentation.

MHC class I (MHC-I) proteins are normally found on all nucleated cells, and present peptide antigens derived from the proteins of pathogens in the cytoplasm of non-phagocytic cells to **cytotoxic T lymphocytes** in the adaptive immune response. You will learn more about T lymphocytes and their role in the adaptive immune response in Section 7.3. MHC-I proteins are also important in the innate immune response, as their absence allows natural killer (NK) cells to identify and destroy infected or damaged cells (Figure 7.2.7).

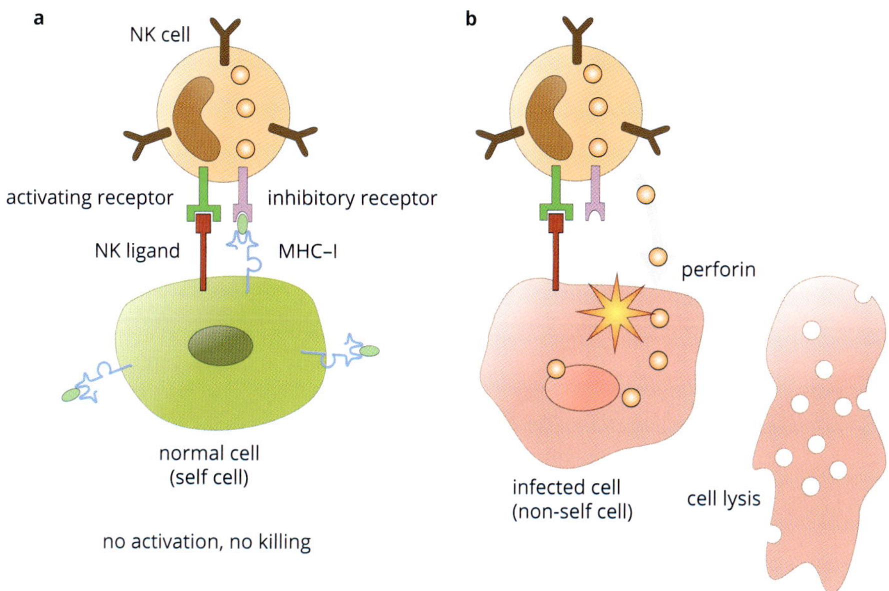

FIGURE 7.2.7 The action of natural killer (NK) cells. (a) The NK cell recognises a normal host cell by the presence of MHC-I and does not elicit an attack. (b) MHC-I is absent from the host cell's surface and the NK cell recognises that the cell is infected or damaged. The NK cell then elicits a response to destroy the infected cell.

MHC class II (MHC-II) proteins can be conditionally expressed on all cells, but are most commonly found on the surface of APCs such as dendritic cells, macrophages and B lymphocytes. This presentation of antigens activates the **helper T lymphocytes** of the adaptive immune response, linking the innate and adaptive immune responses (Figure 7.2.8).

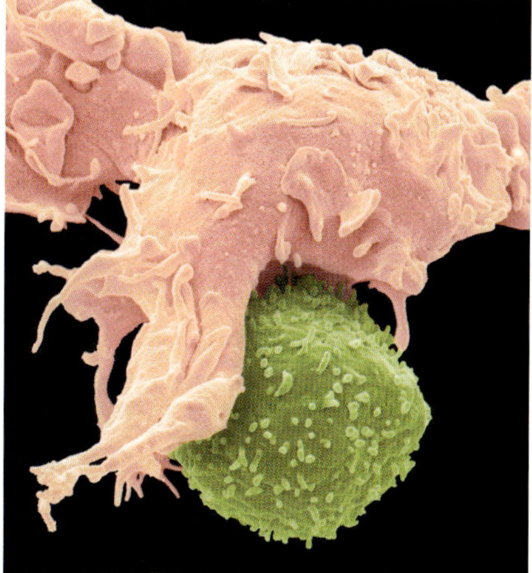

FIGURE 7.2.8 A coloured scanning electron micrograph showing the interaction between a macrophage (pink) and a helper T lymphocyte (yellow).

Summary of innate immune cells

Table 7.2.2 shows some of the leukocytes involved in the innate immune responses and indicates whether they are involved in phagocytosis, antigen presentation, or the release of cytokines that promote inflammation. You will learn more about cytokines later in this section.

Cell type	Function	Cell type	Function
neutrophil (granulocyte)	• phagocytosis • release antimicrobial compounds, such as **defensins** and hydrogen peroxide, that disrupt bacterial and fungal membranes • release cytokines that attract other immune cells and cause inflammation	basophil (granulocyte)	• release **histamine**, which contributes to inflammation and therefore blood vessel dilation • have a limited role in phagocytosis
macrophage	• phagocytosis • antigen presentation • release of cytokines	eosinophil (granulocyte)	• antigen presentation • release cytokines and cytotoxic chemicals • have a limited role in phagocytosis Eosinophils are found in high numbers in parasitic infections.
monocyte	• phagocytosis	**mast cell** (granulocyte)	• play a key role in inflammation, and therefore blood vessel dilation, by releasing histamines • have a limited role in phagocytosis Mast cells stained red/pink with haematoxylin and eosin stain.
dendritic cell	• phagocytosis • antigen presentation Dendritic cells have many grooves that increase their surface area and permit contact with a large number of nearby cells.	natural killer (NK)	• recognise virus-infected and cancerous cells • release cytotoxic chemicals from granules, such as perforin, which punches holes in cell membranes, triggering apoptosis and cell death of virus-infected cells and abnormal cells • release cytokines to attract and activate cells of the adaptive immune system Immunofluorescent light micrograph of natural killer cells; cytotoxin granules (green), nuclei (blue), cytoplasm (red).

TABLE 7.2.2 Some of the leukocytes involved in innate immune responses and their function.

DEFENSIVE MOLECULES

Complement proteins and cytokines are defensive molecules involved in both the innate and adaptive immune responses. (You will learn more about adaptive immune responses in Section 7.3.)

Complement proteins

The **complement proteins** are an array of more than 30 proteins that circulate in the blood and are able to help kill foreign cells. They are found in body fluids in an inactive form, and are activated as part of the non-specific immune response to certain antigens and carbohydrates on the surfaces of some bacteria and parasites.

Activation of complement proteins results in an enzyme-triggered reaction that leads to the **lysis** of the invading pathogens. For example, complement proteins destroy bacteria directly by punching holes in their membranes, causing them to lyse. The release of the bacterial contents attracts phagocytes to the site of infection. Complement proteins activated by antigen–antibody complexes are also involved in specific (adaptive) immune responses.

Complement proteins and cytokines are involved in both the innate and the adaptive immune responses.

Cytokines

Cytokines are small signalling molecules of the immune system and coordinate many aspects of our immune responses. Cytokines can be peptides, proteins or glycoproteins, and are released by body cells in response to cell damage or the presence of pathogens. There are many different cytokines and they trigger a variety of responses, both non-specific and specific. For example, cytokines can promote the proliferation of lymphocytes, induce inflammation and fever, promote antibody responses and activate macrophages. Interferons and chemokines are two different types of cytokines, and they each have different functions.

Interferons

Interferons are a class of cytokines that are produced by, and act on, a host cell infected by a virus. Interferons act in an autocrine manner, activating the infected cells to produce enzymes that break down viral RNA and proteins that block translation. This limits viral replication and release from the cell. Interferons also attract NK cells, which release cytotoxic peptides to kill the virus-infected cell.

Interferons are non-specific and will act against any virus. However, viruses vary widely in their susceptibility to interferons. Many viruses can evade interferon-induced defences and the more virulent viruses may be able to inhibit the production of interferon.

Autocrine refers to a substance secreted by a cell that also has an effect on that cell.

Although an important defence against viruses, interferons also play a smaller role in combating bacterial and parasitic infections. Interferons also regulate the immune system in a number of ways, such as enhancing T lymphocyte activity.

Chemokines

Chemokines are a type of cytokine and act as chemical attractants (or chemo-attractants). Chemokines are important for attracting leukocytes to sites of infection and inflammation.

THE INFLAMMATORY RESPONSE

Inflammation is the accumulation of fluid, plasma proteins and leukocytes that occurs when tissue is damaged or infected, and results in heat, pain, swelling, redness and loss of function. The interaction between leukocytes (especially phagocytes) and pathogens triggers the inflammatory response that results from the production, activation or release of peptides and proteins such as complement proteins and cytokines. Acute inflammation involves phagocytes, and occurs soon after infection as part of the innate immune response, but inflammation can also involve lymphocytes and occur later as part of the adaptive immune response.

A number of steps are involved in the initiation of an inflammatory response to infection (Figure 7.2.9):

1 Bacteria or other pathogens breach the barriers that provide a first line of defence, such as through an open cut or wound in the skin.
2 Injured cells release cytokines (chemokines) that attract neutrophils, and mast cells release histamine, which increases blood vessel dilation and permeability. The dilated, more permeable blood vessels allow leukocytes and fluid containing peptides and proteins such as complement proteins to enter the infected tissue. Platelets release clotting factors at the site of the wound.
3 Neutrophils migrate towards the cytokines and are activated, causing the neutrophils to recruit macrophages and secrete factors, such as defensins and hydrogen peroxide, that degrade and kill pathogens.
4 Macrophages in turn become activated and secrete cytokines and, along with neutrophils, phagocytose pathogens and debris at the site of infection. This may lead to pus, which is fluid containing leukocytes, dead pathogens and cell debris.
5 The inflammatory response continues until the pathogen is eliminated and the wound has healed.

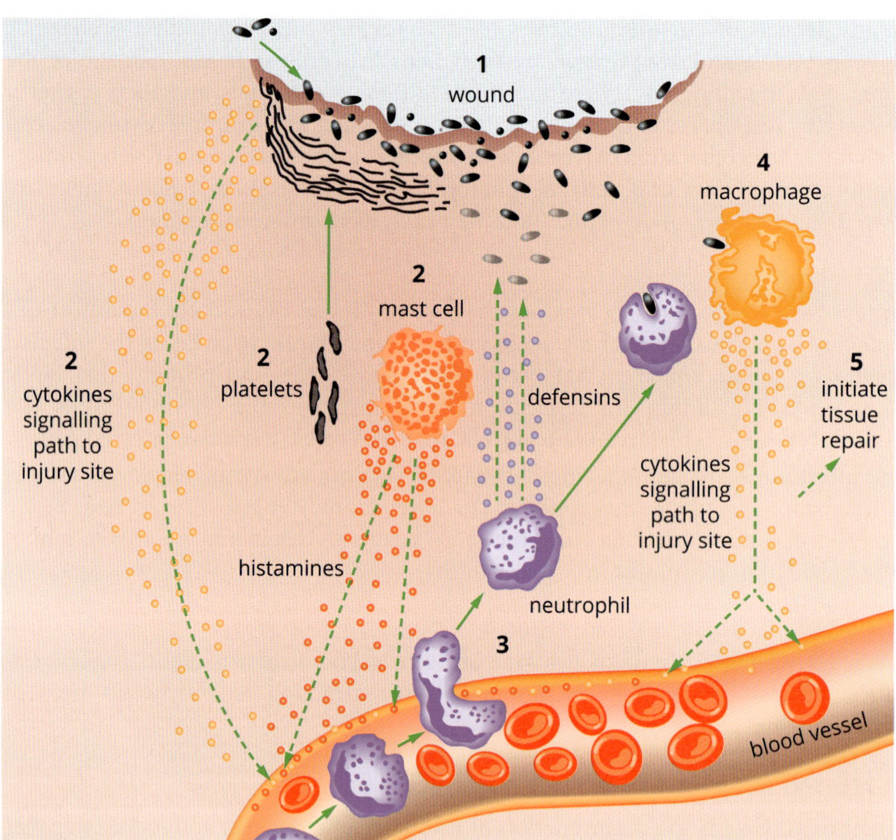

FIGURE 7.2.9 The process of inflammation (cell colours for illustrative purposes only).

Fever

A **fever** is an increase in body temperature that results when the regulated body temperature set point in the hypothalamus of the brain is set to a higher level by inflammatory cytokines. In humans, normal body temperature is around 37 °C. Fever occurs when body temperature is above normal.

Fever slows the replication of bacteria and viruses by shifting the temperature away from their optimal range, and so allows more time for other defences to intervene. Additionally, moderate increases in temperature increase the activity and proliferation of leukocytes, so fever also improves the immune response.

BIOFILE

Plants fight back

Unlike mammals, plants don't have an adaptive immune system, and they lack mobile immune cells that can travel to the site of infection, so every plant cell has to respond to pathogens independently.

Plants have an innate immune system and secrete a number of defensive molecules (Figure 7.2.10). For example, certain plants produce considerable amounts of the insect moulting hormone ecdysone. When parasitic insect larvae eat the leaves, their hormonal balance can be fatally disrupted, with obvious advantages to the plant.

FIGURE 7.2.10 Bread wheat, *Triticum aestivum*, contains small cysteine-rich proteins that act as plant defensins to inhibit the growth of bacteria and fungi.

Eucalypts are under heavy and continuous attack from a vast array of organisms that suck, chew or nibble. Eucalypt leaves contain high levels of toxic compounds, but some insect attackers have adapted to become resistant to these chemicals. Some species of acacias deter insect grazers by producing cyanogenic glycosides (cyanide-generating chemicals that are cytotoxic), killing cells by blocking enzymes involved in cellular respiration.

When these defensive molecules fail to prevent pathogens from infecting the plant, cell-mediated defences can involve self-destruction of infected or damaged cells, which helps to limit a pathogen's access to nutrients and, in turn, limit the pathogen's ability to spread to the rest of the plant. Many plants also produce a hypersensitive response when invaded by parasites such as nematode larvae or bacteria. This response involves the rapid breakdown of cells around the parasite and the release of toxic substances.

7.2 Review

SUMMARY

- Barriers that provide innate resistance to infection include physical, chemical and microbiological barriers.
- Innate immune responses occur when these barriers are breached.
- Innate immune responses are:
  - non-specific—they do not target a specific antigen
  - rapid—they occur within hours
  - present in all animals
  - fixed responses—they do not adapt.
- Innate immune responses do not lead to an immunological memory.
- Leukocytes have pattern recognition molecules called toll-like receptors (TLRs) on their surface, which are able to recognise pathogen-associated molecular patterns (PAMPs).
- Phagocytes are leukocytes that are able to engulf and break down pathogens in a process known as phagocytosis.
- Some phagocytes also act as antigen-presenting cells.
- Defensive molecules include complement proteins and cytokines:
  - Activation of complement proteins results in an enzyme-triggered reaction that leads to the lysis of the invading pathogens.
  - Cytokines are small signalling molecules of the immune system and coordinate many aspects of our immune responses.
- Cytokines include interferon and chemokines:
  - Interferons are produced by virus-infected cells and inhibit viral replication by resulting in the transcription of antiviral genes and expression of antiviral proteins.
  - Chemokines attract white blood cells to the site of infection.
- Inflammation is the accumulation of fluid, plasma proteins and leukocytes that occurs when tissue is damaged or infected. It results in heat, pain, swelling, redness and loss of function.
- Fever is an increase in body temperature that results from the regulated body temperature set point in the hypothalamus of the brain being set to a higher level by inflammatory cytokines. It slows the replication of bacteria and viruses by shifting the temperature away from their optimal range, and improves the immune response by increasing the activity and proliferation of leukocytes.

KEY QUESTIONS

1 Innate immune responses are:
  A specific and delayed
  B non-specific and rapid
  C non-specific and delayed
  D specific and rapid

2 Phagocytes are important cells of the innate immune system. What is the function of some phagocytes that links the innate and adaptive immune systems?

3 The diagram at right shows a natural killer (NK) cell attacking an infected cell. What is the missing inhibitory receptor that indicates this cell is an infected 'non-self' cell and not a healthy 'self' cell?

4 Draw the process of inflammation that occurs when bacteria enter the skin through an open wound. Be sure to label key white blood cells and the molecules they produce in response to the pathogen, and to number the steps involved in the inflammatory response.

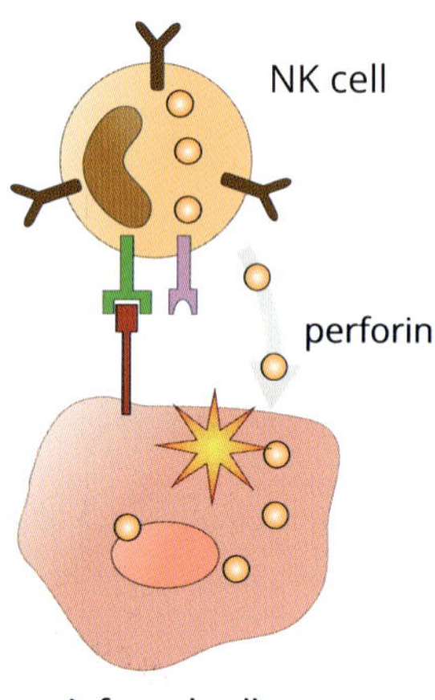

7.3 Adaptive immunity

In Section 7.2, you learnt that if a vertebrate's first line defences are breached by a pathogen, that pathogen is met with a non-specific innate immune response. However, this innate immune response may or may not be successful in eliminating the invader. Fortunately, vertebrates also have **adaptive immunity**, as they have evolved an additional adaptive immune response to pathogens. In this section, you will learn about the adaptive immune response.

FIGURE 7.3.1 B lymphocytes produce antibodies for a specific antigen.

THE NATURE OF THE ADAPTIVE IMMUNE RESPONSE

There are two distinguishing features of the adaptive immune response:

- **specificity**—the ability to recognise and respond exclusively to specific antigens (Figure 7.3.1). On recognising a specific foreign antigen on a pathogen, cells of the adaptive immune system trigger an array of defensive mechanisms that destroy the pathogen.
- **immunological memory**—the ability of cells of the adaptive immune system to 'remember' antigens after primary exposure, and to mount a larger and more rapid response when exposed to the same antigen again.

Lymphocytes: cells of the adaptive immune response

The cells that are crucial to the adaptive immune response are **lymphocytes**. Each lymphocyte has a different receptor for a particular antigen, and is able to proliferate, creating clones of the initial lymphocyte with the specific receptor for the antigen. This is called **clonal selection**.

Lymphocytes are classified as either B lymphocytes or T lymphocytes according to their interaction with the antigen and their response to it. The B and T lymphocyte sub-populations have distinct roles but are both key to the adaptive immune response. Lymphocytes travel through the lymphatic system and become activated when they encounter antigens specific to their receptors. You will learn more about the lymphatic system in Section 7.4.

Lymphocytes are a type of white blood cell (or leukocyte) that are specialised for adaptive immune responses.

B lymphocytes develop in the bone marrow and complete their maturation in the peripheral lymphoid organs and tissues.

T lymphocytes develop in the bone marrow and mature in the thymus.

Mechanisms of adaptive immune responses

There are two mechanisms of adaptive immunity (Figure 7.3.2):

- **humoral immunity**, in which macromolecules, such as complement proteins, and antibodies produced by B lymphocytes, are secreted into the extracellular fluid
- **cell-mediated immunity**, which involves the action of T lymphocytes (which you will learn more about shortly) and phagocytes.

a Humoral immunity B lymphocytes

b Cell-mediated immunity T lymphocytes

FIGURE 7.3.2 (a) B lymphocytes are involved in humoral or antibody-mediated immunity. They originate in the bone marrow and mature in the peripheral lymphoid organs and tissues. (b) T lymphocytes are involved in cell-mediated immunity. They originate in bone marrow and mature in the thymus. Except for plasma cells, the different types of lymphocytes look very similar under a microscope. The only way to know which is which is to identify their different surface proteins.

EXTENSION

Clonal selection theory

An almost infinite number of different antigens exist, and the immune system is able to produce lymphocytes specific to each antigen upon exposure. The clonal selection theory is a scientific theory that explains how lymphocytes are able to produce a large number of antibodies specific to an antigen.

When B and T lymphocytes form, each has a receptor that will react to a single antigen. The clonal selection theory states that a specific antigen will only activate a lymphocyte with a receptor that specifically recognises that antigen. Once activated, this lymphocyte will proliferate into a clone of millions of effector cells dedicated to eliminating the specific antigen that stimulated the immune response (Figure 7.3.3).

Clonal selection theory was developed in 1957 by Sir Frank Macfarlane Burnet, one of Australia's most celebrated scientists.

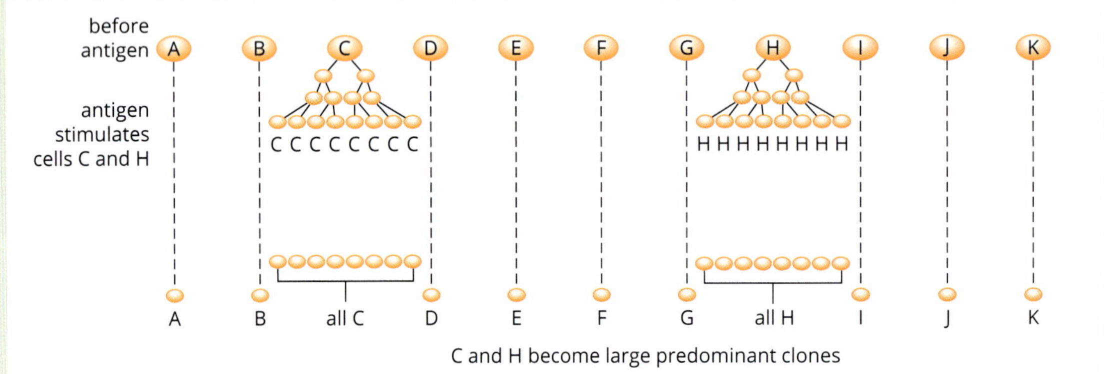

FIGURE 7.3.3 The clonal selection theory and expansion. Lymphocytes that encounter or interact with an antigen, such as lymphocytes C and H, begin to proliferate. This increases the number of lymphocytes with identical receptors, or clones, for the specific antigen that was first encountered.

HUMORAL IMMUNITY

Humoral immunity involves B lymphocytes, which produce specific antibodies against foreign antigens and release them into the blood and lymph (Figure 7.3.4). In medieval times, the term 'humor' referred to body fluids.

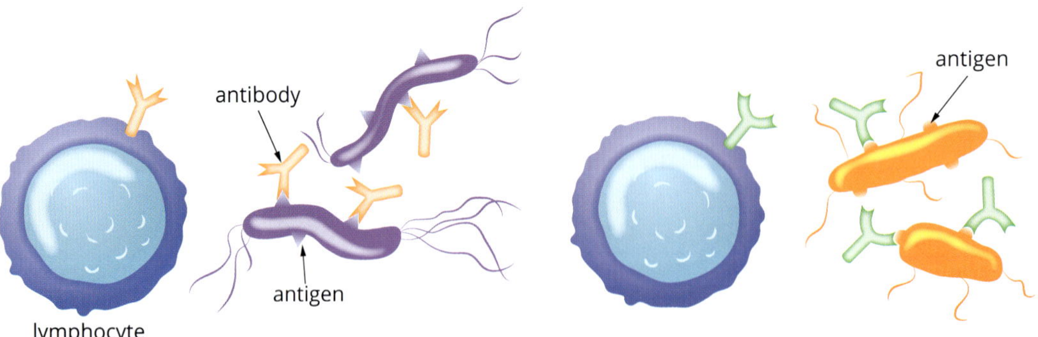

FIGURE 7.3.4 Antibodies specific to a foreign antigen will bind to it, helping to eliminate the invading pathogen.

B lymphocytes

B lymphocytes originate and commence **differentiation** in the bone marrow and complete their maturation in the peripheral lymphoid organs and tissues. At any time there are billions of B lymphocytes circulating in the blood.

When a B lymphocyte meets and binds to a specific antigen, the B lymphocyte can be triggered (or activated) to differentiate and divide (or proliferate). Cytokines released by helper T lymphocytes are also important for helping to activate B lymphocytes. When B lymphocytes are activated, they divide and further differentiate into two types of daughter cells:

- plasma cells
- memory B lymphocytes.

Plasma cells

Activation of B lymphocytes leads to the production of **plasma cells**, which are essentially 'factories' specialising in antibody production (Figure 7.3.5). The antibodies produced are specific to the antigen that activated the B lymphocyte. Plasma cells can produce thousands of antibodies per second.

Memory B lymphocytes

Memory B lymphocytes remain in lymphoid tissues for long periods (even for the lifetime of the animal) and are responsible for the immunity that often follows infection or vaccination. These cells can divide and give rise to plasma cells if secondary exposure to the antigen occurs (Figure 7.3.5).

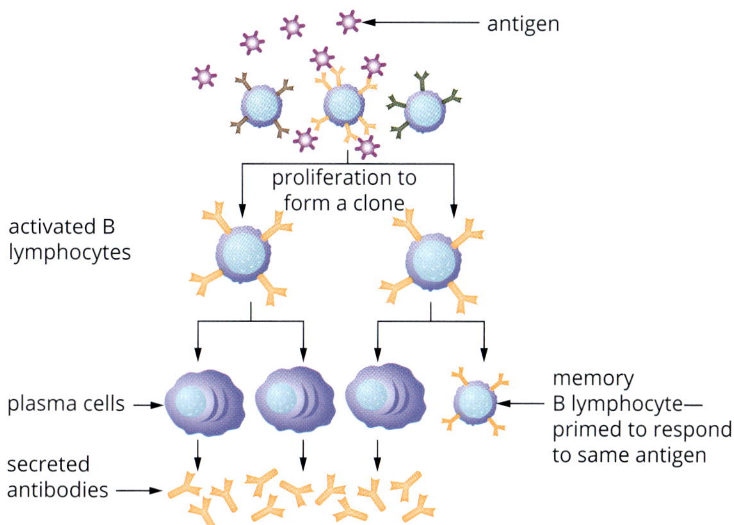

FIGURE 7.3.5 Many B lymphocytes differentiate into plasma cells, which produce and secrete antibodies for immune protection. Others become memory B lymphocytes and are retained in lymph nodes. Helper T lymphocytes are often involved in activating B lymphocytes to produce antibodies.

Antibodies

Antibodies, also known as **immunoglobulins (Ig)**, are produced by B lymphocytes and released into the blood and lymph. Antibodies are proteins that bind to specific antigen molecules.

The basic unit of an antibody molecule is a Y-shaped protein, formed by four polypeptide chains: two long **heavy chains**, and two short **light chains** (Figure 7.3.6). The amino acid sequences that form the top of the 'arms' of the Y-shaped antibody are known as the **variable regions**. It is the variation of these variable regions that allows antibodies to bind to different antigens. The two variable regions are identical antigen-binding sites and attach to identical antigens. The single 'stem' of the Y-shaped antibody is a conserved sequence in all antibodies and is called the **constant region**. The constant region recruits other components of the immune system.

Mature lymphocytes that have not been activated by antigen are said to be 'naive'.

Activated B lymphocytes divide to form antibody-secreting plasma cells or memory B lymphocytes.

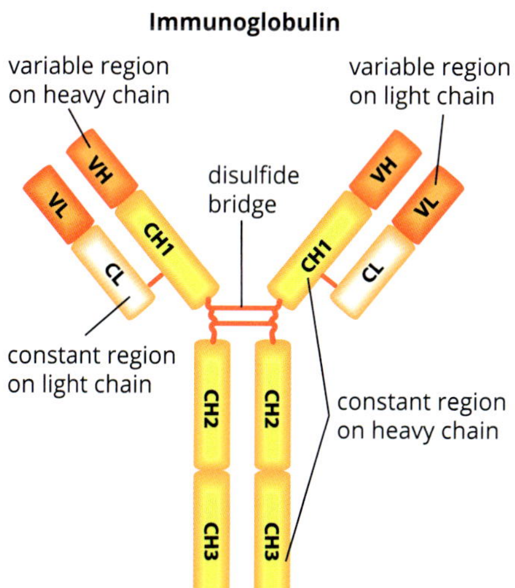

FIGURE 7.3.6 The structure of a basic unit of an antibody. Antibodies have two long heavy (H) chains and two short light (L) chains. Both heavy and light chains have a variable (V) and constant (C) region. Naturally-produced antibody consists of two identical variable regions that are specific for a particular antigen. The constant region is capable of binding to and initiating other immune components, such as the complement proteins.

Antibodies may act singly (monomers), in pairs (dimers) or in groups of five (pentamers). Mammals have five main classes of antibody molecules with different structures and functions (Table 7.3.1).

Class	Half-life in serum	Presence	Functions	Structure
IgG	21 days	blood, lymph and extracellular fluid; most circulating antibodies (>80%); crosses placenta	agglutination, complement activation	
IgM	10 days	blood and lymph; produced early in infection response	agglutination, complement activation	disulfide bond joining chain
IgA	6 days	found in secretions such as tears, saliva and milk	mucosal immunity	joining chain secretory protein
IgD	3 days	blood and lymph; mostly present on B lymphocyte surfaces; small amount in circulation; binds to basophils and mast cells	functions not well understood; possible role in regulating innate immune responses	
IgE	2 days	blood and lymph; attaches to mast cells	involved in allergic reactions	

TABLE 7.3.1 Structure and function of mammalian immunoglobulins.

BIOFILE

Agglutination of red blood cells

If a person is given a blood transfusion with the wrong blood type, antibodies will recognise the transfused blood cells as foreign and will bind to their antigens. This causes clumping (or agglutination) of red blood cells (Figure 7.3.7). Agglutination destroys the red blood cells, which transport oxygen throughout the body, and so can result in severe anaemia and even death.

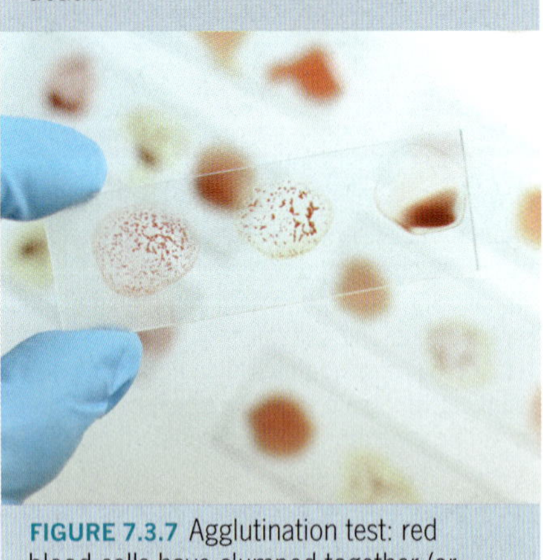

FIGURE 7.3.7 Agglutination test: red blood cells have clumped together (or agglutinated) in these drops of blood on a microscope slide.

Antibody function

Antibodies do not directly destroy pathogens, but carry out several important mechanisms to interfere with the function of the pathogen (Figure 7.3.8):

- **neutralisation** of bacterial toxins: Antibodies bind to bacterial toxins, blocking the action of the toxin.
- neutralisation of pathogens: Antibodies bind to antigens on the surface of the pathogen, which are required for entry into host cells, thus preventing pathogen invasion of host cells.
- **agglutination**: Antibodies bind to antigens on the surface of cells and form **antigen–antibody complexes**, which activate phagocytes and the complement cascade, leading to antigen/cell destruction.
- **precipitation**: Antibodies bind to soluble antigens, causing them to become insoluble and precipitate out of solution.

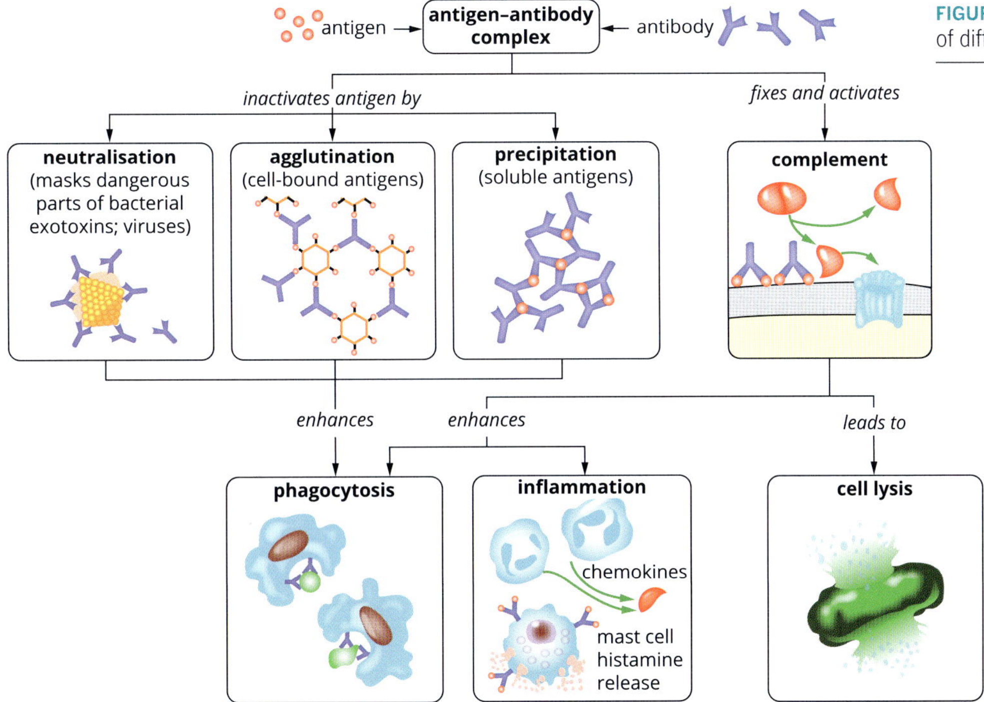

FIGURE 7.3.8 Antibodies function in a number of different ways to help eliminate pathogens.

CELL-MEDIATED IMMUNITY

Unlike humoral immunity, which involves B lymphocytes, cell-mediated immunity is regulated by T lymphocytes (Figure 7.3.9). The response is mediated by the **T cell receptors** (TCR).

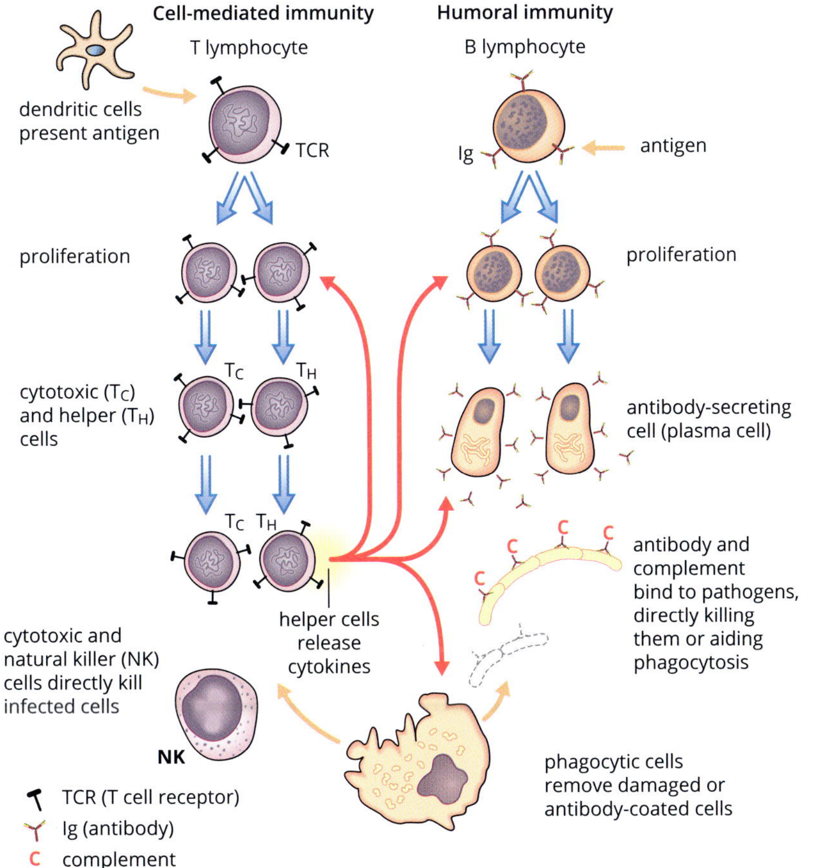

FIGURE 7.3.9 Summary of cell-mediated immunity and humoral responses.

T lymphocytes

Depending on their function, T lymphocytes are classified as helper, cytotoxic or memory T lymphocytes.

Helper T lymphocytes

Helper T lymphocytes do not directly kill pathogens, rather, as their name suggests, they 'help' with immune responses. To assist the immune response, helper T lymphocytes secrete cytokines that promote inflammation, and activate macrophages and B lymphocytes. As mentioned earlier, the activated B lymphocytes go on to become antibody-secreting plasma cells or memory B lymphocytes.

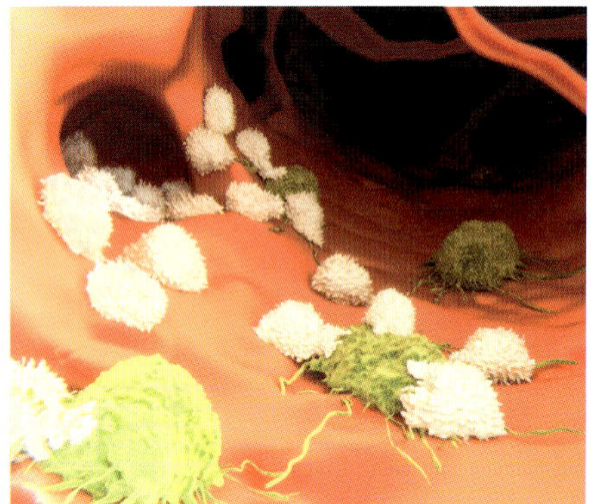

FIGURE 7.3.10 Digital illustration of cytotoxic (white) T lymphocytes attacking migrating cancer cells (yellow).

Cytotoxic T lymphocytes

Cytotoxic T lymphocytes recognise and kill foreign, infected or abnormal host cells by releasing toxic compounds. This includes virus-infected host cells, cancer cells and foreign cells such as those in transplanted tissue (Figure 7.3.10).

Memory T lymphocytes

Memory T lymphocytes are produced after helper and cytotoxic T lymphocytes have been activated during an infection. Once activated, these lymphocytes differentiate into memory T lymphocytes that are antigen-specific. The memory T lymphocytes persist after the infection is resolved, to ensure a prompt response should the same pathogen reinfect the organism.

T cell receptors

T cell receptors (TCRs) are central to the function of T lymphocytes in the adaptive immune response (remember that T lymphocytes are also known as T cells). TCRs are made up of two polypeptide chains. Like antibodies, TCRs have a variable and constant region (Figure 7.3.11). Unlike antibodies, which have two antigen-binding sites, TCRs have only one antigen-binding site.

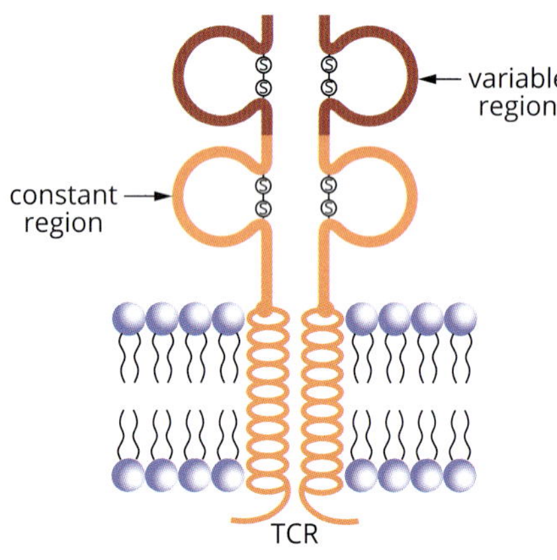

FIGURE 7.3.11 The structure of the T cell receptor (TCR), which is found on helper T and cytotoxic T lymphocytes and binds to fragments of antigen.

TCRs do not bind to antigens on pathogens, as B lymphocyte receptors do; instead, they bind to fragments of antigens that are displayed or presented on the surface of APCs. Receptor binding triggers signal transduction in the T lymphocyte, resulting in proliferation, cytokine release and activation of cytotoxic function.

ANTIGEN RECOGNITION BY T LYMPHOCYTES

T lymphocytes check the antigens of cells they come into contact with in the body, differentiating between cells that belong to the organism (self) and cells that are foreign (non-self). Remember that during their development, lymphocytes that react to self-antigens are normally destroyed. This inability of lymphocytes to respond to self-antigens is known as self-tolerance.

All nucleated cells have surface proteins that present peptide antigens of the proteins being synthesised in that cell. These antigens are presented to cytotoxic T lymphocytes by major histocompatibility complex I (MHC-I) molecules. Antigen-presenting cells (APC) are specialised for presenting antigens. When an APC engulfs a pathogen, the antigens of the pathogen are broken into small peptides in the cell. These antigen fragments bind to MHC-II molecules inside the cell. The antigen–MHC-II complexes then move to the cell surface to present the antigens to helper T lymphocytes. The TCRs on the helper T lymphocytes recognise the antigen–MHC-II complex. Signal transduction in the T lymphocyte leads to activation of the cell, which then proliferates and releases cytokines (Figure 7.3.12).

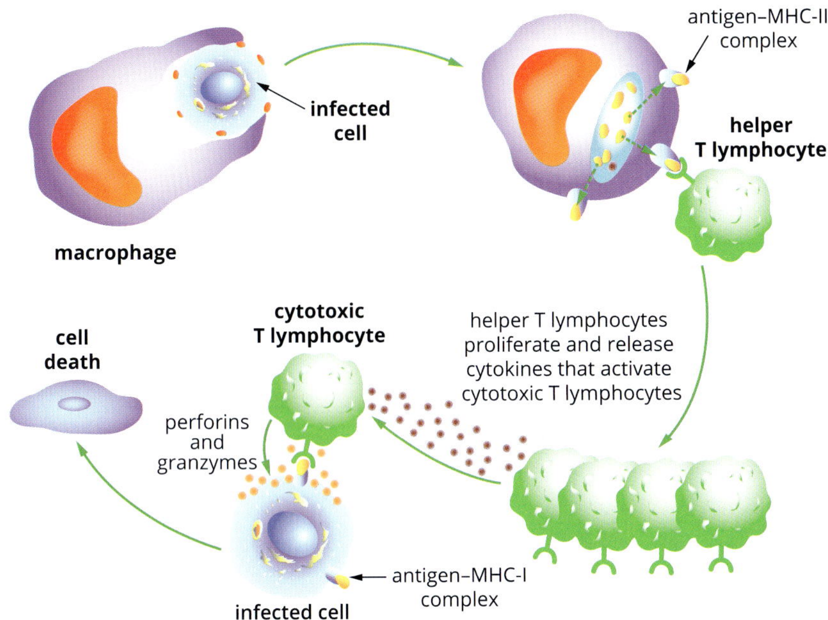

FIGURE 7.3.12 T lymphocytes are activated in cell-mediated immunity. Infected cells are detected and destroyed by cytotoxic T lymphocytes.

IMMUNOLOGICAL MEMORY

The response arising from the first encounter of a T or B lymphocyte with a specific antigen is known as the **primary immune response** (Figure 7.3.13). After the initial exposure, B and T lymphocytes form B and T memory lymphocytes. IgM antibodies are the predominant antibodies produced in a primary response.

The response arising from subsequent encounters with the same antigen is known as the **secondary immune response**. Lymphocyte proliferation and production of antibodies occurs much more quickly during the secondary immune response, because the existing memory lymphocytes, which were produced during the first encounter and which remain for months or years, allow faster proliferation of the required lymphocytes (those with the receptor specific to the antigen). IgG antibodies are the predominant antibodies produced in the secondary response.

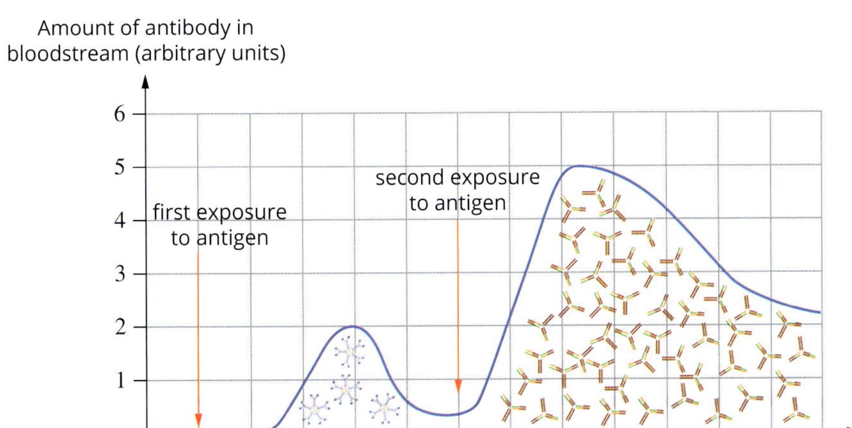

FIGURE 7.3.13 Primary and secondary immune responses after initial and secondary exposure to the same antigen.

EXTENSION

The major histocompatibility complex (MHC)

The major histocompatibility complex (MHC) is a gene region that codes for MHC proteins. Two classes of MHC proteins that are important in antigen presentation are MHC class I (MHC-I) and MHC class II (MHC-II). The two classes differ in structure and function, as they are specialised to present different types of antigens, and so elicit different responses.

MHC-I proteins communicate with cytotoxic T lymphocytes about the proteins being produced within each cell. MHC-I is present on almost every cell in the body, presenting self-antigens and other proteins produced intracellularly, including non-self antigens such as viral antigens produced by virus-infected cells.

If a cytotoxic T lymphocyte reacts to the antigen being presented, it becomes activated and releases toxic peptides (perforin and granzyme) that damage the target cell membrane and induce apoptosis (Figure 7.3.14).

MHC-II is usually only present on specialised antigen-presenting cells such as macrophages and dendritic cells. MHC-II presents antigens that originated extracellularly and have been processed by phagocytosis. Once the antigens are presented, helper T lymphocytes are activated, and secrete cytokines that activate and attract other immune cells to the site of infection (Figure 7.3.15).

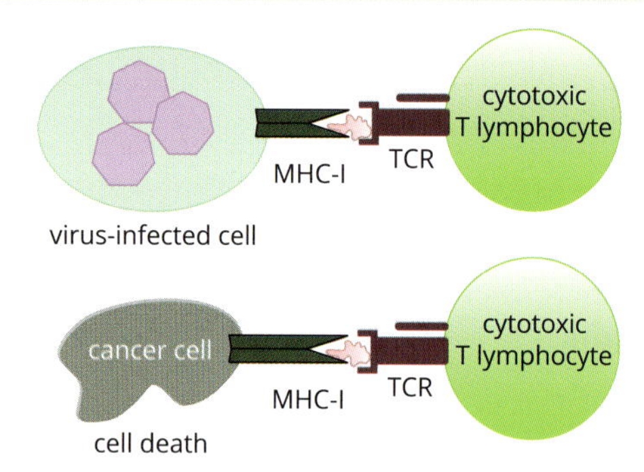

FIGURE 7.3.14 Infected or cancerous cells present antigens on MHC-I to the T cell receptors (TCRs) of cytotoxic T lymphocytes. Cytotoxic T lymphocytes then release cytotoxins that kill the virus-infected cells and cancer cells.

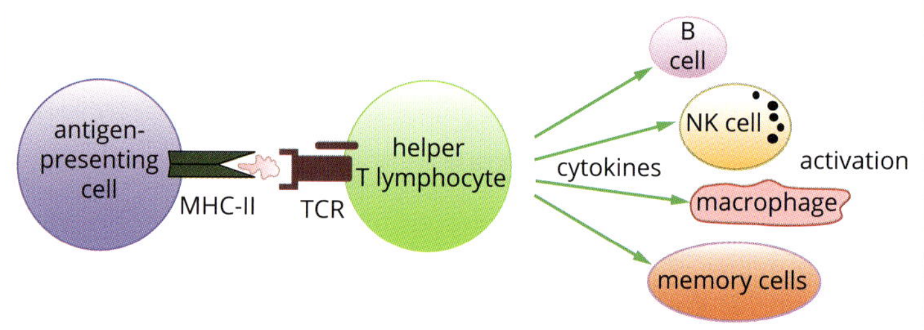

FIGURE 7.3.15 Antigen-presenting cells present antigens on MHC-II to helper T lymphocytes via the T cell receptor (TCR). The CD4 co-receptor assists recognition and signalling. Helper T lymphocytes release cytokines to help activate a range of immune cells.

Transplant rejection

There are several hundred different MHC genes coding for MHC proteins in humans. These different genes result in different amino acid sequences of MHC proteins. T lymphocytes that would trigger an adaptive immune response to your body's own MHC proteins are destroyed in the thymus, but T lymphocytes that respond to non-self MHC proteins remain, and will respond to non-self MHC proteins if exposed to them, as occurs when transplanted donor cells do not have an MHC match with the recipient (Figure 7.3.16a).

However, even when the donor and recipient's MHC is matched, transplant rejection can still occur, because donor (non-self) antigens are still presented by MHC-I on donor cells (Figure 7.3.16b), and the transplant recipient's own antigen-presenting cells can also process donor proteins and present their antigens to T lymphocytes on MHC-II molecules (Figure 7.3.16c).

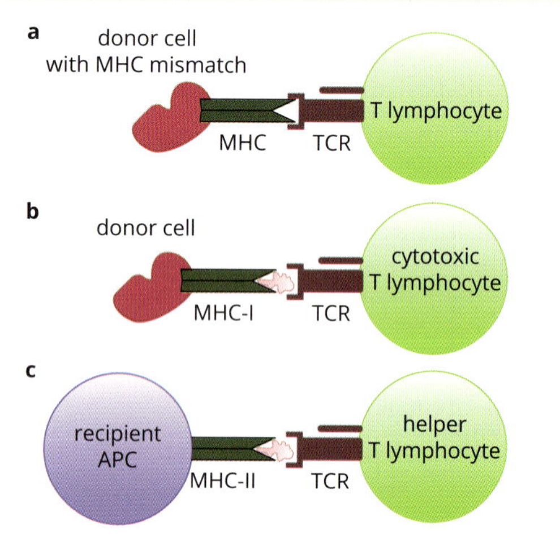

FIGURE 7.3.16 T lymphocyte activation and transplant rejection can be triggered by (a) transplanted donor cells that do not have an MHC match with the recipient, (b) donor cells presenting non-self antigens on MHC-I to cytotoxic T lymphocytes, and (c) recipient antigen-presenting cells (APC) presenting processed donor (non-self) antigens to helper T lymphocytes.

BIOLOGY IN ACTION

Gut microbes boost immunotherapy success in fighting microbes

Immunotherapy is a rapidly growing field of medicine that has shown a lot of promise as a highly effective treatment for cancer. Immunotherapy works by using the patient's immune system to fight cancer cells. There are four main types of immunotherapy: monoclonal antibodies, which are synthetic antibodies designed to destroy, slow the growth, or directly deliver medicine (e.g. chemotherapy) to cancer cells; immune checkpoint inhibitors, which prevent immune cells from being switched off by cancer cells, enabling the immune system to recognise and destroy cancer cells (Figure 7.3.17); cancer vaccines, which trigger the immune system to prevent cancer cell growth or destroy existing cancer cells; and non-specific immunotherapies, which boost the immune system in a non-targeted way to slow or halt cancer cell growth.

In clinical trials, melanoma (skin cancer) patients who received immunotherapy treatments had increased survival rates, with less toxicity and fewer negative side effects than patients who received chemotherapy or radiotherapy. Immunotherapy has also been successful in treating several other types of cancer, including lung, breast and colon. Although this form of treatment holds a lot of promise, the success rates with patients have been varied and there is still much research to be done.

Scientists have recently discovered an unexpected clue as to why some patients may respond well to immunotherapy while others have little to no success. Researchers from the University of Chicago found that introducing the bacteria species *Bifidobacterium* to the digestive systems of mice with melanoma markedly increased their anti-tumour T lymphocyte response.

When mice with the bacterial strain were compared to mice without the bacteria but which were receiving immunotherapy via the drug anti-PD-L1 (an immune checkpoint inhibitor), the researchers found that tumour growth was slowed in both groups. By combining the bacterial treatment with immunotherapy, tumour control was dramatically improved. Another study by researchers at the Institut Gustave Roussy in Paris found that antibiotics reduced the effects of an immunotherapy drug. By replenishing gut microbes in antibiotic-treated and infection-free mice, the anti-cancer effects of the immunotherapy drug were restored.

Further investigation revealed that the *Bifidobacterium* triggered an immune response by interacting with dendritic cells in the intestinal tract. Dendritic cells are antigen-presenting cells that are responsible for detecting potential threats to the immune system and presenting them to the T lymphocytes, thereby triggering an immune response. A genome-wide scan of mice with *Bifidobacterium* also showed upregulation of several genes involved in anti-tumour responses. Both of these studies have demonstrated the important role that a healthy gut microbiome plays in the immune response and the significant implications for the treatment of cancer using immunotherapy.

FIGURE 7.3.17 Some tumour cells switch off T lymphocytes by binding to T cell receptors (a). Immune checkpoint inhibitors, a type of immunotherapy drug, block inhibitory molecules on tumour cells from binding to T cell receptors, allowing T lymphocytes to remain active and infiltrate tumours, keeping them from growing (b). Treatment with immune checkpoint inhibitors will briefly increase the size of a tumour (pseudoprogression) due to T lymphocyte infiltration, before its size is reduced due to tumour cell death (c).

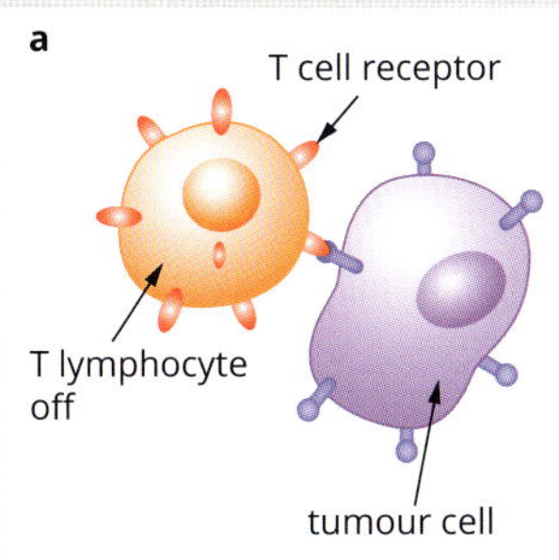

tumour cells turn off activated T lymphocytes when they attach to specific T lymphocyte receptors

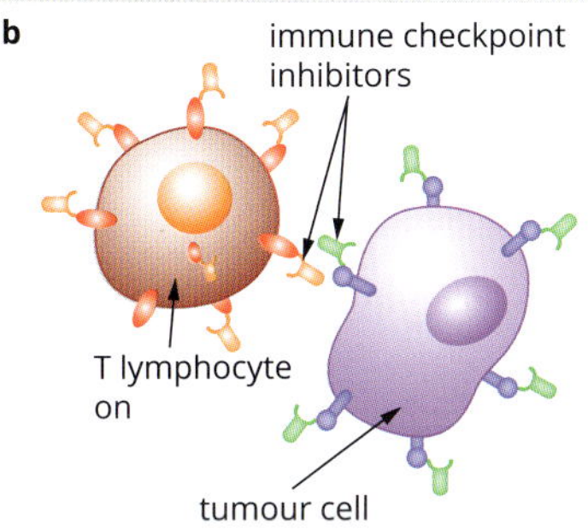

immune checkpoint inhibitors prevent tumour cells from attaching to T lymphocytes so T lymphocytes stay activated

immune checkpoint inhibitors target either T lymphocytes (orange) or tumour cells (green)

c Pseudoprogression (a brief increase in tumour size) results from treatment with immune checkpoint inhibitors.

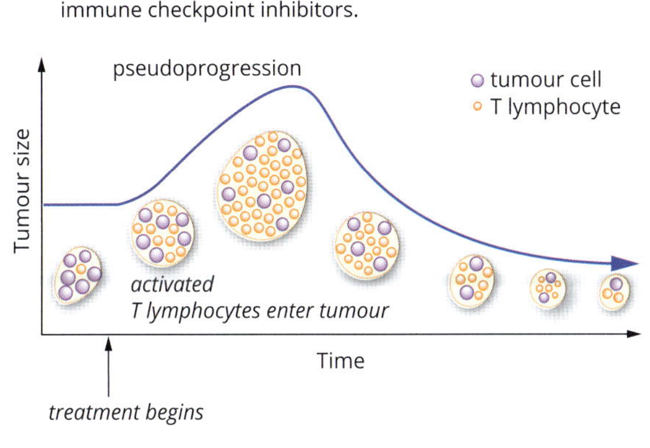

7.3 Review

SUMMARY

- An adaptive immune response is one that is specific to a certain antigen.
- The adaptive immune response in vertebrates is classified as humoral or cell-mediated.
- Humoral immunity involves B lymphocytes, which become activated and proliferate when stimulated by specific antigens or cytokines released by helper T lymphocytes. Activated B lymphocytes become plasma cells that produce antibodies and memory lymphocytes that remain in lymphoid tissues and provide immunological memory.
- Antibodies, also known as immunoglobulins, are proteins that bind to specific antigen molecules.
- Antibodies are Y-shaped proteins that have a constant 'tail' and variable 'arm' regions. The variable regions have antigen-binding sites and the constant region recruits components of the immune system.
- Cell-mediated immunity involves T lymphocytes:
  - Cytotoxic T lymphocytes recognise and kill foreign, infected or abnormal host cells by releasing toxic compounds.
  - Helper T lymphocytes secrete cytokines that promote inflammation, and activate macrophages and B lymphocytes.
  - The major histocompatibility complex (MHC) is important in antigen presentation:
  - MHC-I is expressed on all nucleated cells and presents peptide antigens of proteins being produced within the cell to cytotoxic T lymphocytes.
  - MHC-II is expressed on professional antigen-presenting cells and presents peptides of phagocytosed antigens to helper T lymphocytes.
- Antigen presentation is carried out by antigen-presenting cells (APCs), including dendritic cells, macrophages and B lymphocytes, and involves antigen fragments of a pathogen being presented on the MHC-II of the host cell. Helper T lymphocytes identify and bind to the complex, and become activated to produce and release cytokines.
- Cytotoxic T lymphocytes kill infected cells, which are identifiable by pathogen antigens on MHC-I.
- Memory B and T lymphocytes persist after an infection to enable a larger and faster response upon reinfection with the same pathogen.
- The first infection with a pathogen produces a primary immune response, while reinfection with the same pathogen produces a secondary response due to the presence of memory cells from the primary response (known as immunological memory).

KEY QUESTIONS

1 Briefly describe two key features that make the adaptive immune response different from the innate immune response.

2 What part of an antibody interacts with antigens on a pathogen?

A the constant region
B the disulfide bridge
C the constant region of the heavy chain
D the variable region

3 Which MHC class interacts with helper T lymphocytes, and which interacts with cytotoxic T lymphocytes?

4 Look at the graphs to the right and determine which label on the graphs (A, B, C, D or E) represents:

a the period when the virus has just entered the body and started to multiply (the incubation period)
b the day the patient became infected
c the period the patient felt most ill
d the normal body temperature of 37 °C
e the period when the patient's antibodies destroy the virus.

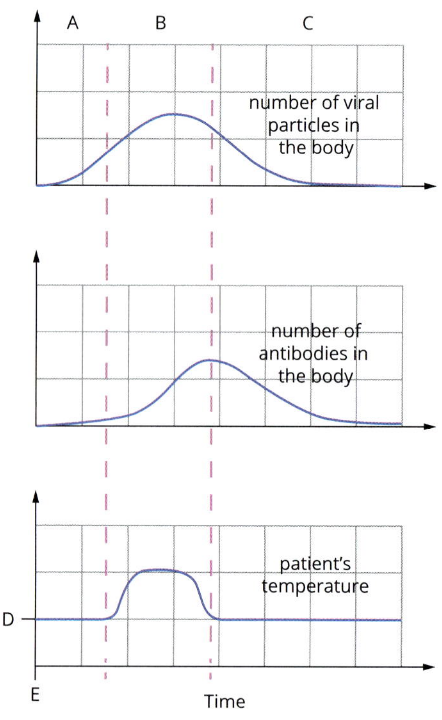

7.4 The lymphatic system

The **lymphatic system** plays a key role in the adaptive immune responses of mammals. It transports immune cells, including antigen-presenting cells, throughout the body, and is where antigen recognition by lymphocytes occurs. In this section, you will learn about the lymphatic system, and how its structures (Figure 7.4.1) are involved in adaptive immune responses.

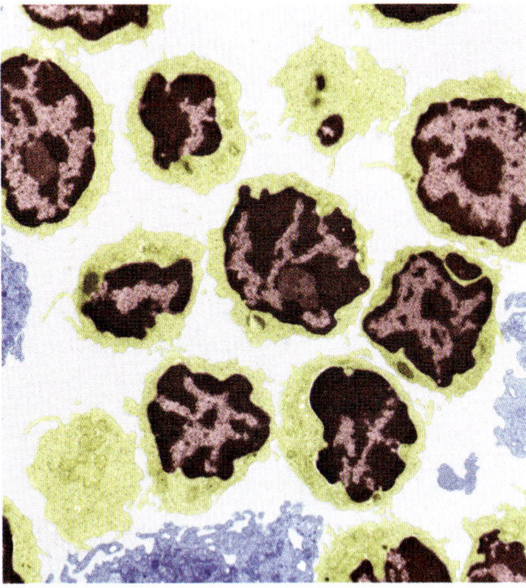

FIGURE 7.4.1 Coloured transmission electron micrograph of a section through a lymph node, showing a variety of lymphocytes (yellow). Lymph nodes are one of the structures of the lymphatic system.

THE ROLE OF THE LYMPHATIC SYSTEM

The mammalian lymphatic system has several roles, including:

- returning fluid that seeps out of the blood vessels into tissues back to the circulatory system
- absorbing and transporting fatty acids and fats from the digestive system
- providing a place for lymphocytes to mature
- transporting lymphocytes and antigen-presenting cells to the lymph nodes, stimulating the adaptive immune response.

The lymphatic system is vital to the immune response. Invading pathogens are transported in the lymph to the lymph nodes, where bacteria, viruses and cancer cells are trapped and destroyed by phagocytes and lymphocytes. This is why your lymph nodes swell up when you have an infection.

THE STRUCTURE OF THE LYMPHATIC SYSTEM

The lymphatic system is made up of lymph, lymphatic vessels and primary and secondary lymphoid organs and tissues (Figure 7.4.2).

When the fluid that surrounds the tissues (or interstitial fluid) is drained into the lymphatic vessels, it is considered **lymph**. Lymph contains immune cells such as lymphocytes and phagocytes.

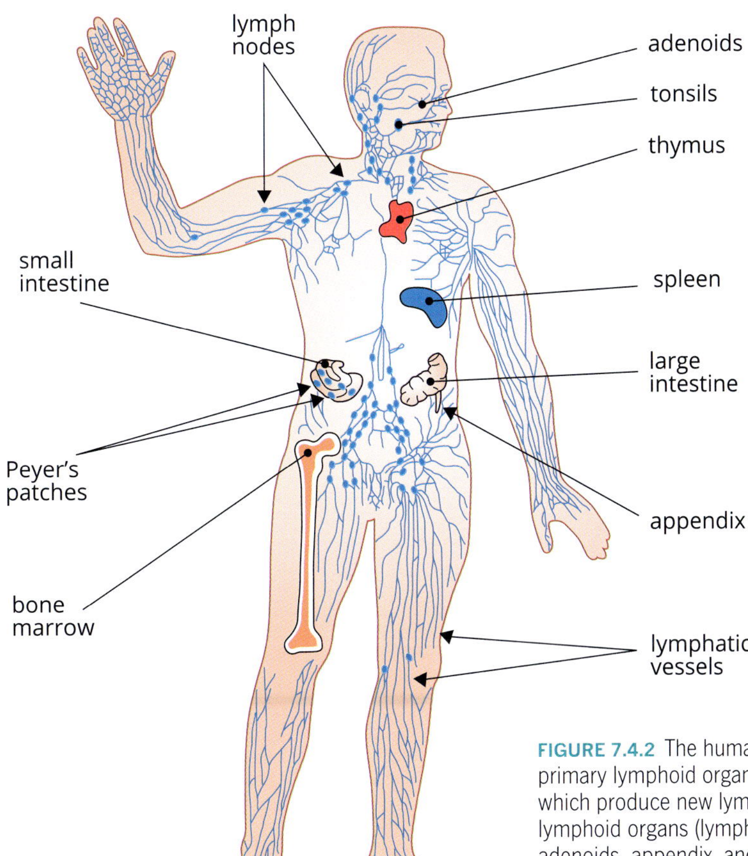

FIGURE 7.4.2 The human immune system is made up of primary lymphoid organs (the thymus and bone marrow), which produce new lymphocytes, and the secondary lymphoid organs (lymph nodes, the spleen, tonsils, adenoids, appendix, and Peyer's patches of the small intestine) in which immune responses occur.

The structure of the lymphatic system is similar to the venous part of the circulatory system. Fine lymphatic capillaries join to form increasingly larger vessels that eventually empty into the large veins near the heart (Figure 7.4.3a). Lymphatic capillaries are widespread, but they are absent from bones and the central nervous system (where excess tissue fluid drains into cerebrospinal fluid). Although blood and lymph capillaries are closed to each other, cells and fluid are able to pass between them through a process called extravasation (Figure 7.4.3b). Some of the larger lymph vessels can contract, but most lymph flow results from the external compression of lymph vessels by muscular activity, such as during movement and breathing. When vessels are compressed, the lymph fluid is forced in one direction because of numerous one-way valves, like those in veins, located along the vessels (Figure 7.4.3c). When a person is inactive (such as standing still or sitting) for a long time, the fluid drainage from tissues decreases and causes swelling. This is especially so in the legs, because fluid drainage must work against gravity.

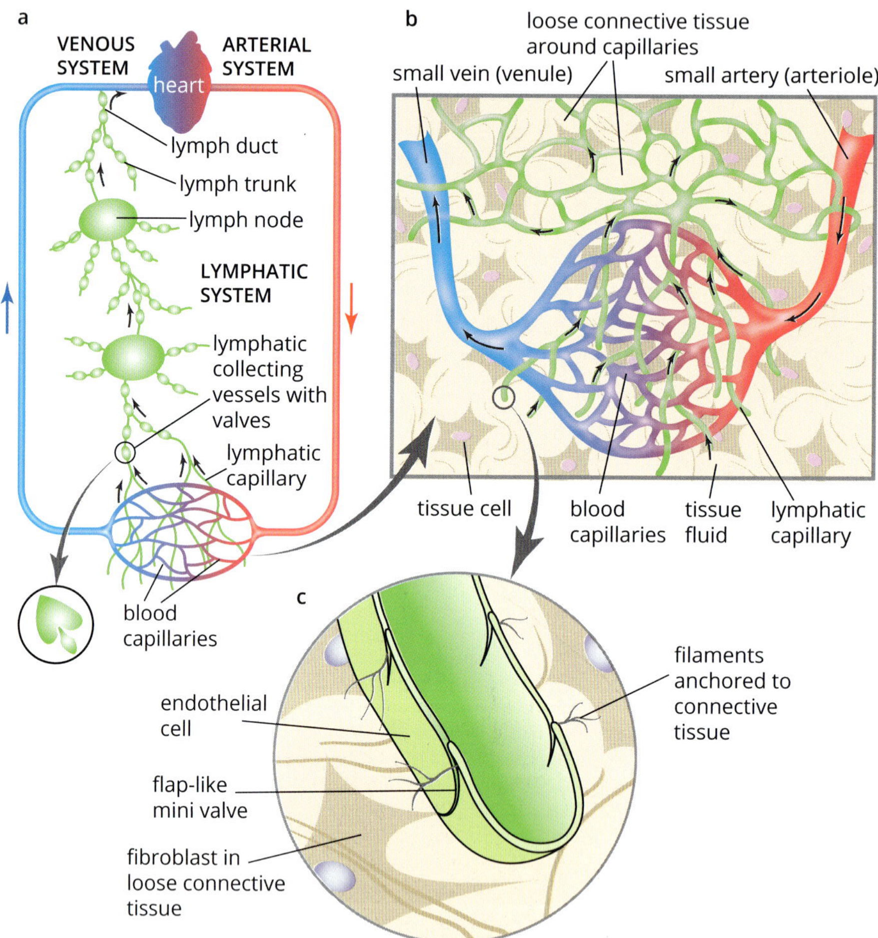

FIGURE 7.4.3 Lymphatic vessels weave through tissue cells and blood capillaries in loose connective tissues of the body. (a) Blood flows from the veins to the heart, then to the lungs to become oxygenated, then through the arteries to tissues. Lymph drains through a lymphatic duct called the thoracic duct, and into a vein called the left subclavian vein, as well as through the right lymphatic duct into the right subclavian vein and the right internal jugular vein. (b) Blood and lymph capillaries are closed to each other, but cells and fluids are able to pass from one vessel to another through a process called extravasation. (c) Lymphatic capillaries are closed-ended tubes in which adjacent endothelial cells overlap each other, forming flap-like mini valves.

Primary lymphoid organs

The **primary lymphoid organs** are bone marrow and the thymus.

Bone marrow contains stem cells from which B and T lymphocytes originate (Figure 7.4.4). B lymphocytes undergo several stages of development in the bone marrow then enter the bloodstream and travel to the spleen and other secondary lymphoid tissues, where they complete their maturation, and also where they become activated after being exposed to antigen.

Immature T lymphocytes travel from the bone marrow to the thymus, where they mature. The thymus is considered a primary lymphoid organ because of its role in the maturation of T lymphocytes. The size of the thymus peaks at puberty, and then gradually shrinks each year, as it becomes replaced by fat (or adipose) tissue. The shrinking of the thymus doesn't have an immediate disastrous effect on immunity, because of the already established pool of peripheral T lymphocytes, but it contributes to the higher risk of infection and cancer that comes with age.

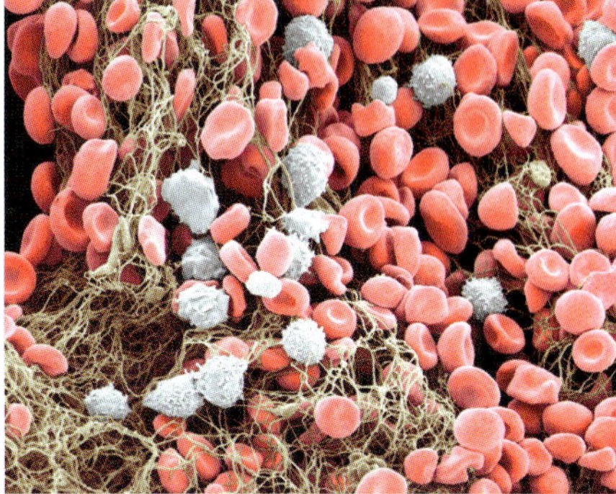

FIGURE 7.4.4 Coloured scanning electron micrograph of a fractured rib. Bone marrow lies between the spongy bone and contains stem cells that give rise to red blood cells (red) and white blood cells such as B and T lymphocytes (grey).

Secondary lymphoid organs and tissues

The **secondary lymphoid organs and tissues** are the lymph nodes, spleen, tonsils, adenoids, appendix and Peyer's patches (Figure 7.4.5). It is in these organs and tissues that adaptive immune responses begin.

Lymphocytes are activated in secondary lymphoid tissues, where they recognise and respond to non-self antigens that are specific to their receptors.

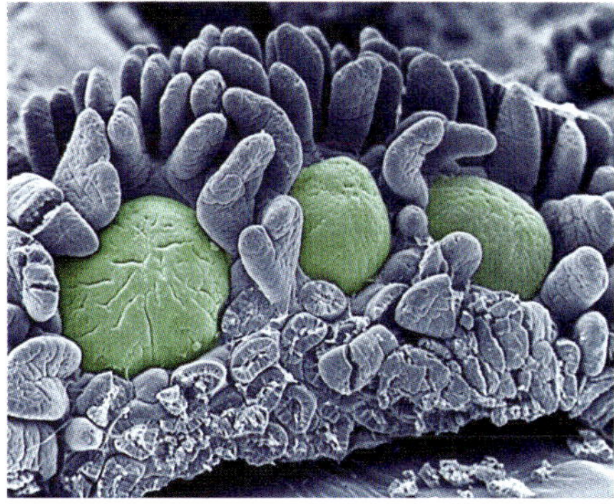

FIGURE 7.4.5 Coloured scanning electron micrograph of Peyer's patches (green) of the small intestine. Peyer's patches defend against infection by supplying lymphocytes to the local intestinal tissue, and are named after the Swiss anatomist Johann Conrad Peyer, who first described them in 1677.

Lymph nodes

Lymph nodes are composed of lymphoid tissue, and are located at regular intervals along the lymphatic system. Lymph passes through lymph nodes on its way back to the bloodstream (Figure 7.4.6). Lymph nodes act as filters, trapping foreign particles, cellular waste, toxins and pathogens.

The structure of lymph nodes maximises the chance of encounters between antigens and immune cells. Some dendritic cells and macrophages are stationed in the lymph nodes, where they phagocytose pathogens, and present the foreign antigens to helper T lymphocytes. Antigen-presenting cells in body tissues also migrate to the lymph nodes after phagocytising pathogens, to present foreign antigens to helper T lymphocytes.

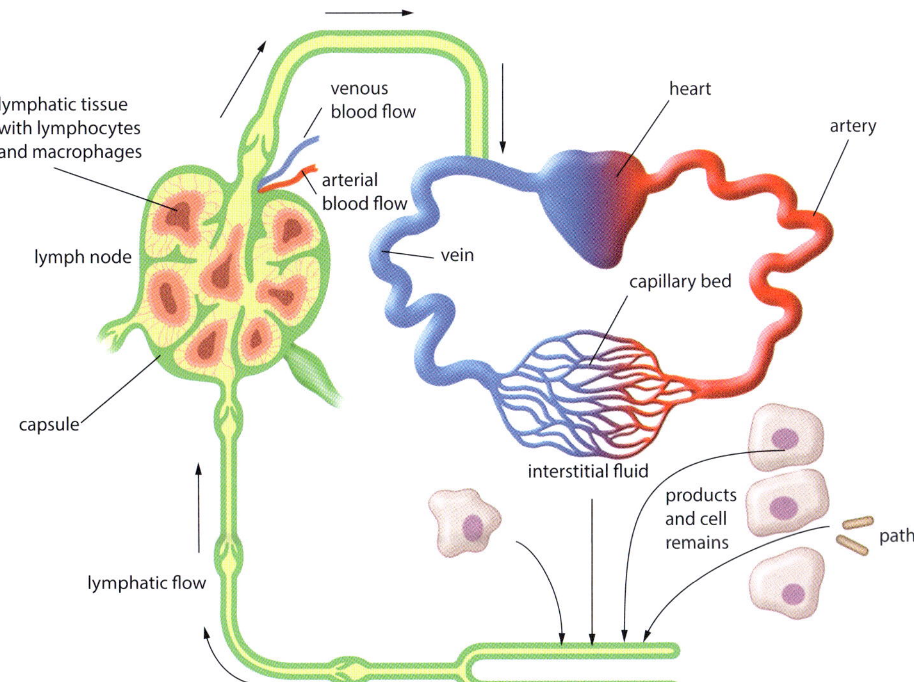

FIGURE 7.4.6 The flow of lymph through the lymphatic system is one-way due to the presence of valves. Lymph nodes act as filters and are important centres of immune cell activity.

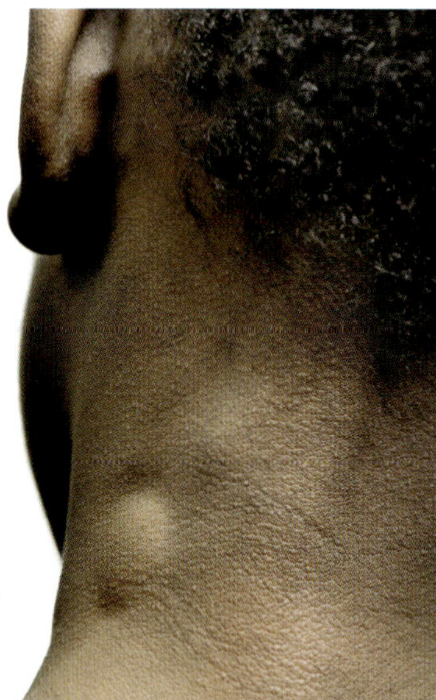

FIGURE 7.4.7 Swollen lymph nodes in a boy's neck. Inflammation of a lymph node (lymphadenitis) is usually a response to an infection. The increased lymph node activity produces more infection-fighting B and T lymphocytes.

B and T lymphocytes interact in the follicles of the lymph nodes. B lymphocytes that identify an antigen undergo clonal expansion and differentiation to plasma cells. Antibody is released into the bloodstream to travel throughout the body. Cytotoxic T lymphocytes are activated, proliferate, and travel through the bloodstream to sites where they are needed.

The size of lymph nodes can expand markedly when cell proliferation is occurring in response to an infection. For example, during a respiratory tract infection, it is common for swollen lymph nodes to occur on the side of the neck (Figure 7.4.7).

Spleen

The spleen's primary function is to control the number of red blood cells in the body by destroying old and defective red blood cells. The spleen also stores up to a quarter of the body's lymphocytes and is a site of B lymphocyte maturation.

If for some reason the spleen needs to be removed, this does not have a disastrous effect on B lymphocyte maturation, because B lymphocytes can still mature in other secondary lymphoid tissues.

BIOFILE

Sentinel lymph nodes

Lymph nodes are filters for antigens and invading microbes, but they can also trap abnormal cells, such as cancer cells that have separated from a primary tumour and travelled in the lymph until reaching a lymph node.

A sentinel lymph node, the first node to which cancer cells are most likely to spread, may be removed and examined under the microscope. The presence of cancer cells in the lymph node indicates that a tumour is malignant.

Tattoos do not cause cancer, but tattoo ink migrates through to the lymph nodes and mimics the appearance of certain types of cancer, making proper diagnosis of some cancers more difficult.

FIGURE 7.4.8 Whether or not tattoos look good on the outside is open to interpretation, but on the inside tattoo ink migrates to your lymph nodes and can make them look cancerous.

7.4 Review

SUMMARY

- The lymphatic system produces lymphocytes and transports them, along with antigen-presenting cells, to the lymph nodes to stimulate adaptive immune responses.
- Primary lymphoid organs (bone marrow and the thymus) are responsible for the production of B and T lymphocytes.
- T lymphocytes mature in the thymus.
- Secondary lymphoid organs and tissues include lymph nodes, spleen, tonsils, adenoids, appendix, and Peyer's patches of the small intestine.
- B lymphocytes develop in the bone marrow and become mature in the peripheral lymphoid organs and tissues.
- Secondary lymphoid organs and tissues are the sites where lymphocytes identify and interact with antigen-presenting cells and are then activated to divide and differentiate.

KEY QUESTIONS

1 Which of the following are the primary lymphoid organs?

A bone marrow and lymph nodes
B bone marrow and spleen
C lymph nodes and spleen
D bone marrow and thymus

2 T lymphocytes mature in:

A bone marrow
B the thymus
C Peyer's patches
D the spleen

3 Explain why the lymphatic system is an important part of the adaptive immune response.

4 Fill in the missing labels A–E for the following diagram of the lymphatic system.

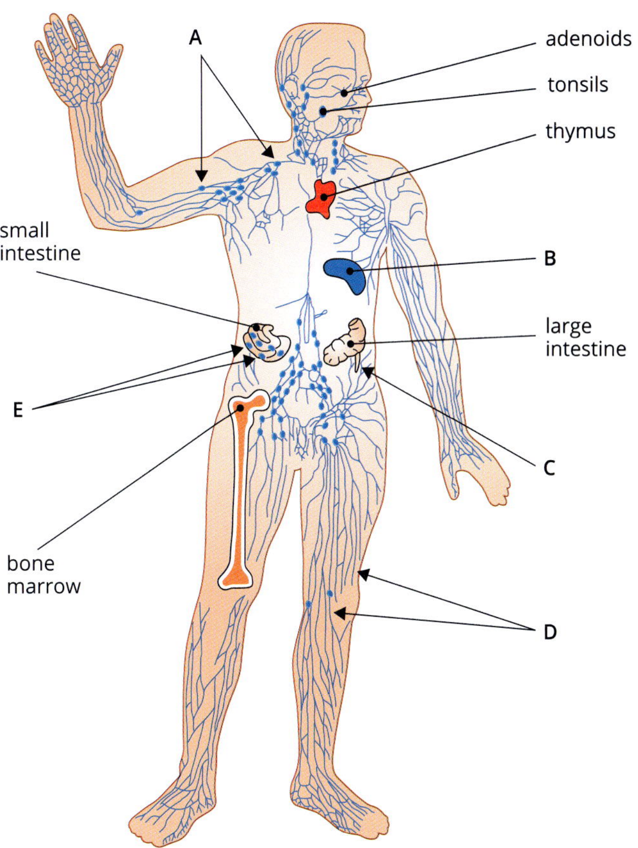

Chapter review

07

KEY TERMS

adaptive immunity
agglutination
allergen
antibody
antigen
antigen–antibody complex
antigen presentation
antigen-presenting cell (APC)
antigenic variation
bacterium (plural bacteria)
cell-mediated immunity
cellular pathogen
chemokine
clonal selection
complement proteins
constant region
cytotoxic T lymphocyte
defensin
differentiation
fever
fungus (plural fungi)
heavy chains
helper T lymphocyte
histamine
human leukocyte antigen
humoral immunity
immunogen
immunoglobulin (Ig)
immunological memory
inflammation
innate immunity
leukocyte
light chain
lymph
lymphatic system
lymphocyte
lysis
lysozyme
major histocompatibility complex (MHC)
mast cell
memory B lymphocyte
memory T lymphocyte
microflora
neutralisation
non-self antigen
oomycete
pathogen
phagocyte
phagosome
plasma cell
precipitation
primary immune response
primary lymphoid organ
prion
protozoan
secondary immune response
secondary lymphoid organs and tissues
self-antigen
self-tolerance
specificity
T cell receptor (TCR)
variable region
viroid
virus

KEY QUESTIONS

1 Which of the following is not one of the initial barriers to infection in animals?

A intact skin

B the action of complement proteins

C lysozymes in saliva

D competition from microflora

2 What are prions and why doesn't the adaptive immune system respond to them?

3 Which of the following is not a role of lymphatic vessels?

A returning fluid that seeps out of the blood vessels into tissues back to the circulatory system

B absorbing and transporting fatty acids and fats from the digestive system

C providing a site for lymphocytes to mature

D transporting lymphocytes and antigen-presenting cells to the lymph nodes

4 Cassava provides a significant proportion of the daily kilojoule intake for over 600 million people throughout the world. In both Africa and India, cassava crops are regularly attacked by the cassava mosaic virus (CMV). One important method of protecting crop production is to use cassava varieties that show a natural resistance to CMV.

Suggest one feature the CMV-resistant varieties of cassava plant could possess which non-resistant plants do not.

5 One way in which the structure of lymph nodes enhances their efficiency is that the nodes have more vessels carrying fluid into the nodes than carrying fluid out.

a i What is the fluid travelling in lymph vessels called?

ii What effect would more vessels leading into the node than out of the node have on the rate of flow?

b One effect of inflammation is to make capillaries more permeable.

i How does the increased permeability of capillaries help the immune response?

ii What effect would the increased permeability of blood vessels have on flow through the lymphatic vessels?

c The lymphatic system is responsible for draining the fluid that leaks from blood vessels. How might the flow of fluid in the lymphatic system be affected by inflammation?

d Explain why it is beneficial to the immune process that lymphocytes accumulate in the lymph nodes.

6 An important function of phagocytes is to destroy bacteria. This is done through the endocytosis of the bacterium, followed by its digestion by lysosomal enzymes. As with all cellular functions, this process is regulated by proteins. Chédiak–Higashi Syndrome is a rare inherited disorder of the immune system in which the proteins that regulate the joining of lysosomes with endosomes are defective.

   a What is the name of the structure formed when a phagosome fuses with a lysosome?

   b The failure of lysosomal breakdown of engulfed bacteria will seriously undermine not only the innate immune response, but also the adaptive immune response. Explain why.

7 Mild infections such as the common cold are a regular experience of children attending childcare. One symptom of these infections is usually a mild fever (up to 39 °C). The usual treatment given to children is a medication such as paracetamol, which reduces their temperature to normal.

   a Explain why reducing the body temperature of a patient with a mild fever may prolong the infection.

   b Patients who have a high fever (over 41 °C) should always be treated to reduce their body temperature. Why would such high temperatures reduce the body's ability to fight off an infection?

8 The first successful human-to-human blood transfusion is reported to have occurred in the 1800s. At that time a blood transfusion was a lottery that might help but could equally make you much worse. This was because the knowledge that humans have blood groups (ABO) was not discovered until 1901 and the idea that transfusions should be matched to the recipient's blood group was not suggested until 1907. It was not until 1939–40 that the Rhesus protein (antigen D) was identified in humans.

   a An individual who presents at a hospital today with serious bleeding will receive a safe, antigen-matched transfusion. Blood group matching is a fairly simple and quick procedure in which antibodies to the blood proteins are mixed with samples of the patient's blood in order to identify the correct blood type for transfusion.

   A patient has presented to the emergency department of the local hospital. The patient needs a blood transfusion. The plates below show the results of the test to determine the patient's blood group.

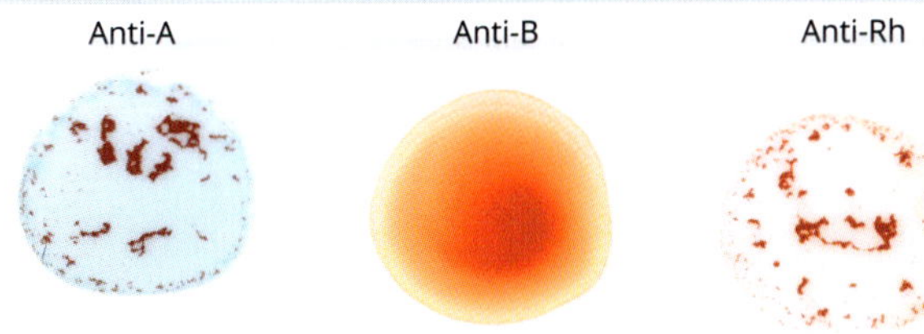

   i What is the patient's blood group?

   ii How do you know?

   iii What blood groups could be used for the patient's transfusion?

   b Occasionally, in hospitals when Rh negative blood is in short supply a man who is Rh negative is given blood that is Rh positive.

   i This procedure can only be done once. Why?

   ii This procedure would not be recommended for a young female patient. Why not?

9 Mammalian antibodies (or immunoglobulins) are generally grouped into five classes. Copy and complete the following table to summarise the role of each class.

Antibody type	Role
IgG	
IgM	
IgA	
IgD	
IgE	

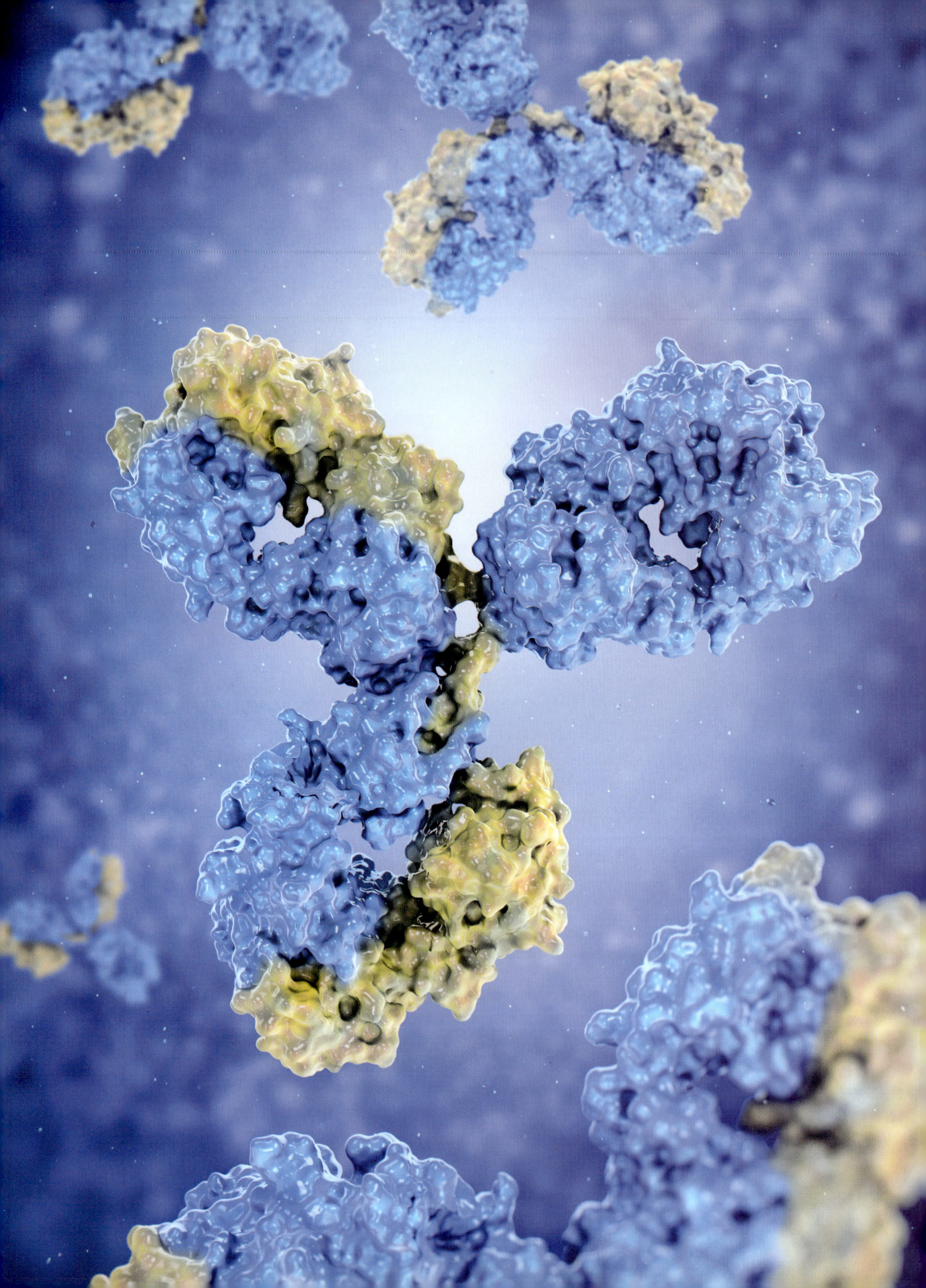

CHAPTER 08 Immunity, immune malfunctions and immunotherapy

Learning outcomes

By the end of this chapter, you will understand the difference between active and passive immunity, as well as natural and artificial means of achieving immunity. You will also be able to explain the role of vaccination programs in the promotion of herd immunity for human populations, describe the deficiencies and malfunctions of the immune system, and have an understanding of the use of monoclonal antibodies in treating cancer.

Key knowledge

- the difference between natural and artificial immunity, and active and passive strategies for acquiring immunity
- vaccination programs and their role in maintaining herd immunity for a particular disease in the human population
- the deficiencies and malfunctions of the immune system as a cause of human diseases including autoimmune diseases (illustrated by multiple sclerosis), immune deficiency diseases (illustrated by HIV) and allergic reactions (illustrated by reactions to pollen)
- the use of monoclonal antibodies in treating cancer.

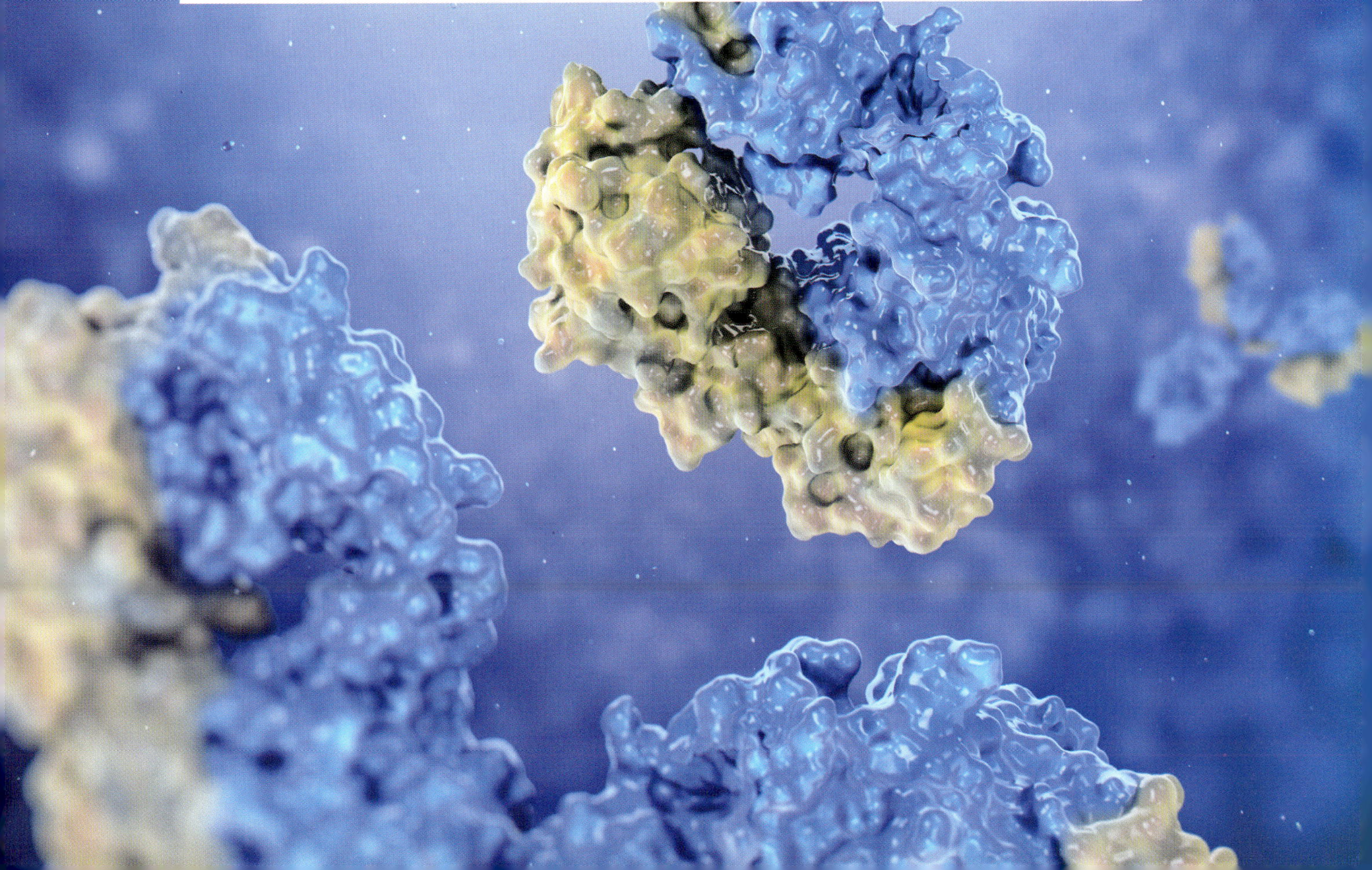

8.1 Immunity

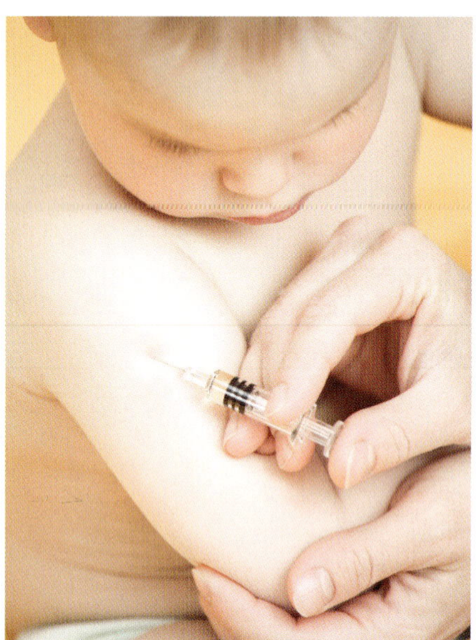

FIGURE 8.1.1 A baby being vaccinated. The vaccine stimulates an adaptive immune response that protects against infection.

In Chapter 7, you learnt about the physical, chemical and microbiological barriers that help prevent infection, as well as the innate and adaptive immune responses that provide immunity against invading pathogens that bypass these first line defences. In this section you will learn that immunity can be active or passive, natural or artificial. You will also learn that vaccination (Figure 8.1.1) is an example of artificial active immunity, and about the role of vaccination programs in creating herd immunity and preventing disease.

TYPES OF IMMUNITY

Immunity is active or passive depending on the origin of the immune response:

- **Active immunity** is protection provided by an individual's own adaptive immune response. This type of immunity takes time to develop, but the memory B and T lymphocytes that result can provide immunological memory that can last for many years, even a lifetime.
- **Passive immunity** is protection provided to an individual by the transfer of antibodies produced by another organism. This type of immunity is immediate, but will only protect the recipient for a limited time because it does not result in immunological memory, and the transferred antibodies degrade over time and are removed from the body.

Table 8.1.1 provides a summary of active and passive immunity.

> Immunological memory is the retention of B and T lymphocytes sensitised to specific antigens. It enables a stronger and more rapid immune response should the same antigens be encountered again.

Active immunity	Passive immunity
• Adaptive immune response to antigen occurs in the individual. • The individual's immune system is activated against the antigen and achieves immunological memory. • Immunity can be maintained by stimulating memory cells, i.e. with booster vaccinations. • Immunity develops over weeks.	• Adaptive immune response occurs in another organism that is exposed to the antigen and antibodies are then transferred to a recipient. • The recipient's immune system is not activated against the antigen and does not achieve immunological memory. • Immunity cannot be maintained. • Immunity is immediate.

TABLE 8.1.1 A summary of the differences between active and passive immunity.

Immunity can develop naturally through exposure to a pathogen, or be induced artificially through purposeful introduction of antigens or antibodies into the body. Both active and passive immunity can arise naturally or artificially.

Natural passive immunity

Natural passive immunity involves the passive transfer of antibodies from mother to foetus through the placenta prior to birth, and from mother to baby through breastfeeding. These maternal antibodies provide protection to the baby for weeks or months, while its own immune system is developing.

Artificial passive immunity

Artificial passive immunity involves an individual receiving, usually by injection of **antiserum**, antibodies produced by another organism. Antiserum is serum that contains specific antibodies. **Serum** is the fluid portion of blood that remains after blood cells and material involved in blood clotting has been removed (Figure 8.1.2). When these transferred antibodies bind to the antigens on the pathogen or toxin, they form an antigen–antibody complex that inhibits the pathogen or toxin before it does much damage.

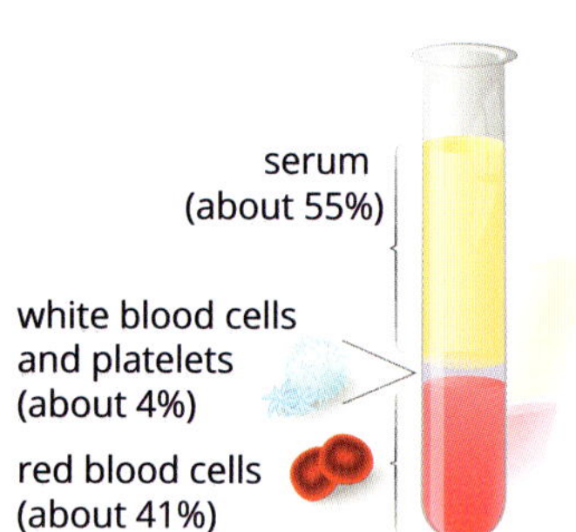

FIGURE 8.1.2 Serum is the fluid portion of blood that remains after blood cells and clotting factors (platelets) have been removed. Antiserum is serum containing specific antibodies injected to treat or protect against disease.

Artificial passive immunisation can be a useful means of treatment of an infection by a pathogen, or a bite or sting by a venomous animal, when death is likely to occur before the primary immune response has had time to develop. For example, the administration of tetanus antiserum protects against tetanus in at-risk patients, such as those with a deep or dirty puncture wound. The antiserum contains antibodies specific for the toxin, called antitoxins, which bind to the tetanus toxin, inhibiting its function. However, introducing antibodies to contain the threat before the person's own adaptive immune response can be mobilised means the protection provided is only temporary, as no immunological memory is formed.

Artificial passive immunisation is also used to supress active immunity when it can be harmful, such as in haemolytic disease of the newborn, which occurs when a mother's natural active immunity causes her immune system to attack the red blood cells of her foetus. The reason this sometimes occurs is because people with the Rhesus blood type (Rh^+) have the Rhesus antigen (or D antigen) on their red blood cells, and people with Rhesus negative (Rh^-) blood type do not (Figure 8.1.3). When a mother who is Rh^- has a baby who is Rh^+, the mother is likely to develop an adaptive immune response to the Rh antigen, as fragments of foetal red blood cells cross the placenta during birth (Figure 8.1.4a).

If the same Rh^- woman has another pregnancy with an Rh^+ baby, her memory cells will trigger the production of antibodies that cross the placenta and damage the baby's blood cells, causing them to develop **haemolytic disease** (from *haemo* meaning blood and *lysis* meaning breakdown). To prevent the mother having an adaptive immune response and producing anti-Rh antibodies during her first pregnancy, she is given a dose of anti-Rh antibodies (Figure 8.1.4b). These administered antibodies will neutralise any foetal Rh antigens before an adaptive immune response by the mother occurs.

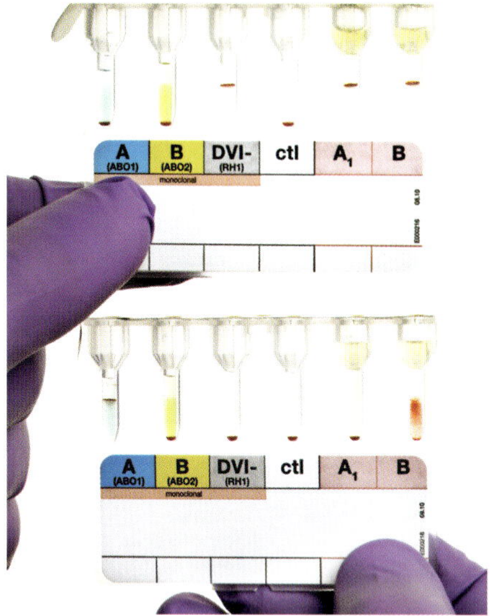

FIGURE 8.1.3 Laboratory tests showing a Rhesus positive result (top), in which there is D antigen present on red blood cells (third vial from the left), and a Rhesus negative result (bottom), in which there is a lack of D antigen.

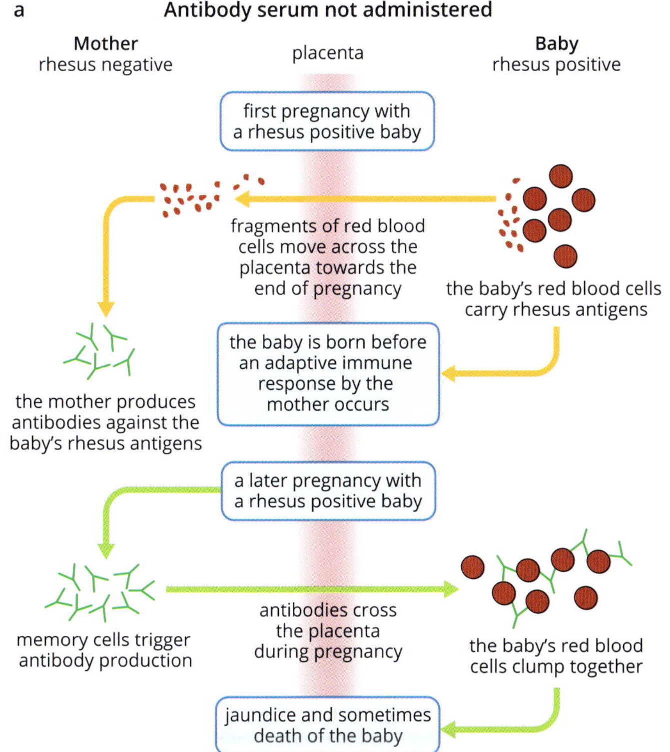

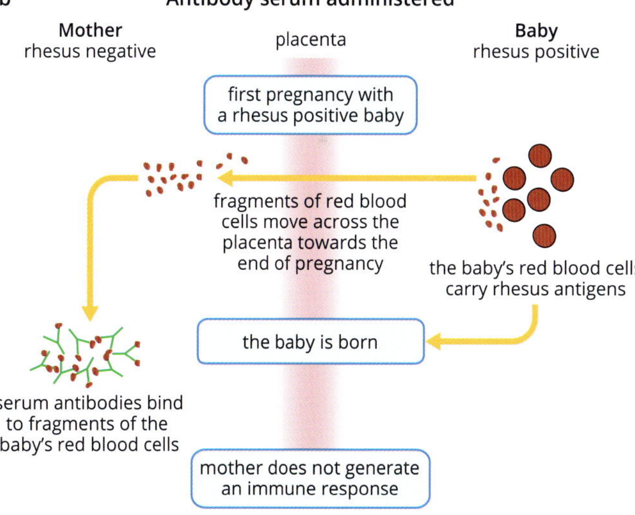

FIGURE 8.1.4 Haemolytic disease of the newborn can be prevented by artificial passive immunisation. A mother who is Rh^- has a baby who is Rh^+. (a) When red blood cell fragments cross the placenta, the mother has an adaptive immune response that makes memory cells, and antibodies against the Rh antigen (called anti-Rh antibodies). In a later pregnancy with a Rh^+ foetus, memory cells trigger the production of antibodies that can cross the placenta and will damage any subsequent Rh^+ foetuses. (b) A dose of anti-Rh antibodies can be administered to the mother to neutralise any foetal Rh antigens before an immune response occurs, protecting any future Rh^+ foetus.

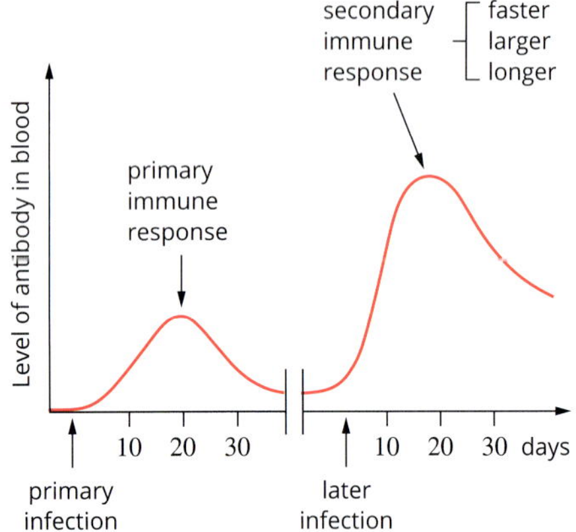

FIGURE 8.1.5 A subsequent infection with the same infectious agent will trigger a secondary immune response. The secondary immune response is faster and stronger than the primary immune response.

Natural active immunity

Natural active immunity develops from the adaptive immune response to a natural infection, and the immunological memory that results. This means that if exposed to the same antigen again in the future, the immune system will recognise it immediately, and a secondary immune response will occur (Figure 8.1.5). Secondary immune responses are much faster and stronger than primary immune responses, and are therefore more likely to minimise disease. For example, if you have had chickenpox, you are unlikely to get it again because your immune system has developed immunological memory specific to the antigens of *Varicella zoster* virus, the virus that causes chickenpox.

Artificial active immunity

Artificial active immunity results from the administration of antigens to induce an adaptive immune response. This artificial technique of inducing an adaptive immune response to produce active immunity is known as **vaccination** (or immunisation), and the material used to induce artificial active immunity is called a vaccine. By administering a specific vaccine usually made of altered, weakened or killed microorganisms, such as bacteria or viruses, or inactivated forms of toxins or proteins, active immunity can be induced.

As with natural active immunity, the primary response to vaccination takes time to develop, and booster vaccines are often needed to stimulate the stronger secondary immune response that provides longer-lasting immunity (Figure 8.1.6).

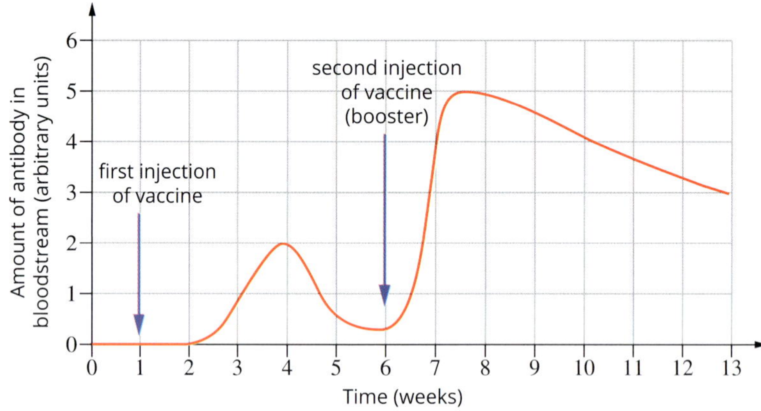

FIGURE 8.1.6 Vaccination induces an adaptive immune response that results in immunological memory. A booster vaccine is often needed to stimulate a stronger secondary response to provide sufficient, longer-lasting protection.

Vaccines need to be highly specific to initiate an adaptive immune response resulting in immunological memory. Increased understanding of microbiology and immunology has led to the development of very safe vaccines that induce the desired immune response with minimal side effects.

Table 8.1.2 provides a summary of examples of the different types of immunity you have just learnt about.

	Active	Passive
Natural	the adaptive immune response of an individual	the passive transfer of antibodies from mother to foetus through the placenta prior to birth, and from mother to baby through breastfeeding
Artificial	vaccination of an individual to stimulate an adaptive immune response	administration of antibodies that have been produced by another organism against a pathogen or toxin

TABLE 8.1.2 Examples of types of immunity.

Live attenuated vaccines

Live attenuated vaccines involve a living microbe that has been weakened in the laboratory, usually through repeated culturing. The advantage of live attenuated vaccines is that a single dose usually provides long-lasting immunity, because the vaccines induce a strong adaptive immune response that produces many types of antibodies directed against multiple antigens. The disadvantages of live attenuated vaccines are that although they are safe for most people, they may cause disease in those with weakened immune systems. Also they may cross the placenta in pregnant women and cause damage to developing foetuses.

Attenuated vaccines are more commonly used for viruses than for bacteria, because bacteria have thousands of genes and thus are much harder to control. Examples of attenuated vaccines include those against measles, mumps, rubella and polio.

Inactivated vaccines

Inactivated vaccines, also known as **killed vaccines**, contain microbes that have been inactivated by heat, radiation or chemical means. The advantages of inactivated vaccines are that they result in the production of many different antibodies, due to the fact they contain many different antigens. Inactivated vaccines can safely be used in people who have weakened immune systems. The disadvantage of inactivated vaccines is that they stimulate a relatively weak immune response compared to live, attenuated vaccines, so they require booster doses to achieve and maintain long-term immunity. Adjuvants can be added to inactivated vaccines to help boost the immune response.

Most vaccines against bacteria are inactivated vaccines. Examples of inactivated vaccines include the inactivated rabies and hepatitis A vaccines.

> Adjuvants are substances that stimulate a stronger immune response against antigens administered at the same time; examples include aluminium phosphate and aluminium hydroxide.

BIOLOGY IN ACTION

The eradication of smallpox

Edward Jenner (1740–1823) was the English doctor who discovered a vaccine for smallpox. Jenner (Figure 8.1.7) had heard of a milkmaid who could not catch smallpox because she had already had cowpox (a mild infection that is never fatal). To test this, in 1796, Jenner infected a young boy with cowpox in the hope of preventing infection with smallpox. He then infected the boy with smallpox, by injecting pus from a smallpox lesion under the boy's skin. The boy did not develop smallpox, providing evidence that inoculating a person with cowpox virus provided immunity against smallpox.

Despite the development of the smallpox vaccine, smallpox persisted in Asia, Africa and South America for many years. In 1966, there was worldwide action to eradicate smallpox, and the last case of epidemic smallpox was registered in 1977 in Somalia, Africa. The only confirmed cases of smallpox after this time were in 1978. They involved a medical photographer called Janet Parker from the University of Birmingham, and her mother. Parker became infected after smallpox travelled through air ducting connecting her office and a laboratory. Although her mother survived, Parker did not.

On 8 May 1980, the World Health Organization announced smallpox had been eradicated. Since then, there have been no vaccinations against smallpox. Laboratories have also greatly improved infection control standards since the time of Parker, and the smallpox virus is now stored safely in only two laboratories (one in Atlanta, USA, and the other in Novosibirsk, Russia). The virus is stored in case there is a need to produce the vaccine again in the future.

Factors that contributed to the eradication of smallpox included the lasting immunity achieved by those who recovered from infection, as well as the fact that the smallpox virus only infected humans and there were no natural reservoirs of infection.

FIGURE 8.1.7 Edward Jenner.

Subunit vaccines

Like inactivated vaccines, **subunit vaccines** do not contain any live microbial components. Unlike inactivated whole-cell vaccines, subunit vaccines contain only parts of microbes selected for their ability to induce an adaptive immune response. Subunit vaccines can contain a fraction of an antigen, a single antigen or multiple antigens, and these antigens can be proteins, detoxified toxins or polysaccharides. Those that contain multiple antigens induce a broader immunity, because they will induce the production of antibodies directed against multiple antigens.

Subunit vaccines share the advantages of inactivated vaccines in that they are safer, more stable, and easier to store than live, attenuated vaccines. However, subunit vaccines also share the disadvantages of inactivated vaccines in that that they require multiple doses and an adjuvant to improve the strength of the immune response.

Subunit vaccines are made by growing the pathogen in the laboratory and chemically extracting the antigens, or by using **recombinant DNA technology**. An example of a recombinant subunit vaccine is one that has been genetically engineered to produce a purified component of the protein coat of the virus that causes foot-and-mouth disease. Recombinant DNA technology can also be used for live vaccines, to genetically modify microbes so that they elicit an immune response but do not cause illness.

Toxoid vaccines, a type of non-recombinant subunit vaccine, use toxins inactivated by formalin (called toxoids) to stimulate an adaptive immune response. Although the toxoid is inactivated, it remains similar enough to the original toxin that the immunological memory for the toxoid is also effective for the toxin. For example, *Clostridium tetani* is a bacterium that produces a neurotoxin that causes tetanus, and the inactivated toxin is used in the vaccine for tetanus (Figure 8.1.9). Another example of a toxoid vaccine is the diphtheria vaccine. Toxoid vaccines often require multiple doses to achieve immunity.

BIOFILE

Human papillomavirus vaccine

Human papillomavirus (or HPV) is a virus that can result in the development of certain types of cancer, including cervical cancer (Figure 8.1.8). Professor Ian Frazer and Dr Jian Zhou from the University of Queensland developed a subunit vaccine for HPV. In 2007, Australia became the first country to roll out a national HPV vaccination program. The program originally only covered girls, but was extended to cover boys in 2013.

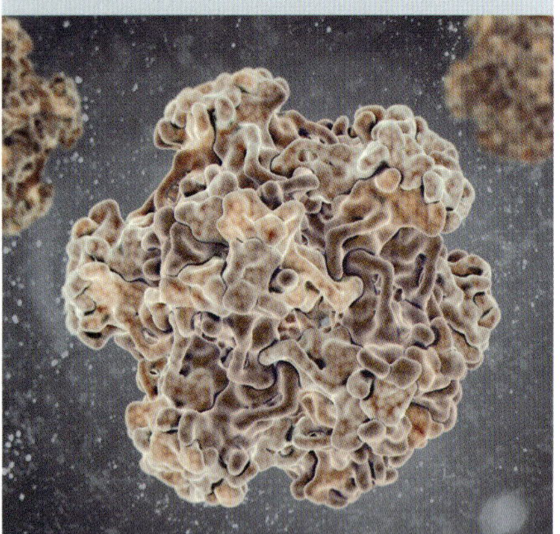

FIGURE 8.1.8 Digital illustration of human papillomavirus.

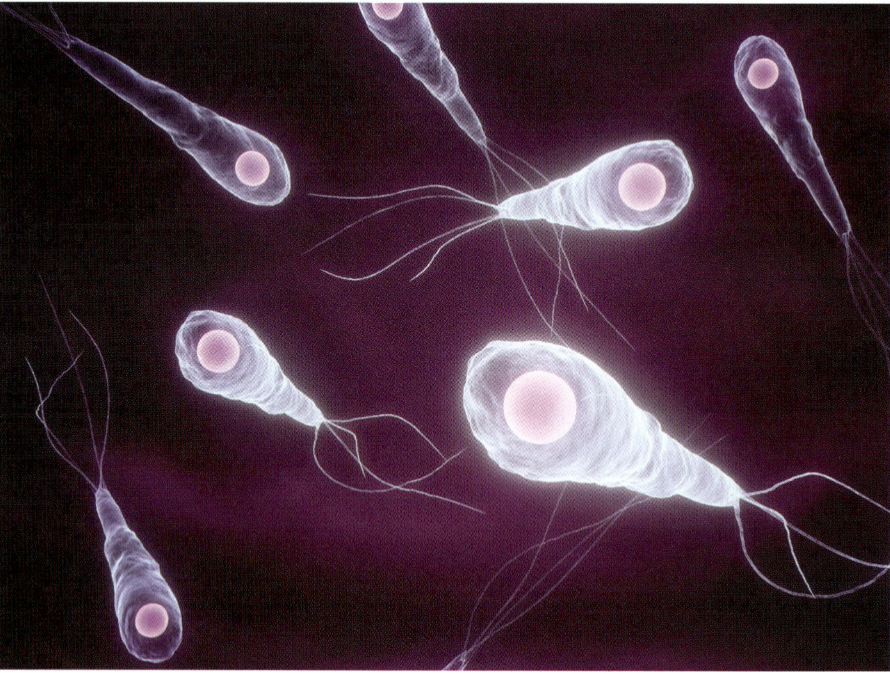

FIGURE 8.1.9 Digital illustration of *Clostridium tetani*, the bacterium that produces a toxin called tetanospasmin. The toxin acts on the central nervous system, causing muscle spasms that can result in convulsions, difficulty breathing and abnormal heart rhythms.

BIOLOGY IN ACTION

Vaccination programs

Vaccination programs aim to decrease the incidence of many diseases, with the aim of ultimately eradicating them (Figure 8.1.10). Through the introduction of vaccination programs, many countries have dramatically reduced new cases of diseases that were common a century ago. For example, in the early 1900s, Australia still experienced epidemics of diseases such as smallpox, polio and measles, but these diseases have either been eliminated (in the case of smallpox) or are now very rare.

Australians enjoy the benefits of vaccination not only because safe and effective vaccines were developed, but also because everyone in Australia receives many vaccinations free of charge as part of the National Immunisation Program (Table 8.1.3). Since the introduction of vaccination for children in 1932, deaths from vaccine-preventable diseases have fallen by 99%, despite a threefold increase in the Australian population over that time.

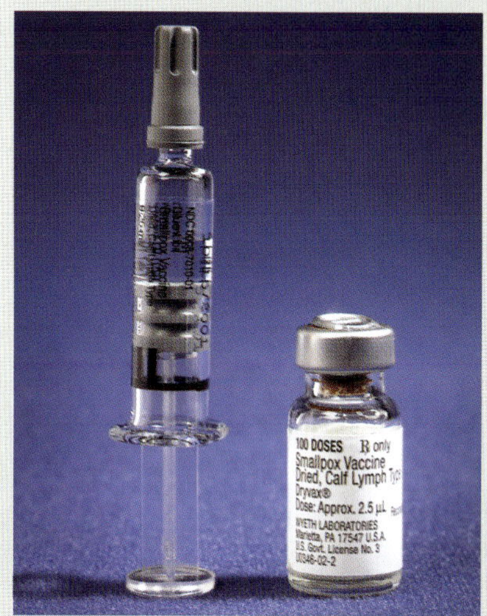

FIGURE 8.1.10 Vaccination helped eradicate smallpox.

Child programs	
Age	**Vaccine**
Birth	• hepatitis B (hepB)
2 months	• hepatitis B, diphtheria, tetanus, acellular pertussis (whooping cough), *Haemophilus influenzae* type b, inactivated poliomyelitis (polio), (hepB-DTP-a-Hib-IPV) • pneumococcal conjugate (13vPVC) • rotavirus
4 months	• hepatitis B, diphtheria, tetanus, acellular pertussis (whooping cough), *Haemophilus influenzae* type b, inactivated poliomyelitis (polio), (hepB-DTP-a-Hib-IPV) • pneumococcal conjugate (13vPVC) • rotavirus
6 months	• hepatitis B, diphtheria, tetanus, acellular pertussis (whooping cough), *Haemophilus influenzae* type b, inactivated poliomyelitis (polio), (hepB-DTP-a-Hib-IPV) • pneumococcal conjugate (13vPVC) • rotavirus b
12 months	• *Haemophilus influenzae* type b, meningococcal C (Hib-MenC) • measles, mumps and rubella (MMR)
18 months	• measles, mumps, rubella and varicella (chickenpox) (MMRV)
4 years	• diphtheria, tetanus, acellular pertussis (whooping cough) and inactivated poliomyelitis (polio) (DTPa-IPV) • measles, mumps and rubella (MMR) (to be given only if MMRV vaccine was not given at 18 months)
School programs	
Age	**Vaccine**
10–15 years	• varicella (chickenpox) • human papillomavirus • diphtheria, tetanus and acellular pertussis (whooping cough) (dTpa)

TABLE 8.1.3 Australian child and school immunisation programs.

Herd immunity

Immunisation is critical, not only for the person immunised, but also for the health of the wider community. For an immunisation program to be successful, a sufficient number of people need to be vaccinated—a phenomenon called **herd immunity** (Figure 8.1.11). The more people who are vaccinated, the less chance there is of an infectious agent spreading throughout a population, because there will be fewer potential carriers. Herd immunity is essential for the protection of those who cannot be vaccinated or who have suppressed immune systems. This includes newborn babies, the elderly, people suffering from an immune disease and people taking immunosuppressant medication.

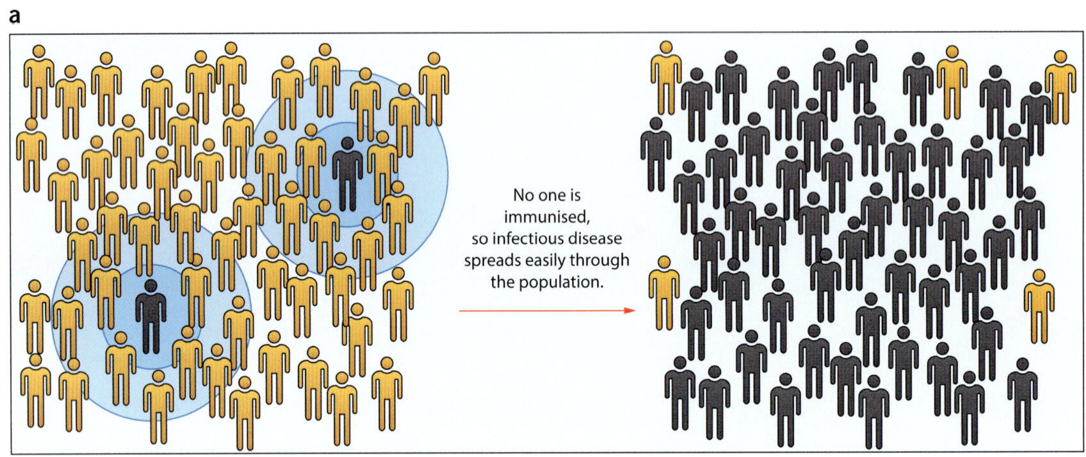

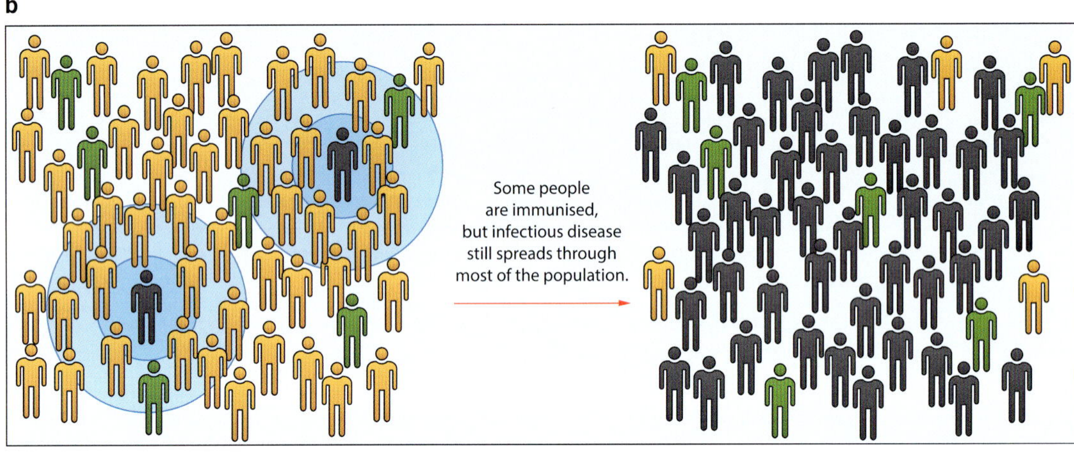

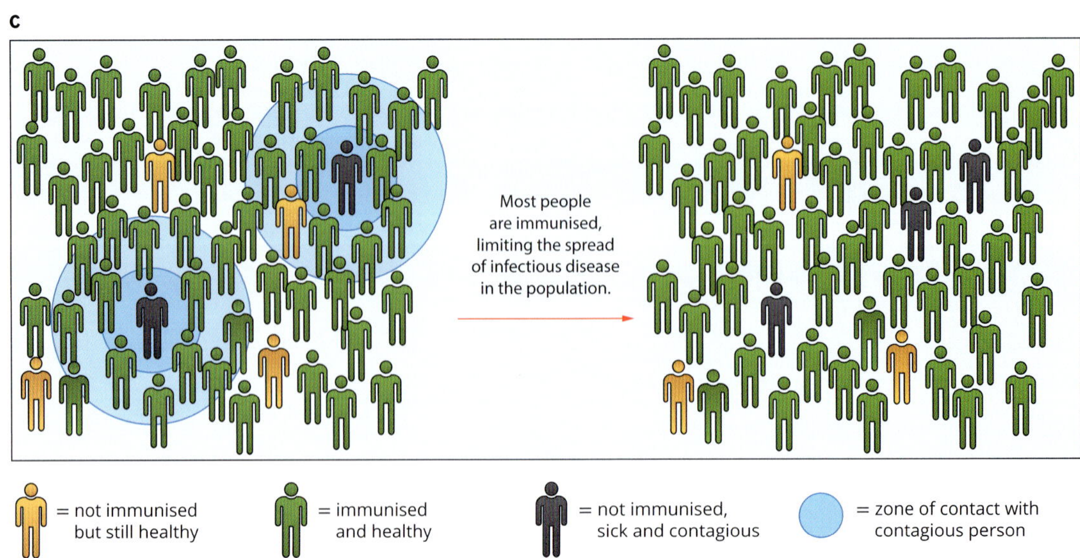

FIGURE 8.1.11 Three diagrams illustrating the effectiveness of herd immunity. (a) With no immunisation in a community, infectious diseases spread easily. (b) With some immunisation in a community, infectious diseases spread less easily. (c) When most of the community is immunised, there are few carriers or infected people and minimal spread of infectious diseases. This is known as herd immunity.

Breakdown of herd immunity: whooping cough

Immunological memory reduces over time, thereby reducing the herd immunity of immunised populations. An example of the breakdown in herd immunity is the recent spike in Australia in cases of whooping cough, a disease caused by the bacterium *Bordetella pertussis*. Although it only causes a persistent cough in adults, approximately 1 in 200 babies under the age of six months who become infected will die. Babies cannot be vaccinated until they are six weeks old and they are not fully protected by this vaccine until about six months of age.

One of the reasons for this breakdown in herd immunity is that not enough people get booster vaccinations. A public education campaign has been implemented to encourage adults to receive a booster vaccination to maintain herd immunity against whooping cough. New parents are offered the booster vaccination when their baby is born, and are encouraged to recommend the vaccination to family and friends who will be in close contact with their baby.

BIOLOGY IN ACTION

Influenza vaccines

Due to their high rate of mutation, influenza (or flu) viruses evolve so rapidly that the immunity developed to one year's strains usually doesn't provide protection against the next year's strains. This is true whether you have caught the flu and developed natural active immunity, or have received a flu vaccine and developed artificial active immunity. The reason the immune system has trouble recognising new strains of influenza viruses is because genetic changes have caused the flu viruses to express different antigens. This change in antigens is called antigenic drift.

To keep up with antigenic drift, new flu vaccines are released every year. Although it is a single injection, each new flu vaccine is a cocktail of vaccines for different influenza strains, and the decision on which strains to vaccinate against is based on what strains are predicted to be the most common in the coming flu season. However, sometimes the most common strains are unexpected, in which case the latest flu vaccine provides little protection.

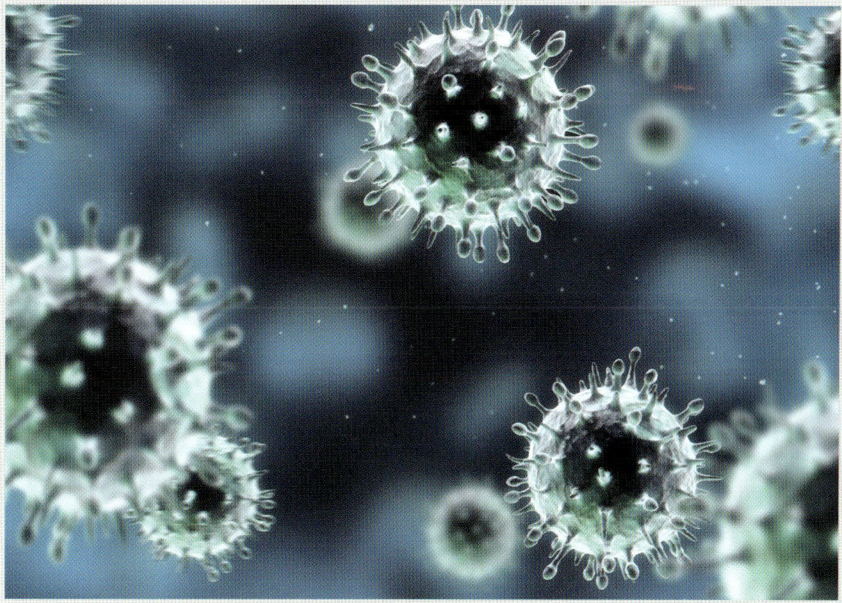

FIGURE 8.1.12 Digital illustration of the H1N1 strain of swine influenza virus.

One of the reasons for new influenza strains being unexpected is the ability of influenza viruses to swap genetic material in a single host. This is known as antigenic shift, because it occurs more rapidly and results in more major changes than antigenic drift. An example of antigenic shift is the H1N1 strain of swine flu virus that was first detected in 2009 (Figure 8.1.12). The reassortment of genes that resulted in H1N1 is thought to have occurred in North American and Eurasian pig herds. The eight RNA strands of H1N1 include one strand from a human influenza strain, two strands from bird (or avian) influenza strains, and five strands from pig (or swine) influenza strains.

8.1 Review

SUMMARY

- Immunity can develop naturally or be induced artificially.
- Passive immunity involves the transfer of antibodies produced in another organism. It does not result in immunological memory and is temporary.
- Active immunity involves the individual's adaptive immune response. It results in immunological memory that can be long-lasting.
- Natural passive immunity is the result of antibodies naturally produced by another organism providing immunity, e.g. maternal antibodies transferred through the placenta and breast milk.
- Artificial passive immunity involves an individual receiving, usually by injection of antiserum, antibodies produced by another organism. The antibodies bind to antigens on the pathogen or toxin, preventing them from causing damage; but may also prevent the body's own adaptive immune response, thereby preventing the development of immunological memory.
- Natural active immunity develops from the adaptive immune response to a natural infection, and the immunological memory that results.
- Artificial active immunity results from the administration of antigens to induce an adaptive immune response, i.e. vaccination. This results in the generation of immunological memory.
- Immunological memory provides a stronger and faster secondary immune response if exposed to the same antigen again.
- Live attenuated vaccines involve a living microbe that has been weakened in the laboratory, usually through repeated culturing.
- Inactivated (or killed) vaccines are composed of microbes inactivated by heat, radiation or chemical means.
- Subunit vaccines contain a fraction of an antigen, a single antigen or multiple antigens, and these antigens can be proteins, detoxified toxins or polysaccharides.
- Toxoid vaccines are non-recombinant subunit vaccines that use toxins inactivated by formalin (called toxoids).
- Vaccination programs are set up by governments in an effort to produce herd immunity.
- Herd immunity is the result of large numbers of people being immune to a pathogen, thus reducing the chance of the pathogen successfully spreading through a population.

KEY QUESTIONS

1 Explain the difference between active and passive immunity.

2 The following graph represents changes in antibody concentrations that occur during a primary and secondary adaptive immune response. Provide appropriate text for labels A, B, C and D.

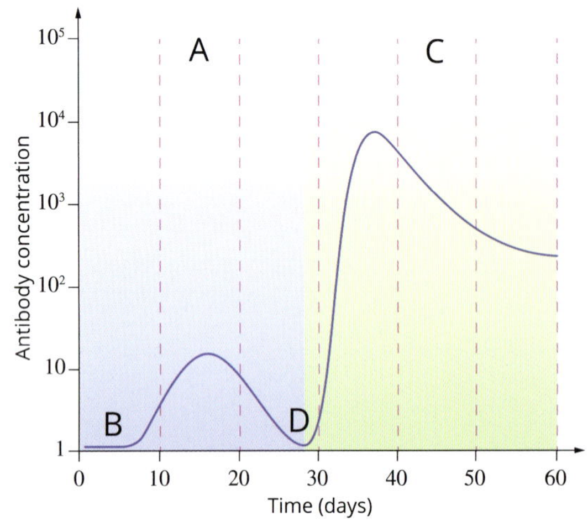

3 Vaccination is an example of:

A artificial passive immunity
B natural active immunity
C artificial active immunity
D natural passive immunity

4 Explain the difference between natural passive immunity and artificial passive immunity. Give examples.

5 Describe herd immunity.

8.2 Diseases of the immune system

In Chapter 7, you learnt that the role of the immune system is to eliminate invading pathogens. However, when the immune system malfunctions and doesn't react to the antigens it should, or it reacts to antigens that are normally harmless, such as those found on pollen grains (Figure 8.2.1), or to self-antigens that are normally tolerated, it can result in mild, moderate or severe illnesses.

In this section, you will learn about hypersensitivity reactions such as allergy, about autoimmune diseases such as multiple sclerosis that result in other types of hypersensitivity reactions, and about causes of a weakened immune system (or immunodeficiency), such as HIV/AIDS.

FIGURE 8.2.1 A digitally enhanced coloured scanning electron micrograph of different types of pollen grain. Pollen can cause an allergic immune response known as hay fever.

HYPERSENSITIVITY REACTIONS

Hypersensitivity reactions occur when the immune system reacts to antigens that pose no threat to the body, causing cell damage and disease. Hypersensitivity reactions are classified into four types, depending on the antigens and immune mechanisms that result in disease:

- type I (or immediate) hypersensitivity
- type II (or cytotoxic) hypersensitivity
- type III (or immune complex) hypersensitivity
- type IV (or delayed-type) hypersensitivity.

Type I hypersensitivity is also known as **allergy**. Autoimmune diseases can result in type II, type III and type IV hypersensitivity reactions. However, autoimmune diseases are not the only causes of these types of hypersensitivity.

Immediate hypersensitivity (type I)

Immediate hypersensitivity reactions (or **allergic reactions**) are due to a rapid and vigorous overreaction of the immune system to antigens that would otherwise be harmless. Antigens that result in type I hypersensitivity reactions are called allergens. Typical allergenic substances include **pollen**, fur, house dust, latex and foods such as peanuts, lobster and monosodium glutamate (MSG). Depending on the particular individual and antigen, the hypersensitivity reaction can range from mild to being a life-threatening reaction known as **anaphylaxis**.

An allergic reaction to pollen is called allergic rhinitis (or hay fever). It is triggered by pollen particles, which carry allergenic antigens on their surfaces (Figure 8.2.2). Grass and tree pollens are the most common cause of hay fever in Australia and New Zealand. Pollen sensitivity has a seasonal pattern of occurrence, as pollen is most abundant in the atmosphere during spring and early summer.

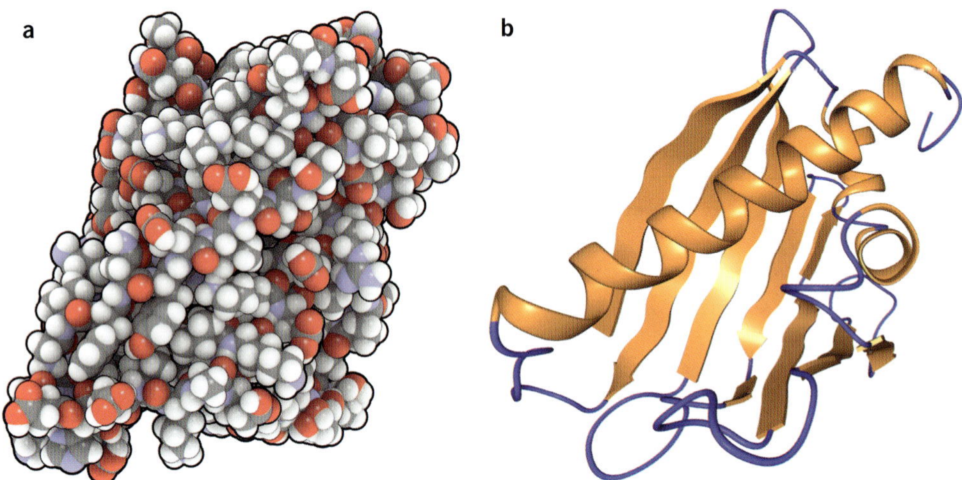

FIGURE 8.2.2 Two different models representing the molecular structure of pollen from a birch tree. (a) A representation showing all the atoms in the molecule, with different coloured balls representing different atoms (carbon (black), oxygen (red) and hydrogen (white)).
(b) A representation of the secondary structures present in the pollen protein. Alpha helices (orange coils), beta-pleated sheets (orange strips) and random coils (blue loops), are easily identified in this model.

Whatever the allergen, mast cell release of histamine is central to immediate hypersensitivity reactions. Allergic reactions are mediated by a specific type of antibody called **immunoglobulin E (IgE)**. IgE is produced by plasma cells and travels in the bloodstream. When IgE comes into contact with mast cells, which are common in epithelial and mucosal tissues, the tail end of the IgE antibody binds to receptors on the cell surface. Upon subsequent exposure to the same allergen, the allergen binds to a pair of adjacent IgE molecules, bridging the gap between (or cross-linking) the two IgE molecules. This binding triggers a signal transduction cascade that causes the mast cells to release histamine (and other mediators of inflammation) from their intracellular vesicles by exocytosis (Figure 8.2.3).

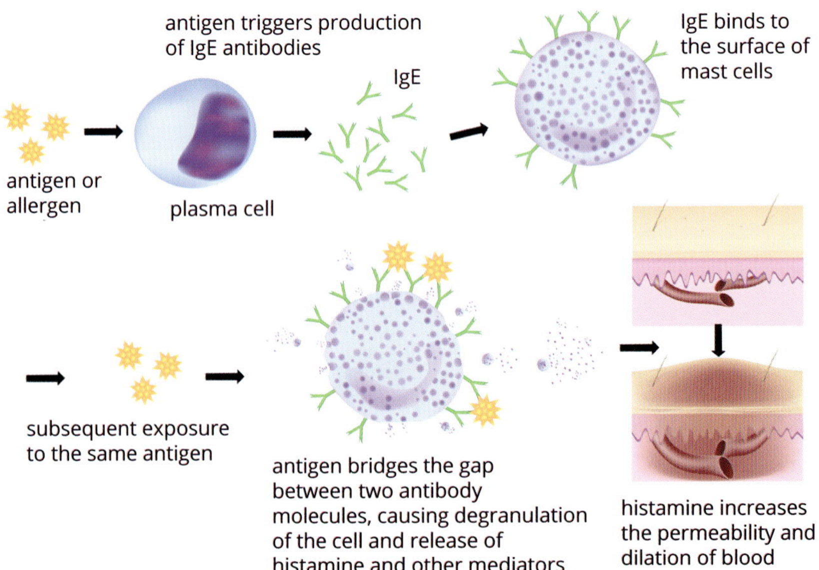

FIGURE 8.2.3 Step 1: Initial exposure to allergens (e.g. pollen) triggers plasma cells to produce IgE molecules specific to the antigen. Step 2: The tail end of the IgE binds to receptors on mast cells. Step 3: Subsequent exposure to the same allergen, causes the antigen to bind to two IgE molecules on a mast cell. Step 4: This binding triggers a signal cascade that causes the release of histamine. Step 5: Histamine binds to receptors on various cells in the body, which produces the classic features of an allergic reaction.

Histamine is an organic nitrogenous compound that binds to specific receptors on various cell types. Histamine causes:

- blood vessel dilation
- a decrease in blood pressure
- an increase in the permeability of blood vessels to immune cells and fluids for a better immune response at the site of antigen contact
- contraction of smooth muscles lining the airways, which can make it more difficult to breathe
- activation of fluid-secreting cells, which results in a runny nose, teary eyes and sneezing, which expels foreign antigens.

An antihistamine is a drug that counteracts the effects of histamine by blocking histamine receptors and therefore suppressing some allergy symptoms.

EXTENSION

Treatment for allergic reactions

Medications

Antihistamines block the effects of histamine by binding to the same receptors as histamine, thereby preventing histamine from binding to them. Other medications used to treat allergies, for example cortisone, suppress the immune system more broadly and reduce the immune response in general.

If the allergic reaction is severe enough to cause anaphylaxis, an immediate intramuscular injection of adrenaline (or epinephrine) is needed (Figure 8.2.4). Adrenaline auto-injectors are commonly known and marketed as EpiPens (Figure 8.2.5). Adrenaline counters the actions of histamine by causing:

- blood vessels to constrict (decreasing swelling and increasing blood pressure)
- muscles in the airways to relax (so the airways open up)
- the heart to beat faster, which increases the blood flow to the heart. (This prevents cardiovascular collapse, which results from a lack of effective blood flow to the heart due to excessive dilation of the blood vessels that results from too much histamine.)

Allergen immunotherapy

Allergen immunotherapy (or desensitisation) is used to treat hypersensitive reactions to particular allergens, such as bee sting toxin. Beginning with extremely small amounts, an allergen is injected multiple times in increasing amounts over a period of months. This causes the formation of specific immunoglobulin G (IgG) antibodies against the allergen. IgG antibodies are a key component of the humoral immune response and are the main immunoglobulins in blood serum.

If IgG antibodies react with the allergen before it can bind to IgE antibodies to cause an allergic response, the allergic response is prevented. During the course of treatment, the individual slowly becomes less and less sensitive, in other words becomes desensitised, to the particular allergen being treated.

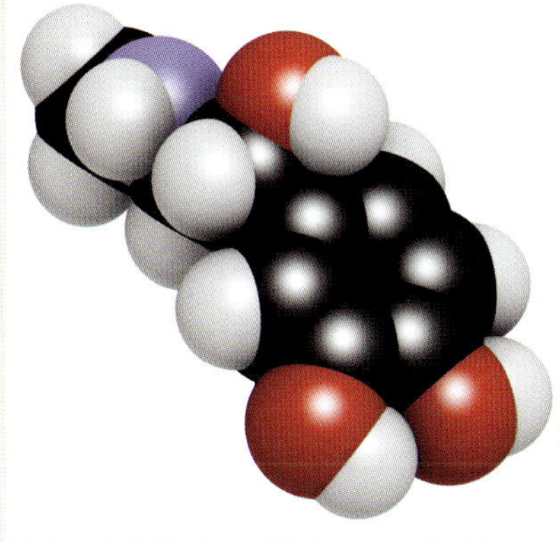

FIGURE 8.2.4 A molecular model of adrenaline (epinephrine), a hormone and neurotransmitter. Atoms: hydrogen (white), carbon (black), oxygen (red), nitrogen (blue).

FIGURE 8.2.5 This girl is using an adrenaline auto-injector to prevent herself going into anaphylactic shock.

Cytotoxic hypersensitivity (type II)

Unlike type I allergic reactions, which involve IgE directed against allergens and usually occur within minutes of exposure to them, type II cytotoxic hypersensitivity reactions involve IgM and IgG antibodies directed against cell surface or extracellular matrix antigens, and can take hours to develop.

An example of type II sensitivity is the antibody-mediated destruction of red blood cells (haemolytic anaemia) that occurs in the newborn when the mother produces antibodies directed against Rhesus antigen on foetal red blood cells (see Section 8.1, page 297). Type II hypersensitivity reactions can also occur as a side effect of taking certain medications. For example, penicillin binds to red blood cells, and if anti-penicillin antibodies are present they bind to the drug and trigger the destruction of the red blood cells to which the penicillin is bound.

Immune complex hypersensitivity (type III)

Like type II cytotoxic hypersensitivity, type III immune complex hypersensitivity reactions also involve IgG, and sometimes IgM. However, the antibodies in type III reactions are directed against soluble antigens, not antigens on the cell surface or extracellular matrix. When the antibodies bind to soluble antigens, they form antigen–antibody immune complexes that can be deposited in tissues, causing inflammation and damage. Type III reactions can also take hours or days to develop.

An example of type III hypersensitivity is serum sickness. It occurs when immune complexes form as a result of a person's antibodies binding to a foreign antigen in serum they have been injected with.

Delayed-type hypersensitivity (type IV)

Unlike type I, II and III hypersensitivity reactions, which are all mediated by antibodies, type IV hypersensitivity is mediated by helper T lymphocytes, which activate macrophages and eosinophils to produce inflammatory responses, and cytotoxic T lymphocytes, which directly attack and kill other cells. As the name suggests, delayed-type hypersensitivity reactions are delayed and take days to develop.

An example of type IV hypersensitivity is the rash caused by coming into contact with poison ivy, which has a lipid-soluble compound called 3-pentadecacatechol (Figure 8.2.6). As this compound is lipid-soluble it can cross the cell membrane. Once inside the cells, it causes new peptides to be produced. These peptides are then delivered to the cell surface, where they are recognised by cytotoxic T lymphocytes.

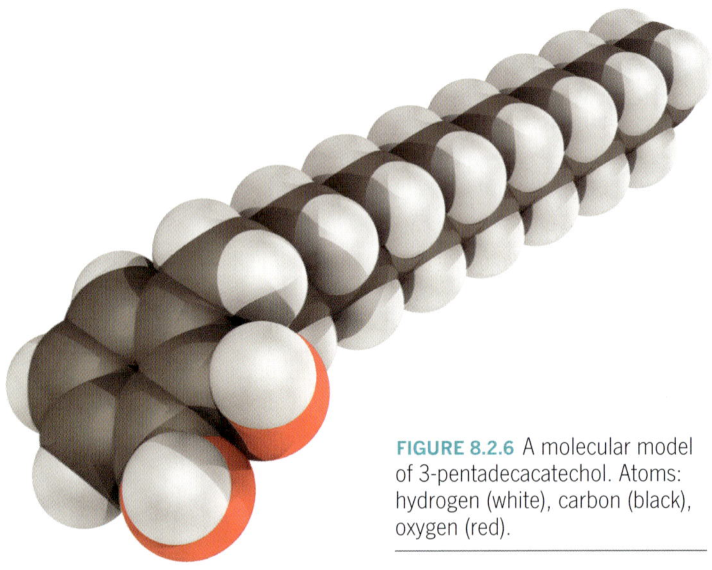

FIGURE 8.2.6 A molecular model of 3-pentadecacatechol. Atoms: hydrogen (white), carbon (black), oxygen (red).

AUTOIMMUNE DISEASES

As you learnt in Chapter 7, normally T and B lymphocytes that are reactive against self-antigens are destroyed. This means that when your immune response is working properly it is directed against non-self antigens, not against self-antigens, and this is known as self-tolerance. **Autoimmune diseases** result from a failure of self-tolerance, which leads to an adaptive immune response directed against specific self-antigens.

Autoimmune diseases result from an adaptive immune response directed against self-antigens.

When autoimmune diseases occur, the cytotoxic T lymphocytes of the adaptive immune response attack the tissues directly; B lymphocytes act indirectly by secreting antibodies. Mast cells are also activated and release histamines, which results in inflammation around the affected tissues.

Particular autoimmune diseases and combinations of autoimmune diseases tend to be inherited and are more common in females. Environmental factors also seem to have an effect in the development of autoimmunity. Researchers are trying to understand the causes and risk factors, as well as the rise of autoimmune diseases in industrialised countries.

Types of autoimmune disease

Autoimmune diseases can be organ-specific or generalised:

- Organ-specific autoimmune diseases are localised to a particular part of the body. For example, multiple sclerosis only affects the brain and spinal cord.
- Generalised autoimmune diseases are those that occur widely throughout the body. For example, systemic lupus erythematosus affects various organs and tissues of the body, such as the joints, skin, kidneys and the brain.

You have already learnt that type II, III and IV hypersensitivity reactions can be caused by a range of things that are unrelated to autoimmune disease. However, autoimmune diseases can also result in these different types of hypersensitivity reactions. You will now learn about some of those diseases.

Autoimmune haemolytic anaemia

Autoimmune haemolytic anaemia is an example of an autoimmune disease that results in a type II hypersensitivity reaction, because it involves antibodies directed against self-antigens (auto-antibodies) on the surface of red blood cells (Figure 8.2.7).

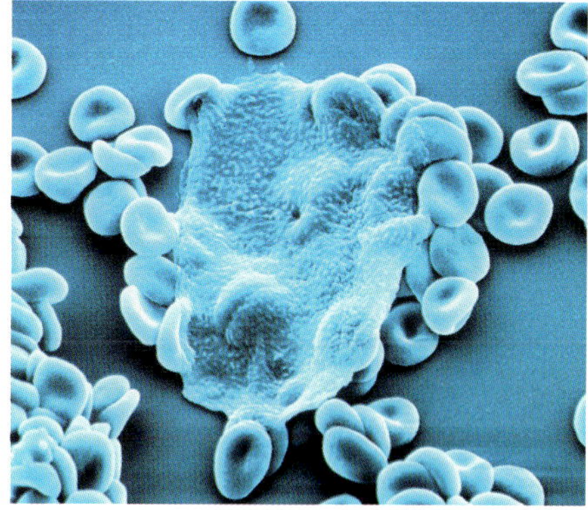

FIGURE 8.2.7 Coloured scanning electron micrograph of a macrophage (centre) engulfing multiple smaller red blood cells around it. Red blood cells recognised by autoantibodies are rapidly phagocytosed by macrophages.

Rheumatoid arthritis

Rheumatoid arthritis is an example of an autoimmune disease that results in a type III hypersensitivity reaction. It is a type III reaction because it involves the deposition of antigen–antibody immune complexes in tissue, which results in inflammation and damage. Rheumatoid arthritis mainly affects the joints. Commonly affected joints are those of the knees and hands (Figure 8.2.8). Rheumatoid arthritis is also thought to result in type IV hypersensitivity reactions, in which T lymphocytes attack an as yet unidentified antigen in the joints.

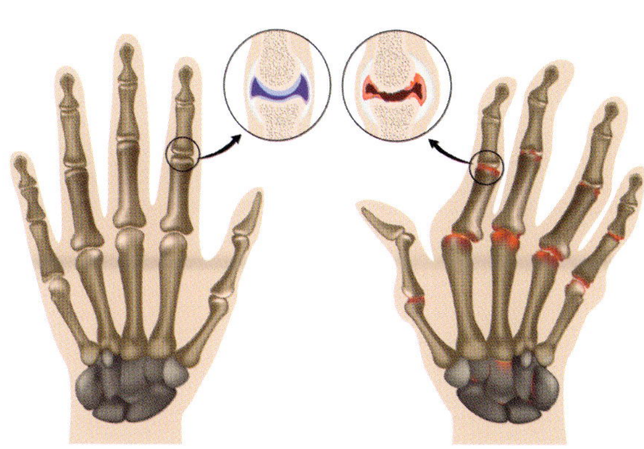

FIGURE 8.2.8 Rheumatoid arthritis causes inflammation in the joints of the hand, and over time can lead to disfigurement.

BIOFILE

Autoimmune diseases of the thyroid

Graves' disease and Hashimoto's disease are both autoimmune diseases that affect the thyroid. In Graves' disease, antibodies bind to a specific receptor on thyroid cells that is normally stimulated by signals from the pituitary gland in the brain. This binding leads to an overactive thyroid gland (hyperthyroidism), because it results in the excessive production of thyroid hormone. Graves' disease can also cause the eyes to bulge (Figure 8.2.9).

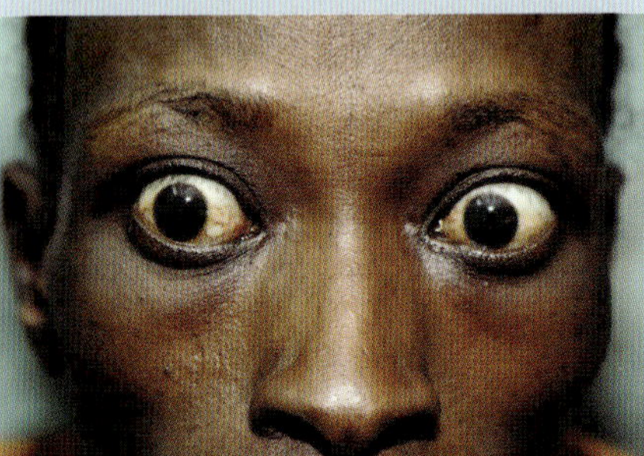

FIGURE 8.2.9 Bulging eyes are a symptom of Graves' disease.

In Hashimoto's disease, T lymphocytes and antibodies attack thyroid cells, which results in an underactive thyroid (hypothyroidism), because not enough thyroid hormone is produced. Both Graves' disease and Hashimoto's disease can result in goitre, which is the swelling of the neck due to an enlarged thyroid gland (Figure 8.2.10). In Graves' disease, goitre results from too much thyroid hormone. In Hashimoto's disease, the pituitary gland in the brain senses the low levels of thyroid hormone, so it produces thyroid-stimulating hormone, which causes the thyroid gland to enlarge.

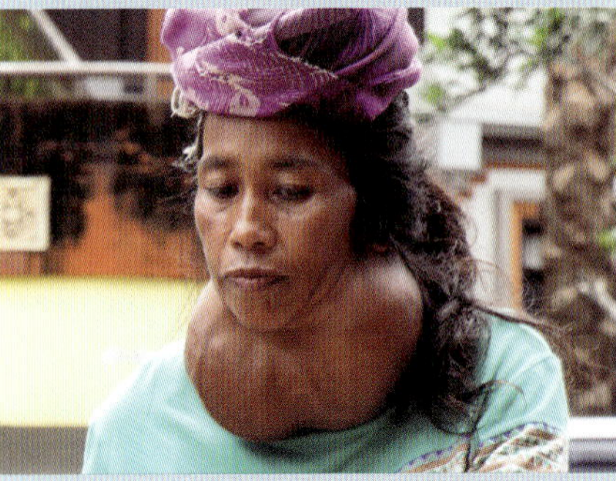

FIGURE 8.2.10 A person with goitre.

Type 1 diabetes

In type 1 diabetes, T lymphocytes attack and destroy beta cells in the pancreas (Figure 8.2.11). Beta cells produce insulin, which regulates the levels of glucose in the blood. People with type 1 diabetes inject insulin to maintain glucose balance. Because the immune reaction is mediated by T lymphocytes, type 1 diabetes is an example of a type IV hypersensitivity reaction.

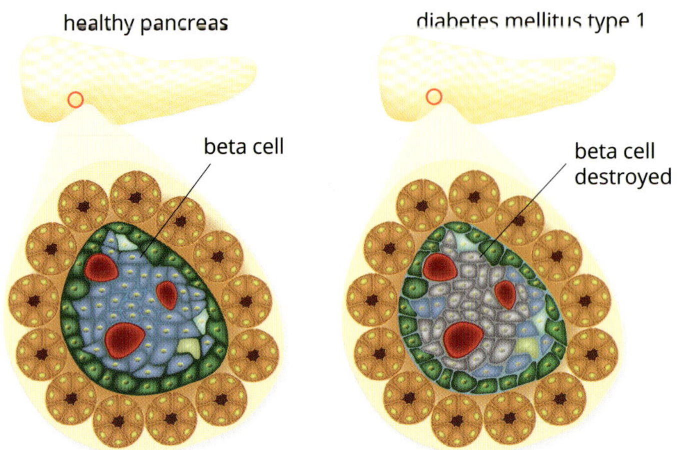

FIGURE 8.2.11 In type 1 diabetes, T lymphocytes attack and destroy beta cells in the pancreas.

Multiple sclerosis

Neurons are the cells of the nervous system that relay signals along their axons. The **central nervous system** is made up of several different types of cell other than neurons. **Oligodendrocytes** are one of these other types of cells, and they produce a substance called **myelin**, which is composed mostly of lipids and some protein. Myelin forms an insulating sheath around the axons. Although some nerves in the peripheral nervous system also have a myelin sheath, multiple sclerosis generally only affects the myelin sheath in the central nervous system (Figure 8.2.12).

In multiple sclerosis, the nerves lose the myelin sheaths insulating them. When this happens, the conduction of signals along the nerve is impaired, and the nerve is eventually damaged.

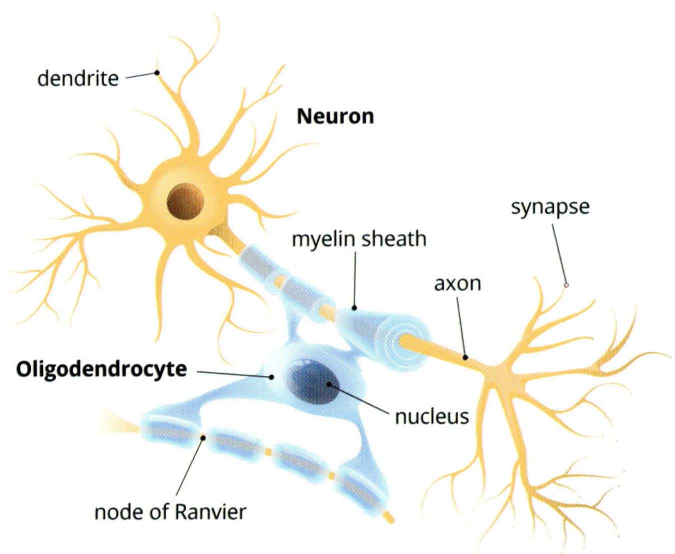

FIGURE 8.2.12 Healthy oligodendrocytes produce multiple myelin sheaths, which surround nerve axons.

It is known that both helper and cytotoxic T lymphocytes are involved in multiple sclerosis, and that plasma cells produce antibodies that target proteins and lipids in the myelin sheath. Mitochondria in oligodendrocytes are damaged, leading to the release of signalling molecules that induce apoptosis in these cells (see Section 6.3, page 250). Macrophages called **microglia** specific to the central nervous system are also involved in oligodendrocyte destruction (Figure 8.2.13). The causes of multiple sclerosis are still not fully understood, but it resembles a type IV hypersensitivity reaction because it is mediated by T lymphocytes and involves the activation of macrophages, which results in inflammation.

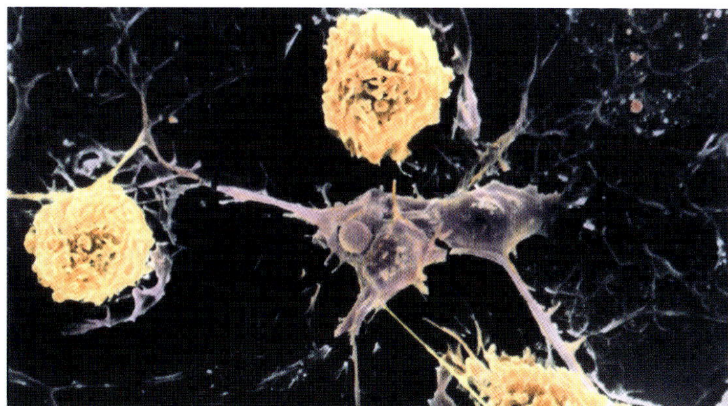

FIGURE 8.2.13 Coloured scanning electron micrograph of microglial cells (yellow), attacking oligodendrocytes (pink).

However it occurs, damage to and death of oligodendrocytes leads to destruction of the myelin sheaths (or **demyelination**). Once the myelin sheaths are damaged, damage also occurs to the nerve axons themselves (Figure 8.2.14).

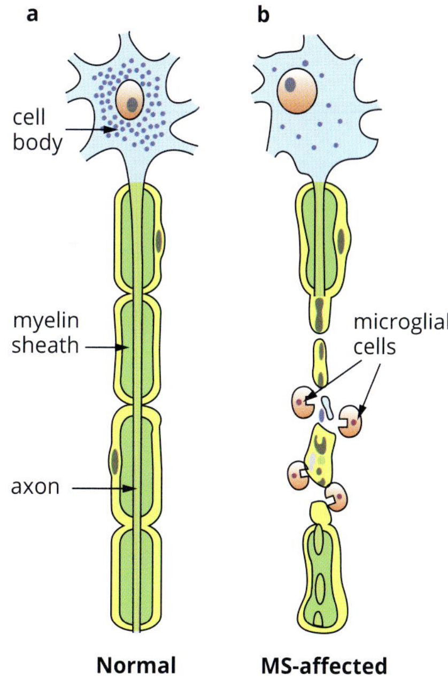

FIGURE 8.2.14 Comparison of the structure and function of myelin sheaths surrounding an axon in (a) a healthy neuron and (b) an MS-affected neuron.

Symptoms vary from person to person and may include visual, motor and sensory problems, such as double vision, tiredness, numbness, muscle weakness, sensitivity to heat, and difficulty with balance and coordination. Symptoms can also include mental problems such as memory lapses, mood swings, depression and epilepsy.

There is no cure for multiple sclerosis. However, certain medications can manage symptoms and delay the progression of the disease. Standard treatments for multiple sclerosis include medications that reduce inflammation (steroids), and that suppress the immune response (immunosuppressants).

IMMUNODEFICIENCY

In contrast to hypersensitivity reactions, which occur when the immune system reacts inappropriately to antigens, **immunodeficiency** occurs when the immune system cannot adequately respond to antigens, or fails to react at all. There are two main forms of immunodeficiency disorders: primary and secondary.

Primary immunodeficiency is **congenital**, meaning that a child is born with the deficiency either as a result of a genetic defect or a developmental abnormality. DiGeorge syndrome is a primary immunodeficiency disease in which the thymus can fail to develop, resulting in impaired production of mature T lymphocytes. In another disease called severe combined immunodeficiency (SCID), B and T lymphocytes do not develop properly or function correctly. SCID is also known as the ‘bubble baby disease’ because people with it need to be isolated from all pathogens present in the environment.

Secondary immunodeficiency is acquired, and results from an external factor, rather than a genetic factor. Secondary immunodeficiency can be temporary or permanent. Temporary immunodeficiency can develop as a result of severe stress or malnutrition. An example of permanent secondary immunodeficiency is **acquired immunodeficiency syndrome (AIDS)**, which results from infection with **human immunodeficiency virus (HIV)**.

HIV/AIDS

HIV is a retrovirus, which is an RNA virus that makes a DNA copy of its genetic information by reverse transcription. After transcription, the viral DNA is inserted into the host cell DNA, and when the DNA is expressed in the host cell, it also results in the production of copies of the virus (Figure 8.2.15).

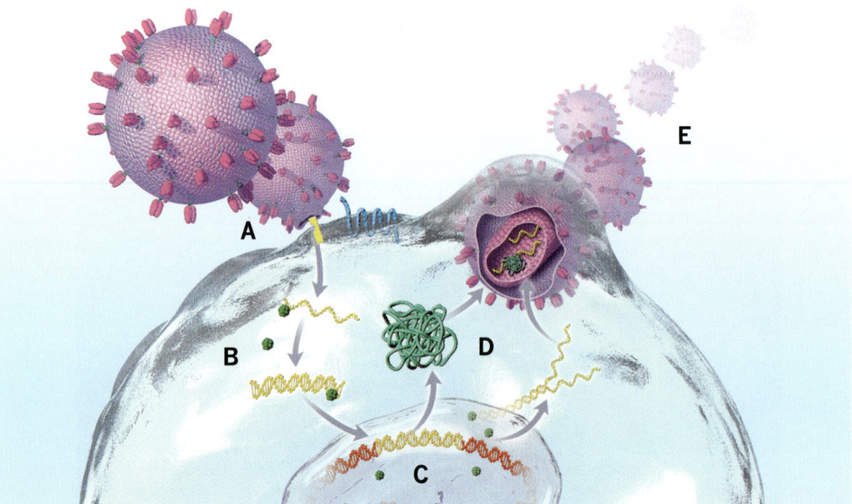

FIGURE 8.2.15 HIV binds to the target host cell (A) and fuses to the cell membrane. It releases viral RNA and enzymes needed for its replication into the cytoplasm. One of these enzymes, called reverse transcriptase, then converts the viral RNA into double-stranded DNA in the cytoplasm (B). This viral DNA enters the nucleus and is integrated into the host's DNA by another viral enzyme called integrase (C). When the viral DNA is expressed, the host cell synthesises the building blocks for a new virus, some of which are processed by another viral enzyme called HIV protease (D). All the components of the new viruses come together before exiting the cell (E).

Although the immune system initially responds to the HIV virus, some copies of the virus survive. One reason for this is the high rate of mutation of HIV, which helps it keep one step ahead of the adaptive immune response. Over time, HIV not only avoids the immune response, it also impairs it. In fact, helper T lymphocytes are the type of cells preferentially infected by HIV (Figure 8.2.16). The destruction of helper T lymphocytes impairs the adaptive immune response because helper T lymphocytes activate both humoral and cell-mediated responses.

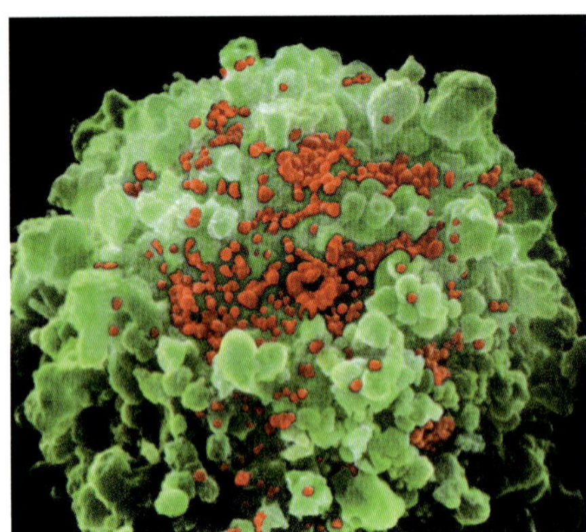

FIGURE 8.2.16 Coloured scanning electron micrograph of a helper T lymphocyte (green) infected with HIV (red), the causative agent of AIDS.

When the number of helper T lymphocytes becomes very low as a result of HIV infection, the adaptive immune system is impaired and AIDS has developed. AIDS makes people susceptible to opportunistic infections, and puts them at greater risk of certain types of cancers. It is these opportunistic infections that usually cause the death of AIDS patients.

HIV transmission

HIV is present in the body fluids of HIV-positive people, including blood, semen, vaginal secretions and breast milk (but not saliva). HIV can be transmitted:

- by unprotected vaginal or anal sex
- by sharing/using contaminated needles
- from HIV-positive mother to foetus during pregnancy, or to baby during birth or by breastfeeding

HIV is not transmitted by:

- saliva, sweat, or tears of HIV-positive people
- mosquitoes or other blood-sucking insects

HIV treatment

There is no cure for HIV, but there are treatments available that extend the life of people infected with the virus. Antiretroviral therapy (ART) involves taking a combination of medicines that prevent the virus from replicating, thereby reducing the amount of HIV in the body, and preventing AIDS from developing. The multiple drugs involved in ART act by blocking:

- the binding and fusion of HIV to the host cells
- reverse transcription (impeding the transcription of the viral genetic material, RNA, to DNA), and
- the enzymes that integrate the viral DNA into the host cell's DNA or are involved in assembly of viral proteins.

8.2 Review

SUMMARY

- Hypersensitivity reactions occur when the immune system overreacts. They are classified into four types:
  - type I (or immediate) hypersensitivity, also known as allergy
  - type II (or cytotoxic) hypersensitivity
  - type III (or immune complex) hypersensitivity
  - type IV (or delayed-type) hypersensitivity
- Antigens that trigger an allergic reaction are called allergens, and include pollens, dust, fur and foods.
- Allergic reactions involve the production of IgE, which attaches to the surface of mast cells. When an allergen cross-links two IgE molecules on a mast cell, it triggers a signal transduction cascade that results in the release of histamine.
- Histamine is a compound that dilates blood vessels, increasing their permeability and promoting inflammation.
- Self-tolerance is the inability to recognise self-antigen.
- Autoimmune diseases result from a failure of self-tolerance, leading to adaptive immune responses against self-antigens. Autoimmune diseases can be organ-specific or generalised.
- Autoimmune diseases can result in type II, type III and type IV hypersensitivity reactions.
- Multiple sclerosis is an organ-specific autoimmune disease that affects the central nervous system. Both helper and cytotoxic T lymphocytes are involved in multiple sclerosis, and plasma cells produce antibodies against the proteins and lipids in the myelin sheath that insulates nerve cell axons. The damage to the myelin sheath impairs the conduction of nerve impulses.
- Immunodeficiency is when the immune system cannot adequately respond to antigens, or fails to react at all.
- Immunodeficiency can be primary or secondary:
  - Primary immunodeficiency is congenital and examples include DiGeorge syndrome and SCID (bubble babies).
  - Secondary immunodeficiency is acquired and can be temporary (malnutrition) or permanent (AIDS).
- AIDS results from untreated infection with HIV, a retrovirus that has such a high rate of mutation that it evolves and evades the immune system. Over time the virus impairs the adaptive immune response by reducing the number of helper T lymphocytes, making infected people susceptible to opportunistic infections and certain types of cancers.

KEY QUESTIONS

1 What type of hypersensitivity reaction are allergic reactions?
   A immediate hypersensitivity (type I)
   B cytotoxic hypersensitivity (type II)
   C immune complex hypersensitivity (type III)
   D delayed-type hypersensitivity (type IV)

2 Explain why type 1 diabetes is an example of a type IV hypersensitivity reaction.

3 Explain the difference between primary and secondary immunodeficiency.

4 Explain how human immunodeficiency virus (HIV) results in acquired immunodeficiency syndrome (AIDS).

8.3 Cancer immunotherapy

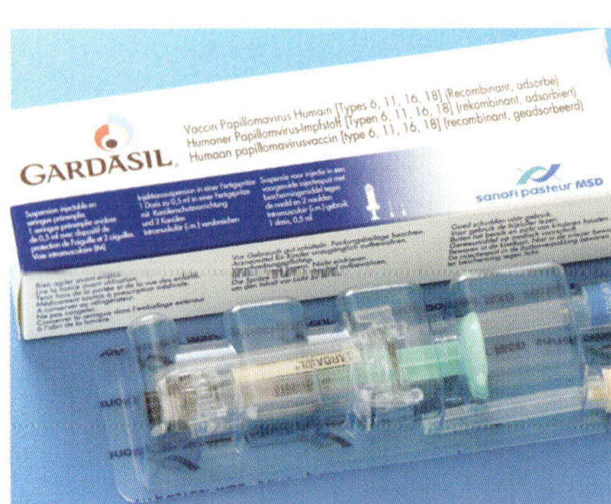

FIGURE 8.3.1 Gardasil is the brand name for a vaccine that provides immunity against certain strains of human papillomavirus (HPV). The vaccine reduces the risk of developing certain types of cancers associated with HPV infection.

Carcinogens are substances that cause damage to cell DNA.

Immunotherapy is a new frontier in the treatment of cancers. It enables more specific (or personalised) medicines than other types of treatments, and improves outcomes while simultaneously decreasing side effects.

In this section, you will learn about cancer and vaccines that can prevent certain types of cancer (Figure 8.3.1). You will also learn about the different types of immunotherapies available for treating cancer, including monoclonal antibody therapy.

CANCER

There are many different types of cancer that affect many different types of cells, but in all cases cancer results from a single abnormal cell that multiplies uncontrollably and spreads throughout the body.

This uncontrolled growth is the result of changes to genes that control how cells grow and divide, and a resistance of these abnormal cells to apoptosis. These genetic changes can be inherited, or develop as a result of damage to cell DNA during one's lifetime.

Substances that can damage cell DNA are called **carcinogens**. Carcinogens can be physical (e.g. radiation), chemical (e.g. asbestos), or biological (e.g. certain viruses; approximately 15–20% of human cancers are thought to be the result of viruses).

Tumours

A **tumour** forms when the number of abnormal cells has increased significantly, forming a clump of cells. Depending on where the tumours form, they may damage or block the normal function of organs and tissues. Not all tumours are cancerous. **Benign tumours** are not cancerous, because their abnormal cells do not invade nearby tissue or spread throughout the body. Although rare, some benign tumours can become cancerous.

Cancer (or malignancy) relates to **malignant tumours**. The cells of malignant tumours are cancerous because they can invade nearby tissue and spread (or metastasise) from the site at which they originate. **Metastasis** occurs when cancer cells break away from the original tumour (or primary tumour), travel through the blood and lymph vessels, and form secondary tumours at other locations (Figure 8.3.2).

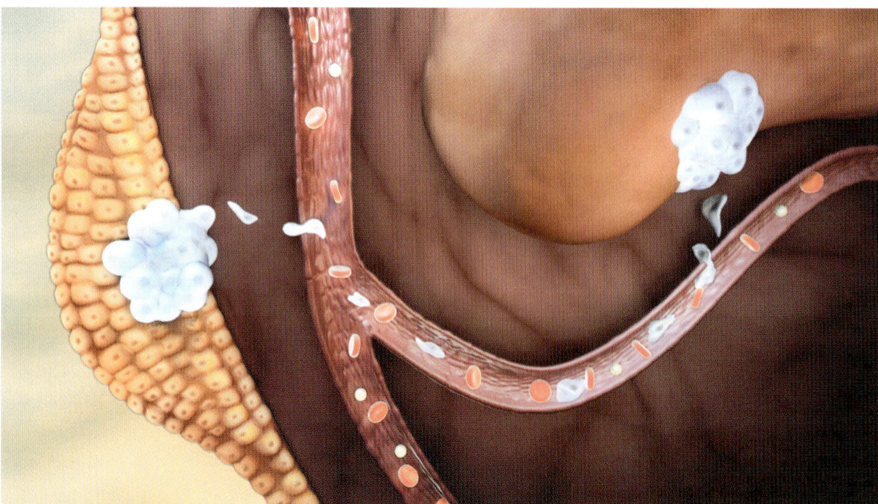

FIGURE 8.3.2 Cancer cells (white) migrating from a tumour and forming a secondary tumour in another organ.

Cancer treatments

Cancer treatments over several decades have included **chemotherapy**, **radiation therapy**, and surgery to remove tumours.

- Chemotherapy involves administering drugs that are **cytotoxic** to cells that multiply rapidly. Although the chemotherapy drugs available today are more specific than those of the past, they are not yet specific enough to avoid damage to healthy cells that also divide rapidly, such as bone marrow and hair follicle cells. This is why chemotherapy has bad side effects.
- Radiation therapy indiscriminately kills cells by damaging their DNA. In cancer treatment, radiotherapy is directed at the cancerous cells, but some damage to surrounding tissue is inevitable.
- Surgery is beneficial in removing solid tumours but it also has its disadvantages. Any surgery takes a toll on the body, and often it is very difficult to ensure that all malignant cells are removed from the body.

The immune response to cancer

People who are older, who use immunosuppressive medications for a long period of time, or who have immunodeficiency, have weakened immune systems and and are at increased risk of developing cancer. However, people do not need to be immunodeficient for their immune system to be ineffective against cancer.

Even people with a normal immune system are unable to control the growth of tumours. This is because tumour cells evade the immune response in a number of ways. They do this by expressing defective MHC-I molecules (so that cytotoxic T lymphocytes cannot detect that the cells are defective), by producing immunosuppressive cytokines, or by releasing enzymes that suppress T lymphocyte responses.

FIGURE 8.3.3 Digital illustration of cytotoxic T lymphocytes (white) attacking a cancer cell (red).

Cytokines are a group of peptides and proteins released from cells that are important in cell signalling, particularly between cells of the immune system.

CANCER IMMUNOTHERAPY

Immunotherapy is any treatment that harnesses the immune system of the patient to fight diseases such as cancer (Figure 8.3.3). Immunotherapy can be non-specific or specific:

- Non-specific immunotherapies stimulate the immune system in general; for example, by the injection of cytokines. Cytokines do not directly target cancer cells, but the stimulation of the immune system can result in a better immune response against cancer cells.
- Specific immunotherapies act on cancer cells by directly stimulating the adaptive immune response against them. Specific immunotherapies include cancer vaccines and monoclonal antibody therapy.

Cancer vaccines

Cancer vaccines stimulate the immune system to attack cancer cells. Some cancer vaccines contain peptides or whole proteins of cancer cells and adjuvants to help stimulate an immune response against them. Sometimes a patient's own immune cells are harvested, exposed to these antigens, and then injected back into the body to produce an immune response. Cancer vaccines have few or no side effects. They can be classified as preventive, therapeutic or personalised.

Preventive cancer vaccines

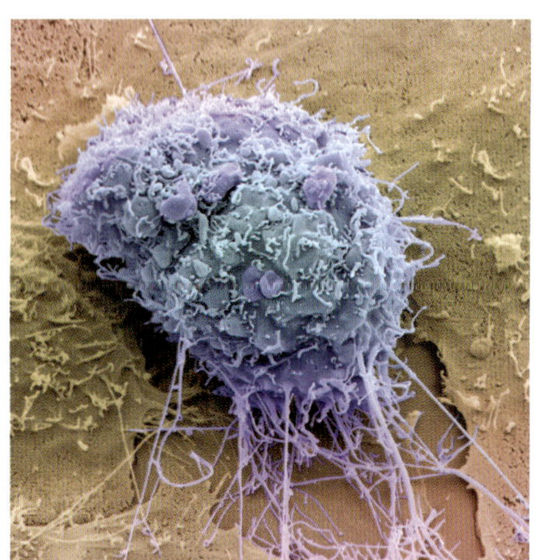

FIGURE 8.3.4 Coloured scanning electron micrograph of a cervical cancer cell.

Preventative cancer vaccines are vaccines directed against viruses that cause cancer. Examples include vaccines for human papillomavirus (HPV), which causes cervical cancer (Figure 8.3.4), and hepatitis B virus (HBV), which causes cancer of the liver. These vaccines introduce specific viral antigens into the body, creating an adaptive immune response that will lead to immunological memory and a stronger and faster response towards the virus if the body is exposed to it.

An example of a preventative cancer vaccine is the HPV vaccine called Gardasil. Professor Ian Frazer and Dr Jian Zhou from the University of Queensland developed Gardasil using a gene for the coat protein of HPV (Figure 8.3.5).

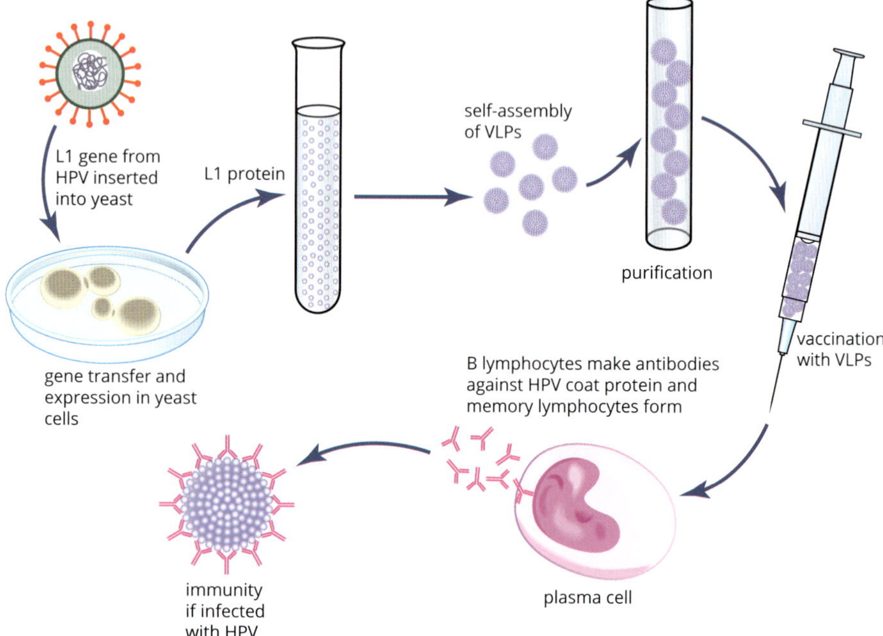

FIGURE 8.3.5 The gene for the HPV coat protein L1 was taken from the virus and inserted into yeast cells, which produce large amounts of the protein. This L1 protein 'self assembles' into particles that look like the virus, and are therefore known as virus-like particles (VLPs). These VLPs do not contain any viral DNA, so they do not cause disease and are not infectious. The VLPs are the antigen used in the Gardasil vaccine.

Therapeutic cancer vaccines

Therapeutic cancer vaccines are given to people who already have cancer. These vaccines are made up of antigens for a specific type of cancer cell (usually proteins or parts of proteins), and often adjuvants are included to help boost the immune response, increasing its ability to identify and destroy cancer cells.

Personalised cancer vaccines

Personalised cancer vaccines are therapeutic vaccines developed for an individual patient. Some involve tumour cells that have been removed from the patient, altered in the laboratory to make them more obvious to the immune system, and then injected back into the patient.

One of the most effective personalised cancer vaccines involves the patient's own tumour and **dendritic cells**, which are antigen-presenting cells of the immune system. This therapy works by presenting the antigen isolated from the patient's tumour sample to the patient's dendritic cells. The activated dendritic cells are then injected back into the patient, where they present antigens to helper T lymphocytes, eliciting an adaptive immune response.

> Antigen-presenting cells present foreign antigens attached to MHC-II molecules on their surface.

Monoclonal antibody therapy

Monoclonal antibodies (mAbs) are antibodies produced by a single clone of a B lymphocyte that is grown in culture to produce a large volume of the same clone. The mAbs produced by the clones are all identical and specific to the same antigen. One of the ways mAbs are used to treat cancer is by targeting specific antigens present on tumour cells (Figure 8.3.6). But they can also be used to target cells of the immune system and direct the immune response in a way that helps destroy tumour cells.

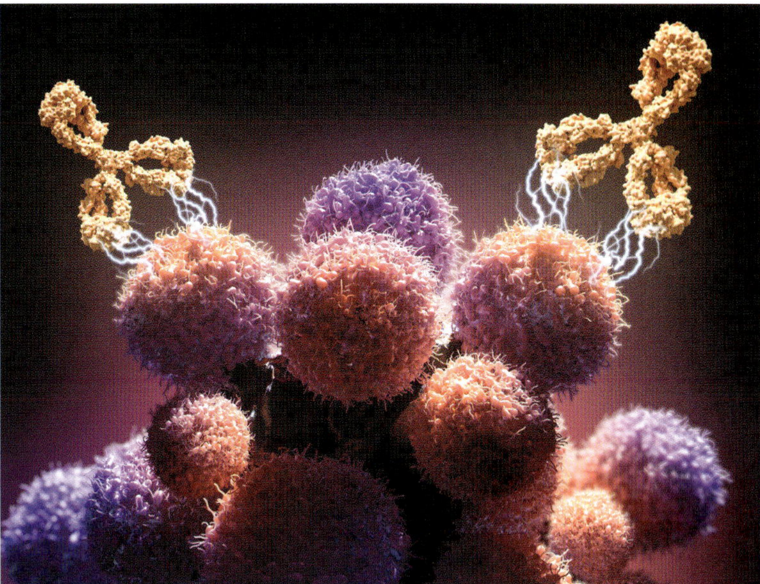

FIGURE 8.3.6 Digital illustration of monoclonal antibodies binding to antigens on cancer cells.

Production of monoclonal antibodies

Figure 8.3.7 illustrates the steps required to make monoclonal antibodies.

- First, mice are injected with a particular antigen, which in the case of cancer therapy is an antigen from a cancer cell.
- This induces the mice's B lymphocytes to produce antibody specific to the antigen, and these B lymphocytes are then isolated from the spleens of the mice.
- In isolation the B lymphocytes only have a limited lifespan, so in order to produce the large quantity of antibodies needed, the isolated B lymphocytes are fused with **myeloma cells**, which are an **immortal cell line**.
- The fusion of the two cells results in a hybridised cell called a **hybridoma**.
- The hybridoma is more stable in tissue culture conditions and the cell secretes multiple copies of the specific antibody (the mAbs), which are then harvested.

Immortal cell lines can continually undergo division without mutation, which would normally occur as a cell ages, and can therefore be cultured for long periods.

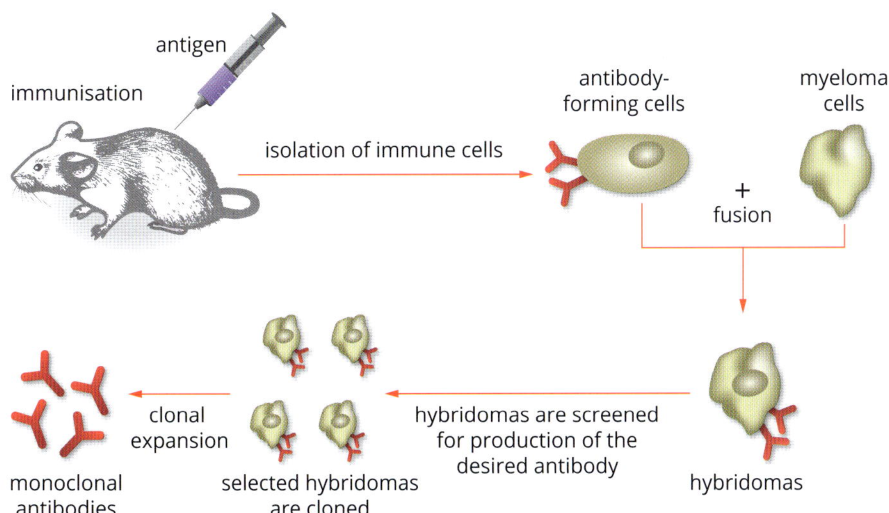

FIGURE 8.3.7 Production of monoclonal antibodies. Mice are injected with an antigen. B lymphocytes sensitised to the antigen are then taken from the mice and fused with myeloma cells. The fused cells, called hybridomas, make antibodies to the antigen, and are grown in culture dishes to produce large quantities of monoclonal antibody specific to the antigen.

Humanised monoclonal antibodies

The first mAbs produced were mouse mAbs made entirely by mouse B lymphocytes, and many still are today. Although these types of mAbs are initially effective when used in human therapy, an immune response is mounted against them once they are identified as foreign (mouse) proteins. Immunological memory is formed, and the adaptive immune response recognises and destroys them faster when the same mAbs are subsequently used again.

To help prevent an immune response directed against them, researchers have been able to replace some components of mouse antibodies with human components using recombinant DNA techniques (Figure 8.3.8). Antibodies with a mixture of mouse and human components are known as **chimeric mAbs**. As mAbs contain more and more human components, they are termed **humanised mAbs**. Some mAbs are now fully human antibodies produced by **transgenic mice**. Although chimeric, humanised and human mAbs are all still produced by mice, these mAbs may be safer and potentially more effective than earlier mAbs.

Transgenic mice have been genetically modified to contain genes from other species.

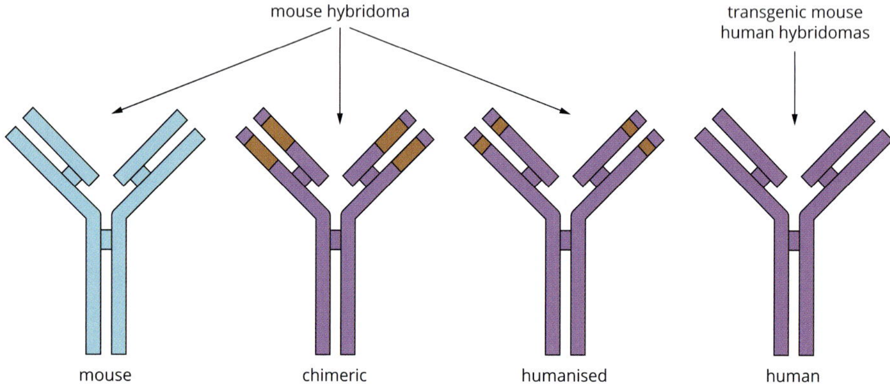

FIGURE 8.3.8 Monoclonal antibodies can be categorised into four types based on their protein composition: animal (most commonly mouse), chimeric (combination of mouse and human), humanised (mostly human) and human.

Conjugated mAbs

Conjugated mAbs are mAbs that have been attached to a chemotherapy drug, a toxin or a **radioactive** particle (Figure 8.3.9a). In this way, conjugated mAbs are used as carriers modified to deliver treatments directly to the specific cancer cells. For example, radioimmunotherapy can be used to treat pancreatic cancer. A radioactive isotope (lead-212) is combined with a specific antibody capable of targeting cancerous cells. This combination of antibody and lead-212 radioisotope is injected intravenously into the body. When it reaches the pancreas, it locks onto the cancerous cells' antigens and the lead-212 destroys the cells by irradiating them. This treatment limits the toxic effects on healthy cells to those located near the cancerous cells.

Bispecific mAbs

Bispecific mAbs are artificially produced using recombinant DNA technology and are used to target cancer cells and activate the immune system simultaneously (Figure 8.3.9b). This type of mAb is used to indirectly activate an adaptive immune response. Naturally produced mAbs have two binding sites, but each site binds to the same antigen. Bispecific mAbs can attach to two different antigens at the same time, because they are composed of parts from two different mAbs and have two different antigen-binding sites (Figure 8.3.10).

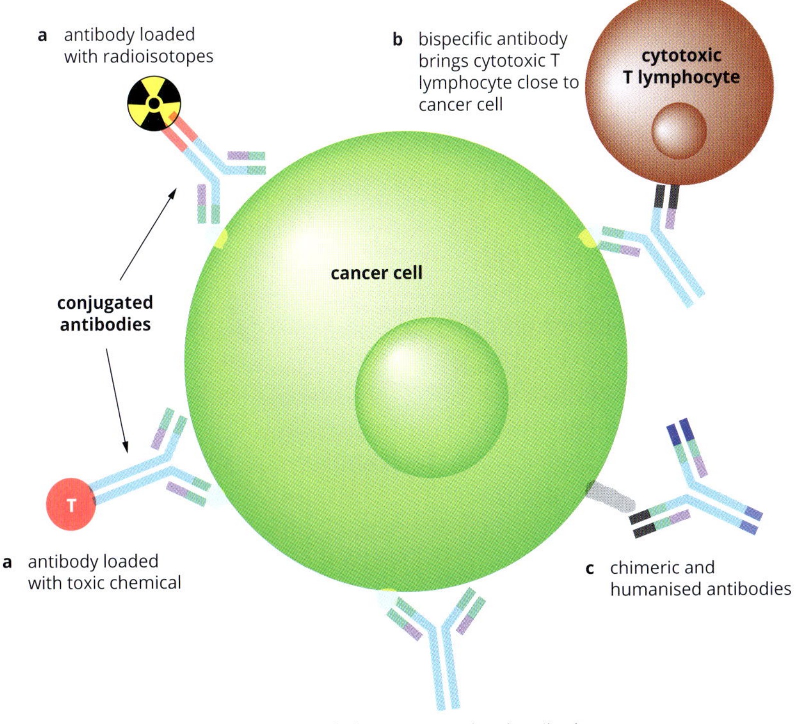

FIGURE 8.3.9 (a) Conjugated antibodies are joined to toxins or radioisotopes to specifically target and kill cancer cells and spare normal body cells. (b) Bispecific antibodies are those engineered to bind two different antigens, in order to bring cytotoxic T lymphocytes close to tumour cells. (c) Chimeric antibodies contain parts from mouse and human antibody, and humanised antibodies contain mostly human parts, but are still produced in mice. (d) Human monoclonal antibodies contain only human components.

An example of a bispecific mAb is Blincyto, which is used to treat some types of acute lymphocytic leukaemia. One part of this mAb attaches to a protein on the surface of the leukaemia cells, while the other part attaches to a protein found on T lymphocytes of the immune system. By binding to both these proteins, the bispecific mAbs are effectively 'identifying' the cancer cells as foreign and 'delivering' them to the immune system.

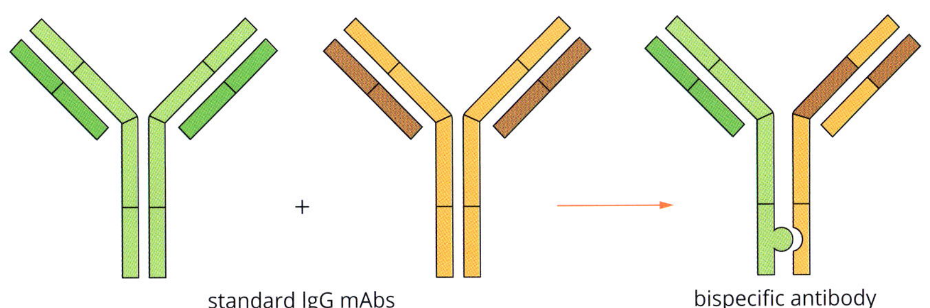

FIGURE 8.3.10 Diagram of a bispecific monoclonal antibody, in which parts of two different antibodies have been fused to form a hybrid that can bind to cancer cells and to T lymphocytes.

BIOFILE

Pembrolizumab

Pembrolizumab is a humanised monoclonal antibody approved for use in Australia to treat metastatic melanoma (Figure 8.3.11). It works by binding to a receptor on T lymphocytes called programmed cell death 1 (or PD-1). PD-1 normally interacts with two ligands on antigen-presenting cells, called PD-L1 and PD-L2. This interaction between PD-1 and the PD-L1 and PD-L2 ligands inhibits T lymphocyte activation and cytokine production. (A ligand is a substance that binds specifically and reversibly to another substance, forming a complex.)

A range of tumour cells, including melanoma cells, have PD-L1 expressed on their surface, so binding of T lymphocyte PD-1 and tumour cell PD-L1 inhibits T lymphocyte responses against the tumour cells (PD-L2 is also expressed on a variety of tumour cells, but it has not been studied as extensively as PD-L1, and its impact on anti-tumour immunity is less clear). The binding of pembrolizumab to PD-1 receptors blocks the inhibition of T lymphocytes, allowing them to become activated and produce inflammatory cytokines. This improves the immune response against tumour cells, but it also results in autoimmune reactions in which healthy cells are damaged.

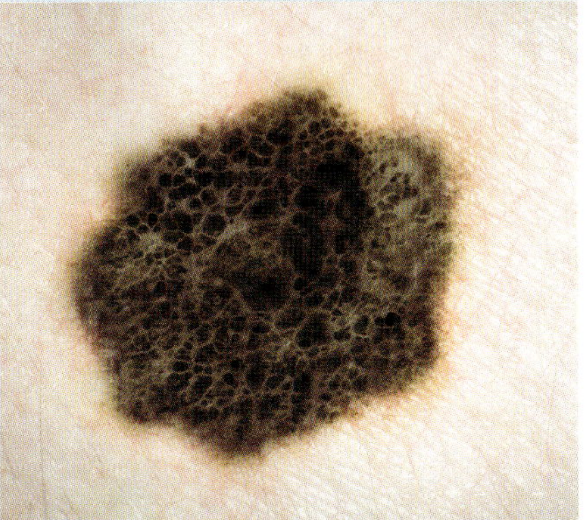

FIGURE 8.3.11 Photograph of melanoma.

BIOLOGY IN ACTION

A cancer vaccine for Tasmanian devils

Tasmanian devils (*Sacrophilus harrisii*) have been plagued for decades by a contagious facial cancer called devil facial tumour disease (DFTD). The cancer originated from a single cancerous Schwann cell in a single Tasmanian devil many years ago, and spread from one Tasmanian devil to another by bites. DFTD has devastated the Tasmanian devil population and threatens the survival of the species.

Contagious cancer is very unusual, and at first it was thought the reason DFTD spread so easily between Tasmanian devils was that they had weakened immune systems, or low genetic variance. Low genetic variance would make it easy for the cancer to spread, because it would mean the immune system of an individual devil wouldn't recognise the cancer cells from another devil as foreign. However, both these hypotheses have been ruled out.

In fact, it appears the cancer cells evade the devil's immune system by destroying their major histocompatibility complex (MHC) molecules, which are vital for immune recognition. Without MHC molecules on the cancer cells, the devil's immune cells do not detect them.

Researchers from the University of Tasmania have shown that a vaccine using killed DFTD tumour cells and adjuvant is able to stimulate a protective adaptive immune response (Figure 8.3.12). The vaccine is currently being trialled, and if successful, it may help bring the Tasmanian devil back from the brink of extinction.

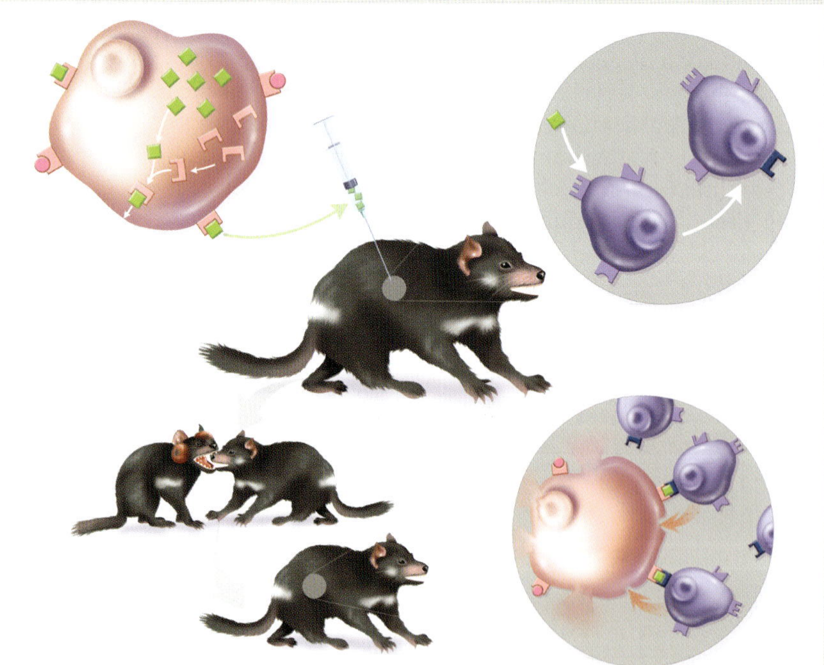

FIGURE 8.3.12 The cancer vaccine for devil facial tumour disease uses tumour-specific antigens from killed tumour cells to stimulate an adaptive immune response involving T lymphocytes.

8.3 Review

SUMMARY

- Cancer occurs when a single rogue (or abnormal) cell multiplies uncontrollably and spreads throughout the body.
- Cancer treatments over the last several decades have included chemotherapy, radiation therapy and surgery. These treatments come with significant side effects.
- Immunotherapy is any treatment that harnesses the immune system of the patient to fight diseases such as cancer.
- Immunotherapies can be non-specific, such as the injection of cytokines, or specific, such as cancer vaccines, personalised immunotherapy or monoclonal antibody therapy.
- Cancer vaccines are made using specific antigens from cancer cells or infectious agents that cause cancer. Like other vaccines, they are administered to a patient to stimulate an immune response that results in the production of an immunological memory.
- Cancer vaccines can be preventative, therapeutic or personalised.
  - Preventative cancer vaccines are directed at viruses that cause cancer such as human papillomavirus (HPV), which causes cervical cancer. Antigens specific to the virus are injected into the body, triggering an adaptive immune response that will lead to immunological memory.
  - Therapeutic cancer vaccines are made up of antigens for a specific type of cancer. These antigens are injected into the body along with adjuvants that help boost the immune response.
  - Personalised cancer vaccines involve cells or substances from the patient, which are removed, altered in a laboratory and then reintroduced by injection. The most effective personalised immunotherapies involve the patient's own tumour and dendritic cells.
- Monoclonal antibody (mAb) therapy involves antibodies produced by a single clone of B lymphocytes that are replicated in culture. mAbs are all identical and specific to the same antigen. Targeting specific cells reduces harm to healthy cells, but identifying the specific antigen in order to create mAbs is a difficult task.
- The first monoclonal antibodies (mAbs) were mouse mAbs made entirely by mouse B lymphocytes. To avoid an immune response against mAbs, chimeric, humanised and human monoclonal antibodies can now be produced using transgenic mice:
  - Chimeric mAbs are a mix of human and mouse components.
  - Humanised mAbs are also a mix but are mostly human.
  - Human mAbs are fully human.
- mAbs can be used as carriers of treatments (drugs, toxins and radioactive particles) for delivery specifically to cancer cells. These types of mAbs are called conjugated mAbs.
- Bispecific mAbs are made up of two different mAbs and have two different binding sites: one is usually for a cancer cell and the other for an immune cell, such as a T lymphocyte. This enables an 'identify' and 'deliver' approach.

KEY QUESTIONS

1 What is the difference between non-specific and specific immunotherapies?

2 Explain the difference between a preventative cancer vaccine and a therapeutic cancer vaccine, including reference to the type of antigen in each vaccine.

3 What is the difference between chimeric, humanised and human monoclonal antibodies?

4 If a monoclonal antibody has a toxin or radioactive substance attached to it, what kind of monoclonal antibody is it?

A bispecific
B conjugated
C chimeric
D humanised

5 Bispecific antibodies are produced artificially. How are these antibodies different from those produced naturally by the immune system?

Chapter review

08

KEY TERMS

acquired immune deficiency syndrome (AIDS)
active immunity
allergy (or allergic reaction)
anaphylaxis
antiserum
artificial active immunity
artificial passive immunity
autoimmune disease
benign tumour
bispecific monoclonal antibody
cancer vaccine
carcinogen
central nervous system
chemotherapy
chimeric monoclonal antibody
congenital
conjugated monoclonal antibody
cytotoxic
demyelination
dendritic cell
haemolytic disease
herd immunity
human immunodeficiency virus (HIV)
humanised monoclonal antibody (humanised mAb)
hybridoma
immortal cell line
immunodeficiency
immunoglobulin E (IgE)
immunotherapy
inactivated vaccine
killed vaccine
live attenuated vaccine
malignant tumour
metastasis
microglia
monoclonal antibody (mAb)
myelin
myeloma cell
natural active immunity
natural passive immunity
oligodendrocyte
passive immunity
pollen
radiation therapy
radioactive
recombinant DNA technology
serum
subunit vaccine
toxoid vaccine
transgenic mouse
tumour
vaccination

KEY QUESTIONS

1 Pregnant women who are Rhesus negative but who may be carrying a Rhesus positive foetus will be given injections in order to protect the foetus from maternal antibodies. These injections make the mother immune to Rhesus proteins. What type of immunity is this?

A artificial passive immunity
B artificial active immunity
C natural passive immunity
D natural active immunity

2 Breastfed babies tend to be healthier than bottle-fed babies. Give a reason why.

3 Chickenpox (varicella) is a disease caused by the *Herpes zoster* virus. Vaccination is available and has been a part of the childhood immunisation schedule in Victoria since 2005. The vaccine is a live virus vaccine that is currently administered at 18 months of age.

a The vaccine contains a live virus. Explain why this virus does not cause the vaccine recipient to develop chickenpox.

b Before the vaccination program it was common for the eldest child in a family to develop chickenpox in their kindergarten or first year of school. Following the eldest child's infection the younger child would then develop the disease, but babies in the family rarely caught the disease and even if they did it was generally very mild. Explain this observation.

c Since the introduction of chickenpox vaccine to the immunisation schedule there has been an increase in the incidence of chickenpox among adults. Why?

4 On occasion the blood bank in Melbourne will advertise for people who have recently recovered from chickenpox to donate blood.
The blood bank takes a blood donation from these individuals and then separates the blood using a centrifuge. The blood serum is collected. The blood cells may be injected back into the donor.
The serum is then purified and some of the proteins are extracted.

a What are these proteins?

b These proteins are given to patients. Which group of patients is most likely to be in need of these proteins?

c Why does the extraction of these proteins not expose the donors of the blood to the possibility of future bouts of chickenpox?

d Explain whether the injection of these proteins gives the recipients long-term immunity.

5 Research in the United States has shown that the common cold costs the American economy around $25 billion each year in lost productivity. The cost of the common cold to Australia would be proportionally similar. The major issue with the common cold is that there are many different strains of the rhinoviruses that cause colds.

a It has been said that every time a person gets a cold it is a new one. Using your understanding of the immune response explain the truth of this statement.

b What information about rhinoviruses is needed before there is any possibility of making a vaccine that will result in immunity to all colds?

6 Diphtheria is a potentially fatal disease caused by either *Corynebacterium diphtheriae* or *C. ulcerans*. Around 10% of individuals infected will die. For this reason diphtheria is one of the diseases for which children are vaccinated. The first vaccinations became available in 1921. By 1929 contacts of people with diphtheria were being vaccinated. School-based vaccination programs began in 1932.

Diphtheria is still a major health issue in some countries. Fewer than 10 cases have occurred in Australia in the last 10 years, all of which have been linked to overseas travel. The last recorded death in Australia was in 2011.

The graph shows the incidence of diphtheria in Australia between 1917 and 1994.

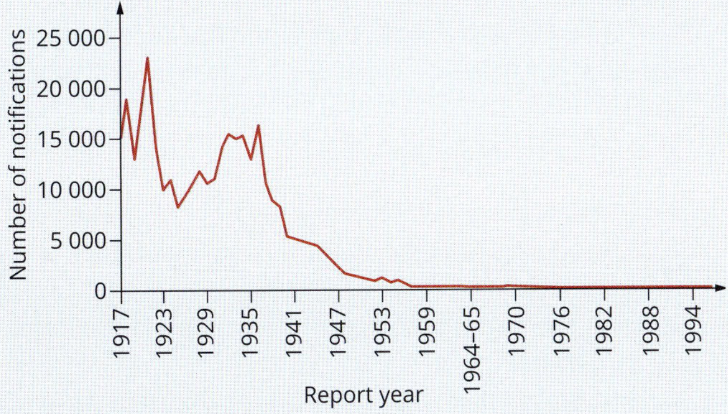

a Using the data in the graph evaluate the effectiveness of the vaccination program in Australia.

b Explain why despite the high incidence of diphtheria in some overseas countries and the ease of international travel only isolated cases and no outbreaks of diphtheria have occurred in Australia.

7 The following graph shows the levels of measles antibodies in a child from birth to seven years of age.

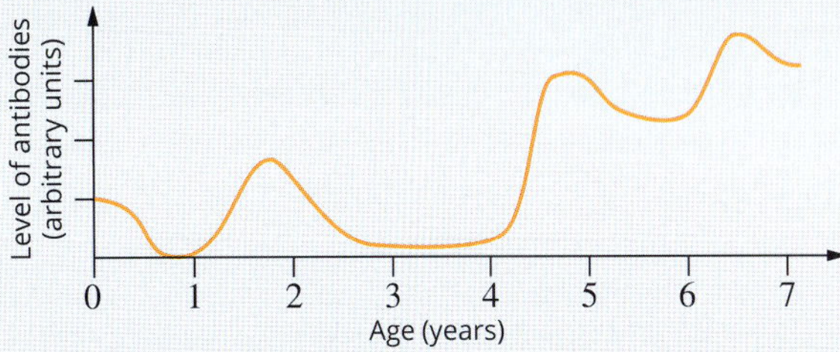

a This child had antibodies against the measles virus at birth. Explain why.

b Why did the antibody levels drop off to zero in the months following birth?

c The child was immunised against measles at one year of age and again at four years of age.

i Explain why antibody production occurs more rapidly and to a higher level after the second vaccination compared with the first vaccination.

ii Why don't the antibody levels drop off to zero after immunisation occurs?

d At six years of age antibody production increases greatly again, but no vaccination has occurred. Explain.

e Can you explain the recent increase in the incidence of measles in children?

8 In 1969, when astronauts returned from landing on the Moon, they were immediately ushered into quarantine for extensive testing before they were allowed to re-enter the community. Explain the reasoning behind this procedure.

9 It is true to say that hypersensitivity:

A always involves histamine

B always involves antibodies

C may lead to an autoimmune disease

D involves all of the above

10 a What is hypersensitivity?

b Describe the four types of hypersensitivity and give an example of each.

11 Consider the life cycle of the malarial parasite (*Plasmodium* sp.) which involves two hosts, as shown in the diagram at right.

- **a** Suggest stages in the life cycle of malaria that have been, or could be, accessible to intervention.
- **b** If a vaccine were to be created to protect people from malaria what information would be needed?
- **c** Explain why the cell-mediated response would be important in the defence against malaria.

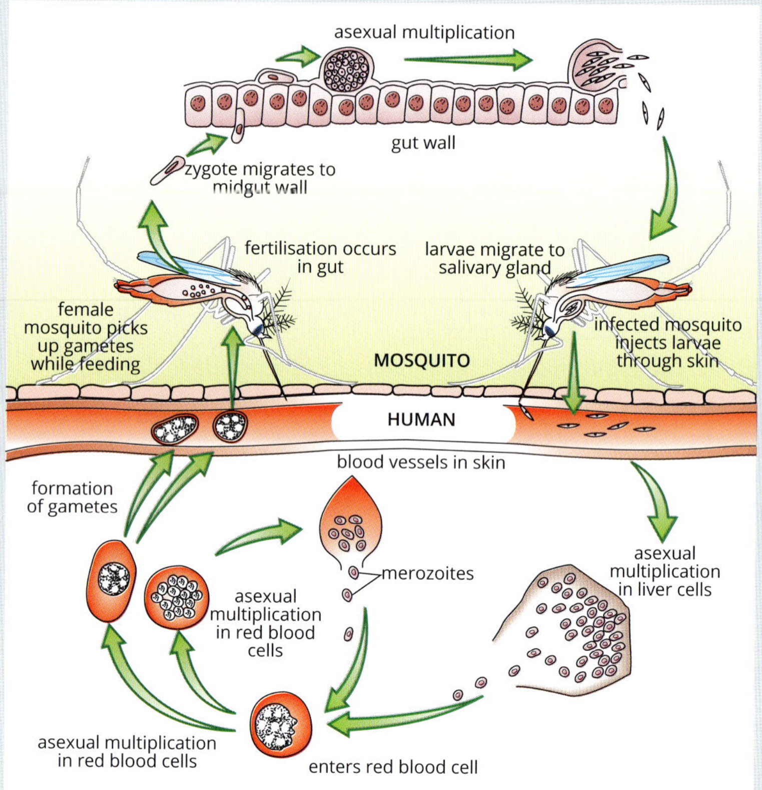

12 Anaphylactic reactions to peanuts have become fairly common Australia. Desensitisation is sometimes used in people who have such severe allergic reactions.

- **a** What class of antibodies is involved in an allergic reaction such as that to peanuts?
- **b** Desensitisation does not suppress the immune system but it does reduce the response to the allergen. Explain how this outcome is achieved.
- **c** Scientists at a biotechnology company have developed a vaccine that they say will immunise people against peanut allergy. They wish to test their vaccine in mice. They have a number of genetically similar laboratory mice that they are going to use for their initial experiments.
  - **i** What is one characteristic that the mice must all have in order to make the experiment valid?
  - **ii** Design an experiment to test the vaccine using the mice as subjects.
  - **iii** What result would provide evidence that this vaccine is successful?
  - **iv** Name the dependent and independent variables in your experiment.
  - **v** One mouse in the trial has an adverse reaction to the vaccine. Should the company abandon the trial at this point? Explain your reasoning.

13 **a** There is ongoing research to find a vaccine to combat HIV. Explain the most likely form that this vaccine will take.

- **b** Propose a reason why clinical trials of a HIV vaccine could raise some ethical issues for researchers.

14 **a** What is the difference between a benign tumour and a malignant tumour?

- **b** Why are malignant tumours more difficult to treat and much more likely to recur?

15 Cancer is characterised by cells that undergo unchecked reproduction, competing with body tissues for space and nutrients.

Leukaemia is a cancer of the white blood cells. In this condition the white blood cells of the bone marrow reproduce prolifically. One treatment for this cancer is radiation therapy to destroy a large percentage of the cancerous cells, followed by a bone marrow transplant. The stem cells of the donor marrow are critical to the patient's recovery potential.

- **a** Suggest why radiation therapy is an important part of the treatment regime for leukaemia.
- **b** What are stem cells? How are they important in such a transplant?
- **c** Explain why it is important that the closest possible tissue match is made between a donor and the recipient for a bone marrow transplant to be successful.

16 Effective immunotherapy is one of the goals of modern research into treatment for cancer.

- **a** What is cancer?
- **b** **i** Briefly describe the three traditional methods of combatting cancer.
  - **ii** Immunotherapy has advantages compared to traditional cancer treatments. Explain why.

17 The antibody shown below is:

A chimeric

B bispecific

C conjugated

D humanised

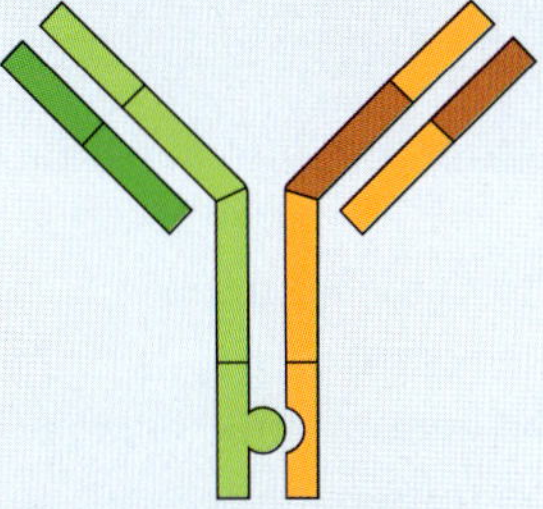

18 The first monoclonal antibodies produced for research were created in mice. These antibodies, however, were not successfully used in humans as they resulted in an immune response.

a Draw and label the parts of an antibody.

b Explain why the mouse antibodies caused an immune response.

c Suggest a possible solution to the problem of immune reaction to the antibodies.

19 Traditional cancer treatments have many side effects. In order to reduce these side effects researchers have sought ways to make anti-cancer drugs more specific to cancer cells. An important advance in this came about with the development of monoclonal antibodies. One group of researchers suspect that monoclonal antibodies can increase the efficacy and reduce the side effects of a commonly used anti-cancer drug by attaching the drug to a monoclonal antibody that has a binding site specific to an antigen expressed only on cancer cells.

a The results of a preliminary trial comparing administration of the drug attached to the antibody with conventional administration are shown in the table below. In both cases there was one round of treatment, which consisted of administration of the drugs once per week for four successive weeks.

i Complete the table by calculating the percentage change in the size of the tumour.

Use the formula:

$$\frac{\text{change in tumour size}}{\text{original tumour size}} \times 100 = \%\text{ change in size}$$

ii Why is it necessary to calculate percentage change in tumour size before analysing the results of the trial?

iii Do the results indicate that further trials of this approach should be undertaken? Explain your reasoning by referring to the data.

iv Suggest a possible explanation for the results observed in patient 5.

b The use of monoclonal antibodies to deliver cancer drugs has been shown to have fewer side effects than conventional methods of drug delivery and that much lower doses of the drug are needed to achieve the same level of response in the patients.

Using your understanding of how monoclonal antibody therapy works, explain these two observations.

Patient number	Antibody (A) or conventional (C) therapy	Tumour size at commencement of treatment	Tumour size after one round of treatment	Change in tumour size	% change in tumour size
1	A	12.5 mm^3	9.6 mm^3		
2	A	23.9 mm^3	15.5 mm^3		
3	A	54.2 mm^3	26.8 mm^3		
4	A	46.8 mm^3	27.9 mm^3		
5	A	53.6 mm^3	56.4 mm^3		
6	C	54.8 mm^3	48.5 mm^3		
7	C	84.1 mm^3	66.9 mm^3		
8	C	36.9 mm^3	30.8 mm^3		
9	C	56.1 mm^3	49.1 mm^3		
10	C	38.9 mm^3	31.5 mm^3		

Tumour response to antibody-boosted or conventional drug treatment.

REVIEW QUESTIONS

How do cells communicate?

Multiple choice questions

1 Hay fever is an overreaction to previously encountered allergens such as animal fur, pollen or dust. The body produces IgE in response to the first contact with these allergens. This antibody then binds to the surface of mast cells. Which is the correct allergic response?

- **A** Allergens bind to the specific binding sites of the IgE antibodies, which then bind to mast cells that migrate via the bloodstream into mucous-producing tissues where they produce the hay fever symptoms.
- **B** Allergens bind to mast cells in the mucus-producing tissue and trigger them to release histamines that cause the allergic symptoms.
- **C** Allergens bind to already-bound IgE antibodies and histamines. The binding of histamines to IgE antibodies causes the hay fever symptoms.
- **D** Allergens bind to pairs of adjacent IgE molecules that are already attached to mast cells. This triggers the release of histamines that cause the hay fever symptoms.

2 Pheromones can be used against insect pests in agriculture as they are natural products that do not leave harmful residues in the environment. Which of the following best describes pheromones?

- **A** They are signalling molecules produced only by insects.
- **B** They are signalling molecules involved in chemical communication between organisms.
- **C** They are a particular type of neurotransmitter.
- **D** They are used to externally digest the plant on which the insect is feeding.

3 Which of the following would it *not* be true to say?

- **A** Complement proteins are part of the innate immune response.
- **B** Complement proteins attract phagocytes to the site of the infection.
- **C** Complement proteins include antibodies.
- **D** Complement proteins can be activated in the absence of antibody–antigen reactions.

4 Which of the following describes paracrine signalling?

- **A** The signalling molecule is carried to the target cell through the bloodstream.
- **B** The signalling molecule acts on the same cell that secretes the molecule.
- **C** The signalling molecule acts on target cells near the cell that secreted it.
- **D** The signalling molecule acts on target cells distant from the cell that produced it.

5 Apoptosis through the intracellular pathway is triggered by:

- **A** the mitochondrion
- **B** the lysosomes
- **C** the endoplasmic reticulum
- **D** the nucleus

6 Natural active immunity is achieved as a result of:

- **A** exposure to live or attenuated vaccines
- **B** infection by particular bacteria or virus
- **C** the administration of antibodies or antitoxin specific to a particular microorganism
- **D** adequate breast-feeding in newborn infants

7 A pathogen shown below enters a human body and antibodies are produced against it.

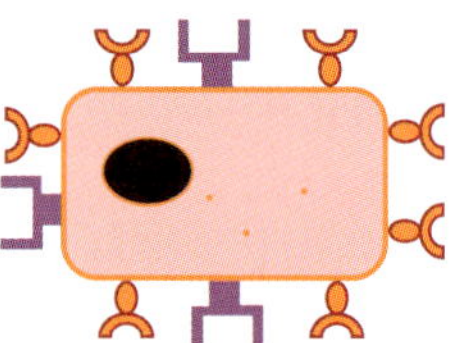

Which antibody would be made in response to the pathogen?

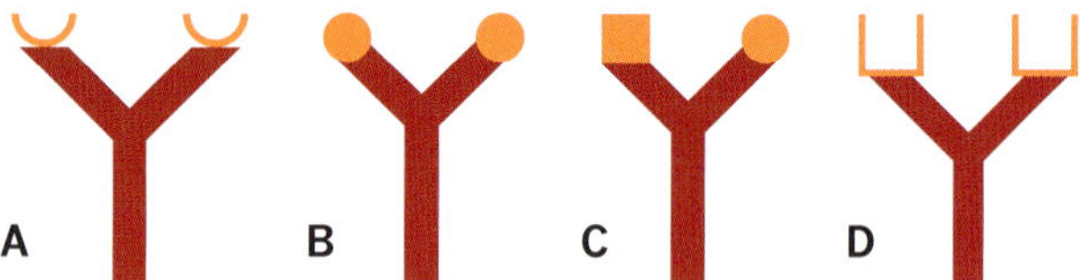

8 Innate immune responses are critical to maintaining the health of an individual because:

- **A** they are specific to the antigens on pathogenic organisms
- **B** the innate response produces antigens, which bind antibodies to mast cells
- **C** they provide immediate and continuous protection against foreign antibodies
- **D** none of the above

Short answer questions

9 Hay fever is an allergic response in which the immune system overreacts to the presence of a previously encountered foreign antigen. The diagram below illustrates the key steps that occur in an allergic reaction.

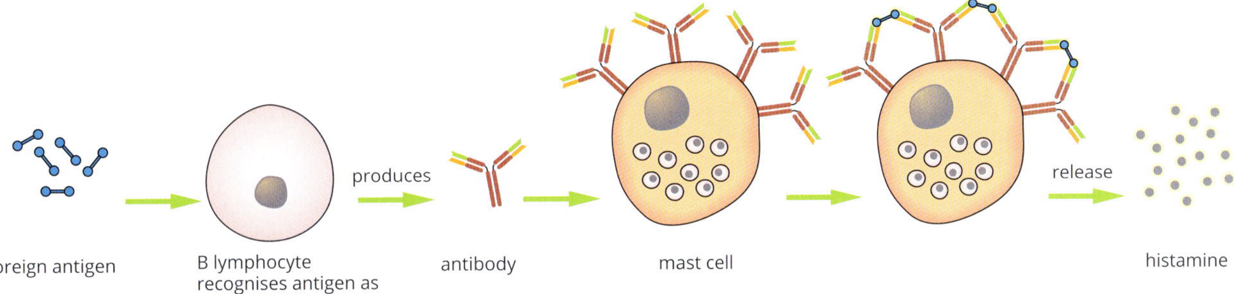

a Define the term 'antigen'.

b Identify the type of antibody involved in allergic reactions.

c Where in the body are mast cells located?

d Describe the event that causes mast cells to release histamine.

e Describe two effects resulting from the release of histamine.

10 Acetylcholine is produced by neurons and binds to receptors on the muscle cell membrane to initiate a sequence of steps that result in muscle contraction. When acetylcholine receptors on muscle cells are blocked by an individual's own antibodies, it results in myasthenia gravis, which is characterised by muscle weakness.

a To which group of diseases does myasthenia gravis belong?

b i To which group of signalling molecules does acetylcholine belong?

ii How would the blocking of acetylcholine receptors on the muscles cause muscle weakness?

c Some patients with myasthenia gravis have been shown to have an abnormally large thymus gland. What role does the thymus gland play in the adaptive immune response?

11 TNF-α (tumour necrosis factor alpha) and IL-6 (interleukin-6) are two important pro-inflammatory cytokines released during an infection. TNF-α and IL-6 are produced by macrophages. One of their functions is to stimulate more macrophages to migrate to the infection site.

a What are cytokines?

b These cytokines are major causes of inflammation during infection. Describe the effects of inflammation on infected tissues.

c Following World War I, a Spanish flu pandemic swept the world. More than 20 million people died. The Spanish flu was unusual because many of the fatalities occurred among young people with healthy immune systems. Modern medical opinion is that most of these deaths were caused by acute respiratory distress syndrome (ARDS) initiated by a 'cytokine storm'.

i TFN-α has been shown to induce apoptosis in lung tissue cells through the death receptor pathway. Explain how the large amounts of TFN-α produced during a cytokine storm leads to the lung damage and death associated with ARDS.

ii Cytokine storms occur as a result of positive feedback between macrophages and the cytokines. This leads to very large concentrations of TNF-α and IL-6. Draw a flow chart to show how a cytokine storm can eventuate.

iii In other situations both TFN-α and IL-6 can act as anti-inflammatories. Explain how the same signalling molecules can cause a different response.

12 The thyroid gland produces the hormone thyroxine, which stimulates cell metabolism. It increases the respiratory rate in cells and so increases the production of both ATP and heat. The thyroid produces thyroxine when it is signalled by a hormone produced by the pituitary gland. This hormone, thyroid stimulating hormone (or TSH) is hydrophilic. The pathway to an increase in cell metabolism is shown below.

pituitary produces TSH
↓
thyroid receptors accept TSH
↓
thyroid produces and releases thyroxine
↓
body cell thyroxine receptors accept thyroxine
↓
body cells increase metabolism

a **i** To which group of hormones does TSH belong?
ii Where are the receptors for TSH? Explain your reasoning.

b One disease associated with the thyroid gland is Hashimoto thyroiditis. In this disease antibodies block the TSH receptors. How might this affect cellular functioning throughout the body? Explain your answer.

c In a different disease called Graves' disease, antibodies attach to the TSH receptors and cause continuous stimulation.
i How would thyroid cells respond to this?
ii What effects would you expect in the body?
iii What kind of disease is Graves' disease?

13 The cone snail (*Conus magnus*) produces a cocktail of different toxins with which it subdues its prey. These toxins are currently the target of considerable medical research as they have numerous effects on both the nervous system and the immune system.

One toxin has been purified and analysed. This toxin is now made artificially and is marketed as a drug called Ziconotide. This drug acts as a painkiller by blocking calcium channels and inhibiting nerve conduction. Explain how Ziconotide can act as a painkiller.

14 A number of cases of measles have been reported to authorities in recent times. Measles is a preventable disease caused by a virus. In 2013 there were 96 000 deaths worldwide from measles. It is the disease with the highest mortality rate of all vaccine-preventable diseases. Most people in Australia are vaccinated against it in childhood.

The vaccination schedule requires children to be given one injection at 12 months and a second injection at 18 months. The antibody response to the vaccinations is shown in the graph below.

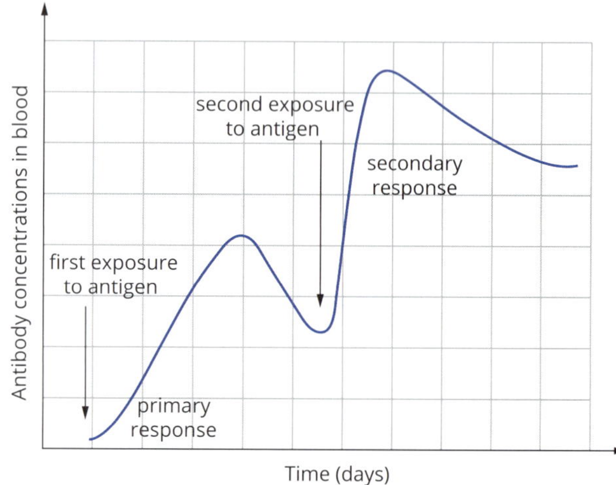

a Explain why the secondary response is so much greater than the primary response.

b T lymphocytes play a significant role in the adaptive immune response of the body against viruses such as the one that causes measles. Describe the role of T cells in immunity.

c In February 2016 Department of Health statistics showed that 93.58% of 5-year-old children in Victoria were fully immunised against measles. In some areas the immunisation rate is as low as 73%. Why is this statistic of such concern?

15 Tetanus is a serious, often fatal disease caused by the bacterium *Clostridium tetani*. The most serious of the symptoms are caused by the toxin produced by the bacteria. The toxin enters neurons in the central nervous system where it blocks the release of the neurotransmitters glycine and GABA. These neurotransmitters stimulate neural pathways that inhibit the contraction of muscles.

a Explain what effect the toxin will have on muscle behaviour.

b **i** A vaccination is available for tetanus. It involves injection of extremely minute amounts of the toxin. Explain how the injection of this minute amount of toxin gives protection from tetanus.

ii Does this produce active or passive immunity? Provide evidence to support your answer.

iii It is recommended that everyone has a booster vaccination for tetanus every 10 years. Why is this necessary?

c An individual who steps on a rusty nail is at significant risk of developing tetanus if they have never been vaccinated. Such individuals are given an injection of a preparation that has been created in horses. This preparation will protect the person against tetanus.

i What is the active constituent of the injection?

ii Explain why this is an example of passive immunity and why it would not give long-term protection.

16 Like animals, plants produce signalling molecules. These molecules are called phytohormones. Auxins are one group of phytohormones. One particular auxin is indole acetic acid (IAA). IAA is the hormone responsible for phototropism.

a What is positive phototropism?

b **i** Where is the site of production of IAA?

ii How does the IAA travel to the site where it is active?

c One mechanism by which auxin causes its action is shown in the flow diagram below. In the case of phototropism, the kinases at step 7 activate enzymes that soften the cell wall, making it more flexible.

Identify the three stages of signal transduction in the action of IAA shown.

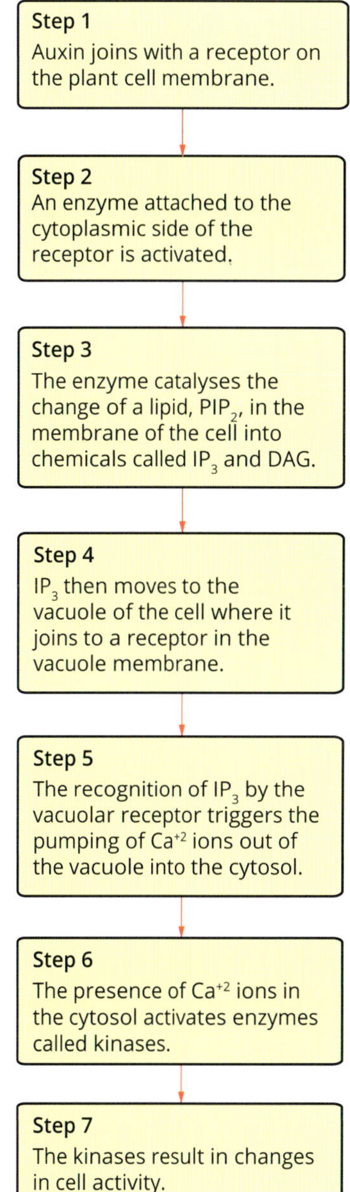

d In the case of phototropism, what is the stimulus and the response of the plant:

i at a cellular level?

ii at a whole organism level?

17 RAS is a family of proteins that are involved in cell signalling. When RAS proteins are switched on they promote cell division and growth. Mutations in the genes that encode the RAS proteins can result in these genes being permanently switched on. Normal RAS proteins are also important in the initiation of cell apoptosis.

- **a** Explain how mutations in RAS genes lead to cancer.
- **b** Another cause of cancer in humans is the overproduction of certain proteins. One human breast cancer variant is caused by a mutation that results in the overproduction of the HER2 protein. This protein forms part of a receptor found on the surface of cell membranes. This receptor is involved in the signalling pathway that results in cell division. Suggest how overproduction of HER2 leads to cancer of the breast.
- **c** One important treatment for HER2-positive breast cancer is a drug called trastuzumab, marketed in Australia as Herceptin. This medication is a monoclonal antibody. It binds to the external domain of the HER2 receptor.
  - **i** What are monoclonal antibodies?
  - **ii** Herceptin is a humanised monoclonal antibody. What does this mean?
  - **iii** Why do scientists humanise monoclonal antibodies?
- **d** A drug company has invented a new monoclonal antibody that it claims will be effective against HER2-positive breast cancers that have been shown to be resistant to current therapies. The company is ready to commence human trials of its proposed treatment.
  - **i** Describe an experiment that could be used to test the effectiveness of the new treatment.
  - **ii** What results would suggest that the drug company's claims are justified?
  - **iii** What are some issues that the researchers could face in performing human trials?
  - **iv** Many human trials enlist only a few individuals. How does that affect the validity of the results?

18 The lymphatic system has a number of important functions.

- **a** List the major functions of the lymphatic system.
- **b** How do lymph nodes assist the efficiency of the adaptive immune response?
- **c** How do primary and secondary lymphoid tissues differ?

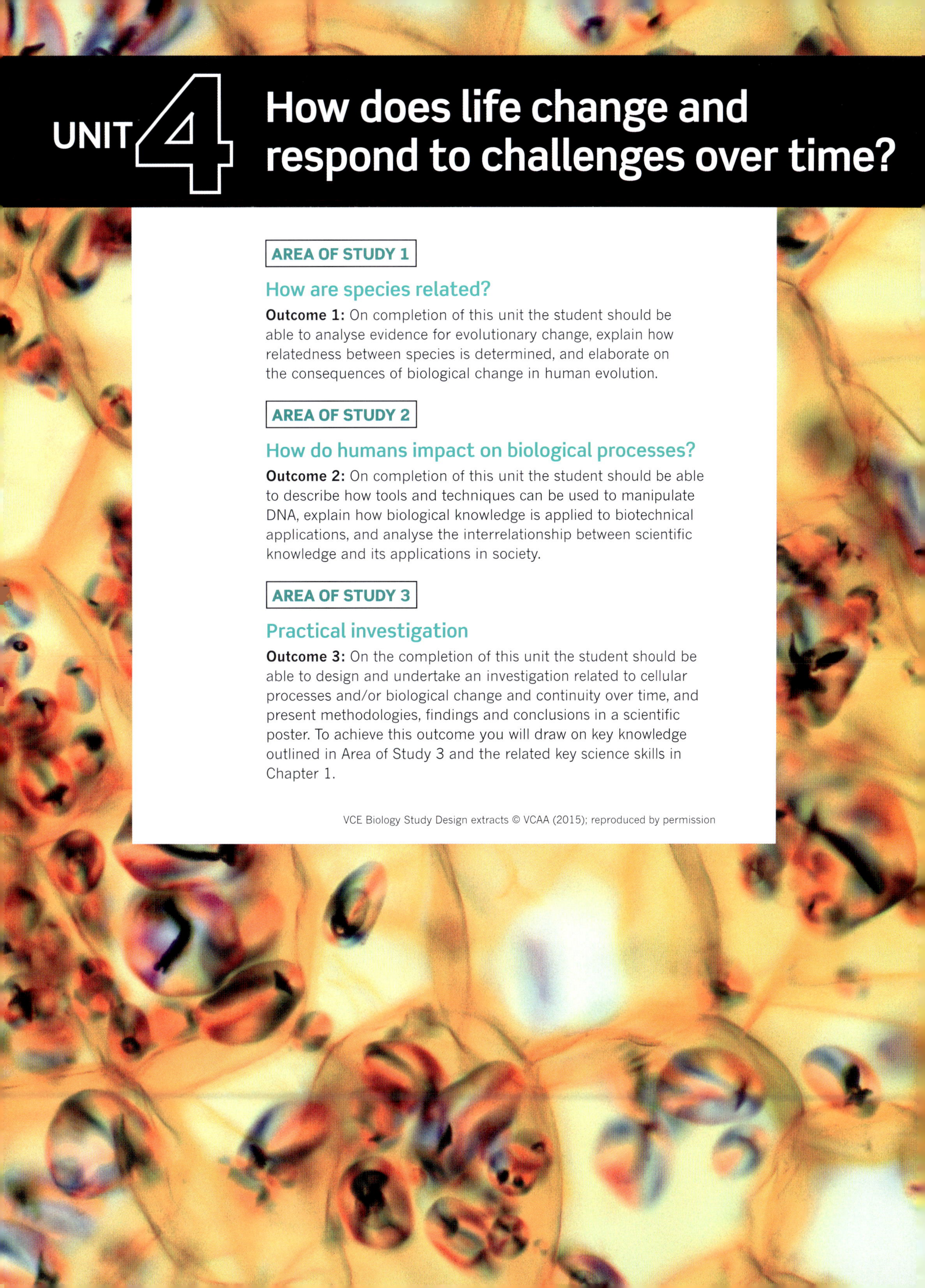

How does life change and respond to challenges over time?

AREA OF STUDY 1

How are species related?

Outcome 1: On completion of this unit the student should be able to analyse evidence for evolutionary change, explain how relatedness between species is determined, and elaborate on the consequences of biological change in human evolution.

AREA OF STUDY 2

How do humans impact on biological processes?

Outcome 2: On completion of this unit the student should be able to describe how tools and techniques can be used to manipulate DNA, explain how biological knowledge is applied to biotechnical applications, and analyse the interrelationship between scientific knowledge and its applications in society.

AREA OF STUDY 3

Practical investigation

Outcome 3: On the completion of this unit the student should be able to design and undertake an investigation related to cellular processes and/or biological change and continuity over time, and present methodologies, findings and conclusions in a scientific poster. To achieve this outcome you will draw on key knowledge outlined in Area of Study 3 and the related key science skills in Chapter 1.

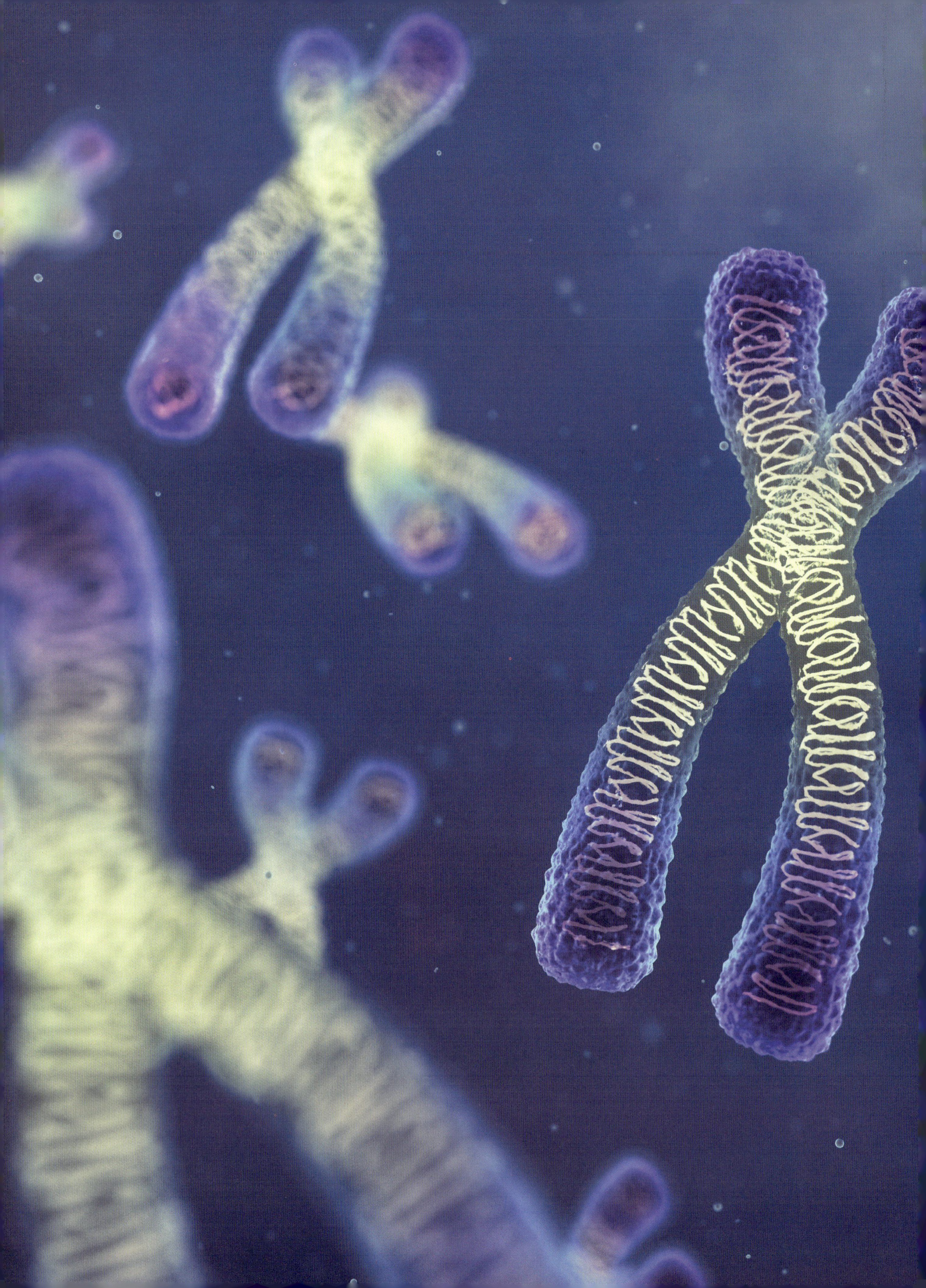

Changes in the genetic make-up of a population

Learning outcomes

To some extent, all physical characteristics of an organism are controlled by their genetic make-up. Individuals of the same species tend to have strong similarities in their genetic make-up, but are not identical.

This chapter focuses on how new genetic differences are introduced into the gene pool of a population and how the likelihood of particular variations persisting in the population may change over time. The driving forces of environmental pressures are examined and natural selection is identified as the mechanism of evolution. The differences between natural and artificial selection are explored, as are the benefits and limitations of selective breeding.

Key knowledge

- the qualitative treatment of the causes of changing allele frequencies in a population's gene pool including types of mutations (point, frameshift, block) as a source of new alleles, chromosomal abnormalities (aneuploidy and polyploidy), environmental selection pressures on phenotypes as the mechanism for natural selection, gene flow, and genetic drift (bottleneck and founder effects) and the biological consequences of such changes in terms of increased or reduced genetic diversity
- processes of evolution including through the action of mutations and different selection pressures on a fragmented population and subsequent isolating mechanisms (allopatric speciation) that prevent gene flow
- the manipulation of gene pools through selective breeding programs.

9.1 Changing allele frequencies

FIGURE 9.1.1 Flower colour (phenotype) in Hydrangeas (*Hydrangea macrophylla*) varies depending on the acidity of the soil in which they grow. Plants grown in slightly acidic soil turn pink and flowers grown in basic soil are blue.

> When a gene has more than one allele it is said to be polymorphic. Human blood groups are an example of a polymorphic trait.

> Because diploid cells contain two copies of each chromosome, an individual can have two alleles for a trait. Homozygotes have two copies of the same allele. Heterozygotes have two different alleles.

As you learnt in Chapter 3, the complete set of genes or DNA make up the genome of a species, and a species is defined by its particular genome. Yet the genome of individuals within that species will vary, depending on their unique combination of alleles and their non-coding portion of DNA. As a population changes over time through births, deaths and migration, the percentage of individuals with particular alleles will also change. In this section, you will learn about some of the mechanisms that change allele frequencies in a population.

ALLELE FREQUENCIES IN A GENE POOL

A gene is a sequence of DNA nucleotides that codes for a particular characteristic or **trait**. Slight variations in the code of a gene can result in different forms of that trait. These variations of a gene are called **alleles**. The various combinations of alleles in an individual make up its **genotype**. The genotype, together with the environment, determine the observable traits (or **phenotype**) of that individual (Figure 9.1.1). Although they share the same genome, the individuals within a species are not genetically identical because they possess different combinations of alleles.

Traits that are controlled by one gene (monogenic) tend to have discontinuous (or discrete) variation. Polygenic traits are controlled by multiple interacting genes. This gene interaction results in continuous variation, as shown by dog coat colour in Figure 9.1.2.

FIGURE 9.1.2 Coat colour in dogs is a polygenic trait controlled by different combinations of alleles. The number of black or brown dogs of this breed will depend on allele frequencies in the population.

All the alleles possessed by an entire population, which may potentially be passed on to the next generation, make up that population's **gene pool**. The relative proportion of a particular allele in a population is referred to as the **allele frequency**, and is often expressed as a percentage or as a decimal (e.g. 25% or 0.25). The following equation shows you how to calculate an allele frequency.

$$\text{allele frequency} = \frac{2(\text{number of homozygotes}) + (\text{number of heterozygotes})}{2(\text{total number of individuals})} \times 100$$

When calculating the allele frequency of a particular trait, it is important to remember that homozygotes with the trait will have two copies of the allele, heterozygotes will only have one copy.

For a variety of reasons, allele frequencies within a gene pool often change over time. For example, new alleles can occur as the result of genetic mutations, and the frequency of particular alleles can alter as a result of changing environmental pressures.

MUTATION

All genetic variation between species and between individuals of the same species is a result of **mutation**. Mutations are changes in DNA. It is important to remember that mutation means *change*, not the introduction of a fault or disease. Mutations can have a beneficial or harmful effect, or no effect at all, on the organism.

Mutations can occur randomly during **replication**. Mutations may affect a single gene, multiple genes or may involve entire chromosomes. They occur spontaneously or as a result of **mutagens**—factors that induce mutation. Common mutagens include different forms of radiation. For example, UV radiation can cause mutations in skin cells that result in skin cancers. Most mutations are detected and repaired by enzymes. Those that cannot be repaired fall into one of three categories: neutral, beneficial or harmful.

- Neutral mutations have no effect on survival.
- Beneficial mutations increase the likelihood of survival.
- Harmful mutations decrease the likelihood of survival.

Somatic mutations occur in body cells and only affect that individual. **Germline mutations** are heritable because they affect gametes and can therefore be passed on to offspring. A germline mutation may bring a new allele into a gene pool, potentially influencing the allele frequencies.

POINT MUTATIONS

Genetic sequences are read in sets of three nucleotides. On the template DNA this set of three nucleotides is called a triplet. In mRNA sequences it is referred to as a codon. These triplet sequences code for specific amino acids. However, using the four nucleotides, there are 64 possible codons but only 20 amino acids. As you learnt in Section 3.2, this means that most of the amino acids are coded for by more than one codon.

A mutation that alters, adds or removes a single nucleotide from a sequence of DNA or RNA is called a **point mutation** (Figure 9.1.3). Point mutations typically only affect a single gene. Point mutations include substitution and frameshift mutations.

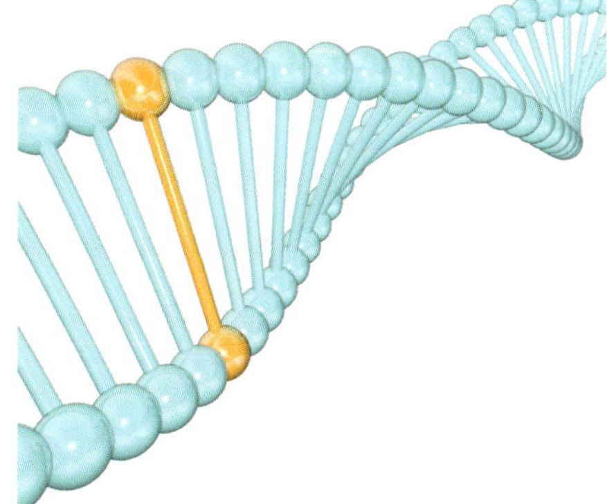

FIGURE 9.1.3 A point mutation involves just one nucleotide in RNA (or nucleotide pair in DNA).

Substitution mutations

A **substitution mutation** is a point mutation in which one nucleotide is replaced by another type of nucleotide. There are different types of substitution mutations including:

- silent mutations
- missense mutations
- nonsense mutations.

Silent mutations

A **silent mutation** occurs when a substitution results in a new codon that still codes for the same amino acid. For example, the DNA triplet CCA is transcribed into the mRNA codon GGU, which codes for the amino acid glycine (gly). If the last nucleotide in the sequence is substituted for any other, the codon will still code for glycine and will not have any effect on the final polypeptide. Below are the normal and a mutated nucleotide sequences of this silent mutation.

Original mRNA sequence:	AUG AAG GAG CGU UUC GGU AUU CAG
Amino acid sequence:	Met – Lys – Glu – Arg – Phe – Gly – Ile – Gln
Mutated mRNA sequence:	AUG AAG GAG CGU UUC GGG AUU CAG
Amino acid sequence:	Met – Lys – Glu – Arg – Phe – Gly – Ile – Gln

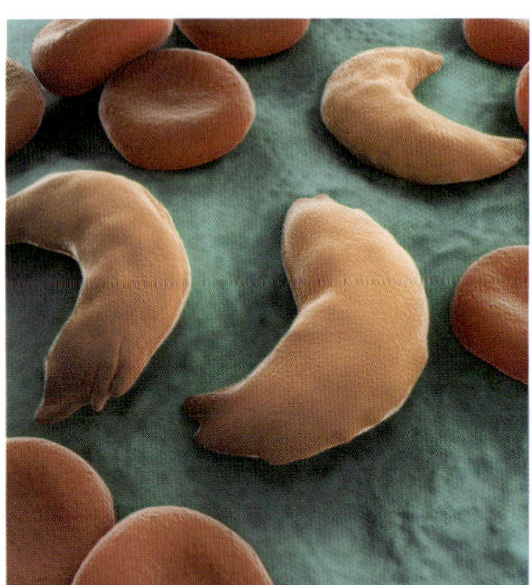

FIGURE 9.1.4 Sickle-cells are characteristically crescent-shaped compared to normal round red blood cells.

Missense mutations

Substitution mutations that result in an amino acid replacement are said to be **missense mutations**. Missense mutations still produce a protein. Whether this altered protein can function properly or not depends on the importance of the amino acid that was replaced.

Normal and sickle-cell haemoglobins differ due to a missense mutation in which glutamic acid is replaced by valine in the beta chain of the protein. The distorted shape of the sickle-cell haemoglobin, called haemoglobin S, affects the overall shape of red blood cells, causing them to become sickle- or crescent-shaped, as shown in the scanning electron micrograph in Figure 9.1.4.

The altered shape of haemoglobin S also reduces its ability to carry oxygen effectively. The fragile nature of sickle cells, their awkward shape and their limited oxygen-carrying abilities result in anaemia and oxygen-deficiency diseases in people who are homozygous for haemoglobin S. Interestingly, people who are heterozygous for haemoglobin S have an increased resistance to malaria.

This resistance has resulted in significant differences in allele frequencies of haemoglobin S in different populations. Sickle-cell has much higher frequencies in populations in sub-Saharan Africa, and small areas of the Mediterranean, Middle East and India as a result of the higher incidence of malaria compared to northern European countries.

Below are the normal and the mutated sickle-cell nucleotide sequences.

Original mRNA sequence:	AUG AAG GAG CGU UUC GGU AUU CAG
Amino acid sequence:	Met – Lys – Glu – Arg – Phe – Gly – Ileu – Gln
Mutated mRNA sequence:	AUG AAG GUG CGU UUC GGU AUU CAG
Amino acid sequence:	Met – Lys – Val – Arg – Phe – Gly – Ileu – Gln

Nonsense mutations

When a substitution mutation results in the creation of a stop codon (UAA, UAG or UGA), it is classified as a **nonsense mutation**, because no other amino acids will be added after this point. Nonsense mutations can have severe effects, particularly if the mutation occurs early in the sequence.

Thalassaemia is an example of a nonsense mutation in which codon 17, of a 147-codon sequence, codes for 'stop' instead of 'lysine' (AAG is changed to UAG). Thalassaemia is, like sickle-cell, a mutation of haemoglobin. It results in oxygen-deficiency disease. The allele frequency of thalassaemia varies greatly among different populations. Like sickle cell, thalassaemia increases resistance to the parasite that causes malaria and so tends to have higher frequencies in populations with Mediterranean, Indian, Middle Eastern and Asian descent.

Below are the normal and the mutated thalassaemia nucleotide sequences.

Original mRNA sequence:	AUG AAG GAG CGU UUC GGU AUU CAG
Amino acid sequence:	Met – Lys – Glu – Arg – Phe – Gly – Ileu – Gln
Mutated mRNA sequence:	AUG UAG GAG CGU UUC GGU AUU CAG
Amino acid sequence:	Met–stop

Frameshift mutations

Frameshift mutations involve one or two nucleotides being either added or removed from a nucleotide sequence, altering every codon in that sequence from that point onwards. These mutations can have significant effects on the polypeptide because as every codon is altered so too is every amino acid they code for after the point of mutation. This results in the loss of functional protein, as it is likely that the resulting polypeptide would be completely different. In cases where the frameshift mutation creates a STOP codon earlier in the sequence, the resulting polypeptide will be shorter.

There are two types of frameshift mutation:

- A nucleotide insertion adds one or two new nucleotides into the sequence and pushes the rest of the nucleotides back one or two places.
- A nucleotide deletion removes one or two nucleotides and pulls all the following nucleotides forwards by one or two places.

Original mRNA sequence:	AUG AAG GAG CGU UUC GGU AUU CAG
Amino acid sequence:	Met – Lys – Glu – Arg – Phe – Gly – Ile – Gln
Insertion mRNA sequence:	AUG AAG GAG CGU AUU CGG UAU UCA G
Amino acid sequence:	Met – Lys – Glu – Arg – Asn – Arg – Tyr – Ser
Deletion mRNA sequence:	AUG AAG GAG CGU UUG GUA UUC AG
Amino acid sequence:	Met – Lys – Glu – Arg – Leu – Val – Phe

BLOCK MUTATIONS

Mutations that affect large sections of a chromosome, typically multiple genes, are called **block mutations** (and sometimes called chromosomal mutations). These types of mutations usually occur during meiosis in eukaryotic cells. They can also be caused by mutagens such as radiation. When a gene is disrupted by the mutation, the effects are serious, even lethal. There are five main forms of block mutations:

- duplication
- deletion
- inversion
- insertion
- translocation.

Duplication mutations

Duplication mutations involve the replication of a section of a chromosome that results in multiple copies of the same genes on that chromosome (Figure 9.1.6). There can be thousands of repeats. This often increases gene expression, which can be harmful or beneficial depending on the gene involved.

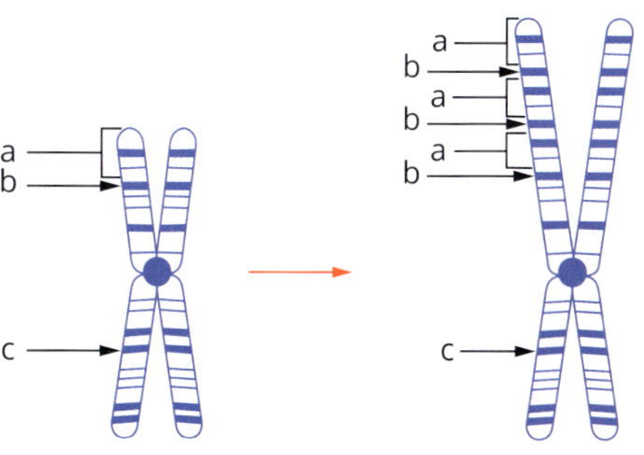

FIGURE 9.1.6 Chromosomal duplication mutations result in multiple repetitions of a sequence of DNA.

Deletion mutations

Deletion mutations remove sections of a chromosome (Figure 9.1.7). Deletions lead to disrupted or missing genes, which can have serious effects on growth and development. Chromosomal deletions are often fatal.

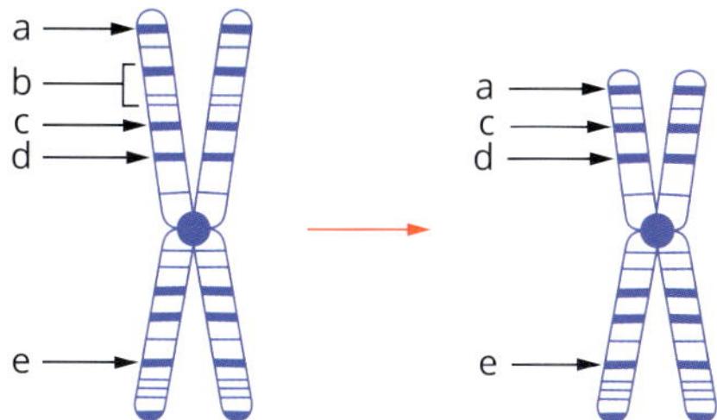

FIGURE 9.1.7 Chromosomal deletion mutations involve the loss of large sequences of DNA from the chromosome (sometimes whole genes).

BIOFILE

A beneficial mutation

An example of a beneficial substitution point mutation is one that involves the *ApoA-1* gene. This gene codes for a protein (apolipoprotein A-1) that is normally involved in the transport of cholesterol and phospholipids to the liver, where they are then redistributed or broken down and excreted. One of the mutated forms of the protein, ApoA-1 Milano, involves a substitution of the amino acid arginine (arg) for cysteine (cys). This mutated protein acts as an antioxidant, reducing cholesterol deposition in arteries, and thereby significantly decreasing the risk of cardiovascular disease.

The mutant form of the ApoA-1 protein (Figure 9.1.5) was first identified in Milan, and so the mutated gene was named after this city. Further investigation, including blood tests of an entire Italian village, traced the origin of the mutation to a single man. The 3.5% frequency of the gene in that village population can be attributed to the descendants of this one man.

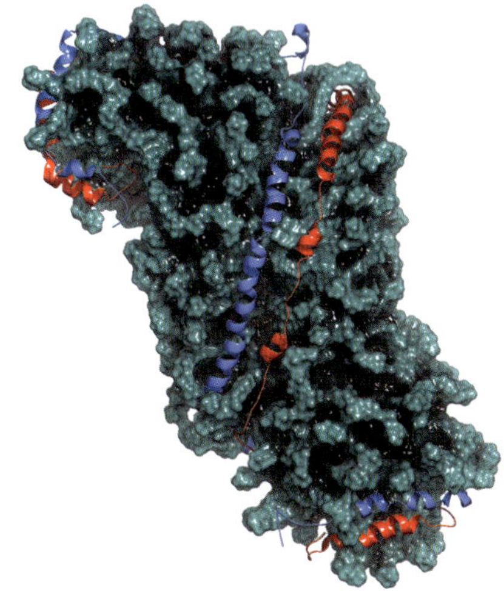

FIGURE 9.1.5 ApoA-1 Milano is a mutated form of a protein that can reduce cholesterol levels in the human blood stream. ApoA-1 Milano is caused by a beneficial point mutation.

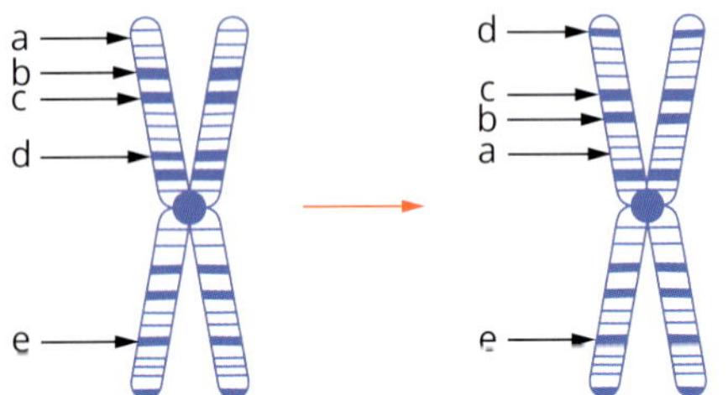

FIGURE 9.1.8 Chromosomal inversion mutations involve a broken section of the sequence rotating 180° before reattaching.

Inversion mutations

During an **inversion mutation**, a section of the sequence breaks off the chromosome, rotates 180° and reattaches to the same chromosome (Figure 9.1.8). Inversions may involve as few as two bases or they may involve several genes.

Insertion mutations

An **insertion mutation** occurs when a section of one chromosome breaks off and attaches to a different chromosome (Figure 9.1.9). In eukaryotes, the effects of this type of mutation depend on whether the cell retains two copies of every gene. During meiosis, if the chromosome now containing the insertion is separated from the chromosome in which the material originated, some gametes may have two copies of the genes in the inserted section, while others will be missing them entirely (Figure 9.1.10).

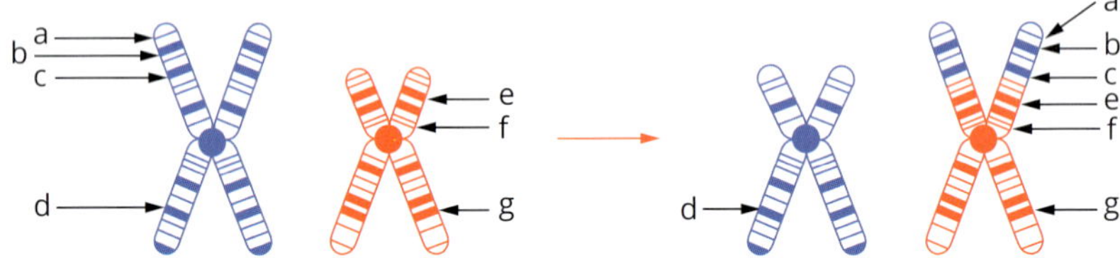

FIGURE 9.1.9 Chromosomal insertion mutations involve a sequence breaking off one chromosome and attaching to another.

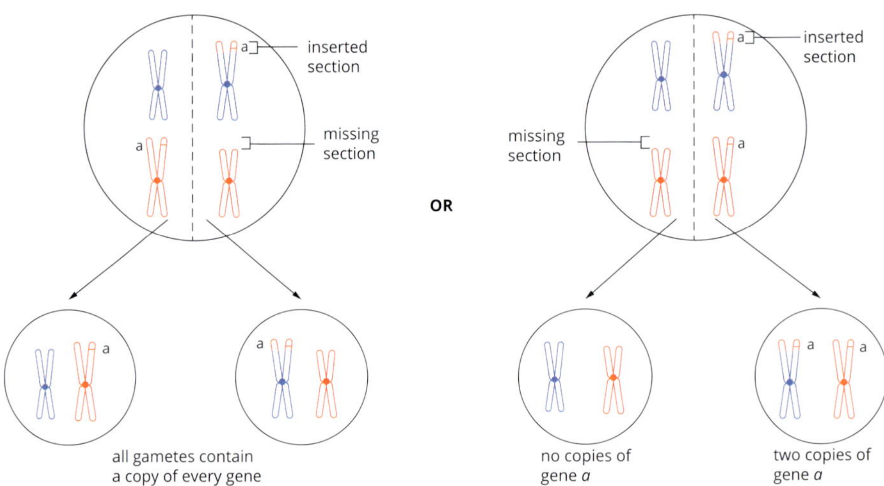

FIGURE 9.1.10 As a result of random assortment in meiosis, a chromosomal insertion mutation may lead to gametes with one, two or no copies of the inserted region.

Translocation mutations

In **translocation mutations**, a whole chromosome or a segment of a chromosome becomes attached to or exchanged with another chromosome or segment. For example, sections from two **non-homologous chromosomes** may break off at the same time. They may reattach to the other chromosome, swapping genetic material (Figure 9.1.11). Translocations typically interrupt normal gene regulation and are the cause of some forms of cancer.

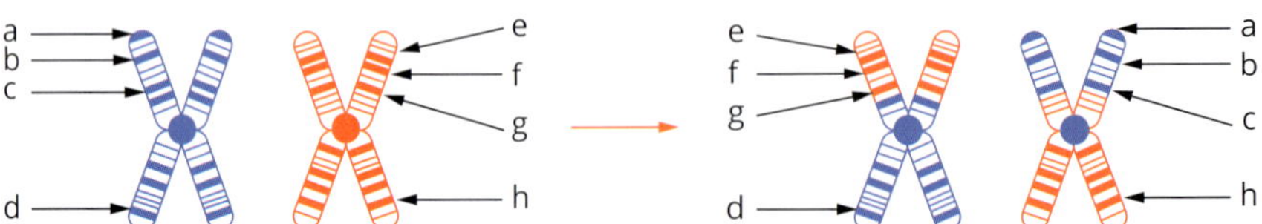

FIGURE 9.1.11 Chromosomal translocation mutations can involve two different chromosomes exchanging segments.

CHROMOSOMAL ABNORMALITIES

When a mutation involves whole chromosomes, or the number of chromosomes, it is termed a **chromosomal abnormality**. This type of mutation is easily detected with a **karyotype**, which is a technique of staining and photographing chromosomes to help classify them and detect chromosomal mutations. There are two main forms of chromosomal abnormalities: aneuploidy and polyploidy.

Aneuploidy

Aneuploidy is the presence of an abnormal number of a particular chromosome, either an extra chromosome (called a **trisomy** when there are three copies of one chromosome) or a missing chromosome. It is usually caused by **non-disjunction** during meiosis. This is when two **homologous chromosomes** do not separate during the first division of meiosis. Aneuploidy results in gametes with an incorrect haploid number, and so results in an abnormal diploid number, which often leads to miscarriage in humans. Aneuploidy can also occur in plants, such as maize, resulting in sterility.

Aneuploidy of human sex chromosomes, X and Y, can result in a number of different conditions, some of which are summarised in Table 9.1.1.

Condition	Genotype	Gender	Incidence (approx.)	Common characteristics
Triple X syndrome	XXX	female	1:1000	no observable characteristics, although typically very tall
Turner syndrome	XO (a single X chromosome) (Figure 9.1.12)	female	1:2500	short stature, skin from neck fused to head and typically infertile
Klinefelter syndrome	XXY	male	1:650	typically sterile, often of tall stature
48, XXXY syndrome	XXXY	male	1:50 000	tall stature, testicular dysfunction and intellectual impairment
48, XXYY	XXYY	male	1:18 000–40 000	sterility, intellectual impairment, tall stature and developmental delays
XYY syndrome	XYY	male	1:1000	rapid physical growth, but usually appear as 'normal' males

TABLE 9.1.1 Types of aneuploidy in sex chromosomes in humans.

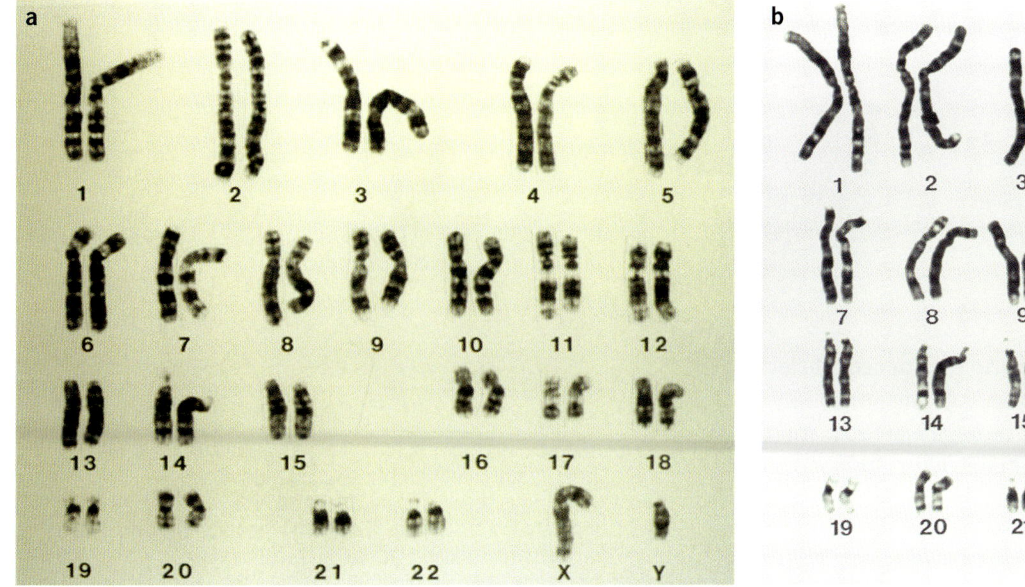

FIGURE 9.1.12 (a) A normal human karyotype. (b) This human karyotype clearly shows that the individual only has a single sex chromosome, an X chromosome, and therefore has Turner syndrome.

EXTENSION

Autosomal aneuploidy

When an abnormal gamete carrying two copies of the same chromosome is fertilised by a normal haploid gamete, it results in trisomy, a cell with three copies of that chromosome. Patau syndrome is caused by trisomy 13 and Down syndrome by trisomy 21.

Errors in mitosis can also result in trisomy. This can occur at various stages of embryonic development. Some people with mild Down syndrome are mosaics. Some of their cells carry three copies of chromosome 21 and other cells are normal. The severity of the condition will depend on which tissues (e.g. brain) are most affected.

Patau syndrome

A person with Patau syndrome can be male or female. They typically present with intellectual disabilities; a small skull, ears and mouth; small, wide hands with short fingers; harelip and a cleft palate. They also commonly have heart defects and seldom survive past the age of one year old. The karyotype in shown in Figure 9.1.13 is that of a male with Patau syndrome. The X and Y chromosomes are clearly visible, as are all three copies of chromosome 13.

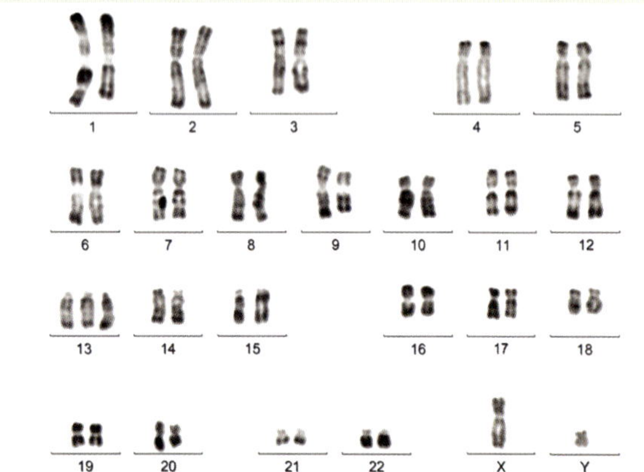

FIGURE 9.1.13 An X and Y chromosome as well as three copies of chromosome 13, seen in this karyotype, indicate a male with Patau syndrome.

Down syndrome

People with Down syndrome also experience intellectual disabilities, although this can vary greatly in severity. They tend to be shorter than average with poor muscle tone and distinctive facial features. Although people with Down syndrome tend to have lowered immunity and an increased risk of heart diseases, with appropriate treatment and healthy lifestyles many live into their 50s and beyond.

There are two genetic mechanisms that lead to Down syndrome. In about 96% of cases, the condition is non-familial; that is, there is no history of the syndrome in the family of the Down individual. For the remaining 4% of cases there is a family history of the condition, and it is called familial Down syndrome. In these families, the incidence of people with the condition may be as high as 20%.

In people with non-familial Down syndrome, the extra copy of chromosome 21 usually results from an error in meiosis during gamete formation. The frequency of these non-disjunction events increases with the age of the parent, especially the mother. For example, the chance of a 35-year-old mother having a Down syndrome child is three times higher than that of a 30-year-old mother. By the age of 45, the risk is even higher.

Familial Down syndrome is caused by the translocation of the long arm of chromosome 21 onto another chromosome, often 14 or 15. The short arm of chromosome 21 does not contain any functional genes, so it is never transcribed. After a few cell divisions it is lost from the cell. The combining of chromosome 21 with another results in an overall diploid number of 45 instead of 46. This individual does not have Down syndrome, but is a carrier. The inheritance of this type of Down syndrome is shown in Figure 9.1.14.

If a gamete containing the translocated chromosome is fertilised, it may result in a child with Down syndrome, a spontaneous abortion or another carrier depending on the other chromosomes present. Unlike non-familial Down syndrome, the age of the mother does not influence incidence. Translocation of chromosome 21 is just as likely to happen in males as females.

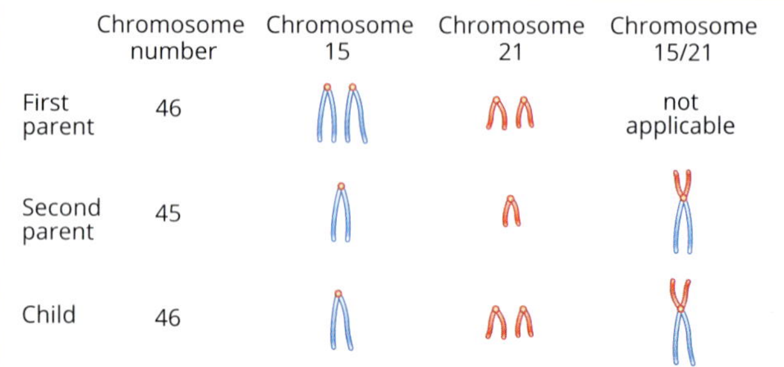

FIGURE 9.1.14 The chromosomes of parents and their child with familial Down syndrome. Chromosome 21 is shown in red and chromosome 15 is shown in blue. The child received two normal copies of chromosome 21 (one from each parent) and a third copy of 21 in a 15/21 translocation (from the second parent). The second parent is a carrier of Down syndrome because they possess the 15/21 translocation chromosome but only possess two copies of chromosome 21, rather than three.

Polyploidy

Most eukaryotic cells are **diploid**; that is, they have two complete sets of chromosomes. For sexually reproducing organisms, the ovum provides one set of chromosomes and the sperm provides the other set. The gametes themselves are typically **haploid**, to ensure that fertilisation restores the correct number of chromosomes to the zygote.

However, errors during meiosis can result in diploid gametes. If one of these gametes is fertilised, it will result in a zygote with more than the usual two sets of chromosomes. **Polyploidy** is the condition that results from cells and organisms that contain more than two full sets of chromosomes; that is, more than two of every chromosome in the set (Figure 9.1.15).

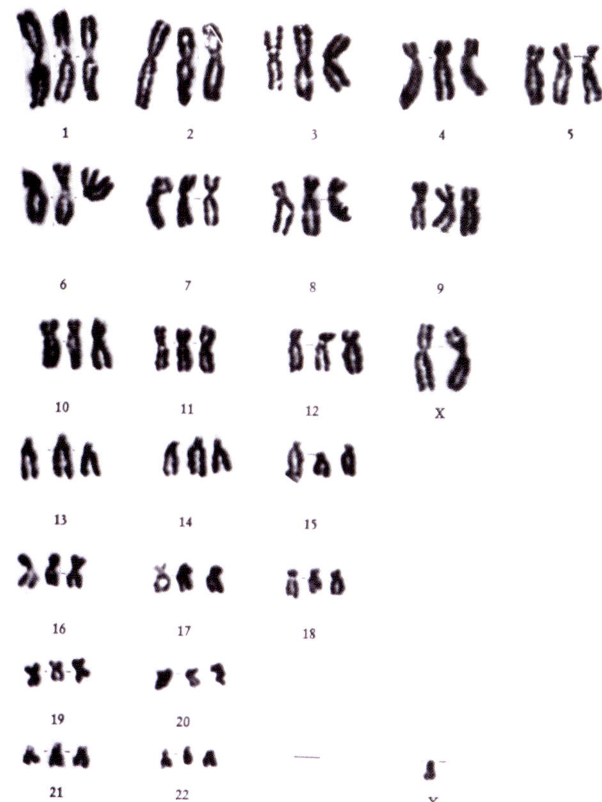

FIGURE 9.1.15 Three copies of all autosomal chromosomes, plus two X and one Y chromosome shown in this karyotype, indicates a male with triploidy.

In humans, polyploid zygotes do not survive. However, polyploidy can also arise from errors during mitosis and produce groups of polyploid somatic cells. Having a few cells with an abnormal chromosome number may not affect health. A liver, for example, may function normally but have patches of polyploid cells. The karyotype shown in Figure 9.1.15 is of a male with 3 copies of each chromosome; known as triploidy.

Polyploid animals include some insects, earthworms and tree frogs. However, polyploidy is more common in plants than animals because many plants can survive by asexual reproduction (Figure 9.1.16). A triploid (3N) organism, resulting from the fusion of one haploid gamete and one diploid gamete, is typically sterile or has low fertility because of problems with chromosome pairing during meiosis and gamete formation. However, a triploid plant could survive by vegetative reproduction or by spontaneous duplication of all chromosomes to 6N. Tetraploid organisms can result from two diploid gametes fusing. A tetraploid organism can produce viable offspring if it breeds with another tetraploid, but not with a diploid individual.

Some banana varieties are triploid; cultivated cotton and potatoes are examples of tetraploid (4N) organisms; bread wheats are hexaploid (6N); and strawberries are often octoploid (8N).

A difference in the number of sets of chromosomes is a form of postzygotic isolation, which can lead to speciation. **Speciation** is the evolution of new species from an ancestral species. You will learn more about the mechanisms of speciation in Section 9.2.

FIGURE 9.1.16 Strawberries are often polyploid. Through asexual reproduction they can possess up to eight copies of each chromosome.

ENVIRONMENTAL PRESSURE

There is always **variation** between individuals within a population. This is due to different combinations of alleles that have arisen from random mating, independent assortment and recombination during gamete formation, and also from mutations. These factors, and the individual differences in gene expression and environmental factors, can lead to differences in phenotypes.

Should a particular phenotype give an individual a survival advantage, that phenotype, and the genes and alleles that control it, is more likely to persist in the population. The conditions or factors that influence allele frequency in a population are known as **selection pressures**. Selection pressures, together with mutation, are the driving force of **evolution**. Selection pressures can be natural environmental pressures or artificial pressures brought about by humans through selective breeding.

NATURAL SELECTION

FIGURE 9.1.17 Drought can create competition for resources such as water and food. This selection pressure can trigger changes in allele frequencies in a population.

Natural selection is the influence of environmental pressures on allele frequency in a population. These environmental selection pressures affect the survival and reproduction of an organism. Individuals with the most advantageous phenotypes have an increased chance of producing fertile offspring.

Examples of environmental selection pressures include:

- climatic conditions such as extreme temperature changes and drought
- competition for resources such as the availability of food and water (Figure 9.1.17), as well as competition for shelter
- mate availability
- predator abundance.

Environmental pressures influence allele frequencies of a gene pool because of a number of factors:

- Variation—There are genetic differences between individuals of a population.
- Reproduction—Organisms can reproduce and alleles are heritable. The offspring are genetically similar (sexual reproduction) or genetically identical (asexual reproduction) to the parents.
- Survival—Not all individuals survive long enough to reproduce and produce offspring.
- Environmental selection pressures—Some phenotypes are better suited to the environmental conditions and give the individual a survival advantage over those of a different phenotype.

When it comes to survival, some phenotypes (traits) have a high **adaptive value** and give the individual an advantage over individuals with phenotypes of lower adaptive value. This concept is often referred to as 'the survival of the fittest'. Having an advantageous trait means the individual is more likely to survive to reproduce and pass their alleles on to the next generation.

Alleles of the advantageous trait tend to increase in frequency in the gene pool, while alleles of the less advantageous trait tend to decrease. Advantageous traits of high adaptive value may persist in the population until all individuals possess them (100% allele frequency). Over time, the population evolves and adapts to its environment.

FIGURE 9.1.18 The key physical characteristics (traits) of the thorny devil (*Moloch horridus*) are a result of increasing frequencies of alleles that provide phenotypes with high adaptive value.

Thorny devils (Figure 9.1.18), for example, have a number of physical adaptations that enable them to thrive in the very arid ecosystems of central Australia. Their mottled camouflage colouring and hard spikes have high adaptive value because these features reduce the likelihood of predation. Thorny devils also have highly textured skin, which allows capillary action to collect any moisture in their environment and channel it directly into their mouths.

Gene flow

Gene pools can change when new individuals join the population from a different gene pool or when some individuals leave a population. Such migration of individuals can result in **gene flow**.

When gene flow exists between two different populations, the gene pools may remain fairly similar. When gene flow is not possible between populations, the gene pools are said to be isolated. Genetic isolation is another key factor in the process of evolution and you will find out more about it in Section 9.2.

> Seed dispersal in plants, interbreeding between different populations and migration can all result in gene flow.

Genetic drift

Allele frequencies in a gene pool may also change randomly over time as a result of chance events. This is called **genetic drift**. Genetic drift is more clearly seen in small populations with little to no gene flow, as the random death of one individual can significantly alter the allele frequencies. Generally, in small populations genetic drift results in the loss of genetic diversity over time as alleles are lost from the gene pool.

Genetic drift can occur when populations decrease for a period of time (a bottleneck effect) or in small founding populations (the founder effect).

Bottleneck effect

The number of individuals in a population can be drastically and quickly reduced as a result of a random event, often a natural disaster. Human activities, such as hunting and land clearance have also greatly and quickly reduced the number of individuals in wild populations of plants and animals. The phenotype of an individual is unlikely to significantly increase its chances of surviving a natural disaster, such as a volcanic eruption, tidal wave or landslide. The individuals that survive will do so by chance. The allele frequencies of the remaining population are unlikely to reflect those of the original population.

The **bottleneck effect** describes the impact on the remaining population (Figure 9.1.19). Because of the reduced population size, the possible reproductive pairings are limited, which leads to high levels of inbreeding. Inbreeding results in reduced variation in the population and an increase in the numbers of homozygous individuals. The smaller the population, the greater the bottleneck effect of genetic drift. Alleles may be lost from the gene pool immediately after the natural disaster or be 'bred out' in only a few generations. The lowered variation may make the population more vulnerable to environmental change.

FIGURE 9.1.19 The mountain pygmy possum (*Burramys parvus*) has a declining population and reduced genetic diversity. This population bottleneck is a result of bushfires, and habitat loss from land clearance and development around Mt Buller.

EXTENSION

Different theories of adaptation over time

Over time, a number of different theories have been proposed to explain the existence of all life on Earth. However, today, the most widely accepted explanation is the theory of evolution, the theory that new forms or species of life have evolved over time. The modern theory of evolution is the result of the combined works of Jean Baptiste Lamarck, Charles Darwin, Alfred Russel Wallace and Gregor Johann Mendel, beginning in the 18th century.

Jean Baptiste Lamarck (1744–1829)

Lamarck was a French naturalist and the first scientist to publish a reasoned theory of evolution. In France, he is regarded as the 'father of evolution'. He developed the theory of inheritance of acquired characteristics as a mechanism to explain how organisms changed over time. Lamarck argued that a particular trait was enhanced or diminished within the lifetime of an individual, depending on its use. The modified trait was inherited by offspring.

According to this theory, if, for example, a short-necked giraffe had to stretch to reach leaves, its neck would became longer over its lifetime. The offspring would have slightly longer necks as a result. In this way, giraffes' necks would continue to stretch until they reached their modern-day length.

Lamarck was developing his theory before the understanding of genetics. His theory was flawed as it did not explain some situations, such as why the offspring of an amputee were born with all their limbs.

FIGURE 9.1.20 Charles Darwin was a key scientist in the development of the theory of evolution.

Charles Darwin (1809–1882)

Darwin (Figure 9.1.20) was an Englishman who sailed as a naturalist on the *HMS Beagle*, collecting specimens while visiting the Galapagos Archipelago, South America, New Zealand and Australia, as well as many other places. Based on his findings during these voyages and the research he performed on domestic pigeons, he developed the theory of evolution by natural selection.

Darwin's theory was an improvement on Lamarck's theory but could not explain how characteristics were inherited.

Alfred Wallace (1823–1913)

Wallace was another Englishman who travelled extensively. He collected specimens in the Amazon and the region of Indo-Malaya. While travelling, he independently came up with the same idea of evolution by natural selection as Darwin (which spurred Darwin on to publish his work, *On the Origin of Species*).

Gregor Mendel (1822–1884)

Mendel was an Austrian monk who studied agriculture and botany. In his garden, he experimented comprehensively with the inheritance of traits in pea plants. His findings became the basis of modern genetics. However, he only became famous after his work on plant genetics was rediscovered in 1900, almost 20 years after his death.

BIOFILE

Founder effect in Tasmania

In Tasmania, many sufferers of Huntington's disease can trace their ancestry to a woman who migrated from Britain in the 19th century.

Founder effect

The **founder effect** occurs when a small group of individuals from a larger population move to a new location and establish a new population. If a small portion of a population becomes separated from the original population, their smaller gene pool is unlikely to reflect the allele frequency of the original population. Like the bottleneck effect, there is increased inbreeding in the 'founder population' and lower variation.

In the new environment, the environmental pressures on the founder population are likely to be different from those experienced by the original population. These differences in environmental pressures drive further changes in allele frequencies and, ultimately, evolution.

9.1 Review

SUMMARY

- Natural variation exists between individuals of the same species because many genes have multiple alleles that are present in the population in different frequencies.
- Allele frequencies are a measure of how common a particular allele is in the gene pool of a population, and are typically expressed as percentages or decimals.
- New alleles, genes and chromosomes are created through mutation.
- Mutations may have a beneficial effect, a harmful effect or no effect at all on the individual.
- Point mutations include:
  - substitution—The mutation replaces one nucleotide in the sequence for another nucleotide.
  - addition—The mutation introduces a new nucleotide into the sequence, resulting in a frameshift.
  - deletion—The mutation removes a nucleotide from the sequence, resulting in a frameshift.
- Substitution mutations appear as three types:
  - silent mutations—The new codon (triplet) still codes for the same amino acid.
  - missense mutations—The new codon codes for a different amino acid (the effects of which may vary depending on the new amino acid and the resulting functionality).
  - nonsense mutations—The new codon is a stop codon and shortens the amino acid chain (which may have severe effects).
  - Addition or deletion mutations are also called frameshift mutations because they alter every triplet from that point onwards in that gene. They typically have more severe effects than substitution mutations.
- Block mutations, or chromosomal mutations, typically involve multiple genes. Types include:
  - duplication mutations—A section of a chromosome is repeated multiple times on that chromosome.
  - deletion mutations—Entire genes are cut from the chromosome.
  - inversion mutations—A large section of a chromosome is removed and rotated 180° before being reinserted, so that the sequence is reversed.
  - insertion mutations—A whole chromosome or section of a chromosome is added to a different chromosome.
  - translocation mutations—Sections from two non-homologous chromosomes are swapped.
- Chromosomal abnormalities are mutations that involve whole chromosomes, or the number of chromosomes. The two main forms are:
  - aneuploidy—The cell or individual has more than or fewer than two copies of a particular chromosome, e.g. trisomy 21 (Down syndrome).
  - polyploidy—The cell or individual has more than two copies of every chromosome, e.g. triploid organisms have three copies of every chromosome.

- Natural selection is the influence of environmental pressures on allele frequencies of a population, which occurs because of genetic variation between individuals, and the survival and reproduction of those individuals with favourable phenotypes (traits):
  - Phenotypes that are better suited to environmental pressures have higher adaptive values than those that are less suited.
  - Individuals with alleles associated with the phenotype are more likely to survive and reproduce.
  - These alleles are more likely to persist in the gene pool and increase in frequency over time.
- Gene flow is the movement of alleles into and out of a gene pool. It can occur when different populations interbreed or individuals migrate between populations.
- Genetic drift is the change in allele frequencies in a population due to random events. This is most likely to affect small populations.
  - The bottleneck effect is seen when an event, often a natural disaster, significantly reduces the size of a population and thus its genetic diversity. It also therefore increases inbreeding, further reducing variation. Genetic drift can remove alleles from the gene pool altogether.
  - The founder effect is seen when a small group of individuals breaks away from the main population and colonises a new habitat. The founding individuals will not necessarily represent the allele frequencies seen in the original population. Further changes in allele frequencies in the founding population can arise from new environmental pressures.

KEY QUESTIONS

1. Which two factors influence phenotype?
2. A single gene with two alleles controls flower colour in pea plants. Purple flowers are dominant to white. In a population of 150 individuals, 80 are homozygous for purple flowers, 40 are homozygous for white flowers and the remaining 30 individuals are heterozygous. Calculate the allele frequency for the purple allele.
3. Why are frameshift mutations more significant than substitution point mutations?
4. Define 'natural selection'.
5. Describe three key factors that contribute to natural selection.
6. How might gene flow occur between two populations?
7. Would genetic drift have more impact on a small population or a large one?

9.2 Processes of evolution

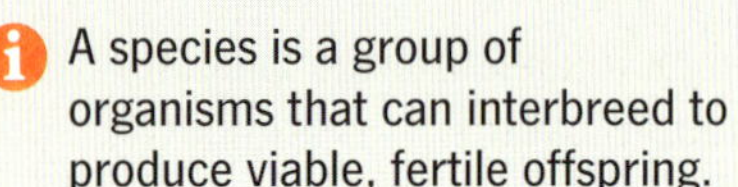

A species is a group of organisms that can interbreed to produce viable, fertile offspring.

Evolution is the change in the genetic composition of populations over time. This can be observed as changes in allele frequencies in a population over time. Genetic change in populations of an ancestral species can lead to new species (Figure 9.2.1). In this section, you will learn that natural selection acting through environmental pressures is the main driving force of evolution.

FIGURE 9.2.1 There are close to one million known insect species on Earth, with new species being discovered every year. The diversity seen in insect species is largely due to different selection pressures over a long period of time.

FIGURE 9.2.2 Koalas are able to interbreed to produce viable fertile offspring and thus are considered to be the same species.

PROCESSES OF EVOLUTION

As you learnt in the previous section, natural selection is a driving force of evolution. It causes alleles to increase or decrease in frequency, depending on their adaptive value. Environmental conditions change over time and so too do the environmental selective pressures on a population. Allele frequencies change in response to natural selection and, eventually, the population may change so significantly that it is deemed a new species. As you also learnt in the previous section, in small populations genetic drift can result in evolutionary change.

It is important to remember that populations and species evolve, not individuals. It is changes in the gene pool that are evidence of evolution, not mutations within an individual.

The relationship between changing environmental pressures and evolution of a species can be seen in ecosystems that have experienced little change over time. If the environment remains unchanged, the traits of organisms will continue to have high adaptive value, and there will be little change to allele frequencies over time.

There are some cases in which the definition of a species has been challenged. For example, there are genes present in the human genome that suggest that *Homo neanderthalensis* and *Homo sapiens* have successfully interbred in the past.

DEFINING SPECIES

A **species** is defined as a group of individuals that are genetically similar enough to produce fertile **viable offspring** when interbreeding (Figure 9.2.2). A species can also be thought of as a gene pool that is isolated from the gene pools of other species. **Genetic isolation** of one species from another can be a result of one or more mechanisms. These mechanisms can be classified into before reproduction (prezygotic) or after reproduction (postzygotic).

Prezygotic isolating mechanisms

Prezygotic isolating mechanisms are those that typically prevent individuals from different populations from interbreeding; in other words, they prevent fertilisation from occurring in the first place. Commonly, prezygotic isolating mechanisms prevent mating. Less common forms of prezygotic isolating mechanisms work after breeding takes place; these prevent gametes from fusing and forming a zygote.

There are a number of prezygotic isolating mechanisms:

- Geographical (spatial) isolation—Populations are separated by physical and geographical barriers, such as oceans, deserts, mountain ranges and glaciers. For example, the southern boobook (*Ninox boobook*) (Figure 9.2.3a) is an Australian owl that is genetically distinct from the New Zealand owl, morepork (*Ninox novaeseelandiae*) (Figure 9.2.3b). One reason that they are genetically isolated is that the Tasman Sea separates them.
- Ecological isolation (or **niche** partitioning)—Populations occupy different niches within the same ecosystem. For example, *Eucalyptus baxteri* (brown stringybark) and *E. verrucata* (Mt Abrupt stringybark) are closely related species that grow side-by-side in the Grampians, in Victoria. *E. verrucata* grows on upper slopes on rocky sites and *E. baxteri* occurs on lower slopes on deeper soils. The two species are usually reproductively isolated but sometimes their flowering times overlap and neighbouring trees will interbreed. **Hybrids** (the offspring of two different species) are fertile, but are generally found only along the border (ecotone) between the two species. The obvious boundary between species can be seen in Figure 9.2.4.

FIGURE 9.2.3 Australia's southern boobook (a) is closely related to the New Zealand morepork (b). The two species are separated by a geographical barrier.

FIGURE 9.2.4 *Eucalyptus verrucata* is a small, hardy shrub that grows on shallow soil and rocky sandstone slopes. *Eucalyptus baxteri* occurs on deeper soils where the bedrock has weathered more. The sharp boundary between the two species can be seen at Mirranatwa Gap, in the southern Grampians. The boundary corresponds with a distinct change in the texture of the sandstone and soil depth.

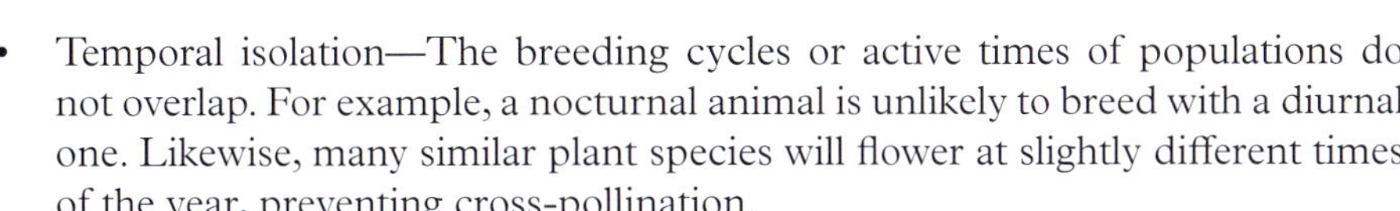

- Temporal isolation—The breeding cycles or active times of populations do not overlap. For example, a nocturnal animal is unlikely to breed with a diurnal one. Likewise, many similar plant species will flower at slightly different times of the year, preventing cross-pollination.
- Behavioural isolation—This occurs when mating calls and courtship rituals, for example, are highly specific. This isolating mechanism is only possible in animals. An example is mate attraction to different types of vocal signals, such as bird songs or frog calls, which are unique to species. Behavioural isolation is often the result of sexual selection, which you will learn more about below.
- Structural or morphological isolation—The reproductive organs of different species are physically incompatible and individuals are unable to mate. For example, a sparrow could not breed with an albatross. For more similar species, even slight differences can prevent mating, such as the different breeding pheromones produced by different moth species.
- Gamete mortality—Egg and sperm fail to fuse in fertilisation. For example, the sperm of one species may not be able to 'find' the egg of another without the appropriate signalling molecules, or the conditions of the female reproductive tract of one species may not sustain the sperm of another species. Pollen may not germinate on the style of the 'wrong' flower due to a chemical barrier, preventing sperm from reaching an egg.

Sexual selection

Sexual selection is one way in which evolution has resulted in behaviour acting as a reproductive or prezygotic isolating mechanism. Most animals exhibit some level of sexual selection in which at least one gender selects their mate based on specific traits. Although it may appear that mates are chosen on the basis of an irrelevant characteristic, the chosen traits are often indicators of good health, strength and fitness or high adaptive value. The alleles of these mates will then be inherited by offspring. Sexual selection is very common in birds. For example, barn swallows (*Hirundo rustica*) select mates on the basis of the length of tail streamers, which indicate health and fitness. Another example is the bowerbird, which selects a mate on the basis of the showiness and particular structure and colour of the bower it builds by retrieving objects from the habitat (Figure 9.2.5), again indicating the bird's health and fitness.

Animals may compete with members of their own sex for mates of the opposite sex. Animals such as sea lions, antelope and kangaroos, come into direct physical conflict over mates, usually resulting in a single male winning the right to mate with a large number of females. These conflicts ensure that the individuals with the 'fittest' phenotypes are the ones that are most likely to produce more offspring and healthier offspring after mating. In this way, the 'fitter' alleles are more likely to be inherited by the next generation and increase the frequency of those favourable alleles in the gene pool over time.

BIOLOGY IN ACTION

The cost of being dangerous: extinction rates higher than speciation rates in poisonous amphibians

FIGURE 9.2.6 A red-backed poison dart frog (*Ranitomeya reticulate*) from the Amazon rainforest in Peru and Ecuador.

Species often evolve defences such as camouflage, warning colouration, toxicity or mimicry of toxic species to escape predation. The evolution of these traits can lead to increased rates of speciation as the prey avoids predation and is able to occupy new niches. An evolutionary hypothesis, known as 'escape-and-radiate', has long been thought to explain such evolutionary relationships between prey and predator; prey species that are constrained by predation will evolve new defences that enable them to 'escape'. Once they have 'escaped', the prey species are free to 'radiate', adapting to new niches and, over time, evolving into new species.

A study has found that this hypothesis may not explain all predator–prey relationships, as it does not take into account extinction rates. Researchers from the University of Liverpool examined speciation and extinction rates in amphibians and found that they varied with the type of defence mechanism the species used.

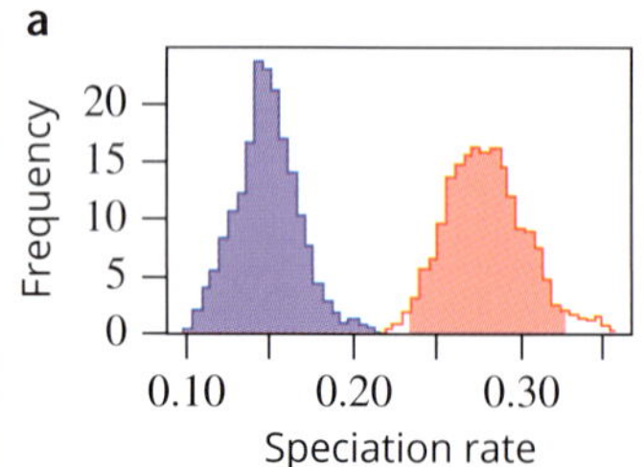

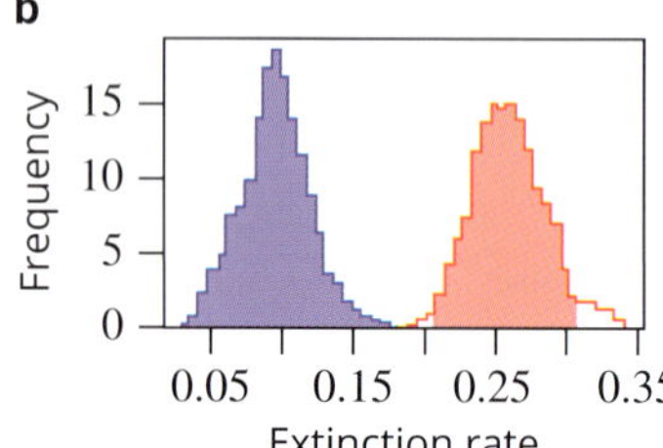

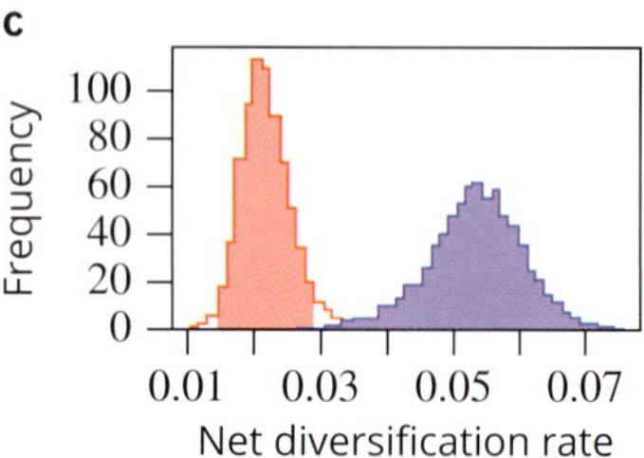

FIGURE 9.2.7 Estimates of rates of speciation (a), extinction (b) and diversification (c) in non-toxic (blue) and toxic (red) amphibian species.

FIGURE 9.2.5 (a) Satin bowerbirds (*Ptilonorhynchus violaceus*) are found in wet forests in the south-east of Australia and are attracted to blue objects. (b) Great bowerbirds (*Chlamydera nuchalis*) occur in the more tropical ecosystems of northern Australia and are attracted to white, green and red. Although the species are also geographically isolated, this difference in mate attraction is an example of sexual selection and behavioural prezygotic isolation.

Species that were poisonous, such as the poison dart frog (Figure 9.2.6), had much higher rates of speciation—twice the rate of non-toxic species—supporting the escape-and-radiate hypothesis (Figure 9.2.7a). However, the researchers also found that the extinction rate of toxic species was up to three times higher than that of non-toxic species (Figure 9.2.7b), resulting in a net loss of toxic species (i.e. lower net diversification rate) (Figure 9.2.7c).

Many toxic species are brightly coloured (conspicuous colouration) to warn predators that they are dangerous. Predators learn from these signals and avoid brightly coloured prey. Mimicry of bright, warning colouration is an anti-predator defence used by some amphibian species. In order to understand the role of conspicuous colouration in speciation and extinction rates, researchers looked at species that were brightly coloured, but not toxic. They found that these species had similar rates of speciation but lower rates of extinction than their toxic counterparts, indicating that bright colouration did not increase extinction risk. Species that had non-conspicuous colouration (e.g. camouflage and non-conspicuous colour variations) had lower rates of speciation than conspicuously coloured amphibians (both toxic and non-toxic) and extinction rates equal to those of non-toxic conspicuous species (Figure 9.2.8).

These findings show that non-toxic species with bright colouration are more likely to be evolutionarily successful, with rapid rates of diversification and speciation, yet a low risk of extinction. This indicates that toxicity comes at an evolutionary cost; the researchers suggest that toxic species might avoid predation but might be vulnerable to other threats, such as infectious disease.

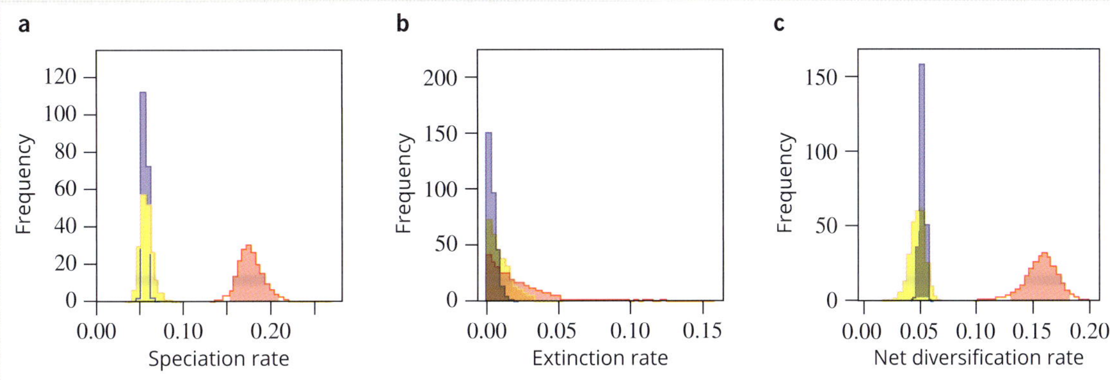

FIGURE 9.2.8 Estimates of rates of speciation (a), extinction (b) and diversification (c) in amphibian species with different colouration: conspicuous (red), camouflage (blue) and polymorphic (colour variants) (yellow).

Postzygotic isolating mechanisms

FIGURE 9.2.9 The zony is a hybrid individual resulting from the interbreeding of a zebra and a donkey.

Postzygotic isolating mechanisms are those that typically prevent a zygote of two different species from developing into a fertile adult. The offspring resulting from interbreeding between individuals from different species are called hybrids (Figure 9.2.9).

Hybrid inviability is mechanism of reproductive isolation in which the sperm from one species does successfully fertilise the egg of another species to form a hybrid zygote, but the hybrid zygote has unmatched chromosomes. As a result, normal embryonic development cannot proceed because of the lack of homologous pairs in the zygote. The zygote does not typically survive long.

Sometimes the zygote survives and undergoes cell division but the offspring does not develop fully and will not reach adulthood. This is known as **reduced hybrid viability**. Most hybrids that do develop into adulthood are sterile; that is, they are incapable of producing offspring themselves (Figure 9.2.10). **Hybrid sterility** usually results from problems during gamete formation. One of the most well-known hybrids is the mule (Figure 9.2.11). As the offspring of a female horse ($2n = 64$) and a male donkey ($2n = 62$), the mule has 63 chromosomes in total. Hybrids such as the mule do not have homologous pairs of chromosomes because their genetic material came from different species. Without homologous pairs, meiosis cannot proceed normally and the gametes, if any are formed at all, cannot interact correctly in order for fertilisation to occur. Mules, like all hybrid offspring, cannot reliably reproduce.

> The interbreeding of different species rarely produces viable offspring. The chromosomes of different species are incompatible because they contain different genes. In other words, the chromosomes are not homologous and the alleles are not for the same genes.

FIGURE 9.2.11 A mule (left) is the offspring of a donkey (right) and a horse. The donkey and horse are separate species, but they can mate to produce viable offspring (the mule). However, the mule is sterile and cannot produce offspring of its own.

FIGURE 9.2.10 An experiment was conducted to crossbreed a western grey kangaroo and an eastern grey kangaroo (shown). The resulting male joeys were all sterile.

In some situations, the first generation of hybrids is semi-fertile and can occasionally produce offspring when reproducing with another hybrid or with one of the parental species. However, the second generation is typically sterile. This form of postzygotic isolating mechanism is called **hybrid breakdown**.

ALLOPATRIC SPECIATION

Speciation is the evolution of new species from an ancestral species. The most common form of speciation is allopatric speciation.

Allopatric speciation occurs when a population becomes divided by a geographical barrier (Figure 9.2.12). The spatial isolation prevents individuals of the separated sub-populations from interbreeding.

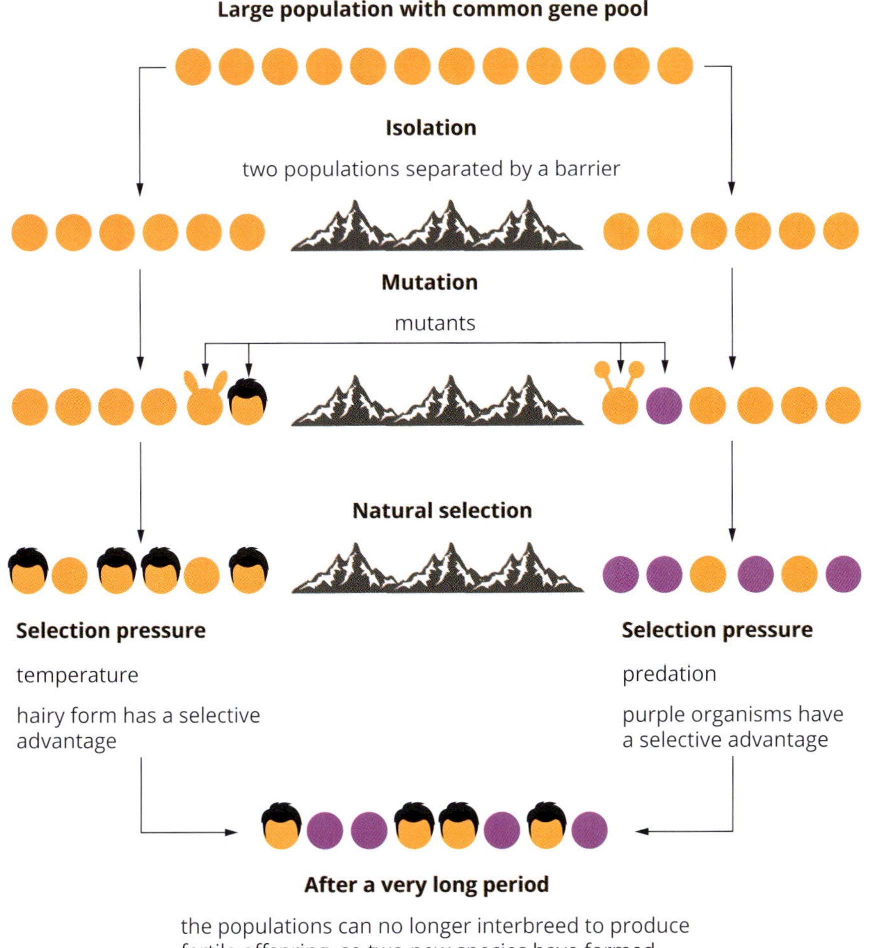

FIGURE 9.2.12 Geographic isolation leads to allopatric speciation.

Over time, different environmental selection pressures and or genetic drift drive change in the allele frequencies of the two sub-populations. Eventually, the two sub-populations may diverge genetically and morphologically, to the point where they can no longer successfully interbreed, should they come into contact again; that is, they may become two distinct species.

Australia can be divided into a number of distinct regions of biodiversity that are isolated geographically by water, mountain ranges and deserts. For example, the south-west corner of Western Australia has a very high number of **endemic** species, both plant and animal, due to its geographic isolation by the central arid zone. Tasmania also has high levels of endemism as a result of its separation from the Australian mainland by Bass Strait (Figure 9.2.14).

BIOFILE

Lions and tigers and ligers

A liger (Figure 9.2.13) is the hybrid offspring of a male lion ($2n = 38$) and a female tiger ($2n = 38$). Ligers also have a diploid number of 38. However, due to differences in the genes in the two species, and despite having the same number of chromosomes, meiosis is rarely successful in ligers. Ligers typically experience hybrid sterility or hybrid breakdown.

Interestingly, the liger tends to be larger than both its parents and is claimed to be the largest cat on Earth. All ligers have been bred in captivity because the natural ranges of tigers and lions do not overlap. The two species are naturally geographically isolated in the wild.

FIGURE 9.2.13 This liger is the hybrid offspring of a Siberian tiger and a lion.

FIGURE 9.2.14 Tasmania has a high number of unique species that are not found on the mainland, such as the leatherwood (*Eucryphia lucida*).

BIOFILE

Allopatric speciation in the coast banksia?

The coast banksia, *Banksia integrifolia*, is one of the largest and most common banksias on Australia's east coast. It grows as far north as Mackay in Queensland and as far south as Port Phillip Bay in Victoria. Over this extensive geographical range, populations vary in the shape of their fruits and particularly in their leaves.

Four forms have been identified. Plant taxonomists have recognised three of these forms as varieties or subspecies of *Banksia integrifolia*. The fourth form, *Banksia aquilonia*, was originally identified as another subspecies but is now recognised as a related but distinct species. Geographically, *B. aquilonia* is separated from the *B. integrifolia* subspecies by more than 200 km.

A molecular study using a DNA fingerprinting technique (called amplified fragment length polymorphism or AFLP technique) confirmed that the three *Banksia integrifolia* subspecies are genetically distinct from one another and different also from *Banksia aquilonia*. The technique also confirmed that plants with a leaf shape intermediate between subspecies are the result of gene flow between populations. The distributions and characteristics of the species and subspecies of coast banksias are summarised in Table 9.2.1.

FIGURE 9.2.15 (a) Coast banksia, *Banksia integrifolia*. (b) The distribution of the different forms of coast banksia.

Species and subspecies	Distribution	Form
B. integrifolia subsp. *compar*	scattered distribution, southern Queensland and north-eastern NSW, coastal	large glossy adult leaves; seedling leaves elliptical, small straight side teeth
B. integrifolia subsp. *monticola*	south-eastern Queensland and north-eastern NSW, montane regions above 650 m in altitude	narrow adult leaves; seedling leaves obovate (widest at the top), large teeth with curved sides
B. integrifolia subsp. *integrifolia*	southern Queensland, NSW, Victoria, coastal	leaves shorter than *B. integrifolia monticola*; seedling leaves obovate, curved
B. aquilonia	north-eastern Queensland, coastal	long narrow adult leaves; largest fruits, fringe of stiff hairs on the midrib, underside of the leaf

TABLE 9.2.1 Distributions and characteristics of species and subspecies of coast banksias.

EXTENSION

Other forms of speciation

All forms of speciation require isolating mechanisms to prevent gene flow between populations. Different isolating mechanisms result in different speciation processes.

Peripatric speciation

Peripatric speciation is similar to the process of allopatric speciation, but instead involves a small population on the edge of the range of a larger population. When this small population becomes spatially isolated from the main population, the founder effect (page 344) can enable low frequency alleles to become more common in only a few generations. The new population typically neighbours the original population and experiences similar environmental pressures. However, genetic drift occurs more quickly in smaller populations and so the new founding population becomes genetically diverse and evolves into a new species.

It is possible that the brown stringybark (*Eucalyptus baxteri*) and Mt Abrupt stringybark (*E. verrucata*) species diverged in this way. Individuals on the edge of the forest ecosystem may have been pushed out to the rocky outcrops, typical of the Grampians, with shallower soil than found on the lower slopes. The difference in environmental conditions would have posed different pressures on individuals. Those with alleles that enabled them to survive in shallow soil would have been more likely to reproduce, and so those alleles would have increased in frequency. With enough time, this and other differences in environmental pressures would have resulted in speciation.

Sympatric speciation

Speciation can occur even when geographical barriers do not isolate populations. Sympatric speciation is more common in plants than in animals, and there are two ways in which it can occur.

Most habitats are made up of microhabitats, small areas with highly specific environmental conditions. When part of a population occupies a microhabitat, the difference in selective pressures can be enough to drive sympatric speciation. In this situation, the two populations become genetically isolated by temporal or behavioural isolating mechanisms.

Sympatric speciation in plants is more commonly caused through polyploidy (page 341). If a mutation causes polyploidy in an individual, the polyploid plant can still reproduce successfully through self-pollination or vegetative reproduction. However, a polyploid plant is unlikely to successfully reproduce with its diploid counterparts and so sympatric speciation by polyploidy is instantaneous.

Parapatric speciation

Parapatric speciation is very rare. This form of speciation occurs where populations are not geographically isolated, but where there is significant variation in habitat conditions within the range of the original population. Gene flow is possible between the two populations, but, in a large population with a large range, individuals are more likely to breed with nearby individuals. Mating is not entirely random. Slight differences in environmental pressures from one end of the range to the other can result in localised variation in allele frequencies within the population. Over time, this may result in two distinct species.

9.2 Review

SUMMARY

- Evolution is change in the genetic composition of populations over time. Allele frequencies change in response to natural selection, depending on the adaptive value of the related phenotypes, or by genetic drift, changes by chance in small populations.
- A species is a group of individuals that can produce viable, fertile offspring through interbreeding, and has a gene pool that is isolated from the gene pools of other species.
- Populations may also evolve into different species as a result of having become fragmented through various genetic isolating mechanisms.
- Isolating mechanisms can be prezygotic, preventing fertilisation of gametes of different species, or postzygotic, preventing production of viable offspring after fertilisation.
- Prezygotic isolating mechanisms include:
  - geographical isolation—physical barriers
  - ecological isolation—niche partitioning within an ecosystem
  - temporal isolation—differences in reproductive timing
  - morphological isolation—differences in reproductive structures
  - behavioural isolation (animals only)—differences in behaviours attracting mates, usually the result of sexual selection
  - gamete mortality—no fertilisation by gametes.
- Postzygotic isolating mechanisms include:
  - hybrid inviability—the hybrid zygote does not survive
  - reduced hybrid viability—the hybrid zygote survives but does not develop fully into a new individual
  - hybrid sterility—the hybrid organism develops but is incapable of reproducing
  - hybrid breakdown—the hybrid cannot reliably produce viable offspring.
- Speciation is the evolution of new species from an ancestral species. The new species are genetically different enough from the ancestral species that they can no longer produce viable offspring should they interbreed.
- Speciation involves either genetic isolation between diverging populations or single species gradually changing over time, with differences in mutations, selective pressures and genetic drift changing allele frequencies and traits in the sub-populations.
- Different types of speciation involve different genetic isolating mechanisms.
- The most common form of speciation is allopatric speciation, in which a population becomes divided by a geographical barrier. In this type of speciation, spatial isolation is the mechanism preventing gene flow, leading to genetic isolation.

KEY QUESTIONS

1 Using the information in the figure below, if you attempted to crossbreed a morepork owl with a southern boobook owl what could you assume about their offspring?

Australia's southern boobook (a) is closely related to the New Zealand morepork (b). The two species are separated by a geographical barrier.

2 Allopatric speciation is characterised by the prevention of gene flow between populations by what type of prezygotic isolating mechanisms? Give an example.

3 In hybrid inviability, why does the zygote fail to develop?

4 Why is genetic isolation an important step in speciation?

5 Other than genetic isolation, what two other factors are required for speciation to occur?

9.3 Selective breeding

Evolution through natural selection is an ongoing and, as the name implies, natural process. In addition, humans have been manipulating allele frequencies in the gene pools of populations for thousands of years through deliberate selection of particular individuals. The process by which humans decide which individuals may breed and leave offspring to the next generation is called **artificial selection** or **selective breeding**. In this section you will learn about some of the potential advantages and disadvantages of selective breeding.

FIGURE 9.3.1 Dairy cows, such as these Holstein Friesians, are selectively bred for their high milk yields.

INCREASING THE FREQUENCY OF DESIRED TRAITS

Artificial selection has led to improved agricultural crops and the domestication of animals for food or other uses (Figure 9.3.1). Darwin, for example, kept and bred pigeons, gathering information from stockbreeders. While developing his theory of natural selection, he observed the success of such artificial selection in producing new types of pigeons.

Through selective breeding, humans choose individual organisms with desirable traits and deliberately interbreed them to increase the allele frequency of those desired traits in the gene pool. This allows certain extreme forms to reproduce while preventing others from reproducing.

There are four basic steps that apply to all forms of selective breeding:

1 Determine the desired trait.
2 Interbreed parents who show the desired trait.
3 Select the offspring with the best form of the trait and interbreed these offspring.
4 Continue this process until the population reliably reproduces the desired trait.

All modern crops and livestock were developed by genetic manipulation of plant and animal species through this process of selective breeding. However, new molecular technologies are being used to alter the characteristics of organisms in a more targeted and specific way, and more quickly than by traditional breeding. These new methods, called genetic engineering, also can allow the exchange of genes between organisms that are sexually incompatible and normally cannot interbreed. New forms of plants and animals developed in this way are referred to as **genetically modified organisms (GMOs)**. You will learn more about GMOs in Chapter 14.

Selective breeding in plants

Most selective breeding of plants is done to produce higher-quality food. Typically, seeds are collected from the individuals with the largest or most numerous grains, fruits, nuts or other part of the plant that will be eaten. Those seeds are planted and the new generation of plants is cross-pollinated with other individuals with similar traits. The resulting plants produce larger, more nutritious or more aesthetically pleasing food products.

Maize, or corn (*Zea mays*), is one of the most widely grown crops in the world. It is thought that maize was selectively bred from a wild grass of the genus *Teosinte*. Modern maize has significantly larger cobs with many more rows of much larger kernels compared to the ancestral *Teosinte*. The higher-yielding modern maize provides more food for people than the ancestral form (Figure 9.3.2).

FIGURE 9.3.2 These varieties of maize, grown in Mexico, are among some of the first that were cultivated from the wild grass *Teosinte* thousands of years ago.

Many other food crops, such as tomatoes, potatoes and bread wheats, have also been modified by selective breeding to have higher yields, as well as greater resistance to common diseases.

Selective breeding in animals

Just as crops have been selectively bred for desired traits, so too have many animal species. In agriculture, sheep have been selected for the quality and quantity of the wool they grow, dairy cows have been selected for the milk they produce, and beef cattle for their muscle mass.

When a species has a variety of traits, different traits may be useful in different situations. A single wild species can be the original source of a great variety of different breeds. For example, it is widely accepted that all domestic dogs were selectively bred from a wolf species (Figure 9.3.3). Today there are hundreds of domesticated dog breeds, some of which would be unlikely to survive in the wild. Examples include soft-mouthed, strong swimming dogs such as Labradors, which were bred for duck hunting, and sheepdogs, bred for their intelligence. Humans have also artificially selected for a wide variety of traits in chickens, horses and many other domestic animals to produce gene pools that consistently produce the desired traits.

FIGURE 9.3.3 Dogs have been selectively bred for particular traits, such as size, coat colour, speed, protectiveness, strength or endurance.

EXTENSION

Selective breeding of edible Australian plants

Bush tucker

One staple food of desert Aboriginal peoples was the maloga bean (*Vigna lanceolata*). It has an edible root and small beans that can be eaten raw. Another related species is *Vigna radiata*, the wild mung bean. Aboriginal peoples did not make use of the bean's small black seeds, but in more recent times the wild mung been has been selectively bred with other plants to produce the cultivated mung bean.

The cultivated mung bean has a seed that can be green or black, and is more than double the size of the wild form (see Figure 9.3.4). The cultivated plant also grows upright, instead of being a wiry creeper.

FIGURE 9.3.4 Seeds of wild mung bean (top) and the larger green seeds of the cultivated variety (bottom).

Conservation and use of wild genes

Indigenous wild plants are of interest to modern plant breeders. Wild plants are a source of genes that may be used for crop improvement. Crosses between wild and cultivated mung beans, for example, may produce hybrid forms best adapted to Australian soils and climates.

To conserve the genetic variation in wild populations for future use, native species that are the wild relatives of agricultural crops are collected and stored in a seed bank.

The CSIRO Centre for Plant Biodiversity Research has established a significant collection of another type of native pea, *Glycine*, which is related to the cultivated soybean. An Australian species of *Glycine* contains resistance to the leaf rust fungus, a trait needed for protecting soybean.

Any new and useful allele discovered by screening collections of wild plants for a particular gene can be incorporated into existing agricultural varieties through standard plant breeding techniques.

Functional genomics

Functional genomics aims to identify genes that determine particular functions. For example, it could be used to determine the genes that allow plants to grow under drier conditions, in saline soils or with resistance to fungal diseases. Once a gene is identified, collections of wild plants can be screened for natural variants of the desired phenotype (for example, drought resistance). Any new and useful alleles discovered by screening can be incorporated into existing agricultural variants through plant breeding by cross-pollination. Also, once identified, unwanted genes can be switched off and desired genes from other species can be introduced through genetic engineering.

POTENTIAL DETRIMENTAL EFFECTS OF SELECTIVE BREEDING

Most selective breeding requires similar individuals to interbreed. Although this increases the allele frequency of the desired trait, it also decreases the frequency of all other alleles for this trait. This reduces the genetic variation within the gene pool and increases the incidence of homozygosity. Furthermore, genes do not exist in isolation, but are carried on chromosomes with other characteristics. Gene linkage means that selecting for one allele may result in the selection of a number of other traits. For example, 'fluffy' cocker spaniels get sore eyes due to extra lashes on the inside of the eyelid. The success of selective breeding of both plants and animals may be limited by the presence of undesirable linked alleles (Figure 9.3.5).

Genes that are located closely on a chromosome are said to be linked.

Selective breeding reduces resistance to environmental change

A population with low genetic variation is one in which all the individuals are very similar. As long as the alleles in the gene pool have a high adaptive value for the environmental conditions, the species will persist. However, should the environmental conditions and the resulting selective pressures change, it is unlikely that the same alleles will still have the same adaptive values. A single disease could potentially wipe out entire populations if none of the individuals is resistant.

FIGURE 9.3.5 Farmers must select wheat varieties with high grain production, as well as strong, moderate-length stems. Some varieties of wheat have long stems that cannot support the weight of the grains. This causes the plants to fall over, making it difficult to harvest the grain.

Selective breeding reduces biodiversity

Selectively bred species are also replacing wild varieties, reducing biodiversity as a whole. This, combined with low genetic variation within the selectively bred populations, puts global food security at great risk, especially with climate change. The loss of a crop species such as sorghum (Figure 9.3.6), maize or rice, could easily lead to mass starvation without wild varieties to fall back on. This possibility has led to the construction of food arks and seed banks around the world, where the seeds of both heirloom and modern varieties are stored. Svalbard Global Seed Vault is funded by the Norwegian government and stores seeds for more than 4000 essential food crops from around the world, including valuable seed stock from Australia.

FIGURE 9.3.6 *Sorghum* is a genus of grasses that includes many species and subspecies, some of which are native to Australia. Grain sorghum, *Sorghum bicolor* subsp. *bicolor*, which originated in Africa, is most commonly grown as a food source worldwide. It is important to conserve other wild species of *Sorghum* as potential sources of novel genotypes if needed for crop breeding in the future.

Selective breeding can increase genetic abnormalities

Selective breeding can also increase genetic abnormalities. Many purebred dogs, for example, have congenital problems including hip dysplasia, deafness and an increased risk of cancers, heart diseases and neurological diseases. Other less detrimental traits, such as an underbite (Figure 9.3.7), are rare in wild dog populations but relatively common in domesticated breeds.

Many of these problems are recessive conditions. This means that an individual needs two copies of the same allele for the condition to be present. Inbreeding and small gene pools greatly increase the frequency of particular alleles. This, in turn, increases the likelihood of homozygous recessive conditions.

FIGURE 9.3.7 An underbite is the result of a malformation of the lower jaw in this boxer. It is not often seen in wild wolves, dingoes or foxes. However, an underbite is common in many domestic dog breeds.

BIOFILE

Musk ox

The musk ox (*Ovibos moschatus*) (Figure 9.3.8) is a large Arctic mammal, prized for its thick wool-like fleece called qiviut. Qiviut is a prized luxury item in North America and musk ox meat is considered a lean alternative to beef. Unregulated hunting of the musk ox led to the near extinction of the species in the late 1800s. Conservation efforts have allowed the species to survive. Hunting restrictions were introduced and musk ox from the surviving populations were relocated to repopulate regions where the animals had died out.

FIGURE 9.3.8 The musk ox (*Ovibos moschatus*).

In the 1950s, the Musk Ox Farm Project was set up in Alaska in an attempt to domesticate the animals. Thirty-three individuals were captured from wild populations and selectively bred for domestication. Several other musk ox farms appeared after this time, greatly reducing the reliance on hunting. The domesticated individuals have been kept as livestock to create sustainable farms, in which qiviut is combed out of the living adults.

In many regions of the Artic Circle, domesticated musk ox were released into the wild where the native populations had been hunted to extinction. In the 1970s, one farm in Northern Quebec closed due to poor profits and 54 musk ox were released into the wild. Slowly, the native population of musk ox has increased to over 1000 adults, all descending from domesticated individuals.

EXTENSION

Domestication of the silver fox

Silver foxes are a natural colour variant of the red fox (*Vulpes vulpes*) and range from blue-grey to black in colour (Figure 9.3.9). They were prized for their unusual colour and hunted for their fur.

FIGURE 9.3.9 These foxes are both red foxes, *Vulpes vulpes*. The silver colour (a) is a natural colour variant of the red fox (b).

In 1959, Russian scientist Dmitri K. Belyaev began an experiment in which he selectively bred silver foxes for 'tameness' (Figure 9.3.10). He found that within eight to ten generations the foxes showed clear signs of domestication, wagging their tails when people approached. However, they no longer resembled their wild ancestors. Instead, the domestic foxes had floppy ears, short or curly tails and their fur had changed considerably in colour and texture. The genes related to 'tameness' are carried on chromosomes with many other characteristics.

The selection of this one trait affected the inheritance of other alleles. By artificially selecting the tamest foxes, the allele frequencies for other traits changed, resulting in domesticated foxes with different phenotypes to the original population.

FIGURE 9.3.10 Dmitri Belyaev with his partially domesticated foxes.

9.3 Review

SUMMARY

- Evolution can occur through natural selection or artificial selection.
- Selective breeding is the traditional form of artificial selection. In selective breeding, humans select desired traits and interbreed organisms with these traits.
- There are four basic steps that apply to all forms of selective breeding:
  - Determine the desired trait.
  - Interbreed parents who show the desired trait.
  - Select the offspring with the best form of the trait and interbreed these offspring.
  - Continue this process until the population reliably reproduces the desired trait.
- More recent molecular technologies have allowed for faster development of genetically modified organisms and for the transfer of DNA between species that cannot interbreed normally.
- Agricultural plants are typically bred for high yield and high resistance to common diseases. Animals are often bred for high quality traits and products (such as wool and milk), or for personality traits (such as loyalty in pets).
- Selectively bred populations tend to have low genetic variation and high homozygosity, meaning that:
  - they are more susceptible to environmental change
  - biodiversity may be reduced if selectively bred populations replace wild populations and varieties
  - an increase in the incidence of genetic abnormalities can occur.

KEY QUESTIONS

1 Identify the steps of artificial selection that would lead to the production of large corn cobs.

2 Explain how artificial selection has changed with the development of new technology.

3 Explain why genetic conditions, such as hip dysplasia in dogs, are more common in selectively bred populations than in 'wild' populations.

4 Selective breeding:
  A reduces biodiversity
  B reduces resistance to environmental change
  C increases genetic abnormalities
  D all of the above

Chapter review

09

KEY TERMS

adaptive value
allele
allele frequency
allopatric speciation
aneuploidy
artificial selection
block mutation
bottleneck effect
chromosomal abnormality
deletion mutation
diploid
duplication mutation
endemic
evolution
founder effect
frameshift mutation
gene flow
gene pool
genetic drift
genetic isolation
genetically modified organism
genotype
germline mutation
haploid
homologous chromosomes
hybrid
hybrid breakdown
hybrid inviability
hybrid sterility
insertion mutation
inversion mutation
karyotype
missense mutation
mutagen
mutation
natural selection
niche
non-disjunction
non-homologous chromosomes
nonsense mutation
phenotype
point mutation
polyploidy
postzygotic isolating mechanism
prezygotic isolating mechanism
reduced hybrid viability
replication
selection pressure
selective breeding
sexual selection
silent mutation
somatic mutation
speciation
species
substitution mutation
thalassaemia
trait
translocation mutation
trisomy
variation
viable offspring

KEY QUESTIONS

1 Which of the following is/are the source of new genetic material?

A gene flow
B natural selection
C genetic drift
D mutation

2 Which sort of mutation involves the movement of DNA between non-homologous chromosomes?

A duplication
B deletion
C inversion
D translocation

3 A sequence of DNA is GGA TTA CCG TCT. It undergoes a mutation that changes the sequence to GGA CTA CCG TCT. What type of mutation is this?

A a frameshift mutation
B a block mutation
C a point mutation
D a deletion mutation

4 On which of the following does natural selection act directly?

A the genotype
B the entire gene pool
C the phenotype
D each allele

5 Natural selection acting over time on a population of wild mung beans is likely to result in which of the following?

A genetic drift
B mutations
C extinction
D a change in the relative frequencies of alleles

6 Choose the best answer. What mechanism(s) is/are likely to maintain reproductive isolation between two different but related species of frogs living in the same marsh?

A The frogs breed at different times of the year.
B The mating calls of the frogs are different and they only respond to their own call.
C Hybrids between the two frogs are sterile.
D All of the above are possible mechanisms.

7 The coyote (*Canis latrans*) and the grey wolf (*Canis lupus*) are closely related species and both have a diploid number of 78. The ranges of the two animals overlap. Hybrids have been identified in the wild. The hybrids seem to express more coyote genes than wolf genes, but are bigger and more able to withstand harsh conditions than pure-breed coyotes. Breeding experiments in which female coyotes were crossed with male wolves resulted in viable offspring but showed reduced fertility, with two out of three pregnancies failing. Hybrids, however, are able to reproduce with a lower success rate than pure breeds of either species.

Genetic studies have shown that the eastern coyote population is almost certainly the result of such matings in the past (more than 200 years ago). Which of the following is the best description of this situation?

A prezygotic isolation
B hybrid inviability
C hybrid sterility
D reduced hybrid viability

8 If a hybrid offspring was formed between a red kangaroo (*Macropus rufus*, $n = 10$) and a grey kangaroo (*Macropus giganteus*, $2n = 16$), what would be the diploid number of the offspring?

A 18
B 36
C 13
D 28

9 Consider the following four populations:

1 a large population experiencing large environmental changes
2 a small population experiencing large environmental changes
3 a large population in a stable environment
4 a small population in a stable environment

The rate of evolution from fastest to slowest in these populations could be expected to be:

A 3, 4, 1, 2
B 4, 2, 3, 1
C 1, 2, 3, 4
D 2, 1, 4, 3

10 Sickle-cell anaemia is a hereditary disease in which the haemoglobin of red blood cells is abnormally formed. The condition results in rapid destruction of red blood cells and severe anaemia. Individuals suffering from sickle-cell disease usually do not survive childhood. The disease is controlled by a single gene with the alleles Hb^S (sickle-cell disease) and Hb^A (normal haemoglobin). The varying degrees of the disease are determined by the combinations of alleles in individuals.

Normal haemoglobin	Sickle-cell trait	Sickle-cell disease
Hb^AHb^A	Hb^AHb^S	Hb^SHb^S

Heterozygotes show greater resistance to the mosquito-borne parasite *Plasmodium falciparum*, which causes malaria, than do normal individuals.

a Sickle-cell disease and sickle-cell trait are relatively common in Africa, India and the Mediterranean, where malaria is prevalent. Given that the homozygous normal genotype develops into a generally healthier individual, explain the steps that would have occurred in human populations originally colonising one of these regions that resulted in a change in the frequency of the Hb^S allele.

b Explain how the variation in these alleles is maintained in human populations in the Mediterranean.

c Describe how variation in the phenotypes is likely to be affected in Australia. Explain.

11 A study involving 23 832 individuals in Lagos, Nigeria, examined the incidence of the blood groups A, B, AB and O in the population. The data on the incidence of the various alleles is shown in the table below:

Blood group	Genotype	Number of individuals
O	I^OI^O	12 700
A	I^AI^A	467
	I^AI^O	4871
B	I^BI^B	403
	I^BI^O	4523
AB	I^AI^B	868

Use the following formula, to calculate the frequency of the I^O allele in the population.

$$\text{allele frequency} = \frac{2(\text{number of homozygotes}) + (\text{number of heterozygotes})}{2(\text{total number of individuals})} \times 100$$

12 There are a number of different antigens found on the surface of red blood cells in humans. The expression of these antigens is genetically controlled. One such antigen is called Duffy. Individuals either have the protein (Duffy positive) or lack it (Duffy negative).

One of the major causes of death in equatorial countries is malaria. Malaria is caused by the invasion of the body's red blood cells by a plasmodium. There are two major species of plasmodium: *Plasmodium falciparum* and *P. vivax*.

Map A, below, shows the distribution of the Duffy negative phenotype. Map B shows the distribution of *P. falciparum* and *P. vivax*.

a It has been suggested that *P. vivax* uses the Duffy protein to assist its entry into the red blood cells of its host. How do the distributions of malaria caused by *P. vivax* and the Duffy antigen support this hypothesis?

b The allele causing the Duffy negative phenotype is likely to have arisen as a result of either a nonsense mutation or a frameshift mutation. Explain how each of these would have caused the Duffy negative phenotype.

c A few individuals with the Duffy negative trait currently exist in Indonesia. Assuming no vaccine is developed against malaria, predict what is likely to happen to the frequency of this allele over the next few hundred years. Explain your reasoning.

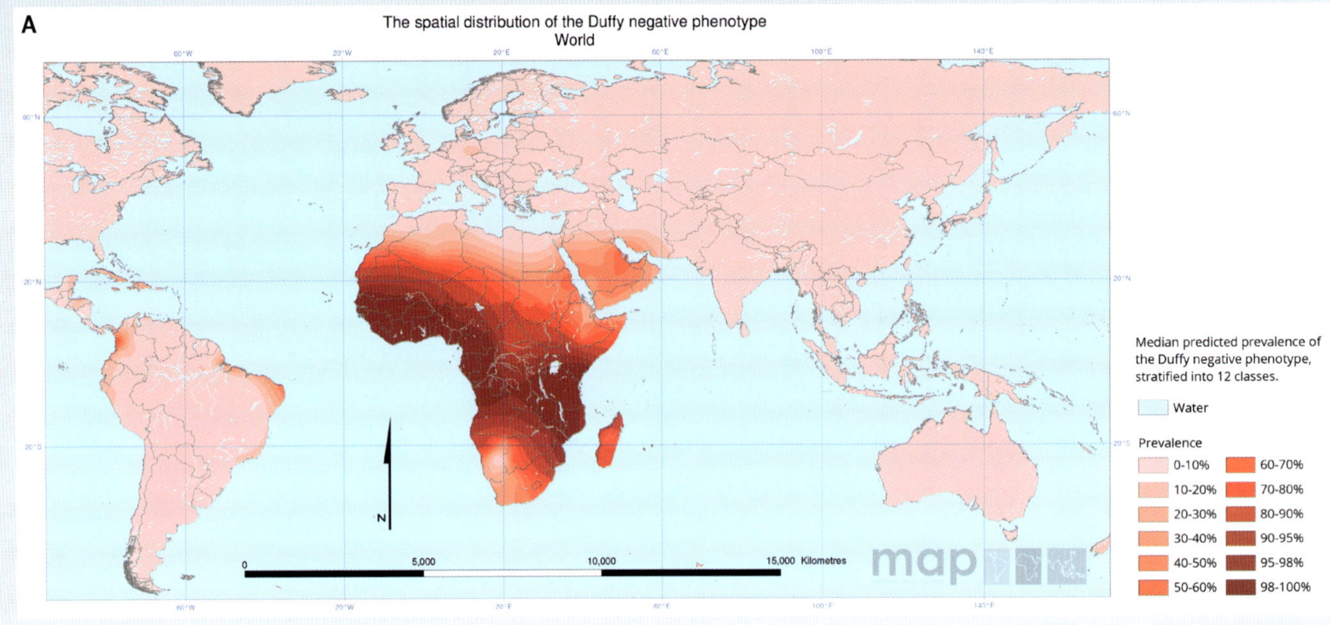

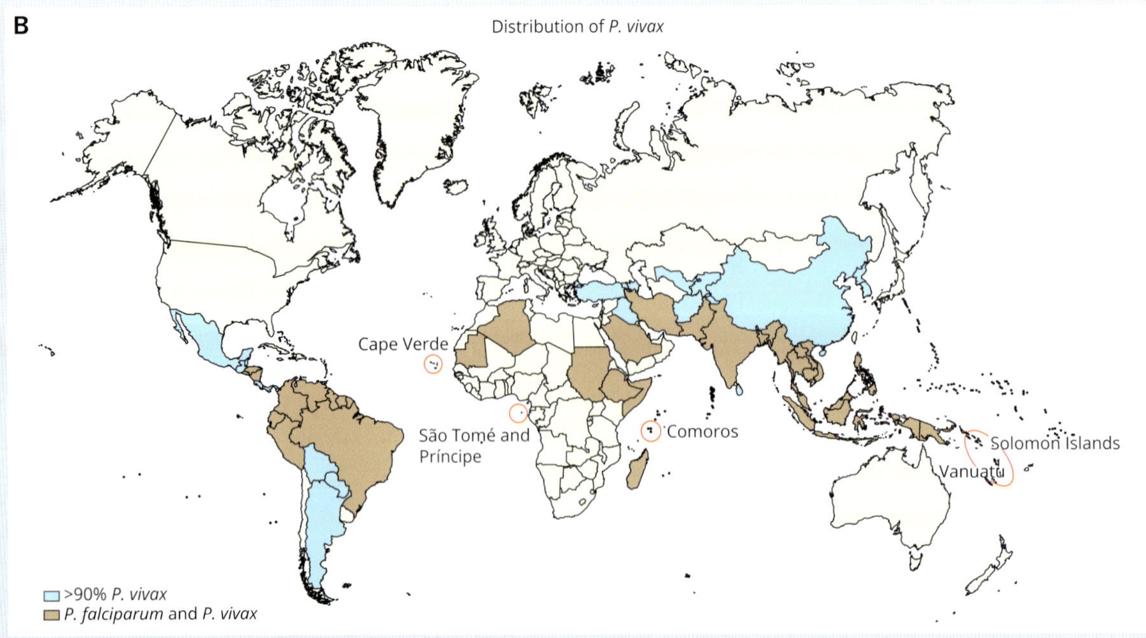

13 a Recent studies of human and chimpanzee genomes have shown that populations of chimpanzees living near each other show greater genetic variation than human populations spread on different continents. Explain how this supports the hypothesis that the human population experienced a genetic bottleneck about 75 000 years ago when Mt Toba, a supervolcano in Sumatra, erupted.

b The genetic bottleneck is one hypothesis to explain the lack of human genetic diversity. Another is that human diversity is low because we are the result of a series of founder populations. Explain why the descendants of founder populations have low diversity.

14 A child was born with some abnormalities. A chromosomal analysis was performed as a part of the testing to identify the cause of these abnormalities. There were some changes identified to two chromosome pairs. The normal chromosomes are shown on the left in the diagram below. Each coloured section is one gene and each letter represents an exon of that gene. (Only the areas of the chromosomes involved are shown.)

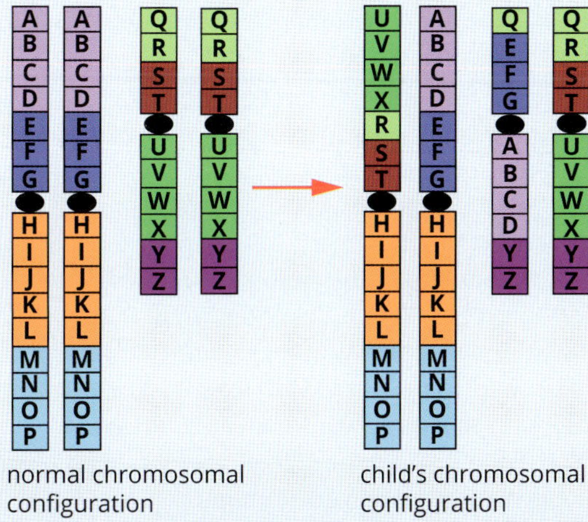

The analysis showed that two translocations had occurred between the chromosomes of these pairs. The first swapped ABCD with UVWX and the second swapped EFG with RST as shown on the right in the diagram above.

a Explain how each of the translocations has contributed to the abnormalities in the phenotype of the child.

b Despite the abnormalities the child grew up and went on to try to start a family of their own. The couple wish to know if the abnormal chromosomal arrangement means that they will be unable to have children. You are the geneticist advising them.

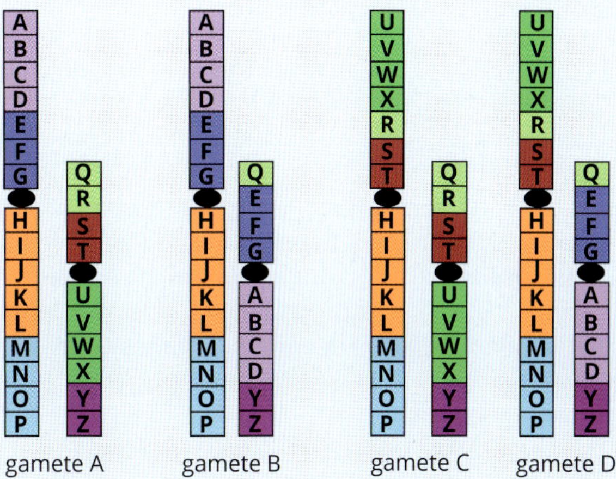

The diagram above shows the possible gametes of the partner with the translocation. The other partner has a normal chromosomal arrangement. Use this information to advise the couple.

15 Rabbit calicivirus is a disease that was introduced into the Australian mainland in 1995 and Tasmania in 1997. The purpose of the introduction was to reduce the number of wild rabbits as they had reached plague levels and were causing land degradation. Initially, millions of rabbits died but by 2005 numbers were again rising as the rabbits developed resistance to the virus.

a Explain how resistance to calicivirus has developed. Ensure you refer to allele frequencies in your answer.

b i Is the action of calicivirus on the Australian rabbit population an example of artificial or natural selection? Explain your reasoning.

ii Could this be considered an example of selective breeding? Why or why not?

c The Department of Primary Industries has a plan to introduce a new and more virulent strain of calicivirus into Tasmania in 2016. Explain if this is likely to be more successful in eradicating the rabbit from Tasmania.

16 *Staphylococcus aureus*, also known as 'golden staph', is a species of bacterium responsible for serious illness in humans. Symptoms include skin rash, aches and pains, fever, boils and ulcers. Complications of *S. aureus* infections can result in limb amputations and even death. *S. aureus* is very contagious, and is easily passed from person to person through skin contact. Hospitals regularly report outbreaks of this infection, which is usually treated effectively with a course of the antibiotic methicillin. In 1996, a child in Japan diagnosed with a *S. aureus* infection failed to recover. He was diagnosed as suffering from MRSA, methicillin-resistant *S. aureus*. He was subsequently treated with the more potent antibiotic vancomycin, but again with no success. Finally the child was prescribed an experimental antibiotic, arbekacin, and eventually recovered.

a Suggest why *S. aureus* is relatively common in hospitals.

b Describe the steps that have resulted in the population of vancomycin-resistant bacteria.

c Outline one advantage and one disadvantage of developing more and more potent antibiotics.

17 Below are shown the karyotypes of a normal individual and one with a chromosomal mutation.

A normal karyotype

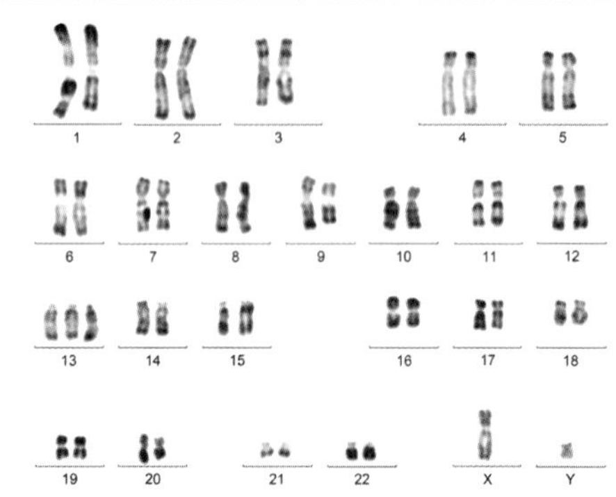

B karyotype with chromosomal abnormalities

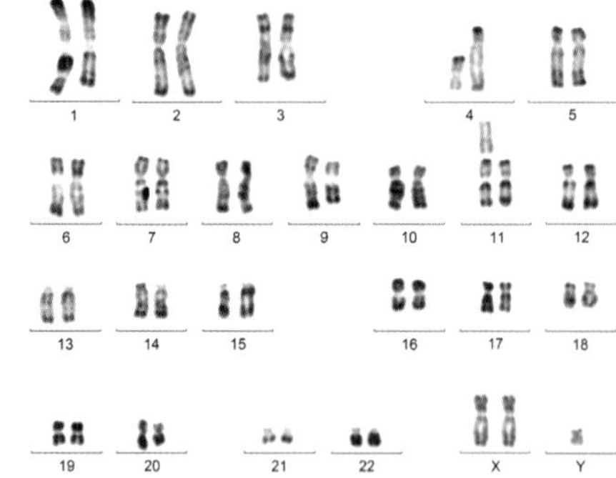

a Identify the sex of the individuals in the karyotypes.

b Identify the abnormalities in karyotype B.

18 Speciation can take tens to hundreds of thousands of years or it can happen very rapidly. An example of rapid speciation has occurred over the last 150 years in the United States. In the early 1900s three species of related wild flowers from the genus *Tragopogon* (*T. dubius*, *T. pratensis* and *T. porrifolius*) were introduced into America from Europe. In all three species $n = 12$. Initially these were separate populations, but eventually their ranges began to overlap and interactions between the individuals of different species occurred and hybrids were formed. These hybrids were sterile.

a Explain why the hybrids were sterile.

b In the 1950s scientists noticed that two varieties of the 'hybrids' (now called *T. miscellus* and *T. mirus*) were reproducing sexually. Each of the variants could reproduce sexually with individuals of the same variant but not with the sterile hybrids or any of the original three species. Examination of the chromosomes of these plants showed that they had double the number of chromosomes of the sterile hybrids.

- **i** How might these hybrids have acquired the ability to reproduce sexually?
- **ii** Why are they now classified as new species?
- **iii** How many chromosomes would these new species possess?

19 A typical example of a breeding program in agriculture is for increased egg production in chickens. At the start of a particular breeding program, the average number of eggs per hen per year in a flock was 125. Hens that produced the most eggs per year were chosen as the female parents of the next generation. Roosters used in the program were the offspring of high-yielding hens. The average number of eggs per hen per year increased from 125 to 230 over 15 years, but the rate of increase was slower in subsequent years.

a Explain how the breeding program is an example of artificial selection by humans.

b What possible reasons might account for the slow down in the increase in egg production?

c What are the possible negative effects on chickens of increasing egg production?

20 One iconic Australian animal is the corroboree frog. There are two species: the southern corroboree frog (*Pseudophryne corroboree*) and the northern corroboree frog (*Pseudophryne pengilleyi*). They are quite small, measuring between 2.5 and 3 cm in length. The two species are closely related but still have distinct differences in colour, mating calls and skin chemistry. Both breed in damp marshy areas. Both species are seriously endangered.

(A) The southern corroboree frog (*Pseudophryne corroboree*).
(B) The northern corroboree frog (*Pseudophryne pengilleyi*).

As seen in the map below, the ranges of the two species do not overlap.

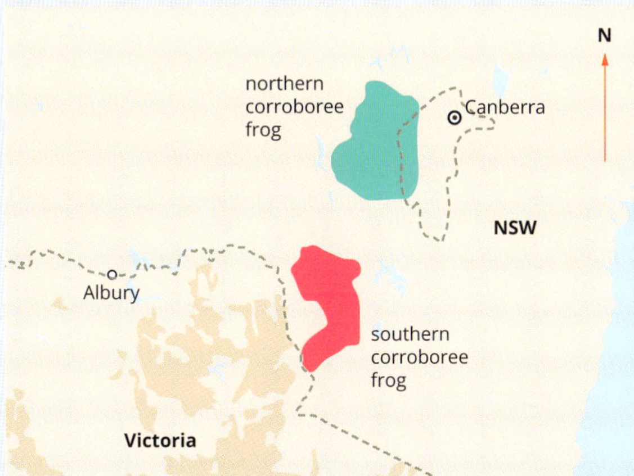

Distribution of the northern and southern corroboree frogs.

a The two species have a recent common ancestor. Describe the processes that could have resulted in the formation of the two species of frog.

b The northern corroboree frog is divided into two genetically distinct populations, both of which contain few individuals. How might this affect the viability of these populations over the next decade?

c There is a captive breeding program at Melbourne Zoo for the northern corroboree frog. It has been suggested that individuals from the two separate populations of the northern corroboree frog should be interbred. How would this help to increase the chances of the survival of this species?

21 Even though not all dogs can interbreed with each other, currently all domestic dogs are considered to be a single species, *Canis lupus familiaris*. This is because alleles can be spread between different breeds by dogs of mixed breed.

If a situation were to occur in which all breeds of dog died out (perhaps due to a disease) except for Jack Russell terriers and Irish wolf hounds, should the two breeds still be considered the same species? Justify your position.

Irish wolfhound (left) and Jack Russell terrier (right).

22 Cancer is not usually transmissible between individuals but at least one known exception to this exists. Devil facial tumour disease is passed between individual Tasmanian devils when they bite each other. Tasmanian devils are aggressive and frequently bite each other when competing for food or even while mating. The facial tumour eventually makes eating impossible and the devil dies from starvation. Ecologists fear that the devil facial tumour will result in the extinction of the Tasmanian devil, so considerable research has been done to try to solve the problem. Two promising lines of research are being investigated.

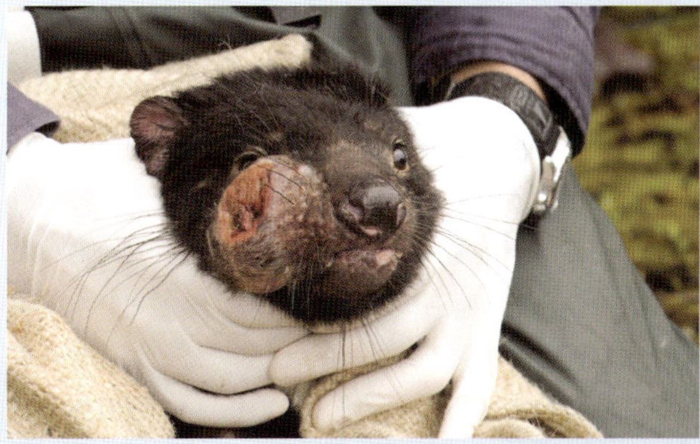

It has been shown that less aggressive devils are much less likely to be infected with the devil facial tumour and that members of the population of Tasmanian devils in north-west Tasmania have what appears to be some level of genetic resistance to the disease. It has been suggested that selective breeding of devils for resistance, gentility or both traits could help to stop the Tasmanian devil becoming extinct.

a Explain how selective breeding in a captive population of Tasmanian devils could be used to help increase the survival chances of the species.

b Are there any possible negative consequences for the gene pool of the Tasmanian devil from such a program?

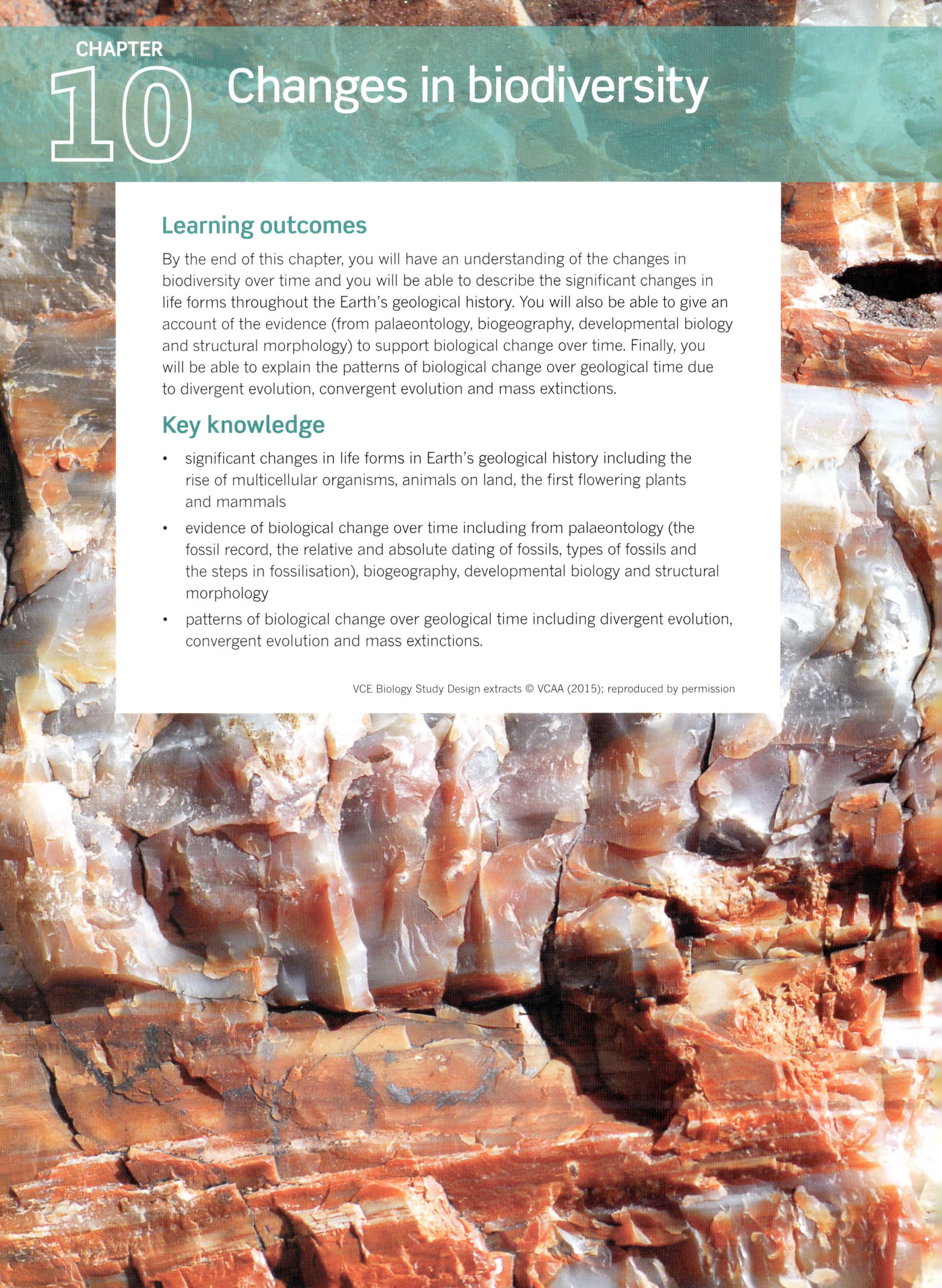

CHAPTER 10

Changes in biodiversity

Learning outcomes

By the end of this chapter, you will have an understanding of the changes in biodiversity over time and you will be able to describe the significant changes in life forms throughout the Earth's geological history. You will also be able to give an account of the evidence (from palaeontology, biogeography, developmental biology and structural morphology) to support biological change over time. Finally, you will be able to explain the patterns of biological change over geological time due to divergent evolution, convergent evolution and mass extinctions.

Key knowledge

- significant changes in life forms in Earth's geological history including the rise of multicellular organisms, animals on land, the first flowering plants and mammals
- evidence of biological change over time including from palaeontology (the fossil record, the relative and absolute dating of fossils, types of fossils and the steps in fossilisation), biogeography, developmental biology and structural morphology
- patterns of biological change over geological time including divergent evolution, convergent evolution and mass extinctions.

10.1 Significant changes in biodiversity over time

The conditions on Earth have always determined the variety of living organisms that can exist. Earth's atmosphere has changed remarkably over time. The early Earth, which formed 4.6 billion years ago, was volcanically very active and the atmosphere was very different from that of today (Figure 10.1.1). Conditions allowed the formation of biomolecules, and cellular life evolved, with the earliest prokaryotes experiencing an atmosphere lacking in oxygen. Climates changed over time, and there have been multiple ice ages and hot, dry periods. The movement of continents has greatly affected the distribution of seas and area of land, with major consequences for organisms.

In this section, you will learn about the significant changes in life forms throughout the Earth's geological history, including the evolution of multicellular organisms, land animals, the first flowering plants and mammals.

FIGURE 10.1.1 When Earth formed 4.6 billion years ago from a hot mix of gases and solids, it had almost no atmosphere.

FIGURE 10.1.2 (a) Geological layers can be seen at Fossil Cove, Tasmania, Australia. (b) Ammonite fossils embedded in rock.

GEOLOGICAL TIME SCALE OF EARTH

The history of Earth and evolving life can be chronologically followed using the **geological time scale**, which covers events that have occurred on Earth from its formation to the present time.

The geological time scale is constructed using the order of rocks laid down in a sedimentary rock sequence (a relative time scale in which the oldest rocks are at the bottom), and the fossilised remains of ancient animals and plants within the rock strata (Figure 10.1.2). Today geologists also use techniques such as radiometric dating to determine directly the age of rocks.

The geological time scale is divided into many subdivisions. The largest of these subdivisions is the **eon**. Eons are subdivided into **eras**, which are further subdivided into **periods**, and into still smaller subdivisions called **epochs** (Table 10.1.1).

Relative duration	Era	Period	Epoch	Age (mya)	Plant life	Animal life
Phanerozoic eon	Cenozoic	Quaternary	Holocene	0.01	modern plants	• evolution of humans
			Pleistocene	2.58		
		Neogene	Pliocene	5.33	angiosperms dominate forests and grasslands	• mammals diversify, including primates • whales appear in oceans
			Miocene	23.03		
		Palaeogene	Oligocene	33.9	angiosperms continue to dominate	• many primate groups appear
			Eocene	56.0	angiosperms continue to dominate	• mammals continue to diversify
			Palaeocene	66.0	angiosperms continue to dominate	• mammals, birds and pollinating insects diversify
	Mesozoic	Cretaceous		145.0	angiosperms become dominant	• dinosaurs become extinct • mammals diversify or further develop • birds appear • first primates
		Jurassic		201.3	conifers abundant, first angiosperms	• age of reptiles, some flying reptiles
		Triassic		252.2	conifer trees dominate forests	• first mammals • first dinosaurs • reptiles dominate land • amphibians decline
	Palaeozoic	Permian		298.9	early seed plants develop, including cycads and early conifers	• reptiles diversify • familiar insects develop • many land vertebrates • many invertebrate sea life become extinct
		Carboniferous		358.9	first large swamp forests of vascular land plants	• insects become more common • first reptiles
		Devonian		419.2	tree-like vascular land plants, including lycopods; ferns appear	• fishes and coral reefs common
		Silurian		443.8	first small vascular land plants, many algae	• many coral reefs, shells • first animals on the land—amphibians and invertebrates
		Ordovician		485.4	types of large algae found as fossils	• many invertebrates • first vertebrates—fishes—found
		Cambrian		541.0	more types of algae appear	• animals with bodies protected by shells • first fishes appear
Proterozoic eon		Ediacaran		635	some algae	• soft-bodied animals • a few fossils found of animals with jelly-like bodies
				2500	• first animal traces • multicellular life develops in shallow warm seas • fossils rarely found, due to the age of the rocks and the soft fragile bodies of these organisms	
Archaean eon				4000	• bacteria (prokaryotes) abundant • fossilised and living stromatolites are still found on Earth today • oldest known sedimentary rocks, and oldest 'fossil' remains—chemical traces of living things	
Hadean time				4600	• solidification of the Earth from a ball of molten rock	

Note: The Proterozoic eon, Archaean eon and Hadean time are collectively known as Precambrian time. Hadean time is not a geological era/eon/period.

TABLE 10.1.1 Geological time scale in millions of years ago (mya).

THE PRECAMBRIAN TIME

The period of Earth's formation is referred to as the **Precambrian time**. Precambrian is not a true geological eon, era, period or epoch, and geologists often refer to it as Precambrian time. The Precambrian is divided into three parts: the **Hadean**, the **Archaean** eon and the **Proterozoic** eon. It is believed that life on Earth first appeared in the Archaean eon.

The Hadean (~4600–4000 mya)

The Earth was formed during the Hadean (4.6–4 billion years ago). Earth was once a molten ball under the bombardment of meteors from space, but as the planet started to cool, molten material solidified into rock, creating a rocky terrain (Figure 10.1.3). At the same time, clouds formed, producing enormous volumes of rainwater that further cooled the crust and formed oceans.

FIGURE 10.1.3 Computer artwork showing how the surface of the Earth may have appeared beneath its clouds during the Hadean, when massive volcanoes and lava fields still dominated the landscape.

The Hadean is technically not a geological period as there is no rock on Earth that is this old. The estimation of when the Earth was formed is based on the age of lunar rocks obtained from the Apollo 16 mission in 1972, which were dated to be approximately 4.5 billion years old. The age of the Moon is considered to be a good reference point for the age of the Earth because it is thought that the Moon was formed from debris left from a collision between the early Earth and a Mars-size planetoid.

Earth's geological history began when solid rock formed on the Earth. The oldest rocks on Earth are approximately 3.8 billion years old. It is believed that the formation of solid rock on Earth most likely happened earlier than 3.8 billion years ago, but evidence for this is lacking due to erosion and plate tectonics, which have probably destroyed all of the solid rocks older than 3.8 billion years. The advent of a rock record roughly marks the beginning of the Archaean eon.

BIOLOGY IN ACTION

Changing times!

Evidence of life on Earth is preserved in the rock record. However, the microfossil record only extends to approximately 3.5 billion years ago. As well as rock records, zircons are used to detect evidence of life. Zircons are heavy, durable minerals that can capture and preserve traces of their immediate environment, acting like tiny time capsules. Carbon taken from inside zircons has been tested and found to have an estimated range in age of up to nearly 4.4 billion years (Figure 10.1.4).

Recently, scientists believe that they have found ancient microorganisms trapped inside zircons (detrital zircons) obtained from Jack Hills, Western Australia. In a study, scientists found that carbon within a zircon, is approximately 4.1 billion years old. If the study results are verified, they would be evidence that life on Earth emerged approximately 300 million years earlier than currently thought.

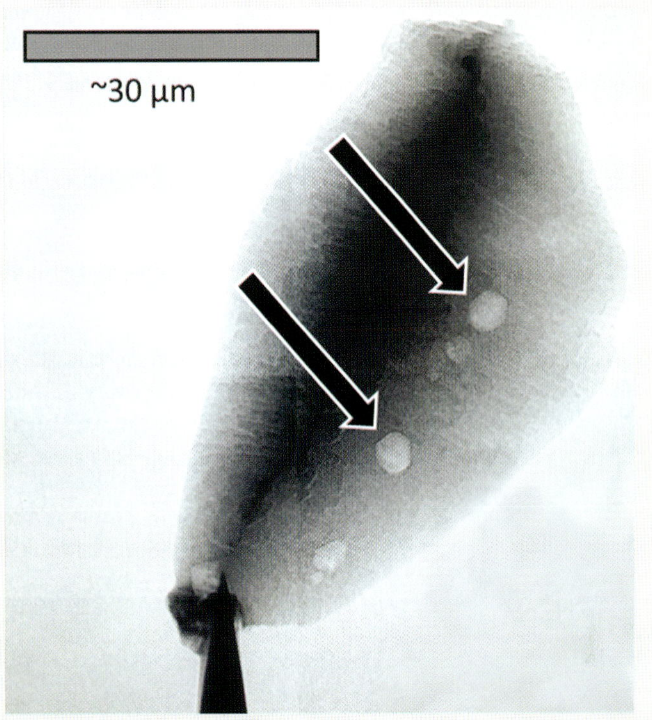

FIGURE 10.1.4 Transmission X-ray image of the zircon showed the presence of pure carbon (graphite) in two locations, shown by the black arrows. The carbon was tested to have an estimated range in age of up to nearly 4.4 billion years.

Archaean eon (4000–2500 mya)

The Archaean eon spanned about 1.5 billion years. The atmosphere at the time was likely composed of mainly methane, ammonia and other gases. There was no oxygen gas on Earth.

It was early in the Archaean eon that life first appeared on Earth and the initial life forms were mainly bacteria (prokaryotes). The oldest **fossils** are microfossils of bacteria that date to roughly 3.5 billion years ago when **stromatolites**, layered rocks that form when certain prokaryotes bind thin films of sediment together, became common in the rock record (Figure 10.1.5a). Stromatolites have been found in early Archaean rocks of South Africa and Western Australia. The early prokaryotes that led to stromatolite formation were Earth's sole inhabitants for more than 1.5 billion years (Figure 10.1.5b). Stromatolites increased in abundance throughout the Archaean, but began to decline during the Proterozoic eon. Stromatolites can still be found in Western Australia today (Figure 10.1.5c).

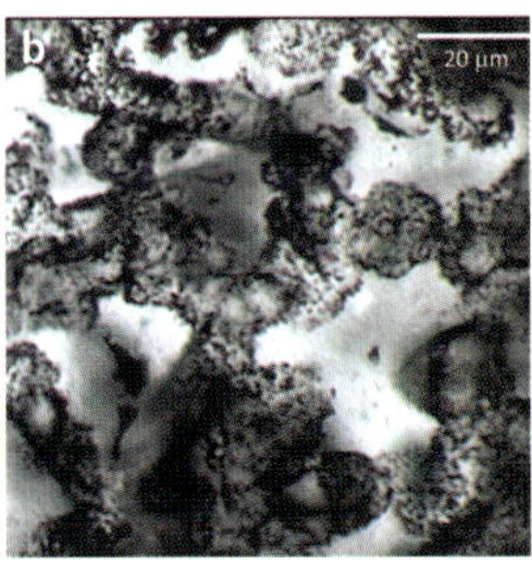

FIGURE 10.1.5 (a) Section through a rock with layers of a fossilised stromatolite. (b) Microfossil of bacteria that lived 3.4 billion years ago found in sandstone at the base of Strelley Pool, Pilbara region, Western Australia. (c) Stromatolites can still be found today at Shark Bay, Western Australia.

Proterozoic eon (2500–541.0 mya)

During the Proterozoic eon, large continental landmasses were formed by the convergence of smaller landmasses. From this eon, there are abundant fossils of bacteria. Some of these bacteria, similar to modern cyanobacteria, evolved the process of photosynthesis, which releases oxygen, and by about 2 billion years ago the atmosphere was rich in oxygen. The build-up of oxygen in the atmosphere caused many early anaerobic life forms to become extinct because oxygen was toxic to them. However, oxygen in the atmosphere allowed for the evolution of aerobic and more complex life forms. The earliest eukaryotes, resembling single-celled green algae, are known from fossils that are 1.4 billion years old. Multicellular algae (red algae and green algae) and the first animals evolved towards the end of the Proterozoic eon.

The end of the Proterozoic eon: the Ediacaran period (~635–541 mya)

The Ediacaran period, at the end of the Proterozoic eon, reveals the earliest evidence of a diversity of multicellular animals. These animal fossils are collectively called the **Ediacaran fauna** after the Ediacara Hills in the Flinders Ranges of South Australia where the fauna was first found. The fauna is found worldwide, and has been recorded from all continents. The fossils are of small, soft-bodied sea creatures that resemble modern sea jellies and segmented worms (Figure 10.1.6). They include representatives of all the major phyla of invertebrate animals.

FIGURE 10.1.6 Traces of fossilised sea jellies can be seen in rippled sandstone in the Flinders Ranges, South Australia.

Theories on how life came about

At some point in time, organic molecules accumulated to form the first self-replicating life form. We know that life depends on elements including carbon, hydrogen, nitrogen, oxygen, phosphorus and sulfur, and that water is vital. Several theories about how life first began are currently being investigated, including the possibility that organic molecules arrived on Earth by meteorite, which may have given a jumpstart to life (Figure 10.1.7). This theory, however, does not answer the question about how those organic molecules originally came to be. The leading theories are described below.

FIGURE 10.1.7 These meteorite fragments landed in Murchison, Victoria, in 1969. The meteorite contained more than 80 amino acids, which are the building blocks of proteins.

The primordial soup theory

The primordial soup theory suggests that amino acids were the result of gases from Earth's atmosphere and molecules in the vast oceans being energised and changed by lightning strikes and ultraviolet light. This theory was proposed independently by both the Russian scientist Aleksandr Oparin and English geneticist John Haldane, and is occasionally referred to as the Oparin–Haldane theory.

Many scientists have attempted to conduct experiments to replicate the conditions of early Earth and produce amino acids, most famously the American scientists Stanley Miller and Harold Urey in 1953. The Miller–Urey experiment produced organic compounds, including amino acids (Figure 10.1.8).

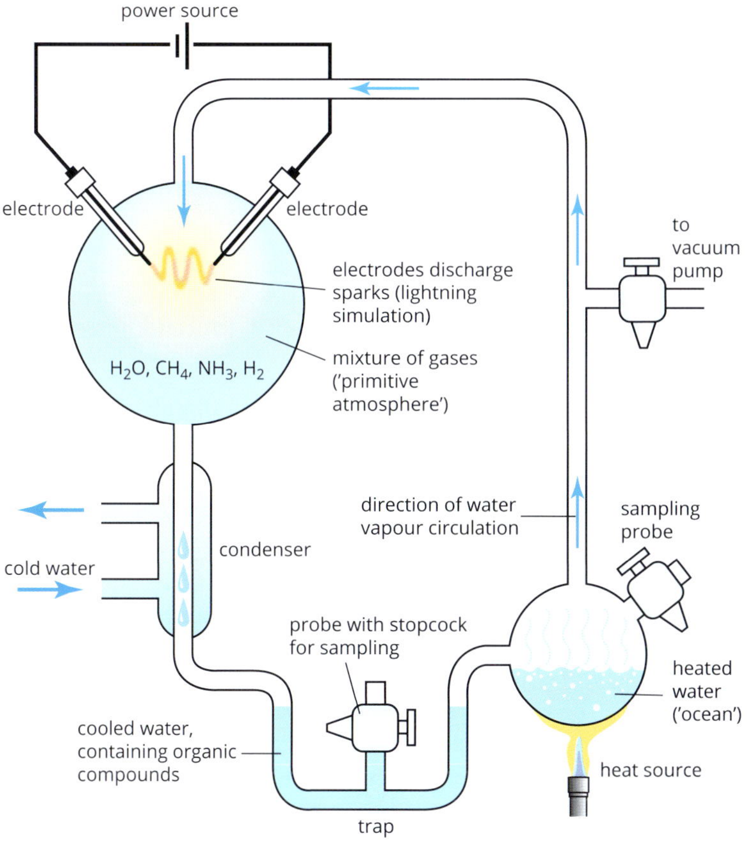

FIGURE 10.1.8 The Miller–Urey experiment.

As our understanding of the ancient atmosphere and weather has developed, it has become apparent that the significance of the Miller–Urey experiment lies in proving that organic molecules can be made from inorganic molecules.

The theory that life evolved near hydrothermal vents

Hydrothermal vents occur in the depths of the ocean where sunlight cannot penetrate (Figure 10.1.9). Although sunlight is the ultimate source of energy for life on the surface of the Earth, this energy cannot penetrate to where the extremophiles survive. Instead these single-celled organisms must source their energy from organic molecules around hydrothermal vents.

Archaea are a domain of prokaryotes that are able to live in extreme environments (hot, acidic, salty). Modern species of Archaea live and thrive near hydrothermal vents and, as they represent an ancient lineage, many scientists view hydrothermal vents as a window into the past.

The water around hydrothermal vents can be extremely hot, and the depths at which these vents are located are under extremely high pressures. These conditions are just right to combine the chemicals present into organic molecules. Similar conditions can occur elsewhere, such as inside volcanic craters.

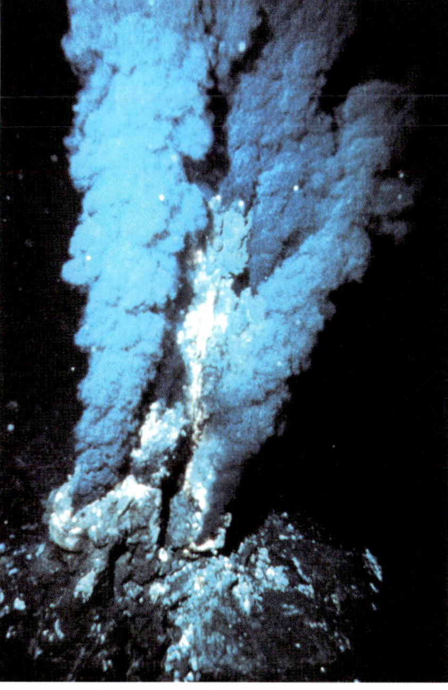

FIGURE 10.1.9 Hydrothermal vents found on the Atlantic floor. Many hydrothermal vents support ecosystems based on bacteria that can utilise the dissolved minerals that come out from the vents. These ecosystems are some of the few types on Earth not directly dependent on the Sun's energy.

The RNA world theory

All living cells on Earth contain the following three biomolecules, each serving a different critical function:

- proteins—made of amino acids; they have structural and catalytic functions
- DNA—double-stranded nucleic acid that carries information from one generation to the next
- RNA—single-stranded nucleic acid that carries information and has metabolic functions.

In the early 1980s, scientists discovered that some RNAs can act as catalysts for chemical reactions. These catalytic RNAs are known as ribozymes, and earned their discoverers Sidney Altman and Thomas Cech the 1989 Nobel Prize in Chemistry.

The discovery of ribozymes supported a theory that RNAs were the first molecules able to store information, replicate and catalyse reactions. Due to the instability of RNA, DNA may have eventually evolved to become the dominant genetic material.

Life began in the seas: evolution of cellular organisms

In the steps leading to the evolution of the cell, replicating organic molecules became enclosed in vesicles, surrounded by a biological membrane. The membrane provided an internal environment that was different from the external environment and in which metabolic processes could develop.

Over time cells also became more structurally complex, with the evolution of membrane-bound, specialised organelles that are characteristic of eukaryotes. The endosymbiotic theory (see Chapter 5, page 186) states that mitochondria and chloroplasts of eukaryotes evolved from bacterial symbionts that were engulfed by a host bacterial cell. The nuclear membrane probably developed from infolding of the bacterial cell membrane, enclosing the bacterial DNA. Through the process of natural selection, metabolic processes became more complex and eventually multicellular organisms evolved.

BIOFILE

The Burgess Shale

The Burgess Shale in the Canadian Rocky Mountains is a fossil site from the Cambrian of significance because it provides a rare glimpse of soft-bodied animals. The Burgess Shale was deposited in the ocean near an underwater algal reef shelf, where occasional undersea landslides buried animals living there. The fine mud prevented decay, preserving the soft parts of the animals. As a result, the Burgess Shale provides a large record of extinct soft-bodied organisms. The deposit also preserved parts of many shelled animals that are not normally seen in other fossil records, including the legs and antennae of trilobites and the setae (hair-like structures) of brachiopods.

THE PALAEOZOIC ERA

The Palaeozoic (meaning 'ancient life') was a time of great change for Earth. By the early Palaeozoic Earth's landmasses had formed two supercontinents (Gondwana and Laurasia), which later came together to form one massive continent (**Pangaea**). These movements of land masses greatly affected the climate, the land and sea environments and hence the evolution of organisms.

Cambrian period (541.0–485.4 mya)

Fossil evidence from 542 million years ago shows a dramatic increase in the number and complexity of life forms in the oceans. This is known as the 'Cambrian explosion'. Fossils include worms, jellyfish, brachiopods and arthropods, the most common being trilobites (Figure 10.1.10). The number and diversity of fossils is significantly more than in the Ediacaran period due to the emergence of organisms with hard exoskeletons, which are more readily preserved.

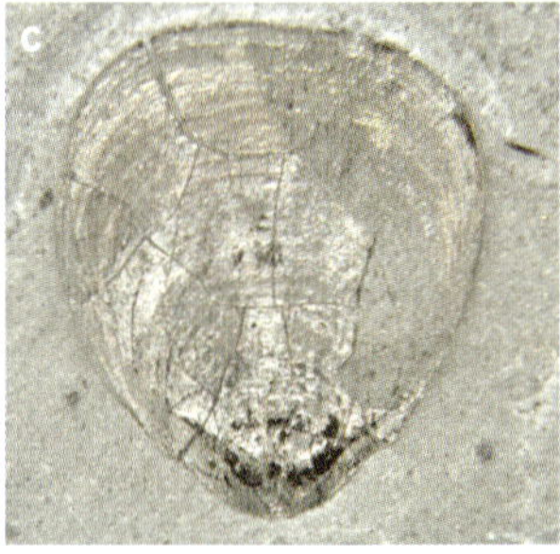

FIGURE 10.1.10 Fossils from the Cambrian period. (a) Onychophoran worm *Hallucigenia sparsa*, from Burgess Shale Formation, Canada. (b) Trilobite fossil *Xystridura saintsmithi* from Mount Isa, Queensland, Australia. (c) Brachipod *Lingulella waptaensis*, from the Burgess Shale, Canada.

Ordovician period (485.4–443.8 mya)

During the Ordovician period, seas were widespread and rich in algae. Trilobites diversified and were quite abundant in this period, together with cephalopods and early corals (Figure 10.1.11). The Ordovician period also provides evidence of the first vertebrates, the jawless armoured fishes (ostracoderms) (Figure 10.1.12).

The end of the period was marked by a major ice age and the second-largest mass extinction of the Palaeozoic era.

BIOFILE

Arthropods in the Ordovician?

In 2002, there was a discovery of arthropod trackways in Cambrian–Ordovician eolian sandstone, in south-eastern Ontario, Canada. It suggested that animals may have ventured onto land early in the Cambrian; however, more evidence is required as dating of the sandstone is still not certain. These animals were arthropods, and resembled centipedes about the size of crayfish. They probably didn't live on land, and may have come ashore to mate or evade predators.

FIGURE 10.1.11 Animals from the Ordovician period: (a) trilobite *Onnia* sp., (b) sea stars *Lapworthura miltoni*, (c) brachiopods *Dalmanella* sp. and *Orthis* sp. (larger shells), and cephalopods (d) nautiloid fossil and (e) illustration of a nautiloid.

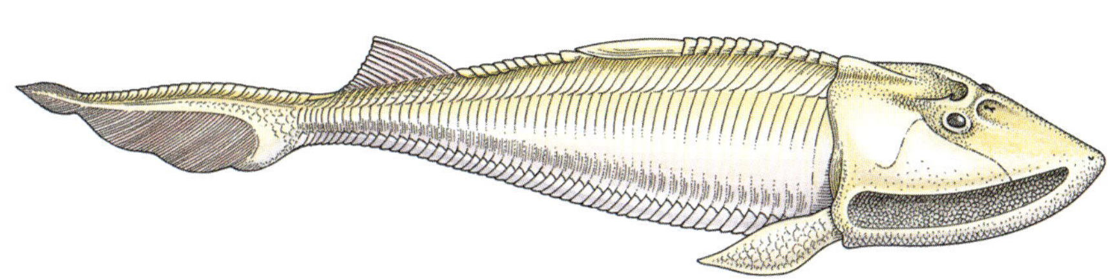

FIGURE 10.1.12 *Cephalaspis*, an armoured fish that belonged to the group of jawless vertebrates called ostracoderms.

Silurian period (443.8–419.2 mya)

The earliest evidence of life on land is from terrestrial rocks of the Silurian period. The first known air-breathing animals were arthropods. Millipedes, centipedes and the earliest arachnids also first appear in the Silurian. The oldest known land plants are from the Late Silurian. These are small, spore-bearing, vascular plants such as *Cooksonia* (Figure 10.1.13). *Cooksonia* had an aerial stem but lacked roots and leaves. They had xylem and phloem tissue that allowed the transport of water and nutrients. It may be that non-vascular plants such as liverworts and mosses evolved on land earlier, but there is no definitive evidence of them until much later. Terrestrial fungi are also recorded as hyphae and spores from the Silurian.

FIGURE 10.1.13 *Cooksonia*, named after the Australian palaeobotanist Isabel Cookson, is probably the earliest-known vascular plant. (a) Model of *Cooksonia* from the Smithsonian National Museum of Natural History exhibit hall. (b) *Cooksonia* fossil.

Devonian period (419.2–358.9 mya): the first land vertebrates

The emergence of land-based plants and animals resulted in organic matter being deposited into the barren soils, promoting further colonisation of the land.

The vegetation of the Devonian period would have only been a few centimetres high and would have been spread by spores. The Devonian period is often referred to as the 'age of fishes' due to the diversification of fishes. In the sea, jawed fishes evolved, along with armoured placoderms, ray-finned and lobe-finned fishes and early sharks (Figure 10.1.14).

FIGURE 10.1.14 Illustration of *Dunkleosteus*, a member of the class Placodermi. Placoderm fishes were distinguished by having heavily-armoured heads and a scaleless body.

> A tetrapod is a vertebrate (which includes amphibians, reptiles and mammals) with four limbs. Animals that had four-limbed ancestors, such as snakes and whales, are also known as tetrapods.

One group of fleshy-finned Devonian fishes developed such sturdy fins that they were able to support their weight at the edge of the water. These animals would give rise to the first terrestrial vertebrates, the **tetrapods** (meaning four-footed) (Figure 10.1.15). The earliest tetrapods include amphibians.

FIGURE 10.1.15 (a) Illustration of a skeleton of an early tetrapod, *Ichthyostega* and (b) an illustration of how *Ichthyostega* might have looked.

Carboniferous period (358.9–298.9 mya)

The Carboniferous period (the coal age) is characterised by abundant terrestrial plant life. Spore-bearing plants (including large lycophytes and sphenophytes) with woody stems, roots and leaves developed tree forms. Forests of these large photosynthesising trees absorbed significant amounts of carbon dioxide from the atmosphere, reducing it to a historically low level. Burial of these trees formed coal beds, the source of fossil fuel mined today. New forest habitats influenced the diversification of animal species, including giant insects such as dragonflies (Figure 10.1.16).

During the Carboniferous period, the tetrapods were losing their amphibian-like bodies in favour of the long snouts and more agile limbs of early reptiles. They were spending more time out of the water adapting to the terrestrial environment. An amniotic egg evolved and supported reproduction on land; scaly skin evolved and protected animals from dehydration.

FIGURE 10.1.16 Artwork of a *Hylonomus* reptile and its prey, a giant dragonfly (*Meganeura monyi*), in a forest during the Carboniferous period.

Permian period (298.9–252.1 mya)

At some point during the Permian period, the landmasses joined to form the supercontinent known as Pangaea, which was characterised by polar ice and deserts (Figure 10.1.17). The remainder of the Earth's surface was one massive ocean.

The fossil evidence shows that a mass extinction event occurred at the end of the Permian period in which up to 90% of all species were wiped out.

FIGURE 10.1.17 Pangaea supercontinent.

BIOFILE

Australian coal

Even though it is true that there are large Carboniferous coal deposits across the world, Carboniferous coal is not significant in Australia. Most coal mined in Victoria is brown coal, which originated in the Palaeogene and Neogene periods (also known as the Tertiary). Queensland and NSW coal is black and is mostly Permian in origin with some Jurassic, Triassic and Cretaceous deposits.

Although the term 'Tertiary period' is no longer part of official terminology, it is still in common usage and refers to the Palaeogene and Neogene periods combined; that is, from 66 million to 2.58 million years ago.

THE MESOZOIC ERA: THE AGE OF THE DINOSAURS

During the Mesozoic era, there was reduced competition following the mass extinction of the Permian period, which opened up **ecological niches**. An ecological niche is the role and position a species has in its environment: how it meets its needs for food and shelter, how it survives, and how it reproduces. The climate was warm and humid, and with these favourable conditions life diversified rapidly and some organisms grew to enormous sizes.

Triassic period (252.1–201.3 mya)

A warmer, drier climate eliminated the polar ice caps and the harsh conditions of the Permian and provided environments that favoured seed plants, such as cycads and gingko trees, as well as ferns. Herbivores grew in number and in size as plant life flourished (Figure 10.1.18).

FIGURE 10.1.18 Artwork of animals that existed in the Triassic period.

Reptiles were the dominant vertebrates in the Triassic period, and included pterosaurs, crocodiles and the earliest dinosaurs. Mammal-like therapsids are thought to have given rise to the true mammals in the Late Triassic. The early mammals were small, insectivorous, nocturnal, hairy and homeothermic (Figure 10.1.19).

FIGURE 10.1.19 Evidence of the first mammals is the genus *Morganucodon*, a 10 cm long weasel-like animal whose fossils were first found in caves in Wales and around Bristol (UK).

Jurassic period (201.3–145.0 mya)

During the Jurassic period the dinosaurs thrived on a warm, forested Earth, giving rise to the first gigantic sauropod and theropod dinosaurs, including bipedal forms such as *Tyrannosaurus rex* (Figure 10.1.20). Sauropod dinosaurs are related to crocodiles and birds. The oldest bird fossil from this period, *Archaeopteryx* (~150 mya), is an early bird with a bony tail and teeth like a dinosaur, and the wishbones and feathers of a modern bird (Figure 10.1.21). Mammals also began to diversify by the early Jurassic (Figure 10.1.22).

Pangaea broke up during the Jurassic into two landmasses: **Gondwana** in the south and **Laurasia** in the north, separating northern and southern flora and fauna.

FIGURE 10.1.20 *Tyrannosaurus rex.*

FIGURE 10.1.21 *Archaeopteryx* fossil. The bones are surrounded by feathers (rippled areas). The head is at centre left, with the wings at upper left and centre right. The legs are at lower centre, and the tail is at lower left.

FIGURE 10.1.22 An illustration of a typical Jurassic ecosystem with three different species from the order Docodonta, a group of now-extinct early mammals.

Cretaceous period (145.0–66.0 mya)

By the middle of the Cretaceous period, dinosaur diversity had reached its peak. The supercontinents Gondwana and Laurasia had started to break up into the continents we know today. Most of the modern forms of bony fishes are found in the fossil record of the end of the Cretaceous.

By the Late Cretaceous small primitive **marsupials** and insectivores were abundant and widespread.

The dawn of the flowering plants

The terrestrial environments in which early Cretaceous animals lived were still dominated by ferns, seed ferns, cycads and conifers; but **angiosperms** (flowering plants) also became part of the flora at this time, about 135 million years ago. Angiosperms diversified rapidly, and by the end of the Cretaceous were by far the most diverse group of terrestrial plants. The first angiosperms co-evolved with thriving insect populations, which pollinated the flowers (Figure 10.1.23).

FIGURE 10.1.23 Earliest flowering plants from the Cretaceous period: (a) *Leefructus mirus* and (b) fossil of *Archaefructus* sp.

BIOFILE

The K–Pg boundary

Originally known as the K–T boundary, marking the end of the Cretaceous period and beginning of the Tertiary period of the Cenozoic era, the K–Pg boundary has been renamed the Cretaceous–Palaeogene boundary.

This point in time is considered significant due to an extinction event that saw the end of the dinosaurs and many other life forms, and has been dated by a section of rock in El Kef, Tunisia. This section of rock highlights the moment when an asteroid hit the Yucatan Peninsula creating the Chicxulub crater and causing massive ecological changes. It contains mineralogical evidence of the asteroid, such as increased iridium content; environmental evidence, such as evidence of tsunamis with ocean sand being significant distances from oceans; and the last of many species being identified in the fossil record.

FIGURE 10.1.24 The K–T boundary, now called the K–Pg boundary. The white line, which is level with the hammer, is iridium-rich boundary clay. Below the line, the rocks are Cretaceous, and above, Tertiary. Iridium is not normally found in Earth's crust but is more commonly associated with meteorites. The high concentration of iridium here is associated with a meteorite impact 65 million years ago that wiped out the dinosaurs as well as many other species.

At the end of the Cretaceous period, about 65 million years ago, another extinction event occurred, possibly due to the Chicxulub asteroid crashing into Earth. Many early dinosaurs became extinct at this time. However, some of them survived and evolved into the birds that we know today.

THE CENOZOIC ERA: THE RISE OF BIRDS AND MAMMALS

The Cenozoic era marked the shift towards the life forms we know today. The extinction of the giants of the Cretaceous period (dinosaurs, pterosaurs and marine reptiles) enabled the smaller species to diversify. Evidence suggests that 50% of plant and animal families and 76% of species were lost at the time of the K–Pg boundary.

Palaeogene period (66.0–23.03 mya)

Mammals began to take advantage of the niches left by the extinction of the dinosaurs and evolved into many new species. Early primates (prosimians) were common but were eventually replaced by true monkeys and apes.

During the Palaeogene period, birds became abundant, as new species of plant life evolved that served as shelter and food.

The continents continued to move to new latitudes, affecting ocean currents and climates and the environments of plants and animals. In what is now Australia, rainforest vegetation was more widespread than it is today.

Neogene period (23.03–2.58 mya)

The Neogene period saw further major continental movement, with what is now Australia well-separated from Antarctica (Figure 10.1.25). Modern forms of mammals, birds and flowering plants continued to evolve. The first hominins (human-like) fossils are 6–7 million years old, and the first of the species of the genus *Homo*, *Homo habilis*, dates from around 2.5 million years ago (Figure 10.1.26).

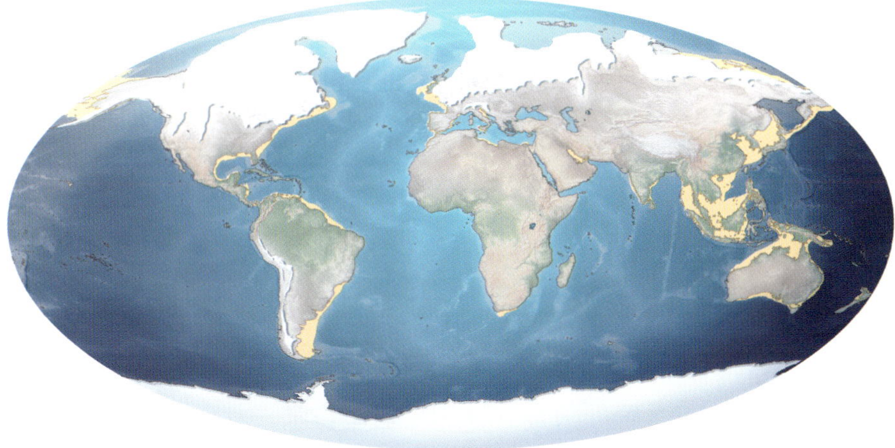

FIGURE 10.1.25 Illustration showing how the continents would have looked in the Neogene period.

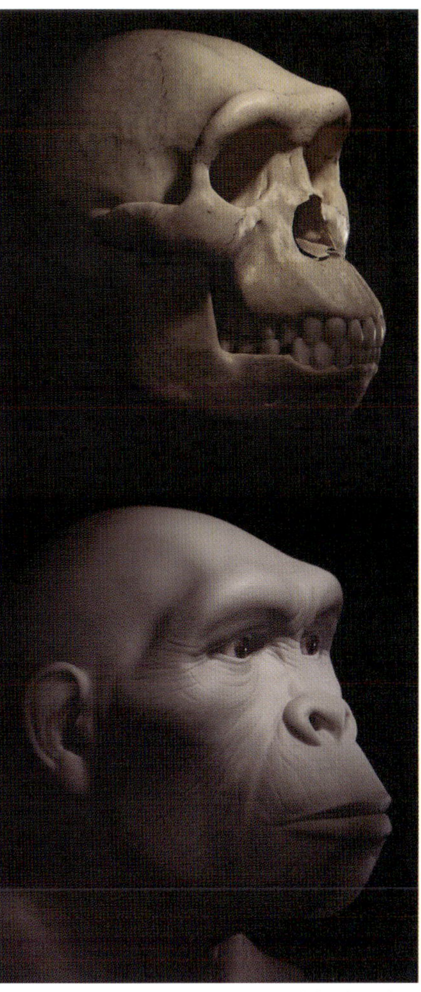

FIGURE 10.1.26 *Homo habilis* skull (top) and illustration of *Homo habilis* (bottom).

Animals across the globe included giant forms—the megafauna. In the Northern Hemisphere there were mammoths, sabre-toothed tigers and ground sloths, while in the Southern Hemisphere there were massive lizards and marsupials (Figure 10.1.27).

FIGURE 10.1.27 Illustration of a herd of mammoths (*Mammuthus columbi*) being attacked by sabre-tooth cats (*Smilodon fatalis*).

FIGURE 10.1.28 Reconstruction of a Neanderthal (*Homo neanderthalensis*) based on the La Chapelle-aux-Saints fossils.

Quaternary period (2.58 mya–today)

The Quaternary period includes two epochs: the Pleistocene epoch (2.58–0.01 mya) and the Holocene epoch (0.01 mya–present). During the Quaternary period climates fluctuated from cold, dry glacial periods and low sea-levels to warmer interglacial periods and higher sea levels. The Holocene epoch (up to today) is an interglacial period of warm conditions.

In what is now Victoria, Australia, the bedrock was in place but the landscape continued to change. The climate started to become drier, which caused the rainforests to shrink in size. Plants and animals that were better-suited to dry conditions started to spread. Giant marsupials such as diprotodon, enormous goannas and large flightless birds that were distantly related to ducks and geese roamed in Australia. The megafauna also shared the landscape with more familiar animals such as kangaroos and emus.

A major extinction event, the Ice Age extinction, occurred worldwide at the end of the Pleistocene epoch. Neanderthals, *Homo neanderthalensis*, became extinct (Figure 10.1.28).

During glacial periods, when water was frozen in ice sheets, sea levels would drop and expose land bridges for species to cross. The Holocene epoch is relatively short in geological time, and does not include significant changes in species, but it does include significant movement of species across continents. Among these species walked the first modern humans.

BIOLOGY IN ACTION

The Cradle of Humankind

A world heritage site that is located approximately 50 km north-west of Johannesburg, South Africa (Figure 10.1.29), is the world's greatest source of hominin fossils, as well as some of the oldest, with some dating back as far as 3.5 million years ago (see Chapter 12).

The first major discovery of a hominin from this region was in 1924 when anatomist Raymond Dart found the skull of a juvenile primate among a box of fossil-bearing rocks. Now known as the Taung Child after the town where it was discovered, it was the first evidence of the hominin species *Australopithecus africanus*. Since 1924, other hominins including *Paranthropus robustus*, *Australopithecus sediba*, *Homo ergaster* and *Homo naledi* have been found in Africa.

The discovery of *Australopithecus sediba* in 2010 is one of the most exciting discoveries. *Australopithecus sediba* is estimated to be nearly 2 million years old, and shares significant traits with the genus *Homo*, indicating their relationship.

The most recent discovery was in September 2015, when Lee Berger of the University of Witwatersrand, in collaboration with National Geographic, announced the discovery of *Homo naledi*. Besides shedding light on the origins and diversity of our genus, *H. naledi* also appears to have intentionally placed bodies of its dead in a remote cave chamber, a behaviour previously thought limited to *Homo sapiens*.

FIGURE 10.1.29 Entrance to Maropeng Exhibition Centre, the Cradle of Humankind World Heritage site in Gauteng South Africa.

10.1 Review

SUMMARY

- The geological time scale is constructed using the relative order of rocks in a sedimentary rock sequence, the fossilised remains of ancient animals and plants within the rock strata, and direct dating of rocks using radiometric techniques.
- The largest geological subdivision of the geological time scale is an eon. Eons are subdivided into eras, which are further subdivided into periods, and into still smaller subdivisions called epochs.
- In the Archaean eon:
  - Life first appeared on Earth. Earth's initial life forms were prokaryotes (bacteria). Stromatolites and other early prokaryotes were Earth's sole inhabitants for more than 1.5 billion years.
- In the Proterozoic eon:
  - Single-celled and multicellular eukaryotes such as algae (red algae and green algae) appeared.
  - The Ediacaran period includes the earliest evidence of multicellular animals called the Ediacaran fauna. These animals were small, soft-bodied sea creatures that resemble modern sea jellies and segmented worms.
- In the Palaeozoic era:
  - Cambrian period—a dramatic increase in the number and complexity of marine life forms, including animals with exoskeletons.
  - Ordovician period—the emergence of the first vertebrates, the jawless armoured fishes (ostracoderms).
  - Silurian period—the first known air-breathing animals were arthropods: millipedes and centipedes. The earliest arachnids also appeared. Small vascular plants colonised swampy land.
  - Devonian period—jawed marine fishes evolved, along with armoured placoderms, ray-finned and lobe-finned fishes and early sharks. One group of finned fishes developed sturdy fins that they were able to support their weight at the edge of the water and led to the evolution of tetrapods.
  - Carboniferous period—evolution of reptiles from amphibian-like ancestors. Formation of forests dominated by tree forms of spore-bearing vascular plants including lycophytes and sphenophytes.
  - Permian period—one massive continent called Pangaea formed, reptiles diversified, and a mass extinction event occurred in which up to 90% of species became extinct.
- In the Mesozoic era:
  - Triassic period—reptiles were the dominant vertebrates. The archosaur reptiles, had diversified into pterosaurs, crocodiles and the earliest dinosaurs. Evidence of the earliest mammals also emerged. Plants included cycads, ferns and Gingko-like trees.
  - Jurassic period—dinosaurs thrived on a warm, forested Earth giving rise to the first gigantic sauropod and theropod dinosaurs. Mammals had begun to diversify. The oldest bird fossils are from the Jurassic.
  - Cretaceous period—dinosaur diversity had reached its peak, small primitive marsupials, and insectivores were abundant and widespread. Angiosperms (flowering plants) evolved.
- In the Cenozoic era:
  - Palaeogene period—mammals evolved into many new species, giving rise to new placental and marsupial species. Birds became abundant and primates evolved.
  - Neogene period—the first of the *Homo* species, *Homo habilis*, evolved.
  - Quaternary period—includes the Pleistocene and Holocene epochs. Climates and sea-levels fluctuated from cold, dry glacial periods and low sea-levels to warm, wetter interglacial periods and higher sea-levels. Some animal species evolved to become giants—the megafauna, which later became extinct.

KEY QUESTIONS

1 a How is the geological time scale constructed?
  b What are the divisions in the geological time scale?

2 What are stromatolites?

3 In which eon did multicellular animals and plants first appear?

4 a What were the first air-breathing animals on land?
  b Which group of organisms gave rise to reptiles?
  c When did the first mammals emerge?

5 a Based on known fossils, what are the earliest land (terrestrial) plants?
  b Name the period with the oldest land plants.
  c When did the first angiosperms appear?

6 Explain the defining features of life forms that can be used to distinguish between the Precambrian and the Palaeozoic era.

7 State which natural event defines the boundary between the Cretaceous and Neogene periods.

10.2 Evidence for evolution

Evolution is a process of change. The modern theory of evolution states that all living organisms share a common origin that dates back to 3.8–4.1 billion years ago. In Section 10.1 you learnt that the environment and its inhabitants have changed dramatically since life began on Earth. The earliest organisms were bacteria and, over a long period of time, very different groups of organisms diverged from these early forms of life. Some groups became extinct, while others changed over time to become the types of organisms that we see today.

In this section, you will learn about the evidence for biological change over time, including evidence from palaeontology (the fossil record, types of fossils, the steps in fossilisation and the relative and absolute dating of fossils) (Figure 10.2.1), biogeography, structural morphology and developmental biology.

FIGURE 10.2.1 Palaeontologists studying 'Lyuba', one of the best-preserved woolly mammoths (*Mammuthus primigenius*). Lyuba was a female calf that died about 40 000 years ago at the age of about one month. Woolly mammoths became extinct about 10 000 years ago.

FIGURE 10.2.2 Macrophotograph of a fossilised midge insect (family: Chironomidae) found embedded in Baltic amber. The insect is related to present-day non-biting midges, and more distantly to mosquitoes. It is about 40 million years old and from the Late Eocene epoch.

PALAEONTOLOGY

Palaeontology involves the study of ancient life represented by fossils. Fossils are the preserved remains, impressions or traces of organisms found in rocks, amber (fossilised tree sap), coal deposits, ice or soil (Figure 10.2.2). Preserved remains are usually hard structures that are not easily destroyed or are slow to decompose, such as bone, shell, wood, leaves, pollen and spores. The **fossil record** refers to the total number of fossils that have been discovered, providing evidence of the evolution of living organisms through geological time. Fossils tell palaeontologists and palaeobotanists (scientists who study either animal or plant fossils) about the kinds of organisms that lived in the past, what they looked like, and where and when they lived, allowing us to put a time scale on evolution.

Fossilisation process

Fossilisation is the preservation of the hardened remains or traces of organisms in rocks.

The chances of an organism becoming fossilised after death are small. Soft-bodied organisms are unlikely to be preserved, because soft body parts decay readily or are subject to predation and scavenging. Fossilised parts of plants are commonly wood and leaves made up of cellulose, which does not decay readily, and spores and pollen, which are even more resistant to decay.

Fossilisation has a chance of occurring when an organism is buried by sediments. This reduces the chance of decay, due to lack of oxygen for decomposer microorganisms, and hides the organism from scavengers. When sediments of sand, silt or mud in the sea, a lake or slow-flowing stream accumulate over the organism, the organism is preserved. The weight of many layers of sediments squeezes out the water between the particles of sand, silt or mud. As the deposit deepens, the temperature increases and soft sediments become solid rock—sandstone, siltstone, mudstone or shale (a mixture of clay and silt) (Figure 10.2.3).

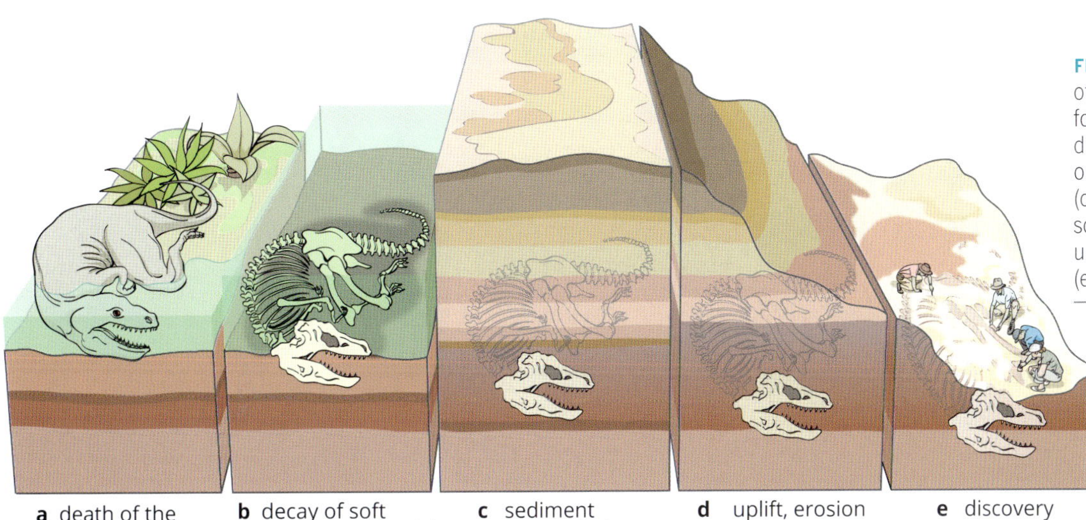

FIGURE 10.2.3 The sequence of events leading to the formation and subsequent discovery of a fossil. (a) The organism dies and (b) is buried. (c) Sediments accumulate and solidify to rock. (d) Subsequent uplift, erosion and exposure (e) lead to its discovery.

Sediments accumulate in bodies of water such as seas, estuaries and lakes, hence a large proportion of fossils are found where ancient bodies of water existed. Fossil shells formed in this way in Tasmania. In the Carboniferous and Early Permian periods (about 280 million years ago) a marine gulf formed the Tasmanian Basin. The basin filled with mud and silt washed down from glaciated uplands. The sediments formed layers of mudstone, siltstone, sandstone and some limestone. The basin later became a larger plain, with lakes and freshwater streams that deposited other sediment layers. Fossil shells, fishes (including lungfish) and amphibians have been found in siltstone and sandstone at Mt La Perouse in Tasmania (Figure 10.2.4).

Organisms on land are less likely to be preserved than those that live in aquatic environments. For example, plants that grow along river banks or the edge of swamps, where sediments can trap leaves, fruits and seeds, are more likely to be fossilised than plants that grow only on rocky outcrops. Delicate plant parts such as flowers are rarely fossilised, although some are preserved by being buried rapidly, for example by ash from an erupting volcano. For these reasons the fossil record is biased towards certain sorts and parts of organisms and certain environmental conditions, which consequently limits evidence of past life and our understanding of it.

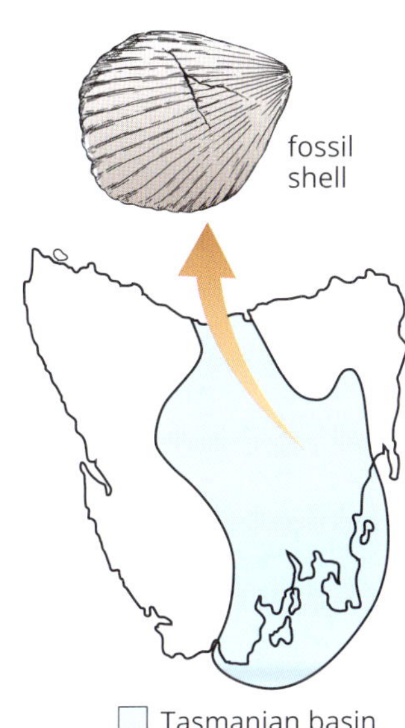

FIGURE 10.2.4 Fossil shells, fishes (including lungfish) and amphibians about 280 million years old have been found at Mt La Perouse in Tasmania.

Types of fossils

The four main types of fossils are impression fossils, mineralised fossils, trace fossils and mummified organisms.

FIGURE 10.2.5 Impression fossil of *Equisetum* (horsetail). Sandstone block split open to show a rare fossil imprint of a terminal bud from an *Equisetum* horsetail plant from the Triassic period.

Impression fossils

Impression fossils are left when the entire organism decays but the shape or impression of the external or internal surface remains (Figure 10.2.5). In some rocks such as limestone the fossils retain their three-dimensional shape, but in rocks (e.g. shales) or in coal deposits that are physically compressed, fossils are flattened. Impression fossils include the internal surface of a shell, tree trunks and plant leaves. If the vacant space of the mould is later filled with foreign material, a three-dimensional 'sculpture' of the organism is formed; this is called a **cast fossil**.

Mineralised fossils

Mineralised fossils occur when minerals replace the spaces in structures of organisms such as bones. Minerals may eventually replace the entire organism, leaving a replica of the original fossil (e.g. petrified wood). This process is known as **mineralisation** or **petrification**. Minerals can include opal, pyrite and silica (Figure 10.2.6).

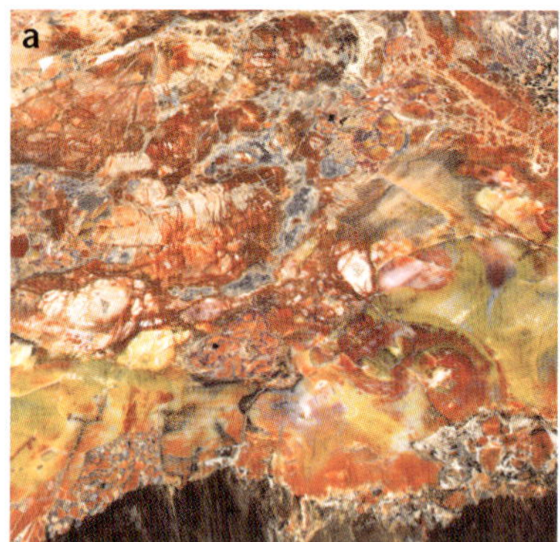

FIGURE 10.2.6 Examples of mineralised fossils. (a) Petrified wood from the Petrified Forest, Arizona, USA. This fossil is from the Late Triassic, when the forest was rapidly buried under volcanic ash. (b) Rock containing a fossil of a *Phareodus* sp. fish from the Eocene epoch. (c) Coloured scanning electron micrograph of fossilised diatoms, single-celled planktonic algae. Diatoms have a wall of silica that provides protection and support, which is readily fossilised. These diatoms are from the Miocene epoch.

Trace fossils

Trace fossils (ichnofossils) are the preserved evidence of an animal's activity or behaviour, without containing parts of the organism. Transient impressions or footprints would be considered trace fossils, as would casts of burrows or even coprolites (fossilised faeces) (Figure 10.2.7).

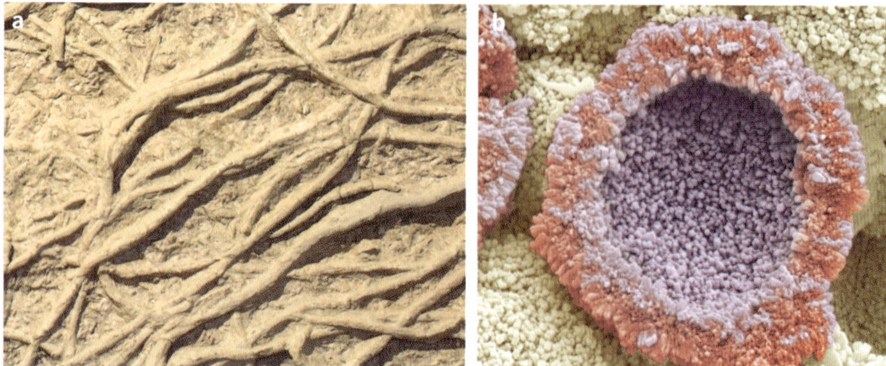

FIGURE 10.2.7 Examples of trace fossils. (a) Fossil worm burrows (*Arthrophycus* sp.) from the early Silurian. (b) Coloured scanning electron micrograph of a section through a coprolite from a dinosaur.

BIOFILE

Fossil footprints

Trace fossils of footprints are formed when an organism steps into soft mud. The impression is then covered with loose sand so that the footprint is filled. The sand in the footprint eventually consolidates and is compacted into sandstone. Finally, when the rock is split open along the bedding surface, the original footprint is revealed.

FIGURE 10.2.8 Fossils appear when rock slowly forms around objects buried in mud. As the rock forms, the shape and anatomy of buried animals and plants can be preserved, including tracks such as these footprints.

Mummified organisms

Mummified organisms are those that have been trapped in a substance under conditions that reduce decay and so are changed little. Examples include insects trapped in amber, leaves that still contain carbon (Figure 10.2.9), animals frozen in ice or trapped in a peat bog (known as 'bog body') (Figure 10.2.10). Fossil mummified animals, including humans, can have hair and skin preserved in a dehydrated state, while limbs and occasionally entire bodies are preserved in peat bogs and tar pits.

FIGURE 10.2.9 This leaf from the coal deposits in Anglesea, Victoria, is a mummified fossil and still contains carbon. It was preserved in layers of mud sandwiched between the layers of coal. Many of the fossils are rainforest plants related to those that survive today in the wet tropics of Queensland. The environment at Anglesea must have once been wet and warm.

FIGURE 10.2.10 Mummified body of Tollund Man, dated 220–40 BCE. This well-preserved body of an adult man was discovered in 1950 in a bog at Tollund Mose in Jutland, Denmark.

BIOFILE

Dinosaur Cove

Dinosaur Cove, on the southern coast of Victoria, is famous for its fossil deposits. An ancient stream flowed through the site 106 million years ago, depositing soft sand and mud, which turned to rock. Dinosaur bones were trapped in these sediments. Dr Tom Rich and Dr Pat Vickers-Rich found and described small, bipedal dinosaurs (hypsilophodontids). At the time that these dinosaurs lived near the cove, Australia was far south and connected to Antarctica. Although not frigidly cold, winter was a long period of darkness. The hypsilophodontids had large optic lobes in their brains, meaning that they could probably see well in the dark!

FIGURE 10.2.11 Cretaceous bird tracks on a slab of sandstone found at Dinosaur Cove, southern Victoria, Australia.

Dating fossils

The age of a fossil is almost as important as the physical details of the fossil because it gives a time scale of evolution. The age of a fossil can be determined by relative dating or by absolute dating methods.

Relative dating

Relative dating is based on stratigraphy. **Stratigraphy** is the study of the relative positions of the rock strata (singular: stratum), or layers, some of which contain fossils. The lowest stratum is the oldest and the upper strata are progressively younger. The age of a fossil is estimated relative to the known age of the layers of rock above and below the layer in which the fossil is found (Figure 10.2.12). For example, if a layer containing fossils lies below rock that is dated at 200 million years old, then the fossils must be at least that age or older. Relative dating can be difficult in areas where rock layers have been eroded away, or where rocks have been buckled, moved or reburied, altering the original sequence of strata.

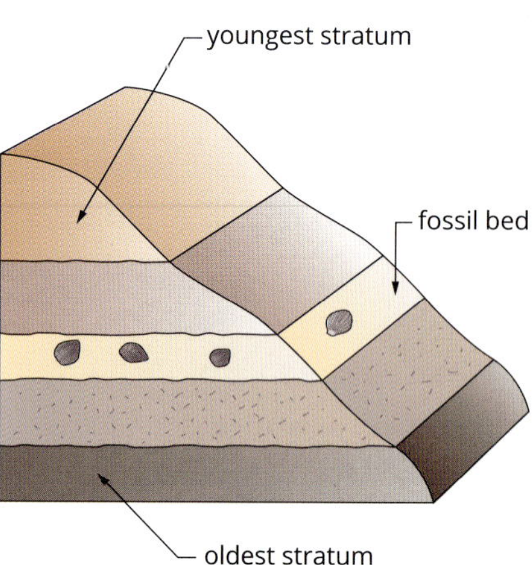

FIGURE 10.2.12 Relative dating assumes that rock strata (layers) are laid down in the order of the formation—the bottom stratum is the oldest, the top the youngest.

An **index fossil** (sometimes known as an indicator fossil) is a fossil used to define and identify geologic periods. Sometimes the only way to age a fossil bed is by using index fossils together with stratigraphy. Index fossils are commonly found fossils from similar sites for which an absolute age has been determined. For example, in Europe the same type of ammonoids (extinct molluscs) are found in different regions. A species of ammonoid fossil is called an index fossil because it indicates that the rocks at each locality are of similar age (Figure 10.2.13).

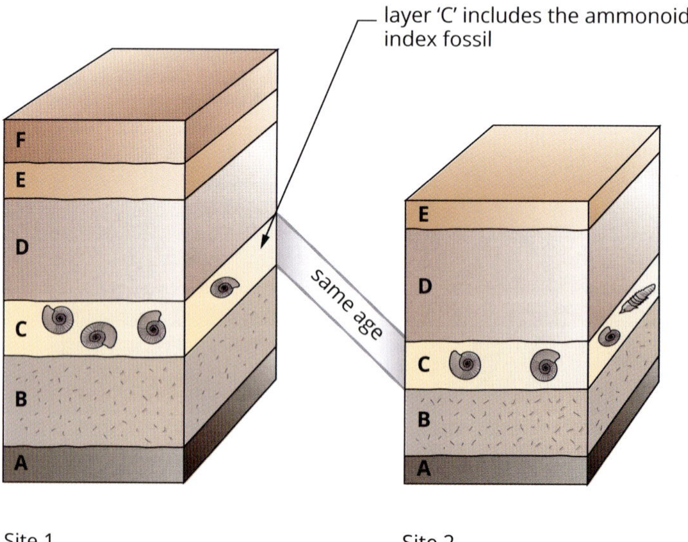

FIGURE 10.2.13 Stratigraphic comparison of sites in different parts of the world provides evidence of the relative age of particular strata. The ammonoid (mollusc) fossils of known age at site 1 are the same as at site 2, so the two strata are assumed to be the same age (even though site 1 has an additional, younger layer, F, at the top). The ammonoid is an index fossil for the age of all the other fossils in the fossil layer at site 2.

EXTENSION

Two fossil sites of inland Australia

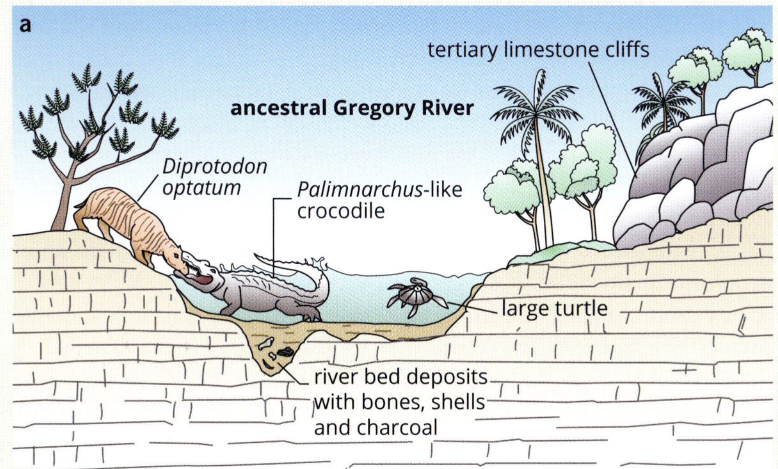

FIGURE 10.2.14 (a) Formation of fossils in the ancestral Gregory River, about 1 million years old. (b) Giant wombat *Diprotodon optatum*.

Riversleigh in north-west Queensland

Riversleigh and the nearby Gregory River contain one of the great fossil sites in the world. Extinct animals are preserved in limestone of various ages, dating back to more than 30 million years (Oligocene period). The fossils at this site reveal the story of ancient rainforest animals and plants that once lived in inland Australia. Fossil remains of the extinct marsupial *Diprotodon optatum* have been uncovered from the ancient bed of the Gregory River. These animals were probably killed by crocodiles as they came to drink from the river, and their bones accumulated as fossils in layers of sands and gravels (Figure 10.2.14). These particular fossils are approximately 1 million years old (from the Late Pleistocene).

Koonwarra in Gippsland, Victoria

At Koonwarra in Gippsland, Victoria, scientists have uncovered a great diversity of organisms trapped in the bed of an ancient lake (Figure 10.2.15). The fossils are preserved in mudstone 115 million years old. Many of the fossils are fishes and the fossil site is named the 'Koonwarra fish beds'. One plant fossil is a leaf of *Ginkgo*. Today, the *Ginkgo* (*G. biloba*) is native to China, but trees of this genus were probably once widespread in the world, as evidenced by the 115 million-year-old fossil from Koonwarra.

FIGURE 10.2.15 (a) Fossil fish and (b) *Ginkgo* leaves, 115 million years old, found at Koonwarra, Victoria.

Absolute dating

Absolute dating provides a more precise estimate of age, although it does not mean that it provides an exact date. Radiometric dating, thermoluminescence and electron spin resonance are all methods of absolute dating that are used to determine the age of fossils.

- Radiometric dating is a quantitative technique used to determine the proportion of particular radioactive elements, **isotopes**, within rocks around the fossil or sometimes within the fossil. Radioactive elements decay into different forms (e.g. uranium to lead, rubidium to strontium) at rates that are constant for a particular element. The rate of decay of the element is independent of the nature of the rocks or the environmental conditions to which they are exposed, so they act as accurate clocks. The half-life of a radioactive element is the time taken for half the element to decay, and can be used to calculate the age of the rock in which it is contained (Figure 10.2.16). Particular isotopes are used depending on the time scale involved. For example, the half-life of radioactive carbon-14 allows an estimate of the age of carbon-bearing materials to be calculated up to about 58 000–62 000 years of age. The method of radioactive carbon dating is limited to samples not older than 60 000 years, because by that age there is very little carbon-14 left. For samples older than 60 000 years, potassium-40, which is found in volcanic rock, can be used. As the volcanic rock cools, its potassium-40 decays into argon-40 with a half-life of 1.25-billion-years.

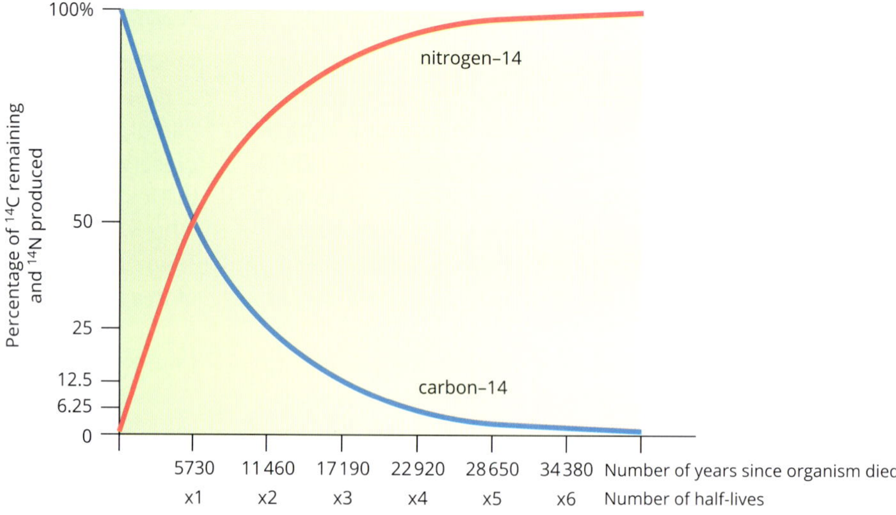

FIGURE 10.2.16 Carbon-14 (^{14}C) dating. The age of a fossil can be determined by measuring the proportion (percentage) of ^{14}C to ^{12}C in a sample. When the fossilised organism was alive, its ^{14}C to ^{12}C ratio was constant (the same as the atmosphere). From the time the organism died the amount of ^{14}C reduced because it decays at a known rate to nitrogen-14 (^{14}N). The ^{14}C decays by half every 5730 years (its half-life), as shown in the graph.

- Thermoluminescence is a technique that can be used to date objects such as pottery, cooking hearths and fire-treated tools up to 500 000 years old, older than is possible with radiocarbon dating. Thermoluminescence is the emission of light from a mineral when it is heated. The amount of light emitted is proportional to the amount of radiation an object has absorbed—the older the object the more light it emits. The intensity of the light can be calibrated to reveal how much time has passed since the object was last heated or burnt in a fire. This technique is used to date artifacts related to human evolution.

- Electron spin resonance (ESR) is used to date calcium carbonate in limestone, coral, fossil teeth, molluscs and egg shells. Palaeoanthropologists have used ESR mostly to date samples from the last 300 000 years. Unlike thermoluminescence dating, the sample is not destroyed with the ESR method and allows samples to be dated more than once.

Electron spin resonance is a spectroscopic technique that detects atoms with orbitals containing unpaired electrons.

Information from fossils

Although fossils provide an indication of the appearance and structure of an organism, other information can be gained or inferred from examining fossils. For example, animal fossils have been found with young in the womb or inside eggs or guarding eggs (Figure 10.2.17). If young are fossilised next to adults, it is likely that the animal parented the young for a period (Figure 10.2.18). If large numbers of organisms are fossilised together, it could be surmised that they lived in herds. The contents of the animal's last meal may even be preserved in the stomach region of the fossil.

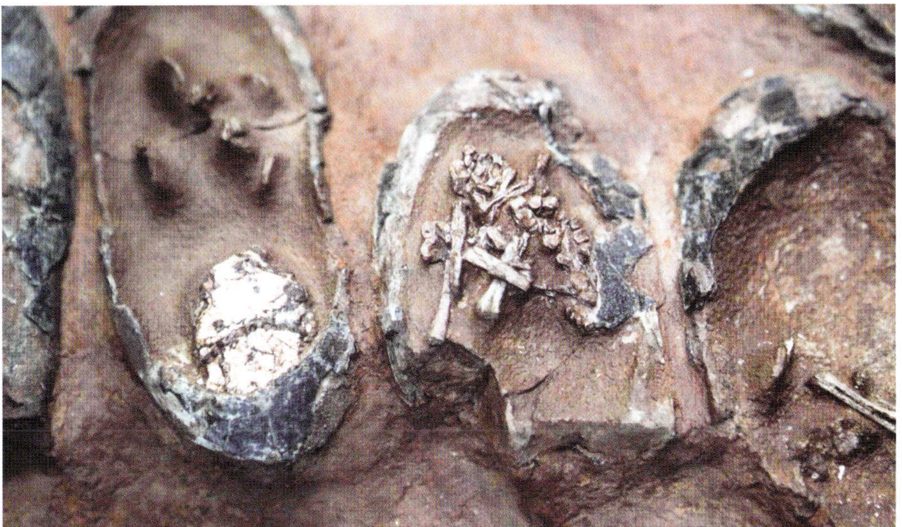

FIGURE 10.2.17 A nest of fossilised dinosaur eggs with remains of dinosaurs inside. The eggs are estimated to be at least 65 million years old.

FIGURE 10.2.18 This dinosaur nest with 15 one-year-old *Protoceratops andrewsi* discovered in the Djadokhta formation in the Gobi Desert, central Asia, suggests these animals were growing together with some sort of parental care.

BIOLOGY IN ACTION

Ancient DNA extracted from fossils of extinct giant kangaroo and wallaby

Ancient DNA has been extracted from the fossils of a giant short-faced kangaroo (*Simosthenurus occidentalis*) and a giant wallaby (*Protemnodon anak*) that inhabited Australia about 40 000–50 000 years ago (Figure 10.2.19). This research was conducted by scientists from the University of Adelaide's Australian Centre for Ancient DNA (ACAD) and provides us with new insight into the evolutionary past of Australia's megafauna.

Remains of the giant short-faced kangaroo and giant wallaby were discovered inside a cave in Mount Cripps, Tasmania. The cool, dry conditions of the cave preserved the remains, allowing the researchers to extract the DNA from the ancient bones and reconstruct part of the mitochondrial genomes of both species (Figure 10.2.20). This is the first glimpse into the DNA of the Australian megafauna; previous attempts to sequence genetic material were unsuccessful due to the poor preservation of the specimens. Well-preserved megafauna specimens are rarely found in Australia due to the harsh climate and age of the remains (39 000–52 000 years). The specimens in this study are the oldest Australian fossils from which DNA has been sequenced.

Information from the mitochondrial genome of the giant wallaby revealed that it is a close relative of the living genus *Macropus*, which includes grey and red kangaroos and the tammar wallaby. DNA evidence confirmed that the short-faced kangaroo belongs to the extinct subfamily Sthenurinae. Although the short-faced kangaroo has no living descendants, its closest living relative is the endangered banded hare-wallaby, the only surviving member of the ancient subfamily, Lagostrophinae. The banded-hare wallaby is now restricted to islands off the coast of Western Australia and is the last representative of a once-diverse lineage of ancient kangaroos.

Ancient DNA gives us insight into the past, but it also strengthens our understanding of the evolutionary history and biology of living species. This knowledge has applications in conservation and environmental management, as we learn from the past to predict the future.

FIGURE 10.2.19 The short-faced kangaroo (*Simosthenurus occidentalis*). Ancient DNA has been sequenced from fossil remains of this species that were discovered in a cave in Tasmania.

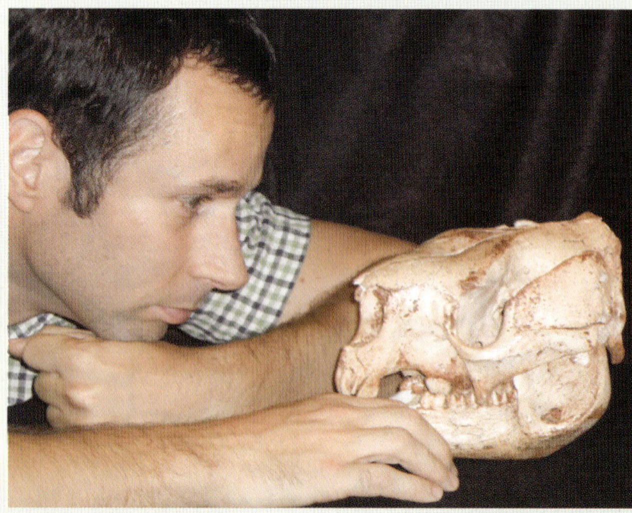

FIGURE 10.2.20 Researcher Dr Bastien Llamas with a skull of the extinct giant short-faced kangaroo (*Simosthenurus occidentalis*). Scientists extracted and sequenced ancient DNA from this species to understand its evolutionary relationship with living and extinct fauna.

EXTENSION

Transition fossils

There are notable examples of fossils that suggest significant stages in evolution, such as the transition from aquatic to land vertebrates. These fossils are interpreted as intermediate or transitional.

Tiktaalik

A 375-million-year-old fossil of the fish *Tiktaalik roseae* was discovered in Arctic Canada in 2004. It had gills and scales typical of a fish, but also limb-like fins, ribs, a flexible neck, and a crocodile-shaped head suggestive of an early tetrapod (Figure 10.2.21). *Tiktaalik* is seen as evidence of the period when aquatic vertebrates began moving ashore. *Acanthostega* is a one of the oldest known tetrapod fossils with recognisable limbs (Figure 10.2.22).

FIGURE 10.2.21 Artwork depicting *Tiktaalik*.

FIGURE 10.2.22 Artwork depicting *Acanthostega*, an early tetrapod. This animal was among the first vertebrates to evolve limbs and be able to move onto land. It is known from the Late Devonian period.

Archaeopteryx

Archaeopteryx is a fossil early bird from the Jurassic (see Section 10.1), first found in Germany in 1860. *Archaeopteryx* possesses similarities to non-avian dinosaurs such as a long, feathered tail and small teeth. However, unlike non-avian dinosaurs, *Archaeopteryx* also has flight feathers and wings, just like a modern bird. The discovery of a fused clavicle bone in *Archaeopteryx* confirmed the relationship between birds and dinosaurs, as they are the only two groups to have this anatomical feature (Figure 10.2.23).

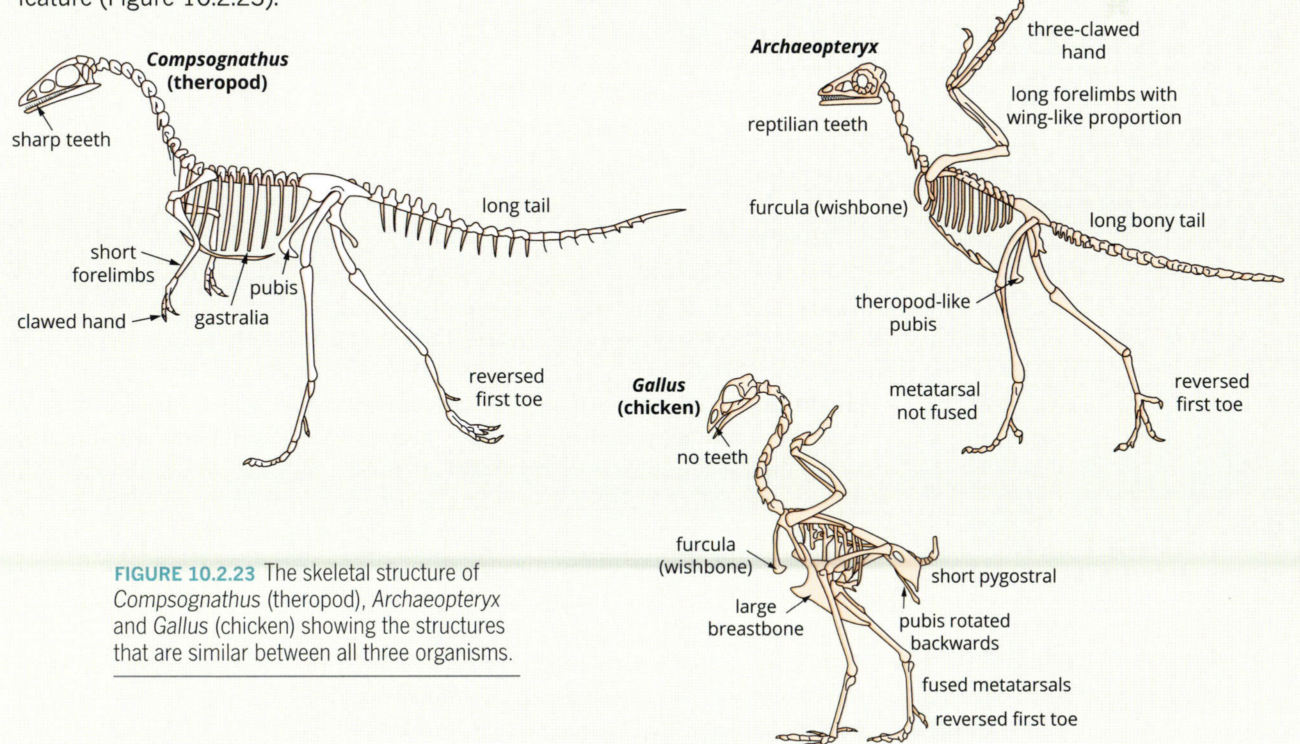

FIGURE 10.2.23 The skeletal structure of *Compsognathus* (theropod), *Archaeopteryx* and *Gallus* (chicken) showing the structures that are similar between all three organisms.

BIOGEOGRAPHY

Biogeography is the study of the distribution of organisms, and is another type of evidence of evolution and the past. Early biogeographers were interested in understanding how similar animals and plants in different parts of the world, separated by vast distances and huge oceans, came about. For example, they sought to understand why there were marsupial possums in Australia and South America, and whether or not the ostrich from Africa, the rhea from South America and the emu from Australia are related.

BIOFILE

Flightless birds

Ratites is a family of large flightless birds that includes ostriches, emus and New Zealand's extinct giant moa. It was once believed that these animals shared a common ancestor that was flightless, and that speciation occurred due to continental drift. However, recent evidence using DNA has shown that ratites do not share this single hypothetical flightless ancestor. Rather they probably evolved from more than one different airborne ancestor.

Patterns of evidence

When distributions of different groups of organisms in the world are mapped, patterns become evident that give clues to the evolutionary histories of the groups and of Earth itself. Observations by **naturalists** such as Alfred Russel Wallace led to the division of the world into a number of **biogeographic regions**. Continental biogeographic regions include the 'Australasian region', the 'Ethiopian region' (Africa) and the 'Neotropical region' (Central and South America). Marine biogeographic regions include the tropical, temperate and Arctic zones, which differ in climate (Figure 10.2.24). Each region is recognised as having a unique set of related organisms, suggesting that the patterns are a result of evolution.

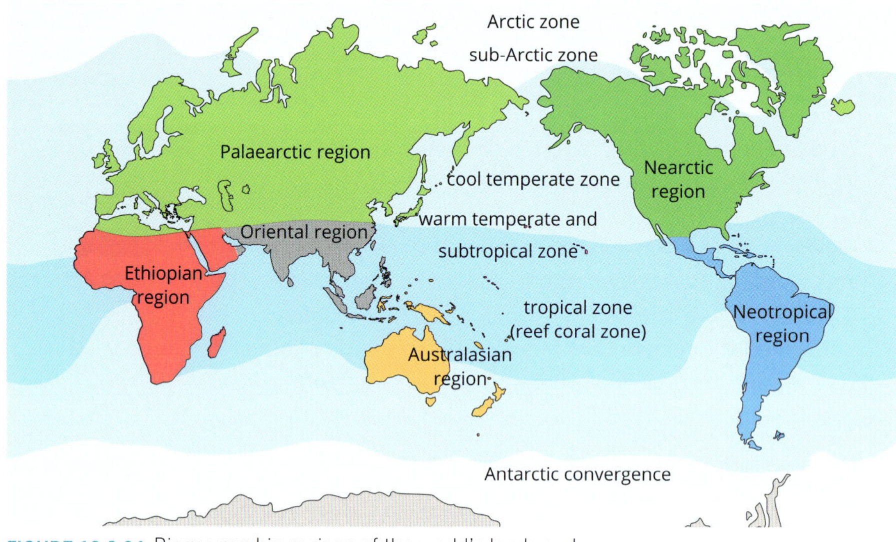

FIGURE 10.2.24 Biogeographic regions of the world's lands and oceans.

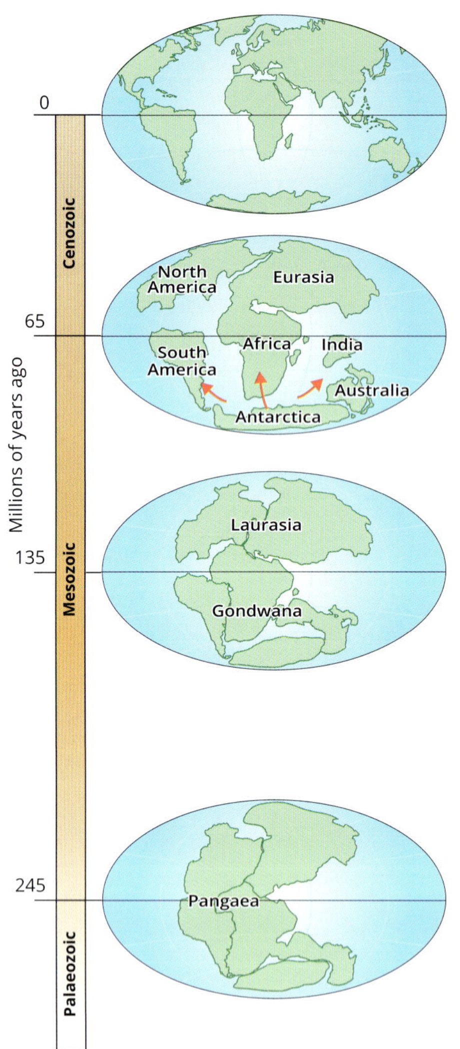

FIGURE 10.2.25 Pangaea and its break-up as crustal plates carrying continents and smaller pieces of land drifted apart.

Continental drift

Scientists noticed that the shape of continents appeared to fit together as if they had been 'cut' from a single larger continent, known today as Pangaea (Figure 10.2.25). The west coast of Africa, for example, seemed to fit well against the east coast of South America. The implications of this were to become important evidence supporting the theory of evolution. If landmasses had at some point been joined, then it follows that organisms on continents separated today might share a common ancestry. Although the theory of continental drift had been proposed in the early 20th century, it was only confirmed much later by geological evidence and the study of plate tectonics. Australia, for example, once part of the southern supercontinent Gondwana, drifted from Antarctica and continues today to move northwards at a rate of approximately 3.7 cm per year for the western side and 5.6 cm per year for the eastern side.

Geologists have been able to calculate the rate at which plates are moving and work backwards to determine the point at which the land masses were joined. This information can help to understand how and when groups of organisms evolved in the different parts of the world, and how they may be related to another.

BIOFILE

Alfred Russel Wallace and Wallace's line

Alfred Russel Wallace independently proposed the theory of evolution by natural selection at about the same time as Charles Darwin. Wallace travelled through the Malay region collecting and studying the distributions of plants and animals. Based on his observations and those of other naturalists, he recognised that the world could be divided up into a number of biogeographic regions. He mapped out a line, now known as Wallace's line, which is the boundary between the Australasian biogeographic region and the Oriental biogeographic region. This boundary marks the point where there is a difference in species on either side of the line. To the west of the line, all of the species are similar or derived from species that are found on the Asian mainland, such as the streaked weaver bird. To the east of the line, there are many species that are of Australian descent, such as the Australian sulphur-crested cockatoo.

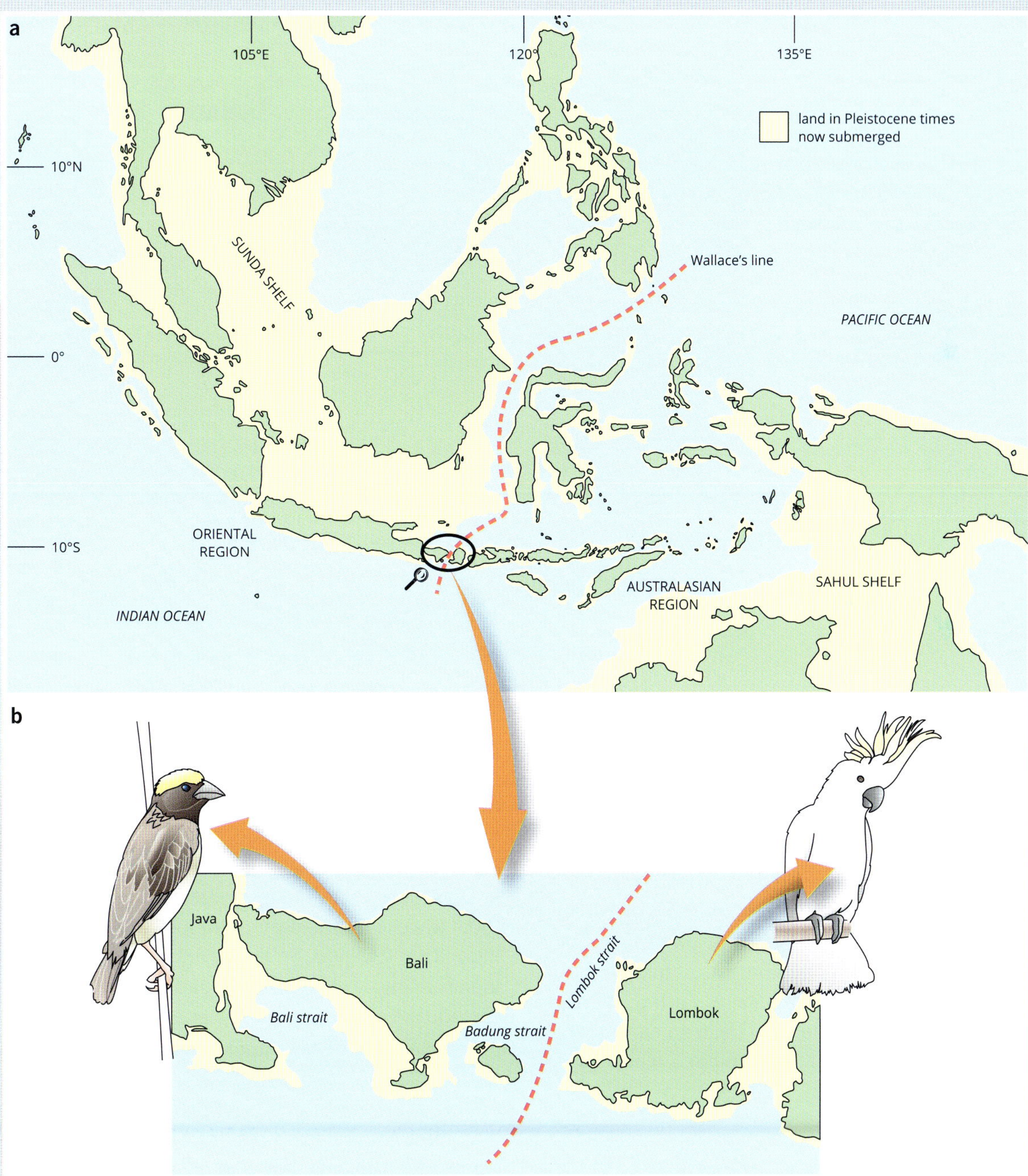

FIGURE 10.2.26 (a) Wallace's line (named after the biogeographer Alfred Russel Wallace) marks approximately the boundary between the Oriental and Australasian biogeographic regions, where plates have collided. (b) The streaked weaver bird (*Ploceus manyar*) lives on Bali while the Australian sulphur-crested cockatoo (*Cacatua sulphurea*) lives on Lombok. These two islands are separated by Wallace's line.

EXTENSION

Plate tectonics

Continental drift

The theory of continental drift was first developed by the geologist Alfred Wegner in 1912. However, his ideas were not accepted until the early 1960s when geological evidence of magnetic reversals and sea-floor spreading were documented.

It is now established that the crustal plates of Earth's surface move relative to one another, like ice floes (Figure 10.2.27). Australia, for example, rides on the top of the Indo-Australian plate, like a log embedded in an ice floe. The Indo-Australian plate also carries with it the southern part of New Guinea and surrounding parts of the Indian, Pacific and Southern oceans.

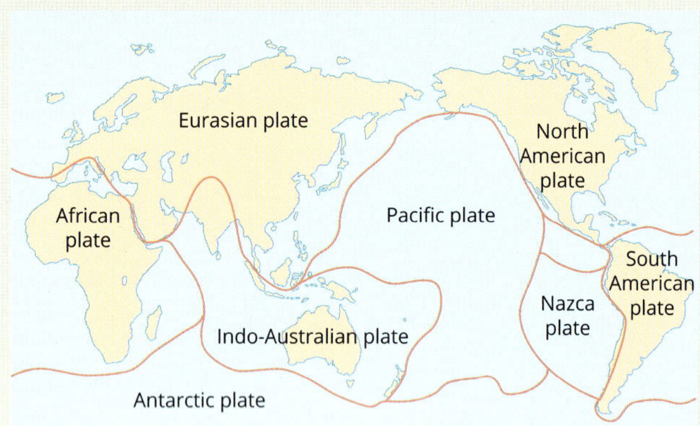

FIGURE 10.2.27 The major tectonic plates of the Earth.

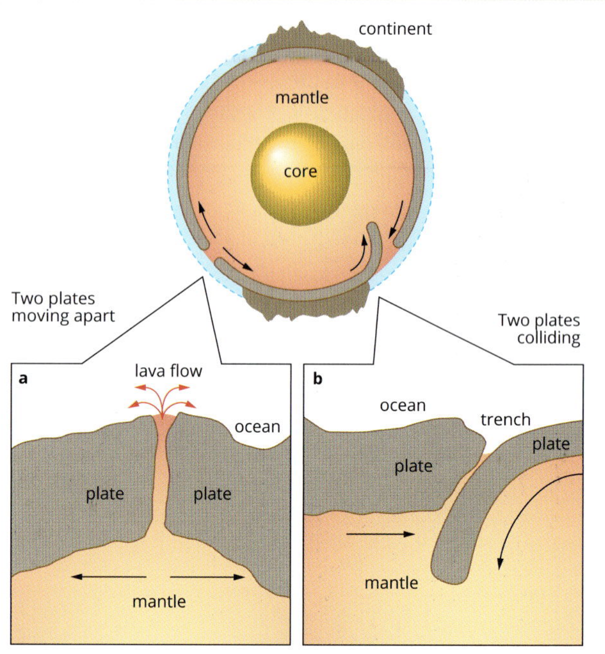

FIGURE 10.2.28 The plates of the Earth's crust float on the underlying mantle. Oceans and continents ride on top of these drifting plates. (a) Where two plates move apart, new material upwells as a lava flow. (b) Where two plates collide head-on, one slides under the other, returning material to the mantle.

Where two plates move apart (Figure 10.2.28a), material below the Earth's crust upwells as a massive flow of lava. This new material forms a ridge marking the edge of two separating plates. Two plates may slide past one another or they may collide (Figure 10.2.28b). When they collide head-on, one plate may thrust under the other, forming a deep trench. Such trenches are the deepest parts of the Earth's surface, the deepest plunging to more than 11 kilometres below sea level in the western Pacific Ocean. Collisions may crumple the edge of the plate and create mountains. Earthquakes are also a result of plate movement. In 1906 an earthquake destroyed most of the city of San Francisco, which straddles a plate boundary.

Significance of plate movement

With our increasing understanding of the Earth's crust and plate movement (plate tectonics), it is becoming possible to predict the occurrence and size of earthquakes. Knowledge such as this may help to avert human tragedies. The most destructive earthquakes on record include the deaths of 830 000 people in China in 1556, 70 000 in Portugal in 1755 and 50 000 people in Iran in 1990. The disastrous earthquake near Sumatra in December 2004 tragically resulted in a tsunami (tidal wave) that killed more than 200 000 people in Indonesia, Thailand, Sri Lanka and adjacent areas of the Indian Ocean. In March 2011, Japan suffered a severe earthquake due to the movement of the Pacific plate.

The theory of plate tectonics provides an explanation for the evolution of different groups of organisms. An important consequence of plate movement is that, throughout Earth's history, the positions of the continents have changed and oceans have been opened up or been lost. At least twice, the continents have come together to form one or two supercontinents. They have then drifted apart, carrying plants and animals that had evolved by that time to new latitudes.

STRUCTURAL MORPHOLOGY

If you compare the human body to that of a chimpanzee, you can see a striking resemblance in structure. The same applies to many other species, and this is more than a coincidence. Studying the **structural morphology** of species, or their body structures, gives an insight into the relationships between species. The field of comparing the structure of organisms is referred to as **comparative morphology** or comparative anatomy.

Homologous features

Features of organisms that have a fundamental similarity based on common ancestry are called **homologous features**. Often homologous features evolve different functions, but their similar structures provides evidence that the organisms shared a **common ancestor** from which they diverged over time. This is known as **divergent evolution**.

Mutations in the genetic sequences regulating the length of the bones in a limb can result in the limb being used in different ways. Close examination of tetrapod forelimbs, for example, shows that the same series of bones is present in each, but the genetic sequence has been modified so that different results occur.

For example, the forelimbs of all mammals, including humans, cats, whales, and bats, show the same arrangement (with different lengths) of bones from the shoulder to the tips of the digits, even though these appendages have very different functions: lifting, walking, swimming and flying (Figure 10.2.30)

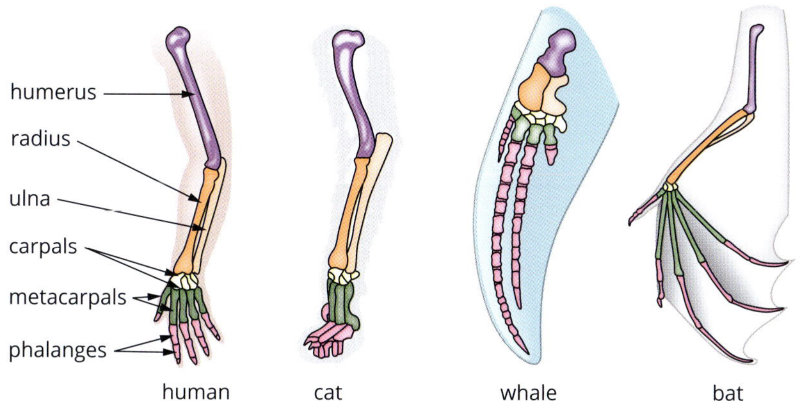

FIGURE 10.2.30 Even though they have become adapted for different functions, the forelimbs of all mammals are constructed from the same basic skeletal elements: one large bone (purple), attached to two smaller bones (orange and tan), attached to several small bones (gold), attached to several metacarpals (green), attached to approximately five digits, each of which is composed of phalanges (pink).

BIOFILE

The dugong

Based on comparisons of anatomy, the dugong (*Dugong dugong*) has the characteristics of a mammal. This is evidence of its evolutionary relationships. It is the only surviving species in its family (Dugongidae) and genus (*Dugong*), with its closest relatives being sea cows and manatees, which are classified in a separate family. The unique characters of these marine mammals evolved as they adapted to an aquatic life. The dugong lives in shallow seas in the Indo-Pacific region and is listed today as an endangered species.

FIGURE 10.2.29 The dugong (*Dugong dugong*) is a marine mammal.

Homologous features are evident in all groups of organisms. For example, the seeds of cycads, *Ginkgo*, conifer trees and flowering plants show a variety of shapes and sizes but they have the same basic structure. The seeds of most conifers are winged and blown about by the wind, whereas the seeds of an acacia lack a wing but have a tough outer coat and a coloured nutritious appendage to attract ants that do the job of dispersal. Despite the variation among seeds, these plants all reproduce by seeds, and it can be argued that they have evolved from a common ancestral group (Figure 10.2.31).

FIGURE 10.2.31 Homologous structures in plants. (a) The seed of a fir tree has a large wing that allows it to be carried by the wind. (b) *Acacia* seeds are small and easily carried by ants. (c) Dandelion seeds have a light parachute attached that easily catches the wind.

Analogous features

FIGURE 10.2.32 (a) The wings of a kingfisher allow the bird to fly and dive fast. This enables it to catch food, increasing its chances of survival. (b) A butterfly's wings allow the insect to travel between flowers in order to drink nectar.

Organisms can also show similarity that is not due to common ancestry. Flying animals such as butterflies and birds have wings (Figure 10.2.32). Fishes and dolphins have swimming appendages and a streamlined body for moving through water. Many burrowing animals have powerful feet. Each of these features can be seen as an adaptation to a particular lifestyle.

Anatomical structures that are found in different groups of organisms, such as wings in birds and butterflies, are described as **analogous features**; that is, they serve the same function but have evolved independently. Analogous features may evolve because unrelated organisms have experienced similar selective pressures. Thus, when biologists attempt to work out evolutionary relationships, they must distinguish between homologous and analogous features. Identifying features as analogous provides evidence of a convergent pattern of evolution but not evidence of descent from a common ancestor. **Convergent evolution** occurs when similar features evolve independently in unrelated groups of organisms.

Vestigial structures

Some organisms possess structures that have little or no function. These structures are often remnants of organs that had a function in an ancestral species but have become reduced in size over time and have ceased to be used. Such structures are referred to as vestigial organs or **vestigial structures**.

Examples of vestigial structures include pelvic bones in whales and pythons (Figure 10.2.33); the coccyx, ear muscles, wisdom teeth and inner eyelid in humans; and the reduced eyes of certain blind cavefish and salamanders. Structures such as the wings of flightless birds can be considered vestigial for flight, but in many cases, such as that of the ostrich, the reduced wings provide a new function of temperature regulation.

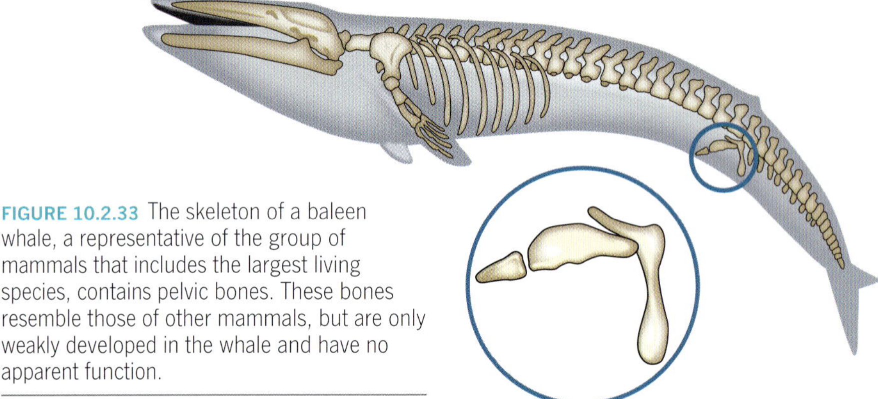

FIGURE 10.2.33 The skeleton of a baleen whale, a representative of the group of mammals that includes the largest living species, contains pelvic bones. These bones resemble those of other mammals, but are only weakly developed in the whale and have no apparent function.

BIOFILE

Homologous holes

By comparing skeletons, it was discovered that all dinosaurs and birds have a hole in their hip socket, known as an acetabulum, which allows the legs to descend straight down from the hips. No other four-legged vertebrate has this feature. The presence of an acetabulum is an example of a homologous feature in birds and dinosaurs—evidence that they shared a common ancestor.

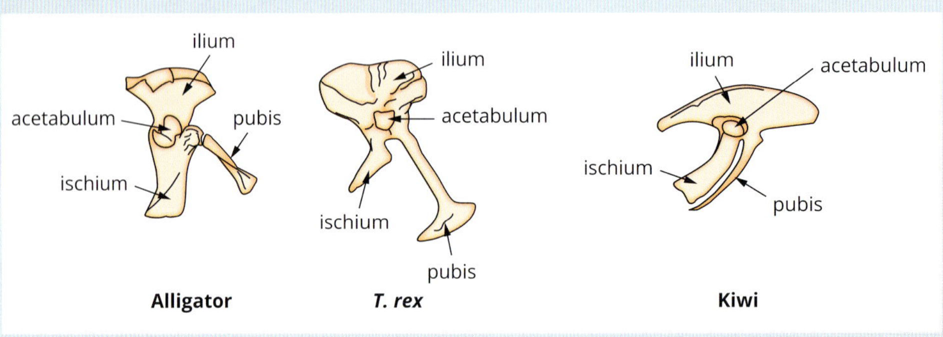

FIGURE 10.2.34 The acetabulum in the alligator, *Tyrannosaurus rex* and the kiwi, a bird.

DEVELOPMENTAL BIOLOGY

The comparison of organisms is complicated by the way organisms change substantially between life stages. The metamorphosis of a tadpole to a frog or a caterpillar to a butterfly are examples of this (Figure 10.2.35). The field of biology that studies the process of how an organism changes from a zygote to the form of an adult is called **developmental biology**. To study evolutionary relationships the same stage of development is compared between organisms.

FIGURE 10.2.35 Leopard lacewing (*Cethosia cyane euanthes*) life stage from caterpillar (top), to pupa (middle) and adult butterfly (bottom).

Comparative embryology

Homologous structures can sometimes be seen in the embryo of species but not in the adult form. When two gametes, an egg and sperm fuse, a zygote and eventually an embryo forms. The development of an embryo is controlled by a series of master genes that organise the position and rate of growth of cells (see Chapter 11). Organisms that shared a common ancestor often have similar master genes. This means that the embryos will pass through similar stages of development. A human embryo, for example, passes through a stage in which it has gill slits like those of a fish (Figure 10.2.36). **Comparative embryology** is therefore an important source of evidence of evolutionary relationships.

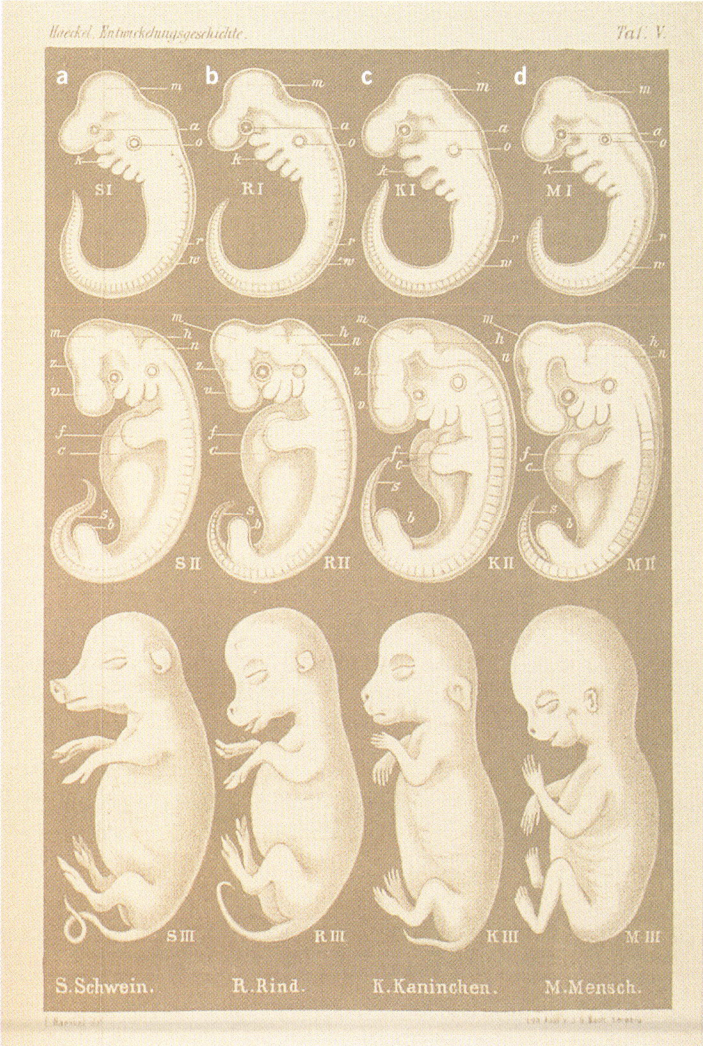

FIGURE 10.2.36 The embryos of four vertebrates in three comparative stages of development are remarkably similar and are used as evidence of common ancestry: (a) pig, (b) cow, (c) rabbit and (d) human. These drawings by the zoologist Ernst Haeckel were published in the 19th century. Haeckel admitted to a bit of artistic licence in his drawings (he curled the tails of some embryos and darkened the eye in a drawing of a chicken embryo, not shown here). Despite this, modern studies confirm his general observation that embryonic structures in early stages of vertebrate development are very similar; for example, all vertebrates at some stage develop pharyngeal slits, which later take on different functions in the different organisms.

10.2 Review

SUMMARY

- The fossil record is the record of the occurrence and evolution of living organisms through geological time as inferred from fossils.
- Palaeontology is the study of ancient life preserved in rocks and ancient sediments. These preserved remains include evidence of the structure of living organisms and how, where and when they lived.
- Fossils are the preserved remains, impressions or traces of organisms found in rocks, amber (fossilised tree sap), coal deposits, ice or soil.
- There are four main types of fossils.
  - Impression fossils are left when the entire organism decays but the shape or impression of the external or internal surface remains.
  - Mineralised fossils occur when minerals replace the spaces in structures of organisms such as bones, and minerals may eventually replace the entire organism leaving a replica of the original fossil.
  - Trace fossils (ichnofossils) are preserved evidence of an animal's activity or behaviour, without containing parts of the organism (e.g. footprints).
  - Mummified organisms are those that have been trapped in a substance and changed little.
- The process of forming impression fossils (fossilisation) involves:
  1. death of the organism
  2. burial of the organism by sediments
  3. the weight of many layers of sediments squeezing out water between the particles of sand, silt or mud
  4. soft sediments become solid rock—sandstone, siltstone, mudstone or shale (a mixture of clay and silt) as the deposit deepens, and pressure and temperature increase.
- Dating of fossils can be determined by:
  - relative dating based on stratigraphy, which places the age of a fossil according to, or relative to, the known age of layers or strata of rock above and below the layer of rock in which the fossil is found
  - using index fossils. Sometimes the only way to age a fossil bed is by using index fossils together with stratigraphy. Index fossils are commonly found fossils from similar sites for which an absolute age has been determined
  - absolute dating using radiometric methods to measure the proportions of particular naturally occurring radioactive isotopes (e.g. the ratio of carbon-14 to carbon-12)
  - thermoluminescence and electron spin resonance, two other methods used in absolute dating.

KEY QUESTIONS

1 What are fossils?

2 What types of organisms or parts of organisms are most likely to be fossilised? Why?

3 Describe some of the different ways in which organisms may be preserved relatively intact.

4 Explain the difference between an impression fossil, a cast fossil and a trace fossil.

5 Describe the following techniques used to date fossils and rock. Outline the applications or limitations of each.
- **a** relative dating
- **b** indicator fossils
- **c** absolute dating using radiocarbon dating

6 The half-life of carbon-14 is 5730 years.
- **a** If a fossil sample originally included 1.0 g of carbon-14, how much would be left after 5730 years?
- **b** How much would be left after 11 460 years?
- **c** How many years would it take for the amount to be 0.125 g?

7 Define 'biogeography'.

8 Define 'continental drift' and outline the biological evidence that supports this theory.

9
- **a** What are homologous structures?
- **b** Use an example of animals and an example of plants to explain the significance of homologous structures in terms of evidence of evolutionary relationships.

10 How are analogous features different from homologous features? Describe an example.

11 Identify the following as analogous or homologous features:
- **a** the wings of butterflies and the wings of birds
- **b** the flippers of whales and the fins of fishes
- **c** the arms of humans and the flippers of seals

12 Hearing is quite a complex process. In humans, and other mammals, it involves an outer membrane (the tympanic membrane or eardrum) and three bones (incus, malleus and stapes), also known as the auditory ossicles. The ear bones can transfer vibrations from the tympanic membrane to an inner fluid-filled chamber called the cochlea.

- Besides giving information on the appearance and structure of an organism, other information such as behaviour of the organism can be gained or inferred from examining fossils.
- Transitional fossils are any fossilised organisms that show intermediate traits and evidence of major change, such as animals moving from aquatic to terrestrial habitats.
- The fossil record is not a complete record of all past life because the chance of a fossil forming is small. Often only hard parts of organisms are preserved and only under certain environmental conditions. The fossil record is biased in favour of organisms that lived in shallow-water sediments.
- Biogeography is the study of the distribution of organisms.
- The world is divided into a number of continental biogeographic regions and marine zones, each with a unique set of animals and plants.
- Some patterns of distribution are best explained by continental drift—the theory that continents ride on crustal plates that gradually move.
- The geographic distributions and evolutionary age of organisms can be correlated with the time of separation of land masses.
- Structural morphology (also known as comparative morphology or comparative anatomy) is the study of the form and structure of organisms.
- Homologous features are those that have a fundamental similarity based on common ancestry. Often these evolve different functions. Homologous features between different organisms are evidence of evolutionary descent from a common ancestor (divergent evolution).
- Analogous features are those shared by different species that have the same function but have evolved separately due to similar selective pressures. Comparative anatomy can reveal that analogous features are structurally different.
- A vestigial structure is a reduced structure with no apparent function but is evidence of evolutionary relationship.
- Developmental biology is the field of biology that studies the process of growth and development of a zygote to an adult organism.
- Embryonic comparisons show that general features of groups of organisms appear early in development. More specialised features, which distinguish the members of a group, appear later in development.

The cochlea is lined with sensors that detect the vibrations in the fluid and transfer the information, via nerve signals, to the brain.

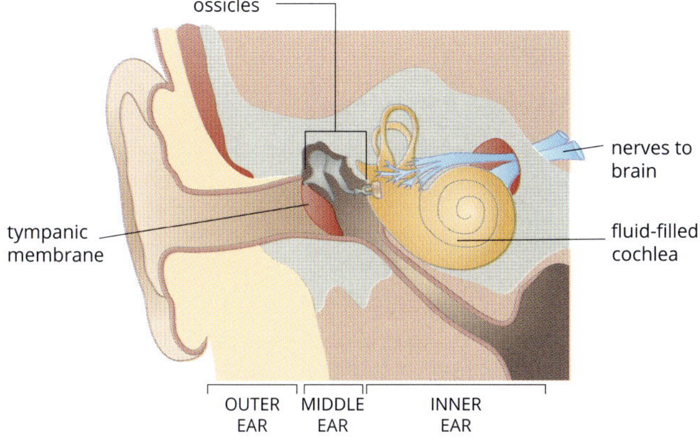

The mammalian incus and malleus evolved from parts of the jaw of fish and the stapes evolved from the hyomandibula bone. In fish this bone helps to support the gills.

Reptiles have a single bone, the columella, which evolved from the hyomandibula bone. The columella transfers the sound from the tympanic membrane to the fluid-filled inner ear.

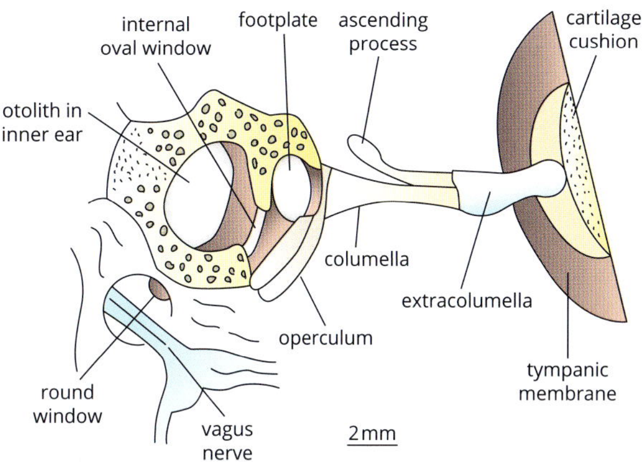

Explain whether the bones of the middle ear in mammals and reptiles are homologous or analogous.

13 What are vestigial structures? Give an example.

14 Outline how comparative embryology provides evidence for evolution. Give an example.

10.3 Patterns of biological change over time

FIGURE 10.3.1 Harlequin ladybirds (*Harmonia axyridis*) vary in colour and the number of spots.

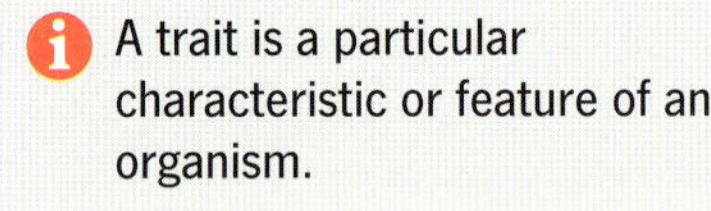

A trait is a particular characteristic or feature of an organism.

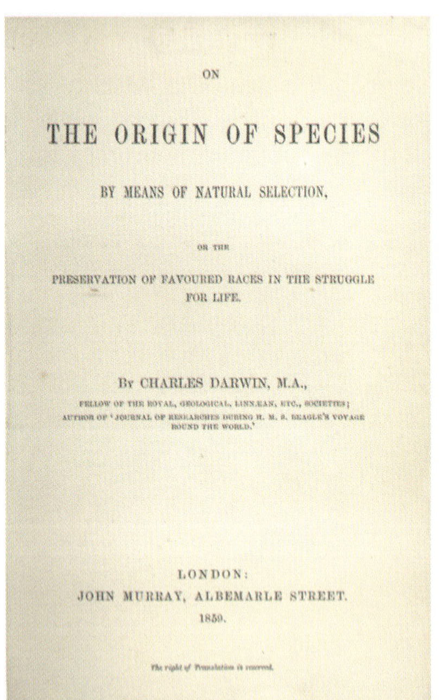

ON

THE ORIGIN OF SPECIES

BY MEANS OF NATURAL SELECTION,

OR THE

PRESERVATION OF FAVOURED RACES IN THE STRUGGLE FOR LIFE.

By CHARLES DARWIN, M.A.,

FELLOW OF THE ROYAL, GEOLOGICAL, LINNÆAN, ETC., SOCIETIES;
AUTHOR OF 'JOURNAL OF RESEARCHES DURING H. M. S. BEAGLE'S VOYAGE ROUND THE WORLD.'

LONDON:
JOHN MURRAY, ALBEMARLE STREET.
1859.

The right of Translation is reserved.

FIGURE 10.3.3 Charles Darwin published his book *On the Origin of Species by Means of Natural Selection, or the Preservation of Favoured Races in the Struggle for Life* in 1859.

In Section 10.2 you learnt that populations that become separated may diverge over time and eventually become different enough to be recognised as different species. We can recognise that these descendant species are related because they will share similar structures that have diverged to have different functions, called homologous features (e.g. the fish swim bladder and the land vertebrate lung are homologous). However, unrelated species may evolve similar features independently, called analogous features, as adaptations to a similar environment (e.g. the eyes of an octopus and a human are analogous).

In this section, you will learn about patterns of biological change over geological time, including divergent evolution, convergent evolution and mass extinctions.

MECHANISMS OF EVOLUTION

The celebrated naturalist Charles Darwin argued that species were not created in their present forms but had evolved from ancestral species. He also proposed a mechanism for evolution, which he termed natural selection, based on two key observations:

1 Members of a population often vary in their inherited traits (Figure 10.3.1).
2 All species produce more offspring than their environment can support, and most of these offspring fail to survive and reproduce (Figure 10.3.2).

Based on his two key observations, Darwin drew two inferences:

1 Individuals whose inherited traits give them a higher probability of surviving and reproducing in a given environment tend to leave more offspring than other individuals.
2 This unequal ability of individuals to survive and reproduce will lead to the accumulation of favourable traits in the population over generations.

Collectively, these proposals make up Darwin's theory of evolution by natural selection, also referred to as **Darwinian theory** (Figure 10.3.3).

FIGURE 10.3.2 The green turtle (*Chelonia mydas*) lays between 100 and 200 eggs in a single clutch, and in each season she may lay up to 8 clutches. However, not all the eggs will hatch, and most hatchlings do not survive to adulthood.

When two populations are separated by a permanent barrier such as continental drift, mountain uplift or river redirection, gene flow between the two populations is prevented. Different selection pressures and different mutations in the two populations cause the two populations to become genetically different. Eventually, the separated populations may accumulate different characteristics, become reproductively isolated and be recognised as two new species. This is an example of divergent evolution and allopatric speciation (see Chapter 9, page 351).

Allopatric speciation occurs when a species is geographically separated into two groups that are isolated from each other and diverge over time.

BIOFILE

Allopatric speciation of snapping shrimps

The Isthmus of Panama in central America, only arose approximately 3 million years ago. As a consequence of the Isthmus, two populations of snapping shrimp became separated with one population isolated on the Pacific Ocean side and the other on the Atlantic Ocean side. A study carried out by Nancy Knowlton and her colleagues of the Smithsonian Tropical Research Institute found that the shrimps on each side of the isthmus appeared almost identical to one another—they had once been members of the same population. However, when she put males and females from different sides of the isthmus together, they snapped aggressively instead of courting. The populations of snapping shrimp divided by the Isthmus had diverged and are reproductively isolated through their different courting behaviours. Biologists now recognise them as separate species, named *Alpheus nuttingi* (Atlantic) and *Alpheus millsae* (Pacific).

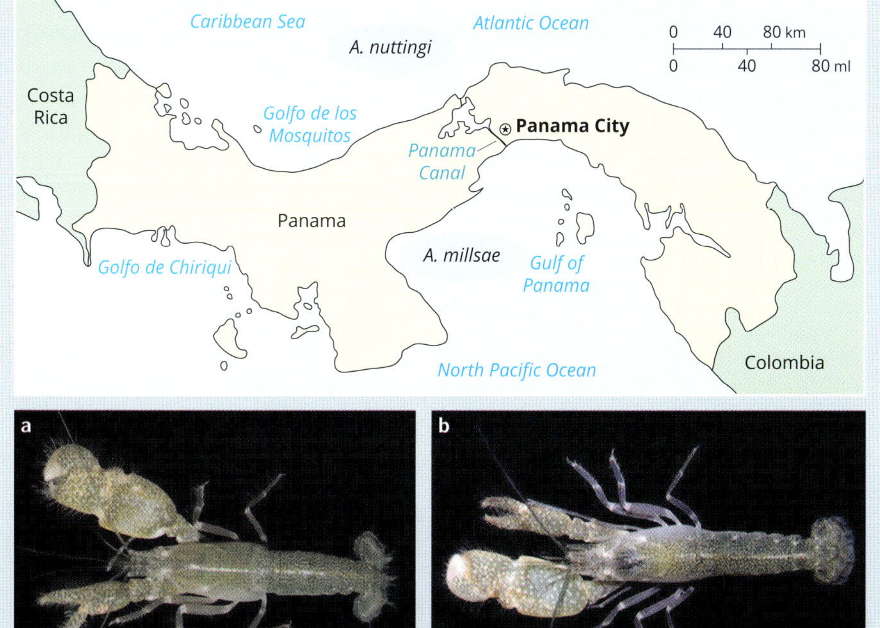

FIGURE 10.3.4 The Isthmus of Panama, a geographic barrier, and the distribution of snapping shrimp and other organisms isolated on either side. *Alpheus nuttingi* (a) *Alpheus millsae* (b).

Genetic drift occurs when a change in the genetic make-up of a population is caused by chance rather than natural selection. Genetic drift is associated with small founder populations or populations that experience a large reduction in numbers (bottlenecks).

DIVERGENT EVOLUTION

Divergent evolution is the evolution of two different species (or populations) from a common ancestral species (or population) (Figure 10.3.5). As time passes, natural selection or genetic drift may lead to the divergence of species or populations. Isolated species and populations accumulate genetic differences and their homologous features may become different, with different functions (e.g. the human arm and a bat's wing; see Section 10.2, page 397).

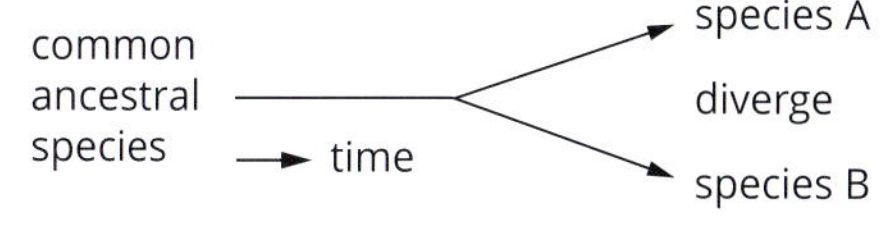

FIGURE 10.3.5 Divergent evolution.

FIGURE 10.3.6 Map of the Galapagos Islands, Ecuador.

Charles Darwin's finches are a classic example of divergent evolution. Darwin collected specimens of finches from the different Galapagos Islands when he sailed on the voyage of *HMS Beagle* (1831–36). The Galapagos Archipelago is an isolated group of volcanic islands about 1000 kilometres off the coast of South America in the eastern Pacific (Figure 10.3.6). When Darwin returned to England, the ornithologist John Gould examined the specimens and recognised them as related but as distinct species. Darwin had collected 13 different species of finch, with every island in the archipelago home to a number of species. Each species has a particular beak, body size and feeding behaviour that is advantageous for the conditions on the island on which they are found.

Adaptive radiation

Adaptive radiation is the rapid divergent evolution of a large number of related species from a single common ancestor. Adaptive radiation results from rapid speciation after organisms evolve different adaptations in response to new conditions and opportunities. This can occur following changes to the environment (e.g. extinction of competitors) or colonisation of a new environment where vacant ecological niches are available. Adaptive radiation can result in a wide diversity of species each with unique adaptations to their environment. Darwin's finches, which have beaks that are adapted to different food types, are an example of adaptive radiation (see Section 11.3).

Australia is well-known for its unique marsupials, a subclass of mammal. The family Macropodidae alone includes 10 genera and 65 species (with some recent extinctions) including kangaroos, wallabies, wallaroos, quokkas, pademelons and tree-kangaroos (Figure 10.3.7). Macropods diverged from a common marsupial ancestor approximately 53 million years ago, with modern kangaroos radiating from about 25 million years ago. For each genus today there are multiple species, with modern species adapted to a browsing or grazing lifestyle, with teeth and digestive system specialised for feeding on plant material. This example of adaptive radiation, like most, occurred as multiple divergences over millions of years.

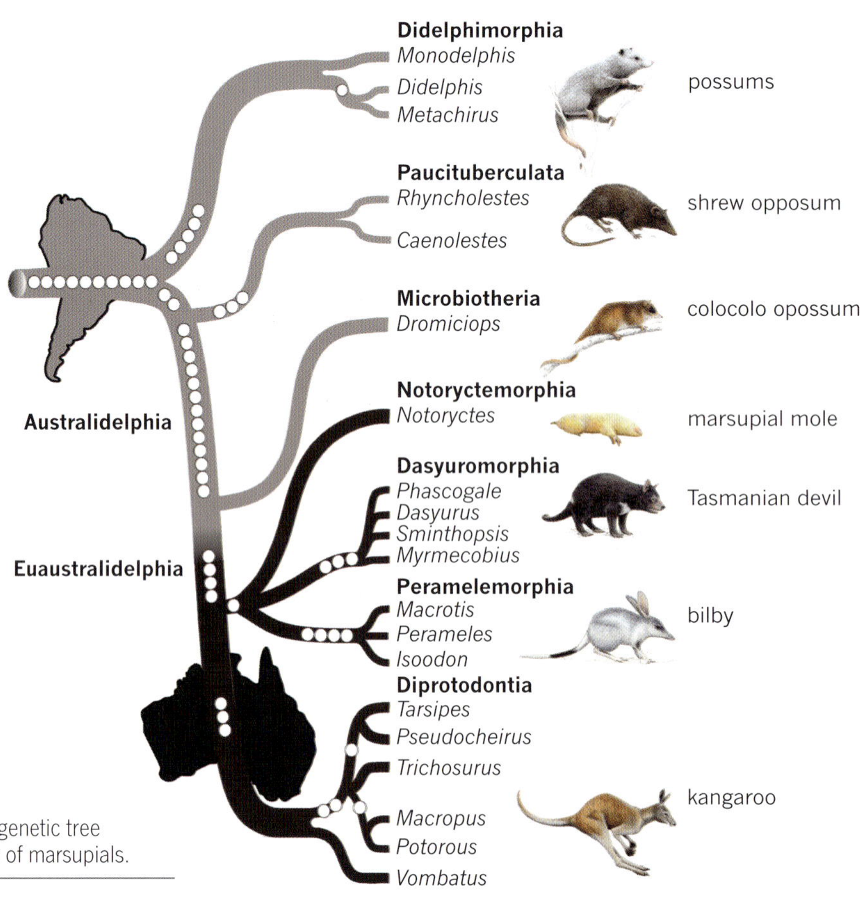

FIGURE 10.3.7 Phylogenetic tree showing the evolution of marsupials.

COEVOLUTION

Species that interact closely exert selection pressures on each other. Both species also experience similar environmental conditions. In such situations, **coevolution** can be seen, with the two species evolving together in a reciprocal response to selection pressures.

New variations of flowers appear through mutation, and these may be more likely to survive and produce seeds. As a result, some pollinators will be more suited to these flowers, and will therefore evolve alongside the flowers (Figure 10.3.8). Coevolution can also be observed in predator–prey relationships. When predators pick off the weaker prey, stronger individuals are left to reproduce. The next generation of predators will need to be stronger and faster to keep up with the strengthening features of the prey.

FIGURE 10.3.8 *Lonicera gracilipes*, an early spring flower, produces nectar collected almost exclusively by andrenid bees (*Andrena (Stenomelissa) lonicerae*). This is achieved by the long narrow corolla tube of the flower allowing the long tongue of the andrenid bees to reach the nectar but preventing pollinators with a short tongue from doing so.

CONVERGENT EVOLUTION

Convergent evolution is the evolution through natural selection of similar features in unrelated groups of organisms (Figure 10.3.9). Features with similar functions but which have evolved from different ancestral structures are termed analogous (see Section 10.2, page 398). Unrelated species that have adapted to a particular environment in similar ways are said to have converged, or become more alike.

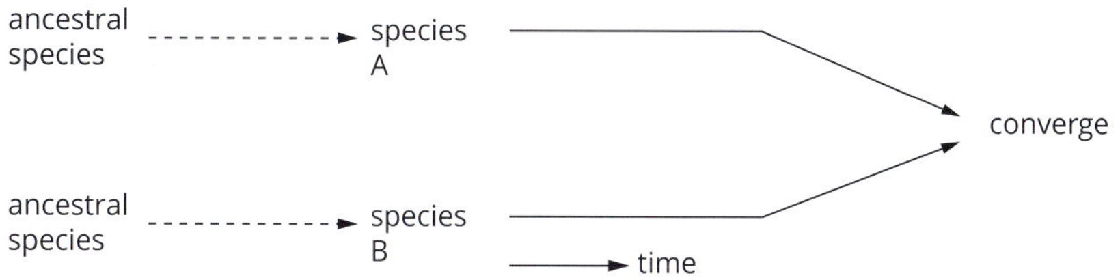

FIGURE 10.3.9 Convergent evolution.

The Australian marsupial sugar glider (*Petaurus breviceps*) and the American placental flying squirrel (*Glaucomys* sp.) have both developed large membranes between their fore and hind limbs that enable them to glide quite successfully (Figure 10.3.10). This is an example of two unrelated species that have converged due to similar environments and lifestyles despite different origins.

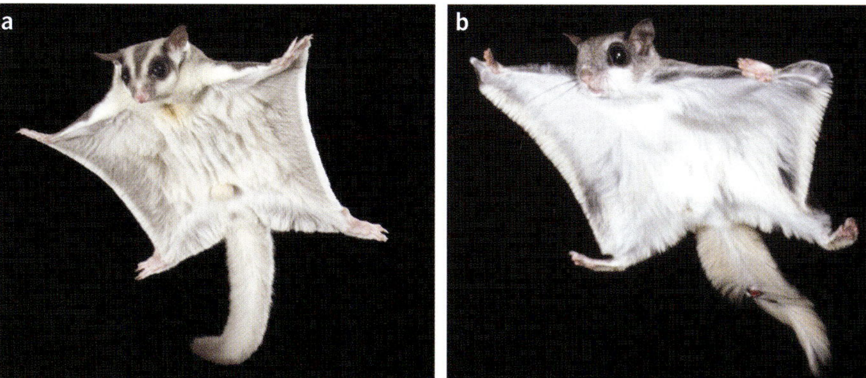

FIGURE 10.3.10 (a) The Australian marsupial sugar glider (*Petaurus breviceps*), and (b) the American flying squirrel (*Glaucomys* sp.) have many features in common, although they are not closely related. They are an example of convergent evolution.

The eyes of octopuses and vertebrates are a striking example of convergent evolution. Even though octopuses and vertebrates both have similar eye structures, the eyes have different origins (Figure 10.3.11).

The leaves of many Australian native shrubs (sclerophylls) in desert heathlands are hard, rigid, small and adapted to dry conditions. They resemble one another even though they are unrelated (e.g. pea family, banksia family, acacia family).

FIGURE 10.3.11 Eye of a giant octopus (*Enteroctopus* sp.).

BIOLOGY IN ACTION

Desert lizards

Dr Jane Melville is a research scientist at Museum Victoria. She is a herpetologist who specialises in the evolution of lizards. She has done extensive field-work in desert environments, comparing species within and between different lizard families based on DNA and morphological and behavioural characteristics.

Divergent Australian species

In outback Australia, the bearded dragon (*Pogona vitticeps*) and the earless pebble dragon (*Tympanocryptis cephalus*) are two related species classified in the same family (Agamidae). But they have diverged over time, look distinctive and behave quite differently (Figure 10.3.12a,b). The bearded dragon is a large animal with short limbs and spines. It is found over large areas of arid and semi-arid Australia, where it perches on tree limbs, stumps and fence posts. When threatened, it extends its spiny beard, opens its mouth and will make lunging movements towards its attacker. In contrast, the earless pebble dragon is a small animal with short limbs and tail. It is found in the stony deserts of Australia and is a stone mimic. When threatened, it freezes and crouches down so that its head and body resemble a stone and its tail a dry twig.

The Australian bearded dragon and the earless pebble dragon have different adaptations to living in Australian desert environments and are an example of divergent evolution.

Stone mimics on two continents

Jane Melville has also studied lizards in stony deserts in the south-west of the United States of America. The round-tailed horned lizard, *Phrynosoma modestum* (Figure 10.3.12c), looks remarkably like the Australian earless pebble dragon *Tympanocryptis cephalus* (family Agamidae) (Figure 10.3.12b), but it is in a different family (Iguanidae). The round-tailed horned lizard is also a stone mimic. When disturbed it freezes, closes its eyes and takes up a pose so that it resembles a small rock. Like the Australian earless pebble dragon, it will curve its back into a hump. Its dark side-markings resemble the shadows of a stone, and the tail banding makes the tail look like a dead twig.

Despite more than 100 million years of evolutionary separation and living on different continents, the Australian pebble dragon and American round-tailed horned lizard show significant convergent evolution in body shape and behaviour—adaptations that enable them to survive in hot, dry conditions and to live on stony desert plains.

FIGURE 10.3.12 (a) The Australian bearded dragon (*Pogona vitticeps*), (b) the Australian earless pebble dragon (*Tympanocryptis cephalus*, family Agamidae), and (c) the American round-tailed horned lizard (*Phrynosoma modestum*, family Iguanidae). The Australian earless pebble dragon (b) and the American round-tailed horned lizard (c) are both stone mimics, with similar adaptations to living in stony deserts in different parts of the world.

EXTINCTION

Species that fail to adapt to environmental changes or to compete for limited resources can die out. This loss of a species or groups of species is called extinction.

Background extinctions

The average rate of natural loss of species is called the **background extinction**. Extinction can occur as a result of changes in the physical environment or of changes in the ecological interactions between species, such as the arrival of a new predator or competitor. The average life of a species varies, depending on the type of organism, but is generally a few million years. Based on the fossil record, some marine animals appear to have existed for 5–10 million years, while mammals tend to last only 1 million years. The coelacanth (*Latimeria chalumnae*) is considered a living fossil because coelacanths as a group date back to the Devonian period of 400 million years ago (Figure 10.3.13).

FIGURE 10.3.13 The coelacanth (*Latimeria chalumnae*) inhabits steep rocky shores in the western Pacific and Indian Oceans, living at depths of 150–700 metres. An adult coelacanth may reach a length of two metres.

Mass extinctions

Throughout the fossil record, there is evidence of some significant mass extinction events. **Mass extinctions** are large-scale extinctions following disruptive changes to the global climate, or loss of sea or land due to the shifting of continents (plate tectonics). Whether these extinctions were caused by global warming or cooling, drowning with sea-level rise, asteroid impacts, volcanic eruptions or perhaps even disease, the aftermath of such large extinction events leads to changes in selection pressures that affect the number of surviving species.

Mass extinctions, like the one that wiped out most of the dinosaurs at the end of the Cretaceous period, led to 'empty' ecological niches. Remaining species may have taken advantage of the sudden availability of resources and reduced competition or predation. Over the course of many generations, natural selection will result in populations becoming more suited to new environments, and new species evolve. Major impacts can also result in surviving populations being small, and thus genetic drift in founding populations can lead to rapid evolution (see Chapter 11). Periods of significant diversification and adaptive radiation often follow mass extinction events.

BIOFILE

Drooping mistletoe

The drooping mistletoe, *Amyema pendulum*, is a flowering plant (family Loranthaceae) that is parasitic on eucalypts and acacias in Victoria. The leaves of the mistletoe hang vertically and closely resemble those of its host, particularly the drooping leaves of a eucalypt tree. The similarity of the leaves of the parasite and its host is a case of convergent evolution in response to harsh sunlight and dry conditions.

FIGURE 10.3.14 The drooping mistletoe, *Amyema pendulum*, parasitic on a eucalypt.

Five mass extinction events are evident from the fossil record, and in each more than 50% of hard-bodied marine species became extinct (Figure 10.3.15). For example, during the Permian period (299–252 million years ago), all of the continents came together and shallow continental seas were gradually lost. This large land mass (called Pangaea) caused reduced rainfall, temperature extremes, harsh conditions and the death of many species including the extinction of marine trilobites. The fossil evidence is particularly well documented as hard-bodied organisms fossilise well, and the environmental conditions of shallow seas are ideal for the fossilisation process.

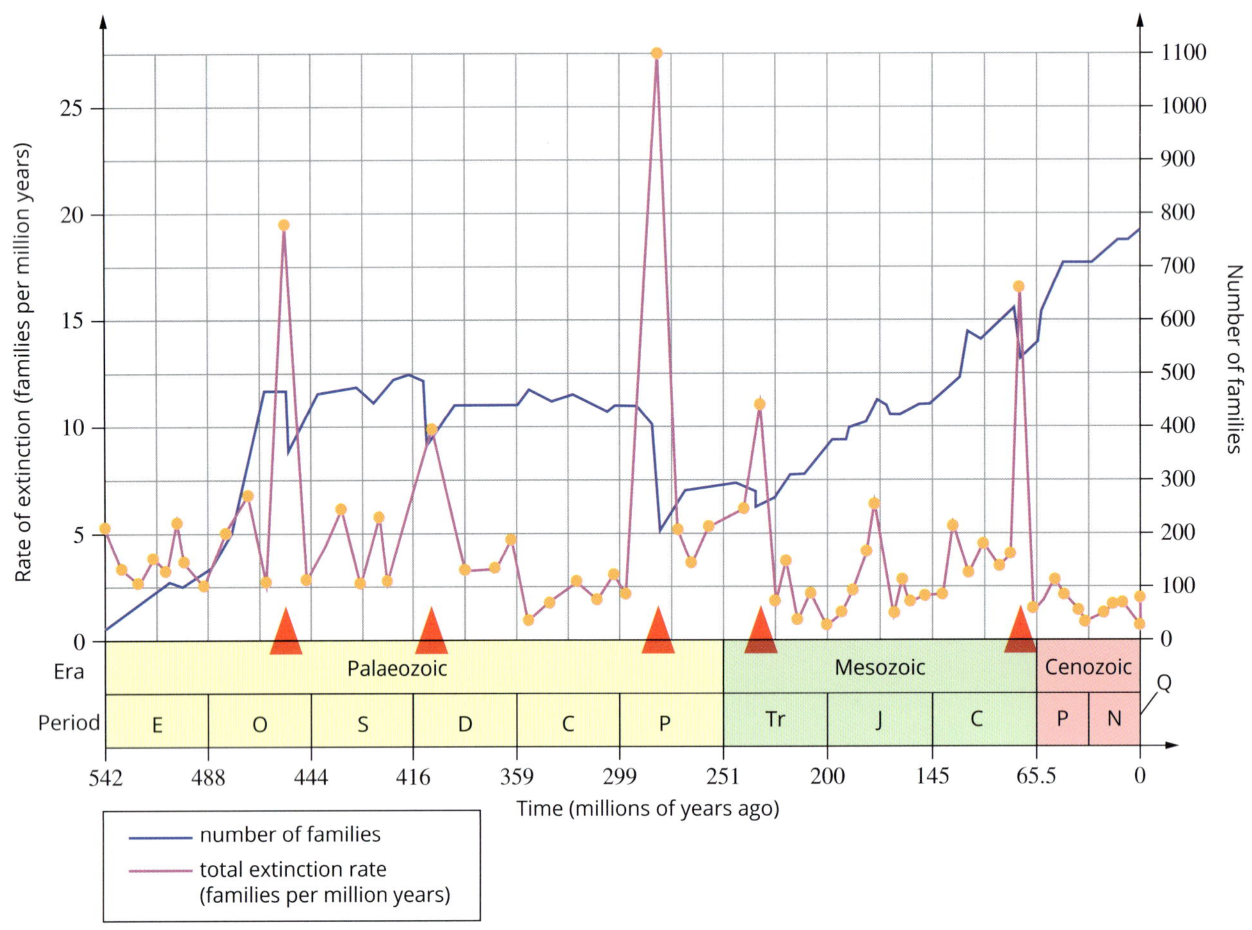

FIGURE 10.3.15 The five generally recognised mass extinction events, indicated by red arrowheads, represent peaks in the extinction rate of marine animal families (red line and left vertical axis). These mass extinctions interrupted the overall increase in the number of marine animal families over time (blue line and right vertical axis).

Recent extinctions

In recent times, humans have also been the cause of species extinctions at an accelerating rate. Since European settlement in Australia, dozens of Australian mammal species have been lost through a combination of factors, including the introduction of cats, foxes and the toxic cane toad. Plants, birds and other animal species have also become extinct as a result of land clearing, habitat loss and altered fire regimes.

10.3 Review

SUMMARY

- Natural selection is the process in which individuals that have certain inherited traits survive and reproduce at higher rates than other individuals because of those traits, and so pass on those traits.
- Genetic drift occurs when a change in the genetic make-up of a population (usually small) is caused by chance rather than natural selection.
- Allopatric speciation occurs when populations are geographically separated, leading to divergent evolution of the isolated populations and eventually new species.
- Divergent evolution is the splitting of an ancestral species (or population) into two different species (populations). The homologous features shared by the species (populations) become different over time (diverge), taking on different functions.
  - Adaptive radiation is a form of divergent evolution in which a common ancestor gives rise to a number of new species rapidly, particularly when a change in the environment opens new environmental niches.
- Coevolution occurs when two interacting species evolve together in a reciprocal response to selection pressures.
- Convergent evolution is the independent development of analogous features in unrelated species (or populations). The species (or populations) become alike over time (converge).
- Species do not last forever but become extinct over time. Background extinction is the average rate of natural loss of species. Extinction can occur as a result of changes in the physical environment or of changes in the interaction between species, such as the arrival of a new predator or competitor.
- Mass extinctions are large-scale extinctions following disruptive changes to global climates and land masses. Five mass extinctions are evident from the fossil record, and in each more than 50% of hard-bodied marine species became extinct.
- When a mass extinction occurs, remaining species may be able to take advantage of the changed environment and available resources. Over the course of many generations, natural selection will result in populations adapting to the various new environments, and new species evolve.
- In recent times, humans have caused an acceleration in species extinctions above the natural background rate.

KEY QUESTIONS

1. Briefly explain Darwin's theory of evolution.
2. Give some examples of conditions that can result in allopatric speciation.
3. Copy and complete the following table, summarising different patterns of evolution.

Pattern of evolution	Definition	Arrow diagram showing change over time	Example
divergent			
convergent			

4. Complete the following sentences.
Adaptive radiation is a form of ___________ evolution. It occurs when several species evolve from a single ___________ ancestor. Adaptive radiation is a result of rapid ___________ that can occur following changes to the ___________. Vacant ecological ___________ provide new opportunities and organisms may ___________ in different ways to different conditions and diversify.

5. There is striking similarity between the eye of a vertebrate, such as a shark, and that of an octopus.
   - **a** What is the evidence that these eyes are not homologous?
   - **b** What is the likely reason for the evolution of a similar type of eye in octopuses and vertebrates?
6. **a** Define 'mass extinction'.
   - **b** Explain how a mass extinction can result in new species evolving.

continued overleaf

10.3 Review *continued*

7 Katydids (also known as bush crickets) are a large group of insects found across the world. There are known to be over 6400 species grouped into nearly 50 genera, which belong to the family Tettigoniidae.

Researchers in South America were investigating hearing in an endemic species of katydid. They found that this particular species has solved the hearing problem in a very interesting manner. The katydid has its hearing organ on its legs. The hearing organ has an outer membrane attached to a stiff inner structure that vibrates in time with the outer membrane. The stiff structure taps on a fluid-filled chamber, which is lined with sensory cells that sense the vibrations and pass the information to the brain of the insect.

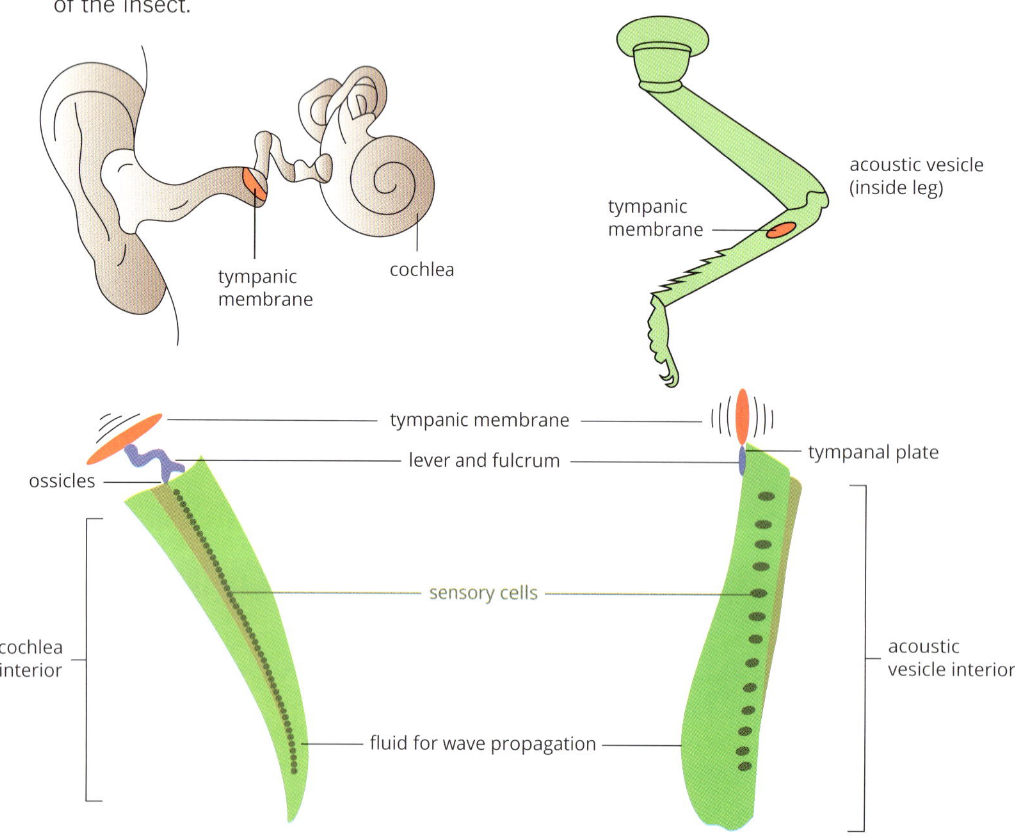

This method of hearing is the same as that seen in mammals.

a Hearing in mammals and katydids is achieved using similar processes, despite the fact that the structures involved are not considered to be homologous, explain why.

b What type of evolution is being shown in this situation? Explain your reasoning.

Chapter review

KEY TERMS

absolute dating (using radiometric methods)
adaptive radiation
analogous features
angiosperm
Archaean
background extinction
biogeographic regions
biogeography
cast fossil
coevolution
common ancestor
comparative embryology
comparative morphology
convergent evolution
Darwinian theory
developmental biology
divergent evolution
ecological niche
Ediacaran fauna
eon
epoch
era
fossil
fossil record
fossilisation
geological time scale
Gondwana
Hadean
homologous features
impression fossil
index fossil
isotopes
Laurasia
marsupial
mass extinctions
mineralisation (or petrification)
mineralised fossil
mummified organism
naturalist
palaeontology
Pangaea
period
Precambrian time
Proterozoic
relative dating
stratigraphy
stromatolite
structural morphology
tetrapod
trace fossil (ichnofossil)
vestigial structures

KEY QUESTIONS

1 Which of the following statements is it true to say of the diagram below of the history of the Earth?

Phanerozoic												Proterozoic	Archaean	Hadean
Cenozoic			Mesozoic			Palaeozoic								
Quaternary	Neogene	Palaeogene	Cretaceous	Jurassic	Triassic	Permian	Carboniferous	Devonian	Silurian	Ordovician	Cambrian			
↑ Q														↑ H

A Point Q occurred 2.6 million years ago and Point H occurred 4000 million years ago.
B Point Q occurred 100 000 years ago and Point H occurred 4000 million years ago.
C Point Q occurred 26 million years ago and Point H occurred 4500 million years ago.
D Point Q occurred 2.6 million years ago and Point H occurred 4500 million years ago.

2 Which of the following statements about carbon-14 dating is incorrect?
A It measures the rate of decay from carbon-12 to carbon-14.
B It requires organic matter to be present in the fossil.
C It is limited to dating fossils that are less than about 50 000 years old.
D It is an absolute measure of dating fossils.

3 Carbon-14 (^{14}C) decays to nitrogen-14 (^{14}N) with a half-life of approximately 5730 years. If a sample of material contained 10 000 atoms of ^{14}C 30 000 years ago, what is the approximate number of ^{14}C atoms that it will contain today?
A 5000
B 1250
C 312
D 156

4 Stromatolites are found today in the shallow, warm waters of Shark Bay in Western Australia. Which of the following statements is false?

A A stromatolite is a type of domed rock formed by layers of growth of cyanobacteria that trap sediments.

B A stromatolite is an accumulation of primitive eukaryotic marine fossils.

C Ancient stromatolites are fossil evidence of the early evolution of prokaryotes.

D Stromatolites represent evidence of early prokaryotic organisms.

5 The coelacanth (*Latimeria chalumnae*) is sometimes described as a living fossil. Coelacanths first appear in the fossil record 400 million years ago. Until 1938 they were thought to have become extinct around 65 million years ago in the K–Pg extinction event. Which of the following is most likely to explain the longevity of the coelacanth?

A They did not compete with each other for food.

B They did not have any predators.

C They produced many offspring.

D Their environment was stable and they were well adapted to it.

6 The long-nosed bat (*Leptonycteris yerbabuenae*) lives in areas near the border of Texas and Mexico. Its main food source is nectar and pollen produced by agave plants (*Agave parryi*). The bats have specialised feeding structures that allow them to reach the nectar. As they eat the nectar the bats become covered in pollen, which they then transfer to another agave plant thereby pollinating it. Many of these plants not only have flowers that are especially shaped for the feeding structures of the bats but they also flower at night when the bats are most active.

Leptonycteris yerbabuenae feeding on *Agave parryi*.

The relationship between the agave plant and the bats has occurred as a result of:

A convergent evolution

B divergent evolution

C parallel evolution

D coevolution

Use the following diagram to answer Questions 7 and 8.

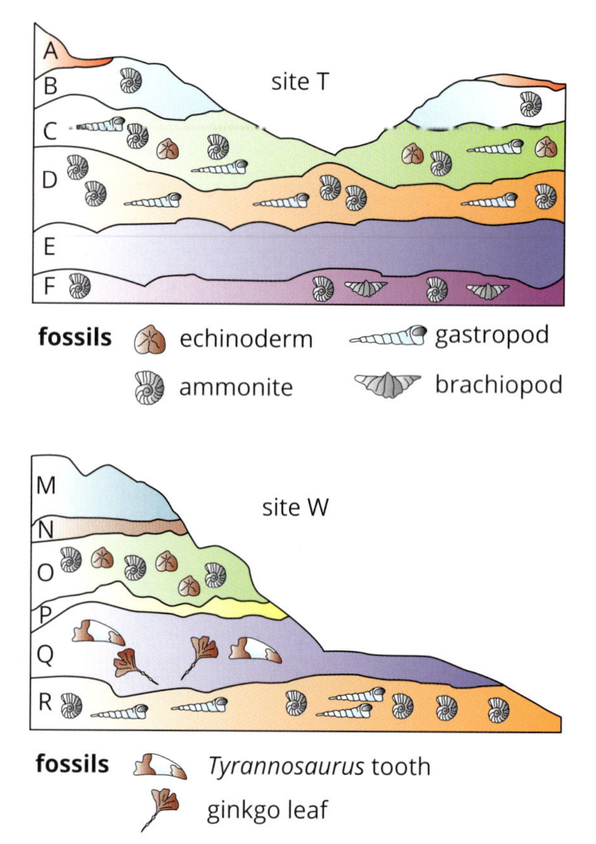

7 Fossils are found in sedimentary rocks. Barring major tectonic upheavals, it can be assumed that the layers have been laid down in order, with the oldest on the bottom.

One method of dating the layers of rock is to use indicator fossils (also called index fossils).

a i Which of the fossils observed at site T has the potential to be a good indicator fossil?

ii Explain your choice.

b Consider the two sites.

i Which layer is the most ancient?

ii How do you know?

c Stratum E at site T is sedimentary but there is no evidence that it contains any fossils. Why might there have been no fossils found in that stratum?

d At site W, stratum R contains ammonites and gastropods while stratum Q contains *Tyrannosaurus* teeth and *Ginkgo*. What does this suggest about the respective environments?

8 A method of absolute dating used for rock strata is radioactive dating. This method can only be used for igneous rocks. In order to date rocks by this method the amounts of a radioactive material in the rock and the decay products are measured. First the half-life for the radioactive material must be calculated.

a **i** What is a half-life?

ii One radioactive material (the parent) and its decay product (the daughter) used for this sort of dating is shown in the graph below. What is another radioactive material and its daughter?

iii A decay curve is generated for the radioactive material in order to determine the age of the rock. Use the graph to determine the half-life of potassium-40 (^{40}K).

b The rocks in strata N and P at site W were determined to be igneous in nature. They were both analysed to determine the percentage of ^{40}K left in the rock.

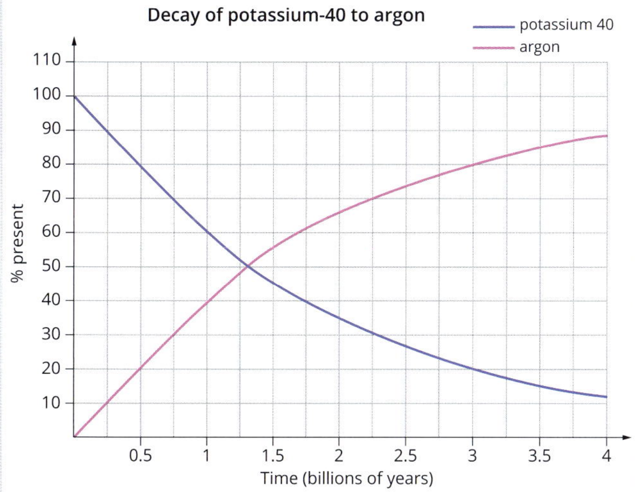

Use the graph to determine the age of the two strata if the rocks of:

i stratum N contains 90% ^{40}K

ii stratum P contains 75% ^{40}K.

c What is the likely age of stratum O?

9 **a** For each of the following fossils identify whether ^{14}C dating would be a suitable method to obtain an absolute date for the fossil:

i an insect preserved in amber that formed sometime during the Palaeogene

ii a dinosaur footprint

iii a hand axe used by a member of *Homo neanderthalensis* during the last days before they became extinct

iv an Egyptian mummy

b For each fossil for which ^{14}C dating is not suitable, explain why.

10 Suggest some reasons why the fossil record is incomplete.

11 The history of the Earth is divided into eons, eras, periods and epochs.

a Which of these spans the shortest length of time?

b The smallest subdivisions of the geological time scale are only seen during the Cenozoic. Explain why.

c Organise the following events from most ancient to most modern:

- adaptive radiation of mammals
- dinosaurs become extinct
- *Homo sapiens* evolve
- angiosperms evolve
- the great extinction event (up to 90% of all species died out)
- first algae

12 Consider the simple model of speciation illustrated.

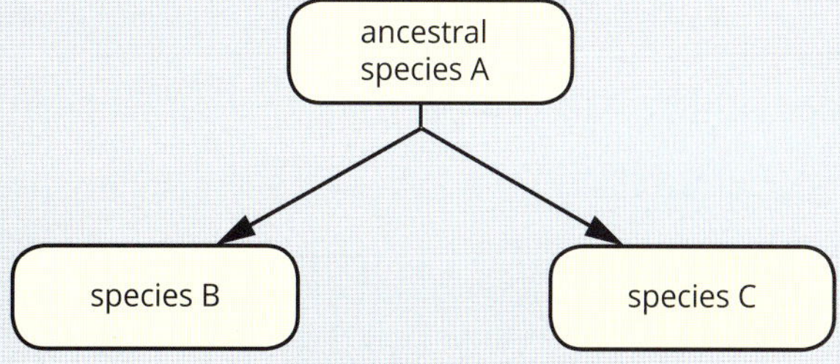

a Use your understanding of variation, isolating mechanisms and natural selection to explain how an ancestral species might evolve into two new species.

b What main criterion is applied to determine whether organisms belong to the same species or not?

c Suggest a mechanism that maintains reproductive isolation between two different but related species of:

i song birds

ii eucalypts in a stand of trees in a forest.

13 The red or lesser panda (*Ailurus fulgens*) was once thought to be closely related to the giant panda (*Ailuropoda melanoleuca*), both species occurring in China. Although members of the order Carnivora, both animals eat bamboo. However, genetic comparison has shown that red pandas are more closely related to racoons and giant pandas are related to bears.

a Suggest reasons for the traditional view that the red and giant pandas were related.

b Explain whether the similarity of each of the following pairs of animals is a result of divergent or convergent evolution:

i red panda and giant panda

ii red panda and raccoon

c Explain how DNA sequencing can be used to assess the relationship between different species of organisms.

14 Tree creepers are a group of birds native to New Guinea and Australia. Currently seven species, divided into two genera (*Cormobates* and *Climacteris*), have been described. These small birds are insectivorous. They forage for food on trees, feeding on insects living on or under the bark. This feeding habit gives the group its name.

White-throated treecreeper *Cormobates leucophaea*.

Tree creepers are a forest and woodland bird. In Australia, populations can become separated by areas of desert.

There are two species of *Cormobates*: *Cormobates leucophaea* (white-throated treecreeper), which is currently found in both northern Australia and New Guinea, and *Cormobates placens* (Papuan treecreeper), which is confined to New Guinea.

a **i** Using the theory of evolution by natural selection, explain how these two species could have evolved.

ii What kind of evolution is involved?

b There are also treecreepers in Europe. These birds look very similar to the Australian and New Guinean varieties and occupy the same niche. The European and Australasian treecreepers all belong to the order Passeriformes but within that order are as distantly related as it is possible to be and still be in the same order.

Eurasian treecreeper.

i Explain how the European and Australasian treecreepers have come to look and behave so similarly.

ii What is the name given to this type of evolution?

15 The wings of bats and birds are both homologous and analogous. Explain how they can be both.

16 Whales have vestigial pelvic bones.

a What is meant by a 'vestigial structure'?

b What do the vestigial pelvic bones suggest about the evolution of whales, and to which animal group they are related?

17 Cane toads (*Rhinella marina*, previously classified as *Bufo marinus*) were introduced into Australia in the 1930s to eat cane beetles that attacked sugar cane in central and northern Queensland. This introduction has proved to be an ecological disaster. The toads did not eat the beetles, but ate many small native animals, including frogs and small mammals, and native predators that eat the cane toads frequently die because cane toads produce a very toxic chemical in their skin.

Cane toads are large. On average their bodies are over 10 cm in length (not including limbs) and they can weigh in excess of 160 g.

Researchers assessing the ecosystems that have been invaded by the fast-spreading toads have noticed an interesting phenomenon in some snakes. Snakes commonly feed on frogs and toads. Two species of snake are particularly susceptible to cane toad poison, the red-bellied black snake (*Pseudechis porphyriacus*) and the green tree snake (*Dendrelaphis punctulatus*). The researchers have observed that in areas invaded by cane toads the populations of both of these snakes are showing an average increase in body size and decrease in mouth gape (how wide the mouth can be opened).

Red-bellied black snake (*Pseudechis porphyriacus*).

Green tree snake (*Dendrelaphis punctulatus*).

Snakes have a generational time of about three years so between 20 and 25 generations would have occurred since the introduction of the toads.

a Suggest how larger body size and smaller mouths could be an advantage in an environment containing cane toads.

b Is this an example of natural selection? Explain your reasoning.

18 Ammonites are an extinct group of molluscs. The last known species died out in the K–Pg event. Ammonites were cephalopods like octopus and squid but their closest living relative is the nautilus.

Nautilus.

a i What was the K–Pg event?

ii What is one suggested cause of this event?

b The picture below shows fossils of two different species of ammonite.

Asteroceras confusum (large ammonite) and *Promicroceras planicosta* (small ammonite).

Other than size, how might palaeontologists have determined that these two ammonites are different species?

c Why might size be a poor feature to use to determine that two individuals are from different species?

19 Concentrations of atmospheric gases have changed over geological time. One view of these changes is represented on the graph below. Although there is consensus over general trends in atmospheric changes, there is no agreement of the small details.

a Why is there a lack of consensus about the details of atmospheric conditions in the early Earth?

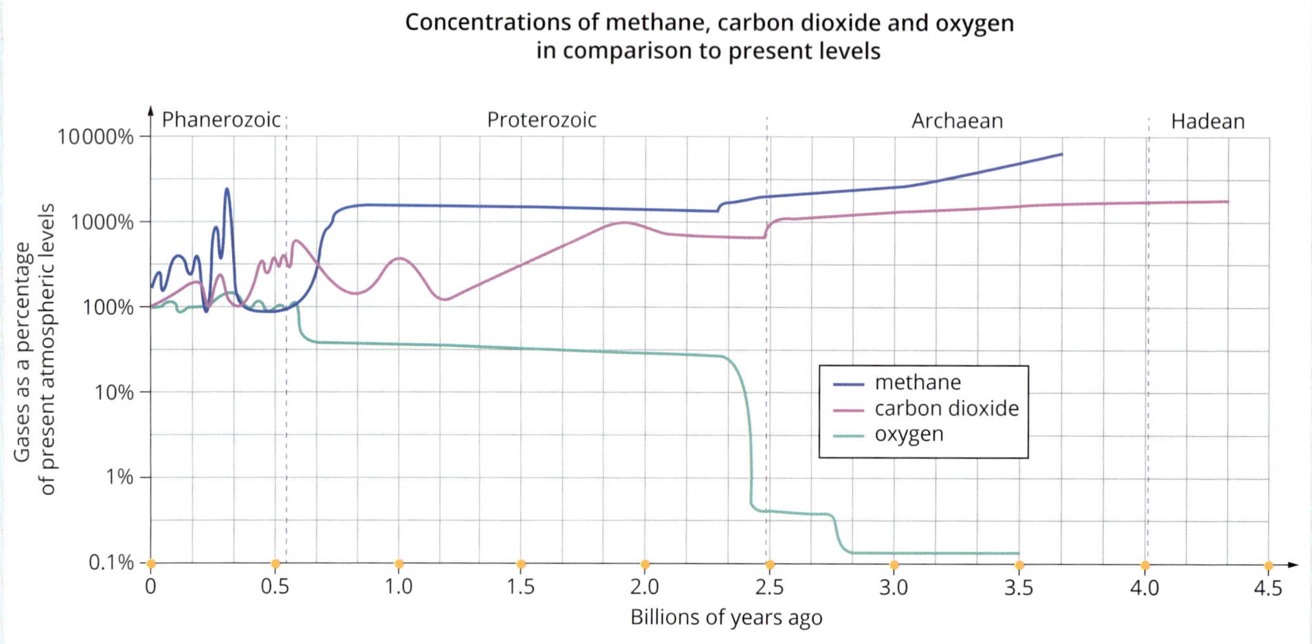

b The graph above uses a logarithmic scale so care needs to be taken with reading values.

The table below shows the concentration of the three gases in the atmosphere today.

Gas	Concentration in parts per million
methane	1.8
carbon dioxide	398
oxygen	210 000

Use the graph to determine the oxygen concentration (in parts per million) 3 billion years ago.

c What evidence can be seen in the graph that supports the contention that photosynthetic organisms arose approximately 2.8 billion years ago?

d The Great Oxygenation Event saw a rapid rise in oxygen levels to levels approaching modern values.

i When did this occur?

ii Which organisms were responsible for the Great Oxygenation Event?

CHAPTER

11 Determining relatedness between species

Learning outcomes

Humans have long contemplated where we came from. Of the many forms of evidence that support evolution, molecular homology is the most recent. Comparing similar molecules, such as genes or proteins, helps us determine which organisms are more closely related. It also allows us to understand how chance mutations in key master regulatory genes that regulate developmental processes can result in large changes in the body structure of an organism. Such significant changes in the structure of an organism may lead to the rapid evolution of new species.

By the end of this chapter, you will have an understanding of how molecular homology provides evidence of relatedness between species and how phylogenetic trees are used to show relatedness between species. You will also be able to describe how new phenotypes can arise from changes in the expression of a few master regulatory genes.

Key knowledge

- molecular homology as evidence of relatedness between species including DNA and amino acid sequences, mtDNA (the molecular clock) and the DNA hybridisation technique
- the use of phylogenetic trees to show relatedness between species
- the evolution of novel phenotypes arising from chance events within genomes, specifically sets of genes that regulate developmental processes and lead to changes in the expression of a few master genes found across the animal phyla, as demonstrated by the expression of gene *BMP4* in beak formation of the Galapagos finches and jaw formation of cichlid fish in Africa.

11.1 Molecular homology

Before molecular techniques were available, structural (or morphological) and functional similarities were the main evidence used to determine relatedness. Humans, chimpanzees, gorillas and orangutans each have forelimbs with hands and five digits (fingers). These morphological similarities can be used as evidence to support the theory that humans are related to these great apes and are descended from a common ancestor.

However, anatomical similarities can also be due to convergent evolution of organisms that did not share a recent common ancestor. A shark and a dolphin, for example, have a similar shape suited to living and swimming in water (Figure 11.1.1). Despite an overall similarity of appearance due to living in similar environments and experiencing similar selection pressures, sharks and dolphins have very different DNA sequences, indicating that they are not closely related. Similar features can result from shared ancestry (homologous) or from independent evolutionary paths (analogous) (see Chapter 10).

In this section, you will learn how amino acid and DNA sequences are used as evidence of relatedness between species, how comparison of mitochondrial DNA (mtDNA) sequences can be used to indicate the time that has passed since the existence of a shared common ancestor, and how the DNA hybridisation technique is used to compare DNA molecules of different species.

> Homologous features are similar because the organisms share a common ancestor. Homologous features are retained through divergent evolution (divergence from a common ancestor).

> Analogous features have similar structure and function but evolved independently and do not share a recent common ancestor (e.g. bird and bat wings). This is a result of convergent evolution.

FIGURE 11.1.1 Although the shark (a) and the dolphin (b) appear to have similar features, they have very different nucleotide sequences in their DNA. This indicates that they are not closely related and the anatomical similarities between these species are due to convergent evolution rather than shared ancestry.

DETERMINING RELATEDNESS BETWEEN SPECIES USING MOLECULES

If species have a very similar set of proteins, chromosomes or DNA sequences, it is evidence that they shared a recent common ancestor. 'Recent', in evolutionary terms, may be hundreds of thousands, even millions, of years.

All living organisms on Earth once shared a common ancestor. If two populations become isolated from each other, they will accumulate different mutations in their DNA. As time passes, the sequence of nucleotides in their DNA becomes more different and what was once similar DNA gradually diverges (Figure 11.1.2). The more mutations that accumulate in the DNA sequences between two species, the more time will have passed since the two species diverged from their common ancestor. For example, there are more differences between the DNA sequence coding for the cartilage protein of a frog and a dog, than between the DNA sequence coding for the cartilage protein of a frog and a toad. This is because a dog and a frog have a more distant common ancestor than a frog and a toad and have therefore had more time to accumulate genetic changes.

Changes in nucleotide sequences are caused by mutations. When a cell copies its DNA, it may make errors. Usually these errors are repaired before mitosis occurs. Occasionally these errors are not repaired and become a permanent part of the genome. This is a mutation. If these mutations occur within the germ line cells, then they can be passed on to the next generation (see Chapter 9).

Mutations occur regularly as a species evolves—like 'molecular clockwork'. As time passes, mutations accumulate in DNA, resulting in genetic differentiation and divergence of species.

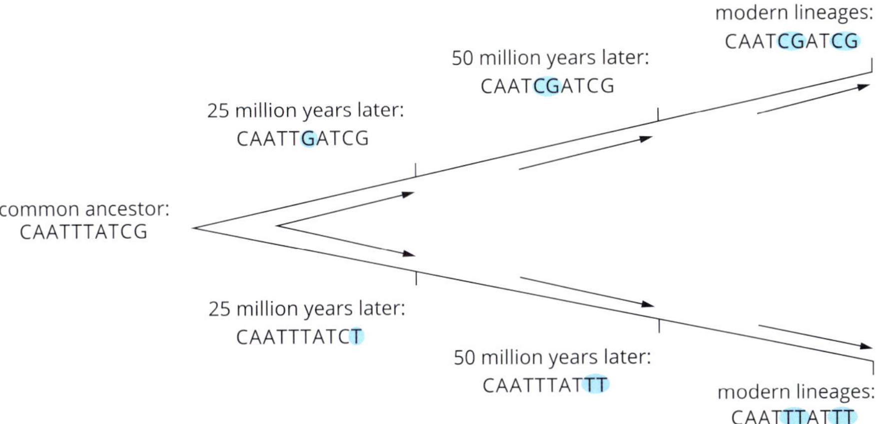

FIGURE 11.1.2 When species diverge, they start accumulating different mutations in their nucleotide sequences. The more time that passes, the more mutations they accumulate.

DIFFERENCES IN AMINO ACID SEQUENCES

As two species diverge from a common ancestor, they accumulate different mutations in their DNA and start accumulating differences in the amino acid sequences of their proteins. The more time that has passed since the two species diverged from the common ancestor, the more differences there are between their amino acid sequences (Figure 11.1.3).

Histone H1 (residues 120–180)

```
human KKASKPKKAASKAPTKKPKATPVKKAKKKLAATPKKAKKPKTVKAKPVKASKPKKAKPVK
mouse KKAAKPKKAASKAPSKKPKATPVKKAKKKPAATPKKAKKPKVVKVKPVKASKPKKAKTVK
  rat KKAAKPKKAASKAPSKKPKATPVKKAKKKPAATPKKAKKPKIVKVKPVKASKPKKAKPVK
  cow KKAAKPKKAASKAPSKKPKATPVKKAKKKPAATPKKTKKPKTVKAKPVKASKPKKTKPVK
chimp KKASKPKKAASKAPTKKPKATPVKKAKKKLAATPKKAKKPKTVKAKPVKASKPKKAKPVK
      *** ********** ************** ****** **** ** ********** * **
```

non-conserved amino acids: conservative; non-conservative; semi-conservative

FIGURE 11.1.3 Amino acid alignment of part of the mammalian histone protein family, H1. Amino acids for which there has been no change are indicated by an asterisk. Conservative amino acid substitutions do not cause a change in the protein. Semi-conservative amino acid substitutions result in an amino acid being replaced by one that is similar in structure but has different biochemical properties. This may lead to a change in the protein. Non-conservative mutations result in an amino acid being replaced by one that is very different and this usually leads to major changes in the protein and its function. The organisms that share a more recent common ancestor have fewer differences in their amino acid sequences (e.g. mouse and rat compared to rat and cow).

Sometimes point mutations, such as nucleotide substitutions, insertions or deletions, in the DNA sequence may not cause a difference in the amino acid sequence. This is because the genetic code is degenerate, which means more than one codon codes for the same amino acid; for example, GUU, GUC, GUA and GUG all code for valine. Consequently, differences in amino acids accumulate more slowly than differences in DNA.

Not all DNA mutations lead to changes in amino acids and proteins. This is because one DNA triplet codes for more than one amino acid.

Even when a point mutation leads to a change in an amino acid, the mutation may not lead to a change in phenotype. The effect an amino acid substitution has on the function and structure of a protein depends on how biochemically similar the substituted amino acid is to the original amino acid in properties such as size, hydrophobicity and charge (Figure 11.1.4). When an amino acid is substituted for another with biochemically similar properties (e.g. glycine for alanine), it is called a **conservative substitution** and does not cause a change in the protein. A **semi-conservative substitution** leads to the replacement of an amino acid with one with a similar shape but different biochemical properties (e.g. alanine for cysteine), possibly leading to a change in protein structure and function. A **non-conservative substitution** results in a substitution with a very different amino acid with different biochemical properties, which often leads to major changes in the protein and its function (e.g. cysteine for arginine).

The order of the nucleotides in the DNA indicates the evolutionary relationship between species more accurately than the amino acid sequence. All mammals, for example, produce milk containing the protein casein, suggesting that all mammal species have the same gene for this protein. However, if the DNA sequence that makes up this gene is compared, slight differences might be observed. For this reason, and because it is now technically easier and cheaper to analyse nucleic acids than proteins, DNA comparisons are the preferred type of data.

Table 11.1.1 shows the number of amino acid differences in the cytochrome *c* molecule between humans and selected organisms, thereby comparing the relatedness of humans to each of these species. For example, there are 51 amino acid differences in the proteins sequences of cytochrome *c* between yeast and humans. This indicates a low level of relatedness. Cytochrome *c* is an important part of the electron transport chain and is essential to an organism's survival.

	Human	Rhesus monkey	Horse	Donkey	Sheep	Dog	Yeast
Human	0						
Rhesus monkey	1	0					
Horse	12	11	0				
Donkey	11	10	1	0			
Sheep	10	9	3	2	0		
Dog	11	10	6	5	3	0	
Yeast	51	51	51	50	50	49	0

TABLE 11.1.1 The number of amino acid differences in the cytochrome c molecule of different organisms.

FIGURE 11.1.4 The properties of amino acids, including the hydrophobicity (non-polar or polar side chains), charge (acidic or basic) and size, influence how amino acid substitutions affect protein structure and function.

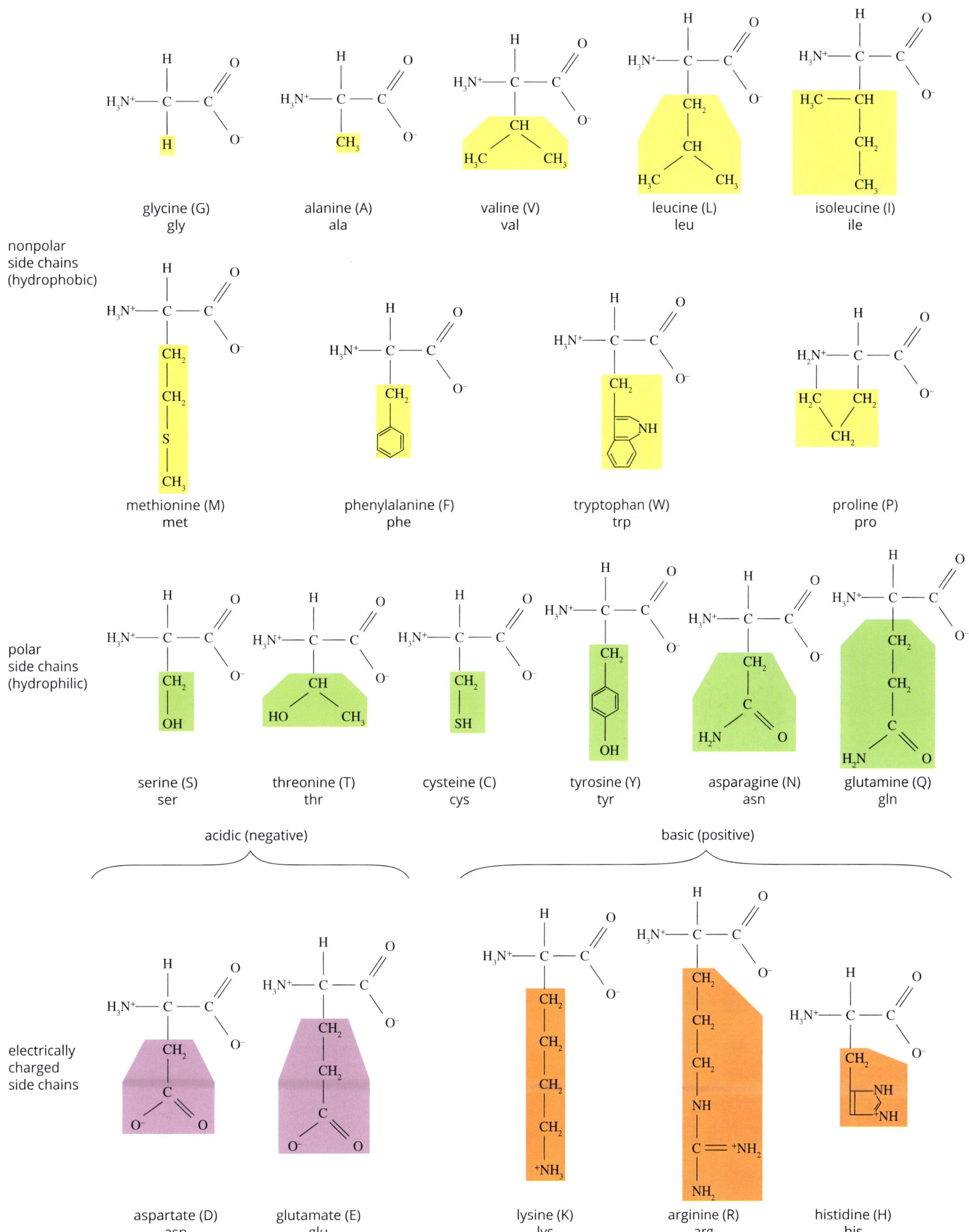

EXTENSION

Chromosomes in common

Homologies between chromosomes of different species can also be used as evidence of relatedness between species. To analyse condensed chromosomes, they are first stained with a dye called Giemsa stain. Some regions of the DNA take up more stain than other regions and appear as dark horizontal bands (Figure 11.1.5). The pattern of bands reflects the structure of the chromosome.

Chromosomes can be identified by three features:

- their length
- the location of the centromere
- the dark sections of the chromosome (banding).

Organisms that shared a recent common ancestor will have many chromosomes in common. A common ancestor is evident when comparing length and banding of human and chimpanzee chromosomes. Although humans have 46 chromosomes and chimpanzees 48, there are many similarities in the banding of the chromosomes of the two species (Figure 11.1.6). The difference in the number of chromosomes can be accounted for by comparing human chromosome 2 with two smaller chimpanzee chromosomes, which have the same banding pattern. Other apes that diverged from a common ancestor to the chimpanzees also have these two chromosomes (Figure 11.1.6). This indicates that in a common ancestor the two smaller chromosomes may have joined together, giving rise to one of the early human ancestors, and leading to the divergence of the two lineages.

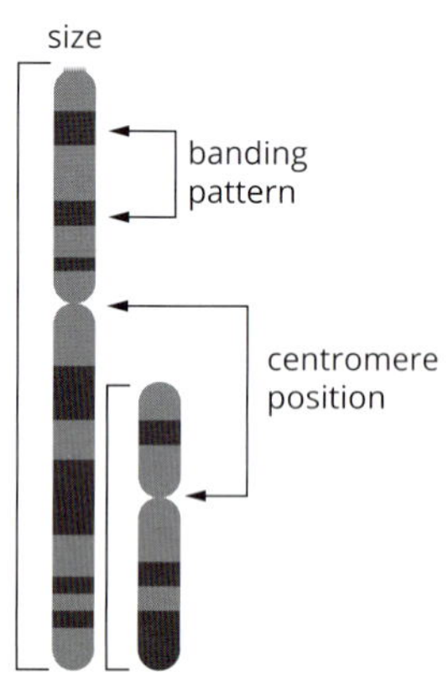

FIGURE 11.1.5 When preparing a karyotype, scientists compare the number and size of the chromosome, the banding pattern and the position of the centromere.

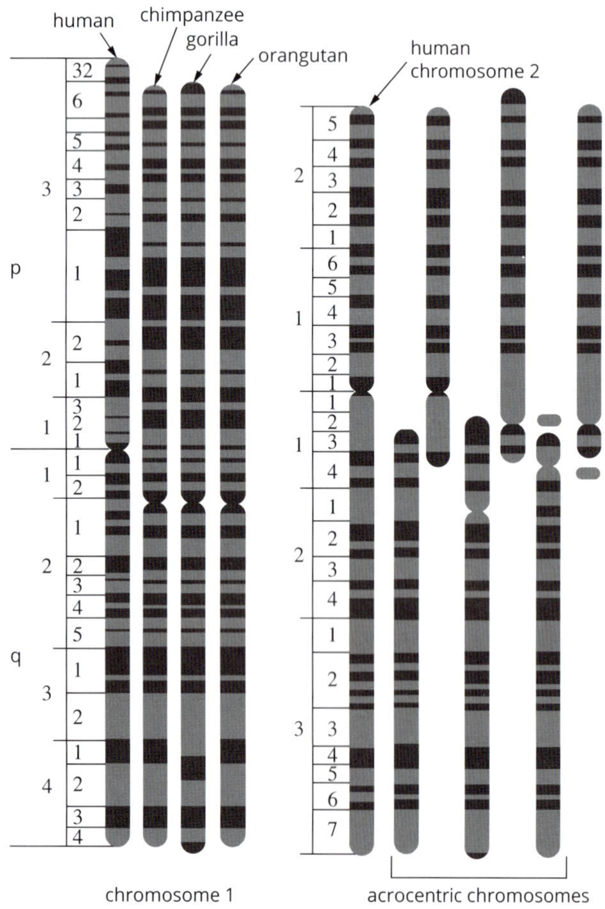

a

FIGURE 11.1.6 A comparison between the karyotypes of humans (a), chimpanzees (b), gorillas (c) and orangutans (d) reveals many similarities between these species. Human chromosome 2 has similar banding patterns to two chromosomes of the chimpanzee, gorilla and orangutan. (Acrocentric chromosomes are chromosomes with centromeres near the end.)

MOLECULAR CLOCKS

A **molecular clock** is a technique that uses the rate of accumulation of mutations in DNA to calculate how long ago organisms diverged from one another. The molecular clock hypothesis is the basis of this technique and was first proposed in the 1960s by Emile Zuckerkandl and Linus Pauling. This hypothesis states that changes in DNA and proteins are constant over evolutionary time and across different lineages.

The change in DNA over time is also known as the **mutation rate** and can be expressed as the number of nucleotide changes that occur every million years. The molecular clock hypothesis can be applied by calculating the rate of mutation of a region of DNA, along with the number of differences between the DNA of two organisms, and using this information to estimate how long ago they diverged (Figure 11.1.7). The more unique mutations each species accumulates, the more time has passed since they shared a common ancestor.

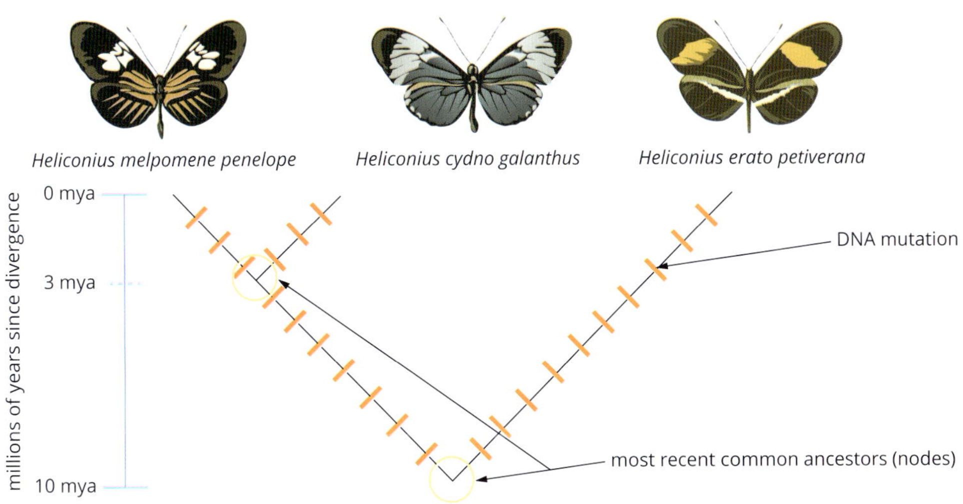

FIGURE 11.1.7 This diagram illustrates the evolutionary relationships between three butterfly species. Each mutation in the DNA sequences used to construct this evolutionary tree is represented by a red line. The time scale (millions of years) over which the divergence occurred is to the left of the tree. The mutation rate can be calculated from the number of mutations that have accumulated over the evolutionary time period. The most recent common ancestors are represented by the nodes (branching points).

In order to estimate how long ago two lineages diverged, the molecular clock is calibrated using evidence from the fossil record. Techniques such as radiometric dating and stratigraphy (the study of rock layers) are used to date fossils (see Chapter 10). The molecular clock for a particular gene can then be calibrated by comparing the number of differences in DNA sequences with the dates of evolutionary branch points known from the fossil record of similar organisms.

Limitations of molecular clocks

The molecular clock is a useful **phylogenetic** tool but it does have some limitations. One of its major limitations is the assumption that the rate of genetic change is constant and therefore accurately represents evolutionary time (Figure 11.1.8). Although we know that genetic difference represents evolutionary distance and that both these factors are positively correlated with time (i.e. the greater the difference between organisms, the more time has passed since they last shared an ancestor), the rate of genetic change over time is not constant. This means that genetic change is not an accurate measure of time from which we can determine exact dates of lineage divergence.

In order for genetic changes to occur at a constant rate, those changes (mutations) need to be neutral or not affected by natural selection. This needs to be considered when applying a molecular clock to genetic data. Any DNA regions that code for the phenotype of the organism (i.e. its structure or function) are under natural selection and will change according to outside selection pressures. Therefore the mutation rate of proteins and protein-coding DNA (genes) will not be constant.

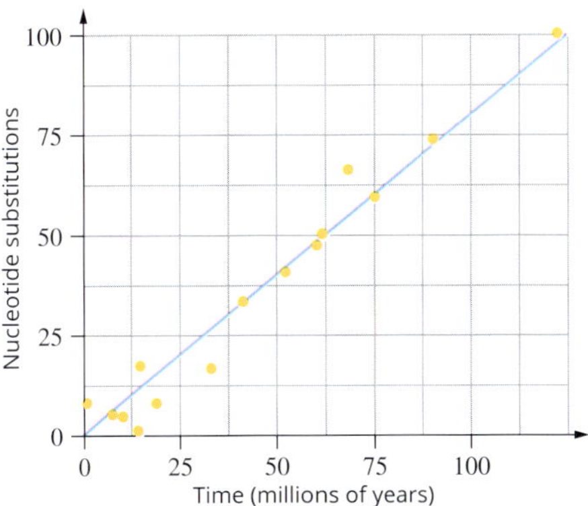

FIGURE 11.1.8 The constant rate of genetic change (nucleotide substitution) over time (millions of years) assumed under the molecular clock hypothesis. Regions of neutral genetic variation undergo relatively constant rates of genetic change; however, this does not apply to many coding regions of DNA.

Some sections of DNA mutate more frequently than others: that is, at a faster rate. This means there are different molecular clocks within an organism, ticking at different speeds in different regions of DNA. Genes that are essential to an organism's survival, such as those that code for cytochrome *c*, very rarely accumulate mutations and the gene sequence is therefore highly conserved (mostly unchanged) throughout evolution (Figure 11.1.9). Any variations to the sequence of these essential genes may result in these proteins losing function and the organism dying. For example, changes in cytochrome *c* may cause the electron transport chain to fail, preventing the formation of ATP. As a result, the organism will die. Sections of DNA that are not essential to the survival of an organism accumulate mutations at a faster rate. As a consequence, there are many molecular clocks within each organism that run at different rates.

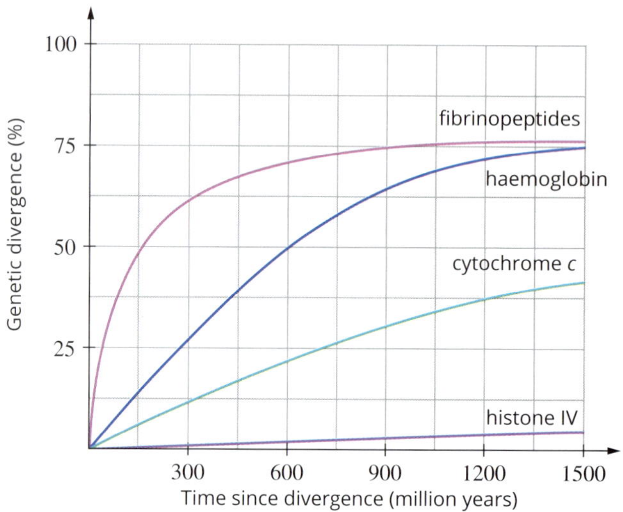

FIGURE 11.1.9 Different proteins mutate at different rates. This graph shows the rate of genetic divergence (%) of different proteins over evolutionary time (millions of years since divergence). The graph shows the rapid mutation rate (% genetic divergence) of fibrinopeptides compared to the slower rates of mutation of haemoglobin, cytochrome *c* and histone IV. The proteins with the slower rates of mutation accumulation are highly conserved, indicating that they are involved in functions that are essential to the survival of the organism.

The molecular clock is also limited when looking at very recent or ancient timescales. When looking at recent timescales, it is less likely that enough time has passed to generate evolutionarily meaningful fixed differences in the sequences of different populations. Instead, alternative alleles that may be present in both populations will lead to an overestimation of evolutionary distance. Over ancient timescales, single sites in the sequence will have changed multiple times and this is known as saturation. Because we can only know of those changes that can be observed today, a molecular clock will underestimate the divergence that has occurred.

Mitochondrial DNA as a molecular clock

Mitochondrial DNA is inherited via the maternal line, while nuclear DNA is passed on from both parents (biparentally inherited).

Genetic material is not just found in the nucleus of a cell. Mitochondria found in eukaryotic cells have their own genome (mitochondrial DNA or mtDNA). In humans, mtDNA contains 37 genes that code for 2 ribosomal RNAs, 22 transfer RNAs and 13 proteins.

MtDNA is unique in that it is passed through the maternal line of sexually reproducing organisms; that is, from mothers to their offspring (Figure 11.1.10). A father's mtDNA is not passed on to his offspring.

Mutations in mtDNA accumulate over time just as they do in nuclear DNA. However, because mtDNA does not have the same repair mechanisms as nuclear DNA, the rate of mutation in mtDNA is usually higher than in nuclear DNA (there are also some highly conserved regions). For this reason, mtDNA can be used as a molecular clock in relatively closely related species, while nuclear DNA is used to compare older lineages. An added advantage of using mtDNA is that it is easier to obtain high yields of DNA because most cells contain many mitochondria.

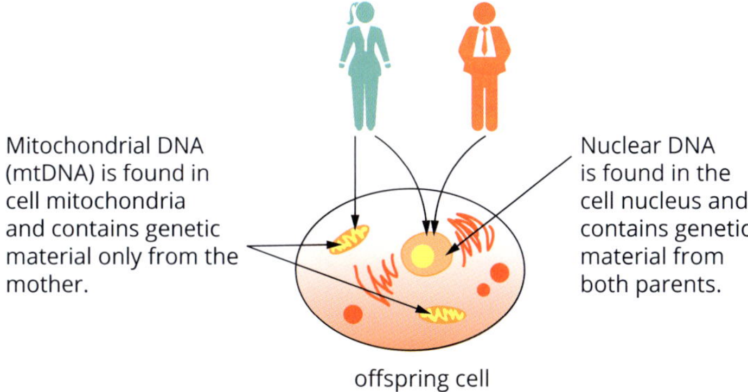

FIGURE 11.1.10 When a sperm and egg fuse, the sperm contributes nuclear DNA only. The egg contributes nuclear DNA as well as the mtDNA of the mitochondria in its cytoplasm. Therefore, mtDNA is only inherited through the maternal line; nuclear DNA is inherited from both parents.

BIOFILE

Mitochondrial Eve

Mitochondrial Eve is the most recent common female ancestor of all present day humans. As mitochondrial DNA (mtDNA) is inherited from the mother (maternally inherited), there is no recombination, making it possible to trace unbroken maternal lines. In 1987, researchers from The University of California, Berkeley, discovered that the mtDNA of all living humans originated from one woman who lived in Africa approximately 140 000 to 200 000 years ago (Figure 11.1.11).

Although Mitochondrial Eve represents an unbroken female lineage from one woman until the present day, she was not the only female alive at the time or the earliest female. Other women who came before her or were alive at the same time have not had a continuous female lineage that persisted to the present day. This is because if a woman doesn't have daughters who also have daughters to pass on her mtDNA, her mitochondrial lineage will die out. It is for this reason that the most recent common ancestor (matrilineal or otherwise) will continually shift over time as lineages end and others carry on.

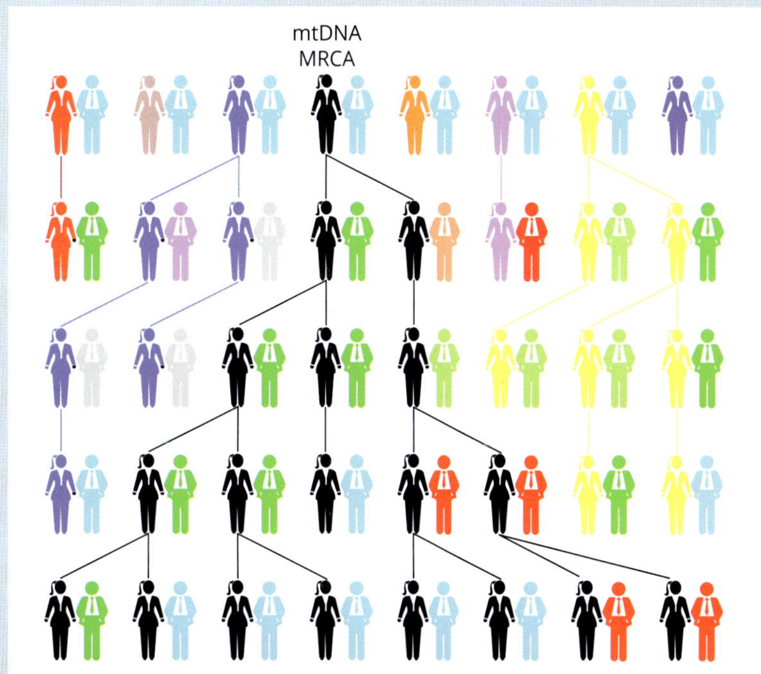

FIGURE 11.1.11 Tracing mitochondrial lineages over five generations. The coloured figures represent extinct mitochondrial lineages, while the black figures represent the female line that is directly descended from the most recent common mitochondrial ancestor (MRCA).

TECHNIQUES FOR MEASURING GENETIC DIFFERENCES BETWEEN ORGANISMS

DNA hybridisation technique

DNA hybridisation is a technique that can be used to determine the level of similarity between sections of DNA of two species. When a small section of double-stranded DNA is heated gently, the hydrogen bonds between complementary bases break and the two strands separate. As the two strands are cooled, the complementary bases match up again and the DNA becomes double stranded once more. If DNA from two different organisms is gently heated to obtain single strands, mixed and then cooled, sections with similar nucleotide sequences will form hybrid double-stranded DNA, containing one strand from each organism (Figure 11.1.12).

The following process would be followed to compare, for example, the region of DNA that produces casein in humans and cows:

1. The casein gene is isolated from the nuclear DNA of each species using **gene probes** (small complementary sequences with an attached dye).
2. Both the human and cow double-stranded DNA are heated to 95 °C to break the hydrogen bonds between complementary bases and separate the individual strands of DNA.
3. The individual strands of DNA of the two species are mixed together and allowed to cool. Where the nucleotide bases of the human and cow genes are complementary, hydrogen bonds will form, creating a strand of **hybridised** DNA.
4. The level of similarity is measured by reheating the hybrid DNA molecule. The temperature needed to separate half of these molecules is recorded as the melting temperature or thermal stability.

For any two species, the more complementary pairs the two strands have in common, the more hydrogen bonds will form and the more strongly the two strands of DNA will be bound together. This means that more heat will be needed to separate them. Therefore:

- a high separation temperature indicates there are relatively more complementary nucleotides between the genes of two species
- a low separation temperature indicates there are fewer complementary nucleotides between the genes of two species.

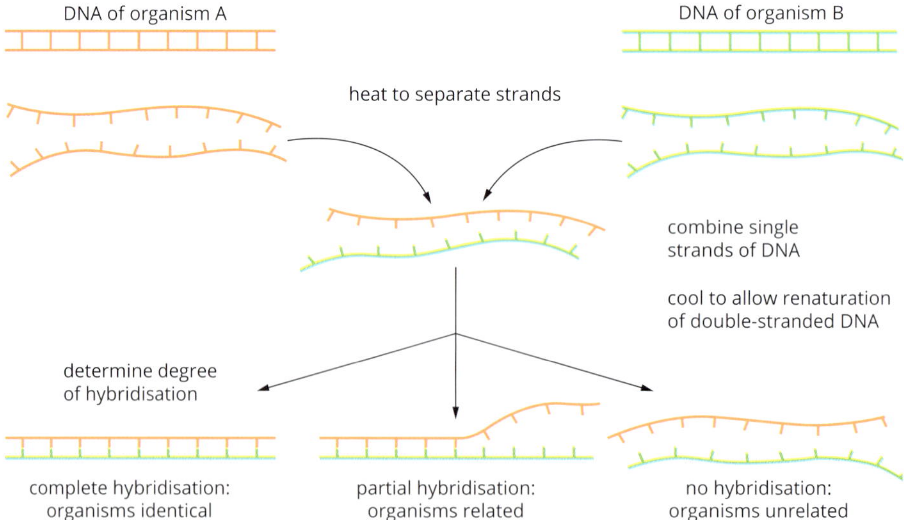

FIGURE 11.1.12 The steps involved in DNA hybridisation to determine the level of similarity. When the strands are reheated, the strands that are completely hybridised will require a higher temperature to separate than the partially hybridised strands of DNA.

Sequencing DNA

DNA hybridisation only provides an overall measure of the degree of difference in the DNA sequences of different species. However, modern DNA sequencing techniques provide a more accurate measure of sequence differentiation by allowing a direct comparison of the nucleotide sequences. The exact number of nucleotide differences between species can then be determined (Figure 11.1.13).

a

	1	2	3	4	5	6	7	8	9	10	11
(eucalypt) *Corymbia*	C	C	C	–	T	C	T	T	T	T	T
(mung bean) *Vigna*	C	T	C	T	T	T	T	T	T	C	A
(soya bean) *Glycine*	C	T	C	T	T	T	T	T	A	C	G
(pine tree) *Pinus*	C	C	T	–	C	C	C	C	C	C	C

b

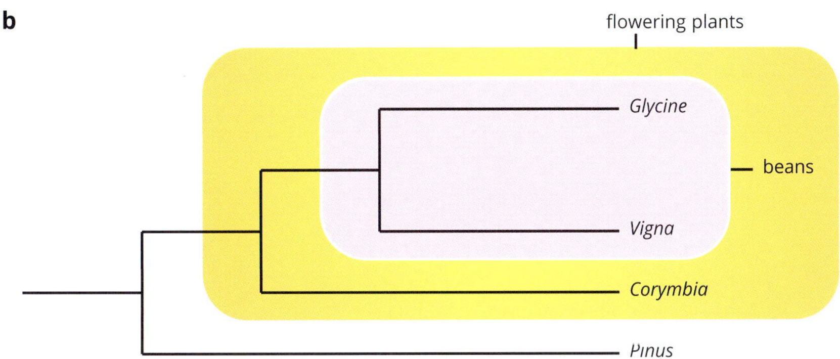

FIGURE 11.1.13 (a) Alignment of part of the sequence of the gene coding for ribosomal RNA (base positions numbered 1–11) of three flowering plants (a eucalypt, *Corymbia*, and two beans, *Vigna* and *Glycine*) and a pine tree (*Pinus*). It shows that the three flowering plants have bases in common (highlighted blue). Furthermore, the two beans have even more bases in common (highlighted green), indicating their high similarity and close relationship. (b) These DNA sequence differences and their evolutionary relationships are represented in a phylogenetic tree.

11.1 Review

SUMMARY

- If two species have a similar set of proteins or DNA sequences, it is evidence that they shared a recent common ancestor.
- Changes in nucleotide sequences are caused by mutations.
- Mutations accumulate throughout the genome over time.
- A mutation to a DNA sequence may not necessarily cause a change to the amino acid sequence because the genetic code is degenerate. However, as more differences in the nucleotide sequence of DNA occur due to mutations, more differences in the amino acid sequence occur.
- Even when a point mutation leads to a change in amino acid, the mutation may not lead to a change in phenotype. This is called a conservative mutation and involves a change from one amino acid to a different amino acid with biochemically similar properties.
- A non-conservative mutation results in a change to a very different amino acid, which often leads to biochemical changes.
- Some genes accumulate mutations faster than others.
- Conserved genes accumulate mutations slowly.
- The more mutations accumulated in the DNA sequences of two species, the more time has passed since they shared a common ancestor. This is the principle of a molecular clock.
- Mitochondrial DNA (mtDNA) is passed through the maternal line.
- MtDNA does not have the same repair mechanisms as nuclear DNA. This means mutations can accumulate at a faster rate than in nuclear DNA, making mtDNA a useful molecular clock for species that diverged recently in evolutionary time.
- Most cells contain many mitochondria and therefore copies of mtDNA are easier to obtain for analysis than nuclear DNA.
- DNA hybridisation involves the heating of DNA from two organisms so that they become single stranded, and mixing them to allow the complementary strands of the two organisms to bond when cooled. The more similar the individual strands of DNA of the two species, the more hydrogen bonds will form between their complementary nitrogen bases, and the greater the amount of heat required to separate them. The formation of many hydrogen bonds means the DNA strands of the two organisms are very similar and share a recent common ancestor; a relatively high temperature will be required to separate the hybridised DNA strands.

KEY QUESTIONS

1 Which of the following species has accumulated the most mutations from the initial sequence: CAATTATCG?

A CATTTATCG
B CAAATAACG
C CTATTTACG
D CATTTGTGC

2 Which of the following is correct?

A The number of differences in the amino acid sequence of a polypeptide would be identical to the number of differences in the DNA sequence of the coding gene.
B The DNA code is degenerate.
C A single change in the DNA will always cause a change in the amino acid sequence.
D The DNA code is grouped into codons.

3 The table below shows the number of differences in the nucleotide sequence between each of the organisms. List the organisms in order from the one with the most distant common ancestor to the horse to the one with the most recent common ancestor to the horse.

Human	0							
Monkey	1	0						
Dog	13	12	0					
Horse	17	16	10	0				
Donkey	16	15	8	1	0			
Pig	13	12	4	5	4	0		
Rabbit	12	11	6	11	10	6	0	
Yeast	66	65	66	68	67	67	67	0
	Human	**Monkey**	**Dog**	**Horse**	**Donkey**	**Pig**	**Rabbit**	**Yeast**

4 Looking at the evolutionary tree below, which two species of butterfly diverged most recently? Circle the point on the tree where they diverged from their common ancestor. Approximately how long ago did these two species diverge from one another?

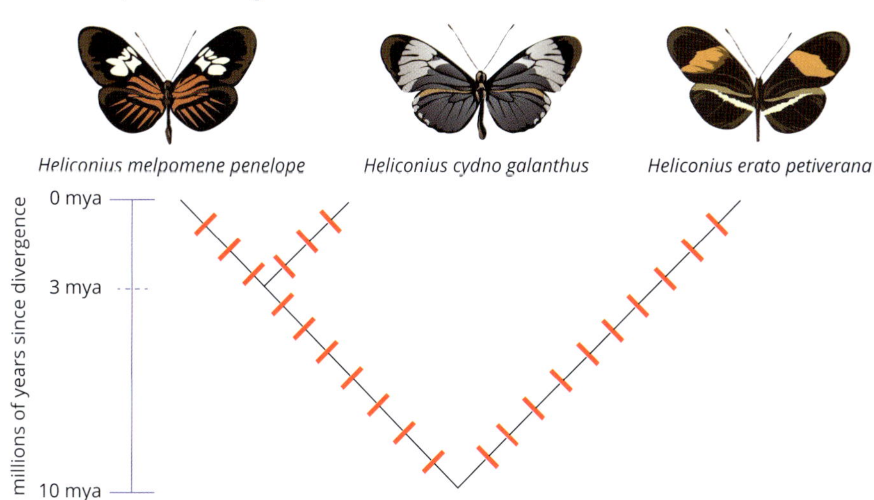

5 Looking at the evolutionary tree in Question 4, how many mutations has *Heliconius erato petiverana* accumulated since diverging from the most recent common ancestor of the other butterfly species? Over what period of time did this occur? What then is the mutation rate (per million years) of the sequence of DNA used to construct this evolutionary tree?

6 Offspring share the same mitochondrial DNA as their:

A father
B paternal grandmother
C mother
D maternal grandfather

7 A small section of DNA from a monkey was hybridised with DNA from species A, B and C. The hybridised DNA was then reheated to determine the temperature at which it became single stranded. Use the graph at right to determine which species is most closely related to the monkey and explain why.

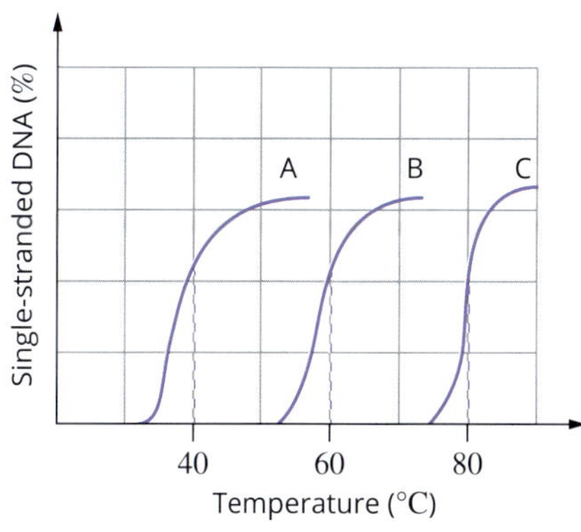

11.2 Phylogenetic trees

Phylogenetic trees (or phylogenies, also known as evolutionary trees) are branching diagrams that depict the evolutionary relationships between different groups of organisms. They are constructed using homologous features, both morphological and molecular, to reveal the branching history of common ancestry between groups of organisms. As information regarding the true evolutionary history of an organism is mostly unknown, scientists use evidence from the morphology and DNA or RNA sequences of living species to reconstruct their evolutionary past. A phylogenetic tree represents an evolutionary hypothesis.

Today, most phylogenetic trees are built using DNA or RNA sequence data, and organisms are grouped on the basis of the similarity of their nucleotide sequences. Using this technology, we have gained remarkable insight into the evolutionary history of life on Earth. Phylogenetic trees based on molecular characters (DNA or RNA nucleotides) can be used to compare any organisms, even if they seem to have very few characteristics in common (Figure 11.2.1). This is because the molecular characters on which the phylogenetic tree is based are the same for all organisms; however, the different mutation rates of different DNA or RNA regions needs to be taken into account when constructing phylogenies. For this reason, phylogenetic trees are usually constructed using sequence alignments of the same gene in different organisms.

In this section, you will learn about the different types of phylogenetic trees and how they are applied to gain an understanding of evolutionary relationships.

FIGURE 11.2.1 Phylogenetic trees based on molecular characters can be used to compare any organisms, even if they seem to have very few characteristics in common.

COMPARATIVE MORPHOLOGY AND CLASSIFICATION

Phylogeny is the evolutionary history of lineages as they diverge from a common ancestor over time. The Swedish naturalist Carl Linnaeus (1707–1778) established the modern biological system of classification. In this system, now known as the **Linnaean system of classification**, organisms are organised into a hierarchy of groups (taxa; singular **taxon**) reflecting their evolutionary relationships: domains, kingdoms, phyla, subphyla, classes, orders, families, genera, and species (Figure 11.2.2). The science of classifying organisms based on their shared characteristics and the evolutionary relationships inferred from this is called **taxonomy**.

A taxon (plural taxa) is a group of organisms that form an evolutionary unit. Biologists (specifically taxonomists) describe these groups based on the Linnaean system of classification (e.g. species, genera, families, etc.).

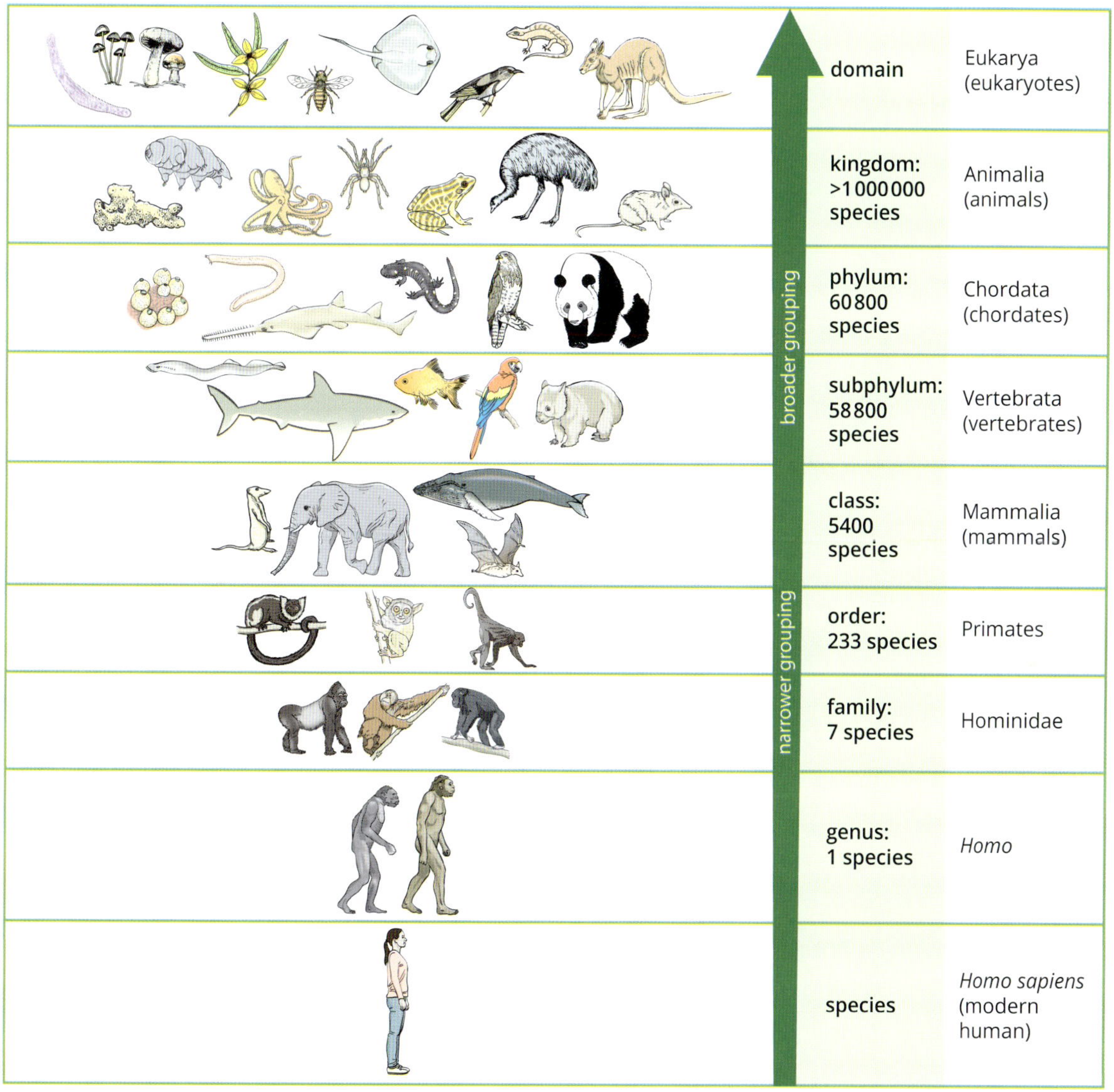

broader grouping ↑ / narrower grouping

Category	Example
domain	Eukarya (eukaryotes)
kingdom: >1 000 000 species	Animalia (animals)
phylum: 60 800 species	Chordata (chordates)
subphylum: 58 800 species	Vertebrata (vertebrates)
class: 5400 species	Mammalia (mammals)
order: 233 species	Primates
family: 7 species	Hominidae
genus: 1 species	*Homo*
species	*Homo sapiens* (modern human)

FIGURE 11.2.2 Categories in the Linnaean system of classification, which is still used today to classify organisms on the basis of their evolutionary relationships. In this figure, the biological classification of humans (*Homo sapiens*) is used as an example.

As you learnt in Chapter 10, organisms were originally grouped on the basis of their physical similarities. This is known as comparative morphology (Chapter 10, page 397). If organisms had features in common, it was hypothesised that they had the same evolutionary origin. Although this practice reflected the best knowledge of phylogeny and biological classification at the time, relying on morphological features alone for phylogenetic classification does have some limitations. For example, some organisms with a common ancestor have lost morphological features or diverged so much that morphological comparisons are not always possible; other organisms that have been genetically separated for a long period of time still share remarkably similar morphology, which means many species are not recognised, and, due to convergent evolution, similar features are not necessarily evidence of evolutionary relationships (see Section 11.1).

As you learnt in the previous section, comparison of DNA sequences is now possible, allowing the discovery of phylogenetic relationships that were previously unknown, and the re-classification of organisms into more accurate taxonomic groups. Molecular data, combined with morphological and ecological information, continues to strengthen our understanding of evolution and refine the system of biological classification (Figure 11.2.3).

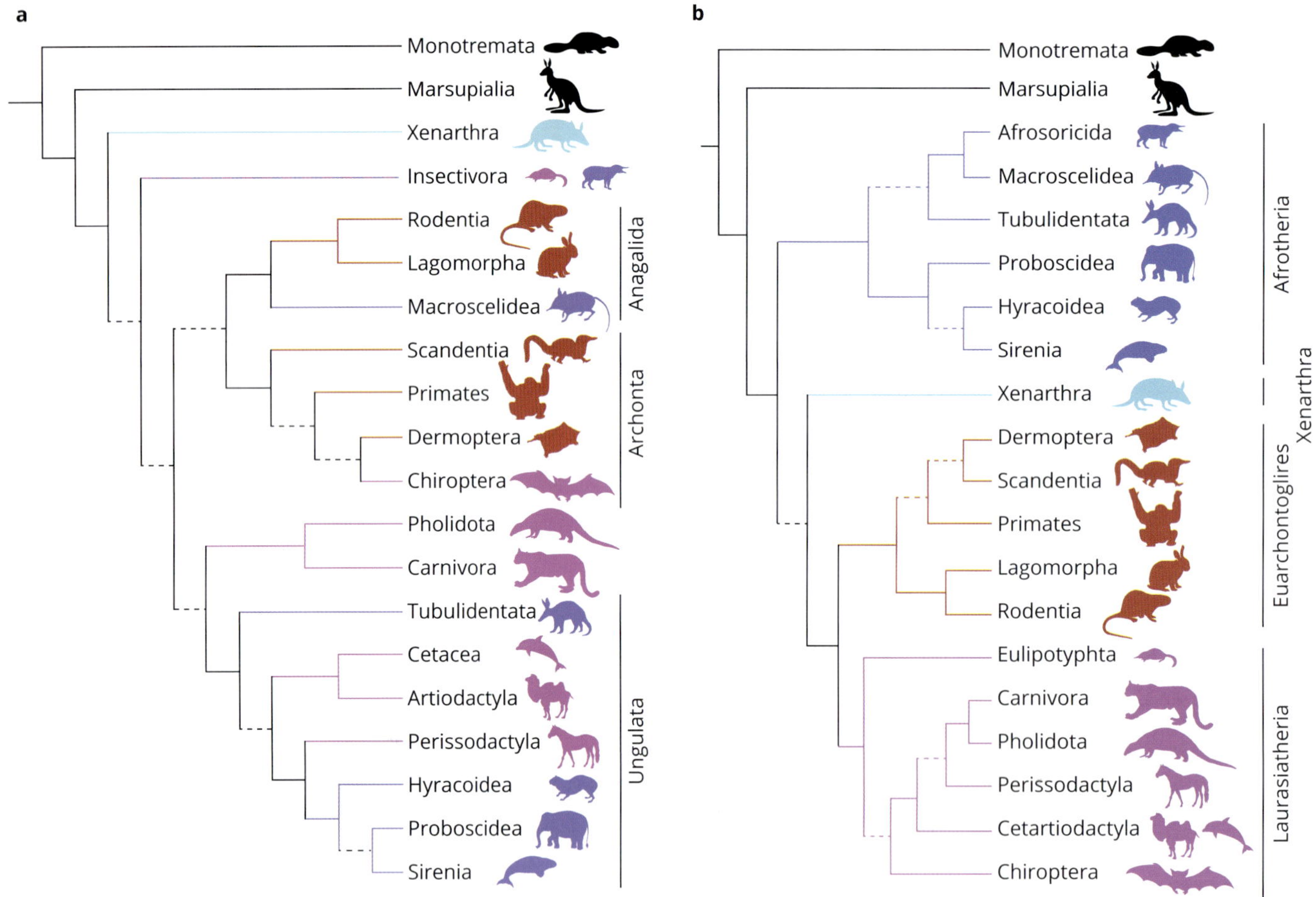

FIGURE 11.2.3 Two phylogenetic trees representing the evolutionary relationships of placental mammals. Tree (a) was constructed using morphological data, and tree (b) was constructed using molecular data. The trees are similar in appearance but the positioning of some taxa differs. The colours represent the four major groups of placental mammals. Marsupials and monotremes are included as a comparison (outgroup). By using a combination of molecular and morphological data our understanding of evolutionary relationships is strengthened.

BUILDING PHYLOGENETIC TREES

Phylogenetic trees are diagrams that show the evolutionary relationships between different groups of organisms. The groups compared in a phylogenetic tree can be different species, genera, phyla or any other level of Linnaean classification.

A phylogenetic tree is built by placing taxa in a branching sequence, according to their shared biological characteristics (e.g. morphological or molecular). A simple example using morphological characters is seen in Figure 11.2.4. By assessing the characters that different organisms share, the evolutionary relationships between taxa can be hypothesised. Starting with the most shared character, which is assumed to be the most ancestral, taxa are added to the tree sequentially, ending with the least shared character at the top of the tree (Figure 11.2.4). Each branch in the tree represents a change in character state (e.g. DNA mutation) from the last common ancestor (Figure 11.2.5). The greater the number of nucleotide differences between sequences or taxa, the greater the distance between them in the tree, reflecting their evolutionary relationships. Most phylogenetic trees are now built using computational methods to generate more complex trees from large datasets.

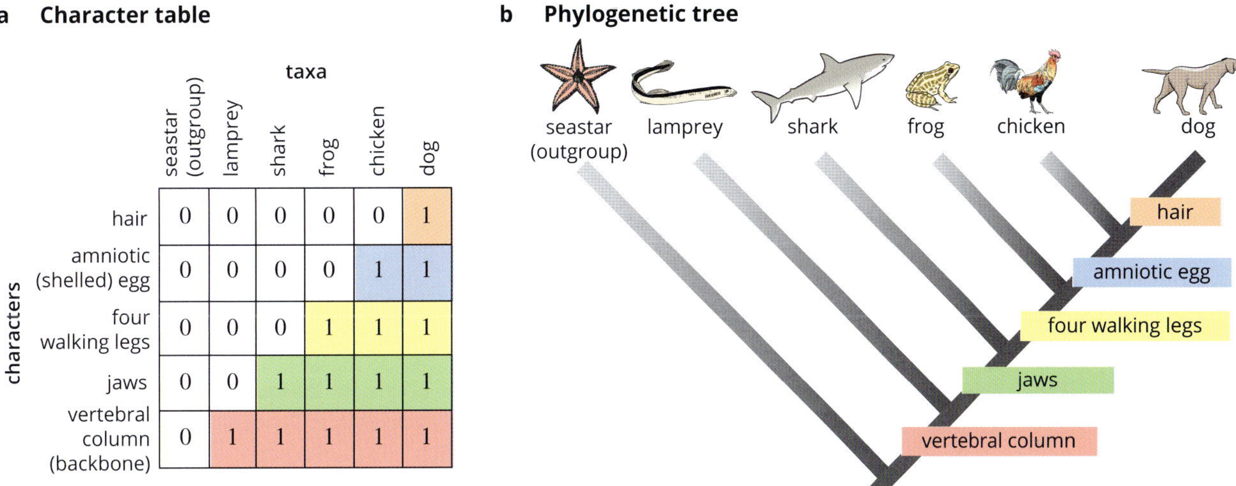

characters	seastar (outgroup)	lamprey	shark	frog	chicken	dog
hair	0	0	0	0	0	1
amniotic (shelled) egg	0	0	0	0	1	1
four walking legs	0	0	0	1	1	1
jaws	0	0	1	1	1	1
vertebral column (backbone)	0	1	1	1	1	1

FIGURE 11.2.4 Building a simple phylogenetic tree using morphological characters. (a) A character table lists different morphological characters and their presence (1) or absence (0) in each taxon is indicated. (b) Each taxon is sequentially added to the tree according to the presence of shared characters, which are assumed to reflect when they appeared in evolutionary history. The more ancestral the character, the more likely it is to be present in a greater number and variety of organisms (e.g. all taxa except the most ancestral group (lancelet) have a vertebral column, so this character is assumed to be the most ancestral and placed at the base of the tree).

FIGURE 11.2.5 Phylogenetic trees can be built using DNA sequence data to estimate the evolutionary distance between taxa. DNA mutations (in this case the ancestral nucleotide C has been substituted for G, highlighted bold) are represented by branches in the tree; the greater the number of nucleotide differences between taxa, the greater the distance between them in the phylogenetic tree.

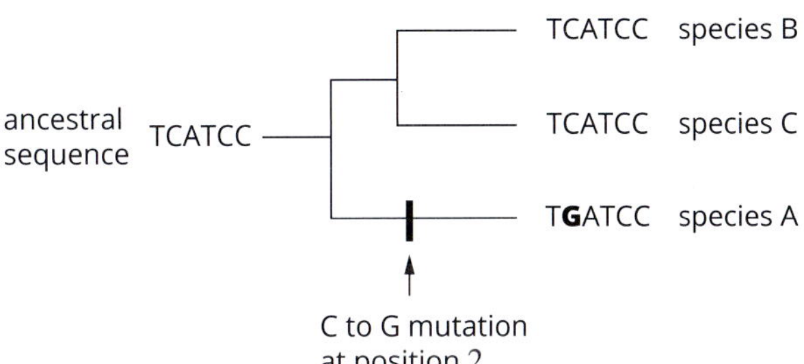

One method that taxonomists use to construct phylogenetic trees is to draw the simplest possible tree to represent evolutionary relationships. This is known as **maximum parsimony** and is based on the principle of parsimony (also known as the theory Occam's Razor), which states that the simplest explanation is the most likely to be correct. In the case of phylogenetic trees, the preferred tree is the one that requires the fewest evolutionary changes to explain the variation (genetic or morphological) observed, as it is the most parsimonious (shows maximum parsimony) (Figure 11.2.6). Although maximum parsimony is a popular method, there are also several other methods used to construct phylogenetic trees, such as maximum likelihood and minimum evolution.

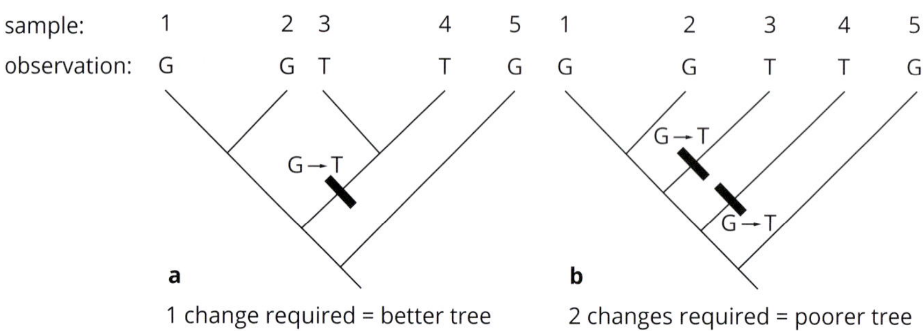

FIGURE 11.2.6 Using maximum parsimony to construct phylogenetic trees means selecting the simplest tree to explain the observed variation and evolutionary relationships between organisms. Tree (a) requires only one change to explain the nucleotide change (G to T), and tree (b) requires two evolutionary changes. According to the principle of parsimony, tree (a) provides a simpler, better explanation of the evolutionary relationships.

Parts of phylogenetic trees

As the name suggests, a phylogenetic tree is shaped like a tree with branches and leaves that extend from the **root** or ancestral **lineage** (Figure 11.2.7). Each line on the tree is called a **branch** and represents the evolutionary path from a common ancestor. The end of each branch contains a scientific name and is called a **leaf**. The point at which two branches diverge is called a **node** (or branch point) and represents the last common ancestor that the diverging groups shared. An **outgroup** species is sometimes included to assess the evolutionary relationships of those in the ingroup (the taxa of focus) relative to more distantly related taxa.

A lineage is all the species that are descendants of a common ancestor.

An outgroup is a taxonomic group that is closely related to the other groups (ingroups) but less closely related than any single one of the ingroups is to each other. An outgroup has a common ancestor with the ingroups that is older than the common ancestor of the ingroups.

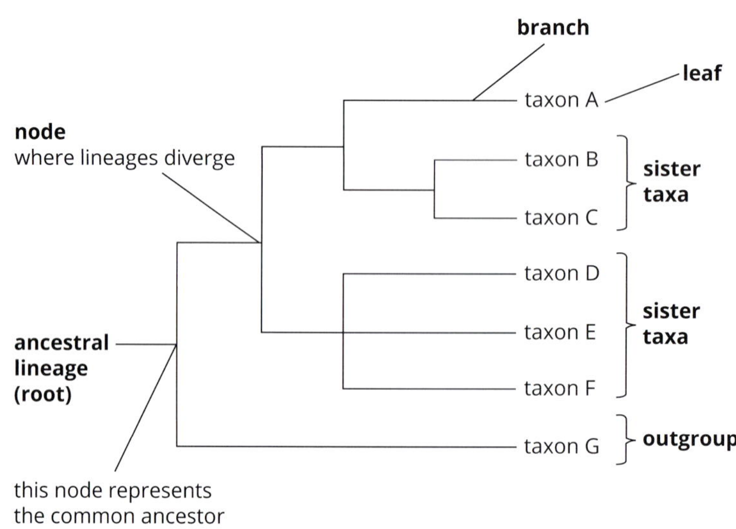

FIGURE 11.2.7 The components of a phylogenetic tree. In this tree, there is a root (an ancestral lineage) from which all of the taxa in the tree have diverged. The branches represent the evolutionary paths (lineages) that each taxon has taken. Each of the nodes represents the divergence point of the taxa that have split off from it. Sister taxa (e.g. taxa B and C) are those that are the most closely related to one another. An outgroup (taxon G) has been included for comparison.

Pairs of taxa grouped together are called **sister taxa** and are the most closely related relative to other taxa in the tree. The most closely related (most recently diverged) taxa have the shortest branch lengths between them. In Figure 11.2.7, you can see that the closest relative of taxon C is taxon B, as that is the closest taxon following the branch from taxon C. Because no other taxa are more closely related in the tree, taxa B and C are sister taxa. The next closest relative to taxon C is taxon A. Even though taxon D is listed under taxon C and they appear close to each other, if you follow the branch lengths between them, you can see that they are actually quite distantly related.

Each section of the phylogenetic tree is called a **clade**. A clade is a group of organisms that includes an ancestor and all the descendants of that ancestor (also called a monophyletic group; shown in Figure 11.2.8). The order in which clades diverged from their common ancestor is represented by the order of the branching points (or nodes), with the oldest branching point closest to the root of the tree (the bottom in vertical trees; the left side in horizontal trees) and the most recent branching point closest to the tips of the tree (the top in vertical trees; the right side in horizontal trees; shown in Figure 11.2.9). As phylogenetic trees can have many different configurations, the order of the taxa at the tips of the tree does not always represent the order in which they diverged.

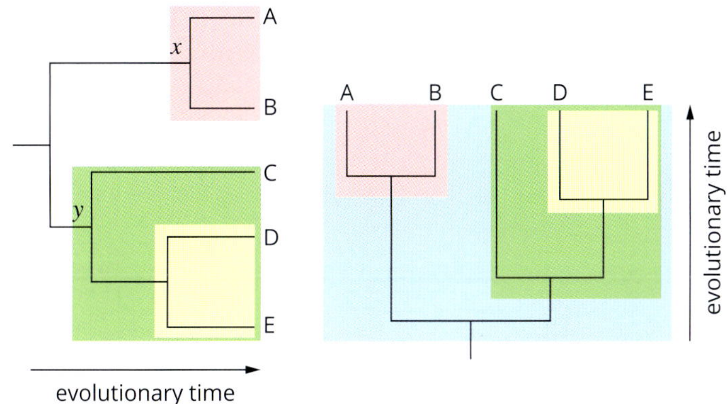

FIGURE 11.2.8 Phylogenetic trees consist of groups of taxa called clades. A clade includes a common ancestor and all its descendants. Each coloured square in these two trees represents a clade.

The more closely related the taxa, the shorter the branch lengths between them will be. Shorter branch lengths in phylogenetic trees indicate that there have been fewer evolutionary splits (divergence points) between the taxa and they are therefore less divergent.

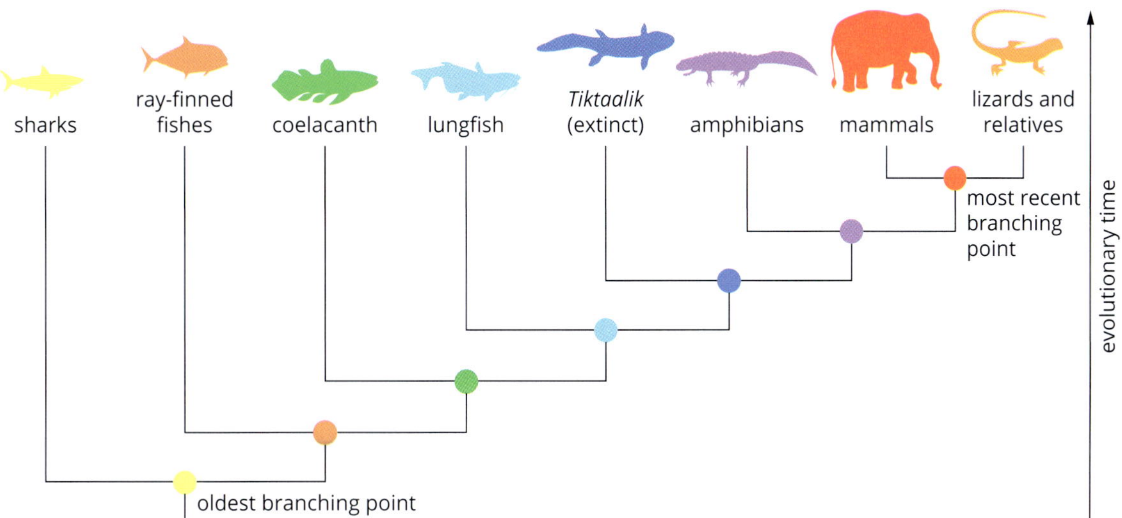

FIGURE 11.2.9 The order of the branching points (or nodes), from root to the tips of the tree (bottom to top in a vertical tree; left to right in a horizontal tree), represents the order in which clades diverged from one another.

There are three different ways in which taxa can be grouped within a phylogenetic tree (Figure 11.2.10):

- Monophyletic groups (one tribe) include a common ancestor and all of its descendants (a clade). This grouping is the only taxonomically viable group and is the basis of evolutionary biology. (A group that does not form a monophyletic group is not a taxonomic group.) For example, reptiles and birds are a monophyletic group (Figure 11.2.11). All the members of a monophyletic group can be removed from the tree with a single 'cut' (Figure 11.2.10).
- Paraphyletic groups (beside or surrounding the tribe) include a common ancestor and only some of its descendants (Figure 11.2.10). Although not taxonomically accurate groupings, paraphyletic groups are useful for describing subsets of evolutionary groups. For example, dinosaurs are a paraphyletic group. Although birds are also dinosaurs (both groups descended from a common ancestor), dinosaurs and reptiles are almost always referred to as a separate evolutionary group from birds (Figure 11.2.11).
- Polyphyletic groups (many tribes) include multiple descendants but do not include a common ancestor (Figure 11.2.10). This grouping is rarely used, but may group taxa on the basis of shared characteristics, such as the ability of birds and mammals to maintain their body temperature (Figure 11.2.11).

A monophyletic group contains a common ancestor and all its descendants. This group is also called a clade and is the only true taxonomic group.

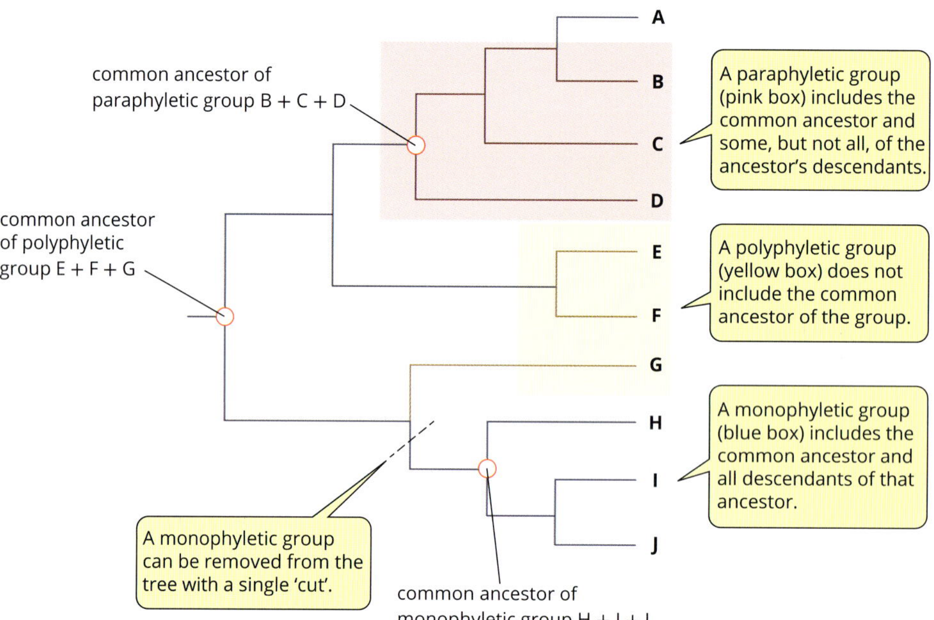

FIGURE 11.2.10 The three types of groupings within phylogenetic trees: monophyletic, paraphyletic and polyphyletic. Monophyletic groups are the only true taxonomic groups.

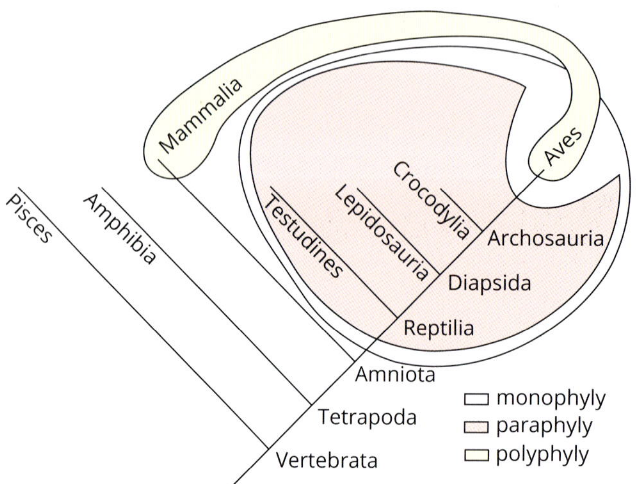

FIGURE 11.2.11 Monophyly, paraphyly and polyphyly represented in a phylogenetic tree. The blue group includes reptiles (Reptilia) and birds (Aves) and so is monophyletic (includes all descendants of a common ancestor).The pink group includes reptiles but excludes birds, so it is paraphyletic. Mammals (Mammalia) and birds can be grouped as warm-blooded (more properly known as endothermic) animals (in yellow), but this group does not contain the common ancestor of these two descendants and so is polyphyletic.

Different forms of phylogenetic trees

Phylogenetic trees come in two main forms: rooted trees (Figure 11.2.12a–d) and unrooted trees (Figure 11.2.12e, f). Both these tree types can also have scaled or unscaled branches. These trees are referred to as **cladograms** if they are unscaled (Figure 11.2.12a, b) and **phylograms** if they are scaled (Figure 11.2.12c, d). The branch lengths of a cladogram are not proportional to the amount of evolutionary divergence (nucleotide changes) between taxa. Phylogram branch lengths are proportional to the number of nucleotide changes that have occurred in the lineage during divergence from the common ancestor and the total length of all the branches connecting two taxa in a phylogram is equal to the evolutionary divergence between them. Both scaled and unscaled trees can be calibrated to a timescale (Figure 11.2.12a, b), and so provide a visual representation of when, in evolutionary time, lineages diverged from one another.

Phylogenetic trees may be presented horizontally (Figure 11.2.12a–d) or vertically, and branches may be diagonal (Figure 11.2.12a, c) or square (Figure 11.2.12b, d). These different tree types and configurations may appear quite different even when representing the same evolutionary relationships. The branch points and their positions relative to one another (order is not important) are the parts of the tree that tell you the most about the evolutionary relationships of the taxa and should remain the same even if the format of the tree is changed (Figure 11.2.12 and Figure 11.2.13 on page 438).

Phylogenetic trees can be drawn in many different ways: vertical or horizontal; rectangular or circular; rooted or unrooted; diagonal or horizontal branches; and scaled or unscaled branches. Trees that look quite different can represent the same evolutionary relationships.

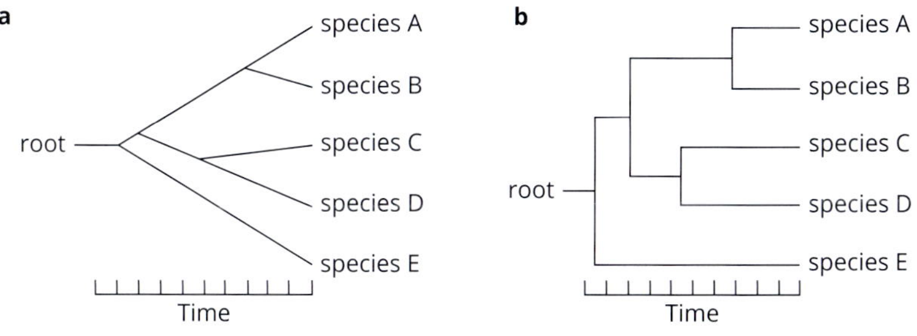

Rooted trees with scaled branches (phylogram)

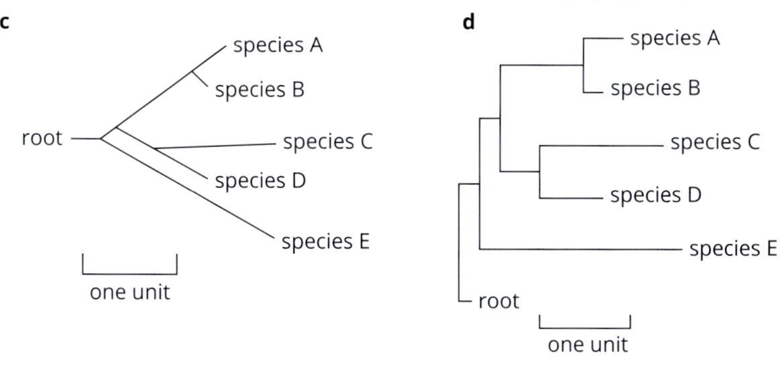

e Unrooted tree with scaled branches (phylogram)

f Unrooted tree with unscaled branches (cladogram)

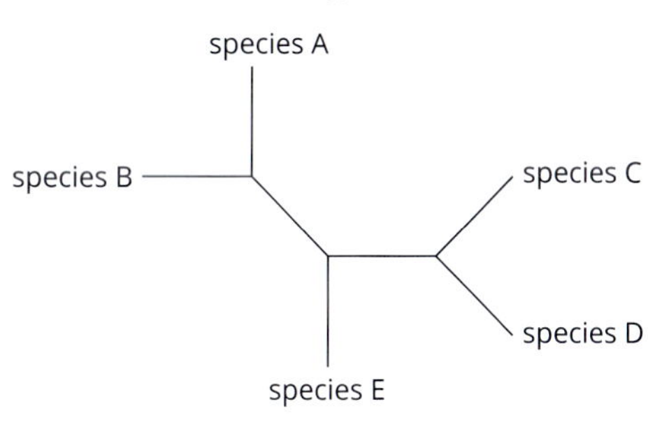

FIGURE 11.2.12 Phylogenetic trees can be depicted in many different ways to show the same evolutionary relationships. Two different ways of drawing rooted trees with unscaled branches (cladograms) (a, b) and scaled branches (phylograms) (c, d). The unscaled trees in (a) and (b) have been calibrated to a timescale. Two unrooted phylogenetic trees, one with scaled branches (e) and the other with unscaled branches (f) are also shown. All of these trees present the same evolutionary relationships in different ways.

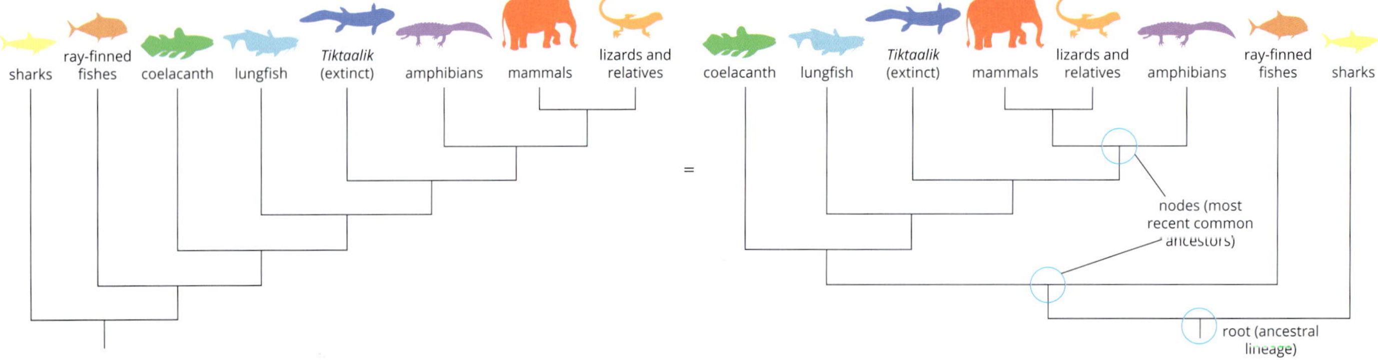

FIGURE 11.2.13 The same evolutionary relationships may be presented in different ways in phylogenetic trees. In these trees, the order of the taxa is different but the configuration of the branches connecting them remains the same.

> The order of branching points (nodes) from root to tip represents the order of evolutionary divergence of taxa.

Rooted trees

Rooted trees are drawn with the first branches coming from the base or 'trunk' of the tree. These may be drawn vertically (Figure 11.2.14), horizontally or in a circular format. The root represents the hypothesised common ancestor of all taxa in the tree and the branches represent the evolutionary path of those taxa over time. In comparison, unrooted trees depict the evolutionary relationships between taxa in the tree but do not show their evolutionary path from a common ancestor.

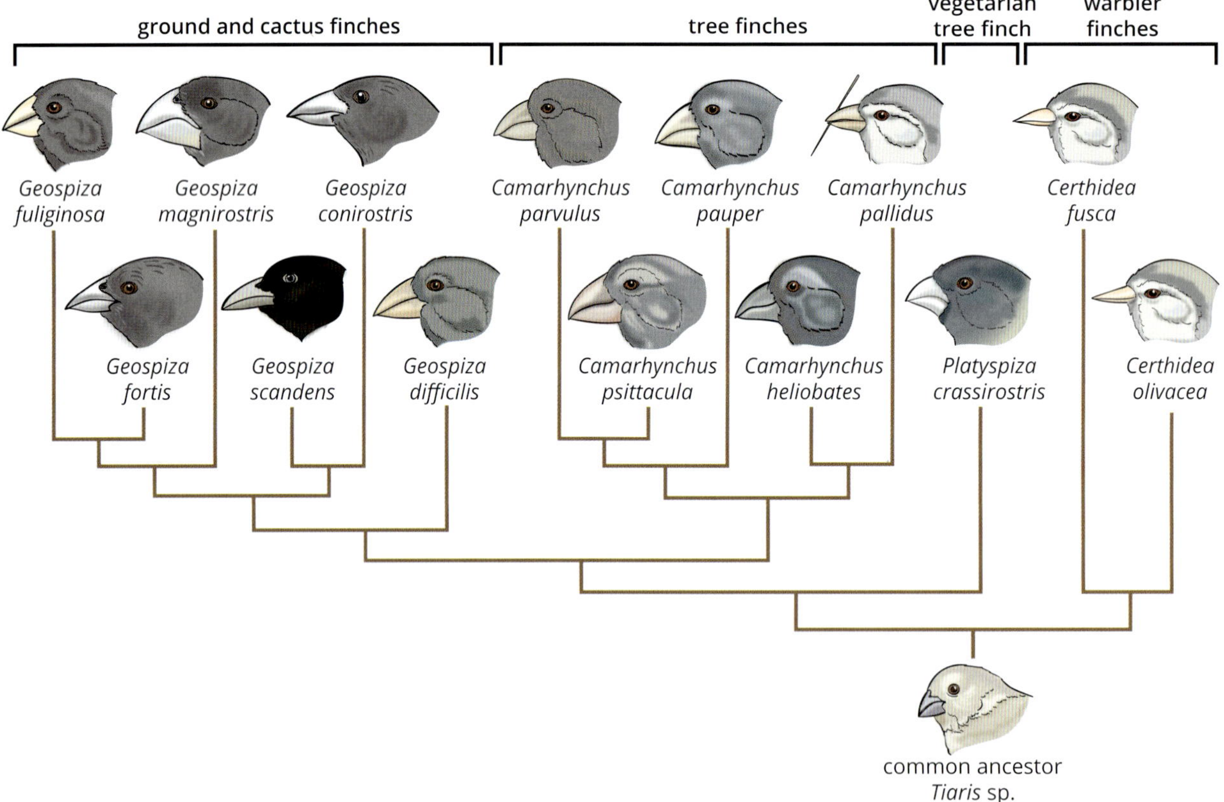

FIGURE 11.2.14 A phylogenetic tree of the evolutionary relationships between Darwin's finches. This tree is rooted to the most recent common ancestor of Darwin's finches, the genus *Tiaris*. The tree has scaled branches, indicating the amount of genetic divergence between the taxa.

In order to root the tree to a common ancestor, an organism that is known to be related to the main group of interest is included in the phylogenetic analysis as an outgroup. The outgroup (warbler finches in Figure 11.2.14) provides a comparison point to assess where the main group of organisms (ingroup) sits in relation to other closely related taxa. The outgroup must be related to the ingroup in order to make meaningful comparisons, but not be more closely related than organisms in the ingroup are to one another. An outgroup can be selected by aligning DNA sequences to determine how closely related they are.

In phylograms the lengths of the branches connecting two organisms indicates the amount of genetic divergence between them. The time that has passed since the organisms shared a common ancestor can also be represented by the branch length, by applying a molecular clock (Section 11.1).

Polytomies

Occasionally there will be nodes with more than two descendent lineages (like a garden rake). These nodes are called **polytomies** (Figure 11.2.15). Polytomies occur where there is not enough information to distinguish the order of evolution or when rapid speciation has occurred after adaptive radiation in a new environment (Chapter 10, page 404). If rapid speciation has occurred, then all the daughter lineages will be closely related.

taxon A
taxon B
taxon C
taxon D
taxon E
taxon F
taxon G
This branch point forms a polytomy: an unresolved pattern of divergence.

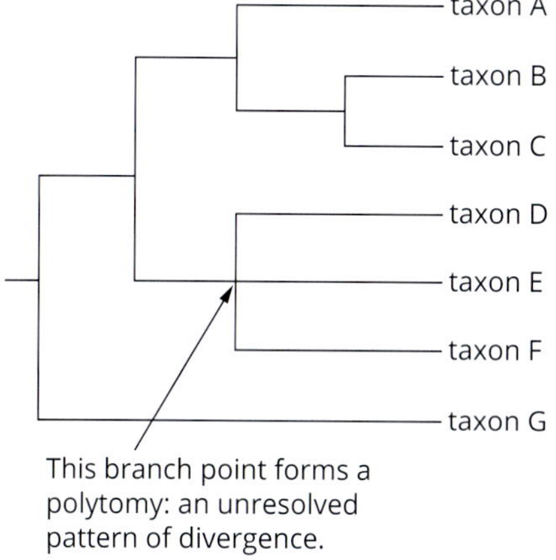

FIGURE 11.2.15 A phylogenetic tree with a polytomy between taxa D, E and F. Polytomies are formed when the evolutionary relationships between three or more taxa cannot be resolved due to lack of information or rapid speciation events.

Unrooted phylogenetic trees

Unrooted phylogenetic trees do not include an ancestral root. Because the ancestor is not defined in unrooted trees, they often have a radial layout. These trees only indicate the relationship between the different leaf nodes without indicating which node is the most ancestral. Like the rooted phylogram, unrooted trees can also have scaled branches where the distance between each leaf (along the branch lengths) is an indication of the amount of genetic divergence between the two groups of organisms. A greater distance suggests more time has passed since they shared a common ancestor (Figure 11.2.16).

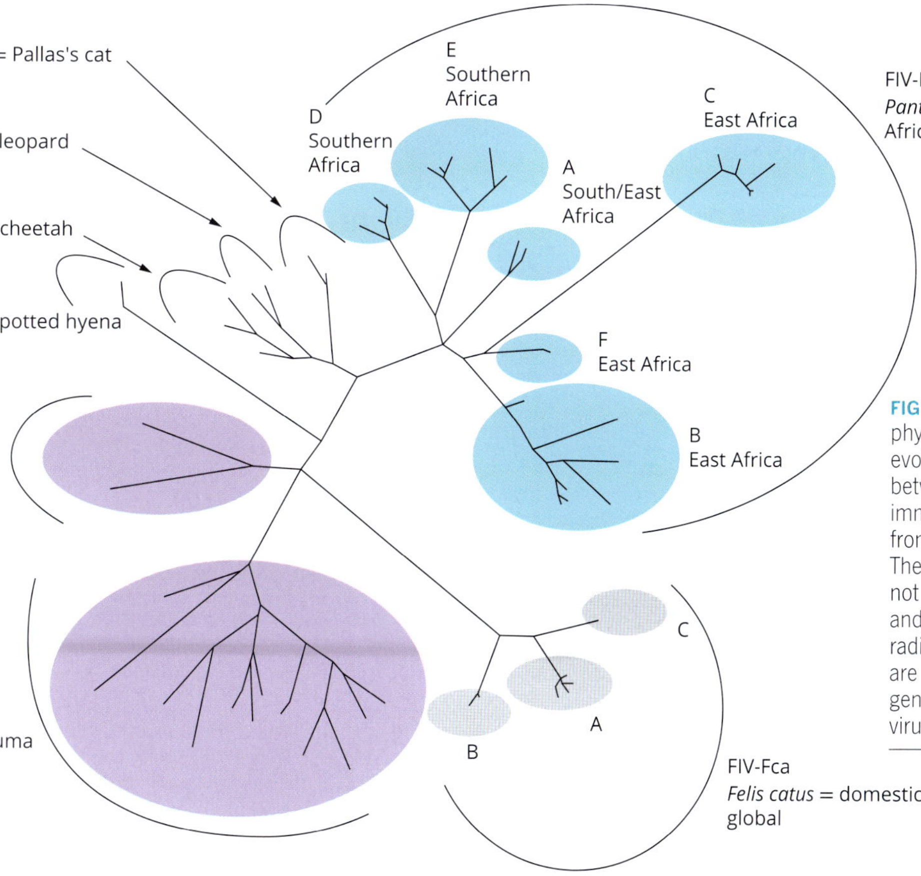

FIV-Oma
Otocolobus manul = Pallas's cat
Asia
FIV-Ppa
Panthera pardus = leopard
Africa
FIV-Aju
Acinonyx jubatus = cheetah
Africa
FIV-Ccr
Crocuta crocuta = spotted hyena
Hyaenidae
Africa
E
Southern
Africa
D
Southern
Africa
C
East Africa
A
South/East
Africa
F
East Africa
B
East Africa
FIV-Ple
Panthera leo = lion
Africa
FIV-Pco Subtype A
Puma concolor =
puma
North America
FIV-Pco Subtype B
Puma concolor = puma
North (NA)
Central (CA)
South America (SA)
C
A
B
FIV-Fca
Felis catus = domestic cat
global

FIGURE 11.2.16 An unrooted phylogenetic tree representing evolutionary relationships between subtypes of the feline immunodeficiency virus (FIV) from seven carnivore species. The common ancestor has not been defined in this tree and so, without a root, it has a radial configuration. Branches are scaled to represent the genetic distance between the virus subtypes.

BIOFILE

The evolutionary uncertainty of ray-finned fish

Phylogenetic modelling of fish species reveals that there is uncertainty around the evolutionary relationships of a group of ray-finned fish known as the Percomorphs (highlighted in yellow in Figure 11.2.17). You might have noticed that the section of the phylogenetic tree (clade) with the Percomorphs looks different from the rest of the tree. Instead of each node (branch point) having two lineages, the Percomorph clade looks like a rake, with many lineages. This feature of a phylogenetic tree is called a polytomy.

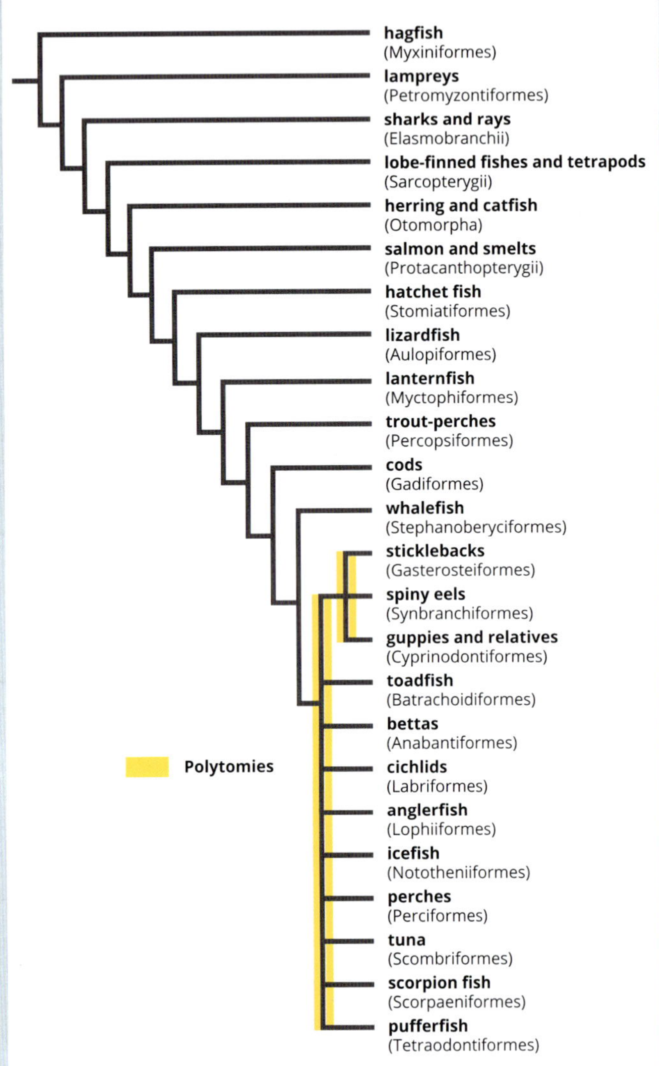

FIGURE 11.2.17 Polytomies in this phylogenetic tree (highlighted in yellow) indicate that the evolutionary relationships of Percomorph fish are uncertain.

In this tree, there are two polytomies (highlighted in yellow), one with three lineages (sticklebacks, spiny eels and guppies) and the other with nine lineages (toadfish, bettas, cichlids, anglerfish, icefish, perches, tuna, scorpion fish and pufferfish, Figure 11.2.18). Polytomies indicate that there is not enough information to resolve the evolutionary relationships of lineages. The tree tells us that these lineages are closely related but we can't be sure of the order in which they diverged from one another. From the phylogenetic tree in Figure 11.2.17, we know that sticklebacks, spiny eels and guppies are closely related, but we can't tell which is most closely related to which. Further study and additional data can help scientists resolve polytomies and gain more certainty about evolutionary relationships.

FIGURE 11.2.18 (a) Spiny eel and (b) scorpion fish

BIOLOGY IN ACTION

Applied phylogeny: the origin of SARS

In 2002 and 2003, there was worldwide panic about a previously unknown virus infecting humans and causing severe acute respiratory syndrome (SARS). Symptoms of the disease include fever, lethargy, sore throat, cough and shortness of breath. The outbreak started in southern China and quickly spread to Hong Kong, Canada, Singapore, Taiwan and Vietnam via international travellers. Between November 2002 and July 2003, the SARS outbreak caused more than 8000 people to become ill and 774 deaths.

Viruses evolve rapidly, accumulating small changes in their DNA and RNA, forming new lineages. Much research was carried out to determine the origin of the SARS virus to improve understanding of the disease and work towards minimising its impact. Scientists suspected that the virus originated from another species and had jumped hosts. There were two suspects: a cat-like mammal called a civet and Chinese horseshoe bats that were found infected at a live animal market in China. Scientists used DNA hybridisation techniques (see Section 11.1) to determine the number of differences in the RNA sequence of the viral genome. These differences were used to construct a phylogenetic tree for the virus using the molecular clock of the virus strains (Figure 11.2.19). The phylogenetic analysis determined that the SARS virus originated in bats and then mutated, enabling it to jump hosts and infect civets. The civet and human SARS viruses were found to be very similar and so the civet SARS is the most likely source of the human SARS virus.

Knowing the origin of a virus helps doctors understand emerging diseases and enables scientists to predict the infection patterns of future outbreaks.

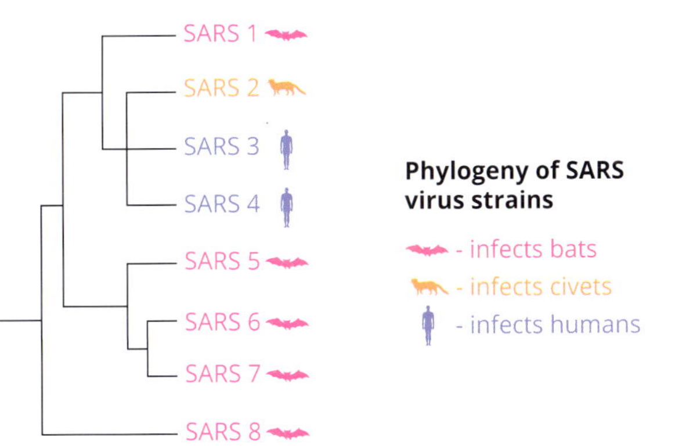

FIGURE 11.2.19 The phylogenetic tree of SARS is an indication of how and when the SARS virus evolved and the host animals in which they are found.

11.2 Review

SUMMARY

- Phylogenetic trees show the evolutionary relationship between different groups of organisms based on morphological and molecular homology.
- The branches and nodes of phylogenetic trees indicate common ancestry between organisms.
- A clade is a group of organisms that includes an ancestor and all the descendants of that ancestor.
- Groups in phylogenetic trees can be described as monophyletic, paraphyletic or polyphyletic depending on their evolutionary relationships.
- Monophyletic groups are the only taxonomically viable group because they contain a common ancestor and all its descendants (a clade).
- Rooted phylogenetic trees can be used to indicate the length of time that has passed since organisms shared a common ancestor.
- Unrooted phylogenetic trees do not include an ancestral root, and only indicate the relationship between the different leaf nodes.

KEY QUESTIONS

1 What is a phylogenetic tree and what does it represent?

2 From the phylogenetic tree below, which animal is most closely related to the pig?

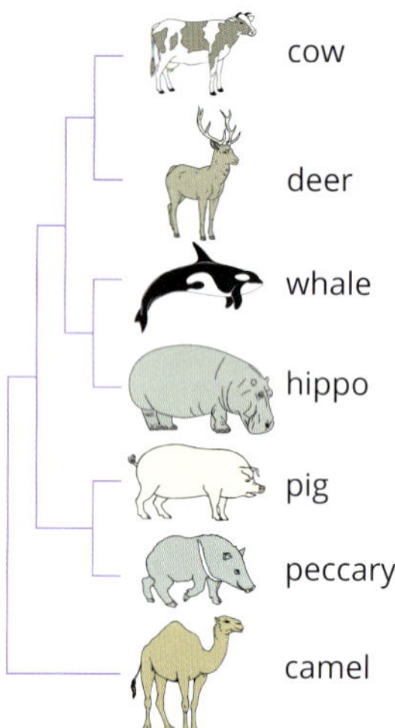

3 Which species is most closely related to species A?

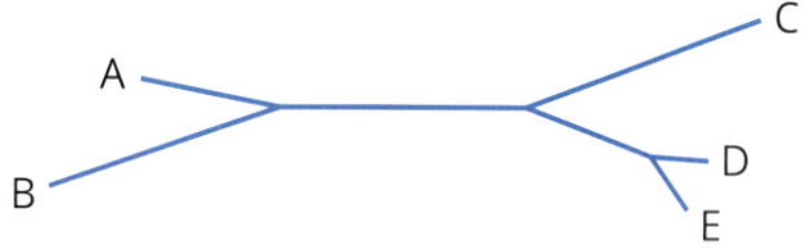

A B
B C
C D
D E

4 Number (1–7) the nodes of the phylogenetic tree in the order in which they diverged.

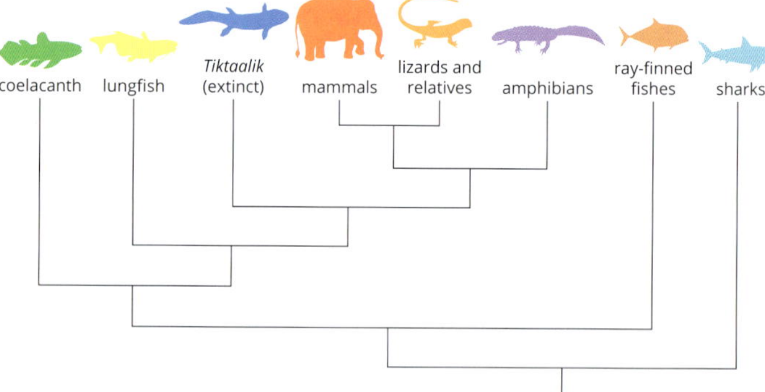

5 List four forms of phylogenetic tree.

6 Why is an outgroup included in some phylogenetic trees?

7 Which phylogenetic grouping (monophyletic, paraphyletic or polyphyletic) is the only taxonomically viable group and why?

11.3 Developments in evolutionary biology

Evolution is an ongoing process and, contrary to what was once thought, it is not always a slow gradual process. Evolution can occur at different rates and for different reasons and is affected by chance events. At various times, evolutionary change has occurred over a very short time, resulting in rapid speciation. In fact, small changes in gene sequence or gene regulation can lead to major phenotypic changes when they affect genes that control the development of an organism from a zygote into an adult.

Evolutionary developmental biologists study developmental processes and how changes in these processes lead to new phenotypes (Figure 11.3.1).

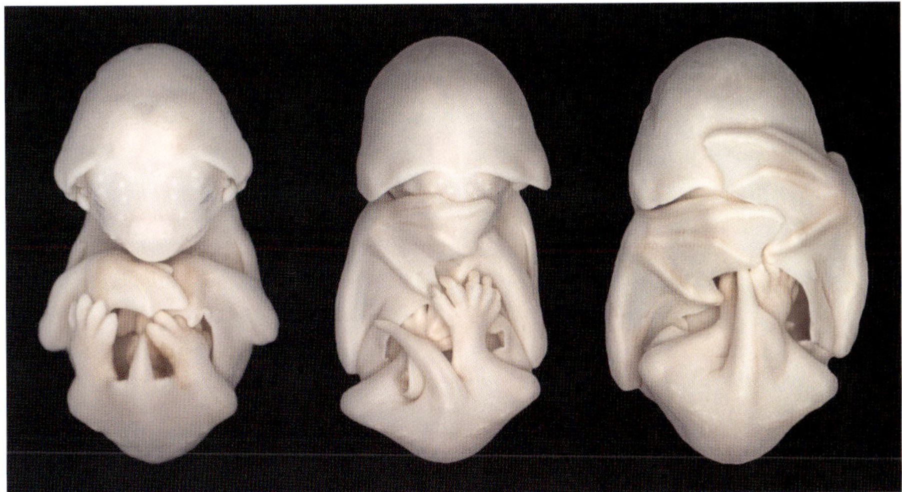

FIGURE 11.3.1 Studying the genes involved in the development of bat wings gives scientists an understanding of how the mutation of master genes can cause new species to evolve. In this image, it can be observed that the fingers in the front paws of a bat embryo grow more rapidly than the rest of the structure and become wings.

RAPID EVOLUTION

There are instances of evolution being a gradual process but comparisons of fossils suggests that this is not always the case. Fossil evidence suggests that some lineages have evolved rapidly in a relatively short period (see adaptive radiation, Chapter 10, page 404). Many of the major groups of animals first appeared in the fossil records during the Cambrian period (541–485 mya) when new species evolved adaptations that were suited to different niches in the environment.

Rapid evolutionary change and species diversity can result from the reorganisation of a genome and changes in the level of gene expression, which can lead to major structural changes. This rapid change in gene expression is a result of changes in master regulatory genes, which control developmental processes, including embryonic development.

MASTER REGULATORY GENES

All embryonic cells have identical copies of every gene. **Master regulatory genes** control the development of these embryonic stem cells into different cell types that will result in the structures of an organism, such as eyes and legs. There are relatively few master regulatory genes, but they are essential for the correct embryonic development of organisms. They affect downstream structural genes either directly or by controlling other regulatory genes. Master regulatory genes are able to switch other genes 'on' or 'off'. By studying embryological development, scientists have been able to improve our understanding of how master regulatory genes evolve and function.

Mutations in master regulatory genes can cause major phenotypic changes, such as segments of the body to be repeated, resulting in new structures that might give rise to new species in a relatively short period of evolutionary time (Figure 11.3.2).

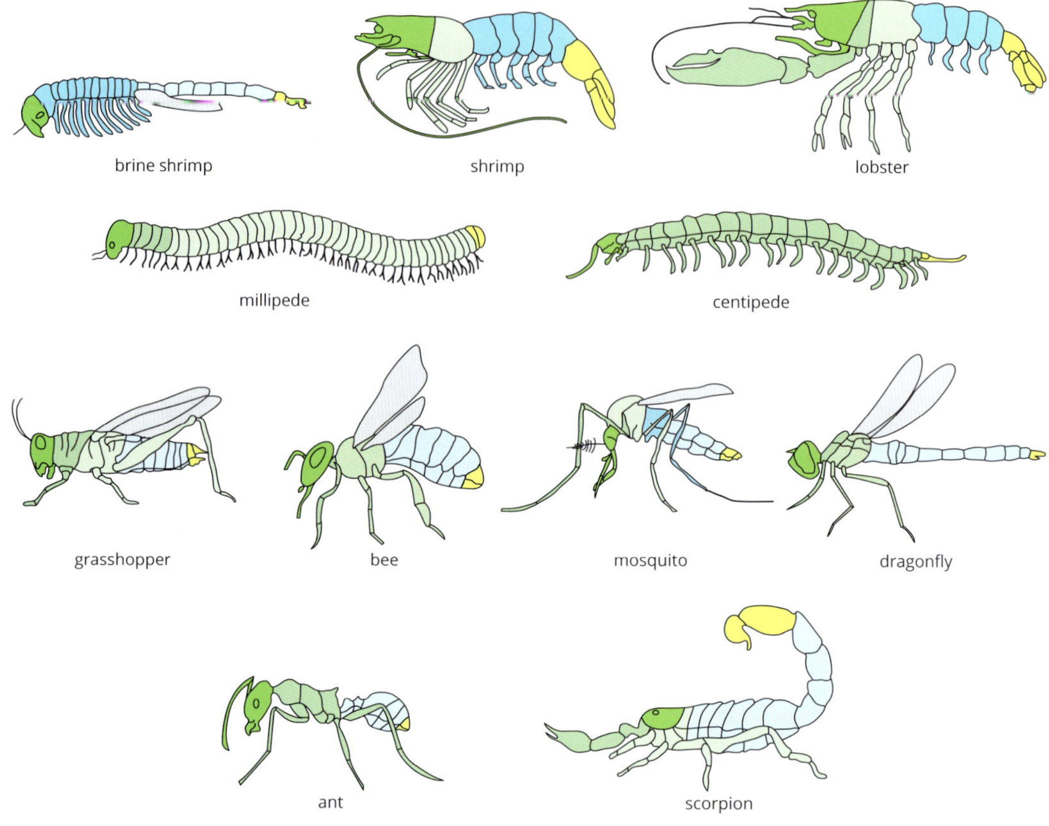

FIGURE 11.3.2 A single mutation in a master regulatory gene can increase the number of segments in the body of an organism. This can result in very different looking structures, from a shrimp's tail to a scorpion's tail.

Master regulatory genes play a vital role in embryonic development by controlling the rate, timing and spatial pattern of the expression of other genes.

Humans share master regulatory genes with other animals. One group of regulatory genes, for example, controls the embryological development of the eye. Small mutations in the master regulatory genes of ancestral animals would have led to the evolution of the complex human eye. Therefore, the evolution of complex new forms can occur by chance following mutations in pre-existing master regulatory genes. Researchers have found that the number of master regulatory genes in which mutations can occur without fatal consequences for the organism are surprisingly small.

MASTER REGULATORY GENES CONTROL THE EXPRESSION OF STRUCTURAL GENES

Master regulatory genes control the expression of other regulatory and structural genes during embryonic development. They control not only the rate and timing of their expression but also the spatial pattern of their expression.

Changes in rate and timing

Sometimes master regulatory genes change the timing and/or rate of expression of a structural gene. This is known as **heterochrony** and it causes the expression of a gene to be slowed down or sped up. This affects the timing of the embryological development of the organism, and can lead to structural changes.

Bat wings, for example, are essentially mammalian paws with really long fingers and skin stretched between them. At some stage during the bat embryo's development, the rate of finger growth must have been increased relative to the development of the rest of the bat's body. This means the bone-growing genes in the bat's fingers were more active than the bone-growing genes in the bat's leg. This makes the bat's fingers proportionately longer than the rest of its bones. This does not happen in other mammals and has resulted from a mutation in the master genes of the bat's ancestral line (Figure 11.3.3).

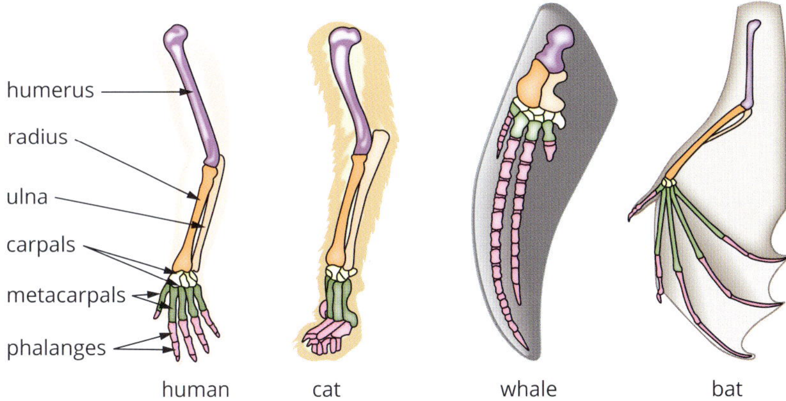

FIGURE 11.3.3 Comparison of homologous bones in the forelimbs of several vertebrates is evidence of how the master genes control bone cell growth. The colour coding in this diagram identifies the shape and position of the same bones in each organism. The limbs and bones are drawn to the same scale.

Changes in spatial pattern

The ***Hox* genes** (a type of **homeotic gene**) are an example of master regulatory genes that affect the spatial pattern of expression of other genes. This group of genes determines the body plan along the head-to-tail axis during embryonic development. The expression of these master regulatory genes during the development of an organism determines the structures that will be formed on a given position. In the fruit fly, these genes determine the position of the abdomen, thorax and head, and the location of the antenna and legs. As an embryo develops, the *Hox* genes direct messages to groups of embryonic cells to, for example, 'grow into a head' or 'develop into an eye'. A mutation in a single *Hox* gene can lead to a fully functional leg growing where the antenna should be (Figure 11.3.4). This means a *Hox* gene has control over the genes that are involved in the development of all the cell types in both leg and antenna.

Hox genes (a subset of homeotic genes) are a group of master regulatory genes that control the development of body segments in the embryo.

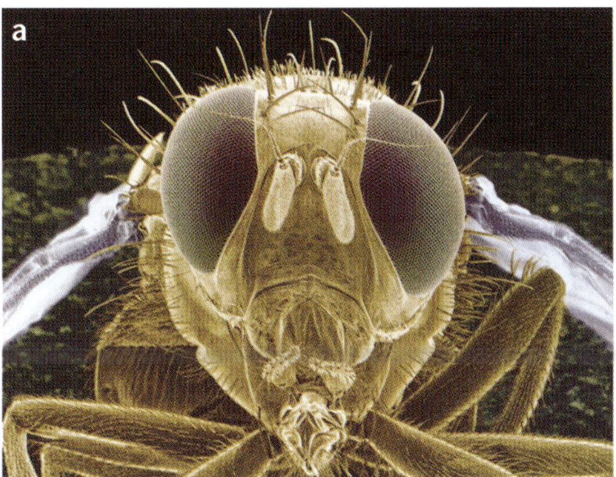

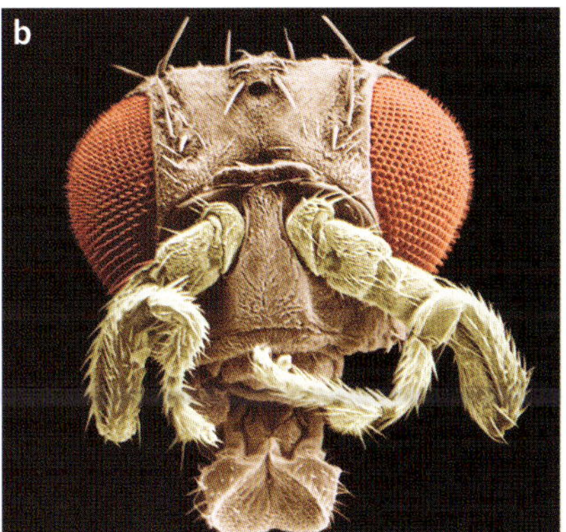

FIGURE 11.3.4 A mutation in a *Hox* master gene can cause a leg to form instead of an antenna. *Hox* genes are involved in the correct placement of body parts. (a) Wild-type *Drosophila* and (b) antennapedia *Hox* mutant *Drosophila* with legs instead of antennae.

Hox genes code for transcription factors that have the role of specifying how different structures in an embryo will be arranged. These transcription factors act by regulating the expression of other regulatory genes or the expression of structural genes. In this way, the *Hox* genes cause the embryonic cells to differentiate. It can be said that they are high in the developmental gene hierarchy (Figure 11.3.5).

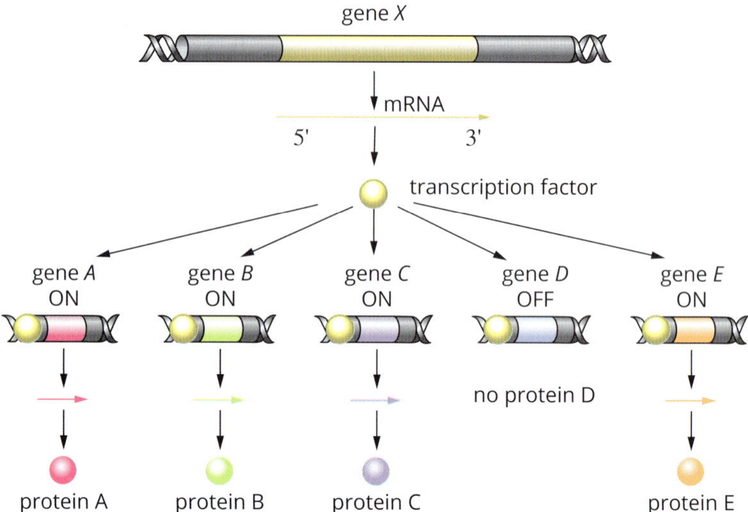

FIGURE 11.3.5 The *Hox* master gene (gene *X*) produces a transcription factor protein that changes the way other genes are transcribed and translated. The *Hox* gene can be described as higher in the gene hierarchy. In this example, genes *A–E* are lower in the gene hierarchy.

Hox genes are highly **conserved** across organisms that are very distant in evolutionary terms. They were first discovered in the fruit fly *Drosophila*, and homologous genes have since been found in vertebrates. As with other highly conserved genes, they are required for basic cellular functions, and mutations in the sequence of *Hox* genes would lead to non-viable forms.

EVOLUTION THROUGH CHANGES IN MASTER REGULATORY GENES

An understanding of master genes has allowed scientists to explain possible mechanisms for the explosion of biodiversity that occurred at various times in the Earth's history. Single mutations in these master genes allow many new phenotypes to develop in a short time.

We will now consider two examples of how chance mutations in master regulatory genes led to the evolution of novel phenotypes.

Cichlid fish in East Africa

Cichlids are a large family of brightly coloured freshwater fish that show enormous diversity in size and appearance. There are about 1650 species found worldwide, with new species still being discovered. Cichlids are found in the Americas and Asia, but Africa is home to the greatest number of species (at least 1600). Here, they are found in a variety of lakes and streams around the central Lake Tanganyika, East Africa. Studies of the surrounding landscape suggest that the lake has been through periods of flooding and drought over millions of years. As a result, populations of cichlid fish from Lake Tanganyika have been periodically isolated and exposed to different selection pressures, leading to the rapid diversification of the cichlid species (Figure 11.3.6). This is an example of adaptive radiation.

When studying cichlid fish from lakes in different geographic locations, scientists were initially puzzled about how the fish species in the lakes were so similar to each other in their feeding adaptations despite being isolated from one another. It seemed that different species of cichlids have responded to environmental pressures independently in exactly the same way, allowing them to adapt to similar feeding niches in the different lakes. Lakes that were far apart had cichlids that had adapted to eat snails in the same way (by developing short stout jaws that allowed them to crush shells) and cichlids that had adapted suction-feeding structures to feed on algae in the same way (with elongated, slender jaws, as shown in Figure 11.3.6).

FIGURE 11.3.6 Cichlid fish and their feeding niches in the three African lakes with the highest species diversity: Lake Malawi, Lake Tanganyika and Lake Victoria. Despite being isolated from each other, the fish have responded to environmental pressures in a similar way, resulting in similar jaw morphology and feeding habits.

Scientists considered that the jaw similarities between fish living in different lakes could be due to sharing a common ancestor. However, molecular studies suggested that species of fish with the same type of jaw had migrated in opposite directions at different times. This implied that jaw adaptations occurred after they shared a common ancestor, in different lineages separately. If this was to be explained through mutations in structural genes it meant that they had to have had exactly the same mutations in exactly the same structural genes controlling jaw growth. This was considered highly unlikely.

An alternative hypothesis considered the existence of a master regulatory gene that controlled the rate of gene expression of genes involved in the development of length, width and depth of the cichlid's jaw. A random mutation in this regulatory master gene affecting the rate or timing (heterochrony) of its expression could lead to a series of adjustments in the rate and timing at which other regulatory genes and structural genes were expressed during development. This hypothesis seemed more plausible as it would imply that only one or few mutations would result in the characteristics that these different species of fish had in common.

It was later found that the ***BMP4*** **(bone morphogenetic protein number 4) gene** is the heterochronic master regulatory gene controlling the cichlid's jaw phenotype. The *BMP4* gene regulates the development of cartilage and the development of cells that become muscle cells in an embryo. Cichlids with high levels of BMP4 protein present during their development have robust, heavy jaws with strong muscles. Cichlids with low levels of BMP4 protein develop longer slender jaws (Figure 11.3.7).

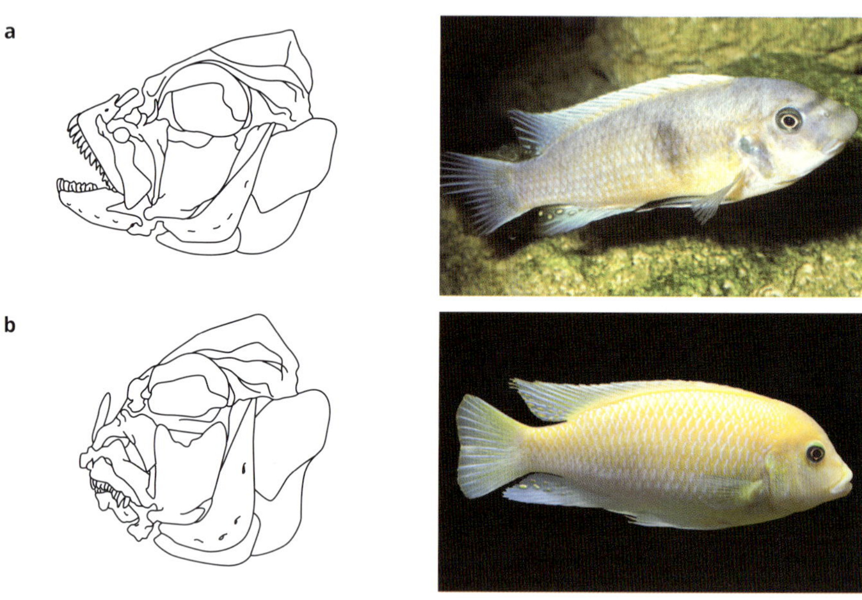

FIGURE 11.3.7 Adaptive radiation in the jaw structures of cichlid fish have led to a high diversity of feeding habits in lakes in East Africa. Similar selective pressures led to cichlid species in different lakes sharing strikingly similar morphology and feeding niches. Specialised biting species are characterised by short, robust jaws and closely spaced teeth (a), while species that feed by sucking are characterised by long, slender jaws and widely spaced teeth (b). It is thought that mutations in a master regulatory gene known as *BMP4* resulted in changes in the cichlid jaw structure.

Mutations in master regulatory genes cause significant changes to the phenotype of organisms. If these changes lead to selective advantages for the organism, rapid speciation is likely to occur.

Mutations in the *BMP4* gene explained the similarities in cichlid jaws present in different species in distant lakes in East Africa. As the fish migrated away from the central Lake Tanganyika, they were exposed to different selection pressures. A single or a few random mutations in the master *BMP4* gene caused the fish to develop different jaw shapes and sizes. Natural selection then ensured the survival of the fish with the most appropriate jaw phenotype. With the same selection pressures in different lakes, the fish responded by evolving similar shaped jaws. Over time, continued morphological change between the different feeding forms resulted in reproductive isolation and speciation, leading to the high species diversity in these lakes today.

Galapagos Islands

When Darwin first examined the variety of birds found across the Galapagos Islands, he had no idea that most of them were finches and belonged to the same family. It was only when he returned to England that John Gould, the famous ornithologist (bird specialist), demonstrated that these species were slight variations of each other. Some, such as the warbler finch, had narrow pointed beaks; others, such as the large ground finch, had strong wide beaks. Darwin used John Gould's observations of beak size and length, along with his own records of the types of food available on each island, to develop his theory of evolution via natural selection.

In the Galapagos Islands, each island has a different environment with different foods available, from cacti to large seeds. These environments provided a variety of selection pressures for birds that migrated between the islands. For example, if a warbler finch, which has a slender beak, migrated to an island with only large seeds and few insects, it would struggle to find enough food to eat, resulting in selection pressure and adaptation to the local environment over time. Table 11.3.1 shows some of the Galapagos finches that diverged from a common ancestor and adapted to their particular environments.

Type of finch	Genus and species	Bill type	Environment / Food type
ground finches	• six species in genus *Geospiza*, including:	• crushing bills, species specialise in eating different sized food	• widespread • live in coastal areas and low-lands, feeding on the ground
	• *G. magnirostris*	• largest finch, with large bill	• seeds
	• *G. scandens*	• longer more pointed bill	• cactus
tree finches	• five species in genus *Camarhynchus*, including:	• grasping bills, one parrot-like	• live in forests, feeding in trees
	• *C. pallidus*	• woodpecker finch uses tools—twigs and cactus spines	• insects, grubs
	• *C. parvulus*	• small tree finch, small bill	• insects
vegetarian tree finch	• only one species in genus • *Platyspiza crassirostris* • absent from outlying islands	• a large bird with a very heavy bill used to pull buds from plants	• lives in forests and feeds on plants, eating fruit, buds and soft seeds
warbler finch	• two species in genus, including:		• widespread species occurring on all of the islands
	• *Certhidea olivacea*	• slender bill	• only feeds on insects
		• searches for food among leaves and branches	• sometimes catches insects when flying

TABLE 11.3.1 Different species of finches that have resulted from adaptive radiation to different feeding niches on the Galapagos Islands.

Roles of BMP4 *and* CaM *in beak formation*

In a bird embryo, the length, width and muscle strength of the upper and lower beak are controlled by a number of genes. The expression of each gene must be controlled to ensure that the correct type and amount of proteins are produced at the right time in an embryo's development.

The master regulatory gene *BMP4* is involved in beak development in finches. It is expressed at the beginning of beak development and, being a master regulatory gene, controls the rate at which other genes are expressed. High levels of BMP4 protein during beak development cause a wide deep beak to be produced, similar to that of the large ground finches (Figure 11.3.8). If BMP4 protein is limited during embryonic development, the finches will develop small slender beaks, like that of the cactus finch (Figure 11.3.8).

A second master gene, the *CaM* gene, is responsible for the production of the calcium-binding protein calmodulin (CaM). Calmodulin regulates the activity of transcription enzymes by binding to them, changing the rate at which other genes are expressed. Differences in the level of CaM in finches' beaks can result in a length difference of up to 14%. Cactus-feeding finches produce large amounts of CaM during their embryonic development, resulting in longer beaks (Figure 11.3.8). Their long, thin beaks enable them to punch holes in the cactus fruit and eat the fleshy pulp inside. In an experiment, chicken embryos were treated with CaM. Chickens exposed to high levels of CaM developed elongated slender beaks.

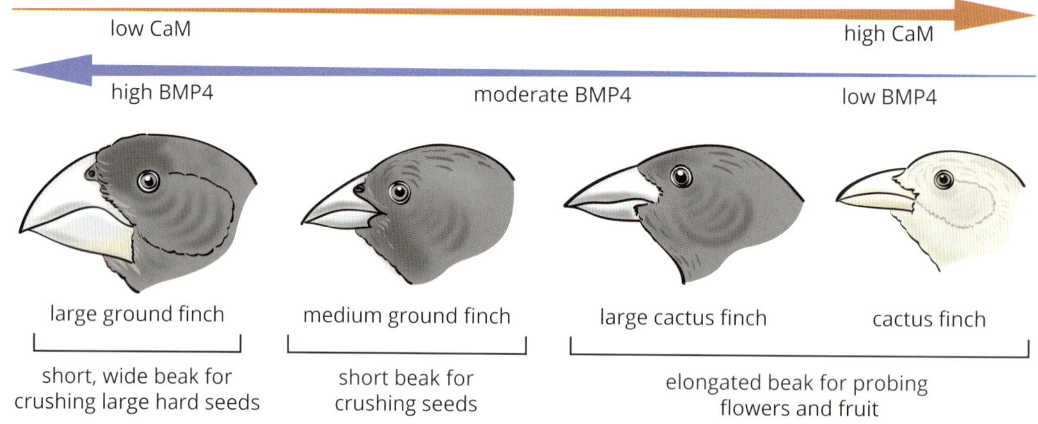

FIGURE 11.3.8 The shape of finch beaks is dependent on the amount of the regulatory proteins BMP4 and CaM present during their development. The production of these proteins is controlled by master regulatory genes. Mutations in these genes can lead to significant changes in the beak structure of finches.

BIOLOGY **IN ACTION**

Research sheds light on the role of *Hox* genes in the development of firefly lanterns

Beetles from the family Lampyridae, also known as fireflies or lightning bugs, are well known for their striking glowing abdomens (Figure 11.3.9). Adult fireflies use this light to attract mates or prey, while the larvae (called glow worms) use it as a deterrent for predators. The light emitted by fireflies is a product of a chemical reaction, known as bioluminescence, which occurs in a specialised organ in their lower abdomen. Although this phenomenon has long fascinated scientists, it is only recently that the evolutionary history and genetic mechanisms behind the firefly's lantern are beginning to be understood.

FIGURE 11.3.9 Firefly (*Photinus* sp.) with bioluminescent abdomen. The glowing abdomen of these insects is used to attract mates or prey.

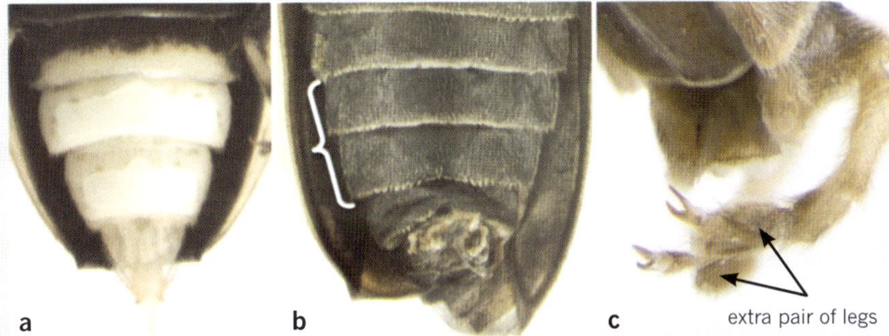

FIGURE 11.3.10 The effects of the down-regulation of *Hox* genes, *abd-A* and *Abd-B* on the development of the lantern organ and abdomen of the firefly (*Photuris* sp.). (a) Control individuals have normal development. (b) Down-regulation of *Hox* genes causes disrupted lantern organ development, excessive pigmentation and (c) an extra pair of legs in place of genitalia.

Researchers at the University of Arizona have found that the development of the lantern structures of fireflies is controlled by master regulatory genes, the *Hox* genes. Two *Hox* genes, *Abdominal-A* (*abd-A*) and *Abdominal-B* (*Abd-b*) were found to play significant roles in the development of lantern organs in fireflies. The researchers discovered this by down-regulating the functions of each of the genes in larvae of the firefly and observing the effects on the phenotypes of the adults. They found that the *Abd-B* gene in particular, was crucial for the correct development of the lantern organ. Severe abdominal deformities, including extra legs and excess pigmentation, were observed in the fireflies that had this gene down-regulated (Figure 11.3.10). The researchers concluded from these observations, that the *Abd-B* gene regulates the formation of the lantern organ and also represses abdominal pigmentation in order to develop a translucent abdomen, which improves light emission. The *abd-A* gene is thought to have secondary effects on the correct functioning of the lantern of fireflies.

Hox genes are well known for their role in determining the body plan during embryonic development. Because they play such a crucial role in the development and viability of organisms, sequences of these genes are usually highly conserved (have very few changes throughout evolutionary history). Until now, it was not known that *Hox* genes could develop the ability to regulate entirely new and complex structures, while still retaining the ability to regulate the development of other body structures. This research sparks many more questions about the role *Hox* genes have played in the evolution of unique and fascinating biological structures.

11.3 Review

SUMMARY

- Evolution does not always occur at a slow, gradual pace.
- Mutations in master regulatory genes give rise to novel phenotypes and may lead to rapid speciation if those phenotypes are advantageous.
- Master regulatory genes control the process of embryonic development.
- Some master regulatory genes can change the rate and timing of gene expression during embryonic development. This causes the expression of a gene to be sped up or slowed down.
- Some master regulatory genes can change the arrangement of body structures during embryonic development.
- *Hox* genes are a type of master regulatory gene that controls the arrangement of the body plan along the head-to-tail axis during embryonic development.
- Master regulatory genes are at the top of the gene hierarchy, controlling the expression of other genes.
- Master regulatory genes are highly conserved across different species because they are so important for correct development and biological functioning.
- Cichlid fish in East Africa are an example of adaptive radiation.
  - The *BMP4* (bone morphogenetic protein number 4) gene regulates the development of cartilage and muscular cell development in the jaws of a cichlid fish.
  - Mutation of the *BMP4* gene causes variations in the size and shape of the jaw of a cichlid fish.
- Darwin's finches on the Galapagos Islands are also an example of adaptive radiation.
  - The *BMP4* gene and *CaM* gene control the size and shape of a finch's beak.
  - BMP4 protein controls the width of the beak. The more BMP4 present during embryonic development, the wider the beak will be.
  - CaM protein controls the length of the beak. If it is present in large amounts during the bird's embryonic development, it will develop a longer beak.

KEY QUESTIONS

1. Complete the following statement by filling in the gaps.
   Mutations in ___________ regulatory genes are thought to cause changes in ___________ genes. These changes can result in major ___________ and functional changes and may lead to ___________.
2. Why are mutations in master regulatory genes relatively rare?
3. Select the most likely explanation for the evolutionary development of bat wings.
   A A bat's wing results from the finger bones growing faster than the rest of the bat's body.
   B A bat has a different set of genes associated with limb development that are not found in any other organism.
   C A bat's wing results from the finger bones continually growing throughout its life.
   D A bat's wing is homologous to a bird's wing.
4. What is the main function of *Hox* genes and why are these genes considered to be at the top of the gene regulation hierarchy?
5. Which of the following statements about cichlid evolution is not correct?
   A The diversity of cichlid fishes is an example of adaptive radiation.
   B The evolution of different cichlid fishes in East Africa is best explained by gradual accumulation of mutations.
   C Cichlid fishes are found in East Africa.
   D The different-sized jaws of cichlid fishes are a result of different amounts of BMP4.
6. How do the master regulatory genes *BMP4* and *CaM* influence the development of finch beaks?

Chapter review

KEY TERMS

BMP4 (bone morphogenetic protein number 4) gene
branch
clade
cladogram
conservative substitution
conserved
gene probe
heterochrony
homeotic gene
Hox genes
hybridised
leaf
lineage
Linnaean system of classification
master regulatory gene
maximum parsimony
molecular clock
mutation rate
node
non-conservative substitution
outgroup
phylogenetic tree
phylogenetics
phylogeny
phylogram
polytomy
root
semi-conservative substitution
sister taxa
taxon (plural taxa)
taxonomy

KEY QUESTIONS

1 The melting temperature of DNA from a single species is 86 °C. A DNA hybridisation experiment was performed between three species A, B and C. The results of the experiment are shown in the table below.

DNA mix	Melting temperature (°C)
A–B	78
A–C	72
B–C	81

Which cladogram is generated from this information?

A

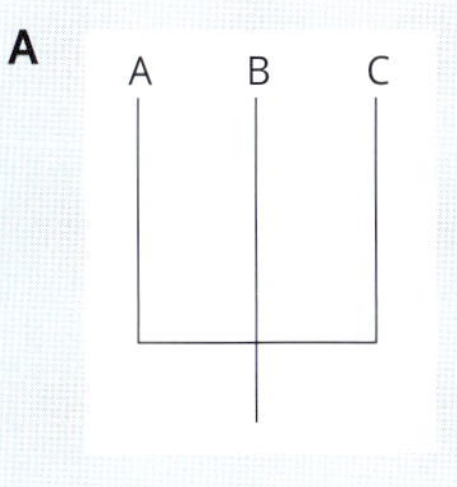

B

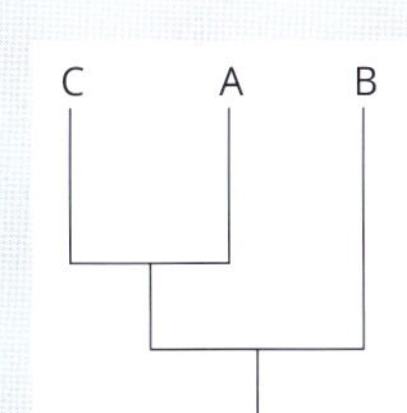

C

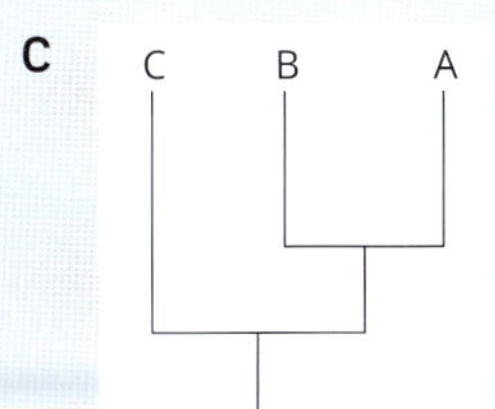

D

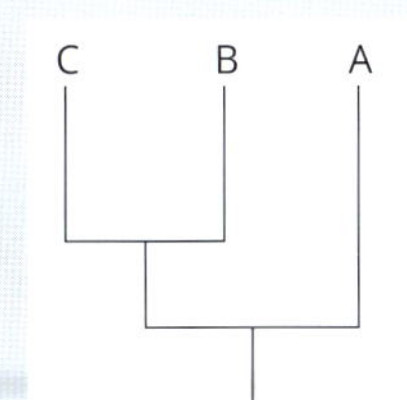

2 Which graph would best illustrate the assumed rate of the molecular clock?

A

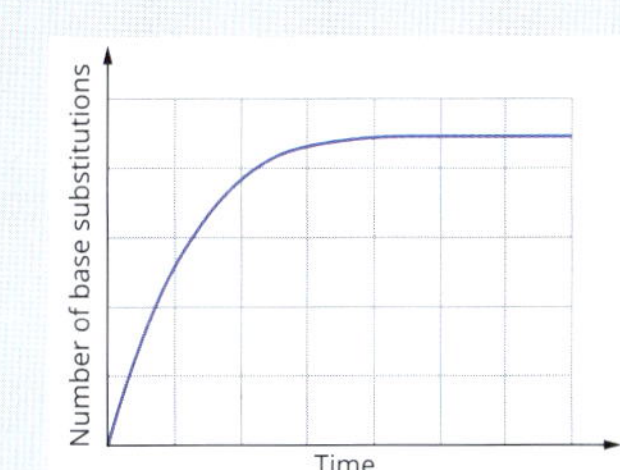

B

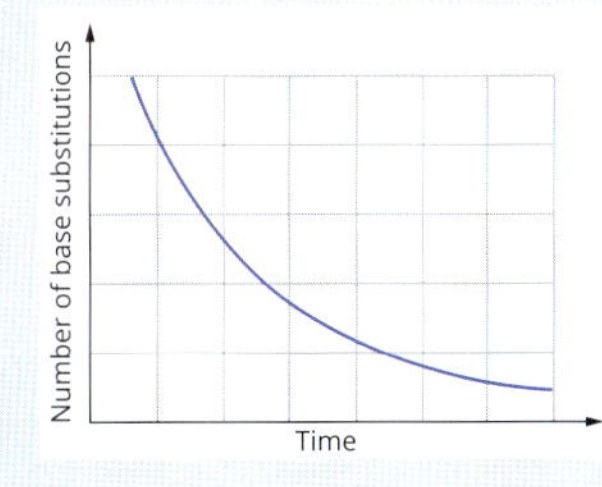

C

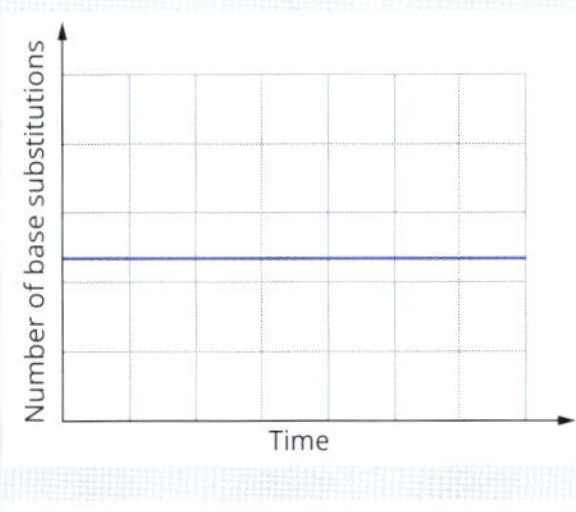

D

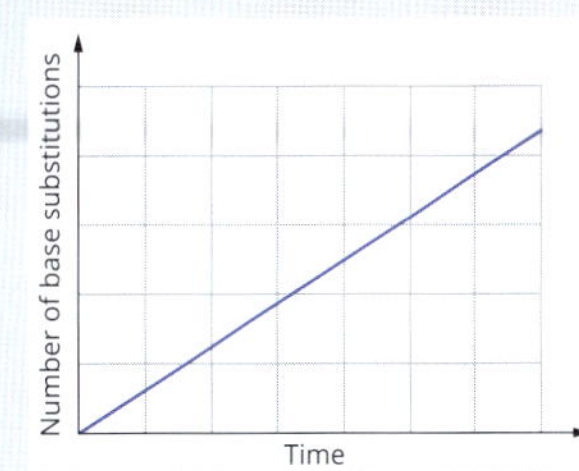

3 Which of the following statements best describes the phylogenetic tree shown below?

A scaled and rooted
B unscaled and rooted
C scaled and unrooted
D unscaled and unrooted

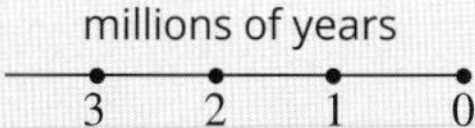

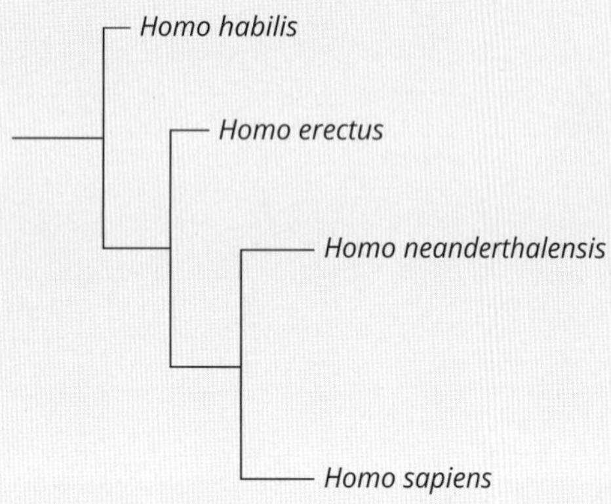

4

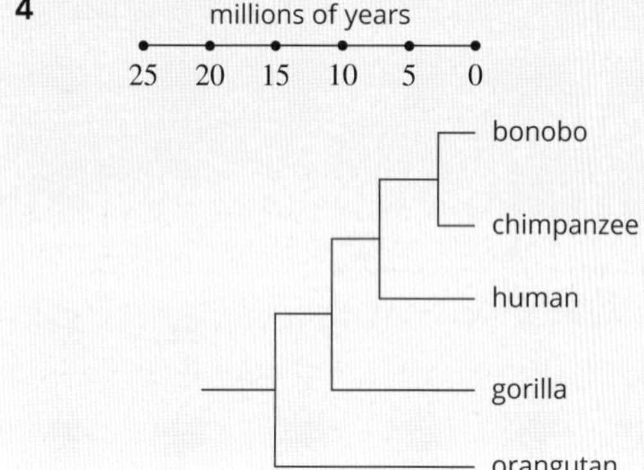

According to the phylogenetic tree of selected primates shown above, which of the following statements is true?

A Humans are more closely related to chimpanzees than they are to bonobos.
B The most recent common ancestor of gorillas and bonobos lived about 15 million years ago.
C The most recent common ancestor of bonobos and chimpanzees lived about 3 million years ago.
D Humans and orangutans last shared a common ancestor around 10 million years ago.

5 It has been established that human mitochondrial DNA is maternally inherited. Using this understanding, scientists have postulated that all living humans are related to a single female (Mitochondrial Eve), who lived between 140 and 200 thousand years ago. There is a general misconception that this means that all humans are descended from only one female. Explain why this interpretation of the situation is incorrect.

6 Changes in the genetic code can result from point mutations—insertions, deletions or changes in single nucleotides. Usually these mutations do not have an effect on the phenotype of the organism.

a **i** Describe a situation in which a change in the DNA sequence does not change the primary structure of the protein.

ii What name is given to this type of mutation?

b On some occasions, mutations in the DNA can cause changes in amino acids, possibly leading to changes in the structure and function of the protein product of a gene. Sometimes a change in the primary structure of the protein does not change its function. This situation arises when the substituted amino acid is of a similar shape and has similar properties to the original amino acid.

The properties and structures of the 20 amino acids are shown on the page opposite.

A DNA mutation leads to a substitution of the amino acid arginine.

i Identify an amino acid that would result in a conservative change to the polypeptide structure.

ii Identify an amino acid that would result in a non-conservative change to the polypeptide structure.

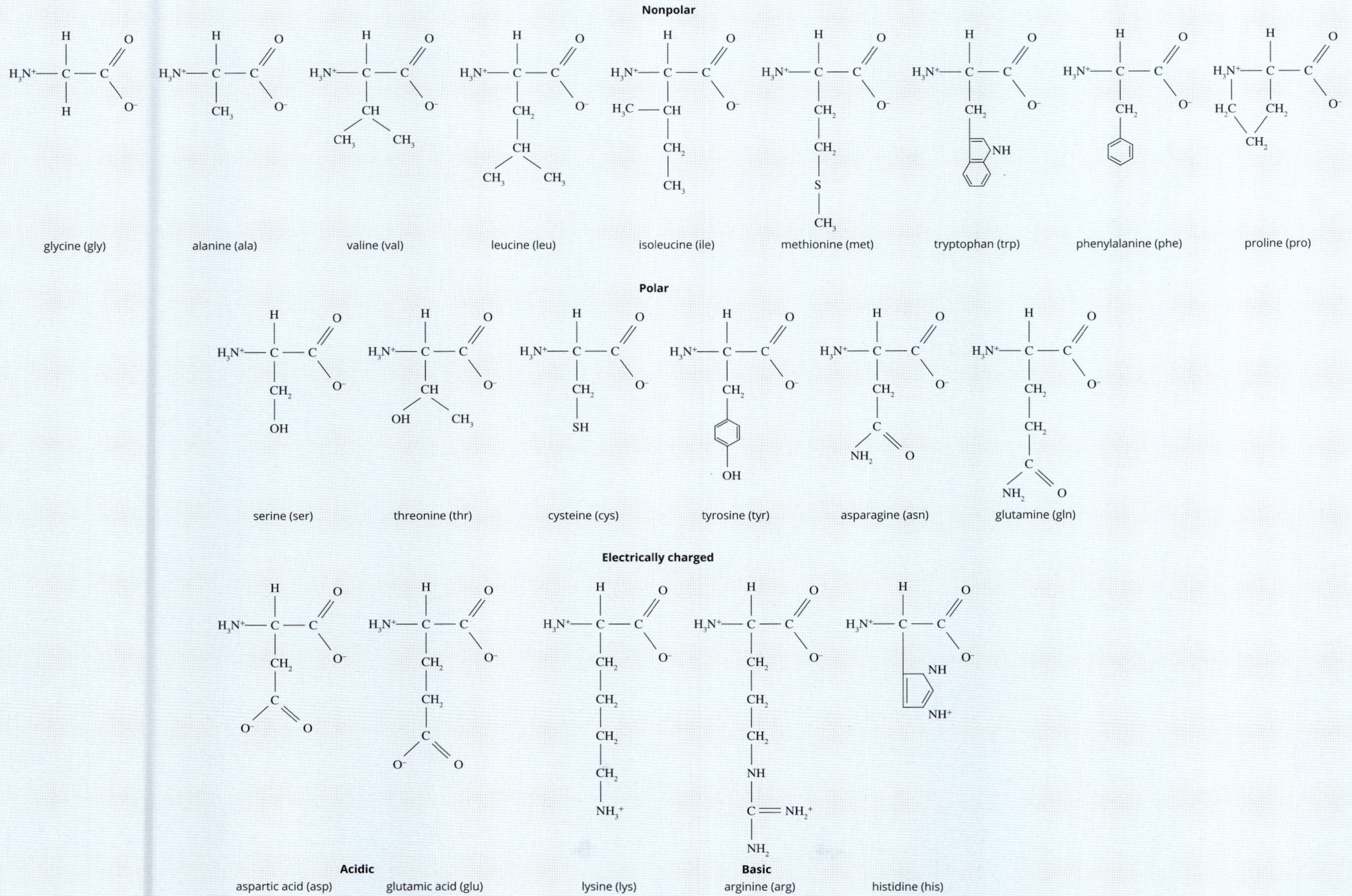

Nonpolar

glycine (gly)

alanine (ala)

valine (val)

leucine (leu)

isoleucine (ile)

methionine (met)

tryptophan (trp)

phenylalanine (phe)

proline (pro)

Polar

serine (ser)

threonine (thr)

cysteine (cys)

tyrosine (tyr)

asparagine (asn)

glutamine (gln)

Electrically charged

Acidic

aspartic acid (asp)

glutamic acid (glu)

Basic

lysine (lys)

arginine (arg)

histidine (his)

7 The tyrosine-related protein 1 gene (*TYRP1*) is a gene that codes for an enzyme that is involved in the production of melanin in the skin, hair and eyes in humans. The gene is also found in many other vertebrates where it also codes for melanin production. Consider the table below, which shows the amino acid sequence for part of the protein. The first column shows the sequence in humans. The other columns show the differences between the sequence in humans and the other species. Empty boxes mean that the amino acid is the same as that in a human.

a **i** Based only on the data given, which organism is most closely related to humans? Justify your opinion.

ii Based only on the data given, which organism is most distantly related to humans? Justify your opinion.

b *Gallus gallus* is a chicken and *Takifugu rubripes* is a pufferfish. Most evolutionary data indicates that chickens have a more recent common ancestor with humans than do pufferfish, but a comparison of the TYRP1 protein of these two organisms and the human protein indicates that pufferfish are more closely related than chickens to humans. Why wouldn't scientists rewrite the evolutionary tree based on this data?

Amino acid position	Species							
	Homo sapiens **(human)**	***Capra hircus*** **(goat)**	***Canis lupus familiaris*** **(dog)**	***Ovis aries*** **(sheep)**	***Mus musculus*** **(mouse)**	***Gallus gallus*** **(chicken)**	***Xenopus laevis*** **(frog)**	***Takifugu rubripes*** **(pufferfish)**
	leu						ser	ala
	ile						leu	
	ser							
	phe		leu			gln	ser	ala
280	asn							
	ser							
	val					ile		ile
	phe							
	ser							
	gln					thr		arg
	trp							
	arg							
	val							
	val					leu		
290	cys							
	asp			glu				
	ser						phe	
	leu					ile	val	val
	glu							
	asp				glu			
	try							
	asp						glu	
	thr					ser	ser	
	leu							
300	gly							
	thr							
	leu					ile	ile	val
	cys							
	asn							

8 The diagram below is a cladogram illustrating the evolutionary relationships between various groups of organisms.

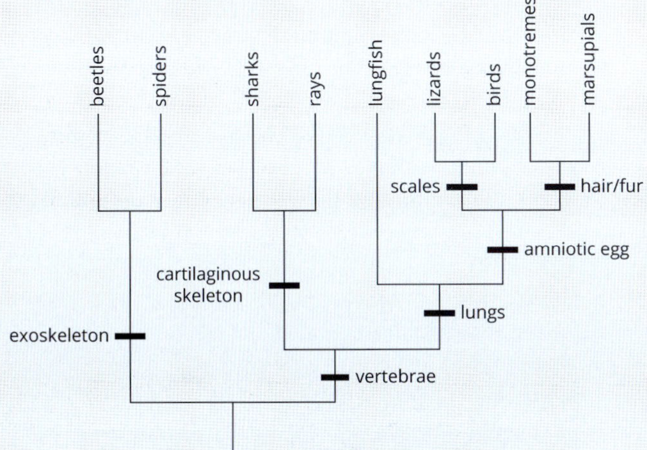

a What extra information could be drawn from the diagram if it had been drawn as a phylogram?

b i Are lungfish more closely related to rays or lizards? Explain.

ii What characteristic arose in the common ancestor of lizards, birds, monotremes and marsupials and is thus present in only that clade today?

c For each group, identify whether it is an example of a monophyletic, paraphyletic or polyphyletic grouping:

i lungfish, sharks and rays

ii lizards, birds, monotremes and marsupials

iii birds, lizards and beetles.

9 The ratite birds have long interested palaeontologists and biologists as they provide something of a mystery. Members of the ratite group (which includes emus and ostriches) are unable to fly. The mystery to be solved was how, if ratites never flew, did they become distributed across the world.

The first hypothesis suggested for their distribution was that they had a common ancestor that lived on the southern supercontinent, Gondwana. This resulted in the formation the many ratite species when Gondwana broke apart over a period of 85 million years.

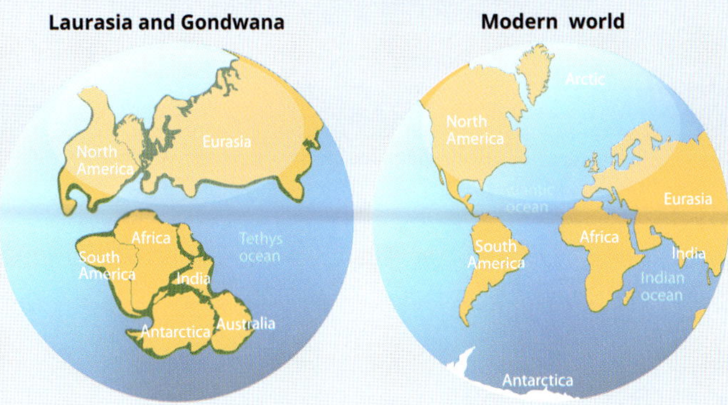

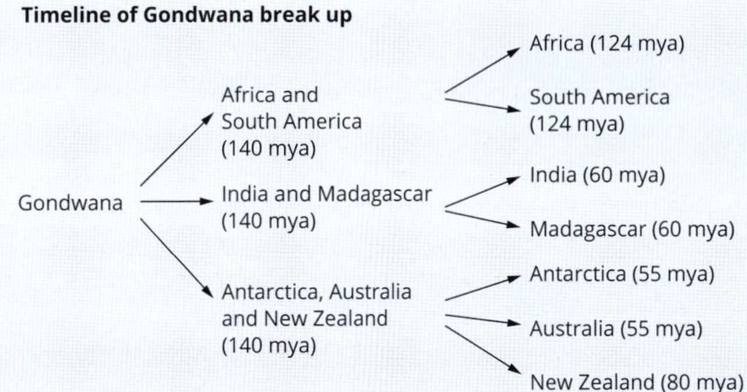

a Explain how the breakup of Gondwana could have resulted in the formation of species as different as the ostrich and the kiwi.

b The phylogenetic tree for the ratites was hypothesised, on the basis of their morphology, to be as shown below.

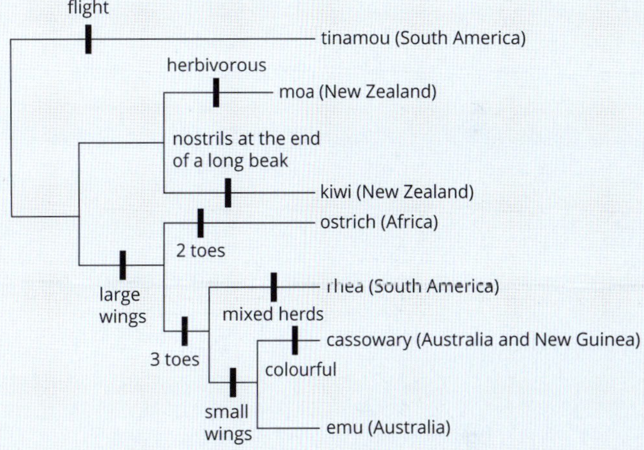

i Which group could be considered to be an outgroup in the tree above?

ii Consider the phylogenetic tree above and the timeline of the continents breaking away from Gondwana. Identify one piece of evidence from the tree that supports the hypothesis that the ratites developed from populations that became isolated when Gondwana broke apart.

iii Identify one piece of evidence in the phylogenetic tree above that refutes the hypothesis that ratites developed as a result of the breakup of Gondwana.

Question continued overleaf

c One extinct member of the ratite group that has not been included in the phylogenetic tree is the elephant bird of Madagascar. Madagascar is an island found off the east coast of Africa. There are no ratites in India and no evidence that they ever lived there, so one hypothesis is that Madagascar was colonised from Africa by rafting. (An organism gets trapped on a log or something similar and is transported across the ocean to an island.)

- **i** Why is it relevant that there are no ratites in India?
- **ii** Considering all of the data presented so far, which of the ratites in the phylogenetic tree on the previous page should be most closely related to the elephant bird?

d In the last few years DNA analysis has advanced to the stage that it has become possible to sequence DNA from fossils of the Moa, which has been extinct for around 600 years, and the elephant bird, which has been extinct for around 250 years. The results of these analyses along with the analysis of the extant (living) species of ratites led to a complete revision of the ratite phylogenetic tree, along with the development of an entirely new hypothesis to explain their distribution.

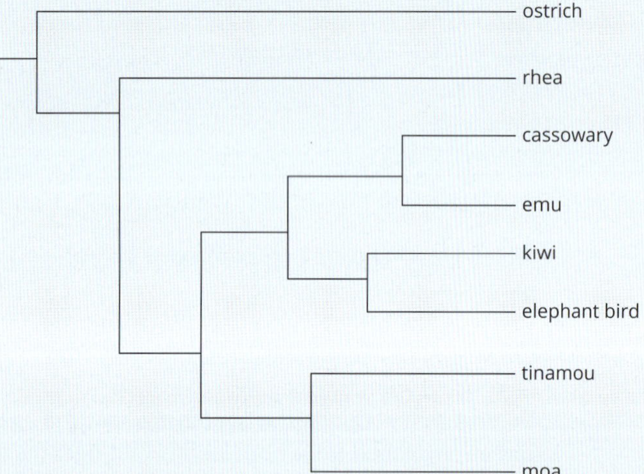

- **i** Explain how evidence from the DNA analysis disagrees with the theory that the ratite distribution occurred due to the separation of Gondwanan populations of an ancestral flightless ratite bird.
- **ii** It has now been hypothesised that the ratite ancestors were all flighted birds whose ancestral populations flew to where they are found today and that the loss of flight occurred independently on several different occasions in this group. Explain the type of evolution that this represents.

10 In the past, the primary method taxonomists used to work out the evolutionary relationships between organisms was to investigate the morphology of the organisms. Important characteristics were compared and scored in a table called a character matrix. Observe the insects below.

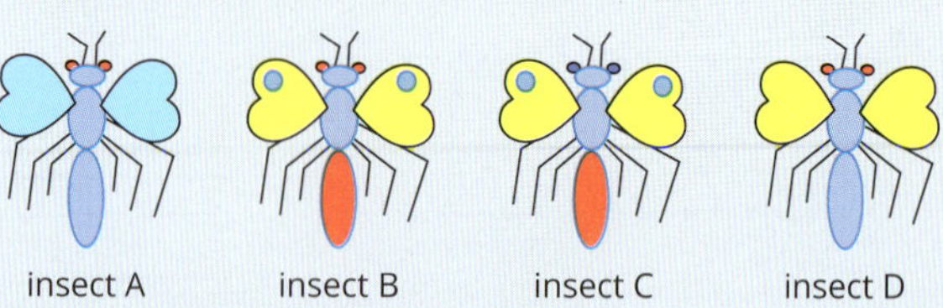

a Complete the character matrix for the insects. Use '0' for ancestral traits and '1' for derived traits. Assume that insect A is an outgroup for these species.

Trait	A	B	C	D
yellow wings				
wing spots				
red abdomen				
blue eyes				

b Draw a cladogram to show the evolutionary relationship between the insects. Include significant traits on the diagram.

11 In the past phylogenetic trees could only be generated using morphology. Modern taxonomists tend to rely on scoring the differences in DNA between two species in order to determine how closely related they are. Scientists were studying the relationships between a number of species in the genus *Conus*. This is a genus in the phylum Mollusca. All members of the genus have cone-shaped shells.

Conus marmoreus shell

Cytochrome *c* oxidase I (COI) is an enzyme found in many species, including the four species of *Conus* below.

Species	
Conus marmoreus	M
Conus chaldeus	C
Conus omaria	O
Conus magnus	MG

The grids below show the base sequence for part of a gene coding for COI. The grids contain 21 triplet nucleotides of the COI gene of all four species and have been coloured to make observing and counting differences easier.

Species	1			2			3			4			5			6			7			8			9			10			11		
M	G	G	T	C	A	A	C	A	A	A	T	C	A	T	A	A	A	G	A	T	A	T	C	G	G	G	A	C	A	T	T	A	T
C	G	G	T	C	A	A	C	A	A	A	T	C	A	T	A	A	A	G	A	C	A	T	T	G	G	G	A	C	A	T	T	G	T
O	A	C	T	A	A	T	C	A	T	A	A	G	G	A	C	A	T	G	G	G	G	A	C	A	T	T	A	T	A	T	A	T	T
MG	G	G	T	C	A	A	C	A	A	A	T	C	A	T	A	A	A	G	A	T	A	T	C	G	G	G	A	C	A	T	T	G	T

Species	12			13			14			15			16			17			18			19			20			21		
M	A	T	A	T	T	T	T	A	T	T	T	G	G	A	A	T	G	T	G	A	T	C	T	G	G	T	T	T	G	G
C	A	T	A	T	T	T	T	A	T	T	T	G	G	G	A	T	A	T	G	G	T	C	C	G	G	T	C	T	G	G
O	T	T	A	T	T	T	G	G	T	A	T	G	T	G	G	T	C	C	G	G	G	T	T	A	G	T	T	G	G	G
MG	A	C	A	T	T	C	T	A	T	T	T	G	C	A	A	T	A	T	C	A	T	C	G	G	G	A	C	T	A	G

a Complete the following table by counting the number of nucleotide differences between pairs of the species. Each pair only needs to be counted once so shaded squares do not need to be filled.

Species	*Conus marmoreus* (M)	*Conus chaldeus* (C)	*Conus omaria* (O)	*Conus magnus* (MG)
Conus marmoreus (M)				
Conus chaldeus (C)				
Conus omaria (O)				
Conus magnus (MG)				

b Use the data obtained from the DNA sequences to generate a possible cladogram for these species.

c An alternative to sequencing DNA and counting the number of nucleotide differences is DNA hybridisation, a technique that can estimate the degrees of genetic differentiation between the DNA of two organisms. A DNA hybridisation experiment was performed between the four members of *Conus* using the COI gene.

- **i** Hybridised DNA from which pair of species would have the lowest dissociation temperature?
- **ii** Explain why.

12 The X and Y chromosomes pair during meiosis due to small homologous areas at either end of the chromosomes (coloured blue). Most areas of the two chromosomes are not homologous (coloured grey in the Y and pink in the X) and contain different genes. An important region that is found only on the Y chromosome is the SRY region (coloured yellow), which determines gender in humans.

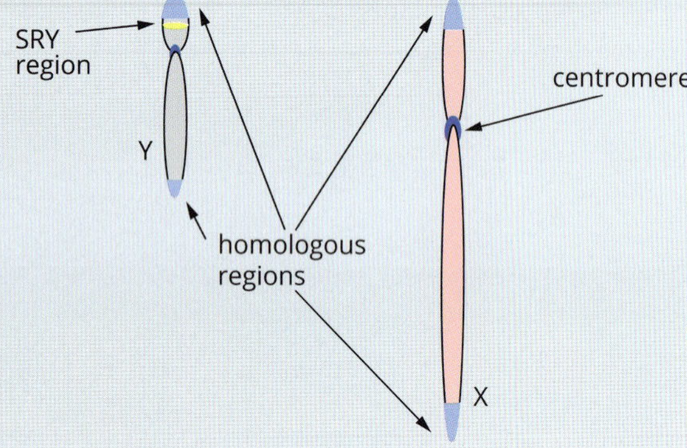

The SRY region of the Y chromosome contains the master regulatory genes responsible for controlling the expression of the genes that lead to the development of a male body plan in humans.

a Explain how master regulatory genes control the body plan of an organism.

In rare cases there is a translocation of the SRY region of the Y chromosome onto the X chromosome.

b Draw and label the chromosome pair after the translocation.

c **i** If a child of this male was to be XX, how would the translocation affect it?

ii If the child of this male was to be XY, how would the translocation affect it?

CHAPTER

12 Human changes over time

Learning outcomes

The quest for answers about our origins and the evolutionary paths that led us to where we are today is more exciting than ever, with rapidly advancing DNA technology and the discovery of new fossils and archaeological artefacts. Our evolutionary story is a dynamic one, changing as new information comes to light or new ideas are found to have greater explanatory power.

In this chapter you will learn about what it is that makes us human: our shared and defining characteristics from other primates, hominoids and hominins. You will also examine the characteristics of some of our known ancestors and come to understand patterns in hominin evolution, from the genera *Australopithecus* and *Paranthropus* to our own genus, *Homo.* You will learn about the most defining characteristic that sets humans apart from other animals, our rich culture, and discover how this evolved from the earliest hominins to present-day humans. Finally, the competing theories of our origins will be discussed, along with the evolutionary challenges we now face as the dominant species on Earth.

Key knowledge

- shared characteristics that define primates, hominoids and hominins
- major trends in hominin evolution from the genus *Australopithecus* to the genus *Homo*, including structural, functional and cognitive changes and the consequences for cultural evolution
- the human fossil record as an example of a classification scheme that is open to interpretations that are contested, refined or replaced when new evidence challenges them or when a new model has greater explanatory power, including whether *Homo sapiens* and *Homo neanderthalensis* interbred and the placement of *Homo denisovans* into the *Homo* evolutionary tree.

12.1 Defining humans

Homo sapiens is the Latin term for our species. The term translates as 'wise man' and refers to anatomically and behaviourally modern humans. *Homo sapiens* is one of the most widespread, adaptable and influential species to have ever existed. Under the Linnaean system of biological classification, *Homo sapiens* is a eukaryote and a member of the animal kingdom.

Among animals, humans are mammals, with the characteristics of body hair and the ability to suckle young (Figure 12.1.1). Humans are also classified as primates, having a grasping hand, bicuspid teeth, a short nose and well-developed eyes and brain. Within the Primate order, humans belong to the same family (**Hominidae**) as the great apes, which include orangutans, gorillas, chimpanzees and bonobos. All **hominids** lack a tail and have similar skeletal and skull features. As well as sharing many anatomical and behavioural features with all of the great apes, modern humans share 98.8% of their DNA with chimpanzees and bonobos, our closest living relatives. Although modern humans are very similar to chimpanzees and bonobos, we did not directly evolve from them or any other living primate. DNA and fossil evidence tells us that we last shared a common ancestor with chimpanzees and bonobos approximately 6–8 million years ago (Figure 12.1.2).

Chimpanzees and bonobos are our closest living relatives but we did not directly evolve from them. Our last shared common ancestor existed approximately 6–8 million years ago.

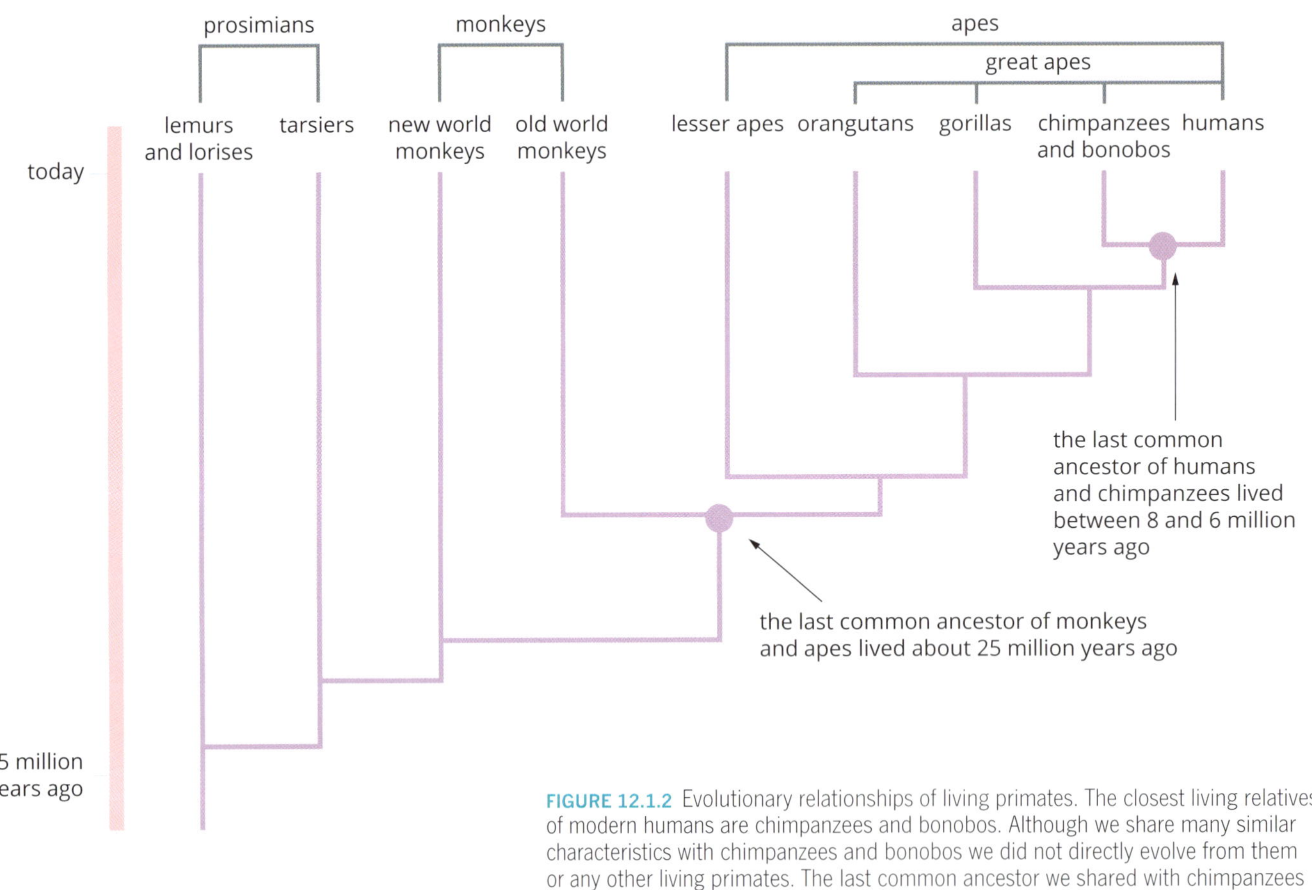

FIGURE 12.1.2 Evolutionary relationships of living primates. The closest living relatives of modern humans are chimpanzees and bonobos. Although we share many similar characteristics with chimpanzees and bonobos we did not directly evolve from them or any other living primates. The last common ancestor we shared with chimpanzees and bonobos lived about 6–8 million years ago. The lineage that led to our species includes many fossils of human-like species that walked upright.

FIGURE 12.1.1 A 'tree of life' depicting the evolutionary relationships of humans with other organisms.

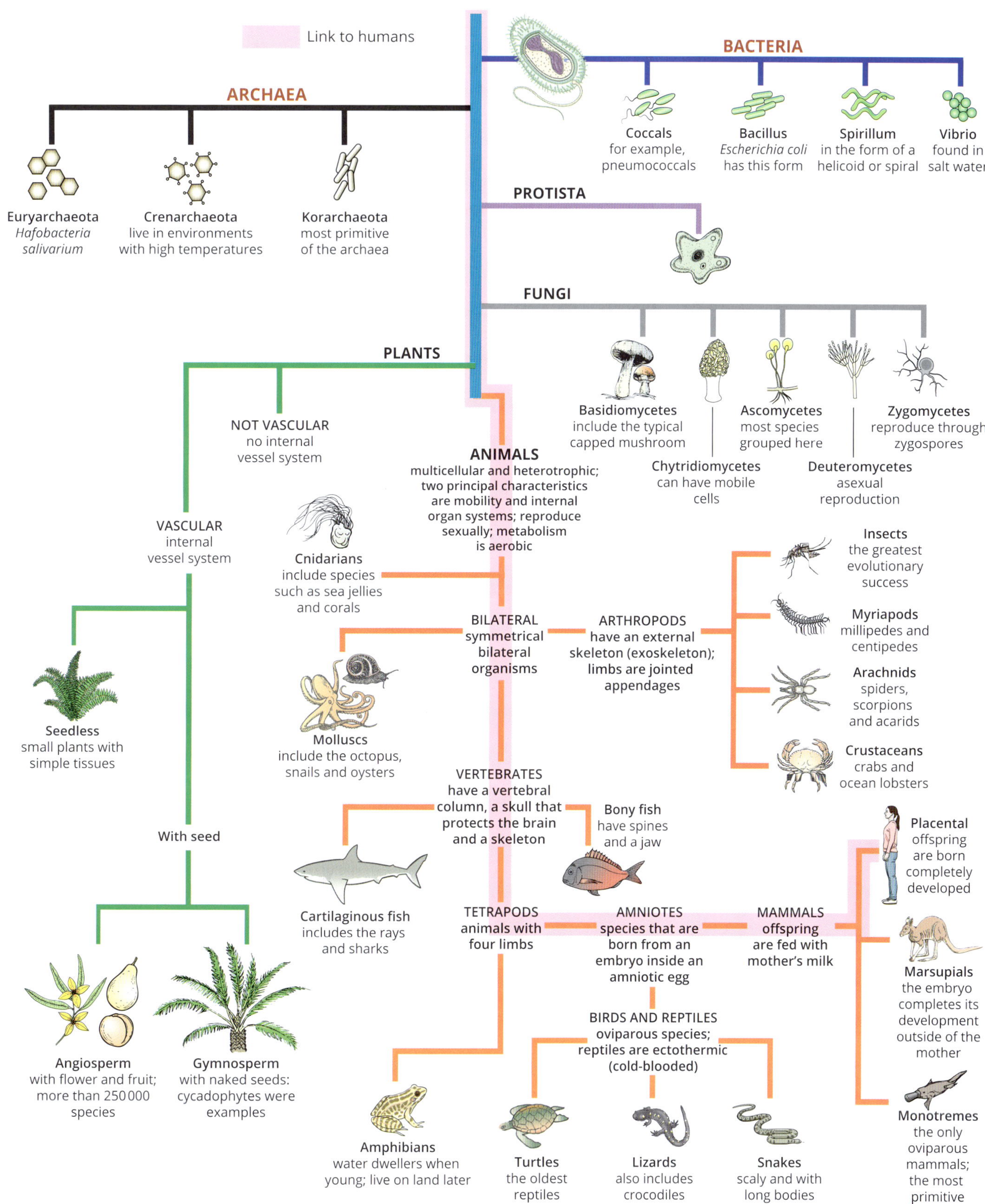

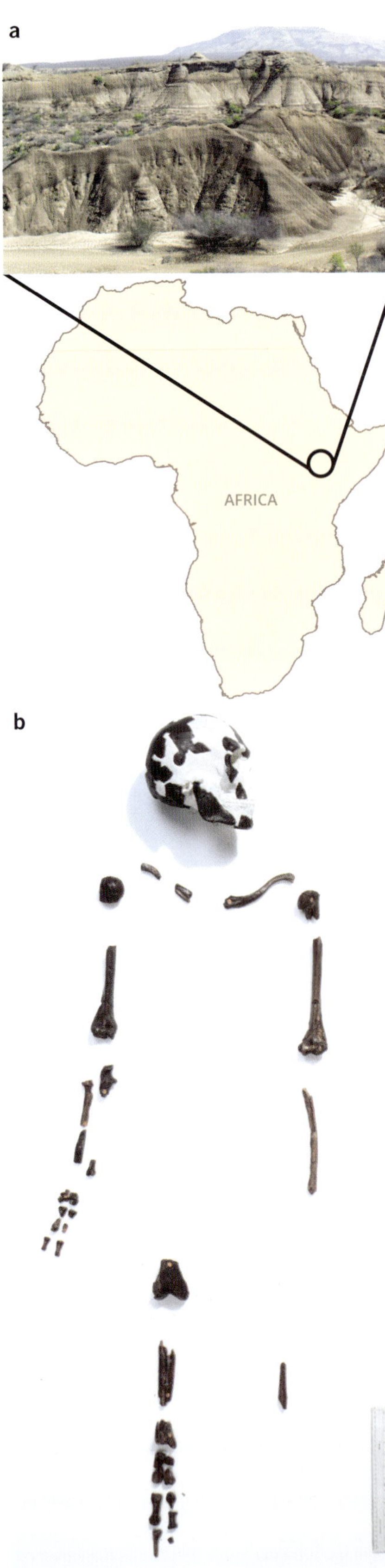

FIGURE 12.1.3 (a) The Omo Kibish formation in Ethiopia, Africa, is the site of the earliest *Homo sapiens* fossils found. These fossils date back to 195 000 years ago. (b) Omo 1, the oldest anatomically modern *Homo sapiens* specimen discovered in Omo Kibish in 1967.

There are various fossilised human-like species (**hominins**) that have been discovered, all characterised by walking upright. *Sahelanthropus* is the oldest known ancestor of modern humans, with skulls found in central Africa that date to approximately 7 million years ago. Some scientists think that *Sahelanthropus* may represent a common ancestor of humans and chimpanzees, but this is still debated in the scientific community. Species of the extinct genus *Australopithecus* are the closest relatives to our genus, ***Homo***. *Australopithecus* had a forward-jutting face, a small brow and a pronounced ridge above the eyes, but hands and teeth that were similar to ours. *Homo* diverged from an australopithecine approximately 2.8 million years ago. The genus *Homo* is characterised by a larger brain and includes 9–15 extinct species known to date (the identities of some fossil specimens are uncertain). Our species, *Homo sapiens*, is the only living one. Of the extinct species, *Homo habilis* is the oldest, found in Africa. A later species, *Homo erectus*, which had a larger brain than *Homo habilis*, spread into South-East Asia about 1.6 million years ago, possibly surviving to about 300 000 years ago.

Fossil and DNA evidence is continuing to uncover new information and helping us to tell our story; but many questions remain concerning the origins of modern humans and fossil species.

HOMO SAPIENS: MODERN HUMANS

Time range: 200 000 years ago–present day

Geographic range: worldwide

Our species, *Homo sapiens*, has been around for approximately 200 000 years, with fossil and genetic evidence placing the earliest of our species in Africa. The oldest fossil record of our species was found in a region called Omo Kibish in Ethiopia and has been dated to 195 000 years ago (Figure 12.1.3). Two specimens, known as Omo 1 and Omo 2, were recovered from this region between 1967 and 1974. Omo 1 has anatomically modern human morphology with some primitive traits, representing a transition from **archaic** *Homo sapiens* to early modern *Homo sapiens* (Figure 12.1.3). Omo 2 has more primitive features than Omo 1, such as a sloping forehead and more robust build, leading researchers to suggest that it belonged to a population that was transitional between *Homo heidelbergensis* and *Homo sapiens*. As both specimens have been dated to approximately the same time of 195 000 years ago, they may have co-existed. Omo Kibish and the surrounding regions are of great **archaeological** significance as this area is the source of many important findings regarding the origin of our species. For this reason, Ethiopia is currently thought to be the site where some of the first modern *Homo sapiens* lived.

Physical characteristics

Modern *Homo sapiens* can generally be characterised by a leaner, more agile (but less strong) **bipedal** build than our **predecessors**. Our pelvis is narrower and deeper than that of our ancestors, and our skulls have a high braincase and short base, with rounding at the back indicating reduced neck muscles. Our brow ridge is much flatter than that of our predecessors, and we have squarer eye sockets, smaller faces and reduced canine teeth (Figure 12.1.4).

Homo sapiens has one of the largest, most complex brains of all the hominin species. There has been a trend throughout hominin evolution for larger braincases, with brain size almost tripling in the last two million years. The braincase capacity of *Homo habilis* was approximately 600 cm^3, while *Homo neanderthalensis* had a braincase capacity of 1600 cm^3, the largest known brain size of all hominins (Figure 12.1.4). The average braincase capacity of present-day humans is about 1100–1300 cm^3, the second largest of the hominins. Although the braincase capacity of Neanderthals is larger than that of *Homo sapiens*, researchers have suggested that Neanderthal brains were specialised for vision and movement, with less brain volume devoted to social interactions and problem solving.

Homo sapiens have a complex larynx, which has allowed the development of remarkably complex languages. Our hands have a well-developed precision grip, which permits fine motor control rather than simply having opposable thumbs.

The cranial capacity of hominin species has tripled in the last two million years. *Homo sapiens* and *Homo neanderthalensis* have the largest cranial capacity of all hominins and their increased brain size and complexity allowed the development of sophisticated tools, innovation, social structures and cultures.

The earliest evidence of anatomically modern humans (*Homo sapiens*) was found in Omo Kibish in Ethiopia. The fossils were dated to 195 000 years ago.

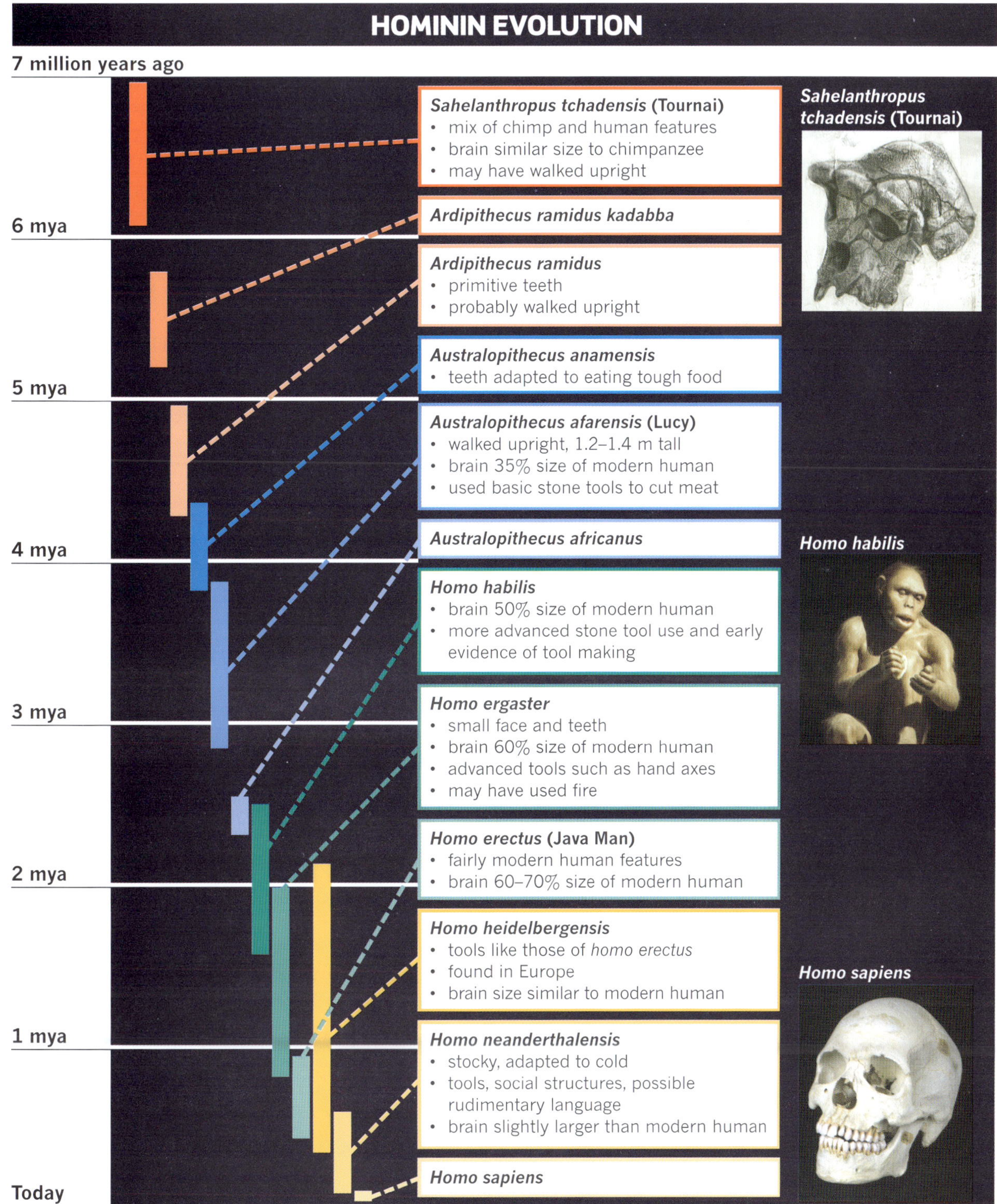

FIGURE 12.1.4 The evolution of the physical and behavioural characteristics of hominin species over the last 7 million years.

Behavioural characteristics

The complex brains of *Homo sapiens* enable much higher levels of communication and interaction between individuals. Our brains also allow for a much better understanding of, and consequent manipulation of, the living and non-living components of our environment compared to any other species. Complex behavioural traits are characteristics that distinguish modern *Homo sapiens* from earlier populations. The behaviour of modern *Homo sapiens* is characterised by the ability to plan, use abstract thinking, ritual and symbolism (e.g. in art, ornamentation and music), large prey capture and advanced tool use. The behavioural complexity of *Homo sapiens* led to the development of diverse cultures around the world.

BIOFILE

Hominins, hominids and hominoids

The taxonomic classification of humans has changed over time, with some confusion arising in regard to the terms 'hominin', 'hominid' and 'hominoid.' Humans belong to the subfamily Homininae and hence are hominins. Homininae also includes chimpanzees and gorillas, as well as extinct *Homo* species, *Australopithecus* and *Paranthropus*. We are also part of a larger group, the family Hominidae, which includes orangutans, hence the term hominid. Finally, the superfamily Hominoidea includes humans, the great apes and the lesser apes (gibbons), hence the term hominoid. So, depending on the group to which you are referring, make sure you get the term correct!

Cultural characteristics

Tool making and communication have had a significant impact on the cultural characteristics of *Homo sapiens*. We have a wide and varied diet, comprising plant and animal sources, and an ability to use fire for cooking and other purposes. The use of tools has allowed the development of complex shelters to protect against the elements, and larger dwellings to accommodate larger groups, promoting the development of social relationships and cooperative living. The establishment of communal living in villages, towns and now cities, brought about agriculture and resource trading. Modern humans have created societal structures that allow communities to work effectively and efficiently together. Our species has also made remarkable advances in developing art, music, dance, jewellery, rituals and complex world views. (In Section 12.2, you will learn more about the cultural evolution of *Homo sapiens*.)

BIOFILE

Tool use

Tool making was once thought to be a defining characteristic of the genus *Homo*, but basic stone tool use has now been found associated with *Australopithecus*. Australopithecines created some of the earliest tools, using sticks and rocks to help them capture and eat small prey. *Homo habilis* made more sophisticated tools, striking flakes off either side of stones. Such tools have been dated as being up to 2.6 million years old.

Tool use has had an important role in shaping culture in *Homo sapiens*. Some of the earliest examples of complex tools that represent modern human culture have been found in Border Cave in South Africa. These tools have been dated to 44 000 years ago and show evidence of complex behaviours and cultural practices, such as using weapons for large prey capture and beads for adornment (Figure 12.1.5).

Although *Homo sapiens* may not have developed the first tools, our larger brains enabled the development of very specialised tools and the use of tools to make other, more complex tools. The invention and manipulation of tools by modern humans has resulted in incredible feats in engineering and demonstrates the remarkable ingenuity of our species.

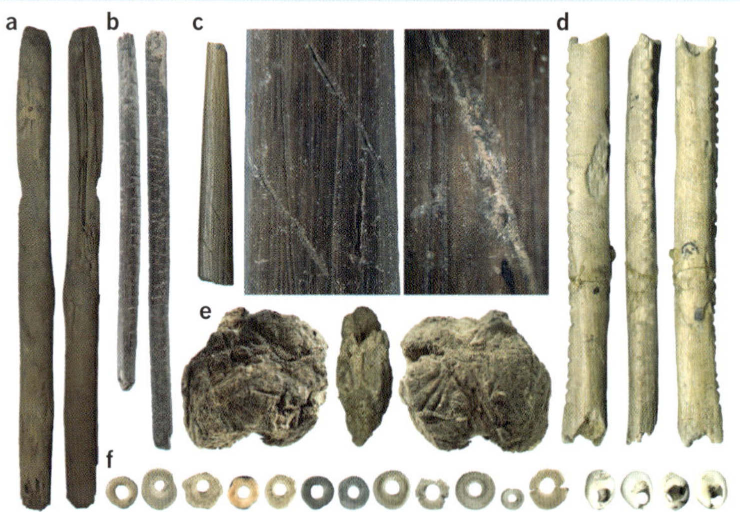

FIGURE 12.1.5 Tools found in Border Cave in South Africa. These are some of the oldest examples of complex tools, dated to 44 000 years ago, and may represent one of the earlier examples of modern human culture. The artefacts recovered from the cave included (a) digging sticks, (b) poison applicators, (c) an arrow point made of bone, (d) bones with notches carved into them, (e) beeswax mixed with resin and (f) beads made from ostrich eggs and marine shells.

GENUS *HOMO*

Fossils that have been classified as belonging to the genus *Homo*, the genus to which we belong, were **gracile**; that is, they have a relatively lightweight skeleton and a more upright bipedal stance (slender build). *Homo* fossils all have a relatively large braincase and reduced jaw size.

There are currently 10–16 species that are recognised within the genus *Homo*; the reason for this range is that some are thought to be **subspecies** or not distinct enough to warrant recognition as separate species. The number and evolutionary relationships of species changes as new evidence from fossils and DNA is uncovered.

Fossil and DNA evidence indicate that different *Homo* species existed at the same time in the same regions (Figure 12.1.6). There is evidence to suggest that some species interbred and gave rise to **hybrid species**.

FIGURE 12.1.6 Evolutionary relationships of currently recognised species of hominins, starting from the earliest known hominins, who appeared in the fossil record approximately 7 million years ago, to modern humans, the only living hominin species. Dates show when the species are first thought to have appeared.

BIOFILE

Homo naledi

A recent discovery of ancient skeletons in a cave in South Africa has changed the story of human evolution and given rise to a new species of hominin: *Homo naledi*. Partial skeletons of 15 specimens were found in the Rising Star Cave near Johannesburg, South Africa. *Homo naledi* has similar morphological features to modern humans, but also has many primitive features such as a small skull and hands and arms specialised for climbing (Figure 12.1.7). Researchers have not yet been able to date the remains but, based on some of the primitive physical features, estimate that *Homo naledi* may be one of the earliest *Homo* species. Given that *Homo naledi* also displays some modern traits, it is possible that it represents a transitional lineage between *Australopithecus* and *Homo* (Figure 12.1.8).

The bones of *Homo naledi* were found scattered on the floor of the cave and buried in shallow sediment. The odd positioning of the bones, piled together on the floor of the remote cave chamber, has led researchers to believe that *Homo naledi* may have deliberately placed the bodies there, using the cave as a burial chamber. This finding has significant implications for our understanding of the evolution of such complex behaviour, which was previously thought to have evolved much later.

The discovery of *Homo naledi* represents one of the greatest palaeoarchaeological finds in history, adding pieces to the puzzle of human evolution while raising even more questions that we are yet to answer.

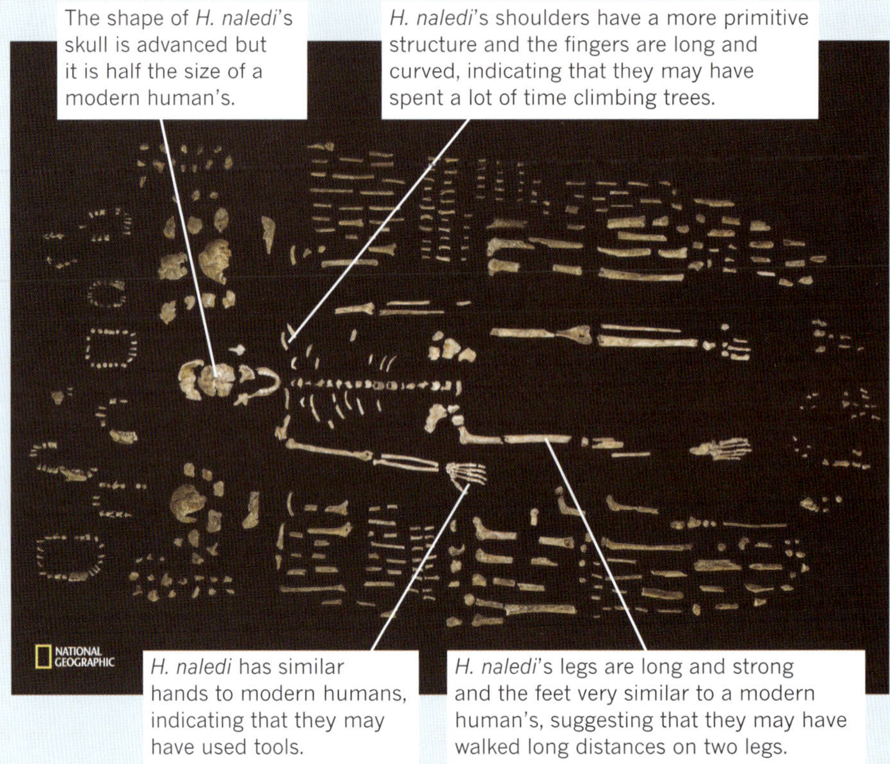

FIGURE 12.1.7 Partial skeleton of the newly discovered hominin species, *Homo naledi*. The specimen was found in a South African cave. It has features similar to modern humans as well as some primitive features retained from our more distant ancestors.

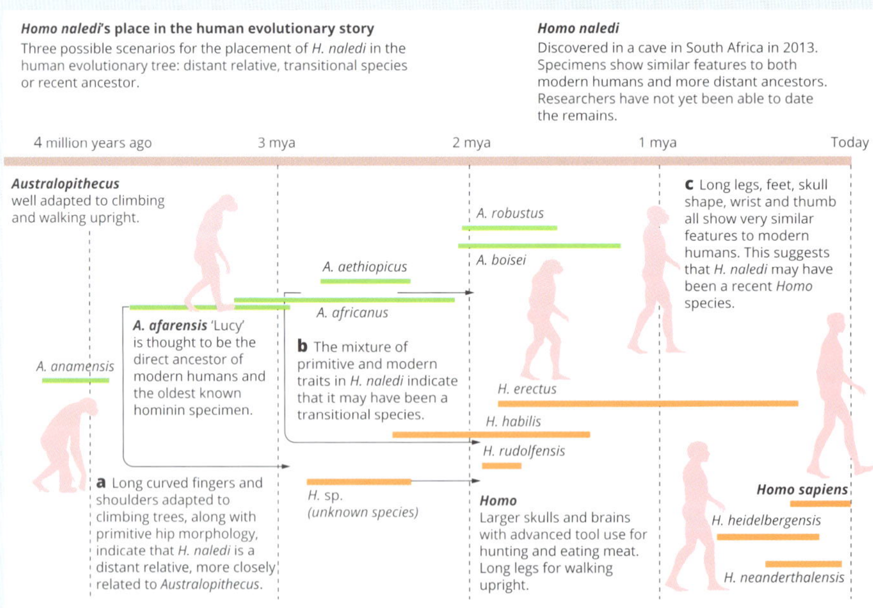

FIGURE 12.1.8 The evolutionary story of *Homo naledi* is currently uncertain. Based on the mix of primitive and modern physical features observed in *Homo naledi* specimens, researchers have suggested three possible scenarios for their origin: at the time of *Australopithecus* (a), between *Australopithecus* and *Homo* (b), or a more recent *Homo* origin (c).

Homo floresiensis

Time range: 17 000–95 000 years ago

Geographic range: Indonesia

Nicknamed 'the hobbit', *Homo floresiensis* is one of the most recent additions to the human family tree. It **coexisted** with *Homo sapiens* throughout its entire known time range and was the most recent *Homo* species to become extinct. Discovered in Indonesia on the island of Flores in 2003, *Homo floresiensis* was the smallest known member of the genus *Homo* (Figure 12.1.9). With a stature of just over one metre tall, and a small brain, oversized teeth and feet, this human relative may have suffered from 'island dwarfism'—an evolutionary process that results from isolation on an island with limited resources and selection pressures from predation.

Homo floresiensis made and used stone tools, successfully hunted pygmy elephants, coped with giant Komodo dragons, and most likely used fire.

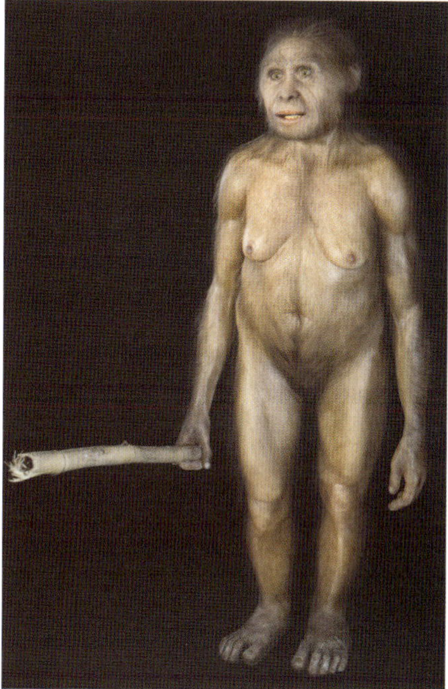

FIGURE 12.1.9 Reconstruction of *Homo floresiensis* based on skeletal remains found in Indonesia in 2003. This hominin appears to have dwarfism and stood just over one metre tall.

Homo denisovans

Time range: 41 000–125 000 years ago (still unconfirmed)

Geographic range: Russia to South-East Asia

Homo denisovans (Denisovan or Denisova hominin) is known only from a finger bone and two teeth that were discovered in Denisova cave in Siberia, Russia. The finger bone that was discovered belonged to a young female who lived approximately 30 000–50 000 years ago. The bone was broad and robust, suggesting that the species had a very robust build, possibly similar to that of the Neanderthals.

DNA from the fossils was well preserved due to the cold, dry conditions of the cave in which the bones and teeth were found. Analysis of mitochondrial DNA (mtDNA) from the finger bone revealed that Denisovans were closely related to Neanderthals and modern humans but genetically distinct.

Fossils from Neanderthals and modern humans have also been found in Denisova cave and evidence from molecular studies suggests that **interbreeding** occurred between Denisovans and both these species. Mitochondrial DNA shows that the Denisovans may have interbred with Neanderthals, with 17% similarity between Denisovan and Neanderthal genomes. Nuclear DNA studies have also shown that 3–5% of the DNA of present-day modern humans from Melanesian and Aboriginal Australian populations is shared with Denisovans, indicating that Denisovans also interbred with modern humans (see Figure 12.1.10).

With such little evidence available, much about this species is still yet to be discovered.

Evidence from DNA and fossils has shown that *Homo sapiens*, *Homo neanderthalensis* and *Homo denisovans* all coexisted and interbred, leaving genetic evidence of these extinct species in the DNA of modern humans.

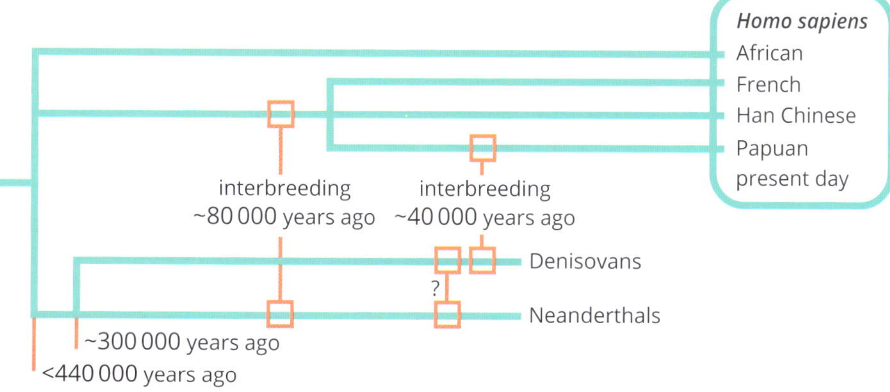

FIGURE 12.1.10 The evolutionary relationship of Denisovans with *Homo sapiens* and Neanderthals. DNA evidence from a finger bone suggests that Denisovans interbred with both Neanderthals and modern humans.

Homo neanderthalensis

Time range: 40 000–400 000 years ago

Geographic range: Europe and south-western to central Asia

Homo neanderthalensis (Neanderthal) is our closest extinct human relative (Figure 12.1.11). Neanderthals differed from *Homo sapiens* in a variety of ways: they had large faces with angled cheek bones, large noses for coping with cold, dry air, and chunkier, shorter builds suited to colder climates.

FIGURE 12.1.11 Reconstruction of a Neanderthal family based on fossil evidence. It is thought that Neanderthals lived in small family groups and had complex social structures and language.

Neanderthals did not use complex tools but they did use fire and lived with family groups in shelters. It is thought that they had complex social structures and were possibly the first human species to have language. Neanderthals were the first species to wear clothes and jewellery, have burial rituals and display symbolic behaviour. Their brains were just as big as ours, relative to body size, and sometimes even larger.

Approximately 1–4% of non-African modern human mitochondrial DNA is shared with Neanderthals, suggesting that interbreeding occurred between these two species after modern humans left Africa.

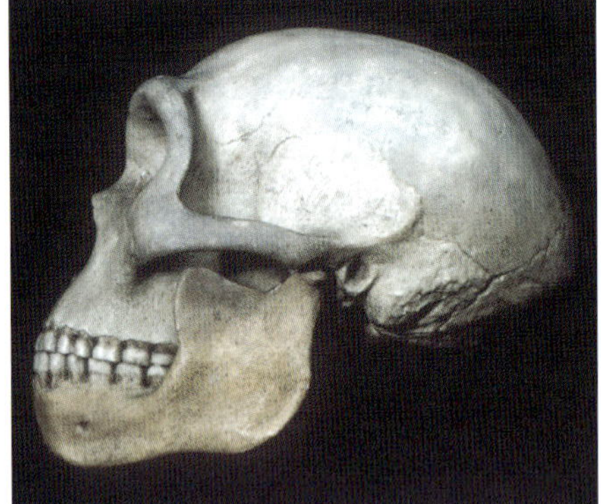

FIGURE 12.1.12 Skull of *Homo heidelbergensis* displaying its characteristic prominent brow ridge.

Homo erectus is considered to be a direct ancestor of *Homo sapiens*.

Homo heidelbergensis

Time range: 200 000–700 000 years ago

Geographic range: Europe, eastern and southern Africa and possibly China

Found in Europe, Africa and possibly Asia, *Homo heidelbergensis* is thought to be the first of the human species to use fire, the first to build shelters and the first to routinely hunt large animals. *Homo heidelbergensis* had a very large brow ridge and a larger braincase than its older relatives (Figure 12.1.12), and was the first of the early humans to successfully live in colder climates. Its short, stocky build would have been beneficial in conserving heat in cold conditions.

Homo erectus

Time range: 143 000–1.89 million years ago

Geographic range: North, eastern and southern Africa and west and east Asia

Homo erectus (Figure 12.1.13) was the earliest *Homo* species to expand out of Africa. They had similar body proportions and behaviours to *Homo sapiens*. From fossils found throughout Africa and Asia, scientists have composed a picture of a species that looked after the sick, old and young, as well as being the first to construct tools using stones and other materials (rather than just using the stones as tools) and the first to cook food. Possibly the longest lived of the human species, *Homo erectus* survived for about nine times as long as we have been around.

FIGURE 12.1.13 Model of a *Homo erectus* based on fossil evidence.

Homo rudolfensis

Time range: 1.8–1.9 million years ago

Geographic range: Eastern Africa

Originally thought to be *Homo habilis*, *Homo rudolfensis* (Figure 12.1.14) had a significantly larger braincase, much larger than the largest *Homo habilis* skull. Found in eastern Africa, *Homo rudolfensis* also had a longer face and larger molar and premolar teeth, and was more like the australopithecines (see *Australopithecus* section on page 472).

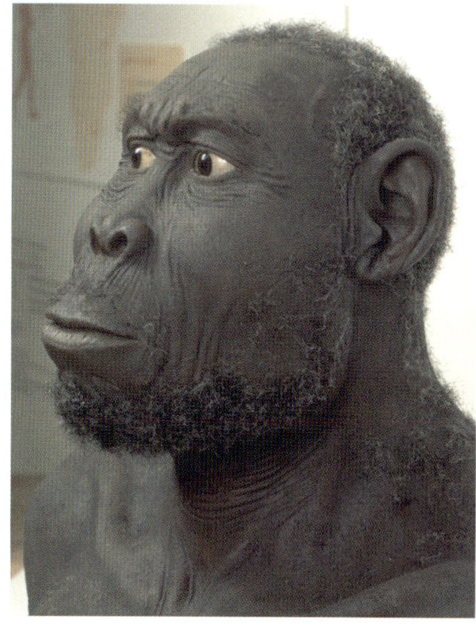

FIGURE 12.1.14 Reconstruction of *Homo rudolfensis* from fossil evidence.

Homo habilis

Time range: 1.4–2.4 million years ago

Geographic range: Eastern and southern Africa

Homo habilis was found only in Africa, and is the earliest known *Homo* species. It had a much larger braincase than its predecessors and has been associated with some of the earliest tools. *Homo habilis* was so-named to represent what was thought to be the first evidence of making stone tools. Although the name stands, the role of 'handy man' has since been taken by an earlier species. *Homo habilis* commonly used tools to break open bones and feed on the bone marrow (Figure 12.1.15). *Homo habilis* had more ape-like features than later species, with longer arms and jaws that project forwards.

FIGURE 12.1.15 *Homo habilis* was thought to be one of the first hominins to use tools. Although we now know that earlier species also used basic tools, *Homo habilis* frequently used stone tools to break open bones.

GENUS *PARANTHROPUS*

Time range: 1.2–2.7 million years ago

Geographic range: South-eastern Africa

Fossils of the genus *Paranthropus* are found in south-eastern Africa and are characterised by their large teeth and powerful jaws. They were bipedal and stood at around 1.3–1.4 metres tall with a muscular build. *Paranthropus* species were prevalent at the time when some species of the *Homo* genus existed, but it is thought that *Paranthropus* species were more specialised and less adaptable than *Homo* species. This lack of adaptability in changing environments may have led to their extinction.

- The face shape of *Paranthropus robustus* (~1.2–1.8 million years ago) was wide and angular to accommodate the powerful jaw muscles. They also demonstrated significant **sexual dimorphism**, with males being taller and heavier than females.

The sagittal crest is a ridge of bone running lengthwise along the midline of the top of the skull. The presence of this ridge of bone indicates that the jaw muscles are exceptionally strong.

- *Paranthropus boisei* (~1.2–2.3 million years ago) had an even wider face than *Paranthropus robustus*, as well as a strong sagittal crest on top of the skull (Figure 12.1.16). This species had very large molars, approximately four times the size of a modern human's, with the thickest dental enamel of any known early human.
- *Paranthropus aethiopicus* (~2.3–2.7 million years ago) possessed similar features for chewing, but few fossils have been found to ascertain other characteristics.

FIGURE 12.1.16 Digital reconstruction of *Paranthropus boisei* from skull specimens.

GENUS *AUSTRALOPITHECUS*

Time range: 2–4.2 million years ago

Geographic range: Eastern and southern Africa

Australopithecine skeletons, all found in eastern Africa, show that they walked upright on a regular basis but still climbed trees. The genus *Homo* is related to *Australopithecus*.

- *Australopithecus afarensis* (~2.95–3.85 million years ago) is better known as 'Lucy's species' due to the key fossil named Lucy, found in Hadar, eastern Africa (Figure 12.1.17). More than 300 individuals have been identified from fossils, making it the best-known of the early hominin species, as well as being one of the longest-lived, having survived more than 900 000 years. *Australopithecus afarensis* had both ape and human features: a flat nose, a projecting chin, a small braincase, long arms and long, curved fingers adapted for climbing trees.
- *Australopithecus garhi* (~2.5 million years ago) and *Australopithecus africanus* (~2.1–3.3 million years ago) had larger brains than Lucy and longer femurs, suggesting they took longer strides when walking upright.
- *Australopithecus anamensis* (~3.9–4.2 million years ago) was also similar to Lucy, but was likely the size of a chimpanzee.

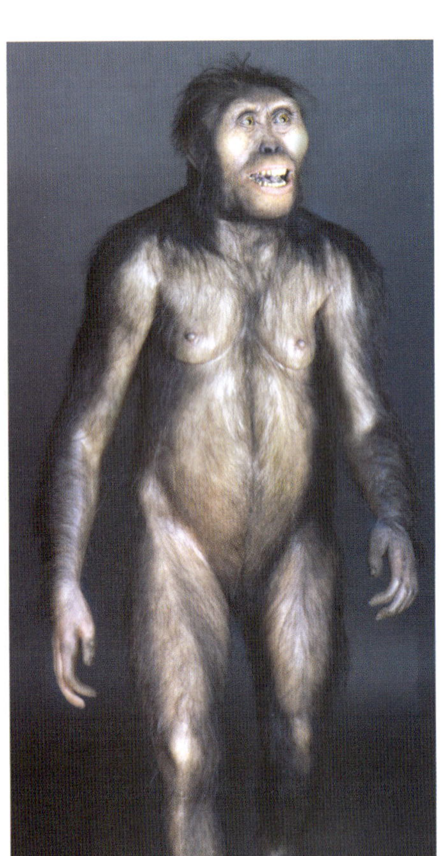

FIGURE 12.1.17 Reconstruction of 'Lucy', *Australopithecus afarensis*.

BEFORE *AUSTRALOPITHECUS*

The earliest hominins predating the australopithecines were significant in standing upright (bipedalism). Bipedalism is thought to be most likely a development related to a savannah environment in Africa that emerged due to drier, warmer climates. The large, open areas of the savannah favoured those who could move quickly over the land, see above the tall grasses, and were not reliant on trees.

- *Ardipithecus ramidus* (~4.4 million years ago) demonstrated features of bipedalism and tree climbing, with little sexual dimorphism evident from the teeth.
- *Ardipithecus kadabba* (~5.2–5.8 million years ago) is known from only a very small sample of skull, tooth and other partial skeletal remains. These fossils are able to provide a lot of information, demonstrating bipedalism, body and brain sizes similar to those of chimpanzees, and canines similar to those of later hominins.

- *Orrorin tugenensis* (~5.8–6.2 million years ago) was also similar in size to chimpanzees, was able to walk on two legs (but it is unknown how often), and possessed small teeth with thick enamel, similar to those of modern humans.
- *Sahelanthropus tchadensis* (~6–7 million years ago) is the oldest fossil species of hominins to be considered ancestral to the lineage that eventually led to *Homo* (Figure 12.1.18). Discovered in west-central Africa, this species is the oldest to possess a **foramen magnum** that is located further forwards than in apes or any other primates except humans. The foramen magnum is the opening for the spinal cord through the skull from the brain. The position of the foramen magnum demonstrates an upright stance that is indicative of time spent on two legs. All this is known from just 9 cranial fossils.

FIGURE 12.1.18 A reconstruction of *Sahelanthropus tchadensis*, the oldest known fossil leading to the lineage of *Homo*.

TRENDS IN HUMAN EVOLUTION

The fossil remains of our human ancestors demonstrate a few significant changes in physical, mental and cultural traits. Their arms got shorter, their legs longer, the pelvis modified and the foramen magnum moved forwards as they spent less time in trees and more time walking upright. Bodies became leaner and taller with less hair as climates became warmer and drier. As their diets moved away from plant material and towards a more omnivorous diet, teeth became smaller, and the increase in cooking food further allowed for a reduction in jaw muscles. This led to an increased capacity for the skull to get larger and accommodate a larger brain (Figure 12.1.19). Culturally, the increased control our ancestors had over the environment enabled progressively less movement, more building of permanent shelters and the gradual development of technology, rituals and societal structures.

Today *Homo sapiens* is one of the most successful species on Earth. Our ability to adapt readily to changing environments and the technology we have developed to support our survival has led us to become one of the most populous, widespread and influential species to ever exist on Earth.

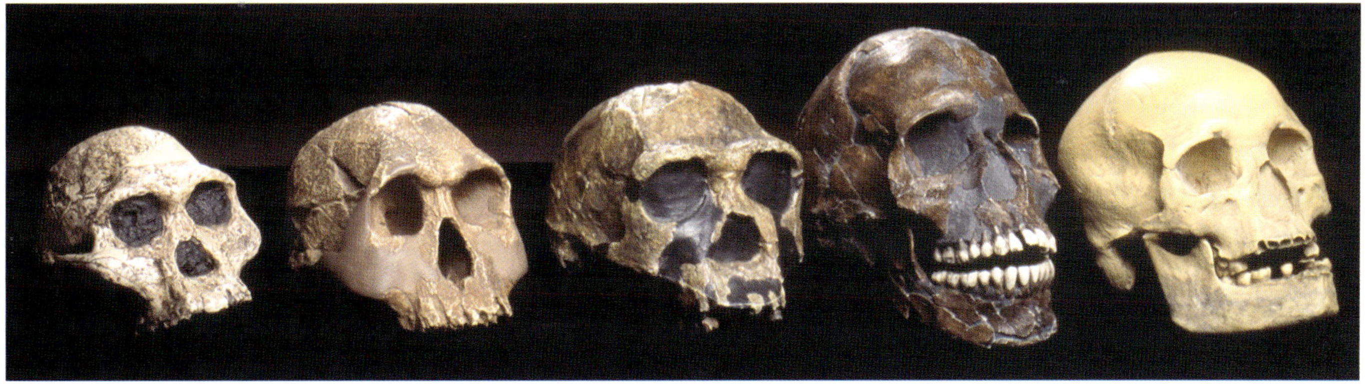

FIGURE 12.1.19 The evolution of skull morphology in *Homo*. Increases in skull size led to larger brains and rapid advances in the development of tools and cultural evolution. From left to right: *Australopithecus africanus* (2.1–3.3 million years ago), *Homo habilis* (1.4–2.4 million years ago), *Homo erectus* (0.14–1.89 million years ago), a modern human (*Homo sapiens*) from Qafzeh in Israel (approximately 92 000 years old), and a Cro-Magnon human (*Homo sapiens*) from France (approximately 22 000 years old).

12.1 Review

SUMMARY

- Modern humans (*Homo sapiens*) are the only living species belonging to the genus *Homo*.
- Humans belong to the subfamily Homininae and hence are called hominins. We are part of a larger group that includes the great apes, the family Hominidae, hence the term hominid. The superfamily Hominoidea includes humans, the great apes and the lesser apes, and are all called hominoids.
- Our closest living relatives are chimpanzees and bonobos, but we did not directly evolve from them or any other living primate. Our last shared common ancestor existed around 6–8 million years ago.
- The oldest fossil record of our species was found in a region called Omo Kibish in Ethiopia and has been dated to 195000 years ago.
- There has been a trend throughout hominin evolution for larger braincases, with brain size tripling in the last two million years.
- *Homo sapiens* has the largest, most complex brain of any of the hominins. Neanderthals had a slightly larger cranial capacity, but it is thought that their brains were more specialised for vision and movement than for problem solving.
- Behaviourally, modern *Homo sapiens* are characterised by the ability to plan, abstract thinking, ritual, symbolism (e.g. in art, ornamentation and music), large prey capture and advanced tool use.
- Tool making and communication have had a significant impact on the cultural characteristics of our species.

KEY QUESTIONS

1 Use the terms below to show how *Homo sapiens* is classified under the Linnaean system of classification.

Chordata *Homo sapiens* Hominidae Animalia Primates Eukarya Mammalia *Homo*

Domain: ______________________

Kingdom: ______________________

Phylum: ______________________

Class: ______________________

Order: ______________________

Family: ______________________

Genus: ______________________

Species: ______________________

- *Homo sapiens* underwent a cultural 'explosion' approximately 50 000 years ago with rituals, burials, the wearing of clothes and more complex hunting techniques becoming more evident in the fossil record.
- The number of hominin species known is changing with new discoveries and evidence from fossils and DNA. Currently there is evidence of 10–16 *Homo* species, but many of these are thought to be subspecies or not distinct enough to warrant recognition as separate species.
- Our closest extinct relative is *Homo neanderthalensis*. *Homo sapiens* coexisted with Neanderthals and there is evidence of interbreeding between the two species. DNA and fossil evidence also suggests that both these species coexisted and interbred with *Homo denisovans*.
- *Homo erectus* was the first of our relatives to walk upright all the time, and spread into South-East Asia.
- *Homo habilis* is the oldest known species of the genus *Homo*.
- *Paranthropus* is a genus of hominin that existed at the same time as some species of the genus *Homo*.
- The genus *Homo* is related to members of the genus *Australopithecus*.
- *Australopithecus*, including the fossil Lucy, represents a turning point in the evolution of bipedalism, which is thought to be an adaptation to the African savannah environment.

2 When did humans last share a common ancestor with chimpanzees?

3 Discuss how the trend for increasing cranial capacity in hominins has influenced the evolution of these species.

4 Which two hominin species were thought to have interbred with *Homo sapiens*?

A *Homo habilis*
B *Homo neanderthalensis*
C *Homo denisovans*
D *Homo floresiensis*

5 Complete the following sentences using the terms below:

Hominoidea Homininae hominids hominins hominoids Hominidae

Humans belong to the subfamily __________ and hence are __________. We are part of a larger group that includes the great apes, the family __________, hence the term __________. The superfamily __________ includes humans, the great apes and the lesser apes, hence the term __________.

6 Name three significant physical changes *Homo sapiens* has undergone throughout evolution.

12.2 Patterns in hominin evolution

The human fossil record is an example of a classification scheme that is open to interpretations that are contested, refined or replaced when new evidence challenges them or when a new model has greater explanatory power. This has occurred numerous times, sometimes with little impact and at other times completely overturning our idea of our ancestry. With new discoveries of fossils and archaeological artefacts, along with rapid advances in DNA technology and bioinformatics, our family tree is being discovered in more detail (Figure 12.2.1), and our understanding of our relationships continues to grow. Most recently, evidence has been discovered that supports the theory that *Homo sapiens* and *Homo neanderthalensis* interbred. A new fossil species of *Homo*, *Homo floresiensis*, was discovered in 2004 in Indonesia; another, *Homo denisovans*, was discovered in Russia in 2008. In 2015 palaeoanthropologists described yet another member of the genus *Homo*: *Homo naledi* from South Africa (see Section 12.1). **Palaeoanthropology** is a branch of anthropology that involves the study of fossil hominins, and contributes to our knowledge of human evolution.

Our knowledge of the human fossil record and human evolution are evolving as new information comes to light.

FIGURE 12.2.1 Models of our hominin ancestors based on reconstructions of fossils. Back row, from left to right: *Homo ergaster*, two male Neanderthals (*Homo neanderthalensis*) carrying dead animals with a Neanderthal female and child in between, a Cro-Magnon (*Homo sapiens*) hunter throwing a spear; middle row, left to right: a female *Homo habilis*, a male *Homo georgicus* wearing fur, *Australopithecus africanus*, *Paranthropus boisei*, *Sahelanthropus tchadensis*; front row, left to right: *Homo georgicus* (female with throwing stone), Lucy and Lucien (*Australopithecus afarensis*).

CULTURAL EVOLUTION

One of the world's foremost evolutionary biologists Richard Dawkins famously declared, "What lies at the heart of every living thing is not a fire, not warm breath, not a 'spark of life'. It is information, words, instructions." What sets humans apart from all other species is our cultural evolution.

Culture is the accumulated knowledge passed on to the next generation by verbal, written or symbolic communication. A very important step for the human species was the development of language and the ability to record information. The study of the relationships of different languages, together with information from skeletal remains, artefacts and DNA, provides evidence of the evolution of the different cultures of *Homo sapiens* living today in different geographic regions.

As humans became increasingly able to communicate their discoveries from one generation to the next, their culture also changed over time. This is called **cultural evolution**. As culture is passed on, it is modified by each generation and is subject to influences such as new discoveries about the environment. Like gene flow, the transmission of cultural information can also be interrupted by geographic barriers: over time cultures that were geographically isolated evolved independently and became increasingly different, resulting in the cultural diversity in *Homo sapiens* that we see today.

The most significant time for early cultural development of *Homo sapiens* is thought to have occurred approximately 50 000 years ago (although evidence of earlier culture is beginning to emerge). It is around this time that rituals, burials, clothes and more complex hunting techniques become evident in the fossil record (Figure 12.2.2).

Culture is the accumulated knowledge that is passed on from generation to generation and evolves with changing environments, information and time.

FIGURE 12.2.2 Timeline of the major events in the cultural evolution of *Homo*.

Tools

FIGURE 12.2.3 A flint hand axe used by *Homo neanderthalensis* in France.

The first use of primitive stone tools is dated from approximately 3.3 million years ago for *Australopithecus*. The earliest evidence of stone tool use by *Homo* dates back to 2.6 million years ago in Ethiopia. Sharp stone flakes were used for cutting through animal skin and meat. There is evidence of more advanced craftsmanship of stone tools, from flake tools to axes, by *Homo erectus* and *Homo habilis* approximately 1.6 million years ago.

Throughout evolutionary time, and with the increasing brain size of *Homo* species, tools became more complex and their uses more varied (Figure 12.2.3). The earliest evidence of the emergence of modern behaviour and culture in *Homo* is from tools found in Africa from approximately 300 000–400 000 years ago. Blades, bone tools, specialised weapons for hunting, and evidence of long-distance travel and trade signify the beginnings of cultural development in humans. The sophisticated use of tools was crucial to the evolution of human culture for hunting, adornment, ritual, clothes making, agriculture, writing, building and many other important advances. Tool use marked the beginning of cultural evolution in *Homo*.

Art

Evidence of the use of ochre (clay in shades of yellow, brown, red and purple) and pierced shells for decoration date back to 100 000–200 000 years ago in sub-Saharan Africa. These indicate that culture, in the form of ritual and symbolic expression, may have begun to emerge much earlier than the cultural explosion that was thought to have occurred 50 000 years ago.

More complex forms of symbolic expression are evident in cave paintings and carvings that became prolific about 35 000–40 000 years ago. It has long been thought that cave art first developed in Europe, with cave paintings dated to 35 000 years ago found in caves in France and Spain. Recent research has suggested that paintings found in a cave in Sulawesi, an Indonesian island, date back to 39 900 years ago, making them the earliest known forms of artwork (Figure 12.2.4). In Australia, scientists are attempting to date Aboriginal rock art that may prove to be some of the oldest in the world. The skeleton of Mungo Man, unearthed on the edge of Lake Mungo, in New South Wales, is evidence of a ceremonial burial 40 000 years ago (see Biofile on page 480).

FIGURE 12.2.4 The earliest known forms of art created by modern humans are found in caves in Indonesia. Paintings of outlines of human hands on cave walls on the Indonesian island of Sulawesi have been dated to 39 900 years ago.

BIOLOGY IN ACTION

Lascaux Caves

Some of the most spectacular examples of Palaeolithic cave art can be seen in the World Heritage-listed Lascaux Caves in south-western France. The cave paintings are thought to be 17 300 years old and mostly depict large animals, such as horses, deer and aurochs (extinct species of wild cattle) that once lived in the area (Figure 12.2.5), along with some human figures and abstract symbols. The cave walls have more than 2000 figures that were painted using red, yellow and black mineral pigments (ochre) mixed with water, clay or animal fat.

FIGURE 12.2.5 Cave paintings of Palaeolithic bulls and horses on the walls of the famous Lascaux Caves in south-western France. These paintings are some of the most spectacular cave art in the world and are thought to be 17 300 years old.

Many interpretations of what the images mean have been proposed. Some scholars hypothesise that the depictions of animals represent hunting success, while others suggest that the images are spiritual in nature. There has also been the suggestion that some of the dot paintings on the walls of the cave represent star charts while others function as a lunar calendar.

The cave paintings were discovered in 1940 and soon attracted up to 1200 visitors per day for many years. The exposure of the caves to excess carbon dioxide and light, along with changes in air circulation, led to problems with lichen growing on the walls and so the caves were closed to visitors in 1963. In 2001 an infestation of white mould began growing throughout the caves after the air conditioning system was changed and in 2007 a black fungus began growing on the cave walls. In 2008 the caves were closed for three months; not even scientists or preservationists were permitted to enter. One person was allowed to enter the caves once per week to measure climatic conditions inside. An International Scientific Committee has been established to assess measures that can be taken to eliminate the fungus and determine how much human access, if any, should be permitted to the World Heritage-listed caves.

Religion and ritual

> Tools, art, ritual and religion all played a significant role in the development of communication and the evolution of human culture.

The development of religion and elaborate ritual marks one of the most significant transitions to modern human behaviour, and the emergence of an important element of human culture. Evidence of the beginning of religious belief and practice is found in intentional burial markers. This has been observed as early as 130 000 years ago in *Homo sapiens* and *Homo neanderthalensis*, with groups of specimens found buried together or deliberately placed in cave pits.

Ritual burials with symbolic objects, such as jewellery and tools, and the use of ochre to cover the dead are first evident 100 000 years ago in Israel. Ritual burial practices represent empathy and concern for what happens to the living after death. The placement of symbolic objects with the dead indicates a belief in the afterlife and is evidence of spiritual belief by *Homo sapiens* from at least 100 000 years ago.

FIGURE 12.2.6 The skeleton of the 'Young Prince' from the Arene Candide cave in Italy shows evidence of burial rituals from 23 500 years ago. The 15-year-old male was covered in a layer of red ochre and buried with shell jewellery, deer antler and a long flint tool in his hand. A wound below his chin was covered in yellow ochre before his burial.

Evidence of religion and elaborate ritual became more prevalent in the archaeological record approximately 50 000 years ago, during the cultural explosion of *Homo sapiens*. One of the most well preserved **Palaeolithic** burial sites is the Arene Candide cave in Italy. Of the 19 remains found in the cave, the most famous is of a 15-year-old male known as the 'Young Prince' who lived approximately 23 500 years ago. He was found buried several metres below the surface of the cave, covered in a layer of red ochre, with shell jewellery, deer antler and a long flint tool in his hand (Figure 12.2.6).

Throughout human history, symbolic images have become increasingly important in the transmission of information and the evolution of human culture. Symbols continue to play an integral role in human communication and culture today.

BIOFILE

Lake Mungo

Lake Mungo is a dry lake located in south-western New South Wales, Australia. It is part of a World Heritage-listed region because of the many significant archaeological remains that have been found there. Three bodies have been discovered in the sediment deposits of the lake, dated to between 25 000 and 50 000 years ago. These bodies are known as Mungo Lady (or Lake Mungo 1), Mungo Man (or Lake Mungo 3) (Figure 12.2.7) and Lake Mungo 2.

The remains of Mungo Lady and Mungo Man have been dated to 40 000 years old, which makes them the second oldest anatomically modern humans to be found east of India. There is evidence that Mungo Lady's body was cremated and covered in ochre (mineral pigment), making her one of the earliest examples of cremation and burial rituals in the world. Mungo Man's hands were found placed in his lap, also indicating ritualistic practices in placing the body for burial.

Lake Mungo is a significant place for indigenous Australians and the story of the migration of modern humans throughout the world.

FIGURE 12.2.7 Mungo Man is one of the oldest anatomically modern human specimens found in Australia, dated to approximately 40 000 years ago. His hands were placed in his lap before burial, indicating deliberate placement of the body and ritual surrounding death and burial. This is one of the earliest examples of ritual burial practices in the world.

The Neolithic Revolution (agricultural revolution)

Throughout the last 12 000 years there is evidence of human settlements, the establishment of permanent dwellings and the beginning of agriculture. The transition from a hunter-gatherer lifestyle to agriculture is known as the **Neolithic Revolution** and was one of the most significant advances in human evolution. Permanent dwellings and food supplies supported a growing population and encouraged the development of complex cultural and social structures.

The earliest evidence of the transition to agriculture comes from the Fertile Crescent (also known as the Cradle of Civilisation) in the Middle East, approximately 10 000–12 000 years ago. It was here that the first domesticated plants were grown and used as food crops, and the first domesticated animals were bred as modern cows, sheep, horses and goats. As the civilisation developed, technology and communication became more sophisticated, leading to the evolution of writing and education systems, the development of trade and political systems, the division of labour and the refinement of art and architecture. Property ownerships also led to a hierarchical society, class conflicts and the establishment of armies. Some of the most important innovations in human history came from this early civilisation: the invention of the wheel, agriculture and the first cereal crops (Figure 12.2.8), written language (Figure 12.2.9), mathematics and astronomy.

FIGURE 12.2.8 A Neolithic grindstone used to process grain.

FIGURE 12.2.9 Clay tablet from Assyria engraved with cuneiform, one of the earliest forms of writing.

BIOFILE

Cro-Magnon: European early modern humans

Archaeological deposits in Europe from the Upper Palaeolithic period (approximately 10 000 to 40 000 years ago) tell the story of the first early modern *Homo sapiens* in Europe. These people are called Cro-Magnon or European early modern humans. Cro-Magnon had a robust build, a short, wide face, a prominent chin and a slightly larger cranial capacity than people of today. Skeletons have been found with evidence of long-term injuries, indicating that Cro-Magnon may have cared for sick and injured group members. It is thought that Cro-Magnon may have constructed semi-permanent dwellings, with evidence from a series of huts made from mammoth bones that date back approximately 15 000 years in a Ukrainian village. Evidence of their culture has been found in sophisticated tools, jewellery, clothing, cave paintings and the use of ochre for colouring objects (Figure 12.2.10).

FIGURE 12.2.10 Reconstruction of a Cro-Magnon man painting on a cave wall.

THE ORIGIN OF MODERN HUMANS

Homo sapiens have spread throughout the world, but their origins are still debated. Palaeoanthropologists conclude that *Homo sapiens* first evolved in Africa and that much of human evolution occurred on that continent, given the evidence of fossil hominins in Africa from as far back as 6 million years ago.

Scientists currently agree that there are 10–16 or more fossil species of *Homo*, but there is disagreement surrounding how, when and where each species evolved, how they are related and whether they died out or evolved into a new species. Fossil evidence puts *Homo* in Africa from 2.5 million years ago; but some populations of *Homo erectus* migrated northwards, and arrived in Asia and South-East Asia 1.8–2 million years ago, in Europe 1–1.5 million years ago and other parts of the world much later. Our species, *Homo sapiens*, is thought to have left Africa between 60 000 and 125 000 years ago, arriving in Europe and Asia 35 000 to 45 000 years ago, Australia 50 000 years ago and the Americas 12 000 to 20 000 years ago (Figure 12.2.11 on page 482), but exact migration dates are speculative. Of the theories that abound, three models have stood the test of time and much debate: the **Multiregional evolution (continuity) model**, the **Out of Africa (replacement) model** and the **Assimilation (partial replacement) model**.

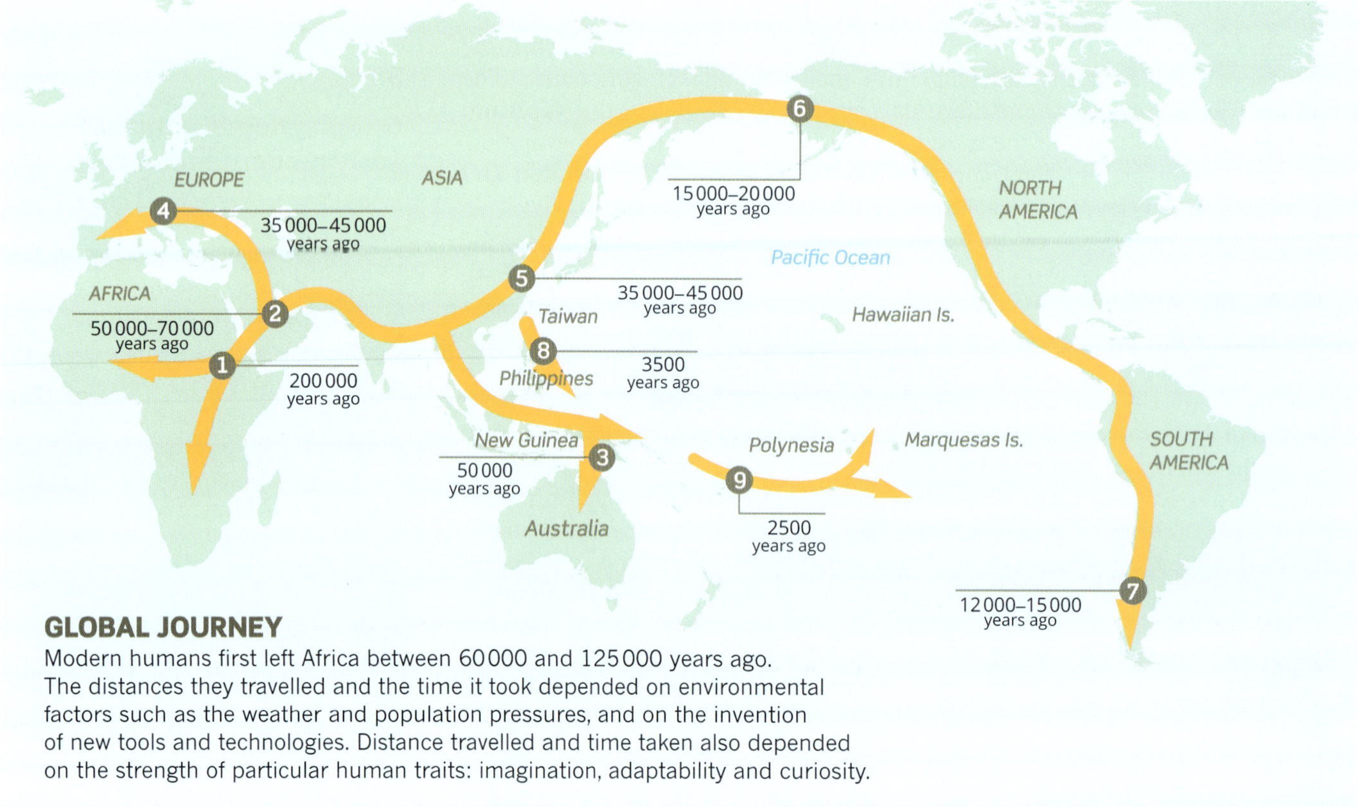

FIGURE 12.2.11 The journey of modern humans throughout the world. Fossil and DNA evidence supports the theory that all modern humans originated in Africa and began to migrate to other parts of the world between 60 000 and 125 000 years ago.

> The three dominant theories of present-day human origins all support an African origin but have different views on the pattern of migration and interbreeding that occurred in other parts of the world.

Multiregional evolution (continuity) model

Several scenarios have also been proposed that include the interbreeding of *Homo sapiens* with archaic human species in Africa prior to migration. The Multiregional theory proposes significant migration of *Homo erectus* across Africa, Asia and Europe for the last 1.8 million years. It is argued that isolation of the populations resulted in the divergence of gene pools, traits and behaviour, but occasional contact ensured some gene flow was maintained and led to concurrent evolution of all groups as one recognisable species, *Homo sapiens* (Figure 12.2.12a). There is speculation that inbreeding could have occurred on any number of occasions but that, over time, genetic drift has led to the DNA of the other species being lost from the genome of *Homo sapiens*.

Supporting evidence

- Physical variations in modern *Homo sapiens* indicate a relatively long time period since migration from an African ancestor.
- Similarities exist between modern humans and extinct species found in the same region.
- Living humans show little genetic diversity, consistent with regular gene flow across all regions.

> The Out of Africa theory is the most widely accepted model for the evolution of modern humans.

Out of Africa (replacement) model

The Out of Africa theory suggests that after *Homo erectus* left Africa, populations became isolated and diverged into different species (e.g. *Homo neanderthalensis* and *Homo heidelbergensis*). All living modern humans evolved from a single common ancestor in Africa about 200 000 years ago. Migrations of *Homo sapiens* from Africa occurred some time between 60 000 and 125 000 years ago, although stone artefacts found in the Arabian Desert suggest it may be as early as 160 000 years ago.

As modern humans (*Homo sapiens*) spread throughout the world, they displaced all other human species (Figure 12.2.12b); *Homo heidelbergensis* was displaced in Africa and Europe, *Homo erectus* was displaced in Asia and *Homo neanderthalensis* was displaced in Europe. Evidence also points towards several migration events, the earliest about 130 000 years ago along the southern coastlines into Australia and Papua New Guinea. There was a later migration about 50 000 years ago along the Nile River Valley northwards into Europe and Asia, and possible additional migration events between and after these events.

The most extreme version of this model suggests competition between *Homo sapiens* and other human species without interbreeding, although analysis of the Neanderthal genome in 2010 supports theories that includes some interbreeding, at least between *Homo neanderthalensis* and *Homo sapiens*.

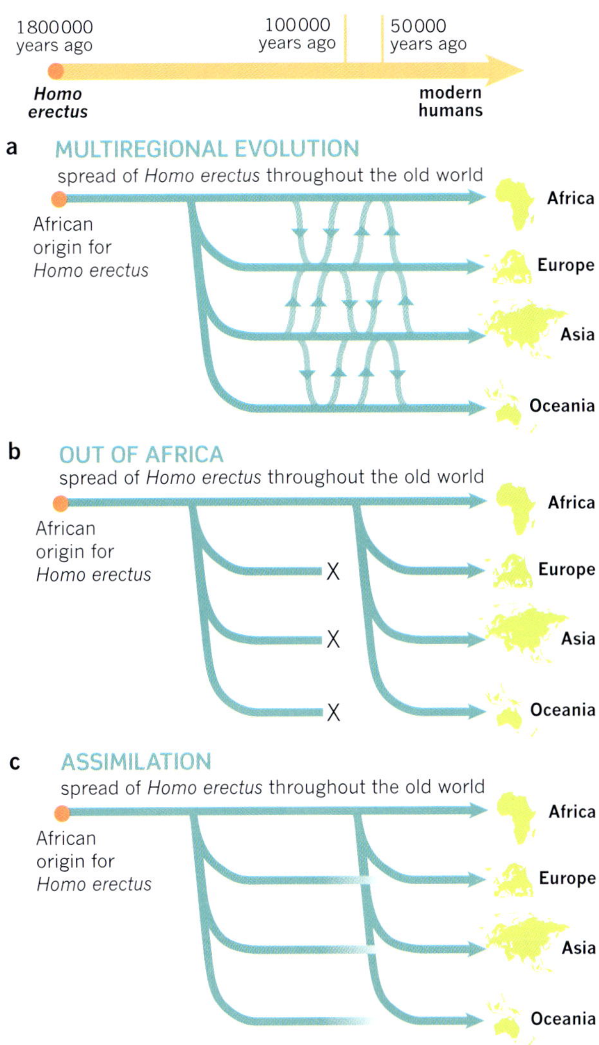

FIGURE 12.2.12 The three dominant theories of the origin of all living humans: (a) the Multiregional evolution model, (b) the Out of Africa model and (c) the Assimilation model. All three theories support an African origin of modern humans but vary in the pattern of migration and interbreeding that occurred in other parts of the world.

Supporting evidence

- The oldest fossil evidence of *Homo sapiens* has been found in Africa.
- Living *Homo sapiens* show little genetic diversity, suggesting a relatively recent emergence.
- Analysis of the DNA of living humans can be used to map the movement of humans and the approximate timeline on the basis of the estimated rates of mutation. This analysis points to Africa as the point of origin.
- DNA analysis of more than 1000 unrelated humans from different populations shows that present-day humans belong to three genetic groups: Africans, Eurasians (people native to Europe, the Middle East and southwest Asia) and East Asians (people native to Japan, South-East Asia, the Americas and Oceania). Differences between these groups have been found to be mostly due to genetic drift during periods of isolation. The African group has the most genetic diversity, indicating that they are the source population, which supports the Out of Africa theory.

Assimilation (partial replacement) model

The Assimilation model is a newer hypothesis that combines elements of the Out of Africa and Multiregional evolution models. This model proposes that all modern humans had an African origin and that when people migrated out of Africa there was occasional interbreeding with archaic humans who were already living in other parts of the world, resulting in hybrid populations (assimilation) (Figure 12.2.12c).

Supporting evidence

- The abrupt appearance of modern humans in Europe approximately 40 000–45 000 years ago (Cro-Magnon people) suggests that modern humans originated in Africa and then migrated to Europe and Asia.
- DNA and fossil evidence supports the theory of interbreeding between Neanderthals and modern humans. It is proposed that hybrid populations of Neanderthals and Cro-Magnon may have evolved into modern Europeans.
- Skeletal remains that have characteristics of both archaic and modern humans have been found.
- DNA evidence shows that interbreeding between *Homo* species from Asia, Europe and Africa has been occurring over the past 600 000 years.

Living in the Anthropocene

Scientists have dubbed the geological era in which we exist the **Anthropocene** or the 'Age of Humans'. The Anthropocene is defined as the time when human activities began to have significant impacts on the Earth's systems on a global scale. For the Anthropocene to be officially recognised as a true geological time period, a whole range of specialists must agree upon a start date on the basis of evidence of the impact of humans on the Earth (e.g. atmospheric carbon dioxide levels). Some argue for the beginning of modern humans, others for the advent of agriculture; some suggest the Industrial Revolution (late 18th century) was the beginning of significant human-induced environmental damage, and yet others suggest the nuclear age from the mid-1940s is more appropriate.

No matter when it began, the Anthropocene is a significant concept. People live on every continent and the total population is growing at a rapid rate, placing demands on the environment for resources. The exponential growth in the human populations has led most scientists to agree that human activities have had significant long-term detrimental effects on ecosystems, biodiversity and Earth's climate system. Because of these effects, we are now experiencing Earth's sixth major extinction: the greatest rate of species loss since the mass extinction of the dinosaurs at the end of the Cretaceous period.

Climate change caused by increasing levels of carbon dioxide in the atmosphere from human activities is expected to result in further biodiversity loss, and a rise in sea levels (Figure 12.2.13). The cascade of events that is likely to occur because of climate change presents one of the greatest threats to modern humans. Although changing climates are not unique to the Anthropocene, the rate of change we are now experiencing is unprecedented. From the beginnings of life on Earth, organisms have faced significant changes in climate, and it is these changes that have shaped the species inhabiting our planet. Their ability to survive determined their fate. A number of scientists suggest that some of the most significant evolutionary changes in human history have occurred during times of climatic instability. The consequences of our modification and exploitation of our environment now threatens our survival. Our unique ability to reflect upon our impact on the Earth and modify our behaviour and technology to adapt to a changing environment may now be more crucial than ever to the survival of our species.

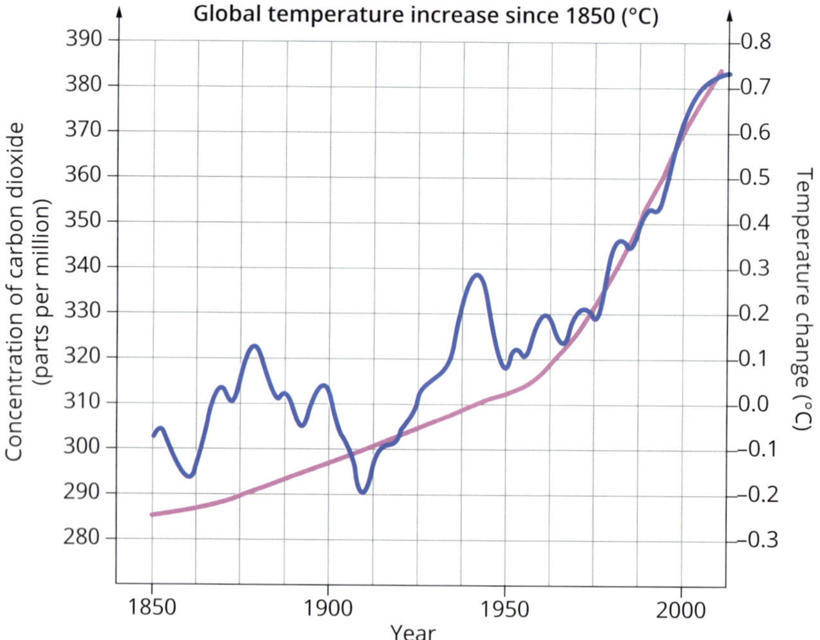

FIGURE 12.2.13 Relationship between increasing global temperature (°C) since 1850 (blue line) and increasing concentration of carbon dioxide (parts per million) in Earth's atmosphere (purple line).

12.2 Review

SUMMARY

- The human fossil record is open to interpretations that are contested, refined or replaced when new evidence challenges them or when a new model has greater explanatory power.
- Culture is the accumulated knowledge passed on to the next generation by verbal, written or symbolic communication and this evolves over time.
- Tools were (and still are) crucial to the evolution of human culture for hunting, adornment, ritual, clothes making, agriculture, writing, building and many other important advances. Tool use marked the beginning of cultural evolution in *Homo*.
  - 300 000–400 000 years ago (Africa): The earliest evidence of the emergence of modern behaviour and culture in *Homo* is from tools found in Africa.
  - 100 000–200 000 ago (Africa): Culture, in the form of ritual and symbolic expression, is evident from the use of ochre and decorative pierced shells.
  - 130 000 years ago (*Homo sapiens* and *Homo neanderthalensis*, Africa): Signs of ritual and religion are evident in intentional burials, with groups of individuals found buried together or deliberately placed in cave pits.
  - 100 000 years ago (Israel): Evidence of spiritual belief is found in ritual burials with symbolic objects, such as jewellery and tools, and the use of ochre to cover the dead.
  - 50 000 years ago: Evidence of religion and elaborate ritual became more prevalent in the archaeological record from approximately 50 000 years ago.
  - 35 000–40 000 years ago (Europe, Indonesia): Complex forms of symbolic expression are evident in cave paintings and carvings.
  - 10 000–12 000 years ago (the Fertile Crescent, Middle East): The transition from a hunter-gatherer lifestyle to agriculture began and is known as the Neolithic Revolution. This was one of the most significant advances in human evolution.
- *Homo sapiens* spread to all corners of the Earth, but their origins are still debated. Evidence puts hominins in Africa from 6 million years ago, and *Homo* in Africa from 2.5 million years ago, in Asia 1.8–2 million years ago, in Europe 1–1.5 million years ago and other parts of the world much later.
- There are three main theories to explain the origins of *Homo sapiens*: the Out of Africa theory, the Multiregional theory and the Assimilation theory.
  - The Out of Africa theory is currently the most widely accepted model for the origin of modern humans. It suggests that all living modern humans evolved from a single common ancestor in Africa about 200 000 years ago and spread throughout the rest of the world, replacing other hominin species. Migrations of modern humans from Africa occurred at some time between 60 000 and 90 000 years ago.
  - The Multiregional theory proposes significant migration of *Homo erectus* across Africa, Asia and Europe, with concurrent evolution of all groups into *Homo sapiens*. The Multiregional theory suggests that there was interbreeding and gene flow between various populations in Africa and Eurasia.
  - The Assimilation theory proposes that all living humans had an African origin and migrated out of Africa, occasionally interbreeding with archaic humans, resulting in hybrid populations (assimilation).
- The Anthropocene is defined as the time when human activities began to have significant impacts on Earth's systems on a global scale.
- Most scientists agree that human activities have had significant long-term detrimental effects on ecosystems, biodiversity and Earth's climate system.
- Our unique ability to modify and adapt to changing environments may be more significant than ever for our species' survival in the face of the current rapid rate of population growth, biodiversity crisis and climate change.

continued overleaf ➤

12.2 Review *continued*

KEY QUESTIONS

1. Give three reasons why our understanding of human evolution might change.
2. What is culture and how does culture evolve over time?
3. Why was tool use important to the cultural evolution of the genus *Homo* and our species *Homo sapiens*?
4. What are some of the first sources of evidence for religious belief and ritual of early *Homo sapiens*?
5. Complete the following statements about the Neolithic Revolution:

   The Neolithic Revolution, also known as the ___________, began approximately ___________ and marked the beginning of the transition from a ___________ lifestyle to ___________. This allowed people to establish ___________ and have a more reliable ___________, supporting a growing ___________. The evolution of these early civilisations led to significant changes in human ___________. Some of the most important ___________ in human history came from this time, such as ___________ and ___________.
6. What are the three main theories for the origin of present-day humans?
7. What is the Anthropocene?
8. What are five major impacts on Earth's environment that are caused by human activities?

Chapter review

KEY TERMS

Anthropocene
archaeological
archaic
Assimilation (partial replacement) model
bipedal
coexisted
cultural evolution
culture
foramen magnum
gracile
hominid
Hominidae
hominin
Homo
hybrid species
interbreeding
Multiregional evolution (continuity) model
Neolithic Revolution
Out of Africa (replacement) model
palaeoanthropology
Palaeolithic
predecessors
sexual dimorphism
subspecies

KEY QUESTIONS

1 The family Hominidae includes which groups of primates?

A Old World monkeys, gibbons, gorillas, chimpanzees and orangutans
B lemurs, gibbons, gorillas, chimpanzees, orangutans and humans
C African great apes including gorillas, chimpanzees and humans
D only humans

2 The hominid fossil record is fragmentary and confusing and there are a of number species that have been identified from very limited fossil evidence. Such fossils are classified by their structures into the genus that has other better-studied fossils to which they are most similar both in structure and time. Some of the less well-known hominin species are listed. Which of these is likely to be the most ancient?

A *Paranthropus walkeri*
B *Australopithecus sediba*
C *Ardipithecus kaddaba*
D *Homo naledi*

3 Consider the evolutionary tree below, which includes the genera *Homo*, *Paranthropus*, *Australopithecus* and *Sahelanthropus*.

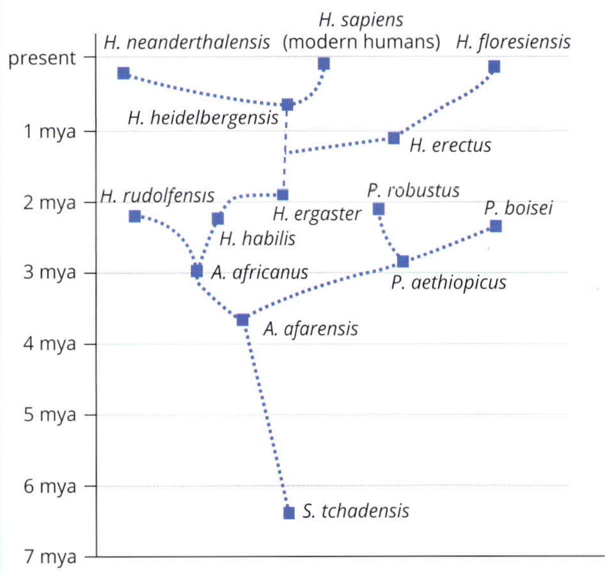

According to the evolutionary tree, which of the following is true?

A *Homo erectus* gave rise to modern humans.
B *Homo habilis* is a direct descendent of *Homo ergaster*.
C *Australopithecus afarensis* lived 4.5 million years ago.
D The most recent common ancestor of *Homo erectus* and *Homo heidelbergensis* lived about 1.5 million years ago.

4 The diagram below shows one model of human spread.

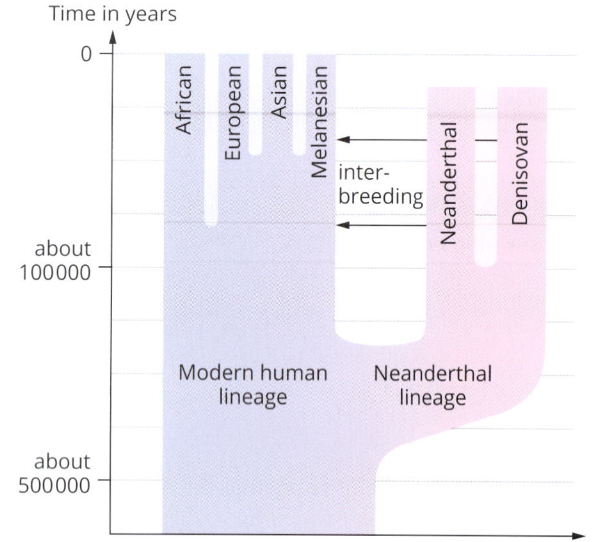

The model represented in this diagram is closest to:

A the Assimilation model
B the Out of Africa model
C the Multiregional evolution model
D none of the above

5 Culture includes all of the following except:

A art
B religion
C brain size
D knowledge

6 The first evidence of culture in hominins was:

A evidence of the use of ochre
B the appearance of stone tools
C cave paintings
D evidence of cooking fires

7 Which of the following statements about culture is true?

A Human culture has had no effect on biological evolution.
B Culture is a mental phenomenon, so no evidence of past culture can be found.
C Culture has the ability to influence biological evolution.
D The only species that has exhibited evidence of cultural evolution is *Homo sapiens*.

8 Which of the following statements about culture is true?

A It is learned behaviour that is passed from one generation to the next through processes that do not involve genes.
B It is all of the ideas and knowledge held by a population of humans.
C Culture is subject to faster change than biology.
D All of the above.

9 Examine the table below. For each feature of humans, place an X in the relevant boxes to identify whether the feature is common to all primates, hominids or is only observed in hominins.

Feature	Primates	Hominids	Hominins
binocular vision			
foramen magnum in central position			
opposable thumbs			
S-shaped spine			
teeth of four types			
grasping hands			
skeletal flexibility (e.g. shoulder rotation)			
broad chest			
no tail			
brains larger than 800 cm^3			
flat faces			
tool use			
fingernails and toenails			
large head and short neck to femur			
long period of juvenile development			

10 **a** Three skulls are shown below. Look carefully at the skulls.

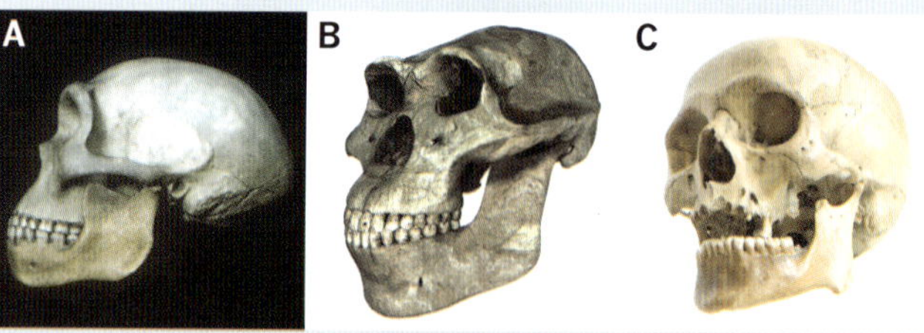

i List the skulls in order from oldest to youngest.
ii Describe the features of the skull that support your contention.

b Bipedalism is one defining feature of modern humans and their ancestors. The fossil record leading to modern humans shows a series of developments in skeletal structure that make bipedal locomotion more efficient. What was the reason for this development?

11 Anthropologists have various theories about the evolutionary tree that led to *Homo sapiens*. The two such evolutionary trees shown below include members of the genera *Homo*, *Paranthropus*, *Australopithecus* and *Sahelanthropus*.

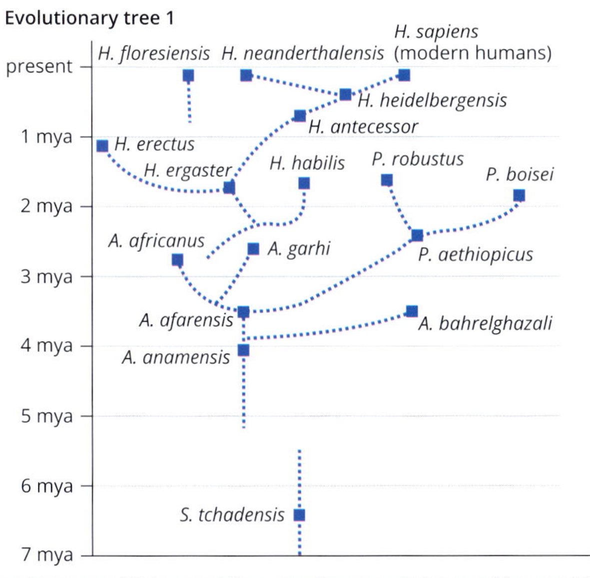

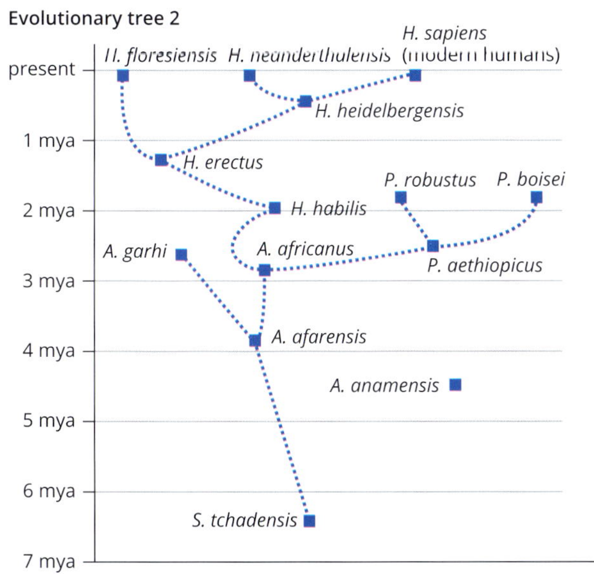

a Identify one point of agreement and one point of difference between the two evolutionary trees.

b Why are anthropologists unable to agree on a single view of human evolution?

c When anthropologists discover a new hominid fossil they must first decide whether the fossil should be classified as a hominin and then if it is a hominin they must decide where it belongs in the hominin family tree. What is a hominin?

12 Ancient fossils are unlikely to contain any usable DNA, so all the anthropologists can use to decide the classification of the fossil is its morphology. In 2003 a new fossil hominin was discovered on the island of Flores in Indonesia. One view of this fossil hominin is that it represents a previously undescribed new species of hominin that was a direct descendant of *Homo erectus*. This new species has been given the name *Homo floresiensis*. The skulls of *Homo floresiensis* and *Homo erectus* are shown below.

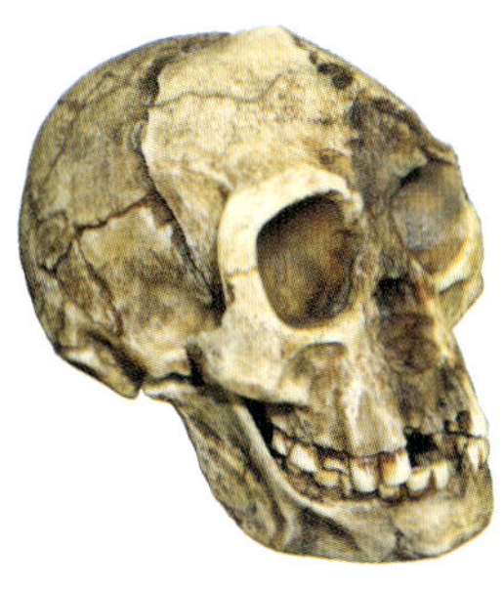
Homo floresiensis

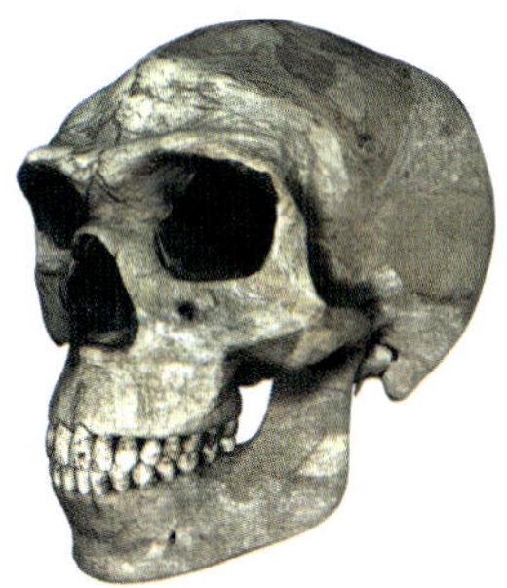
Homo erectus

a i Describe two features of the skulls which indicate that these two species are closely related.

ii There are some anthropologists who argue that *Homo floresiensis* is not a new species but is a variant of *Homo sapiens*. Describe one feature of the skull that would support this idea.

b Anthropologists examined the skull of *Homo floresiensis* and inferred that it had a significantly smaller brain than *Homo sapiens*.

i What is the difference between an observation and an inference?

ii What observation about the skull did the anthropologists make that led them to infer that *Homo floresiensis* had a smaller brain than *Homo sapiens*?

13 Sections of DNA from two archaic species of human, Neanderthals and Denisovans, have been sequenced and compared to the genome of modern humans. This research has shown that Neanderthals in Europe and the Middle East and Denisovans in Asia interbred with modern humans.

Only modern humans descended from European populations have Neanderthal DNA in their genomes and only populations descended from native Australians, Melanesians and some native South-East Asian groups, such as the Manobo of The Philippines, have Denisovan DNA.

a There is a significant number of scientists who argue that the classification of Neanderthals and Denisovans should be *Homo sapiens denisovan* and *Homo sapiens neanderthalensis*. Why might they be suggesting this?

b Mitochondrial DNA from Neanderthals and modern humans have both been fully sequenced. Modern human mtDNA shows no relationship between Neanderthals and modern humans. How does this support the idea that they are separate species?

14 a The development of agriculture allowed humans to live in much larger groups and to allow specialisation of tasks within the large groups. Explain how agriculture resulted in these developments.

b Some of the major cultural changes of humans would not have been able to occur without the existence of earlier biological changes. Give two examples that support this contention.

REVIEW QUESTIONS

How are species related?

Multiple choice questions

1 A species of moth known as Morgan's sphinx moth (*Xanthopan morganii*) has an extremely long tongue that has evolved over many thousands of generations. This moth exploits a food resource unavailable to other moths. It uses its long tongue to collect nectar from the flower of a plant called Darwin's orchid (*Angraecum* sp.). This plant has a very tubular flower so only the long tongue of the moth can reach the nectar. In the process of collecting the nectar the moth collects pollen and then it carries it to other orchids, which it pollinates. Over time both the flower tubes and the moth tongues have become longer.

This relationship between the moth and the orchid is best described as an example of:

A natural selection
B allopatric speciation
C coevolution
D convergent evolution

2 Mass extinction events have been a feature of Earth's biological history. The largest of these extinction events is known to have occurred at the end of which period?

A Devonian
B Ordovician
C Cretaceous
D Permian

3 The family Felidae includes many species of felines. Three species in this family are *Catopuma badia* (species A), *Catopuma temminckii* (species B) and *Profelis aurata* (species C). Together these are known as golden cats. An outgroup of this clade is *Panthera leo* (species D).

Assuming that these species are classified correctly, which of these phylogenetic trees most accurately shows the relationships between the species?

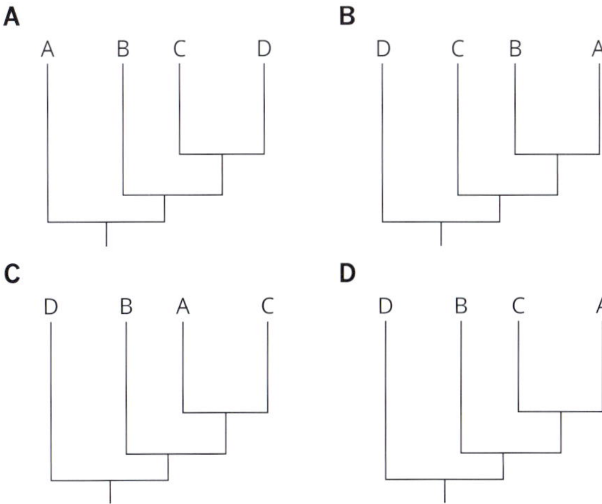

4 Radioactive dating methods are frequently used to give absolute dates to fossils. One radioactive dating method used to date fossils directly uses the carbon-14 isotope. The graph below shows the decay of carbon-14 to nitrogen-14.

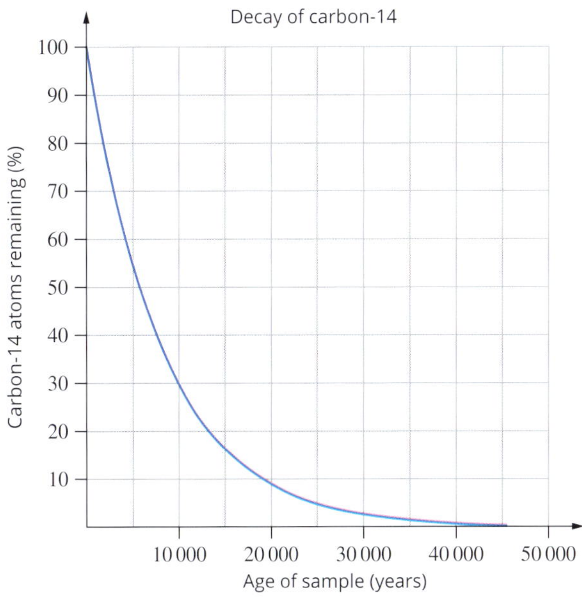

A fossil is found in which the percentage of carbon-14 remaining is 60%. What is the approximate age of the fossil?

A 10 000 years
B 8000 years
C 6000 years
D 4000 years

5 A molecular clock is used to measure the rate of evolutionary change in a group of related organisms. The molecular clock uses the known mutation rate of particular genes common to the groups under investigation to estimate when the groups diverged from each other. This information is then used to construct scaled phylogenetic trees. In order to be used as a molecular clock, a gene needs several features. Which of these is the most important?

A It has a recent origin.
B It is being acted upon by a strong selection pressure.
C It is positioned close to a telomere.
D It has a constant mutation rate.

6 Stratigraphic correlation is used to work out the relative ages of sedimentary strata at different sites. Strata containing the same index fossils are assumed to be the same age. It is also assumed that lower layers are older than upper layers.

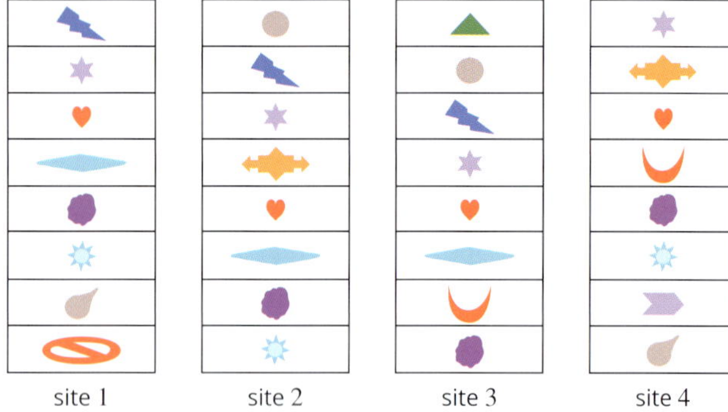

Consider the four sites illustrated. Which one of the following statements is accurate?

A Site 3 contains the youngest stratum and site 4 the oldest.
B Site 1 contains both the youngest and the oldest strata.
C All strata at the same depth are the same age.
D Site 1 contains the oldest stratum and site 3 the youngest.

7 There are two wild species of banana: *Musa acuminata* and *Musa balbisiana*. Today there are many varieties of this popular fruit. These original wild varieties have 22 chromosomes. The most popular variety of bananas grown in Australia today is the Cavendish. It accounts for more than 95% of all production. Cavendish bananas have been around since before 1850. The Cavendish banana has a chromosome count of 33. The development of this variety of banana is most likely due to:

A polyploidy
B hybridisation between *Musa acuminata* and *Musa balbisiana*
C genetic engineering
D self-fertilisation by a member of *Musa acuminata*

Short answer questions

8 Birds and mammals share a common ancestor. The ancestor of these two classes was a terrestrial vertebrate. Today some species in both taxa spend large amounts of time in an aquatic environment. Two such species are penguins and seals. The hypothesised evolution of both of these groups is shown in the phylogenetic tree below.

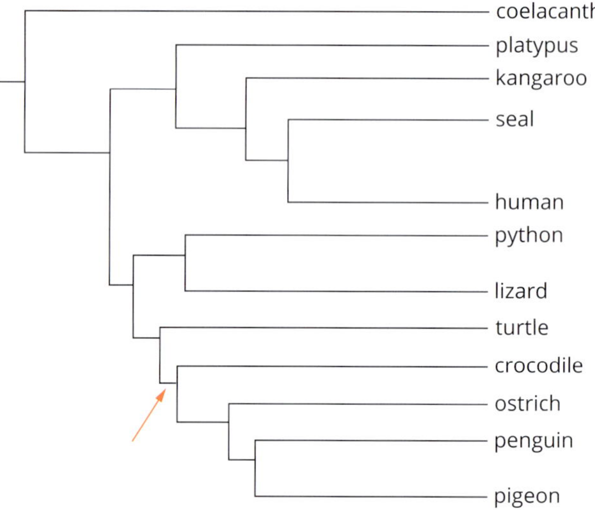

The last common ancestor of penguins and seals was a vertebrate that lived some time between 305 and 340 million years ago. Evidence suggests that, like modern reptiles, this organism was ectothermic.

a Penguins and seals are both endothermic. What kind of evolution has resulted in the two groups having this characteristic? Explain.

b Why is there so much uncertainty about when the last common ancestor of penguins and seals lived?

c Seals and penguins both have forelimbs modified as flippers. These flippers can be considered to be both homologous and analogous. Explain how this can be the case.

d On the phylogenetic tree above the red arrow points to a common ancestor called an archosaur, which palaeontologists believe gave rise to a number of later lineages. Identify all of the members of the monophyletic group that includes the archosaurs.

e Identify a polyphyletic group from the tree. Explain how this group is polyphyletic.

9 *Lymnaea* is a genus of mollusc having snail-like shells. Members of *Lymnaea* have shells that coil either left or right. In most individuals, development of this trait is purely genetic, with offspring showing a phenotype that is identical to the maternal phenotype. Occasionally, and seemingly at random, an environmental factor influences the outcome and an individual with a genetically right-coiling shell grows a shell that coils left and vice versa.

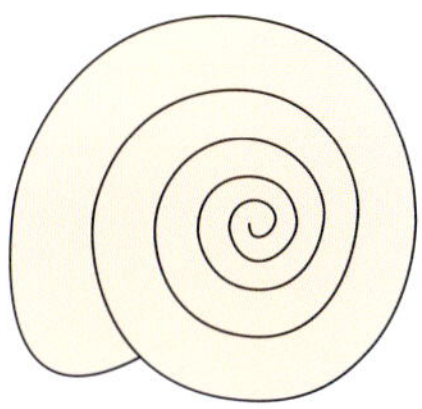
left-coiled shell

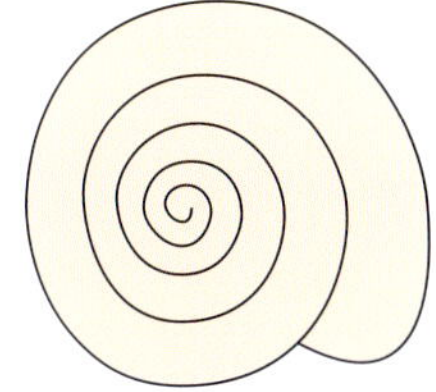
right-coiled shell

a As a result of physical incompatibility individuals with shells that coil in the opposite direction are unable to mate.
   i If the environmental effect were to disappear the two populations would find mating very difficult. What name is given to this sort of isolation?
   ii Without the occasional environmental effect on shell growth this incompatibility could lead to speciation of the two groups. Explain how this could occur.

b Why is speciation unlikely as long as the environmental effect continues to create individuals that are genetically of one form but phenotypically the other form?

10 Wheat (*Triticum* sp.) is one of the world's major food crops. A disease that causes severe crop losses is wheat leaf rust. This disease is caused by three species of fungus, all of which belong to the genus *Puccinia*. Leaf rust fungus is found in all major wheat-growing areas, including Australia. Leaf rust can cause crop losses of up to 20% when infection occurs. Cost of *Puccinia* infections to farmers in Australia is about $127 000 000 annually. This has resulted in significant research into methods of fighting the disease.

Resistant strains of wheat have been identified and selective breeding has been used to enhance the offspring of these strains. One such strain, called Norm, has been shown to have strong resistance to leaf rust fungus.

a Around 46 different genes have been shown to be associated with resistance to the leaf rust fungus. Scientists have selectively bred varieties of wheat to increase the frequency of the alleles conferring most resistance in wheat plants. Selective breeding for particular traits can have unexpected effects on the phenotypes of the organisms that have been subject to this process. Explain why.

b The identification of Norm as highly resistant to leaf rust means that many farmers are deciding to plant this variety of wheat. Some scientists consider the exclusive use of one strain to be a very dangerous practice that could lead to worldwide famine. Do you agree with the scientists? Explain why or why not.

c i Are these resistant populations of wheat likely to maintain their resistance to leaf rust fungus over the long term? Explain.
   ii What kind of evolution is being demonstrated by the wheat and the leaf rust fungus?

11 The genus *Equus* includes several species, both extinct and extant (still living). The extant species includes two species of horse: Przewalski's horse (*Equus przewalskii*) and the domestic horse (*Equus caballus*). Przewalski's horse is listed as endangered. In 1945, 13 individuals were held in captivity. All current members of the species are descendants of 9 members of this captive population. The last wild member of *Equus przewalskii* was seen in 1965, shortly after which it was determined that it was extinct in the wild. Since then, a conservation and breeding programme based on the original 9 members of the captive population has been so successful that they were re-introduced into their natural habitat, the steppes of Mongolia. The wild population now numbers over 400 and the captive population is in excess of 1500.

The phylogenetic tree for the genus *Equus* is shown below.

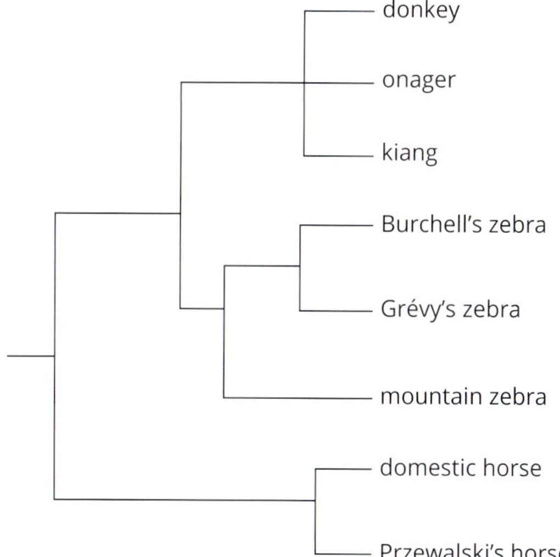

a The population of Przewalski's horse from which the current population descended was quite small in number. Does this represent a genetic bottleneck or a founder population? Explain.

b Despite the considerable increase in the numbers of Przewalski's horse, it is still considered to be endangered. Explain why this is the case.

c The nearest living relative of Przewalski's horse is the domestic horse. Przewalski's horses have 66 chromosomes and domestic horses have 64. Despite this difference, matings between domestic horses and Przewalski's horses produce fertile offspring.

i How might the change in chromosome numbers have come about?

ii Przewalski's horse and the modern domestic horse are very closely related. The relationship is so close that some taxonomists classify them as subspecies of the same species: *Equus ferus*. What is a subspecies?

iii Why might there be so much argument about whether Przewalski's horse and the domestic horse are separate species or subspecies of *Equus ferus*?

12 While on his famous trip on the HMS *Beagle*, Darwin observed finches on the Galapagos Islands. These islands contain 13 different species. On Cocos Island there is a fourteenth species of finch that has been shown to be related to the Galapagos finches. The Galapagos finches are hypothesised to be most closely related to the tanager finches of Central and South America.

The map below shows the relationship between the islands and the mainland of South America.

The hypothesised relationships between the various finches is shown in the phylogenetic tree. Note: Vegetarian finches are also known as herbivorous finches.

Genetic sequences show that finches with similar feeding styles tend to be closely related.

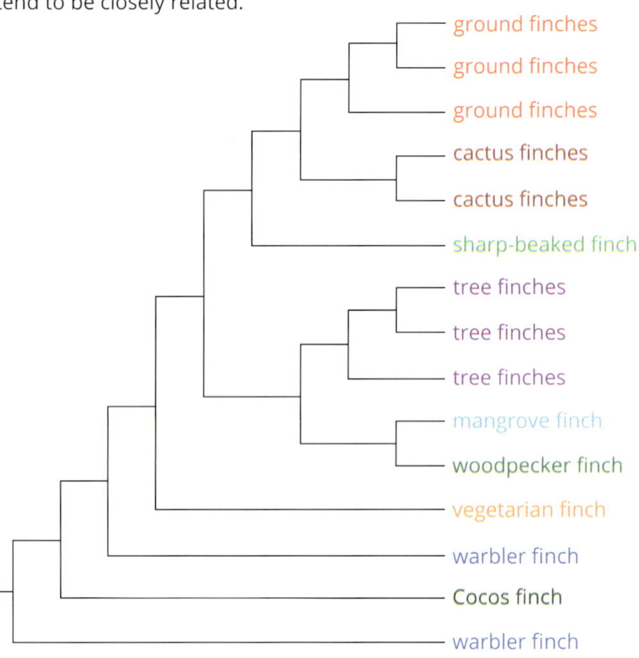

a The finches of the Galapagos Islands show considerable genetic diversity. What does this suggest about the size of the founder population?

b It is thought that the first finches arrived on the Galapagos about 2 million years ago. Since that time the environment has changed considerably. There has been volcanic activity, which added 14 islands to the chain. There have also been changes in sea level, isolating some islands, and climate has changed from lush and tropical to cool and dry. How might the changes in the Galapagos Islands have resulted in speciation of the original population of finches to arrive there?

c DNA studies were used to establish that the Cocos Island finches are closely related to the Galapagos finches.

i What is the name of the DNA manipulation used to identify relationships between species?

ii Describe the process used to establish the relationship between two species.

d According to the phylogenetic tree shown, which of the Galapagos groups is most closely related to the Cocos Island finches?

13 The fossil record has been an important source of evidence to support the theory of evolution.

Shown below is a series of fossil skeletons showing the evolution of modern whales.

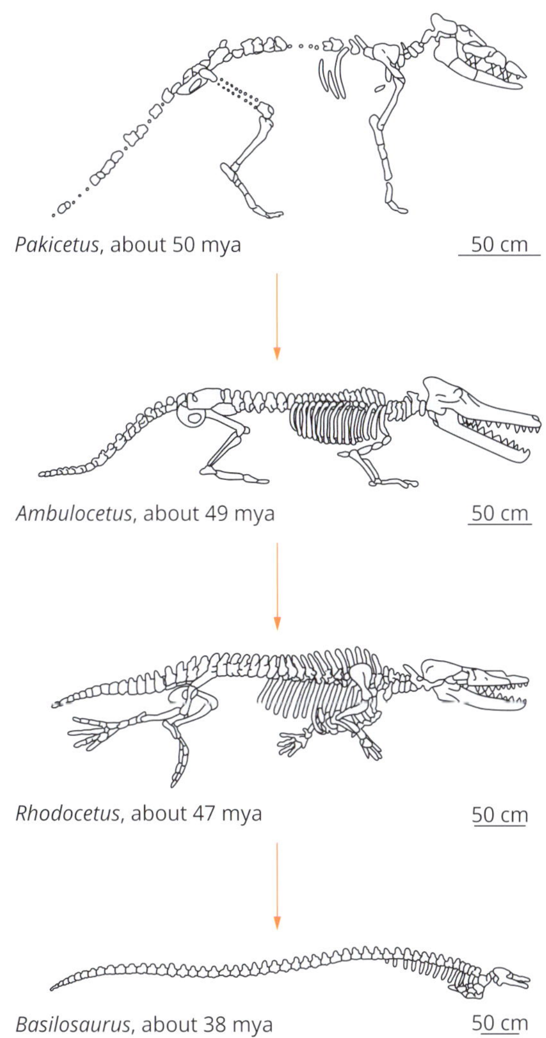

a Explain how this group of fossils provides evidence for evolution.

b Describe how the age of the *Ambulocetus* fossil was determined.

c *Dorudon* was another whale ancestor. Its skeleton is shown below.

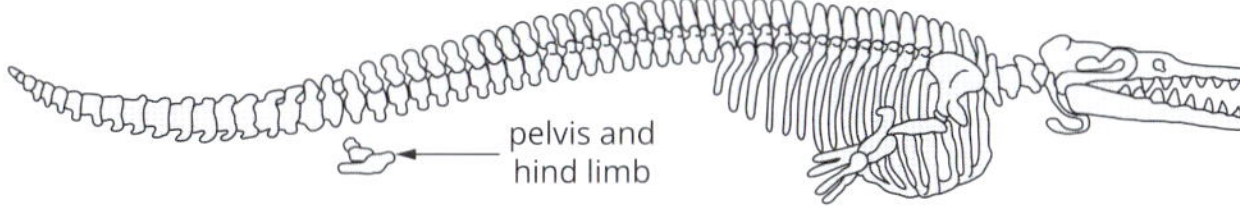

Dorudon was a contemporary of one of the above organisms. Which of the fossil series is this most likely to be? Provide reasoning to support your answer.

14 The body plan of insects is that they have a head, a thorax and an abdomen. The thorax is formed from three embryonic segments: T1, T2 and T3. In most insects each segment grows one pair of legs and segments T2 and T3 also grow a pair of wings. Flies, however, have only one pair of wings, which grow on T2. It is the *Ultrabithorax* (*Ubx*) gene that controls this situation. It is a master regulator gene that encodes a transcription factor that represses the formation of wings on T3.

a i What is a transcription factor?

ii What is the likely effect of a non-lethal mutation in the *Ubx* gene in a fly?

b *PAX-6* is a gene found in a large variety of animals. It is a transcriptional factor that encodes a protein that binds to DNA. It has two binding sites and its binding controls the development of sensory organs, especially eyes. This gene has been highly conserved over time. The mouse and human form are identical. In fact, the genes are so similar in most eyed organisms that the human gene when inserted into the genome of otherwise sightless fruit flies (*Drosophila melanogaster*) results in the formation of eyes.

The human eye is quite different in structure from the compound eye of the fruit fly. Explain how *PAX-6* can result in eye formation in both humans and fruit flies.

15 Taxonomists use the features of skeletons, and for hominins associated cultural materials, to develop cladograms and phylogenetic trees for species to show how they may be related. In order to develop their cladogram, the scientist will first look at the various traits of the group of organisms and develop a character matrix. Such a matrix is shown below for one group of hominins.

Species	Opposable big toe	Large brow ridge	Recognisable tool making associated	Relatively small teeth	Taller than 1.5 m	Cranial capacity normally greater than 1200 cm^3	Sophisticated ritual burial practices
1	1	1	0	0	0	0	0
2	0	1	0	0	0	0	0
3	0	1	1	0	0	0	0
4	0	1	1	1	0	0	0
5	0	0	1	1	1	1	0
6	0	0	1	1	1	1	1

a Using the information in the matrix, develop a cladogram for the species shown.

b The species used to develop this matrix are listed below:
Homo neanderthalensis, *Australopithecus* sp., *Homo habilis*, *Homo sapiens*, *Ardipithecus* sp., *Homo erectus*

Using your knowledge of hominin evolution, match each species in the matrix to its correct name.

c The search for hominin fossils continues across the world. There have been some significant finds in many places, for example on the island of Flores and in the Denisovan caves.
Consider a hypothetical group of anthropologists working in China who have found a group of fossilised hominin bones. As is common, some of the bones are missing and some show signs of animal activity such as teeth marks.
Why would many fossilised bones show signs of animal activity?

d The bones found by the researchers had a number of significant features. The cranial capacity was around 600 cm^3. The length of the femur was around 30% shorter than the femur of a modern human. The brow ridges were pronounced and the zygomatic arches were extremely large. The teeth were large and a number were distinctly pointed. The foramen magnum was positioned more forwards than in chimpanzees and gorillas but very slightly more to the rear of the skull than in the australopithecines. The arms bones were significantly shorter than the leg bones. No bones from the feet or ankles were found.

i If you were presented with these bones, where would you put the hominin on your cladogram? Give reasons.

ii What other inferences could you make about this hominin?

16 The controlled use of fire is thought to have occurred about 1.7 million years ago, but it may have been as recent as 400 000 years ago. It was an important cultural advance. It increased the chances of survival of the hominins who mastered its use. The use of fire not only enhanced social relationships, it assisted with cultural advances and may have played a role in hominin biological evolution.

a Discuss how fire has influenced human cultural evolution and how it may have had an effect on biological evolution.

b Like many inferences about the ancient past, the exact date when fire was mastered is a matter of debate among palaeontologists and anthropologists. Discuss why there is often little consensus about matters that relate to the ancient past.

CHAPTER 13

DNA manipulations: PCR, gel electrophoresis, DNA transformations

Learning outcomes

By the end of this chapter, you will have developed an understanding of the molecular tools and techniques used to manipulate DNA molecules for particular purposes. You will have learnt how recombinant plasmids are created and then used as vectors in the process of bacterial transformation. You will also have an understanding of the other molecular tools and techniques explored in this chapter, including gel electrophoresis and the polymerase chain reaction (PCR).

Key knowledge

- the use of enzymes including endonucleases (restriction enzymes), ligases and polymerases
- amplification of DNA using the polymerase chain reaction
- the use of gel electrophoresis in sorting DNA fragments, including interpretation of gel runs
- the use of recombinant plasmids as vectors to transform bacterial cells.

13.1 DNA manipulation

To work with DNA, it is necessary to have more than a few DNA molecules. The polymerase chain reaction (PCR) is a technique used for DNA amplification—it makes millions of identical copies of a piece of DNA. Using PCR, forensic scientists can amplify the DNA in traces of blood left at the scene of a crime, or a particular gene from a sample of DNA.

Gel electrophoresis is a method used to separate and visualise nucleic acids and proteins according to their size (Figure 13.1.1). This is usually performed after PCR to either confirm that the correct DNA fragment was amplified, or to identify DNA fragments present in the sample (DNA profiling). It is also used to separate DNA fragments to be used in gene manipulation processes.

In this section, you will learn about PCR and gel electrophoresis.

FIGURE 13.1.1 A scientist loads DNA samples into a gel electrophoresis chamber.

A microsatellite is a short repeated sequence of nucleotides found at a defined locus on a chromosome. Variation in the number of repeats between individuals makes microsatellites useful in DNA profiling.

DNA AMPLIFICATION

Many DNA manipulation techniques require a large quantity of DNA to work with. However, sometimes only a very small sample of DNA is available for scientists. For example, only trace samples of DNA may be left at a crime scene or extracted from fossils of extinct species (Figure 13.1.2), and only small samples can be removed for medical tests and in embryonic or foetal DNA screening for genetic disorders. In these cases, DNA amplification is required to increase the amount of the target DNA sample so that it is large enough to be used or analysed in other techniques and processes.

DNA amplification uses the **polymerase chain reaction** (PCR) to create a large quantity of DNA that is identical to the initial trace sample. DNA amplification enables the rapid and accurate replication of target DNA, resulting in millions of copies being produced in a matter of hours. The term 'target DNA' is used to describe a particular region of a DNA molecule that a scientist intends to study or manipulate (e.g. a specific gene or a **microsatellite**, which is a variable region of the genome used for DNA profiling).

FIGURE 13.1.2 A scientist extracts fossilised DNA from a Neanderthal (*Homo neanderthalensis*) bone. The DNA will be amplified for further DNA analysis.

Primers determine the start and end points of a nucleotide sequence to be amplified.

POLYMERASES

As the name indicates, polymerase chain reaction (PCR) is based on the action of polymerases. **Polymerases** are enzymes that catalyse the formation of polymers, in particular nucleic acids. There are two different groups of polymerases:

- DNA polymerase
- RNA polymerase.

DNA polymerase

DNA polymerase acts to assemble DNA. This enzyme uses each strand of the DNA double helix as a template for building the new DNA strands, following the complementary base pairing rules. Each double-stranded DNA copy is therefore identical to the original. In the initial steps of DNA replication of prokaryotic and eukaryotic cells the DNA needs to be unzipped. In cells, a **helicase** enzyme unwinds and unzips the DNA molecule. DNA polymerase can only attach to the end of a nucleotide chain that is base paired with the template strand, so a short string of nucleotides called a **primer** binds to the DNA template strand. DNA polymerase then attaches and moves along the strand, continuously adding bases in the 5' to 3' direction until it reaches the end of the molecule.

Scientists mimic this process in a test tube for techniques such as PCR and DNA sequencing. In PCR, the strands of the double-stranded DNA are separated by heating rather than with a helicase enzyme. To be amplified, each double-stranded DNA sequence needs two primers. The primers are made in a laboratory to have a nucleotide sequence that is complementary to each end of the DNA sequence to be amplified, enabling them to target particular regions of a nucleotide sequence within a gene, chromosome or genome. The DNA polymerase enzyme used in PCR has particular properties.

Taq polymerase

Taq polymerase is the DNA polymerase that is most commonly used in PCR. It was originally extracted from the **thermophilic** bacterium *Thermus aquaticus*. The heat-resistant properties of Taq polymerase make it extremely useful in DNA manipulation techniques such as PCR.

RNA polymerase

RNA polymerase acts to assemble RNA. In cells, RNA polymerases synthesise mRNA, rRNA and tRNA by transcription of genes. RNA polymerase does not require a primer to **anneal** to the DNA template strand to start the synthesis of RNA. It has a subunit that unwinds the DNA, allowing it to add bases, one at a time in the 5' to 3' direction. RNA polymerase adds nucleotides until a 'stop' sequence is reached. A single-stranded piece of RNA is created. RNA polymerase works much more slowly than DNA polymerase. RNA polymerase is used in specialised laboratory methods to study transcription and RNA amplification.

Reverse transcriptase

Reverse transcriptase is a DNA polymerase that synthesises single-stranded DNA using single-stranded RNA as a template. This is the reverse of the usual transcription process in which DNA is transcribed into RNA. Reverse transcriptase is used in the laboratory to produce DNA molecules that can be amplified by PCR for further analysis. It is also used to make **complementary DNA** (**cDNA**) from mRNA that has already had the introns spliced out. The cDNA is then inserted into bacterial plasmids so that the protein coded for by the mRNA can be made in the laboratory, as you will learn in Section 13.2.

BIOFILE

Retroviruses and reverse transcriptase

Retroviruses, such as the human immunodeficiency virus (HIV), are a natural source of reverse transcriptase. Retroviruses have an RNA genome. Reverse transcriptase copies the viral RNA genome into double-stranded DNA, which is then inserted into the host cell chromosome. The host cell transcribes and translates the viral genes to produce viral proteins. Retroviral genomes may become a permanent part of the host genome. They are then called endogenous retroviruses. These sequences may account for up to 9% of the human genome.

BIOFILE

Thermus aquaticus

A field trip to Yellowstone National Park in the 1960s radically altered the course of molecular genetics research. Thomas Brock, a bacteriologist from the University of Wisconsin-Madison, found bacteria in water taken from a hot spring. He named the new species *Thermus aquaticus* (Latin for 'hot water').

Enzymes are normally denatured if heated to temperatures of 95 °C for more than a few seconds. For *T. aquaticus* to survive in the hot springs, its enzymes, including DNA polymerase, need to tolerate these high temperatures. Therefore, the DNA polymerase from *T. aquaticus* ('Taq polymerase') has proved to be an ideal enzyme for use in PCR.

FIGURE 13.1.3 A scientist obtains a sample of *Thermus aquaticus* from a hot spring.

Number of PCR cycles (n)	Number of double-stranded copies of original DNA (2^n)
0	1
1	2
2	4
3	8
4	16
5	32
6	64
7	125
8	256
9	512
10	1024
20	1048576
30	1073741824

TABLE 13.1.1 Exponential growth in number of target DNA molecules during PCR.

THE POLYMERASE CHAIN REACTION

The polymerase chain reaction (PCR) is a method of amplifying specific target sequences of DNA. The DNA polymerase Taq polymerase is used in this process, as it resists the changes in temperature required during PCR. Each strand of the DNA acts as a template for a new copy of itself.

PCR is carried out in cycles. Each cycle doubles the amount of DNA, resulting in exponential growth. In other words, it is a chain reaction that creates billions of copies of target DNA, and it does so in only a few hours. Table 13.1.1 shows how each PCR cycle increases the amount of target DNA exponentially. After 30 cycles, there are over one billion copies.

For this reaction to be carried out, a PCR mixture is required, containing:

- DNA, including the target DNA to be amplified (the sample) (Figure 13.1.4)
- free nucleotides, to build new DNA strands
- a heat-resistant DNA polymerase (usually Taq polymerase), to elongate the new DNA strands by adding the free nucleotides
- two DNA primers complementary to the ends of the target DNA, to specify the start and finish of the DNA fragment to be amplified. The two primers are synthetic, single-stranded DNA molecules up to 30 bases in length. They are specifically created to bind through hydrogen bonding with complementary base pairs on either end of the DNA to be amplified.

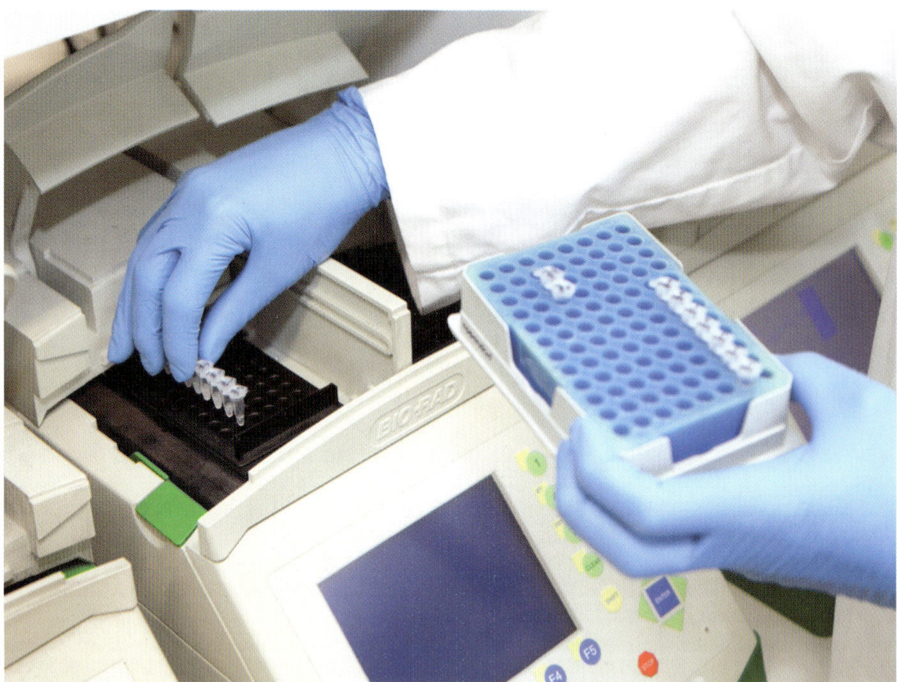

FIGURE 13.1.4 A scientist loads DNA samples into a thermocycler to be amplified by PCR.

BIOFILE

Trace samples

The process of PCR requires as few as one or two cells for DNA amplification. Scientists at the Victoria Forensic Science Centre have shown that merely touching an object deposits sufficient material for successful DNA amplification. In handling keys, opening a door or driving a car, the cellular material deposited by a criminal provides ample DNA for analysis following PCR.

Steps in the polymerase chain reaction

The PCR mixture is placed in a **DNA thermocycler**, which alters the temperature in pre-programmed steps. Each PCR cycle involves three steps (Figure 13.1.5):

1. Denaturation: The sample is heated to 95 °C to break the hydrogen bonds between the two strands of double-stranded DNA to obtain single strands of DNA.
2. Annealing: The temperature is reduced to 50–60 °C. This allows the primers to anneal (or bind) to complementary sequences on opposite strands at each end of the target DNA sequence.
3. Extension: The temperature is increased to 72 °C. This allows Taq polymerase to attach to the primers on the DNA strands. The Taq polymerase moves along each strand, adding free nucleotides to form double-stranded DNA.

This three-step cycle of heating and cooling is repeated up to 50 times to ensure there is sufficient target DNA produced to work with.

Annealing of primers to their complementary nucleotide sequences may take from 10 to 30 seconds depending on the length and base composition of the primer.

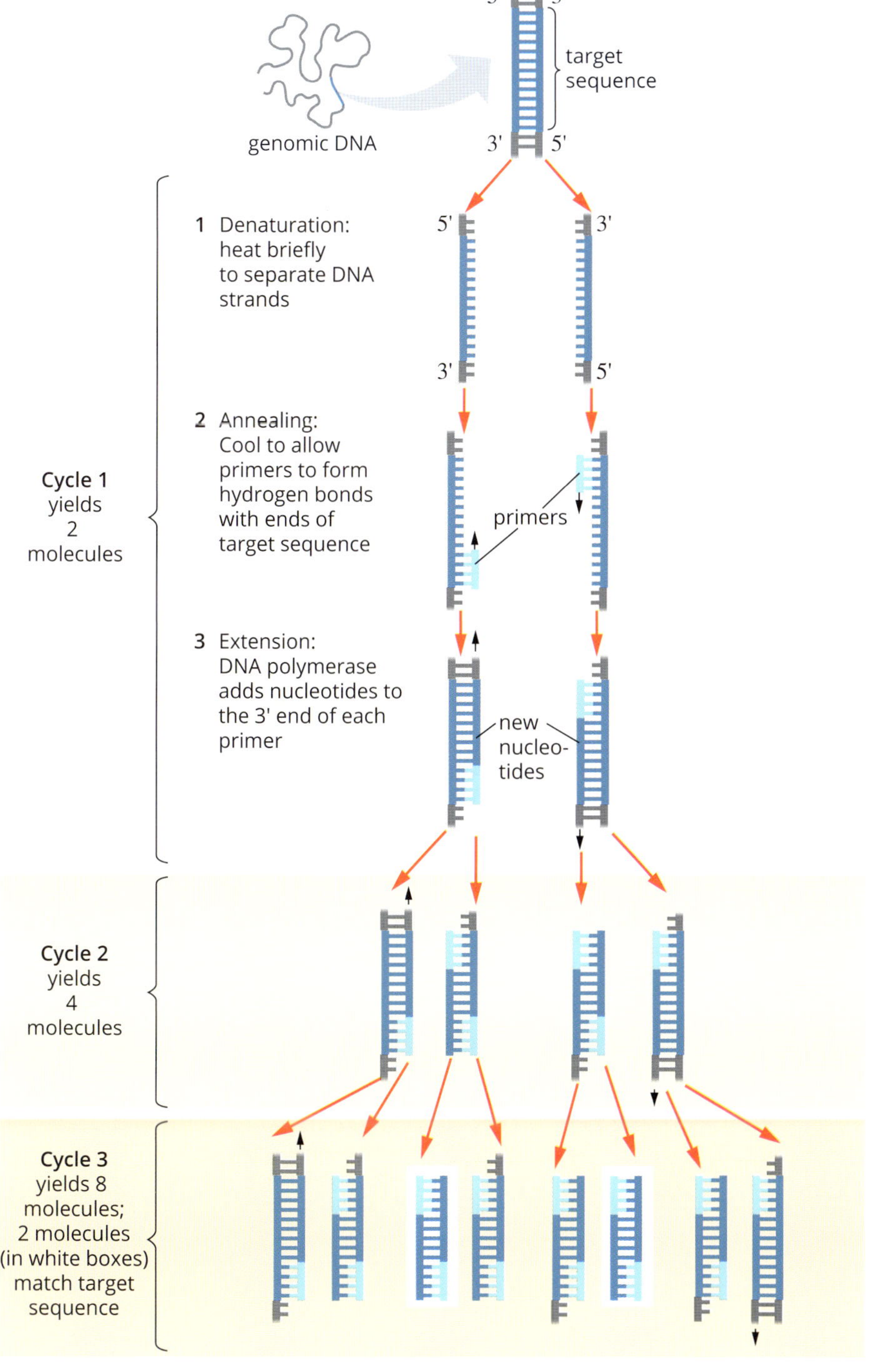

FIGURE 13.1.5 The three steps of PCR: denaturation, annealing and extension. Each PCR cycle increases the amount of target DNA exponentially.

DNA SEPARATION BASED ON FRAGMENT SIZE

Gel electrophoresis is commonly performed after DNA amplification. Gel electrophoresis allows scientists to separate out the different DNA fragments present in a sample based on their size. This information can be used to match fragments from other samples, as in DNA profiling, or to isolate a particular fragment for further use in another technique, such as DNA recombination and bacterial transformation.

> In addition to separating DNA and RNA, gel electrophoresis can be used to separate proteins of different sizes.

Gel electrophoresis

Gel electrophoresis is a technique for separating fragments of nucleic acids (DNA and RNA). When an electric current is applied to the gel, the negatively charged DNA molecules in the gel move towards the positive terminal. Small DNA molecules move faster than large ones, causing them to separate based on their size.

> DNA is negatively charged due to the negative charge on the phosphate group in each nucleotide.

Gel electrophoresis is used to compare DNA fragments for a number of applications such as:

- DNA screening
- confirming the correct gene has been amplified in PCR
- identifying DNA fragments obtained by restriction enzyme digestion.

The following steps and Figure 13.1.6 outline the process of gel electrophoresis:

1. An electrophoresis gel is prepared. It has a jelly-like texture and is usually composed of agarose (a purified form of agar). The gel is rectangular and contains small wells (holes) at one end.
2. The gel is placed into a gel electrophoresis chamber with the wells situated at the negative terminal of the chamber.
3. Each DNA sample is loaded into one of the wells within the gel.
4. A **DNA ladder** containing DNA fragments of known length is also run on the gel for comparison with the samples. This allows the length of the sample DNA fragments to be estimated.
5. The gel is placed in an electrophoresis bath where it is covered with a controlled pH solution that contains ions to conduct an electric current.
6. A power source is attached to the electrophoresis bath and switched on. The electrical current causes the negatively charged DNA fragments to migrate through the gel towards the positive terminal of the chamber.
7. Smaller fragments move faster through the gel, so they migrate further through the gel than larger fragments in a given period of time. This sorts the fragments by length.
8. The DNA fragments are detected by applying a stain that binds to DNA. This can be done with a fluorescent stain (which may be included in the gel or added after) or with methylene blue stain (which is added after running the gel). Fluorescent stains are viewed with ultraviolet light.

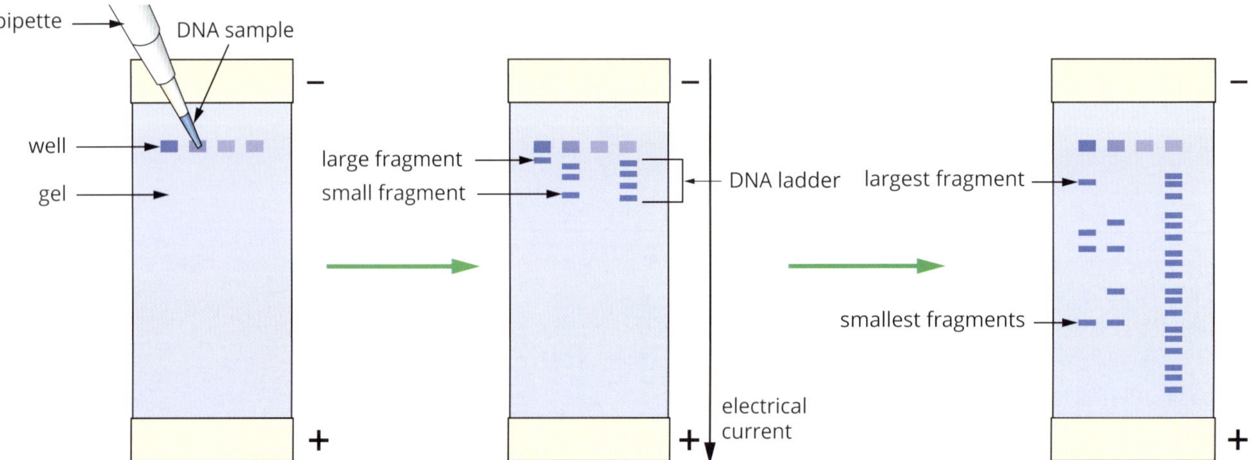

FIGURE 13.1.6 The process of gel electrophoresis, showing two DNA samples being loaded into wells and migration of the DNA fragments through the wells based on their size. The last well contains a DNA ladder for comparing and estimating the fragment length of the two DNA samples.

If the gel is stained with methylene blue, the DNA fragments are visible by eye as blue bands. If a fluorescent stain is used, the DNA in the gel is observed on an ultraviolet (UV) light box (Figure 13.1.7). This is because the fluorescent stain, which binds to the DNA in the gel, emits a bright colour when exposed to UV light. Bright coloured bands are observed wherever there is DNA present in the gel.

The size of the DNA fragment bands of various samples can be determined by comparing them against a DNA ladder, which contains DNA fragments of known length, measured in base pairs. The DNA ladder is run on the gel next to the samples to determine the approximate length of each fragment (Figure 13.1.8).

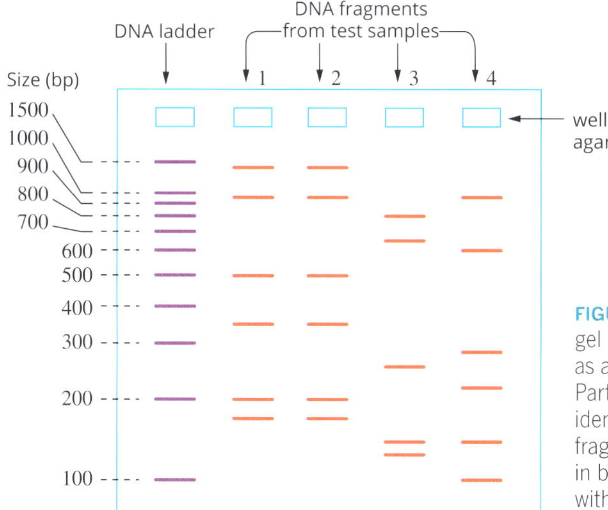

FIGURE 13.1.8 The results from gel electrophoresis can be shown as a diagram that can be analysed. Particular fragments can be identified and compared. Each fragment length can be measured in base pairs (bp) by comparing it with the DNA ladder.

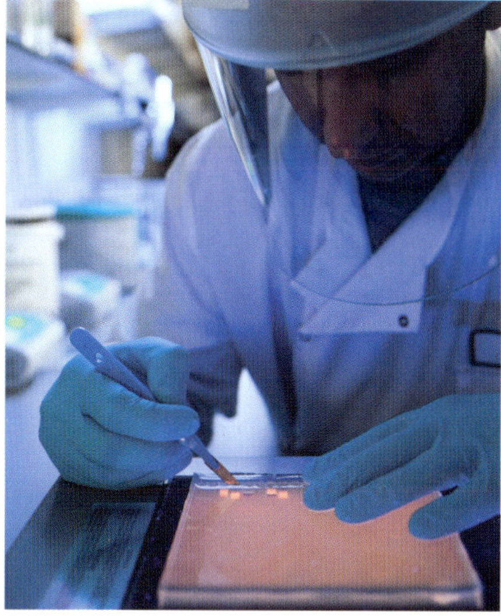

FIGURE 13.1.7 A scientist cuts a sample from an electrophoresis gel. The gel, which was stained with ethidium bromide, a fluorescent DNA stain, shows the DNA fragments as pink fluorescent bands.

Smearing of bands in the gel can indicate overloading of the wells or the degradation of the DNA or protein.

BIOLOGY IN ACTION

Surveillance of waterborne RNA viruses

Poor sanitation and faecal contamination of the water supply with viruses, bacteria and protists is a major health problem in many regions globally. The World Health Organization provides guidelines on water safety and promotes improved methods of monitoring microbial contamination in drinking water. Outbreaks of poliovirus and hepatitis A and E virus infections may be due to contaminated water. Traditional methods of identifying viruses, by growing them in cells, take several weeks and are not successful for all viruses. Molecular methods such as PCR offer faster detection and the potential to detect multiple viruses in one analysis. Researchers are working to develop such methods. Viruses such as poliovirus and hepatitis viruses have an RNA genome, so a technique called RT-PCR is used. First, reverse transcriptase (RT) is used to copy the viral RNA into DNA, and then the polymerase chain reaction (PCR) amplifies the DNA.

Primers specific for different viral genes are used in the PCR. Multiple sets of primers can sometimes be used in one assay, allowing the detection of several viruses at once (Figure 13.1.9). If a virus is present in the water sample, its RNA will be copied to DNA and amplified. The DNA resulting from the reaction can be visualised by the technique of gel electrophoresis.

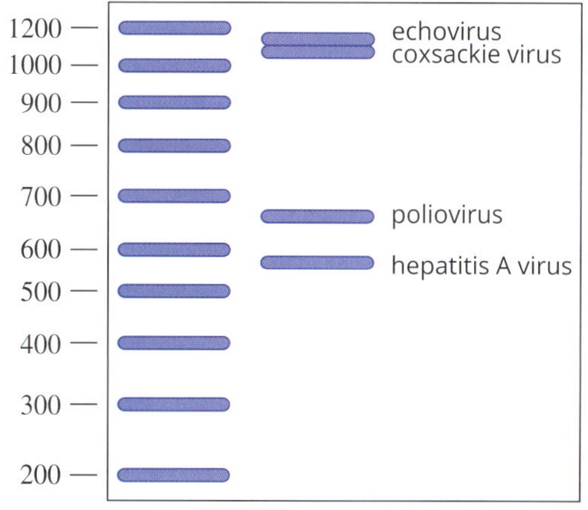

FIGURE 13.1.9 Illustration of gel electrophoresis results obtained in water sampling studies using RT-PCR and multiple primer sets to detect four different RNA viruses. DNA standards ranging in size from 200 to 1200 base pairs (bp) are on the left. Amplified viral genes were detected as bands of the following size: hepatitis A virus, 589 bp; poliovirus, 671 bp; coxsackie virus, 1084 bp; echovirus, 1128 bp.

Combining molecular tools to detect mutations

Mutations are usually discovered because of the effect they have on the individual carrying them. If an individual has the symptoms of cystic fibrosis (CF) then the cystic fibrosis transmembrane conductance regulator (*CFTR*) gene would be analysed. The *CFTR* gene codes for a large membrane protein of the same name, which regulates chloride ion movement across cell membranes. The *CFTR* gene is very large and many different mutations can cause disease. The most common mutation, called the ΔF508 mutation, is a deletion of 3 base pairs, leading to deletion of the amino acid phenylalanine (single letter code F) from position 508 of the protein. Families with a history of cystic fibrosis may wish to undergo screening for this mutation in parents, unborn foetuses or newborns.

PCR and electrophoresis can be used to detect the mutant allele by:

1 isolating DNA from the individual. The DNA can come from a mouth swab of an adult or from the amniotic fluid surrounding an unborn child.
2 using PCR primers that are complementary to the DNA sequences on either side of the site of the ΔF508 mutation to amplify the DNA.
3 comparing the amplified DNA molecules by gel electrophoresis. For a normal allele the amplified region is 98 base pairs long. In a ΔF508 mutant allele the amplified region is 95 base pairs long (Figure 13.1.10a). A DNA ladder is run next to the samples to enable identification of the normal and mutant alleles based on their size (Figure 13.1.10b).

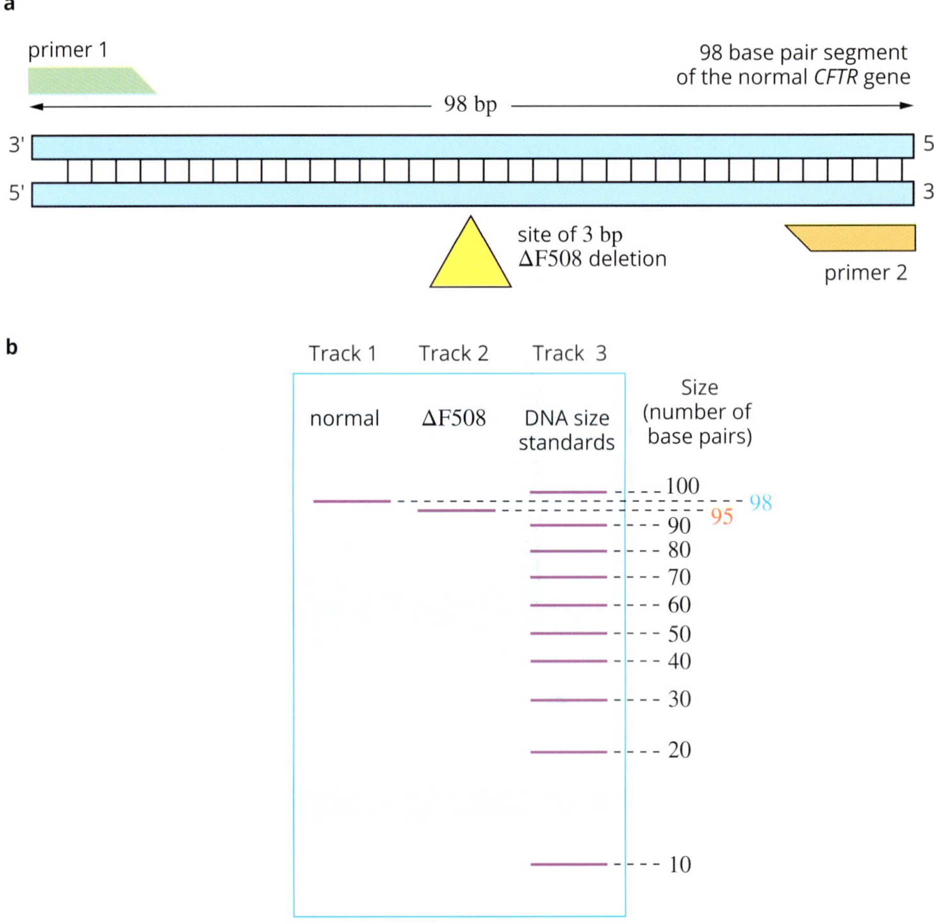

FIGURE 13.1.10 (a) PCR primers that span the site of the ΔF508 mutation are chosen to test for the presence of the CF mutation. (b) Diagram of an electrophoresis gel showing PCR products of an individual carrying only the normal allele of 98 bp (track 1) and an individual carrying only the ΔF508 mutation of 95 bp (track 2). The DNA ladder (size standards) in track 3 allows the size of the bands in tracks 1 and 2 to be determined.

Identification of carriers

The following example shows how these molecular techniques are used to identify family members carrying the mutation.

A couple had a healthy daughter but their next child, a boy, failed to grow as expected, had bowel problems, lung congestion and salty-tasting skin. These symptoms led to a diagnosis of cystic fibrosis (Figure 13.1.11a). The couple were not aware of the incidence of CF in their families, yet a genetic counsellor suggested that they might be carriers of an allele for the disease. To develop cystic fibrosis, you need two copies of the mutant allele (that is, you need to be homozygous for the mutant allele). If you have one copy of the normal allele and one copy of the mutant allele (a heterozygous carrier) you are healthy but able to pass the mutant allele to your children. There is a 25% chance of two carriers having an affected child.

When the couple fell pregnant with a third child they chose to have genetic tests for the common ΔF508 mutation to find out whether they were carriers and whether this child could develop CF. DNA samples from all family members were taken (cheek swabs, or amniotic fluid for the unborn child) and analysed by PCR and gel electrophoresis. The results, shown in Figure 13.1.11, reveal that each parent had one normal allele (the 98 bp fragment) and one copy of the ΔF508 mutation (the 95 bp fragment). That is, they were both heterozygous carriers. The daughter inherited a normal allele from each parent, so was healthy. The son inherited a mutant allele from each parent, so was homozygous for the mutant allele and developed CF. The unborn child inherited one normal allele and one copy of the mutant allele; he or she is a heterozygous carrier, but should not show symptoms of cystic fibrosis (Figure 13.1.11b). This child should be informed that they are a carrier so they can choose to ask their future partners to be tested before having children.

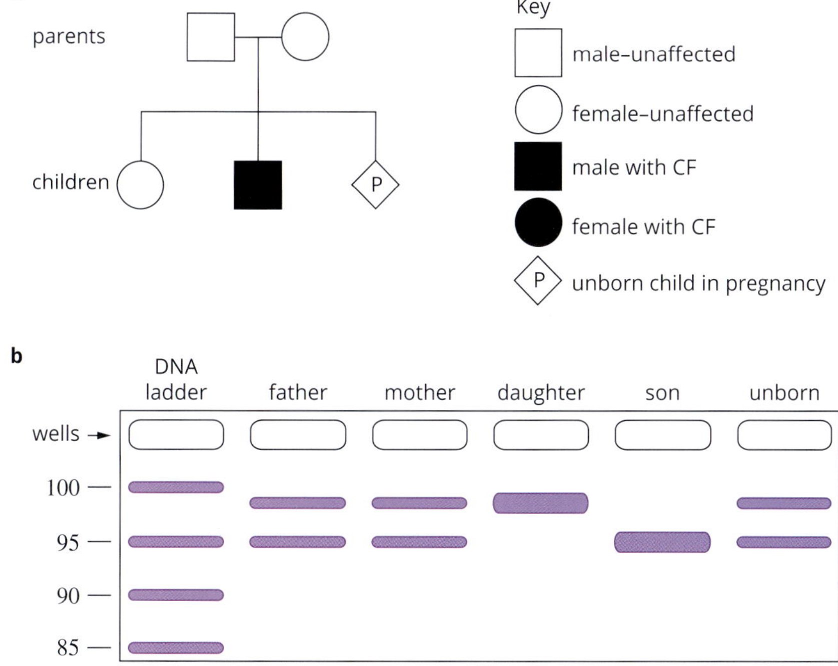

FIGURE 13.1.11 (a) Pedigree chart showing the appearance of cystic fibrosis in the son of two unaffected parents. The first-born daughter is healthy. The parents want to know whether the unborn child (sex unknown) carries the CFTR mutation. (b) Gel electrophoresis results for the DNA test for the ΔF508 mutation. DNA from each family member was amplified by PCR with primers for the ΔF508 mutation region. The amplified DNA was analysed by gel electrophoresis alongside DNA size markers (DNA ladder). The unborn child (child 3) has one copy of the normal allele (98 bp) and one copy of the ΔF508 mutant allele (95 bp).

EXTENSION

Genome sequencing: from gel electrophoresis to next generation sequencing

The entire human genome (~20 000 genes) and the genomes of many other species have now been sequenced. Whole-genome sequencing became faster and affordable through advances in electrophoresis methods and development of fluorescence labelling and detection systems.

Genome sequencing is the process of obtaining the order of the bases, A, C, G and T, in every gene of an organism. For gene sequencing, a DNA sample is placed in a reaction with normal nucleotides, modified nucleotides that terminate the growing DNA strand, and DNA polymerase. The modified nucleotides carry a fluorescent tag, with a different colour for each of the four bases. The reaction generates many fragments that vary in size by one nucleotide, and end with a fluorescent base. The sample is run on a gel and the order of the bases is 'read' by the fluorescent colour emitted by each band (Figure 13.1.12a). The early large polyacrylamide slab gels were replaced with more efficient capillary gel electrophoresis. The fluorescent nucleotides are read by a laser to produce a chromatogram, with each peak representing a band on the gel and hence the base sequence (Figure 13.1.12b).

Faster and cheaper new methods, called next-generation sequencing, take place on chips rather than in test tubes and gels, and use PCR as part of the technique (Figure 13.1.12c). Genome sequences are made publicly available and scientists worldwide can search the databases, for example, to compare genes from different cell types and species, research gene regulation and study pathogen evolution. Studying genomes provides a greater understanding of genetic diversity, evolution, diseases, development and heredity.

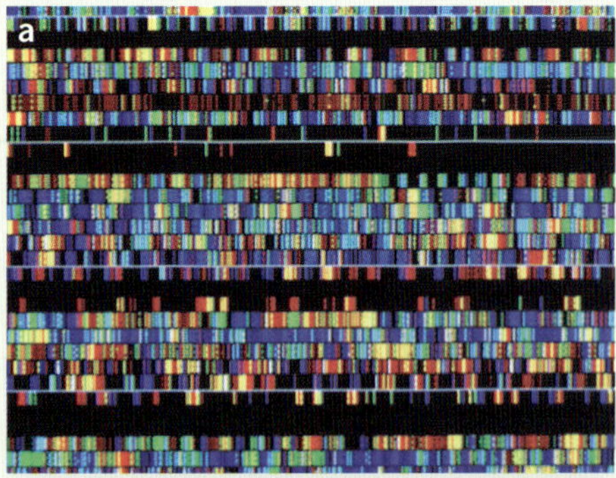

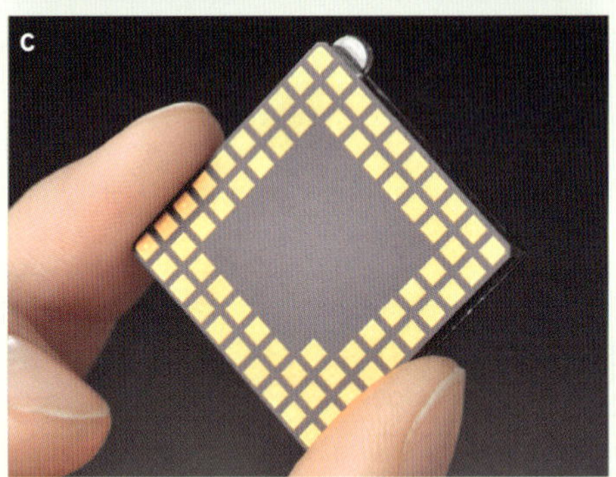

FIGURE 13.1.12 Genome sequencing. (a) Computer screen display of a human DNA sequence as a series of coloured bands on gel electrophoresis. Each row contains the DNA from one sample run through a gel. (b) The order of nucleotides as they run through the gel is translated into a chromatogram for DNA sequence analysis. (c) A next-generation gene-sequencing chip.

BIOLOGY IN ACTION

BOLD barcodes

A DNA barcode is a short sequence of nucleotides that uniquely identifies a species. It is obtained by using the polymerase chain reaction (PCR) and DNA sequencing (Figure 13.1.13). Sequences are submitted to online databases such as BOLD, the Barcode of Life Database, or other similar databases. There are global DNA barcoding projects underway to catalogue all of life, including bees, butterflies, mosquitoes, fungi, mammals and plants. Scientists and non-scientists alike can access these sequences for research.

To compare and identify species, you need a gene sequence that is present in all organisms but differs slightly between different groups or species. For eukaryotes, the mitochondrial gene CO1 (cytochrome oxidase subunit 1) is often used. Genes in plant plastids help to further identify plant species. For bacteria, a gene for ribosomal RNA can be used.

Examples of barcoding projects include:

- species identification and diversity; for example, barcoding organisms of the Great Barrier Reef, CSIRO scientists barcoding Australian fish species
- tracking pathogenic and non-pathogenic bacterial populations
- food authentication—Is it shark and chips? Is the beef burger really a horse burger?
- monitoring wildlife crime, such as illegal trade in protected and endangered species
- ecology and evolution; for example, seed identification and seed banking of Australian *Acacia* (wattle) genus for conservation and restoration of biodiversity.

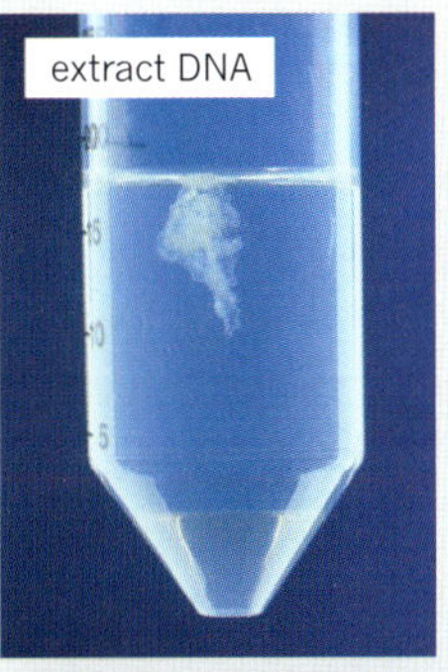

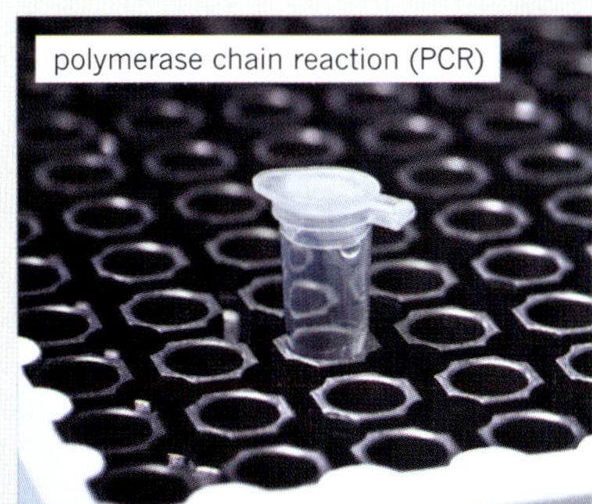

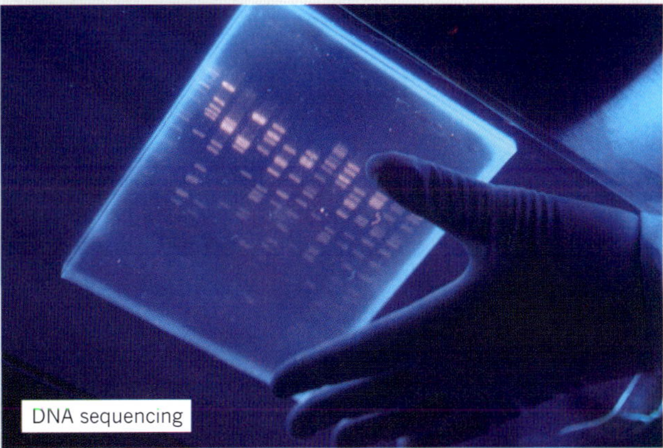

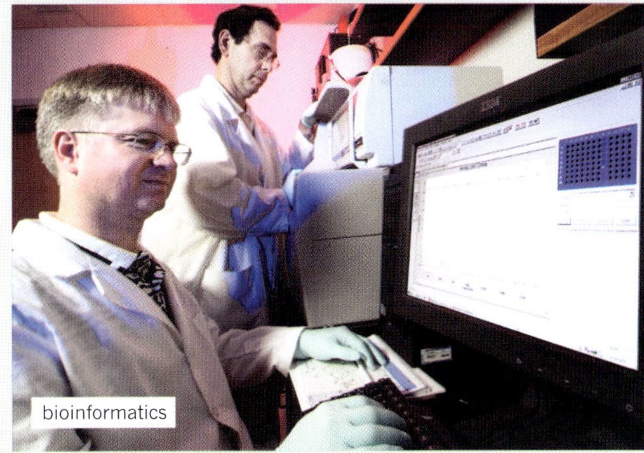

FIGURE 13.1.13 The process of obtaining a DNA barcode.

13.1 Review

SUMMARY

- DNA is often found or extracted in trace amounts and requires amplification to produce an adequate sample for scientists to work with.
- DNA amplification uses the polymerase chain reaction (PCR) to rapidly increase the amount of DNA (creating identical copies of the DNA sample).
- Polymerases are enzymes that create polymers of nucleic acids (DNA and RNA). DNA polymerase acts to assemble DNA molecules, and RNA polymerase acts to assemble RNA molecules.
  - Taq polymerase is a heat-resistant DNA polymerase used in PCR.
- PCR is a technique that amplifies DNA exponentially to create millions of identical copies of the DNA from a very small sample. The sample plus two primers, free nucleotides and Taq polymerase are added to a test tube and the test tube is placed in a thermocycler. The thermocycler alters the temperature in pre-programmed stages to enable a three-step process to be carried out:
  - denaturation—separation of the two strands of DNA at 95 °C
  - annealing—joining at 50–60 °C of a primer to each strand to create a starting point for Taq polymerase to copy
  - extension—Taq polymerase adds free nucleotides to the DNA to create new, identical DNA at 72 °C.
- The three steps of the PCR cycle are repeated many times (up to 50 cycles) to create large quantities of the DNA sample.
- Gel electrophoresis is a technique that separates fragments of negatively charged DNA by length.
  - Each DNA sample is loaded into a different well in an agarose gel. The wells are located at the negative end of a gel electrophoresis chamber.
  - A DNA ladder is also added to a well.
  - The agarose gel is submerged in a buffer solution and the electricity is switched on.
  - Negatively charged DNA migrates through pores in the gel towards the positive terminal. Smaller fragments move faster and therefore further through the gel than longer fragments in a given time.
  - The electricity is switched off and the gel is viewed to identify the number and size of DNA fragments. DNA is visible as bands in a stained gel. The location of a band directly correlates with its specific length so visible fragments can be identified and measured by comparing their position to the DNA ladder.

KEY QUESTIONS

1 **a** What is meant by amplifying a piece of DNA?
  b What do the letters PCR stand for?
  c Draw a simple, labelled flow diagram to summarise the key steps in the process of PCR.
  d What is the role of the enzyme Taq polymerase? Why does it have to be heat resistant?
  e If you started a PCR reaction with one DNA molecule, how many molecules would you have after 30 cycles of amplification?

2 Which one or more of the following would be a suitable target DNA for PCR?
  A a single gene
  B a genome
  C a short variable region within a chromosome
  D a microsatellite
  E a whole chromosome

3 Describe the function of a polymerase enzyme in cells and name two types used in cells.

4 **a** What is the function of the enzyme reverse transcriptase?
  b When is it used in molecular techniques?

5 You have a single-stranded RNA molecule with the sequence 3'-AAUUGCGCA-5'. If you place it in a test tube with nucleotides and reverse transcriptase, what sequence would be made on the complementary strand?

6 **a** What is gel electrophoresis?
  b What does this technique reveal about the DNA fragments being tested?

7 A DNA ladder is used in gel electrophoresis. Explain what a DNA ladder is and its purpose.

8 Gel electrophoresis was conducted with a DNA ladder (standard) and two DNA samples (A and B). The diagram illustrates the resulting gel.

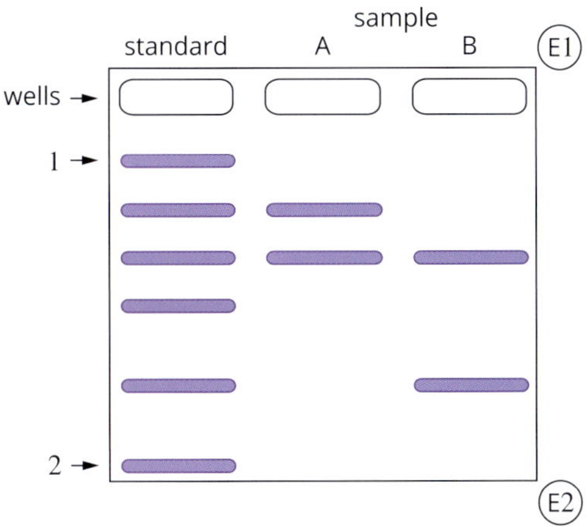

a What do the arrows marked 1 and 2 indicate?

b E1 and E2 represent the electrodes. Which one is the positive electrode?

c Explain the direction in which DNA migrates through the gel when an electric current is applied.

d How many DNA fragments are in:

i the standard (DNA ladder)?

ii sample A?

iii sample B?

e You have been given information about the DNA ladder. The sizes of the fragments are 600, 500, 400, 300, 200 and 100 base pairs. What are the sizes of the fragments in samples A and B?

13.2 Bacterial transformation

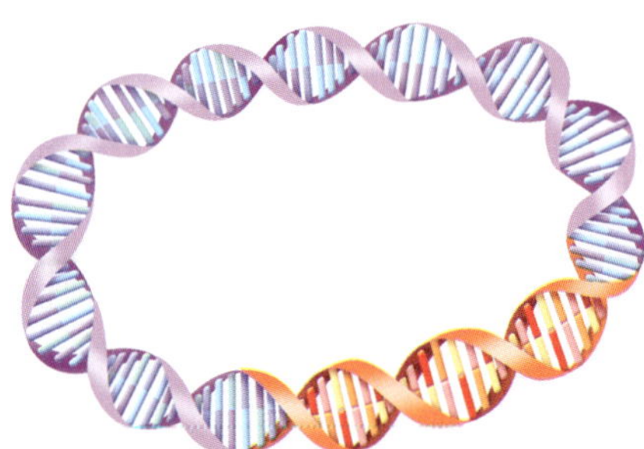

FIGURE 13.2.1 An illustration representing a recombinant plasmid, showing the gene of interest (orange) combined with the plasmid DNA (purple).

Plasmids are small, circular DNA molecules found in bacterial cells. They are often used as vectors (carriers) when scientists move target DNA from one organism to another. Genes can be inserted into plasmids, which can then be incorporated into bacterial cells in a process known as **bacterial transformation**. The inserted gene can then be replicated through the self-replicating properties of the plasmid and bacterial cells and the gene can express the proteins for which it codes.

In this section, you will learn how restriction enzymes and ligases are used to create recombinant DNA (Figure 13.2.1), using plasmids as vectors. You will also learn how these plasmids are then incorporated into bacterial cells for replication.

RESTRICTION ENZYMES (ENDONUCLEASES)

DNA molecules are far too long for biologists to work with in their entirety. The discovery and isolation of **restriction enzymes**, also known as **endonucleases**, has enabled scientists to cut DNA into smaller, more usable fragments and isolate particular regions of interest, such as a single gene.

Restriction enzymes are a large group of enzymes that occur naturally in bacteria. They form part of a bacterial cell's defence system, targeting foreign DNA that may enter the cell, such as the DNA of **bacteriophages**. The restriction enzymes cut up foreign DNA into smaller fragments, destroying it and preventing it from replicating.

Each restriction enzyme targets a specific sequence of nucleotides, usually four to six base pairs in length. This sequence is called a **recognition site**. Every time a restriction enzyme passes its recognition site, it breaks the phosphodiester backbone once on each DNA strand. As a result, the DNA molecule is cut up into fragments of different lengths.

Restriction enzymes of bacterial cells do not cut their own DNA. Bacteria have an enzyme called methylase that adds a methyl group to a specific nucleotide within the recognition site of the restriction enzymes made by the bacterium. This blocks the restriction enzymes from binding to and cutting the bacterium's own DNA.

Restriction enzymes cleave the phosphodiester (sugar–phosphate backbone) bonds in double-stranded DNA.

Over 800 restriction enzymes have been isolated from bacteria and these enzymes recognise and cut more than 100 recognition sites.

BIOFILE

NotI

NotI is an example of a restriction enzyme. It has been sourced from the bacterium *Nocardia otitidis* and has a recognition site of eight base pairs. This means it cuts a DNA molecule less frequently than restriction enzymes that have shorter recognition sites. NotI is used by scientists to cut DNA into large genomic fragments for use in other techniques and to create genomic libraries

Types of restriction enzymes

There are two types of restriction enzyme, which cut DNA differently:

- sticky-end restriction enzymes
- blunt-end restriction enzymes.

Sticky-end restriction enzymes

Sticky-end restriction enzymes leave DNA fragments with overhanging ends. They cut the DNA backbone at a different location on each strand within the recognition site (Figure 13.2.2). This results in a staggered cut, leaving two fragments with exposed bases or 'sticky ends'. The exposed bases are then able to form complementary base pairs through hydrogen bonding with nucleotides of other DNA molecules that have complementary sticky ends.

EcoRI is an example of a sticky-end restriction enzyme extracted from *Escherichia coli*. It cuts the recognition site GAATTC between the G and A nucleotides on each strand. The sequence on one strand is GAATTC and on the complementary strand it is CTTAAG, the same sequence when read backwards. This is called a palindromic sequence. EcoRI cuts at a different location on each strand, so sticky end fragments are produced.

The base-pairing ability of sticky ends allows DNA from very different species to ligate, forming recombinant DNA molecules.

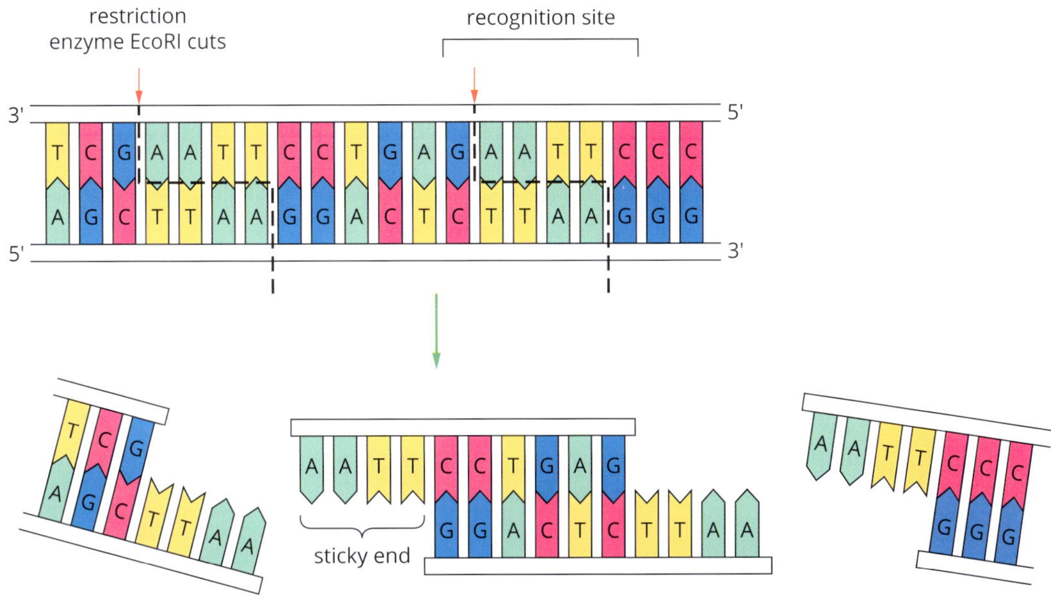

FIGURE 13.2.2 The sticky-end restriction enzyme EcoRI cuts the DNA between the G and the A of its specific recognition site, GAATTC, creating fragments of DNA with sticky ends.

Blunt-end restriction enzymes

Blunt-end restriction enzymes leave clean-cut ends by cutting the sugar–phosphate backbone on both strands of the DNA molecule at the same location within the recognition site (Figure 13.2.3).

HaeIII is an example of a blunt-end restriction enzyme, which is extracted from the bacterium *Haemophilus aegyptius*. It cuts the recognition site GGCC between the G and C nucleotides. These two bases are in the exact same location on either strand of the DNA molecule, resulting in a straight cut through both stands that leaves two fragments with blunt ends.

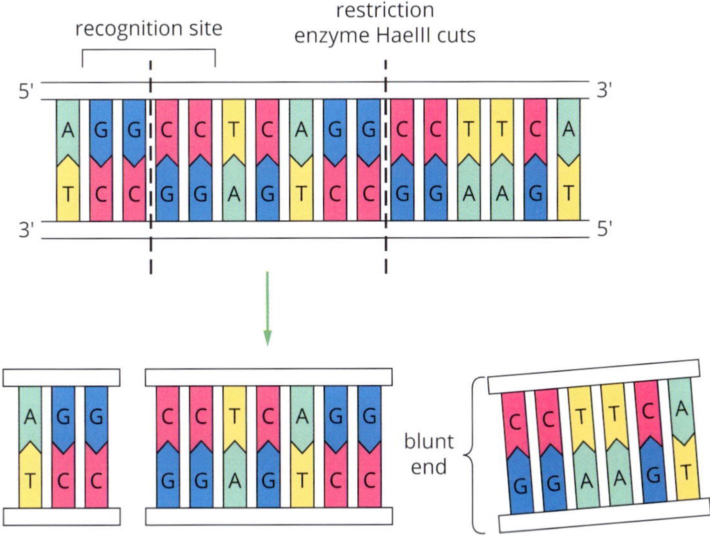

FIGURE 13.2.3 The blunt-end restriction enzyme HaeIII cuts the DNA between the G and the C of its specific recognition site, GGCC, creating fragments of DNA with blunt ends.

BIOFILE

Naming restriction enzymes

The first three letters of the name of a restriction enzyme identify the bacterial species from which they were isolated. The fourth letter refers to the particular strain of bacteria. A roman numeral is also included if more than one restriction enzyme has been isolated from this bacterial strain. For example, the restriction enzyme EcoRI comes from *Escherichia coli* strain RY13, and was the first restriction enzyme isolated from that strain of *E. coli*. Other frequently used enzymes are BamHI, the first restriction enzyme found in *Bacillus amyloliquefaciens* H strain, and HindIII, the third restriction enzyme isolated from *Haemophilus influenzae* D strain.

Restriction enzymes as tools to identify polymorphisms and mutations

> Haemoglobin is a protein with quaternary structure. It consists of four polypeptide chains, two alpha (α) globin and two beta (β) globin polypeptides, each with an iron-containing haem group that binds oxygen. The α-globin and β-globin polypeptides are encoded by different genes.

Small variations in DNA sequence occur within a population. These are called **polymorphisms**. To be considered a polymorphism rather than a mutation, the least common allele has to have a frequency of 1% or more in a population.

A mutation changes a previously normal allele to a new rare (abnormal) variant, different from the predominant allele within a population. For example, individuals with sickle-cell anaemia have a mutation in the β-globin gene. β-globin is a component of haemoglobin, the oxygen carrying protein in your red blood cells.

A common mutation occurring in the β-globin gene is a single base change from A to T, resulting in a single amino acid substitution in the protein—a missense mutation (Figure 13.2.4). This changes the structure of the β-globin polypeptide, causing haemoglobin to clump, and reduces the amount of oxygen carried. The red blood cells take on a sickle shape and tend to clog and rupture the small blood vessels, the capillaries.

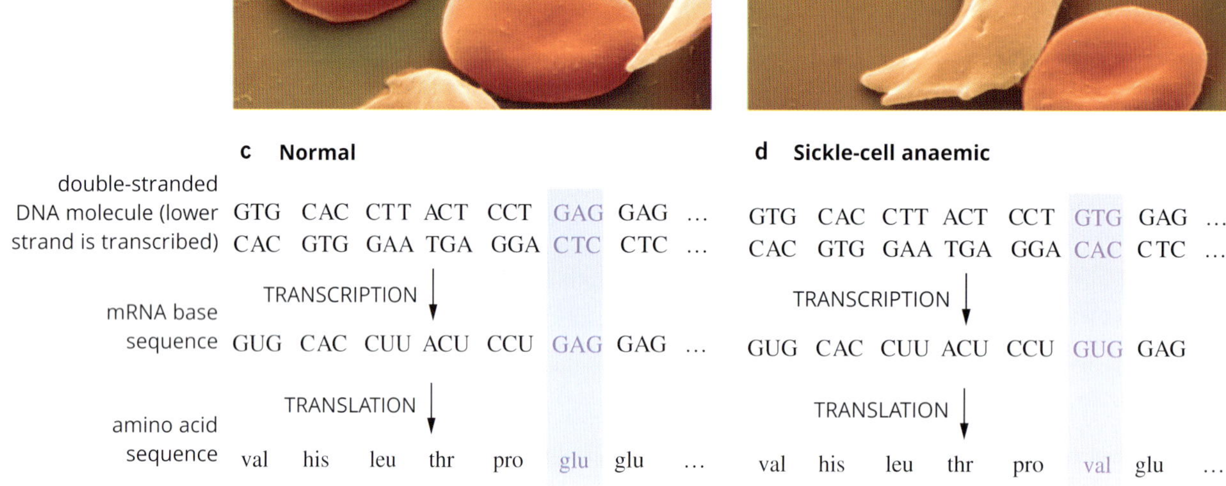

FIGURE 13.2.4 Viewed under the electron microscope there are obvious differences between the (a) normal and (b) sickle-cell red blood cell. Comparisons of parts of the gene and amino acid sequences of β-globin from (c) normal and (d) individuals with sickle-cell anaemia. The DNA sequence is transcribed to mRNA which is then translated into protein (amino acid sequence). A single base change in the DNA (A replaced with T) leads to a change in one amino acid in the protein (glutamate replaced with valine).

Molecular tools are used to identify individuals carrying this mutation. By chance, the mutation occurs at a restriction enzyme site. The base change eliminates the recognition site for MstII (Figure 13.2.5a). To detect the mutation, DNA is extracted from individuals, the region of DNA containing the recognition site for MstII is amplified by PCR, and then the PCR products are incubated with the MstII restriction enzyme. The normal allele will be cut at the MstII recognition site. The mutant allele will not be cut. The difference in the size of the DNA fragments is identified by gel electrophoresis (Figure 13.2.5b).

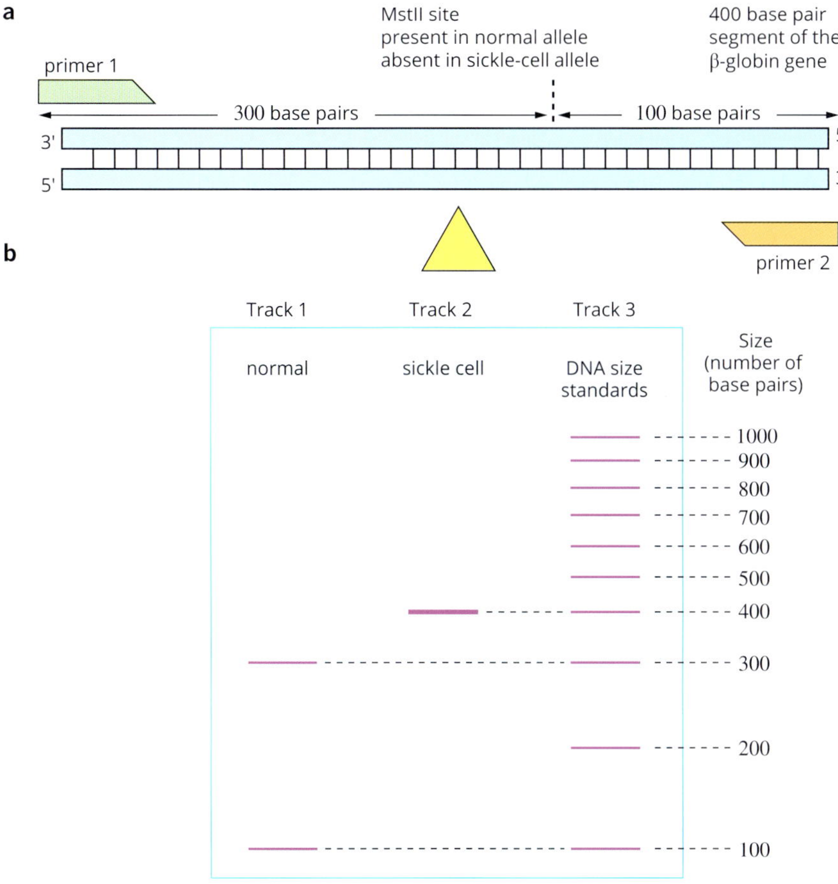

FIGURE 13.2.5 (a) A portion of the β-globin gene is shown. The position of the PCR primers define a 400 bp region. An MstII site is present in the normal allele, resulting in fragments of 300 and 100 bp when the PCR product is cut. The MstII site is missing in the sickle-cell allele, resulting in a 400 bp PCR product. (b) Diagram of an electrophoresis gel showing the DNA fragments produced when the DNA of an individual carrying two copies of the normal allele (track 1) or an individual carrying two copies of the sickle-cell allele (track 2) is amplified by PCR and exposed to the restriction enzyme MstII. The DNA size standards in track 3 allow the size of the bands in tracks 1 and 2 to be determined.

LIGASES

Ligases are a group of enzymes that join fragments of DNA or RNA in a process called **ligation**. **DNA ligase** is used to join fragments of DNA, and **RNA ligase** is used to join fragments of RNA. The role of DNA ligase in a cell is to join segments of newly replicated DNA and to repair breaks in DNA molecules.

Assuming they were cut with the same restriction enzymes, DNA ligase can join DNA fragments extracted from different organisms or different species, as DNA has a universally consistent molecular structure. Depending on the characteristics of the DNA ends to be joined (sticky or blunt ends), the conditions of the ligation reaction (incubation time and temperature) need to be adjusted to ensure efficient DNA ligation is achieved.

Ligases join two DNA fragments and create a phosphodiester bond between them.

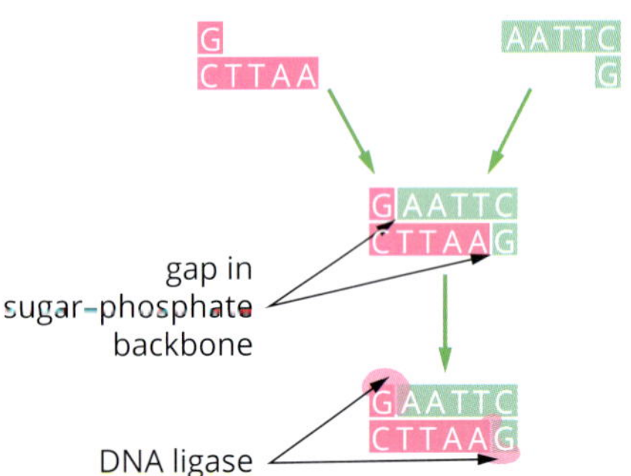

FIGURE 13.2.6 Two sticky-end DNA fragments come together by complementary base pairing, and then DNA ligase permanently links the sugar–phosphate backbone.

Ligation of sticky-end fragments

Ligation to join sticky-end fragments is specific because the exposed bases of sticky-end fragments first bind by complementary base pairing. Complementary bases are attracted by weak hydrogen bonds that hold them together. After this, the ligase joins the fragments (Figure 13.2.6) by creating a phosphodiester bond between the 3' OH end and 5' phosphate end of the adjoining nucleotides. This technique makes recombinant DNA and is used in processes such as **gene cloning**.

Ligation of blunt-end fragments

Ligation of blunt-end fragments is random. Any two fragments can join if they come in contact and the DNA ligase joins them. For this reason, blunt end fragments are more difficult to use in DNA manipulation processes that require the joining of specific fragments. However, sometimes blunt ends are unavoidable. For instance, a blunt-end enzyme might be the only type available to cut out your target gene without damaging the gene itself. Using DNA ligase, scientists are able to attach short linker DNA fragments onto blunt-end DNA to create sticky ends.

PLASMIDS

In Chapter 3 you learnt about the structure of DNA and chromosomes. In eukaryotic cells, DNA forms long, linear chromosomal strands containing thousands of genes. In prokaryotic cells, DNA forms double-stranded circular chromosomes.

In addition to the circular chromosome, bacterial cells also contain small circular pieces of double-stranded DNA called **plasmids**. Plasmids replicate independently of the chromosome. As bacterial cells do not have a nucleus, both chromosomal and plasmid DNA are located in the cytosol (Figure 13.2.7).

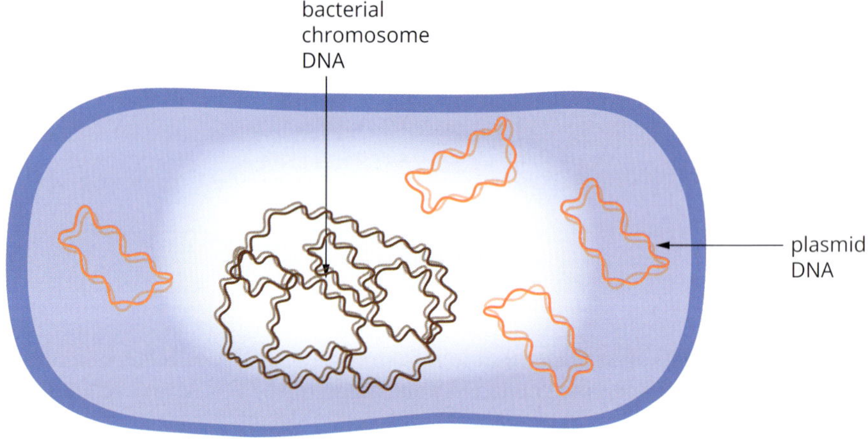

FIGURE 13.2.7 Bacterial cells contain a circular bacterial chromosomal DNA (brown) and small circular pieces of DNA called plasmids (red).

RECOMBINANT DNA

When DNA from two different sources is joined together, the resulting molecule is called **recombinant DNA**. Scientists create recombinant DNA to clone (make multiple copies of) a particular gene. Following this step, they may also produce large quantities of the protein expressed by the cloned gene. For example, an insulin-coding gene may be cloned and then incorporated into bacteria to produce large quantities of the protein insulin to be used as treatment for diabetes.

Using plasmids as vectors

When scientists create recombinant DNA, they often use a plasmid as the vector. They insert target DNA into the plasmid, producing a **recombinant plasmid**. The plasmid is then placed in a bacterial cell, where the self-replicating system of the plasmid and cell replicates the plasmid genes. Each bacterial cell containing the plasmid will produce the protein products of the genes in the plasmid (including those of the target DNA). For example, if a recombinant plasmid containing the insulin-coding gene is placed in a bacterial cell, the cell will produce insulin.

Plasmids are used as vectors when creating recombinant DNA for the following reasons:

- Their small size makes them easy to manipulate in a laboratory.
- Plasmids carry a range of restriction enzyme sites. A plasmid containing the appropriate recognition sites is chosen to suit your needs. For example, if your gene of interest was cut from a chromosome with EcoRI, you would use a plasmid with a single EcoRI site.
- Recombinant plasmids self-replicate independently once they are placed inside a host bacterial cell and at a faster rate than their bacterial host's chromosomal DNA. This is of vital importance in manufacturing vast quantities of proteins.

To enable the identification of cells that have incorporated the recombinant plasmid, the plasmids used as vectors need to have particular characteristics including:

- an antibiotic-resistance gene
- a gene that can be easily identified—such as a gene that produces coloured or fluorescent proteins.

One example is a plasmid containing the ***lacZ* gene** (Figure 13.2.8). Restriction enzyme sites for inserting the gene of interest are located within the *lacZ* gene. If the gene insertion is successful it will disrupt the *lacZ* gene. We will see how this is used in bacterial selection later in this section. The plasmid also contains the antibiotic resistance gene amp^R, which encodes resistance to ampicillin.

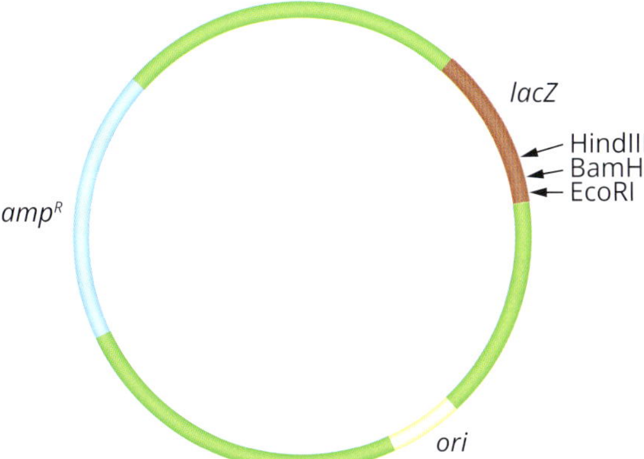

FIGURE 13.2.8 An example of a plasmid used for recombinant DNA and bacterial transformation showing the gene for ampicillin resistance (amp^R), the origin of replication (*ori*) and the *lacZ* gene, which creates blue colonies when grown on agar containing an indicator called X-gal. Sites for three restriction enzymes, HindIII, BamHI and EcoRI, lie within the *lacZ* gene.

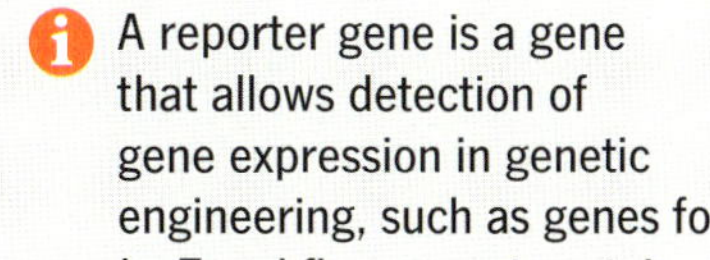

A reporter gene is a gene that allows detection of gene expression in genetic engineering, such as genes for *lacZ* and fluorescent proteins.

BIOFILE

Green fluorescent protein (GFP)

Green fluorescent protein (GFP) is a protein first isolated from jellyfish. The *GFP* gene can be inserted into plasmids to be used in bacterial transformation. Bacterial cells that have incorporated the plasmid vector fluoresce green under UV light.

Genes that code for fluorescent proteins are often attached to another gene that is being investigated. When this gene of interest is expressed, the fluorescent protein is also produced, so it sends an obvious 'visual report' that the gene is being expressed. Hence they are referred to as reporter genes. Other reporter genes and proteins include luciferase, an enzyme from fireflies that produces a yellow fluorescent product, and a red fluorescent protein from a coral.

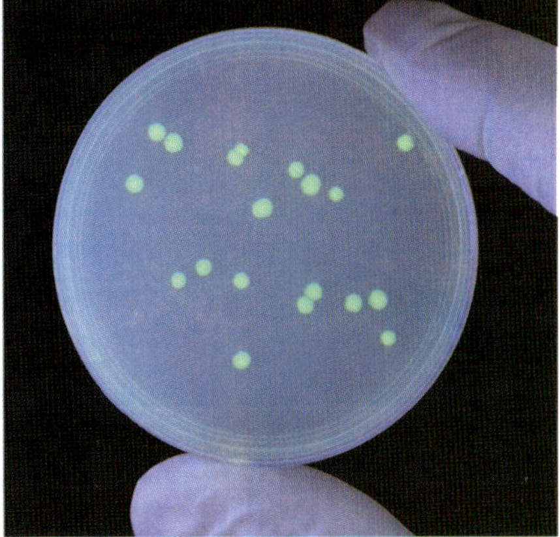

FIGURE 13.2.9 Transformed bacterial colonies can be identified as they contain the *GFP* gene and fluoresce under UV light.

Creating recombinant DNA

The process of creating a recombinant plasmid is outlined below and in Figure 13.2.10:

1 The target DNA is cut out using a sticky-end restriction enzyme and then isolated.
2 The bacterial plasmid is cut by the same restriction enzyme used to cut out the target DNA. The plasmid and the target DNA now have the same sticky ends with exposed bases that are complementary to each other.
3 The target DNA and plasmids are placed together. Some plasmids will simply close back up (known as non-recombinant plasmids), while other plasmids will incorporate the target DNA by complementary base pairing (known as recombinant plasmids).
4 DNA ligase is added to rejoin the sugar–phosphate backbone of the DNA.

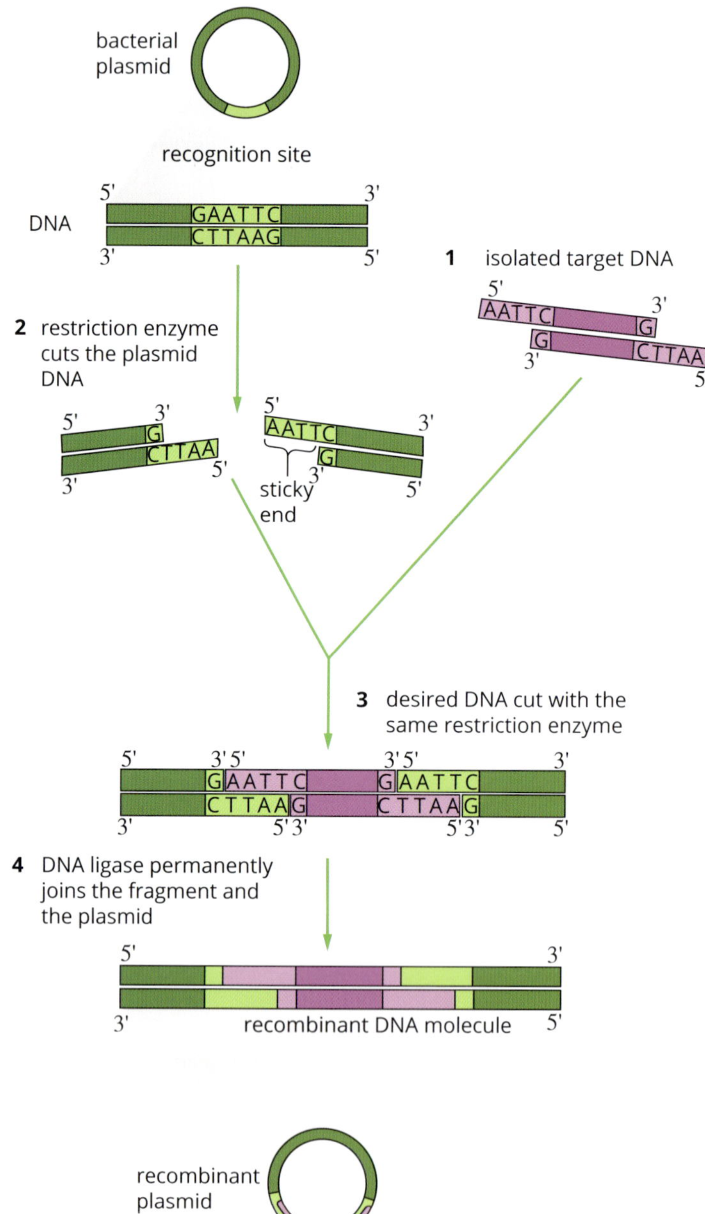

FIGURE 13.2.10 A recombinant plasmid is created by joining a target DNA fragment and a plasmid that have both been cut with the same sticky-end restriction enzyme. They are joined using DNA ligase.

Complementary DNA

To produce a eukaryotic protein in a bacterial cell, complementary DNA (cDNA) is used as the target DNA. cDNA is DNA that has been copied from mature mRNA and contains only exons. cDNA is synthesised using the reverse transcriptase enzyme.

Reverse transcriptase

In Section 13.1, you learnt that reverse transcriptase is an enzyme with the ability to make cDNA from mRNA (Figure 13.2.11). This is useful because mature mRNA has already had the introns spliced out. Prokaryotic cells are unable to splice out introns. Reverse transcriptase allows the synthesis of DNA from mature RNA in a test tube (*in vitro*).

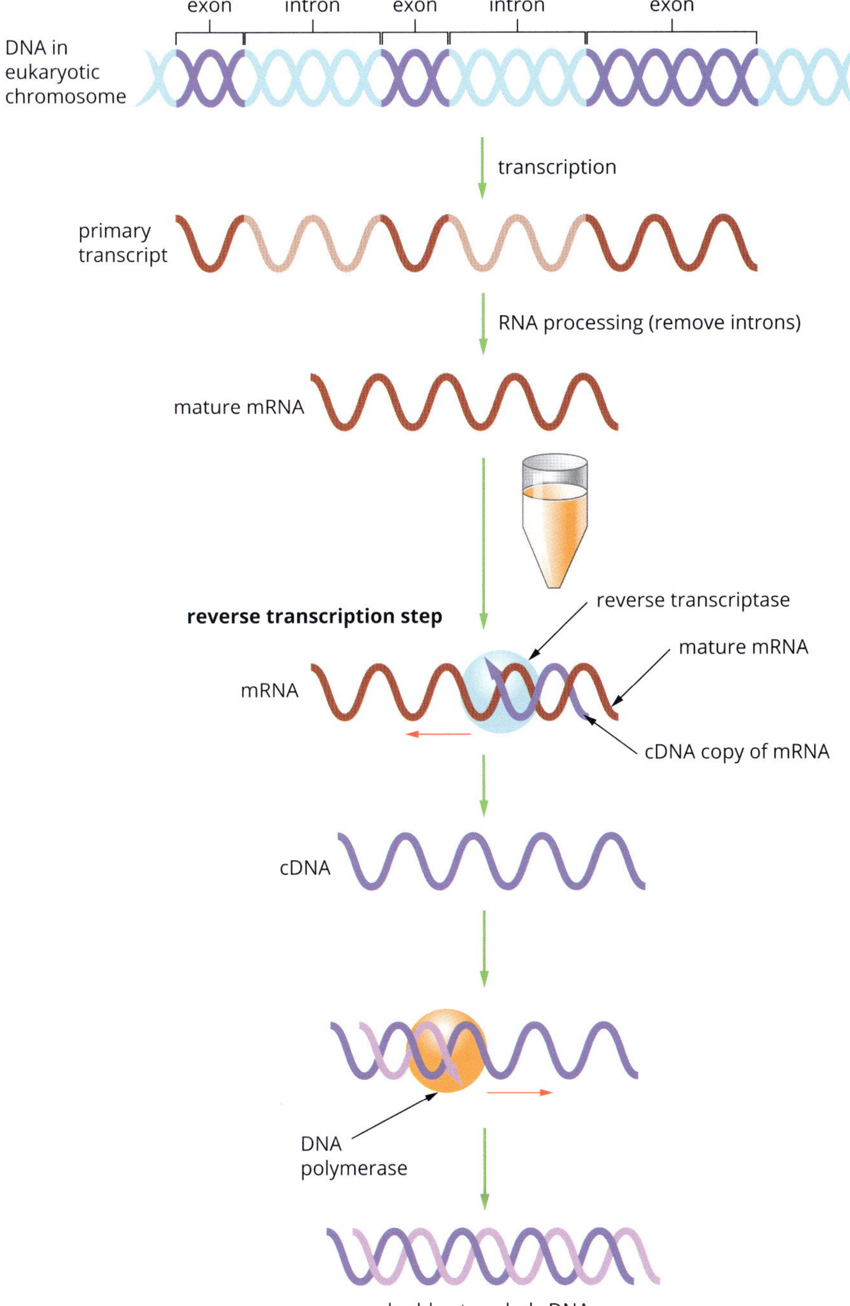

FIGURE 13.2.11 The process of creating cDNA from mRNA using reverse transcriptase. Once the strand of cDNA is produced, DNA polymerase is used to make the cDNA double stranded.

BIOFILE

Insulin from animals

Insulin was previously extracted from the pancreas of other animals, such as pigs and cattle, for the treatment of type 1 diabetes. This was an expensive and time-consuming method that also involved the risk of an allergic reaction to the foreign molecule and potential for contracting diseases. Porcine (pig) and bovine (cattle) insulin are similar, but not identical, to human insulin. Their biological activity is not as effective as human insulin, so it is preferable to use the human hormone. Recombinant human insulin became available for treatment in the 1980s.

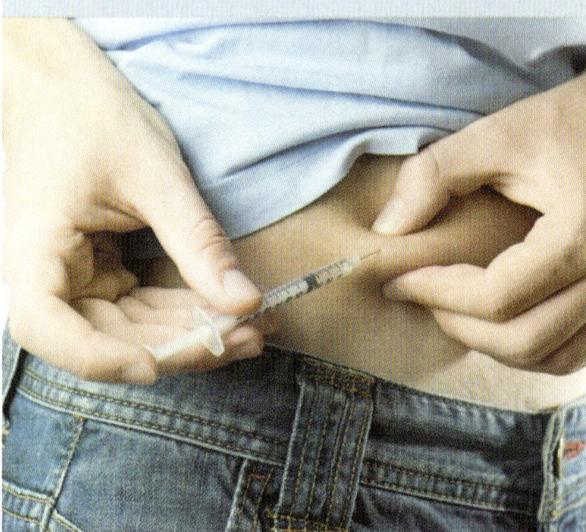

FIGURE 13.2.12 Woman giving herself an injection of insulin.

When cDNA is inserted into a plasmid, which in turn is then incorporated into a bacterial cell, the protein encoded by this cDNA will be expressed (Figure 13.2.13). This method is now commonly used to produce vast quantities of therapeutic proteins such as human insulin, growth hormone and cytokines.

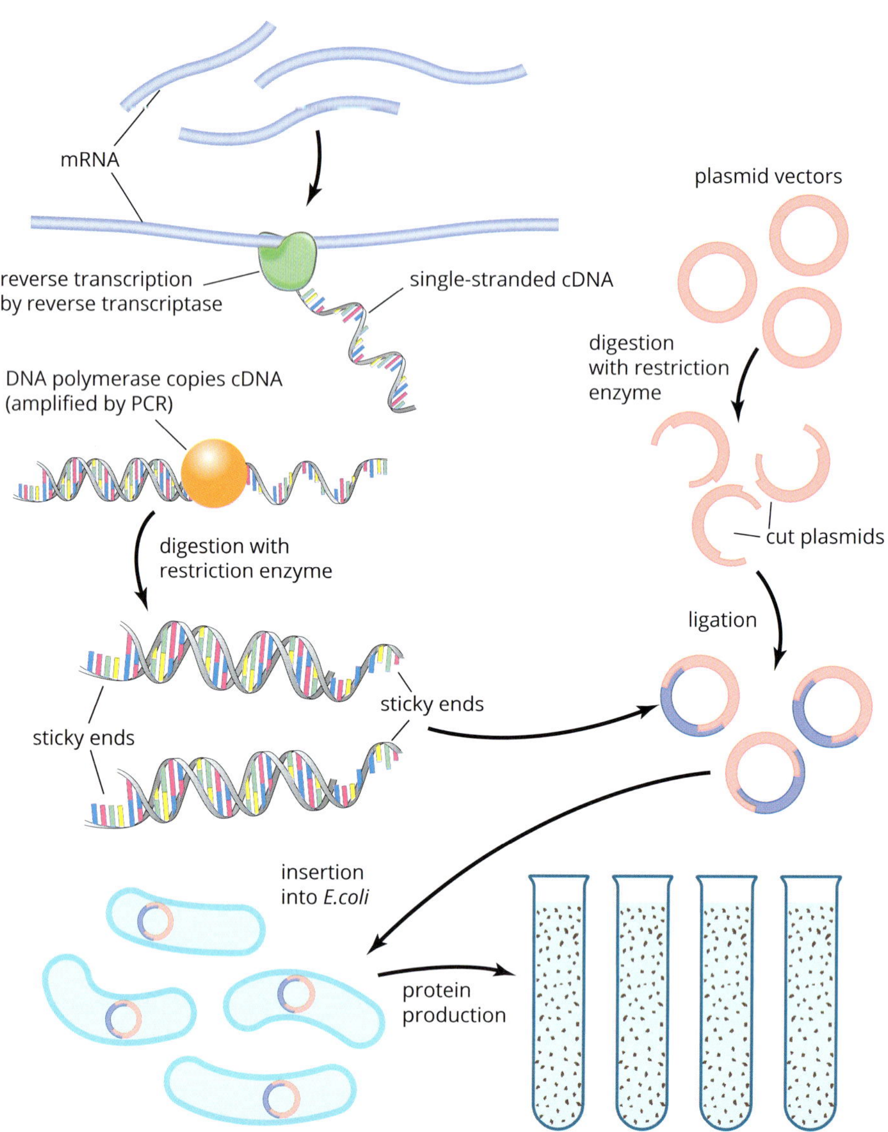

FIGURE 13.2.13 cDNA is used as target DNA to create a eukaryote protein product within a bacterial cell. The cDNA is treated to create sticky ends so that it can be inserted into a plasmid, creating a recombinant plasmid. This plasmid is then incorporated into a bacterial cell, which expresses the protein product.

Regulatory genes in recombinant DNA

Regulatory genes may be included in plasmids for the purpose of controlling the expression of the target gene that is inserted into the plasmid. The regulatory gene is turned on by an **inducer** molecule, for example a sugar such as lactose or arabinose, or by metal ions such as iron, copper or zinc. Once the regulator is transcribed, the target gene can be transcribed and translated. Inducers are important in regulating gene expression, particularly when the aim of the recombinant DNA technology is protein production, and when a gene is being expressed and studied in plant and animal models.

TRANSFORMING BACTERIAL CELLS

Cells that have had foreign DNA incorporated into them are said to be **transformed**. For example, when a foreign plasmid is incorporated into a bacterial cell, the bacterial cell is then 'transformed' because it can express a new gene and therefore has a new characteristic.

Two methods of artificial bacterial transformation are used: heat shock and electroporation.

- Heat shock involves placing bacterial cells and a mixture of recombinant and non-recombinant plasmids in an ice-cold solution containing calcium ions and then rapidly increasing the temperature to disrupt the plasma membrane of the bacterial cells. The plasmids can then penetrate the membrane and enter the bacteria.
- In electroporation, the bacterial cells and a mixture of recombinant and non-recombinant plasmids are subjected to an electrical current that alters the plasma membrane. Again, the plasmids are then able to enter the bacteria.

Very few of the bacterial cells will be transformed with recombinant plasmids. Some will take up the non-recombinant plasmids (plasmids without the target DNA) and others will not be transformed at all or will die from the heat shock or electroporation treatment (Figure 13.2.14).

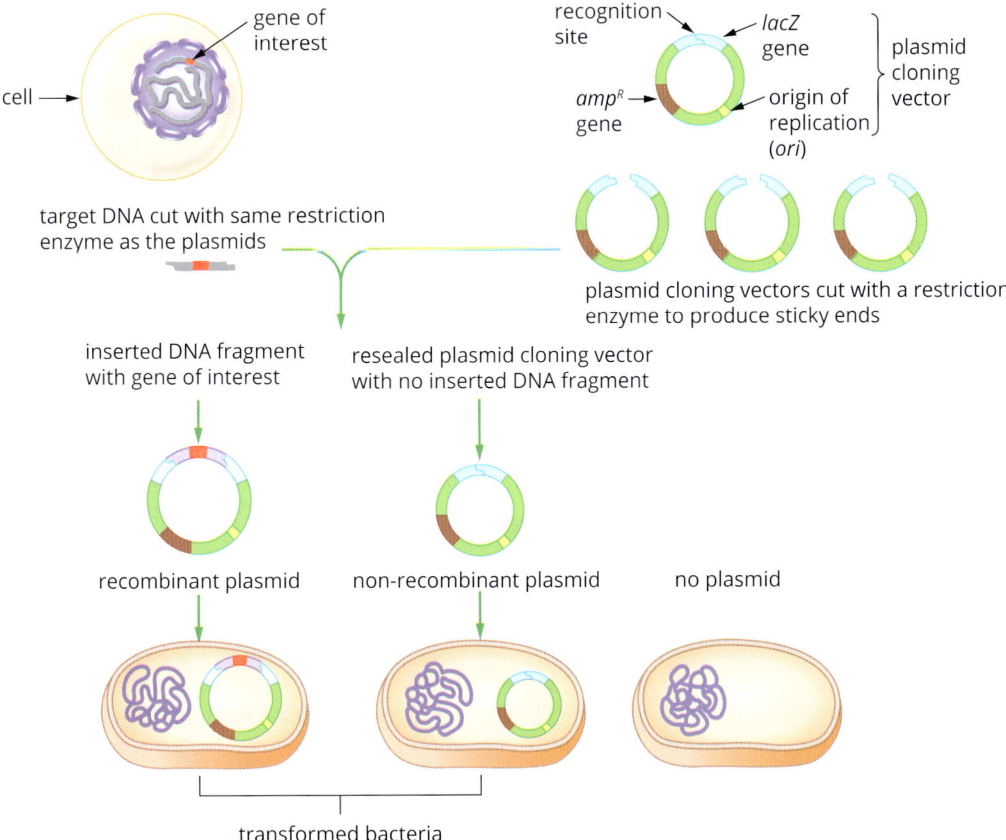

FIGURE 13.2.14 The process of bacterial transformation involves creating recombinant plasmids and then inserting them into bacterial cells. Some will successfully take up the recombinant plasmid but others will either contain a non-recombinant plasmid or no plasmid at all.

BIOLOGY IN ACTION

Natural transformers

Bacterial transformation is used in crop biotechnology to transfer genes into plants and thus introduce a desirable trait, such as increased productivity, salt tolerance, insect resistance or herbicide resistance. Such plants are called transgenic. The bacterium *Agrobacterium tumefaciens* is commonly used as it has the ability to infect plant tissue and incorporate specific parts of its plasmid into the host plant's DNA.

The study and use of *A. tumefaciens* has a long history. This common soil bacterium was identified as the cause of the tumours of crown gall disease over 100 years ago. A plasmid that can move from the bacterium into plant cells was later identified as the key agent. This is a natural transformation of plant cells by a bacterial plasmid. The plasmid in *A. tumefaciens* is called the Ti or tumour-inducing plasmid (Figure 13.2.18). It has genes that direct the movement of the plasmid from bacterium into plant cells and for insertion into the chromosomes of the plant. It also has genes for growth-promoting plant hormones that cause the tumours.

This plasmid is used for genetic engineering. The genes that cause tumour growth are removed while the genes that ensure transfer from bacterium to plant cell remain. A foreign gene inserted into this plasmid can then be transferred into plant cells. Genes for traits such as insect resistance, herbicide resistance, and tolerance to salinity and frost are all being used in crop biotechnology.

The foreign gene is inserted into the plasmid then *A. tumefaciens* is transformed by the methods described earlier. Transformed bacteria and plant material are mixed (Figure 13.2.19a). *A. tumefaciens* inserts a section of its plasmid (containing the desired gene) into the plant cell, which is then incorporated into the plant cell DNA. This plant tissue is cultured (Figure 13.2.19b) to create a transgenic plant that expresses the foreign gene. In Australia, transgenic cotton is created using bacterial transformation to confer insect resistance. You will learn more about the production of transgenic plants such as cotton in Chapter 14.

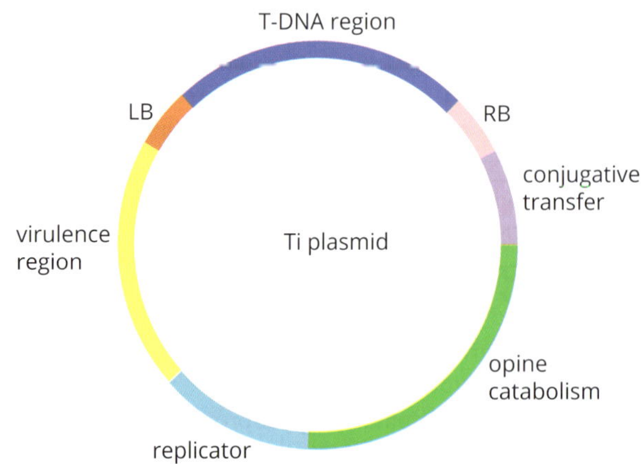

FIGURE 13.2.18 The Ti plasmid from *Agrobacterium tumefaciens*. The T-DNA region carries the genes that cause tumours in crown gall disease. This region is removed and replaced with a foreign gene, making a recombinant plasmid.

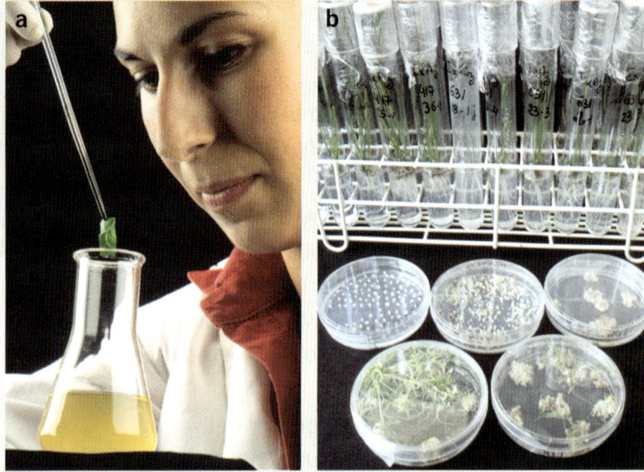

FIGURE 13.2.19 (a) Plant material is exposed to *Agrobacterium* to allow transfer of recombinant plasmid from bacteria to plant cells. (b) The plants carrying the new gene are cultured and selected in the laboratory before release for field testing of the new characteristics acquired by gene transfer.

13.2 Review

SUMMARY

- Restriction enzymes are enzymes that cut DNA at particular recognition sites.
  - Sticky-end restriction enzymes leave fragments with overhanging ends that have exposed bases.
  - Blunt-end restriction enzymes cut DNA to leave flat-ended fragments.
- Ligase enzymes permanently join fragments of DNA or RNA together in a process called ligation. DNA ligase joins DNA fragments, RNA ligase joins RNA fragments.
- Plasmids are small, circular pieces of double-stranded DNA found in bacterial cells. They replicate independently of the bacteria's chromosomal DNA.
- Recombinant plasmids are plasmids that have had target DNA inserted into them. The same sticky-end or blunt-end restriction enzyme is used to cut both the targeted gene and the plasmid, and then DNA ligase is used to permanently join the two together.
- Plasmids with antibiotic resistance are generally used to enable identification of bacterial transformation later on, as only bacterial cells containing these plasmids will survive when grown in cultures containing the antibiotic.
- A gene that produces an identifiable phenotype, such as a coloured product or fluorescence, may be used as a reporter gene to identify transformed cells.
- Disruption of a reporter gene by insertion of the target gene can indicate successful bacterial transformation, as the gene no longer functions. For example, if the *lacZ* gene is disrupted by insertion of a target gene, the bacterial cell cannot break down the X-gal indicator present in the agar medium. Colonies appear white instead of blue.
- Positive expression of a reporter gene, such as green fluorescent protein, is also used to indicate successful bacterial transformation.
- There are three steps in bacterial transformation:
  - Gene uptake: Bacterial cells are induced to take up the recombinant plasmids either by heat shock or electroporation methods. Very few bacterial cells will be transformed (and some of these will contain non-recombinant plasmids).
  - Selection of transformed bacteria: The bacteria are grown in the presence of an antibiotic. Bacterial cells that have been transformed will survive, as the gene for antibiotic resistance is located on the plasmid.
  - Identification of transformed colonies: Transformed bacteria containing recombinant plasmids are identified by the reporter gene in the recombinant plasmid. For example, if the *lacZ* gene is the identifying gene, bacterial colonies will be white rather than blue as the *lacZ* gene will be disrupted by the insertion of the target DNA, and will no longer be functional.

KEY QUESTIONS

1. a Restriction enzymes are a basic molecular tool in gene technology. What is a restriction enzyme and what can it do?
   b Describe the difference between sticky ends and blunt ends produced by restriction enzymes.
2. Define the following terms and give an example of each:
   a gene cloning
   b recombinant DNA
3. What is a plasmid? Describe the role played by plasmids in gene cloning.
4. Outline the purpose of DNA ligase in recombinant DNA technology.
5. Draw a flow diagram outlining the different techniques and processes involved in insulin production. Use diagrams where possible to assist your explanations.
6. a Explain what is meant by 'genetic transformation'.
   b List three protein products manufactured using genetic transformation and outline their importance in medicine or agriculture.
7. Give two features of a plasmid that are important for identifying cells containing a recombinant plasmid.
8. Describe how antibiotics are used to select transformed bacteria.
9. Three restriction enzymes and their recognition sites are illustrated below.

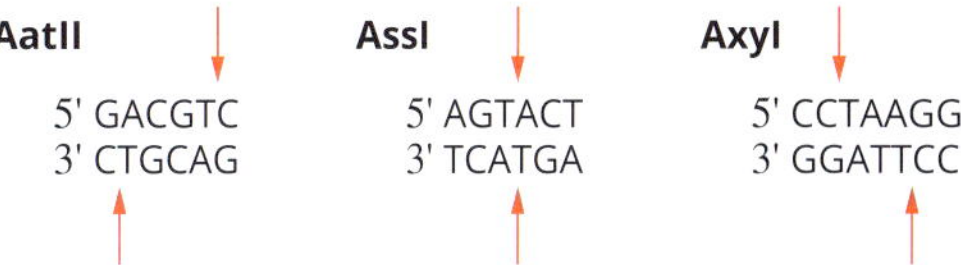

   Use the information given to determine whether the enzyme leaves sticky or blunt ends.
   a AatII cuts between T and C on each strand.
   b AssI cuts between T and A on each strand.
   c AxyI cuts between C and T on each strand.

Chapter review

13

KEY TERMS

anneal
bacterial transformation
bacteriophage
blunt-end restriction enzyme
complementary DNA (cDNA)
DNA amplification
DNA ladder
DNA ligase
DNA polymerase
DNA thermocycler
endonuclease
gel electrophoresis
gene cloning
helicase
inducer
lacZ gene
ligase
ligation
microsatellite
plasmid
polymerase
polymerase chain reaction
polymorphism
primer
recognition site
recombinant DNA
recombinant plasmid
restriction enzyme
reverse transcriptase
RNA ligase
sticky-end restriction enzyme
Taq polymerase
thermophilic
transformed

KEY QUESTIONS

1 Reverse transcriptase is used to make:

A copied DNA
B complementary DNA
C constructed DNA
D none of the above

2 Some students are doing an experiment involving bacterial transformation. Bacteria were incubated with plasmids containing resistance to the antibiotic ampicillin and then grown on agar plates. A plate that will have only transformed bacteria growing will have which of the following?

A nutrient agar only
B nutrient agar and ampicillin
C plain agar with ampicillin
D nutrient agar, ampicillin and penicillin

3 Genes such as the *lacZ* gene can be used as reporter genes. Which of the following are reporter genes used to determine?

A that a bacterium has absorbed a plasmid
B that a plasmid has accepted the gene of interest
C that a bacterium has absorbed a plasmid containing the gene of interest
D none of the above

4 Duchenne muscular dystrophy (DMD) is a genetic disease caused by the deletion of part of the sequence of the dystrophin gene, which is located on the X chromosome. The dystrophin protein is very large. The normal gene is 2 220 390 base pairs long and contains many exons and introns. The total length of the coding sequences (the exons) is 11 055 nucleotides.

a How many amino acids are in the normal protein?

A young woman who has a family history of DMD is about to start trying to have a family. Her partner's family has no history of the condition.

b The first step the young woman takes is to be tested for the condition. How could she have the genetic change and not know?

c The relevant sections of the X chromosomes of the young woman were cut using an appropriate restriction enzyme and then run on an electrophoresis gel. A set of DNA standards was also run. The gel is shown below.

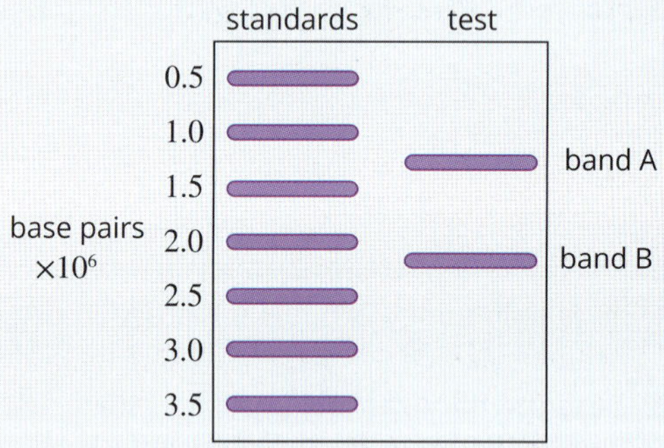

i The wells have not been draw on the diagram. Indicate where the wells should have been drawn.

ii Explain whether the woman has the DMD mutation.

iii What are the approximate lengths of the pieces of DNA represented by bands B and A?

d The woman decides to use IVF and preimplantation testing to become pregnant. The doctors harvest eight eggs, five of which are then fertilised with her partner's sperm. The embryos are then tested for the DMD allele using gel electrophoresis.

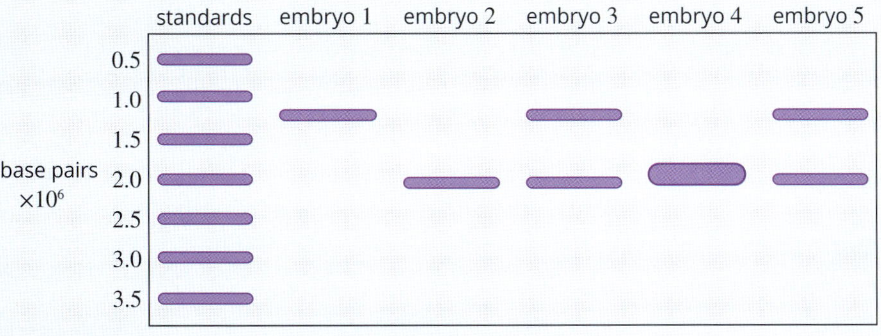

- **i** Explain which of the embryos is/are male.
- **ii** Explain which of the embryos is/are most suitable for implantation.

5 Bacteria are commonly genetically engineered to produce human proteins. The DNA sequence for the gene of one of these proteins is shown below.

```
5' – GCT TCT TCC CGT GCA TAT AGA TAC TCT GAA ACA CTG TGC GGC GGT GAA CTG
3' – CGA AGA AGG GCA CGT ATA TCT ATG AGA CTT TGT GAC ACG CCG CCA CTT GAC

... many base pairs ...  CTG TGC ACC TAT TGT GCT ACT CCC GCA AAG TCC GAA TAG TAG GCT TCT
                         GAC ACG TGA ATA ACA CGA TGA GGG CGT TTC AGG CTT ATC ATC CGA AGA

CGC TGC TCC CGT GCT TCT CGC GTA TGT CCG – 3'
GCG ACG AGG GCA CGA AGA GCG CAT ACA GGC – 5'
```

A restriction enzyme is used to cut the gene from the human genome. Four possible enzymes have recognition sequences and cutting sites as shown.

enzyme 1	enzyme 2	enzyme 3	enzyme 4
GC│TTCT CGAA│GA	C│TGTGC GACAC│G	TCC│CGT AGG│GCA	CTG│TGC GAC│ACG

The DNA sequence is quite long, so only the beginning and end are shown along with a section before and after the gene. The start and stop triplets are underlined.

- **a** Explain which of the restriction enzymes would be most suitable to cut out the gene so that it can be inserted into the bacterium that will produce the protein.
- **b** A mutation can occur that changes the base indicated with the arrow from a T to a C. One way to identify individuals who have this mutation is to cut the DNA with a restriction enzyme and run the DNA on an electrophoresis gel. Explain why enzyme 4 is the most appropriate to use for this purpose.
- **c** This mutation runs in one particular family. John and Sarah are members of the family and decide to be tested. John turns out to be normal and Sarah is heterozygous (one normal and one mutant allele).
  - **i** If enzyme 4 is used, how many DNA bands will result from the cutting of John's DNA?
  - **ii** If enzyme 4 is used, how many DNA bands will result from the cutting of Sarah's DNA?
  - **iii** Re-draw and complete the picture of the electrophoresis of the DNA of John and Sarah after cutting with enzyme 4.
  - **iv** Show the positions of the positive and negative terminals on the electrophoresis set below, and explain why you placed them in those positions.

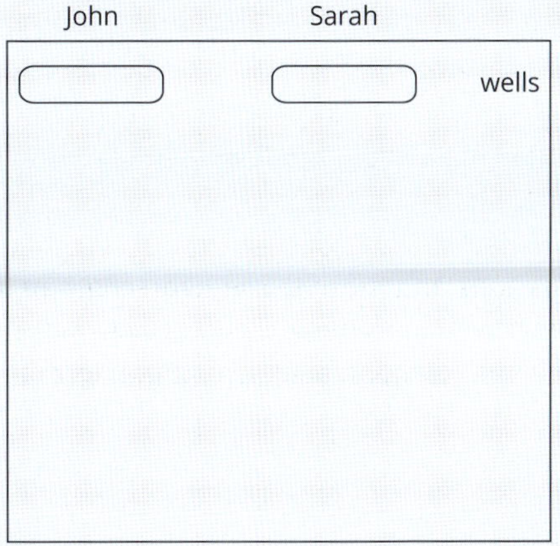

6 Before any gene can be inserted into bacteria in order to make proteins for human use, the number of copies of the gene must be increased. In order to do this a process called PCR is used.

a i For what do the letters PCR stand?

ii What is the role of the 'P' in the process?

iii What is the source of the 'P' used in this process? Why is that particular source used?

b PCR is semi-conservative replication. What does this mean?

c One particular PCR machine uses the following sequence: heat to 94 °C for 1 minute, cool to 56 °C for 1.5 minutes, then heat to 72 °C for 1.5 minutes.

i Describe what is happening at each stage.

ii How long would it take to obtain 8000 copies of the target DNA?

7 Members of a family with a history of cystic fibrosis underwent genetic testing to determine whether they carried the common ΔF508 mutation. DNA samples obtained from cheek cells were analysed by PCR using primers specific for the ΔF508 region, followed by gel electrophoresis. The normal allele yields a 98 bp DNA fragment. The mutant allele yields a 95 bp DNA fragment.

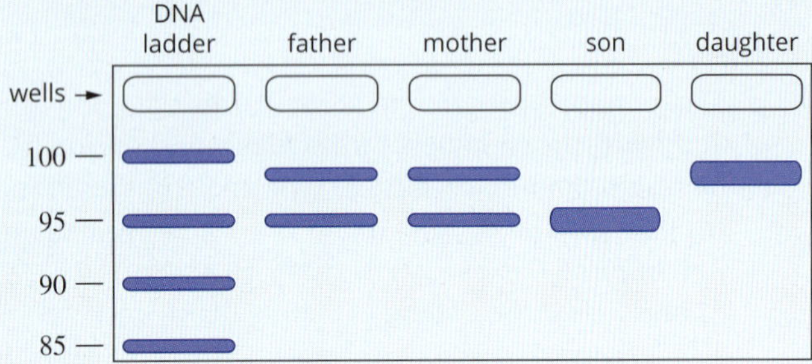

a How is PCR able to identify the allele responsible for cystic fibrosis?

b Describe the purpose of gel electrophoresis in this type of genetic testing.

c The parents are carriers of cystic fibrosis. Explain how the PCR and gel electrophoresis results show this.

d What does the genetic test show about the son?

e The daughter gets a cold and chest infection every winter. Is this likely to be related to the lung congestion seen in cystic fibrosis?

8 A plasmid of total length 1000 bp and a segment of a linear chromosome are being used to make recombinant DNA (diagram below, top). The DNA was cut with the restriction enzyme Tat1, which leaves sticky ends. The cutting sites are indicated by arrows. The resulting fragments were run on a gel (diagram below, bottom). The purpose of the experiment is to insert a segment of the chromosome into the plasmid for gene cloning.

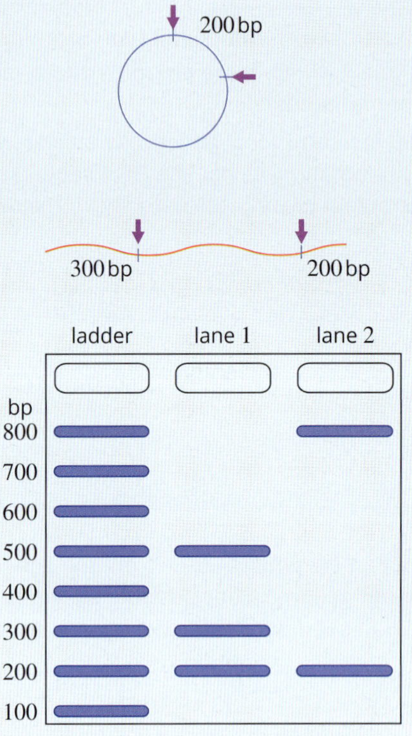

a Which lane on the gel has the fragments from the plasmid digestion? Explain your choice.

b What length is the starting chromosomal DNA? Explain your answer.

c What enzyme will be needed to make a recombinant plasmid using the large chromosome fragment?

CHAPTER

14 Biological knowledge and society

Learning outcomes

By the end of this chapter, you will have an understanding of how gene technologies can be applied to producing copies of DNA for research and the production of human proteins for the treatment of disease. You will also have an understanding of how DNA profiling can be used to identify individuals, how embryos and adults can be screened for genetically inherited traits and the issues that can arise from the use of these technologies.

You will also learn about identifying emerging diseases, using scientific knowledge to identify pathogens and the methods used to treat those affected by them as well as to control their spread. You will encounter rational drug design as a way of producing drugs that precisely target a molecule or pathogen.

Key knowledge

- techniques that apply DNA knowledge (specifically gene cloning, genetic screening and DNA profiling) including social and ethical implications and issues
- the distinction between genetically modified and transgenic organisms, their use in agriculture to increase crop productivity and to provide resistance to insect predation and/or disease, and the biological, social and ethical implications that are raised by their use
- strategies that deal with the emergence of new diseases in a globally connected world, including the distinction between epidemics and pandemics, the use of scientific knowledge to identify the pathogen, and the types of treatments
- the concept of rational drug design in terms of the complementary nature (shape and charge) of small molecules that are designed to bind tightly to target biomolecules (limited to enzymes) resulting in the enzyme's inhibition and giving rise to a consequential therapeutic benefit, illustrated by the Australian development of the antiviral drug Relenza as a neuraminidase inhibitor
- the use of chemical agents against pathogens including the distinction between antibiotics and antiviral drugs with reference to their mode of action and biological effectiveness.

14.1 Applications of DNA technologies

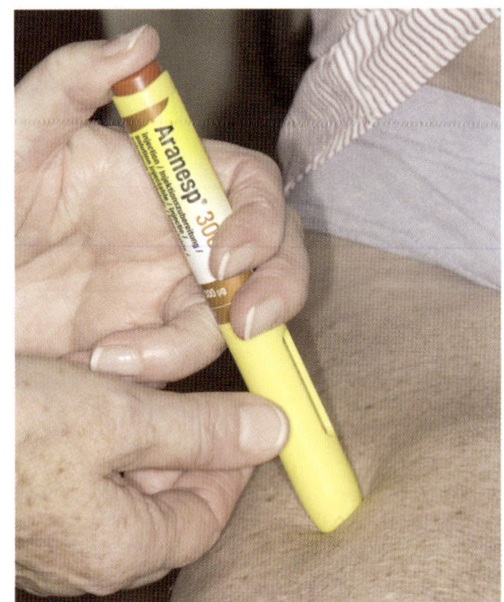

FIGURE 14.1.1 Recombinant DNA technology is routinely used to produce hormones for therapeutic purposes. Recombinant erythropoietin (EPO), for example the brand Aranesp is used to treat severe anaemia caused by chemotherapy and chronic kidney failure.

Gene cloning, DNA profiling and genetic screening are all powerful gene technologies that use DNA manipulation techniques (Figure 14.1.1). In this section, you will learn how these techniques can be used to detect and treat genetic variations, and some of the associated implications and issues that arise. Scientific knowledge can bring great benefits to society; but it can also bring great challenges, as societies consider if or how new technologies should be implemented, and the possible biological, social, economic, moral and ethical implications of their use.

GENE CLONING

Gene cloning allows scientists to produce exact copies of a gene of interest. Gene cloning is very different from the cloning of whole organisms. In gene cloning the end product is many copies of a specific gene.

Techniques of gene cloning

Two techniques can be used for gene cloning: an *in vivo* and an *in vitro* method. The *in vivo* technique involves the use of restriction enzymes, ligases and **vectors** to incorporate the desired gene into the DNA of a living organism, where this gene will replicate. Because the DNA code is universal, the original gene taken from one organism will express the same protein in the host organism. In the *in vitro* gene cloning technique, the polymerase chain reaction (PCR) is used to produce multiple copies of the specific gene in a solution.

Studies that are *in vivo* are 'within the living', such as when cells are studied in a living organism.

Studies that are *in vitro* are 'in glass' or in a dish or test tube, such as when cells are removed from the organism and studied in a culture dish.

Applications of gene cloning

There are many benefits of gene cloning that affect the individual and the community. Applications of gene cloning include research into genes and gene expression, determining DNA sequences and locations, the production of recombinant proteins, the use of gene therapy and the production of transgenic organisms.

Research into genes and gene expression

Gene cloning technology allows scientists to obtain the complete DNA sequence of different species. This allows the genetic code of different species to be compared and their evolutionary relationships to be refined.

Gene cloning also allows scientists to examine how genes are regulated by the environment or by master genes and how problems with gene expression can lead to disease. The methods enable the mapping of genes associated with disease, such as the *BRCA1* gene associated with breast cancer, with the view to earlier detection and the development of specific treatments.

Production of recombinant proteins

In Chapter 13, you learnt that the production of recombinant proteins is achieved by introducing recombinant DNA into bacteria or eukaryotes and allowing them to synthesise the protein. The main types of proteins produced by this technology are hormones, cytokines, enzymes and vaccines for human therapeutic purposes. This is much safer and more effective than using proteins purified from other organisms, such as the purification of insulin from pigs and growth hormone from human pituitary glands, as was done in the past. Many recombinant proteins are also being developed for research and veterinary medicine. Examples of human proteins that are now produced in bacteria, yeast or animal cell culture are shown in Table 14.1.1.

BIOFILE

Scanning whole genomes

DNA sequence variation is being studied on a large scale in populations. Genome-wide association studies (GWAS) search for gene variants, often single base changes called single nucleotide polymorphisms (SNPs) that are associated with various traits. A GWAS will screen for thousands of SNPs at once. SNPs associated with age-related macular degeneration were the first to be identified. These SNPs can then be used for screening individuals to identify those at risk of developing the disorder.

Product	Type of protein	Action and use
hepatitis B vaccine	viral coat protein	induces immune response to hepatitis B virus, to prevent infection by hepatitis B virus
human insulin	hormone	treatment for diabetes
human growth hormone	hormone	promotes bone and tissue growth; treatment for some forms of dwarfism and other growth problems
interleukin-2	cytokine	treatment for cancer
tissue plasminogen activator	enzyme	breaks blood clots; treatment of heart attacks
erythropoietin	hormone	increases red blood cell production; treatment of anaemia
atrial natriuretic factor	hormone	dilates blood vessels; treatment of high blood pressure
colony stimulating factor	cytokine	promotes white blood cell production and differentiation; treatment of leukaemia
human papilloma virus vaccine	viral capsid protein	induces immune response against papilloma virus; protects against cervical cancer
interferon-beta	cytokine	reduces relapse rate in multiple sclerosis
glucocerebrosidase	lysosomal enzyme	treatment of Gaucher's disease, a lysosomal storage disease

TABLE 14.1.1 Examples of human proteins produced by recombinant DNA technology in bacteria or eukaryotic cells.

Implications of recombinant DNA technologies for society

All technologies impact on people's lives and social structures. Social implications may affect people's financial position, lifestyle and reproductive decisions. Biological implications relate to the organisms used in and affected by the technology, its safety and short-term or lasting changes to human biology. Ethical, philosophical, moral and religious issues are also part of the impacts of biotechnologies.

Some of the implications of gene technologies for making therapeutic recombinant proteins are listed in Figure 14.1.2. Think about other issues you consider to be important in the use of DNA technology.

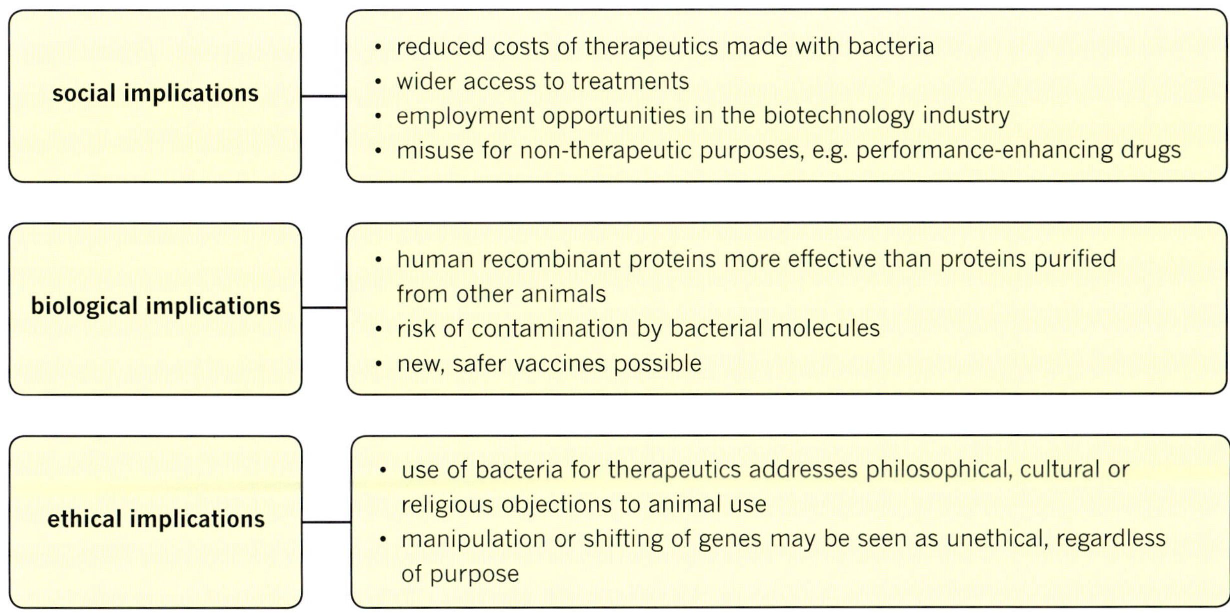

FIGURE 14.1.2 Implications of recombinant DNA technology and recombinant protein production.

BIOLOGY IN ACTION

Recombinant human erythropoietin

Red blood cell production is essential for maintaining oxygen homeostasis. A drop in oxygen supply to tissues (hypoxia) normally triggers the release of the hormone erythropoietin (also known as EPO) from the kidneys. EPO promotes red blood cell production in the bone marrow to restore the oxygen-carrying capacity of blood and its delivery to tissues (Figure 14.1.3).

In chronic kidney disease not enough EPO is made by the kidneys, resulting in low red blood cell counts and anaemia. Recombinant human erythropoietin for the medical treatment of this disease is produced in cultured mammalian cells. A copy of the human EPO gene is inserted into a plasmid, which is introduced into mammalian host cells.

Erythropoietin is a glycoprotein and must have the correct carbohydrates attached to the protein chain in order to function properly. Bacteria cannot do this, therefore mammalian cells must be used for making the recombinant protein (Figure 14.1.4). Recombinant erythropoietin has also been developed for veterinary use, such as recombinant feline EPO for cats with chronic kidney disease.

Because EPO promotes red blood cell production and oxygen-carrying capacity, it has been used by athletes seeking an advantage. EPO has been at the centre of sports doping scandals in recent years, particularly in endurance sports such as cycling, long-distance running and the triathlon. It has also been used in horse racing. The World Anti-doping Agency (WADA) works with drug testing laboratories to develop and validate tests that can distinguish between EPO produced naturally in the athlete and pharmaceutical EPO.

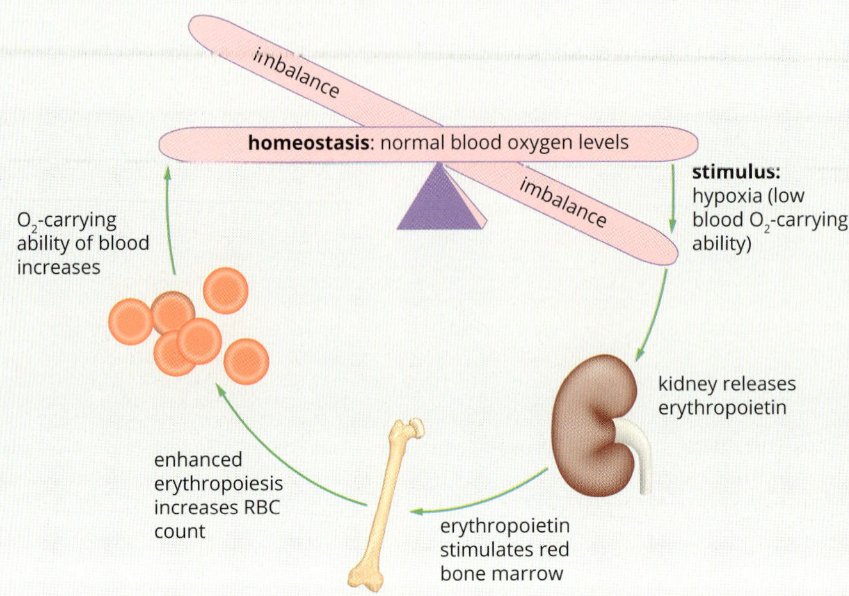

FIGURE 14.1.3 Erythropoietin is a hormone released mainly by the kidney to maintain oxygen homeostasis. It promotes red blood cell production (erythropoiesis) in the bone marrow.

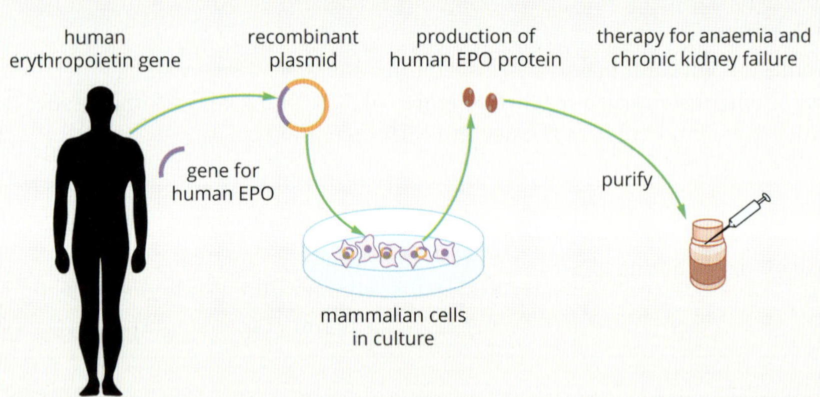

FIGURE 14.1.4 Recombinant human erythropoietin is produced in cultured mammalian cells. It is used to treat anaemia and chronic kidney disease.

Gene therapy

Gene therapy is an application of cloning technology in medicine. Gene therapy refers to the insertion of a gene into an individual's cells or tissues to correct or replace defective gene function that leads to disease.

Some methods of gene therapy involve removing cells from the body, such as bone marrow stem cells, inserting genes into these cells and then returning the cells to the patient. Other forms of gene therapy deliver a gene directly into the affected tissue in the body using a suitable vector. For example, a copy of the normal gene coding for the membrane protein that is faulty in cystic fibrosis can be inserted into some of the patient's lung cells. The expression of the normal target gene allows a functional protein to be produced, relieving the symptoms of the disease.

Specialised vectors are used to deliver genes into eukaryotic cells for gene therapy.

- **Viral vectors** naturally insert genes into cells (Figure 14.1.5). Adeno-associated virus (AAV) is one example of a viral vector. It is a small non-pathogenic virus that can enter a wide variety of human cells and is able to insert DNA into a defined region of a chromosome. The virus is easily modified to carry a specific gene, and usually does not induce an immune response.
- **Liposomes** are very small phospholipid vesicles that can diffuse across plasma membranes or enter cells by endocytosis. DNA is inserted into the liposome and carried into the cell, where it is released. Liposomes can also be used for delivery of drugs to cells. They may be safer gene delivery vehicles than viruses.

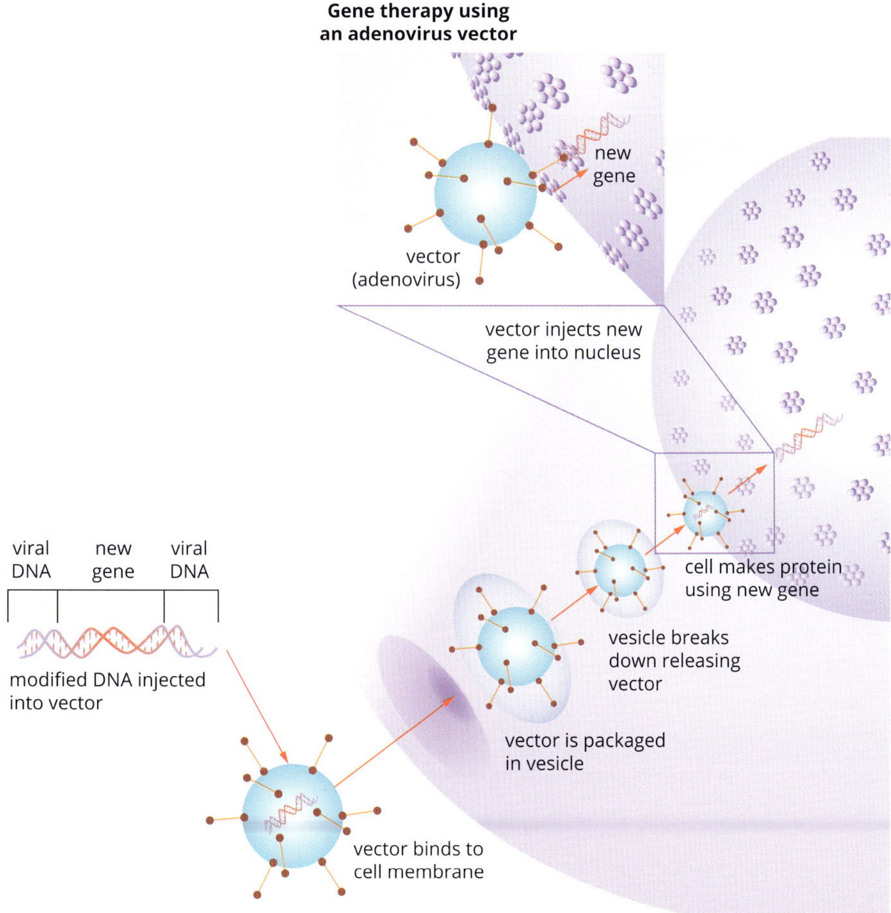

FIGURE 14.1.5 Illustration of a viral vector delivering recombinant DNA to a eukaryotic cell nucleus.

Biological issues have arisen in gene therapy trials. The death of a participant in an early clinical trial prevented further trials for some time while scientists investigated the cause. Some people experience an unexpected adverse immune reaction to the viral vector. In other gene therapy trials, the target gene was inserted into a vital gene, disrupting its function and causing additional health problems in the patient. There have been cases of contamination with an infectious virus when viral vectors have been used.

A range of social and ethical issues arise from gene therapy; some of these are listed in Figure 14.1.6. A big question is whether gene therapy should be permitted for germ line cells (gametes), which would mean the genetic change could be inherited, or should remain limited to somatic (body) cells so that the genetic change is not passed on to offspring.

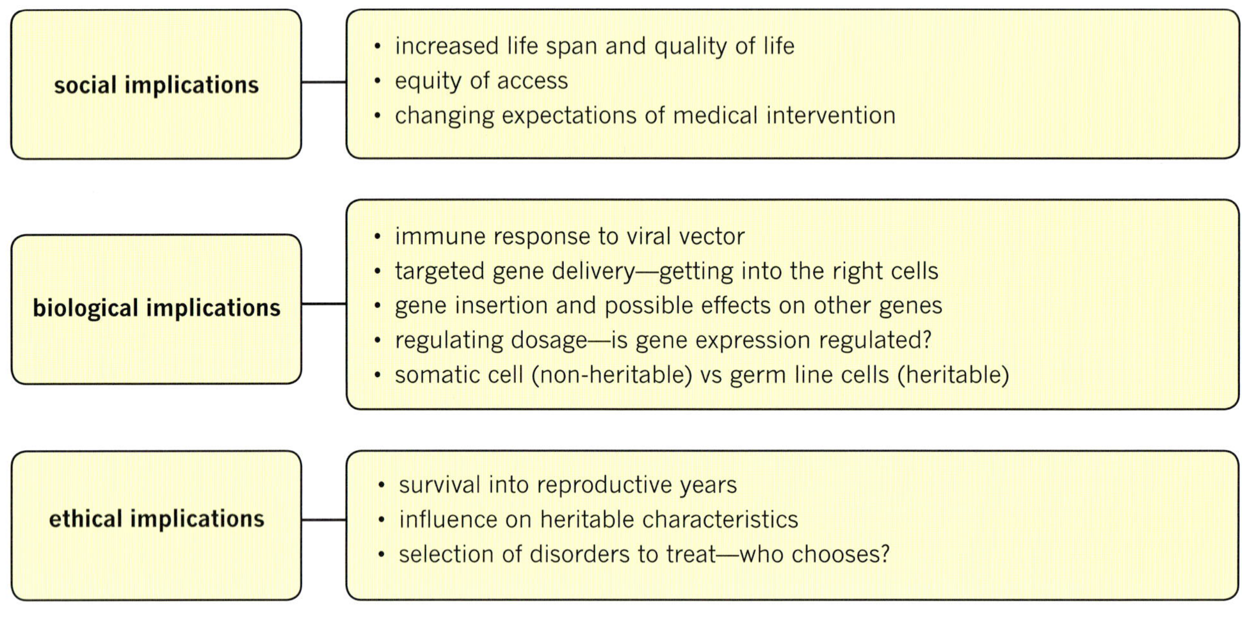

FIGURE 14.1.6 Some of the issues arising from the use of gene therapy.

Production of transgenic organisms

The scope of gene cloning for research and agricultural purposes is large. The desired genes from one species can be introduced into another species to give benefits such as pest resistance and enhanced yield production. These organisms are known as transgenic organisms and will be discussed in Section 14.2.

EXTENSION

Gene therapy

Gene therapy has been a work in progress since the first trial in 1990. Delivering genes effectively and safely into eukaryotic cells is more difficult than the transformation of bacteria. Biomedical researchers continue to investigate and improve on the methods for gene delivery, including safer and more effective vectors, better targeting of cells, and more control over where the gene is inserted and how it is regulated. The methods that seem most effective are illustrated in Figure 14.1.7.

Gene therapy may become more mainstream medicine if the hurdles outlined in Table 14.1.2 can be consistently overcome.

Gene therapy clinical trials using different approaches continue for several debilitating genetic disorders:

- Recombinant adeno-associated virus is used to deliver the gene for blood clotting factor IX to treat haemophilia B, the dystrophin gene to treat muscular dystrophy, and the CFTR gene to treat cystic fibrosis.
- Liposomes carrying a recombinant plasmid with the CFTR gene are inhaled with the aid of a nebuliser (the device used to inhale asthma medication) to deliver the gene directly to lung cells.
- Stem cell biology and molecular biology are combined for thalassaemia gene therapy. Stem cells are removed from the patient's bone marrow, the normal beta-globin gene is delivered to these cells by a suitable vector and the modified cells are cultured to increase the population before they are returned to the patient.

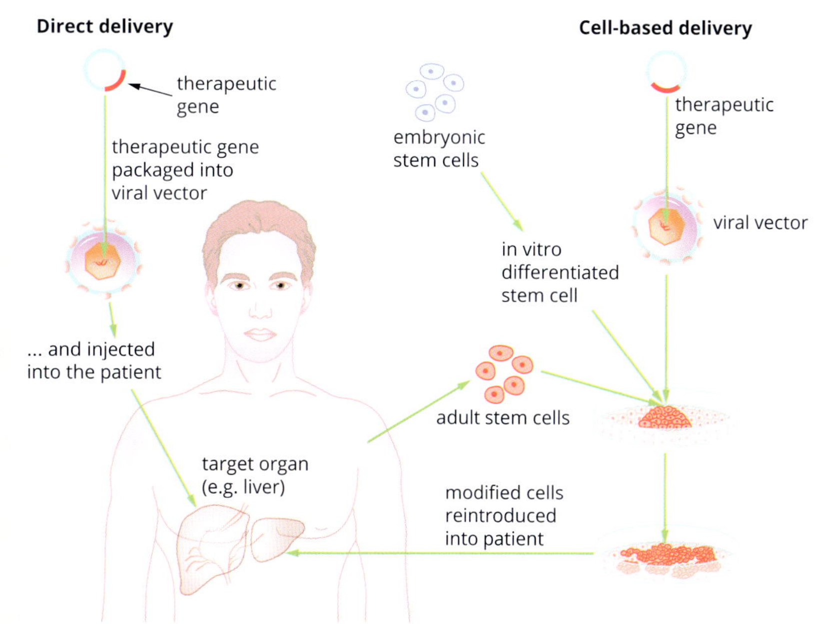

FIGURE 14.1.7 Recombinant DNA and cell biology methods are used to deliver genes to cells for gene therapy.

The biological hurdles	The approach developed or under investigation
ensuring viral vectors are safe	• self-inactivating vectors—removal of genes for viral replication • viruses that do not cause a large immune response
non-viral vectors can deliver the gene	• liposome vector that enters the cell by endocytosis or diffusion across plasma membrane
targeting specific cells	• delivery method to specific organs, e.g. inhalation into lungs • membrane receptors for specific viral-vector docking • insertion of gene into stem cell before differentiation to specialised cells
targeting specific or safe regions of the genome	• DNA sequences included in the vector to recognise and insert into non-coding regions of a chromosome
gene regulation—expressing the gene only when needed	• insert regulatory sequences with the gene

TABLE 14.1.2 Technical factors that may limit the success of gene therapy.

DNA PROFILING

In 1984, an English geneticist named Alec Jeffreys developed a technique he called DNA fingerprinting to distinguish between individuals on the basis of variable regions of their DNA. Today, the more commonly used method is called **DNA profiling**. DNA profiling can be used to identify one individual from any other individual. It is often used in forensics to identify the perpetrator of a crime. DNA profiling can also be used to identify bodies after disasters or to confirm if a child is genetically related to a parent.

DNA profiling relies on an individual's unique DNA. The non-coding sections of the DNA, those that do not code for proteins, can vary widely between individuals. These variations are called polymorphisms. DNA profiling uses the differences between a number of polymorphic sections to identify individuals. In particular, short, repeated sections of between two and six bases, called **short tandem repeats (STRs)**, are examined. STRs are also referred to as microsatellites.

There are thousands of STR loci throughout the human genome and 13 STRs are generally used in DNA profiling. The same STR sequence occurs on each member of a homologous pair of chromosomes, and, by chance, the number of repeats at each locus may be the same or different (Figure 14.1.8). For example, at one location a person might have 25 repeats, while another person might have 45 repeats at the same site.

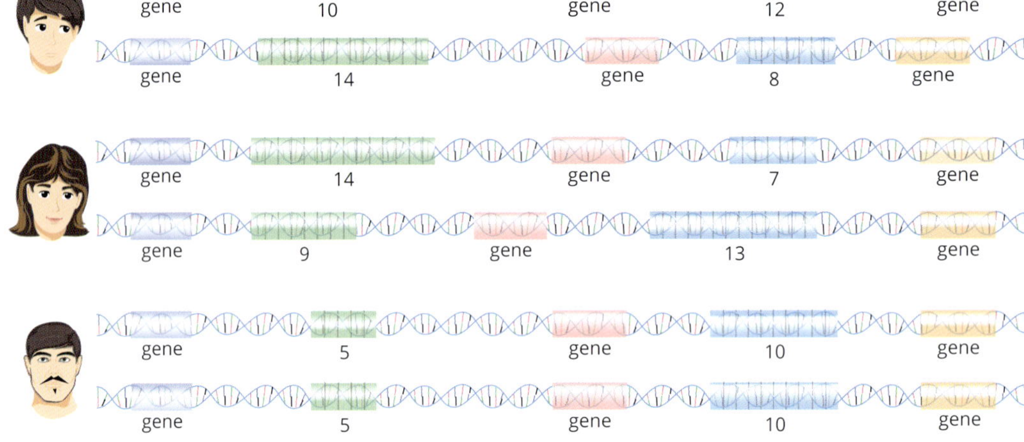

FIGURE 14.1.8 An example of the variation that can be seen in the STRs in three individuals for two STR sites. As chromosomes occur as homologous pairs, each person has two copies of each STR, which may vary in length.

BIOFILE

The FBI's CODIS

The 13 STR sites used in DNA profiling are part of the CODIS system developed by the FBI in the United States of America. CODIS stands for Combined DNA Index System. Using 13 STRs gives an extremely high probability that the DNA profile is unique and that the only perfect match of all 13 DNA sites will be with DNA from the same person. DNA profiling also includes regions on the X and Y chromosome for sex determination.

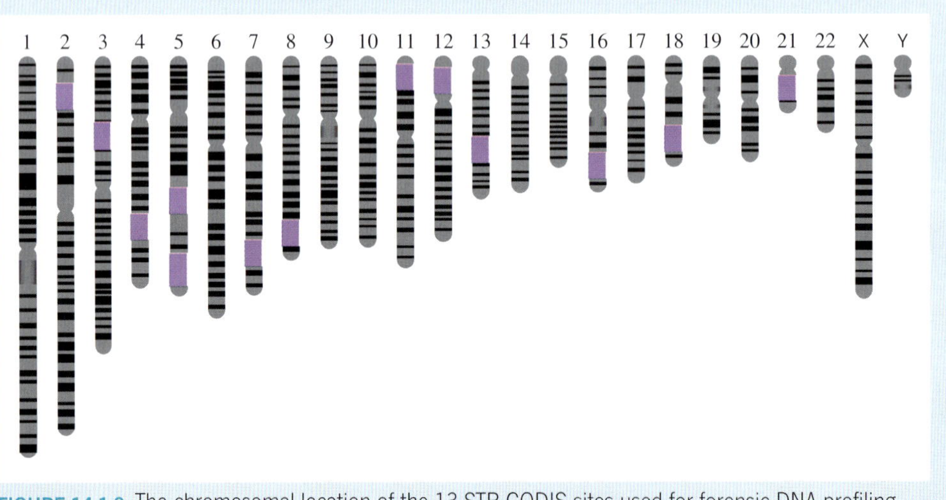

FIGURE 14.1.9 The chromosomal location of the 13 STR CODIS sites used for forensic DNA profiling.

Techniques involved in DNA profiling

If, for example, a small amount of blood (containing white blood cells), semen or another type of DNA sample is found at a crime scene, the following steps will be carried out to determine the DNA profile and match it to a perpetrator:

- DNA is extracted from the sample using restriction enzymes.
- The STRs are amplified using PCR, with specific primers for each STR. This produces a much larger sample to test.
- Differences in the size of the STRs can be detected by standard gel electrophoresis or by capillary electrophoresis, a rapid, automated method. In capillary electrophoresis the DNA fragments move in a thin tube under the influence of an electric field. The smaller the size of the fragment, the faster it moves through the capillary tube. As each fragment moves through the tube a laser detector registers a peak on a graph (Figure 14.1.10).
- The printout of the STR analysis taken from the crime scene is compared with that obtained from the DNA of a suspect or suspects.
- The DNA sample is matched to that of a suspect if the lengths of the particular STRs at all sites are the same.

When the lengths of 13 STRs from two DNA samples match perfectly, the chance that the two samples are from different people is hundreds of billions to one.

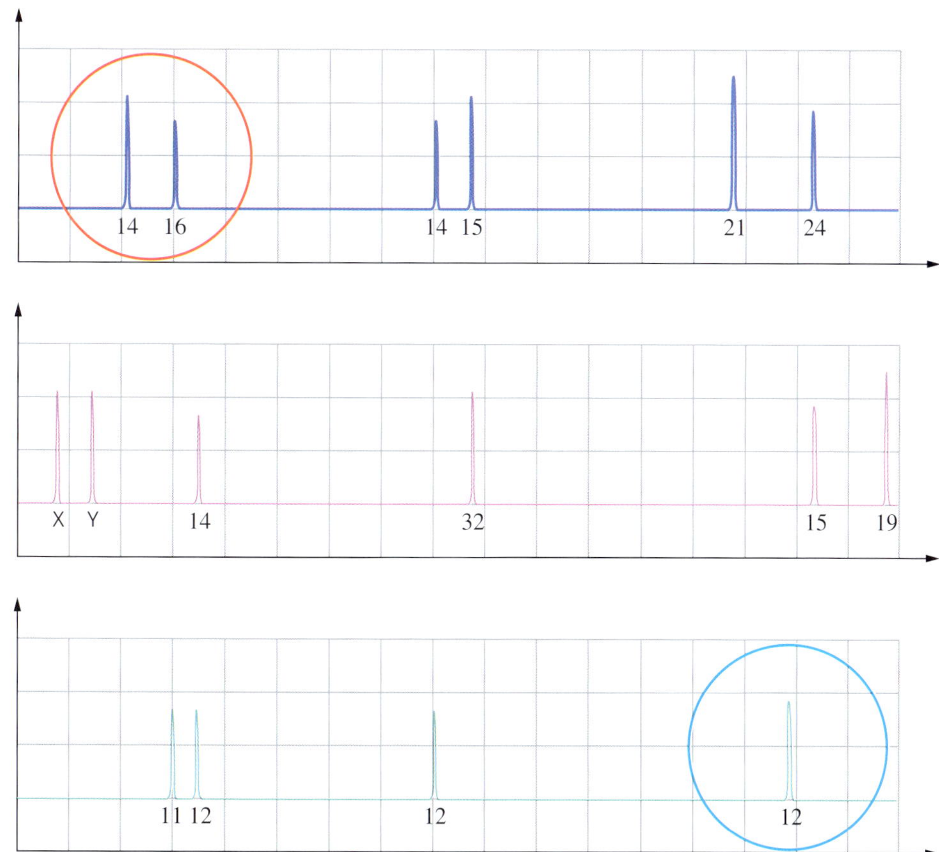

FIGURE 14.1.10 DNA profiling printout obtained from the sample of one individual. Ten different regions (nine different STRs and the sex chromosome markers) have been analysed. We have 2 copies of each STR, one on each homologous (paired) chromosome so most STRs appear as a pair of peaks on the graph (the same number of repeats on each chromosome appears as a single peak). The numbers below the peaks indicate the number of repeats at that locus. For example, the first two peaks (circled in red) show this person has 14 repeats and 16 repeats of that STR on that homologous pair of chromosomes, while the last peak (circled in blue) indicates 12 repeats of that STR on each homologous chromosome. The X and Y chromosome markers show this sample is from a male.

Issues related to DNA profiling

It is unlikely that a person will be incorrectly identified by DNA profiling if the procedure is carried out accurately. However, there is the possibility of incorrectly identifying an individual through DNA profiling. For example, foreign DNA may contaminate a sample at a crime scene or at the laboratory where the sample is tested. Failure to properly clean equipment could cause a sample from one suspect to contaminate a sample for another test.

Privacy is a contentious issue related to DNA profiling. In Victoria, DNA samples cannot be obtained from a person unless they give their permission. They can, however, be ordered to do so if there is strong evidence that they may have committed the crime under investigation and if the DNA profile could help to confirm or deny their guilt. These DNA samples must be destroyed if the person is not guilty or is not charged. However, in some countries, the DNA may be kept for up to 10 years. This has enabled the identification of criminals who have committed crimes in unsolved cases that occurred before DNA profiling technology was developed. It has also resulted in the exoneration of people who have been wrongly accused.

Storing DNA after a person has served their sentence for a crime may be seen as unethical, as the person has already been punished for the particular crime and is no longer considered a criminal. Others are in favour of the creation of a 'bank' of DNA samples, provided by everyone in the community, which could be used to solve crimes and perhaps trace the remains of unidentified missing persons. Opponents of a DNA 'bank' argue that there would be potential for these samples to be stolen or used unethically.

BIOFILE

Tsar rediscovered using DNA

In July 1918, Tsar Nicholas II of Russia, the Tsarina Alexandra, their five children, Olga, Tatyana, Maria, Anastasia and Alexis, three female servants and the royal physician were executed by a Bolshevik firing squad in the town of Etkaterinburg. Historical accounts indicate that two of the children's bodies were burned, although others claim that Anastasia escaped execution. The remaining bodies were thrown into a shallow grave and sulfuric acid poured over them.

In 1991 two amateur historians, Gely Ryabov and Alexander Avdonin, discovered nine skeletons in a grave near Etkaterinburg. The remains were tested to find out whether they came from the Tsar and his family. DNA extracted from bone tissue samples was amplified by PCR. The first step in the analysis was to identify the sex of the skeletons. This was done using PCR of a gene that is found on the Y chromosome. This indicated that there were two males and seven females.

Using DNA profiling of the bone samples, it was possible to conclude with certainty that five of the skeletons were those of two parents and their three daughters. But these could have been the remains of any family. To establish the identity of the bones, comparison to the DNA from a related person was needed. The evidence came when the DNA profiles of the skeletons were compared with those generated from known relatives of the Tsar (George, brother of Nicholas II, whose remains were exhumed from a crypt in St Petersburg) and Tsarina (Prince Philip, husband of Queen Elizabeth II). The presence of common bands in the DNA profiles of the five bodies, Prince Philip and George indicated that all of the individuals were relatives. Indeed, it was estimated that the probability that the bones found in Etkaterinburg are the remains of the Tsar and his family is approximately 99 999 out of 100 000.

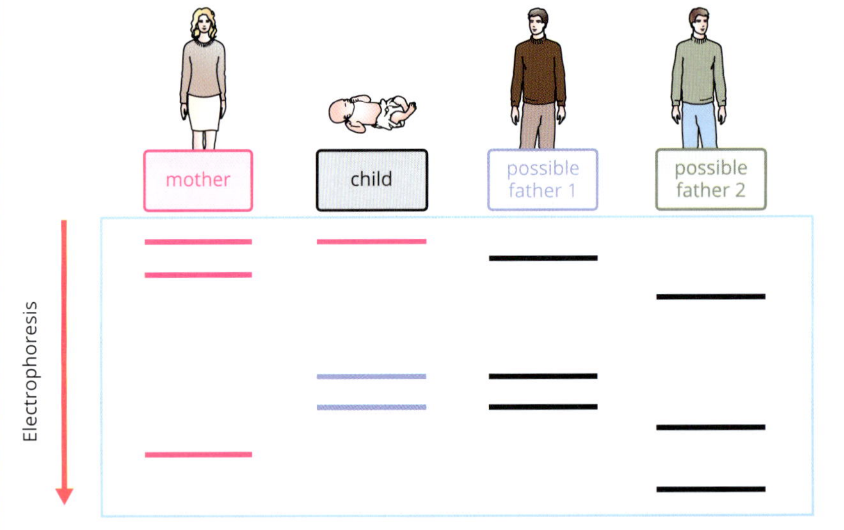

FIGURE 14.1.11 Family members share common bands in DNA profiles, although the combination of bands in each individual is unique.

Most recently, DNA profiles have been used to identify key features of an individual's appearance. The presence of alleles for eye, skin and hair colour allows investigators to narrow down the list of suspects. Occasionally the suspect's ancestry can also be determined. This is a developing field of science and the analysis may not yet be reliable.

BIOFILE

Applications of DNA profiling

DNA profiling is used for other applications such as genealogy, biogeographical population comparisons, historical population migration patterns and evolutionary relationships. For these purposes the DNA sequences used for comparisons include STRs (different sites from those used for crime scene analysis), mitochondrial DNA, Y chromosome genes and single nucleotide polymorphisms or SNPs.

GENETIC SCREENING

Genetic screening is used to detect abnormalities or changes to genes. Screening is used to confirm or rule out suspected genetic conditions, to determine if a person carries a faulty gene that can be passed on to their offspring or whether a person is at risk of developing a disease.

Genetic screening in embryos

Preimplantation genetic diagnosis involves testing embryos created through ***in vitro* fertilisation (IVF)** (Figure 14.1.12) before implanting them into the mother. To do this, a single cell is taken from an embryo that is about 8 cells in size. This cell multiplies to produce several cells and then the DNA of the cells is tested. This is performed when parents are carriers of genetic conditions such as cystic fibrosis and are at risk of passing this on to their offspring. Parents then choose which embryo to implant in the mother's uterus.

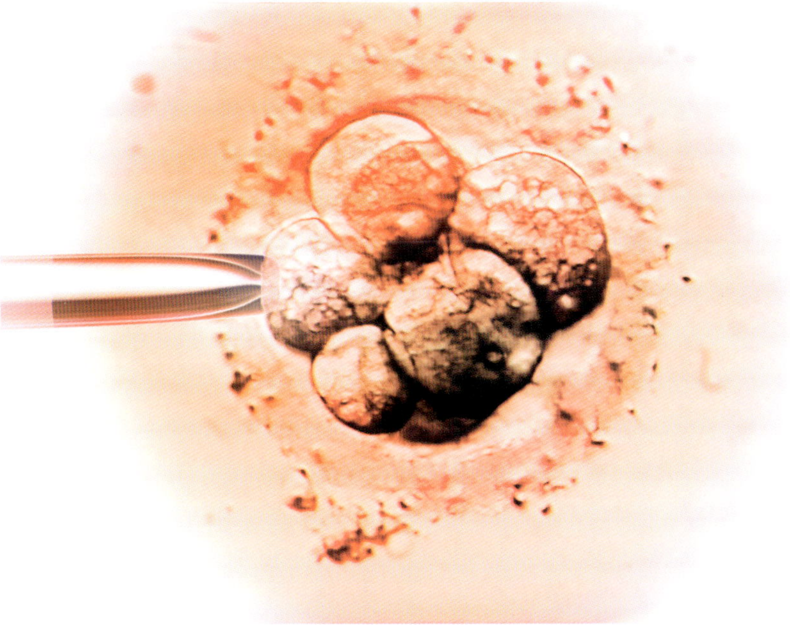

FIGURE 14.1.12 Coloured light micrograph of a three-day-old embryo about to undergo preimplantation genetic diagnosis. The micropipette (left) will remove a single cell for DNA analysis.

Genetic screening in foetuses

If a woman is already pregnant and there is a risk of the child having a genetic abnormality, she may choose to have genetic screening by either **amniocentesis** or **chorionic villus sampling**. These techniques are invasive tests and increase the possibility of miscarriage.

Chorionic villus sampling (CVS) can be done earlier in the pregnancy (9–14 weeks) than amniocentesis and takes a sample of the cells from the placenta using a needle (Figure 14.1.13a on page 538). The placenta is made up of cells produced by the embryo and therefore contains the same genetic information as the embryo itself.

Amniocentesis is performed at between 14 and 20 weeks after conception. A small sample of fluid surrounding the foetus is removed (Figure 14.1.13b). This fluid contains cells from the foetus and these cells are grown and tested for genetic abnormalities.

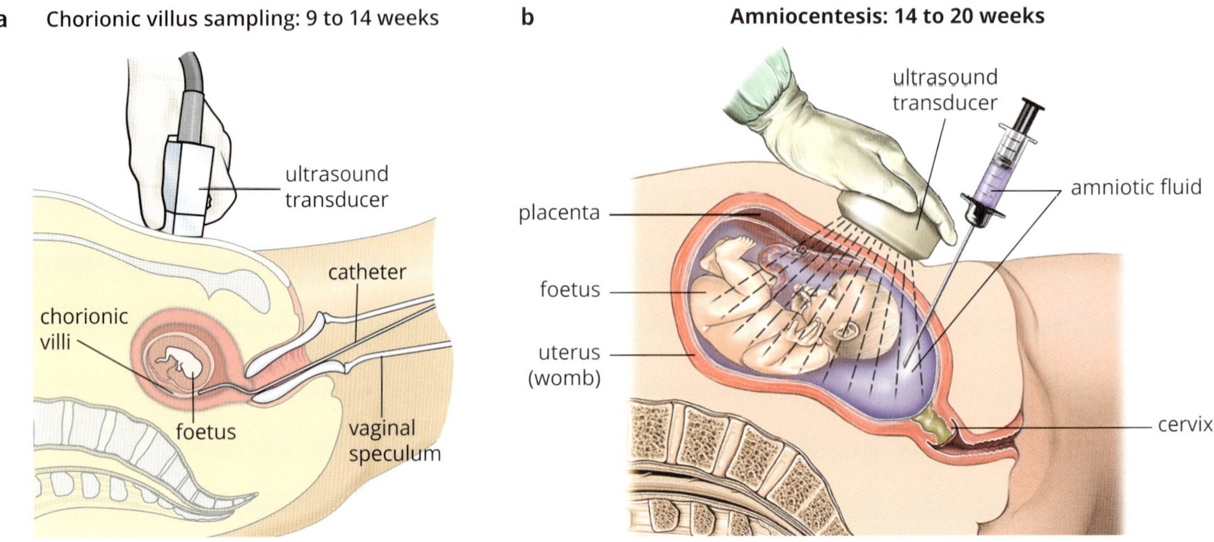

FIGURE 14.1.13 (a) Chorionic villus sampling and (b) amniocentesis.

Postnatal genetic screening

The parents of every newborn baby in Australia can choose to have their child genetically screened for medical conditions such as phenylketonuria (PKU), cystic fibrosis, hypothyroidism and some other rare metabolic disorders. All of these conditions have a genetic basis and many of their negative effects can be reduced if treated immediately. To perform the test, a small drop of blood is collected from the heel of the baby soon after birth and this drop is placed on a card called a Guthrie card (Figure 14.1.14). The DNA in the white blood cells is used to analyse the genes affected in these diseases.

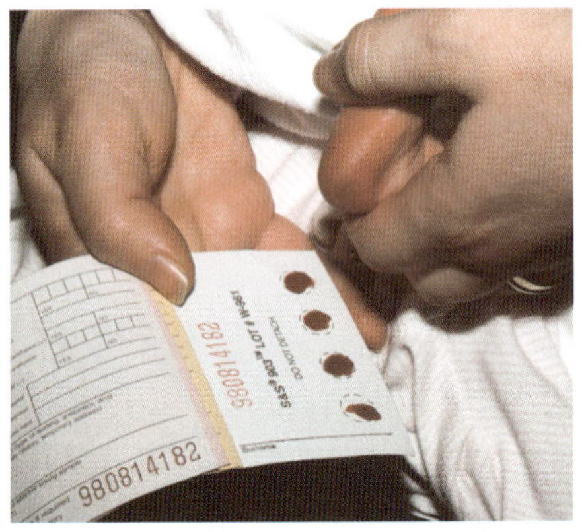

FIGURE 14.1.14 The heel of a newborn is pricked to collect blood on a Guthrie card for genetic screening. Guthrie cards are stored and can be used at a later time if needed for genetic screening.

Many other genetic diseases, such as achondroplasia (a disorder of bone growth), are not routinely tested for in genetic screening because there is no treatment that will stop these conditions developing.

Adults may also choose genetic screening to find if they carry changes or mutations in their genes that may either increase the likelihood of developing that disease or passing it on to their offspring.

DNA techniques used in genetic screening

The methods employed for detecting different types of mutation use the key tools and techniques of PCR, restriction enzymes and gel electrophoresis. You saw examples of these applications is Chapter 13 for detecting the beta-globin allele that causes sickle cell anaemia, and a mutation for cystic fibrosis.

Other methods include the use of DNA probes. A probe is a short sequence of DNA that is complementary to the gene or mutation of interest. A fluorescent tag can be attached to the probe so that when it binds to the DNA its location can be detected. This is how fluorescence *in situ* hybridisation (FISH) detects mutations on chromosomes (Figure 14.1.15).

Examples of methods used in genetic screening are listed in Table 14.1.3.

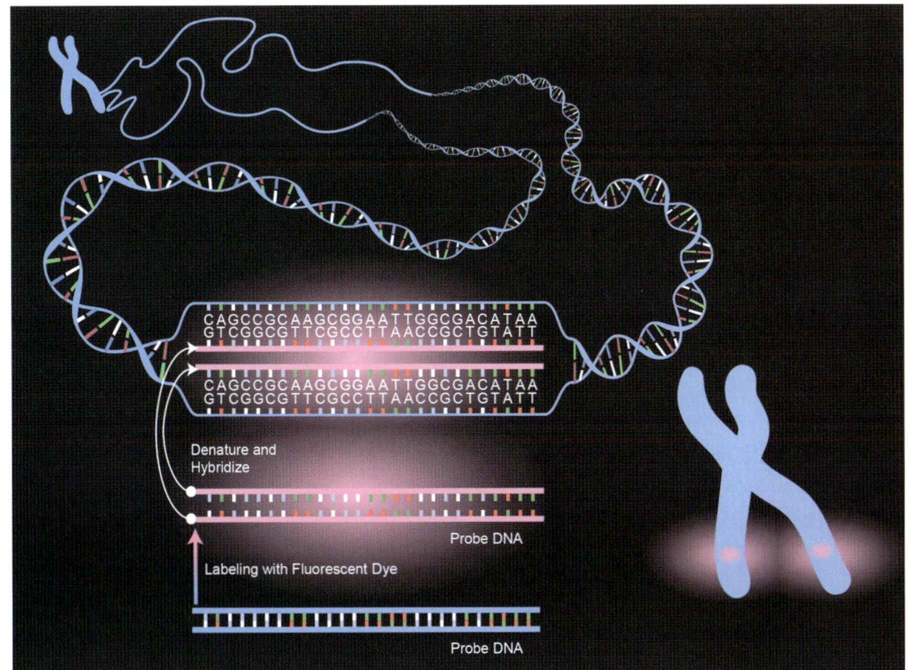

FIGURE 14.1.15 Fluorescence *in situ* hybridisation may be used to detect mutations such as those causing Fragile X syndrome, Di George syndrome and extra copies of whole chromosomes, as in Patau syndrome. This diagram illustrates how a fluorescent DNA probe binds by complementary base pairing to the target sequence on a chromosome.

Disease examples	Type of mutation	Methods used	Detection
Huntington's disease	CAG repeat mutation	• PCR • gel electrophoresis	different length of allele on gel
• phenylketonuria (PKU) • cystic fibrosis	point mutation occurs in restriction enzyme recognition site	• PCR • restriction enzymes • gel electrophoresis	specific fragments detected on gel
• Fragile X syndrome • Di George syndrome • Patau syndrome	• duplication mutation • deletion mutation • trisomy (3 copies of chromosome)	• fluorescence *in situ* hybridisation (FISH) • fluorescence microscopy	mutation located on chromosomes

TABLE 14.1.3 Examples of methods used in genetic screening.

BIOFILE

Phenylketonuria

Phenylketonuria (PKU) is one of the genetic conditions routinely screened for in newborns. It is a disease caused by a number of possible mutations in the a gene located on chromosome 12. This gene codes for an enzyme called phenylalanine hydroxylase (PAH), which catalyses the conversion of the amino acid phenylalanine to tyrosine. Tyrosine is necessary to make some hormones, neurotransmitters and melanin, a skin pigment molecule. The mutation causes the loss of a functioning enzyme, so phenylalanine is not converted to tyrosine. In PKU, phenylalanine builds up in the blood and tyrosine is not produced. If too much phenylalanine builds up in the blood, it will cause permanent brain damage. A newborn found to have PKU can be placed on a diet that contains tyrosine and is low in phenylalanine. This diet will limit the build up of phenylalanine and so prevent the damage to the child's brain.

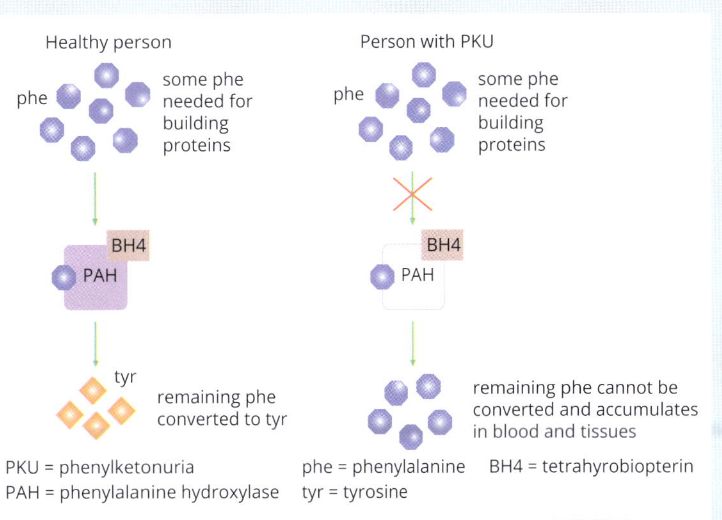

FIGURE 4.1.16 Phenylketonuria. Normal phenylalanine hydroxylase (PAH) enzyme converts phenylalanine (phe) to tyrosine (tyr). In phenylketonuria (PKU) the mutant enzyme does not work and phe builds up in the blood.

Issues related to genetic screening

There is a broad range of potential issues associated with any form of genetic testing, some of which are listed below (Figure 14.1.17).

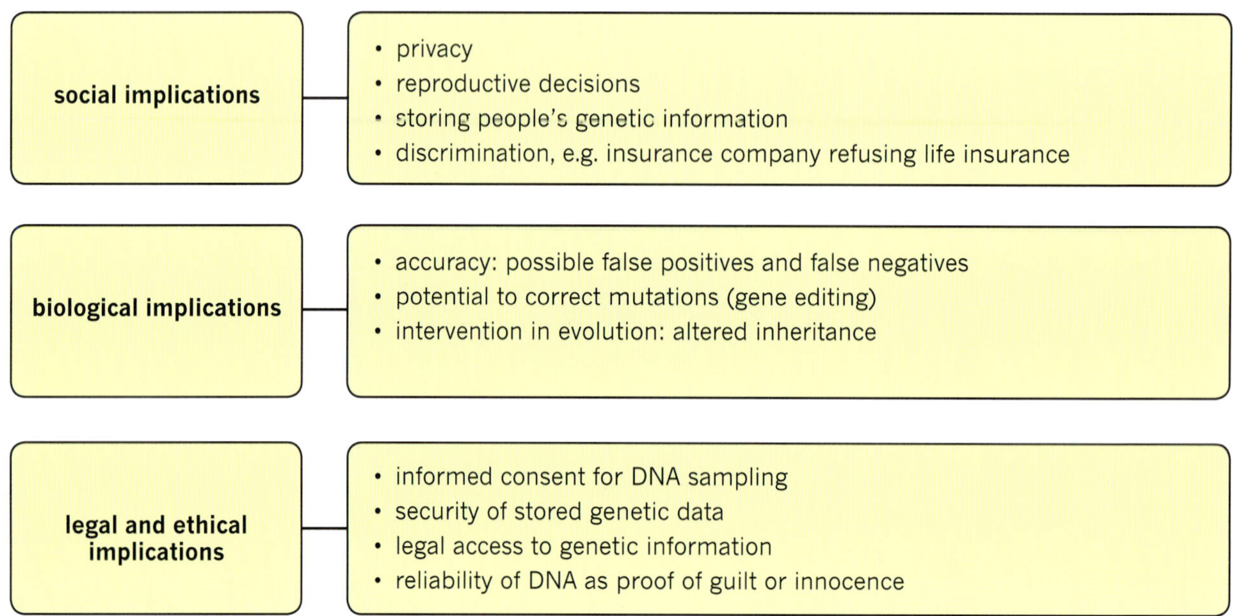

FIGURE 14.1.17 Summary of issues arising from DNA profiling and genetic screening.

Preimplantation genetic diagnosis identifies genetic abnormalities before the embryo is implanted. At this stage, the embryo is a collection of undifferentiated cells. Many people feel that choosing not to implant an affected embryo is a choice that reduces potential suffering for both the potential child and its parents. Others believe that at fertilisation the embryo is uniquely human and has the same right to continue on its path to being a person as any other child. Some argue that removing embryos that are less than perfect will bring about a society that lacks compassion and acceptance of difference.

If a foetus is found to have a genetic abnormality, a decision has to be made to continue the pregnancy or to terminate it; that is, the mother would undergo an abortion. The difficult decision requires balancing the loss of the pregnancy with the knowledge or prediction of the suffering and challenges for the child if it were born. The decision will be informed by the severity of the genetic disorder and the capacity of the family to accommodate any special needs of the child. Genetic counsellors may assist by providing information on the disorder, discuss issues related to passing on the genetic defect to other potential children and help the parents come to a decision consistent with their religious, ethical and moral principles.

Genetic testing after birth or in older individuals can result in other issues. In the case of testing for PKU, for example, there is very little risk or pain associated with the test and the benefits of early treatment are large. The Guthrie cards with blood samples are securely stored indefinitely in Victoria, but this varies across states. Privacy is also an issue; parents may choose to give permission for the samples to be used in research projects for which individual identification is removed. Access to the samples may be legally protected by legislation in some states because they contain personal, identifiable, genetic information. However, the genetic information in the samples is the property of the child who cannot, at the time of testing, give consent for it to be used.

Many people argue that there is a potential for misuse of this stored genetic information in the future. For example, unless there is legislation to prevent discrimination, an insurance company may refuse life insurance to a person carrying disease-associated alleles detected by genetic screening. The legislation relating to DNA data varies across states and countries. Laws may change in response to new understanding of the technology and improved techniques, experience in particular legal cases, or changes in political or social attitudes.

Like many other procedures, genetic testing can have errors. Parents may sue individual doctors or companies for the cost of raising their child because the doctors failed to identify the child had a genetic defect.

Individuals who have genetic testing and find they carry a particular allele for a disease may choose not to share this information with others to whom they are related. If this happens then family members may become ill or give birth to babies who are affected by the disease without knowing they were at risk.

BIOFILE

Couple sues Royal Children's Hospital

In 2015 a couple sued a Melbourne hospital because it had failed to identify that their first child had a condition called Fragile X syndrome before they had their second child. Fragile X syndrome is characterised by behavioural, emotional, developmental and physical features such as anxiety, intellectual disabilities and hyperflexible joints.

The hospital performed genetic testing on the first child but did not test for Fragile X, even though the child displayed characteristics for this. Following the birth of their second child, who also displayed these characteristics, it was discovered that both children had the genetic disorder. Both children would need care and medical treatment for the rest of their lives.

The couple is also considering bringing legal action against a relative who knew they were a carrier of the condition but did not tell the rest of their family. The couple said that if they had known they were carriers for this condition they would not have chosen to have children, or would have chosen to use IVF to select unaffected embryos or to have used donor eggs.

The mutation responsible for Fragile X syndrome is a duplication of 3 bases, CGG, more than 200 times in the *FMR1* gene on the X chromosome. One way to detect the mutation is by fluorescence *in situ* hybridisation, which can identify the abnormality on the X chromosome.

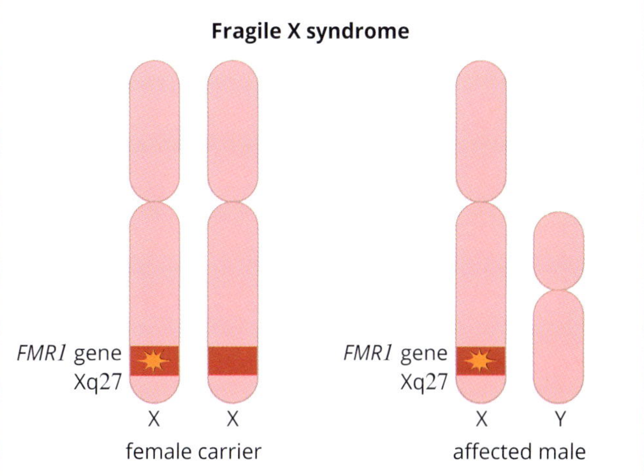

FIGURE 14.1.18 Fragile X is a sex-linked mutation causing a number of health and behavioural issues. As females carry two X chromosomes they are less often affected than males who only carry one X chromosome, because females may have one normal *FMR1* gene to counteract the effect of the mutation.

14.1 Review

SUMMARY

- Gene cloning is the copying of individual genes for sequencing, research and the production of recombinant proteins.
- Therapeutic recombinant human proteins, made from recombinant DNA in bacteria, yeast or animal cell cultures, include hormones, cytokines, enzymes and vaccines.
- Gene therapy replaces the defective gene that causes disease with a normal gene.
- Viral vectors and liposomes are used to deliver recombinant DNA to target cells for gene therapy.
- DNA profiling compares variable short tandem repeat (STR) regions of the genome for identification of individuals.
- Genetic screening identifies disease-associated mutations in foetuses, newborns or adults.
- Issues of DNA technology applications include cost and scale of production of therapeutic proteins; ethics of gene transfer between species; safety of gene transfer; accuracy of genetic tests; storage, privacy and disclosure of genetic information; reproductive choices; genetic selection.

KEY QUESTIONS

1 Describe two reasons why scientists clone genes.

2 a What types of human proteins are commonly produced by recombinant DNA technology?

b Suggest an advantage of this method of production compared to a traditional approach.

3 Draw up a table to list a positive and a negative aspect of recombinant DNA technology for making human proteins. Include social, ethical and biological issues.

4 The following diagram illustrates an application of DNA technology. Name and describe the method illustrated. Give an example of its application.

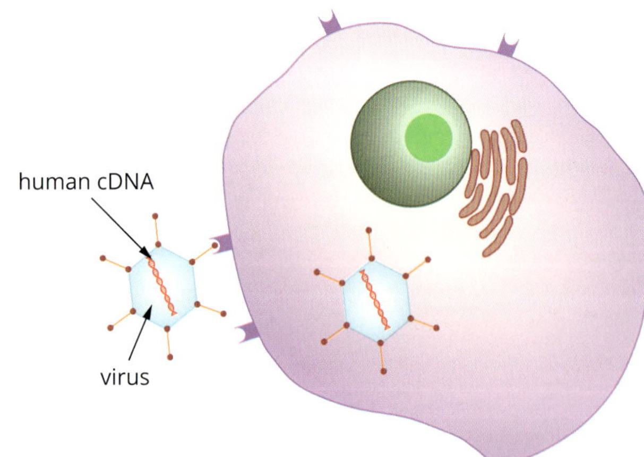

5 The following diagram represents STR regions of DNA used to identify a person who died in a natural disaster. Three people who were looking for a missing sibling submitted DNA for comparison.

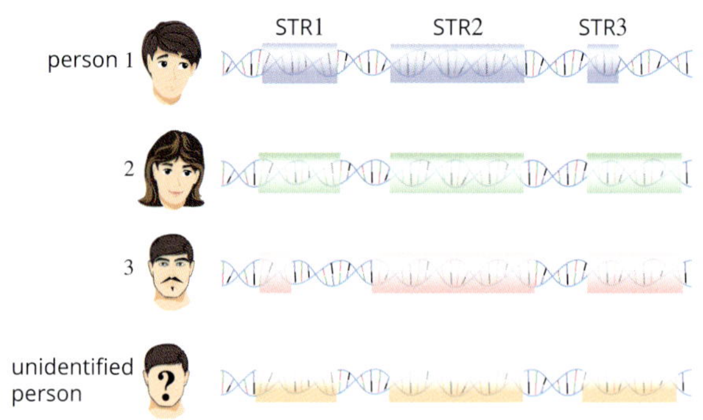

a What is an STR? List the steps involved in determining a DNA profile by STR analysis.

b The following illustration represents gel electrophoresis of the STR analysis. Which person is most likely the sibling of the unidentified person? Explain your choice.

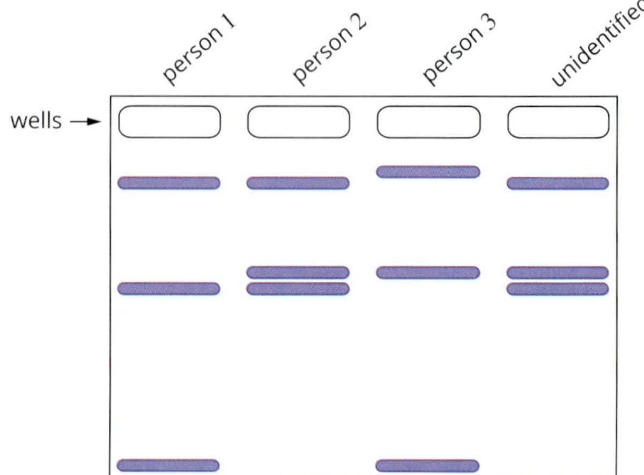

6 a Explain how DNA profiling can be used to help determine the guilt or innocence of a murder suspect found with blood stains on their clothes.
   b Explain why a sample may need to be tested in more than one laboratory.

7 Genetic screening may be performed on blood taken from a newborn baby whose family carries the inherited disease phenylketonuria.
   a What would the genetic screening test be looking for?
   b Describe the methods that may be used to perform the genetic screening.

8 In a chart or table list some of the social, biological and ethical issues of genetic screening. Indicate whether you consider them to be positive or negative aspects of the technology application.

9 Recombinant human proteins may be produced in eukaryotic cells, such as cultured mammalian cells or yeast, rather than bacteria. Suggest a reason for this method of producing recombinant proteins.

10 Gardasil is a recombinant vaccine against the capsid protein of human papilloma virus. The viral capsid protein is expressed or produced in large-scale yeast cultures. The viral protein is then purified and used for vaccination. Use a flow chart or diagram to outline a method that can be used to create yeast cells that can produce a viral capsid protein.

11 Explain why DNA profiling can discriminate between:
   a a twin brother and sister
   b a mother and daughter
   but NOT
   c identical twins.

14.2 Genetically modified and transgenic organisms

FIGURE 14.2.1 Genetically modified cotton on the left shows its insect resistance compared to the non-modified cotton on the right, which has been ravaged by insects.

Humans have used selective breeding to produce animals and plants with more useful or more attractive characteristics for tens of thousands of years. They chose those animals or plants that expressed the characteristics they wanted to conserve and selectively bred them together, hoping that their offspring would show even more of these characteristics. In the past, selective breeding could only utilise characteristics that already existed in the genetic pool of a species. We now have the knowledge and skills to use recombinant DNA techniques and to transfer genes from one species to another to produce organisms with DNA combinations never seen before (Figure 14.2.1). This could lead to many benefits, but it may also lead to questions regarding whether the impacts of this manipulation are socially or ethically acceptable.

GENETICALLY MODIFIED ORGANISMS

Over the last few decades techniques were developed that allowed for the alteration of an organism's genome and for the transfer of genes from one organism to another. Because the DNA code is universal, almost any gene transferred from one organism to another will express the protein that it expressed in the original organism. This means that a desirable characteristic seen in one animal or plant could be transferred to another organism lacking this characteristic.

An organism that has had its genome altered in this way is considered to be a genetically modified organism (GMO). An organism that has had genes from another species inserted is additionally called a **transgenic organism**. The gene that came from another organism is called a transgene. Organisms may have their genome modified by directed mutation (mutagenesis) or by newer technologies called **gene editing**. These methods may change the genes of an individual without introducing a new gene, so it would be genetically modified but not a transgenic organism.

TRANSGENIC ANIMALS

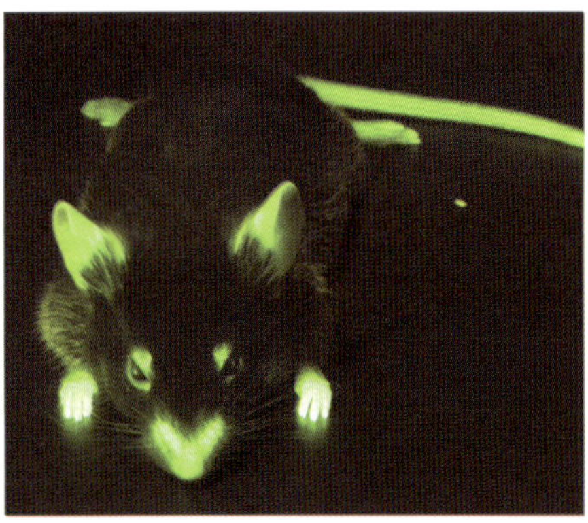

FIGURE 14.2.2 This transgenic mouse has been genetically modified so that it carries the gene for green fluorescent protein (GFP). This mouse will be used in scientific research. The GFP gene is linked to the gene being studied, and so acts as a reporter gene for the expression and location of the protein of interest.

Transgenic animals have been used in scientific research (Figure 14.2.2), medicine, pharmaceutical production and agriculture. In research, mice, rats and rabbits are used as living models of biological processes in healthy and disease states. Scientists can study disease progression and/or potential treatments. For example, scientists studying motor neuron disease have identified a mutation in the gene for the enzyme superoxide dismutase 1 (SOD1) that is associated with some inherited cases of this disease. To study the disease in an animal model, scientists produced a transgenic mouse expressing the SOD1 enzyme mutation to establish links between the mutation and disease symptoms, and investigate treatments.

Scientists may also produce knock-out mouse models in which gene technology is used to disable or knock-out expression of a particular gene. The phenotype (appearance, behaviour and biological function) of the animal gives the first indications of the function of the protein encoded by the gene. Structural and regulatory genes are studied in this way.

In agriculture, transgenic sheep and cows are used for improved fertility, meat production, milk quality and yield, and wool quality and yield. The use of genetically modified farm animals has not expanded to the extent it has for GM plants, perhaps because of detrimental effects of some modifications in animals. For example, genes that promote growth may also cause altered skeletal growth, arthritis, and heart and kidney problems.

To date, GM animals are not approved for human consumption in Australia. Transgenic fish have been approved in the USA (see the Biofile on page 551) but they are unlikely to be approved for sale in Australia in the near future.

Some farm animals are being used for the production of therapeutic proteins such as antibodies that are difficult to make in bacteria and cultured cells. This process has been referred to as 'pharming', combining the words farming and pharmaceutical. The products are released into blood or milk and they can be readily extracted from there.

Spider silk protein (Figure 14.2.3) is an example of a potentially useful product made in transgenic goats. The gene for spider 'dragline' silk has been put into the genome of goats, along with regulatory genes so that it is expressed in the milk. Spider silk is of great interest for its extraordinary strength and flexibility. Potential applications include a biopolymer for artificial ligaments and tendons, bandages, biodegradable bottles and tough bulletproof clothing.

Medical researchers are also exploring ways to modify genes in insect vectors of disease (see Biology in Action: 'GM mosquitoes', page 546) and modify pig embryos to make their organs suitable for transplantation into humans, a procedure called **xenotransplantation**.

FIGURES 14.2.3 Researchers are untangling the mystery of what makes spider silk so super strong. The key to the silk's strength, which exceeds that of steel, is its cross-linked beta-pleated sheet structure.

Genetically modified cells

Scientists regularly produce genetically modified cells and cell lines to study normal and abnormal cellular processes and to expand the range of cell-based therapies. One area of application is cancer immunotherapy. In Section 8.3 you learnt about the use of monoclonal antibodies for cancer immunotherapy. Not only are transgenic mice being used to produce fully human monoclonal antibodies, but genetically modified lymphocytes are also being tested. Recall that T lymphocytes use their specific T cell receptor to recognise and attack cancer cells. Scientists at the Peter MacCallum Cancer Centre in Melbourne are investigating ways to boost the number of specific T cells in cancer patients. T cells are removed from the patient and genetically modified with genes coding for receptors that make the T cells better able to target the cancer cells. The genetically modified T cells are grown *in vitro* and then returned to the patient to attack the cancer (Figure 14.2.4).

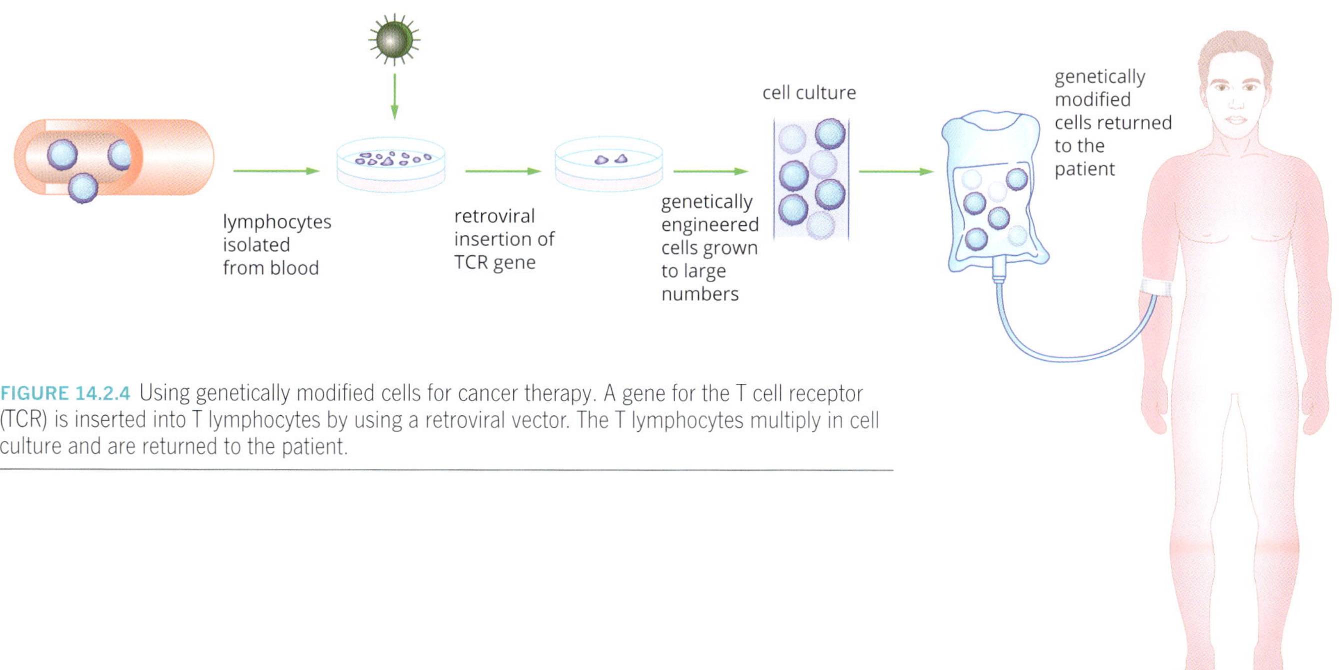

FIGURE 14.2.4 Using genetically modified cells for cancer therapy. A gene for the T cell receptor (TCR) is inserted into T lymphocytes by using a retroviral vector. The T lymphocytes multiply in cell culture and are returned to the patient.

BIOLOGY IN ACTION

GM mosquitoes

Genetically modified mosquitoes are being used for disease control. Some mosquito species are vectors of disease-causing viruses or protozoa. Mosquitoes of the *Aedes* genus are vectors for several disease-causing viruses, including the yellow fever virus, the dengue virus and the Zika virus. A biotechnology company has developed genetically modified mosquitoes that carry a dominant lethal gene for the purpose of reducing the population of mosquitoes carrying the viruses.

Males carrying the lethal gene are released into the wild, where they mate with normal 'wild type' females and pass the lethal gene on to their offspring. The offspring die as larvae. The DNA used to make the genetically modified mosquitoes also has a reporter gene for red fluorescent protein, enabling scientists to easily identify the adults and larvae carrying the lethal gene (Figure 14.2.5).

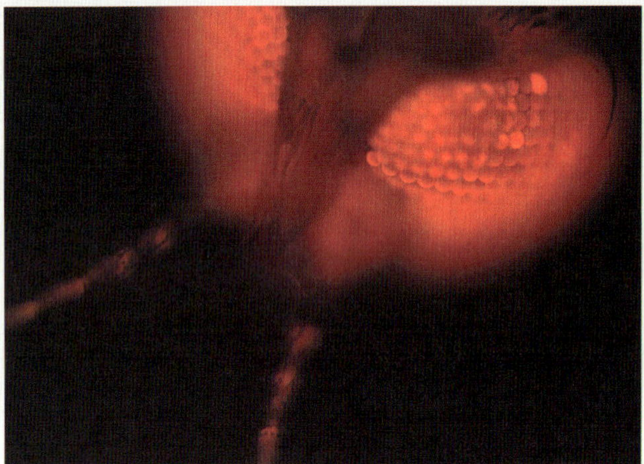

FIGURE 14.2.5 The gene for red fluorescent protein has been linked to a lethal gene to enable researchers to identify genetically modified mosquitoes. These insects are released into the wild to transmit the lethal gene to their offspring for control of virus-carrying mosquito populations.

Field trials using genetically modified *Aedes aegypti*, the vector of dengue virus, have been conducted in the Cayman Islands, Brazil and Panama where dengue fever is a widespread and serious health problem. The recent outbreaks and rapid spread of Zika virus, which is also transmitted by *Aedes* mosquitoes, has prompted trials of these genetically modified mosquitoes in areas of Brazil affected by outbreaks of Zika virus.

Another approach to mosquito control is the release of sterile insects. Mosquitoes of the *Anopheles* genus transmit the malaria parasite *Plasmodium*. Male *Anopheles* mosquitoes have been genetically modified with genes expressed in the testes that cause the males to be unable to make sperm, so they are sterile. Female *Anopheles* mosquitoes mate only once, so mating with a sterile male limits population growth. The genes causing the sterility are linked to a reporter gene for green fluorescent protein for easy identification of the genetically modified insects (Figure 14.2.6). The aim of the research is to reduce the populations of mosquitoes that carry and transmit *Plasmodium* sp. and thus reduce the incidence of malaria, a serious health problem in many developing countries.

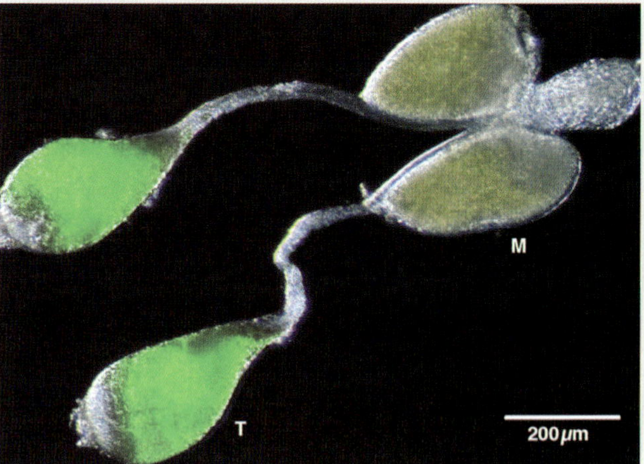

FIGURE 14.2.6 The internal reproductive organs of a genetically modified male *Anopheles gambiae* mosquito. The testes (T), where sperm cells develop, are fluorescent green due to the expression of a green fluorescent protein (GFP) that is linked to the genes causing sterility. The male accessory glands (M) that produce seminal secretions are not expressing GFP.

This genetic approach to controlling insect vector populations specifically targets the particular vector species. It has advantages over the large-scale use of insecticide, which impacts ecosystems by killing many other insect species. Disadvantages include the development and implementation costs of these highly specialised methods. Thorough risk assessments and compliance with regulations concerning the release of biological controls and genetically modified organisms must be completed in many countries.

TRANSGENIC CROPS

Transgenic crops are used in agriculture to increase crop productivity, provide resistance to insect predation and prevent disease. Several genetically modified crops have been developed or are grown in Australia. For example, insect-resistant GM cotton has been grown since 1996, and herbicide-tolerant GM canola was approved for commercial production in Victoria in 2008. In Australia, the Office of the Gene Technology Regulator (OGTR) assesses all GM animals or plants before research, agricultural and commercial use.

Techniques for producing transgenic plants

Transferring a gene into plant cells can be limited by the presence of the cell wall. The introduction of foreign genes into plants is usually done by using a biological vector. One method utilises *Agrobacterium tumefaciens*, a soil bacterium that is able to naturally transfer a plasmid into plant cells (Figure 14.2.7). *Agrobacterium* normally causes crown gall disease because it carries a plasmid with genes that cause the growth of a tumour. A recombinant plasmid (the vector), carrying a desired gene from a different species but lacking the tumour-inducing genes, is introduced into *Agrobacterium* cells. When the transformed *Agrobacterium* is cultured with plant cells, the recombinant plasmid is transferred into the plant cells. These transformed plant cells are then grown in tissue culture into new plants for transplanting into the field as a transgenic crop.

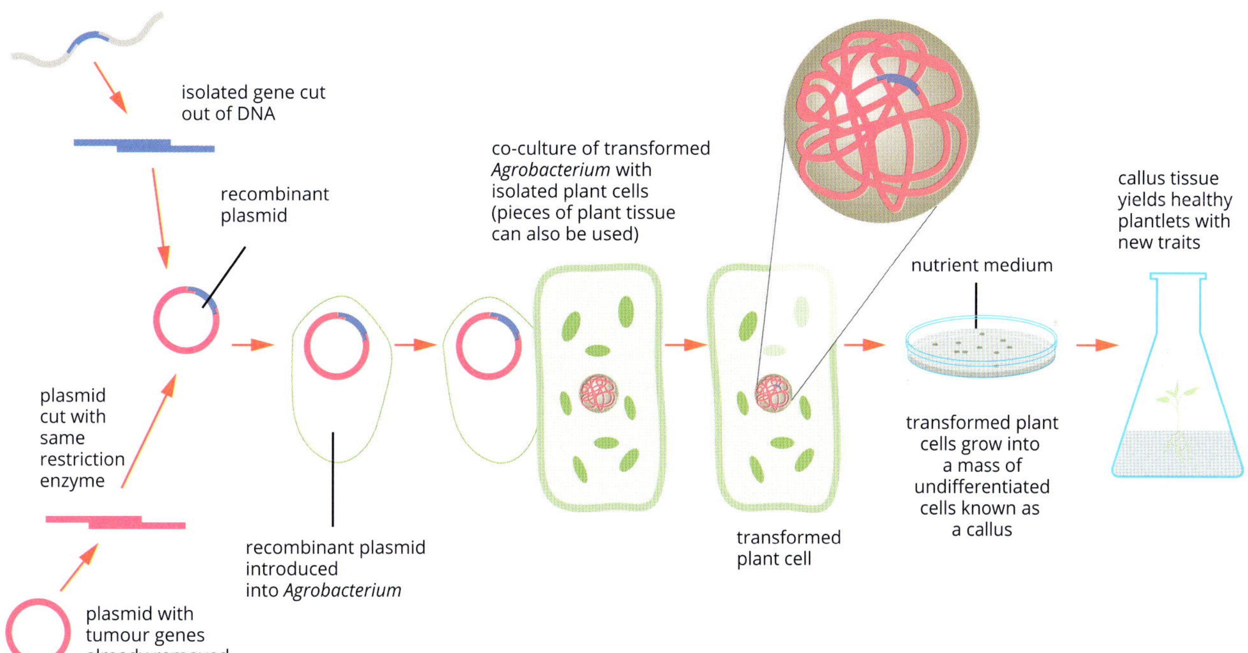

FIGURE 14.2.7 The use of *Agrobacterium tumefaciens* in gene cloning and the production of genetically modified and transgenic plants. Relative sizes of the plasmid, bacterium and plant cell are not to scale.

Salt-tolerant wheat

Soil salinity is a major problem for Australian agriculture. A high level of sodium salts in the soil leads to osmotic water loss from roots and other tissues in which salt accumulates. Cells are stressed due to the altered ratios of sodium and potassium ions in cells. Salt-tolerant plants protect themselves from the effects of salinity by preventing sodium entry into cells, storing the salt in the vacuole, or pumping the sodium out of the cells. Molecular biologists have found the genes that control these features of salt-tolerant plants.

FIGURE 14.2.8 Australian scientists have produced wheat plants that can grow in salty soil.

To increase crop productivity, Australian scientists from the University of Adelaide introduced a gene from a salt-tolerant Australian native plant into wheat plants. This greatly improved the grain yield of wheat grown on salty soils without affecting grain yield in normal soil (Figure 14.2.8). The salt-tolerant gene codes for a protein that removes sodium from the leaves, allowing water to move normally from the roots to the leaves. This increases the geographical range that can be used for wheat production in Australia and other countries facing salinity problems, which is becoming increasingly important as the global population grows.

Bt cotton

Cotton is a plant that attracts many insect pests. To protect the cotton crops, they are sprayed with insecticides up to four times before the crop is harvested. This high use of insecticides impacts the populations of both harmful and beneficial insects, and of the animals that feed on them. Insecticides may also have an impact on human health. In addition, insecticides are expensive.

Bt cotton is a transgenic crop that has been modified to contain two genes from the soil bacterium *Bacillus thuringiensis*. Expression of these genes produces proteins in the cotton plant that kill the main caterpillar pest of cotton by disrupting its digestive system. In Australia, almost all cotton grown is Bt cotton and this has reduced the use of pesticides dramatically. This decreases the environmental impacts of pesticides and saves the farmers many dollars. Australian regulators have reported no adverse effects over 15 years of Bt cotton use in Australia. Cotton seed oil extracted from Bt cotton can be sold without GM labelling as the extraction processes separate the oil from the plant's proteins and nucleic acids, therefore the oil does not have any GM components.

BIOFILE

Editing genes

Another way to genetically modify an organism is to directly edit the genes in the organism's cells. New gene editing technologies, especially a method called CRISPR, have become popular. The method uses a bacterial enzyme called Cas9 with a guide RNA that is made to match the gene sequence to be edited. When the guide RNA finds its target gene, the enzyme cuts the DNA so that the gene is disabled, cut out or replaced with another gene.

The technique is proving to be a very accurate, cheap and easy tool and has been used to edit cancer genes and explored as a way to remove HIV genes from HIV/AIDS patients. It has the potential for gene therapy to replace faulty genes, and as a method for making genetically modified crops without the need for plasmids. Scientists from around the world held a summit in 2015 to discuss the safety, applications, social impacts and ethics of using the technique, as concerns were raised by reports of gene editing in human embryos.

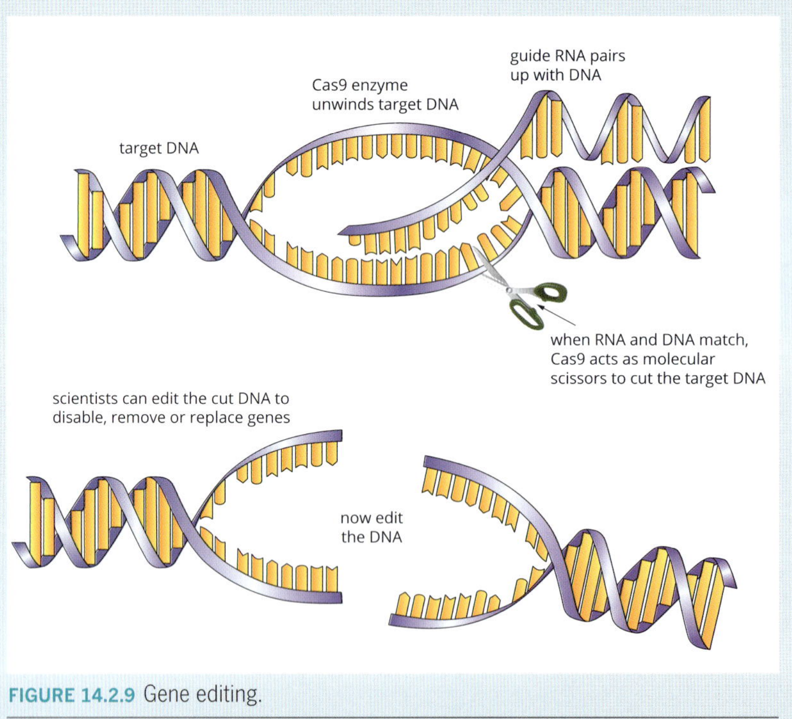

FIGURE 14.2.9 Gene editing.

Golden rice

Vitamin A is a coenzyme that is essential for healthy eyes and the immune system. Diets poor in vitamin A contribute to the deaths of many children in developing countries each year. Vitamin A deficiency can lead to preventable blindness in children and increases the risk of common infections such as measles and diarrhoea.

Rice is a staple food for millions of people around the world. Traditional white rice varieties grown in countries such as Vietnam, Bangladesh and the Philippines are low in vitamin A. Golden rice is a transgenic rice crop that contains pro-vitamin A (Figure 14.2.10). Golden rice is produced when two plant genes and one bacterial gene are inserted into the white rice genome. These genes switch on a biochemical pathway in the rice plant that sends vitamin A to the rice grains rather than the leaves, as would occur in non-modified rice.

Australian researchers are exploring other genetic modifications to improve the ability of rice plants to use nitrogen for improved growth and reduce the need for nitrogen fertiliser. They also aim to make rice grains richer in elements such as iron, zinc and phosphorus, a process called biofortification.

FIGURE 14.2.10 Golden rice is yellow due to the presence of pro-vitamin A, also known as beta-carotene.

ISSUES RELATED TO GENETICALLY MODIFIED ORGANISMS

GMOs are controversial for a number of reasons. Debates surrounding their biological, social and ethical implications are common in the scientific and general media. Some of the issues surrounding GMOs are discussed below.

Biological implications

A number of potential problems have been raised:

- Biological impact on the health of the organism carrying a foreign gene: Crops approved for commercial growth appear relatively unaffected by a transgene (apart from the new trait). Transgenic animals may experience more adverse effects of transgenes that affect growth rates.
- Danger of altering own DNA by eating GMOs: One argument against GMOs is the claim that eating these foods will change our own DNA in unpredictable ways. There is no reason to think this is the case as animals have been eating the DNA of plants and other animals for millions of years and this DNA does not result in any change to the core genome of the consumer.
- Uncontrollable pest plant species: Crops that have been modified for herbicide or insect resistance may breed with other plants, producing a hybridised pest species that farmers may not be able to control. Plants do not usually pollinate different species, so this is only a potential concern for cross-pollination with wild relatives, such as between crop canola and wild canola. Uncontrollable growth would be a concern if the genes code for rapid growth and there was a selective advantage to having the genes in a wild environment.
- Cross pollination between GM crops and non-GM crops: Pollen distributed by wind and insects can transfer genes between neighbouring plots of the same species. Cross pollination may happen in both directions and so will have an impact on both types of crop.
- Genetically altered animals may escape and interbreed or compete with natural populations with unforeseen consequences. This may be a potential concern if there are wild relatives of the GM animals (animals usually don't breed with unrelated species) and there is a selective advantage to the presence of the transgene in a wild population; for example, GM salmon escaping from aquaculture enclosures adjacent to the ocean and breeding with wild salmon.
- Loss of biodiversity: As more farmers use genetically modified crops or animals, there will be fewer crops grown that vary from these, reducing the genetic pool of some species. If disease or environmental change occurs, there could be widespread and catastrophic effects on food production. However, this is already an issue in modern agriculture where monocultures and commercial practices have reduced genetic diversity. GM technology may enhance the trend or it could be used to re-introduce characteristics lost from wild relatives.
- Most of the biological concerns apply equally to livestock and crops that have been selected and bred by traditional breeding methods as wells as by GM technology.

Social implications

- Solve malnutrition and hunger: GM crops could improve the nutrition and yield of foods already grown by farmers. This could lead to improved health and result in greater availability of food to some of the poorest people in the world who, at the same time, would continue to use their traditional and well-established farming practices and therefore maintain their livelihoods. Although this is good in theory, some people suggest that the best way to end hunger and malnourishment is a political solution to inequitable global food distribution rather than reliance on any agricultural technology, including GMOs.
- Create more social inequality: Many of these technologies are expensive but also have the potential to bring about great benefits. Who should be able to benefit from these technologies and could limiting access through patents and pricing bring about greater social inequalities? These concerns do not appear to have limited the uptake of GM technologies, and many developing countries have embraced GM food and fibre crops while some developed countries oppose their introduction.
- Consumer choice, including the rights of the individual to choose to eat GM crops: In many countries, but not all of them, manufacturers are required by law to label whether GMOs have been used in the product.
- Do the biotechnology companies that create the GMOs provide equal access to the technology and at an affordable price for farmers?
- The interplay between environmental impacts of GMO use and social consequences: For example, reduced pesticide use saves money, which may be invested in education; decreased disease load due to GM-insect vector control improves the health of the whole community, enabling greater investment in economic, educational and social development.
- Changes to business practices: For example, changes in herbicide and insecticide usage affect sales, labour requirements and employment.

Ethical implications

As stated earlier, genetically modified organisms, including transgenic crops, violate the ethical, philosophical and religious principles of some people.

Some of the ethical concerns identified are listed below.

- Patents:
  - If companies control the rights to the genome of GM crops, they also control the prices of the seeds. Farmers have to spend money each year to buy seeds for their crops. For example, the BT cotton seeds lose their efficacy after one generation and the farmer is therefore forced to purchase new seed each year. This may limit the farmer's ability to make a living.
  - Cross contamination of a GM crop with a non-GM crop occurs naturally as pollen and seeds are carried by wind or animals. There is the potential for companies owning patents for GMOs to sue farmers whose crops become contaminated by non-GMOs for breaching commercial agreements. The opposite can also happen when organic crops become contaminated with GM crops and the farmer loses the ability to be certified as organic.
- Animal rights:
  - Some people believe that producing transgenic animals violates the fundamental rights of an animal.
  - Some consider that using transgenic animals as a source of human proteins or organs is immoral and against animal rights. Animals are already a source of food for humans, so the judgement to be made is whether adding a gene changes these issues.

- Some people consider that introducing human genes into animals, for example to produce pharmaceuticals, makes the animal closer to humans and deserving of the rights and protections given to humans. A question to consider is whether one gene can change the identity of an organism.
- Animals produced through genetic modification may have characteristics that are seen as beneficial to humans and agricultural productivity but may in fact be detrimental to the organism's welfare. For example, some transgenic pigs grow very quickly and this affects their heart and joints, and they cannot be as active as normal pigs.

• Intervention in evolution:
- GM technology expands our ability to interfere with and override normal evolutionary processes.

Figure 14.2.11 summarises the social, biological and ethical implications arising out of the use of genetically modified and transgenic organisms.

social implications	• increased food supply, nutritional content and food quality • expanded range for growth of agricultural species • access to the technology; social equality/inequality • labelling and consumer choice • patents and pricing; control of access by biotechnology companies • costs to farmers
biological implications	• safety of consuming GMOs • cross-pollination between GM plants and wild plants • cross-pollination between GM and non-GM crops • viability of transgenic organisms in the wild • health of GMOs • genetic variation in agriculture
ethical implications	• violation of animal rights • human self-interest overrides ethical treatment of other organisms • intervention in evolutionary process

FIGURE 14.2.11 Summary of issues arising from the use of genetically modified and transgenic organisms.

BIOFILE

Genetically modified salmon

Recently the US government cleared the way for genetically modified Atlantic salmon to be used for human consumption. A gene from another salmon species, along with a promoter sequence from a fish called a pout, means that the transgenic salmon eat all through the year, not only when the water temperature is warm. This increases the growth rates of these fish dramatically and means they are ready for harvest much sooner than non-modified Atlantic salmon. The eggs of the GM salmon are treated to create infertile adult fish (99% of the adults are reported to be sterile), thus reducing the chances of interbreeding with wild salmon if they escape from their pens. This will be the first genetically modified animal of any type to be cleared for human consumption in the USA.

FIGURE 14.2.12 Transgenic salmon cleared for human consumption.

14.2 Review

SUMMARY

- Genetically modified organisms are organisms with modifications made to one or more genes.
- Transgenic organisms carry a gene from a different organism.
- Genetically modified animals are used in research, disease control, medicine and biomolecule production.
- *Agrobacterium tumefaciens* and plasmid transfer is a well established method of transferring genes into plant cells.
- Transgenic plants are used in agriculture, providing varieties that resist insect attack, are herbicide resistant or have improved yield or nutritional content.
- A range of social, ethical and biological issues arise from the application of genetically modified organisms.

KEY QUESTIONS

1 **a** What does 'genetic modification' of an organism mean? Include an example in your answer.
  b Describe a successful application of genetic modification in agriculture in recent years.

2 Describe a genetically modified organism and compare it to a transgenic organism.

3 Which Australian regulatory body oversees the development, use and commercial or medical introduction of genetically modified organisms?

4 Research the kinds of foods that are commonly genetically modified. The internet will be a useful resource. Compile an inventory of your pantry and/or of supermarket shelf items to find which pre-packaged food products contain GM foods.

5 Describe how genetic modification can be useful as a tool to fight vector-borne disease, such as a disease carried and transmitted by an insect vector.

6 Give an example of how genetic modification may be used to change a characteristic in:
  a an animal
  b a crop plant.
  List any potential benefits or disadvantages to the organism, the environment and/or to society.

7 The following diagram illustrates two model organisms used in research and the molecular procedures being used to alter a genetic characteristic. State whether the resulting organism is genetically modified, transgenic or both.

a

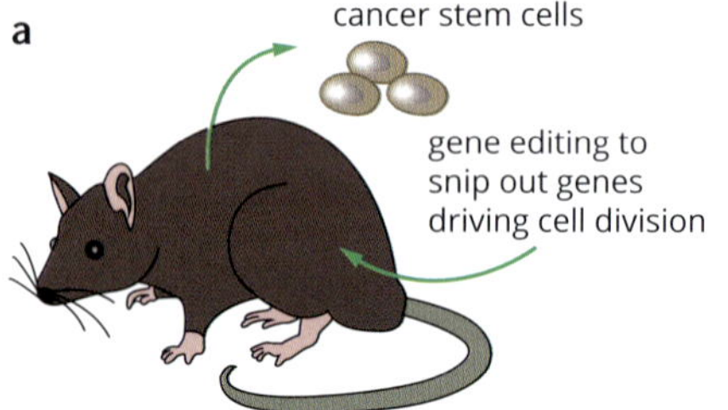

b

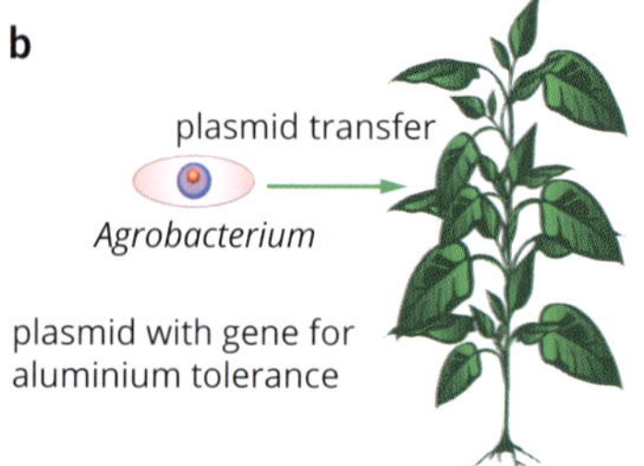

8 In your notebook draw up a table (like the table below) to identify what you consider to be the key issues surrounding the use of genetically modified organisms. List the points in categories that you consider to be positive or negative aspects of the technology.

	Pros/Positives	Cons/Negatives
social		
biological		
ethical		

9 From the cases described in this section, identify one example of:
  a a genetic modification that leads to a reduced environmental impact
  b a modification that has potential commercial opportunity for increased production or a new product
  c a modification that impacts on both the viability of an animal and the benefit to human health.

14.3 Strategies to deal with emerging new diseases

There have been a number of significant diseases that have plagued the world over many centuries (Figure 14.3.1). As world populations become larger and humans come into closer contact with other animals, new diseases have emerged. More recently, with rapid transport between countries and bigger cities bringing people into closer contact, new and rapidly spreading diseases have proved globally challenging, and strategies have been implemented to prevent the spread of these fatal diseases. Controlling these diseases will require a global effort.

The study and surveillance of newly emerging and re-emerging diseases aims to find ways to predict, prevent and respond to outbreaks of disease. Many of the diseases of local and global importance are infectious diseases. **Emerging infectious diseases** may be defined as:

- new or previously unrecognised diseases
- diseases that have increased in incidence, **virulence** (the ability of a pathogen to cause disease) or geographic range over the past 20 years
- diseases that may increase in the near future.

Organisations that monitor global disease, such as the US Centers for Disease Control and Prevention and the World Health Organization, generate priority lists of emerging diseases to direct global surveillance and collection of data. They work on early detection, containment (preventing the spread to other areas) and early treatment. As an island nation, Australia is highly susceptible to the introduction of new disease-causing agents. Australia has a strong emphasis on **biosecurity** to prevent, respond to, and recover rapidly from pests and diseases, to help secure our agriculture, biodiversity and human health.

Although scientific knowledge has increased over the decades, allowing us to identify, diagnose and treat more diseases, other factors have led to new infectious diseases becoming a real challenge for governments and medical personnel around the world.

FIGURE 14.3.1 During the bubonic plagues, so many people died in Europe and elsewhere that bodies were placed into mass graves.

THE EMERGENCE OF NEW DISEASES

Infectious diseases can be caused by viruses, bacteria, parasites or other pathogens. New diseases can emerge due to mutations in a pathogen. They may also emerge as pathogens adapt to a new host under different selection pressures or circumstances. Mutations may occur when a pathogen moves from one host to another and the new mutation may increase the ability of the pathogen to move into a wider range of host organisms, including humans (Table 14.3.1). When a disease passes from another animal to a human host it is known as a **zoonotic** disease. Many of the emerging and re-emerging diseases are zoonotic infections. The World Health Organization and experts in the field expect that the next pandemic is likely to be a zoonotic disease.

New or previously unrecognised diseases	• HIV/AIDS: other primate → human • SARS: bat → human • MERS: camel → human • Hendra virus: bat → horse → human • Zika virus: mosquito → human • vCJD prion BSE: cattle → human
Increased in incidence, virulence or range over past 20 years	• Ebola: bat → human • dengue virus • West Nile virus • cholera • MRSA • *Clostridium difficile*
May increase in the near future	• influenza • antibiotic-resistant bacteria • cholera • dengue virus • prion diseases • non-infectious diseases: diabetes, obesity, Alzheimer's

TABLE 14.3.1 Examples of emerging diseases of human populations. Some infectious diseases, such as dengue fever, may increase in incidence due to climate change. Some non-infectious diseases may increase due to lifestyle factors and demographic change.

An **epidemic** is the rapid spread of a disease to a large number of people. When the spread of the disease reaches global proportions, it is known as a **pandemic**. There are many factors that interplay and influence the emergence and spread of diseases. Some of these factors are:

- human demographics
- human behaviour
- changes in farming practices and food production
- uncontrolled or inappropriate use of antimicrobials
- lack of sanitation and poor hygiene.

An epidemic is the sudden increase in the number of cases of a disease above what is normally expected in that population in that area.

A pandemic is an epidemic that has spread over several countries or continents, usually affecting a large number of people.

Human demographics

Over the centuries, there has always been movement of populations either due to war or for socio-economic reasons. In any migration, there is the potential to introduce pathogens to a new area. Pathogens that were contained in a small area may now be exposed to new hosts, who might not have the same immunity against them. Today, with people moving between continents in short periods of time, emergent diseases have the potential to spread rapidly around the globe.

Bubonic plague

The bubonic plague historically caused devastation to human populations and remains a re-emerging disease in several parts of the world. Sometimes simply called the plague, it is caused by the bacterium *Yersinia pestis*. This bacterium infects rodents and is transmitted from rodent to rodent by fleas. It is transmitted to humans when a flea carrying the bacterium bites a susceptible person. Once bacteria reach the lungs, they become airborne and highly contagious (Figure 14.3.2). Symptoms appear 7–10 days after infection. The first pandemic plague recorded occurred in the 6th century, and is believed to have been brought to Europe from Africa by the fleas on rats in trade ships. There have been several epidemics and pandemics of the plague throughout the centuries, which have claimed the lives of millions of people.

The incidence of plague has dramatically declined, largely due to improved living standards. However it continues to appear in several countries, mainly in Africa (Figure 14.3.3). The bacterium resides in wild rodent populations and can transmit the pathogen to rodents living in human communities. Australia is fortunate to be the only continent that does not have infected rodent populations. Prevention and control measures are put in place where affected rodent populations are identified, and early detection and treatment of people exposed to the bacterium enables rapid recovery. Australia's strong quarantine regulations help to keep *Yersinia pestis* infections out of the country.

FIGURE 14.3.2 (a) A mask used by 16th-century Venetian doctors to protect themselves against the highly contagious bubonic plague. The beak was filled with herbs to prevent the 'bad' air from being inhaled. (b) An illustration demonstrating the full outfit worn by doctors attending to patients suffering from the bubonic plague. The mask and outfit acted as protective wear so the doctor too did not become infected.

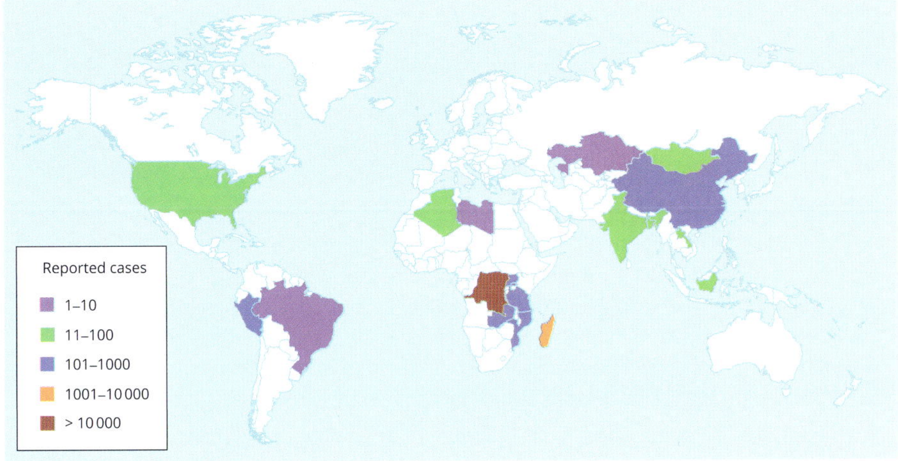

FIGURE 14.3.3 World Health Organization data on plague cases for the years 2000–2009.

Influenza

From 1918 to 1920, approximately 50 million people were killed worldwide during an influenza (flu) pandemic. The influenza virus was highly contagious and deadly, even to healthy young adults. The disease was spread by air as the pathogen was breathed or coughed out of the infected person's lungs. World War I was occurring during the same period and infected soldiers spread the disease throughout Europe, the USSR and the USA in three deadly waves. It was dubbed the Spanish flu, although it did not originate in Spain; the origins of the flu are still debated.

Flu virus circulates in populations of migratory birds, domestic poultry, pigs and humans. In humans, a new variant emerges almost every year and infects large numbers of people; this is called seasonal flu. New strains evolve from the genetic mixing or reassortment of genes. This occurs when two or more different viruses infect the same human or other host animal (see Biology in Action: Influenza vaccines, Chapter 8, page 303). New strains resulting from reassortment have new combinations of the surface antigens **haemagglutinin** (H) and **neuraminidase** (N), and may be more virulent. New strains arising in birds or pigs that gain the ability to infect human cells are a high risk for pandemic flu.

FIGURE 14.3.4 Influenza virus infects lung tissue. It is spread in water droplets released through coughing and sneezing.

Seasonal flu is often the H3N2 type of virus. The recent 2009 pandemic, an H1N1 type, was also called 'swine' flu because it originated in pigs. It killed about 18 500 people worldwide, mostly in North America.

Strategies for preventing the evolution of new pandemic flu strains include regular gene sequencing of human, bird and pig viruses to identify problem variants, the culling of animals that are infected with dangerous new strains, and limiting the close interaction between humans and their animals. Strategies to contain the virus include hygiene, wearing of face masks to prevent the spread of airborne viral particles (Figure 14.3.4), vaccination and antiviral medication. Vaccination before each flu season limits the number of infected people. When fewer people and other animals are infected with influenza virus, there is a reduced chance of genetic reassortment giving rise to dangerous new strains.

The Victorian Infectious Disease Reference Laboratory in Melbourne is part of the World Health Organization Global Influenza Surveillance and Response System. It analyses influenza viruses circulating in the human population to help determine which viral strains should be included in the flu vaccine each year.

BIOFILE

Reconstructing the Spanish flu virus

In an attempt to understand more about the Spanish flu virus and why it was so deadly, the body of a person killed by the virus was exhumed from permafrost in Alaska. The Spanish flu had killed the majority of the people living in the village at the time. Genome sequencing of the virus was completed in 2005 and it was identified as H1N1 influenza A, originating in birds. Researchers extracted the viral RNA, grew active virus in cultured cells, and infected mice and monkeys with the reconstructed virus (Figure 14.3.5). The virus was highly virulent, killing the first experimental mice in 6 days.

In a normal response to the flu, the immune system wanes over a period of time. However, when monkeys were infected with the reactivated Spanish flu virus, their immune systems went into overdrive and did not switch off. Large quantities of cytokines were produced, resulting in uncontrolled inflammation. Similar responses were observed in the H5N1 avian flu, an epidemic that appeared in Asia in 1997. Researchers are using this information to investigate possible treatments.

The reactivated virus may help to identify the mutations that produce a virulent virus so that they can be identified in new variants of seasonal flu, and help prevent another pandemic.

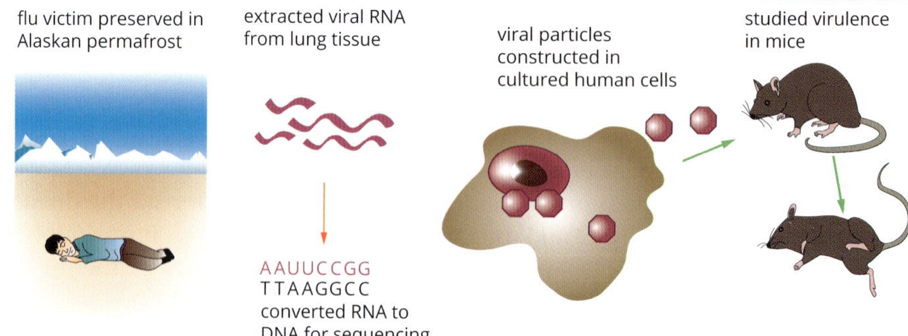

FIGURE 14.3.5 How the Spanish flu virus was reconstructed in the laboratory and tested in animal models.

Human behaviour

Human behaviour also has an impact on the emergence and re-emergence of diseases, including sexually transmitted diseases.

HIV

Human immunodeficiency virus (HIV) is the virus that causes acquired immune deficiency syndrome (AIDS). HIV was first recognised as a new disease in 1981, when five young men in Los Angeles presented with a rare case of lung infection and a range of other infections. The men's immune systems had shut down due to the virus infecting and destroying helper T lymphocytes (Figure 14.3.6).

HIV appears to have a zoonotic origin, as a genetically similar virus is found in some African monkeys. Researchers suggest that there were two independent transmissions from animal to human. These transmissions might have occurred between pet monkeys and their owners or following the slaughter of a monkey by a person, perhaps for bush meat. In either case, the bodily fluids of the monkeys would have infected the humans.

HIV is spread through contact with infected bodily fluids and can occur through sexual contact, using contaminated hypodermic needles, and by blood transfusions and organ transplants. Some people present with flu-like symptoms within two weeks; in others, the virus incubation period lasts many years and they show no symptoms. Therefore, a person may not be aware that they are infected if they are not tested.

From the initial epidemic in the 1980s, people began to modify their behaviour, including following safe sex practices to reduce the incidence of transmission. However, a couple of decades later, the numbers of infections are once again climbing. HIV has infected 78 million people since the pandemic began and has killed 39 million worldwide. A suite of antiviral drugs is available for keeping HIV infection under control. So, unlike in the early days of the pandemic, people with access to medications may live for many years with the infection. However, social, political and economic factors contribute to unequal access to these drugs globally.

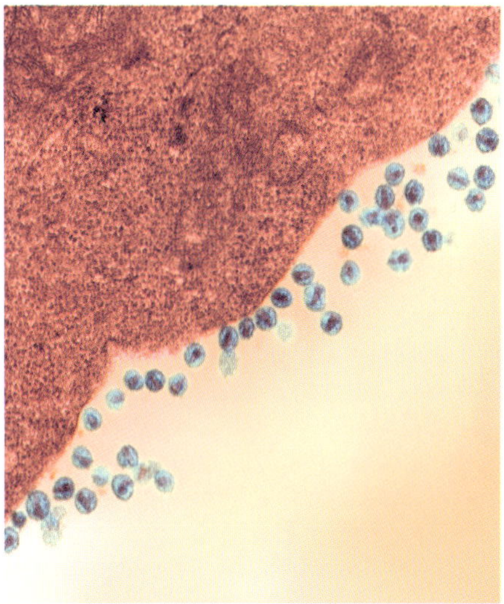

FIGURE 14.3.6 HIV infecting a cell. Transmission electron micrograph of a section showing human immunodeficiency virus (HIV) particles (virions, round) on the surface of a cultured cell. HIV attacks helper T lymphocytes (specialised white blood cells), which are crucial in the body's immune system.

Kuru

Kuru is a disease of the nervous system caused by prion proteins that cause normal proteins to change shape and form abnormal protein clusters in the brain, impeding normal function and resulting in the death of neurons (Figure 14.3.7). These changes are similar to those of bovine spongiform encephalopathy (BSE) affecting cows, scrapie affecting sheep and Creutzfeldt–Jakob disease (CJD), another human prion disease. Death occurs approximately one year after the onset of symptoms, which include coordination problems, tremors and pain.

An epidemic of Kuru began in the 1920s, having emerged in a tribe in Papua New Guinea due to a cannibalistic practice they carried out. During funeral rites, the women would eat the brain of the deceased. The infectious prion protein in the brain was then transmitted to a new host. During the 1950s the ritual was banned, leading to a decline in the disease. However, the incubation period can be as long as 50 years and cases continue to be reported.

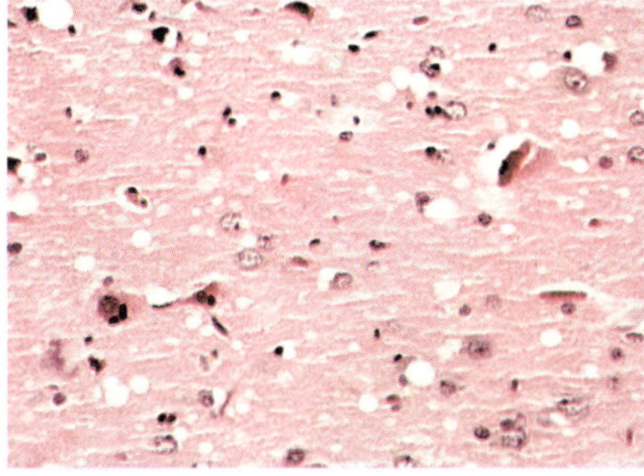

FIGURE 14.3.7 A slice of brain tissue showing the spongy appearance of spongiform encephalopathy disease such as CJD or Kuru. The white holes are gaps where neurons have died.

Changes in farming practices and food production

Farming practices and food production have changed over the centuries to adapt to an increasing population and to globalisation. A greater demand for meat has created pressures on the farming industry. Throughout the world, dense farming practices have developed and this has occurred in close proximity to human populations. This has allowed diseases in animals to transfer to humans more easily. These new diseases can have severe effects as the population has no previous resistance to them.

BIOFILE

The Fore People and cannibalism

The Fore People of Papua New Guinea (Figure 14.3.8) were once known for their cannibalism. A debilitating disease, Kuru, was spread through communities when they ate the brains of affected individuals during funeral rituals, thus transmitting the infectious prions.

In most people, the prions form fibrils in the brain, interrupting its normal function and causing neuron death. Recent research has discovered that some members of the Fore People are resistant to Kuru. Molecular biologists tracked this down to mutations in the prion gene. Transgenic mice carrying this mutated prion gene were completely protected from disease when infected with prions. This is a beneficial mutation.

There are now hopes that further studies using these mice will aid in understanding the progression of sporadic prion diseases such as Creutzfeldt–Jakob disease and the new variant, vCJD, which is acquired by eating beef from cattle affected by prion disease (see Mad cow disease, right).

The appearance and spread of the beneficial prion mutations in the Fore population is an example of evolution in action.

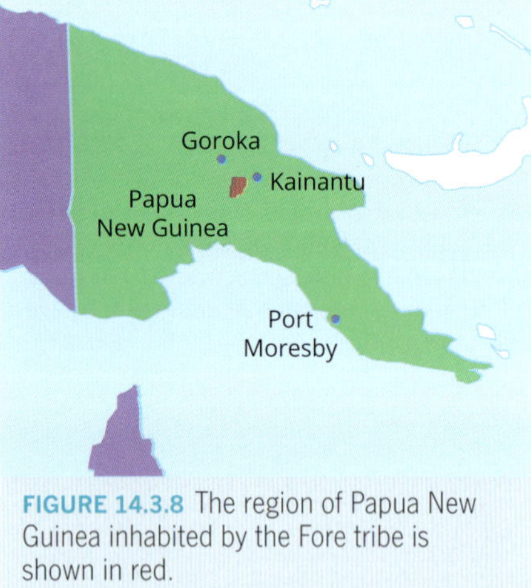

FIGURE 14.3.8 The region of Papua New Guinea inhabited by the Fore tribe is shown in red.

Mad cow disease

Bovine spongiform encephalopathy (BSE) or mad cow disease is caused by an infectious prion protein and is a fatal neurological disease in cows. The first documented case of BSE in the UK was recorded in 1986 and the incidence grew rapidly. At the height of the epidemic in 1993, up to 1000 new cases were reported weekly. Over 180 000 cattle were affected. BSE is caused by abnormal prion proteins that misfold; these prions then cause normal prion proteins to misfold and aggregate. This damages the brain tissue.

In the 1990s, a new human disease appeared, variant Creutzfeldt–Jakob disease (vCJD). It was acquired by eating products from BSE-infected cattle. Transmission occurred because the abnormal prion proteins are not denatured or destroyed when cooked, as occurs with most proteins.

A farming practice that contributed to the rapid and extensive spread of BSE in the UK was feeding meat and bone meal to farm animals. Meat and bone meal is the ground up remains of farm animals, and was used to increase the protein content of animal feed. It is possible that BSE started when this meal contained tissue from sheep with scrapie (a prion disease). Controlling the spread initially involved animal culls. The practice of feeding mammalian protein to cattle was banned and stringent slaughter practices, testing (Figure 14.3.9) and surveillance was introduced. Australia's thorough screening and tight importation regulations on animal products has kept the country free of BSE.

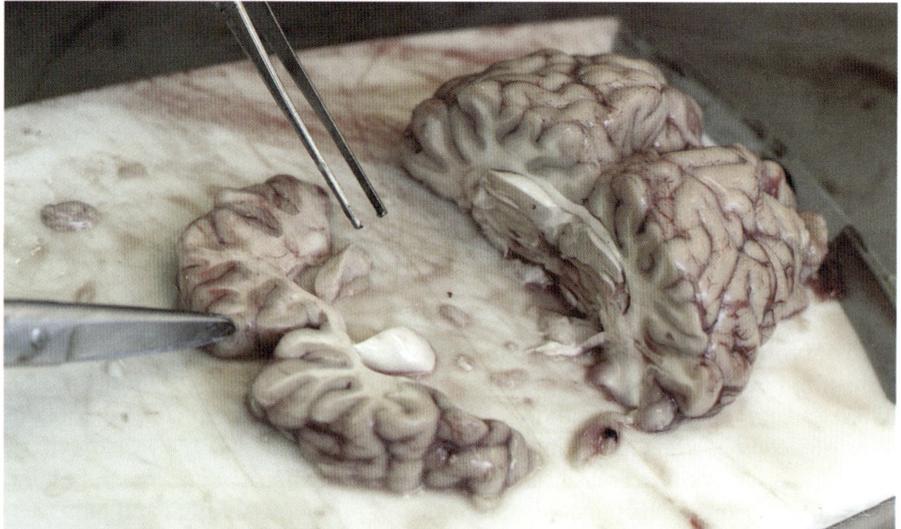

FIGURE 14.3.9 A vet slicing a cow brain to look for signs of BSE and other diseases. Tissue showing signs of damage must not enter the food chain.

Uncontrolled use of antimicrobials

The overuse and mismanagement of **antimicrobial drugs** against pathogens such as bacteria, viruses and parasites has led to a global issue that has the potential to lead to a major crisis. Although **antimicrobial resistance** is a natural phenomenon, it has been made worse by excessive use of antimicrobial drugs, particularly in medicine and agriculture. This has resulted in the emergence of some drug-resistance in pathogens. Resistance in pathogens can develop in several ways and two examples are given.

Antibiotic-resistant pathogens ('super bugs')

Antibiotics act against bacteria but are not effective against viruses. Antibiotics target specific biological pathways that disrupt cellular function and kill or slow the growth of the bacteria. There is the chance that a population of bacteria will become resistant to that particular antibiotic when treated with it. This can occur because some bacteria in a population already carry resistance genes; they are not killed by the antibiotic and pass on the resistance genes. Alternatively, new spontaneous mutations may occur in the bacterial DNA, making the bacteria resistant to the antibiotic. These bacteria are then able to multiply rapidly without competition from sensitive strains, and produce cloned offspring that carry the resistance gene. Bacterial cells also exchange plasmid DNA between each other in a process known as **horizontal gene transfer**. Plasmids carrying a drug resistance gene can be acquired in this way.

Transmission of resistant bacteria can be indirect, through food and water, or through direct contact with an infected person or animal. Methicillin-resistant *Staphylococcus aureus* (MRSA) and vancomycin-resistant *Enterococcus* (VRE) are examples of bacteria that are now resistant to top-line antibiotics, making them almost impossible to treat. They are frequently acquired in hospitals.

Antimicrobial resistance is a natural phenomenon. Microorganisms compete with each other for space and resources. Fungi and some bacteria produce antimicrobial compounds that damage other microbes. For example, the *Penicillium* fungus releases penicillin, and the soil bacterium *Bacillus subtilis* produces the antibiotic bacitracin. In response, microorganisms have evolved mechanisms to resist these molecular attacks. Bacterial antibiotic resistance genes are often located on plasmids, which are readily transferred between bacteria.

Malaria

Malaria is a disease that has been present since ancient times. Despite repeated and ongoing attempts to eliminate the parasite and its vector, malaria continues to emerge and is likely to spread as the climate warms. Approximately half the world's population is at risk of malaria, and there were 214 million cases in 2015. The parasite *Plasmodium falciparum* has developed an intricate relationship with two hosts, the *Anopheles* mosquito and primates, including humans. The parasite is carried by the female mosquito and transmitted to humans when they are bitten by an infected mosquito. The disease may also be passed between humans during blood transfusions. In humans, the parasite replicates inside red blood cells, causing them to burst. Symptoms include fever, vomiting and headaches.

The antimalarial drug chloroquine was discovered in 1934 and used in the 1940s as it was a safe and highly effective drug. However, in the late 1950s chloroquine-resistant parasites emerged in Asia. A mine along the border of Thailand and Cambodia attracted many workers from neighbouring regions. As well as the environmental changes that occurred from the construction of the mine, which favoured the breeding of mosquitoes, workers were given inadequate doses of chloroquine. A mutation in the DNA of the parasite allowed the parasite to survive in the presence of chloroquine and spread to infect the many workers in the region, and then further afield as the workers migrated (Figure 14.3.10).

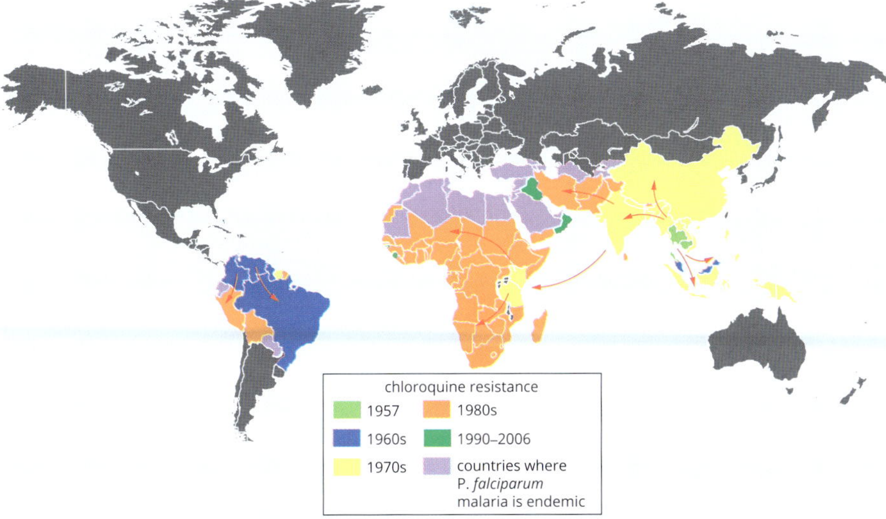

FIGURE 14.3.10 The history of chloroquine-resistant *P. falciparum*.

Artemisinin, a compound from the sweet wormwood plant (*Artemisia annua*), has been recently identified as a very potent antimalarial compound, earning the Chinese scientist who discovered it the 2015 Nobel Prize in Medicine. Sweet wormwood has been used in Chinese traditional medicine for many years. It is now a key component of combination therapy for malaria and is effective against drug-resistant malaria.

Lack of sanitation and poor hygiene

Outbreaks of various diseases are common in developing countries or in isolated communities in developed countries. These places often lack the infrastructure for reliable sanitation amenities, such as clean running water, effective sewerage systems and adequate health services.

Cholera

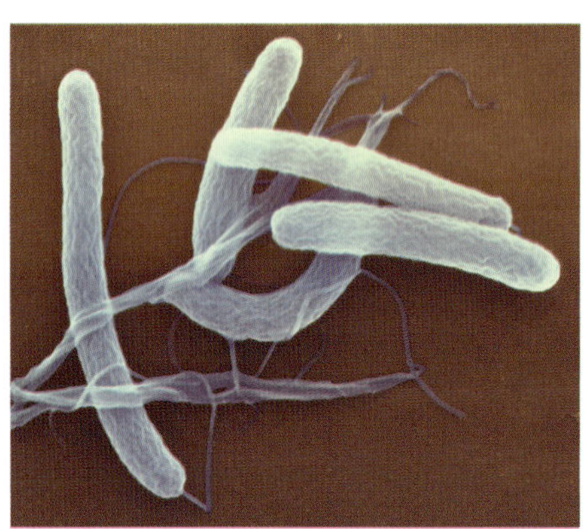

FIGURE 14.3.11 Coloured scanning electron micrograph of *Vibrio cholerae*.

The cholera epidemic that occurred in Haiti in the wake of the devastating 2010 earthquake is the worst cholera epidemic in recent history, with more than 700 000 cases and almost 9000 deaths. It was a surprise as Haiti had not experienced a cholera epidemic in more than a century. The epidemic was caused by the toxigenic strain of *Vibrio cholerae* (strain O1, Figure 14.3.11). Cholera affects the gastrointestinal tract, causing watery diarrhoea and vomiting, dehydration and in some cases death.

The origin of the Haitian epidemic was traced to a camp of Nepalese army personnel who came to the country to aid the relief effort. Epidemiology and scientific methods were used to trace the source and identify the infectious agent. DNA fingerprinting and sequencing matched the epidemic strain to a *V. cholerae* strain previously identified in Bangladesh, rather than strains that already existed at low levels in Latin America. The *V. cholerae* bacterium was most likely introduced by the Nepalese soldiers, who had recently been in a region experiencing a cholera outbreak. Contaminated sewage from the camp was released into the local river. The communities downstream used this water for washing and cooking, because the water supply network had been damaged and was still being repaired.

Despite improved sanitation and repairs to water infrastructure new infections continue to occur. Reasons why cholera continues to be a problem in Haiti some time after the initial cause of the epidemic include:

- limited sanitation infrastructure
- limited access to adequate medical services
- gaps in the water quality systems, limited chlorination
- limits to the alert and coordination systems
- displacement of peoples.

A different strain of *V. cholerae*, which emerged in 1992–93, was responsible for a cholera epidemic in a large area along the Bay of Bengal. Scientific studies of this *V. cholerae* strain (O139) are providing clues as to how bacteria evolve as new variants with the potential to cause epidemics.

Multiple factors

It is frequently the case that multiple factors contribute to an epidemic or pandemic. Existing pathogens may emerge in new locations and among populations that lack the ability to recognise and deal with it early enough to prevent it spreading.

Ebola

In 2014, an epidemic of the Ebola virus occurred in Western Africa, affecting over 11 000 people in multiple countries. Ebola was first identified in 1976 and, although this was the largest and longest outbreak, Ebola had affected small numbers of people previously at least 26 times, mainly in equatorial African countries.

Ebola is an often-fatal virus that is carried by the fruit bat, in which it remains silent. Ebola entered into the human population when local people ate bat meat. Ebola is spread through bodily fluids and has a relatively short incubation period of as little as two days.

The 2014 epidemic was an example of an old virus in a new context. Equatorial African countries with a history of local Ebola outbreaks had the experience to recognise it early, the laboratory facilities to identify it and the facilities to contain an outbreak. West African countries, however, had none of this experience. Several other factors contributed to its spread: years of poverty and civil unrest, poor transport and road systems, a highly mobile population and cultural practices, especially burial practices and the continued use of traditional healers. In addition, the affected countries had a limited number of healthcare workers and health facilities, and they lacked the knowledge to deal with Ebola. Hospitals lacked the equipment and facilities to handle the numbers of infected people. A delay in the foreign aid response, mistrust by locals towards the health workers and reluctance to change practices allowed the contagious disease to spread more than it might have otherwise.

BIOFILE

Identification of the Ebola virus in 1976

In 1976, approximately 100 people died in a remote village in The Democratic Republic of Congo. The deaths were initially believed to be due to an outbreak of yellow fever. Two vials containing the blood sample from an infected patient were sent from the village to a laboratory in Belgium for identification. The vials were packaged in a regular thermos and carried in the hand luggage of a passenger. On arrival, it was discovered that one of the vials had broken in transit and its contents were mixed through the ice. It was extremely fortunate that the pathogen did not enter the environment and infect the passengers or the scientists, as it would have inadvertently caused multiple infections and a possible pandemic.

If the dangers of the pathogen that was contained in the package had been known, tighter measures would have been implemented, as they are in the transport of samples today.

The laboratory methods available in 1976 to identify viruses included transmission electron microscopy of the blood samples, which revealed a very unusual and large new virus. Soon, fluorescent antibody tests were available to identify the virus in blood.

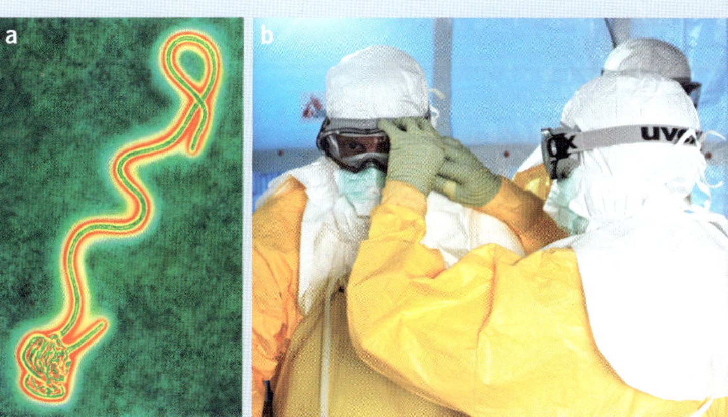

FIGURE 14.3.12 (a) Coloured transmission electron micrograph of the Ebola virus. (b) The personal protective equipment that healthcare workers were issued during the 2014 Ebola outbreak in Sierra Leone.

Today the virus is identified by an assay that detects either the viral antigen or antibodies present in the blood of people who have been exposed to the virus (called an enzyme-linked immunosorbent assay or ELISA). The virus is also detected by viral RNA genome sequencing.

EMERGING DISEASES IN AGRICULTURE AND WILDLIFE

Wildlife and agricultural species are at risk of new diseases emerging from pathogens already in the environment and from introduced species. Australian biosecurity agencies and research organisations conduct intensive surveillance to identify potential pathogens, prevent their entry into the country, and prevent them spreading throughout vulnerable habits and farmlands. Some examples of emerging diseases of importance to agriculture and wildlife biodiversity are listed in Table 14.3.2.

New or previously unrecognised diseases	• Hendra virus in horses • bat lyssavirus • devil facial tumour disease • cattle BSE • bee *Varoa* mite • amphibian chytrid fungus • sugar cane orange rust • Coral Sea fan fungus	
Increased in incidence, virulence or range over the last 20 years	• foot-and-mouth disease virus in cattle • blue tongue virus in sheep • wheat stem rust fungus • jarrah dieback *Phytophthora cinnamomi* • loggerhead turtle fungus	

TABLE 14.3.2 Examples of emerging diseases affecting Australian wildlife and agricultural plants and animals.

Exotic species originate in another country. Native species have not coevolved with them and so often do not have any defence against exotic pathogenic species.

Infectious fungal diseases, for example, are emerging globally as major threats to whole ecosystems and to particular plants and animals. Species extinctions are occurring as organisms have not evolved any natural defences against the new diseases (Figure 14.3.13). Human activity, such as travel within and between countries, bushwalking and mountain biking through affected areas, and the import and export of agricultural products, increases the speed at which these pathogens move around the globe.

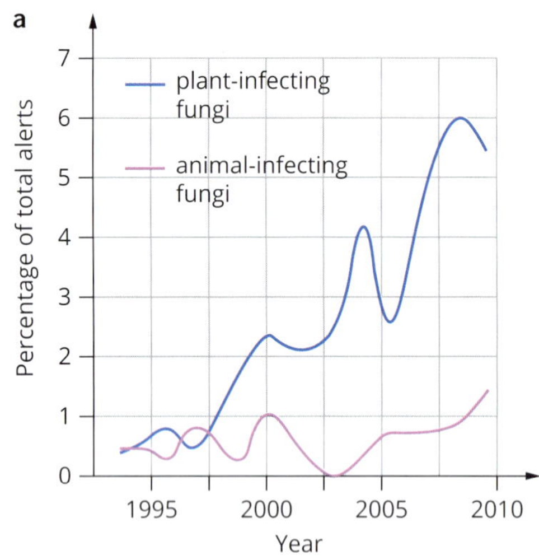

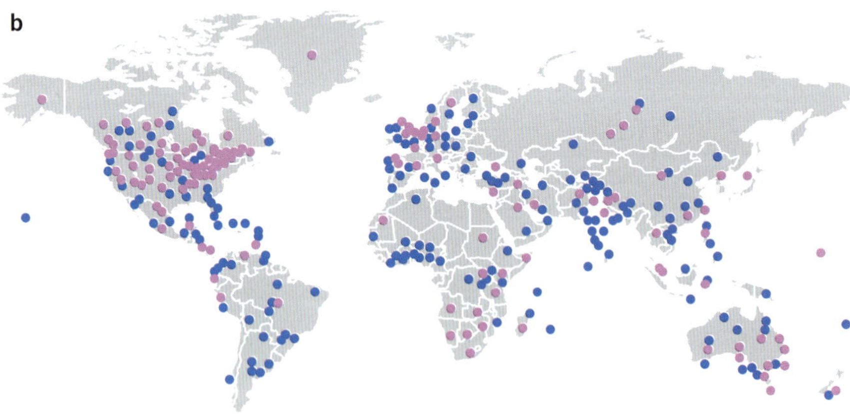

FIGURE 14.3.13 Data showing the global rise in emerging fungal infections of animals and plants (a) and the location of the reported diseases (b).

A significant fungal disease of plants in many parts of the world, and now Australia (including Victoria), is dieback, caused by *Phytophthora* (Figure 14.3.14a). Although it has been in the country since the 1930s, it has recently spread to regions that were previously unaffected, and poses a significant threat to Australian biodiversity. *Phytophthora cinnamomi* penetrates the roots and stem to get nutrients, clogging the vascular tissue and thus starving the plant. Other species affect commercially important fruit crops (Figure 14.3.14b). Strategies to control dieback range from simple steps for everyone visiting an affected area, such as washing shoes/boots after bushwalking (Figure 14.3.14c), to aerial spraying of large affected areas with potassium phosphonate.

FIGURE 14.3.14 The exotic fungus *Phytophthora* affects many plant species in Australia. (a) *Phytophthora cinnamomi* causes root rot or dieback disease in Jarrah forests in Western Australia. (b) *Phytophthora palmivora* causes fruit rot in papaya. (c) A forest worker washes boots to avoid carrying dieback fungus outside an affected area.

IDENTIFYING DISEASE

To control a fast-spreading disease, the cause of the disease must be identified and a test for the infection developed that does not rely on diagnosing the symptoms, especially for diseases with long incubation periods.

Tests can be developed to identify the presence of the pathogen's proteins or DNA. Typically, tissue or blood samples are tested. Proteins, such as viral coat proteins and bacterial toxins, can be detected by gel electrophoresis and the use of antibodies that recognise these molecules. When testing for a pathogen's genetic material, PCR is commonly used to amplify the genetic material (DNA, or RNA in the case of RNA viruses). The amplified DNA is then used to identify genes for toxins, analyse DNA fingerprints of variable repeat sequences, analyse single nucleotide polymorphisms (SNPs), or undergo full genome sequencing.

Another approach for animal diseases is to look for the presence of antibodies against the pathogen. A test called an **enzyme-linked immunosorbent assay (ELISA)** can be performed. An ELISA identifies the presence of specific proteins, such as antibodies. It uses an antigen from the pathogen to bind to antibodies present in the blood, which would have been produced by the immune system if the pathogen were present in the animal.

The development of rapid molecular techniques, databases of protein and gene sequences and bioinformatics has made identification of pathogens much faster, enabling a more rapid response and a greater chance of control and containment of disease outbreaks.

Controlling disease and containment

Infectious diseases pass from one person to another. Controlling the spread relies on preventing the pathogen coming into contact with more people or preventing infection in those people. Disease control can be aided in the following ways:

- Prevention: Infection can be reduced if people understand how the disease is spread and are educated in simple hygiene and prevention methods. These can include hand washing, disinfection, sterilisation and use of personal protective equipment, such as gloves and face masks.
- Isolation and quarantine: Infected people in hospitals may be separated from other patients. Governments may restrict the movement of people, animals, plants and biological products into and out of the affected countries. Airports may have medical personnel testing people who show symptoms on arrival. Schools and other communal places may also be closed until the epidemic or pandemic has slowed.
- Control carriers: For example, cattle and sheep infected with mad cow disease (BSE) and scrapie, and ducks and chickens infected with avian flu are destroyed. Bans on the consumption and commercialisation of meat and egg products may also be put in place.
- Eradication of vectors: Insect vectors such as mosquitoes and midges may be eliminated, or repelled, to control malaria and other vector-borne diseases.
- Vaccination: Vaccines often take some time to develop but are effective in preventing future infections.

- Education: Informing people of the cause of the disease, the ways in which the disease is transmitted and the reasons for the measures used to control it helps to prevent the spread. This may be difficult if control measures, such as quick burial of the dead, are not customary.
- Response plans: Plans by governments or communities need to be put in place to instigate and enforce all of the above controls for both existing infectious diseases and outbreaks of new diseases. Governments often work together to share information and control the spread of diseases.

Figure 14.3.15 summarises the key steps in the containment and control of infectious disease.

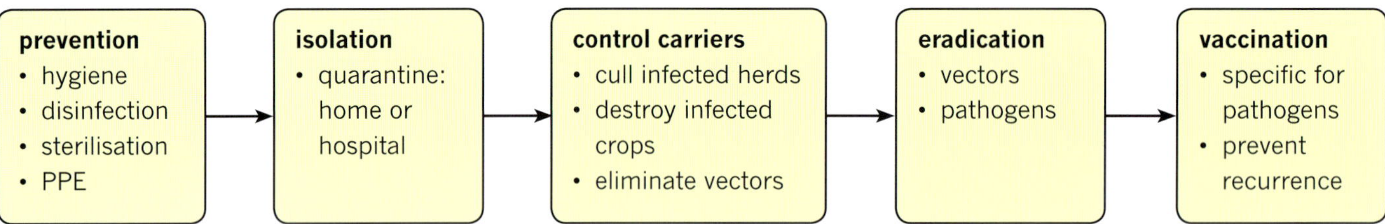

FIGURE 14.3.15 Key steps in the containment and control of infectious diseases.

BIOFILE

Eradicating smallpox

The World Health Organization Smallpox Eradication Programme ran from 1966 to 1980. Smallpox, caused by the *Variola* virus is highly contagious. The global programme involved centres for vaccination, surveillance, prevention and containment of epidemics. The last identified natural case of smallpox was reported in Somalia in 1977.

Stocks of the virus still exist in two secure laboratories in the event of a future need to make another vaccine. Some remain concerned about keeping any stocks of the pathogenic virus due to the risk of it getting into the hands of bioterrorists.

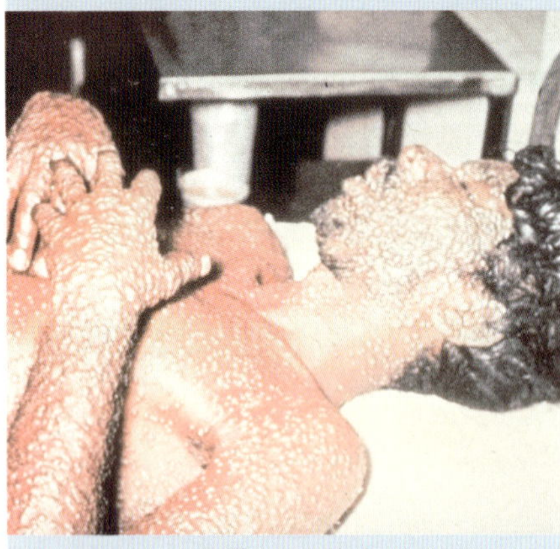

FIGURE 14.3.16 This man's body is covered with smallpox lesions.

TREATMENT

The treatment for diseases will vary depending on their cause and effects (Table 14.3.3). Different types of treatments are explored further in Section 14.4.

- Viral diseases, such as Ebola, are treated with fluids, antiviral drugs and pain-relieving medication, but there is little else that can currently be done.
- Retroviruses, such as HIV, are treated with antiretroviral drugs that minimise the viral load in infected people. This limits the effects of the virus on the immune system and reduces the chance of it spreading to other people.
- Bacterial diseases are treated with antibiotics. However, antibiotic resistance is becoming a concern. It is essential that antibiotics are given only for bacterial infections, and that once prescribed the entire course of antibiotics is taken.
- Prion diseases, such as vCJD, currently have no treatment. Antibodies against prions are currently being tested as potential treatments for prion diseases.
- Combination therapies use more than one drug at the same time, and include drugs that work in different ways. Multiple lines of attack makes it harder for the pathogen to evolve resistance.

Treat symptoms	Medication for type of pathogen	Drug for specific pathogen
• rehydration • fever reduction • pain relief	• antibiotics for bacteria • antivirals for viruses • antimycotics for fungi • antihelminths for worms	• Relenza for influenza virus • artemisinin for *P. falciparum* malaria

TABLE 14.3.3 Treatment depends on the nature of the pathogen and the symptoms.

BIOLOGY IN ACTION

Bats, horses and Hendra

A new disease struck racehorses and their handlers on a Queensland property in 1994. Within just a few days 14 horses and their owner had died. Government departments and the CSIRO Australian Animal Health Laboratory in Geelong moved into action to identify the mystery disease. The culprit was the Hendra virus (Figure 14.3.17).

The natural host of the Hendra virus is the Australia fruit bat, also called the flying fox. The fruit bat species most associated with the Hendra virus is the black flying fox that resides in Queensland. The virus has not been found in Victoria's grey-headed flying foxes, so the risk of this disease in Victoria is low. However, it is possible that a horse incubating the virus could be brought into Victoria and transfer the virus to other horses and to human handlers.

Exactly how the Hendra virus is transmitted from bats to horses is not fully understood. It is likely that horses are infected by ingesting or inhaling the virus in droplets of fluid secretions from bats, when horse properties overlap with bat roosts. The virus is transmitted from horse to horse through direct contact and infectious body fluids (Figure 14.3.18).

The virus can move from horse to human through contact with infected mucus and/or blood or other body fluids. Hendra virus does not appear to be transmitted from person to person. Horse owners and vets need to be aware of the signs of Hendra infection, which include fever, increased heart rate and breathing difficulty, and seek treatment early. Prevention by immunisation is best.

CSIRO has developed a vaccine against Hendra virus, using a viral coat protein. Equivac HeV is for immunisation of horses. This is the most effective way to prevent the spread of Hendra virus and protect horses, horse owners and vets.

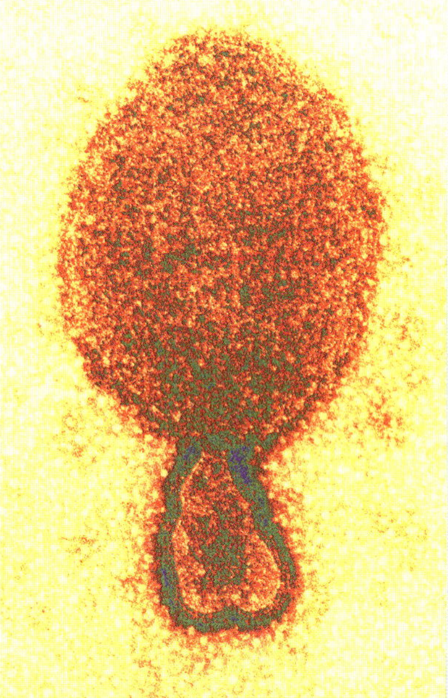

FIGURE 14.3.17 Coloured electron micrograph of Hendra virus. Hendra virus is an RNA virus in the genus *Henipavirus*. Henipaviruses reside in bats.

a amount of virus

- bat food supply
- bat population size
- frequency of bat visiting
- rate of virus shedding
- virus excretion

b horse exposure

- pasture quality
- horses feed beneath trees
- horse breeding and rearing
- horse behaviour

c horse infection

- horse health
- innate immunity
- adaptive immunity
- route of exposure

FIGURE 14.3.18 Risk factors for the development of Hendra virus infection by horses. The concentration of virus in the environment is affected by the quantity of virus shed by bats and the survival of the virus out of the bat host (a). Exposure of horses to the virus is affected by several factors (b). The effectiveness of innate and adaptive immunity determines whether horse exposure leads to infection (c).

14.3 Review

SUMMARY

- The emergence of new and recurring infectious disease poses a challenge to global health.
- An epidemic is the sudden increase in the number of cases of a disease in a localised area.
- A pandemic is an epidemic that has spread over several countries or continents.
- Human demographics and behaviours, mobility, cultural traditions, changes in lifestyle, sanitation and general hygiene contribute to the emergence and spread of disease.
- Agricultural practices and procedures may contribute to the introduction and spread of infectious disease in animals and crops.
- Overuse and inappropriate application of antimicrobial drugs has contributed to the evolution of resistant pathogens.
- Zoonotic diseases pose a threat to populations in close contact with natural host animals.
- Control of zoonotic and vector-borne diseases, such as insect vectors carrying parasites, require control of the natural host, the vector and the pathogen.
- Control of disease outbreaks requires detection, identification and containment of the affected population.

KEY QUESTIONS

1. What is meant by the term 'emerging disease'?
2. How can an organism acquire the ability to become pathogenic and cause disease?
3. Distinguish between an epidemic and a pandemic.
4. What is a zoonotic disease? Give an example of this type of disease.
5. Give an example of an emerging disease influenced by factors a–g. Then research to find another example of a disease influenced by each factor.
   - **a** human demographics, demographic change and/or mobility
   - **b** human behaviour
   - **c** farming practices
   - **d** overuse of an antimicrobial agent
   - **e** poor sanitation
   - **f** limited social, transport or health infrastructure
   - **g** close association between wildlife and domestic animals
6. Why does the influenza virus repeatedly emerge each year, and sometimes cause a pandemic?
7. How is scientific knowledge applied to dealing with the emergence of new diseases?
8. Discuss how government agencies and research organisations can help prevent the emergence of new diseases that may impact on agriculture and wildlife biodiversity.

14.4 Chemical agents against pathogens and rational drug design

The spread of viruses and bacteria outside the body may be limited by using chemical agents such as disinfectants. Inside the body, specific drugs may be required to treat the pathogen.

A drug is a molecule that alters the functioning of an organism. Drugs interact with target molecules in the body. Some drugs increase the activity of a target molecule, while others block it (Figure 14.4.1; this drug is discussed in more detail on pages 573 and 574). Drugs can also be used to restrict or kill pathogens. Antibiotics can treat bacterial infections. Drugs called **antivirals** are specific to viral infections.

In the past, many drugs, such as the antibiotic penicillin, were discovered by chance or by large-scale screening of naturally occurring substances to see if they had effects on a range of diseases. Since the 1980s, rational drug design has become more common.

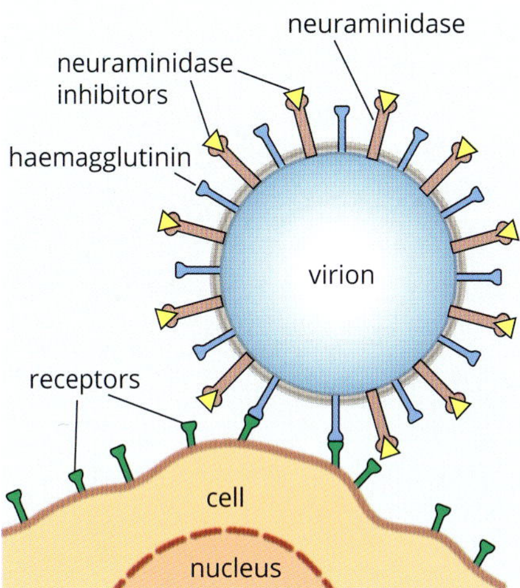

FIGURE 14.4.1 The drug Relenza (yellow) blocks the active site of neuraminidase on a flu virus. The drug plays a key role in preventing the virus exiting the host cell.

CHEMICAL AGENTS

One way of controlling pathogens and their spread is to reduce the number of pathogens in the outside environment. Chemicals are one way of doing so. **Disinfectants** are used to kill pathogens on surfaces such as door handles and hospital equipment. **Antiseptics** are used to kill pathogens on the body. By reducing the number of pathogens, the chance that an infection will occur is reduced.

Disinfectants and antiseptics are non-specific antimicrobial agents, that is, they inactivate or destroy most biological agents (bacteria, viruses, fungi). Some common antiseptics and disinfectants and their modes of action are listed in Table 14.4.1. Bacteria with waxy surfaces and microbes that form spores and cysts may be resistant to the effects of these compounds. Disease-causing prions are resistant to these chemicals and to strong acids, including stomach acid.

TABLE 14.4.1 Examples of widely used antiseptics and disinfectants and how they kill microbes.

Disinfectants	Antiseptics	Modes of action
• alcohols: ethanol, isopropanol • chlorine • formaldehyde • phenol compounds • hydrogen peroxide • copper sulfate	• alcohols: ethanol, isopropanol • some detergents • phenol compounds • hydrogen peroxide • chlorhexidine (in hand wash) • iodine (in Betadine) • quaternary ammonium compounds	• denature proteins • dissolve lipids; disrupt cell membranes • act as oxidising agent to alter or destroy proteins • act as alkylating agent or fixative, linking proteins and destroying function

BIOFILE

Death by hospital

In the early 1800s, one of the greatest causes of death was admission to hospital. Doctors knew nothing of bacteria or viruses and did not know how infections occurred. A man called Joseph Lister argued that doctors were somehow transferring disease from one patient to another. He was not listened to by his colleagues, who thought his ideas were ridiculous.

Lister conducted trials in a hospital ward and showed that the washing of hands by doctors and nurses after seeing each patient significantly decreased the number of patients that became ill. He also suggested that operations should be performed in sterile environments and persuaded surgeons to operate under a mist of antiseptic to reduce the number of pathogens that might infect the wounds (Figure 14.4.2). You might also notice that the surgeons are wearing their everyday clothes while operating, a practice that no longer occurs in order to limit the number of microbes entering the operating theatre.

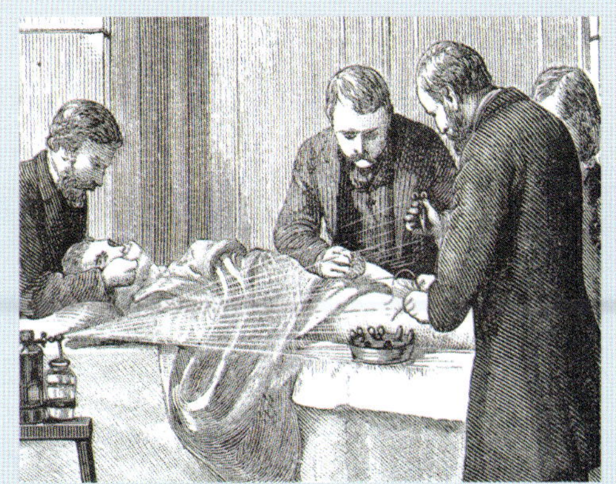

FIGURE 14.4.2 Joseph Lister using antiseptics during an operation.

Bacteria and viruses are very different pathogens. When infection occurs in humans, specific drugs are able to treat the pathogen. Antibiotics treat bacterial infections. Antivirals treat viral infections. These drugs may affect a specific pathogen or a common feature of a group of pathogens.

Antibiotics

Many bacteria are pathogens and can be treated using drugs called antibiotics. Penicillin, discovered in 1928, was the first commercial antibiotic. When it was first mass produced, it was often described as a 'magic bullet' because diseases that had once been fatal could now be easily cured. Since the 1930s, many other antibiotics have been discovered and developed. Many antibiotics are naturally occurring molecules produced by fungi or bacteria. These natural molecules may then be extracted and chemically modified to improve their effectiveness. Alternatively, new antibiotic compounds have been discovered through trial and error chemical screening, or by design to fit a particular pathogen.

To be successful, antibiotics ideally kill bacteria without damaging the cells of the organism being treated. Therefore the mechanism of action should target biochemical pathways and molecules specific to the microbe. Different classes of antibiotic target bacterial cell walls, ribosomes, enzymes for DNA and RNA synthesis, protein synthesis and metabolic pathways (Figure 14.4.3).

Some antibiotics slow bacterial growth; they are bacteriostatic. Others kill the bacteria; they are bactericidal. These effects may depend on the concentration of antibiotic used.

BIOFILE

Adverse effects of antibiotics

Although antibiotics specifically target bacterial structures and proteins, some antibiotics unfortunately cause damage to eukaryotic cells, particularly when used at high concentrations. Damage to the sensory hair cells in the ear, leading to deafness, and disruption of kidney tubules and tendons may be caused by some antibiotics. This may be due to cross-reaction of the drug with eukaryotic membranes and enzymes.

Recent studies have identified a potential new mechanism for some of the adverse effects of antibiotics: alteration of mitochondrial function. Mitochondria most likely evolved from bacteria by endosymbiosis, and indeed mitochondria share structural and molecular similarities with bacteria, including related ribosomes. Antibiotics that inhibit protein synthesis in bacteria also inhibit the mitochondrial ribosomes. Other antibiotic effects observed in mitochondria include decreased energy generation and excessive production of reactive oxygen species (free radicals), which damage proteins and lipids in the cell.

A positive side to these effects is the potential use of antibiotics to inhibit mitochondrial activity in cancer stem cells and thus slow their growth.

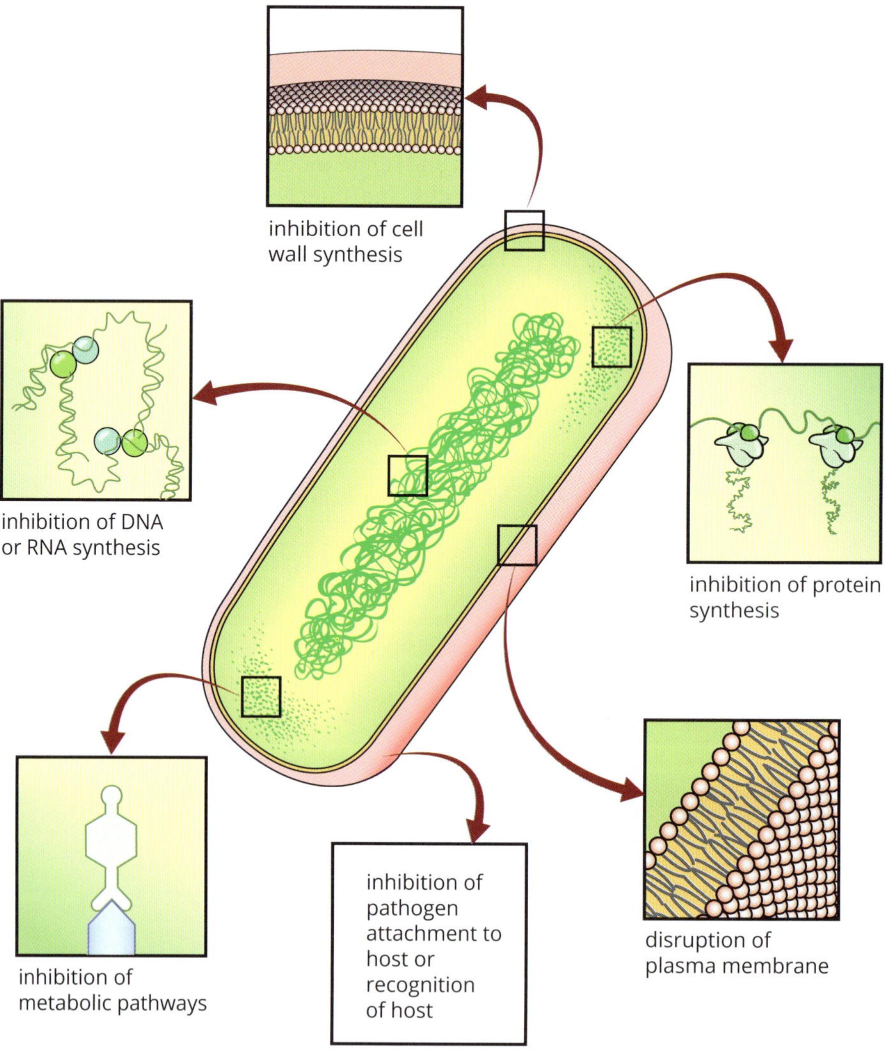

FIGURE 14.4.3 Mechanisms of antibiotic action.

BIOFILE

Antibiotics for plants

Plants, like animals, are susceptible to bacterial, fungal and viral infections. Although plants have an arsenal of antimicrobial compounds to defend them against pathogens, sometimes pathogens overcome these barriers. Spraying with antimicrobial chemicals, such as copper compounds, may be sufficient to control an infection. In some places, when agricultural commercially valuable crops are infected, antibiotics may be used.

FIGURE 14.4.4 Fruit tree damaged by fire blight disease.

Apple crops in New Zealand are affected by the highly contagious disease fire blight, which is caused by the bacterium *Erwinia amylovora*. Affected parts of the plant may appear as though they have been scorched by fire; fruits and leaves may be blackened, dried, shrivelled and cracked. Some orchard owners in NZ treat crops by spraying with the antibiotic streptomycin. In other countries fire blight is treated by injecting the antibiotic oxytetracycline into trees. This is a costly exercise, so is less common than chemical sprays. Antibiotics are not approved for use as a crop spray in Australia. There is also concern that the widespread use of antibiotics as sprays would speed the development of resistant bacteria in the environment.

Australia does not have fire blight, and orchardists want the restriction on the importation of New Zealand apples and pears to remain in place because of the high biosecurity risk. This remains a major concern for Australian fruit growers.

One group of antibiotics, including penicillin, inhibits the synthesis of a bacterial cell wall component called **peptidoglycan**, which is composed of amino acids and sugars linked into a mesh-like substance. These antibiotics prevent the bacteria from fully developing their cell walls, causing the cell walls to weaken and the bacteria to die. Bacteria with a thick layer of peptidoglycan in the cell wall are more susceptible to the effects of these antibiotics than bacteria with little peptidoglycan in their cell wall. Animals do not produce this peptidoglycan and so these antibiotics have no effect on their cells.

Another group of antibiotics, called sulfonamide drugs, act as competitive inhibitors in the metabolic pathway for folic acid (folate) production. Folate is a B vitamin essential for DNA and RNA synthesis in all cells. Animal cells can absorb this vitamin, which is obtained in the diet, through their cell membranes, but the thick bacterial walls will not allow folic acid to cross. Without the ability to make their own folic acid, the bacteria die.

Other antibiotics are inhibitors of transcription, blocking mRNA synthesis (e.g. actinomycin). Protein synthesis may be blocked by antibiotics that interfere with the ribosomes (e.g. tetracycline), or that block transfer RNA (e.g. puromycin). Bacterial ribosomes differ from eukaryotic ribosomes, and the enzymes for RNA synthesis and DNA replication are specific in prokaryotes and eukaryotes. Most antibiotics inhibit only the bacterial molecules.

The antibiotics that inhibit peptidoglycan production in the bacterial cell wall are called the beta-lactam antibiotics; they have a beta-lactam ring in their chemical structure. Bacteria that produce an enzyme called beta-lactamase can destroy the beta-lactam ring, making them resistant to these antibiotics.

Antiviral drugs

Viruses are non-cellular pathogens that must be inside living cells to replicate (see Figure 7.1.15, page 265 to review viral structures). Viruses may be either RNA or DNA viruses. RNA viruses tend to have a higher rate of mutation than DNA viruses. Viruses can survive for a limited time outside their host cells. This extracellular form of a virus is called a **virion**. A virion consists of genetic material in the form of DNA or RNA, surrounded by a protective protein coat called a **capsid**.

Once the virion has penetrated the host cell, it disassembles, freeing its genetic material to transcribe and translate new viral proteins. Some viruses, called enveloped viruses, are surrounded by a lipid envelope that is picked up when they bud from the host cell in which they replicated. The envelope contains viral proteins that are used as recognition molecules for binding to, and entering, target host cells. Because replicating viruses are found inside cells, they are difficult to destroy without damaging the host cell. A virus uses the host ribosomes for protein synthesis and may use host cell enzymes for nucleic acid replication.

Good targets for antiviral drugs are the capsid proteins, envelope proteins, and the DNA/RNA polymerase enzymes encoded by viral genes. Drugs that cross-react with host cell enzymes cause serious side effects. Antiviral drugs have only been in use since 1960, but a broad range has already been developed.

Antiviral drugs can work by several possible methods (Figure 14.4.5):

- preventing the virus from entering the cell by binding to receptors that allow the virus to enter
- inhibiting enzymes that catalyse reproduction of the virus genome
- blocking transcription and translation of viral proteins
- preventing the viruses from leaving the cell, and so preventing the infection of other cells

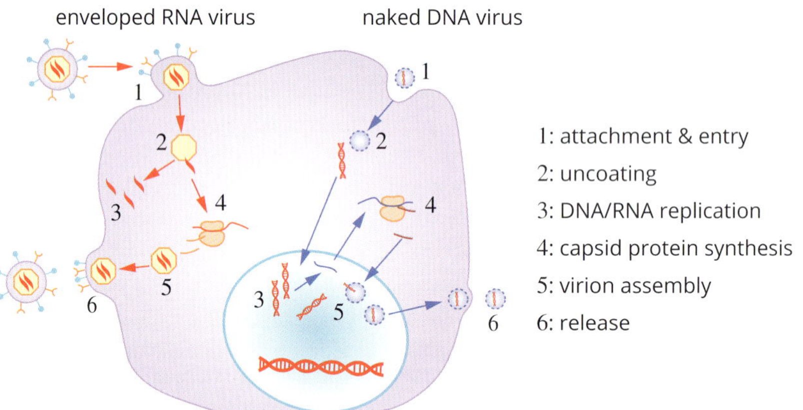

FIGURE 14.4.5 Targets for antiviral drug development. Any one of the key steps 1–6 in virus replication is a potential target for designing an antiviral drug. The illustration represents simplified life cycles for an enveloped RNA virus and a naked (non-enveloped) DNA virus.

DNA viruses

DNA viruses have a genome of DNA. Some DNA viruses use the host DNA polymerase while others have a gene for DNA polymerase, which may be a target for antiviral drugs. DNA viruses include the *Herpes simplex* virus that causes cold sores and *Varicella zoster* virus that causes chicken pox. The drug acyclovir lessens the symptoms by blocking the production of the *Herpes* DNA polymerase and therefore blocking replication of the virus. Acyclovir comes as an ointment that is applied to cold sores (Figure 14.4.6). It is also used in tablet form for other *Herpes* viruses, such as *Herpes zoster*, which causes shingles.

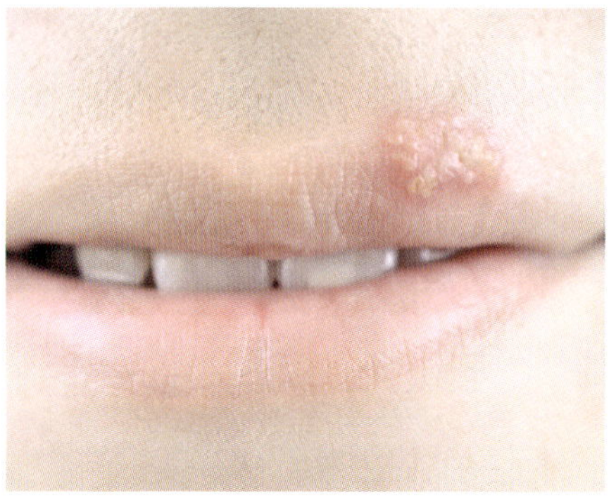

FIGURE 14.4.6 Cold sores caused by *Herpes simplex* virus often appear around the mouth, nose and eyes. Acyclovir (aciclovir) specifically targets *Herpes* virus replication.

RNA viruses

RNA viruses have a genome of RNA. RNA viruses have their own RNA replication enzymes, which may be a target for antiviral drugs. Examples of RNA viruses are those that cause influenza, mumps, hepatitis C, dengue fever and Ebola. Experimental drugs targeting viral RNA, to block translation, are being trialled against dengue virus and hepatitis C virus.

Some RNA viruses use an enzyme called reverse transcriptase to produce DNA from their RNA. These viruses are called **retroviruses**, and drugs used to treat them are called **antiretroviral drugs**. These viruses also have an enzyme called integrase, which inserts, or integrates, the retroviral DNA into a host chromosome where it stays permanently, making such viruses very difficult to treat. In addition to the antiviral targets mentioned above, antiretroviral drugs can work by specifically inhibiting the reverse transcriptase and integrase enzymes (Figure 14.4.7).

The human immunodeficiency virus (HIV) is an example of a retrovirus. The use of a number of antiretroviral drugs together, often called a 'cocktail' of drugs, has been shown to be more effective against HIV than the use of a single antiretroviral drug. A combination of drugs, called combination antiretroviral therapy or highly active antiretroviral therapy, has improved the health and life expectancy of people with HIV/AIDS and reduced their chances of transmitting the virus to others. The cocktail includes inhibitors of reverse transcriptase, DNA polymerase, and the enzyme that allows the virus to exit from the cell.

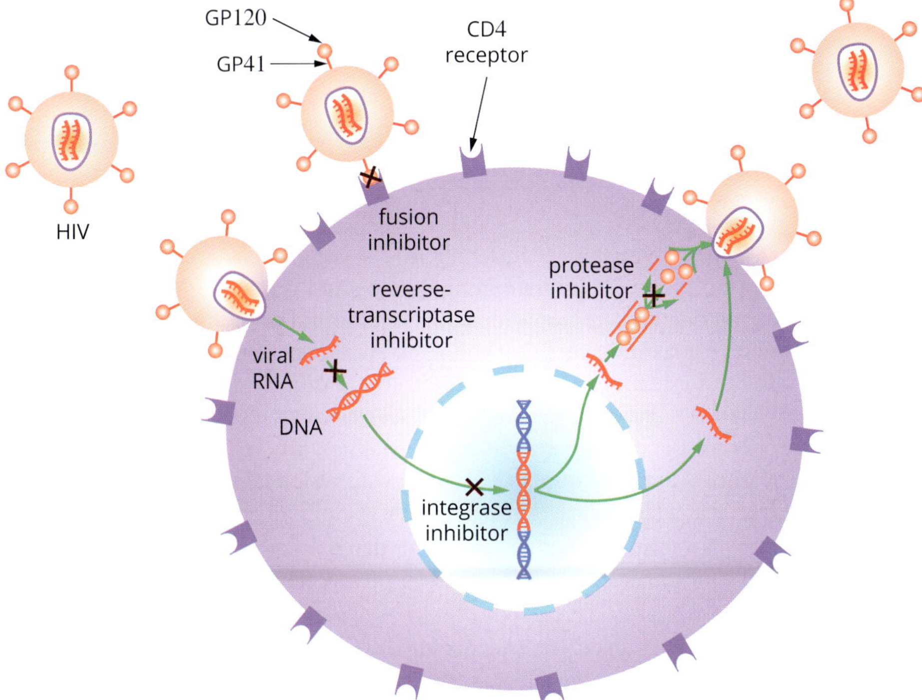

FIGURE 14.4.7 Targets for antiretroviral drug development, illustrating HIV.

DRUG RESISTANCE

There is an increasing concern that the antibiotics humans have relied on for so long to fight bacterial infections may be losing their effect. Bacteria that are exposed to antibiotics can develop a resistance to them. The development of drug resistance has become a huge problem, limiting the usefulness of all these drugs. For example, there was a time when penicillin killed more than 97% of all *Staphylococcus aureus*, 'golden staph' bacteria, which can cause blood infection (sepsis). Now in Australia 80–90% of these bacteria are resistant to penicillin and about 20% are resistant to a higher-level antibiotic, methicillin. *S. aureus* normally resides on the skin, and in the nasal passage and mouth, but causes infection when it breaches the epithelial layer. Infection is strongly linked with hospital visits. In addition, the extensive use of antibiotics as growth-promoting agents in livestock is driving the development of antibiotic resistance in species common in agricultural animals and the environment.

Bacteria can resist antibiotics in a variety of ways, as illustrated in Figure 14.4.8. For example, bacteria may reduce the intake of the drug into the cell, alter the target molecule to which the drug attaches, pump the drug out of the cell or enzymatically deactivate the drug. If these resistance properties are present in members of the bacterial population, the bacteria possessing them will survive the drug treatment and go on to become the dominant population.

BIOFILE

Australian Group on Antimicrobial Resistance (AGAR)

A group of medical scientists called the Australian Group on Antimicrobial Resistance (AGAR) tests and gathers information on the level of antibiotic resistance in bacteria causing important and life-threatening infections, particularly *Staphylococcus aureus* and *Enterococcus*. To increase global awareness of this problem, and to help educate the population on appropriate uses of antibiotics, the World Health Organization runs World Antibiotic Awareness Week each year.

FIGURE 14.4.9 The World Health Organization supports World Antibiotic Awareness Week.

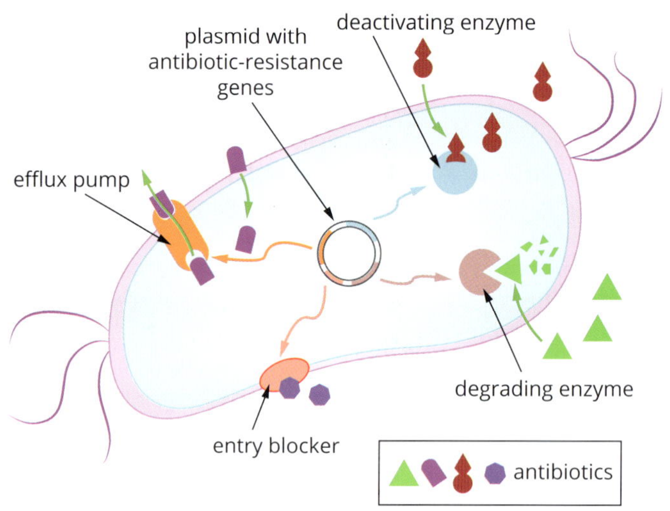

FIGURE 14.4.8 Mechanisms of bacterial resistance to antibiotics.

Genes for drug resistance may already exist in bacteria or arise from spontaneous mutations. Resistance genes are passed on through cell replication, as well as during bacterial conjugation. This is a natural process of horizontal gene transfer in which plasmids carrying the resistance genes are transferred between bacteria of the same or different species. In addition to methicillin-resistant *S. aureus* (MRSA), vancomycin-resistant *Enterococcus* (VRE) and multidrug-resistant tuberculosis (*Mycobacterium tuberculosis*) are major health concerns. Although scientists are working to develop new antibiotics, there is a concern that a strain of bacteria will emerge that will be resistant to all antibiotics.

Resistance can also evolve to antiviral drugs, especially in patients with compromised immune systems. Once this happens, new forms of treatment may be necessary.

Resistance to drugs used for protozoan diseases, such as those used to treat malaria, has already arisen, as you learnt in Section 14.3. The use of combination therapies, with multiple lines of attack on different parts of the microbe's replication cycle, is proving a more effective way to kill microbes and prevent the evolution of resistance.

Resistance to antifungal treatments is also an emerging problem. Drugs that work against fungal infections are called antimycotics. The yeast *Candida*, which causes skin infections and thrush, is a serious problem when it gets into the bloodstream. Resistance of *Candida* to antimycotics is a growing problem in the hospital setting and in immunocompromised patients, such as those undergoing chemotherapy or with HIV/AIDS.

BIOFILE

The last line of defence

Recently in mainland China, bacteria have been found in pigs and humans that are resistant to the 'last line of defence' antibiotic polymyxins. The resistance has come about because antibiotics are widely used in agriculture in China and a mutation occurring in a gene in plasmids of common bacteria has allowed these bacteria to survive when polymyxins are present. Unfortunately, these mutated plasmids can be passed between species of bacteria and this will increase the number of different species that will have this resistance.

THE CONCEPT OF RATIONAL DRUG DESIGN

Rational drug design is a targeted approach to designing new drugs. Rational drug design involves analysing the structure of a pathogen or disease-causing molecule and using this information to design a drug that will mimic or block the action of the disease-causing agent.

Designed drugs have complementary shapes and charges to the active sites of the pathogen or molecule they are targeting. Using methods such as **X-ray crystallography** and other sophisticated imaging techniques, the detailed structure of the active site or receptor-binding site on a molecule can be identified. It is then a case of finding an existing drug from databases or manufacturing a new drug that has a complementary shape to the active site or receptor site. The interaction of the drug with the target molecule can then be tested in the laboratory.

Figure 14.4.10 summarises the process of rational drug design.

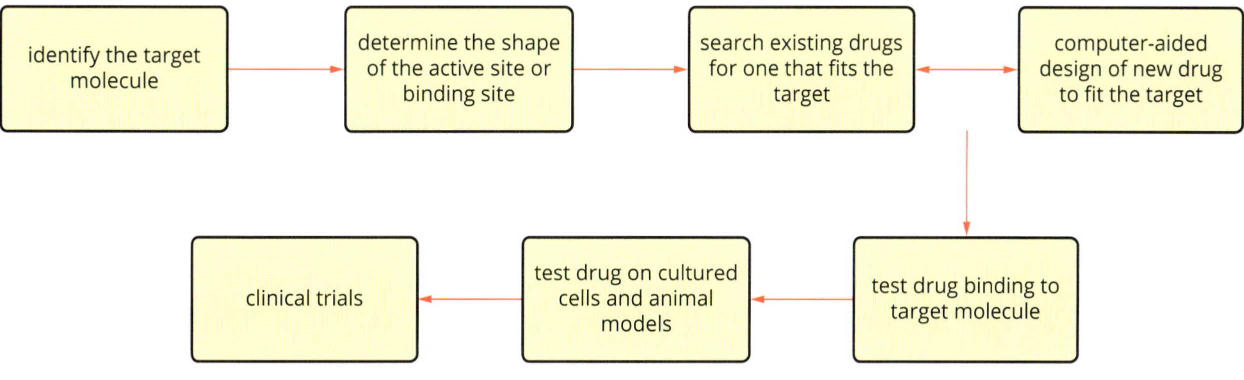

FIGURE 14.4.10 How to design a drug (rationally).

Relenza

A drug developed at CSIRO in Australia called zanamivir, and marketed as Relenza, was one of the first antiviral drugs to be produced using rational drug design. Relenza targets a protein on the surface of the flu virus. It is used both to treat the early stages of influenza and as a **prophylaxis**, to prevent influenza. Relenza has been approved for use in more than 50 countries.

Influenza (flu) is a viral disease that affects up to 500 million people each year. Unlike many other viral diseases such as measles and mumps, where a lifelong immunity can be acquired, the flu virus can cause repeated infections. This is due to random mutations in the virus, causing the antigens on the protein coat to change (see the discussion of antigenic shift and antigenic drift in Chapter 7, page 265). As a result, new versions of the vaccine must be produced each year. For a drug to be effective against a number of strains of the virus, it is necessary to find a common structure between them.

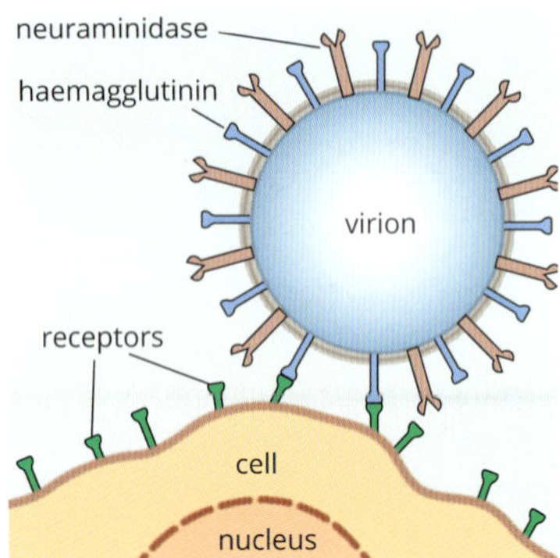

1. New virion being released from a cell is bound by haemagglutinin (blue cup) to a receptor (green).

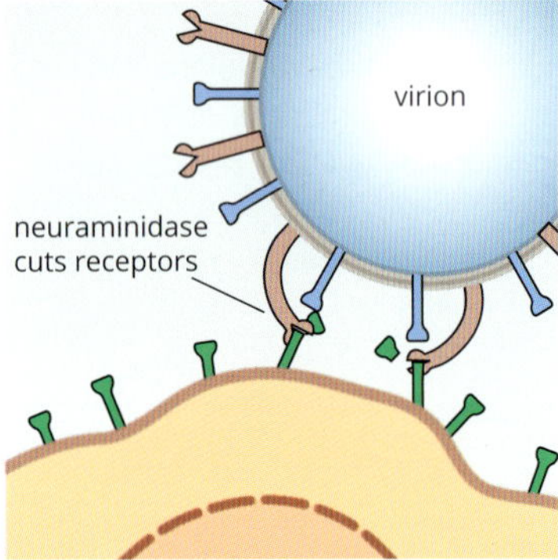

2. Neuraminidase cuts the receptors to let the virus escape and infect another cell.

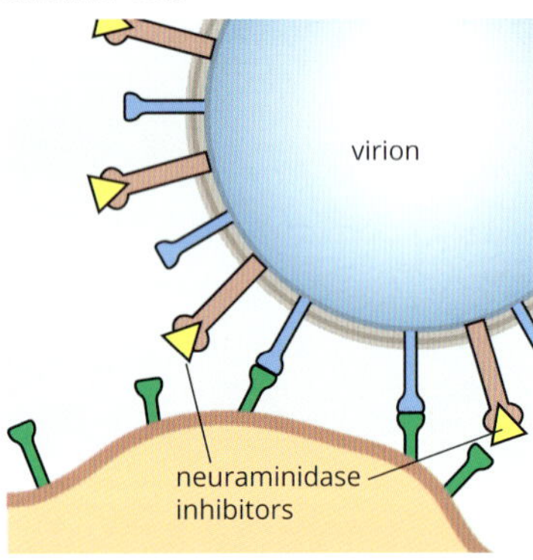

3. Relenza (yellow) blocks the active site of neuraminidase.

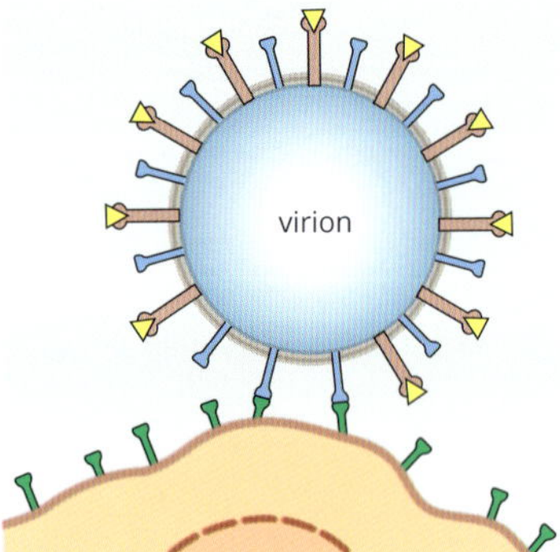

4. Now the virus cannot escape from the cell surface.

FIGURE 14.4.11 Mechanism of Relenza action.

Scientists targeted neuraminidase, one of the envelope proteins of influenza A viruses. When newly made influenza virions are budding from an infected cell, they are connected to the host cell membrane by an attachment between haemagglutinin on the virus and a cell receptor molecule. Neuraminidase is an enzyme that cuts this molecular connection, allowing new virions to escape from the cell and go off to infect more cells. Blocking the action of neuraminidase would therefore stop the release of new viruses. Using X-ray crystallography, scientists focused on the active site of neuraminidase. Unlike other proteins on the virus, this site remains constant in different strains of the flu virus.

Relenza has a complementary shape to the active site of neuraminidase. Also, the active site of neuraminidase has a negatively charged area and Relenza has a slightly positive charge. These features allow the drug to attach to the active site and act as a competitive inhibitor (Figure 14.4.11). When Relenza binds to the active site, neuraminidase does not catalyse the reaction that releases viral particles from the infected cells. This means there are fewer virions to infect other cells. The body's immune system is then usually able to destroy the remaining virions.

Since the development of Relenza, many other drugs have been developed through rational drug design to treat a wide range of illnesses. For example, drugs that treat HIV have been developed to target the active sites of viral reverse transcriptase, proteases, or the entry and exit of the virus from the cell.

Other rationally designed drugs

Rational drug design is not just for treating pathogens and infectious diseases. Drugs to treat heart disease, cancer, inflammatory diseases and mental health problems are all within the scope of drugs being designed specifically to target molecules on the affected cells.

Here are a few examples:

- Gleevec was designed to treat chronic myeloid leukaemia. It is a competitive inhibitor of a cancer-causing protein because it is complementary to the active site of an abnormal enzyme that leads to uncontrolled cell division.
- Selective serotonin re-uptake inhibitors (SSRIs), which are used to treat depression, were the first psychotropic drugs developed by rational drug design. They target the serotonin transporter protein.
- Anti-inflammatory drugs were designed to inhibit the enzyme COX-2, which is involved in inflammation and pain. The aim is to avoid the side effects that other anti-inflammatory drugs have on the cardiovascular and gastrointestinal systems.

Summary of rational drug design

The more scientists know about the structure of molecules and how molecules interact and function within cells, the more precisely targeted drugs can be. The expansion of genomics and proteomics aids the process. Yet this powerful approach can raise issues for society to consider. In being more specific and a faster approach to new drug development, rational drug design may speed up the advance of personalised medicine, or attempt to enhance intelligence or sports performance, for example. Society as a whole, rather than scientists and medical practitioners alone, should be involved in discussion and decisions about whether these uses are ethical or equitable.

14.4 Review

SUMMARY

- Antiseptics and disinfectants control the number of pathogenic organisms or agents in the environment and on surfaces.
- Antibiotics are natural or synthetic molecules that slow bacterial growth (bacteriostatic) or kill bacteria (bactericidal).
- Antibiotics affect different aspects of the bacterium: cell wall, cell membrane, protein synthesis, DNA or RNA synthesis or metabolic pathways.
- Antiviral drugs target proteins for virus entry into cells, replication within cells or exit from cells.
- Antiretroviral drugs are specific for retroviruses, such as HIV; they inhibit specific enzymes such as reverse transcriptase.
- Bacteria, protists, fungi and viruses develop resistance to antimicrobial drugs.
- Drug resistance arises by mutation and transfer of these new gene variants.
- Rational drug design specifically develops a drug to be complementary to the shape and/or charge on a target molecule; the drug fits the active site or binding site to prevent its action.
- Relenza was developed by rational drug design to block an enzyme on the influenza virus.

KEY QUESTIONS

1. Explain how antiseptics and disinfectants differ from antibiotics.
2. List several ways in which antibiotics slow down or kill bacteria.
3. List some ways in which antiviral drugs may limit the spread of a virus.
4. Describe the fire blight infection, including the infectious agent, the type of organisms affected and how it can be controlled. Discuss any issues of this disease for the Australian fruit industry.
5. Describe the cause of cold sores and how they may be treated.
6. HIV/AIDS is caused by a retrovirus.
   - **a** How is this type of virus different from other viruses?
   - **b** If you were designing a drug to treat HIV/AIDS, which molecule(s) would be a good target?
7. How does antibiotic resistance arise and spread through bacterial populations?
8. List some mechanisms in bacteria for antibiotic resistance.
9. Describe the aim and process of rational drug design.
10. Outline the development of Relenza, its molecular target, its purpose and how it works.

Chapter review

KEY TERMS

amniocentesis
antibiotic
antimicrobial drug
antimicrobial resistance
antiretroviral drug
antiseptic
antiviral
biosecurity
capsid
chorionic villus sampling
disinfectant
DNA profiling
emerging infectious disease
enzyme-linked immunosorbent assay (ELISA)
epidemic
gene editing
gene therapy
genetic screening
haemagglutinin
horizontal gene transfer
in vitro fertilisation (IVF)
liposome
neuraminidase
pandemic
peptidoglycan
prophylaxis
rational drug design
retrovirus
short tandem repeat (STR)
transgenic organism
vector
viral vector
virion
virulent/virulence
xenotransplantation
X-ray crystallography
zoonotic

KEY QUESTIONS

1 *Drosophila melanogaster* is commonly used for genetic research. One particular mutation results in the deletion of a section of DNA 200 bp long from one particular gene. The gene was extracted from a fly that is homozygous for the mutant gene and the same gene was extracted from a fly that is homozygous for the wild type (normal) version of the gene. Both versions of the gene were amplified using PCR. The PCR products were then run on gel electrophoresis. Which gel most accurately shows the PCR products?

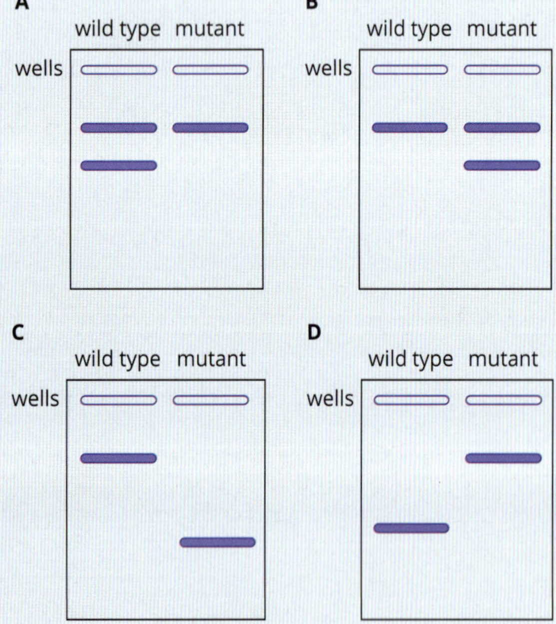

2 The use of bactericidal hand-washes has become very common. These hand-washes are types of:

A antiseptics
B disinfectants
C antibiotics
D antivirals

3 One result of a genetic application was the Flavr Savr tomato. Tomatoes have a short shelf life due the effects of an enzyme called polygalacturonase. This enzyme catalyses the breakdown of the cell walls of the tomato, causing the tomatoes to become soft and unappetising. To slow down this process, the sequence of the polygalacturonase gene was determined and an antisense gene was produced. The antisense gene has a complementary nucleotide sequence to the polygalacturonase gene. The antisense gene was inserted into the tomatoes. When the antisense gene is transcribed, the mRNA produced is complementary to the mRNA for the polygalacturonase gene, so the two mRNAs join to form double-stranded mRNA. Double-stranded mRNA cannot be translated, so the enzyme is not formed and the cell walls are not broken down. The Flavr Savr tomatoes can be considered to be:

A only transgenic
B only genetically modified
C both genetically modified and transgenic
D none of the above

4 Arsenic contamination of soil is a serious problem in some countries. The arsenic contaminates groundwater and drinking wells. The flow chart at right illustrates a process used to insert a gene that enables plants to absorb arsenic from the soil. In your notebook write the terms that describe steps A–E.

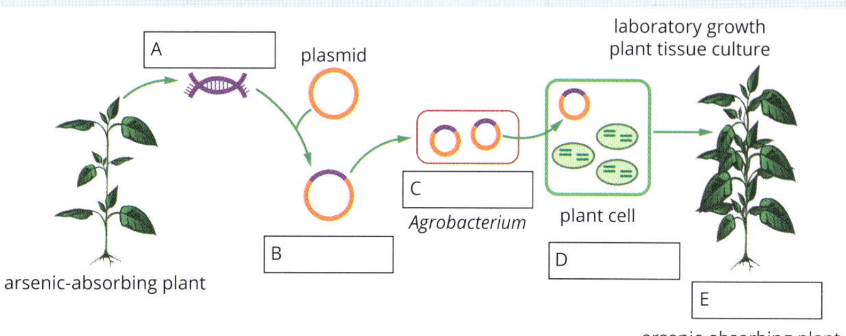

5 MRSA has become a serious problem in most major hospitals around the world.

a What is MRSA?

b Why is it an issue?

c Explain the relationship between MRSA and the inappropriate use of antibiotics.

6 Q fever is an example of a zoonotic disease. It is caused by a bacterium called *Coxiella burnetii*. In most people it causes a flu-like illness, but in some individuals it can cause liver or heart disease. One pathway for infection is shown below.

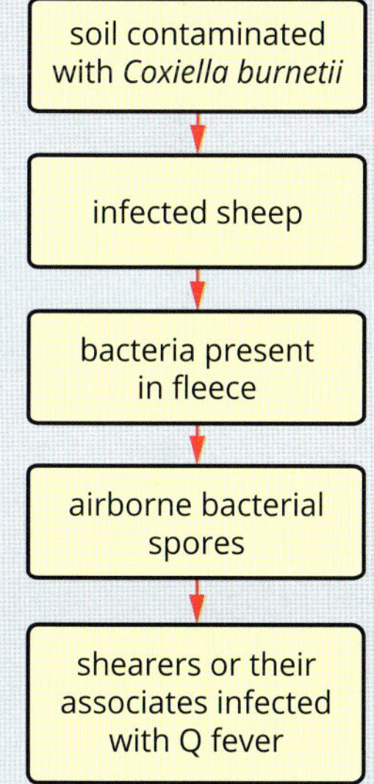

a What is a zoonotic disease?

b Identify three possible practices that could reduce the chance of infection with the bacterium.

7 Huntington's disease is a genetic condition. It results in significant and continuing neurological decline leading to severe problems with cognition, emotional control and movement. Many sufferers develop severe depression. This condition is caused by a single gene that codes for a protein called huntingtin. This protein is involved in signal transduction and synapse formation in neurons. The gene has been shown to be close to the telomere on the short arm of chromosome 4.

The normal gene for the huntingtin protein contains a section in which the triplet CAG is repeated between 11 and 29 times. In people with Huntington's disease, the CAG repeats number from 40 to over 80. Individuals having only one copy of the gene coding for the defective protein will develop Huntington's disease. No person has ever been found who is homozygous for the condition. Most people having the mutated gene do not show symptoms until they are in their forties or later; however, the greater the number of repeats the earlier the symptoms appear.

a Many people with Huntington's disease in their families wish to be tested to determine whether they have the mutated gene. How might such a test be performed?

b There is no treatment for Huntington's disease. It will result in serious mental deterioration. It is ultimately fatal. It is a dominant condition so having one allele means having the condition is certain. If a child has the condition then the parent with the condition in their family must also have the mutated allele and will develop Huntington's disease. This raises issues associated with testing for the condition.

i Describe an ethical issue associated with testing for Huntington's disease.

ii Describe a social or financial issue associated with testing for Huntington's disease.

8 WAGR syndrome is a genetic condition caused by deletion of several genes from chromosome 11. It results in a number of symptoms. The condition gets its name from the four most serious: *W*ilms tumour (a cancer of the kidneys), *A*niridia (lack of the iris in the eyes leading to vision loss), *G*enital and urinary tract abnormalities, mental *R*etardation.
When the WAGR syndrome is suspected, FISH is used to confirm the diagnosis.
  a What does FISH stand for?
  b Normally when FISH is used two gene probes are used: one for a sequence known to be on the same chromosome as the target DNA and the other for the target DNA. The first probe allows the researcher to clearly identify the correct chromosome. When the probe joins to its target it fluoresces. This fluorescence can be observed under a microscope. What is a probe?
  c **i** In order to perform FISH, what must the geneticists know?
  ii If the individual has normal chromosomes, how many glowing spots would be visible on the chromosomes?
  iii If the individual has WAGR, how many glowing spots would be visible on the chromosomes?

9 At a small country hospital three babies were born on one night. This stretched the resources of the hospital to such an extent that normal procedures failed and the babies were not labelled with their mother's name. In order to ensure the correct babies were taken home by the correct parents DNA testing was performed.
Homologous chromosomes come in pairs. Each member of the pair has alleles for the same genes. STRs also come in allelic forms. The number of repeats on both members of a homologous pair can be different.
A STR on chromosome 6 that has between 7 and 20 ATTG repeats was investigated in order to match the parents with their babies. The results for the couples and the babies are shown below.

Couple one		Couple two		Couple three	
Mother	Father	Mother	Father	Mother	Father
11, 14	7, 12	14, 20	12, 18	18, 20	11, 18

Baby one	Baby two	Baby three
12, 20	11, 20	12, 14

  a Match each baby with its correct parents.
  b Explain how you matched the couples with their children.
  c Figure 14.1.10 shows one way of analysing a series of STRs. It shows the analysis of 10 sites. Some sites have two peaks and others only one. Explain why this is the case.

10 Strawberries are a very fragile fruit. If they are exposed to freezing temperatures the fruit becomes soft and unappealing. This results in considerable economic loss to strawberry farmers. Scientists have been searching for a way to make strawberries more resistant to frost. One approach that is still in laboratory trials is to genetically engineer strawberries.
Arctic flounder live in near-freezing waters but their blood does not freeze because they make a protein that acts as an antifreeze. Scientists have cut this gene from the genome of the Arctic flounder and inserted it into a plasmid. They have also inserted into the plasmid a gene that makes epidermal cells produce a blue pigment.
Agrobacterium tumefaciens is used to modify the strawberry cells. When the strawberry cells have been successfully modified they produce both the antifreeze protein and the blue pigment. Blue frost-resistant strawberries are produced.

  a The resulting strawberry plants are genetically modified. Are they also transgenic? Explain.
  b Blue strawberries are unlikely to have large-scale consumer appeal so when the strawberries finally move to field trials this gene will not be in the modified strawberries. Why is it being used in this early stage of research?
  c Draw a flow chart of the steps needed to make the strawberries frost resistant.
  d If the frost-resistant strawberries ever get to the stage of field trials, applications will need to be made to the government department that oversees the control of genetically modified organisms. Why can't scientists just plant genetically modified crops without any oversight?

11 Classify the following as prevention or cure.
  a placing an antiseptic on a fresh cut
  b taking a course of antibiotics for a bacterial chest infection
  c having a vaccination against cholera before going overseas
  d washing hands after going to the toilet
  e isolating an exotic bird in quarantine after it has been found in luggage by customs officials

12 Zika virus is a mosquito-transmitted virus that has been linked to an increase in birth defects in South America. It has been postulated that it is a cause of an increase in the number of babies born with microcephaly (a small skull and brain).

There is currently no vaccine or treatment for Zika virus, but considerable research efforts are being directed towards the development of both.

Preliminary research indicates that the virus uses a receptor on cell surfaces called AXL to attach to cells and gain entry. This receptor is one in a family that is very important in cell signalling.

a **i** Draw a labelled diagram showing a hypothetical AXL receptor with the virus attached to it.

ii What can you say about the shape of the protein the virus uses to attach to the cell and the shape of the AXL receptor?

b Design a drug that will block the virus from attaching to and entering the cell. Draw its shape.

c How might the spread of Zika virus be limited even in the absence of a vaccine or treatment?

13 Ghrelin is a hormone that helps in the regulation of body weight through its action in promoting hunger. It acts on receptors in the brain. The stimulation of these receptors causes a feeling of hunger and an increase in food intake.

When ghrelin was first discovered it was thought that blocking the ghrelin receptors in the brain might be a viable treatment for obesity. (For a number of reasons this has proved to not be the case.)

If blocking the ghrelin receptors had proved to be a valid treatment for people with obesity, how would doctors have gone about designing a suitable drug?

14 A medical student studying the impact of antibiotics on pathogenic bacteria set up the following experiment. Three sterilised nutrient agar plates were prepared. Plate A remained sealed with nothing added to it. Plate B was exposed to bacterial spores, then sealed. Plate C was treated with three drops of a common antibiotic and then exposed to the bacteria in the same way as plate B, then it too was sealed. All three plates were incubated at 37 °C for 24 hours and then examined for the growth of bacterial colonies. The results are set out in the following diagram.

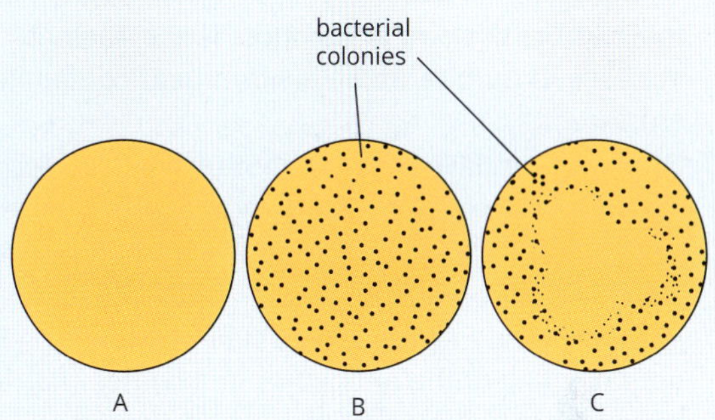

a Suggest a hypothesis that the medical student was testing.

b **i** State which agar plate was the negative control and which was the positive control.

ii Explain the significance of the positive control in this experiment.

c Explain whether or not the experimental results support the hypothesis.

UNIT 4 • Area of Study 2

REVIEW QUESTIONS

How do humans impact on biological processes?

Multiple choice questions

1 Which of the following statements most correctly describes how antibiotic resistance has developed in bacteria?

- **A** Colonies of antibiotic-resistant bacteria have developed as a result of the bacteria undergoing mutations in response to the antibiotics.
- **B** Some bacteria have a pre-existing allele that allows them to break down antibiotics. These bacteria increase in number in an environment containing antibiotics.
- **C** Bacteria that have undergone mutations in response to antibiotics pass that resistance to other bacteria by horizontal transfer. These bacteria then increase in number.
- **D** Humans create antibiotic resistance in bacteria by not taking all of the antibiotics in a course that has been prescribed by a doctor.

2 Which step in the process of PCR best describes annealing?

- **A** separating the DNA strands
- **B** binding the primers
- **C** adding the polymerase
- **D** building the complementary DNA strands

3 The figure below is a DNA profiling printout obtained from one individual. Ten regions have been analysed, 9 STRS and the sex chromosome markers. In the centre of the figure there is a peak labelled as 32.

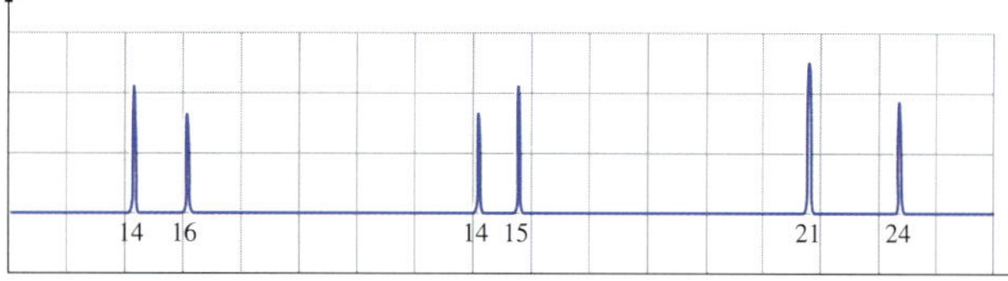

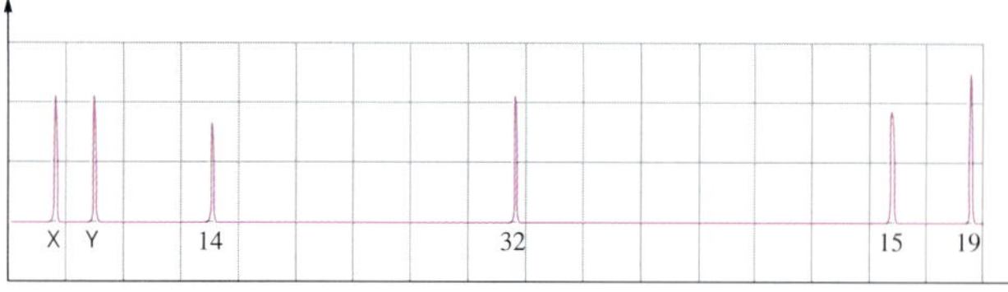

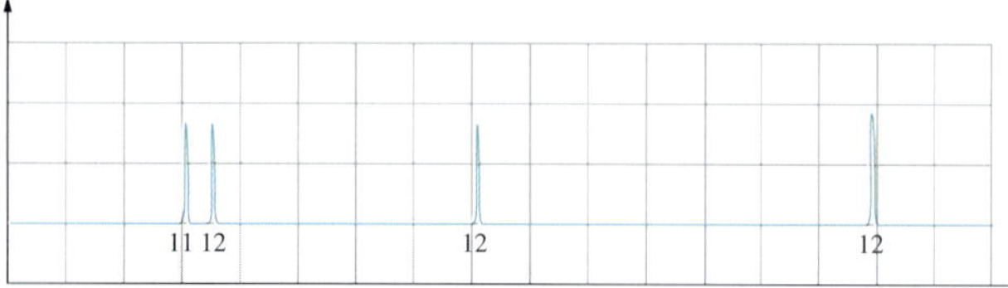

Which of the following best describes why there is only one peak?

- **A** The individual has one chromosome with that STR and the STR is 32 units in length.
- **B** The individual has two chromosomes with that STR and the STR is 32 units in length.
- **C** The individual has two chromosomes with that STR and the STR is 16 units in length.
- **D** The peak represents an STR on the Y chromosome and the person tested was male.

4 Students were performing an experiment in which they were transforming *E. coli* bacteria to make them resistant to the antibiotic kanamycin. The experiment involved the use of four plates. Two were plain nutrient agar plates and two were nutrient agar and kanamycin. One plain nutrient plate and one of the kanamycin plates were exposed to untreated *E. coli*. A second batch of *E. coli* was incubated with a plasmid containing the kanamycin-resistance gene and then heat shocked. These bacteria were then divided between the remaining plates—one nutrient only and the other with kanamycin. The plates were then incubated at 37 °C for 24 hours. Unfortunately, when the students came to check their plates they realised that they had forgotten to label them. Using your understanding of the processes involved, determine which of the following alternatives correctly identifies each of the plates.

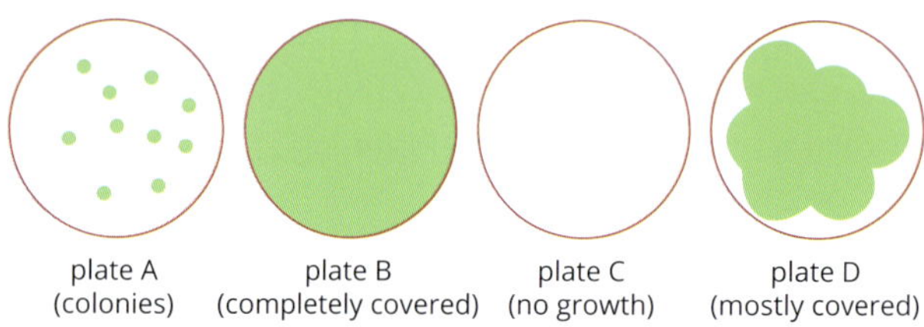

	Untreated *E. coli*		Transformed *E. coli*	
	Nutrient agar only	Nutrient agar and kanamycin	Nutrient agar only	Nutrient agar and kanamycin
A	plate B	plate C	plate D	plate A
B	plate A	plate B	plate C	plate D
C	plate D	plate A	plate B	plate C
D	plate C	plate D	plate A	plate B

5 Antiviral medications provide treatments to reduce the seriousness of viral infections. Which of the following best describes how do they do this?

- **A** They block the receptors used by the virus to attach to and enter the cells.
- **B** They block transcription in the infected cells.
- **C** They prevent the virus particles from leaving the cell.
- **D** All of the above.

6 Different drugs are appropriate for different pathogens. Canestan fights fungi, penicillin fights bacteria, Relenza is effective against influenza virus. The plasmodium which causes malaria is a protozoan. Which would be the most effective treatment for malaria?

- **A** Canestan
- **B** penicillin
- **C** Relenza
- **D** None of the above.

7 Rational design of a drug to combat a virus involves targeting a particular viral protein. Which of the following characteristics would be the most useful in designing a drug to combat the virus?

- **A** The protein is used by the virus in the early stages of its reproduction.
- **B** The protein is used by the virus to attach to its host cells.
- **C** The protein is used by the virus in the later stages of its reproduction.
- **D** The protein is common to several strains of the virus.

Short answer questions

8 The 'Ever-Open Convenience Store' had experienced a number of robberies. The police were keen to catch the offender, who brandished a gun during each robbery. The police had five suspects, but were unable to gather sufficient evidence to clearly identify the perpetrator.

The robber wore rubber gloves, a mask, concealing clothing and a balaclava. After the fourth robbery the police found the little finger ripped from a pair of rubber gloves. The piece of glove was carefully collected and sent to the forensic science laboratory to be tested for DNA. Such material will contain a very small amount, if any, of DNA.

- **a** What is the source of the DNA found inside the glove?
- **b** Such small amounts of DNA are not suitable for preparing a DNA profile. How will the forensic scientists acquire enough DNA to create a DNA profile? Draw a flow chart describing the process.
- **c** DNA from the crime scene was collected and amplified. A DNA profile was made using the amplified DNA from the crime scene and DNA from each of the five suspects. The profile is shown at the bottom of the page.
  - **i** The standards are 1000 kb, 2000 kb, 4000 kb, 5000 kb, 7000 bp and 10 000 bp. What is the size of the band indicated by the red arrow?
  - **ii** Why are standards needed?
  - **iii** Explain which suspect best matches the crime scene sample.
  - **iv** Does a match mean that the suspect performed the crime?

9 Myotonic dystrophy is a serious disease that causes wastage of muscles. It can affect cardiac muscle, resulting in heart problems. The most severe form of the disease is caused by a mutation in the *DMPK* gene, which is found on the long arm of chromosome 19. It is caused by a CTG trinucleotide repeat. In most people there are between 5 and 37 repeats but in individuals with myotonic dystrophy the number of repeats exceeds 50. It is often an adult-onset disease and has an autosomal dominant pattern of inheritance. This means that if the allele is inherited it is certain that the disease will develop, but the person may not know until later in life.

- **a** Explain how electrophoresis could be used to identify whether an individual has the mutated allele.
- **b** Before a person can undergo genetic testing they must spend some time discussing associated issues with a counsellor. What are some issues that could be associated with genetic testing for myotonic dystrophy?

10 Kuru is a disease that was once common in the highlands of New Guinea. It has been established that it is caused by a prion. Like mad cow disease in cattle and Creutzfeldt–Jakob disease (vCJD) in humans, this prion builds up in neurons, causing plaques that eventually destroy the cells, resulting in compromised neurological function.

Researchers studying the problem in New Guinea have discovered that there are some individuals who are highly resistant to the misfolding of their proteins into the prion form. Study of these individuals has established that they possess a mutated protein.

Further study of this protein is needed as it may lead to a treatment or cure for both Kuru and vCJD.

In order to study this protein, a large and readily available pure supply is needed, so the scientists wish to introduce the gene for the protein into bacteria, which will then produce a constant supply for research.

a Assume that the amino acid sequence of only part of the protective protein has been identified. Using this information, mRNA for the protein has been extracted from human cells. This will be used to make the gene to insert into the bacterium.

i How will the gene be produced from the mRNA?

ii Why is it better to use mRNA in this case rather than DNA?

b Once a functional copy of the gene has been created, many copies will be required. To do this a plasmid that can be inserted into a cell is needed. A plasmid containing the *lacZ* and ampicillin (an antibiotic) resistance genes is obtained. (These are made commercially today and would just be bought from a biological supplier.)

i What does the *lacZ* gene do in bacteria normally?

ii What does it do in the transformed bacteria?

iii Why is the ampicillin resistance gene included?

c Once the plasmid is obtained, the gene for the Kuru-protective protein must be inserted into the plasmid. The sequences shown below are for the coding strand.

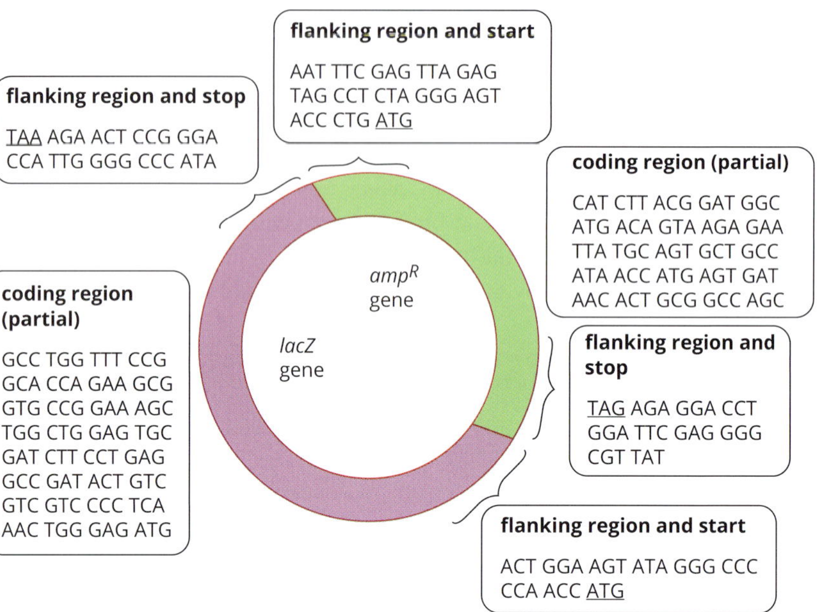

Many plasmids are cut open with a restriction enzyme and are incubated with the gene of interest. You have four restriction enzymes you could use. The enzymes have cutting sites as shown. The slash (/) indicates the cutting site. Enzymes 1 and 2 create sticky ends and enzymes 3 and 4 create blunt ends.

Enzyme	1	2	3	4
Cutting site	GGG/CCC	CTT/CCT	GA/TACT	GAA/AGC

i What is the difference between sticky ends and blunt ends?

ii Explain which enzyme should be used to cut the plasmid.

d Once the plasmid has been cut it should be incubated with the gene to allow the gene to be incorporated into it. The bacteria will then be mixed with the plasmids and a proportion of the bacteria will be transformed.

i Name the enzyme needed to incorporate the gene for the protein of interest into the plasmid.

ii How will the proportion of bacteria that is transformed be increased?

11 Severe Combined Immune Deficiency (SCID) is an inherited disease. The most common form of the disease has been linked to mutations in the gene *IL2RG*, which is found on the X chromosome, so this form of the disease is called SCID-X1. As a result of the mutation the body lacks the ability to make natural killer (NK) cells, T cells and B cells, effectively resulting in a total absence of the adaptive immune response and a depleted innate response to viruses. Typically, individuals die from viral infections within the first year of life. As girls need to inherit the trait from both parents and boys with SCID-X1 do not survive to maturity, this condition is only seen in boys.

The first trials of a new treatment for SCID-X1 began in 1999. Bone marrow stem cells were collected from the patients and the active form of *IL2RG* was inserted into the nuclei of the cells using a viral vector.

a **i** What is a vector in this context?

ii Why is a virus used as a vector?

b By 2002, 20 boys had been treated but only 18 of them actually developed the ability to make effective lymphocytes. Long-term monitoring of the patients showed that eventually 5 of the boys developed leukaemia and, of those 5, one eventually died. Further research into the causes of the leukaemia discovered that the cancer developed because the *IL2RG* gene inserted into the chromosomes by the virus was inserted in such a way as to activate a cancer-causing gene.

How might the insertion of a virus cause a cancer-causing gene to become activated?

c Gene therapy trials have recommenced following identification of the cause of the leukaemia and subsequent modification of the vector to better regulate its insertion into the chromosomes. Even so, there are issues associated with the use of this therapy, especially because the recipients are all infants.

What issues may arise in gene therapy of infants that do not arise when the patients are adults?

12 Antithrombin is a plasma protein. Its function is to stop blood clots forming in inappropriate places. People with a mutation in the gene for antithrombin production will easily develop a thrombosis (blood clot) and will generally require hospitalisation. Formation of blood clots in the brain and heart can cause death. Like many blood-clotting diseases, antithrombin deficiency is treated by injecting required amounts of the protein. The challenge for doctors is the supply of the protein. In order to create a steady supply of antithrombin, goats have been engineered to produce the protein in their milk.

The process is summarised below:

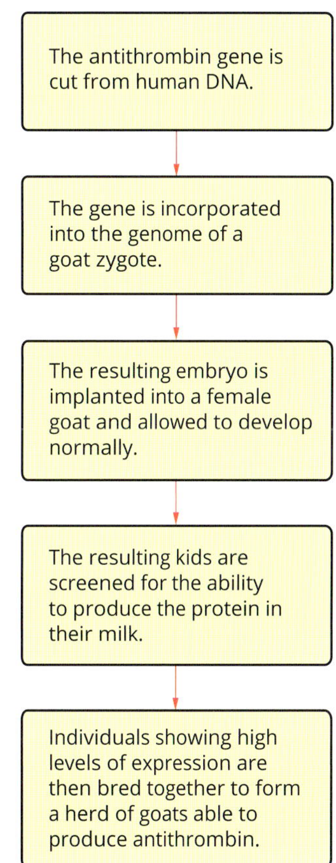

The goats are both genetically modified and transgenic.

a Explain the difference between being genetically modified and transgenic. Give examples to illustrate your understanding.

b Are there any possible drawbacks to using animals to make human proteins?

c Identify an ethical issue associated with the use of animals for the production of human pharmaceuticals.

13 Rational drug design is the new area of development in pharmaceuticals. It is based on learning all it is possible to know about the cause of a disease, whether its cause is a pathogen or a malfunction of a body system due to genetics or lifestyle. New and returning diseases are an important target for rational drug design. One disease that has been studied extensively is HIV. It has been discovered that about 1% of Caucasian people are highly resistant to infection by the virus. The resistance of these people to infection has provided an important line of research.

It was determined that HIV enters a cell by attaching to a receptor called CD4. This is an important receptor for many cell-signalling pathways and so is not a suitable target for an anti-HIV drug. However, further investigation showed that HIV also needs a second receptor (a co-receptor) that usually works with CD4. This receptor is called CCR5. In most people this is a transmembrane protein that assists HIV to enter the cell. The resistant individuals have a mutation in the gene for CCR5 that causes the protein to be much shorter so that it is totally within the cell and is thus not able to help HIV enter the cell. People heterozygous for the mutation are resistant to HIV and those homozygous for it are highly resistant. These people show no ill effects from having this mutation.

a **i** Describe a possible reason why this molecule would make an ideal target for a possible drug against HIV.

ii Describe the design process to develop such a drug.

b Are there likely to be any issues associated with the trialling of an anti-HIV drug?

14 Imagine you a member of a group of scientists who have been working on finding a treatment for Ebola. Your team has used rational drug design techniques to design a drug that will block the ability of Ebola to enter and infect cells. You are still in the initial stages of trialling and have decided that the initial tests will take place in mice. You have 200 genetically similar mice at your disposal.

a **i** Design an experiment to test the effectiveness of your drug in slowing the progress of the disease.

ii In your answer make sure you identify the dependent and independent variables.

b What results would indicate that trials should move on to the next stage with this drug?

c What kinds of safety precautions would be needed in performing your trials?

Glossary

5' cap (five-prime cap) A special nucleotide that is added to the 5' end of primary transcripts in eukaryotes. The process is known as mRNA capping and it functions to make stable, mature mRNA that is ready to undergo translation.

A

abscisic acid A plant hormone involved in many plant development processes.

absolute dating (using radiometric methods) A direct quantitative method of determining the age of a rock or object using radioactivity.

accuracy The ability to obtain an exact (or true) value.

acquired immune deficiency syndrome (AIDS) A syndrome in which helper T lymphocytes are destroyed and are in sufficiently low numbers that the immune response to infection is impaired. Can result from malignancy or HIV/AIDS.

action potential The reversal of the resting potential difference across a plasma membrane (between the intracellular and extracellular voltage), typically of a neuron.

activation In proteins, activation is the modification of the 3D shape of a precursor molecule to turn it into a functional molecule; for example, of enzymes.

activation energy The energy that is required to start a biochemical reaction.

active immunity Immunity that involves an individual's own adaptive immune response, through B and T lymphocytes.

active site The specific site of the enzyme that binds the substrate and where catalysis occurs.

adaptive immunity An immune response that is specific to a particular antigen; only present in vertebrates.

adaptive radiation A process of rapid evolution and divergence of species from a common ancestor in response to new environmental conditions.

adaptive value A measure of how well suited a particular phenotype is to a particular environmental condition and the likelihood of an individual with the allele that codes for the phenotype surviving and reproducing successfully. Alleles with high adaptive values tend to persist and increase in frequency in gene pools compared to those with low adaptive values.

adenine (A) A nitrogenous base (a purine) that occurs in nucleotides of DNA and RNA.

aerobic respiration The breakdown of glucose in the cell using oxygen, the energy from which is used to synthesise ATP. Aerobic respiration is comprised of glycolysis, the Krebs cycle and the electron transport chain.

agglutination The process in which antibodies bind to antigens on the surface of cells and form antigen–antibody complexes that clump together and activate phagocytes and the complement cascade, which leads to antigen/cell destruction.

aim A statement describing in detail what will be investigated.

allele Different forms of a gene. Different alleles of a single gene will have differences in their genetic code, resulting in different forms of the phenotype. For example, the gene that codes for hairline shape in humans has two alleles—straight and widow's peak.

allele frequency The relative proportion of a particular allele in a gene pool. Typically presented as a decimal or percentage of the allele of that gene in the gene pool.

allergen An antigen that elicits an allergic response.

allergy (or allergic reaction) The rapid and vigorous overreaction of the immune system to antigens called allergens. Allergic reactions involve the production of IgE by B lymphocytes and the release of histamine by mast cells.

allopatric speciation The divergent evolution of two new species from an ancestral species resulting from separation by a geographical barrier. The lack of gene flow between populations and the slight variations in environmental conditions of the two habitats drives natural selection in slightly different ways until the two populations become genetically distinct.

allosteric site A site on an enzyme other than the active site to which an effector molecule binds.

alpha helix A coiled secondary protein structure within a polypeptide chain stabilised by hydrogen bonds between adjacent amino acids.

amine group (amino group) A $-NH_2$ group.

amino acid The monomer of polypeptides. All amino acids contain an amine group at one end of the molecule and a carboxyl group at the other end.

amino-acid-derived hormones Small signalling molecules derived from the amino acids tyrosine and tryptophan. They include thyroid hormones (hydrophobic) and catecholamines (hydrophilic).

amniocentesis Technique to take a sample of amniotic fluid surrounding a foetus during pregnancy for genetic testing.

amplification The process of increasing the strength of a cellular signal, typically by increasing the number of molecules involved at each step of a transduction cascade. For example, when a peptide-based hormone binds to a plasma receptor, it triggers the production of many second messenger molecules. Each second messenger molecule then triggers the activation of proteins in the cascade, with each protein triggering the activation of the next protein in the sequence.

anabolic reaction Biochemical reactions in which larger molecules are made from smaller molecules; requires an input of energy to build new bonds.

analogous features Features (e.g. organ or structure) in different organisms that have the same function but have evolved independently and not from a common ancestor. Analogous features are the result of convergent evolution. Analogous features may evolve because unrelated organisms have experienced similar selective pressures.

anaphylaxis A severe allergic reaction that can be life-threatening.

aneuploidy A condition in which an individual does not have the correct diploid number of chromosomes. As a result of non-disjunction of homologous chromosomes during meiosis, individuals typically have an extra copy (trisomy) or only one copy of a particular chromosome. Triple X syndrome is an example of a human disorder caused by aneuploidy and is characterised by three copies of the X chromosome.

angiosperm A flowering, fruit-bearing plant.

anneal The joining together of DNA or RNA fragments by complementary base pairing.

Anthropocene The time period when human activities began to have significant impacts on Earth's systems on a global scale. The Anthropocene is not yet recognised as a true geological time period.

antibiotic A substance, produced by a microorganism or synthesised, that inhibits the growth of a type of bacteria.

antibody Also known as immunoglobulins, antibodies are proteins produced by plasma cells that are highly selective for, and bind to, specific antigen molecules

anticodon The three nucleotides on a transfer RNA (tRNA) molecule that join to the codons on mRNA by complementary base pairing during the process of translation.

antigen A substance that reacts with antibodies and T lymphocyte receptors; antigens that induce an immune response are immunogens.

antigen–antibody complex A specific chemical interaction between an antibody (immunoglobulin) molecule and an antigen molecule.

antigen presentation The presentation of antigens by antigen-presenting cells.

antigen-presenting cell (APC) A cell that uses MHC-II on its surface to present foreign antigens to helper T lymphocytes to elicit an adaptive immune response. Examples include dendritic cells and macrophages.

antigenic variation The mechanism of changing surface antigens, usually to avoid detection or an immune attack. Employed by certain protozoans such as *Plasmodium* sp.

antimicrobial drug A compound that inhibits the growth of microorganisms, including bacteria, fungi, protists and viruses.

antimicrobial resistance The ability of a microbe to survive in the presence of an antimicrobial drug.

antiparallel Running in opposite directions, with one strand in the 5' to 3' direction and the other in the 3' to 5' direction (referring to the two strands in DNA molecules).

antiretroviral drug A drug that specifically inhibits replication of retroviruses.

antiseptic A non-specific chemical that is used to kill disease-causing organisms on body surfaces.

antiserum A serum containing specific antibodies.

antiviral A drug that inhibits replication of a virus by blocking entry or exit from the cell, or blocking viral replication enzymes.

apoptosis Regulated and programmed cell death. The controlled destruction of the cell does not spill the contents and does not trigger an immune response.

apoptosome A large protein formed in the process of apoptosis. It is made up of cytochrome *c* and a protein called Apaf-1.

Archaean A period of time on Earth between 4000 and 2500 million years ago

archaeological Relating to the scientific study of human history or prehistory and their culture from the analysis of fossil remains and artefacts.

archaic Very old or ancient.

artificial active immunity Active immunity resulting from the administration of antigens, such as through vaccination.

artificial passive immunity The administration, usually by injection, of antibodies produced by another organism to provide an immediate, specific immune response.

artificial selection A process of changing the allele frequencies of a population through human intervention. It is a form of selective breeding. Phenotypes that are selected for may not necessarily be better suited to the environment, but may be desired by humans and so the alleles that produce the desired phenotypes are increased in frequency in the population. For example, dairy cows are artificially selected for high milk yields.

Assimilation (partial replacement) model A model to explain the origins of present day humans. This model proposes that all living humans had an African origin and when they migrated out of Africa, there was occasional interbreeding with archaic humans that were already living in other parts of the world, resulting in hybrid populations.

ATP (adenosine triphosphate) The energy-carrying molecule of the cell that provides energy for cellular processes. ATP releases energy when its terminal phosphate bond is hydrolysed.

autocrine signalling A type of chemical signalling in which the signalling molecule is received by the same cell or cell type that secreted it.

autoimmune disease Disease in which there is a failure of tolerance and an adaptive immune response is directed against a self-antigen, causing T lymphocytes to attack tissues directly and B lymphocytes to produce antibodies against the self-antigen. Autoimmune diseases can be organ-specific or generalised.

autotroph Living organism capable of synthesising all of its own food by photosynthesis or chemosynthesis.

auxin A plant hormone that is released from the growing tip and stimulates growth. Synthetic auxins are used as herbicides.

axon A long projection of a neuron that transmits the action potential. Often wrapped with myelin sheath.

B

B lymphocyte (or B cell) Lymphocytes that when stimulated produce large quantities of antibodies specific to a particular antigen. They are responsible for the humoral immune response and include both memory and plasma cells.

background extinction The average rate of natural loss of species.

bacteria All prokaryotes not members of the domain Archaea.

bacterial transformation The incorporation of DNA from another organism into a bacterial cell.

bacteriophage A virus that infects bacteria.

bar graph A graph that shows the value of the dependent variable by the length of the horizontal bar; the categories are labelled up the *y*-axis.

base One of five nitrogenous chemicals present in the nucleotides of nucleic acids (DNA or RNA). The five bases are adenine, guanine, cytosine, thymine (DNA only) and uracil (RNA only).

Bcl-2 A family of proteins that regulate apoptosis.

benign tumour A mass of abnormal (but not cancerous) cells. They are not cancerous because they do not invade nearby tissue or spread throughout the body through a process of metastasis.

beta-pleated sheet A secondary protein structure stabilised by hydrogen bonds between different regions of a polypeptide chain that create pleat-like formations.

biochemical pathway (metabolic pathway) A sequence of biochemical reactions, each catalysed by a specific enzyme in which the product of one reaction becomes the substrate of the next.

biogeographic region An area inhabited by a unique set of animals and plants, indicating a common history or environment, for example the Australasian region or neotropical region.

biogeography The study of the distribution of species and ecosystems in geographic space and through (geological) time.

biomacromolecule A large molecule formed by joining together many monomers to form a chain. Examples include proteins and polysaccharides.

biomolecule A molecule involved in the maintenance or metabolism of living organisms.

biosecurity Measures to detect, respond rapidly to and recover from pests and diseases, including introduced species, to protect agricultural production and wildlife biodiversity.

bipedal An animal that walks on two legs.

bispecific monoclonal antibody (bispecific mAb) Monoclonal antibody (mAb) constructed through molecular technology to have two different binding sites: one for a cancer cell and one for an immune cell. Bispecific mAbs 'identify' and 'deliver' cancer cells to the immune system.

bleb A protrusion or bulge of the plasma membrane of a cell.

block mutation A type of mutation that typically involves entire genes or multiple neighbouring genes on a chromosome. There are five types of block mutations: duplication, chromosomal deletion, inversion, chromosomal insertion and translocation.

blunt-end restriction enzyme A restriction enzyme that leaves clean-cut ends because it cuts both strands of the DNA molecule at the same location within the recognition site.

***BMP4* gene** The gene that regulates the development of cartilage and muscular cell development in the jaws of a cichlid fish.

bottleneck effect The resulting impact when a large portion of a population is removed from the habitat by chance, typically as a result of a natural disaster. The effect of genetic drift is more significant on the smaller population, as the remaining gene pool has reduced diversity.

branch Line on a phylogenetic tree that represents the evolutionary path from a common ancestor (lineage).

C

C3 plant A plant that uses the most common form of photosynthesis in plants, in which the first carbon-containing product is a 3-carbon compound.

Calvin cycle A cyclic biochemical pathway in the light-independent reactions of photosynthesis, in which carbon from carbon dioxide becomes fixed in the synthesis of carbohydrate.

cancer A group of diseases characterised by uncontrolled cell division.

cancer vaccine Vaccines made up of cancer cells, parts of cells, or pure antigens that stimulate the immune system to prevent or fight cancer cells.

capsid The protein coat of a virus.

carbohydrate A class of organic compounds made from three elements: carbon, hydrogen and oxygen. Includes monosaccharides, disaccharides and polysaccharides. Carbohydrates are also known as saccharides.

carboxyl group A -COOH group containing a carbonyl and hydroxyl group.

carcinogen A substance that damages cell DNA. A carcinogen can be physical, chemical or biological.

carrier protein A type of protein found on the plasma membrane that is involved in facilitated diffusion and active transport.

cascade A multi-step process in which each step must occur in a set order with each step triggering the next in the sequence.

caspase One of a group of enzymes involved in protein and DNA cleavage. Caspases are involved in apoptosis.

cast fossil A three-dimensional 'sculpture' of an organism formed by materials such as silica or phosphate filling the vacant space in an impression or fossil mould.

catabolic reaction Biochemical reaction in which there is a breakdown of macromolecules into smaller molecules; releases energy.

catalyse To increase the rate of a reaction.

catalytic power The ability or potential of an enzyme to increase the rate of a biochemical reaction compared to the reaction occurring without the enzyme present.

catecholamines Hydrophilic amino-acid-derived hormones; examples include adrenaline and dopamine.

cell Cells are the smallest structural and functional units from which all living things are built.

cell-mediated immunity An immune response that is mediated by T lymphocytes. Compare with *humoral immunity*.

cell wall An external structure that surrounds the plasma membrane for structural support and protection. Composed of cellulose (in plants) or murein (in bacteria).

cellular pathogens Cellular organisms that are a source of non-self antigens and cause disease. They include bacteria, protozoa, oomycetes, fungi, several types of worms and arthropods.

cellular respiration (1) General—the complete breakdown of glucose to provide energy in cells. (2) Specifically—refers to the second aerobic stage that occurs in the mitochondria and produces 36–38 molecules of ATP per molecule of glucose.

central nervous system The primary organs of the nervous system; in vertebrates, the brain and spinal cord.

centriole A small structure composed of microtubules and involved in cell division or in the formation of cell structures such as flagella and cilia.

channel protein A type of protein found on the plasma membrane that is involved in facilitated diffusion.

chaperonin (chaperone protein) A group of proteins that ensure the correct folding of newly synthesised proteins into their tertiary structures.

chemical group A group of covalently linked atoms, such as an amino group or hydroxyl group, that has a characteristic chemical behaviour.

chemokines Cytokines that attract white blood cells to the site of infection.

chemotherapy The use of one or more chemicals administered to a person for the treatment of cancer.

chimeric monoclonal antibody (chimeric mAb) An antibody made of mouse and human molecular components.

chlorophyll The green pigment found in chloroplasts in plants, and within some prokaryotic cells, that absorbs light energy for photosynthesis.

chloroplast An organelle that uses light energy, carbon dioxide and water to produce glucose.

chorionic villus sampling Technique to sample membranes surrounding a foetus for genetic testing in early pregnancy.

chromosomal abnormality A change to the number of chromosomes, or the genetic composition of an individual or multiple chromosomes, which can lead to disease.

cilia An organelle involved in movement (of the cell or things around the cell).

***cis* face** The side of the Golgi apparatus facing the nucleus.

cisternae Flattened sac-like membranes found in the Golgi apparatus and endoplasmic reticulum.

clade A group of organisms that includes an ancestor and all descendants of that ancestor.

cladogram A branching diagram representing the evolutionary relationships between taxa. The branches of a cladogram are scaled (the lengths of the branches do not represent evolutionary distance).

clonal selection The theory that in a group of lymphocytes, a specific antigen will activate only the lymphocyte that has a receptor that specifically recognises it. This lymphocyte will proliferate into clones of itself.

coding strand The strand of DNA that has the same nucleotide base sequence as the mRNA strand produced by transcription (uracil in the mRNA in place of thymine in the DNA).

codon Basic unit of the genetic code. A sequence of three nucleotides on mRNA that codes for a particular amino acid, or indicates the beginning or end of translation.

coenzyme A small organic molecule that combines with an enzyme and is necessary for its activity.

coevolution The evolution of two species in response to one another such as those between pollinator and flowering plant or between parasite and host.

coexisted Existed at the same time and place.

cofactor A chemical component such as a metal ion or coenzyme that is required for the proper function of proteins.

cohesive (cohesion) describes the attraction between particles of the same type, such as water molecules.

column graph A graph that shows the value of the dependent variable by the height of the column; the categories are labelled across the *x*-axis.

common ancestor An organism from which two or more species diverged. Also known as shared ancestor.

comparative embryology The branch of embryology that compares embryos of different species to show how animals are related.

comparative morphology The analysis of the body plan and structures of organisms in order to understand the relationships between organisms and their morphological features. It formed the basis of the Linnaean system of biological classification and the study of taxonomy.

competitive inhibition The inhibition of an enzyme due to a molecule that binds to the active site of the enzyme, preventing the substrate from binding.

complement proteins Proteins that are able to kill foreign cells by lysis. There are over 30 different complement proteins that are activated in response to antigen–antibody complexes, antigens and carbohydrates on the surfaces of some bacteria and parasites.

complementary base pairing The pairing in DNA and RNA molecules of the nitrogenous bases between two strands. In DNA adenine always pairs with thymine, and cytosine always pairs with guanine.

complementary DNA (cDNA) Double-stranded DNA that contains no introns; copied from mRNA by the enzyme reverse transcriptase.

complex polysaccharide A polysaccharide that has different monosaccharide subunits in the same molecule, such as murein found in the cell walls of bacteria. Also known as complex carbohydrate.

compound A molecule containing two or more different elements.

concentration gradient The difference in concentration of a solute between one region and another; for example, across a membrane.

condensation polymerisation The reaction in which monomers are joined to create a polymer by the removal of water; a bond is formed between them.

conformational change A change in the spatial (three-dimensional) arrangement of atoms in a macromolecule such as a protein or nucleic acid.

congenital A condition present from birth either as a result of a genetic defect or developmental abnormality.

conjugated monoclonal antibody (conjugated mAb) A monoclonal antibody (mAb) that has been attached to a drug, toxin or radioactive particle in order to deliver the treatment specifically to cancer cells.

conjugated protein A protein that contains a non-protein (prosthetic) group.

conservative substitution A change in the nucleotide sequence of DNA or RNA that leads to the replacement of one amino acid with a functionally similar one, and having little or no impact on the protein synthesised.

conserved DNA sequence that has remained unchanged throughout evolution. Conserved genes are often important for basic cellular functions and therefore viable forms of life. Because of their functional importance, these genes rarely accumulate mutations.

constant region The region of antibody molecules that remains the same and interacts with receptors on the body's cells.

constitutive gene A gene or protein that is always expressed or active.

continuous variable A variable that can have any number value within a given range.

control group The experimental conditions of the control group are identical to those of the experimental group, except that the variable of interest (the independent variable) is also kept constant.

controlled variables The variables that are kept constant during the investigation.

convergent evolution The evolution of similar features in unrelated groups of organisms (i.e. they do not share a recent common ancestor). For example, wings in birds and bats look similar and have the same function but evolved independently.

cortisol A (lipid-based) hormone produced in the adrenal cortex and belonging to a group of hormones called glucocorticoids.

crista (pl. cristae) The folded structures of the inner membrane of a mitochondrion.

cultural evolution Cultural change over time via the transmission of information from generation to generation. The evolution of culture is influenced by the environment and new information from it.

culture The accumulated knowledge passed on to the next generation by verbal, written or symbolic communication.

cyanobacteria Photosynthesising prokaryotic organisms with a similar structure to chloroplasts.

cytochrome *c* A component of the electron transport chain in mitochondria.

cytokine One of a group of peptides and proteins released from cells that are important in cell signalling, particularly between cells of the immune system.

cytokinin A plant growth hormone that promotes cell division or cytokinesis in plant roots and shoots.

cytosine (C) A nitrogenous base (a pyrimidine) that occurs in nucleotides of DNA and RNA.

cytoskeleton A network of microtubules and microfilaments that provides a supporting framework for cells and attachments for various organelles.

cytotoxic A substance or process that is toxic to a cell, and can cause death of that cell.

cytotoxic T lymphocyte T lymphocyte that is stimulated by cytokines to bind to antigen–MHC I complexes on infected host cells and release cytotoxic compounds that destroy the infected cells.

D

Darwinian theory The theory of biological evolution by natural selection developed by Charles Darwin.

death receptor Specific receptors on the outer surface of cells that will bind to cytokines and initiate the extrinsic pathway of apoptosis.

defensins Molecules active against bacteria, fungi and certain viruses.

degenerate More than one codon may code for a particular amino acid.

deletion mutation When a point mutation: a type of mutation that involves the loss of one or two nucleotides in a sequence. This type of point mutation causes a frameshift mutation. When a block mutation: multiple genes are cut from a chromosome.

demyelination Damage to the myelin sheath that surrounds nerve cell axons, which limits or stops the ability of the nerve to transmit electrical impulses.

denature (or denaturation) An irreversible change in the tertiary structure of a protein, as a result, for example, of heating the protein above a critical temperature.

dendrite The branched projections of a neuron that receive signals from other neurons across synaptic junctions.

dendritic cell A type of antigen-presenting cell.

deoxyribonucleic acid (DNA) A double-stranded nucleic acid that contains the genetic code in its sequence of bases. DNA is found in all organisms, and most viruses, in chromosomes, as well as in mitochondria and chloroplasts.

deoxyribose The five-carbon sugar molecule found in DNA. Deoxyribose is derived from ribose but lacks an oxygen molecule; it has a hydrogen atom rather than a hydroxyl group.

dependent variable A variable that may change in response to a change in the independent variable, and is measured or observed.

dephosphorylation A reaction that involves the removal of a covalently coupled phosphate group from another molecule.

developmental biology The study of the processes by which an organism develops from a zygote to its adult form.

differentiation The modification of the structure and function of a cell that occurs during its development.

diffusion The passive movement of a solute from a region of higher concentration to a region of lower concentration.

dinucleotide Two nucleotides joined together through a condensation polymerisation reaction that joins the phosphate of one nucleotide to the 3' end of the other nucleotide's sugar molecule. Water is removed in the process.

dipeptide Two amino acids joined by a peptide bond through a condensation reaction, which results in the production of a dipeptide and a molecule of water.

diploid A condition in which a cell has two (di-) sets of chromosomes, typically one set inherited from each parent in sexually reproducing organisms. The chromosomes with the same genes exist in homologous pairs. Denoted as $2n$.

disaccharide A carbohydrate consisting of two linked monosaccharide molecules.

discrete variables Separate or distinct values that can be counted.

disinfectant A substance that inhibits the growth of disease-causing microorganisms, or kills them. A disinfectant is generally toxic to human cells, but if diluted sufficiently can be used as an antiseptic.

divergent evolution The evolution of two or more different species from a common ancestral species.

DNA See *deoxyribonucleic acid.*

DNA amplification The process of creating millions of identical copies of a DNA sample using the polymerase chain reaction (PCR).

DNA ladder DNA standards; a set of DNA molecules of known size used as a 'molecular ruler' on gel electrophoresis to determine the size of other DNA molecules.

DNA ligase An enzyme that joins together fragments of DNA by forming a phosphodiester bond between the 3'-hydroxyl and 5'-phosphate of adjacent nucleotides.

DNA polymerase An enzyme that catalyses the formation of polymers of DNA by linking nucleotides into a chain by complementary base pairing with a template strand.

DNA profiling Technique to produce an individual's unique pattern of DNA bands on a gel. Produced by analysing short tandem repeat (STR) regions of the genome.

DNA thermocycler The machine used in the polymerase chain reaction that alters the temperature in pre-programmed steps.

double helix The double-stranded, coiled structure of a DNA molecule.

duplication mutation A block mutation that involves the repetition of a section of a chromosome, usually containing multiple genes. The duplication may involve thousands of repeats, significantly lengthening the chromosome.

E

ecological niche The distribution and ecological role of a species in its environment; how it meets its needs for food and shelter, how it survives, and how it reproduces.

Ediacaran fauna Assemblage of soft-bodied animals preserved as fossil impressions in marine sediments from the Proterozoic eon at the close of Precambrian time; found on all continents but named after the Ediacaran Hills, a fossil site in South Australia.

effector cell A cell that responds to signalling molecules. For example, a cell in which a signal transduction pathway activates an enzyme and causes a metabolic change in the cell.

electrons Negatively charged subatomic particles that usually occupy the orbit surrounding the nucleus of an atom.

emerging infectious disease Disease that is newly identified and has increased in incidence over recent years, or that may increase in the near future.

endemic A term used to describe a species as being native to a specific area or region. For example, koalas are endemic to Australia.

endocrine signalling A mode of transmission for signalling molecules that act on cells that are far from the cell that secretes them.

endocrine system The animal body system that is responsible for the production of hormones.

endocytosis The process in which a cell takes in a substance or particle from outside the cell by engulfing it within the plasma membrane to form a vesicle, bringing it into the cell.

endonuclease An enzyme, also called a restriction enzyme, that occurs naturally in bacteria and can cut DNA at a particular site (a recognition site); used in genetic engineering.

endosymbiosis The theory of endosymbiosis proposes that eukaryotic organelles such as mitochondria and chloroplasts were once single-cell organisms that were engulfed by cells and lived in symbiosis with them, eventually evolving into a unified cell.

enzyme A protein molecule that acts as a biological catalyst. Enzymes speed up rates of reactions that would otherwise take place much more slowly. Their action is often specific to only one type of reaction.

enzyme-linked immunosorbent assay (ELISA) A method to detect the presence of antigen in a sample, or antibodies in blood or serum.

enzyme–substrate complex The complex that forms when an enzyme binds to a substrate.

eon One of several subdivisions of geologic time enabling cross-referencing of rocks and geologic events from place to place. Eons are the largest subdivisions.

epidemic The sudden increase in the number of cases of a disease above what is normally expected in that population in that area.

epoch One of several subdivisions of geologic time enabling cross-referencing of rocks and geologic events from place to place. Epochs are the smallest subdivisions.

era One of several subdivisions of geologic time enabling cross-referencing of rocks and geologic events from place to place. Eons are larger subdivisions than eras; eras may be divided into periods and epochs.

error The difference between the true value and the measured value.

ethene A gaseous plant hormone involved in promoting fruit ripening.

eukaryote An organism with cells that contain a nucleus and other membrane-bound organelles.

evolution A change in the inheritable traits of a population (or species) over successive generations. See *genetic drift* and *natural selection*.

exocytosis The movement of materials out of a cell via a vesicle. The vesicle fuses with the plasma membrane, and the vesicle contents are released out of the cell.

exon The region of a gene that codes for a protein.

experimental group Controlled (fixed) variables are kept constant, a single experimental (independent) variable is changed and the dependent variable is measured to determine any effect of the change.

exponential relationship Variables that are exponentially proportional to each other will produce a curved trend line when graphed.

extracellular matrix The non-cellular fluid and fibres surrounding a cell, found between the cells of tissues.

F

facilitated diffusion A process in which molecules are moved down a concentration gradient across the plasma membrane by carrier proteins or channel proteins.

FAD (flavin adenine dinucleotide) A coenzyme that functions as an electron carrier in metabolic reactions.

feedback inhibition Occurs when a product produced late in a biochemical pathway acts as the inhibitor of an enzyme acting earlier in the pathway.

fermentation Stage in the breakdown of glucose to yield energy for the production of ATP, which follows glycolysis when there is no oxygen present. Produces either lactic acid (in most animals) or alcohol in the form of ethanol (in most plants and microorganisms).

fever An increase in body temperature that results from the regulated body temperature set point in the hypothalamus of the brain being set to a higher level by inflammatory cytokines, to slow the replication of bacteria and improve the adaptive immune response.

fibrous protein A type of protein that forms long fibres and provides structural support to cells and tissues.

first-hand data The measurements or observations that you collect during your investigation.

flagellum (pl. flagella) An organelle involved in movement (of the cell or things around the cell).

fluid mosaic model Describes the structure of the plasma membrane in which phospholipids and unanchored proteins are free moving, giving the membrane fluidity. Anchored proteins are scattered throughout, giving the membrane a matrix pattern.

foramen magnum The hole at the base of the skull through which the spinal cord connects to the brain.

fossil The preserved remains, impressions or traces of organisms found in rocks, amber (fossilised tree sap), ice or soil.

fossil record The record of the evolution of organisms through geological time based on information from fossils.

fossilisation The process of preservation of the hardened remains, impressions or traces of organisms in rocks.

founder effect Occurs when a small portion of a population disperses to a new location and becomes genetically isolated from the main population. The allele frequencies of the founding population are completely dependent on those of the specific individuals that were relocated, and therefore may be significantly different from those of the original population.

frameshift mutation A type of gene mutation that involves the insertion or deletion of one or two nucleotides, affecting every triplet (and codon) in that gene from the point of mutation.

fungus (pl. fungi) Non-phototrophic eukaryotes that have rigid cell walls made from chitin; includes moulds, yeasts, mushrooms and toadstools.

G

gel electrophoresis A technique used for separating fragments of DNA, or different proteins, based on their molecular weight (or length). Fragments migrate through a gel at rates that are dependent on their length and charge.

gene A specific sequence of nucleotides that codes for a particular protein or RNA molecule. It is the unit of heredity.

gene cloning The production of identical copies of a gene.

gene editing The modification of genes by removal, substitution or alteration by mutation, without necessarily introducing a foreign gene.

gene expression The process that leads to the transformation of the information stored in a gene into a functional gene product (usually a protein or RNA molecule).

gene flow The movement of alleles between individuals of different populations; includes the dispersal of pollen and seeds in plants.

gene pool All the alleles possessed by members of a population, which may potentially be passed to the next generation.

gene probe A section of DNA with a base sequence complementary to a particular gene, to which it base pairs. When labelled with a fluorescent dye or radioactive marker, it is used to find a particular DNA sequence within a DNA fragment.

gene regulation Processes that control gene expression, turning genes on or off.

gene therapy The replacement of faulty genes by genetic engineering techniques.

genetic code The linear sequence of three nucleotides in DNA or RNA that determines, or codes for, the sequence of amino acids in a protein.

genetic drift Random changes to allele frequencies in a gene pool as the result of a chance event. This has a more significant impact on smaller populations, as the chance death of one individual could eliminate an allele from the gene pool.

genetic isolation The prevention of gene flow between two populations.

genetic screening Genetic testing of samples from embryos, foetuses, newborns or adults to identify disease-associated alleles.

genetically modified organism (GMO) An organism with a genetic modification (GM) made by transfer of specific genes from another organism (transgenic) or by gene editing techniques.

genome The complete set of genes or DNA in an organism.

genotype The genetic composition of an individual. Contrast with *phenotype*.

geological time scale Time scale of events that have occurred on Earth from its formation to the present time.

germline mutation A mutation that can affect gamete formation and can therefore be inherited by offspring.

gibberellin A class of plant hormones that promote stem elongation and germination.

globular protein Type of protein that is folded and coiled to form a compact spherical shape. It has a tertiary or quaternary structure specific to its function; for example, enzymes.

glucose A common six-carbon monosaccharide carbohydrate, or hexose. It is the product of photosynthesis and the substrate for respiration.

glycolysis The biochemical pathway in which glucose is broken down to pyruvate. The first stage of cellular respiration.

Golgi apparatus (also known as Golgi body, Golgi complex) An organelle composed of a stack of cisternae in which proteins are assembled and then packaged in vesicles for exocytosis.

Gondwana A supercontinent of the past that formed when the process of plate tectonics united the land masses of the Southern Hemisphere and included present-day Africa, Madagascar, India, Australia, Antarctica and South America.

gracile Slender, thin build.

granum (pl. grana) A stack of flattened discs composed of the thylakoid membranes found in chloroplasts.

growth hormone A peptide-based hormone secreted by the pituitary gland that stimulates growth in plants and animals.

guanine (G) A nitrogenous base (a purine) that occurs in nucleotides of DNA and RNA.

H

Hadean The earliest time in the history of the Earth from its formation (around 4600 million years ago) until the date of the oldest known rocks (about 3800 million years ago).

haemagglutinin An envelope protein on the influenza virus used to attach to cells for entry. Abbreviated as H in virus name, as in H7N9.

haemolytic disease A condition in which red blood cells are abnormally broken down and removed from circulation. The causes of this include hereditary conditions or immune-mediated causes such as infection or autoimmune disease. Symptoms include fatigue, dizziness and shortness of breath.

haploid The number of chromosomes in a cell in which there is only one copy of every chromosome present, rather than homologous pairs of chromosomes. Gametes are typically haploid so that fertilisation restores the diploid number of the organism. Denoted as *n*.

heat capacity The amount of heat required to raise the temperature of a substance by one degree Celsius.

heavy chains The polypeptide chain that forms the 'stem' of a Y-shaped antibody molecule.

helicase A type of enzyme that unwinds and unzips the two strands of a DNA molecule.

helper T lymphocyte Helper T lymphocytes bind to antigen-MHC II complexes on antigen-presenting cells and activate B lymphocytes to secrete antibodies, macrophages to phagocytose, and cytotoxic T cells to kill infected cells.

herd immunity Phenomenon in which vaccination of a large proportion of a population provides protection from a pathogen to non-immune or non-vaccinated individuals.

heterochrony Refers to changes by a master gene in the rate or timing of a gene's expression during embryonic development.

histamine An organic compound involved in inflammatory responses and allergic reactions, which causes surface blood vessels to dilate and become more permeable to immune cells and fluids. Common hay fever symptoms such as runny nose and eyes and sneezing are the result of histamine action and are aimed at flushing out allergens.

homeotic gene A gene that determines the type or location of a body part during an organism's development.

hominid Member of the family Hominidae, which includes humans, chimpanzees, gorillas and orangutans.

Hominidae Family of primates, which includes humans, chimpanzees, gorillas and orangutans.

hominin Member of the subfamily Homininae, which includes humans, chimpanzees, gorillas and extinct species of *Homo*, *Australopithecus*, *Paranthropus* and *Ardipithecus*.

Homo The genus of anthropoid mammals of which humans (*Homo sapiens*) are the only living species; includes numerous extinct species and subspecies, such as Neanderthals (*H. neanderthalensis*), *H. erectus*, *H. ergaster* and *H. habilis*.

homologous chromosomes Matching pairs of chromosomes in a diploid organism. Homologous chromosomes carry the same genes in the same loci.

homologous features Structures that have a common evolutionary origin, which is evident in the underlying fundamental similarities in their structure. Homologous features are found in different organisms and may have evolved different functions (e.g. a human hand and a bat wing) as a result of divergent evolution from a common ancestor. DNA sequences or proteins can also be homologous.

horizontal gene transfer Transfer of genes between cells, sometimes of different species, such as transfer of plasmids between bacteria. Contrasts with inheritance of genes from parent to daughter cell through cell division (vertical gene transfer).

hormone A molecule that regulates the growth or activity of those cells capable of responding to it (target cells). Hormones are produced by specialised cells within an organism.

***Hox* genes** A subset of homeotic genes; master regulatory genes that control the arrangement of body structures during embryonic development.

human immunodeficiency virus (HIV) The retrovirus that if left untreated is responsible for causing AIDS.

human leukocyte antigen Another name for major histocompatibility complex in humans.

humanised monoclonal antibody (humanised mAb) A monoclonal antibody that is mostly comprised of human molecular components.

humoral immunity An immune response involving B lymphocytes that produce specific antibodies against foreign antigens.

hybrid breakdown A form of postzygotic isolation between different species in which any first-generation hybrid offspring cannot reliably produce viable second-generation offspring when interbred with other hybrids.

hybrid species It is the result of mixing, through sexual reproduction, two individuals of different breeds, varieties, species or genera.

hybrid inviability A form of postzygotic isolation between different species where any fertilised hybrid zygotes do not develop properly and do not reach birth/germination.

hybrid sterility A form of postzygotic isolation between different species where hybrid offspring that may survive to reproductive maturity are incapable of reproducing.

hybridised A double-stranded DNA molecule formed from two strands of DNA that are from different species.

hybridoma The product of the fusion of an immortal cell line with a B lymphocyte to produce an immortal B lymphocyte; used in the production of antibodies.

hydrophilic Polar ions and molecules that dissolve easily in water; means 'water loving'.

hydrophobic Non-polar molecules that are relatively insoluble in water; means 'water fearing'.

hypothesis A possible explanation to a research question that can be used to make predictions that can often be tested experimentally.

I

immortal cell line A cell line that can continually undergo division without the mutations that would normally occur as a cell ages, and can therefore be cultured for long periods.

immunodeficiency An inadequate response by the immune system to the presence of antigens. Immunodeficiency diseases can be acquired (e.g. AIDS) or congenital (e.g. DiGeorge syndrome).

immunogens Antigens that elicit an immune response.

immunoglobulin (Ig) Alternate name for an antibody; a type of protein produced by B lymphocytes in an immune response to the presence of a particular antigen, to which the immunoglobulin binds.

immunoglobulin E (IgE) A type of antibody that mediates allergic reactions.

immunological memory The ability of lymphocytes of the adaptive immune system to 'remember' antigens after primary exposure, and to mount a larger and more rapid response when exposed to the same antigen again.

immunotherapy Any treatment that harnesses the immune system of the patient to fight diseases; for example, monoclonal antibody therapy.

impression fossil A type of fossil where the impression of the external or internal surface of the organism is preserved.

in vitro Occurring in a culture dish, test tube or other location outside the living organism (compared to 'in vivo', a process taking place in a living organism).

in vitro fertilisation (IVF) Fertilisation of eggs with sperm in a laboratory dish; an assisted reproduction technology.

inactivated vaccine Vaccines made from inactivated (killed) forms of pathogens. Inactivation destroys the pathogen's ability to replicate, but keeps it 'intact' so it can be recognised by the immune system.

independent variable The variable that is altered during an experiment to test its effect on another variable (the dependent variable). Also called experimental variable.

index fossil A fossil that is used to define and identify geologic periods.

induced To promote or activate. When a gene is induced its transcription is activated.

inducer A molecule that regulates gene expression.

inducible operon An operon that is under the control of regulatory proteins (transcription factors).

infer To deduce something from evidence or reasoning.

inference Something that is inferred. See *Infer*.

inflammation (or inflammatory response) A protective response triggered by damaged tissue or invading pathogens, that leads to increased blood flow and migration of white blood cells to the site of damage/infection. It results in heat, pain, swelling, redness and loss of function.

inhibition The binding of another component and/or change in the shape of a molecule that prevents the molecule from functioning; a common method of regulating enzymes. It can be reversible or irreversible.

innate immunity Innate immunity non-specifically protects against a wide variety of pathogens. It consists of physical, chemical and microbiological barriers that provide resistance to infection, and an innate immune response to infection that involves phagocytes and defensive molecules.

inorganic compound Any compound that does not include carbon. However, oxides, carbonates, bicarbonates, carbides and cyanides are usually also considered to be inorganic compounds.

insertion mutation When a point mutation: a type of mutation that involves the addition of one or two nucleotides into the existing sequence. When a block mutation: occurs as a result of one part of a chromosome breaking off and attaching to another chromosome.

insulin In vertebrates, a peptide-based hormone secreted by beta cells in the pancreas. It controls the concentration of glucose in the blood.

integral protein Protein that is partially or fully embedded within the plasma membrane.

interbreeding The mating of two different species.

interferon A type of cytokine important in antiviral immunity. Interferons are produced by virus-infected cells to inhibit viral replication by resulting in the transcription of antiviral genes and the expression of antiviral proteins; have a lesser role in bacterial and parasitic immune responses. They regulate the immune response in a number of ways, such as enhancing T lymphocyte activity.

interleukin A type of cytokine produced by leukocytes for regulating immune responses.

intron Section of DNA that does not code for proteins and is spliced during mRNA processing in eukaryotes.

inverse relationship A mathematical relationship in which one variable increases as the other decreases.

inversion mutation A type of block mutation that involves the reversal of a sequence on a chromosome.

ion channel A transmembrane protein that forms a pore in the plasma membrane to allow the passage of charged particles (ions) across the membrane.

isotopes One of two or more atoms that have the same atomic number (the same number of protons) but a different number of neutrons; for example, carbon-12 and carbon-14.

K

karyotype A description of the number, size and shape of chromosomes observed in an individual.

killed vaccine See *inactivated vaccine*

Krebs cycle A cyclic biochemical pathway that metabolises acetyl coenzyme A, producing carbon dioxide, ATP, NADH and $FADH_2$. The Krebs cycle is one of the three major biochemical pathways that constitute aerobic respiration.

L

***lac* operon** It contains the genes that code for three proteins involved in the metabolism of lactose in *E. coli* and a few other bacteria. It is an inducible operon and its genes are transcribed only when lactose is available. Also known as lactose operon.

***lac* repressor** A DNA-binding protein that binds to the operator site of the *lac* operon inhibiting the transcription of the genes in it.

lacI A regulatory gene that codes for the *lac* repressor.

lactose A carbohydrate, specifically a disaccharide, made up of galactose and glucose and found in milk.

***lacZ* gene** A gene in the *lac* operon that codes for beta galactosidase; used in recombinant plasmids for detecting transformed bacteria.

Laurasia A supercontinent of the past formed by the land masses of the Northern Hemisphere near the end of the Palaeozoic era.

leaf The branch tip on a phylogenetic tree where the scientific name of the taxon is.

leukocytes White blood cells; includes phagocytes and lymphocytes.

ligase An enzyme that joins together two molecules or fragments of molecules.

ligation The process of joining two fragments of DNA using a DNA ligase enzyme.

light chains The short polypeptide chains that form the 'arms' of a Y-shaped antibody molecule.

light-dependent reaction The reactions in photosynthesis in which light captured by chlorophyll is used to split water to produce oxygen, and ATP and NADPH for use in the light-independent reactions.

light-independent reaction The reactions in photosynthesis in which the energy in ATP and NADPH from the light-dependent reactions is used to fix carbon into carbohydrates. These reactions are part of a biochemical pathway called the Calvin cycle.

light saturation curve A plateau-shaped graph of the rate of photosynthesis versus light intensity. The point at which the curve plateaus is the point at which a further increase in light intensity brings about no further increase in photosynthesis.

line graph A type of graph that is useful for representing continuous quantitative data.

lineage All the species that are descendants of a common ancestor.

linear relationship A mathematical relationship in which variables are directly proportional to each other and produce a straight trend line when graphed.

Linnaean system of classification A hierarchical system of classifying organisms that is aimed at reflecting their relationships.

lipid An organic compound that consists of three elements: hydrogen, carbon and oxygen. Lipids are readily soluble in non-polar solvents but not in

liposome Small phospholipid vesicle that can diffuse across plasma membranes or enter cells by endocytosis. It is used as a carrier for genes or drugs across the plasma membrane into cells.

live attenuated vaccine Vaccine that uses a weakened form of the disease-causing agent to stimulate an immune response, but which doesn't cause disease.

loaded coenzyme The form of an enzyme that has a proton, electron or chemical group to donate.

lymph A colourless fluid that contains white blood cells, bathes tissues, and travels through the lymphatic system, draining into the bloodstream.

lymphatic system The body system that transports immune cells including antigen-presenting cells throughout the body, and is where antigen recognition by lymphocytes occurs; important for adaptive immune responses in mammals.

lymphocytes A type of leukocyte involved in adaptive immune responses; includes B and T lymphocytes.

lysis The destruction of a cell, usually by rupturing the cell membrane.

lysosome An organelle vesicle containing digestive enzymes used in the digestion of waste and foreign material.

lysozyme An antibacterial enzyme present in body secretions such as saliva and tears. It disrupts the bacterial cell wall.

M

macrophage A type of large white blood cell that is responsible for engulfing and digesting foreign matter in the body, as well as damaged cells or the remnants of apoptosis.

major histocompatibility complex (MHC) proteins A group of major histocompatibility complex proteins on the surface of cells, involved in antigen presentation to T cells. MHC proteins are also known as human leukocyte antigens.

malignant tumour Mass of cancer cells that can invade nearby tissue and spread throughout the body through a process of metastasis.

marsupial A subclass of mammals characterised by a pouch for carrying the young, which are born immature and complete their development in the pouch.

mass extinctions Large-scale worldwide extinctions evident in the fossil record and caused by major disruptive changes to global climate and the shifting of continents.

mast cell An immune cell containing granules of histamine. This cell mediates allergic responses by binding IgE–allergen complexes and releasing histamines.

master regulatory gene A gene that controls the expression of two or more other genes in the development of an embryo.

maximum parsimony A method for constructing the simplest possible phylogenetic tree to explain evolutionary relationships. This model is based on the principle of parsimony (also known as Occam's razor), which states that the simplest explanation is the most likely to be correct.

mean The average value of a set of values, calculated by dividing the sum of the values by the number of values.

median The value in the middle of an ordered list of values.

memory B lymphocytes B lymphocytes activated against a specific antigen that remain in the lymphoid tissues for a long time, and permit a faster and more effective secondary immune response if the same antigen is encountered again.

memory T lymphocytes T lymphocytes activated against a specific antigen that remain in the lymphoid tissues for a long time, and permit a faster and more effective secondary immune response if the same antigen is encountered again.

meniscus The curved upper surface of liquid in a tube, caused by surface tension. A meniscus can be concave (as in water in a glass tube) or convex (as in mercury in a thermometer).

messenger RNA (mRNA) RNA molecule transcribed from DNA in the nucleus, which passes into the cytoplasm and binds to a ribosome. At the ribosome, mRNA is translated into a polypeptide.

metabolism The total of all chemical processes that take place in an organism.

metastasis The process by which cancer cells break away from the original (or primary) tumour, travel through the blood and lymph vessels, and form secondary tumours at other locations.

microflora Microorganisms that colonise particular sites; normal microflora do not usually cause disease.

microglia Macrophage of the nervous system.

microsatellite A short repeated sequence of nucleotides found at a defined locus on a chromosome. The number of repeats varies between individuals and so are useful in DNA profiling.

mineralisation (or petrification) A process of fossilisation in which minerals replace the spaces in the structures of organisms such as bones.

mineralised fossil Fossil in which minerals replace the spaces in the structure of the organism such as bone, Minerals may eventually replace the entire organism, leaving a replica of the original fossil.

missense mutation A type of substitution mutation that results in a different amino acid in the sequence.

mitochondrial matrix The aqueous solution enclosed by the inner membrane of the cristae of mitochondria. It is the site of the reactions of the Krebs cycle.

mitochondrion (pl. mitochondria) An organelle in eukaryotic cells consisting of folded membrane structures called cristae. It is where the Krebs cycle and the electron transport chain in aerobic respiration occur.

mode The value that appears most often in a data set.

molecular clock The estimated rate of mutation in a region of DNA. It is used to estimate the rate of evolutionary change.

monoclonal antibody (mAb) Antibody produced by a single clone of B lymphocytes grown in culture. The antibodies produced by the clone are identical and specific to the same antigen.

monomer A smaller subunit of a larger unit (called a polymer); examples include amino acids and nucleotides.

monosaccharide A type of carbohydrate consisting of single sugars. They include glucose, fructose and galactose.

Multiregional evolution (continuity) model A model used to explain the origins of modern humans. This theory proposes significant migration of *Homo erectus* across Africa, Asia and Europe for the last 1.8 million years. Isolation between the populations resulted in the divergence of biology and behaviour, but occasional contact ensured gene flow was maintained and led to concurrent evolution of all groups into *Homo sapiens*.

mummified organism A type of fossil in which the organism is fully preserved and may include features such as skin, fur and organs.

mutagen Any substance or condition that causes mutation. Some chemicals and radiation are common types of mutagens.

mutation A permanent change in a genetic sequence, including changes to the nucleotide sequence of DNA or chromosomal arrangement. Mutations can have a beneficial effect, a harmful effect or no effect at all on the survival ability of the individual.

mutation rate The rate at which genetic mutations occur over time.

myelin A substance that forms a layer around the nerve axons (myelin sheath), which enables the conduction of nerve impulses to occur along the axons.

myeloma cell Myeloma is a cancer of B lymphocytes. Myeloma cells are B lymphocytes that proliferate uncontrollably.

N

NAD$^+$ (nicotinamide adenine dinucleotide) A coenzyme that functions as an electron carrier during cellular respiration.

NADP$^+$ (nicotinamide adenine dinucleotide phosphate) A coenzyme that functions as an electron carrier during photosynthesis.

natural active immunity Active immunity induced as a result of survival of a natural infection.

natural passive immunity Passive transfer of antibodies from mother to foetus through the placenta prior to birth, and from mother to baby through breastfeeding.

natural selection A mechanism of evolution. Phenotypes with high adaptive values are 'selected for' because they are better suited to the environmental conditions. Individuals who are better adapted are more likely to survive and reproduce successfully and so are more likely to pass their alleles with high adaptive values on to the next generation.

naturalist A person who is an expert or interested in botany or zoology (the natural world), especially in the field.

Neolithic Revolution The time period when human cultures transitioned from hunter-gatherer lifestyles to permanent settlements centred around agriculture. Also known as the Agricultural Revolution.

neuraminidase Enzyme used by influenza virus to release new virus particles from the host cell. Abbreviated as N in virus name, as in H1N1. The drug Relenza blocks the active site of neuraminidase.

neuron A nerve cell, including its various processes and attachments. It is the fundamental unit of the nervous system in animals.

neurotransmitter A group of signalling molecules produced by neurons and used to carry a signal across synapses between cells.

neutralisation The binding of neutralising antibodies to toxins or antigens on the surface of pathogens that inhibits their action or ability to enter cells.

niche Role or place in an ecosystem.

nitrogen fixation A process in which nitrogen from the atmosphere is incorporated into the tissue of certain plants.

node The point at which two branches in a phylogenetic tree diverge (also known as branch point). The node represents the last common ancestor that the two diverging taxa shared.

nominal variable A categorical variable in which there is no inherent order. Nominal variables can be counted but not ordered.

non-competitive inhibition The inhibition of an enzyme due to an inhibiting molecule binding to an allosteric site on the enzyme. This causes a conformational change in the active site of the enzyme that prevents substrate from binding, or otherwise prevents a catalytic reaction from proceeding even if substrate is bound.

non-conservative substitution A change in the nucleotide sequence of DNA or RNA that leads to the replacement of one amino acid with a functionally different one, resulting in biochemical changes.

non-disjunction The failure of homologous pairs of chromosomes to separate during metaphase I of meiosis. Non-disjunction results in aneuploidy because two of the gametes formed will have two copies of the chromosome, while the other two gametes will be missing that chromosome entirely.

non-homologous chromosomes Chromosomes that contain alleles for different types of genes.

non-self antigens Antigens that do not belong to an organism's own cells.

nonsense mutation A type of substitution mutation that results in the creation of a stop codon within the coding sequence.

nuclear pore A channel in the nuclear envelope that is used in the transportation of molecules between the nucleus and the cytoplasm.

nucleic acid The genetic material of all organisms (DNA and RNA) that controls cellular activities, and is made up of monomer units called nucleotides.

nucleolus An organelle in the nucleus involved in the synthesis of ribosomal RNA and the formation of incomplete ribosomes.

nucleotide Monomer, or building block, of the nucleic acids DNA and RNA. Consists of a phosphate, a sugar and a nitrogenous base.

nucleus An organelle that contains genetic information (used for the synthesis of proteins) and directs the activities of the cell.

O

observation Closely monitoring something or someone.

oestrogen A steroid hormone produced mainly in the ovaries, that initiates the development of secondary sex characteristics and controls the ovarian cycle.

oligodendrocyte A cell in the central nervous system that produces myelin, which forms myelin sheaths around the axons of neurons.

oomycetes Fungus-like pathogens of plants with branching hyphae (haustoria) that penetrate living cells and absorb nutrients, or release enzymes that digest cytoplasm into molecules that can be absorbed.

operator The segment of DNA that is the binding site of the transcription factor.

operon A unit of DNA under the regulation of a single promoter that codes for several proteins.

ordinal variable A categorical variable in which there is an inherent order; they can be counted and ordered.

organelle Any of the specialised structures in a cell, such as the nucleus, the Golgi apparatus, mitochondria and vacuoles.

organic compound Any chemical substance containing carbon, once thought to come from living organisms. Common organic compounds are proteins, carbohydrates and lipids. However, oxides, carbonates, bicarbonates, carbides and cyanides are usually not considered to be organic compounds.

osmosis The movement of free water molecules through a semipermeable membrane from a region of higher free water concentration to a lower free water concentration.

osmotic gradient The difference in concentration between solutions on either side of a semipermeable membrane.

osmotic pressure The measure of the tendency of a solution to take in water by osmosis.

Out of Africa (replacement) model A model to explain the origins of modern humans. This model proposes that all living modern humans evolved from a single common ancestor in Africa about 200 000 years ago and as they spread throughout the world they displaced all other human species.

outgroup A taxonomic group that is closely related to the other groups (ingroups) but less closely related than any single one of the ingroups is to each other. It has a common ancestor with the ingroups that is older than the common ancestor of the ingroups. It is included in phylogenetic trees for comparison to the group of interest.

outlier A value that lies outside the main group of data of which it is a part.

P

palaeoanthropology A branch of anthropology that involves the study of fossil hominins, contributing to our knowledge of human evolution.

Palaeolithic Also known as the Stone Age, this was the cultural period of the development and use of stone tools, beginning approximately 2.6 million years ago to approximately 10 000 years ago.

palaeontology The study of ancient life preserved as fossils in rocks and ancient sediments.

pandemic An epidemic that has spread over several countries or continents, usually affecting a large number of people.

Pangaea The supercontinent of enormous land mass that formed when Laurasia (northern land mass) and Gondwana (southern land mass) united by the end of the Palaeozoic (225 million years ago).

paracrine signalling A form of cell-to-cell communication where signalling molecules travel in the interstitial fluid between cells.

passive immunity Immunity provided by the transfer of antibodies produced in another organism.

passive transport The process of moving molecules and other substances across the plasma membranes without the expenditure of energy.

pathogen An organism that can produce disease in another organism; includes many micro-organisms and parasites.

peer-reviewed Other scientists have checked the information and have agreed that it is appropriate for publication.

peptide A polypeptide that consists of fewer than 50 amino acids.

peptide and protein hormones Hydrophilic signalling molecules that are peptides, such as insulin, or proteins, such as growth hormone and follicle stimulating hormone.

peptidoglycan A component of bacterial cell walls, composed of sugars and amino acids. Gram-positive bacteria have a thick layer of peptidoglycan. Gram-negative bacteria have a thin layer. Penicillin-type antibiotics inhibit its synthesis.

period One of several subdivisions of geological time enabling cross-referencing of rocks and geologic events from place to place. Eons and eras are larger subdivisions than periods while periods themselves may be divided into epochs.

peripheral protein A type of protein attached to the outer surface of the phospholipids or the integral proteins in the plasma membrane. They do not penetrate the hydrophobic centre of the plasma membrane.

peroxisome An organelle that contains catalase. Peroxisomes provide a compartment in which oxidation of harmful materials takes place.

petrification See *mineralisation*.

phagocytes Cells capable of engulfing pathogens or foreign particles to destroy them.

phagocytosis The engulfment of solid materials in which the plasma membrane surrounds the material, forming a vacuole (phagosome) and allowing the substance into the cell.

phagosome A vacuole in a cell's cytoplasm that contains a phagocytosed particle.

phenotype The observable trait; expression of a genotype in an individual for a particular trait. The dominance of the alleles and the environmental conditions influence the phenotype of an individual. For example, nutrient availability may influence the pigment synthesis in flower petals or hair follicles in animals.

pheromone A group of signalling molecules that are excreted from the body and diffuse through the air to elicit a response from another individual, typically of the same species.

phosphodiester bond The bond that joins nucleotides into a chain of DNA and RNA by linking the phosphate group of one nucleotide and the sugar of another.

phospholipid Essential component of the plasma membrane, comprised of a hydrophilic phosphate head and a hydrophobic fatty acid tail.

phosphorylation The process of adding a phosphate group to a molecule. Phosphorylation is a common way of activating enzymes and other functional molecules.

photosynthesis The process used by plants, algae and some prokaryotes in which nutrients are produced from carbon dioxide, water and light energy.

phylogenetic tree A diagram that represents the evolutionary relationships between different species, but does not show evolutionary distance. It may be rooted or unrooted.

phylogenetics The study of the evolutionary history and relationships of groups of organisms. Biologists use information from inherited characteristics such as morphology and DNA to determine the relatedness of species or populations of organisms and produce phylogenies or phylogenetic trees.

phylogeny Evolutionary relationships of organisms, usually represented by a branching tree diagram (phylogenetic tree).

phylogram A branching diagram representing the evolutionary relationships between taxa. The branches of a phylogram are scaled (i.e. the lengths of the branches represent evolutionary distance).

phytohormone Signalling molecule produced by plants to regulate growth and development. Also referred to commonly as plant hormone.

pie chart A circular diagram divided into sections, with each section representing the value of one set of data as a proportion of the total data set; useful for presenting qualitative and categorical data.

pinocytosis The engulfment of liquid substances, in which the plasma membrane surrounds the substance, forming a vesicle and allowing the substance into the cell.

plasma cells Activated B lymphocytes that produce large quantities of the same type of antibody.

plasmid Small, circular pieces of double-stranded DNA found in bacterial cells. Plasmids replicate independently of the bacteria's chromosomal DNA and are used in genetic engineering for creating recombinant DNA.

plastid A large organelle with a double membrane found in plant cells. Plastids contain their own DNA.

point mutation A type of gene mutation that typically only affects a single nucleotide. Types of point mutations include substitution and frameshift mutations.

pollen Granules containing the male gametes of flowers. Pollen is the allergen that causes hay fever.

poly-A tail A long tail of adenine (A) nucleotides (100–250) that is added to the end of mRNA during processing. This increases the stability of the mRNA.

polymer A large molecule that is made up of many repeating smaller molecules strung together (e.g. nucleic acids, carbohydrates and proteins).

polymerase A group of enzymes that catalyse the formation of polymers, in particular the formation of nucleic acid polymers by complementary base pairing with a template strand.

polymerase chain reaction A laboratory technique used to amplify (make millions of copies of) a piece of DNA in a short period of time.

polymorphism Genetic variation within a population. The least common allele has to have a frequency in a population of 1% or more to be considered polymorphism rather than mutation.

polynucleotide A polymer of nucleotides joined together through a condensation polymerisation reaction. Can refer to DNA or RNA.

polypeptide A polymer of amino acids joined by peptide bonds through a condensation polymerisation reaction.

polyploidy Describes the number of sets of chromosomes in a cell in which every chromosome has more than two copies. It is denoted xn, where x is the number of copies of chromosomes, such as $3n$ (triploid) or $6n$ (hexaploid) wheat varieties.

polysaccharide A polymer consisting of many repeating units of monosaccharides (e.g. glucose, fructose, galactose) or disaccharides (e.g. sucrose, lactose) linked by glycosidic bonds. They range in structure from linear to highly branched. Examples include starch, glycogen, cellulose and chitin.

polytomy A node in a phylogenetic tree that indicates three or more lineages evolving from a common ancestor.

porin A transmembrane protein, commonly found in the membranes of prokaryotes, mitochondria and chloroplasts, that acts as a channel through which molecules can diffuse.

postsynaptic neuron A neuron to which an electrical impulse is transmitted across a synaptic cleft by the release of neurotransmitters from the axon terminal of a presynaptic neuron.

postzygotic isolating mechanism A process that stops successful gene flow between different species by causing reproductive failures after fertilisation. Common forms of postzygotic isolating mechanisms are hybrid inviability, reduced hybrid viability, hybrid sterility and hybrid breakdown.

Precambrian time The earliest time of Earth history, the geological era from 4.6 billion to 541 million years ago, from the time that the Earth's crust formed, including the oldest fossils of prokaryotic life (sometime between 3.8 billion to 4.1 billion years old), up to the time of the oldest fossils of marine animals (541 million years old).

precipitation (in immunity) One of the mechanisms used by antibodies to interfere with the function of the pathogens by binding to soluble antigens, causing them to become insoluble and precipitate out of solution.

precision The ability to consistently obtain the same value.

predecessor Person who came before (preceded) another. In terms of human evolution, predecessors are ancestors.

prezygotic isolating mechanism A process that stops successful gene flow between different species by preventing fertilisation. Common forms of prezygotic isolating mechanisms include spatial, temporal, ecological, structural and behavioural isolation.

primary immune response The immune response to an antigen that has been encountered for the first time.

primary lymphoid organs The major organs of the lymphatic system: the bone marrow and the thymus.

primary source A source that includes first-hand information, such as the results of an original experiment.

primary structure The linear sequence of amino acids in the polypeptide chain of a protein.

primer A short strand of DNA or RNA that is able to bind or anneal to single-stranded DNA to create a region where DNA polymerase can join and initiate DNA synthesis.

principle A principle is usually more specific than a theory. See *theory*.

prion A small protein particle that, when its shape is altered due to mutation, causes protein aggregation and is toxic to neurons. Prions are the cause of spongiform encephalopathy diseases, BSE in cattle and CJD in humans.

processed data Data that has been mathematically manipulated in some way.

prokaryote A type of organism with a simple cellular structure, lacking a nucleus and other membrane-bound organelles; for example, bacteria.

promoter Upstream region of a gene (a specific DNA sequence) to which RNA polymerase attaches, initiating transcription.

prophylaxis Drug or other measure taken to prevent the onset of disease.

prosthetic group An non-protein compound that is involved in protein structure or function. A protein with a prosthetic group is known as a conjugated protein.

protein An organic compound consisting of one or more long chains of amino acids connected by peptide bonds, and has a distinct and varied three-dimensional structure.

protein synthesis The production of a protein through the processes of gene expression which, in eukaryotes, comprises transcription, RNA processing and translation.

proteome The entire set of protein products of the genome.

proteomics The study of proteomes, including the structure, function and interactions of proteins.

Proterozoic Latter part of the Precambrian era, from about 2.5 billion to 541 million years ago, characterised by the appearance of prokaryotes (bacteria), marine algae and the first animals.

proton A positively charged subatomic particle that forms part of the nucleus of an atom.

protozoan A unicellular, eukaryotic organism that may have multiple stages in a complete life cycle and may replicate within the cells of its host.

pseudopodia A temporary protrusion of the cytoplasm of a cell (such as an amoeba or a white blood cell) that functions especially as an organ of locomotion or in taking up food or other particulate matter.

purine A nitrogenous base that has a double ring structure (e.g. adenine and guanine). Each purine base pairs with a specific pyrimidine base (cytosine, thymine, uracil).

pyrimidine A nitrogenous base that has a single ring structure (e.g. cytosine, thymine, and uracil). Each pyrimidine base pairs with a specific purine base (adenine and guanine).

Q

qualitative data Data collected about categorical variables.

quantitative data Data collected about numeric variables.

quaternary structure Two or more polypeptide chains joined as a single functional protein.

R

radiation therapy Treatment of disease using ionising radiation, especially of cancer to destroy cancerous cells.

radioactive Emitting radiation (a form of energy from the nucleus of an unstable atom).

random coil A polymer conformation with the monomers orientated randomly. Adjacent monomers are bonded together.

random selection A form of sampling in which subjects are randomly selected to participate in a study.

range The difference between the highest and lowest values.

rational drug design Directed chemical and computer-aided design of a drug based on knowledge of the shape and charge of the target molecule.

raw data The data you record in your logbook.

receptor A molecule in a cell membrane that binds and responds to specific molecules such as hormones and neurotransmitters, triggering a response.

receptor-mediated endocytosis A method of transport of specific substances into a cell. A receptor in the plasma membrane binds to a molecule, triggering its entry into the cell via a vesicle formed from the plasma membrane.

recognition site The short sequence of DNA bases recognised and cut by a restriction enzyme; also called a restriction site.

recombinant DNA DNA that has been genetically engineered by joining fragments of DNA from two or more different organisms.

recombinant DNA technology Technology that combines DNA molecules from two or more sources in cell cultures in the laboratory to create a new DNA molecule or genetic sequence.

recombinant plasmid A plasmid containing a foreign gene that has been inserted by the use of restriction enzymes and DNA ligase.

reduced hybrid viability A form of postzygotic isolation in which offspring of different species (hybrids) survive to birth or germination but do not reach reproductive maturity and are therefore incapable of reproducing.

regulatory gene A gene that codes for transcription factors (which in turn control gene expression at the transcription stage).

relative dating Method of dating geological deposits based on the relative order of layers (strata) and, if present, the fossils within those layers. It is assumed that the deepest layer is the oldest and the uppermost layer is the youngest.

reliability The ability to consistently reproduce results.

repeat trials Collecting multiple data sets by performing an experiment again after the initial test.

replication (1) Mechanism by which DNA can be copied. (2) Experimentation carried out on duplicate sets at the same time.

repressed Describes a gene that is inhibited and cannot be transcribed.

research question A statement that defines what is being investigated.

response In cellular communication, the response is the action or change in functionality of a cell that occurs as a result of a specific stimulus. For example, the release of neurotransmitters into a synapse is the response to an action potential reaching the synaptic terminal of a neuron.

restriction enzyme A type of enzyme, also called an endonuclease, that occurs naturally in bacteria and can cut DNA at a particular site (a recognition site); used in genetic engineering.

retrovirus An RNA virus that uses reverse transcriptase to copy its RNA genome into DNA for integration into the chromosome of a host.

reverse transcriptase A type of polymerase enzyme used by retroviruses to copy their RNA genome into DNA; used in genetic engineering to copy messenger RNA (mRNA) into complementary DNA (cDNA).

ribonucleic acid (RNA) A nucleic acid that is a single strand made up of a sequence of ribose sugars and bases (adenine, cytosine, guanine and uracil) linked by phosphodiester bonds. There are three forms: messenger RNA (mRNA), ribosomal RNA (rRNA) and transfer RNA (tRNA).

ribose A five-carbon sugar molecule found in RNA as a component of RNA nucleotides.

ribosomal RNA (rRNA) A nucleic acid synthesised in the nucleolus that forms part of a ribosome.

ribosome A non-membrane bound organelle, made of RNA and protein, which acts as the site of translation in the process of protein synthesis.

risk assessment A systematic way of identifying the potential risks associated with an activity.

RNA See *ribonucleic acid*.

RNA ligase A ligase enzyme that joins together fragments of RNA.

RNA polymerase An enzyme that catalyses the synthesis of RNA, using an existing strand of DNA as a template.

RNA processing The removal of introns from the primary transcript produced in transcription. The exons are joined to form mRNA, ready for translation. This stage of gene expression occurs only in eukaryotes.

root The root of a phylogenetic tree represents the common ancestor of all the taxa in the tree.

rough endoplasmic reticulum (RER) vvA large organelle comprised of layers of cisternae with ribosomes studding the surface. It is the site of protein synthesis and modification.

S

safety data sheet A document that contains important information about the possible hazards in using a substance and how the substance should be handled and stored.

scatterplot A graph in which two variables are plotted as points. Used when looking to see if there is a correlation or relationship between two quantitative variables.

scientific method The experimental approach to the study of science that involves formulating a hypothesis, designing and performing an experiment to test the hypothesis, and analysing whether the results support or refute the hypothesis.

second messenger A group of small, non-protein molecules that are produced inside a cell when a signalling molecule binds to a receptor on the outer surface of the plasma membrane. The production of second messengers is a form of signal transduction.

second-hand data Data you have not collected yourself.

secondary immune response The immune response to an antigen that has previously been encountered and which elicited a primary immune response. The process activates memory cells and so is faster and more effective that the primary response.

secondary lymphoid organs and tissues The organs and tissues of the lymphatic system in which adaptive immune responses initiate: the lymph nodes, spleen, tonsils, adenoids and appendix.

secondary source A resource that interprets primary documents, written after the event by a person who was not a witness to the event.

secondary structure The folding or coiling of the polypeptide chains in proteins due to hydrogen bonds. The main forms are the alpha helix structure, beta-pleated sheets and random coils.

secretory protein A protein synthesised for export out of the cell.

secretory vesicle Vesicle that buds from the Golgi apparatus and contains material that is to be secreted out of the cell by exocytosis.

selection pressure An environmental factor that affects the survival and reproductive success of an individual based on their particular phenotype.

selective breeding The artificial selection of individuals with desired traits to be interbred. Humans selectively breed both plant and animal species, creating specific strains or breeds.

self-antigens An organism's own antigens, which are normally tolerated (do not elicit an immune response).

self-tolerance The inability of the adaptive immune system to respond to self-antigen.

semi-conservative substitution A change in the nucleotide sequence of DNA or RNA that leads to the replacement of one amino acid with one that is similar in structure but has different biochemical properties.

serum The fluid portion of blood that remains after blood cells and material involved in blood clotting has been removed.

sexual dimorphism Marked differences in the physical appearance of males and females of the same species (in addition to differences in sexual organs).

sexual selection The difference in the ability of individuals to acquire mates. It typically involves contests between males or choice by females and leads to the selection of characteristics relating to mate attraction. Individuals that possess the desired characteristic are more likely to mate and pass on their desired alleles to the next generation. The desired trait, such as number of eye spot feathers in peacocks, is often an indication of overall health and fitness and other alleles of high adaptive value.

short tandem repeat (STR) Regions of non-coding DNA with 4–6 base pair repeated sequences. Used in DNA profiling.

signal transduction The process of transmitting a signal into or out of a cell, or changing the form of the signal. Examples are the production of second messengers inside a cell when signalling molecules bind to a cell surface receptor and the conversion of an electrical signal (action potential) to a chemical signal (neurotransmitters) in neural pathways.

signalling molecule A molecule, such as a neurotransmitter or hormone, that is involved in chemical communication between cells.

silent mutation A type of substitution mutation that results in a different codon that codes for the same amino acid as the original sequence. This type of mutation has no effect on the individual.

sister taxa A pair of taxa grouped together in a phylogenetic tree (the closest relative of any given taxon).

smooth endoplasmic reticulum (SER) A continuous membrane system that forms a series of flattened sacs within the cytoplasm of eukaryotic cells. It is not associated with ribosomes and is involved in the synthesis of lipids.

solute A substance that is dissolved in a solvent to form a solution.

solvent The substance in which a solute dissolves to form a solution.

somatic mutation A mutation that occurs in somatic, or non-gamete, cells of an organism. These types of mutations may affect the individual, but cannot be passed on to offspring. Cancer is a form of somatic mutation.

speciation The formation of new species following a lineage splitting event. Speciation may result from geographic, anatomical, physiological or behavioural barriers to breeding, leading to divergence over evolutionary time, or may be rapid as a result of adaptive radiation.

species A group of individuals that are able to interbreed to produce viable, fertile offspring.

specificity The ability to recognise and respond exclusively to specific antigens.

spliceosome Enzyme that removes the introns from the primary transcript to create mRNA during RNA processing (in eukaryotes).

splicing The removal of introns from the primary transcript produced in transcription. The exons are joined to form mRNA, ready for translation.

steroid hormone Lipid hormones based on cholesterol that are hydrophobic.

sticky-end restriction enzyme A type of restriction enzyme that makes a staggered cut in DNA to leave fragments with overhanging (or 'sticky') ends. The exposed bases of these sticky ends are then able to form complementary base pairs with nucleotides of other DNA molecules that have been cut with the same restriction enzyme.

stimulus A chemical or physical change that activates a receptor molecule in a cell and generates a response, such as the production of a hormone.

stimulus–response model An explanation of the mechanism by which an organism, organ or cell changes its behaviour or physiology as a consequence of changes in its internal or external environment. At the cell level, the stimulus–response model involves three basic steps: reception, transduction and cellular response.

stoma (pl. stomata) A pore structure bordered by two guard cells found in plants (often more abundant on the underside of leaves) that allows exchange of gases between the outside and inside of the leaf. The stomata open to allow carbon dioxide to enter and oxygen to be released, and close to reduce water loss.

stratigraphy The study of the relative positions of layers of rock (strata), some of which contain fossils. The lowest stratum is the oldest and upper strata are progressively younger.

stroma (pl. stromata) The fluid matrix part of a chloroplast in which the light-independent reactions occur.

stromatolite A layered rock that forms when certain marine prokaryotes bind thin films of sediment together; includes fossil and present-day rocks.

structural gene A gene that codes for proteins and RNA molecules that are not involved in gene regulation (e.g. enzymes).

structural morphology The study of the form and structure of organisms.

subspecies Populations within a species that show genetic differences across a geographic range.

substitution mutation A type of point mutation in which individual nucleotides, typically one or two, are replaced by different nucleotides. Substitution mutations can result in silent, missense or nonsense mutations.

substrate A molecule that is acted on by an enzyme.

subunit vaccine Vaccine that contains one or more antigens that stimulate an adaptive immune response.

symbiosis (adj. symbiotic) A close long-term relationship between individual organisms of two different species, usually to the advantage of both.

synapse The point of communication between two cells, where at least one of the cells is a neuron. It includes the membrane of the presynaptic neuron, the synaptic gap and the membrane of the postsynaptic cell (which may be, for example, a neuron, muscle cell or gland cell).

synaptic terminal The structure on the tip of an axon terminal of a neuron that forms the presynaptic component of a synapse. The synaptic terminal is where neurotransmitters are produced and secreted from the neuron.

T

T cell receptor A molecule found on the surface of T lymphocytes that is responsible for recognising fragments of antigen as peptides bound to major histocompatibility complex (MHC) proteins. It is made up of two polypeptide chains that have a variable and a constant region and only one antigen-binding site.

T lymphocyte (or T cell) A type of lymphocyte that originates in the bone marrow and matures in the thymus, and is responsible for cell-mediated immune responses. See *cytotoxic T lymphocyte* and *helper T lymphocyte*.

Taq polymerase A type of heat-resistant DNA polymerase that is widely used in PCR.

TATA box A name given to a common sequence of bases in eukaryotic genes, which is TATAAA, that codes for the promoter region.

taxon (pl. taxa) A biological group classified on the basis of their shared characteristics and evolutionary relationships.

taxonomy The science of the classification of organisms into hierarchical groupings based on their shared characteristics and evolutionary relationships.

template strand A strand of DNA or RNA used as a template for building a complementary strand of a precise nucleotide sequence.

tertiary structure The structure in proteins created by further folding as a result of bonds forming between the R groups of the amino acids, leading to greater stability than the folding in secondary structures.

testosterone A steroid hormone produced in male vertebrates and responsible for the development of various sex characteristics. It is produced by the testes in mammals and to a lesser extent by the ovaries.

tetrapod A vertebrate (which includes amphibians, reptiles and mammals) with four limbs. Animals that had four-limbed ancestors (e.g. snakes and whales) are also known as tetrapods.

thalassaemia A disease caused by a nonsense mutation that affects the formation of haemoglobin molecules in humans.

theory When, after many experiments, a hypothesis has been supported by all the results so far, it is referred to as a theory or principle.

thermophilic Of or relating to an organism that favours high temperatures (a thermophile).

thylakoid The internal membranes of a chloroplast in which chlorophyll molecules are located.

thylakoid lamella (pl. thylakoid lamellae) The sheet-like thylakoid membranes between the grana in a chloroplast.

thymine (T) A nitrogenous base (pyrimidine) found in the nucleotides of DNA.

thyroid hormones Hydrophobic amino-acid-derived hormones; for example, thyroxine.

toxoid vaccine A type of non-recombinant subunit vaccine that uses toxins inactivated by formalin to stimulate an adaptive immune response.

trace fossil (or ichnofossil) Preserved evidence of an animal's activity or behaviour, such as footprints, without containing parts of the organism.

trait A particular characteristic or feature of an organism.

***trans* face** The side of the Golgi apparatus facing the plasma membrane.

transcription Process by which a base sequence in DNA is used to produce a base sequence in RNA.

transcription factor Proteins that control gene expression at the transcription stage by binding to DNA sequences close to the promoter region of a gene or to the RNA polymerase to induce or repress the expression of specific genes.

transfer RNA (tRNA) An RNA molecule that brings a specific amino acid to a ribosome so it can be joined to other amino acids during translation.

transformed A bacterium that has incorporated DNA from another organism into its own or taken up a plasmid containing foreign DNA.

transgenic mouse Mouse that has been genetically modified to contain genes from other species.

transgenic organism An organism with a genetic modification (GM) made by transfer of specific gene(s) from another organism.

translation The process in which the base sequence of a mRNA molecule is used to produce the amino acid sequence of a polypeptide.

translocation mutation A type of block mutation that involves sections of two different chromosomes switching positions.

transmembrane protein A functional protein, often composed of more than one polypeptide molecule, which spans the entire thickness of the plasma membrane. Hydrophilic hormone receptors, ion channels and ion pumps are all examples of transmembrane proteins.

transport vesicle Vesicle that buds from the rough endoplasmic reticulum and contains materials that are to be transported to the Golgi apparatus.

triplet Sequence of three nucleotides in DNA that carries the genetic information for the sequence of amino acids in a protein. Each triplet usually codes for one amino acid.

trisomy An abnormality in which there are three copies of a particular chromosome in a cell. Down syndrome is characterised by trisomy 21; that is, three copies of chromosome 21.

tumour An abnormal growth of cells resulting from uncontrolled cell division or failure of programmed cell death. It may be benign or malignant.

U

uncertainty The range of values within which the true value of a measured quantity probably occurs. Uncertainty is caused by random and systematic errors.

unloaded coenzyme The form of a coenzyme that is free to accept a proton, electron or chemical group.

uracil (U) A nitrogenous base (purine) found in the nucleotides of RNA. It forms a base pair with adenine.

V

vaccination The technique of artificially inducing an adaptive immune response by administering (usually by injection) a vaccine usually made of altered, weakened or killed microorganisms, or inactivated forms of toxins or antigens.

vacuole An organelle involved in storage in plant cells.

valid See *validity*.

validity How strong (or sound) your results are.

variable A factor or condition that can change during your experiment.

variable R group A side chain found on an amino acid.

variable region The region of an antibody molecule that varies between different antibodies and allows them to interact with different antigens.

variation Differences between individuals in a population in terms of their phenotypes, genotypes or specific genetic sequences.

vector (1) In infectious disease: object or organism that transfers a parasite from one host to another. (2) In molecular biology: a vehicle used to transfer foreign DNA into a cell (e.g. a plasmid, virus or liposome).

vestigial structure A remnant structure of an organism that has lost all or most of its original function in the course of evolution.

viable offspring Members of the next generation who survive to maturity and are able to reproduce successfully.

viral vector Virus used as a vehicle to transfer genes (foreign DNA) into cells; used in gene therapy, vaccine production and research.

virion A complete mature virus particle that is metabolically inert and is in the transmission (infectious) phase.

viroids Infectious agents of plants that are a type of self-cleaving RNA enzyme (or ribozyme); composed of short, circular strands of RNA that lack a protein coat.

virulence (adj. virulent) The ability of a pathogen to cause disease.

virus An infectious agent composed of genetic material (DNA or RNA) enclosed in a protein coat, and sometimes also a lipoprotein envelope; is only able to multiply in a host cell.

WXYZ

X-ray crystallography Technique using X-ray diffraction to determine the structure of molecules, such as proteins and nucleic acids.

xenotransplantation Organ, tissue or cell transplant from one species to another, such as from pig into human.

zoonotic Infectious disease transmitted from non-human animal to human.

Index

Note: Page numbers in **bold** refer to the main discussion of key terms in the text.

Acknowledgements

We thank the following for their contributions to our textbook:

COVER: **Alamy Stock Photo**, Noriyuki Otani. **123RF**: pp.95bl, 152b, 200b, 296b, 397cr; alycan, p.563t; Visarute Angkatavanich, p.415rt; antonprado, pp.1,2; Micha Baraski, p. 375t5; Roberto Biasini, pp.72br, 94; blueringmedia, pp. 75tl, 75bl; 75br; Ekaterina Borisova, p.448rb; Charles Brutlag, p.23tr; Lynne Carpenter, p.23br; Hans Christiansson, p.207t; luk Cox, p.260t; Andrea Danti, p.81; pan demin, p.430lb; Designpics, p.406b; designua, pp.75cl, 111t, 116bl, 116br, 229bl, 248br, 279r, 280r, 310rt, 310rb; elnavegante, p.461; dirk ercken, p.348l; Nicolas Fernandez, p.375t6; Juan Gartner, p.261r; Deyan Georgiev, p.30tl; Ben Goode, p. 342t; gozzoli, p.355t; Jeff Grabert , p.358tl; Kevin Griffin, p.440rb; Tim Hester, p.527; p.562tc; Hypedesk, p.82t; culig, pp. 83lt,162-3; iqoncept, p.335; Eric Isselee, p.418l; vladislav ivantsov, p.488bc; jeanchen, p.398lt; Helen Jobson, p.382b; Sutisa Kangvansap, p.399r; kasto, p.30br; Sebastian Kaulitzki, p.300r; Andrey Khrobostov, p.228t4; Valerii Kirsanov, p.358tr; Kjuuurs, p.350bl; Vit Krajicek , p.118; Lightspring, p.259; Erik Lam, p.356t; Peter Lamb, p.105br; Anton Lebedev, pp.86lt, 86lc, 86lb, 87tl, 87tr, 126b; Steven Lovegrove, p.406t; Dana Mihaela Matei, p.378b; Ivan Mikhaylov, p.41b; miloszg , p.262lc; molekuul, p.337br; natursports, p.375t1; Derrick Neill, p.379t; Sergey Nivens, p. 133; nobeastsofierce, pp.221, 323-330, 332-3, 491-6; Golkin Oleg, p.418t; Jarun Ontakrai, p.280bl; peterwaters, p.230lt; petervick167, pp.354r, 347rb; prill, p.397br; reddogs, p.357bl; rioblanco, p.556t; Bruce Rolff, pp. 497, 580-4; David Smith, pp.366-7; snapgalleria, p.75cr; soleg, p.247cc; Luca Alarico Sperotti, p.458br; srum, p.103b;tempusfugit, p.334t; Leah-Anne Thompson, p.352tl; Akhararat Wathanasing, p.27lc; Geoffrey Whiteway, p.32bl; Wong Hock Weng, p.410t; zuzanaa, p.263tr. **AAP**: p.125b, AP, pp.468t, 480r; AP/Aqua Bounty Farms, p.551; Julian Smith, p.254. **Alamy Stock Photo**: Agencja Fotograficzna Caro, p.500; Stephen Barnes/ Business, p.19; Sabena Jane Blackbird, p.489rt; blickwinkel, pp.91tl, 397bl; BSIP SA, p.9l, 320, 314t; B Christopher, p.473t; domonabikebali, p.310lb; FLPA, p.406c; Frans Lanting, p.491c; Medicimage, p.310lt; MShieldsPhotos, p.32br; Nature's Geometry by NanoArt , p.262lb; Roberto Nistri, p.448rt; Friedrich Saurer, p.489rb; Martin Shields, p.515r; The Natural History Museum, pp.374bl, 375brb, 464b, 491l; Duncan Usher, p.402tl; World History Archive, p.506b; Robert Wyatt, p.349tr. **Alexbateman**: p.35b. **Anker, Arthur**: pp.403cl, 403cr. **ATOR (Arc-team Open Research)**: p.472t. **Aubert, Maxime**: p.478b. **Australian Government, Department of Health**: Adapted from: 'Vaccine preventable diseases in Australia, 2005 to 2007' by Chiu C1, Dey A, Wang H, Menzies R, Deeks S, Mahajan D, et al. Commun Dis Intell 2010; 34 Supp: S1-S167 p.S65, pp.53b, 51l. **Banwell, Tom**: p.555rt. **BioMed Central**: Catteruccia et al; licensee BioMed Central Ltd. 2009, p.546r. **Boden, Gareth**: p.562tc. **Cengage**: Cengage Advantage, Biology: The Dynamic Science, 3rd Edition 2014 by Peter J. Russell & Paul E. Hertz p.395, pp.519, 520r; Solomon & Martin, Biology 10th Edition 2015, p.319, p.517. **Centers for Disease Control and Prevention**: p.564b. **China Daily**: p.391t. **Clifford, Trevor**: p.40bl. **Commonwealth Scientific and Industrial Research Organisation (CSIRO)**: Electron Microscopy Unit AAHL, p.565t. **Cotton Australia**: Photo by Greg Kauter, courtesy of Cotton Australia, p.544t. **Darwin, Charles**: Origin of Species 1859, 402bl. **d'Errico, Francesco and Backwell, Lucinda**: p.466. **Diagnosis**: Diet: www.diagnosisdiet.com, Suzi Smith, p.168. **DK Images**: p.375c. **Elsevier**: 'Molecules consolidate the placental mammal tree' by Mark S. Springer, Michael J. Stanhope, Ole Madsen, Wilfried W. de Jong, p.432. **Haeckel, Ernst**: Anthropogenie (1874), plates V:, p.399l. **Fergus Photography**: Mark Fergus Photography, p.86br. **Fleagle, John**: p.464t. **Fotolia**: san724, p.187b; elenabsl, p. 171bl; evgenia sh, p. 196tl; Dr_Kateryna, pp.71, 216-20; KAR, p.42; kmiragaya, p.480c; Ruslan Olinchuk, p.549; juanjo tugores, p.548t; Sophia Winters, p.10. **Getty Images**: pp.253c, 270, 272t5; Addenbrookes Hospital/Science Photo Library, p.339r; Ramon Andrade, p.235t; Mauricio Anton, p.381bl; Asklepios Medical Atlas, p.289b; Auscape/UIG, p.372r; Jonathon Blair, p.380r; Massimo Brega, p.14; BSIP, pp.184r, 143t; Dr Jeremy Burgess/Science Photo Library, p.192tl; CDC/Science Photo Library, p.7; Peter Chadwick, p.350tl; Bethany Clarke/ Stringer, p.563b; CNRI, p.200t; Henning Dalhoff, p.312t; Dea/A. Dagli Orti, p.480l; E R Degginger, pp.470lt, 488ba; Wally Eberhart, p.90bc; Jason Edwards, p.343t; Encyclopaedia Britannica , pp.76r, 76l; Equinox Graphics/Science Photo Library, p.110b; ericsphotography, p.9b; Eye of Science, p.205tr; Eye of Science/Science Photo Library, pp. 512tl, 512tr; Mauro Fermariello, p. 28r; Pascal Goetgheluck, p.522br; Peggy Grb/US Department of Agriculture/ Science Photo Library, p.28bl; Steve Gschmeissner, pp.90br, 272t1, 386br; Steve Gschmeissner/Science Photo Library, pp.33b, 247r; Hipersynteza/Science Photo Library, p.231c2; James Jordan Photography, p.451bl; Wolfgang Kaehler, p.471l; Manfred Kage/ Science Photo Library, p.56bl; Paul Kennedy, p.242br; Kennis and Kennis/MSF, p.465tr; Laguna Design, pp.155t, 231c3; Laguna Design/Science Photo Library, p.521tl; Claus Lunau, p.387t; Dr P Marazzi, p.230br; Joe McDonald, p.405bl; Medi-Mation/Science Photo Library, p.237b;Cathlyn Melioan, p.228t2; Peter Menzel/Science Photo Library, p.499; Christian Koch, Microchemicals, p.186; Matyiai Fendrych, Moritz K. Nowack, Marlies Huysmans, Microscopy Core Facility, VIB Gent, p.246tr; Dr Gopal Murti/Science Photo Library, p.246tl; NIBSC/ Science Photo Library, p.312l; Ralph Orlowski, p. 558r; Heiti Paves, p11; Alfred Pasieka/ Science Photo Library, pp.305, Pasieka, p 251bl: Alfred Pasieka, p.384l; Phillipe Plailly, pp.465cr, 470t, 476; Sebastien Plailly, p.469t; Keith R Porter, p.217tr3; Ralph Jr, p.557b; Nicolle R. Fuller, p.234c; Ed Reschke, p.56tl; Science Photo Library, pp.201bl, 203tl, 208, 251br, 289cr, 386cr; 488bb; Science Picture Co, p.470lb; H. Singh/Custom Medical Stock Photo/Science Photo Library, p.234tl; Dr Robert Spicer/Science Photo Library, p.380bl; Sinclair Stammers/Science Photo Library, p.375t4; Volker Steger, p.498b; Ariram Subramaniam/National Institutes of Health, p.272t7; Hans Surfer, p.334r; Ted Thai, p.465br; Universal Images Group, p.252b; Tim Vernon/Science Photo Library, p.315; Dr Simon Walker-Samuel, UCL CABI, p.223r; Mike Werner/Science Photo Library, p.241; Dirk Wiersma, p.478t; Dr John Zajicek/Science Photo Library, p.311t. **Hooke, Robert**: Micrographia (1665), p.73l. **ImageFolk** (was Amana/Corbis): pp.78br, 79bl, 87r, 91tr, 111c, 115t, 123, 116tl, 130bc, 266, 336, 349tl, 371bl, 445bl, 537, 538tr; Tkachev Andrei/ITAR-TASS, p.384r; Anthony Bannister/ Gallo Images, p.146bb; Bettmann, p.4tl; Jonathan Blair, p.402br; Regis Bossu/Sygma, p.471t; John Cancalosi/National Geographic Creative, p.386cc; Nigel Cattlin/Visuals Unlimited, p.194r; CDC/PHIL, p.301; Jurgen Freund/Nature Picture Library, p.184tl; Juan Gaertner/ Science Photo Library p.277t; Steve Gschmeissner/Science Photo Library, p.253t; Hero Images/Hero Images Inc./Hero Images, p.498t; Hoberman Collection, p.407l; Laguna Design/ Science Photo Library, pp.152l, 231c1, 308; Joe McDonald, p.405bc; Dr Gopal Murti/Visuals Unlimited, p.246b; Roberta Olenick/All Canada Photos, p412b; Photo Quest Ltd/Science Photo Library, p.272t3; Louie Psihoyos, p.379bl; Michael Rosenfeld/Science Faction, p.3b; Science Picture Co., p.381tr; Srdjan Suki/epa, p.226l; Dr Alvin Telser/Visuals Unlimited, p.272t6; Tom Till/SuperStock, p.386cl; Steven Vidler, p.415rb; Dave Wattsl/Visuals Unlimited, p.365b. **Italy's Ministry of Infrastruture and Transport**: Ministero delle Infrastrutture e Trasporti - Provveditorato Interregionale per le Opere Pubbliche del Triveneto - Consorzio Venezia Nuova, p.533. **Kohout, Michele**: p.263l. **Ladiges, Pauline**: p.347l.**Low, Tim**: p.356b. **Malaria Atlas Project**: 'The global distribution of the Duffy blood group' by Rosalind E. Howes, Anand P. Patil, Frederic B. Piel, Oscar A Nyangiri, Caroline W Kabaria, Peter W. Gething, Peter A. Zimmerman, Celine Barnadas, Cynthia M. Beall, Amha Gebremedhin, Didier Menard, Thomas N. Williams, David J. Weatherall, Simon I Hay. (2011) Nature Communications 2, Article number: 266., p.362t. **McGraw-Hill Australia Pty Ltd**: McGraw-Hill Education, Australia, p.278t. **MDPI**: 'Emerging Viruses in the Felidae: Shifting Paradigms' by Stephen J. O'Brien, Jennifer L. Troyer, Meredith A. Brown, Warren E. Johnson, Agostinho Antunes, Melody E. Roelke and Jill Pecon-Slattery, Viruses 2012, 4(2), 236-257; doi:10.3390/ v4020236: p.439b. **Museum Victoria**: Reproduced courtesy of Museum Victoria/Rodney Start, p.372l; Reproduced courtesy of Museum Victoria/John Bloomfield, p.374bc. **National Geographic Society**: O. Louis Mazzatent, p.291b. **National Human Genome Research Institute**: p.139br;. Natural History Museum Picture Library: Michael R Long/The Trustees of the Natural History Museum, London, p.392t. **Nature Publishing Group**: 'Fine-tuned Bee-Flower Coevolutionary State Hidden within Multiple Pollination Interactions' by Akira Shimizu, Ikumi Dohzono, Masayoshi Nakaji, Derek A. Roff, Donald G. Miller III, Sara Osato, Takuya Yajima, Shûhei Niitsu, Nozomu Utsugi, Takashi Sugawara & Jin Yoshimura, Scientific Reports 4, Article number: 3988 (2014) doi:10.1038/srep03988, p.405tr. **National Human Genome Research Institute**: Darryl Leja, p.539t. **National Oceanic and Atmospheric Administration (NOAA)**: Kyle Carothers, p.211r. **Natural History Photographic Agency/Photoshot**: De Agostini, p.376br. Newman, Hans: p.31a. **New South Wales Office of Environment and Heritage**: David Hunter, pp.365lt1, 365lt2. **NSW Government, Department of Health**: Adapted from 'Diphtheria in Australia, recent trends and future prevention strategies' by H. Gidding et. al., Communicable Diseases Intelligence, 24(6), 2000. © Commonwealth of Australia. Reproduced with permission., p.323l. **Oxford Designers & Illustrators Ltd**: pp.29t, 260b. **Oxford University Press**: 'Multilocus Species Trees Show the Recent Adaptive Radiation of the Mimetic Heliconius Butterflies' by Kozak KM, Wahlberg N, Neild AFE, Dasmahapatra KK, Mallet J, Jiggins CD. 2015, Systematic Biology Vol 64, 505–524, pp.429c, 423t. **PaDIL**: Dr J Liberato, Dr M Suzuki & Dr A Drenth, p.563c. **Pearson Education Australia**: Campbell Biology 10th Edition (c) 2014 by Jane B. Reece p.414, p.516; Campbell Biology 9th Edition by Jane B. Reece, Michael L. Cain, Steven A Wasserman, Lisa A. Urry, figure 7.17, p.135, p.107; figure 7.19, p.136, p.108bl; figure 5.20, p.83, p.122b; figure 5.23, p.85, p.124t; p.99t; figure 7.10, p. 99b; figure 7.6, p.127, p.96t; figure 7.6, p.128, p.96bl; figure 7.6, p.128, p.97t. **PNAS** (Proceedings of the National Academy of Sciences): 'Potentially biogenic carbon preserved in a 4.1 billion-year-old zircon' by Elizabeth A. Bella, Patrick Boehnkea, T. Mark Harrisona, and Wendy L. Mao, p. 371t; 'Directional selection has shaped the oral jaws of Lake Malawi cichlid fishes' by R. Craig Albertson, J. Todd Streelman, and Thomas D. Kocher, pp. 450lt, 450lb; 'Antipredator defenses predict diversification rates' by Kevin Arbuckle and Michael P. Speed, PNAS 2015 112 (44) 13597-13602; published ahead of print October 19, 2015, pp.348b, 349b. **Popular historia 2/2015**: p.567b. **Public Library of Science**: 'Genetic Analysis of the Roles of BMP2, BMP4, and BMP7 in Limb Patterning and Skeletogenesis' by Bandyopadhyay A, Tsuji K, Cox K, Harfe BD, Rosen V, Tabin CJ (2006) PLoS Genet 2(12): e216. doi:10.1371/journal.pgen.0020216, p.247cb. **RCSB Protein Data Bank (PDB)**: 'Molecule of the Month' feature by David S. Goodsel, H.M. Berman et al. (2000) Nucleic Acids Research, 28: 235-242; www.rcsb.org, p.141. **Royal Ontario Museum**: With permission of the Royal Ontario Museum and Parks Canada © ROM, p. 374br. **Royal Society Publishing**: 'The function of Hox and appendage-patterning genes in the development of an evolutionary novelty, the Photuris firefly lantern' by Matthew S. Stansbury, Armin P. Moczek. Published 19 March 2014, p.451t. **Science Photo Library**: pp.29br, 31lt, 37, 80c, 92r, 113br, 164b, 165t, 205br, 265r, 290l; AMI Images, p.268c; AMI Images/Dartmouth college, Louisa Howard, pp.557t, 560; Mauricio Anton, p.389tr; Stephen Ausmus/US Department of Agriculture, p.33r; Juergen Berger, p.264l; Bildagentur-Online/Tschanz, p.555rb; Ian Boddy, p.307b; Dr Tony Brain, p.263br; Massimo Brega, The Lighthouse, p.522bl; BSIP, Beranger, p.34lb; Ron Bonser, p.569; Wladimir Bulgar, p.520l; Chris Butler, p. 368r; Dr Jeremy Burgess, pp.130br, 407r; Nicoletta Casartelli and Dr Olivier Schwartz, Institute Paseur, p.248bl; CDC, p.561r; Martyn F. Chillmaid, pp.29bl, 31b, 31c; Carlos Clarivan, p.185cr; Jaime Chirinos, p.376t; Christian Jegou Publiphoto Diffusion, pp.393c, 377t 378t; Clouds Hill Imaging Ltd, p.264br; CNRI, pp.92l, 113bl, 341t; Kevin Curtis, p.506c; Elizabeth Daynes, p.472l; Maurizio de Angelis, p.317t; Deagostini/UIG, p.376bl; Department of Clinical Cytogenetics, Addenbrookes Hospital, p.339l; Doncaster and Bassetlaw Hospitals, p.297t; Dan Dunkley, pp.58, 503tr; A. Dowsett, Health Protection Agency, p.561l; John Durham, p.4tr; Gunilla Elam, pp.130bl, 314b; S. Entressangle/E.Daynes, p.382t; Equinox graphics, p. 64tl; Eye of Science, pp.77tr, 89bl, 146bt; Mauro Fermariello, p. 33tlb; Simon Fraser, p.538br; Dr David Furness, pp.77tl, 79tr; James Gathany/CDC, p.507b; Pascal Goetgheluck, pp.79bl, 473b; Steve Gschmeissner, pp.77br, 247t, 247ca, 262lt, 268t, 268b, 271b, 272t4, 278t, 298tr,316tl; Gary Hincks, p.377b; Dorit Hockman, p.443; Hossler, Custom Medical Stock Photo, p.79tc; Louise Hughes, pp.78bc; 217tr2; ISM, p. 201br; Jacopin/BSIP, pp.230tr, 510; Manfred Kage, p.309r; Kallista Images/Custom Medical Stock Photo, pp.91br, 124bl, 209b; Nancy Kedersha, p.272t8; Dr Alexey Khodjakov, p.80r; Russell Kightley, pp.72lc, 72lb, 72lt, 79br; James King-Holmes, p.171tr, 265bl, 506, 507tr; Laguna Design, pp.13b, 174t, 240lt, 240lb, 248t; Martin Land, p.375t2; Dr Kari Lounatmaa, p.78bl; Claus Lunau, pp.12, 381tl; Dr P. Marazzi, pp. 249b, 538; Dr Charles Mazel/Visuals Unlimited., INC, p.544b; Medimage, p.110t; Gilles Mermet, p.386t; Microscape, pp.217tl, 217tr1; Dr Brad Mogen/Visuals Unlimited, p.521tr; Profs P. Motta & T. Naguro, pp.77tc, 77bl, 77bc, 78tl, 78tr; Profs P.M. Motta, S. Makabe, T. Naguro, p.78tc; Dr Gopal Murti, p.137c; Walter Myers, p.370; National Science Foundation, p.393t; J. G. Paren, p.171t; Alfred Pasieka, pp. 9t, 124br, 129tla, 129tlb, 129tlc, 137br; Phillipe Plailly, pp. 253b, 507c, 521b; Power and Syred, p. 90lb; Philippe Psaila, p. 355b; Revi, ISM, p.293; J.C. Revy, ISM, p.507tl; p Rona/OAR/National Undersea Research Program/NOAA, p.373r; Dr Lothar Schermelleh, p.73t; SCI-COMM Studios, p.35l; Science Stock Photography, p.386bl; Science VU/Dr F Rudolph Turner, Visuals Unlimited, p.445br; Scubazoo, p.397bc; Silkeborg Museum, Denmark/Munoz-Yague, p.387b; Sputnik, p. 358br; Sinclair Stammers, p.546l; Volker Steter/ Nordstar - 4 Million Years of Man, p.481br; James Stevenson, p.252t; Prof Miodrag Stojkovic, p.17; Tek Image, p.33tla; Trevor Clifford Photography, pp.30bl, 106bl; Javier Trueba/MSF, p.22r; Prof Philippe Vago, ISM, pp.340l, 364t, 364b; Richard Wehr, p.89br; Carol & Mike Werner/Visuals Unlimited, Inc., p.98t; Dr Keith Wheeler, p.80l; Dirk Wiersma, p.481l; Dr Torsten Wittman, p.34lt; Zephyr, p.251tr. **Shutterstock**: Potapov Alexander, p.346t; Alila Medical Media, pp.262r, 271t, 277b, 306b, 309b; Allocricetulus, p.31d; Anneka, p.224l; attem, p.27lt; Australis Photography, p.319br; Bitcyte, p.117t; BMCL, p.346l; Bork, p.30tr; Rob Byron, p.170t; catolla, p.4b; Rich Carey, p.185tc; Katarina Christenson, pp.415lt, 414bl; Crevis, p.430lt; Ethan Daniels, p.185tr; Dariush M, p.417;Designua, pp.167b, 235b, 236l,237t, 283t, 457bl; Dmussman, p.365tr; dotshock, p.38t; ellepigrafica, pp.271c, 276; Everett - Art, p.479; Everett Historical, p.343b; extender_01, p.116tr; Jacek Fulawka, p.430br; Juan Gaertner, pp.222, 282t; gillmar, p.357t; Jens Goepfert, p.28tl; Richard Griffin, p.209t; Gypsytwitcher, pp.347rt, 354l; Rob Hainer, p.422brc; Dieter Hawlan, p.405br; hfuchs, p.414tr; llaszlo, p.185tl; Ipatov, p.350rt; iofoto, p.290r; Tania Izquierdo, p.578; Natalie Jean, p.193br; Jianghaistudio, p.440rt; JKlingebiel, p.422brd; Monica Johansen, p.371br; Matej Kastelic, p.22l; Kaspri, p.23tl; Sebastian Kaulitzki, pp.210t, 303; KentaStudio, p.27lb; D. Kucharski K. Kucharska, p.264t; Travis Klein, p. 415lb; Georgios Kollidas, p. 299; Levent Konuk, p.571t; David Lade, pp.350rb, 368lt; Lebendkulturen.de, p.27tr; Janelle Lugge, p.342b; magnetix, pp.104t, 122t; mangostock, p.572l; Matusciac Alexandru, p.545t; Jamilia Marini, p.111b; Meletios, p.244br; mihalec, p.228t1; Miles Aways Photography, p.210b; molekuul.be, pp.306tl, 306tr, 307t;_Mopic, p.100; Claudia Naerdemann, p.350r; Dmitry Naumov, p.296t; NH, p.207bl; Kathie Nichols, p.89tr; Antonio Jorge Nunes, p.422bra; Mizina Oksana, p.206; Sari ONeal, p.398lb; optimarc, p.89cr; Jamen Percy, p.430rt; Catalin Petolea, p.275; photong, p.30tc; Quick Shot, p.231rb; Alexander Raths, p.518l; Jane Rix, p.368bb; Ronsmith, p.38b; Deborah Lee Rossiter, p.232t; RTimages, p.26l; Andrei S, p.229t; Science Photo Library, p.165b; sevenke, p. 228t3; Viorel Sima, p.562tb; Sozaijiten, p.193bl; stoonn, p.357br; Tefl, p.227t; toeytoey, p.272t2; Stephen van Horn, p.192b; vetpathologist, p.172bl; vitstudio, pp.294-5, 300l; vovan, pp.182-3; wang song, p.226tr; Bogdan Wankowicz, p.228t5; Tony Wear, p.230lb; Y.F. Wong, p.422brb; yuris, p.341b; Serg Zastavkin, p.358l; zimowa, p.91bl; Bildagentur Zoonar GmbH, p.264bc. **Sinaeur Associates Inc**: p.436t. **Smithsonian Institute, National Museum of Natural History**: p.375brt. **South Australian Museum**: Michael Lee/The Museum Board of South Australia 2016, p.392b. **Sozaijiten**: pp.70, 331. **Splettstoesser, Thomas**: p.571b. **Tait, Alan**: p.388l. **Taylor & Francis**: 'Molecular Biology of the Cell. 4th edition. (c) 2002 Garland Science' by Alberts B, Johnson A, Lewis J, et al., pp.83lb, 95t. **The New York Botanical Garden**: Photograph courtesy of The New York Botanical Garden, p.242bl. **The University of Chicago**: Al Neander, p.379br. **The University of Melbourne**: Pauline Ladiges, p.389bl, 387cr; Dr Andrew Drinnan/School of Botany, p.389br. **UNC School of Medicine**: Max Englund, p.74l. **Victorian Curriculum and Assessment Authority (VCAA)**: Selected extracts from the VCE Physics Study Design (2017-2021) are copyright Victorian Curriculum and Assessment Authority (VCAA), reproduced by permission. VCE® is a registered trademark of the VCAA. The VCAA does not endorse this product and makes no warranties regarding the correctness or accuracy of its content. To the extent permitted by law, the VCAA excludes all liability for any loss or damage suffered or incurred as a result of accessing, using or relying on the content. Current VCE Study Designs and related content can be accessed directly at www.vcaa.vic.edu.au pp.1, 2, 70, 71, 133, 163, 183, 221, 259, 295, 331, 333, 267, 417, 461, 497, 527. **Wacey, David**: p.371bc. **Weizmann Institute of Science**: Noam Sobel, pp. 231rt, 231rc. **Western Australian Museum**: p.3t. **World Health Organisation**: p.555l.

Every effort has been made to trace and acknowledge copyright. However, should any infringement have occurred, the publishers tender their apologies and invite copyright owners to contact them.